Statistical models

Statistical models

A. C. Davison

Swiss Federal Institute of Technology,
Lausanne

CAMBRIDGE
UNIVERSITY PRESS

PUBLISHED BY THE PRESS SYNDICATE OF THE UNIVERSITY OF CAMBRIDGE
The Pitt Building, Trumpington Street, Cambridge, United Kingdom

CAMBRIDGE UNIVERSITY PRESS
The Edinburgh Building, Cambridge CB2 2RU, UK
40 West 20th Street, New York, NY 10011-4211, USA
477 Williamstown Road, Port Melbourne, VIC 3207, Australia
Ruiz de Alarcón 13, 28014 Madrid, Spain
Dock House, The Waterfront, Cape Town 8001, South Africa

http://www.cambridge.org

First published 2003

Printed in the USA

Typeface Times 10/13 pt   *System* LATEX 2_ε [TB]

A catalogue record for this book is available from the British Library

ISBN 0 521 77339 3 hardback

Contents

Appendix A. Practicals   696

Preface

A statistical model is a probability distribution constructed to enable inferences to be drawn or decisions made from data. This idea is the basis of most tools in the statistical workshop, in which it plays a central role by providing economical and insightful summaries of the information available.

This book is intended as an integrated modern account of statistical models covering the core topics for studies up to a masters degree in statistics. It can be used for a variety of courses at this level and for reference. After outlining basic notions, it contains a treatment of likelihood that includes non-regular cases and model selection, followed by sections on topics such as Markov processes, Markov random fields, point processes, censored and missing data, and estimating functions, as well as more standard material. Simulation is introduced early to give a feel for randomness, and later used for inference. There are major chapters on linear and nonlinear regression and on Bayesian ideas, the latter sketching modern computational techniques. Each chapter has a wide range of examples intended to show the interplay of subject-matter, mathematical, and computational considerations that makes statistical work so varied, so challenging, and so fascinating.

The target audience is senior undergraduate and graduate students, but the book should also be useful for others wanting an overview of modern statistics. The reader is assumed to have a good grasp of calculus and linear algebra, and to have followed a course in probability including joint and conditional densities, moment-generating functions, elementary notions of convergence and the central limit theorem, for example using Grimmett and Welsh (1986) or Stirzaker (1994). Measure is not required. Some sections involve a basic knowledge of stochastic processes, but they are intended to be as self-contained as possible. To have included full proofs of every statement would have made the book even longer and very tedious. Instead I have tried to give arguments for simple cases, and to indicate how results generalize. Readers in search of mathematical rigour should see Knight (2000), Schervish (1995), Shao (1999), or van der Vaart (1998), amongst the many excellent books on mathematical statistics.

Solution of problems is an integral part of learning a mathematical subject. Most sections of the book finish with exercises that test or deepen knowledge of that section, and each chapter ends with problems which are generally broader or more demanding.

Real understanding of statistical methods comes from contact with data. Appendix A outlines practicals intended to give the reader this experience. The practicals themselves can be downloaded from

```
http://statwww.epfl.ch/people/~davison/SM
```

together with a library of functions and data to go with the book, and errata. The practicals are written in two dialects of the S language, for the freely available package R and for the commercial package S-plus, but it should not be hard for teachers to translate them for use with other packages.

Biographical sketches of some of the people mentioned in the text are given as sidenotes; the sources for many of these are Heyde and Seneta (2001) and

`http://www-groups.dcs.st-and.ac.uk/~history/`

Part of the work was performed while I was supported by an Advanced Research Fellowship from the UK Engineering and Physical Science Research Council. I am grateful to them and to my past and present employers for sabbatical leaves during which the book advanced. Many people have helped in various ways, for example by supplying data, examples, or figures, by commenting on the text, or by testing the problems. I thank Marc-Olivier Boldi, Alessandra Brazzale, Angelo Canty, Gorana Capkun, James Carpenter, Valérie Chavez, Stuart Coles, John Copas, Tom DiCiccio, Debbie Dupuis, David Firth, Christophe Girardet, David Hinkley, Wilfred Kendall, Diego Kuonen, Stephan Morgenthaler, Christophe Osinski, Brian Ripley, Gareth Roberts, Sylvain Sardy, Jamie Stafford, Trevor Sweeting, Valérie Ventura, Simon Wood, and various anonymous reviewers. Particular thanks go to Jean-Yves Le Boudec, Nancy Reid, and Alastair Young, who gave valuable comments on much of the book. David Tranah of Cambridge University Press displayed exemplary patience during the interminable wait for me to finish. Despite all their efforts, errors and obscurities doubtless remain. I take responsibility for this and would appreciate being told of them, in order to correct any future versions.

My long-suffering family deserve the most thanks. I dedicate this book to them, and particularly to Claire, without whose love and support the project would never have been finished.

Lausanne, January 2003

1

Introduction

Statistics concerns what can be learned from data. Applied statistics comprises a body of methods for data collection and analysis across the whole range of science, and in areas such as engineering, medicine, business, and law — wherever variable data must be summarized, or used to test or confirm theories, or to inform decisions. Theoretical statistics underpins this by providing a framework for understanding the properties and scope of methods used in applications.

Statistical ideas may be expressed most precisely and economically in mathematical terms, but contact with data and with scientific reasoning has given statistics a distinctive outlook. Whereas mathematics is often judged by its elegance and generality, many statistical developments arise as a result of concrete questions posed by investigators and data that they hope will provide answers, and elegant and general solutions are not always available. The huge variety of such problems makes it hard to develop a single over-arching theory, but nevertheless common strands appear. Uniting them is the idea of a *statistical model*.

The key feature of a statistical model is that variability is represented using probability distributions, which form the building-blocks from which the model is constructed. Typically it must accommodate both random and systematic variation. The randomness inherent in the probability distribution accounts for apparently haphazard scatter in the data, and systematic pattern is supposed to be generated by structure in the model. The art of modelling lies in finding a balance that enables the questions at hand to be answered or new ones posed. The complexity of the model will depend on the problem at hand and the answer required, so different models and analyses may be appropriate for a single set of data.

Charles Robert Darwin (1809–1882) was rich enough not to have to earn his living. His reading and studies at Edinburgh and Cambridge exposed him to contemporary scientific ideas, and prepared him for the voyage of the Beagle (1831–1836), which formed the basis of his life's work as a naturalist — at one point he spent 8 years dissecting and classifying barnacles. He wrote numerous books including *The Origin of Species*, in which he laid out the theory of evolution by natural selection. Although his proposed mechanism for natural variation was never accepted, his ideas led to the biggest intellectual revolution of the 19th century, with repercussions that continue today. Ironically, his own family was in-bred and his health poor. See Desmond and Moore (1991).

Examples

Example 1.1 (Maize data)   Charles Darwin collected data over a period of years on the heights of *Zea mays* plants. The plants were descended from the same parents and planted at the same time. Half of the plants were self-fertilized, and half were cross-fertilized, and the purpose of the experiment was to compare their heights. To

1

| | Height (eighths of an inch) | | |
Pot	Crossed	Self-fertilized	Difference
I	188	139	49
	96	163	−67
	168	160	8
II	176	160	16
	153	147	6
	172	149	23
III	177	149	28
	163	122	41
	146	132	14
	173	144	29
	186	130	56
IV	168	144	24
	177	102	75
	184	124	60
	96	144	−48

Table 1.1 Heights of young *Zea mays* plants, recorded by Charles Darwin (Fisher, 1935a, p. 30).

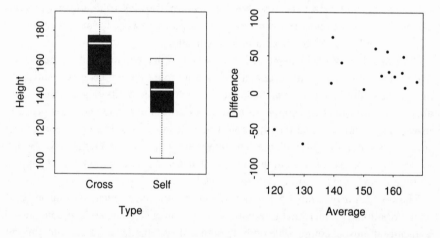

Figure 1.1 Summary plots for Darwin's *Zea mays* data. The left panel compares the heights for the two different types of fertilization. The right panel shows the difference for each pair plotted against the pair average.

this end Darwin planted them in pairs in different pots. Table 1.1 gives the resulting heights. All but two of the differences between pairs in the fourth column of the table are positive, which suggests that cross-fertilized plants are taller than self-fertilized ones.

This impression is confirmed by the left-hand panel of Figure 1.1, which summarizes the data in Table 1.1 in terms of a *boxplot*. The white line in the centre of each box shows the median or middle observation, the ends of each box show the observations roughly one-quarter of the way in from each end, and the bars attached to the box by the dotted lines show the maximum and minimum, provided they are not too extreme.

Cross-fertilized plants seem generally higher than self-fertilized ones. Overlaid on this systematic variation, there seems to be variation that might be ascribed to chance: not all the plants within each group have the same height. It might be possible, and for some purposes even desirable, to construct a mechanistic model for plant growth that could explain all the variation in such data. This would take into account genetic variation, soil and moisture conditions, ventilation, lighting, and so forth, through a vast system of equations requiring numerical solution. For most purposes, however, a deterministic model of this sort is quite unnecessary, and it is simpler and more useful to express variability in terms of probability distributions.

If the spread of heights within each group is modelled by random variability, the same cause will also generate variation between groups. This occurred to Darwin, who asked his cousin, Francis Galton, whether the difference in heights between the types of plants was too large to have occurred by chance, and was in fact due to the effect of fertilization. If so, he wanted to estimate the average height increase. Galton proposed an analysis based essentially on the following model. The height of a self-fertilized plant is taken to be

$$Y = \mu + \sigma\varepsilon, \tag{1.1}$$

where μ and σ are fixed unknown quantities called *parameters*, and ε is a random variable with mean zero and unit variance. Thus the mean of Y is μ and its variance is σ^2. The height of a cross-fertilized plant is taken to be

$$X = \mu + \eta + \sigma\varepsilon, \tag{1.2}$$

where η is another unknown parameter. The mean height of a cross-fertilized plant is $\mu + \eta$ and its variance is σ^2. In (1.1) and (1.2) variation within the groups is accounted for by the randomness of ε, whereas variation between groups is modelled deterministically by the difference between the means of Y and X. Under this model the questions posed by Darwin amount to:

- is η non-zero?
- Can we estimate η and state the uncertainty of our estimate?

Galton's analysis proceeded as if the observations from the self-fertilized plants, $Y_1, \ldots, Y_{15}$, were independent and identically distributed according to (1.1), and those from the cross-fertilized plants, $X_1, \ldots, X_{15}$, were independent and identically distributed according to (1.2). If so, it is natural to estimate the group means by $\overline{Y} = (Y_1 + \cdots + Y_{15})/15$ and $\overline{X} = (X_1 + \cdots + X_{15})/15$, and to compare $\overline{Y}$ and $\overline{X}$. In fact Galton proposed another analysis which we do not pursue.

In discussing this experiment many years later, R. A. Fisher pointed out that the model based on (1.1) and (1.2) is inappropriate. In order to minimize differences in humidity, growing conditions, and lighting, Darwin had taken the trouble to plant the seeds in pairs in the same pots. Comparison of different pairs would therefore involve these differences, which are not of interest, whereas comparisons within pairs would depend only on the type of fertilization. A model for this writes

$$Y_j = \mu_j + \sigma\varepsilon_{1j}, \quad X_j = \mu_j + \eta + \sigma\varepsilon_{2j}, \quad j = 1, \ldots, 15. \tag{1.3}$$

The parameter μ_j represents the effects of the planting conditions for the jth pair, and the ε_{gj} are taken to be independent random variables with mean zero and unit

Francis Galton (1822–1911) was a cousin of Darwin from the same wealthy background. He explored in Africa before turning to scientific work, in which he showed a strong desire to quantify things. He was one of the first to understand the implications of evolution for homo sapiens, *he invented the term regression and contributed to statistics as a by-product of his belief in the improvement of society via eugenics. See Stigler (1986).*

Ronald Aylmer Fisher (1890–1962) was born in London and educated there and at Cambridge, where he had his first exposure to Mendelian genetics and the biometric movement. After obtaining the exact distributions of the t *statistic and the correlation coefficient, but also having begun a life-long endeavour to give a Mendelian basis for Darwin's evolutionary theory, he moved in 1919 to Rothamsted Experimental Station, where he built the theoretical foundations of modern statistics, making fundamental contributions to likelihood inference, analysis of variance, randomization and the design of experiments. He wrote highly influential books on statistics and on genetics. He later held posts at University College London and Cambridge, and died in Adelaide. See Fisher Box (1978).*

	Stress (N/mm^2)				
950	900	850	800	750	700
225	216	324	627	3402	12510+
171	162	321	1051	9417	12505+
198	153	432	1434	1802	3027
189	216	252	2020	4326	12505+
189	225	279	525	11520+	6253
135	216	414	402	7152	8011
162	306	396	463	2969	7795
135	225	379	431	3012	11604+
117	243	351	365	1550	11604+
162	189	333	715	11211	12470+
$\bar{y}$ 168	215	348	803	5636	9828
s 33	43	58	544	3864	3355

Table 1.2 Failure times (in units of 10^3 cycles) of springs at cycles of repeated loading under the given stress (Cox and Oakes, 1984, p. 8). + indicates that an observation is right-censored. The average and estimated standard deviation for each level of stress are $\bar{y}$ and s.

variance. The μ_j could be eliminated by basing the analysis on the $X_j - Y_j$, which have mean η and variance $2\sigma^2$.

The right panel of Figure 1.1 shows a *scatterplot* of pair differences $x_j - y_j$ against pair averages $(y_j + x_j)/2$. The two negative differences correspond to the pairs with the lowest averages. The averages vary widely, and it seems wise to allow for this by analyzing the differences, as Fisher suggested.   ∎

Both models in Example 1.1 summarize the effect of interest, namely the mean difference in heights of the plants, in terms of a fixed but unknown parameter. Other aspects of secondary interest, such as the mean height of self-fertilized plants, are also summarized by the parameters μ and σ of (1.1) and (1.2), and $\mu_1, \ldots, \mu_{15}$ and σ of (1.3). But even if the values of all these parameters were known, the distributions of the heights would still not be known completely, because the distribution of ε has not been fully specified. Such a model is called *nonparametric*. If we were willing to assume that ε has a given distribution, then the distributions of Y and X would be completely specified once the parameters were known, giving a *parametric model*. Most of this book concerns such models.

The focus of interest in Example 1.1 is the relation between the height of a plant and something that can be controlled by the experimenter, namely whether it is self- or cross-fertilized. The essence of the model is to regard the height as random with a distribution that depends on the type of fertilization, which is fixed for each plant. The variable of primary interest, in this instance height, is called the *response*, and the variable on which it depends, the type of fertilization, is called an *explanatory variable* or a *covariate*. Many questions arising in data analysis involve the dependence of one or more variables on another or others, but virtually limitless complications can arise.

Example 1.2 (Spring failure data) In industrial experiments to assess their reliability, springs were subjected to cycles of repeated loading until they failed. The failure 'times', in units of 10^3 cycles of loading, are given in Table 1.2. There were 60 springs divided into groups of 10 at each of six different levels of stress.

Figure 1.2 Failure times
(in units of 10^3 cycles) of
springs at cycles of
repeated loading under the
given stress. The left
panel shows failure time
boxplots for the different
stresses. The right panel
shows a rough linear
relation between log
average and log variance
at the different stresses.

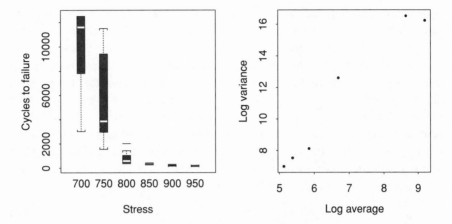

As stress decreases there is a rapid increase in the average number of cycles to failure, to the extent that at the lowest levels, where the failure time is longest, the experiment had to be stopped before all the springs had failed. The observations are *right-censored*: the recorded value is a lower bound for the number of cycles to failure that would have been observed had the experiment been continued to the bitter end. A right-censored observation is indicated as, say, 11520+, indicating that the failure time would be greater than 11520.

Let us represent the jth number of cycles to failure at the kth loading by y_{lj}, for $j = 1, \ldots, 10$ and $l = 1, \ldots, 6$. Table 1.2 shows the average failure time for each loading, $\overline{y}_{l.} = 10^{-1} \sum_j y_{lj}$, and the sample standard deviation, s_l, where the sample variance is $s_l^2 = (10 - 1)^{-1} \sum_j (y_{lj} - \overline{y}_{l.})^2$. The average and variance at the lowest stresses underestimate the true values, because of the censoring. The average and standard deviation decrease as stress increases.

The boxplots in the left panel of Figure 1.2 show that the cycles to failure at each stress have the marked pattern already described. The right panel shows the log variance, $\log s_l^2$, plotted against the log average, $\log \overline{y}_{l.}$. It shows a linear pattern with slope approximately two, suggesting that variance is proportional to mean squared for these data.

Our inspection has revealed that:

(a) failure times are positive and range from $117\text{–}12510 \times 10^3$ or more cycles;
(b) there is strong dependence between the mean and variance;
(c) there is strong dependence of failure time on stress; and
(d) some observations are censored.

To proceed further, we would need to know how the data were gathered. Do systematic patterns, of which we have been told nothing, underlie the data? For example, were all 60 springs selected at random from a larger batch and then allocated to the different stresses at random? Or were the ten springs at 950 N/mm^2 selected from one batch, the ten springs at 900 N/mm^2 from another, and so on? If so, the apparent dependence on stress might be due to differences among batches. Were all measurements made with the same machine? If the answers to these and other such questions were un-satisfactory, we might suggest that better data be produced by performing another experiment designed to control the effects of different sources of variability.

Suppose instead that we are provisionally satisfied that we can treat observations at each loading as independent and identically distributed, and that the apparent dependence between cycles to failure and stress is not due to some other factor. With (a) and (b) in mind, we aim to represent the failure time at a given stress level by a random variable Y that takes continuous positive values and whose probability density function $f(y; \theta)$ keeps the ratio (mean)^2/variance constant. Clearly it is preferable if the same parametric form is used at each stress and the effect of changing stress enters only through θ. A simple model is that Y has exponential density

$$f(y; \theta) = \theta^{-1} \exp(-y/\theta), \quad y > 0, \theta > 0, \tag{1.4}$$

whose mean and variance are θ and θ^2, so that (mean)^2 = variance. We can express systematic variation in the density of Y in terms of stress, x, by

$$\theta = \frac{1}{\beta x}, \quad x > 0, \beta > 0, \tag{1.5}$$

though of course other forms of dependence are possible.

Equations (1.4) and (1.5) imply that when $x = 0$ the mean failure time is infinite, but it decreases to zero as stress x increases. Expression (1.4) represents the random component of the model, for a given value of θ, and (1.5) the systematic component, which determines how mean failure time θ depends on x. ∎

In Examples 1.1 and 1.2 the response is continuous, and there is a single explanatory variable. But data with a discrete response or more than one explanatory variable often arise in practice.

Example 1.3 (Challenger data) The space shuttle Challenger exploded shortly after its launch on 28 January 1986, with a loss of seven lives. The subsequent US Presidential Commission concluded that the accident was caused by leakage of gas from one of the fuel-tanks. Rubber insulating rings, so-called 'O-rings', were not pliable enough after the overnight low temperature of $31°F$, and did not plug the joint between the fuel in the tanks and the intense heat outside.

There are two types of joint, nozzle-joints and field-joints, each containing a primary O-ring and a secondary O-ring, together with putty that insulates both rings from the propellant gas. Table 1.3 gives the number of primary rings, r, out of the total $m = 6$ field-joints, that had experienced 'thermal distress' on previous flights. Thermal distress occurs when excessive heat pits the ring — 'erosion' — or when gases rush past the ring —- 'blowby'. Blowby can occur in the short gap after ignition before an O-ring seals. It can also occur if the ring seals and then fails, perhaps because it has been eroded by the hot gas. Bench tests had suggested that one cause of blowby was that the O-rings lost their resilience at low temperatures. It was also suspected that pressure tests conducted before each launch holed the putty, making erosion of the rings more likely.

Table 1.3 O-ring thermal distress data. r is the number of field-joint O-rings showing thermal distress out of 6, for a launch at the given temperature (°F) and pressure (pounds per square inch) (Dalal *et al.*, 1989).

Flight	Date	Number of O-rings with thermal distress, r	Temperature (°F) x_1	Pressure (psi) x_2
1	21/4/81	0	66	50
2	12/11/81	1	70	50
3	22/3/82	0	69	50
5	11/11/82	0	68	50
6	4/4/83	0	67	50
7	18/6/83	0	72	50
8	30/8/83	0	73	100
9	28/11/83	0	70	100
41-B	3/2/84	1	57	200
41-C	6/4/84	1	63	200
41-D	30/8/84	1	70	200
41-G	5/10/84	0	78	200
51-A	8/11/84	0	67	200
51-C	24/1/85	2	53	200
51-D	12/4/85	0	67	200
51-B	29/4/85	0	75	200
51-G	17/6/85	0	70	200
51-F	29/7/85	0	81	200
51-I	27/8/85	0	76	200
51-J	3/10/85	0	79	200
61-A	30/10/85	2	75	200
61-B	26/11/86	0	76	200
61-C	21/1/86	1	58	200
61-I	28/1/86	—	31	200

Figure 1.3 O-ring thermal distress data. The left panel shows the proportion of incidents as a function of joint temperature, and the right panel shows the corresponding plot against pressure. The x-values have been jittered to avoid overplotting multiple points. The solid lines show the fitted proportions of failures under a model described in Chapter 4.

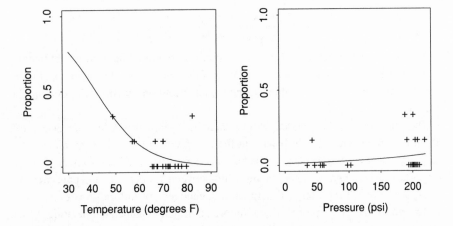

Table 1.3 shows the temperatures x_1 and test pressures x_2 associated with thermal distress of the O-rings for flights before the disaster. The pattern becomes clearer when the proportion of failures, r/m, is plotted against temperature and pressure in Figure 1.3. As temperature decreases, r/m appears to increase. There is less pattern in the corresponding plot for pressure.

Years of smoking t	Daily cigarette consumption d						
	Nonsmokers	1–9	10–14	15–19	20–24	25–34	35+
15–19	10366/1	3121	3577	4317	5683	3042	670
20–24	8162	2937	3286/1	4214	6385/1	4050/1	1166
25–29	5969	2288	2546/1	3185	5483/1	4290/4	1482
30–34	4496	2015	2219/2	2560/4	4687/6	4268/9	1580/4
35–39	3512	1648/1	1826	1893	3646/5	3529/9	1336/6
40–44	2201	1310/2	1386/1	1334/2	2411/12	2424/11	924/10
45–49	1421	927	988/2	849/2	1567/9	1409/10	556/7
50–54	1121	710/3	684/4	470/2	857/7	663/5	255/4
55–59	826/2	606	449/3	280/5	416/7	284/3	104/1

Table 1.4 Lung cancer deaths in British male physicians (Frome, 1983). The table gives man-years at risk/number of cases of lung cancer, cross-classified by years of smoking, taken to be age minus 20 years, and number of cigarettes smoked per day.

For these data, the response variable takes one of the values $0, 1, \ldots, 6$, with fairly strong dependence on temperature and possibly weaker dependence on pressure. If we assume that at a given temperature and pressure, each of the six rings fails independently with equal probability, we can treat the number of failures R as binomial with denominator m and probability π,

$$\Pr(R = r) = \frac{m!}{r!(m-r)!}\pi^r(1-\pi)^{m-r}, \quad r = 0, 1, \ldots, m, \ 0 < \pi < 1. \quad (1.6)$$

One possible relation between temperature x_1, pressure x_2, and the probability of failure is $\pi = \beta_0 + \beta_1 x_1 + \beta_2 x_2$, where the parameters β_0, β_1, and β_2 must be derived from the data. This has the drawback of predicting probabilities outside the range [0, 1] for certain values of x_1 and x_2. It is more satisfactory to use a function such as

$$\pi = \frac{\exp(\beta_0 + \beta_1 x_1 + \beta_2 x_2)}{1 + \exp(\beta_0 + \beta_1 x_1 + \beta_2 x_2)},$$

so $0 < \pi < 1$ wherever $\beta_0 + \beta_1 x_1 + \beta_2 x_2$ roams in the real line. It turns out that the function $e^u/(1 + e^u)$, the logistic distribution function, has an elegant connection to the binomial density, but any other continuous distribution function with domain the real line might be used.

The night before the Challenger was launched, there was a lengthy discussion about how the O-rings might behave at the low predicted launch temperature. One approach, which was not taken, would have been to try and predict how many O-rings might fail based on an estimated relationship between temperature and pressure. The lines in Figure 1.3 represent the estimated dependence of failure probability on x_1 and x_2, and show a high probability of failure at the actual launch temperature. When this is used as input to a probability model of how failures occur, the probability of catastrophic failure for a launch at 31°F is estimated to be as high as 0.16. To obtain this estimate involves extrapolation outside the available data, but there would have been little alternative in the circumstances of the launch. ∎

Example 1.4 (Lung cancer data) Table 1.4 shows data on the lung cancer mortality of cigarette smokers among British male physicians. The table shows the man-years

Figure 1.4 Lung cancer
deaths in British male
physicians. The figure
shows the rate of deaths
per 1000 man-years at
risk, for each of three
levels of daily cigarette
consumption.

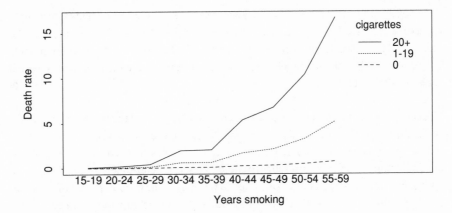

Figure 1.4 Lung cancer deaths in British male physicians. The figure shows the rate of deaths per 1000 man-years at risk, for each of three levels of daily cigarette consumption.

at risk and the number of cases with lung cancer, cross-classified by the number of years of smoking, taken to be age minus twenty years, and the number of cigarettes smoked daily. The man-years at risk in each category is the total period for which the individuals in that category were at risk of death.

As the eye moves from top left to the bottom right of the table, the figures suggest that death rate increases with increased total cigarette consumption. This is confirmed by Figure 1.4, which shows the death rate per 100,000 man-years at risk, grouped by three levels of cigarette consumption. Data for the first two groups show that death rate for smokers increases with cigarette consumption and with years of smoking. The only nonsmoker deaths are one in the age-group 35–39 and two in the age-group 75–79.

In this problem the aspect of primary interest is how death rate depends on cigarette consumption and smoking, and we treat the number of deaths in each category as the response. To build a model, we suppose that the death rate for those smoking d cigarettes per day after t years of smoking is $\lambda(d, t)$ deaths per man-year. Thus we may imagine deaths occurring at random in the total T man-years at risk in that category, at rate $\lambda(d, t)$. If deaths are independent point events in a continuum of length T, the number of deaths, Y, will have approximately a Poisson density with mean $T\lambda(d, t)$,

$$\Pr(Y = y) = \frac{\{T\lambda(d, t)\}^y}{y!} \exp\{-T\lambda(d, t)\}, \quad y = 0, 1, 2, \ldots. \quad (1.7)$$

One possible form for the mean deaths per man-year is

$$\lambda(d, t) = \beta_0 t^{\beta_1} \left(1 + \beta_2 d^{\beta_3}\right), \quad (1.8)$$

based on a deterministic argument and used in animal cancer mortality studies. In (1.8) there are four unknown parameters, and power-law dependence of death rate on exposure duration, t, and cigarette consumption, d. We expect that all the parameters β_r are positive. The background death-rate in the absence of smoking is given by $\beta_0 t^{\beta_1}$, the death-rate for nonsmokers. This represents the overall effect of other causes of lung cancer.

Expressions (1.7) and (1.8) give the random and systematic components for a simple model for the data, based on a blend of stochastic and deterministic arguments. An increasingly important development in statistics is the use of very complex models for real-world phenomena. Stochastic processes often provide the blocks with which such models are built.   ■

There is an important difference between Example 1.4 and the previous examples. In Example 1.1, Darwin could decide which plants to cross and where to plant them, in Example 1.2 the springs could be allocated to different stresses by the experimenter, and in Example 1.3 the test pressure for field joints was determined by engineers. The engineers would have no control over the temperature at the proposed time of a launch, but they could decide whether or not to launch at a given temperature. In each case, the allocation of treatments could in principle be controlled, albeit to different extents. Such situations, called *controlled experiments*, often involve a random allocation of treatments — type of fertilization, level of stress or test pressure — to units — plants, springs, or flights. Strong conclusions can in principle be drawn when randomization is used — though it played no part in Examples 1.1 or 1.3, and we do not know about Example 1.2.

In Example 1.4, however, a new problem rears its head. There is no question of allocating a level of cigarette consumption over a given period to individuals — the practical difficulties would be insuperable, quite apart from ethical considerations. In common with many other epidemiological, medical, and environmental studies, the data are *observational*, and this limits what conclusions may be drawn. It might be postulated that propensities to smoking and to lung cancer were genetically related, causing the apparent dependence in Table 1.4. Then for an individual to stop smoking would not reduce their chance of contracting lung cancer. In such cases data of different types from different sources must be gathered and their messages carefully collated and interpreted in order to put together an unambiguous story.

Despite differences in interpretation, the use of probability models to summarize variability and express uncertainty is the basis of each example. It is the subject of this book.

Outline

The idea of treating data as outcomes of random variables has implications for how they should be treated. For example, graphical and numerical summaries of the observations will show variation, and it is important to understand its consequences. Chapter 2 is devoted to this. It deals with basic ideas such as parameters, statistics, and sampling variation, simple graphs and other summary quantities, and then turns to notions of convergence, which are essential for understanding variability in large samples and generating approximations for small ones. Many statistics are based on quantities such as the largest item in a sample, and order statistics are also discussed. The chapter finishes with an account of moments and cumulants.

Variation in observed data leads to uncertainty about the reality behind it. Uncertainty is a more complicated notion, because it entails considering what it is reasonable to infer from the data, and people differ in what they find reasonable. Chapter 3 explains one of the main approaches to expressing uncertainty, leading to the construction of confidence intervals via quantities known as pivots. In most cases these can only be approximate, but they are often exact for models based on the normal distribution, which are then described. The chapter ends with a brief account of Monte Carlo simulation, which is used both to appreciate variability and to assess uncertainty.

In some cases information about model parameters θ can be expressed as a density $\pi(\theta)$, separate from the data y. Then the prior uncertainty $\pi(\theta)$ may be updated to posterior uncertainty $\pi(\theta \mid y)$ using Bayes' theorem

Thomas Bayes
(1702–1761) was a
nonconformist minister
and also a mathematician.
His theorem is contained
in his *Essay towards*
solving a problem in the
doctrine of chances, found
in his papers after his
death and published in
1764.

$$\pi(\theta \mid y) = \frac{\pi(\theta)f(y \mid \theta)}{f(y)},$$

which converts the conditional density $f(y \mid \theta)$ of observing data y, given that the true parameter is θ, into a conditional density for θ, given that y has been observed. This Bayesian approach to inference is attractive and conceptually simple, and modern computing techniques make it feasible to apply it to many complex models. However many statisticians do not agree that prior knowledge can or indeed should always be expressed as a prior density, and believe that information in the data should be kept separate from prior beliefs, preferring to base inference on the second term $f(y \mid \theta)$ in the numerator of Bayes' theorem, known as the likelihood.

Likelihood is a central idea for parametric models, and it and its ramifications are described in Chapter 4. Definitions of likelihood, the maximum likelihood estimator and information are followed by a discussion of inference based on maximum likelihood estimates and likelihood ratio statistics. The chapter ends with brief accounts of non-regular models and model selection.

Chapters 5 and 6 describe some particular classes of models. Accounts are given of the simplest form of linear model, of exponential family and group transformation models, of models for survival and missing data, and of those with more complex dependence structures such as Markov chains, Markov random fields, point processes, and the multivariate normal distribution.

Chapter 7 discusses more traditional topics of mathematical statistics, with a more general treatment of point and interval estimation and testing than in the previous chapters. It also includes an account of estimating functions, which are needed subsequently.

Regression models describe how a response variable, treated as random, depends on explanatory variables, treated as fixed. The vast majority of statistical modelling involves some form of regression, and three chapters of the book are devoted to it. Chapter 8 describes the linear model, including its basic properties, analysis of variance, model building, and variable selection. Chapter 9 discusses the ideas underlying the use of randomization and designed experiments, and closes with an account of mixed effect models, in which some parameters are treated as random. These two chapters are largely devoted to the classical linear model, in which the responses are supposed normally distributed, but since around 1970 regression modelling has greatly broadened. Chapter 10 is devoted to nonlinear models. It starts with an account of likelihood estimation using the iterative weighted least squares algorithm, which subsequently plays a unifying role and then describes generalized linear models, binary data and loglinear models, semiparametric regression by local likelihood estimation and by penalized likelihood. It closes with an account of regression modelling of survival data.

Bayesian statistics is discussed in Chapter 11, starting with discussion of the role of prior information, followed by an account of Bayesian analogues of procedures developed in the earlier chapters. This is followed by a brief overview of Bayesian computation, including Laplace approximation, the Gibbs sampler and Metropolis–Hastings algorithm. The chapter closes with discussion of hierarchical and empirical Bayes and a very brief account of decision theory.

Likelihood is a favourite tool of statisticians but sometimes gives poor inferences. Chapter 12 describes some reasons for this, and outlines how conditional or marginal likelihoods can give better procedures.

The main links among the chapters of this book are shown in Figure 1.5.

Notation

The notation used in this book is fairly standard, but there are not enough letters in the Roman and Greek alphabets for total consistency. Greek letters generally denote parameters or other unknowns, with α largely reserved for error rates and confidence levels in connection with significance tests and confidence sets. Roman letters X, Y, Z, and so forth are mainly used for random variables, which take values x, y, z.

Probability, expectation, variance, covariance, and correlation are denoted $\Pr(\cdot)$, $\mathrm{E}(\cdot)$, $\mathrm{var}(\cdot)$ $\mathrm{cov}(\cdot, \cdot)$, and $\mathrm{corr}(\cdot, \cdot)$, while $\mathrm{cum}(\cdot, \cdot, \cdots)$ is occasionally used to denote a cumulant. We use $I(A)$ to denote the indicator random variable, which equals 1 if the event A occurs and 0 otherwise. A related function is the Heaviside function

$$H(u) = \begin{cases} 0, & u < 0, \\ 1, & u \geq 0, \end{cases}$$

whose generalized derivative is the Dirac delta function $\delta(u)$. This satisfies

$$\int \delta(y - u)g(u)\, du = g(y)$$

for any function g.

The Kronecker delta symbols δ_{rs}, δ_{rst}, and so forth all equal unity when all their subscripts coincide, and equal zero otherwise.

We use $\lfloor x \rfloor$ to denote the largest integer smaller than or equal to x, and $\lceil x \rceil$ to denote the smallest integer larger than or equal to x.

The symbol $\equiv$ indicates that constants have been dropped in defining a log likelihood, while $\doteq$ means 'approximately equals'. The symbols $\sim$, $\overset{\cdot}{\sim}$ $\overset{\text{ind}}{\sim}$, and $\overset{\text{iid}}{\sim}$ are

Figure 1.5 A map of the main dependencies among chapters of this book. A solid line indicates strong dependence and a dashed line indicates partial dependence through the given subsections.

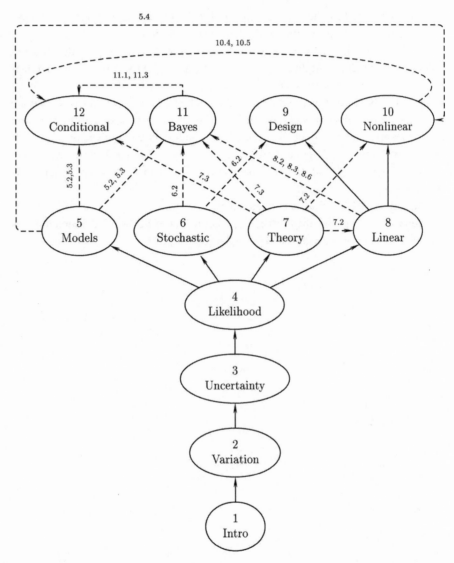

shorthand for 'is distributed as', 'is approximately distributed as', 'are independently distributed as', and 'are independent and identically distributed as', while $\stackrel{D}{=}$ means 'has the same distribution as'. $X \perp Y$ means 'X is independent of Y'. We use $\stackrel{D}{\longrightarrow}$ and $\stackrel{P}{\longrightarrow}$ to denote convergence in distribution and in probability. To say that $Y_1, \ldots, Y_n$ are a random sample from some distribution means that they are independent and identically distributed according to that distribution.

We mostly reserve Z for standard normal random variables. As usual $N(\mu, \sigma^2)$ represents the normal distribution with mean μ and variance σ^2. The standard normal cumulative distribution and density functions are denoted Φ and ϕ. We use $c_\nu(\alpha)$, $t_\nu(\alpha)$, and $F_{\nu_1, \nu_2}(\alpha)$ to denote the α quantiles of the chi-squared distribution, Student t distribution with ν degrees of freedom, and F distribution with ν_1 and ν_2 degrees of freedom, while $U(0, 1)$ denote the uniform distribution on the unit interval. Almost everywhere, z_α is the α quantile of the $N(0, 1)$ distribution.

The data values in a sample of size n, typically denoted $y_1, \ldots, y_n$, are the observed values of the random variables $Y_1, \ldots, Y_n$; their average is $\overline{y} = n^{-1} \sum y_j$ and their sample variance is $s^2 = (n - 1)^{-1} \sum (y_j - \overline{y})^2$.

We avoid boldface type, and rely on the context to make it plain when we are dealing with vectors or matrices; a^T denotes the matrix transpose of a vector or matrix a. The identity matrix of side n is denoted I_n, and 1_n is a $n \times 1$ vector of ones. If θ is a $p \times 1$ vector and $\ell(\theta)$ a scalar, then $\partial \ell(\theta)/\partial \theta$ is the $p \times 1$ vector whose rth element is $\partial \ell(\theta)/\partial \theta_r$, and $\partial^2 \ell(\theta)/\partial \theta \partial \theta^\mathsf{T}$ is the $p \times p$ matrix whose (r, s) element is $\partial^2 \ell(\theta)/\partial \theta_r \partial \theta_s$.

The end of each example is marked thus: ■

Exercise 2.1.3 denotes the third exercise at the end of Section 2.1, Problem 2.3 is the third problem at the end of Chapter 2, and so forth.

2

Variation

The key idea in statistical modelling is to treat the data as the outcome of a random experiment. The purpose of this chapter is to understand some consequences of this: how to summarize and display different aspects of random data, and how to use results of probability theory to appreciate the variation due to this randomness. We outline the elementary notions of statistics and parameters, and then describe how data and statistics derived from them vary under sampling from statistical models. Many quantities used in practice are based on averages or on ordered sample values, and these receive special attention. The final section reviews moments and cumulants, which will be useful in later chapters.

2.1 Statistics and Sampling Variation

2.1.1 Data summaries

The most basic element of data is a single observation, y — usually a number, but perhaps a letter, curve, or image. Throughout this book we shall assume that whatever their original form, the data can be recoded as numbers. We shall mostly suppose that single observations are scalar, though sometimes they are vectors or matrices.

We generally deal with an ensemble of n observations, $y_1, \ldots, y_n$, known as a *sample*. Occasionally interest centres on the given sample alone, and if n is not tiny it will be useful to summarize the data in terms of a few numbers. We say that a quantity $s = s(y_1, \ldots, y_n)$ that can be calculated from $y_1, \ldots, y_n$ is a *statistic*. Such quantities may be wanted for many different purposes.

Location and scale

Two basic features of a sample are its typical value and a measure of how spread out the sample is, sometimes known respectively as *location* and *scale*. They can be summarized in many ways.

Example 2.1 (Sample moments) Sample moments are calculated by putting mass n^{-1} on each of the y_j, and then calculating the mean, variance, and so forth. The simplest of these sample moments are

$$\overline{y} = \frac{1}{n} \sum_{j=1}^{n} y_j = \frac{1}{n}(y_1 + \cdots + y_n) \quad \text{and} \quad \frac{1}{n} \sum_{j=1}^{n}(y_j - \overline{y})^2;$$

we call the first of these the *average*. In practice the denominator n in the second moment is usually replaced by $n - 1$, giving the *sample variance*

$$s^2 = \frac{1}{n-1} \sum_{j=1}^{n}(y_j - \overline{y})^2. \tag{2.1}$$

The denominator $n - 1$ is justified in Example 2.14.

Here $\overline{y}$ and s have the same dimensions as the y_j, and are measures of location and scale respectively.   ∎

Potential confusion is avoided by using the word *average* to refer to a quantity calculated from data, and the words *mean* or *expectation* for the corresponding theoretical quantity; this convention is used throughout this book.

Example 2.2 (Order statistics)   The *order statistics* of $y_1, \ldots, y_n$ are their values put in increasing order, which we denote $y_{(1)} \leq y_{(2)} \leq \cdots \leq y_{(n)}$. If $y_1 = 5$, $y_2 = 2$ and $y_3 = 4$, then $y_{(1)} = 2$, $y_{(2)} = 4$ and $y_{(3)} = 5$. Examples of order statistics are the *sample minimum* $y_{(1)}$ and *sample maximum* $y_{(n)}$, and the lower and upper *quartiles* $y_{(\lceil n/4 \rceil)}$ and $y_{(\lceil 3n/4 \rceil)}$. The lowest quarter of the sample lies below the lower quartile, and the highest quarter lies above the upper quartile.

$\lceil u \rceil$ denotes the smallest integer greater than or equal to u.

Among statistics that can be based on the $y_{(j)}$ are the *sample median*, defined as

$$\text{median}(y_j) = \begin{cases} y_{((n+1)/2)}, & n \text{ odd,} \\ \frac{1}{2}\left(y_{(n/2)} + y_{(n/2+1)}\right), & n \text{ even.} \end{cases} \tag{2.2}$$

This is the centre of the sample: equal proportions of the data lie above and below it.

All these statistics are examples of *sample quantiles*. The pth sample quantile is the value with a proportion p of the sample to its left. Thus the minimum, maximum, quartiles, and median are (roughly) the $0, 1, 0.25, 0.75$ and 0.5 sample quantiles. Like the median (2.2) when n is even, the pth sample quantile for non-integer pn is usually calculated by linear interpolation between the order statistics that bracket it.

Another measure of location is the average of the central observations of the sample. Suppose that p lies in the interval $[0, 0.5)$, and that $k = pn$ is an integer. Then the $p \times 100\%$ *trimmed average* is defined as

$$\frac{1}{n - 2k} \sum_{j=k+1}^{n-k} y_{(j)},$$

which is the usual average $\overline{y}$ when $p = 0$. The 50% trimmed average ($p = 0.5$) is defined to be the median, while other values of p interpolate between the average and the median. Linear interpolation is used when pn is non-integer.

The statistics above measure different aspects of sample location. Some measures of scale based on the order statistics are the *range*, $y_{(n)} - y_{(1)}$, the *interquartile range* and the *median absolute deviation*,

$$\mathrm{IQR} = y_{(\lceil 3n/4 \rceil)} - y_{(\lceil n/4 \rceil)}, \quad \mathrm{MAD} = \mathrm{median}\{|y_i - \mathrm{median}(y_j)|\}.$$

These are, respectively, the difference between the largest and smallest observations, the difference between the observations at the ends of the central 50% of the sample, and the median of the absolute deviations of the observations from the sample median. One would expect the range of a sample to grow with its size, but the IQR and MAD should depend less on the sample size and in this sense are more stable measures of scale. ∎

It is easy to establish that the mapping $y_1, \ldots, y_n \mapsto a + by_1, \ldots, a + by_n$ changes the values of location and scale measures in the previous examples by $m, s \mapsto a + bm, bs$ (Exercise 2.1.1); this seems entirely reasonable.

Bad data

The statistics described in Examples 2.1 and 2.2 measure different aspects of location and of scale. They also differ in their susceptibility to bad data. Consider what happens when an error, due perhaps to mistyping, results in an observation that is unusual compared to the others — an *outlier*. If the 'true' y_1 is replaced by $y_1 + \delta$, the average changes from $\bar{y}$ to $\bar{y} + n^{-1}\delta$, which could be arbitrarily large, while the sample median changes by a bounded amount — the most that can happen is that it moves to an adjacent observation. We say that the sample median is *resistant*, while the average is not. Roughly a quarter of the data would have to be contaminated before the interquartile range could change by an arbitrarily large amount, while the range and sample variance are sensitive to a single bad observation. The large-sample proportion of contaminated observations needed to change the value of a statistic by an arbitrarily large amount is called its *breakdown point*; it is a common measure of the resistance of a statistic.

Example 2.3 (Birth data) Table 2.1 shows data extracted from a census of all the women who arrived to give birth at the John Radcliffe Hospital in Oxford during a three-month period. The table gives the times that women with vaginal deliveries — that is, without caesarian section — spent in the delivery suite, for the first seven of 92 successive days of data.

The initial step in dealing with data is to scrutinize them closely, and to understand how they were collected. In this case the time for each birth was recorded by the midwife who attended it, and numerous problems might have arisen in the recording. For example, one midwife might intend 4.20 to mean 4.2 hours, but another might mean 4 hours and 20 minutes. Moreover it is difficult to believe that a time can be known as exactly as 2 hours and 6 minutes, as would be implied by the value 2.10. Furthermore, there seems to be a fair degree of rounding of the data. In fact the data collection form was carefully prepared, and the midwives were trained in how to compile it, so the data are of high quality. Nevertheless it is important always to ask how the data were collected, and if possible to see the process at work.

Ideally the statistician assists in deciding what data are collected, and how.

Woman	Day 1	2	3	4	5	6	7
1	2.10	4.00	2.60	1.50	2.50	4.00	2.00
2	3.40	4.10	3.60	4.70	2.50	4.00	2.70
3	4.25	5.00	3.60	4.70	3.40	5.25	2.75
4	5.60	5.50	6.40	7.20	4.20	6.10	3.40
5	6.40	5.70	6.80	7.25	5.90	6.50	4.20
6	7.30	6.50	7.50	8.10	6.25	6.90	4.30
7	8.50	7.25	7.50	8.50	7.30	7.00	4.90
8	8.75	7.30	8.25	9.20	7.50	8.45	6.25
9	8.90	7.50	8.50	9.50	7.80	9.25	7.00
10	9.50	8.20	10.40	10.70	8.30	10.10	9.00
11	9.75	8.50	10.75	11.50	8.30	10.20	9.25
12	10.00	9.75	14.25		10.25	12.75	10.70
13	10.40	11.00	14.50		12.90	14.60	
14	10.40	11.20			14.30		
15	16.00	15.00					
16	19.00	16.50					

Table 2.1 Seven successive days of times (hours) spent by women giving birth in the delivery suite at the John Radcliffe Hospital. (Data kindly supplied by Ethel Burns.)

The average of the $n = 95$ times in Table 2.1 is $\overline{y} = 7.57$ hours. The variance of the time spent in the delivery suite can be estimated by the sample variance, $s^2 = 12.97$ squared hours. The minimum, median, and maximum are 1.5, 7.5 and 19 hours respectively, and the quartiles are 4.95 and 9.75 hours. The 0.2 and 0.4 trimmed averages, 7.48 and 7.55 hours, are similar to $\overline{y}$ because there are no gross outliers. ∎

Shape

The shape of a sample is also important. For example, the upper tails of annual income distributions are typically very fat, because a few individuals earn enormously more than most of us. The shape of such a distribution can be used to assess inequality, for example by considering the proportion of individuals whose annual income is less than one-half the median. Since shape does not depend on location or scale, statistics intended to summarize it should be invariant to location and scale shifts of the data.

Example 2.4 (Sample skewness) One measure of shape is the *standardized sample skewness*,

$$g_1 = \frac{n^{-1} \sum_{j=1}^{n} (y_j - \overline{y})^3}{\left\{ (n-1)^{-1} \sum_{j=1}^{n} (y_j - \overline{y})^2 \right\}^{3/2}}.$$

If the data are perfectly symmetric, $g_1 = 0$, while if they have a heavy upper tail, $g_1 > 0$, and conversely. For the times in the delivery suite, $g_1 = 0.65$: the data are somewhat skewed to the right. ∎

Example 2.5 (Sample shape) Measures of shape can also be based on the sample quantiles. One is $(y_{(\lceil 0.95n \rceil)} - y_{(\lceil 0.5n \rceil)})/(y_{(\lceil 0.5n \rceil)} - y_{(\lceil 0.05n \rceil)})$, which takes value one for a symmetric distribution, and is more resistant to outliers than is the sample skewness.

For the times in the delivery suite, this is 1.43, again showing skewness to the right. A value less than one would indicate skewness to the left. ∎

It is straightforward to show that both these statistics are invariant to changes in the location and scale of $y_1, \ldots, y_n$.

Graphs

Graphs are indispensable in data analysis, because the human visual system is so good at recognizing patterns that the unexpected can leap out and hit the investigator between the eyes. An adverse effect of this ability is that patterns may be imagined even when they are absent, so experience, often aided by suitable statistics, is needed to interpret a graph. As any plot can be represented numerically, it too is a statistic, though to treat it merely as a set of numbers misses the point.

This can lead to inter-ocular trauma.

Example 2.6 (Histogram) Perhaps the best-known statistical graph is the *histogram*, constructed from scalar data by dividing the horizontal axis into disjoint bins — the intervals $I_1, \ldots, I_K$ — and then counting the observations in each. Let n_k denote the number of observations in I_k, for $k = 1, \ldots, K$, so $\sum_k n_k = n$. If the bins have equal width δ, then $I_k = [L + (k-1)\delta, L + k\delta)$, where L, δ, and K are chosen so that all the y_j lie between L and $L + K\delta$. We then plot the proportion n_k/n of the data in each bin as a column over it, giving the probability density function for a discretized version of the data.

The upper left panel of Figure 2.1 shows this for the birth data in Table 2.1, with $L = 0$, $\delta = 2$, and $K = 13$; the *rug* of tickmarks shows the data values themselves. As we would expect from Examples 2.4 and 2.5, the plot shows a density skewed to the right, with the most popular values in the range 5–10 hours. To increase δ would give fewer, wider, bins, while decreasing δ would give more, narrower, bins. It might be better to vary the bin width, with narrower bins in the centre of the data, and wider ones at the tails. ∎

Example 2.7 (Empirical distribution function) The *empirical distribution function* (EDF) is the cumulative probability distribution that puts probability n^{-1} at each of $y_1, \ldots, y_n$. This is expressed mathematically as

$$n^{-1} \sum_{j=1}^{n} H(y - y_j), \tag{2.3}$$

where the distribution function that puts mass one at $u = 0$, that is,

$$H(u) = \begin{cases} 0, & u < 0, \\ 1, & u \geq 0, \end{cases}$$

is known as the Heaviside function. The EDF is a step function that jumps by n^{-1} at each of the y_j; of course it jumps by more at values that appear in the sample several times.

The upper right panel of Figure 2.1 shows the EDF of the times in the delivery suite. It is more detailed than the histogram, but perhaps conveys less information about the

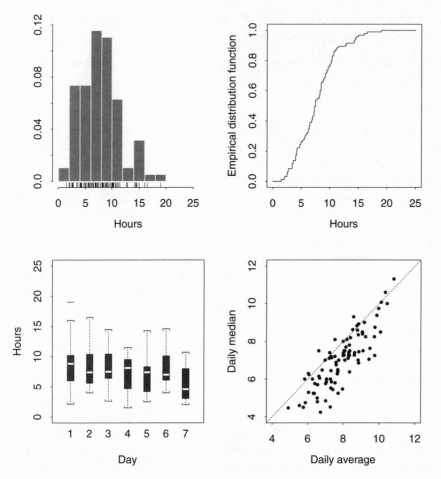

Figure 2.1 Summary
plots for times in the
delivery suite, in hours.
Clockwise from top left:
histogram, with rug
showing values of
observations; empirical
distribution function;
scatter plot of daily
average hours against
daily median hours, for all
92 days of data, with a
line of unit slope through
the origin; and boxplots
for the first seven days.

shape of the data. Which is preferable is partly a matter of taste, and depends on the
use to which they will be put.   ∎

Example 2.8 (Scatterplot) When an observation has two components, $y_j = (u_j, v_j)$, a *scatter plot* is a plot of the v_j on the vertical axis against the u_j on the
horizontal axis. An example is given in the lower right panel of Figure 2.1, which
shows the median daily time in the delivery suite plotted against the average daily
time, for the full 92 days for which data are available. As most points lie below the
line with unit slope, and as the slope of the point cloud is slightly greater than one,
the medians are generally smaller and somewhat more variable than the averages. The
average and sample variance of the medians are 7.03 hours and 2.15 hours squared;
the corresponding figures for the averages are 7.90 and 1.54.   ∎

Example 2.9 (Boxplot) Boxplots are usually used to compare related sets of data.
An illustration is in the lower left panel of Figure 2.1, which compares the hours in
the delivery suite for the seven different days in Table 2.1. For each day, the ends
of the central box show the quartiles and the white line in its centre represents the

daily median: thus about one-half of the data lie in the box, and its length shows the interquartile range IQR for that day. The bracket above the box shows the largest observation less than or equal to the upper quartile plus 1.5IQR. Likewise the bracket below shows the smallest observation greater than or equal to the lower quartile minus 1.5IQR. Values outside the brackets are plotted individually. The aim is to give a good idea of the location, scale, and shape of the data, and to show potential outliers clearly, in order to facilitate comparison of related samples. Here, for example, we see that the daily median varies from 5–10 hours, and that the daily IQR is fairly stable. ■

It takes thought to make good graphs. Some points to bear in mind are:

- the data should be made to stand out, in particular by avoiding so-called *chart-junk* — unnecessary labels, lines, shading, symbols and so forth;
- the axis labels and caption should make the graph as self-explanatory as possible, in particular containing the names and units of measurement of variables;
- comparison of related quantities should be made easy, for example by using identical scales of measurement, and placing plots side by side;
- scales should be chosen so that the most important systematic relations between variables are at about 45° to the axes;
- the *aspect ratio* — the ratio of the height of a plot to its width — can be varied to highlight different features of the data;
- graphs should be laid out so that departures from 'standard' appear as departures from linearity or from random scatter; and
- major differences in the precision of points should be indicated, at least roughly.

Nowadays it is easy to produce graphs, but unfortunately even easier to produce bad ones: there is no substitute for drafting and redrafting each graph to make it as clear and informative as possible.

2.1.2 Random sample

So far we have supposed that the sample $y_1, \ldots, y_n$ is of interest for its own sake. In practice, however, data are usually used to make inferences about the system from which they came. One reason for gathering the birth data, for example, was to assess how the delivery suite should be staffed, a task that involves predicting the patterns with which women will arrive to give birth, and how long they are likely to stay in the delivery suite once they are there. Though it is not useful to do this for births that have already occurred, the data available can help in making predictions, provided we can forge a link between the past and future. This is one use of a statistical model.

The fundamental idea of statistical modelling is to treat data as the observed values of random variables. The most basic model is that the data $y_1, \ldots, y_n$ available are the observed values of a *random sample of size n*, defined to be a collection of n independent identically distributed random variables, $Y_1, \ldots, Y_n$. We suppose that each of the Y_j has the same cumulative distribution function, F, which represents the population from which the sample has been taken. If F were known, we could in

Perception experiments have shown that the eye is best at judging departures from 45°.

Or sometimes a simple random sample.

principle use the rules of probability calculus to deduce any of its properties — such as its mean and variance, or the probability distribution for a future observation — and any difficulties would be purely computational. In practice, however, F is unknown, and we must try to infer its properties from the data. Often the quantity of central interest is a nonrandom function of F, such as its mean or its p quantile,

$$E(Y) = \int y \, dF(y), \quad y_p = F^{-1}(p) = \inf\{y : F(y) \geq p\}; \tag{2.4}$$

We use $dF(y)$ to accommodate the possibility that F is discrete. If it bothers you, take $dF(y) = f(y) \, dy$.

these are the population analogues of the sample average and quantiles defined in Examples 2.1 and 2.2. Often there is a simple form for F^{-1} and the infimum is unnecessary. Other population quantities such as the interquartile range, $F^{-1}(\frac{3}{4}) - F^{-1}(\frac{1}{4})$, are defined similarly.

Example 2.10 (Laplace distribution)   A random variable Y for which

$$f(y; \eta, \tau) = \frac{1}{2\tau} \exp\left(-|y - \eta|/\tau\right), \quad -\infty < y < \infty, \ -\infty < \eta < \infty, \tau > 0, \tag{2.5}$$

is said to have the Laplace distribution. As $f(\eta + u; \eta, \tau) = f(\eta - u; \eta, \tau)$ for any u, the density is symmetric about η. Its integral is clearly finite, so $E(Y) = \eta$, and evidently its median $y_{0.5} = \eta$ also. Its variance is

$$\mathrm{var}(Y) = \frac{1}{2\tau} \int_{-\infty}^{\infty} (y - \eta)^2 \exp\left(-|y - \eta|/\tau\right) dy = \tau^2 \int_0^{\infty} u^2 e^{-u} \, du = 2\tau^2,$$

Pierre-Simon Laplace (1749–1827) helped establish the metric system during the French Revolution but was dismissed by Napoleon 'because he brought the spirit of the infinitely small into the government' — presumably Bonaparte was unimpressed by differentiation. Laplace worked on celestial mechanics, published an important book on probability, and derived the least squares rule.

as follows after the substitution $u = (y - \eta)/\tau$ and integration by parts; see Exercise 2.1.3. Integration of (2.5) gives

$$F(y) = \begin{cases} \frac{1}{2} \exp\{(y - \eta)/\tau\}, & y \leq \eta, \\ 1 - \frac{1}{2} \exp\{-(y - \eta)/\tau\}, & y > \eta, \end{cases}$$

so

$$F^{-1}(p) = \begin{cases} \eta + \tau \log(2p), & p < \frac{1}{2}, \\ \eta - \tau \log\{2(1 - p)\}, & p \geq \frac{1}{2}, \end{cases}$$

the interquartile range is

$$F^{-1}\left(\frac{3}{4}\right) - F^{-1}\left(\frac{1}{4}\right) = \eta + \tau \log 2 - (\eta - \tau \log 2) = 2\tau \log 2,$$

and the median absolute deviation is $\tau \log 2$ (Exercise 2.1.5).   ∎

Quantities such as $E(Y)$, $\mathrm{var}(Y)$ and $F^{-1}(p)$ are called *parameters*, and as their values depend on F, they are typically unknown. If F is determined by a finite number of parameters, θ, the model is *parametric*, and we may write $F = F(y; \theta)$, with corresponding probability density function $f(y; \theta)$. Ignorance about F then boils down to uncertainty about θ.

It is natural to use sample quantities for inference about model parameters. Suppose that the data $Y_1, \ldots, Y_n$ are a random sample from a distribution F, that we are interested in a parameter θ that depends on F, and that we wish to use the statistic

We use the term probability density function to mean the density function for a continuous variable, and the mass function for a discrete variable, and use the notation $f(y; \theta)$ in both cases.

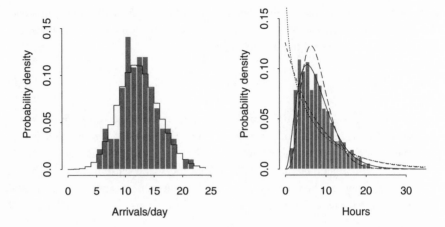

Figure 2.2 Comparisons of 92 days of delivery suite data with Poisson and gamma models. The left panel shows a histogram of the numbers of arrivals per day, with the PDF of the Poisson distribution with mean $\theta = 12.9$ overlaid. The right panel shows a histogram of the hours in the delivery suite for the 1187 births, with the PDFs of gamma distributions overlaid. The gamma distributions all have mean $\kappa/\lambda = 7.93$ hours. Their shape parameters are $\kappa = 3.15$ (solid), 0.8 (dots), 1 (small dashes), and 5 (large dashes).

$S = s(Y_1, \ldots, Y_n)$ to make inferences about θ, for example hoping that S will be close to θ. Then we call S an *estimator* of θ and say that the particular value that S takes when the observed data are $y_1, \ldots, y_n$, that is, $s = s(y_1, \ldots, y_n)$, is an *estimate* of θ. This is the usual distinction between a random variable and the value that it takes, here S and s.

Siméon Denis Poisson (1781–1840) learned mathematics in Paris from Laplace and Lagrange. He did major work on definite integrals, on Fourier series, on elasticity and magnetism, and in 1837 published an important book on probability.

Example 2.11 (Poisson distribution) The Poisson distribution with mean θ has probability density function

$$\Pr(Y = y) = f(y; \theta) = \frac{\theta^y}{y!} e^{-\theta}, \quad y = 0, 1, 2, \ldots, \quad \theta > 0. \tag{2.6}$$

This discrete distribution is used for count data. For example, the left panel of Figure 2.2 shows a histogram of the number of women arriving at the delivery suite for each of the 92 days of data, together with the probability density function (2.6) with $\theta = 12.9$, equal to the average number of arrivals over the 92 days. This distribution seems to fit the data more or less adequately. ∎

Example 2.12 (Gamma distribution) The gamma distribution with scale parameter λ and shape parameter κ has probability density function

$\Gamma(\kappa)$ is the *gamma function*; see Exercise 2.1.3 for some of its properties.

$$f(y; \lambda, \kappa) = \frac{\lambda^\kappa y^{\kappa-1}}{\Gamma(\kappa)} \exp(-\lambda y), \quad y > 0, \quad \lambda, \kappa > 0. \tag{2.7}$$

This distribution has mean κ/λ and variance κ/λ^2.

When $\kappa = 1$ the density is exponential, for $0 < \kappa < 1$ it is L-shaped, and for $\kappa > 1$ it falls smoothly on either side of its maximum. These shapes are illustrated in the right panel of Figure 2.2, which shows the hours in the delivery suite for the 1187 births that took place over the three months of data. In each case the mean of the density matches the data average of 7.93 hours; the value $\kappa = 3.15$ of the shape parameter was chosen to match the variance of the data by solving simultaneously the equations $\kappa/\lambda = 7.93$, $\kappa/\lambda^2 = 12.97$. Evidently the solid curve gives the best fit of those shown.

It is important to appreciate that the parametrization of F is not carved in stone. Here it might be better to rewrite (2.7) in terms of its mean $\mu = \kappa/\lambda$ and the shape parameter κ, in which case the density is expressed as

$$\frac{1}{\Gamma(\kappa)} \left(\frac{\kappa}{\mu}\right)^{\kappa} y^{\kappa-1} \exp(-\kappa y/\mu), \quad y > 0, \quad \mu, \kappa > 0, \tag{2.8}$$

with variance μ^2/κ. As functions of y the shapes of (2.7) and (2.8) are the same, but their expression in terms of parameters is not. The range of possible densities is the same for any 1–1 reparametrization of (κ, λ), so one might write the density in terms of two important quantiles, for example, if this made sense in the context of a particular application. The central issue in choice of parametrization is directness of interpretation in the situation at hand. ∎

Example 2.13 (Laplace distribution)   To express the Laplace density (2.5) in terms of its mean and variance η and $2\tau^2$, we set $\tau^2 = \sigma^2/2$, giving

$$\frac{1}{\sqrt{2}\sigma} \exp(-\sqrt{2}|y - \eta|/\sigma) \quad -\infty < y < \infty, \quad -\infty < \eta < \infty, \sigma > 0.$$

Its shape as a function of y is unchanged, but the new formula is uglier. ∎

2.1.3 Sampling variation

If the data $y_1, \ldots, y_n$ are regarded as the observed values of random variables, then it follows that the sample and any statistics derived from it might have been different. However, although we would expect variation over possible sets of data, we would also expect to see systematic patterns induced by the underlying model. For instance, having inspected the lower left panel of Figure 2.1, we would be surprised to be told that the median hours in the delivery suite on day 8 was 15 hours, though any value between 5 and 10 hours would seem quite reasonable. From a statistical viewpoint, data have both a random and a systematic component, and one common goal of data analysis is to disentangle these as far as possible. In order to understand the systematic aspect, it makes sense to ask how we would expect a statistic $s(y_1, \ldots, y_n)$ to behave on average, that is, to try and understand the properties of the corresponding random variable, $S = s(Y_1, \ldots, Y_n)$.

Example 2.14 (Sample moments)   Suppose that $Y_1, \ldots, Y_n$ is a random sample from a distribution with mean μ and variance σ^2. Then the average $\overline{Y}$ has expectation and variance

$$\mathrm{E}(\overline{Y}) = \mathrm{E}\left(\frac{1}{n}\sum_{j=1}^{n} Y_j\right) = \frac{n}{n}\mathrm{E}(Y_j) = \mu,$$

$$\mathrm{var}(\overline{Y}) = \mathrm{var}\left(\frac{1}{n}\sum_{j=1}^{n} Y_j\right) = \frac{1}{n^2}\sum_{j=1}^{n}\mathrm{var}(Y_j) = \frac{\sigma^2}{n},$$

because the Y_j are independent identically distributed random variables. Thus the expected value of the random variable $\overline{Y}$ is the population mean μ.

To find the expectation of the sample variance $S^2 = (n-1)^{-1} \sum_j (Y_j - \overline{Y})^2$, note that

$$
\begin{aligned}
\sum_{j=1}^{n} (Y_j - \overline{Y})^2 &= \sum_{j=1}^{n} \{Y_j - \mu - (\overline{Y} - \mu)\}^2 \\
&= \sum_{j=1}^{n} (Y_j - \mu)^2 - 2 \sum_{j=1}^{n} (Y_j - \mu)(\overline{Y} - \mu) + \sum_{j=1}^{n} (\overline{Y} - \mu)^2 \\
&= \sum_{j=1}^{n} (Y_j - \mu)^2 - 2n(\overline{Y} - \mu)^2 + n(\overline{Y} - \mu)^2 \\
&= \sum_{j=1}^{n} (Y_j - \mu)^2 - n(\overline{Y} - \mu)^2.
\end{aligned}
$$

As

$$
\begin{aligned}
E\{(n-1)S^2\} &= nE\{(Y_j - \mu)^2\} - nE\{(\overline{Y} - \mu)^2\} \\
&= n\sigma^2 - n\sigma^2/n \\
&= (n-1)\sigma^2,
\end{aligned}
$$

we see that S^2 has expected value σ^2. This explains our use of the denominator $n-1$ when defining the sample variance s^2 in (2.1): the expectation of the corresponding random variable equals the population variance.

The birth data of Table 2.1 have $n = 95$, and the realized values of the random variables $\overline{Y}$ and S^2 are $\overline{y} = 7.57$ and $s^2 = 12.97$. Thus $\overline{y}$ has estimated variance $s^2/n = 12.97/95 = 0.137$ and estimated standard deviation $0.137^{1/2} = 0.37$. This suggests that the underlying 'true' mean μ of the population of times spent in the delivery suite by women giving birth is close to 7.6 hours. ∎

Example 2.15 (Birth data)   Figure 2.2 suggests the following simple model for the birth data. Each day the number N of women arriving to give birth is Poisson with mean θ. The jth of these women spends a time Y_j in the delivery suite, where Y_j is a gamma random variable with mean μ and variance σ^2. The values of these parameters are $\theta \doteq 13$, $\mu \doteq 8$ hours and $\sigma^2 \doteq 13$ hours squared. The average time and median times spent, $\overline{Y} = N^{-1} \sum Y_j$ and M, vary from day to day, with the lower right panel of Figure 2.1 suggesting that $E(M) < E(\overline{Y})$ and $\mathrm{var}(M) > \mathrm{var}(\overline{Y})$, properties we shall see theoretically in Example 2.30. ∎

Much of this book is implicitly or explicitly concerned with distinguishing random and systematic variation. The notions of sampling variation and of a random sample are central, and before continuing we describe a useful tool for comparison of data and a distribution.

2.1.4 Probability plots

It is often useful to be able to check graphically whether data $y_1, \ldots, y_n$ come from a particular distribution. Suppose that in addition to the data we had a random sample $x_1, \ldots, x_n$ known to be from F. In order to compare the shapes of the samples, we could sort them to get $y_{(1)} \leq \cdots \leq y_{(n)}$ and $x_{(1)} \leq \cdots \leq x_{(n)}$, and make a *quantile-quantile* or *Q-Q plot* of $y_{(1)}$ against $x_{(1)}$, $y_{(2)}$ against $x_{(2)}$, and so forth. A straight line would mean that $y_{(j)} = a + b x_{(j)}$, so that the shape of the samples was identical, while distinct curvature would indicate systematic differences between them. If the line was close to straight, we could be fairly confident that $y_1, \ldots, y_n$ looks like a sample from F — after all, it would have a shape similar to the sample $x_1, \ldots, x_n$ which is from F.

Quantile-quantile plots are helpful for comparison of two samples, but when comparing a single sample with a theoretical distribution it is preferable to use F directly in a *probability plot*, in which the $y_{(j)}$ are graphed against the *plotting positions* $F^{-1}\{j/(n+1)\}$. This use of the $j/(n+1)$ quantile of F is justified in Section 2.3 as an approximation to $\mathrm{E}(X_{(j)})$, where $X_{(j)}$ is the random variable of which $x_{(j)}$ is a particular value. For example, the jth plotting positions for the normal and exponential distributions $\Phi\{(x - \mu)/\sigma\}$ and $1 - e^{-\lambda x}$ are $\mu + \sigma \Phi^{-1}\{j/(n+1)\}$ and $-\lambda^{-1} \log\{1 - j/(n+1)\}$. When parameters such as μ, σ, and λ are unknown, the plotting positions used are for standardized distributions, here $\Phi^{-1}\{j/(n+1)\}$ and $-\log\{1 - j/(n+1)\}$, which are sometimes called *normal scores* and *exponential scores*. Probability plots for the normal distribution are particularly common in applications and are also called *normal scores plots*. The interpretation of a probability plot is aided by adding the straight line that corresponds to perfect fit of F.

Example 2.16 (Birth data)   The top left panel of Figure 2.3 shows a probability plot to compare the 95 times in the delivery suite with the normal distribution. The distribution does not fit the largest and smallest observations, and the data show some upward curvature relative to the straight line. The top right panel shows that the exponential distribution would fit the data very poorly. The bottom left panel, a probability plot of the $\log y_j$ against normal plotting positions, corresponding to checking the log-normal distribution, shows slight downward curvature. The bottom right panel, a probability plot of the y_j against plotting positions for the gamma distribution with mean $\bar{y}$ and variance s^2, shows the best fit overall, though it is not perfect.

In the normal and gamma plots the dotted line corresponds to the theoretical distribution whose mean equals $\bar{y}$ and whose variance equals s^2; the dotted line in the exponential plot is for the exponential distribution whose mean equals $\bar{y}$; and the dotted line in the log-normal plot is for the normal distribution whose mean and variance equal the average and variance of the $\log y_j$.   ■

Some experience with interpreting probability plots may be gained from Practical 2.3.

Figure 2.3 Probability plots for hours in the delivery suite, for the normal, exponential, gamma, and log-normal distributions (clockwise from top left). In each panel the dotted line is for a fitted distribution whose mean and variance match those of the data. None of the fits is perfect, but the gamma distribution fits best, and the exponential worst.

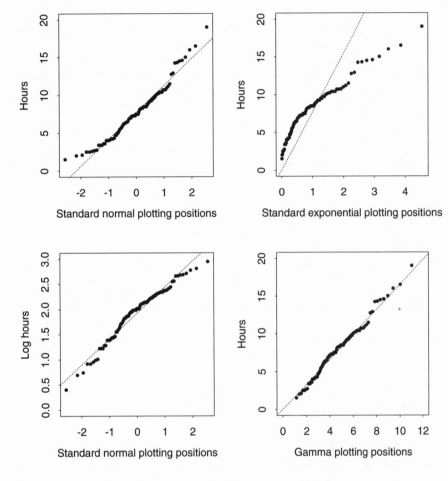

Exercises 2.1

1   Let m and s be the values of location and scale statistics calculated from $y_1, \ldots, y_n$; m and s may be any of the quantities described in Examples 2.1 and 2.2. Show that the effect of the mapping $y_1, \ldots, y_n \mapsto a + by_1, \ldots, a + by_n$, $b > 0$, is to send $m, s \mapsto a + bm, bs$. Show also that the measures of shape in Examples 2.4 and 2.5 are unchanged by this transformation.

2   (a) Show that when δ is added to one of $y_1, \ldots, y_n$ and $|\delta| \to \infty$, the average $\overline{y}$ changes by an arbitrarily large amount, but the sample median does not. By considering such perturbations when n is large, deduce that the sample median has breakdown point 0.5.

A sketch may help.

(b) Find the breakdown points of the other statistics in Examples 2.1 and 2.2.

3   (a) If $\kappa > 0$ is real and k a positive integer, show that the gamma function

$$\Gamma(\kappa) = \int_0^\infty u^{\kappa-1} e^{-u} \, du,$$

has properties $\Gamma(1) = 1$, $\Gamma(\kappa + 1) = \kappa \Gamma(\kappa)$ and $\Gamma(k) = (k - 1)!$. It is useful to know that $\Gamma(\frac{1}{2}) = \pi^{1/2}$, but you need not prove this.

The *mode* of a density f is a value y such that $f(y) \geq f(x)$ for all x.

(b) Use (a) to verify the mean and variance of (2.7).

(c) Show that for $0 < \kappa \leq 1$ the maximum value of (2.7) is at $y = 0$, and find its mode when $\kappa > 1$.

4   Give formulae analogous to (2.4) for the variance, skewness and 'shape' of a distribution
    F. Do they behave sensibly when a variable Y with distribution F is transformed to
    $a + bY$, so $F(y)$ is replaced by $F\{(y-a)/b\}$?

5   Let Y have continuous distribution function F. For any η, show that $X = |Y - \eta|$ has
    distribution $G(x) = F(\eta + x) - F(\eta - x)$, $x > 0$. Hence give a definition of the median
    absolute deviation of F in terms of F^{-1} and G^{-1}. If the density of Y is symmetric about
    the origin, show that $G(x) = 2F(x) - 1$. Hence find the median absolute deviation of the
    Laplace density (2.5).

6   A probability plot in which $y_1, \ldots, y_n$ and $x_1, \ldots, x_n$ are two random samples is called a
    quantile-quantile or *Q-Q plot*. Construct this plot for the first two columns in Table 2.1.
    Are the samples the same shape?

7   The *stem-and-leaf display* for the data 2.1, 2.3, 4.5, 3.3, 3.7, 1.2 is

    1 | 2
    2 | 13
    3 | 37
    4 | 5

    If you turn the page on its side this gives a histogram showing the data values themselves
    (perhaps rounded); the units corresponding to intervals [1, 2), [2, 3) and so forth are to
    the left of the vertical bars, and the digits are to the right. Construct this for the combined
    data for days 1–3 in Table 2.1. Hence find their median, quartiles, interquartile range, and
    range.

8   Do Figures 2.1–2.3 follow the advice given on page 21? If not, how could they be im-
    proved? Browse some textbooks and newspapers and think critically about any statistical
    graphics you find.

2.2 Convergence

2.2.1 Modes of convergence

Intuition tells us that the bigger our sample, the more faith we can have in our
inferences, because our sample is more representative of the distribution F from
which it came — if the sample size n was infinite, we would effectively know F. As
$n \to \infty$ we can think of our sample $Y_1, \ldots, Y_n$ as converging to F, and of a statistic
$S = s(Y_1, \ldots, Y_n)$ as converging to a limit that depends on F. For our purposes there
are two main ways in which a sequence of random variables, $S_1, S_2, \ldots$, can converge
to another random variable S.

Convergence in probability

We say that S_n *converges in probability* to S, $S_n \xrightarrow{P} S$, if for any $\varepsilon > 0$

$$\Pr(|S_n - S| > \varepsilon) \to 0 \quad \text{as} \quad n \to \infty. \tag{2.9}$$

A special case of this is the *weak law of large numbers*, whose simplest form is that
if $Y_1, Y_2, \ldots$ is a sequence of independent identically distributed random variables
each with finite mean μ, and if $\overline{Y} = n^{-1}(Y_1 + \cdots + Y_n)$ is the average of $Y_1, \ldots, Y_n$,
then $\overline{Y} \xrightarrow{P} \mu$. We sometimes call this simply the *weak law*. It is illustrated in the
left-hand panels of Figure 2.4, which show histograms of 10,000 averages of random
samples of n exponential random variables, with $n = 1, 5, 10$, and 20. The individual

Figure 2.4 Convergence in probability and in distribution. The left panels show how histograms of the averages $\overline{Y}$ of 10,000 samples of n standard exponential random variables become more concentrated at the mean $\mu = 1$ as n increases through 1, 5, 10, and 20, due to the convergence in probability of $\overline{Y}$ to μ. The right panels show how the distribution of $Z_n = n^{1/2}(\overline{Y} - 1)$ approaches the standard normal distribution, due to the convergence in distribution of Z_n to normality.

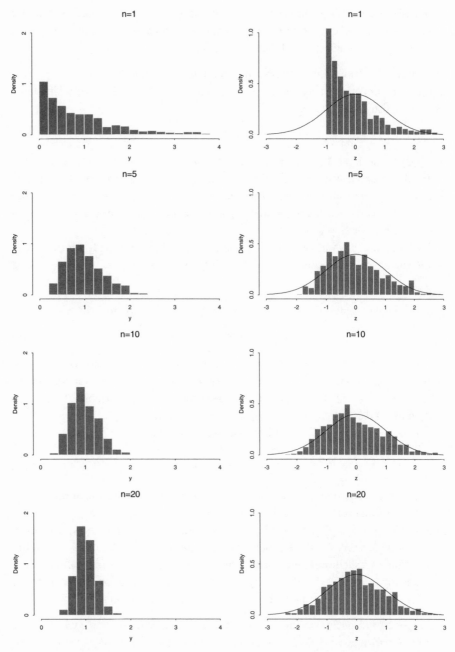

variables have density e^{-y} for $y > 0$, so their mean μ and variance σ^2 both equal one. As n increases, the values of $S_n = \overline{Y}$ become increasingly concentrated around μ, so as the figure illustrates, $\Pr(|S_n - \mu| > \varepsilon)$ decreases for each positive ε.

Statistics that converge in probability have some useful properties. For example, if s_0 is a constant, and h is a function continuous at s_0, then if $S_n \xrightarrow{P} s_0$, it follows that $h(S_n) \xrightarrow{P} h(s_0)$ (Exercise 2.2.1).

An estimator S_n of a parameter θ is *consistent* if $S_n \xrightarrow{P} \theta$ as $n \to \infty$, whatever the value of θ. Consistency is desirable, but a consistent estimator that has poor properties for any realistic sample size will be useless in practice.

Example 2.17 (Binomial distribution)   A binomial random variable $R = \sum_{j=1}^{m} I_j$ counts the numbers of ones in the random sample $I_1, \ldots, I_m$, each of which has a Bernoulli distribution,

$$\Pr(I_j = 1) = \pi, \quad \Pr(I_j = 0) = 1 - \pi, \quad 0 \le \pi \le 1.$$

It is easy to check that $E(I_j) = \pi$ and $\text{var}(I_j) = \pi(1 - \pi)$. Thus the weak law applies to the proportion of successes $\widehat{\pi} = R/m$, giving $\widehat{\pi} \xrightarrow{P} \pi$ as $m \to \infty$. Evidently $\widehat{\pi}$ is a consistent estimator of π. However, the useless estimator $\widehat{\pi} + 10^6/\log m$ is also consistent — consistency is a minimal requirement, not a guarantee that the estimator can safely be used in practice.

Jacob Bernoulli (1654–1705) was a member of a mathematical family split by rivalry. His major work on probability, *Ars Conjectandi*, was published in 1713, but he also worked on many other areas of mathematics.

Each of the I_j has variance $\pi(1 - \pi)$, and this is estimated by $\widehat{\pi}(1 - \widehat{\pi})$, a continuous function of $\widehat{\pi}$ that converges in probability to $\pi(1 - \pi)$.   ∎

Convergence in distribution

We say that the sequence $Z_1, Z_2, \ldots$, *converges in distribution* to Z, $Z_n \xrightarrow{D} Z$, if

$$\Pr(Z_n \le z) \to \Pr(Z \le z) \quad \text{as} \quad n \to \infty \tag{2.10}$$

at every z for which the distribution function $\Pr(Z \le z)$ is continuous. The most important case of this is the *central limit theorem*, whose simplest version applies to a sequence of independent identically distributed random variables $Y_1, Y_2, \ldots$, with finite mean μ and finite variance $\sigma^2 > 0$. If the sample average is $\overline{Y} = n^{-1}(Y_1 + \cdots + Y_n)$, the Central Limit Theorem states that

$$Z_n = n^{1/2}\frac{(\overline{Y} - \mu)}{\sigma} \xrightarrow{D} Z, \tag{2.11}$$

where Z is a standard normal random variable, that is, one having the normal distribution with mean zero and variance one, written $N(0, 1)$; see Section 3.2.1.

The right panels of Figure 2.4 illustrate such convergence. They show histograms of Z_n for the averages in the left-hand panels, with the standard normal probability density function superimposed. Each of the right-hand panels is a translation to zero of the histogram to its left, followed by 'zooming in': multiplication by a scale factor $n^{1/2}/\sigma$. As n increases, Z_n approaches its limiting standard normal distribution.

Example 2.18 (Average)   Consider the average $\overline{Y}$ of a random sample with mean μ and finite variance $\sigma^2 > 0$. The weak law implies that $\overline{Y}$ is a consistent estimator of its expected value μ, and (2.11) implies that in addition $\overline{Y} = \mu + n^{-1/2}\sigma Z_n$, where $Z_n \xrightarrow{D} Z$. This supports our intuition that $\overline{Y}$ is a better estimate of μ for large n, and makes explicit the rate at which $\overline{Y}$ converges to μ: in large samples $\overline{Y}$ is essentially a normal variable with mean μ and variance σ^2/n.   ∎

Example 2.19 (Empirical distribution function)   Let $Y_1, \ldots, Y_n$ be a random sample from F, and let $I_j(y)$ be the indicator random variable for the event $Y_j \le y$. Thus

$I_j(y)$ equals one if $Y_j \leq y$ and zero otherwise. The empirical distribution function of the sample is

$$\widehat{F}(y) = n^{-1} \sum_{j=1}^{n} I_j(y),$$

a step function that increases by n^{-1} at each observation, as in the upper right panel of Figure 2.1. We thought of (2.3) as a summary of the data $y_1, \ldots, y_n$; $\widehat{F}(y)$ is the corresponding random variable.

The $I_j(y)$ are independent and each has the Bernoulli distribution with probability $\Pr\{I_j(y) = 1\} = F(y)$. Therefore $\widehat{F}(y)$ is an average of independent identically distributed variables and has mean $F(y)$ and variance $F(y)\{1 - F(y)\}/n$. At a value y for which $0 < F(y) < 1$,

$$\widehat{F}(y) \xrightarrow{P} F(y), \quad n^{1/2} \frac{\{\widehat{F}(y) - F(y)\}}{[F(y)\{1 - F(y)\}]^{1/2}} \xrightarrow{D} Z, \quad \text{as } n \to \infty, \qquad (2.12)$$

where Z is a standard normal variate. It can be shown that this pointwise convergence for each y extends to convergence of the function $\widehat{F}(y)$ to $F(y)$. The empirical distribution function in Figure 2.1 is thus an estimate of the true distribution of times in the delivery suite. ∎

The alert reader will have noticed a sleight-of-word in the previous sentence. Convergence results tell us what happens as $n \to \infty$, but in practice the sample size is fixed and finite. How then are limiting results relevant? They are used to generate approximations for finite n — for example, (2.12) leads us to hope that $n^{1/2}\{\widehat{F}(y) - F(y)\}/[F(y)\{1 - F(y)\}]^{1/2}$ has approximately a standard normal distribution even when n is quite small. In practice it is important to check the adequacy of such approximations, and to develop a feel for their accuracy. This may be done analytically or by simulation (Section 3.3), while numerical examples are also valuable.

Slutsky's lemma

Evgeny Evgenievich Slutsky (1880–1948) made fundamental contributions to stochastic convergence and to economic time series during the 1920s and 1930s. In 1902 he was expelled from university in Kiev for political activity. He studied in Munich and Kiev and worked in Kiev and Moscow.

Convergence in distribution is useful in statistical applications because we generally want to compare probabilities. It is weaker than convergence in probability because it does not involve the joint distribution of S_n and S. If s_0 and u_0 are constants, these modes of convergence are related as follows:

$$S_n \xrightarrow{P} S \Rightarrow S_n \xrightarrow{D} S, \qquad (2.13)$$

$$S_n \xrightarrow{D} s_0 \Rightarrow S_n \xrightarrow{P} s_0, \qquad (2.14)$$

$$S_n \xrightarrow{D} S \text{ and } U_n \xrightarrow{P} u_0 \Rightarrow S_n + U_n \xrightarrow{D} S + u_0, S_n U_n \xrightarrow{D} S u_0. \qquad (2.15)$$

Devotees of tricky analysis will find references to proofs of (2.13)–(2.15) in Section 2.5.

The third of these is known as *Slutsky's lemma*.

Example 2.20 (Sample variance) Suppose that $Y_1, \ldots, Y_n$ is a random sample of variables with finite mean μ and variance σ^2. Let

$$S_n = n^{-1} \sum_{j=1}^{n} (Y_j - \overline{Y})^2 = n^{-1} \sum_{j=1}^{n} Y_j^2 - \overline{Y}^2,$$

where $\overline{Y}$ is the sample average. The weak law implies that $\overline{Y} \xrightarrow{P} \mu$, and the function $h(x) = x^2$ is continuous everywhere, so $\overline{Y}^2 \xrightarrow{P} \mu^2$. Moreover

$$E\left(Y_j^2\right) = \text{var}(Y_j) + \{E(Y_j)\}^2 = \sigma^2 + \mu^2,$$

so $n^{-1} \sum Y_j^2 \xrightarrow{P} \sigma^2 + \mu^2$ also. Now (2.13) implies that $n^{-1} \sum Y_j^2 \xrightarrow{D} \sigma^2 + \mu^2$, and therefore (2.15) implies that $S_n \xrightarrow{D} \sigma^2$. But σ^2 is constant, so $S_n \xrightarrow{P} \sigma^2$.

The sample variance S^2 may be written as $S_n \times n/(n-1)$, which evidently also tends in probability to σ^2. Thus not only is it true that for all n, $E(S^2) = \sigma^2$, but the distribution of S^2 is increasingly concentrated at σ^2 in large samples. ∎

These ideas extend to functions of several random variables.

Example 2.21 (Covariance and correlation)   The covariance between random variables X and Y is

$$\gamma = E[\{X - E(X)\}\{Y - E(Y)\}] = E(XY) - E(X)E(Y).$$

An estimate of γ based on a random sample of data pairs $(X_1, Y_1), \ldots, (X_n, Y_n)$ is the sample covariance

$$C = \frac{1}{n-1} \sum_{j=1}^n (X_j - \overline{X})(Y_j - \overline{Y}) = \frac{n}{n-1}\left(n^{-1}\sum_{j=1}^n X_j Y_j - \overline{XY}\right),$$

where $\overline{X}$ and $\overline{Y}$ are the averages of the X_j and Y_j. Provided the moments $E(XY)$, $E(X)$ and $E(Y)$ are finite, the weak law applies to each of $n^{-1}\sum X_j Y_j$, $\overline{X}$ and $\overline{Y}$, which converge in probability to their expectations. The convergence is also in distribution, by (2.13), so (2.15) implies that $C \xrightarrow{D} \gamma$. But γ is constant, so (2.14) implies that $C \xrightarrow{P} \gamma$.

The correlation between X and Y,

$$\rho = \frac{E(XY) - E(X)E(Y)}{\{\text{var}(X)\text{var}(Y)\}^{1/2}},$$

is such that $-1 \le \rho \le 1$. When $|\rho| = 1$ there is a linear relation between X and Y, so that $a + bX + cY = 0$ for some nonzero b and c (Exercise 2.2.3). Values of ρ close to ± 1 indicate strong linear dependence between the distributions of X and Y, though values close to zero do not indicate independence, just lack of a linear relation. The parameter ρ can be estimated from the pairs (X_j, Y_j) by the sample *correlation coefficient*,

Also known as the product moment correlation coefficient.

$$R = \frac{\sum_{j=1}^n (X_j - \overline{X})(Y_j - \overline{Y})}{\left\{\sum_{i=1}^n (X_i - \overline{X})^2 \sum_{k=1}^n (Y_k - \overline{Y})^2\right\}^{1/2}}.$$

The keen reader will enjoy showing that $R \xrightarrow{P} \rho$. ∎

Example 2.22 (Studentized statistic)   Suppose that $(T_n - \theta)/\text{var}(T_n)^{1/2}$ converges in distribution to a standard normal random variable, Z, and that $\text{var}(T_n) = \tau^2/n$, where $\tau^2 > 0$ is unknown but finite. Let V_n be a statistic that estimates τ^2/n, with the property that $nV_n \overset{P}{\longrightarrow} \tau^2$. The function $h(x) = \tau/(nx)^{1/2}$ is continuous at $x = 1$, so $\tau/(nV_n)^{1/2} \overset{P}{\longrightarrow} 1$. Therefore

$$Z_n = n^{1/2}\frac{(T_n - \theta)}{\tau} \times \frac{\tau}{(nV_n)^{1/2}} \overset{D}{\longrightarrow} Z \times 1,$$

by (2.15). Thus Z_n has a limiting standard normal distribution provided that nV_n is a consistent estimator of τ^2.

The best-known instance of this is the average of a random sample, $\overline{Y} = n^{-1}(Y_1 + \cdots + Y_n)$. If the Y_j have finite mean θ and finite positive variance, σ^2, $\overline{Y}$ has mean θ and variance σ^2/n. The Central Limit Theorem states that

$$n^{1/2}\frac{(\overline{Y} - \theta)}{\sigma} \overset{D}{\longrightarrow} Z.$$

Consider $Z_n = n^{1/2}(\overline{Y} - \theta)/S$, where $S^2 = (n-1)^{-1}\sum(Y_j - \overline{Y})^2$. Example 2.20 shows that $S^2 \overset{P}{\longrightarrow} \sigma^2$, and it follows that $Z_n \overset{D}{\longrightarrow} Z$.

The replacement of $\mathrm{var}(T_n)$ by an estimate is called studentization to honour W. S. Gossett. Publishing under the pseudonym 'Student' in 1908, he considered the effect of replacing σ by S for normal data; see Section 3.2. ∎

Intuition suggests that bigger samples always give better estimates, but intuition can mislead or fail.

Example 2.23 (Cauchy distribution)   A Cauchy random variable centred at θ has density

$$f(y;\theta) = \frac{1}{\pi\{1 + (y - \theta)^2\}}, \quad -\infty < y < \infty, \quad -\infty < \theta < \infty. \tag{2.16}$$

Although (2.16) is symmetric with mode at θ, none of its moments exist, and in fact the average $\overline{Y}$ of a random sample $Y_1, \ldots, Y_n$ of such data has the same distribution as a single observation. So if we were unlucky enough to have such a sample, it would be useless to estimate θ by $\overline{Y}$: we might as well use Y_1. The difficulty is that the tails of the Cauchy density decrease very slowly. Data with similar characteristics arise in many financial and insurance contexts, so this is not a purely mathematical issue: the average may be a poor estimate, and better ones are discussed later. ∎

2.2.2 Delta method

Variances and variance estimates are often required for smooth functions of random variables. Suppose that the quantity of interest is $h(T_n)$, and

$$(T_n - \mu)/\mathrm{var}(T_n)^{1/2} \overset{D}{\longrightarrow} Z, \quad n\mathrm{var}(T_n) \overset{P}{\longrightarrow} \tau^2 > 0,$$

as $n \to \infty$, and Z has the standard normal distribution. Then we may write $T_n = \mu + n^{-1/2}\tau Z_n$, where $Z_n \overset{D}{\longrightarrow} Z$. If h has a continuous non-zero derivative h' at μ, Taylor series expansion gives

$$h(T_n) = h\big(\mu + n^{-1/2}\tau Z_n\big) = h(\mu) + n^{-1/2}\tau Z_n h'\big(\mu + n^{-1/2}\tau W_n\big),$$

William Sealy Gossett (1876–1937) worked at the Guinness brewery in Dublin. Apart from his contributions to beer and statistics, he also invented a boat with two rudders that would be easy to manoeuvre when fly fishing.

Augustin Louis Cauchy (1789–1857) made contributions to all the areas of mathematics known at his time. He was a pioneer of real and complex analysis, but also developed applied techniques such as Fourier transforms and the diagonalization of matrices in order to work on elasticity and the theory of light. His relations with contemporaries were often poor because of his rigid Catholicism and his difficult character.

where W_n lies between Z_n and zero. As h' is continuous at μ, it follows that $h'(\mu + n^{-1/2}\tau W_n) \overset{P}{\longrightarrow} h'(\mu)$, so (2.15) gives

$$
\begin{aligned}
\frac{n^{1/2}\{h(T_n) - h(\mu)\}}{\tau h'(\mu)} &= \frac{n^{1/2}\{h(T_n) - h(\mu)\}}{\tau h'\left(\mu + n^{-1/2}\tau W_n\right)} \times \frac{h'\left(\mu + n^{-1/2}\tau W_n\right)}{h'(\mu)} \\
&= Z_n \times \frac{h'\left(\mu + n^{-1/2}\tau W_n\right)}{h'(\mu)} \\
&\overset{D}{\longrightarrow} Z
\end{aligned}
$$

as $n \to \infty$. This implies that in large samples, $h(T_n)$ has approximately the normal distribution with mean $h(\mu)$ and variance $\operatorname{var}(T_n)h'(\mu)^2$, that is,

$$
h(T_n) \overset{\cdot}{\sim} N(h(\mu), \operatorname{var}(T_n)h'(\mu)^2). \tag{2.17}
$$

> $\overset{\cdot}{\sim}$ means 'is approximately distributed as'.

This result is often called the *delta method*. Analogous results apply if the limiting distribution of Z_n is non-normal.

Furthermore, if $h'(\mu)$ is replaced by $h'(T_n)$ and τ^2 is replaced by a consistent estimator, S_n, a modification of the argument in Example 2.22 gives

$$
\frac{n^{1/2}\{h(T_n) - h(\mu)\}}{S_n^{1/2}|h'(T_n)|} \overset{D}{\longrightarrow} Z. \tag{2.18}
$$

Thus the same limiting results apply if the variance of $h(T_n)$ is replaced by a consistent estimator. In particular, replacement of the parameters in $\operatorname{var}(T_n)h'(\mu)^2$ by consistent estimators gives a consistent estimator of $\operatorname{var}\{h(T_n)\}$.

Example 2.24 (Exponential transformation)   Consider $h(\overline{Y}) = \exp(\overline{Y})$, where $\overline{Y}$ is the average of a random sample of size n, and each of the Y_j has mean μ and variance σ^2. Here $h'(\mu) = e^\mu$, so $\exp(\overline{Y})$ is asymptotically normal with mean e^μ and variance $n^{-1}\sigma^2 e^{2\mu}$. This can be estimated by $n^{-1}S^2 \exp(2\overline{Y})$, where S^2 is the sample variance.   ∎

Several variables

The delta method extends to functions of several random variables $T_1, \ldots, T_p$; we suppress dependence on n for ease of notation. As $n \to \infty$, suppose that for each r, $n^{-1/2}(T_r - \theta_r) \overset{D}{\longrightarrow} N(0, \omega_{rr})$, that the joint limiting distribution of $n^{-1/2}(T_r - \theta_r)$ is multivariate normal (see Section 3.2.3) and $n\operatorname{cov}(T_r, T_s) \to \omega_{rs}$, where the $p \times p$ matrix Ω whose (r, s) element is ω_{rs} is positive-definite; note that Ω is symmetric. Now suppose that a variance is required for the scalar function $h(T_1, \ldots, T_p)$. An argument like that above gives

$$
h(T_1, \ldots, T_p) \overset{\cdot}{\sim} N\{h(\theta_1, \ldots, \theta_p), n^{-1}h'(\theta)^{\mathrm{T}}\Omega h'(\theta)\}, \tag{2.19}
$$

where $h'(\theta)$ is the $p \times 1$ vector whose rth element is $\partial h(\theta_1, \ldots, \theta_p)/\partial\theta_r$; the requirement that $h'(\theta) \neq 0$ also holds here. As in the univariate case, the variance can be estimated by replacing parameters with consistent estimators.

Example 2.25 (Ratio)   Let $\theta_1 = \mathrm{E}(X) \neq 0$ and $\theta_2 = \mathrm{E}(Y)$, and suppose we are interested in $h(\theta_1, \theta_2) = \theta_2/\theta_1$. Estimates of θ_1 and θ_2 based on random samples

$X_1, \ldots, X_n$ and $Y_1, \ldots, Y_n$ are $T_1 = \overline{X}$ and $T_2 = \overline{Y}$, so the ratio is consistently estimated by T_2/T_1. The derivative vector is $h'(\theta) = (-\theta_2/\theta_1^2, \theta_1^{-1})^{\mathsf{T}}$, and the limiting mean and variance of T_2/T_1 are

$$\frac{\theta_2}{\theta_1}, \quad n^{-1}\begin{pmatrix} -\theta_2/\theta_1^2 & \theta_1^{-1} \end{pmatrix} \begin{pmatrix} \omega_{11} & \omega_{12} \\ \omega_{21} & \omega_{22} \end{pmatrix} \begin{pmatrix} -\theta_2/\theta_1^2 \\ \theta_1^{-1} \end{pmatrix},$$

the second of which equals

$$\left(n\theta_1^2\right)^{-1} \left\{ \omega_{11}\left(\frac{\theta_2}{\theta_1}\right)^2 - 2\omega_{12}\frac{\theta_2}{\theta_1} + \omega_{22} \right\},$$

assumed finite and positive. The variance tends to zero as $n \to \infty$, so we should aim to estimate $n\mathrm{var}(T_2/T_1)$, which is not a moving target.

Examples 2.20 and 2.21 imply that ω_{11}, ω_{22}, and ω_{12} are consistently estimated by $S_1^2 = (n-1)^{-1}\sum(X_j - \overline{X})^2$, $S_2^2 = (n-1)^{-1}\sum(Y_j - \overline{Y})^2$, and $C = (n-1)^{-1}\sum(X_j - \overline{X})(Y_j - \overline{Y})$ respectively. Therefore $n\mathrm{var}(\overline{Y}/\overline{X})$ is consistently estimated by

$$\overline{X}^{-2}\left\{ S_1^2\left(\frac{\overline{Y}}{\overline{X}}\right)^2 - 2C\frac{\overline{Y}}{\overline{X}} + S_2^2 \right\} = \frac{1}{(n-1)\overline{X}^2}\sum_{j=1}^{n}\left(Y_j - \frac{\overline{Y}}{\overline{X}}X_j\right)^2,$$

as we see after simplification.   ∎

Example 2.26 (Gamma shape) In Example 2.12 the shape parameter κ of the gamma distribution was taken to be $\overline{y}^2/s^2 = 3.15$, based on $n = 95$ observations. The corresponding random variable is T_1^2/T_2, where $T_1 = \overline{Y}$ and $T_2 = S^2$ are calculated from the random sample $Y_1, \ldots, Y_n$, supposed to be gamma with mean $\theta_1 = \kappa/\lambda$ and variance $\theta_2 = \kappa/\lambda^2$. We take $h(\theta_1, \theta_2) = \theta_1^2/\theta_2$, giving $h'(\theta_1, \theta_2) = (2\theta_1/\theta_2, -\theta_1^2/\theta_2^2)^{\mathsf{T}}$. The variance of T_1 is θ_2/n, that is, $n^{-1}\kappa/\lambda^2$, and it turns out that

$$\mathrm{var}(T_2) = \mathrm{var}(S^2) = \frac{\kappa_4}{n} + \frac{2\kappa_2^2}{n-1}, \quad \mathrm{cov}(T_1, T_2) = \mathrm{cov}(\overline{Y}, S^2) = \frac{\kappa_3}{n},$$

where $\kappa_2 = \kappa/\lambda^2$, $\kappa_3 = 2\kappa/\lambda^3$, and $\kappa_4 = 6\kappa/\lambda^4$. Thus

$$\mathrm{var}(T_1^2/T_2) \doteq \begin{pmatrix} 2\lambda & -\lambda^2 \end{pmatrix} \begin{pmatrix} \frac{\kappa}{n\lambda^2} & \frac{2\kappa}{n\lambda^3} \\ \frac{2\kappa}{n\lambda^3} & \frac{6\kappa}{n\lambda^4} + \frac{2\kappa^2}{(n-1)\lambda^4} \end{pmatrix} \begin{pmatrix} 2\lambda \\ -\lambda^2 \end{pmatrix}$$

$$= \frac{2\kappa}{n}\left(1 + \frac{n\kappa}{n-1}\right),$$

or roughly $2n^{-1}\kappa(\kappa + 1)$.   ∎

Big and little oh notation: O and o

This can be skipped on a first reading.

For two sequences of constants, $\{s_n\}$ and $\{a_n\}$ such that $a_n \geq 0$ for all n, we write $s_n = o(a_n)$ if $\lim_{n\to\infty}(s_n/a_n) = 0$, and $s_n = O(a_n)$ if there is a finite constant k such that $\lim_{n\to\infty}|s_n| \leq a_n k$. A sequence of random variables $\{S_n\}$ is said to be $o_p(a_n)$ if $(S_n/a_n) \xrightarrow{P} 0$ as $n \to \infty$, and is said to be $O_p(a_n)$ if S_n/a_n is bounded in probability as $n \to \infty$, that is, given $\varepsilon > 0$ there exist n_0 and a finite k such that for all $n > n_0$,

$$\Pr(|S_n/a_n| < k) > 1 - \varepsilon.$$

This gives a useful shorthand for expansions of random quantities.

To illustrate this, suppose that $\{Y_j\}$ is a sequence of independent identically distributed variables with finite mean μ, and let $S_n = n^{-1}(Y_1 + \cdots + Y_n)$. Then the weak law may be restated as $S_n = \mu + o_p(1)$, and if in addition the Y_j have finite variance σ^2, the Central Limit Theorem implies that $\overline{Y} = \mu + O_p(n^{-1/2})$. More precisely, $\overline{Y} \stackrel{D}{=} \mu + n^{-1/2}\sigma Z + o_p(n^{-1/2})$, where Z has a standard normal distribution. Such expressions are sometimes used in later chapters.

$\stackrel{D}{=}$ means 'has the same distribution as'.

Exercises 2.2

1 Suppose that $S_n \stackrel{P}{\longrightarrow} s_0$, and that the function h is continuous at s_0, that is, for any $\varepsilon > 0$ there exists a $\delta > 0$ such that $|x - y| < \delta$ implies that $|h(x) - h(y)| < \varepsilon$. Explain why this implies that

$$\Pr(|S_n - s_0| < \delta) \le \Pr\{|h(S_n) - h(s_0)| < \varepsilon\} \le 1,$$

and deduce that $\Pr\{|h(s_0) - h(S_n)| < \varepsilon\} \to 1$ as $n \to \infty$. That is, $h(S_n) \stackrel{P}{\longrightarrow} h(s_0)$.

2 Let s_0 be a constant. By writing

$$\Pr(|S_n - s_0| \le \varepsilon) = \Pr(S_n \le s_0 + \varepsilon) - \Pr(S_n \le s_0 - \varepsilon),$$

for $\varepsilon > 0$, show that $S_n \stackrel{D}{\longrightarrow} s_0$ implies that $S_n \stackrel{P}{\longrightarrow} s_0$.

3 (a) Let X and Y be two random variables with finite positive variances. Use the fact that $\text{var}(aX + Y) \ge 0$, with equality if and only if the linear combination $aX + Y$ is constant with probability one, to show that $\text{cov}(X, Y)^2 \le \text{var}(X)\text{var}(Y)$; this is a version of the *Cauchy–Schwarz inequality*. Hence show that $-1 \le \text{corr}(X, Y) \le 1$, and say under what conditions equality is attained.
  (b) Show that if X and Y are independent, $\text{corr}(X, Y) = 0$. Show that the converse is false by considering the variables X and $Y = X^2 - 1$, where X has mean zero, variance one, and $\text{E}(X^3) = 0$.

4 Let $X_1, \ldots, X_n$ and $Y_1, \ldots, Y_n$ be independent random samples from the exponential densities $\lambda e^{-\lambda x}$, $x > 0$, and $\lambda^{-1}e^{-y/\lambda}$, $y > 0$, with $\lambda > 0$. If $\overline{X}$ and $\overline{Y}$ are the sample averages, show that $\overline{X}\,\overline{Y} \stackrel{P}{\longrightarrow} 1$ as $n \to \infty$.

5 Show that as $n \to \infty$ the skewness measure in Example 2.4 converges in probability to the corresponding theoretical quantity

$$\frac{\int (y - \mu)^3 dF(y)}{\left\{\int (y - \mu)^2 dF(y)\right\}^{3/2}},$$

provided this has finite numerator and positive denominator. Under what additional condition(s) is the skewness measure asymptotically normal?

6 If $Y_1, \ldots, Y_n \stackrel{\text{iid}}{\sim} N(\mu, \sigma^2)$, show that $n^{1/2}(\overline{Y} - \mu)^2 \stackrel{P}{\longrightarrow} 0$ as $n \to \infty$. Given that $\text{var}\{(Y_j - \mu)^2\} = 2\sigma^4$, deduce that $(S^2 - \sigma^2)/(2\sigma^4/n)^{1/2} \stackrel{D}{\longrightarrow} Z$, where $Z \sim N(0, 1)$. When is this true for non-normal data?

$\stackrel{\text{iid}}{\sim}$ means 'are independent and identically distributed as'.

7 Let R be a binomial variable with probability π and denominator m; its mean and variance are $m\pi$ and $m\pi(1 - \pi)$. The *empirical logistic transform* of R is

$$h(R) = \log\left(\frac{R + \frac{1}{2}}{m - R + \frac{1}{2}}\right).$$

Show that for large m,

$$h(R) \stackrel{\cdot}{\sim} N\left\{\log\left(\frac{\pi}{1-\pi}\right), \frac{1}{m\pi(1-\pi)}\right\}.$$

What is the exact value of $E[\log\{R/(m-R)\}]$? Are the $\frac{1}{2}$s necessary in practice?

8 Truncated Poisson variables Y arise when counting quantities such as the sizes of groups, each of which must contain at least one element. The density is

$$\Pr(Y = y) = \frac{\theta^y e^{-\theta}}{y!(1 - e^{-\theta})}, \quad y = 1, 2, \ldots, \quad \theta > 0.$$

Find an expression for $E(Y) = \mu(\theta)$ in terms of θ. If $Y_1, \ldots, Y_n$ is a random sample from this density and $n \to \infty$, show that $\overline{Y} \stackrel{p}{\longrightarrow} \mu(\theta)$. Hence show that $\widehat{\theta} = \mu^{-1}(\overline{Y}) \stackrel{p}{\longrightarrow} \theta$.

9 Let $Y = \exp(X)$, where $X \sim N(\mu, \sigma^2)$; Y has the log-normal distribution. Use the moment-generating function of X to show that $E(Y^r) = \exp(r\mu + r^2\sigma^2/2)$, and hence find $E(Y)$ and $\mathrm{var}(Y)$.

If $Y_1, \ldots, Y_n$ is a log-normal random sample, show that both $T_1 = \overline{Y}$ and $T_2 = \exp(\overline{X} + S^2/2)$ are consistent estimators of $E(Y)$, where $X_j = \log Y_j$ and S^2 is the sample variance of the X_j. Give the corresponding estimators of $\mathrm{var}(Y)$.

Are the estimators based on the Y_j or on the X_j preferable? Why?

10 The binomial distribution models the number of 'successes' among independent variables with two outcomes such as success/failure or white/black. The multinomial distribution extends this to p possible outcomes, for example total failure/failure/success or white/black/red/blue/.... That is, each of the discrete variables $X_1, \ldots, X_m$ takes values $1, \ldots, p$, independently with probability $\Pr(X_j = r) = \pi_r$, where $\sum \pi_r = 1, \pi_r \geq 0$. Let $Y_r = \sum_j I(X_j = r)$ be the number of X_j that fall into category r, for $r = 1, \ldots, p$, and consider the distribution of $(Y_1, \ldots, Y_p)$.

(a) Show that the marginal distribution of Y_r is binomial with probability π_r, and that $\mathrm{cov}(Y_r, Y_s) = -m\pi_r\pi_s$, for $r \neq s$. Is it surprising that the covariance is negative?

(b) Hence give consistent estimators of positive probabilities π_r. What happens if some $\pi_r = 0$?

(d) Suppose that $p = 4$ with $\pi_1 = (2 + \theta)/4$, $\pi_2 = (1 - \theta)4$, $\pi_3 = (1 - \theta)/4$ and $\pi_4 = \theta/4$. Show that $T = m^{-1}(Y_1 + Y_4 - Y_2 - Y_3)$ is such that $E(T) = \theta$ and $\mathrm{var}(T) = a/m$ for some $a > 0$. Hence deduce that T is consistent for θ as $m \to \infty$.

Give the value of T and its estimated variance when (y_1, y_2, y_3, y_4) equals $(125, 18, 20, 34)$.

2.3 Order Statistics

Summary statistics such as the sample median, interquartile range, and median absolute deviation are based on the ordered values of a sample $y_1, \ldots, y_n$, and they are also useful in assessing how closely a sample matches a specified distribution. In this section we study properties of ordered random samples.

The *r*th *order statistic* of a random sample $Y_1, \ldots, Y_n$ is $Y_{(r)}$, where

$$Y_{(1)} \leq Y_{(2)} \leq \cdots \leq Y_{(n-1)} \leq Y_{(n)}$$

is the ordered sample. We assume that the cumulative distribution F of the Y_j is continuous, so $Y_{(r)} < Y_{(r+1)}$ with probability one for each r and there are no ties.

Density function

To find the probability density of $Y_{(r)}$, we argue heuristically. Divide the line into three intervals: $(-\infty, y)$, $[y, y + dy)$, and $[y + dy, \infty)$. The probabilities that a single observation falls into each of these intervals are $F(y)$, $f(y)dy$, and $1 - F(y)$ respectively. Therefore the probability that $Y_{(r)} = y$ is

> The dy is a rhetorical device so that we can say the *probability* that $Y = y$ is $f(y)dy$.

$$\frac{n!}{(r-1)!\,1!\,(n-r)!} \times F(y)^{r-1} \times f(y)dy \times \{1 - F(y)\}^{n-r}, \qquad (2.20)$$

where the second term is the probability that a prespecified $r - 1$ of the Y_j fall in $(-\infty, y)$, the third the probability that a prespecified one falls in $[y, y + dy)$, the fourth the probability that a prespecified $n - r$ fall in $[y + dy, \infty)$, and the first is a combinatorial multiplier giving the number of ways of prespecifying disjoint groups of sizes $r - 1$, 1, and $n - r$ out of n.

If we drop the dy, expression (2.20) becomes a probability density function, from which we can derive properties of $Y_{(r)}$. For example, its mean is

$$\mathrm{E}\big(Y_{(r)}\big) = \frac{n!}{(r-1)!(n-r)!} \int_{-\infty}^{\infty} yf(y)F(y)^{r-1}\{1 - F(y)\}^{n-r}\,dy \qquad (2.21)$$

when it exists; of course we expect that $\mathrm{E}(Y_{(1)}) < \cdots < \mathrm{E}(Y_{(n)})$.

Example 2.27 (Uniform distribution) Let $U_1, \ldots, U_n$ be a random sample from the uniform distribution on the unit interval,

$$\Pr(U \le u) = \begin{cases} 0, & u \le 0, \\ u, & 0 < u \le 1, \\ 1, & 1 < u; \end{cases} \qquad (2.22)$$

we write $U_1, \ldots, U_n \overset{\text{iid}}{\sim} U(0, 1)$. As $f(u) = 1$ when $0 < u < 1$, $U_{(r)}$ has density

$$f_{U_{(r)}}(u) = \frac{n!}{(r-1)!(n-r)!}u^{r-1}(1-u)^{n-r}, \quad 0 < u < 1, \qquad (2.23)$$

and (2.21) shows that $\mathrm{E}(U_{(r)})$ equals

$$\frac{n!}{(r-1)!(n-r)!} \int_0^1 u\,u^{r-1}(1-u)^{n-r}\,dy = \frac{n!}{(r-1)!(n-r)!}\frac{r!(n-r)!}{(n+1)!}$$

$$= \frac{r}{n+1};$$

the value of the integral follows because (2.23) must have integral one for any r in the range $1, \ldots, n$ and any positive integer n. The expected positions of the n order statistics divide the unit interval and hence the total probability under the density into $n + 1$ equal parts.

It is an exercise to show that $U_{(r)}$ has variance $r(n - r + 1)/\{(n + 1)^2(n + 2)\}$ (Exercise 2.3.1). For large n this is approximately $n^{-1}p(1 - p)$, where $p = r/n$, and hence we can write $U_{(r)} = r/(n + 1) + \{p(1 - p)/n\}^{1/2}\varepsilon$, where ε is a random variable with mean zero and variance approximately one. ∎

Recall that every
distribution function is
right-continuous.

Integrals such as (2.21) are nasty, but a good approximation is often available. Let
$U, U_1, \ldots, U_n \overset{\text{iid}}{\sim} U(0, 1)$ and $F^{-1}(u) = \min\{y : F(y) \geq u\}$. Then

$$\Pr\{F^{-1}(U) \leq y\} = \Pr\{U \leq F(y)\} = F(y),$$

which is the distribution function of Y. Hence $Y \overset{D}{=} F^{-1}(U)$; note that for continuous F the variable $F(Y)$ has the $U(0, 1)$ distribution; $F(Y)$ is called the *probability integral transform* of Y. It follows that $F^{-1}(U_1), \ldots, F^{-1}(U_n)$ is a random sample from F and that the joint distributions of the order statistics $Y_{(1)}, \ldots, Y_{(n)}$ and of $F^{-1}(U_{(1)}), \ldots, F^{-1}(U_{(n)})$ are the same; in fact this is true for general F. Consequently $\mathrm{E}(Y_{(r)}) = \mathrm{E}\{F^{-1}(U_{(r)})\}$. But Example 2.27 implies that $U_{(r)} \overset{D}{=} r/(n + 1) + \{p(1 - p)/n\}^{1/2}\varepsilon$, where ε is a random variable with mean zero and unit variance. If we apply the delta method with $h = F^{-1}$, we obtain

$$\mathrm{E}\big(Y_{(r)}\big) = \mathrm{E}\big\{F^{-1}\big(U_{(r)}\big)\big\} \doteq F^{-1}\big\{\mathrm{E}\big(U_{(r)}\big)\big\} = F^{-1}\{r/(n + 1)\}. \tag{2.24}$$

Hence the plotting positions $F^{-1}\{r/(n + 1)\}$ are approximate expected order statistics, justifying their use in probability plots; see Section 2.1.4.

Several order statistics

The argument leading to (2.20) can be extended to the joint distribution of any collection of order statistics. For example, the probability that the maximum, $Y_{(n)}$, takes value v and that the minimum, $Y_{(1)}$, takes value u, is

$$\frac{n!}{1!(n - 2)!1!} \times f(u)du \times \{F(v) - F(u)\}^{n-2} \times f(v)dv, \quad u < v,$$

and is zero otherwise. Similarly the joint density of all n order statistics is

$$f_{Y_{(1)}, \ldots, Y_{(n)}}(y_1, \ldots, y_n) = n! f(y_1) \times \cdots \times f(y_n), \quad y_1 < \cdots < y_n. \tag{2.25}$$

In principle one can use (2.25) to calculate other properties of the joint distribution of the $Y_{(r)}$, but this can be very tedious. Here is an elegant exception:

Example 2.28 (Exponential order statistics)   Consider the order statistics of a random sample $Y_1, \ldots, Y_n$ from the exponential density with parameter $\lambda > 0$, for which $\Pr(Y > y) = e^{-\lambda y}$. Let $E_1, \ldots, E_n$ denote a random sample of standard exponential variables, with $\lambda = 1$. Thus $Y_j \overset{D}{=} E_j/\lambda$.

The reasoning uses two facts. First, the distribution function of $\min(Y_1, \ldots, Y_r)$ is

$$\begin{aligned}
1 - \Pr\{\min(Y_1, \ldots, Y_r) > y\} &= 1 - \Pr\{Y_1 > y, \ldots, Y_r > y\} \\
&= 1 - \Pr(Y_1 > y) \times \cdots \times \Pr(Y_r > y) \\
&= 1 - \exp(-r\lambda y);
\end{aligned}$$

this is exponential with parameter $r\lambda$. Second, the exponential density has the *lack-of-memory property*

$$\Pr(Y - x > y \mid Y > x) = \frac{\Pr(Y > x + y)}{\Pr(Y > x)} = \frac{\exp\{-\lambda(x + y)\}}{\exp(-\lambda x)} = \exp(-\lambda y),$$

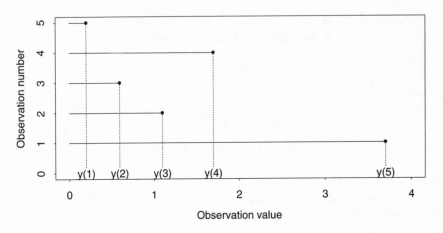

Figure 2.5 Exponential order statistics for a sample of size $n = 5$. The time to $y_{(1)}$ is the time to first event in a Poisson process of rate 5λ, and so it has the exponential distribution with mean $1/(5\lambda)$. The spacing $y_{(2)} - y_{(1)}$ is the time to first event in a Poisson process of rate 4λ, and is independent of $y_{(1)}$ because of the lack-of-memory property. It follows likewise that the spacings are independent and that the rth spacing has the exponential distribution with parameter $(n + 1 - r)\lambda$.

implying that given that $Y - x$ is positive, its distribution is the same as the original distribution of Y, whatever the value of x.

We now argue as follows. Since $Y_{(1)} = \min(Y_1, \ldots, Y_n)$, its distribution is exponential with parameter $n\lambda$: $Y_{(1)} \stackrel{D}{=} E_1/(n\lambda)$. Given $Y_{(1)}$, $n - 1$ of the Y_j remain, and by the lack-of-memory property the distribution of $Y_j - Y_{(1)}$ for each of them is the same as if the experiment had started at $Y_{(1)}$ with just $n - 1$ variables; see Figure 2.5. Thus $Y_{(2)} - Y_{(1)}$ is exponential with parameter $(n - 1)\lambda$, independent of $Y_{(1)}$, giving $Y_{(2)} - Y_{(1)} \stackrel{D}{=} E_2/\{(n - 1)\lambda\}$. But given $Y_{(2)}$, just $n - 2$ of the Y_j remain, and by the lack-of-memory property the distribution of $Y_j - Y_{(2)}$ for each of them is exponential independent of the past; hence $Y_{(3)} - Y_{(2)} \stackrel{D}{=} E_3/\{(n - 2)\lambda\}$. This argument yields the *Rényi representation*

$$Y_{(r)} \stackrel{D}{=} \lambda^{-1} \sum_{j=1}^{r} \frac{E_j}{n + 1 - j}, \tag{2.26}$$

from which properties of the $Y_{(r)}$ are easily derived. For example,

$$E(Y_{(r)}) = \lambda^{-1} \sum_{j=1}^{r} \frac{1}{n + 1 - j}, \quad \mathrm{cov}(Y_{(r)}, Y_{(s)}) = \lambda^{-2} \sum_{j=1}^{r} \frac{1}{(n + 1 - j)^2}, \quad s \geq r.$$

The upper right panel of Figure 2.3 shows a plot of the ordered times in the delivery suite against standard exponential plotting positions or *exponential scores*, $\sum_{j=1}^{r}(n + 1 - j)^{-1} \doteq -\log\{1 - r/(n + 1)\}$. The exponential model fits very poorly.

The argument leading to (2.26) may be phrased in terms of Poisson processes. A superposition of independent Poisson processes is itself a Poisson process with rate the sum of the individual rates, so the period from zero to $Y_{(1)}$ is the time to the first event in a Poisson process of rate $n\lambda$, the time from $Y_{(1)}$ to $Y_{(2)}$ is the time to first event in a Poisson process of rate $(n - 1)\lambda$, and so on, with the times between events independent by definition of a Poisson process; see Figure 2.5. Exercise 2.3.4 gives another derivation. ∎

During the second world war Alfréd Rényi (1921–1970) escaped from a labour camp and rescued his parents from the Budapest ghetto. He made major contributions to number theory and to probability. He was a gifted raconteur who defined a mathematician as 'a machine for turning coffee into theorems'.

Approximate density

Although (2.20) gives the exact density of an order statistic for a random sample of any size, approximate results are usually more convenient in practice. Suppose that r is the smallest integer greater than or equal to np, $r = \lceil np \rceil$, for some p in the range $0 < p < 1$. Then provided that $f\{F^{-1}(p)\} > 0$, we prove at the end of this section that $Y_{(r)}$ has an approximate normal distribution with mean $F^{-1}(p)$ and variance $n^{-1}p(1-p)/f\{F^{-1}(p)\}^2$ as $n \to \infty$. More formally,

$$\sqrt{n}\frac{\{Y_{(r)} - F^{-1}(p)\}f\{F^{-1}(p)\}}{\{p(1-p)\}^{1/2}} \xrightarrow{D} Z \quad \text{as} \quad n \to \infty, \tag{2.27}$$

where Z has a standard normal distribution.

Example 2.29 (Normal median)   Suppose that $Y_1, \ldots, Y_n$ is a random sample from the $N(\mu, \sigma^2)$ distribution, and that $n = 2m + 1$ is odd. The *median* of the sample is its central order statistic, $Y_{(m+1)}$. To find its approximate distribution in large samples, note that $(m+1)/(2m+1) \doteq \frac{1}{2}$ for large m, and since the normal density is symmetric about μ, $F^{-1}(\frac{1}{2}) = \mu$. Moreover $f(y) = (2\pi\sigma^2)^{-1/2}\exp\{-(y-\mu)^2/2\sigma^2\}$, so $f\{F^{-1}(\frac{1}{2})\} = (2\pi\sigma^2)^{-1/2}$. Thus (2.27) implies that in large samples $Y_{(m+1)}$ is approximately normal with mean μ and variance $\pi\sigma^2/(2n)$.   ∎

Example 2.30 (Birth data)   In Figure 2.1 and Example 2.8 we saw that the daily medians of the birth data were generally smaller but more variable than the daily averages. To understand why, suppose that we have a sample of $n = 13$ observations from the gamma distribution F with mean $\mu = 8$ and shape parameter $\kappa = 3$; these are close to the values for the data. Then the average $\overline{Y}$ has mean μ and variance $\mu^2/(n\kappa)$; these are 8 and 1.64, comparable with the data values 7.90 and 1.54. The sample median has approximate expected value $F^{-1}(\frac{1}{2}) = 7.13$ and variance $n^{-1}\frac{1}{2}(1-\frac{1}{2})/f\{F^{-1}(\frac{1}{2})\}^2 = 4.02$, where f denotes the density (2.8); these values are to be compared with the average and variance of the daily medians, 7.03 and 2.15. The expected values are close, but the variances are not; we should not rely on an asymptotic approximation when $n = 13$. The theoretical variance of the median exceeds that of the average, so the sampling properties of the daily average and median are roughly what we might have expected: $\text{var}(M) > \text{var}(\overline{Y})$, and $\text{E}(M) < \text{E}(\overline{Y})$. Our calculation presupposes constant n, but in the data n changes daily; this is one source of error in the asymptotic approximation.   ∎

Vilfredo Pareto (1848–1923) studied mathematics and physics at Turin, and then became an engineer and director of a railway, before becoming professor of political economy in Lausanne. He pioneered sociology and the use of mathematics in economic problems. The Pareto distributions were developed by him to explain the spread of wealth in society.

Expression (2.27) gives asymptotic distributions for *central order statistics*, that is, $Y_{(r)}$ where $r/n \to p$ and $0 < p < 1$; as $n \to \infty$ such order statistics have increasingly more values on each side. Different limits arise for *extreme order statistics* such as the minimum, for which $r = 1$ and $r/n \to 0$, and the maximum, for which $r = n$ and $r/n \to 1$. We discuss these more fully in Section 6.5.2, but here is a simple example.

Example 2.31 (Pareto distribution)   Suppose that $Y_1, \ldots, Y_n$ is a random sample from the Pareto distribution, whose distribution function is

$$F(y) = \begin{cases} 0, & y < a, \\ 1 - (y/a)^{-\gamma}, & y \geq a, \end{cases}$$

where $a, \gamma > 0$. The minimum $Y_{(1)}$ exceeds y if and only if all the $Y_1, \ldots, Y_n$ exceed y, so $\Pr(Y_{(1)} > y) = (y/a)^{-n\gamma}$. To obtain a non-degenerate limiting distribution, consider $M = \gamma n(Y_{(1)} - a)/a$. Now

$$\Pr(M > z) = \Pr\left(Y_{(1)} > \frac{az}{n\gamma} + a\right) = \left(\frac{\frac{az}{n\gamma} + a}{a}\right)^{-n\gamma} \rightarrow e^{-z}$$

as $n \rightarrow \infty$. Consequently $\gamma n(Y_{(1)} - a)/a$ converges in distribution to the standard exponential distribution.

There are two differences between this result and (2.27). First, and most obviously, the limiting distribution is not normal. Second, as the power of n by which $Y_{(1)} - a$ must be multiplied to obtain a non-degenerate limit is higher than in (2.27), the rate of convergence to the limit is faster than for central order statistics. Accelerated convergence of extreme order statistics does not always occur, however; see Example 6.32. ■

Derivation of (2.27)

This may be omitted at a first reading.

Consider $Y_{(r)}$, where $r = \lceil np \rceil$ and $0 < p < 1$ is fixed; hence $r/n \rightarrow p$ as $n \rightarrow \infty$. We saw earlier that $Y_{(r)} \overset{D}{=} F^{-1}(U_{(r)})$, where $U_{(r)}$ is the rth order statistic of a random sample $U_1, \ldots, U_n$ from the $U(0, 1)$ density, and that $U_{(r)} = r/(n+1) + \{p(1 - p)/n\}^{1/2}\varepsilon$, where ε has mean zero and variance tending to one as $n \rightarrow \infty$. Recall that F is a distribution whose density f exists. Hence the delta method gives $\mathrm{E}(Y_{(r)}) \doteq F^{-1}\{r/(n+1)\} \doteq F^{-1}(p)$, and as

$$\mathrm{var}(Y_{(r)}) = \mathrm{var}\{F^{-1}(U_{(r)})\} \doteq \mathrm{var}(U_{(r)}) \times \left\{\frac{dF^{-1}(p)}{dp}\right\}^2$$

and

$$\frac{d}{dp}F\{F^{-1}(p)\} = f\{F^{-1}(p)\}\frac{d}{dp}F^{-1}(p) = 1,$$

we have $\mathrm{var}\{Y_{(r)}\} \doteq p(1 - p)/[f\{F^{-1}(p)\}^2 n]$ provided $f\{F^{-1}(p)\} > 0$.

To find the limiting distribution of $Y_{(r)}$, note that

$$\Pr(Y_{(r)} \leq y) = \Pr\left(\sum_j I_j(y) \geq r\right), \qquad (2.28)$$

where $I_j(y)$ is the indicator of the event $Y_j \leq y$. The $I_j(y)$ are independent, so their sum $\sum_j I_j(y)$ is binomial with probability $F(y)$ and denominator n. Therefore (2.28) and the central limit theorem imply that for large n,

$$\Pr(Y_{(r)} \leq y) \doteq 1 - \Phi\left(\frac{r - nF(y)}{[nF(y)\{1 - F(y)\}]^{1/2}}\right). \qquad (2.29)$$

Now choose $y = F^{-1}(p) + n^{-1/2}z\{p(1 - p)/f\{F^{-1}(p)\}^2\}^{1/2}$, so that

$$F(y) = p + n^{-1/2}z\{p(1 - p)\}^{1/2} + o(n^{-1/2}),$$

and recall that $r = \lceil np \rceil \doteq np$. Then (2.28) and (2.29) imply that, as required,

$$\Pr\left(n^{1/2}\frac{\{Y_{(r)} - F^{-1}(p)\}}{\{p(1-p)/f\{F^{-1}(p)\}^2\}^{1/2}} \le z\right)$$

approximately equals

$$1 - \Phi\left[\frac{np - np - n^{1/2}z\{p(1-p)\}^{1/2}}{\{np(1-p)\}^{1/2}}\right] = 1 - \Phi(-z) = \Phi(z).$$

Exercises 2.3

1   If $U_{(1)} < \cdots < U_{(n)}$ are the order statistics of a $U(0,1)$ random sample, show that $\mathrm{var}(U_{(r)}) = r(n - r + 1)/\{(n + 1)^2(n + 2)\}$. Find $\mathrm{cov}(U_{(r)}, U_{(s)})$, $r < s$ and hence show that $\mathrm{corr}(U_{(r)}, U_{(s)}) \to 1$ for large n as $r \to s$.

2   Let $U_1, \ldots, U_{2m+1}$ be a random sample from the $U(0,1)$ distribution. Find the exact density of the median, $U_{(m+1)}$, and show that $U_{(m+1)} \overset{\cdot}{\sim} N\{\frac{1}{2}, (8m)^{-1}\}$ for large m.

3   Let the $X_1, \ldots, X_n$ be independent exponential variables with rates λ_j. Show that $Y = \min(X_1, \ldots, X_n)$ is also exponential, with rate $\lambda_1 + \cdots + \lambda_n$, and that $\Pr(Y = X_j) = \lambda_j/(\lambda_1 + \cdots + \lambda_n)$.

4   Verify that the joint distribution of all the order statistics of a sample of size n from a continuous distribution with density $f(y)$ is (2.25). Hence find the joint density of the *spacings*, $S_1 = Y_{(1)}$, $S_2 = Y_{(2)} - Y_{(1)}, \ldots, S_n = Y_{(n)} - Y_{(n-1)}$, when $f(y) = \lambda e^{-\lambda y}$, $y > 0$, $\lambda > 0$. Use this to establish (2.26).

5   Use (2.27) to show that $Y_{(r)} \overset{P}{\longrightarrow} F^{-1}(p)$ as $n \to \infty$, where $r = \lceil pn \rceil$ and $0 < p < 1$ is constant.

Consider IQR and MAD (Example 2.2). Show that $\mathrm{IQR} \overset{P}{\longrightarrow} 1.35\sigma$ for normal data and hence give an estimator of σ. Find also the estimator based on MAD.

6   Let N be a random variable taking values $0, 1, \ldots$, let $G(u)$ be the probability-generating function of N, let $X_1, X_2, \ldots$ be independent variables each having distribution function F, and let $Y = \max\{X_1, \ldots, X_N\}$. Show that Y has distribution function $G\{F(y)\}$, and find this when N is Poisson and the X_j exponential.

7   Let M and IQR be the median and interquartile range of a random sample $Y_1, \ldots, Y_n$ from a density of form $\tau^{-1}g\{(y - \eta)/\tau\}$, where $g(u)$ is symmetric about $u = 0$ and $g(0) > 0$. Show that as $n \to \infty$,

$$n^{1/2}\frac{M - \eta}{\mathrm{IQR}} \overset{D}{\longrightarrow} N(0, c),$$

for some $c > 0$, and give c in terms of g and its integral G.

Give c when $g(u)$ equals $\frac{1}{2}\exp(-|u|)$ and $\exp(u)/\{1 + \exp(u)\}^2$.

8   The probability that events in a Poisson process of rate $\lambda > 0$ observed over the interval $(0, t_0)$ occur at $0 < t_1 < t_2 < \cdots < t_n < t_0$ is

$$\lambda^n \exp(-\lambda t_0), \quad 0 < t_1 < t_2 < \cdots < t_n < t_0.$$

By integration over $t_1, \ldots, t_n$, show that the probability that n events occur, regardless of their positions, is

$$\frac{(\lambda t_0)^n}{n!}\exp(-\lambda t_0), \quad n = 0, 1, \ldots,$$

and deduce that given that n events occur, the conditional density of their times is $n!/t_0^n$, $0 < t_1 < t_2 < \cdots < t_n < t_0$. Hence show that the times may be considered to be order statistics from a random sample of size n from the uniform distribution on $(0, t_0)$.

9   Find the exact density of the median M of a random sample $Y_1, \ldots, Y_{2m+1}$ from the uniform
    density on the interval $(\theta - \frac{1}{2}, \theta + \frac{1}{2})$. Deduce that $Z = m^{1/2}(M - \theta)$ has density

$$f(z) = \frac{(2m+1)!}{(m!)^2 m^{1/2}} \left(\frac{1}{4} + \frac{z^2}{m} \right)^m, \quad |z| < \frac{1}{2} m^{1/2},$$

and by considering the behaviour of $\log f(z)$ as $m \to \infty$ or otherwise, show that for large   Stirling's formula implies
m, $Z \overset{\cdot}{\sim} N(0, 1/8)$. Check that this agrees with the general formula for the asymptotic   that $\log m! \sim \frac{1}{2}\log(2\pi) +$
distribution of a central order statistic. $(m + \frac{1}{2})\log m - m$ as
                                                                                                           $m \to \infty$.

2.4 Moments and Cumulants

Calculations involving moments often arise in statistics, but they are generally simpler
when expressed in terms of equivalent quantities known as cumulants.

The *moment-generating function* of the random variable Y is $M(t) = \mathrm{E}(e^{tY})$, provided $M(t) < \infty$. Let

$$M'(t) = \frac{dM(t)}{dt}, \quad M''(t) = \frac{d^2 M(t)}{dt^2}, \quad M^{(r)}(t) = \frac{d^r M(t)}{dt^r}, \quad r = 3, \ldots,$$

denote derivatives of M. If finite, the rth *moment* of Y is $\mu'_r = M^{(r)}(0) = \mathrm{E}(Y^r)$,
giving the power series expansion

$$M(t) = \sum_{r=0}^{\infty} \mu'_r t^r / r!.$$

The quantity μ'_r is sometimes called the rth moment about the origin, whereas $\mu_r =$ The characteristic
$\mathrm{E}\{(Y - \mu'_1)^r\}$ is the rth *moment about the mean*. Among elementary properties of the   function $\mathrm{E}(e^{itY})$, with
moment-generating function are the following: $M(0) = 1$; the mean and variance of $i^2 = -1$ is defined more
Y may be written   broadly than $M(t)$, but as
                                                                                        we shall not need the extra
                                                                                        generality, $M(t)$ is used
$$\mathrm{E}(Y) = M'(0), \quad \mathrm{var}(Y) = M''(O) - \{M'(0)\}^2;$$ almost everywhere in this
                                                                                        book.

random variables $Y_1, \ldots, Y_n$ are independent if and only if their joint moment-
generating function factorizes as

$$\mathrm{E}\{\exp(Y_1 t_1 + \cdots + Y_n t_n)\} = \mathrm{E}\{\exp(Y_1 t_1)\} \cdots \mathrm{E}\{\exp(Y_n t_n)\};$$

and the fact that any moment-generating function corresponds to a unique probability
distribution.

Cumulants

The *cumulant-generating function* or *cumulant generator* of Y is the function $K(t) =$
$\log M(t)$, and the rth *cumulant* is $\kappa_r = K^{(r)}(0) = d^r K(0)/dt^r$, giving the power series
expansion

$$K(t) = \sum_{r=1}^{\infty} t^r \kappa_r / r!, \tag{2.30}$$

provided all the cumulants exist. Differentiation of (2.30) shows that the mean and variance of Y are its first two cumulants

$$\kappa_1 = K'(0) = \frac{M'(0)}{M(0)} = \mu'_1, \quad \kappa_2 = K''(0) = \frac{M''(0)}{M(0)} - \frac{M'(0)^2}{M(0)^2} = \mu'_2 - (\mu'_1)^2.$$

Further differentiation gives higher-order cumulants. Cumulants are mathematically equivalent to moments, and can be defined as combinations of powers of moments, but we shall see below that their statistical interpretation is much more natural than is that of moments.

Example 2.32 (Normal distribution) If Y has the $N(\mu, \sigma^2)$ distribution, its moment-generating function is $M(t) = \exp(t\mu + \frac{1}{2}t^2\sigma^2)$ and its cumulant-generating function is $K(t) = t\mu + \frac{1}{2}t^2\sigma^2$. The first two cumulants are μ and σ^2, and all its higher-order cumulants are zero. The standard normal distribution has $K(t) = \frac{1}{2}t^2$. ∎

The cumulant-generating function is very convenient for statistical work. Consider independent random variables $Y_1, \ldots, Y_n$ with respective cumulant-generating functions $K_1(t), \ldots, K_n(t)$. Their sum $Y_1 + \cdots + Y_n$ has cumulant-generating function

$$\log M_{Y_1 + \cdots + Y_n}(t) = \log E\left\{\exp(tY_1 + \cdots + tY_n)\right\} = \log \prod_{j=1}^{n} M_{Y_j}(t) = \sum_{j=1}^{n} K_j(t).$$

It follows that the rth cumulant of a sum of independent random variables is the sum of their rth cumulants. Similarly, the cumulant-generating function of a linear combination of independent random variables is

$$K_{a+\sum_{j=1}^{n} b_j Y_j}(t) = \log E\left\{\exp(ta + tb_1 Y_1 + \cdots + tb_n Y_n)\right\} = ta + \sum_{j=1}^{n} K_j(b_j t). \tag{2.31}$$

Example 2.33 (Chi-squared distribution) If $Z_1, \ldots, Z_\nu$ are independent standard normal variables, each Z_j^2 has the chi-squared distribution on one degree of freedom, and (3.10) gives its moment-generating function, $(1 - 2t)^{-1/2}$. Therefore each Z_j^2 has cumulant-generating function $-\frac{1}{2}\log(1 - 2t)$, and the χ_ν^2 random variable $W = \sum_{j=1}^{\nu} Z_j^2$ has cumulant-generating function

$$K(t) = -\frac{\nu}{2}\log(1 - 2t) = -\frac{\nu}{2}\sum_{r=1}^{\infty}(-1)^{r-1}\frac{(-2t)^r}{r} = \nu\sum_{r=1}^{\infty}2^{r-1}(r-1)!\frac{t^r}{r!},$$

provided that $|t| < \frac{1}{2}$. Therefore W has rth cumulant $\kappa_r = \nu 2^{r-1}(r-1)!$. In particular, the mean and variance of W are ν and 2ν. ∎

Example 2.34 (Linear combination of normal variables) Let $L = a + \sum_{j=1}^{n} b_j Y_j$ be a linear combination of independent random variables, where Y_j has the normal distribution with mean μ_j and variance σ_j^2. Then L has cumulant-generating function

$$at + \sum_{j=1}^{n} \left\{ (b_j t)\mu_j + \frac{1}{2}(b_j t)^2 \sigma_j^2 \right\} = t \left(a + \sum_{j=1}^{n} b_j \mu_j \right) + \frac{t^2}{2} \left(\sum_{j=1}^{n} b_j^2 \sigma_j^2 \right),$$

corresponding to a $N(a + \sum b_j \mu_j, \sum b_j^2 \sigma_j^2)$ random variable.   ∎

Skewness and kurtosis

The third and fourth cumulants of Y are called its *skewness*, κ_3, and *kurtosis*, κ_4. Example 2.32 showed that $\kappa_3 = \kappa_4 = 0$ for normal variables. This suggests that they be used to assess the closeness of a variable to normality. However, they are not invariant to changes in the scale of Y, and the *standardized skewness* $\kappa_3/\kappa_2^{3/2}$ and *standardized kurtosis* κ_4/κ_2^2 are used instead for this purpose; small values suggest that Y is close to normal.

Some authors define the kurtosis to be $\kappa_4 + 3\kappa_2^2$, in our notation.

The average $\overline{Y}$ of a random sample of observations, each with cumulant-generating function $K(t)$, has mean and variance κ_1 and $n^{-1}\kappa_2$. Expression (2.31) shows that the random variable $Z_n = n^{1/2}\kappa_2^{-1/2}(\overline{Y} - \kappa_1)$, which is asymptotically standard normal, has cumulant-generating function

$$nK\left(n^{-1/2}\kappa_2^{-1/2}t\right) - n^{1/2}\kappa_2^{-1/2}\kappa_1 t,$$

and this equals

$$n\left\{ \frac{t}{n^{1/2}} \frac{\kappa_1}{\kappa_2^{1/2}} + \frac{1}{2}\frac{t^2}{n}\frac{\kappa_2}{\kappa_2} + \frac{1}{6}\frac{t^3}{n^{3/2}}\frac{\kappa_3}{\kappa_2^{3/2}} + \frac{1}{24}\frac{t^4}{n^2}\frac{\kappa_4}{\kappa_2^2} + o\left(\frac{t^4}{n^2}\right) \right\} - n^{1/2}t\frac{\kappa_1}{\kappa_2^{1/2}}.$$

After simplification we find that the cumulant-generating function of Z_n is

$$\frac{1}{2}t^2 + \frac{1}{3}n^{-1/2}t^3\frac{\kappa_3}{\kappa_2^{3/2}} + \frac{1}{24}n^{-1}t^4\frac{\kappa_4}{\kappa_2^2} + o\left(\frac{t^4}{n}\right). \tag{2.32}$$

Hence convergence of the cumulant-generating function of Z_n to $\frac{1}{2}t^2$ as $n \to \infty$ is controlled by the standardized skewness and kurtosis $\kappa_3/\kappa_2^{3/2}$ and κ_4/κ_2^2.

Example 2.35 (Poisson distribution) Let $Y_1, \ldots, Y_n$ be independent Poisson observations with means $\mu_1, \ldots, \mu_n$. The moment-generating function of Y_j is $\exp\{\mu_j(e^t - 1)\}$, so its cumulant-generating function is $K_j(t) = \mu_j(e^t - 1)$ and all its cumulants equal μ_j. As the cumulant-generating function of $Y_1 + \cdots + Y_n$ is $\sum_j \mu_j(e^t - 1)$, the sum $\sum Y_j$ has a Poisson distribution with mean $\sum \mu_j$.

Now suppose that all the μ_j equal μ, say. From (2.31), the cumulant-generating function of the standardized average, $n^{1/2}\mu^{-1/2}(\overline{Y} - \mu)$, is

$$nK\left\{t(n\mu)^{-1/2}\right\} - t(n\mu)^{1/2} = n\mu\left\{e^{t(n\mu)^{-1/2}} - 1\right\} - t(n\mu)^{1/2}$$

$$= n\mu \sum_{r=2}^{\infty} \frac{t^r}{(n\mu)^{r/2}r!}.$$

Thus $\overline{Y}$ has standardized skewness and kurtosis $(n\mu)^{-1/2}$ and $(n\mu)^{-1}$; in general $\kappa_r = (n\mu)^{-(r-2)/2}$ for $r = 2, 3, \ldots$ Hence $\overline{Y}$ approaches normality for fixed μ and large n or fixed n and large μ. ∎

Vector case

A vector random variable $Y = (Y_1, \ldots, Y_p)^{\mathrm{T}}$ has moment-generating function $M(t) = \mathrm{E}(e^{t^{\mathrm{T}}Y})$, where $t^{\mathrm{T}} = (t_1, \ldots, t_p)$. The joint moments of the Y_r are the derivatives

$$\mathrm{E}\big(Y_1^{r_1} \cdots Y_p^{r_p}\big) = \frac{\partial^{r_1 + \cdots + r_p} M(t)}{\partial t_1^{r_1} \cdots \partial t_p^{r_p}}\bigg|_{t=0}.$$

The cumulant-generating function is again $K(t) = \log M(t)$, and the joint cumulants of the Y_r are given by mixed partial derivatives of $K(t)$ with respect to the elements of t. For example, the covariance matrix of Y is the $p \times p$ symmetric matrix whose (r, s) element is $\kappa_{r,s} = \partial^2 K(t)/\partial t_r \partial t_s$, evaluated at $t = 0$.

Suppose that $Y = (Y_1, Y_2)^{\mathrm{T}}$, and that the scalar random variables Y_1 and Y_2 are independent. Then their joint cumulant-generating function is

$$K(t) = \log \mathrm{E}\{\exp(t_1 Y_1 + t_2 Y_2)\} = \log \mathrm{E}\{\exp(t_1 Y_1)\} + \log \mathrm{E}\{\exp(t_2 Y_2)\},$$

because the moment-generating function of independent variables factorizes. But since every mixed derivative of $K(t)$ equals zero, all the joint cumulants of Y_1 and Y_2 equal zero also. This observation generalizes to several variables: the joint cumulants of independent random variables are all zero. This is not true for moments, and partly explains why cumulants are important in statistical work.

Joint derivatives are not needed to obtain first cumulants, which are not joint cumulants.

Example 2.36 (Multinomial distribution) The probability density of a multinomial random variable $Y = (Y_1, \ldots, Y_p)^{\mathrm{T}}$ with denominator m and probabilities $\pi = (\pi_1, \ldots, \pi_p)$, that is $\Pr(Y_1 = y_1, \ldots, Y_p = y_p)$, equals

$$\frac{m!}{y_1! \cdots y_p!} \pi_1^{y_1} \cdots \pi_p^{y_p}, \quad y_r = 0, 1, \ldots, m, \quad \sum_{r=1}^{p} y_r = m;$$

note that $\pi_r \geq 0$, $\sum_r \pi_r = 1$. This arises when m independent observations take values in one of p categories, each falling into the rth category with probability π_r. Then Y_r is the total number falling into the rth category. If $Y_1, \ldots, Y_p$ are independent Poisson variables with means $\mu_1, \ldots, \mu_p$, then their joint distribution conditional on $Y_1 + \cdots + Y_p = m$ is multinomial with denominator m and probabilities $\pi_r = \mu_r / \sum \mu_s$.

The moment-generating function of Y is

$$\mathrm{E}\big(e^{t^{\mathrm{T}}Y}\big) = \sum \frac{m!}{y_1! \cdots y_p!} \pi_1^{y_1} \cdots \pi_p^{y_p} e^{y_1 t_1 + \cdots + y_p t_p} = (\pi_1 e^{t_1} + \cdots + \pi_p e^{t_p})^m;$$

the sum is over all vectors $(y_1, \ldots, y_p)^{\mathrm{T}}$ of non-negative integers such that $\sum_r y_r = m$. Thus $K(t) = m \log(\pi_1 e^{t_1} + \cdots + \pi_p e^{t_p})$. It follows that the joint cumulants of the elements of Y are

$$\kappa_r = m\pi_r,$$
$$\kappa_{r,s} = m\left(\pi_r\delta_{rs} - \pi_r\pi_s\right),$$
$$\kappa_{r,s,t} = m\left(\pi_r\delta_{rst} - \pi_r\pi_s\delta_{rt}[3] + 2\pi_r\pi_s\pi_t\right),$$
$$\kappa_{r,s,t,u} = m\left\{\pi_r\delta_{rstu} - \pi_r\pi_s\left(\delta_{rt}\delta_{su}[3] + \delta_{stu}[4]\right) + 2\pi_r\pi_s\pi_t\delta_{ru}[6] - 6\pi_r\pi_s\pi_t\pi_u\right\};$$

here a Kronecker delta symbol such as δ_{rst} equals 1 if $r = s = t$ and 0 otherwise, and a term such as $\pi_r\pi_s\delta_{rt}[3]$ indicates $\pi_r\pi_s\delta_{rt} + \pi_s\pi_t\delta_{rs} + \pi_r\pi_t\delta_{st}$. The value of $\kappa_{r,s}$ implies that components of Y are negatively correlated, because a large value for one entails low values for the rest. Zero covariance occurs only if $\pi_r = 0$, in which case Y_r is constant. ∎

Exercises 2.4

1 Show that the third and fourth cumulants of a scalar random variable in terms of its moments are
$$\kappa_3 = \mu_3' - 3\mu_1'\mu_2' + 2(\mu_1')^3, \quad \kappa_4 = \mu_4' - 4\mu_3'\mu_1' - 3(\mu_2')^2 + 12\mu_2'(\mu_1')^2 - 6(\mu_1')^4.$$

2 Show that the cumulant-generating function for the gamma density (2.7) is $-\kappa\log(1 - t/\lambda)$. Hence show that $\kappa_r = \kappa(r-1)!/\lambda^r$, and confirm the mean, variance, skewness and kurtosis in Examples 2.12 and 2.26.
   If $Y_1, \ldots, Y_n$ are independent gamma variables with parameters $\kappa_1, \ldots, \kappa_n$ and the same λ, show that their sum has a gamma density, and give its parameters.

3 The Cauchy density (2.16) has no moment-generating function, but its characteristic function is $E(e^{itY}) = \exp(it\theta - |t|)$, where $i^2 = -1$. Show that the average $\overline{Y}$ of a random sample $Y_1, \ldots, Y_n$ of such variables has the same characteristic function as Y_1. What does this imply?

This demands nodding acquaintance with characteristic functions.

2.5 Bibliographic Notes

The idea that variation observed around us can be represented using probability models provides much of the motivation for the study of probability theory and underpins the development of statistics. Cox (1990) and Lehmann (1990) give complementary general discussions of statistical modelling and a glance at any statistical library will reveal hordes of books on specific topics, references to some of which are given in subsequent chapters. Real data, however, typically refuse to conform to neat probabilistic formulations, and for useful statistical work it is essential to understand how the data arise. Initial data analysis typically involves visualising the observations in various ways, examining them for oddities, and intensive discussion to establish what the key issues of interest are. This requires creative lateral thinking, problem solving, and communication skills. Chatfield (1988) gives very useful discussion of this and related topics.

J. W. Tukey and his co-workers have played an important role in stimulating development of approaches to exploratory data analysis both numerical and graphical; see Tukey (1977), Mosteller and Tukey (1977), and Hoaglin *et al.* (1983, 1985, 1991).

John Wilder Tukey (1915–2000) was educated at home and then studied chemistry and mathematics at Brown University before becoming interested in statistics during the 1939–45 war, at the end of which he joined Princeton University. He made important contributions to areas including time series, analysis of variance, and simultaneous inference. He underscored the importance of data analysis, computing, robustness, and interaction with other disciplines at a time when mathematical statistics had become somewhat introverted, and invented many statistical terms and techniques. See Fernholtz and Morgenthaler (2000).

Figure 2.6 Match the sample to the density. Upper panels: four densities compared to the standard normal (heavy). Lower panels: normal probability plots for samples of size 100 from each density.

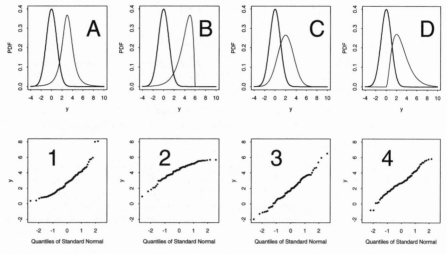

Two excellent books on statistical graphics are Cleveland (1993, 1994), while Tufte (1983, 1990) gives more general discussions of visualizing data. For a brief account see Cox (1978).

Cox and Snell (1981) give an excellent general account of applied statistics.

Most introductory texts on probability and random processes discuss the main convergence results; see for example Grimmett and Stirzaker (2001). Bickel and Doksum (1977) give a more statistical account; see their page 461 for a proof of Slutsky's lemma. See also Knight (2000).

Arnold *et al.* (1992) give a full account of order statistics and many further references.

Most elementary statistics texts do not describe cumulants despite their usefulness. McCullagh (1987) contains forceful advocacy for them, including powerful methods for cumulant calculations. See also Kendall and Stuart (1977), whose companion volumes (Kendall and Stuart, 1973, 1976) overlap considerably with parts of this book, from a quite different viewpoint.

2.6 Problems

Pin the tail on the density.

1 Figure 2.6 shows normal probability plots for samples from four densities. Which goes with which?

2 Suppose that conditional on μ, X and Y are independent Poisson variables with means μ, but that μ is a realization of random variable with density $\lambda^\nu \mu^{\nu-1} e^{-\lambda\mu} / \Gamma(\nu)$, $\mu > 0$, $\nu, \lambda > 0$. Show that the joint moment-generating function of X and Y is

$$\mathrm{E}\left(e^{sX+tY}\right) = \lambda^\nu \{\lambda - (e^s - 1) - (e^t - 1)\}^{-\nu},$$

and hence find the mean and covariance matrix of (X, Y). What happens if $\lambda = \nu/\xi$ and $\nu \to \infty$?

3 Show that a binomial random variable R with denominator m and probability π has cumulant-generating function $K(t) = m \log(1 - \pi + \pi e^t)$. Find $\lim K(t)$ as $m \to \infty$ and

$\pi \to 0$ in such a way that $m\pi \to \lambda > 0$. Show that

$$\Pr(R = r) \to \frac{\lambda^r}{r!}e^{-\lambda},$$

and hence establish that R converges in distribution to a Poisson random variable. This yields the Poisson approximation to the binomial distribution, sometimes called the *law of small numbers*. For a numerical check in the S language, try

```
y <- 0:10;  lambda <- 1; m <- 10; p <- lambda/m
round(cbind(y,pbinom(y,size=m,prob=p),ppois(y,lambda)),digits=3)
```

with various other values of m and λ.

4 (a) Let X be the number of trials up to and including the first success in a a sequence of independent Bernoulli trials having success probability π. Show that $\Pr(X = k) = \pi(1 - \pi)^{k-1}, k = 1, 2, \ldots$, and deduce that X has moment-generating function $\pi e^t/\{1 - (1 - \pi)e^t\}$; hence find its mean and variance. X has the *geometric distribution*.
(b) Now let Y_n be the number of trials up to and including the nth success in such a sequence of trials. Show that

$$\Pr(Y_n = k) = \binom{k-1}{n-1}\pi^n(1 - \pi)^{k-n}, \quad k = n, n+1, \ldots;$$

this is the *negative binomial distribution*. Find the mean and variance of Y_n, and show that as $n \to \infty$ the sequence $\{Y_n\}$ satisfies the conditions of the Central Limit Theorem. Deduce that

$$\lim_{n\to\infty} 2^{1-n} \sum_{k=0}^{n} \binom{k+n-1}{n-1}\frac{1}{2^k} = 1.$$

(c) Find the limiting cumulant-generating function of $\pi Y_n/(1 - \pi)$ as $\pi \to 0$, and hence show that the limiting distribution is gamma.

5 Let $Y_1, \ldots, Y_n$ be a random sample from a distribution with mean μ and variance σ^2. Find the mean of

$$T = \frac{1}{2n(n-1)} \sum_{j \neq k}(Y_j - Y_k)^2,$$

and by writing $Y_j - Y_k = Y_j - \overline{Y} - (Y_k - \overline{Y})$, show that $T = S^2$.

6 Let $Y_1, \ldots, Y_n$ be a random sample from the uniform distribution on the interval $(\theta - \frac{1}{2}, \theta + \frac{1}{2})$. Show that the joint density of the sample maximum and minimum, $Y_{(n)}$ and $Y_{(1)}$, is

$$f_{Y_{(1)}, Y_{(n)}}(u, v) = n(n-1)(v - u)^{n-2}, \quad \theta - \frac{1}{2} < u < v < \theta + \frac{1}{2}.$$

The *sample range* is $R = Y_{(n)} - Y_{(1)}$, and a natural estimator of θ is the *midrange*, $T = (Y_{(n)} + Y_{(1)})/2$. Show that the conditional density of T given R is

$$f(t \mid r; \theta) = (1 - r)^{-1}, \quad 0 < r < 1, \ \theta + \frac{1}{2} - \frac{r}{2} > t > \theta - \frac{1}{2} + \frac{r}{2}.$$

How precisely is θ determined by this density as $r \to 0$ and $r \to 1$?

7 A random variable X with the *Weibull distribution* with index α has distribution function $1 - \exp\{-(x/\lambda)^\alpha\}, x > 0, \lambda, \alpha > 0$. The idea that a system with many similar components will fail when the weakest component fails has led to widespread use of this distribution in industrial reliability.
(a) Suppose that $X_1, \ldots, X_n$ are independent identically distributed continuous non-negative random variables such that as $t \to 0$, the density and distribution functions are asymptotically $at^{\kappa-1}$ and at^α/α respectively, where $a, \alpha > 0$. Let $Y = \min(X_1, \ldots, X_n)$

Waloddi Weibull (1887–1979) was a Swedish engineer who in 1937 published the distribution that bears his name; it is widely used in reliability.

and let $W = (a/\alpha)^{1/\alpha} n^{1/\alpha} Y$. Show that as $n \to \infty$, W has as its limiting distribution the Weibull distribution with index α.

(b) Explain why a probability plot for the Weibull distribution may be based on plotting the logarithm of the rth order statistic against $\log\{-\log(1 - \frac{r}{n+1})\}$, and give the slope and intercept of such a plot. Check whether the data in Table 1.2 follow Weibull distributions.

8   Let $Y_1, \ldots, Y_{2m+1}$ be a random sample from the uniform density

$$f(y) = \begin{cases} \theta^{-1}, & 0 \le y \le \theta, \\ 0, & \text{otherwise.} \end{cases}$$

Derive the density function of the sample median $T = Y_{(m+1)}$ and find its exact mean and variance.

Find the density function of $Z = 2(2m + 3)^{1/2}(Y_{(m+1)} - \theta/2)/\theta$ and use Stirling's formula to show directly that, as $m \to \infty$, Z has asymptotically a standard normal distribution. Deduce that asymptotically $\text{var}(T) \sim 3\text{var}(\overline{Y})$.

9   The *coefficient of variation* of a random sample $Y_1, \ldots, Y_n$ is $C = S/\overline{Y}$, where $\overline{Y}$ and S^2 are the sample average and variance. It estimates the ratio $\psi = \sigma/\mu$ of the standard deviation relative to the mean. Show that

$$\text{E}(C) \doteq \psi, \quad \text{var}(C) \doteq n^{-1}\left(\psi^4 - \gamma_3\psi^3 + \frac{1}{4}\gamma_4\psi^2\right) + \frac{\psi^2}{2(n-1)}.$$

10   If T_1 and T_2 are two competing estimators of a parameter θ, based on a random sample $Y_1, \ldots, Y_n$, the *asymptotic efficiency* of T_1 relative to T_2 is $\lim_{n\to\infty} \text{var}(T_2)/\text{var}(T_1) \times 100\%$. If $n = 2m + 1$, find the asymptotic efficiency of the sample median $Y_{(m+1)}$ relative to the average $\overline{Y} = n^{-1}\sum_j Y_j$ when the density of the Y_j is: (a) normal with mean θ and variance σ^2; (b) Laplace, $(2\sigma)^{-1}\exp\{-|y - \theta|/\sigma\}$ for $-\infty < y < \infty$; and (c) Cauchy, $\sigma/[\pi\{\sigma^2 + (y - \theta)^2\}]$ for $-\infty < y < \infty$.

11   Show that the covariance matrix for the multinomial distribution may be written $m(\text{diag}\{\pi\} - \pi\pi^{\text{T}})$, and deduce that it has determinant zero. Explain why the distribution is degenerate.

12   (a) If X has the $N(\mu, \sigma^2)$ distribution, show that X^2 has cumulant-generating function

$$t\mu^2/(1 - 2t\sigma^2) - \frac{1}{2}\log(1 - 2t\sigma^2).$$

(b) If $X_1, \ldots, X_\nu$ are independent normal variables with variance σ^2 and means $\mu_1, \ldots, \mu_\nu$, show that the cumulant-generating function of $W = X_1^2 + \cdots + X_\nu^2$ is

$$t\delta^2\sigma^2/(1 - 2t\sigma^2) - \frac{\nu}{2}\log(1 - 2t\sigma^2),$$

where $\delta^2 = (\mu_1^2 + \cdots + \mu_\nu^2)/\sigma^2$. The distribution of W/σ^2 is said to be *non-central chi-squared* with ν degrees of freedom and non-centrality parameter δ^2. Show that the moment-generating function of W may be written

$$\exp\left\{-\frac{1}{2}\delta^2 + \frac{1}{2}\delta^2(1 - 2t\sigma^2)^{-1}\right\}(1 - 2t\sigma^2)^{-\nu/2},$$

and that this equals

$$e^{-\delta^2/2}\sum_{r=0}^{\infty}\frac{1}{r!}\left(\frac{\delta^2}{2}\right)^r (1 - 2t\sigma^2)^{-r-\nu/2}. \tag{2.33}$$

Use (2.33) and (3.10) to write down an expression for the density of W.

(c) Hence deduce that (i) $W \overset{D}{=} W_\nu + W_{2N}$, where $W \sim \sigma^2\chi_\nu^2$ independent of $W_{2N} \sim \sigma^2\chi_{2N}^2$, with χ_0^2 taking value 0 with unit probability, and N is Poisson with mean $\delta^2/2$, and (ii) $W \overset{D}{=} (\delta\sigma + Y_1)^2 + Y_2^2 + \cdots + Y_\nu^2$.

3

Uncertainty

In the previous chapter we saw how variation arises in data generated by a model. We now confront a central issue: how to transform knowledge of this variation into statements about the uncertainty surrounding the model parameters. Uncertainty is a more elusive concept than variation and there is more disagreement about how it should be expressed. One important approach described in Chapter 11 uses Bayes' theorem to convert prior knowledge into posterior uncertainty, conditional on the data observed. The route taken below is more common in applications, and is usually known as the *frequentist*, *repeated sampling*, or *classical* approach. The next section describes how uncertainty may be expressed in terms of confidence intervals. In practice confidence intervals are usually approximate, but exact inferences are possible from some central models derived from the normal distribution, and these are described in the following section, followed by a brief summary of methods for prediction. There follows an introduction to the use of simulated data to appreciate both variation and uncertainty, for example in assessing the quality of approximate confidence intervals.

3.1 Confidence Intervals

3.1.1 Standard errors and pivots

In Section 2.2 we saw that many statistics approach limiting distributions in large samples. In practice a sample size is never infinite, but nevertheless these limits may be used to help quantify uncertainty. Suppose that T is an estimator of a parameter ψ based on a random sample $Y_1, \ldots, Y_n$, that its unknown variance $\text{var}(T)$ has form τ^2/n, and that nV is a consistent estimator of τ^2, so $nV \xrightarrow{P} \tau^2$ as $n \to \infty$. Statements of uncertainty about an estimator often involve its standard deviation $n^{-1/2}\tau$, but usually τ is unknown and must be estimated. An estimated standard deviation is known as a *standard error*, so $V^{1/2}$ is a standard error for T.

Example 3.1 (Average) Suppose each of the Y_j has mean μ and variance σ^2. The sample average, $\overline{Y}$, has mean μ and variance σ^2/n. Now $S^2 = (n-1)^{-1} \sum (Y_j - \overline{Y})^2$ is an estimator of σ^2, so $V^{1/2} = n^{-1/2}S$ is a standard error for $\overline{Y}$. ∎

Example 3.2 (Gamma shape) In Example 2.26 we saw that the shape parameter κ of a gamma random sample may be estimated by $T = \overline{Y}^2/S^2$, and that this estimate has variance approximately $2\kappa(\kappa + 1)/n$, which may itself be estimated by $V = 2T(T + 1)/n$. We saw also that for the $n = 95$ observations in Example 2.3, $\overline{y}^2/s^2 = 3.15$. It follows that a standard error for this estimate is

$$\{2 \times 3.15(3.15 + 1)/95\}^{1/2} = 0.52.$$

■

In statements of uncertainty for an unknown parameter ψ, a central role is played by a *pivot* — a function of the data and the parameter whose distribution is known. In the discussion below we must distinguish a generic value of ψ from its true but unknown value ψ_0. The condition that a quantity $Z(\psi_0)$ be pivotal means that for each z $\Pr\{Z(\psi_0) \leq z\}$ is the same for every ψ_0; that is, the distribution of $Z(\psi_0)$ does not depend on ψ_0.

Example 3.3 (Exponential sample) Let $Y_1, \ldots, Y_n$ be a random sample from the exponential distribution $1 - \exp(-y/\psi_0)$, $y > 0$, where $\psi_0 > 0$ is unknown. Then Y_j/ψ_0 has distribution function

$$\Pr(Y_j/\psi_0 \leq u) = \Pr(Y_j \leq u\psi_0) = 1 - \exp(-u),$$

which is known, even though the distribution of Y_j itself is not. Each of the Y_j/ψ_0 has this same distribution, and they are independent, so the distribution of $Z(\psi_0) = \psi_0^{-1} \sum Y_j$ is known, at least in principle. In fact the density of $Z(\psi_0)$ is $z^{n-1} \exp(-z)/(n - 1)!$ for $z > 0$; this is the gamma density (2.7) with parameters $\lambda = 1$ and $\kappa = n$. As n is known, every property of the distribution of $Z(\psi_0)$ may be obtained. ■

Exact pivots are rare, but approximate ones are legion. For example, let $Z(\psi_0) = (T - \psi_0)/V^{1/2}$ be based on a sample of size n, and suppose that the limiting distribution of $Z(\psi_0)$ as $n \to \infty$ is standard normal; the results of Chapter 2 suggest that this will often be the case if T is based on averages. Then if n is large, $Z(\psi_0)$ is roughly standard normal, and so is an approximate pivot. Now

$$\Pr\{Z(\psi_0) \leq z\} = \Pr\left(\frac{T - \psi_0}{V^{1/2}} \leq z\right) \doteq \Phi(z),$$

where Φ is the standard normal distribution function. Then

$$\Pr\left(z_\alpha \leq \frac{T - \psi_0}{V^{1/2}} \leq z_{1-\alpha}\right) \doteq 1 - 2\alpha, \tag{3.1}$$

where z_α is the α quantile of this distribution, that is, $\Phi(z_\alpha) = \alpha$. Equivalently

$$\Pr\left(T - V^{1/2}z_{1-\alpha} \leq \psi_0 \leq T - V^{1/2}z_\alpha\right) \doteq 1 - 2\alpha. \tag{3.2}$$

Hence the random interval whose endpoints are

$$T - V^{1/2}z_{1-\alpha}, \qquad T - V^{1/2}z_\alpha \tag{3.3}$$

contains ψ_0 with probability approximately $(1 - 2\alpha)$, whatever the value of ψ_0. This interval is variously called an approximate $(1 - 2\alpha) \times 100\%$ *confidence interval for*

ψ_0 or a confidence interval for ψ_0 with approximate *coverage probability* $(1 - 2\alpha)$; we call it a $(1 - 2\alpha)$ confidence interval for ψ_0. We regard the interval as random, containing ψ_0 with a specified probability. Conventionally α is a number such as 0.1, 0.05, 0.025, or 0.005, corresponding to 0.8, 0.9, 0.95 and 0.99 confidence intervals for ψ_0; these intervals will be increasingly wide. As $z_\alpha = -z_{1-\alpha}$, (3.3) may be written $T \pm V^{1/2}z_\alpha$. When $1 - 2\alpha = 0.95$, $z_\alpha = -1.96 \doteq -2$, so (3.3) is roughly $T \pm 2V^{1/2}$.

Given a particular set of data, $y_1, \ldots, y_n$, we calculate the confidence interval from (3.3) by replacing T and V with their observed values t and v; this gives $t \pm v^{1/2}z_\alpha$. This interval either does or does not contain ψ_0, though we do not know which in any particular case. We interpret this by reference to a hypothetical infinite sequence of sets of data generated by the same mechanism or experiment that gave the data from which the interval was calculated. We then argue that if the observed data had been selected at random from these sets of data, then the interval actually obtained could be regarded as being selected randomly from a sequence of intervals with the property (3.2), and in this sense it would contain ψ_0 with probability $(1 - 2\alpha)$. With this interpretation, on average 19 out of every 20 confidence intervals with coverage 0.95 will contain ψ_0, and on average 99 out of every 100 intervals with coverage 0.99 will contain ψ_0, and so forth. Such an interval will also contain other values of ψ, but we would like it to be as short as possible on average, so that it does not contain too many of them.

Example 3.4 (Birth data)   We use the data from Example 2.3 to construct a 95% confidence interval for the population mean time in the delivery suite, μ_0 hours, assuming that the times for each day are a random sample $Y_1, \ldots, Y_n$ from the population.

An obvious choice of estimator T is the average, $\overline{Y}$, and we may take V to equal $n^{-1}S^2 = \{n(n-1)\}^{-1}\sum(Y_j - \overline{Y})^2$. In this case a $(1 - 2\alpha) \times 100\%$ confidence interval has endpoints $\overline{Y} \pm n^{-1/2}Sz_\alpha$, and if $(1 - 2\alpha) = 0.95$, then $\alpha = 0.025$ and $z_\alpha = -1.96$. On day 1 there were $n = 16$ deliveries, with average $\overline{y} = 8.77$ and sample variance $s^2 = 18.46$, so a 95% confidence interval for μ_0 based on these data is $\overline{y} \pm n^{-1/2}sz_{0.025} = (6.66, 10.87)$ hours.

The upper left panel of Figure 3.1 shows 95% confidence intervals for μ_0 based on data for each of the first 20 days. The dotted line shows the average time in the delivery suite for all three months of data, which should be close to μ_0. The intervals vary in length and in location, with 18 of them containing the three-month average. We expect about 19 of these 20 intervals to contain the true parameter, and the data seem consistent with this.

The upper right panel illustrates the calculation of the confidence interval from the day 1 data. The horizontal axis shows values of μ, and the diagonal line shows the function $z(\mu) = (8.77 - \mu)/(18.46/16)^{1/2}$. The confidence interval is obtained by reading off those values of μ for which $z(\mu) = z_{0.025}, z_{0.975} = \pm 1.96$, and these are shown by the vertical dashed lines, values of μ between which lie in the interval.

Other values of $\overline{Y}$ and S^2 that might have been observed would give different functions $Z(\mu) = (\overline{Y} - \mu)/(S^2/n)^{1/2}$. The lower right panel shows the observed values $z(\mu)$ of these for each of the first ten days of data. An infinite number of days would induce a probability density for $Z(\mu_0)$, corresponding to the points where the solid

Figure 3.1 Confidence intervals for the mean time in the delivery suite. Upper left: 95% confidence intervals calculated using each of the first 20 days of data, with the average time for three months (92 days) of data (dots). Upper right: $z(\mu) = (\bar{y} - \mu)/(s^2/n)^{1/2}$ as a function of μ for the data from day 1 (diagonal line). The dotted lines show $z_{0.025} = -1.96$ and $z_{0.975} = 1.96$, from which the confidence interval is read off by solving $z(\mu) = \pm 1.96$. Lower right: lines $z(\mu)$ for ten different samples; their intersections $z(\mu_0)$ with the vertical line at μ_0 (blobs) have the standard normal density shown. If μ_0 were different, the density would be translated in the x-direction but remain unchanged, because $Z(\mu_0)$ is a pivot. Lower left: proportion of all 92 95% confidence intervals that include different values of μ. The vertical line (dots) shows the most likely value of μ_0, where the coverage probability should be 0.95, given by the horizontal line (dashes).

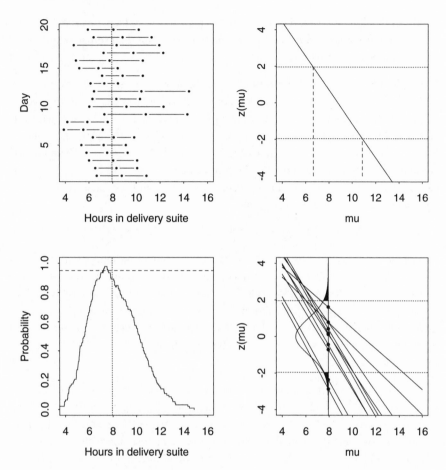

vertical line intersects with the diagonal lines, and this density is illustrated also. If μ_0 was equal to the three-month average of 7.93 hours, we would expect a proportion 0.025 of the blobs at $z(7.93)$ to lie outside ± 1.96. Exact pivotality of $Z(\mu_0)$ would mean that even if μ_0 was not 7.93 hours, so that the density was shifted horizontally, it would not change shape. In fact the normal approximation is not perfect here, as we shall see in Example 3.6.

We can compute the probability that the confidence interval (3.3) contains any value of μ. For μ_0 this should be $(1 - 2\alpha)$, but it will be lower for other values of μ. The lower left panel of Figure 3.1 shows the proportion of the 92 separate daily 95% confidence intervals containing each value of μ. This shows the shape we would expect: values close to the three-month average lie in most of the intervals, while values far from it are rarely covered. The corresponding proportions from an infinite number of days of data are the coverage probabilities

$$\Pr\left(T - z_{1-\alpha} V^{1/2} \leq \mu \leq T - z_\alpha V^{1/2} \middle| \text{true value is } \mu_0 \right).$$

If the approximation (3.2) was perfect, this probability would equal 0.95 when $\mu = \mu_0$, but a poor approximation would give a probability different from 0.95. We would

hope that this function would be as peaked as possible, to reduce the probability that
a value other than μ_0 is contained in the interval: we want the average length of the
intervals to be as short as possible.   ■

Example 3.5 (Binomial distribution)   In opinion polls about the status of the
political parties in the UK, $m = 1000$ people are typically asked about their voting
intentions. Let the number of these who support a particular party be denoted by R,
supposed binomial with probability π. An estimate of π is $\widehat{\pi} = R/m$, and since $\widehat{\pi}$
has variance $\pi(1 - \pi)/m$, the standard error of $\widehat{\pi}$ is $\{\widehat{\pi}(1 - \widehat{\pi})/m\}^{1/2}$. Example 2.17
combined with Slutsky's lemma (2.15) implies that $(\widehat{\pi} - \pi)/\{\widehat{\pi}(1 - \widehat{\pi})/m\}^{1/2}$ converges in distribution to a standard normal variable, and consequently a $(1 - 2\alpha)$
confidence interval for π has endpoints

$$\widehat{\pi} - z_{1-\alpha}\{\widehat{\pi}(1 - \widehat{\pi})/m\}^{1/2}, \quad \widehat{\pi} - z_{\alpha}\{\widehat{\pi}(1 - \widehat{\pi})/m\}^{1/2}.$$

For the two main parties π usually lies in the range 0.3–0.4, so suppose that $\widehat{\pi} = 0.35$,
$m = 1000$, and we want a 95% confidence interval for π, so that $z_{0.975} = -z_{0.025} =$
$1.96 \doteq 2$. Then as $(0.35 \times 0.65/1000)^{1/2} \doteq 0.015$, the interval lies roughly 0.03 on
either side of $\widehat{\pi}$. In percentage terms this is the '3% margin of error' sometimes
mentioned when the results of such a poll are reported. The margin depends little on
$\widehat{\pi}$ because the function $\pi(1 - \pi)$ is fairly flat over the usual range 0.2–0.5 of support
for the main parties.   ■

There are infinitely many confidence intervals with coverage $(1 - 2\alpha)$, because
we can replace $z_{1-\alpha}$ and z_{α} in (3.3) with any pair $z_{1-\alpha_1}$, z_{α_2} such that $\alpha_1, \alpha_2 \geq 0$
and $1 - \alpha_1 - \alpha_2 = 1 - 2\alpha$. The choice $\alpha_1 = \alpha_2 = \alpha$ gives the *equi-tailed* intervals
discussed above, and these are common in practice. Other standard choices are $\alpha_1 =$
$2\alpha, \alpha_2 = 0$ or $\alpha_1 = 0, \alpha_2 = 2\alpha$, which give *one-sided* intervals $(T - V^{1/2}z_{1-2\alpha}, \infty)$
or $(-\infty, T - V^{1/2}z_{2\alpha})$ respectively. These are appropriate when a lower or an upper
confidence bound is required for ψ_0. For example, insurance companies are interested
in upper confidence bounds for potential losses, lower bounds being of little interest.

Complications

In order not to obscure the main points, the discussion above has been deliberately
oversimplified. One complication is that realistic models rarely have just one parameter, so our notion of a pivot must be generalized.

Suppose that in addition to ψ, the model has another parameter λ whose value
is not of interest, and that we seek to construct a confidence interval for ψ_0 using a
pivot $Z(\psi_0)$. Our previous definition must be extended to mean that the distribution
of $Z(\psi_0)$ depends neither on ψ_0 nor on λ. This is a stronger requirement than before
and harder to satisfy.

A second complication is that there may be several possible (approximate) pivots,
so that some basis is needed for choosing the best of them. Obviously we would
like a pivot whose distribution depends as little as possible on the parameters, and
preferably one that is exact, but we should also like short confidence intervals and a
reliable general procedure for obtaining them. We describe some such procedures in

Figure 3.2 Densities of
two approximate pivots
for setting confidence
intervals for the gamma
mean, based on samples
of size $n = 15$ from the
gamma distribution. Left
panel: density estimates
based on 10,000 values of
$Z_1(\mu_0) =$
$n^{1/2}(\overline{Y} - \mu_0)/S$, for shape
parameter $\kappa = 2$ (solid), 3
(dots), 4 (dashes), with
$N(0, 1)$ density (heavy).
Right panel: density of
$Z_2(\mu_0) = \overline{Y}/\mu_0$ for $\kappa = 2$
(line), 3 (dots) 4 (dashes).

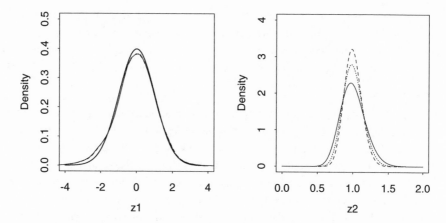

Chapter 4, and return to a general discussion in Chapter 7. The following example illustrates some of the difficulties.

Example 3.6 (Gamma distribution) A random variable Y with gamma density (2.8) may be expressed as $Y = \mu X$, where X has density (2.8) with $\mu = 1$, that is, it has unit mean and shape parameter κ. If $Y_1, \ldots, Y_n$ is a sample from the gamma density with parameters μ_0 and κ, then

$$Z_1(\mu_0) = \frac{\overline{Y} - \mu_0}{\left\{\frac{1}{n(n-1)} \sum (Y_j - \overline{Y})^2\right\}^{1/2}} = \frac{\overline{X} - 1}{\left\{\frac{1}{n(n-1)} \sum (X_j - \overline{X})^2\right\}^{1/2}},$$

and hence the distribution of $Z_1(\mu_0)$ is independent of μ_0. As $n \to \infty$, $Z_1(\mu_0) \xrightarrow{D} N(0, 1)$, giving the confidence interval (3.3), but for any given n the distribution of $Z_1(\mu_0)$ depends on n and on κ. Estimates of this density for $n = 16$ and $\kappa = 2$, 3, and 4 are shown in the left panel of Figure 3.2. The density seems stable over κ, but it is skewed to the left compared to the limiting normal density. Thus although $Z_1(\mu_0)$ appears to be roughly pivotal, values of the normal quantiles z_α might not give good confidence bounds; this would chiefly affect the upper limit.

Another possible pivot here is $Z_2(\mu_0) = \overline{Y}/\mu_0 = \overline{X}$, which turns out to have the gamma density (2.8) with unit mean and shape parameter $n\kappa$. Let $g_\alpha(n\kappa)$ be the α quantile of this distribution. Then

$$1 - 2\alpha = \Pr\{g_\alpha(n\kappa) \leq \overline{Y}/\mu_0 \leq g_{1-\alpha}(n\kappa)\}$$
$$= \Pr\{\overline{Y}/g_{1-\alpha}(n\kappa) \leq \mu_0 \leq \overline{Y}/g_\alpha(n\kappa)\},$$

giving a $(1 - 2\alpha)$ confidence interval $(\overline{y}/g_{1-\alpha}(n\kappa), \overline{y}/g_\alpha(n\kappa))$ based on a sample $y_1, \ldots, y_n$. In practice κ is unknown and must be replaced by an estimate $\widehat{\kappa}$, so $Z_2(\mu_0)$ is also an approximate pivot.

Consider the day 1 data for the delivery suite, for which $n = 16$, $\overline{y} = 8.77$ and suppose $\widehat{\kappa} = 3$. With $\alpha = 0.025$ we find that $g_\alpha(n\widehat{\kappa}) = 0.737$, $g_{1-\alpha}(n\widehat{\kappa}) = 1.302$. This gives 95% confidence interval (6.74, 11.89) hours for μ_0. This interval is longer than that given by the pivot $Z_1(\mu_0)$, (6.66, 10.87), and it is not symmetric about $\overline{y}$.

Densities for $Z_2(\mu_0)$ shown in the right panel of Figure 3.2 depend much more on κ than those for $Z_1(\mu_0)$. Thus here we have a choice between two approximate pivots, one which is close to pivotal but whose distribution can only be estimated, and another which is further from pivotal but whose quantiles are known. ∎

Interpretation

The repeated sampling basis for interpretation of confidence intervals is not universally accepted. The central issue is whether or not hypothetical repetitions bear any relevance to the data actually obtained. One view is that since every set of data is unique, such repetitions would be irrelevant even if they existed, and another basis must be found for statements of uncertainty; see Chapter 11. However it is reassuring that intervals derived from different principles are often similar and sometimes identical for standard problems, and in practice most users do not worry greatly about the precise interpretation of the uncertainty measures they report. The essential point is to provide some assessment of uncertainty, as honest as possible.

Another view is that the repeated sampling interpretation is secure provided the hypothetical data contain the same information, defined suitably, as the original data, but that if the set of hypothetical datasets taken is too large then it is irrelevant to the data actually observed. Thus in the delivery suite example we might argue that as day 1 had 16 arrivals, the relevant hypothetical repetitions are for days with 16 arrivals, because to know the number of arrivals is informative about the precision of any parameter estimate, though not about its value.

3.1.2 Choice of scale

The delta method provides standard errors and limiting distributions for smooth functions of random variables. This poses a problem, however: on what scale should a confidence interval for ψ_0 be calculated? For suppose that h is a monotone function, and that (L, U) is a $(1 - 2\alpha)$ confidence interval for $h(\psi_0)$, that is, $\Pr\{L \leq h(\psi_0) \leq U\} \doteq 1 - 2\alpha$. Then, as

$$\Pr\{h^{-1}(L) \leq \psi_0 \leq h^{-1}(U)\} \doteq 1 - 2\alpha,$$

the interval $(h^{-1}(L), h^{-1}(U))$ is a $(1 - 2\alpha)$ confidence interval for ψ_0. Which of the many possible transformations h should we use? Sometimes the choice is suggested by the need to avoid intervals that contain silly values of ψ, as in the following example.

Example 3.7 (Binomial distribution) Suppose that we want a 95% confidence interval for the support π for a small political party, based on a sample of $m = 100$ individuals. If $\widehat{\pi} = 0.02$, the standard error is $(0.02 \times 0.98/100)^{1/2} = 0.014$, so the 95% interval, roughly $(-0.008, 0.034)$, contains negative values of π.

To avoid this, let us construct an interval for $h(\pi) = \log \pi$ instead, so that $h'(\pi) = \pi^{-1}$. Now $\log \widehat{\pi} = -3.91$, with standard error $\widehat{\pi}^{-1}\{\widehat{\pi}(1 - \widehat{\pi})/m\}^{1/2} = 0.7$. Hence the 95% interval for $\log \pi$ is roughly $-3.91 \pm 1.96 \times 0.7$, and the corresponding interval for π is $(\exp(-3.91 - 1.4), \exp(-3.91 + 1.4)) = (0.005, 0.08)$. The

Table 3.1 Exact mean and variance of variance-stabilized form $Y^{1/2}$ of Poisson random variable.

θ	0.25	0.5	1	2	5	10	20
$E(Y^{1/2})$	0.23	0.44	0.77	1.27	2.17	3.12	4.44
$\text{var}(Y^{1/2})$	0.20	0.31	0.40	0.39	0.29	0.26	0.26

distribution of R/m is too far from normal here to take this interval very seriously, but at least it contains only positive values. ∎

A different approach is to choose a transformation for which $\text{var}\{h(T)\}$ is roughly constant, independent of ψ. Let T be an estimator of ψ, and suppose that $\text{var}(T) = \phi V(\psi)/n$, where ϕ is independent of ψ. The function $V(\psi)$ is called the *variance function* of T. We aim to choose h such that

$$1 \propto \text{var}\{h(T)\} \doteq h'(\psi)^2 \text{var}(T) = h'(\psi)^2 \phi V(\psi)/n,$$

where the approximation results from the delta method. This implies that

$$h(\psi) \propto \int^{\psi} \frac{du}{V(u)^{1/2}}, \tag{3.4}$$

which is called the *variance-stabilizing transformation* for T.

Example 3.8 (Poisson distribution) The mean and variance of the Poisson density (2.6) are both θ, so the average of a random sample of n such variables has mean θ and variance θ/n, giving $V(\theta) = \theta$ and $\phi = 1$. The variance-stabilizing transform is $h(\theta) = \int^{\theta} u^{-1/2}\, du \propto \theta^{1/2}$; the constant of proportionality is irrelevant. The delta method gives $\text{var}(Y^{1/2}) \doteq 0.25$. The exact mean and variance of $Y^{1/2}$ are given in Table 3.1. Variance-stabilization does not work perfectly, but $\text{var}(Y^{1/2})$ depends much less on θ than $\text{var}(Y)$ does.

To apply this to the birth data, we use the 16 arrivals on the first day. To construct a $(1 - 2\alpha)$ confidence interval for the mean arrivals per day, we recall that the Poisson mean and variance both equal θ and suppose that $(Y - \theta)/\theta^{1/2} \sim N(0, 1)$. An estimator of the denominator is $Y^{1/2}$, and taking $(Y - \theta)/Y^{1/2} \sim N(0, 1)$ gives $(Y - Y^{1/2}z_{1-\alpha}, Y - Y^{1/2}z_{\alpha})$ as approximate confidence interval. With $\alpha = 0.025$ and $y = 16$ this yields $(8.2, 23.8)$.

It is better to take $Y^{1/2} \sim N(\theta^{1/2}, 0.25)$, giving $(1 - 2\alpha)$ confidence intervals

$$\left(Y^{1/2} - \frac{1}{2}z_{1-\alpha}, Y^{1/2} - \frac{1}{2}z_{\alpha}\right), \quad \left(\left(Y^{1/2} - \frac{1}{2}z_{1-\alpha}\right)^2, \left(Y^{1/2} - \frac{1}{2}z_{\alpha}\right)^2\right)$$

for $\theta^{1/2}$ and θ. With $\alpha = 0.025$ and $y = 16$ this gives $(9.1, 24.8)$, which is shifted to the right relative to the interval above, and is not symmetric about y. Here the effect of transformation is small, but it can be much larger in other problems. ∎

3.1.3 Tests

The distribution of the pivot $Z(\psi_0)$ implies that some values of ψ are more plausible than others, and we can gauge this using confidence intervals: values of ψ close to the centre of a (say) 95% confidence interval are evidently more plausible than are those that only just lie within it. In some applications a particular value of ψ has special meaning and we may want to assess its plausibility in the light of some data. Given a set of data, a pivot $Z(\psi)$ and a value ψ_0 whose plausibility we wish to establish, one approach is to obtain the observed value of the pivot, $z(\psi_0)$, and then regard the probability $\Pr\{Z(\psi_0) \leq z(\psi_0)\}$ as a measure of the consistency of ψ_0 with the data. The key point is that if ψ_0 was the value of ψ which generated the data, then we would expect $z(\psi_0)$ to be a plausible value for $Z(\psi_0)$, but if not, we would expect $z(\psi_0)$ to be more extreme relative to the known distribution of the pivot.

Example 3.9 (Birth data) If the average time in the delivery suite for 10,000 women at a hospital in Manchester was 6 hours, then we might want to see if this is consistent with the times in Oxford; the Manchester sample is so large that we can treat the 6 hours as fixed. The times for day 1 of the Oxford data seem longer, but how sure can we be?

If ψ_0 for Oxford was equal to 6 hours, then the observed value of $Z(\psi_0)$ for day 1 of the Oxford data,

$$z(\psi_0) = (\bar{y} - \psi_0)/(s^2/n)^{1/2} = (8.77 - 6)/(18.46/16)^{1/2} = 2.58,$$

would be the value of an approximately normal variable. However this seems unlikely: with ψ_0 equal to 6 we get

$$\Pr\{Z(\psi_0) \leq 2.58\} \doteq \Phi(2.58) = 0.995.$$

This is an event which might take place about once in 200 repetitions, and it suggests two possibilities: either the Manchester and Oxford data actually are consistent but an unusual event has occurred, or they are not consistent, and in fact the average time is indeed shorter in Manchester.   ∎

Tests and their relation to confidence intervals are discussed further in Sections 4.5 and 7.3.4.

3.1.4 Prediction

In some applications the focus of interest is the likely value of an as-yet unobserved random variable Y_+, to be predicted using known data y, taken to be a realization of a random variable Y. By analogy with using pivots to make inferences on unknown parameters, it may then be possible to construct a function $Q = q(Y_+, Y)$ whose distribution is independent of the parameters and such that

Prediction intervals are also known as tolerance intervals.

$$\Pr\{q(Y_+, Y) \in R_\alpha\} = \Pr\{l_\alpha(Y) \leq Y_+ \leq u_\alpha(Y)\} = 1 - 2\alpha.$$

Then $(l_\alpha(y), u_\alpha(y))$ is a $(1 - 2\alpha)$ *prediction interval* for Y_+.

Example 3.10 (Location-scale model)   Suppose that Y_+ is to be predicted using an independent random sample $Y_1, \ldots, Y_n$ from a location-scale model. We can write $Y_+ = \eta + \tau \varepsilon_+$ and $Y_j = \eta + \tau \varepsilon_j$, where the εs have common and known density g, say. If $\overline{Y}$ and S^2 are the sample average and variance of $Y_1, \ldots, Y_n$, then the distribution of $Q = (Y_+ - \overline{Y})/S$ depends only on g, and its quantiles q_α may be found numerically. Then

$$\Pr\{q_\alpha \leq (Y_+ - \overline{Y})/S \leq q_{1-\alpha}\} = \Pr(\overline{Y} + Sq_\alpha \leq Y_+ \leq \overline{Y} + Sq_{1-\alpha}) = 1 - 2\alpha,$$

and hence $(\overline{y} + sq_\alpha, \overline{y} + sq_{1-\alpha})$ is an equitailed $(1 - 2\alpha)$ prediction interval for Y_+. ∎

Exercises 3.1

1   Calculate a two-sided 0.95 confidence interval for the mean population time in the delivery suite based on day 2 of the data in Table 2.1. Obtain also lower and upper 0.90 confidence intervals.

2   Let $Y_1, \ldots, Y_n$ be defined by $Y_j = \mu + \sigma X_j$, where $X_1, \ldots, X_n$ is a random sample from a known density g with distribution function G. If $M = m(Y)$ and $S = s(Y)$ are location and scale statistics based on $Y_1, \ldots, Y_n$, that is, they have the properties that $m(Y) = \mu + \sigma m(X)$ and $s(Y) = \sigma s(X)$ for all $X_1, \ldots, X_n, \sigma > 0$ and real μ, then show that $Z(\mu) = n^{1/2}(M - \mu)/S$ is a pivot.

When n is odd and large, g is the standard normal density, M is the median of $Y_1, \ldots, Y_n$ and $S = $ IQR their interquartile range, show that $S/1.35 \xrightarrow{P} \sigma$, and hence show that as $n \to \infty$, $Z(\mu) \xrightarrow{D} N(0, \tau^2)$, for known $\tau > 0$. Hence give the form of a 95% confidence interval for μ.

Compare this interval and that based on using $Z(\mu)$ with $M = \overline{Y}$ and S^2 the sample variance, for the data for day 4 in Table 2.1.

3   If Y is Poisson with large mean θ, then $(Y - \theta)/\theta^{1/2} \stackrel{\cdot}{\sim} N(0, 1)$. Show that the limits of a $(1 - 2\alpha)$ confidence interval for θ are the solutions of the equation $(Y - \theta)^2 = z_\alpha^2 \theta$. Obtain them and compare them with the intervals for the birth data in Example 3.8.

4   Suppose that the unemployment rate π is estimated by sampling randomly from the potential workforce. A total of m individuals are sampled and the number unemployed R is found, giving $\widehat{\pi} = R/m$. How large should m be if $\pi \doteq 0.05$ and a standard error of at most 0.005 is required? What if $\pi = 0.1$?

In some countries such surveys are conducted by telephone interviews with a fixed number of households chosen randomly from the phone book and then asking how many people in the household are eligible for work (not children, retired, ...) and how many are working. Suppose that the total number of people is n, of whom M are eligible to work; suppose that M is binomial with denominator n and probability θ. Of the M, R are eligible to work, so $\widehat{\pi} = R/M$ with M now random. If $n = 12,000$, $\theta = 0.5$ and $\pi = 0.05$, use the delta method to compute a variance for $\widehat{\pi}$. Compute also the variance when $M = 6000$ is treated as fixed. Does the variability of M change the variance by much?

What problems might arise when sampling from the phone book?

5   One way to construct a confidence interval for a real parameter θ is to take the interval $(-\infty, \infty)$ with probability $(1 - 2\alpha)$, and otherwise take the empty set $\emptyset$. Show that this procedure has exact coverage $(1 - 2\alpha)$. Is it a good procedure?

6   A binomial variable R has mean $m\pi$ and variance $m\pi(1 - \pi)$. Find the variance function of $Y = R/m$, and hence obtain the variance-stabilizing transform for R.

7 Let I be a confidence interval for μ based on an estimator T whose distribution is $N(\mu, \sigma^2)$. Show that $\exp(I)$ is a confidence interval for the median of the distribution of $\exp(T)$. How would you compute a confidence interval for its mean, if σ^2 is (i) known and (ii) unknown?

8 If R is binomial with denominator m and probability π, show that

$$\frac{R/m - \pi}{\{\pi(1-\pi)/m\}^{1/2}} \xrightarrow{D} Z \sim N(0, 1),$$

and that the limits of a $(1 - 2\alpha)$ confidence interval for π are the solutions to

$$R^2 - \left(2mR + mz_\alpha^2\right)\pi + m\left(m + z_\alpha^2\right)\pi^2 = 0.$$

Give expressions for them.

In a sample with $m = 100$ and 20 positive responses, the 0.95 confidence interval is $(0.13, 0.29)$. As this interval either does or does not contain the true π, what is the meaning of the 0.95?

9 I am uncertain about what will happen when I next roll a die, about the exact amount of money at present in my bank account, about the weather tomorrow, and about what will happen when I die. Does uncertainty mean the same thing in all these contexts? For which is variation due to repeated sampling meaningful, do you think?

10 Let $Y_1, \ldots, Y_n$ be a random sample from a model in which $Y_j = \theta X_j$, where the X_j are independent with known density g. Show that $\sum Y_j/\theta$ is a pivot, and deduce that a $(1 - 2\alpha)$ confidence interval for θ based on $\sum Y_j$ has form $(\sum Y_j/a, \sum Y_j/b)$, where a and b are known constants.

If $g(x) = e^{-x}$, $x > 0$, is the exponential density, then the 0.025, 0.05, 0.1, 0.5, 0.9, 0.95 and 0.975 quantiles of $\sum X_j$ for $n = 12$ are 6.20, 6.92, 7.83, 11.67, 16.60, 18.21 and 19.68. Use them to give two-sided 0.80 and 0.95 confidence intervals for θ, based on the data in Practical 2.5. Give also upper and lower 0.90 confidence intervals for θ.

3.2 Normal Model

3.2.1 Normal and related distributions

The previous section described an approach to approximate statements of uncertainty, useful in many contexts. We now discuss exact inference for a model of central importance, when the data available form a random sample from the normal distribution. That is, we treat the data $y_1, \ldots, y_n$ as the observed values of $Y_1, \ldots, Y_n$, where the Y_j are independently taken from the normal density

$$f(y; \mu, \sigma^2) = \frac{1}{(2\pi\sigma^2)^{1/2}} \exp\left\{-\frac{1}{2\sigma^2}(y - \mu)^2\right\}, \quad -\infty < y < \infty, \qquad (3.5)$$

Laplace named this the Gaussian density, after Johann Carl Friedrich Gauss (1777–1855), who derived it while writing on the combination of astronomical observations by least squares.

with μ real and σ positive. The normal model owes its ubiquity to the central limit theorem, which, in addition to applying to functions of many observations, may apply to individual measurements themselves. For example, in Example 1.1 it is reasonable to suppose that a plant's height is determined by the effects of many genes, to which an averaging effect may apply, leading to a normal distribution of heights for the population to which the individual belongs, and therefore suggesting the use of normal distributions in (1.1), (1.2), and (1.3). In other situations the simplicity of inference for the normal distribution leads to its use as an approximation even where no such argument applies. Of course it is important to check that the data do appear normally distributed, for example by a normal probability plot (Section 2.1.4).

Before considering inference for the normal sample, we discuss the normal and some related distributions. All are widely tabulated, and their density and distribution functions and quantiles are readily calculated in statistical packages.

See Lindley and Scott (1984) or Pearson and Hartley (1976), for example.

Normal distribution

If we change variable in (3.5) from y to $z = (y - \mu)/\sigma$, we see that the corresponding random variable $Z = (Y - \mu)/\sigma$ has density

$$\phi(z) = (2\pi)^{-1/2} \exp\left(-\frac{1}{2}z^2\right), \quad -\infty < z < \infty; \tag{3.6}$$

this is the density of the *standard normal* random variable Z. The density (3.6) is symmetric about $z = 0$, and $E(Z) = 0$ and $\text{var}(Z) = 1$ (Exercise 3.2.1). Consequently the mean and variance of $Y = \mu + \sigma Z$ are μ and σ^2. We write $Y \sim N(\mu, \sigma^2)$ as shorthand for 'Y has the normal distribution with mean μ and variance σ^2'.

The distribution function corresponding to (3.6),

$$\Phi(z) = (2\pi)^{-1/2} \int_{-\infty}^{z} \exp\left(-\frac{1}{2}u^2\right) du, \tag{3.7}$$

has no closed form, and neither do its quantiles, $z_p = \Phi^{-1}(p)$. Two useful values are $z_{0.025} = -1.96$ and $z_{0.05} = -1.65$. The symmetry of (3.6) about $z = 0$ implies that $z_p = -z_{1-p}$.

The moment-generating function of Y is

$$\begin{aligned}
M(t) &= E(e^{tY}) \\
&= \frac{1}{(2\pi\sigma^2)^{1/2}} \int_{-\infty}^{\infty} \exp\left\{ ty - \frac{1}{2\sigma^2}(y - \mu)^2 \right\} dy \\
&= \frac{1}{(2\pi\sigma^2)^{1/2}} \int_{-\infty}^{\infty} \exp\left\{ \mu t + \sigma^2 \frac{t^2}{2} - \frac{1}{2\sigma^2}(y - \mu - t\sigma)^2 \right\} dy \\
&= \exp\left(\mu t + \sigma^2 t^2/2\right) \int_{-\infty}^{\infty} f(y; \mu + \sigma t, \sigma^2) \, dy \\
&= \exp\left(\mu t + \sigma^2 t^2/2\right), \tag{3.8}
\end{aligned}$$

since for any real t, $f(y; \mu + \sigma t, \sigma^2)$ is just a normal density and has unit integral. We often use variants of this argument to sidestep integration.

The mean and variance of Y can be read off from its cumulant-generating function, $K(t) = \log M(t) = \mu t + \sigma^2 t^2/2$: $\kappa_1 = E(Y) = \mu$ and $\kappa_2 = \text{var}(Y) = \sigma^2$.

Chi-squared distribution

If $Z_1, \ldots, Z_\nu$ are independent standard normal random variables, we say that $W = Z_1^2 + \cdots + Z_\nu^2$ has the *chi-squared distribution on ν degrees of freedom*: we write $W \sim \chi_\nu^2$. The probability density function of W,

Here $\Gamma(\kappa) = \int_0^\infty u^{\kappa-1} e^{-u} \, du$ is the gamma function; see Exercise 2.1.3.

$$f(w) = \frac{1}{2^{\nu/2} \Gamma(\nu/2)} w^{\nu/2 - 1} e^{-w/2}, \quad w > 0, \ \nu = 1, 2, \ldots, \tag{3.9}$$

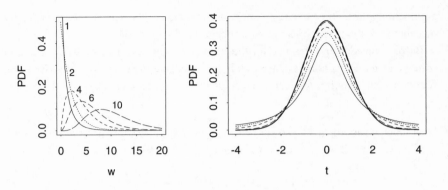

Figure 3.3 Chi-squared and Student t density functions (3.9) and (3.11). Left panel: chi-squared densities with 1 (solid), 2 (dots), 4 (dashes), 6 (larger dashes), and 10 (largest dashes) degrees of freedom. Right panel: t densities with 1 (solid), 2 (dots), 4 (dashes), and 20 (large dashes) degrees of freedom, and standard normal density (heavy solid). The scale is chosen to show the much heavier tails of the t density with few degrees of freedom.

is shown in the left panel of Figure 3.3 for various values of ν. As one would expect from its definition, both the mean and variance of W increase with ν. Its p quantile, denoted $c_\nu(p)$, has the property that $\Pr\{W \leq c_\nu(p)\} = p$. When $\nu = 1$, $W = Z^2$, where $Z \sim N(0, 1)$, so

$$\Pr(W \leq w) = \Pr(-\sqrt{w} \leq Z \leq \sqrt{w}),$$

implying that $c_1(1 - 2p) = z_p^2$.

It is clear from the definition of W that if $W_1 \sim \chi_{\nu_1}^2$ and $W_2 \sim \chi_{\nu_2}^2$ and they are independent, then $W_1 + W_2 \sim \chi_{\nu_1+\nu_2}^2$; evidently this extends to finite sums of independent chi-squared variables. Chi-squared and gamma distributions are closely related: if X has the gamma density (2.7) with parameter λ and shape κ, then $\lambda X \sim \frac{1}{2}\chi_{2\kappa}^2$ (Exercise 3.2.2).

To find the moment-generating function of W, we first find the moment-generating function of Z_j^2, namely

$$\begin{aligned} \mathrm{E}\left(e^{tZ_j^2}\right) &= \frac{1}{(2\pi)^{1/2}} \int_{-\infty}^{\infty} e^{tz^2-z^2/2}\, dz \\ &= (1 - 2t)^{-1/2} \frac{1}{(2\pi)^{1/2}} \int_{-\infty}^{\infty} e^{-u^2/2}\, du \\ &= (1 - 2t)^{-1/2}, \quad t < \frac{1}{2}, \end{aligned} \qquad (3.10)$$

where we have changed variable from z to $u = (1 - 2t)^{1/2}z$. The Z_j^2 are independent and identically distributed, so W has moment-generating function $\{(1 - 2t)^{-1/2}\}^\nu = (1 - 2t)^{-\nu/2}$, differentiation of which shows that the mean and variance of W are ν and 2ν.

Student t distribution

Suppose now that Z and W are independent, that Z is standard normal and W is chi-squared with ν degrees of freedom, and let $T = Z/(W/\nu)^{1/2}$. The random variable T is said to have a *Student t distribution on ν degrees of freedom*; we write $T \sim t_\nu$. Its density is

$$f(t) = \frac{\Gamma\{(\nu + 1)/2\}}{\sqrt{\nu\pi}\,\Gamma(\nu/2)} \frac{1}{(1 + t^2/\nu)^{(\nu+1)/2}}, \quad -\infty < t < \infty, \ \nu = 1, 2, \ldots. \quad (3.11)$$

The right panel of Figure 3.3 shows (3.11) for various values of ν. The distribution of T approaches that of Z for large ν, because the fact that $W/\nu \xrightarrow{P} 1$ as $\nu \to \infty$ implies that $T \xrightarrow{D} Z$; see Example 2.22. The extra variability induced by dividing Z by $(W/\nu)^{1/2}$ spreads out the distribution of T relative to that of Z, by a large amount when ν is small, but by less when ν is large. One consequence of this is that as $\nu \to \infty$ the quantiles of T, denoted $t_\nu(p)$, approach those of Z, that is, $t_\nu(p) \to z_p$. For example, the 0.025 quantiles for $\nu = 2$, 10, and 20 are -4.30, -2.23 and -2.09, while $t_\infty(0.025) = z_{0.025} = -1.96$. The symmetry of (3.11) about $t = 0$ implies that $t_\nu(p) = -t_\nu(1 - p)$.

Not all the moments of T are finite, because the function $t^r f(t)$ is integrable only if $r < \nu$. One simple way to calculate its mean and variance, when they exist, is to use the identities

$$E\{h(Z, W)\} = E_W [E\{h(Z, W) \mid W\}], \qquad (3.12)$$

$$\mathrm{var}\{h(Z, W)\} = E_W [\mathrm{var}\{h(Z, W) \mid W\}] + \mathrm{var}_W [E\{h(Z, W) \mid W\}], \qquad (3.13)$$

which hold for any random variables Z and W; the inner expectation and variance are over the distribution of Z for W fixed (Exercise 3.2.3). If $h(Z, W) = Z/(W/\nu)^{1/2}$ and Z and W are independent, then

$$E\{Z/(W/\nu)^{1/2} \mid W\} = (W/\nu)^{-1/2}E(Z) = 0,$$

$$\mathrm{var}\{Z/(W/\nu)^{1/2} \mid W\} = (W/\nu)^{-1}\mathrm{var}(Z) = (W/\nu)^{-1}.$$

Consequently (3.12) and (3.13) imply that $E(T) = E_W\{Z/(W/\nu)^{1/2}\} = 0$ and

$$\begin{aligned}
\mathrm{var}(T) &= E_W(\nu/W) \\
&= \frac{\nu}{2^{\nu/2}\Gamma(\nu/2)} \int_0^\infty w^{-1} \cdot w^{\nu/2-1} e^{-w/2}\, dw \\
&= \frac{\nu}{2^{\nu/2}\Gamma(\nu/2)} 2^{\nu/2-1}\Gamma(\nu/2 - 1) \\
&= \frac{\nu}{\nu - 2}, \quad \nu = 3, 4, \ldots,
\end{aligned}$$

the first equality following from (3.13), the second from (3.9), the third on noticing that the integrand is proportional to the chi-squared density on $\nu - 2$ degrees of freedom — whose integral must equal one — and the fourth on using the fact that $\Gamma(\kappa + 1) = \kappa\Gamma(\kappa)$, for $\kappa > 0$ (Exercise 2.1.3). The variance of T is finite only if $\nu \geq 3$, and its mean is finite only if $\nu \geq 2$. Setting $\nu = 1$ in (3.11) gives the Cauchy density (2.16), useful for counter-examples.

F distribution

Suppose that W_1 and W_2 have independent chi-squared distributions with ν_1 and ν_2 degrees of freedom respectively. Then

$$F = \frac{W_1/\nu_1}{W_2/\nu_2}$$

has the *F distribution on* v_1 *and* v_2 *degrees of freedom*: we write $F \sim F_{v_1,v_2}$. Its density function is

$$f(u) = \frac{\Gamma\left(\frac{1}{2}v_1 + \frac{1}{2}v_2\right) v_1^{v_1/2} v_2^{v_2/2}}{\Gamma\left(\frac{1}{2}v_1\right) \Gamma\left(\frac{1}{2}v_2\right)} \frac{u^{\frac{1}{2}v_1-1}}{(v_2 + v_1 u)^{(v_1+v_2)/2}}, \quad u > 0, \ v_1, v_2 = 1, 2, \ldots,$$

(3.14)

and its p quantile is denoted $F_{v_1,v_2}(p)$. When $v_1 = 1$, $F = Z^2/(W_2/v_2)$, where $Z \sim N(0, 1)$ is independent of $W_2 \sim \chi^2_{v_2}$, so F then has the same distribution as T^2, where $T \sim t_{v_2}$.

3.2.2 Normal random sample

When a random sample $Y_1, \ldots, Y_n$ is normal, there are compelling reasons to base inference for μ and σ^2 on its average and variance, $\overline{Y}$ and S^2. At the end of this section we shall prove that their joint distribution is given by

We suppose that n is two or more, so $S^2 > 0$ with probability 1.

$$\left. \begin{array}{c} \overline{Y} \sim N(\mu, n^{-1}\sigma^2), \\ (n-1)S^2 \sim \sigma^2 \chi^2_{n-1}, \end{array} \right\} \quad \text{independently.} \qquad (3.15)$$

Another way to express this is

$$\left. \begin{array}{ll} \overline{Y} \overset{D}{=} \mu + n^{-1/2}\sigma Z, & Z \sim N(0, 1), \\ S^2 \overset{D}{=} (n-1)^{-1}\sigma^2 W, & W \sim \chi^2_{n-1}, \end{array} \right\} \quad Z, W \text{ independent.}$$

The studentized form of $\overline{Y}$ may therefore be written

$$\begin{aligned} T &= \frac{\overline{Y} - \mu}{(S^2/n)^{1/2}} \qquad\qquad\qquad (3.16) \\ &\overset{D}{=} \frac{n^{-1/2}\sigma Z}{\{\sigma^2(n-1)^{-1}W/n\}^{1/2}} \\ &= \frac{Z}{\{W/(n-1)\}^{1/2}}, \end{aligned}$$

which has the t distribution with $n - 1$ degrees of freedom.

As the distribution of $T = (\overline{Y} - \mu)/(S^2/n)^{1/2}$ is known, T is an exact pivot, and there is no need for large-sample approximation when a confidence interval is required for μ. That is,

$$\begin{aligned} 1 - 2\alpha &= \Pr\left\{ t_{n-1}(\alpha) \leq \frac{\overline{Y} - \mu}{(S^2/n)^{1/2}} \leq t_{n-1}(1 - \alpha) \right\} \\ &= \Pr\left\{ \overline{Y} - n^{-1/2}St_{n-1}(1 - \alpha) \leq \mu \leq \overline{Y} - n^{-1/2}St_{n-1}(\alpha) \right\}. \end{aligned}$$

As the t distribution is symmetric, the random interval with endpoints

$$\overline{Y} \pm n^{-1/2}St_{n-1}(\alpha) \qquad\qquad (3.17)$$

contains μ with probability exactly $(1 - 2\alpha)$, for all $n \geq 2$. In practice, $\overline{Y}$ and S are replaced by their observed values $\overline{y}$ and s, and the resulting interval has the repeated sampling interpretation outlined in Section 3.1.

Example 3.11 (Maize data) The final column of Table 1.1 contains the differences in heights between $n = 15$ pairs of self- and cross-fertilized plants. Suppose that these differences are a random sample from the $N(\mu, \sigma^2)$ distribution; here μ and σ have units of eighths of an inch, and represent the mean and standard deviation of a population of such differences.

The values of the average and sample variance are $\bar{y} = 20.93$ and $s^2 = 1424.6$. As $t_{14}(0.025) = -2.14$, the 95% confidence interval for μ is $\bar{y} \pm n^{-1/2} s t_{n-1}(\alpha)$, that is, $20.93 \pm (1424.6/15)^{1/2} \times 2.14 = (0.03, 41.84)$ eighths of an inch. This interval suggests that the mean difference in heights is positive; the best estimate of μ is about $2\frac{1}{2}$ inches. However, the value $\mu = 0$ is only just outside the interval, so the evidence for a height difference between the two types of plants is not overwhelming. ∎

A similar argument gives confidence intervals for σ^2. If $(n - 1)S^2 \sim \sigma^2 \chi_{n-1}^2$, then $(n - 1)S^2/\sigma^2 \sim \chi_{n-1}^2$ is another exact pivot. Thus

$$\Pr\left\{ c_{n-1}(\alpha) \leq \frac{(n-1)S^2}{\sigma^2} \leq c_{n-1}(1 - \alpha) \right\} = 1 - 2\alpha,$$

leading to the exact $(1 - 2\alpha)$ confidence interval for σ^2,

$$((n - 1)S^2/c_{n-1}(1 - \alpha), (n - 1)S^2/c_{n-1}(\alpha)). \tag{3.18}$$

Example 3.12 (Maize data) Table 1.1 shows samples of sizes $n_1 = n_2 = 15$ on the heights of plants; the sample variances are $s_1^2 = 837.3$ and $s_2^2 = 269.4$ for the cross- and self-fertilized plants respectively.

If we take $\alpha = 0.025$, then $c_{14}(0.025) = 5.629$ and $c_{14}(0.975) = 26.119$. Hence the 95% confidence interval (3.18) for the variance for the cross-fertilized data is $(14s_1^2/c_{14}(0.975), 14s_1^2/c_{14}(0.025))$, that is, $(449, 2082)$ eighths of inches squared. ∎

The F distribution gives a means to compare the variances of two normal samples. Suppose that S_1^2 and S_2^2 are the sample variances for two independent normal samples of respective sizes n_1 and n_2, and that the variances of those samples are σ^2 and $\psi\sigma^2$. That is, ψ is the ratio of the variances of the samples. Then $(n_1 - 1)S_1^2/\sigma^2$ and $(n_2 - 1)S_2^2/(\psi\sigma^2)$ have independent chi-squared distributions on $n_1 - 1$ and $n_2 - 1$ degrees of freedom, and

$$\Pr\left\{ F_{n_1-1,n_2-1}(\alpha) \leq \frac{S_1^2/\sigma^2}{S_2^2/(\psi\sigma^2)} \leq F_{n_1-1,n_2-1}(1 - \alpha) \right\} = 1 - 2\alpha,$$

or equivalently

$$\Pr\left\{ F_{n_1-1,n_2-1}(\alpha) \frac{S_2^2}{S_1^2} \leq \psi \leq F_{n_1-1,n_2-1}(1 - \alpha) \frac{S_2^2}{S_1^2} \right\} = 1 - 2\alpha.$$

Thus, given two normal random samples whose variances are s_1^2 and s_2^2,

$$\left(F_{n_1-1,n_2-1}(\alpha) s_2^2 / s_1^2, \, F_{n_1-1,n_2-1}(1 - \alpha) s_2^2 / s_1^2 \right) \tag{3.19}$$

is a $(1 - 2\alpha)$ confidence interval for the ratio of variances, ψ. Here the pivot is $\psi S_1^2 / S_2^2$, which has an exact F_{n_1-1, n_2-1} distribution.

Example 3.13 (Maize data) Following on from Example 3.12, we take $\alpha = 0.025$, giving $F_{14,14}(0.025) = 0.336$, $F_{14,14}(0.975) = 2.979$. The 95% confidence interval (3.19) for the ratio of the variances for self- and cross-fertilized plants is $(0.108, 0.958)$. The value $\psi = 1$ is not in this interval, which suggests that the self-fertilized plants are less variable in height than the cross-fertilized ones. ∎

The comparison of variance estimates using F statistics is a crucial ingredient in the analysis of variance, discussed in Section 8.5.

3.2.3 Multivariate normal distribution

The normal distribution plays a central role in inference for scalar data. Its simple properties generalize elegantly to vectors of variables, and these we study now.

One measure of the strength of association between scalar random variables Y_1 and Y_2 is their *covariance*,

$$\text{cov}(Y_1, Y_2) = \text{E}\left[\{Y_1 - \text{E}(Y_1)\} \{Y_2 - \text{E}(Y_2)\}\right].$$

Evidently $\text{cov}(Y_1, Y_1) = \text{var}(Y_1)$, $\text{cov}(Y_1, Y_2) = \text{cov}(Y_2, Y_1)$, and if a and b are constants then $\text{cov}(a + bY_1, Y_2) = b\text{cov}(Y_1, Y_2)$.

In general we may have several random variables. If Y denotes the $p \times 1$ vector $(Y_1, \ldots, Y_p)^\text{T}$ and Z denotes the $q \times 1$ vector $(Z_1, \ldots, Z_q)^\text{T}$, let $\text{E}(Y)$ be the $p \times 1$ vector whose rth element is $\text{E}(Y_r)$. We define the covariance of Y and Z to be the $p \times q$ matrix

$$\text{cov}(Y, Z) = \text{E}\left[\{Y - \text{E}(Y)\} \{Z - \text{E}(Z)\}^\text{T}\right]$$

whose (r, s) element is $\text{cov}(Y_r, Z_s)$. In particular, $\text{cov}(Y, Y) = \Omega$, the $p \times p$ symmetric matrix whose (r, s) element is $\omega_{rs} = \text{cov}(Y_r, Y_s)$; this is called the *covariance matrix* of Y. It is symmetric because $\text{cov}(Y_r, Y_s) = \text{cov}(Y_s, Y_r)$, positive semi-definite because

Or sometimes just the *variance matrix*.

$$\text{var}(a^\text{T}Y) = \text{cov}(a^\text{T}Y, a^\text{T}Y) = a^\text{T}\text{cov}(Y, Y)a = a^\text{T}\Omega a \geq 0$$

for any constant $p \times 1$ vector a, and positive definite unless the distribution of Y is *degenerate*, here meaning that some Y_r is constant or can be expressed in terms of a linear combination of the others (Exercise 3.2.14).

The covariance matrix of the linear combinations $a + B^\text{T}Y$ and $c + D^\text{T}Y$, where a and c are respectively $q \times 1$ and $r \times 1$ constant vectors, and B and D are respectively $p \times q$ and $p \times r$ constant matrices, is

$$\begin{aligned}
\text{cov}(a + B^\text{T}Y, c + D^\text{T}Y) &= \text{E}\left[\{B^\text{T}Y - \text{E}(B^\text{T}Y)\}\{D^\text{T}Y - \text{E}(D^\text{T}Y)\}^\text{T}\right] \\
&= \text{E}\left[B^\text{T}\{Y - \text{E}(Y)\} \{Y - \text{E}(Y)\}^\text{T} D\right] \\
&= B^\text{T}\Omega D.
\end{aligned}$$

When a, b, c, d are constants, $\text{cov}(a + bY_1, c + dY_2) = bd\text{cov}(Y_1, Y_2)$, and thus covariance is not an absolute measure of the association between the variables, because it depends on their units. A measure that is invariant to the choice of units is the *correlation* of Y_1 and Y_2, namely

$$\text{corr}(Y_1, Y_2) = \frac{\text{cov}(Y_1, Y_2)}{\{\text{var}(Y_1)\text{var}(Y_2)\}^{1/2}},$$

some of whose properties were outlined in Example 2.21 and Exercise 2.2.3. Positive correlation between Y_1 and Y_2 indicates that large values of Y_1 and Y_2 tend to occur together, and conversely; whereas negative correlation means that if Y_1 is larger than $\text{E}(Y_1)$, Y_2 tends to be smaller than $\text{E}(Y_2)$. The correlation matrix of a $p \times 1$ vector Y has as its (r, s) element the correlation between Y_r and Y_s, and may be expressed as $\Omega_d^{-1/2}\Omega\Omega_d^{-1/2}$, where Ω_d is the diagonal matrix $\text{diag}(\omega_{11}, \ldots, \omega_{pp})$. The diagonal of $\Omega_d^{-1/2}\Omega\Omega_d^{-1/2}$ consists of ones.

Multivariate normal distribution

A p-dimensional *multivariate normal* random variable $Y = (Y_1, \ldots, Y_p)^{\text{T}}$ with $p \times 1$ vector mean μ and $p \times p$ covariance matrix Ω has density

$$f(y; \mu, \Omega) = \frac{1}{(2\pi)^{p/2}|\Omega|^{1/2}} \exp\left\{-\frac{1}{2}(y - \mu)^{\text{T}}\Omega^{-1}(y - \mu)\right\}; \tag{3.20}$$

we write $Y \sim N_p(\mu, \Omega)$. Here Y, y, and μ take values in $\mathbb{R}^p$. We assume that the distribution is not degenerate, in which case Ω is positive definite, implying amongst other things that its determinant $|\Omega| > 0$.

The moment-generating function of Y is

$$M(t) = \text{E}\left(e^{t^{\text{T}}Y}\right) = \frac{1}{(2\pi)^{p/2}|\Omega|^{1/2}} \int \exp\left\{t^{\text{T}}y - \frac{1}{2}(y - \mu)^{\text{T}}\Omega^{-1}(y - \mu)\right\} dy,$$

where t^{T} is the $1 \times p$ vector $(t_1, \ldots, t_p)$ and $Y = (Y_1, \ldots, Y_p)^{\text{T}}$; the integral is over $y \in \mathbb{R}^p$. To simplify $M(t)$ we write the exponent inside the integral as

$$t^{\text{T}}\mu + \frac{1}{2}t^{\text{T}}\Omega t - \frac{1}{2}(y - \mu - \Omega t)^{\text{T}}\Omega^{-1}(y - \mu - \Omega t).$$

The first two terms of this do not depend on y, so

$$M(t) = \exp\left(t^{\text{T}}\mu + \frac{1}{2}t^{\text{T}}\Omega t\right) \int f(y; \mu + \Omega t, \Omega) \, dy = \exp\left(t^{\text{T}}\mu + \frac{1}{2}t^{\text{T}}\Omega t\right),$$

because for any value of μ, (3.20) is a probability density function. We obtain the moments of Y by differentiation:

$$\text{E}(Y_r) = \frac{\partial M(0)}{\partial t_r} = \mu_r,$$

$$\text{cov}(Y_r, Y_s) = \frac{\partial^2 M(0)}{\partial t_r \partial t_s} - \frac{\partial M(0)}{\partial t_r}\frac{\partial M(0)}{\partial t_s} = \omega_{rs} + \mu_r\mu_s - \mu_r\mu_s = \omega_{rs}.$$

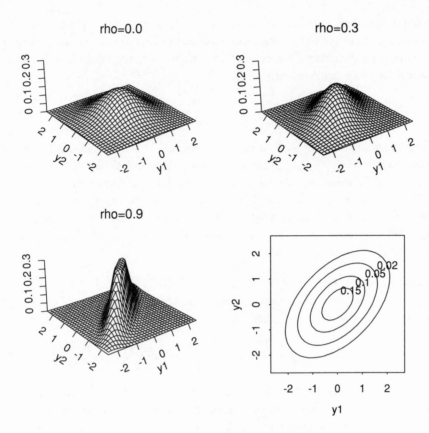

Figure 3.4 The bivariate normal density, with correlation $\rho = 0, 0.3$, and 0.9. The lower right panel shows contours of the density when $\rho = 0.3$; note that they are elliptical. In higher dimensions the contours of equal density are ellipsoids.

The cumulant-generating function of Y is

$$K(t) = \log M(t) = t^{\mathrm{T}}\mu + \frac{1}{2}t^{\mathrm{T}}\Omega t = \sum_{r=1}^{p} t_r \mu_r + \frac{1}{2}\sum_{r=1}^{p}\sum_{s=1}^{p} t_r t_s \omega_{rs}.$$

Thus the first and second cumulants are $\kappa_r = \mu_r$ and $\kappa_{r,s} = \omega_{rs}$, which are respectively the rth element of μ and the (r, s) element of Ω; all higher cumulants are zero.

A special case of (3.20) is the bivariate normal distribution, whose covariance matrix is

$$\begin{pmatrix} \omega_{11} & \omega_{12} \\ \omega_{21} & \omega_{22} \end{pmatrix};$$

the correlation between Y_1 and Y_2 is $\rho = \omega_{12}/(\omega_{11}\omega_{22})^{1/2}$. This density is shown in Figure 3.4 for $\mu = 0$; the effect of increasing ρ is to concentrate the probability mass close to the line $y_1 = y_2$. The corresponding densities for negative ρ are obtained by reflection in the line $y_1 = 0$. When $p = 2$ the contours of constant density are ellipses, but when $p > 2$ they are the ellipsoids given by constant values of $(y - \mu)^{\mathrm{T}}\Omega^{-1}(y - \mu)$.

Marginal and conditional distributions

To study the distribution of a subset of Y, we write $Y^{\mathrm{T}} = (Y_1^{\mathrm{T}}, Y_2^{\mathrm{T}})$, where now Y_1 has dimension $q \times 1$ and Y_2 has dimension $(p - q) \times 1$. Partition t, μ, and Ω conformably, so that

$$t = \begin{pmatrix} t_1 \\ t_2 \end{pmatrix}, \quad \mu = \begin{pmatrix} \mu_1 \\ \mu_2 \end{pmatrix}, \quad \Omega = \begin{pmatrix} \Omega_{11} & \Omega_{12} \\ \Omega_{21} & \Omega_{22} \end{pmatrix},$$

where t_1 and μ_1 are $q \times 1$ vectors and Ω_{11} is a $q \times q$ matrix, t_2 and μ_2 are $(p - q) \times 1$ vectors and Ω_{22} is a $(p - q) \times (p - q)$ matrix, and $\Omega_{12} = \Omega_{21}^{\mathrm{T}}$ is a $q \times (p - q)$ matrix. The moment-generating function of Y is

$$\begin{aligned} \mathrm{E}\big(e^{t^{\mathrm{T}}Y}\big) &= \mathrm{E}\big(e^{t_1^{\mathrm{T}}Y_1 + t_2^{\mathrm{T}}Y_2}\big) \\ &= \exp\big\{ t_1^{\mathrm{T}}\mu_1 + t_2^{\mathrm{T}}\mu_2 + \tfrac{1}{2}\big(t_1^{\mathrm{T}}\Omega_{11}t_1 + 2t_1^{\mathrm{T}}\Omega_{12}t_2 + t_2^{\mathrm{T}}\Omega_{22}t_2\big)\big\}, \end{aligned}$$

from which we obtain the moment-generating functions of Y_1 and Y_2 by setting t_2 and t_1 respectively equal to zero, giving

$$\mathrm{E}\big(e^{t_1^{\mathrm{T}}Y_1}\big) = \exp\big(t_1^{\mathrm{T}}\mu_1 + \tfrac{1}{2}t_1^{\mathrm{T}}\Omega_{11}t_1\big), \ \ \mathrm{E}\big(e^{t_2^{\mathrm{T}}Y_2}\big) = \exp\big(t_2^{\mathrm{T}}\mu_2 + \tfrac{1}{2}t_2^{\mathrm{T}}\Omega_{22}t_2\big).$$

Thus the marginal distributions of Y_1 and Y_2 are multivariate normal also. Note that Y_1 and Y_2 are independent if and only if their joint moment-generating function factorizes, that is,

$$\mathrm{E}\big(e^{t_1^{\mathrm{T}}Y_1 + t_2^{\mathrm{T}}Y_2}\big) = \mathrm{E}\big(e^{t_1^{\mathrm{T}}Y_1}\big)\mathrm{E}\big(e^{t_2^{\mathrm{T}}Y_2}\big), \quad \text{for all } t_1, t_2,$$

which occurs if and only if $\Omega_{12} = \Omega_{21}^{\mathrm{T}} = 0$.

Equivalently and more elegantly, the cumulant-generating function of Y_1 and Y_2 is

$$K(t_1, t_2) = t_1^{\mathrm{T}}\mu_1 + t_2^{\mathrm{T}}\mu_2 + \tfrac{1}{2}\big(t_1^{\mathrm{T}}\Omega_{11}t_1 + 2t_1^{\mathrm{T}}\Omega_{12}t_2 + t_2^{\mathrm{T}}\Omega_{22}t_2\big),$$

and Y_1 and Y_2 are independent if and only if its coefficient in t_1 and t_2, $t_1^{\mathrm{T}}\Omega_{12}t_2$, is identically zero; this is the case if $\Omega_{12} = 0$ but not otherwise.

Thus for normal random variables zero covariance is equivalent to independence. One implication is that if $Y_1, \ldots, Y_n$ is a random sample from the normal distribution with mean μ and variance σ^2, then we can write

$$Y \ \sim \ N_n(\mu 1_n, \sigma^2 I_n).$$

1_n denotes the $n \times 1$ vector of 1s and I_n the $n \times n$ identity matrix.

The conditional distribution of Y_1 given that $Y_2 = y_2$ is (Exercise 3.2.18)

$$N_q\big(\mu_1 + \Omega_{12}\Omega_{22}^{-1}(y_2 - \mu_2), \Omega_{11} - \Omega_{12}\Omega_{22}^{-1}\Omega_{21}\big). \tag{3.21}$$

In the bivariate normal distribution with zero mean and unit variances,

$$N_2\left\{ \begin{pmatrix} 0 \\ 0 \end{pmatrix}, \begin{pmatrix} 1 & \rho \\ \rho & 1 \end{pmatrix} \right\},$$

the conditional mean of Y_1 given $Y_2 = y_2$ is ρy_2, and the conditional variance is $1 - \rho^2$. Thus $\mathrm{var}(Y_1 \mid Y_2 = y_2) \to 0$ as $|\rho| \to 1$. In the lower right panel of Figure 3.4 this conditional density is supported on a horizontal line passing through y_2, and the conditional mean of Y_1 increases with y_2.

Example 3.14 (Trivariate distribution)   Let $Y \sim N_3(\mu, \Omega)$, where

$$
\mu = \begin{pmatrix} 1 \\ 2 \\ 1 \end{pmatrix}, \quad \Omega = \begin{pmatrix} 2 & 0 & 1 \\ 0 & 2 & 1 \\ 1 & 1 & 2 \end{pmatrix}.
$$

The marginal distribution of Y_1 is $N(1, 2)$ and the marginal distribution of $(Y_1, Y_2)^{\mathrm{T}}$ is

$$
N_2 \left\{ \begin{pmatrix} 1 \\ 2 \end{pmatrix}, \begin{pmatrix} 2 & 0 \\ 0 & 2 \end{pmatrix} \right\};
$$

Y_1 and Y_2 are marginally independent.

For the conditional distribution of $(Y_1, Y_2)^{\mathrm{T}}$ given Y_3 we set

$$
\mu_1 = \begin{pmatrix} 1 \\ 2 \end{pmatrix}, \ \mu_2 = (1), \ \Omega_{11} = \begin{pmatrix} 2 & 0 \\ 0 & 2 \end{pmatrix}, \ \Omega_{12} = \Omega_{21}^{\mathrm{T}} = \begin{pmatrix} 1 \\ 1 \end{pmatrix}, \ \Omega_{22} = (2).
$$

Given $Y_3 = y_3$, $(Y_1, Y_2)^{\mathrm{T}}$ is bivariate normal with mean vector and variance matrix

$$
\begin{pmatrix} 1 \\ 2 \end{pmatrix} + \begin{pmatrix} 1 \\ 1 \end{pmatrix} 2^{-1}(y_3 - 1), \quad \begin{pmatrix} 2 & 0 \\ 0 & 2 \end{pmatrix} - \begin{pmatrix} 1 \\ 1 \end{pmatrix} 2^{-1}(1, 1) = \begin{pmatrix} 3/2 & -1/2 \\ -1/2 & 3/2 \end{pmatrix}.
$$

Thus knowledge of Y_3 induces correlation between Y_1 and Y_2 despite their marginal independence. Moreover the conditional variance of Y_1 is smaller than the marginal variance: knowing Y_3 makes one more certain about Y_1. The positive covariance between Y_1 and Y_3 means that if Y_3 is known to exceed its mean, that is, $y_3 > 1$, then the conditional mean of Y_1 exceeds its marginal mean by an amount that depends on the difference $y_3 - 1$.   ∎

Linear combinations of normal variables

Linear combinations of normal random variables often arise. The moment-generating function of the linear combination $a + b^{\mathrm{T}}Y$, where the constants a and b are respectively a scalar and a $p \times 1$ vector, is

$$
\mathrm{E}\left\{ e^{t(a + b^{\mathrm{T}}Y)} \right\} = e^{ta} \exp\left\{ (bt)^{\mathrm{T}}\mu + \frac{1}{2}(bt)^{\mathrm{T}}\Omega(bt) \right\}
$$

$$
= \exp\left\{ t(a + b^{\mathrm{T}}\mu) + \frac{t^2}{2} b^{\mathrm{T}}\Omega b \right\},
$$

and hence $a + b^{\mathrm{T}}Y$ has the normal distribution with mean $a + b^{\mathrm{T}}\mu$ and variance $b^{\mathrm{T}}\Omega b$. This extends to vectors $U = a + B^{\mathrm{T}}Y$, where a is a $q \times 1$ constant vector and B is a $p \times q$ constant matrix. Then U has moment-generating function

$$
\mathrm{E}(e^{t^{\mathrm{T}}U}) = e^{t^{\mathrm{T}}a}\mathrm{E}(e^{t^{\mathrm{T}}B^{\mathrm{T}}Y}) = e^{t^{\mathrm{T}}a}\mathrm{E}(e^{(Bt)^{\mathrm{T}}Y})
$$

$$
= \exp\left\{ t^{\mathrm{T}}a + (Bt)^{\mathrm{T}}\mu + \tfrac{1}{2}(Bt)^{\mathrm{T}}\Omega(Bt) \right\}
$$

$$
= \exp\left\{ t^{\mathrm{T}}(a + B^{\mathrm{T}}\mu) + \tfrac{1}{2}t^{\mathrm{T}}B^{\mathrm{T}}\Omega Bt \right\},
$$

and so U has a multivariate normal distribution with $q \times 1$ mean $a + B^{\mathrm{T}}\mu$ and $q \times q$ covariance matrix $B^{\mathrm{T}}\Omega B$; this is singular and the distribution degenerate unless B has full rank and $q \leq p$. That is, if $Y \sim N_p(\mu, \Omega)$, then

$$a + B^{\mathrm{T}}Y \sim N_q(a + B^{\mathrm{T}}\mu, B^{\mathrm{T}}\Omega B). \tag{3.22}$$

Example 3.15 (Trivariate distribution) In the previous example, consider the joint distribution of $U_1 = Y_1 + Y_2 + Y_3 - 4$ and $U_2 = Y_1 - Y_2 + Y_3$:

$$U = \begin{pmatrix} -4 \\ 0 \end{pmatrix} + \begin{pmatrix} 1 & 1 & 1 \\ 1 & -1 & 1 \end{pmatrix} \begin{pmatrix} Y_1 \\ Y_2 \\ Y_3 \end{pmatrix}.$$

The mean vector and covariance matrix of U are

$$\begin{pmatrix} -4 \\ 0 \end{pmatrix} + \begin{pmatrix} 1 & 1 & 1 \\ 1 & -1 & 1 \end{pmatrix} \begin{pmatrix} 1 \\ 2 \\ 1 \end{pmatrix} = \begin{pmatrix} 0 \\ 0 \end{pmatrix},$$

$$\begin{pmatrix} 1 & 1 & 1 \\ 1 & -1 & 1 \end{pmatrix} \begin{pmatrix} 2 & 0 & 1 \\ 0 & 2 & 1 \\ 1 & 1 & 2 \end{pmatrix} \begin{pmatrix} 1 & 1 \\ 1 & -1 \\ 1 & 1 \end{pmatrix} = \begin{pmatrix} 10 & 4 \\ 4 & 6 \end{pmatrix}.$$

∎

A further consequence of (3.22) follows from the spectral decomposition $\Omega = ELE^{\mathrm{T}}$, where the columns of E are eigenvectors of Ω, L is the diagonal matrix containing the corresponding eigenvalues, and $EE^{\mathrm{T}} = E^{\mathrm{T}}E = I_p$. For positive definite Ω, the elements of L are strictly positive and hence $\Omega^{-1} = EL^{-1}E^{\mathrm{T}}$. We set $U = L^{-1/2}E^{\mathrm{T}}(Y - \mu)$, and note that $U \sim N_p(0, I_p)$, so

$$(Y - \mu)^{\mathrm{T}}\Omega^{-1}(Y - \mu) = (Y - \mu)^{\mathrm{T}}EL^{-1}E^{\mathrm{T}}(Y - \mu) = U^{\mathrm{T}}U \sim \chi_p^2. \tag{3.23}$$

Two samples

Result (3.22) has many uses. For example, suppose that a random sample of size n_1 is available from the $N(\mu_1, \sigma_1^2)$ density and an independent random sample of size n_2 is available from the $N(\mu_2, \sigma_2^2)$ density, and that the focus of interest is the difference of means $\mu_1 - \mu_2$. This is the situation in Example 1.1. Then since (3.15) applies to each sample separately,

$$\begin{pmatrix} \overline{Y}_1 \\ \overline{Y}_2 \end{pmatrix} \sim N_2 \left\{ \begin{pmatrix} \mu_1 \\ \mu_2 \end{pmatrix}, \begin{pmatrix} \sigma_1^2/n_1 & 0 \\ 0 & \sigma_2^2/n_2 \end{pmatrix} \right\},$$

and an application of (3.22) with $a = 0$ and $B^{\mathrm{T}} = (1, -1)$ gives that $\overline{Y}_1 - \overline{Y}_2$ has a normal distribution with mean $\mu_1 - \mu_2$ and variance $n_1^{-1}\sigma_1^2 + n_2^{-1}\sigma_2^2$. To simplify matters, let us suppose that the variances σ_1^2 and σ_2^2 both equal σ^2, in which case

$$\overline{Y}_1 - \overline{Y}_2 \stackrel{D}{=} (\mu_1 - \mu_2) + \sigma\left(n_1^{-1} + n_2^{-1}\right)^{1/2}Z,$$

where $Z \sim N(0, 1)$, and $(n_1 - 1)S_1^2/\sigma^2$ and $(n_2 - 1)S_2^2/\sigma^2$ are independent chi-squared variables with $n_1 - 1$ and $n_2 - 1$ degrees of freedom respectively, so

$(n_1 - 1)S_1^2 + (n_2 - 1)S_2^2 \sim \sigma^2 \chi_{n_1+n_2-2}^2$. Hence the *pooled estimate of* σ^2, S^2, has distribution given by

$$S^2 = \frac{(n_1 - 1)S_1^2 + (n_2 - 1)S_2^2}{n_1 + n_2 - 2} \stackrel{D}{=} \sigma^2 W/(n_1 + n_2 - 2),$$

where $W \sim \chi_{n_1+n_2-2}^2$, independently of $\overline{Y}_1 - \overline{Y}_2$. Consequently the quantity

$$\frac{\overline{Y}_1 - \overline{Y}_2 - (\mu_1 - \mu_2)}{\left\{S^2\left(n_1^{-1} + n_2^{-1}\right)\right\}^{1/2}} \stackrel{D}{=} \frac{Z}{\{W/(n_1 + n_2 - 2)\}^{1/2}} \sim t_{n_1+n_2-2}$$

is a pivot from which confidence intervals for $\mu_1 - \mu_2$ may be determined. The argument parallels that leading to (3.17) and shows that the *two-sample t confidence interval* whose endpoints are

$$(\overline{Y}_1 - \overline{Y}_2) \pm \left\{S^2\left(n_1^{-1} + n_2^{-1}\right)\right\}^{1/2} t_{n_1+n_2-2}(\alpha) \tag{3.24}$$

is a $(1 - 2\alpha)$ confidence interval for $\mu_1 - \mu_2$ based on the two samples. In practice, the random variables in (3.24) are replaced by their observed values, and the resulting interval is given the repeated sampling interpretation.

Example 3.16 (Maize data) For the data in Example 1.1, we have $n_1 = n_2 = 15$, $\overline{y}_1 = 161.5$, $s_1^2 = 837.3$, $\overline{y}_2 = 140.6$ and $s_2^2 = 269.4$. The difference of averages is 20.9 and the pooled estimate of variance is 553.3; note that pooling here ignores the evidence of Example 3.13 that the self-fertilized plants are less variable, that is, $\sigma_2^2 < \sigma_1^2$.

The 0.025 quantile of t_{28} is -2.05, so the two-sample 0.95 confidence interval for $\mu_1 - \mu_2$ is $20.9 \pm 553.3^{1/2}(1/15 + 1/15)^{1/2} \times 2.05 = (3.34, 38.53)$ eighths of an inch. This confidence interval is slightly narrower than that given in Example 3.11, based on differences of pairs of plants, and gives correspondingly stronger evidence for a height difference in mean heights. However, this interval is less appropriate, both because of the pairing of plants in the original experiment, and because of the evidence for a difference in variances.   ∎

If there are two normal samples with unequal variances, $\sigma_1^2 \neq \sigma_2^2$, there is no exact pivot. One fairly accurate approach to confidence intervals for the difference of sample means, $\mu_1 - \mu_2$, is based on the approximate pivot

$$T = \frac{\overline{Y}_1 - \overline{Y}_2 - (\mu_1 - \mu_2)}{\left(S_1^2/n_1 + S_2^2/n_2\right)^{1/2}} \stackrel{\cdot}{\sim} t_\nu, \quad \nu = \frac{\left(S_1^2/n_1 + S_2^2/n_2\right)^2}{S_1^4/\{n_1^2(n_1 - 1)\} + S_2^4/\{n_2^2(n_2 - 1)\}}.$$

The idea of this is to replace the exact variance of $\overline{Y}_1 - \overline{Y}_2$, $\sigma_1^2/n_1^{-1} + \sigma_2^2/n_2^{-1}$, by an estimate, and then to find the t distribution whose degrees of freedom give the best match to the moments of T.

Example 3.17 (Maize data) For the data in Example 1.1, we have $\nu = 22.16$, and $t_\nu(0.025) = -2.07$. Now $s_1^2/n_1 + s_2^2/n_2 = 73.78$, so an approximate 95% confidence interval is $20.9 \pm 2.07 \times 73.78^{1/2}$, that is, $(3.13, 38.74)$. As mentioned before, this interval is more appropriate for these data, but it differs only slightly from the interval in Example 3.16.   ∎

Joint distribution of $\overline{Y}$ and S^2

We now derive the key result (3.15). The most direct route starts from noting that if $Y_1, \ldots, Y_n$ is a random sample from the $N(\mu, \sigma^2)$ distribution, the distribution of $Y = (Y_1, \ldots, Y_n)^{\mathrm{T}}$ is $N_n(\mu 1_n, \sigma^2 I_n)$. We now consider the random variable $U = B^{\mathrm{T}} Y$, where the $n \times n$ matrix B^{T} equals

$$
\begin{pmatrix}
\frac{1}{n^{1/2}} & \frac{1}{n^{1/2}} & \frac{1}{n^{1/2}} & \frac{1}{n^{1/2}} & \cdots & \frac{1}{n^{1/2}} \\
\frac{1}{2^{1/2}} & -\frac{1}{2^{1/2}} & 0 & 0 & \cdots & 0 \\
\frac{1}{6^{1/2}} & \frac{1}{6^{1/2}} & -\frac{2}{6^{1/2}} & 0 & \cdots & 0 \\
\vdots & \vdots & \vdots & \vdots & & \vdots \\
\frac{1}{\{n(n-1)\}^{1/2}} & \frac{1}{\{n(n-1)\}^{1/2}} & \frac{1}{\{n(n-1)\}^{1/2}} & \frac{1}{\{n(n-1)\}^{1/2}} & \cdots & -\frac{n-1}{\{n(n-1)\}^{1/2}}
\end{pmatrix}.
$$

For $j = 2, \ldots, n$, the jth row contains $\{j(j-1)\}^{-1/2}$ repeated $j-1$ times, followed by $-(j-1)\{j(j-1)\}^{-1/2}$ once, with any remaining places filled by zeros. Note that $B^{\mathrm{T}} B = I_n$ and $B^{\mathrm{T}} 1_n = (n^{1/2}, 0, \ldots, 0)^{\mathrm{T}}$, which imply that

$$
U \sim N_n\{(n^{1/2}\mu, 0, \ldots, 0)^{\mathrm{T}}, \sigma^2 I_n\}.
$$

Thus the components of U are independent, and only the first, U_1, has non-zero mean; in fact $U_1 = n^{-1/2} \sum Y_j = n^{1/2}\overline{Y}$, from which we see that $\overline{Y} \sim N(\mu, n^{-1}\sigma^2)$, thus establishing the first line of (3.15). Now

$$
\sum_{j=1}^{n} Y_j^2 = Y^{\mathrm{T}} Y = Y^{\mathrm{T}} B^{\mathrm{T}} B Y = U^{\mathrm{T}} U = \sum_{j=1}^{n} U_j^2 = n\overline{Y}^2 + U_2^2 + \cdots + U_n^2,
$$

which implies that

$$
(n-1)S^2 = \sum_{j=1}^{n} (Y_j - \overline{Y})^2 = \sum_{j=1}^{n} Y_j^2 - n\overline{Y}^2 = U_2^2 + \cdots + U_n^2.
$$

Thus $(n-1)S^2/\sigma^2$ equals the sum of the squares of the $n-1$ standard normal variables $U_2/\sigma, \ldots, U_n/\sigma$, and therefore has the chi-squared distribution with $n-1$ degrees of freedom, independent of U_1, and hence independent of $\overline{Y}$. This establishes the remainder of (3.15).

Exercises 3.2

1 Show that the first two derivatives of $\phi(z)$ are $-z\phi(z)$ and $(z^2 - 1)\phi(z)$. Hence use integration by parts to find the mean and variance of (3.6).

2 If X has density (2.7), show that $2\lambda X$ has density (3.9) with $\nu = 2\kappa$.

3 Let $h(Z, W)$ be a function of two random variables Z and W whose variance is finite, and let $g(W) = E_W\{h(Z, W) \mid W\}$. Show that $h(Z, W) - g(W)$ has mean zero and is uncorrelated with $g(W)$. Hence establish (3.13).

4 Let N be a random variable taking values $0, 1, \ldots$, let $G(u)$ be the probability-generating function of N, and let $X_1, X_2, \ldots$ be independent variables each having moment-generating function $M(t)$. Use (3.12) to show that $Y = X_1 + \cdots + X_N$ has moment-generating function $G\{M(t)\}$, and hence find the mean and variance of Y in terms of those of X and N.
Use (3.12) and (3.13) to find $E(Y)$ and $\mathrm{var}(Y)$ directly.

5 Use (3.6) and (3.9) to derive (3.11).

6 Use (3.9) to derive (3.14).

7 Check carefully the derivations of (3.8) and (3.10).

8 Assuming that the times for each day in Table 2.1 are a random sample from the normal distribution, use the day 2 data to compute (i) a two-sided 0.95 confidence interval for the population mean time in delivery suite and (ii) a 0.95 confidence interval for the population variance. Also give two-sided 0.95 confidence intervals for the difference in mean times for day 1 and day 2, assuming that their variances are (iii) equal and (iv) unequal. Give a 0.95 confidence interval for the ratio of their variances. Repeat (i) and (ii) giving 0.95 upper and lower confidence intervals.

9 If $Z \sim N(0, 1)$, derive the density of $Y = Z^2$. Although Y is determined by Z, show they are uncorrelated.

10 If $W \sim \chi_\nu^2$, show that $E(W) = \nu$, $\text{var}(W) = 2\nu$ and $(W - \nu)/\sqrt{2\nu} \xrightarrow{D} N(0, 1)$ as $\nu \to \infty$.

11 (a) If $F \sim F_{\nu_1, \nu_2}$, show that $1/F \sim F_{\nu_2, \nu_1}$. Give the quantiles of $1/F$ in terms of those of F.
(b) Show that as $\nu_2 \to \infty$, $\nu_1 F$ tends in distribution to a chi-squared variable, and give its degrees of freedom.
(c) If Y_1 and Y_2 are independent variables with density e^{-y}, $y > 0$, show that Y_1/Y_2 has the F distribution, and give its degrees of freedom.

12 Let $f(t)$ denote the probability density function of $T \sim t_\nu$.
(a) Use $f(t)$ to check that $E(T) = 0$, $\text{var}(T) = \nu/(\nu - 2)$, provided $\nu > 1, 2$ respectively.
(b) By considering $\log f(t)$, show that as $\nu \to \infty$, $f(t) \to \phi(t)$.   Recall Stirling's formula.

13 If Y and Z are $p \times 1$ and $q \times 1$ vectors of random variables, show that $\text{cov}(Y, Z) = E(YZ^\mathsf{T}) - E(Y)E(Z)^\mathsf{T}$.

14 Verify that if there is a non-zero vector a such that $\text{var}(a^\mathsf{T}Y) = 0$, either some Y_r takes a single value with probability one or $Y_r = \sum_{s \neq r} b_s Y_s$, for some r, b_s not all equal to zero.

15 Suppose $Y \sim N_p(\mu, \Omega)$ and a and b are $p \times 1$ vectors of constants. Find the distribution of $X_1 = a^\mathsf{T}Y$ conditional on $X_2 = b^\mathsf{T}Y = x_2$. Under what circumstances does this not depend on x_2?

16 Otherwise, or by noting that

$$\sigma^{-1} \int \Phi(a + by)\phi\left(\frac{y - \mu}{\sigma}\right) dy = E_Y \left\{\Pr(Z \leq a + bY \mid Y = y)\right\},$$

where $Z \sim N(0, 1)$, independent of $Y \sim N(\mu, \sigma^2)$, show that

$$\sigma^{-1} \int \Phi(a + by)\phi\left(\frac{y - \mu}{\sigma}\right) dy = \Phi\left\{\frac{a + b\mu}{(1 + b^2\sigma^2)^{1/2}}\right\}. \tag{3.25}$$

17 Let $Y = X_1 + bX_2$, where the X_j are independent normal variables with means μ_j and variances σ_j^2. Show that conditional on $X_2 = x$, the distribution of Y is normal with mean $\mu_1 + bx$ and variance σ_1^2, and hence establish that

$$\int \frac{1}{\sigma_1}\phi\left(\frac{y - \mu_1 - bx}{\sigma_1}\right) \frac{1}{\sigma_2}\phi\left(\frac{x - \mu_2}{\sigma_2}\right) dx = \frac{1}{(\sigma_1^2 + b^2\sigma_2^2)^{1/2}}\phi\left\{\frac{y - \mu_1 - b\mu_2}{(\sigma_1^2 + b^2\sigma_2^2)^{1/2}}\right\}.$$

18 To establish (3.21), show that the variables $X = Y_1 - \Omega_{12}\Omega_{22}^{-1}Y_2$ and Y_2 have a joint multivariate normal distribution and are independent, find the mean of X, and show that its variance matrix is $\Omega_{11} - \Omega_{12}\Omega_{22}^{-1}\Omega_{21}$. Then use the fact that if X and Y_2 are independent, conditioning on $Y_2 = y_2$ will not change the distribution of X, to give (3.21).

19 Let Y have the p-variate multivariate normal distribution with mean vector μ and co-variance matrix Ω. Partition Y^{T} as $(Y_1^{\mathrm{T}}, Y_2^{\mathrm{T}})$, where Y_1 has dimension $q \times 1$ and Y_2 has dimension $r \times 1$, and partition μ and Ω conformably. Find the conditional distribution of Y_1 given that $Y_2 = y_2$ direct from the probability density functions of Y and Y_2.

20 Conditional on $M = m$, $Y_1, \ldots, Y_n$ is a random sample from the $N(m, \sigma^2)$ distribution. Find the unconditional joint distribution of $Y_1, \ldots, Y_n$ when M has the $N(\mu, \tau^2)$ distribution. Use induction to show that the covariance matrix Ω has determinant $\sigma^{2n-2}(\sigma^2 + n\tau^2)$, and show that Ω^{-1} has diagonal elements $\{\sigma^2 + (n-1)\tau^2\}/\{\sigma^2(\sigma^2 + n\tau^2)\}$ and off-diagonal elements $-\tau^2/\{\sigma^2(\sigma^2 + n\tau^2)\}$.

3.3 Simulation

3.3.1 Pseudo-random numbers

Simulation, or the computer generation of artificial data, has many purposes. Among them are:

- to see how much variability to expect in sampling from a particular model. For example, a probability plot for a small sample can be hard to interpret, and in assessing whether any pattern in it is imagined or real it is helpful to compare it with those for sets of simulated data;
- to assess the adequacy of a theoretical approximation. This is illustrated by Figure 2.4, which compares histograms of the average of n simulated exponential variables with the normal density arising from the central limit theorem. The simulations suggest that the approximation is poor when $n \leq 5$, but much improved when $n \geq 20$;
- to check the sensitivity of conclusions to assumptions — for example, how badly do the methods of the previous section fail when the data are not normal? We discuss this in Example 3.24 below;
- to give insight or confirm a hunch, on the principle that a rough answer to the right question is worth more than a precise answer to the wrong question; and
- to provide numerical solutions when analytical ones are unavailable.

The starting point is an algorithm that provides a stream of *pseudo-random* variables, $U_1, U_2, \ldots$, supposed independent and uniformly distributed on the interval $(0, 1)$. These are called *pseudo*-random because although the algorithm should ensure that they seem independent and identically distributed, they are predictable to anyone knowing the algorithm. One important class is the *linear congruential generators* defined by

Some authors call them *quasi-random*.

$$X_{j+1} = (aX_j + c) \bmod M, \quad U_j = X_j/M,$$

for some natural number M, with $a, c \in \{0, 1, \ldots, M - 1\}$; such a generator will repeat with period at most M. The values of M, a and c are chosen to maximize the period and speed of the generator, and the apparent randomness of the output. An example is $M = 2^{48}, a = 5^{17}$ and $c = 1$, giving $M/4$ elements of the set $\{0, \ldots, M - 1\}/M$ in what appears to be a random order.

Not only is it important that the U_j are uniform, but also that they seem independent. One way to do this is to consider k-tuples $(U_j, U_{j+1}, \ldots, U_{j+k-1})$ of successive values as points in the set $(0, 1)^k$, where they should be uniformly distributed; see Practical 3.5. Many of the algorithms in standard packages have been thoroughly tested, but it is wise to store the *seed* X_0 so that if necessary the sequence can be repeated, and to perform important calculations using two different generators. Below we suppose it safe to assume that $U_1, U_2, \ldots$ are independent identically distributed variables from the $U(0, 1)$ distribution (2.22) and refer to them as random rather than pseudo-random.

Inversion

The simplest way to convert uniform variables into those from other distributions is *inversion*. Let F be the distribution function of a random variable, Y, and let $F^{-1}(u) = \inf\{y : F(y) \geq u\}$. If U has the $U(0, 1)$ distribution (2.22), we saw on page 39 that $Y \stackrel{D}{=} F^{-1}(U)$, and that $F^{-1}(U_1), \ldots, F^{-1}(U_n)$ is a random sample from F.

Example 3.18 (Exponential distribution)   The distribution function of an exponential random variable with parameter $\lambda > 0$ is

$$F(y) = \begin{cases} 0, & y \leq 0, \\ 1 - \exp(-\lambda y), & 0 < y, \end{cases}$$

and for $0 < u < 1$ the solution to $F(y) = u$ is $y = -\lambda^{-1} \log(1 - u)$. Therefore a random variable from F is $Y = -\lambda^{-1} \log(1 - U) \stackrel{D}{=} -\lambda^{-1} \log U$, because U and $1 - U$ have the same distribution.   ■

Example 3.19 (Normal, chi-squared and t distributions)   A normal random variable with mean μ and variance σ^2 has distribution function $F(y) = \Phi\{(y - \mu)/\sigma\}$, and therefore $\mu + \sigma \Phi^{-1}(U_1), \ldots, \mu + \sigma \Phi^{-1}(U_n)$ is a normal random sample.

If $Z_1, Z_2, \ldots$ is a stream of standard normal variables, $V = \sum_{j=1}^{\nu} Z_j^2$ is chi-squared with ν degrees of freedom, and $T = Z_{\nu+1}/(V/\nu)^{1/2}$ has the Student t distribution with ν degrees of freedom. Since $Z_j = \Phi^{-1}(U_j)$, V and T are easily obtained.   ■

Pseudo-random variables from other distributions and processes can be constructed using their definitions, though statistical packages usually contain specially-programmed algorithms. One general approach for discrete variables is the *look-up method*. Suppose that Y takes values in $\{1, 2, \ldots\}$ and that we have created a table containing the values of $\Pi_r = \Pr(Y \leq r)$ and $\pi_r = \Pr(Y = r)$. Then inversion amounts to this algorithm:

1   generate $U \sim U(0, 1)$ and set $r = 1$; then
2   while $\Pi_r \leq U$ set $r = r + 1$; and finally
3   return $Y = r$.

The number of comparisons at step 2 can be reduced by sorting the π_r into decreasing order and re-ordering $\{1, 2, \ldots\}$ accordingly. An alternative is to begin searching at a place that depends on U. Each involves initial expense in obtaining and manipulating the π_r's, and as the trade-off between this and the number of comparisons is complicated, fast algorithms for discrete distributions can be complex.

Rejection

Inversion is simple, but to be efficient it requires a fast algorithm for F^{-1}. Another approach is *rejection*. Suppose we wish to generate from an awkward density f, and can easily generate from the uniform distribution and from a density g for which $\sup_y f(y)/g(y) = b < \infty$; note that $b > 1$. The rejection algorithm to generate Y from f is:

Sometimes called the *acceptance-rejection* or *envelope* method.

1   generate X from g and U from the $U(0, 1)$ density, independently;
2   set $Y = X$ if $Ubg(X) \le f(X)$, and otherwise go to 1; finally
3   return Y.

To see why this works, note that the interpretation of $\Pr(X \le a)$ as the area under g to the left of a implies that $(X, Ubg(X))$ is uniformly distributed on the set $\{(x, w) : 0 \le w \le bg(x)\}$, and a value Y is returned only if $Ubg(X) \le f(X)$. For a single pair (X, U), the probability a value Y is returned and is less than y is

$$\Pr\{Ubg(X) \le f(X) \text{ and } X \le y\} = \int_{-\infty}^{y} \Pr\left\{U \le \left.\frac{f(X)}{bg(X)}\right| X = x\right\} g(x)\,dx,$$

$$= \int_{-\infty}^{y} \frac{f(x)}{bg(x)} g(x)\,dx$$

$$= b^{-1} \int_{-\infty}^{y} f(x)\,dx,$$

because U is uniform, independent of X. Hence

$$\Pr(Y \le y \mid \text{value returned}) = \frac{\Pr\{Ubg(X) \le f(X) \text{ and } X \le y\}}{\Pr\{Ubg(X) \le f(X) \text{ and } X \le \infty\}}$$

$$= \int_{-\infty}^{y} f(x)\,dx;$$

the density of Y is indeed f. The probability a value is returned is b^{-1}, so the algorithm is most efficient when b is as small as possible, and the envelope function $bg(x)$ should ensure both this and fast simulation from g.

Example 3.20 (Half-normal density)   A half-normal variable is defined by $Y = |Z|$, where $Z \sim N(0, 1)$. Its density, $f(y) = 2\phi(y)$ for $y > 0$, is shown by the solid line in the left panel of Figure 3.5. The exponential density $g(y) = \lambda e^{-\lambda y}$, declines more slowly than $f(y)$ for large y, and the ratio

$$\frac{f(y)}{g(y)} = \frac{2(2\pi)^{-1/2} e^{-y^2/2}}{\lambda e^{-\lambda y}} = \exp\left\{\lambda y - \frac{1}{2}y^2 + \frac{1}{2}\log\left(\frac{2}{\pi\lambda^2}\right)\right\}$$

is maximized at $y = \lambda$, giving $b = \sup_y f(y)/g(y) = (2/\pi\lambda^2)^{-1/2} e^{\lambda^2/2}$. The function $bg(x)$ with $\lambda = 1$ is shown by the dotted line in the figure.

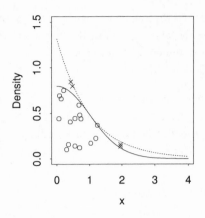

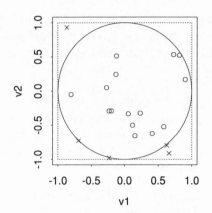

Figure 3.5 Simulation by rejection algorithms. Left panel: half-normal density f (solid) and envelope function bg (dots), with points for which X rejected shown by crosses and those accepted by circles. Right panel: pairs (V_1, V_2) are generated uniformly in the square $[-1, 1] \times [-1, 1]$, but only those in the disk $v_1^2 + v_2^2 \leq 1$ are accepted. They are then transformed into two independent normal variables.

Circles shows pairs $(X, Ubg(X))$ accepted, giving $Y = X$, and crosses show pairs for which X is rejected. These lie in the set $\{(x, w) : f(x) \leq w \leq bg(x)\}$, whose area is $b - 1$, while the area under $bg(x)$ is of course b. The proportion of rejections is minimized by choosing λ to minimize b, and this occurs when $\lambda = 1$, giving $b^{-1} = 0.760$. Whether the resulting algorithm is faster than simply taking $Y = |\Phi^{-1}(U)|$ will depend on the speeds of the functions and the arithmetical operations involved.   ∎

Rejection can be combined with other methods to give efficient algorithms.

Example 3.21 (Normal distribution)   Let Z_1 and Z_2 be two independent standard normal variables. Their joint density is

$$f(z_1, z_2) = \phi(z_1)\phi(z_2) = \frac{1}{2\pi} \exp\left\{-\frac{1}{2}(z_1^2 + z_2^2)\right\}, \quad -\infty < z_1, z_2 < \infty.$$

The polar coordinates of the point (z_1, z_2) in the plane are $r = (z_1^2 + z_2^2)^{1/2}$ and $\theta = \tan^{-1}(z_2/z_1)$, in terms of which $z_1 = r \cos\theta$, $z_2 = r \sin\theta$. The transformation from (z_1, z_2) to (r, θ) has Jacobian

$$\left|\frac{\partial(z_1, z_2)}{\partial(r, \theta)}\right| = \left|\begin{matrix} \cos\theta & \sin\theta \\ -r\sin\theta & r\cos\theta \end{matrix}\right| = r > 0,$$

so the joint density of $R = (Z_1^2 + Z_2^2)^{1/2}$ and $\Theta = \tan^{-1}(Z_2/Z_1)$ is

$$f(r, \theta) = f(z_1, z_2)\left|\frac{\partial(z_1, z_2)}{\partial(r, \theta)}\right| = \frac{1}{2\pi} r \exp\left(-\frac{1}{2}r^2\right), \quad r > 0, \ 0 \leq \theta < 2\pi.$$

Evidently R and Θ are independent, with Θ uniform on the interval $[0, 2\pi)$ and R having distribution $\Pr(R \leq r) = 1 - \exp(-r^2/2)$. Thus if $U_1, U_2 \overset{\text{iid}}{\sim} U(0, 1)$, we can generate Z_1 and Z_2 by setting $Z_1 = R \cos\Theta$, $Z_2 = R \sin\Theta$, where $\Theta = 2\pi U_1$ and $R = (-2 \log U_2)^{1/2}$; this amounts to inversion for R and Θ.

A drawback of this method is that trigonometric functions such as $\sin(\cdot)$ and $\cos(\cdot)$ tend to be slow. It is better to avoid them by using rejection, as follows. We first generate $U_1, U_2 \overset{\text{iid}}{\sim} U(0, 1)$ and set $V_1 = 2U_1 - 1$ and $V_2 = 2U_2 - 1$; (V_1, V_2) is uniformly distributed in the square $[-1, 1] \times [-1, 1]$. If $S = V_1^2 + V_2^2 > 1$, we reject (V_1, V_2) and start again; see the right panel of Figure 3.5. If it is accepted, the point (V_1, V_2) is uniform in the unit disk, S is independent of the angle $\Theta = \tan^{-1}(V_2/V_1)$ by symmetry, and comparison of areas gives $\Pr(S \leq s) = (s\pi)/\pi = s, 0 \leq s \leq 1$, so $S \sim U(0, 1)$; this implies that $R \overset{D}{=} (-2 \log S)^{1/2}$. Furthermore, if (V_1, V_2) has been accepted, then $\cos \Theta = V_1/S^{1/2}$, $\sin \Theta = V_2/S^{1/2}$. Then $Z_1 = R \cos \Theta = V_1(-2S^{-1} \log S)^{1/2}$ and $Z_2 = R \sin \Theta = V_2(-2S^{-1} \log S)^{1/2}$ are independent standard normal variables, and may be obtained without recourse to trigonometric functions. The efficiency of this algorithm is $\pi/4 \doteq 0.785$.

If $Z_1, Z_2 \overset{iid}{\sim} N(0, 1)$, then their ratio $C = Z_2/Z_1$ has a Cauchy distribution. Thus if we want to generate a Cauchy variable, we need only take $R \sin \Theta/(R \cos \Theta) = V_2/V_1$, where (V_1, V_2) lies inside the unit disk. This suggests the *ratio of uniforms* method (Problem 3.7). ∎

It may be hard to find an envelope density $g(y)$ for $f(y)$, leading to a high initialization cost for rejection sampling. If $f(y)$ is log-concave, however, so $h(y) = \log f(y)$ is concave in y, then it turns out to be easy to find an envelope from which quick simulation is possible. To see how, let $f(y)$ be a log-concave density with known support $[y_L, y_U]$, where possibly $y_L = -\infty$ or $y_U = \infty$ or both. Then for any y_1, y_2 in $[y_L, y_U]$,

$$h\{\gamma y_1 + (1 - \gamma)y_2\} \geq \gamma h(y_1) + (1 - \gamma)h(y_2), \quad 0 \leq \gamma \leq 1,$$

and if $h(y)$ is piecewise differentiable, as we henceforth assume, then $h'(y) = dh(y)/dy$ is monotonic decreasing in y, though perhaps $h(y)$ has straight line segments or $h'(y)$ is discontinuous.

Let $y_L \leq y_1 < \cdots < y_k \leq y_U$ and suppose that $h(y_1), \ldots, h(y_k)$ and $h'(y_1), \ldots, h'(y_k)$ are known. If $y_L = -\infty$ we choose y_1 so that $h'(y_1) > 0$. Likewise if $y_U = \infty$, we choose y_k so that $h'(y_k) < 0$. We then define a function $h_+(y)$ by taking the upper boundary of the convex hull generated by the tangents to $h(y)$ at $y_1, \ldots, y_k$; see Figure 3.6. That is,

$$h_+(y) = \begin{cases} h(y_1) + (y - y_1)h'(y_1), & y_L < y \leq z_1, \\ h(y_{j+1}) + (y - y_{j+1})h'(y_{j+1}), & z_j \leq y \leq z_{j+1}, \ j = 1, \ldots, k-1, \\ h(y_k) + (y - y_k)h'(y_k), & z_k \leq y < y_U, \end{cases}$$

where

$$z_j = y_j + \frac{h(y_j) - h(y_{j+1}) + (y_{j+1} - y_j)h'(y_{j+1})}{h'(y_{j+1}) - h'(y_j)}, \quad j = 1, \ldots, k-1,$$

are the values of y at which the tangents at y_j and y_{j+1} intersect; we also set $z_0 = y_L$ and $z_k = y_U$. As the density $g_+(y) \propto \exp\{h_+(y)\}$ consists of k piecewise exponential portions, a variable X with density g_+ may be generated by inversion and then rejection applied. If the X thus generated is rejected, then $h(X)$ and $h'(X)$ can be used to update h_+ and provide a better envelope for subsequent simulation.

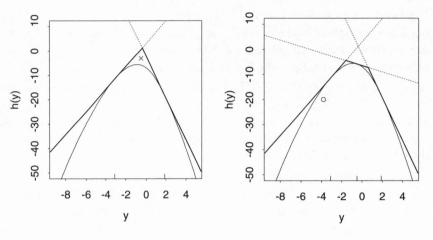

Figure 3.6 Adaptive rejection sampling from log-concave density proportional to $h(y)$ (solid). The left panel shows the initial envelope (heavy), formed as the concave hull of tangents (dotted) to $h(y)$ at $y = -3.1, 1.9$ (rug). The envelope density looks like two exponential densities, back to back, from which a value shown by a cross is generated. This value is rejected but used to update the envelope to that on the right, so the corresponding density has three exponential parts. This time the value generated by rejection sampling (circle) is accepted.

This discussion suggests an *adaptive rejection sampling* algorithm:

1. Initialize by choosing $y_1 < \cdots < y_k$, calculating $h(y_1), \ldots, h(y_k)$ and $h'(y_1), \ldots, h'(y_k), h_+(y)$ and $g_+(y)$. Then
2. generate independent variables X from g_+ and U from the $U(0, 1)$ density. If $U \leq \exp\{h(X) - h_+(X)\}$ then set $Y = X$ and return Y; otherwise
3. replace k by $k + 1$, update $y_1, \ldots, y_k, h(y_1), \ldots, h(y_k)$ and $h'(y_1), \ldots, h'(y_k)$ by adding X, $h(X)$ and $h'(X)$, recompute $h_+(y)$ and $g_+(y)$ and go to 2.

This can be accelerated by using $h(y_1), \ldots, h(y_k)$ and $h'(y_1), \ldots, h'(y_k)$ to add a lower envelope $h_-(y)$ and then accepting X if $U \leq \exp\{h_-(X) - h_+(X)\}$, in which case $h(X)$ need not be computed (Problem 3.12).

Example 3.22 (Adaptive rejection)   To illustrate this we take

$$h(y) = ry - m \log(1 + e^y) - \frac{(y - \mu)^2}{2\sigma^2} + c, \quad -\infty < y < \infty,$$

where $m, \sigma^2 > 0$ and c is the constant ensuring that $\exp\{h(y)\}$ has unit integral; see Example 11.26. As we deal only with ratios of densities we can ignore c below, and Figure 3.6 shows $h(y)$ for $r = 2$, $m = 10$, $\mu = 0$, $\sigma^2 = 1$ and when we set $c = 0$; here $y_L = -\infty$ and $y_U = \infty$.

An initial search establishes that $h'(-3.1) > 0$ and $h'(1.9) < 0$, and the resulting envelope is shown in the left panel. The corresponding density $g_+(y)$ looks like two back-to-back exponential densities, from which it is easy to simulate a value X. This is accepted if $Ug_+(X) < h(X)$, where $U \sim U(0, 1)$. In the event, the value -0.5 is generated but not accepted, and the envelope is updated to that shown in the right panel. A value generated from the new $g_+(y)$ is accepted, terminating the algorithm. Otherwise the envelope would again be updated, and the process repeated. ∎

Applications

Fast, tested generators are available in many statistical packages, so the details can often — but not always — be ignored. Here are two uses of them.

Figure 3.7 Numbers of women in the delivery suite over a week of simulations from the model for the birth data. Also shown are arrival and departure times for the first 25 simulated women.

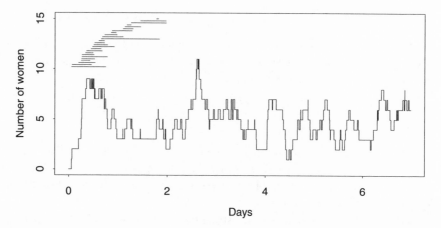

Example 3.23 (Birth data) The data in Example 2.3 were collected in order to assess the workload in the delivery suite. Examples 2.11 and 2.12 suggest that the daily number of arrivals leading to normal deliveries is Poisson with mean about $\lambda = 12.9$, and that each woman remains for a period whose density is roughly gamma with shape $\alpha = 3.15$ and mean $\mu = 7.93$ hours, independent of the others. To simulate t days of data from this model, we generate a Poisson random variable N with mean λ for each day, and then generate N arrival times uniformly through the day. We create departure times by adding a gamma variable with mean μ and shape α to each arrival time; of course a woman may not depart on the day she arrived. We repeat this for each day, and record how many women are present at each arrival and departure.

Figure 3.7 shows a week of simulated workload. Note the initial 'burn-in' period, due to starting with no women present rather than in steady state. The number present has long-run average $12.9 \times 7.93/24 = 4.26$, but it fluctuates widely, with bursts of activity when several women arrive almost together.

Such simulations show the random variation in the process due to the model, but they do not reflect the fact that the model itself is uncertain, because it has been estimated. However it would be easy to change λ, α, and μ, or to replace the gamma by a different distribution, and then to repeat the simulation. This would help assess the effect of model uncertainty.

On leaving the delivery suite, women and their babies go to a ward where midwives give post-natal care. At one stage hospital managers hoped to save money by imposing a rigid demarcation between ward and delivery suite, but this would have been counter-productive. According to hospital guidelines, each woman in the delivery suite should have a midwife with her at all times, so when bursts of activity begin it is essential to be able to call in midwives immediately. It is more expensive to do so from outside, so costs are reduced by allowing easy transfer of workers between ward and suite. ∎

The previous example illustrates a particularly simple queueing system — each 'customer' must be dealt with at once, so there is no queue! More complicated queues arise in many contexts, and *discrete-event simulation* packages exist to help operations researchers estimate quantities such as the average waiting-time.

We now use simulation use to assess properties of a statistical procedure.

Example 3.24 (*t* statistic)   The elements of a random sample $Y_1, \ldots, Y_n$ from the $N(\mu, \sigma^2)$ density may be expressed $Y_j = \mu + \sigma Z_j$, where the Z_j are standard normal variables. The t statistic may be written as

Recall that $\overline{Y}$ and S^2 are the average and sample variance of $Y_1, \ldots, Y_n$.

$$T = \frac{\overline{Y} - \mu}{(S^2/n)^{1/2}} = \frac{n^{1/2}(\mu + \sigma\overline{Z} - \mu)}{\left\{(n-1)^{-1}\sigma^2 \sum_j (Z_j - \overline{Z})^2\right\}^{1/2}} = \frac{n^{1/2}\overline{Z}}{S_Z},$$

say, whether or not the Z_j are normal. When they are, T has a Student t distribution on $n - 1$ degrees of freedom and its quantiles $t_{n-1}(\alpha)$ may be explicitly calculated, leading to the exact $(1 - 2\alpha)$ confidence interval (3.17). How badly does that interval fail when the data are not normal?

Suppose the Z_j have mean zero but distribution F otherwise unspecified. Then the confidence interval (3.17) contains μ with probability

$$\Pr\left\{\overline{Y} - n^{-1/2}St_{n-1}(1 - \alpha) \le \mu \le \overline{Y} - n^{-1/2}St_{n-1}(\alpha)\right\}, \tag{3.26}$$

and this equals

$$\Pr\left\{t_{n-1}(\alpha) \le T \le t_{n-1}(1 - \alpha)\right\} = \Pr\left\{t_{n-1}(\alpha) \le \frac{n^{1/2}\overline{Z}}{S_Z} \le t_{n-1}(1 - \alpha)\right\}$$
$$= p(1 - \alpha, n, F) - p(\alpha, n, F),$$

say, where

$$p(\alpha, n, F) = \Pr\left\{\frac{n^{1/2}\overline{Z}}{S_Z} \le t_{n-1}(\alpha)\right\}.$$

When F is normal, $p(\alpha, n, F) = \alpha$ and (3.26) is $(1 - 2\alpha)$, as it should be.

Given any F, α and n, we estimate $p(\alpha, n, F)$ thus. For $r = 1 \ldots, R$,

- generate $Z_1, \ldots, Z_n \overset{\text{iid}}{\sim} F$;
- calculate $T_r = n^{1/2}\overline{Z}/S_Z$; then
- set $I_r = I\{T_r \le t_{n-1}(\alpha)\}$.

$I\{A\}$ is the indicator of the event A.

Having obtained $I_1, \ldots, I_R$, we compute $\widehat{p} = R^{-1} \sum_r I_r$, whose expectation is

$$E\left(R^{-1}\sum_{r=1}^{R} I_r\right) = E\left[I\left\{\frac{n^{1/2}\overline{Z}}{S_Z} \le t_{n-1}(\alpha)\right\} \middle| Z_1, \ldots, Z_n \overset{\text{iid}}{\sim} F\right] = p(\alpha, n, F).$$

Now $\sum_r I_r$ is binomial with denominator R and probability $p(\alpha, n, F)$, so $\widehat{p}$ has variance $R^{-1}p(\alpha, n, F)\{1 - p(\alpha, n, F)\}$. This can be used to gauge the value of R needed to estimate $p(\alpha, n, F)$ with given precision. For example, if $p(\alpha, n, F) \doteq \alpha = 0.05$, then $R = 1600$ gives standard deviation roughly $\{0.05(1 - 0.05)/1600\}^{1/2} = 0.0054$, and a crude 95% confidence interval for $p(\alpha, n, F)$ is $\widehat{p} \pm 0.01$.

Table 3.2 shows values of $100\widehat{p}$ for various distributions F, using $n = 10$ and $R = 1600$. The second and third columns are for $\alpha = 0.05, 0.95$, while the fourth shows the estimated probability that the confidence interval contains μ; ideally this

Table 3.2 Estimated coverage probabilities $p(\alpha, n, F)$, $p(1 - \alpha, n, F)$, and $p(1 - \alpha, n, F) - p(\alpha, n, F)$, for $\alpha = 0.05$ and 0.025, for 1600 samples of size $n = 10$ from various distributions. The Laplace and mixture densities are $\frac{1}{2}\exp(-|z|)$ and $0.9\phi(z) + 0.1\phi(z/3)/3$, for $z \in \mathbb{R}$, and t_ν denotes the t density on ν degrees of freedom. The 'slash' distribution is that of Z/U, where $Z \sim N(0, 1)$ and $U \sim U(0, 1)$ independently. The estimates have been multiplied by 100 for convenience and have standard errors of about 0.5.

				Target		
F	5	95	90	2.5	97.5	95
Normal	4.9	94.7	88.8	2.6	97.1	94.4
Laplace	4.1	94.9	90.8	2.2	98.1	95.9
Mixture	4.0	94.9	90.9	1.9	98.0	96.1
t_{20}	5.4	95.4	90.1	2.2	97.7	95.5
t_{10}	6.1	93.9	87.8	2.6	97.0	94.4
t_5	4.6	95.3	90.7	2.5	98.1	95.6
t_1 (Cauchy)	2.3	97.3	95.1	0.8	99.1	98.3
Slash	2.6	97.4	94.9	1.3	99.3	98.0
Gamma, $\alpha = 2$	9.7	97.9	88.3	6.3	99.1	92.8

would be $1 - 2\alpha = 0.90$. Columns 5–7 give the same quantities for 95% confidence intervals. The first row is included to check the simulation: it does not hit the target exactly, due to simulation randomness, but it is close. Laplace, mixture, 'slash' and t_ν densities have heavier tails than the normal; the mixture corresponds to $N(0, 1)$ samples that are occasionally contaminated by $N(0, 3^2)$ variables. The results suggest that heavy-tailed data have little effect on the probabilities until the extreme cases $\nu = 1$ and the 'slash' distribution, for both of which the Z_j have infinite mean. Then the intervals are too wide and therefore have too great a chance of containing μ. The gamma distribution is the only asymmetric case, and this shows in the estimated one-tailed probabilities p, though the estimates of $p(1 - \alpha, n, F) - p(\alpha, n, F)$ remain reasonably close to $(1 - 2\alpha)$. Overall the performance of T seems fairly satisfactory unless the data are grossly non-normal.

Simulation timings depend on the computer and language used, as well as the skill of the programmer, so they are often uninformative. Having said this, it took about 20 seconds to obtain each row of the table, using about 25 lines of code in total. This compares very favourably with the time and effort that would be involved in getting such results analytically. ∎

3.3.2 Variance reduction

This section may be skipped on a first reading.

Even though it involves no chemicals or nasty smells, a simulation experiment is nonetheless an experiment, and it may be worth considering how to increase its precision for a given effort. There are numerous ways to do this, but as they all involve extra work on the part of the experimenter, they are only worthwhile when the amount of simulation is large: a reduction from 30 to five seconds matters much less than one from 30 to five days.

Suppose that we wish to estimate properties of a rather awkward statistic $T = t(Y_1, \ldots, Y_n)$ that is correlated with a statistic $W = w(Y_1, \ldots, Y_n)$ with known properties. Then one way to use W is to write $T = W + (T - W) = W + D$, say, work out the relevant properties of the *control variate* W analytically, and use simulation only for the difference D. For example, if moments of W are available explicitly but

		p				
F		0 (Average)	0.1	0.2	0.3	0.4	0.5 (Median)
Normal	$n\text{var}(T)$	1	1.05	1.13	1.23	1.35	1.54
	Correlation	1	0.98	0.95	0.91	0.86	0.81
	Efficiency gain	∞	10.4	4.9	3.1	2.3	1.9
t_5	$n\text{var}(T)$	1.67	1.38	1.37	1.42	1.53	1.73
	Correlation	1	0.93	0.89	0.84	0.80	0.75
	Efficiency gain	∞	2.1	1.4	1.1	1	0.9

Table 3.3 Estimated variances of $p \times 100\%$ trimmed averages in samples of size $n = 21$ from the normal and t_5 distributions.

we want to estimate the variance of T, we write

$$\text{var}(T) = \text{var}(W) + 2\text{cov}(W, D) + \text{var}(D),$$

where only terms involving D need to be estimated by simulation. We then generate R independent samples $Y_1, \ldots, Y_n$ and calculate T, W and D for each, giving (T_r, W_r, D_r), $r = 1, \ldots, R$. Then $\text{var}(T)$ is estimated by

$$V_1 = \text{var}(W) + \frac{2}{R - 1} \sum_{r=1}^{R} (W_r - \overline{W})(D_r - \overline{D}) + \frac{1}{R - 1} \sum_{r=1}^{R} (D_r - \overline{D})^2, \quad (3.27)$$

where the exact quantity $\text{var}(W)$ replaces the sample variance of $W_1, \ldots, W_R$. The usual estimate of $\text{var}(T)$ would be $V_2 = (R - 1)^{-1} \sum_r (T_r - \overline{T})^2$. If $\text{var}(W)$ is a large part of $\text{var}(T)$, then $\text{var}(V_1)$ may be much smaller than $\text{var}(V_2)$, but the *efficiency gain* $\text{var}(V_2)/\text{var}(V_1)$ will depend on the correlation between W and T.

Example 3.25 (Trimmed average) Let $Y_1, \ldots, Y_n$ be a random sample from a distribution F with mean μ and variance σ^2. One estimate of μ is the sample average $\overline{Y}$, but as this is sensitive to bad values it may be preferable to use the $p \times 100\%$ trimmed average

$$T = (n - 2k)^{-1} \sum_{j=k+1}^{n-k} Y_{(j)},$$

where $Y_{(1)} \leq \cdots \leq Y_{(n)}$ are the order statistics of the sample and $k = pn$ is an integer. One measure of the precision of T is its variance, and if we found that $\text{var}(T) < \text{var}(\overline{Y})$ for many different distributions F, we might choose to use T rather than $\overline{Y}$. Given F, $\text{var}(T)$ can in principle be obtained exactly, but as the calculations are tedious it is simpler to simulate.

An obvious control variate is $W = \overline{Y} = n^{-1} \sum_j Y_{(j)}$, which has variance σ^2/n and is perfectly correlated with T if $p = 0$. We simulate as described above, obtaining R values of W_r, T_r and $D_r = T_r - W_r$, and estimate $\text{var}(T)$ using (3.27). Table 3.3 shows values of $n V_1$ for samples of size $n = 21$ from the normal and the t_5 distribution, using various values of p; we took $R = 1000$ replicates. The table also shows the estimated correlation between W and T, and the efficiency gains due to use of control variates, estimated by repeating the experiment 50 times. In practice one would have just one value of V_1 and one of V_2; the repetition here was needed only to find the efficiency gains. These are largest when p is small, and even infinite when $p = 0$, when $W = T$ and $\text{var}(V_2) = 0$. In this case $D = T - W = 0$, and as $\text{var}(W) = \text{var}(\overline{Y})$ is known exactly, V_1 is constant and hence $\text{var}(V_1) = 0$; simulation is then unnecessary. The efficiency gains depend not only on the correlation between W and T, but also on the underlying distribution F.

For normal data, the increase in variance when using T rather than $\overline{Y}$ is modest for $p < 0.3$, and for t_5 data $\text{var}(T) < \text{var}(\overline{Y})$ when $0 < p < 0.5$. This suggests that a lightly trimmed average may be preferable to $\overline{Y}$ for non-normal data and not much more variable than $\overline{Y}$ for normal data, but we would need more extensive results to be sure. ∎

Importance sampling

Another approach to variance reduction is *importance sampling*. The key idea here is that sometimes most of the sampling is unproductive, and then it is better to concentrate on the parts of the sample space where it is most valuable. The idea is often used in Monte Carlo integration. Suppose we want to estimate

$$\psi = \text{E}\{m(Y)\} = \int m(y)g(y)\,dy.$$

The direct approach is to generate $Y_1, \ldots, Y_R$ independently from density g, and to set $\widehat{\psi} = R^{-1} \sum_r m(Y_r)$. This has mean and variance

$$\text{E}(\widehat{\psi}) = \text{E}\{m(Y)\} = \psi, \quad \text{var}(\widehat{\psi}) = \int m(y)^2 g(y)\,dy - \psi^2,$$

but it may be a very poor estimate. For example, if $m(Y) = I(Y \leq a)$ and $\psi = \text{Pr}(Y \leq a)$ is very small, then most of the Y_r will not contribute to $\widehat{\psi}$, and the effort spent in generating them will be wasted. Instead we try simulating from a density h, chosen to concentrate effort in the important part of the sample space; the support

The support of a density f is $\{y : f(y) > 0\}$.

E_h denotes expectation with respect to density h.

of h must include the support of g. The resulting estimator is the *raw importance sampling estimator* $\widehat{\psi}_{\text{raw}} = R^{-1} \sum_r m(Y_r)w(Y_r)$, where $W = w(Y) = g(Y)/h(Y)$ is known as the *importance sampling weight*. The mean and variance of $\widehat{\psi}_{\text{raw}}$ are

$$\text{E}(\widehat{\psi}_{\text{raw}}) = \text{E}_h\{m(Y)w(Y)\} = \int m(y)\frac{g(y)}{h(y)}h(y)\,dy = \int m(y)g(y)\,dy = \psi,$$

$$\begin{aligned}
\text{var}(\widehat{\psi}_{\text{raw}}) &= R^{-1}\text{var}_h\{m(Y)w(Y)\} \\
&= R^{-1}[\text{E}_h\{m(Y)^2 w(Y)^2\} - \text{E}_h\{m(Y)w(Y)\}^2] \\
&= R^{-1}\left\{\int m(y)^2\frac{g(y)}{h(y)}g(y)\,dy - \psi^2\right\}.
\end{aligned}$$

Hence $\widehat{\psi}_{\text{raw}}$ will be a big improvement on $\widehat{\psi}$ if

$$\frac{\text{var}(\widehat{\psi})}{\text{var}(\widehat{\psi}_{\text{raw}})} = \frac{\int m(y)^2 g(y)\,dy - \psi^2}{\int m(y)^2 \frac{g(y)}{h(y)}g(y)\,dy - \psi^2} \tag{3.28}$$

is large. This ratio depends on h, a bad choice of which can make $\widehat{\psi}_{\text{raw}}$ much more variable than is $\widehat{\psi}$. The trick is to choose h well.

z	-3	-2	-1	0	1	2	3
μ_z	-3.15	-2.22	-1.34	-0.62	-0.18	-0.03	-0.002
Efficiency gain	222	19	4.1	1.75	1.19	1.04	1.004

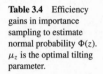

Table 3.4 Efficiency gains in importance sampling to estimate normal probability $\Phi(z)$. μ_z is the optimal tilting parameter.

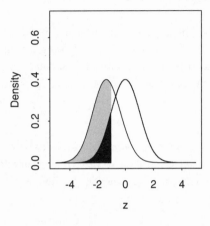
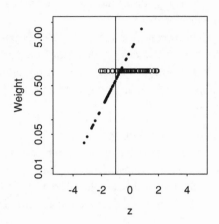

Figure 3.8 Importance sampling for normal tail probability. Left: $N(0, 1)$ density and area $\Phi(z)$ to be estimated (heavy shading), with importance sampling density $N(\mu_z, 1)$, whose lightly shaded area contributes to $\widehat{\psi}_{\text{raw}}$. Right: weights for samples with $R = 50$ from $N(0, 1)$ (circles) and from $N(\mu_z, 1)$ (blobs). The vertical line shows $z = -1$; only points to the left of that line contribute to estimation of $\Phi(z)$.

Example 3.26 (Normal probability)   Marooned on a desert island with only parrots for company, a shipwrecked statistician decides to realize his lifelong ambition of memorizing values of the normal integral $\Phi(z)$; he hopes to make himself more attractive to the statisticienne of his dreams. His statistical tables have been ruined by salt water, but washed up on the beach he finds a programmable solar-powered calculator on which he is able to implement a slow but reliable normal random number generator.

Rather than estimate $\psi = \Phi(z)$ directly, he decides to use importance sampling from the $N(\mu, 1)$ distribution, taking $m(Y) = I(Y \leq z)$, $g(y) = \phi(y)$, and $h(y) = \phi(y - \mu)$. If $Y_1, \ldots, Y_R \overset{\text{iid}}{\sim} g$, then $\widehat{\psi} = R^{-1} \sum I(Y_r \leq z)$ has mean $\Phi(z)$ and variance $R^{-1}\Phi(z)\{1 - \Phi(z)\}$. If he samples from h, the importance sampling estimate is $\widehat{\psi}_{\text{raw}} = R^{-1} \sum_r w(Y_r) I(Y_r \leq z)$, where $w(y) = \phi(y)/\phi(y - \mu) = \exp(\frac{1}{2}\mu^2 - \mu y)$, and it turns out that

$$\text{var}(\widehat{\psi}_{\text{raw}}) = R^{-1}\{\exp(\mu^2)\Phi(z + \mu) - \Phi(z)^2\}. \tag{3.29}$$

Given z, therefore, the optimal value μ_z of μ minimizes $e^{\mu^2}\Phi(z + \mu)$. Table 3.4 shows values of μ_z and the efficiency gain (3.28) for a few values of z. Note how $\mu_z \doteq z$ for $z < 0$, but not for $z > 0$, and how importance sampling becomes increasingly effective as $z \to -\infty$, when almost none of the Y_r contribute to $\widehat{\psi}$. For $z > 0$, most of the observations contribute to $\widehat{\psi}$ and importance sampling gives little improvement.

The panels of Figure 3.8 show the optimal importance sampling distribution when $z = -1$ and the weights obtained in samples of size $R = 50$ from the $N(0, 1)$ and $N(\mu_z, 1)$ distributions. Most of the observations generated from $\phi(y - \mu_z)$ contribute to $\widehat{\psi}_{\text{raw}}$, whereas only a few of those from $\phi(y)$ contribute to $\widehat{\psi}$. The efficiency gain of 4.1 implies that 50 observations from the $N(\mu_z, 1)$ distribution are worth about 200 from the $N(0, 1)$ distribution. The gains are larger when $z \to -\infty$, and combined with the fact that $\Phi(z) = 1 - \Phi(-z)$ should enable our hero to fulfil his ambition before he is rescued. ∎

A difficulty with $\widehat{\psi}_{\mathrm{raw}}$ is that the weights W_r can be very variable, with one or two large ones dominating the rest, leading to the average weight $\overline{W}$ being very different from its expectation $\mathrm{E}_h(W) = 1$. This can be dealt with by rescaling the weights to $W'_r = W_r/\overline{W}$, for which $\overline{W'} = 1$, resulting in the *importance sampling ratio estimator* $\widehat{\psi}_{\mathrm{rat}} = R^{-1} \sum_r W'_r m(Y_r)$. Another approach treats W as a control variate, assuming that the pair $(T, W) = (m(Y)w(Y), w(Y))$ has approximately a bivariate normal distribution, and then estimating the conditional mean of T given $W = 1$. This results in the *importance sampling regression estimator*

$$\widehat{\psi}_{\mathrm{reg}} = \overline{T} + \frac{\sum_r(W_r - \overline{W})(T_r - \overline{T})}{\sum_r(W_r - \overline{W})^2}(1 - \overline{W}), \quad T_r = m(Y_r)w(Y_r);$$

note that $\widehat{\psi}_{\mathrm{raw}} = \overline{T}$. If T and W are positively correlated, the ratio here will be positive, and if $\overline{W} > 1$ the adjustment reduces $\widehat{\psi}_{\mathrm{raw}}$ by an amount that depends on $1 - \overline{W}$. This makes sense because if T and W are positively correlated and $\overline{W} > \mathrm{E}(W) = 1$, then it is likely that $\overline{T} > \mathrm{E}(T)$. Both ratio and regression estimators tend to improve on $\widehat{\psi}_{\mathrm{raw}}$.

Exercises 3.3

1. Show how to use inversion to generate Bernoulli random variables. If $0 < \pi < 1$, what distribution has $\sum_{j=1}^m I(U_j \le \pi)$?

2. Write down algorithms to generate values from the gamma density with small integer shape parameter by (a) direct construction using exponential variables, (b) rejection sampling with an exponential envelope.

3. The Cholesky decomposition of an $p \times p$ symmetric positive matrix Ω is the unique lower triangular $p \times p$ matrix L such that $LL^{\mathrm{T}} = \Omega$. Find the distribution of $\mu + LZ$, where Z is a vector containing a standard normal random sample $Z_1, \ldots, Z_p$, and hence give an algorithm to generate from the multivariate normal distribution.

4. If inversion can be used to generate a variable Y with distribution function F, discuss how to generate values from F conditioned on the events (a) $Y \le y_U$, (b) $y_L < Y \le y_U$. Under what circumstances might rejection sampling be sensible?
   Define Z by setting $Z = j$ when $Y \le y_j$, for $y_1 < \cdots < y_{k-1} < y_k = \infty$. Give an algorithm to generate Z.

5. If X has density $\lambda e^{-\lambda x}$, $x > 0$, show that $\Pr(r - 1 \le X \le r) = e^{-\lambda(r-1)}(1 - e^{-\lambda})$.
   If Y has geometric density $\Pr(Y = r) = \pi(1 - \pi)^{r-1}$, for $r = 1, 2, \ldots$ and $0 < \pi < 1$, show that $Y \stackrel{D}{=} \lceil \log U / \log(1 - \pi) \rceil$. Hence give an algorithm to generate geometric variables.

6. Construct a rejection algorithm to simulate from $f(x) = 30x(1 - x)^4$, $0 \le x \le 1$, using the $U(0, 1)$ density as the proposal function g. Give its efficiency.

7. Verify (3.29).

3.4 Bibliographic Notes

The idea of a confidence interval belongs to statistical folklore, but its mathematical formulation and the repeated sampling interpretation were developed by J. Neyman in the 1930s. Fisher argued strongly against the repeated sampling interpretation and developed his own approaches based on conditioning and fiducial inference. Welsh (1996) gives a thoughtful comparison of these and other approaches to inference.

Inference procedures for normal samples are treated in many basic statistics texts.

Stochastic simulation is a very large topic. In addition to books such as Rubinstein (1981), Fishman (1996), Morgan (1984), Ripley (1987), and Robert and Casella (1999), there is a rapidly growing literature on simulation for stochastic processes, often using Markov chain theory; see the bibliographic notes to Chapter 11.

Jerzy Neyman (1894–1981) was born in Moldavia and studied mathematics at Kharkov University and then statistics in Warsaw and University College London, where he worked on the basis of hypothesis testing with Egon Pearson, on experimental design, and on sampling theory. In 1938 he moved to Berkeley and became a leading figure in the development of statistics in the USA.

3.5 Problems

1 Suppose that $Y_1, \ldots, Y_4$ are independent normal variables, each with variance σ^2, but with means $\mu + \alpha + \beta + \gamma, \mu + \alpha - \beta - \gamma, \mu - \alpha + \beta - \gamma, \mu - \alpha - \beta + \gamma$. Let

$$Z^{\mathrm{T}} = \tfrac{1}{4}(Y_1 + Y_2 + Y_3 + Y_4, Y_1 + Y_2 - Y_3 - Y_4, Y_1 - Y_2 + Y_3 - Y_4, Y_1 - Y_2 - Y_3 + Y_4).$$

Calculate the mean vector and covariance matrix of Z, and give the joint distribution of Z_1 and $V = Z_2^2 + Z_3^2 + Z_4^2$ when $\alpha = \beta = \gamma = 0$. What is then the distribution of $Z_1/(V/3)^{1/2}$?

2 $W_i, X_i, Y_i,$ and $Z_i, i = 1, 2,$ are eight independent, normal random variables with common variance σ^2 and expectations μ_W, μ_X, μ_Y and μ_Z. Find the joint distribution of the random variables

$$T_1 = \frac{1}{2}(W_1 + W_2) - \mu_W, \ T_2 = \frac{1}{2}(X_1 + X_2) - \mu_X,$$

$$T_3 = \frac{1}{2}(Y_1 + Y_2) - \mu_Y, \ T_4 = \frac{1}{2}(Z_1 + Z_2) - \mu_Z,$$

$$T_5 = W_1 - W_2, \ T_6 = X_1 - X_2, \ T_7 = Y_1 - Y_2, \ T_8 = Z_1 - Z_2.$$

Hence obtain the distribution of

$$U = 4\frac{T_1^2 + T_2^2 + T_3^2 + T_4^2}{T_5^2 + T_6^2 + T_7^2 + T_8^2}.$$

Show that the random variables $U/(1 + U)$ and $1/(1 + U)$ are identically distributed, without finding their probability density functions. Find their common density function and hence determine $\Pr(U \le 2)$.

3 Figure 3.9 shows samples of size 100 from densities in which (i) X and Y are independent; (ii) $\mathrm{corr}(X, Y) = -0.7$; (iii) $\mathrm{corr}(X, Y) = 0.7$; (iv) $\mathrm{corr}(X, Y) = 0$. Say which is which and why.

4 (a) Suppose that conditional on η, $Y_1, \ldots, Y_n$ is a random sample from the $N(\eta, \sigma^2)$ distribution, but that η has itself a $N(\mu, \sigma_\eta^2)$ distribution. Show that the unconditional distribution of $Y_1, \ldots, Y_n$ is multivariate normal, with correlation $\rho = \sigma_\eta^2/(\sigma^2 + \sigma_\eta^2)$ between different variables.
(b) Show that

$$W = (\overline{Y} - \mu)/(S^2/n)^{1/2} \stackrel{D}{=} \{1 + n\rho/(1 - \rho)\}^{1/2}T,$$

where $T \sim t_{n-1}$. Hence show that the probability that the usual confidence interval (3.17) contains μ is $1 - 2\Pr\{T \le t_{n-1}(\alpha)(1 + n\sigma_\eta^2/\sigma^2)^{-1/2}\}$ and verify that when $\alpha = 0.025$, $n = 10$ and $\rho = 0.1$, this probability is 0.85, and that when $n = 100$ and $\rho = 0.01, 0.02$, it is 0.84, 0.74.

Figure 3.9 Samples
from bivariate
distributions with
correlations $-0.7, 0, 0.7$;
one sample has
independent components.
Which is which? Why?

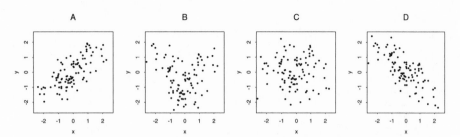

What does this tell you about the assumptions underlying (3.17)?

5   If Z is standard normal, then $Y = \exp(\mu + \sigma Z)$ is said to have the log-normal distribution.
    Show that $E(Y^r) = \exp(r\mu)M_Z(r\sigma)$ and hence give expressions for the mean and variance
    of Y. Show that although all its moments are finite, Y does not have a moment-generating
    function.

6   (a) Let $Y = Z_1$ and $W = Z_2 - \lambda Z_1$, where Z_1, Z_2 are independent standard normal vari-
    ables and λ is a real number. Show that the conditional density of Y given that $W < 0$ is
    $f(y; \lambda) = 2\phi(y)\Phi(\lambda y)$; Y is said to have a *skew-normal distribution*. Sketch $f(y; \lambda)$ for
    various values of λ. What happens when $\lambda = 0$?
    (b) Show that $Y^2 \sim \chi_1^2$.
    (c) Use Exercise 3.2.16 to show that Y has cumulant-generating function $t^2/2 + \log \Phi(\delta t)$,
    where $\delta = \lambda/(1 + \lambda^2)^{1/2}$, and hence find its mean and variance. Show that the standardized
    skewness of Y varies in the range $(-0.995, 0.955)$.

7   For $h(x)$ a non-negative function of real x with finite integral, let

$$C_h = \left\{(u, v) : 0 \leq u \leq h(v/u)^{1/2}\right\}.$$

    (a) By considering the change of variables $(u, v) \to (w = u, x = v/u)$, show that C_h has
    finite area, and that if (U, V) is uniformly distributed on C_h, then $X = V/U$ has density
    $h(x)/\int h(y)\,dy$.
    (b) If $h(x)$ and $x^2h(x)$ are bounded and $a = \sqrt{\sup\{h(x) : -\infty < x < \infty\}}$,

$$b_+ = \sqrt{\sup\{x^2h(x) : x \geq 0\}}, \quad b_- = -\sqrt{\sup\{x^2h(x) : x \leq 0\}},$$

    show that $C_h \subset [0, a] \times [b_-, b_+]$. Hence justify the following algorithm:
    1 Repeat
        • generate $U_1, U_2 \overset{\text{iid}}{\sim} U(0, 1)$;
        • let $U = aU_1$, $V = b_- + (b_+ - b_-)U_2$;
        until $(U, V) \in C_h$.
    2 Return $X = V/U$.
    (c) If $h(x) = (1 + x^2)^{-1}$ on $-\infty < x < \infty$, show that this algorithm gives the method for
    generating Cauchy variables described in Example 3.21.
    (d) If $h(x) = e^{-x}$ on $0 < x < \infty$, show that $a = 1$, $b_- = 0$, and $b_+ = 2/e$, and give the
    algorithm.
    (e) If $h(x) = e^{-x^2/2}$ on $-\infty < x < \infty$, find the values of a, b_- and b_+, and show that X
    is accepted if and only if $V^2 \leq -4U^2 \log U$. Hence give the algorithm.

8   Let R_1, R_2 be independent binomial random variables with probabilities π_1, π_2 and de-
    nominators m_1, m_2, and let $P_i = R_i/m_i$. It is desired to test if $\pi_1 = \pi_2$.
    Let $\widehat{\pi} = (m_1 P_1 + m_2 P_2)/(m_1 + m_2)$. Show that when $\pi_1 = \pi_2$, the statistic

$$Z = \frac{P_1 - P_2}{\sqrt{\widehat{\pi}(1 - \widehat{\pi})(1/m_1 + 1/m_2)}} \xrightarrow{D} N(0, 1)$$

    when $m_1, m_2 \to \infty$ in such a way that $m_1/m_2 \to \xi$ for $0 < \xi < 1$.

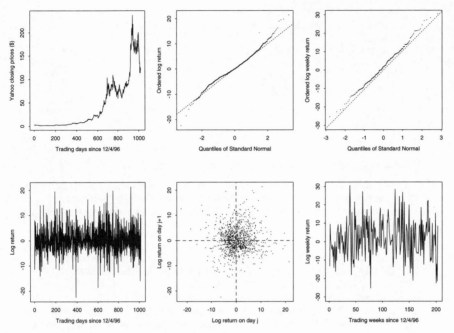

Figure 3.10 Analysis of Yahoo.com share values. Left: share price x_j from 12 April 1996 to 26 April 2000 (above); log daily returns $y_j = 100 \log(x_j/x_{j-1})$ (below). Centre: normal probability plot of y_j (above) and plot of y_{j+1} against y_j (below). Right: normal probability plot of log weekly returns (above); log weekly returns (below).

Now consider a 2×2 table formed using two independent binomial variables and having entries R_i, S_i where $R_i + S_i = m_i$, $R_i/m_i = P_i$, for $i = 1, 2$. Show that if $\pi_1 = \pi_2$ and $m_1, m_2 \to \infty$, then

$$X^2 = (n_1 + n_2)(R_1 S_2 - R_2 S_1)^2 / \{n_1 n_2 (R_1 + R_2)(S_1 + S_2)\} \xrightarrow{D} \chi_1^2.$$

Two batches of trees were planted in a park: 250 were obtained from nursery A and 250 from nursery B. Subsequently 41 and 64 trees from the two groups die. Do trees from the two nurseries have the same survival probabilities? Are the assumptions you make reasonable?

9   If $\overline{Y}$ is the average of a random sample $Y_1, \ldots, Y_n$ from density $\theta^{-1} \exp(-y/\theta)$, $y > 0$, $\theta > 0$, give the limiting distribution of $Z(\theta) = n^{1/2}(\overline{Y} - \theta)/\theta$ as $n \to \infty$. Hence obtain an approximate two-sided 95% confidence interval for θ.
    Show that for large n, $\log(\overline{Y}) \doteq \log \theta + n^{-1/2} Z$, find an approximate mean and variance for $\log \overline{Y}$, and hence give another approximate two-sided 95% confidence interval for θ. Which interval would you prefer in practice?

10 Independent pairs (X_j, Y_j), $j = 1, \ldots, m$ arise in such a way that X_j is normal with mean λ_j and Y_j is normal with mean $\lambda_j + \psi$, X_j and Y_j are independent, and each has variance σ^2. Find the joint distribution of $Z_1, \ldots, Z_m$, where $Z_j = Y_j - X_j$, and hence show that there is a $(1 - 2\alpha)$ confidence interval for ψ of form $A \pm m^{-1/2} B c$, where A and B are random variables and c is a constant.
    Obtain a 0.95 confidence interval for the mean difference ψ given (x, y) pairs (27, 26), (34, 30), (31, 31), (30, 32), (29, 25), (38, 35), (39, 33), (42, 32). Is it plausible that $\psi \neq 0$?

11 The upper left panel of Figure 3.10 shows daily closing share prices x_j for Yahoo.com from 12 April 1996 to 26 April 2000. We define the log daily *returns* $y_j = 100 \log(x_j/x_{j-1})$; y_j is roughly the daily percentage change in price.
    (a) The lower left panel shows the y_j. Does their distribution seem to change with time?
    (b) The upper central panel shows a normal probability plot of the y_j. Do they seem normal to you? If not, describe how they differ from normal variates.

(c) The lower central panel shows a plot of y_{j+1} against y_j. Are successive daily log returns correlated? What would be the implication if they were?

(d) The $n = 1015$ values of y_j have average and variance $\bar{y} = 0.376$ and $s^2 = 25.35$. Is $E(y_j) > 0$?

(e) We can also define the log weekly returns, $w_j = y_{5(j-1)+1} + \cdots + y_{5j}$, whose normal probability plot is shown in the top right panel. Are they normal? They have average and variance 1.878 and 110.07. Is their mean positive?

(f) The data suggest the simple *geometric Brownian motion* model that the stock value at the end of week k is $S_k = s_0 \exp\left(k\mu + \sigma \sum_{j=1}^{k} Z_j\right)$, where the Z_j are a standard normal random sample and s_0 is the initial stock value. If I bought \$100 worth of stock when it was launched and its value on 26 April 2000 was \$4527, give its median predicted value and a 95% prediction interval for its value 400 weeks after launch. Do you find this credible? Under the normal model, how long must I wait before the probability is 0.5 that I am a millionaire?

Remember: past performance is no guide to the future!

12 (a) Check the expressions for z_j for adaptive rejection sampling.

(b) Show that $G_+(y) = \int_{-\infty}^{y} g_+(x)\,dx$ satisfies

$$G_+(z_i) = \frac{\sum_{j=0}^{i} \frac{1}{h'(y_{j+1})} \{\exp h_+(z_{j+1}) - \exp h_+(z_j)\}}{\sum_{j=0}^{k-1} \frac{1}{h'(z_{j+1})} \{\exp h_+(z_{j+1}) - \exp h_+(z_j)\}};$$

let c_k denote the denominator of this expression. Show that a value X from g_+ is generated by taking $U \sim U(0, 1)$, finding the largest z_i such that $G_+(z_i) < U$ and setting

$$X = z_i + \frac{1}{h'(y_{i+1})} \log\left[1 + \frac{h'(y_{i+1})c_k \{U - G_+(z_i)\}}{\exp h_+(z_i)}\right].$$

(c) Let $h_-(y)$ be defined by taking the chords between the points $(y_j, h(y_j))$, for $j = 1, \ldots, k$, and let it be $-\infty$ outside $[y_1, y_k]$. Explain how to use $h_-(y)$ to speed up sampling from f when h is complicated, by performing a pretest based on $\exp\{h_-(X) - h_+(X)\}$. (Gilks and Wild, 1992; Wild and Gilks, 1993)

4

Likelihood

4.1 Likelihood

4.1.1 Definition and examples

Suppose we have observed the value y of a random variable Y, whose probability density function is supposed known up to the value of a parameter θ. We write $f(y; \theta)$ to emphasize that the density is a function of both data and a parameter. In general both y and θ will be vectors whose respective elements we denote by y_j and θ_r. The parameter takes values in a *parameter space* Θ, and the data Y take values in a *sample space* $\mathcal{Y}$. Our goal is to make statements about the distribution of Y, based on the observed data y. The assumption that f is known apart from uncertainty about θ reduces the problem to making statements about what range of values of θ within Θ is plausible, given that y has been observed.

A fundamental tool is the *likelihood* for θ based on y, which is defined to be

$$L(\theta) = f(y; \theta), \quad \theta \in \Theta, \tag{4.1}$$

regarded as a function of θ for fixed y. Our interest in this is motivated by the idea that it will be relatively larger for values of θ near that which generated the data. When Y is discrete we use $f(y; \theta) = \Pr(Y = y; \theta)$, while if Y is continuous, we take $f(y; \theta)$ to be its probability density function. Owing to rounding, the recorded y is always discrete in practice, and occasional minor difficulties can be avoided by taking this into account, as we shall see in Example 4.42. However in constructing (4.1) for continuous Y we almost always use its density function. When $y = (y_1, \ldots, y_n)$ is a collection of independent observations the likelihood is

$$L(\theta) = f(y; \theta) = \prod_{j=1}^{n} f(y_j; \theta). \tag{4.2}$$

Example 4.1 (Poisson distribution)   Suppose that y consists of a single observation from the Poisson density (2.6). Here the data and the parameter are both scalars, and $L(\theta) = \theta^y e^{-\theta} / y!$. The parameter space is $\{\theta : \theta > 0\}$ and the sample space is

Figure 4.1 Likelihoods for the spring failure data at stress 950 N/mm². The upper left panel is the likelihood for the exponential model, and below it is a perspective plot of the likelihood for the Weibull model. The upper right panel shows contours of the log likelihood for the Weibull model; the exponential likelihood is obtained by setting $\alpha = 1$. that is, slicing L along the vertical dotted line. The lower right panel shows the profile log likelihood for α, which corresponds to the log likelihood values along the dashed line in the panel above, plotted against α.

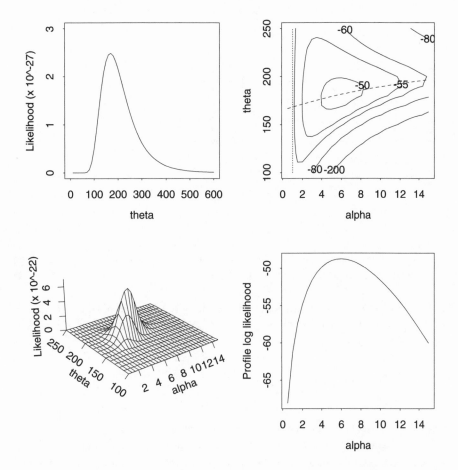

$\{0, 1, 2, \ldots\}$. If $y = 0$, $L(\theta)$ is a monotonic decreasing function of θ, whereas if $y > 0$ it has a maximum at $\theta = y$, and limit zero as θ approaches zero or infinity. ∎

Example 4.2 (Exponential distribution)   Let y be a random sample $y_1, \ldots, y_n$ from the exponential density $f(y; \theta) = \theta^{-1}e^{-y/\theta}$, $y > 0, \theta > 0$. The parameter space is $\Theta = \mathbb{R}_+$ and the sample space the Cartesian product $\mathbb{R}_+^n$. Here (4.2) gives

$$L(\theta) = \prod_{j=1}^{n} \theta^{-1}e^{-y_j/\theta} = \theta^{-n} \exp\left(-\frac{1}{\theta}\sum_{j=1}^{n} y_j\right), \quad \theta > 0. \qquad (4.3)$$

The spring failure times at stress 950 N/mm² in Example 1.2 are

$$225, \ 171, \ 198, \ 189, \ 189, \ 135, \ 162, \ 135, \ 117, \ 162,$$

and the top left panel of Figure 4.1 shows the likelihood (4.3). The function is unimodal and is maximized at $\theta \doteq 168$; $L(168) \doteq 2.49 \times 10^{-27}$. At $\theta = 150$, $L(\theta)$ equals 2.32×10^{-27}, so that 150 is $2.32/2.49 = 0.93$ times less likely than $\theta = 168$ as an explanation for the data. If we were to declare that any value of θ for which

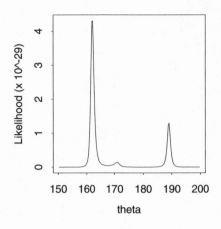 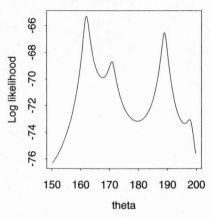

Figure 4.2 Cauchy likelihood and log likelihood for the spring failure data at stress 950N/mm².

$L(\theta) > cL(168)$ was "plausible" based on these data, values of θ in the range $(120, 260)$ or so would be plausible when $c = \frac{1}{2}$.   ∎

Example 4.3 (Cauchy distribution)   The Cauchy density centered at θ is $f(y; \theta) = [\pi\{1 + (y - \theta)^2\}]^{-1}$, where $y \in \mathbb{R}$ and $\theta \in \mathbb{R}$. Hence the likelihood for a random sample $y_1, \ldots, y_n$ is

$$L(\theta) = \prod_{j=1}^{n} \frac{1}{\pi\{1 + (y_j - \theta)^2\}}, \quad -\infty < \theta < \infty.$$

The sample space is $\mathbb{R}^n$ and the parameter space is $\mathbb{R}$.

The left panel of Figure 4.2 shows $L(\theta)$ for the spring data in Example 4.2. There seem to be three local maxima in the range for which $L(\theta)$ is plotted, with a global maximum at $\theta \doteq 162$. We can see more detail in the log likelihood $\log L(\theta)$ shown in the right panel of the figure. There are at least four local maxima — apparently one at each observation, with a more prominent one when observations are duplicated. By contrast with the previous example, for some values of c a "plausible" set for θ here consists of disjoint intervals.   ∎

Example 4.4 (Weibull distribution)   The Weibull density is

$$f(y; \theta, \alpha) = \frac{\alpha}{\theta}\left(\frac{y}{\theta}\right)^{\alpha-1} \exp\left\{-\left(\frac{y}{\theta}\right)^{\alpha}\right\}, \quad y > 0, \quad \theta, \alpha > 0. \qquad (4.4)$$

When $\alpha = 1$ this is the exponential density of Example 4.2; the exponential model is nested within the Weibull model, the parameter space for which is $\mathbb{R}_+^2$, and the sample space for which is $\mathbb{R}_+^n$.

A random sample $y = (y_1, \ldots, y_n)$ from (4.4) has joint density

$$f(y; \theta, \alpha) = \prod_{j=1}^{n} f(y_j; \theta, \alpha) = \prod_{j=1}^{n}\left[\frac{\alpha}{\theta}\left(\frac{y_j}{\theta}\right)^{\alpha-1} \exp\left\{-\left(\frac{y_j}{\theta}\right)^{\alpha}\right\}\right]$$

and hence the likelihood is

$$L(\theta, \alpha) = \frac{\alpha^n}{\theta^{n\alpha}} \left(\prod_{j=1}^{n} y_j \right)^{\alpha-1} \exp\left\{ -\sum_{j=1}^{n} \left(\frac{y_j}{\theta} \right)^{\alpha} \right\}, \quad \theta, \alpha > 0. \quad (4.5)$$

The lower left panel of Figure 4.1 shows $L(\theta, \alpha)$ for the data of Example 4.2. The likelihood is maximized at $\theta \doteq 181$ and $\alpha \doteq 6$, and $L(181, 6)$ equals 6.7×10^{-22}. This is 2.7×10^5 times greater than the largest value for the exponential model. The top right panel shows contours of the log likelihood, $\log L(\theta, \alpha)$. The dotted line indicates the slice corresponding to the exponential density obtained when $\alpha = 1$. The factor 2.5×10^5 gives a difference of $\log(2.7 \times 10^5) = 12.5$ between the maximum log likelihoods. This big improvement suggests that the Weibull model fits the data better. However, if we judge model fit by the maximum likelihood value, the Weibull model is bound to fit at least as well as the exponential, because $\max_{\theta, \alpha} L(\theta, \alpha) \geq \max_{\theta} L(\theta, 1)$, with equality only if the maximum occurs on the line $\alpha = 1$. ∎

The examples above involve random samples, but (4.1) and (4.2) apply also to more complex situations.

Example 4.5 (Challenger data) Consider the data in Table 1.3 on O-ring thermal distress. For now we ignore the effect of pressure, and treat the temperature x_1 at launch as fixed and the number of O-rings with thermal distress as binomial variables with denominator m and probability π, giving

$$\Pr(R = r) = \frac{m!}{r!(m-r)!}\pi^r(1-\pi)^{m-r}, \quad r = 0, 1, \ldots, m.$$

If π depends on temperature through the relation

$$\pi = \frac{\exp(\beta_0 + \beta_1 x_1)}{1 + \exp(\beta_0 + \beta_1 x_1)},$$

then the parameter β_0 determines the probability of thermal distress when $x_1 = 0°F$, which is $e^{\beta_0}/(1 + e^{\beta_0})$. The parameter β_1 determines how π depends on temperature; we expect that $\beta_1 < 0$, since π decreases with increasing x_1.

If the data for the jth flight consist of r_j O-rings with thermal distress at launch temperature x_{1j}, $j = 1, \ldots, n$, and $n = 23$ and $m = 6$, we have

$$\Pr(R_j = r_j; \beta_0, \beta_1) = \frac{m!}{r_j!(m-r_j)!} \left\{ \frac{e^{\beta_0 + \beta_1 x_{1j}}}{1 + e^{\beta_0 + \beta_1 x_{1j}}} \right\}^{r_j} \left\{ \frac{1}{1 + e^{\beta_0 + \beta_1 x_{1j}}} \right\}^{m-r_j}$$

$$= \frac{m!}{r_j!(m-r_j)!} \frac{\exp\{r_j(\beta_0 + \beta_1 x_{1j})\}}{\{1 + \exp(\beta_0 + \beta_1 x_{1j})\}^m}.$$

If the R_j are independent, the likelihood for the entire set of data is

$$L(\beta_0, \beta_1) = \prod_{j=1}^{n} \Pr(R_j = r_j; \beta_0, \beta_1)$$

$$= \prod_{j=1}^{n} \frac{m!}{r_j!(m-r_j)!} \times \frac{\exp\left(\beta_0 \sum_{j=1}^{n} r_j + \beta_1 \sum_{j=1}^{n} r_j x_{1j} \right)}{\prod_{j=1}^{n} \{1 + \exp(\beta_0 + \beta_1 x_{1j})\}^m}. \quad (4.6)$$

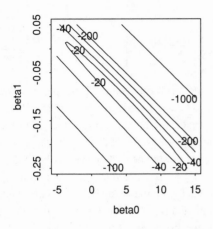 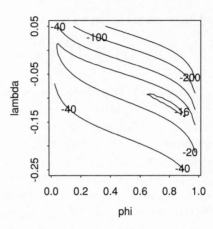

Figure 4.3 Log likelihoods for a binomial model for the O-ring thermal distress data. The probability of thermal distress is taken to be $\psi = \exp(\beta_0 + \beta_1 x_1)/\{1 + \exp(\beta_0 + \beta_1 x_1)\}$. The left panel gives the log likelihood for parameters β_0 and β_1, and the right panel the log likelihood for the probability of thermal distress at 31°F, $\psi = \exp(\beta_0 + 31\beta_1)/\{1 + \exp(\beta_0 + 31\beta_1)\}$ and $\lambda = \beta_1$.

The left panel of Figure 4.3 shows contours of this function, which is largest at $\beta_0 \doteq 5$ and $\beta_1 \doteq -0.1$. However it is difficult to interpret because of the strong negative association between β_0 and β_1: the values of β_1 most plausible for $\beta_0 \doteq 0$ are different from those most plausible when $\beta_0 \doteq 10$.   ∎

Dependent data

In the examples above the data are assumed independent, though not necessarily identically distributed. In more complicated problems the dependence structure of the data may be very complex, making it hard to write down $f(y; \theta)$ explicitly. Matters simplify when the data are recorded in time order, so that y_1 precedes y_2 precedes $y_3, \ldots$. Then it can help to write

This is sometimes called the prediction decomposition.

$$f(y; \theta) = f(y_1, \ldots, y_n; \theta) = f(y_1; \theta) \prod_{j=2}^{n} f(y_j \mid y_1, \ldots, y_{j-1}; \theta). \qquad (4.7)$$

For example, if the data arise from a Markov process, (4.7) becomes

$$f(y; \theta) = f(y_1; \theta) \prod_{j=2}^{n} f(y_j \mid y_{j-1}; \theta), \qquad (4.8)$$

where we have used the Markov property, that given the "present" Y_{j-1}, the 'future', $Y_j, Y_{j+1}, \ldots$, is independent of the 'past', $\ldots, Y_{j-3}, Y_{j-2}$.

Example 4.6 (Poisson birth process)   Suppose that $Y_0, \ldots, Y_n$ are such that, given that $Y_j = y_j$, the conditional density of Y_{j+1} is Poisson with mean θy_j. That is,

$$f(y_{j+1} \mid y_j; \theta) = \frac{(\theta y_j)^{y_{j+1}}}{y_{j+1}!} \exp(-\theta y_j), \qquad y_{j+1} = 0, 1, \ldots, \quad \theta > 0.$$

If Y_0 is Poisson with mean θ, the joint density of data $y_0, \ldots, y_n$ is

$$f(y_0; \theta) \prod_{j=1}^{n} f(y_j \mid y_{j-1}; \theta) = \frac{\theta^{y_0}}{y_0!} \exp(-\theta) \prod_{j=0}^{n-1} \frac{(\theta y_j)^{y_{j+1}}}{y_{j+1}!} \exp(-\theta y_j),$$

so the likelihood (4.8) equals

$$L(\theta) = \left(\prod_{j=0}^{n} y_j!\right)^{-1} \exp\left(s_0 \log\theta - s_1\theta\right), \qquad \theta > 0,$$

where $s_0 = \sum_{j=0}^{n} y_j$ and $s_1 = 1 + \sum_{j=0}^{n-1} y_j$. ∎

4.1.2 Basic properties

It can be convenient to plot the likelihood on a logarithmic scale. This scale is also mathematically convenient, and we define the *log likelihood* to be

$$\ell(\theta) = \log L(\theta).$$

Statements about relative likelihoods become statements about differences between log likelihoods. When y has independent components, $y_1, \ldots, y_n$, we can write

$$\ell(\theta) = \sum_{j=1}^{n} \log f(y_j; \theta) = \sum_{j=1}^{n} \ell_j(\theta), \qquad (4.9)$$

where $\ell_j(\theta) \equiv \ell(\theta; y_j) = \log f(y_j; \theta)$ is the contribution to the log likelihood from the jth observation. The arguments of f and ℓ are reversed to stress that we are primarily interested in f as a function of y, and in ℓ as a function of θ.

To combine the likelihoods for two independent sets of data y and z that both carry information about θ, note that their joint probability density is just the product of their individual densities, and therefore the likelihood based on y and z is the product of the individual likelihoods:

$$L(\theta; y, z) = f(y; \theta)f(z; \theta) = L(\theta; y)L(\theta; z),$$

say, where for clarity the data are an additional argument in the likelihoods.

An important property of likelihood is its invariance to known transformations of the data. Suppose that there are two observers of the same experiment, and that one records the value y of a continuous random variable, Y, while the other records the value z of Z, where Z is a known 1–1 transformation of Y. Then the probability density function of Z is

$$f_Z(z; \theta) = f_Y(y; \theta)\left|\frac{dy}{dz}\right|, \qquad (4.10)$$

where y is regarded as a function of z, and $|dy/dz|$ is the Jacobian of the transformation from Y to Z. As (4.10) differs from (4.1) only by a constant that does not depend on the parameter, the log likelihood based on z equals that based on y plus a constant: the relative likelihoods of different values of θ are the same. This implies that within a particular model f the absolute value of the likelihood is irrelevant to inference about θ. When the maximum value of the likelihood is finite we define the *relative likelihood* of θ to be

$$RL(\theta) = \frac{L(\theta)}{\max_{\theta'} L(\theta')}.$$

This takes values between one and zero, and its logarithm takes values between zero and minus infinity. As the absolute value of $L(\theta)$, or equivalently $\ell(\theta)$, is irrelevant to inference about θ, we can neglect constants and use whatever version of L we wish. Henceforth we use the notation $\equiv$ to indicate that constants have been ignored in defining a log likelihood. However we may not neglect constants if our goal is to compare models from different families of distributions.

Example 4.7 (Spring failure data)   We can compare the Cauchy and Weibull models for the data in Examples 4.2–4.4 in terms of the maximum likelihood value achieved. Under this criterion, the Weibull model, for which the largest log likelihood is about -48, is a much better model than is the Cauchy, for which the maximum log likelihood is about -66. Evidently it makes no sense to add a constant to one of these and not to the other.   ∎

Suppose that the distribution of Y is determined by ψ, which is a 1–1 transformation of θ, so that $\theta = \theta(\psi)$. Then the likelihood for ψ, $L^*(\psi)$, and the likelihood for θ, $L(\theta)$, are related by the expression $L^*(\psi) = L\{\theta(\psi)\}$. The value of L is not changed by this transformation, so the likelihood is invariant to 1–1 reparametrization. We can use a parametrization that has a direct interpretation in terms of our particular problem.

Example 4.8 (Challenger data)   We focus on the probability of thermal distress at 31°F, expressed in terms of the original parameters as

$$\psi = \frac{\exp(\beta_0 + 31\beta_1)}{1 + \exp(\beta_0 + 31\beta_1)}.$$

If we reparametrize L in terms of ψ and $\lambda = \beta_1$, we have $\beta_0(\psi, \lambda) = \log\{\psi/(1 - \psi)\} - 31\lambda$, and $L^*(\psi, \lambda) = L\{\beta_0(\psi, \lambda), \lambda\}$. The plot of the log likelihood $\ell^*(\psi, \lambda) = \log L^*(\psi, \lambda)$ in the right panel of Figure 4.3 is easier to interpret than the plot of $\ell(\beta_0, \beta_1)$ in the left panel, because the plausible range of values for ψ changes more slowly with λ. The contours in the left panel seem roughly elliptical, but those in the right are not. The most plausible range of values for ψ is $(0.7, 0.9)$, throughout which the value of λ is roughly -0.1.   ∎

Interpretation

When there is a particular parametric model for a set of data, likelihood provides a natural basis for assessing the plausibility of different parameter values, but how should it be interpreted? One viewpoint is that values of θ can be compared using a scale such as

$$
\begin{aligned}
1 &\geq RL(\theta) > \tfrac{1}{3}, & &\theta \text{ strongly supported,} \\
\tfrac{1}{3} &\geq RL(\theta) > \tfrac{1}{10}, & &\theta \text{ supported,} \\
\tfrac{1}{10} &\geq RL(\theta) > \tfrac{1}{100}, & &\theta \text{ weakly supported,} \qquad (4.11) \\
\tfrac{1}{100} &\geq RL(\theta) > \tfrac{1}{1000}, & &\theta \text{ poorly supported,} \\
\tfrac{1}{1000} &\geq RL(\theta) > 0, & &\theta \text{ very poorly supported.}
\end{aligned}
$$

Under this *pure likelihood* approach, values of θ are compared solely in terms of relative likelihoods. A scale such as (4.11) is simple and directly interpretable, but as it has the disadvantages that the numbers $\frac{1}{3}$, $\frac{1}{10}$ and so forth are arbitrary and take no account of the dimension of θ, this interpretation is not the most common one in practice. We discuss repeated sampling calibration of likelihood values in Section 4.5.

Exercises 4.1

1. Sketch the Cauchy likelihood for the observations 1.1, 2.3, 1.5, 1.4.
   Show that the distribution function of the *two-parameter Cauchy density*,

   $$f(u; \theta, \sigma) = \frac{\sigma}{\pi\{\sigma^2 + (u - \theta)^2\}}, \quad -\infty < u < \infty, \ \sigma > 0, -\infty < \theta < \infty,$$

   is $F(u) = \frac{1}{2} + \pi^{-1} \tan^{-1}\{(u - \theta)/\sigma\}$. Hence find $\Pr(|Y - \theta| < 20)$ when $\sigma = 1$, and with hindsight explain why the model in Example 4.3 fits poorly.

2. Find the likelihood for a random sample $y_1, \ldots, y_n$ from the geometric density $\Pr(Y = y) = \pi(1 - \pi)^y$, $y = 0, 1, \ldots$, where $0 < \pi < 1$.

3. Verify that the likelihood for $f(y; \lambda) = \lambda \exp(-\lambda y)$, $y, \lambda > 0$, is invariant to the reparametrization $\psi = 1/\lambda$.

4. Show that the log likelihood for two independent sets of data is the sum of their log likelihoods.

5. Let $A_n \subset A_{n-1} \subset \cdots \subset A_1$ be events on the same probability space. Show that

   $$\Pr(A_n) = \Pr(A_n \mid A_{n-1})\Pr(A_{n-1}) = \Pr(A_n \mid A_{n-1}) \cdots \Pr(A_2 \mid A_1)\Pr(A_1)$$

   and hence establish (4.7).

4.2 Summaries

4.2.1 Quadratic approximation

In a problem with one or two parameters, the likelihood can be visualized. However models with a few dozen parameters are commonplace, and sometimes there are many

For example, an image of 512×512 pixels may have a parameter for each pixel.

more, so we often need to summarize the likelihood.

A key idea is that in many cases the log likelihood is approximately quadratic as a function of the parameter. To illustrate this, the left panel of Figure 4.4 shows log likelihoods for random samples of size $n = 5, 10, 20, 40$ and 80 from an exponential

As usual, $\bar{y} = n^{-1} \sum y_j$.

density, $\theta^{-1} \exp(-u/\theta)$, $\theta > 0$, $u > 0$. In each case the sample has average $\bar{y} = e^{-1}$. The panel has two general features. First, the maximum of each log likelihood is at $\theta = e^{-1}$. To see why, note that (4.3) implies that

$$\ell(\theta) = -n \log \theta - \theta^{-1} \sum_{j=1}^{n} y_j = -n \left(\log \theta + \bar{y}/\theta\right),$$

which is maximized when $d\ell(\theta)/d\theta = 0$, that is, when $\theta = \bar{y}$. Now

$$\frac{d^2\ell(\theta)}{d\theta^2} = -n \left(-\frac{1}{\theta^2} + \frac{2\bar{y}}{\theta^3}\right)$$

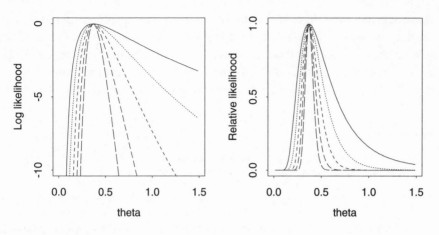

Figure 4.4 Log
likelihoods and relative
likelihoods for
exponential samples with
sample sizes $n = 5, 10,$
$20, 40, 80.$ The curvature
of the functions increases
with n, so the highest
curve in each panel is for
$n = 5$, and the lowest is
for $n = 80$.

takes the value $-n/\overline{y}^2$ at $\theta = \overline{y}$, so $\overline{y}$ gives the unique maximum of ℓ. The value of θ for which L, or equivalently ℓ, is greatest is called the *maximum likelihood estimate*, $\widehat{\theta}$. In this case $\widehat{\theta} = \overline{y}$. For future reference, note that the values of $-n^{-1}d^2\ell(\theta)/d\theta^2$ and its derivative $-n^{-1}d^3\ell(\theta)/d\theta^3$ are bounded in a neighbourhood $\mathcal{N} = \{\theta : |\theta - \widehat{\theta}| < \delta\}$ of $\widehat{\theta}$, provided $\mathcal{N}$ excludes $\theta = 0$.

Second, the curvature of the log likelihood at the maximum increases with n, because the second derivative of ℓ, which measures its curvature as a function of θ, is a linear function of n. The function $-d^2\ell(\theta)/d\theta^2$ is called the *observed information*. In this case its value at $\widehat{\theta}$ is $n/\overline{y}^2 = n/\widehat{\theta}^2$.

The right panel of Figure 4.4 shows the relative likelihoods corresponding to the left panel. The effect of increasing n is that the likelihood becomes more concentrated about the maximum, and so it becomes relatively less and less plausible that each value of θ a fixed distance from $\widehat{\theta}$ generated the data. To express this algebraically, we write the log relative likelihood, $\log RL(\theta)$, as $\ell(\theta) - \ell(\widehat{\theta})$ and expand $\ell(\theta)$ in a Taylor series about $\widehat{\theta}$ to obtain

$$\log RL(\theta) = \ell(\widehat{\theta}) + (\theta - \widehat{\theta})\ell'(\widehat{\theta}) + \frac{1}{2}(\theta - \widehat{\theta})^2\ell''(\theta_1) - \ell(\widehat{\theta}) = \frac{1}{2}(\theta - \widehat{\theta})^2\ell''(\theta_1);$$
$$(4.12)$$

θ_1 lies between θ and $\widehat{\theta}$. We denote differentiation with respect to θ by a prime, thus $\ell'(\theta) = d\ell(\theta)/d\theta$, and so forth; note that $\ell'(\widehat{\theta}) = 0$. Each derivative of ℓ is a sum of n terms. As n increases, we see that the bound on $-n^{-1}\ell''(\theta_1)$ implies that (4.12) will become increasingly negative except at $\theta = \widehat{\theta}$. Hence $RL(\theta)$ tends to zero unless $\theta = \widehat{\theta}$, while $RL(\widehat{\theta}) = 1$ for all n.

To examine the behaviour of the log likelihood more closely, we take another term in the Taylor expansion leading to (4.12), to find that

$$\log RL(\theta) = \frac{1}{2}(\theta - \widehat{\theta})^2\ell''(\widehat{\theta}) + \frac{1}{6}(\theta - \widehat{\theta})^3\ell'''(\theta_2),$$

where θ_2 lies between θ and $\widehat{\theta}$. Now consider what happens, not at a fixed distance from $\widehat{\theta}$, but at $\theta = \widehat{\theta} + n^{-1/2}\delta$. As n increases this corresponds to "zooming in" and examining the region around $\widehat{\theta}$ ever more closely. Now

$$\log RL\left(\widehat{\theta} + n^{-1/2}\delta\right) = \frac{1}{2}\delta^2 n^{-1}\ell''(\widehat{\theta}) + \frac{1}{6}\delta^3 n^{-3/2}\ell'''(\theta_2), \qquad (4.13)$$

and crucially, both $\ell''(\theta)$ and $\ell'''(\theta)$ are linear functions of n. The bound on $-n^{-1}\ell'''(\theta)$ implies that the last term on the right of (4.13) disappears as $n \to \infty$, but the quadratic term becomes $-\frac{1}{2}\delta^2\{-n^{-1}\ell''(\widehat{\theta})\}$, which in this case is $-\frac{1}{2}\delta^2/\overline{y}^2$. Thus in large samples the likelihood close to the maximum is a quadratic function and can be summarized in terms of the maximum likelihood estimate $\widehat{\theta}$ and the observed information $-\ell''(\widehat{\theta})$. One implication of this is that if we restrict ourselves to parameter values that are plausible relative to the maximum likelihood estimate, say those values of θ such that $RL(\theta) > c$, we find $\log RL(\theta) > \log c$. Comparison with (4.13) shows that our range of 'plausible' θ is decreasing with n and has length roughly proportional to $n^{-1/2}$.

This discussion concerns a scalar parameter, but extends to higher dimensions, where $d^2\ell/d\theta^2$ is replaced by the matrix of second derivatives of ℓ.

Whether a quadratic approximation to ℓ is useful depends on the problem. To summarize the log likelihood in Figure 4.2 in such terms would be very misleading, unless a summary was required only very close to the maximum. If feasible, it is sensible to plot the likelihood.

Example 4.9 (Uniform distribution)   Suppose we are presented with a random sample $y_1, \ldots, y_n$ from the uniform density on $(0, \theta)$:

$$f(u; \theta) = \begin{cases} \theta^{-1}, & 0 < u < \theta, \\ 0, & \text{otherwise.} \end{cases}$$

The likelihood is

$$L(\theta) = \prod_j f(y_j; \theta) = \begin{cases} \theta^{-n}, & 0 < y_1, \ldots, y_n < \theta, \\ 0, & \text{otherwise.} \end{cases}$$

It is maximized at $\widehat{\theta} = \max(y_j)$, but $d\ell(\widehat{\theta})/d\theta \neq 0$ and $-d^2\ell(\widehat{\theta})/d\theta^2 = -n/\widehat{\theta}^2 < 0$, and $\ell(\theta)$ becomes increasingly spikey as $n \to \infty$ and is not approximately quadratic near $\widehat{\theta}$ for any n.   ∎

4.2.2 Sufficient statistics

In well-behaved problems and with large samples the likelihood may be summarized in terms of the maximum likelihood estimate and observed information, though Examples 4.3 and 4.9 show that this can fail. A better approach rests on the fact that the likelihood often depends on the data only through some low-dimensional function $s(y)$ of the y_j, and then a suitable summary can be given in terms of this. Thus in Examples 4.2 and 4.9 the likelihoods depend on the data through $(n, \sum y_j)$ and $(n, \max y_j)$ respectively. If we believe that our model is correct, we need only these functions to calculate the likelihoods for any value of θ. These functions are examples of *sufficient statistics*.

Suppose that we have observed data, y, generated by a distribution whose density is $f(y; \theta)$, and that the statistic $s(y)$ is a function of y such that the conditional density of the corresponding random variable Y, given that $S = s(Y)$, is independent of θ. That is,

$$f_{Y|S}(y \mid s; \theta) \tag{4.14}$$

does not depend on θ. Then S is said to be a *sufficient statistic for θ based on Y*, or just a *sufficient statistic for θ*. The idea is that any extra information in Y but not in S is given by the conditional density (4.14), and if this conditional density is free of θ, Y contains no more information about θ than does S. We shall see later that S is not unique.

Definition (4.14) is hard to use, because we must guess that a given statistic S is sufficient before we can calculate the conditional density. An equivalent and more useful definition is via the *factorization criterion*. This states that a necessary and sufficient condition for a statistic S to be a sufficient statistic for a parameter θ in a family of probability density functions $f(y; \theta)$ is that the density of Y can be expressed as

$$f(y; \theta) = g\{s(y); \theta\}h(y). \tag{4.15}$$

Thus the density of Y factorizes into a function g of $s(y)$ and θ, and a function of y, h, that does not depend on θ.

The equivalence of these two definitions is almost self-evident. First note that if S is a sufficient statistic, the conditional distribution of Y given S is independent of θ, that is,

$$f_{Y|S}(y \mid s) = \frac{f_{Y,S}(y, s; \theta)}{f_S(s; \theta)} \tag{4.16}$$

is free of θ. But as S is a function $s(Y)$ of Y, the joint density of S and Y is zero except where $S = s(Y)$, and so the numerator of the right-hand side of (4.16) is just $f_Y(y; \theta)$. Rearrangement of (4.16) implies that if S is sufficient, (4.15) holds with $g(\cdot) = f_S(\cdot)$ and $h(\cdot) = f_{Y|S}(\cdot)$.

Conversely, if (4.15) holds, we find the density of S at s by summing or integrating (4.15) over the range of y for which $s(y) = s$. In the discrete case

$$f_S(s; \theta) = \sum g\{s(y); \theta\}h(y) = g\{s; \theta\} \sum h(y),$$

because the sum is over those y for which $s(y)$ equals s. Therefore the conditional density of Y given S is

$$\frac{f_Y(y; \theta)}{f_S(s; \theta)} = \frac{g\{s(y); \theta\}h(y)}{g\{s; \theta\} \sum h(y)} = \frac{h(y)}{\sum h(y)},$$

Proof in the continuous case would replace the sum here by an integral, but a detailed proof is not simple because all elements of the parametric model must be dominated by a single measure. See for example Theorem 2.21 of Schervish (1995).

which shows that S is sufficient.

Example 4.10 (Bernoulli distribution)   A Bernoulli random variable Y records the 'success' or 'failure' of a binary trial. Thus

$$\Pr(Y = 1) = 1 - \Pr(Y = 0) = \pi, \quad 0 \leq \pi \leq 1,$$

with $Y = 1$ representing success and $Y = 0$ failure. The likelihood contribution from a single trial with outcome $Y = y$ may be written $\pi^y(1-\pi)^{1-y}$, and hence the likelihood for π based on the outcomes of n independent trials is

$$L(\pi) = \prod_{j=1}^{n} \pi^{y_j}(1-\pi)^{1-y_j} = \pi^r(1-\pi)^{n-r},$$

say, where $r = \sum y_j$ is the number of successes in the n trials. The distribution of the corresponding random variable, $R = \sum Y_j$, is binomial with probability π and denominator n, that is,

$$\Pr(R = r) = \binom{n}{r} \pi^r(1-\pi)^{n-r}, \quad r = 0, \ldots, n.$$

Hence the distribution of $Y_1, \ldots, Y_n$ conditional on R is

$$\Pr\left(Y_1 = y_1, \ldots, Y_n = y_n \mid \sum Y_j = r\right) = \frac{1}{\binom{n}{r}},$$

which puts equal probability on each of the $\binom{n}{r}$ permutations of r 1's and $n - r$ 0's. This conditional distribution does not depend on π, so R is sufficient for π, as is intuitively clear.

Although there is no loss of information about π when $Y_1, \ldots, Y_n$ is reduced to R, the original data are more useful for some purposes. For example, if $y_1, \ldots, y_n$ consisted of a sequence of zeros followed by a sequence of ones, we might want to revise our belief that the trials were independent, but we could not know this if only $\sum y_j$ had been reported. ∎

Example 4.11 (Exponential distribution) Suppose that Y_1 and Y_2 are independently exponentially distributed. Then their joint density is

$$\begin{aligned} f(y; \lambda) &= \lambda e^{-\lambda y_1} \cdot \lambda e^{-\lambda y_2}, \quad y_1, y_2 > 0, \\ &= \lambda^2 \exp\{-\lambda(y_1 + y_2)\} \\ &= \lambda^2 \exp(-\lambda s) \cdot 1, \end{aligned}$$

which factorizes into a function of $s = y_1 + y_2$ and the constant 1. Therefore $S = Y_1 + Y_2$ is sufficient, using the factorization criterion (4.15).

To verify this using the original definition (4.14), note that S is a sum of two independent exponential random variables, and so has the gamma distribution with density

$$f(s; \lambda) = \lambda^2 s \exp(-\lambda s), \quad s > 0.$$

Thus the conditional density of Y_1 and Y_2 given that $S = Y_1 + Y_2 = s$ is

$$\frac{f(y_1, y_2; \lambda)}{f(s; \lambda)} = \frac{\lambda^2 \exp\{-\lambda(y_1 + y_2)\}}{\lambda^2 s \exp(-\lambda s)} = \frac{1}{s}, \quad y_1 + y_2 = s > 0.$$

This, the uniform density on $(0, s)$, is free of λ. Thus given the particular value s for the line $Y_1 + Y_2 = s$ on which the point (Y_1, Y_2) lies, the position of (Y_1, Y_2) on the line conveys no extra information about λ. ∎

Example 4.12 (Random sample)   Let $Y_1, \dots, Y_n$ be a random sample of scalar observations from a density $f(y; \theta)$. Now as all the observations are on an equal footing, their order is irrelevant. It follows that the order statistics $Y_{(1)}, \dots, Y_{(n)}$ are sufficient for θ. To see this, note that we saw at (2.25) that the joint density of the order statistics is

$$n! f(y_{(1)}; \theta) \times \cdots \times f(y_{(n)}; \theta), \quad y_{(1)} \leq \cdots \leq y_{(n)}.$$

Hence the conditional density of $Y_1, \dots, Y_n$ given $Y_{(1)}, \dots, Y_{(n)}$ is $1/n!$, provided that $Y_{(1)}, \dots, Y_{(n)}$ is a permutation of $Y_1, \dots, Y_n$, and is zero otherwise. Evidently this conditional density is free of θ, and hence the order statistics are a sufficient statistic of dimension n for θ.

If we are willing to make more specific assumptions about $f(y; \theta)$, we can reduce the data further. For the exponential density, for example, the likelihood is $\theta^{-n} \exp(-\theta^{-1} \sum y_j)$, so it follows that $(N, \sum Y_j)$ is also sufficient for θ. Thus there can be different sufficient statistics for a single model.   ∎

Example 4.13 (Capture-recapture model)   Capture-recapture models are widely used to estimate the sizes of animal populations and survival rates from one year to the next. The idea is to capture animals on a number of separate occasions, to mark them, and to return them to the wild after each occasion. The proportion of marked animals seen on the second and subsequent occasions gives an idea of the quantities of interest. For example, if the population is large and only a small proportion of it is seen on the first occasion, then few of the animals captured next time will already be marked.

Suppose there are three capture occasions (years) labelled 0, 1, and 2, that the probability of survival from one occasion to the next is ψ, and that, for an animal alive in year s, the probability of recapture is λ_s. Then the possible capture histories and their probabilities are

111 $\psi\lambda_1 \times \psi\lambda_2$,	011 $\psi\lambda_2$,
110 $\psi\lambda_1 \times \{1 - \psi + \psi(1 - \lambda_2)\}$,	010 $1 - \psi + \psi(1 - \lambda_2)$,
101 $\psi(1 - \lambda_1) \times \psi\lambda_2$,	001 1,
100 $1 - \psi + \psi(1 - \lambda_1)\{1 - \psi + \psi(1 - \lambda_2)\}$	

where, for example, 110 represents an animal seen in years 0 and 1, but not 2. The probability of being alive and seen in year 1 is $\psi\lambda_1$, and conditional on being alive in year 1, the animal may be dead in year 2, with probability $1 - \psi$, or alive but not seen, with probability $\psi(1 - \lambda_2)$. Without further assumptions we can say nothing about animals with history 000, which we never see.

If animals are assumed independent, the likelihood is a product of such terms, and we notice that, for example, there is a contribution $\psi\lambda_1$ from animals with history 111 or 110, a contribution $\psi\lambda_2$ from animals with history 111 or 011, and so on. Thus the likelihood may be written as

$$(\psi\lambda_1)^{r_{01}} \{\psi(1 - \lambda_1)\psi\lambda_2\}^{r_{02}} \{1 - \psi\lambda_1 - \psi(1 - \lambda_1)\psi\lambda_2\}^{m_0 - r_{01} - r_{02}}$$
$$\times (\psi\lambda_2)^{r_{11}} (1 - \psi\lambda_2)^{m_1 - r_{11}},$$

Table 4.1 Sufficient statistics and probabilities for capture-recapture model.

Year	Number captured	Number first recaptured in year		Number never recaptured
		1	2	
0	m_0	r_{01}	r_{02}	$m_0 - r_{01} - r_{02}$
1	m_1		r_{11}	$m_1 - r_{11}$

Year	Probability first recaptured in year		Probability never recaptured
	1	2	
0	$\psi\lambda_1$	$\psi(1-\lambda_1)\psi\lambda_2$	$1 - \psi\lambda_1 - \psi(1-\lambda_1)\psi\lambda_2$
1		$\psi\lambda_2$	$1 - \psi\lambda_2$

where m_s is the number of animals seen in year s, of whom r_{st} are first seen again in year t. Evidently the quantities m_s and r_{st} are sufficient statistics. We lay out these and the corresponding probabilities in Table 4.1, which is a standard representation for such data. With k occasions the number of individual histories is $2^k - 1$ but the table contains just $\frac{1}{2}(k+2)(k-1)$ elements, so the reduction can be considerable, but more importantly the data structure is clearer in terms of the sufficient statistics. ∎

Minimal sufficiency

Even for a single model, sufficient statistics are not unique. Apart from the possibility that different functions $s(Y)$ might satisfy the factorization criterion, the data themselves form a sufficient statistic. Moreover it is easy to see from (4.15) that any known 1–1 function of a sufficient statistic is itself sufficient. What is unique to each sufficient statistic is the partition that it induces on the sample space.

To see this, we say that two samples Y_1 and Y_2 with corresponding sufficient statistics $S_1 = s(Y_1)$ and $S_2 = s(Y_2)$ are equivalent if $S_1 = S_2$. This evidently satisfies the three properties of an equivalence relation:

- reflexivity, Y is equivalent to itself;
- symmetry, Y_1 is equivalent to Y_2 if Y_2 is equivalent to Y_1; and
- transitivity, Y_1 is equivalent to Y_3 whenever Y_1 is equivalent to Y_2 and Y_2 is equivalent to Y_3.

Therefore the sample space is partitioned by the relation into equivalence classes, corresponding to each of the distinct values that S can take. Unlike the sufficient statistic itself, this partitioning is invariant under 1–1 transformation of S. By the factorization criterion it has the property that the conditional density of the data Y given that Y falls into a particular equivalence class is independent of the parameter, and hence is called a *sufficient partition*. Such a partition has the property that if we are told into which of its equivalence classes the data fall, we can reconstruct the log likelihood up to additive constants. A mathematical discussion of sufficiency would be in terms of sufficient partitions rather than sufficient statistics. However it is more natural to think in terms of sufficient statistics, and we mostly do so.

As sufficient statistics are not unique, we can choose which to use. The biggest reduction of the data is obtained by taking a sufficient statistic whose dimension is as small as possible, that is, a *minimal sufficient statistic*. A sufficient statistic is said to be minimal if it is a function of any other sufficient statistic. This corresponds to the coarsest sufficient partition of the sample space, while the data generate the finest sufficient partition. To find a minimal sufficient statistic, we return to the likelihood. Suppose that the likelihoods of two sets of data, y and z, are the same up to a constant. Then $L(\theta; y)/L(\theta; z)$ does not depend on θ, and the partition that this equivalence relation generates is minimally sufficient. Thus a minimal sufficient statistic is obtained by examining the likelihood to see on what functions of the data it depends.

Example 4.14 (Exponential distribution)   In Example 4.11 the sample space into which (Y_1, Y_2) falls is $\mathbb{R}_+^2$, and this is partitioned by the lines $y_1 + y_2 = s$, $s > 0$, each of which corresponds to an equivalence class.

In order to find a minimal sufficient statistic, note that the likelihood based on data y_1, y_2 is $\lambda^2 \exp\{-\lambda(y_1 + y_2)\}$, whereas the likelihood based on $x_1, \ldots, x_m$ would be $\lambda^m \exp\{-\lambda(x_1 + \cdots + x_m)\}$ The ratio of these would be independent of λ only if $m = 2$ and $x_1 + x_2 = y_1 + y_2$. Hence a minimal sufficient statistic is (N, S), the number of observations in the sample, and their sum. Usually N is chosen without regard to λ, and S alone is regarded as minimal sufficient. ∎

Example 4.15 (Poisson birth process)   We saw in Example 4.6 that the likelihood based on data $y_0, \ldots, y_n$ from such a process is

$$L(\theta) = \left(\prod_{j=0}^{n} y_j!\right)^{-1} \exp\left(s_0 \log \theta - s_1 \theta\right), \qquad \theta > 0,$$

where $s_0 = \sum_{j=0}^{n} y_j$ and $s_1 = 1 + \sum_{j=0}^{n-1} y_j$. The factorization criterion shows that a sufficient statistic is (S_0, S_1), but equally so is (S_0, Y_n), since $S_1 = S_0 + 1 - Y_n$. Evidently either of these is also minimal sufficient. ∎

Example 4.16 (Logistic regression)   Suppose that independent binomial random variables R_j have denominators m_j and probabilities π_j, where

$$\pi_j = \frac{\exp(\beta_0 + \beta_1 x_{1j})}{1 + \exp(\beta_0 + \beta_1 x_{1j})}, \quad j = 1, \ldots, n,$$

and the x_{1j} are known constants. The likelihood is (4.6), and on applying the factorization criterion we see that a minimal sufficient statistic for (β_0, β_1) is $S = (\sum R_j, \sum R_j x_{1j})$. Although the m_j, x_{1j}, and n are needed to calculate the likelihood, they are non-random and not included in S. ∎

Exercises 4.2

1   Find the maximum likelihood estimate and observed information in Example 4.1. Find also the maximum likelihood estimate of $\Pr(Y = 0)$.

2 Find maximum likelihood estimates for θ based on a random sample of size n from the densities (i) $\theta y^{\theta-1}$, $0 < y < 1, \theta > 0$; (ii) $\theta^2 y e^{-\theta y}$, $y > 0, \theta > 0$; and (iii) $(\theta + 1)y^{-\theta-2}$, $y > 1, \theta > 0$;

3 Plot the likelihood for θ based on a random sample $y_1, \ldots, y_n$ from the density

$$f(x;\theta) = \begin{cases} 1/(2c), & \theta - c < x < \theta + c, \\ 0, & \text{otherwise}, \end{cases}$$

where c is a known constant. Find a maximum likelihood estimate, and show that it is not unique.

4 In the discussion following (4.13), show that if the log likelihood was exactly quadratic and we agreed that values of θ such that $RL(\theta) > c$ were 'plausible', the range of plausible θ would be $\widehat{\theta} \pm \{2 \log c / \ell''(\widehat{\theta})\}^{1/2}$.

5 Data are available from n independent experiments concerning a scalar parameter θ. The log likelihood for the jth experiment may be summarized as a quadratic function, $\ell_j(\theta) \doteq \widehat{\ell}_j - \frac{1}{2} J_j(\widehat{\theta}_j)(\theta - \widehat{\theta}_j)^2$, where $\widehat{\theta}_j$ is the maximum likelihood estimate and $J_j(\widehat{\theta}_j)$ is the observed information. Show that the overall log likelihood may be summarized as a quadratic function of θ, and find the overall maximum likelihood estimate and observed information.

6 In a *first-order autoregressive process*, $Y_0, \ldots, Y_n$, the conditional distribution of Y_j given the previous observations, $Y_1, \ldots, Y_{j-1}$, is normal with mean αy_{j-1} and variance one. The initial observation Y_0 has the normal distribution with mean zero and variance one. Show that the log likelihood is proportional to $y_0^2 + \sum_{j=1}^{n}(y_j - \alpha y_{j-1})^2$, and hence find the maximum likelihood estimate of α and the observed information.

7 Find a minimal sufficient statistic for θ based on a random sample $Y_1, \ldots, Y_n$ from the Poisson density (2.6).

8 Let $Y_1, \ldots, Y_n$ be a random sample from the $N(\mu, \sigma^2)$ distribution.
   (a) Use the factorization criterion to show that $(\sum Y_j, \sum Y_j^2)$ is sufficient for (μ, σ^2). Say, giving your reasons, which of the following are also sufficient: (i) $(\overline{Y}, S^2)$; (ii) $(\overline{Y}^2, S)$; (iii) the order statistics $Y_{(1)} < \cdots < Y_{(n)}$.
   (b) If $\sigma^2 = 1$, show that the sample average is minimal sufficient for μ.
   (c) Suppose that μ equals the known value μ_0. Show that $S = \sum(Y_j - \mu_0)^2$ is a minimal sufficient statistic for σ^2, and give its distribution. Show that S is a function of the minimal sufficient statistic when both parameters are unknown.

9 Find the minimal sufficient statistic based on a random sample $Y_1, \ldots, Y_n$ from the gamma density (2.7).

10 Use the factorization criterion to show that the maximum likelihood estimate and observed information based on $f(y;\theta)$ are functions of data y only through a sufficient statistic $s(y)$.

11 Verify that the relation 'y_1 is equivalent to y_2' if $L(\theta; y_1)/L(\theta; y_2)$ is independent of θ is an equivalence relation and that the corresponding partition is sufficient. Deduce that the likelihood itself is minimal sufficient.

4.3 Information

4.3.1 Expected and observed information

In a model with log likelihood $\ell(\theta)$, the *observed information* is defined to be

$$J(\theta) = -\frac{d^2\ell(\theta)}{d\theta^2}.$$

When $\ell(\theta)$ is a sum of n components, so too is $J(\theta)$, because (4.9) implies that

$$J(\theta) = -\frac{d^2\ell(\theta)}{d\theta^2} = -\frac{d^2}{d\theta^2}\sum_{j=1}^{n}\ell_j(\theta) = \sum_{j=1}^{n} -\frac{d^2\log f(y_j;\theta)}{d\theta^2}. \qquad (4.17)$$

We saw in Section 4.2.1 that when the log likelihood is roughly quadratic, the relative plausibility of parameter values near the maximum likelihood estimate is determined by the observed information. High information, or equivalently high curvature, will pin down θ more tightly than if the observed information is low. The amount of information is typically related to the size of the dataset, a fact useful in planning experiments. Before we conduct an experiment it is valuable to assess what information there will be in the data, to see if the proposed sample is large enough. Otherwise we may need more data or a more informative experiment. Before the experiment is performed we have no data, so we cannot obtain the observed information. However we can calculate the *expected* or *Fisher information*,

$$I(\theta) = \mathrm{E}\left\{-\frac{d^2\ell(\theta)}{d\theta^2}\right\},$$

which is the mean information the data will contain when collected, if the model is correct and the true parameter value is θ.

If the data are a random sample, (4.17) implies that $I(\theta) = ni(\theta)$, where $i(\theta)$ is the information from a single observation,

$$i(\theta) = \mathrm{E}\left\{-\frac{d^2\log f(Y_j;\theta)}{d\theta^2}\right\}.$$

When θ is a $p \times 1$ vector, the information matrices are

$$J(\theta) = -\frac{\partial^2\ell(\theta)}{\partial\theta\partial\theta^{\mathrm{T}}}, \quad I(\theta) = -\mathrm{E}\left\{\frac{\partial^2\ell(\theta)}{\partial\theta\partial\theta^{\mathrm{T}}}\right\};$$

these are symmetric $p \times p$ matrices whose (r, s) elements are respectively

For a $p \times 1$ vector θ we use $\partial\ell/\partial\theta$ to denote the $p \times 1$ vector whose rth element is $\partial\ell/\partial\theta_r$, and $\partial^2\ell/\partial\theta\partial\theta^{\mathrm{T}}$ to denote the $p \times p$ matrix whose (r, s) element is $\partial^2\ell/\partial\theta_r\partial\theta_s$.

$$-\frac{\partial^2\ell(\theta)}{\partial\theta_r\partial\theta_s}, \quad \mathrm{E}\left\{-\frac{\partial^2\ell(\theta)}{\partial\theta_r\partial\theta_s}\right\}.$$

Example 4.17 (Binomial distribution)   The likelihood for a binomial variable R with denominator m and probability of success $0 < \pi < 1$ is $L(\pi) = \binom{m}{r}\pi^r(1 - \pi)^{m-r}$, so $\ell(\pi) \equiv r\log\pi + (m - r)\log(1 - \pi)$ and

$$J(\pi) = -\frac{d^2\ell(\pi)}{d\pi^2} = \frac{r}{\pi^2} + \frac{m - r}{(1 - \pi)^2},$$

given an observed value r of R. Before the experiment has been performed the value of r is unknown, and we replace it by the corresponding random variable R. In this case $J(\pi)$ too is random, and

$$
\begin{aligned}
I(\pi) &= \mathrm{E}\{J(\pi)\} \\
&= \mathrm{E}\left\{\frac{R}{\pi^2} + \frac{m-R}{(1-\pi)^2}\right\} \\
&= \frac{m\pi}{\pi^2} + \frac{m(1-\pi)}{(1-\pi)^2} = \frac{m}{\pi(1-\pi)},
\end{aligned}
$$

since $\mathrm{E}(R) = m\pi$. The expected information $I(\pi)$ increases linearly with m and is symmetric in π, for $0 < \pi < 1$. ∎

Example 4.18 (Normal distribution) The density function of a normal random variable with mean μ and variance σ^2 is (3.5), so the log likelihood for a random sample $y_1, \ldots, y_n$ is

$$
\ell(\mu, \sigma) \equiv -\frac{n}{2}\log\sigma^2 - \frac{1}{2\sigma^2}\sum_{j=1}^{n}(y_j - \mu)^2.
$$

Its first derivatives are

$$
\frac{\partial\ell}{\partial\mu} = \sigma^{-2}\sum(y_j - \mu), \quad \frac{\partial\ell}{\partial\sigma^2} = -\frac{n}{2\sigma^2} + \frac{1}{2\sigma^4}\sum(y_j - \mu)^2,
$$

and the elements of the observed information matrix $J(\mu, \sigma^2)$ are given by

$$
\frac{\partial^2\ell}{\partial\mu^2} = -\frac{n}{\sigma^2}, \quad \frac{\partial^2\ell}{\partial\mu\partial\sigma^2} = -\frac{n}{\sigma^4}(\overline{y} - \mu), \quad \frac{\partial^2\ell}{\partial(\sigma^2)^2} = -\frac{n}{2\sigma^4} + \frac{1}{\sigma^6}\sum(y_j - \mu)^2.
$$

On replacing y_j with Y_j and taking expectations, we get

$$
I(\mu, \sigma^2) = \begin{pmatrix} n/\sigma^2 & 0 \\ 0 & n/(2\sigma^4) \end{pmatrix}, \tag{4.18}
$$

because $\mathrm{E}(Y_j) = \mu$ and $\mathrm{E}\{(Y_j - \mu)^2\} = \sigma^2$. ∎

4.3.2 Efficiency

Suppose that we might adopt one of two sampling schemes, and we wish to see which is most efficient in the sense of needing least data to pin down the parameter to a given range. One way to do this is to compare the information in each likelihood. If θ is scalar, the *asymptotic efficiency* of sampling scheme A relative to sampling scheme B is

$$
\frac{I_A(\theta)}{I_B(\theta)}, \tag{4.19}
$$

where $I_A(\theta)$ and $I_B(\theta)$ are the expected information quantities for schemes A and B. In simple random samples (4.19) equals $n_A i_A(\theta)/\{n_B i_B(\theta)\}$, where n_A and n_B observations are used by the sampling schemes. The information from both schemes is equal if

$$
\frac{n_B}{n_A} = \frac{i_A(\theta)}{i_B(\theta)} \tag{4.20}
$$

and we see that $i_A(\theta)/i_B(\theta)$ can be interpreted as the number of observations an observer using scheme B would need in order to get the information in a single observation sampled under scheme A, when the parameter value is θ. Expression (4.19) is called the asymptotic efficiency because this use of the information rests on the quadratic likelihoods usually entailed by large samples.

Example 4.19 (Poisson process)   Over short periods the times at which vehicles pass an observer on a country road might be modelled as a Poisson process of rate λ vehicles/hour. Observer A decides to estimate λ by counting how many cars pass in a period of t_0 minutes. Observer B, who is more diligent, records the times at which they pass.

The total number of events, N, when a Poisson process of rate λ is observed for a period of length t_0 has the Poisson distribution with mean λt_0. Hence A bases her inference on the likelihood

$$L_A(\lambda) = \frac{(\lambda t_0)^N}{N!} e^{-\lambda t_0}, \qquad \lambda > 0,$$

for which the observed and expected information quantities are

$$J_A(\lambda) = N/\lambda^2, \qquad I_A(\lambda) = t_0/\lambda,$$

since $E(N) = \lambda t_0$.

The times between events in a Poisson process of rate λ have independent exponential distributions with density $\lambda e^{-\lambda u}$, $u > 0$. Therefore if observer B records cars passing at times $0 < t_1 < \cdots < t_N < t_0$, his likelihood is

$$\lambda e^{-\lambda t_1} \times \lambda e^{-\lambda(t_2-t_1)} \times \cdots \times \lambda e^{-\lambda(t_N-t_{N-1})} \times e^{-\lambda(t_0-t_N)},$$

where the final term corresponds to observing no cars in the interval (t_N, t_0). Thus B bases his inference on

$$L_B(\lambda) = \lambda^N e^{-\lambda t_0},$$

for which the observed and expected information quantities are the same as those for A. Thus the efficiency of A relative to B is $I_A(\lambda)/I_B(\lambda) = 1$: no information is lost by recording only the number of cars. This is because $L_A(\lambda) \propto L_B(\lambda)$; under either sampling scheme, the statistic N is sufficient for λ.

Inference for Poisson processes is discussed in Section 6.5.1.   ∎

Example 4.20 (Censoring)   A widget has lifetime T, but trials to estimate widget lifetimes finish after a known time c when the vice president for widget testing has a tea break. The available data are the observed lifetime $Y = \min(T, c)$, and $D = I(T \le c)$, where D indicates whether T has been observed. If $T > c$ then T is said to be *right-censored*: we know only that its value exceeds c.

$I(\cdot)$ is the indicator function of the event '$\cdot$'.

If T has density and distribution functions $f(t; \theta)$ and $F(t; \theta)$, the likelihood contribution from (Y, D) is

$$f(Y; \theta)^D \{1 - F(c; \theta)\}^{1-D},$$

so the likelihood for a random sample of data $(y_1, d_1), \ldots, (y_n, d_n)$ is

$$\prod_{j=1}^{n}[f(y_j;\theta)^{d_j}\{1 - F(y_j;\theta)\}^{1-d_j}] = \prod_{\text{uncens}} f(y_j;\theta) \times \prod_{\text{cens}}\{1 - F(c;\theta)\},$$

where the first product is over uncensored data, and the second is over censored data.

The likelihood for a random sample with exponential density $f(u;\lambda) = \lambda e^{-\lambda u}$, $u > 0, \lambda > 0$, and distribution $F(u;\lambda) = 1 - e^{-\lambda u}, u > 0$, is

$$\prod_{\text{uncens}} \lambda e^{-\lambda y_j} \times \prod_{\text{cens}} e^{-\lambda c} = \exp\left(\sum_{j=1}^{n} d_j \log\lambda - \lambda \sum_{j=1}^{n} y_j\right).$$

The observed information is $J(\lambda) = \sum d_j/\lambda^2$, which decreases as $\sum d_j$ decreases: if n is known, there is information only in observations that were seen to fail. To find the expected information $I_c(\lambda)$ when there is censoring at c, note that

$$\mathrm{E}\left(\sum_{j=1}^{n} D_j\right) = n\mathrm{Pr}(Y \le c) = n(1 - e^{-\lambda c}),$$

so that $I_c(\lambda) = n(1 - e^{-\lambda c})/\lambda^2$. By letting $c \to \infty$ we can obtain the expected information when there is no censoring, $I_\infty(\lambda) = n/\lambda^2$. Therefore the relative efficiency when there is censoring at c is

$$\frac{I_c(\lambda)}{I_\infty(\lambda)} = \frac{n(1 - e^{-\lambda c})/\lambda^2}{n/\lambda^2} = 1 - e^{-\lambda c}.$$

This equals the proportion of uncensored data, which is unsurprising, as we saw above that censored observations do not contribute to $J(\lambda)$. As one would anticipate, the loss of information becomes more severe as c decreases.

Inference for censored data is discussed in Sections 5.4 and 10.8. ∎

When θ is a $p \times 1$ vector, we replace (4.19) by the ratio

$|C|$ is the determinant of the $p \times p$ matrix C.

$$\left\{\frac{|I_A(\theta)|}{|I_B(\theta)|}\right\}^{1/p},$$

which preserves the interpretation of efficiency given at (4.20) in terms of numbers of observations. This is an overall measure of the efficiency of the schemes, but often in practice one may want to compare the efficiency of estimation for a single component of θ, say θ_r. For reasons to be given in Section 4.4.2, the appropriate measure is then $I_B^{rr}(\theta)/I_A^{rr}(\theta)$, where $I_A^{rr}(\theta)$ is the (r, r)th element of the inverse matrix $I_A(\theta)^{-1}$.

Example 4.21 (Rounding) What information is lost when the sample 2.71828, 3.14159, . . . is rounded to 2.7, 3.1, . . .? Let Y denote a real-valued continuous random variable with distribution function $F(y;\theta)$. In recording the data, Y is rounded to X, the nearest multiple of δ. Thus $X = k\delta$ if $(k - \frac{1}{2})\delta \le Y < (k + \frac{1}{2})\delta$, an event with probability

$$\pi_k(\theta) = F\left\{\left(k + \frac{1}{2}\right)\delta;\theta\right\} - F\left\{\left(k - \frac{1}{2}\right)\delta;\theta\right\}.$$

Table 4.2 Efficiency (%) of likelihood inference when $N(0, \sigma^2)$ data are rounded to the nearest δ.

δ/σ	0.001	0.01	0.1	0.2	0.5	1	1.5	2	3
Overall efficiency	100	100	99.9	99.5	97.0	88.9	77.9	64.0	37.5
Efficiency for μ	100	100	99.9	99.7	98.0	92.3	84.2	75.5	54.2
Efficiency for σ^2	100	100	99.8	99.3	96.0	85.5	72.0	54.2	25.9

The density of a single rounded observation may be written $\prod_k \pi_k(\theta)^{I(X=k\delta)}$, so the log likelihood for θ based on X is

$$\ell(\theta) = \sum_{k=-\infty}^{\infty} I(X = k\delta) \log \pi_k(\theta).$$

On differentiation we find that

$$\frac{\partial^2 \ell(\theta)}{\partial \theta_r \partial \theta_s} = \sum_{k=-\infty}^{\infty} I(X = k\delta) \left\{ \frac{1}{\pi_k} \frac{\partial^2 \pi_k}{\partial \theta_r \partial \theta_s} - \left(\frac{1}{\pi_k} \frac{\partial \pi_k}{\partial \theta_r} \right) \left(\frac{1}{\pi_k} \frac{\partial \pi_k}{\partial \theta_s} \right) \right\},$$

and as $\sum_k \pi_k(\theta) = 1$ for all θ and $\mathrm{E}\{I(X = k\delta)\} = \pi_k(\theta)$, the (r, s) element of the expected information matrix for a random sample $X_1, \ldots, X_n$ is

$$n \sum_{k=-\infty}^{\infty} \frac{1}{\pi_k(\theta)} \frac{\partial \pi_k(\theta)}{\partial \theta_r} \frac{\partial \pi_k(\theta)}{\partial \theta_s}. \tag{4.21}$$

For concreteness, suppose that Y is normally distributed with mean μ and variance σ^2, in which case $\pi_k(\mu, \sigma^2) = \Phi(z_{k+1}) - \Phi(z_k)$ and

$$\frac{\partial \pi_k}{\partial \mu} = -\frac{1}{\sigma} \{\phi(z_{k+1}) - \phi(z_k)\}, \quad \frac{\partial \pi_k}{\partial \sigma^2} = -\frac{1}{2\sigma^2} \{z_{k+1}\phi(z_{k+1}) - z_k\phi(z_k)\}, \quad (4.22)$$

where $z_k = \sigma^{-1}\{(k - \frac{1}{2})\delta - \mu\}$. With $\mu = 0$ it turns out that the expected information may be written as

$$n \begin{pmatrix} \sigma^{-2} I_{\mu\mu}(\delta/\sigma) & 0 \\ 0 & (4\sigma^4)^{-1} I_{\sigma\sigma}(\delta/\sigma) \end{pmatrix},$$

where the elements are given by substituting (4.22) into (4.21). On comparing this with (4.18), we see that the overall efficiency for the two parameters is $\{I_{\mu\mu}(\delta/\sigma)I_{\sigma\sigma}(\delta/\sigma)/2\}^{1/2}$, while the efficiencies for μ and σ^2 separately are $I_{\mu\mu}(\delta/\sigma)$ and $\frac{1}{2}I_{\sigma\sigma}(\delta/\sigma)$. Table 4.2 shows that these are remarkably high even with quite heavy rounding. When $\delta = \sigma = 1$, rounding Y to X gives a discrete distribution with almost all its probability on the seven values $-3, -2, \ldots, 3$, but a sample $x_1, \ldots, x_{100}$ of such values gives almost the same efficiency as 89 of the corresponding ys: the overall loss of efficiency is only 11%. If the data are rounded to the equivalent of one decimal place, $\delta = 0.1\sigma$, there is effectively no information lost. with $\delta = 1.5\sigma$ or more the loss is more dramatic, particularly for estimation of σ, and with $\delta = 3\sigma$ the data are almost binary.

Although suggestive, these results should be regarded with caution for two reasons. First, they apply to large samples, and the efficiency loss might be different in small samples. Second, they rest on the assumption that the multinomial likelihood based on the x_j is used, but in practice the rounded data would usually be treated as continuous and inference based on the (incorrect) log likelihood $\sum_j \log f(x_j; \theta)$. Practical 4.1 considers the effect of this. ∎

Exercises 4.3

1 (a) Show that the log likelihood for a random sample from density (2.7) is

$$\ell(\lambda, \kappa) = n\kappa \log \lambda + (\kappa - 1) \sum \log y_j - \lambda \sum y_j - n \log \Gamma(\kappa),$$

deduce that the observed information is

$$J(\lambda, \kappa) = n \begin{pmatrix} \kappa/\lambda^2 & -1/\lambda \\ -1/\lambda & d^2 \log \Gamma(\kappa)/d\kappa^2 \end{pmatrix},$$

and find the expected information $I(\lambda, \kappa)$.

(b) Suppose that we write $\lambda = \kappa/\mu$, where μ is the distribution mean. Find the log likelihood in terms of μ and κ, and show that $J(\mu, \kappa)$ is random and $I(\mu, \kappa) = n \text{diag}\{2\kappa/\mu^2, d^2 \log \Gamma(\kappa)/d\kappa^2 - 1/\kappa\}$.

2 Check the details of Example 4.19.

3 $Y_1, \ldots, Y_n$ are independent normal random variables with unit variances and means $E(Y_j) = \beta x_j$, where the x_j are known quantities in $(0, 1]$ and β is an unknown parameter. Show that $\ell(\beta) \equiv -\frac{1}{2} \sum (y_j - x_j \beta)^2$ and find the expected information $I(\beta)$ for β.

A sketch may help. Suppose that $n = 10$ and that an experiment to estimate β is to be designed by choosing the x_j appropriately. Show that $I(\beta)$ is maximized when all the x_j equal 1. Is this design sensible if there is any possibility that $E(Y_j) = \alpha + \beta x_j$, with α unknown?

4 Use (4.21) and (4.22) to give expressions for the quantities $I_{\mu\mu}(\delta/\sigma)$ and $I_{\sigma\sigma}(\delta/\sigma)$ in Example 4.21. Show that $I_{\mu\sigma}(\delta/\sigma) = 0$ when $\mu = 0$.

5 Find the expected information for θ based on a random sample $Y_1, \ldots, Y_n$ from the geometric density

$$f(y; \theta) = \theta(1 - \theta)^{y-1}, \quad y = 1, 2, 3, \ldots, \quad 0 < \theta < 1.$$

A statistician has a choice between observing random samples from the Bernoulli or geometric densities with the same θ. Which will give the more precise inference on θ?

6 Suppose a random sample $Y_1, \ldots, Y_n$ from the exponential density is rounded down to the nearest δ, giving δZ_j, where $Z_j = \lfloor Y_j/\delta \rfloor$. Show that the likelihood contribution from a rounded observation can be written $(1 - e^{-\lambda\delta})e^{-Z\lambda\delta}$, and deduce that the expected information for λ based on the entire sample is $n\delta^2 \exp(-\lambda\delta)\{1 - \exp(-\lambda\delta)\}^{-2}$. Show that this has limit n/λ^2 as $\delta \to 0$, and that if $\lambda = 1$, the loss of information when data are rounded down to the nearest integer rather than recorded exactly, is less than 10%. Find the loss of information when $\delta = 0.1$, and comment briefly.

4.4 Maximum Likelihood Estimator

4.4.1 Computation

The maximum likelihood estimate of θ, $\widehat{\theta}$, is a value of θ that maximizes the likelihood, or equivalently the log likelihood. Suppose $\psi = \psi(\theta)$ is a 1–1 function of θ. Then in terms of ψ the likelihood is

$$L^*(\psi) = L^*\{\psi(\theta)\} = L(\theta),$$

so the largest values of L^* and L coincide, and the maximum likelihood estimate of ψ is $\widehat{\psi} = \psi(\widehat{\theta})$. This simplifies calculation of maximum likelihood estimates, as we can compute them in the most convenient parametrization, and then transform them to the scale of interest.

Often, though not invariably, $\widehat{\theta}$ satisfies the *likelihood equation*

$$\frac{\partial \ell(\widehat{\theta})}{\partial \theta} = 0. \tag{4.23}$$

If θ is a $p \times 1$ vector, (4.23) is a $p \times 1$ system of equations that must be solved simultaneously for the components of $\widehat{\theta}$. We check that $\widehat{\theta}$ gives a local maximum by verifying that $-d^2 \ell(\widehat{\theta})/d\theta^2 > 0$, or in the vector case that the observed information matrix $J(\theta) = -d^2 \ell(\theta)/d\theta d\theta^{\mathsf{T}}$ is positive definite at $\widehat{\theta}$. If there are several solutions to (4.23), in principle we find them all, check which are maxima, and then evaluate $\ell(\theta)$ at each local maximum, thereby obtaining the global maximum. If there are numerous local maxima, as in Figure 4.2, doubt is cast on the usefulness of summarizing $\ell(\theta)$ in terms of $\widehat{\theta}$ and $J(\widehat{\theta})$, but many log likelihoods can be shown to be strictly concave. Then a local maximum is also the global maximum, so there is a unique maximum; moreover if there is a solution to (4.23), it is unique and gives the maximum.

Example 4.22 (Normal distribution) The likelihood equation for a random sample $y_1, \ldots, y_n$ from the normal distribution with mean μ and variance σ^2 is (Example 4.18)

$$\begin{pmatrix} \frac{\partial \ell(\mu, \sigma^2)}{\partial \mu} \\ \frac{\partial \ell(\mu, \sigma^2)}{\partial \sigma^2} \end{pmatrix} = \begin{pmatrix} \sigma^{-2} \sum (y_j - \mu) \\ -\frac{n}{2\sigma^2} + \frac{1}{2\sigma^4} \sum (y_j - \mu)^2 \end{pmatrix} = \begin{pmatrix} 0 \\ 0 \end{pmatrix}.$$

The first of these has the sole solution $\widehat{\mu} = \overline{y}$ for all values of σ^2, and $\ell(\widehat{\mu}, \sigma^2)$ is unimodal with maximum at $\widehat{\sigma}^2 = n^{-1} \sum (y_j - \overline{y})^2$. At the point $(\widehat{\mu}, \widehat{\sigma}^2)$, the observed information matrix $J(\mu, \sigma^2)$ is diagonal with elements $\mathrm{diag}\{n/\widehat{\sigma}^2, n/(2\widehat{\sigma}^4)\}$, and so is positive definite. Hence $\overline{y}$ and $n^{-1} \sum (y_j - \overline{y})^2$ are the sole solutions to the likelihood equation, and therefore are the maximum likelihood estimates.

If we wish to estimate the mean of $\exp(Y)$, which is $\psi = \exp(\mu + \sigma^2/2)$, then rather than reparametrize in terms of ψ and μ, say, and maximizing directly, we use the earlier results on transformations to see that the maximum likelihood estimate of ψ is $\widehat{\psi} = \exp(\widehat{\mu} + \widehat{\sigma}^2/2)$. ∎

In most realistic cases (4.23) must be solved iteratively, and often variants of the Newton–Raphson algorithm can be used. Given a starting-value $\theta^{\dagger}$, we expand (4.23) by Taylor series about $\theta^{\dagger}$ to obtain

$$0 = \frac{\partial \ell(\widehat{\theta})}{\partial \theta} \doteq \frac{\partial \ell(\theta^{\dagger})}{\partial \theta} + \frac{\partial^2 \ell(\theta^{\dagger})}{\partial \theta \partial \theta^{\mathsf{T}}} (\widehat{\theta} - \theta^{\dagger}). \tag{4.24}$$

On rearranging (4.24) we obtain

$$\widehat{\theta} \doteq \theta^{\dagger} + J(\theta^{\dagger})^{-1} U(\theta^{\dagger}), \tag{4.25}$$

where $U(\theta) = \partial \ell(\theta)/\partial \theta$ is called the *score statistic* or *score vector*, and $J(\theta)$ is the observed information (4.17). In the vector case $\widehat{\theta}$, $\theta^{\dagger}$ and $U(\theta^{\dagger})$ are $p \times 1$ vectors and

$J(\theta^\dagger)$ is a $p \times p$ matrix. The log likelihood is usually maximized in a few iterations of (4.25), using $\widehat{\theta}$ from one iteration as $\theta^\dagger$ for the next. In doubtful cases it is wise to try several initial values of $\theta^\dagger$.

The iteration (4.25) gives $\widehat{\theta}$ in one step if $\ell(\theta)$ is actually quadratic, so convergence is accelerated by choosing a parametrization in which $\ell(\theta)$ is as close to quadratic as possible. Often it helps to transform components of θ to take values in the real line, for example removing the restrictions $\lambda > 0$ and $0 < \pi < 1$ by maximizing in terms of $\log \lambda$ and $\log\{\pi/(1 - \pi)\}$. This also avoids steps that take $\widehat{\theta}$ outside the parameter space. Another simple trick is to use a variable step-length in (4.25). We replace $J(\theta^\dagger)^{-1}U(\theta^\dagger)$ by $cJ(\theta^\dagger)^{-1}U(\theta^\dagger)$, choose c to maximize ℓ along this line, then recalculate U and J, and try again. Many standard models are readily fitted with a few lines of code in statistical packages, but fitting more adventurous models may involve writing special programs.

Example 4.23 (Weibull distribution) The log likelihood for a random sample from the Weibull density (4.4) is

$$\ell(\theta, \alpha) = n \log \alpha - n \log \theta + (\alpha - 1) \sum_{j=1}^{n} \log \left(\frac{y_j}{\theta}\right) - \sum_{j=1}^{n} \left(\frac{y_j}{\theta}\right)^\alpha,$$

the score function is

$$U(\theta, \alpha) = \begin{pmatrix} \partial\ell/\partial\theta \\ \partial\ell/\partial\alpha \end{pmatrix} = \begin{pmatrix} -n\alpha/\theta + \alpha\theta^{-1}\sum(y_j/\theta)^\alpha \\ n/\alpha + \sum\log(y_j/\theta) - \sum(y_j/\theta)^\alpha \log(y_j/\theta) \end{pmatrix},$$

and the likelihood equation (4.23) cannot be solved analytically. The observed information matrix $J(\theta, \alpha)$ is

$$\begin{pmatrix} \alpha(\alpha + 1)/\theta^2 \sum(y_j/\theta)^\alpha - n\alpha\theta^{-2} & \theta^{-1}\sum[1 - (y_j/\theta)^\alpha\{1 + \alpha\log(y_j/\theta)\}] \\ \theta^{-1}\sum[1 - (y_j/\theta)^\alpha\{1 + \alpha\log(y_j/\theta)\}] & n/\alpha^2 + \sum(y_j/\theta)^\alpha\{\log(y_j/\theta)\}^2 \end{pmatrix},$$

and to obtain maximum likelihood estimates we would iterate (4.25) until it converged. Suitable starting-values could be obtained by setting $\alpha^\dagger = 1$, in which case $\theta^\dagger = \overline{y}$. If trouble arose in using (4.25), it would be sensible to write the problem in terms of $\psi = (\log\theta, \log\alpha)^\mathsf{T}$, and iterate based on $\widehat{\psi} = \psi^\dagger + J(\psi^\dagger)^{-1}U(\psi^\dagger)$.

In this case a two-dimensional maximization can be avoided by noticing that for fixed α the unique maximum likelihood estimate of θ is

$$\widehat{\theta}_\alpha = \left(n^{-1}\sum_{j=1}^{n} y_j^\alpha\right)^{1/\alpha}.$$

The dashed line in the upper right panel of Figure 4.1 shows the curve traced out by $\widehat{\theta}_\alpha$ as a function of α. The value of ℓ along this curve, the *profile log likelihood for* α,

$$\ell_\mathrm{p}(\alpha) = \max_\theta \ell(\theta, \alpha) = \ell(\widehat{\theta}_\alpha, \alpha),$$

is shown in the lower right panel of the figure. This function is unimodal, and from it we see that $\widehat{\alpha} \doteq 6$. More precise estimates are obtained maximizing $\ell_\mathrm{p}(\alpha)$ numerically over α, to obtain $\widehat{\alpha}$ and hence $\widehat{\theta} = \widehat{\theta}_{\widehat{\alpha}}$. ∎

A variant of the Newton–Raphson method, *Fisher scoring*, replaces $J(\theta^\dagger)$ in (4.25) with the expected information $I(\theta^\dagger)$. This is useful when $J(\theta^\dagger)$ is badly behaved — for example, not positive definite — but typically (4.25) works well. It has the advantage that it can be implemented in an automatic way using numerical first and second derivatives of $\ell(\theta)$. In simple problems where minimizing programming time is more important than saving computing time it is generally simplest to maximize the log likelihood directly using a packaged routine.

4.4.2 Large-sample distribution

Thus far we have treated the maximum likelihood estimate as a summary of a likelihood based on a given sample $y_1, \ldots, y_n$, rather than as a random variable. Evidently, however, we may consider its properties when samples are repeatedly taken from the model. Suppose we have a random sample $Y_1, \ldots, Y_n$ from a density $f(y; \theta)$ that satisfies the regularity conditions:

- the true value θ^0 of θ is interior to the parameter space Θ, which has finite dimension and is compact;
- the densities defined by any two different values of θ are distinct;
- there is a neighbourhood $\mathcal{N}$ of θ^0 within which the first three derivatives of the log likelihood with respect to θ exist almost surely, and for $r, s, t = 1, \ldots, p$, $n^{-1}\mathrm{E}\{|\partial^3 \ell(\theta)/\partial\theta_r \partial\theta_s \partial\theta_t|\}$ is uniformly bounded for $\theta \in \mathcal{N}$; and
- within $\mathcal{N}$, the Fisher information matrix $I(\theta)$ is finite and positive definite, and its elements satisfy

$$I(\theta)_{rs} = \mathrm{E}\left\{ \frac{\partial\ell(\theta)}{\partial\theta_r} \frac{\partial\ell(\theta)}{\partial\theta_s} \right\} = \mathrm{E}\left\{ -\frac{\partial^2\ell(\theta)}{\partial\theta_r\partial\theta_s} \right\}, \quad r, s = 1, \ldots, p.$$

  We shall see below that this implies that $I(\theta)$ is the variance matrix of the score vector.

Some cases where these conditions fail are described in Section 4.6. If they do hold, the main results below also apply to many situations where the data are neither independent nor identically distributed.

At the end of this section we establish two key results. First, as $n \to \infty$ there is a value $\widehat{\theta}$ of θ such that $\ell(\widehat{\theta})$ is a local maximum of $\ell(\theta)$ and $\Pr(\widehat{\theta} \to \theta^0) = 1$; this is a strongly consistent estimator of θ. Second,

$$I(\theta^0)^{1/2}(\widehat{\theta} - \theta^0) \xrightarrow{D} Z \quad \text{as} \quad n \to \infty, \tag{4.26}$$

where Z has the $N_p(0, I_p)$ distribution. The first holds very generally, but the second requires smoothness of certain log likelihood derivatives. The condition $n \to \infty$ can often be replaced by $I(\theta^0) \to \infty$.

Another way to express (4.26) is to say that for large n, $\widehat{\theta} \overset{\cdot}{\sim} N(\theta^0, I(\theta^0)^{-1})$, and this explains our definition of asymptotic relative efficiency for components of vector parameters, on page 113: we compare asymptotic variances of two different estimators of θ^0.

Figure 4.5 Repeated sampling likelihood inference for the exponential mean. The upper left panel shows the functions $\log RL(\theta)$ for ten random samples of size $n = 10$ from the exponential distribution with mean $\theta^0 = 1$; the dashed line shows θ^0. The lower left panel shows a histogram of 5000 maximum likelihood estimates $\widehat{\theta}$, together with their approximate normal density. The upper right panel shows a probability plot of 5000 replicates of $W(\theta^0) = -2 \log RL(\theta^0)$ against quantiles of the χ_1^2 distribution. The lower right panel shows the construction of a 95% confidence region for the value of θ using ten observations from the spring failure data. The region is the set of all θ such that $\log RL(\theta) \geq -\frac{1}{2} c_1(0.95)$, where $c_1(0.95)$ is the 0.95 quantile of the χ_1^2 distribution; the dotted horizontal line shows $\frac{1}{2} c_1(0.95)$ and the limits of the region are the dashed vertical lines.

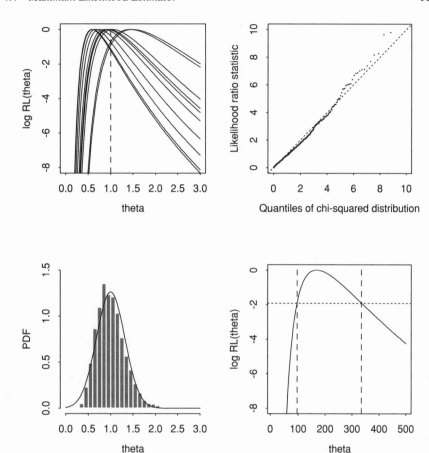

We illustrate (4.26) with random samples of size $n = 10$ from the exponential distribution with true mean $\theta^0 = 1$. As we saw in Section 4.2.1, the log likelihood for a random sample $y_1, \ldots, y_n$ is $\ell(\theta) \equiv -n(\log \theta + \overline{y}/\theta)$, and the maximum likelihood estimate is $\widehat{\theta} = \overline{y}$. The observed information and expected information are $J(\theta) = n(2\overline{y}/\theta^3 - 1/\theta^2)$ and $I(\theta) = n/\theta^2$. The upper left panel of Figure 4.5 shows the log relative likelihoods for ten such samples. Each curve is asymmetric about its maximum, so the distribution of $\widehat{\theta}$ is skewed; see the lower left panel. The density of $\widehat{\theta}$ is roughly normal with mean $\theta^0 = 1$ and variance $I(\theta^0)^{-1} = 1/10$, but this is a poor approximation. In fact $\overline{Y}$ has an exact gamma density with shape parameter 10 and unit mean.

On replacing $I(\theta^0)$ in (4.26) by $I(\widehat{\theta})$, we obtain the approximation

$$\widehat{\theta} \ \dot\sim \ N_p(\theta^0, V), \tag{4.27}$$

where $V = I(\widehat{\theta})^{-1}$ is the inverse expected information. Provided (4.26) is true, replacement of $I(\theta^0)$ by $I(\widehat{\theta})$ or $J(\widehat{\theta})$ is justified by the fact that both converge in probability to $I(\theta^0)$, so we can apply Slutsky's lemma (2.15). The main use of (4.27) is to construct confidence regions for components of θ^0.

Scalar parameter

If θ is scalar, (4.27) boils down to

$$I(\widehat{\theta})^{1/2}(\widehat{\theta} - \theta^0) \;\dot\sim\; N(0, 1).$$

Thus $I(\widehat{\theta})^{1/2}(\widehat{\theta} - \theta^0)$ is an approximate pivot from which to find confidence intervals for θ^0. That is,

z_α is the α quantile of the standard normal distribution.

$$1 - 2\alpha = \Pr\{z_\alpha \le I(\widehat{\theta})^{1/2}(\widehat{\theta} - \theta^0) \le z_{1-\alpha}\}$$
$$= \Pr\{\widehat{\theta} - z_{1-\alpha}I(\widehat{\theta})^{-1/2} \le \theta^0 \le \widehat{\theta} - z_\alpha I(\widehat{\theta})^{-1/2}\},$$

giving the $(1 - 2\alpha)$ *confidence interval for* θ^0,

$$\left(\widehat{\theta} - z_{1-\alpha}I(\widehat{\theta})^{-1/2}, \widehat{\theta} - z_\alpha I(\widehat{\theta})^{-1/2}\right). \tag{4.28}$$

The corresponding interval using the observed information $J(\widehat{\theta})$,

$$\left(\widehat{\theta} - z_{1-\alpha}J(\widehat{\theta})^{-1/2}, \widehat{\theta} - z_\alpha J(\widehat{\theta})^{-1/2}\right), \tag{4.29}$$

is easier to calculate than (4.28) because it requires no expectations, and moreover its coverage probability is often closer to the nominal level. Both intervals are symmetric about $\widehat{\theta}$.

Example 4.24 (Spring failure data) We reconsider the exponential model fitted to the data of Example 4.2, for which $n = 10$ and $\widehat{\theta} = \overline{y} = 168.3$. For this model $I(\widehat{\theta}) = J(\widehat{\theta}) = n/\overline{y}^2$, so the 95% confidence intervals (4.28) and (4.29) for the true mean both equal $\overline{y} \pm z_{0.025}n^{-1/2}\overline{y}$, that is, (64.0, 272.6). ∎

Example 4.25 (Cauchy data) To see the quality of these confidence intervals, we take samples of size n from the Cauchy density (2.16), for which

$$\ell(\theta) \equiv -\sum_{j=1}^{n}\log\{1 + (y_j - \theta)^2\}, \quad J(\theta) = 2\sum_{j=1}^{n}\frac{1 - (y_j - \theta)^2}{\{1 + (y_j - \theta)^2\}^2}, \quad I(\theta) = \frac{1}{2}n;$$

we take $\theta^0 = 0$. The basis of (4.28) and (4.29) is large-sample normality of $Z_I = I(\widehat{\theta})^{1/2}(\widehat{\theta} - \theta^0)$ and $Z_J = J(\widehat{\theta})^{1/2}(\widehat{\theta} - \theta^0)$, and to assess this we compare Z_I and Z_J with a standard normal variable Z. Symmetry of the Cauchy density about θ^0 implies that Z_I and Z_J are distributed symmetrically about the origin, so the left panel of Figure 4.6 compares quantiles of $|Z_J|$ with those of $|Z|$ in a half-normal plot (Practical 3.1), for 5000 simulated Cauchy samples of size $n = 15$. Evidently the distribution of Z_J is close to normal; its empirical 0.9, 0.95, 0.975 and 0.99 quantiles are 1.34, 1.76, 2.08 and 2.55, compared with 1.28, 1.65, 1.96 and 2.33 for Z. With $\alpha = 0.025$, (4.29) has estimated coverage probability 0.93, close to the nominal 0.95. The right panel shows that Z_I has heavier tails than Z_J; the coverage probability for (4.28) with $\alpha = 0.025$ is 0.91. Use of observed information is preferable, but the large-sample approximations seem accurate enough for practical use even with $n = 15$.

Just one of the 5000 log likelihoods had two local maxima, compared to 36 for 5000 samples with $n = 10$; the rest appeared unimodal. Thus $\widehat{\theta}$ was almost invariably the sole solution to the likelihood equation. ∎

Figure 4.6 Inference based on observed and expected information in samples of $n = 15$ Cauchy observations. Left: half-normal plot of $|Z_J| = J(\widehat{\theta})^{1/2}|\widehat{\theta} - \theta^0|$; the dotted line shows the ideal, so Z_J is slightly heavier-tailed than normal. Right: comparison of $|Z_I| = I(\widehat{\theta})^{1/2}|\widehat{\theta} - \theta^0|$ with $|Z_J|$. $|Z_I|$ has heavier tails.

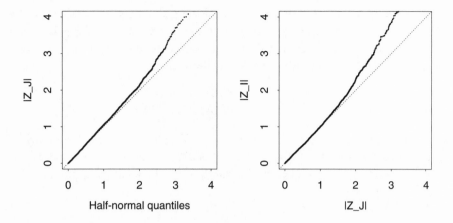

Vector parameter

When θ is a vector, confidence sets for the rth element of θ^0, θ_r^0, may be based on the fact that the corresponding maximum likelihood estimator, $\widehat{\theta}_r$, has approximately the $N(\theta_r^0, v_{rr})$ distribution, where v_{rr} is the (r, r) element of $V = I(\widehat{\theta})^{-1}$ or $J(\widehat{\theta})^{-1}$. This gives intervals (4.28) and (4.29), but with $\widehat{\theta}$, $I(\widehat{\theta})^{-1}$, and $J(\widehat{\theta})^{-1}$ replaced by $\widehat{\theta}_r$, v_{rr}, and the (r, r) element of $J(\widehat{\theta})^{-1}$.

Example 4.26 (Normal distribution)   In Examples 4.18 and 4.22 we saw that the maximum likelihood estimates of the mean and variance of the normal distribution based on a random sample $y_1, \ldots, y_n$ are $\widehat{\mu} = \overline{y}$ and $\widehat{\sigma}^2 = n^{-1} \sum(y_j - \overline{y})^2$, and that the expected information matrix is $\mathrm{diag}\{n/\sigma^2, n/(2\sigma^4)\}$. Hence $V = \mathrm{diag}\{n^{-1}\widehat{\sigma}^2, n^{-1}2\widehat{\sigma}^4\}$, and the $(1 - 2\alpha)$ confidence intervals for μ and σ^2 based on the large-sample results above are

$$\overline{y} \pm n^{-1/2}\widehat{\sigma} z_\alpha, \quad \widehat{\sigma}^2 \pm (2/n)^{1/2}\widehat{\sigma}^2 z_\alpha.$$

$s^2 = n\widehat{\sigma}^2/(n-1)$ is the unbiased estimate of σ^2.

The asymptotic approximation gives an interval for μ with the same form as the exact interval, $\overline{y} \pm n^{-1/2}s t_{n-1}(\alpha)$, but with s replaced by $\widehat{\sigma}$ and the t quantile replaced by the corresponding normal quantile. Provided that $n > 20$ or so, these alterations will typically have little effect on the interval. Larger samples are needed for the interval for σ^2 to be good, because normal approximation to the distribution of $\widehat{\sigma}^2$ is poorer than to the distribution of $\widehat{\mu}$. ■

The use of (4.27) to give confidence regions for the whole of θ rests on the fact that (4.27) entails $(\widehat{\theta} - \theta^0)^T V^{-1}(\widehat{\theta} - \theta^0) \overset{\cdot}{\sim} \chi_p^2$. Hence an approximate $(1 - 2\alpha)$ confidence region is

$$\{\theta : (\widehat{\theta} - \theta)^T V^{-1}(\widehat{\theta} - \theta) \le c_p(1 - 2\alpha)\};$$

an ellipsoid centred at $\widehat{\theta}$, with shape determined by the elements of V and volume determined by $c_p(1 - 2\alpha)$. Another version replaces $I(\widehat{\theta})^{-1}$ with $J(\widehat{\theta})^{-1}$.

Example 4.27 (Challenger data) Examples 4.5 and 4.8 discuss a model for the Challenger data, where the probability of O-ring thermal distress depends on the launch temperature. The maximum likelihood estimates for this model are $\widehat{\beta}_0 = 5.084$ and $\widehat{\beta}_1 = -0.116$, and the inverse observed information is

$$J(\widehat{\beta}_0, \widehat{\beta}_1)^{-1} = \begin{pmatrix} 9.289 & -0.142 \\ -0.142 & 0.00220 \end{pmatrix},$$

yielding standard errors $9.289^{1/2} = 3.048$ and $0.00220^{1/2} = 0.0469$. The estimated correlation of $\widehat{\beta}_0$ and $\widehat{\beta}_1$, $-0.142/(9.289 \times 0.00220)^{1/2}$, equals -0.993, and we see that the matrix $J(\widehat{\beta}_0, \widehat{\beta}_1)$ is close to singular. In view of the left panel of Figure 4.3 this is not surprising.

A joint 95% confidence region for (β_0, β_1) is the ellipsoid given by

$$(\beta_0 - 5.084, \beta_1 + 0.116) J(\widehat{\beta}_0, \widehat{\beta}_1) \begin{pmatrix} \beta_0 - 5.084 \\ \beta_1 + 0.116 \end{pmatrix} \le c_2(0.95) = 5.99.$$

■

Often we focus on a scalar parameter $\psi = \psi(\theta)$, estimated by $\widehat{\psi} = \psi(\widehat{\theta})$. To approximate the variance of $\widehat{\psi}$ we apply the delta method (2.19), giving

$$\psi(\widehat{\theta}) \doteq \psi(\theta^0) + \frac{\partial \psi(\theta^0)}{\partial \theta^{\mathrm{T}}}(\widehat{\theta} - \theta^0).$$

Consequently

$$\mathrm{var}\{\psi(\widehat{\theta})\} \doteq \frac{\partial \psi(\theta^0)}{\partial \theta^{\mathrm{T}}} \widehat{\mathrm{var}(\theta)} \frac{\partial \psi(\theta^0)}{\partial \theta} \doteq \frac{\partial \psi(\widehat{\theta})}{\partial \theta^{\mathrm{T}}} J(\widehat{\theta})^{-1} \frac{\partial \psi(\widehat{\theta})}{\partial \theta},$$

where $\partial \psi(\widehat{\theta})/\partial \theta$ is the $p \times 1$ vector of derivatives of ψ evaluated at $\widehat{\theta}$. Thus an approximate $(1 - 2\alpha)$ confidence interval for ψ is

$$\psi(\widehat{\theta}) \pm z_\alpha \{\partial \psi(\widehat{\theta})/\partial \theta^{\mathrm{T}} J(\widehat{\theta})^{-1} \partial \psi(\widehat{\theta})/\partial \theta\}^{1/2}. \tag{4.30}$$

Example 4.28 (Challenger data) One quantity of particular interest is the probability of failure at $31°\mathrm{F}$, $\psi = e^{\beta_0 + 31\beta_1}/(1 + e^{\beta_0 + 31\beta_1})$. Its maximum likelihood estimate and derivatives are

$$\widehat{\psi} = \frac{e^{\widehat{\beta}_0 + 31\widehat{\beta}_1}}{1 + e^{\widehat{\beta}_0 + 31\widehat{\beta}_1}} = 0.816, \qquad \frac{\partial \psi}{\partial \beta_0} = \psi(1 - \psi), \qquad \frac{\partial \psi}{\partial \beta_0} = 31\psi(1 - \psi).$$

The 95% confidence interval (4.30) for ψ is $0.816 \pm 1.96 \times 0.242 = (0.34, 1.29)$. As this contains values greater than one it is less than satisfactory, so we need a better approach, such as the one described in Section 4.5.2. ■

Consistency of $\widehat{\theta}$

We now obtain the key convergence results for maximum likelihood estimation of a scalar, subject to the regularity conditions on page 118.

Let $h : \mathbb{R} \to \mathbb{R}$ be convex. Then for any real x_1, x_2,

$$h\{\pi x_1 + (1 - \pi)x_2\} \leq \pi h(x_1) + (1 - \pi)h(x_2), \quad 0 \leq \pi \leq 1.$$

If X is a real-valued random variable, then *Jensen's inequality* says that $\mathrm{E}\{h(X)\} \geq h\{\mathrm{E}(X)\}$, with equality if and only if X is degenerate.

Let $Y_1, \ldots, Y_n$ be a random sample from a density $f(y; \theta)$, where θ is scalar with true value θ^0, and let $\overline{\ell}(\theta) = n^{-1} \sum \log f(Y_j; \theta)$. Now

$$
\begin{aligned}
\mathrm{E}\{\overline{\ell}(\theta) - \overline{\ell}(\theta^0)\} &= \mathrm{E}\left[\log\left\{\frac{f(Y_1; \theta)}{f(Y_1; \theta^0)}\right\}\right] \\
&\leq \log \mathrm{E}\left\{\frac{f(Y_1; \theta)}{f(Y_1; \theta^0)}\right\} \\
&= \log \int \frac{f(y; \theta)}{f(y; \theta^0)} f(y; \theta^0)\, dy = 0,
\end{aligned}
\tag{4.31}
$$

where we have applied Jensen's inequality to the convex function $-\log x$. The inequality is strict unless the density ratio is constant, so that the densities are the same, and according to our regularity conditions this may occur only if $\theta = \theta^0$. As $n \to \infty$, the weak law of large numbers applies to the average $\overline{\ell}(\theta) - \overline{\ell}(\theta^0)$, which converges in probability to

$$\int \log\left\{\frac{f(y; \theta)}{f(y; \theta^0)}\right\} f(y; \theta^0)\, dy = -D(f_\theta, f_{\theta^0}),$$

say. This is negative unless $\theta = \theta^0$. The quantity $D(f, g) \geq 0$ is known as the *Kullback–Leibler discrepancy* between f and g; it is minimized when $f = g$. In fact this convergence is almost sure, that is, $\overline{\ell}(\theta) - \overline{\ell}(\theta^0)$ converges to $-D(f_\theta, f_{\theta^0})$ with probability one. This shores up our earlier informal discussion of Figure 4.4, for we see that if $\theta \neq \theta^0$, then

$$\ell(\theta) - \ell(\theta^0) \sim nD(f_\theta, f_{\theta^0}) \to -\infty$$

with probability one as $n \to \infty$.

Now for any $\delta > 0$, $\overline{\ell}(\theta^0 - \delta) - \overline{\ell}(\theta^0)$ and $\overline{\ell}(\theta^0 + \delta) - \overline{\ell}(\theta^0)$ converge with probability one to the negative quantities $-D(f_{\theta^0 - \delta}, f_{\theta^0})$ and $-D(f_{\theta^0 + \delta}, f_{\theta^0})$. Hence for any sequence of random variables $Y_1, \ldots, Y_n$ there is an n' such that for $n > n'$, $\overline{\ell}(\theta)$ has a local maximum in the interval $(\theta^0 - \delta, \theta^0 + \delta)$. If we let $\widehat{\theta}$ denote the value at which this local maximum occurs, then $\Pr(\widehat{\theta} \to \theta^0) = 1$ and $\widehat{\theta}$ is said to be a *strongly consistent* estimate of θ^0. This implies $\widehat{\theta} \xrightarrow{P} \theta^0$, so $\widehat{\theta}$ is consistent in our usual, weaker, sense.

As this proof does not require $f(y; \theta)$ to be smooth it is very general. It says nothing about uniqueness of $\widehat{\theta}$, merely that a strongly consistent local maximum exists, but if $\ell(\theta)$ has just one maximum, then $\widehat{\theta}$ must also be the global maximum. A more delicate argument is needed when θ is vector, because it is then not enough to consider only the two values $\theta^0 \pm \delta$.

Solomon Kullback (1907–1994) was born and educated in New York. He had careers in the US Defense Department and then at George Washington University. His main scientific contribution is to information theory. Richard Arthur Leibler (1914–) has spent much of his life working in the US defense community. Their definition of information was published in 1951.

Asymptotic normality of $\widehat{\theta}$

To prove asymptotic normality of $\widehat{\theta}$, we assume that $\widehat{\theta}$ satisfies the likelihood equation and consider the score statistic, $U(\theta) = d\ell(\theta)/d\theta$. Its mean and variance are

$$E\{U(\theta)\} = \sum_{j=1}^{n} E\left\{\frac{d\log f(Y_j;\theta)}{d\theta}\right\} = n\,E\{u(\theta)\},$$

$$var\{U(\theta)\} = \sum_{j=1}^{n} var\left\{\frac{d\log f(Y_j;\theta)}{d\theta}\right\} = n\,var\{u(\theta)\},$$

where $u(\theta) = d\log f(Y_j;\theta)/d\theta$ is the score function for a single random variable. Provided the order of differentiation and integration may be interchanged, the mean of $u(\theta)$ is

$$E\{u(\theta)\} = \int \frac{d\log f(y;\theta)}{d\theta} f(y;\theta)\,dy = \int \frac{df(y;\theta)}{d\theta}\,dy = \frac{d}{d\theta}\int f(y;\theta)dy = 0,$$
(4.32)

because $f(y;\theta)$ has integral one for each value of θ. Furthermore

$$0 = \frac{d}{d\theta}\int \frac{d\log f(y;\theta)}{d\theta} f(y;\theta)\,dy$$

$$= \int \frac{d^2\log f(y;\theta)}{d\theta^2} f(y;\theta)dy + \int \left\{\frac{d\log f(y;\theta)}{d\theta}\right\}^2 f(y;\theta)\,dy,$$

and so

$$var\{u(\theta)\} = E\{u(\theta)^2\} = -\int \frac{d^2\log f(y;\theta)}{d\theta^2} f(y;\theta)dy = i(\theta),$$
(4.33)

the expected information from a single observation.

Now both $U(\theta^0)$ and $J(\theta^0) = -\sum d^2\ell(\theta^0)/d\theta^2$ are sums of n independent random variables, and $E\{U(\theta^0)\} = 0$, $var\{U(\theta^0)\} = I(\theta^0) = ni(\theta^0)$, while $E\{J(\theta^0)\} = I(\theta^0) = ni(\theta^0)$. Hence the central limit theorem (2.11) and the weak law of large numbers imply that

$$I(\theta^0)^{-1/2}U(\theta^0) \xrightarrow{D} Z, \quad I(\theta^0)^{-1}J(\theta^0) \xrightarrow{P} 1,$$
(4.34)

where Z has the standard normal distribution.

If the log likelihood is sufficiently smooth to allow Taylor series expansion, then $\widehat{\theta}$ satisfies the likelihood equation

$$0 = U(\widehat{\theta}) \doteq U(\theta^0) + \frac{d^2\ell(\theta^0)}{d\theta^2}(\widehat{\theta} - \theta^0),$$

rearrangement of which gives

$$\widehat{\theta} - \theta^0 \doteq J(\theta^0)^{-1}U(\theta^0),$$

where $J(\theta^0)$ is the observed information and we require that the missing terms of the Taylor series are asymptotically small enough to be ignored. If so,

$$
\begin{aligned}
I(\theta^0)^{1/2}(\widehat{\theta} - \theta^0) &\doteq I(\theta^0)^{1/2} J(\theta^0)^{-1} U(\theta^0) \\
&= I(\theta^0)^{1/2} J(\theta^0)^{-1} I(\theta^0)^{1/2} \times I(\theta^0)^{-1/2} U(\theta^0) \\
&\xrightarrow{D} Z,
\end{aligned}
$$

by (4.34) and Slutsky's lemma (2.15). Replacement of $I(\theta^0)$ by $I(\widehat{\theta})$ or $J(\widehat{\theta})$ is justified by the fact that both converge in probability to $I(\theta^0)$ as $n \to \infty$.

This argument is generalized to vector θ by interpreting the score as a $p \times 1$ vector and the information quantities as $p \times p$ matrices, with Z having a $N_p(0, I_p)$ distribution.

Exercises 4.4

1. In Example 4.23, show that $\widehat{\alpha}$ is the solution of the equation

$$
\widehat{\alpha} = \left\{ \frac{\sum_j y_j^{\widehat{\alpha}} \log y_j}{\sum_j y_j^{\widehat{\alpha}}} - n^{-1} \sum_j \log y_j \right\}^{-1}.
$$

2. If the log likelihood for a $p \times 1$ vector of parameters is $\ell(\theta) = a + b^{\mathsf{T}}\theta - \frac{1}{2}\theta^{\mathsf{T}}C\theta$, where the constants a, b and C are respectively scalar, a $p \times 1$ vector, and a $p \times p$ symmetric positive definite matrix, show that the score statistic can be written $b - C\theta$. Find the observed information $J(\theta)$, and show that $\widehat{\theta}$ is attained in one step of (4.25) from any initial value of θ.

3. The *Laplace* or *double exponential* distribution has density

$$
f(y; \mu, \sigma) = \frac{1}{2\sigma} \exp\left(-|y - \mu|/\sigma\right), \qquad -\infty < y < \infty, \quad -\infty < \mu < \infty, \sigma > 0.
$$

Sketch the log likelihood for a typical sample, and explain why the maximum likelihood estimate is only unique when the sample size is odd. Derive the score statistic and observed information. Is maximum likelihood estimation regular for this distribution?

4. Eggs are thought to be infected with a bacterium *salmonella enteriditis* so that the number of organisms, Y, in each has a Poisson distribution with mean μ. The value of Y cannot be observed directly, but after a period it becomes certain whether the egg is infected ($Y > 0$) or not ($Y = 0$). Out of m such eggs, r are found to be infected. Find the maximum likelihood estimator $\widehat{\mu}$ of μ and its asymptotic variance. Is the exact variance of $\widehat{\mu}$ defined?

5. If $Y_1, \ldots, Y_n$ is a random sample from density $\theta^{-1} e^{-x/\theta}$, show that the maximum likelihood estimator $\widehat{\theta}$ has an asymptotic normal distribution with mean θ and variance θ^2/n. Deduce that an approximate $(1 - 2\alpha)$ confidence interval for θ is

z_α is the α quantile of the standard normal distribution.

$$
\frac{\widehat{\theta}}{1 + z_\alpha n^{-1/2}} \geq \theta \geq \frac{\widehat{\theta}}{1 + z_{1-\alpha} n^{-1/2}}.
$$

Show that $\widehat{\theta}/\theta$ is an exact pivot, having the gamma distribution with unit mean and shape parameter $\kappa = n$. Hence find an exact confidence interval for θ, and compare it with the approximate one when $n = 10$ and $\widehat{\theta} = 100$.

6. If $Y_1, \ldots, Y_n \overset{\text{iid}}{\sim} N(\mu, c\mu^2)$, where c is a known constant, show that the minimal sufficient statistic for μ is the same as for the $N(\mu, \sigma^2)$ distribution. Find the maximum likelihood estimate of μ and give its large-sample standard error. Show that the distribution of $\overline{Y}^2/S^2$ does not depend on μ.

4.5 Likelihood Ratio Statistic

4.5.1 Basic ideas

Suppose that our model is determined by a parameter θ of dimension p, whose true but unknown value is θ^0, and for which the maximum likelihood estimate is $\widehat{\theta}$. Then provided the model satisfies the conditions for asymptotic normality of the maximum likelihood estimator given in the previous section, in large samples the *likelihood ratio statistic*

$$W(\theta^0) = -2\log RL(\theta^0) = 2\{\ell(\widehat{\theta}) - \ell(\theta^0)\} \tag{4.35}$$

has an approximate chi-squared distribution on p degrees of freedom under repeated sampling of data from the model. That is, as $I(\theta^0) \to \infty$,

$$W(\theta^0) \xrightarrow{D} \chi_p^2, \tag{4.36}$$

so $W(\theta^0) \overset{\cdot}{\sim} \chi_1^2$ when θ is scalar. In practice this result is used to generate approximations for finite samples. It is illustrated in the upper right panel of Figure 4.5, which compares 5000 simulated values of $W(\theta^0)$, based on exponential samples of size $n = 10$, with quantiles of the χ_1^2 distribution. Here $p = 1$, $W(\theta^0) = 2n\{\overline{Y}/\theta^0 - 1 - \log(\overline{Y}/\theta^0)\}$, and $\theta^0 = 1$. This approximation seems better than that for $\widehat{\theta}$.

To establish (4.36), we note that $d\ell(\widehat{\theta})/d\theta = 0$ and make a Taylor series expansion of $W(\theta^0)$, giving

$$
\begin{aligned}
W(\theta^0) &= 2\{\ell(\widehat{\theta}) - \ell(\theta^0)\} \\
&\doteq 2\left\{ \ell(\widehat{\theta}) - \ell(\widehat{\theta}) - (\theta^0 - \widehat{\theta})^{\mathrm{T}} \frac{\partial\ell(\widehat{\theta})}{\partial\theta} - \frac{1}{2}(\theta^0 - \widehat{\theta})^{\mathrm{T}} \frac{\partial^2\ell(\widehat{\theta})}{\partial\theta\,\partial\theta^{\mathrm{T}}}(\theta^0 - \widehat{\theta}) \right\} \\
&= (\widehat{\theta} - \theta^0)^{\mathrm{T}} J(\widehat{\theta})(\widehat{\theta} - \theta^0) \\
&\doteq (\widehat{\theta} - \theta^0)^{\mathrm{T}} I(\theta^0)(\widehat{\theta} - \theta^0),
\end{aligned}
$$

and the limiting normal distribution for $\widehat{\theta}$ at (4.26) and the relation (3.23) linking this to the chi-squared distribution yield (4.36).

Expression (4.36) shows that $W(\theta^0)$ is an approximate pivot which may be used to provide confidence regions for θ^0. For if $W(\theta^0) \overset{\cdot}{\sim} \chi_p^2$, then

$$\Pr\{W(\theta^0) \le c_p(1 - 2\alpha)\} \doteq 1 - 2\alpha,$$

$c_p(\alpha)$ denotes the α quantile of the χ_p^2 distribution.

and hence values of θ for which $W(\theta) \le c_p(1 - 2\alpha)$ may be regarded as 'plausible' at the $(1 - 2\alpha)$ level. Equivalently, the set

$$\left\{ \theta : \ell(\theta) \ge \ell(\widehat{\theta}) - \frac{1}{2}c_p(1 - 2\alpha) \right\} \tag{4.37}$$

is a $(1 - 2\alpha)$ confidence region for the unknown θ^0. We use $(1 - 2\alpha)$ here for consistency with our earlier discussion of confidence intervals.

These 'plausible' sets of θ based on $W(\theta^0)$ under repeated sampling have the same form as those for the pure likelihood approach described at the end of Section 4.1.2, since the condition $RL(\theta) \geq c$ is equivalent to $W(\theta) \leq -2 \log c$. Here however the constant $-2 \log c$ is replaced by $c_p(1 - 2\alpha)$, chosen with respect to the approximate distribution of $W(\theta^0)$ under repeated sampling. Often α is taken to be 0.05, 0.025 or 0.005, values that correspond to regions containing θ^0 with approximate probabilities 0.9, 0.95 and 0.99.

Example 4.29 (Spring failure data) The likelihood ratio statistic for the exponential model in Example 4.2 is $W(\theta) = 2n\{\bar{y}/\theta - 1 - \log(\bar{y}/\theta)\}$. As $c_1(0.95) = 3.84$, a 95% confidence region for θ based on $W(\theta)$ is the set

$$\{\theta : 2n\{\bar{y}/\theta - 1 - \log(\bar{y}/\theta)\} \leq 3.84\}.$$

This set is found by plotting the log likelihood and reading off the values of θ for which $\ell(\theta) \geq \ell(\widehat{\theta}) - \frac{1}{2} \times 3.84$. The lower right panel of Figure 4.5 shows this region, $(96, 335)$, which is not symmetric about the maximum likelihood estimate $\bar{y} = 168.3$.

We saw in Example 4.24 that the 95% confidence interval for θ based on the asymptotic normal distribution of $\widehat{\theta}$, $(64, 273)$, is symmetric about $\widehat{\theta}$. The difference between intervals based on $W(\theta)$ and $\widehat{\theta}$ would vanish in sufficiently large samples, but it can be important to capture the asymmetry of $\ell(\theta)$ when n is small or moderate. Regions defined by (4.37) need not be connected, unlike those based on normal approximation to the distribution of $\widehat{\theta}$, which may be problematic when $\ell(\theta)$ is multimodal. ■

When θ is vector, confidence regions for θ^0 can in principle be obtained from (4.37) through contour plots of ℓ. This seems infeasible when p exceeds three. We discuss one resolution of this in the next section.

4.5.2 Profile log likelihood

In the previous section we treated all elements of θ equally, but in practice some are more important than others. We write $\theta^{\mathrm{T}} = (\psi^{\mathrm{T}}, \lambda^{\mathrm{T}})$, where ψ is a $p \times 1$ vector of *parameters of interest*, and λ is a $q \times 1$ vector of *nuisance parameters*. Our enquiry centres on ψ, but we cannot avoid including λ in the model. We may wish to check whether a particular value ψ^0 of ψ is consistent with the data, or to find a plausible range of values for ψ, but in either case the value of λ is irrelevant or of at most secondary interest. The division into ψ and λ may change in the course of an investigation.

Two models are said to be *nested* if one reduces to the other when certain parameters are fixed. Thus a model with parameters (ψ^0, λ) is nested within the more general model with parameters (ψ, λ); the corresponding parameter spaces are $\{\psi^0\} \times \Lambda$ and $\Psi \times \Lambda$, where $\psi^0 \in \Psi$. Under the more restrictive model the value of λ that maximizes the log likelihood $\ell(\psi^0, \lambda)$ is $\widehat{\lambda}_{\psi^0}$, whereas the overall maximum likelihood estimate, $(\widehat{\psi}, \widehat{\lambda})$, maximizes ℓ over both parameters. Of course, $\ell(\widehat{\psi}, \widehat{\lambda}) \geq \ell(\psi^0, \widehat{\lambda}_{\psi^0})$.

Example 4.30 (Weibull distribution) The Weibull density (4.4) has two parameters α and θ, and reduces to the exponential density when $\alpha = 1$. In terms of our general discussion we set $\alpha = \psi$ and $\lambda = \theta$, with $\psi^0 = 1$, $\Lambda = \mathbb{R}_+$, and $\Psi = \mathbb{R}_+$. Then the vertical dotted line in the upper right panel of Figure 4.1 corresponds to $\{\psi^0\} \times \Lambda$, while the entire upper right quadrant of the plane is $\Psi \times \Lambda$. Evidently the likelihood reaches its maximum away from the exponential submodel. The maximum likelihood estimates under the submodel are (1, 168), while overall they are roughly (6, 181); the difference of log likelihoods is 12.5.   ∎

A natural statistic with which to compare two nested models is the log ratio of maximized likelihoods,

$$W_{\mathrm{p}}(\psi^0) = 2\{\ell(\widehat{\psi}, \widehat{\lambda}) - \ell(\psi^0, \widehat{\lambda}_{\psi^0})\}. \tag{4.38}$$

This is sometimes called the *generalized likelihood ratio statistic* because it generalizes (4.35), but as (4.38) is the version almost invariably used in practice we shall refer to both simply as likelihood ratio statistics. At the end of this section we show that for regular models (4.36) generalizes to

$$W_{\mathrm{p}}(\psi^0) \xrightarrow{D} \chi_p^2. \tag{4.39}$$

That is, even though nuisance parameters are estimated, the likelihood ratio statistic has an approximate chi-squared distribution in large samples.

Often the parameter of interest, ψ, is scalar or has much smaller dimension than the nuisance parameter, λ, and we wish to form a confidence region for its true value ψ^0 regardless of λ. To do so we use the *profile log likelihood*,

$$\ell_{\mathrm{p}}(\psi) = \max_{\lambda} \ell(\psi, \lambda) = \ell(\psi, \widehat{\lambda}_\psi),$$

where $\widehat{\lambda}_\psi$ is the maximum likelihood estimate of λ for fixed ψ. The above result for $W_{\mathrm{p}}(\psi^0)$ implies that confidence regions for ψ^0 can be based on ℓ_{p} for regular models. A $(1 - 2\alpha)$ confidence region for ψ^0 is the set

$$\left\{ \psi : \ell_{\mathrm{p}}(\psi) \geq \ell_{\mathrm{p}}(\widehat{\psi}) - \frac{1}{2} c_p(1 - 2\alpha) \right\}. \tag{4.40}$$

This is our primary approach to finding confidence regions from likelihoods. It often yields good approximations to standard intervals.

When ψ is scalar we define the *signed likelihood ratio statistic*

This is sometimes called the *directed deviance statistic*.

$$Z(\psi^0) = \mathrm{sign}(\widehat{\psi} - \psi^0)[2\{\ell(\widehat{\psi}, \widehat{\lambda}) - \ell(\psi^0, \widehat{\lambda}_{\psi^0})\}]^{1/2}.$$

The relation between the normal and chi-squared distributions implies that $c_1(1 - 2\alpha) = z_\alpha^2 = z_{1-\alpha}^2$, so

$$1 - 2\alpha \doteq \Pr\{W_{\mathrm{p}}(\psi^0) \leq c_1(1 - 2\alpha)\}$$
$$= \Pr\{Z(\psi^0) \leq z_{1-\alpha}\} - \Pr\{Z(\psi^0) \leq z_\alpha\},$$

and $Z(\psi^0)$ may be regarded as having an approximate standard normal distribution and is an approximate pivot on which inference for ψ^0 may be based; when $p = 1$, a different way of writing (4.40) is

$$\{\psi : z_\alpha \leq Z(\psi) \leq z_{1-\alpha}\}. \tag{4.41}$$

We have briefly mentioned the effect of reparametrization on likelihood. If ψ is of central interest, inference should be invariant to *interest-preserving transformations*, under which $\psi, \lambda \mapsto \eta(\psi), \zeta(\psi, \lambda)$, where the map $\psi \to \eta$ is one-one for each value of ψ, and so too is the map $\lambda \mapsto \zeta$. For such a reparametrization, $\ell_p(\eta) = \ell_p(\psi)$, so $W_p(\psi)$ is invariant; so too is $Z(\psi)$ apart from a possible change in sign.

Example 4.31 (Normal distribution) The log likelihood for a normal sample $y_1, \ldots, y_n$ is

$$\ell(\mu, \sigma^2) \equiv -\frac{1}{2}\left\{ n \log \sigma^2 + \frac{1}{\sigma^2}\sum_{j=1}^{n}(y_j - \mu)^2 \right\}.$$

To use the profile log likelihood to find a confidence region for μ, we set $\psi = \mu$, $\lambda = \sigma^2$, and note that for fixed μ, the maximum likelihood estimate of σ^2 is

$$\begin{aligned}
\widehat{\sigma}_\mu^2 &= n^{-1} \sum (y_j - \mu)^2 \\
&= n^{-1}\left\{ \sum (y_j - \overline{y})^2 + n(\overline{y} - \mu)^2 \right\} \\
&= \frac{n-1}{n}s^2\left\{ 1 + \frac{t(\mu)^2}{n-1} \right\},
\end{aligned}$$

where $t(\mu) = (\overline{y} - \mu)/(s^2/n)^{1/2}$ is the observed value of the t statistic (3.16) and $s^2 = (n-1)^{-1}\sum (y_j - \overline{y})^2$. Thus the profile log likelihood for μ is

$$\ell_p(\mu) = \ell\big(\mu, \widehat{\sigma}_\mu^2\big) \equiv -\frac{n}{2}\log[s^2\{1 + t(\mu)^2/(n-1)\}],$$

and as the overall maximum likelihood estimate of μ is $\widehat{\mu} = \overline{y}$, $t(\widehat{\mu}) = 0$ and

$$W_p(\mu) = n \log\left\{ 1 + \frac{T(\mu)^2}{n-1} \right\}, \quad Z(\mu) = \text{sign}(\overline{Y} - \mu)\left[n \log\left\{ 1 + \frac{T(\mu)^2}{n-1} \right\} \right]^{1/2},$$

whose values are large when $T(\mu) = (\overline{Y} - \mu)/(S^2/n)^{1/2}$ is large, that is, when μ differs from $\overline{Y}$ in either direction. Evidently the confidence interval (4.40) has the form $T(\mu)^2 \leq c$ and may be written $\overline{Y} \pm n^{-1/2}Sc^{1/2}$. The usual $(1 - 2\alpha)$ confidence interval, based on the exact distribution of $T(\mu)$, sets $c^{1/2}$ to be a quantile of the Student t distribution, $t_{n-1}(1 - \alpha)$. For $n = 15$ and $\alpha = 0.025$, $t_{n-1}(1 - \alpha) = 2.14$, while the value of $c^{1/2}$ from (4.40) is 2.05. This close agreement is not surprising, as Taylor series expansion shows that $W_p(\mu) \doteq nT(\mu)^2/(n-1)$, $T(\mu)^2$ has the $F_{1,n-1}$ distribution, and the F_{1,v_2} distribution approaches the χ_1^2 distribution when $v_2 \to \infty$.

The lower left panel of Figure 4.7 shows $z(\mu) = \text{sign}(\overline{y} - \mu)w_p(\mu)^{1/2}$ for the differences between cross- and self-fertilized plant heights in Table 1.1, for which $n = 15$, $\overline{y} = 20.93$, and $s^2 = 1424.6$. The function $z(\mu)$ differs only slightly from the straight line $t(\mu) = (\overline{y} - \mu)/(s^2/n)^{1/2}$. The inner dotted lines at $z_\alpha, z_{1-\alpha} = \pm 1.96$ lead to the confidence set (4.41), here (1.23, 40.63), shown by the inner vertical dashed lines. This is only slightly narrower than the exact interval (0.03, 41.84) obtained by solving $t(\mu) = \pm t_{14}(0.025)$; this interval is shown by the outer dotted and dashed lines.

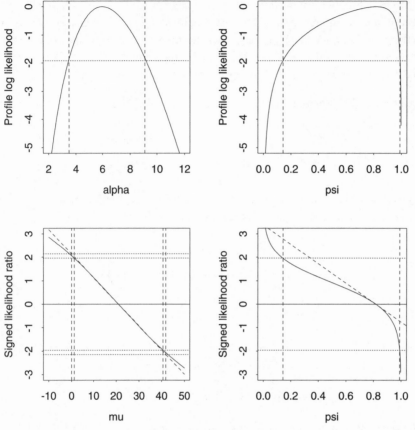

Figure 4.7 Inference from likelihood ratio statistics. Top left and right: profile log likelihoods for the shape parameter of the Weibull model for the springs failure data, and for the probability, ψ, of O-ring thermal distress at 31°F for the Challenger data. The dashed vertical lines show 95% confidence intervals based on the approximate distribution of the likelihood ratio statistic, that is, the set of ψ such that $\ell_p(\psi) \geq \ell_p(\widehat{\psi}) - \frac{1}{2}c_1(0.95)$, with the horizontal dotted line at $-\frac{1}{2}c_1(0.95)$. Bottom left and right: signed likelihood ratio statistics for the maize data and the Challenger data probability ψ. The solid curves are $Z(\mu)$ and $Z(\psi)$, and the dotted horizontal lines are at $z_\alpha, z_{1-\alpha} = \pm 1.96$; the dashed vertical lines show 95% confidence intervals. The dashed diagonal line in the right panel shows $(0.816 - \psi)/0.242$ and corresponds to using approximate normality of $\widetilde{\psi}$ to set a confidence interval. The dashed diagonal line in the left panel shows the Student t quantity $t(\mu)$, with the outer dotted lines showing $\pm t_{14}(0.025)$, from which the t confidence interval shown by the outer dashed lines is read off.

In practice the exact interval would be used, but such results build confidence in use of (4.40) and (4.41) when there is no exact interval.   ∎

Example 4.32 (Weibull distribution)   For the data in Example 4.4, we saw that the difference of maximized likelihoods for the Weibull and exponential models is roughly 12.5, and so $W_p(\alpha^0) = 2\{\ell(\widehat{\theta}, \widehat{\alpha}) - \ell(\widehat{\theta}_{\alpha^0}, \alpha^0)\} \doteq 25$. If $\alpha^0 = 1$ was the true value for α, (4.39) implies that the distribution of $W_p(\alpha^0)$ would be approximately χ_1^2. However the 0.95 and 0.99 quantiles of this distribution are respectively $c_1(0.95) = 3.84$ and $c_1(0.99) = 6.635$, and a value as large as 25 is very unlikely to arise by chance. Thus the Weibull model fits the data appreciably better than the exponential one.

A 95% confidence region for the true value of α based on the profile log likelihood is the set of α such that $\ell_p(\alpha) \geq \ell_p(\widehat{\alpha}) - \frac{1}{2} \times 3.84$; we read this off from the top left panel of Figure 4.7 and obtain (3.5, 9.2). As we would expect, this interval does not contain $\alpha = 1$.   ∎

Example 4.33 (Challenger data)   Examples 4.5, 4.8, and 4.27 concern likelihood analysis of a binomial model for the data in Table 1.3. Our model is that at temperature x_1 and pressure x_2, the number of O-rings suffering thermal distress is binomial with denominator $m = 6$ and probability

$$\pi(\beta_0, \beta_1, \beta_2) = \frac{\exp(\beta_0 + \beta_1 x_1 + \beta_2 x_2)}{1 + \exp(\beta_0 + \beta_1 x_1 + \beta_2 x_2)}.$$

Apart from a constant, the corresponding log likelihood is

$$\beta_0 \sum_{j=1}^{n} r_j + \beta_1 \sum_{j=1}^{n} r_j x_{1j} + \beta_2 \sum_{j=1}^{n} r_j x_{2j} - m \sum_{j=1}^{n} \log\{1 + \exp(\beta_0 + \beta_1 x_{1j} + \beta_2 x_{2j})\}.$$

We maximize this first as it is, then with β_2 held equal to zero, and then with both β_1 and β_2 held equal to zero, and obtain -15.05, -15.82 and -18.90. To check whether there is a pressure effect when temperature is included, we calculate the corresponding likelihood ratio statistic, $2 \times \{-15.05 - (-15.82)\} = 1.54$. This is smaller than the 0.95 quantile of the χ_1^2 distribution, $c_1(0.95) = 3.84$, so any pressure effect is slight. Assuming no pressure effect, the likelihood ratio statistic for no temperature effect is $2 \times \{-15.82 - (-18.90)\} = 6.16$, which we again compare to the χ_1^2 distribution. But $\Pr(\chi_1^2 \geq 6.16) = 0.013$, so 6.16 is unlikely to occur by chance if the true value of β_1 is zero: there seems to be a temperature effect.

The focus in this problem is the probability of thermal distress at temperature $x_1 = 31°F$, and if there is an effect of temperature but not of pressure this probability is $\psi = \pi(\beta_0, \beta_1, 0)$, for which we would like confidence intervals. In Example 4.28 we saw how to apply the delta method to $\widehat{\psi}$, but it gave the unsatisfactory 95% confidence interval $(0.34, 1.29)$.

The upper right panel of Figure 4.7 shows the profile log likelihood $\ell_p(\psi)$. A 95% confidence interval based on this is $(0.14, 0.99)$; unlike intervals based on normal approximation to $\widehat{\psi}$, this is guaranteed to be a subset of $(0, 1)$. The panel below shows the signed likelihood ratio statistic, which is far from a straight line because the profile log likelihood is far from quadratic in ψ. The dashed diagonal line shows how the interval based on the normal distribution of $\widehat{\psi}$ contains values outside the interval $[0, 1]$; an interval symmetric about $\widehat{\psi}$ is wholly inappropriate. ∎

In both the preceding examples the profile log likelihood is asymmetric. Particularly in the second example, the profile log likelihood or equivalently $W_p(\psi)$ or $Z(\psi)$, provide better confidence intervals than normal approximation to the distribution of the maximum likelihood estimate.

4.5.3 Model fit

So far we have supposed that the model is known apart from parameter values, but this is rarely the case in practice and it is essential to check model fit. Graphs play an important role in this, with variants of probability plots (Section 2.1.4) particularly useful. A more formal approach is to nest the model in a larger one, and then to assess whether the expanded model fits the data appreciably better. If its log likelihood is $\ell(\psi, \lambda)$ and the original model restricts ψ to ψ_0, the two may be compared using a likelihood ratio statistic. The usefulness of this approach depends on the expanded model: if it is uninteresting, so too will be the comparison. We have already seen an application of this in Example 4.33.

Example 4.34 (Generalized gamma distribution)   A random variable Y with the generalized gamma distribution has density function

$$f(y; \lambda, \alpha, \kappa) = \frac{\alpha \lambda^{\kappa} y^{\alpha\kappa-1}}{\Gamma(\kappa)} \exp(-\lambda y^{\alpha}), \quad y > 0, \quad \lambda, \alpha, \kappa > 0. \tag{4.42}$$

This arises on supposing that for some α, Y^{α} has a gamma distribution, and reduces to the gamma density (2.7) when $\alpha = 1$, to the Weibull density (4.4) with $\theta = \lambda^{-1/\alpha}$ when $\kappa = 1$, and to the exponential density when $\alpha = \kappa = 1$; it is a flexible generalization of these models. In terms of our general discussion $\psi = \alpha$, with $\psi^0 = 1$, and $\lambda = (\kappa, \lambda)^{\mathsf{T}}$.

When applied to the data in Table 2.1, the maximized log likelihoods are -250.65 for the generalized gamma model, -251.12 for the gamma model, and -251.17 for the Weibull model. The likelihood ratio statistic for comparison of the gamma and generalized gamma densities is $2 \times \{-250.65 - (-251.12)\} = 0.94$, to be treated as χ_1^2. There is no evidence that (4.42) fits better than the gamma density, which fits about equally as well as the Weibull density.   ∎

One useful approach in this context is a *score test*. Suppose that ψ and λ have dimensions $p \times 1$ and $q \times 1$, and let $I_{\lambda\psi} = \mathrm{E}(-\partial^2 \ell / \partial\lambda \partial\psi^{\mathsf{T}})$, and so forth. The idea is that if the restricted model is adequate, then the maximized log likelihood $\ell(\psi^0, \widehat{\lambda}_{\psi^0})$ will not increase sharply in the ψ-direction, so its gradient $\partial\ell(\psi, \lambda)/\partial\psi$ evaluated at $(\psi^0, \widehat{\lambda}_{\psi^0})$ should be modest. We show at the end of this section that

$$\frac{\partial\ell(\psi^0, \widehat{\lambda}_{\psi^0})}{\partial\psi} \;\dot\sim\; N_p\big(0, I_{\psi\psi} - I_{\psi\lambda}I_{\lambda\lambda}^{-1}I_{\lambda\psi}\big),$$

implying that if the simpler model is adequate, then

$$S = \frac{\partial\ell(\psi^0, \widehat{\lambda}_{\psi^0})}{\partial\psi^{\mathsf{T}}}\big(I_{\psi\psi} - I_{\psi\lambda}I_{\lambda\lambda}^{-1}I_{\lambda\psi}\big)^{-1}\frac{\partial\ell(\psi^0, \widehat{\lambda}_{\psi^0})}{\partial\psi} \;\dot\sim\; \chi_p^2, \tag{4.43}$$

where S is evaluated at $(\psi^0, \widehat{\lambda}_{\psi^0})$. When $p = 1$ the signed square root of S should have an approximate standard normal distribution. The statistic S is asymptotically equivalent to the likelihood ratio statistic $W_{\mathrm{p}}(\psi^0)$, but is more convenient because it involves maximization only under the simpler model. Expected information quantities may be replaced by observed information quantities.

Example 4.35 (Spring failure data)   We illustrate the score test by checking whether $\alpha = 1$ for the spring failure data. In terms of our general discussion, $\psi = \alpha$, with $\psi^0 = 1$, and $\lambda = \theta$. The score and observed information are given in Example 4.23. When $\alpha = 1, \widehat{\theta} = \overline{y} = 168.3$. At $(\widehat{\theta}, 1)$, we have $\partial\ell(\theta, \alpha)/\partial\alpha = 9.64$ and $(J_{\alpha\alpha} - J_{\alpha\theta}J_{\theta\theta}^{-1}J_{\theta\alpha})^{-1} = 0.097$, so S takes value 8.99. Compared to the χ_1^2 distribution this gives strong evidence that $\alpha \neq 1$.   ∎

Chi-squared statistics

Sometimes it is useful to assess fit without a specific alternative in mind. One approach is to group the data and to use a *chi-squared statistic*.

Suppose we have n independent observations that fall into categories $1, \ldots, k$, with Y_i denoting the number of observations in category i. The probability that a single observation falls into this category is π_i, where $0 < \pi_i < 1$ and $\sum_{i=1}^{k} \pi_i = 1$, but as $\pi_k = 1 - \pi_1 - \cdots - \pi_{k-1}$, the parameter space is the interior of a simplex in k dimensions, that is, the set

$$\left\{ (\pi_1, \ldots, \pi_k) : \sum_{i=1}^{k} \pi_i = 1, \ 0 < \pi_1, \ldots, \pi_k < 1 \right\} \tag{4.44}$$

of dimension $k - 1$. The model whose fit we wish to assess is that category i has probability $\pi_i(\lambda)$, where $\sum_i \pi_i(\lambda) = 1$ for each λ and the parameter λ has dimension p. This is multinomial with probabilities $\pi_1, \ldots, \pi_k$ and denominator n; see Example 2.36. We suppose that there is a 1–1 mapping between $\pi = (\pi_1, \ldots, \pi_{k-1})^{\mathsf{T}}$ and (ψ, λ), and that setting $\psi = \psi_0$ corresponds to the restricted model $\pi(\lambda) = (\pi_1(\lambda), \ldots, \pi_{k-1}(\lambda))^{\mathsf{T}}$. Thus our model of interest restricts π to a p-dimensional subset of (4.44), where $p < k - 1$, and is nested within the full multinomial model with $k - 1$ parameters.

Given data $y_1, \ldots, y_k$, the likelihood under the general model is

$$L(\pi) = \frac{n!}{y_1! \cdots y_k!} \pi_1^{y_1} \times \cdots \times \pi_k^{y_k}, \quad \sum_{i=1}^{k} \pi_i = 1, \ 0 < \pi_1, \ldots, \pi_k < 1,$$

where $\sum_i y_i = n$, so the log likelihood is

$$\ell(\pi) \equiv \sum_{i=1}^{k-1} y_i \log \pi_i + y_k \log(1 - \pi_1 - \cdots - \pi_{k-1}), \tag{4.45}$$

resulting in score vector and observed information matrix with components

$$\frac{\partial \ell(\pi)}{\partial \pi_i} = \frac{y_i}{\pi_i} - \frac{y_k}{1 - \pi_1 - \cdots - \pi_{k-1}}, \tag{4.46}$$

$$-\frac{\partial^2 \ell(\pi)}{\partial \pi_i d\pi_j} = \begin{cases} \frac{y_i}{\pi_i^2} + \frac{y_k}{(1-\pi_1-\cdots-\pi_{k-1})^2}, & i = j, \\ \frac{y_k}{(1-\pi_1-\cdots-\pi_{k-1})^2}, & i \neq j, \end{cases}$$

where i and j run over $1, \ldots, k - 1$. Manipulation of the likelihood equations shows that the maximum likelihood estimators are $\widehat{\pi}_i = Y_i/n$ (Exercise 4.5.4). The expected information matrix involves $\mathrm{E}(Y_i)$, which may be calculated by noting that if we regard an observation in category i as a 'success', Y_i is the number of successes out of n independent trials, so its marginal distribution is binomial with denominator n and probability π_i and mean $n\pi_i$; see Example 2.36. The expected information is the

$(k-1) \times (k-1)$ matrix

$$I(\pi) = n \begin{pmatrix} 1/\pi_1 + 1/\pi_k & 1/\pi_k & \cdots & 1/\pi_k \\ 1/\pi_k & 1/\pi_2 + 1/\pi_k & \cdots & 1/\pi_k \\ \vdots & \vdots & \ddots & \vdots \\ 1/\pi_k & 1/\pi_k & \cdots & 1/\pi_{k-1} + 1/\pi_k \end{pmatrix}, \qquad (4.47)$$

and it is straightforward to verify that its inverse is

$$I(\pi)^{-1} = n^{-1} \begin{pmatrix} \pi_1(1-\pi_1) & -\pi_1\pi_2 & \cdots & -\pi_1\pi_{k-1} \\ -\pi_2\pi_1 & \pi_2(1-\pi_2) & \cdots & -\pi_2\pi_{k-1} \\ \vdots & & \ddots & \vdots \\ -\pi_{k-1}\pi_1 & -\pi_{k-1}\pi_2 & \cdots & \pi_{k-1}(1-\pi_{k-1}) \end{pmatrix};$$

this is unsurprising, because $\widehat{\pi}_i = Y_i/n$. Provided none of the π_i equals zero or one, the usual large-sample properties of maximum likelihood estimates are satisfied as $n \to \infty$, and in particular $\widehat{\pi}$ has a limiting normal distribution.

We now return to the restricted model, whose log likelihood is

$$\ell(\lambda) = \ell\{\pi(\lambda)\} \equiv \sum_{i=1}^{k-1} y_i \log \pi_i(\lambda) + y_k \log\{1 - \pi_1(\lambda) - \cdots - \pi_{k-1}(\lambda)\},$$

maximization of which gives the maximum likelihood estimator $\widehat{\lambda}$. The first and second derivatives of $\ell(\lambda)$ are

$$\frac{\partial \ell(\lambda)}{\partial \lambda_r} = \sum_{i=1}^{k-1} \frac{\partial \pi_i}{\partial \lambda_r} \frac{\partial \ell(\pi)}{\partial \pi_i},$$

$$\frac{\partial^2 \ell(\lambda)}{\partial \lambda_r \partial \lambda_s} = \sum_{i=1}^{k-1} \frac{\partial^2 \pi_i}{\partial \lambda_r \partial \lambda_s} \frac{\partial \ell(\pi)}{\partial \pi_i} + \sum_{i=1}^{k-1} \sum_{j=1}^{k-1} \frac{\partial \pi_i}{\partial \lambda_r} \frac{\partial \pi_j}{\partial \lambda_s} \frac{\partial^2 \ell(\pi)}{\partial \pi_i \partial \pi_j},$$

and as $E\{\partial \ell(\pi)/\partial \pi_i\} = 0$, the expected information for λ is the $p \times p$ matrix

$$I(\lambda) = \frac{\partial \pi^{\mathrm{T}}}{\partial \lambda} E\left\{-\frac{\partial^2 \ell(\pi)}{\partial \pi \partial \pi^{\mathrm{T}}}\right\} \frac{\partial \pi}{\partial \lambda^{\mathrm{T}}} = \frac{\partial \pi^{\mathrm{T}}}{\partial \lambda} I(\pi) \frac{\partial \pi}{\partial \lambda^{\mathrm{T}}},$$

where $\partial \pi^{\mathrm{T}}/\partial \lambda$ is the $p \times (k-1)$ matrix of partial derivatives of the π_i with respect to the λ_r, and $I(\pi)$ is given by (4.47); see Problem 4.2. Thus provided $\partial \pi^{\mathrm{T}}/\partial \lambda \neq 0$, the parameter λ has a large-sample normal distribution under the restricted model, and the general results in Section 4.5.2 imply that the likelihood ratio statistic used to compare the two models satisfies

$$W = 2 \sum_{i=1}^{k} y_i \log\left\{\frac{\widehat{\pi}_i}{\pi_i(\widehat{\lambda})}\right\} = 2 \sum_{i=1}^{k} y_i \log\left\{\frac{y_i}{n\pi_i(\widehat{\lambda})}\right\} \overset{\cdot}{\sim} \chi^2_{k-1-p}$$

if the simpler model is true. We may write $W = 2 \sum O_i \log(O_i/E_i)$, where $O_i = y_i$ and $E_i = n\pi_i(\widehat{\lambda})$ are the ith observed and expected values under the fitted model; We take $0 \log 0 = \lim_{y \downarrow 0} y \log y = 0$.

Karl Pearson (1857–1936) was a leader of the English biometrical school, which applied statistical ideas to heredity and evolution. His energy was astonishing: he practised law and wrote books on history and religion as well as the classic 'The Grammar of Science' and over 500 other publications. He coined the terms 'standard deviation', 'histogram' and 'mode'. He invented the correlation coefficient and also the chi-square test. He feuded with Fisher, who pointed out that Pearson gave P too many degrees of freedom. The statistic P is sometimes denoted X^2 or χ^2.

as $\sum \pi_i(\widehat{\lambda}) = 1$, it is true that $\sum E_i = \sum O_i = n$. Taylor series expansion shows that $W \doteq \sum (O_i - E_i)^2 / E_i$ (Exercise 4.5.5), leading to *Pearson's statistic*,

$$P = \sum_{i=1}^{k} \frac{\{y_i - n\pi_i(\widehat{\lambda})\}^2}{n\pi_i(\widehat{\lambda})};$$

this too has an approximate χ^2_{k-1-p} distribution if the simpler model is true.

Both W and P provide checks on the adequacy of the restricted multinomial compared to the most general multinomial possible, which requires only that the probabilities sum to one. The approximate distributions of W and P apply when there are large counts, and experience suggests that the chi-squared approximations are more accurate if most of the fitted values exceed five. Though asymptotically equivalent to W, P behaves better in small samples because it does not involve logarithms.

Example 4.36 (Birth data) Figure 2.2 shows the Poisson density with mean $\widehat{\theta} = 12.9$ fitted to the numbers of daily arrivals for the delivery suite data. How good is the fit? Here $p = 1$ parameters are estimated under the Poisson model. With the $n = 92$ daily counts split among the $k = 13$ categories $[0, 7.5), [7.5, 8.5), \ldots, [18.5, \infty)$, the values for O and E are

O	6	3	3	8	13	10	11	11	8	6	4	4	5
E	5.23	4.37	6.26	8.08	9.48	10.19	10.11	9.32	8.01	6.46	4.91	3.52	6.07

and P takes value 4.39, to be treated as a χ^2_{11} variable. As $\Pr(\chi^2_{11} \geq 4.39) \doteq 0.96$, the Poisson model fits very well, perhaps surprisingly so.

A minor problem here is that $\widehat{\theta}$ is obtained from the original data rather than from the data grouped into the k categories. However the maximum likelihood estimate from the grouped data is 12.89, so the fit is hardly affected at all. Use of the parameter estimate from the ungrouped data increases the degrees of freedom for the test, because slightly fewer than p degrees of freedom must be subtracted from the $k - 1$. The estimates will usually be similar unless the grouping is very coarse. ∎

Example 4.37 (Two-way contingency table) Suppose that each of n individuals chosen at random from a population is classified according to two sets of categories. The first corresponds to the r rows of the table, and the second to the c columns; there are $k = rc$ cells indexed by (i, j), $i = 1, \ldots, r$, $j = 1, \ldots, c$. Such a setup is known as an $r \times c$ *contingency table* or *two-way contingency table*. The top part of Table 4.3 shows an example in which 422 people have been cross-classified according to presence or absence of the antigens 'A' and 'B' in their blood. There are 202 people without either antigen, 179 with antigen 'A' but not 'B', and so forth. This is the simplest cross-classification, a 2×2 *table*.

Suppose that there are y_{ij} individuals in the (i, j) cell, so $\sum_{i,j} y_{ij} = n$. If the individuals are independently chosen at random from a population in which the proportion in cell (i, j) is π_{ij}, the joint density of the cell counts Y_{ij} is multinomial with

Table 4.3 Blood groups in England (Taylor and Prior, 1938). The upper part of the table shows a cross-classification of 422 persons by presence or absence of antigens 'A' and 'B', giving the groups 'A', 'B', 'AB', 'O' of the human blood group system. The lower part shows genotypes and corresponding probabilities under one- and two-locus models. See Example 4.38 for details.

		Antigen 'B'		
		Absent	Present	Total
Antigen 'A'	Absent	'O': 202	'B': 35	237
	Present	'A': 179	'AB': 6	185
Total		381	41	422

	Two-locus model		One-locus model	
Group	Genotype	Probability	Genotype	Probability
'A'	$(AA; bb), (Aa; bb)$	$\alpha(1 - \beta)$	$(AA), (AO)$	$\lambda_A^2 + 2\lambda_A\lambda_O$
'B'	$(aa; BB), (aa; Bb)$	$(1 - \alpha)\beta$	$(BB), (BO)$	$\lambda_B^2 + 2\lambda_B\lambda_O$
'AB'	$(AA; BB), (Aa; BB),$	$\alpha\beta$	(AB)	$2\lambda_A\lambda_B$
	$(AA; Bb), (Aa; Bb)$			
'O'	$(aa; bb)$	$(1 - \alpha)(1 - \beta)$	(OO)	λ_O^2

denominator n and probabilities π_{ij}, that is,

$$\frac{n!}{y_{11}! y_{12}! \cdots y_{rc}!} \pi_{11}^{y_{11}} \pi_{12}^{y_{12}} \cdots \pi_{rc}^{y_{rc}}, \quad y_{ij} = 0, \ldots, n, \quad \sum_{i,j} y_{ij} = n,$$

where $0 < \pi_{ij} < 1$ and $\sum_{i,j} \pi_{ij} = 1$. The log likelihood is

$$\ell(\pi) \equiv \sum_{i,j} y_{ij} \log \pi_{ij}, \quad 0 < \pi_{ij} < 1, \quad \sum_{i,j} \pi_{ij} = 1;$$

there are $rc - 1$ parameters because of the constraint that the probabilities sum to one. The preceding general results imply that estimated proportion of the population in cell (i, j) is the sample proportion in that cell, that is, $\widehat{\pi}_{ij} = y_{ij}/n$, so the maximized log likelihood is $\sum_{i,j} y_{ij} \log(y_{ij}/n)$.

Often the question arises whether the row and column classifications are independent. If so, and if the proportion of the population in row category i is α_i, and that in column category j is β_j, then $\pi_{ij} = \alpha_i \beta_j$. As $\sum_i \alpha_i = \sum_j \beta_j = 1$, this model has $p = (r - 1) + (c - 1)$ parameters. The log likelihood is $\sum_{i,j} y_{ij} \log(\alpha_i \beta_j)$, and to maximize it subject to the constraints on the α_i and β_j we use Lagrange multipliers ζ and η and seek extremal points of

$$\ell^*(\alpha, \beta, \zeta, \eta) = \sum_{i,j} y_{ij} \log(\alpha_i \beta_j) + \zeta \left(\sum_i \alpha_i - 1 \right) + \eta \left(\sum_j \beta_j - 1 \right).$$

We find that $\widehat{\alpha}_i = y_{i.}/n$ and $\widehat{\beta}_j = y_{.j}/n$, where $y_{i.} = \sum_j y_{ij}$ and $y_{.j} = \sum_i y_{ij}$; these are respectively the observed proportions of observations in the ith row and jth column categories. The fitted value in cell (i, j) is $n\widehat{\alpha}_i \widehat{\beta}_j = y_{i.} y_{.j}/n$, and the maximized log likelihood is $\sum_{i,j} y_{ij} \log(\widehat{\alpha}_i \widehat{\beta}_j)$.

The likelihood ratio statistic for comparing the independence model with the more general model is

$$W = 2 \sum_{i,j} \left\{ y_{ij} \log \left(\frac{y_{ij}}{n} \right) - y_{ij} \log \left(\frac{y_{i\cdot} y_{\cdot j}}{n^2} \right) \right\} = 2 \sum_{i,j} y_{ij} \log \left(\frac{n y_{ij}}{y_{i\cdot} y_{\cdot j}} \right),$$

and when the independence model is true, the approximate distribution of W is χ^2_{k-1-p}; here $k - 1 - p = rc - 1 - \{(r - 1) + (c - 1)\} = (r - 1)(c - 1)$.

In this case Pearson's statistic may be expressed as

$$P = \sum_{i,j} \frac{(y_{ij} - y_{i\cdot} y_{\cdot j}/n)^2}{y_{i\cdot} y_{\cdot j}/n},$$

with an approximate $\chi^2_{(r-1)(c-1)}$ distribution when the categorizations are independent. ∎

Example 4.38 (ABO blood group system) The most important classification of human blood types is into the four groups 'A', 'B', 'AB', and 'O', corresponding to presence or absence of the antigens 'A' and 'B'; 'AB' refers to the presence of both and 'O' to their absence. In a set of data shown in Table 4.3, the frequencies of these groups were 179, 35, 6, and 202.

According to a model thought credible until the 1920s, the blood group of a person is controlled by two loci $(1; 2)$ on a pair of chromosomes, one chromosome being inherited from each parent. At the loci they independently inherit alleles $(x_1; y_1)$ from their mother and $(x_2; y_2)$ from their father, where x_1 and x_2 are one of a or A, and y_1 and y_2 are one of b or B. Thus their genotype $(x_1 x_2; y_1 y_2)$ is any one of $(aa; bb), \ldots, (AA, BB)$, and they have the antigen 'A' only if allele A is present; similarly for antigen 'B'. In fact $(Aa; Bb)$ is indistinguishable from $(aA; bB)$ and so forth, so under this model there are nine genotypes shown in the second column of the lower part of Table 4.3. Since the loci are independent, the probabilities that a person randomly taken from the population will have blood groups 'A', 'B', 'AB' and 'O' may be written as $\alpha(1 - \beta)$, $(1 - \alpha)\beta$, $\alpha\beta$, and $(1 - \alpha)(1 - \beta)$, where α and β are the probabilities that they have antigens 'A' and 'B'.

An alternative model posits a single locus at which three alleles, A, B, and O may appear, A and B conferring the respective antigens, and O conferring nothing. If λ_A, λ_B and λ_O denote the probabilities that a parent has the three alleles on one chromosome, and if the population is in equilibrium, then the probabilities that the child has blood types 'A', 'B', 'AB' and 'O' are

$$\pi_A = \lambda_A^2 + 2\lambda_A \lambda_O, \ \pi_B = \lambda_B^2 + 2\lambda_B \lambda_O, \ \pi_{AB} = 2\lambda_A \lambda_B, \ \pi_O = \lambda_O^2.$$

where $\lambda_O = 1 - \lambda_A - \lambda_B$.

Under the two-locus model, Example 4.37 implies that the maximum likelihood estimates of α and β are the corresponding sample proportions, $\widehat{\alpha} = 185/422 = 0.438$ and $\widehat{\beta} = 41/422 = 0.097$. The fitted values, 213.97, 167.03, 23.02, 17.97, are rather far from 202, 179, 35, 6. The values for W and P are 17.66 and 15.73, to be treated as χ^2_{k-1-p} if the two-locus model is adequate; here $k - 1 - p = 4 - 1 - 2 = 1$. As $c_1(0.95) = 3.84$, the fit is poor.

Under the single-locus model, the log likelihood is

$$179 \log\left(\lambda_A^2 + 2\lambda_A\lambda_O\right) + 35 \log\left(\lambda_B^2 + 2\lambda_B\lambda_O\right) + 6 \log(2\lambda_A\lambda_B) + 202 \log\left(\lambda_O^2\right),$$

where $\lambda_O = 1 - \lambda_A - \lambda_B$, and maximization in terms of $(\log \lambda_A, \log \lambda_B)$ gives $\widehat{\lambda}_A = 0.252$, $\widehat{\lambda}_B = 0.050$. The fitted values for the blood groups are 205.85, 174.99, 30.54, and 10.62, and the values of W and P are 3.17 and 2.82. The single-locus model is much better supported by the data.   ∎

Derivations of (4.39) and (4.43)

This may be skipped at a first reading.

In the regular case when the model is correct and the true values of the $p \times 1$ and $q \times 1$ vectors ψ and λ are ψ^0 and λ^0, we denote the score vector and observed and expected information matrices by

$$U(\psi^0, \lambda^0) = \begin{pmatrix} U_\psi \\ U_\lambda \end{pmatrix}, \quad J(\psi^0, \lambda^0) = \begin{pmatrix} J_{\psi\psi} & J_{\psi\lambda} \\ J_{\lambda\psi} & J_{\lambda\lambda} \end{pmatrix}, \quad I(\psi^0, \lambda^0) = \begin{pmatrix} I_{\psi\psi} & I_{\psi\lambda} \\ I_{\lambda\psi} & I_{\lambda\lambda} \end{pmatrix},$$

where, for example, U_λ is the $q \times 1$ vector $\partial\ell/\partial\lambda$, $J_{\lambda\psi}$ is the $q \times p$ matrix $-\partial^2\ell/\partial\lambda\partial\psi^{\mathrm{T}}$, and and $I_{\lambda\psi} = \mathrm{E}(-\partial^2\ell/\partial\lambda\partial\psi^{\mathrm{T}})$, evaluated at (ψ^0, λ^0). The components of U are $O_p(n^{1/2})$, those of J are $O_p(n)$, and those of I are $O(n)$.

To establish (4.43), we expand the likelihood equations $U(\widehat{\psi}, \widehat{\lambda}) = 0$ and $\partial\ell(\psi^0, \widehat{\lambda}_{\psi^0})/\partial\lambda = 0$ about (ψ^0, λ^0), giving

$$\begin{pmatrix} U_\psi \\ U_\lambda \end{pmatrix} = J(\psi^0, \lambda^0)\begin{pmatrix} \widehat{\psi} - \psi^0 \\ \widehat{\lambda} - \lambda^0 \end{pmatrix} + o_p\left(n^{1/2}\right)$$

$$= I(\psi^0, \lambda^0)\begin{pmatrix} \widehat{\psi} - \psi^0 \\ \widehat{\lambda} - \lambda^0 \end{pmatrix} + o_p\left(n^{1/2}\right),$$

$$U_\lambda = J_{\lambda\lambda}(\widehat{\lambda}_{\psi^0} - \lambda^0) + o_p\left(n^{1/2}\right) = I_{\lambda\lambda}(\widehat{\lambda}_{\psi^0} - \lambda^0) + o_p\left(n^{1/2}\right).$$

Thus

$$\widehat{\lambda}_{\psi^0} - \lambda^0 = I_{\lambda\lambda}^{-1}U_\lambda + o_p\left(n^{-1/2}\right) = \widehat{\lambda} - \lambda^0 + I_{\lambda\lambda}^{-1}I_{\lambda\psi}(\widehat{\psi} - \psi^0) + o_p\left(n^{-1/2}\right).$$

Taylor series expansion gives

$$\frac{\partial\ell(\psi^0, \widehat{\lambda}_{\psi^0})}{\partial\psi} = U_\psi - I_{\psi\lambda}(\widehat{\lambda}_{\psi^0} - \lambda^0) + o_p\left(n^{-1/2}\right) = U_\psi - I_{\psi\lambda}I_{\lambda\lambda}^{-1}U_\lambda + o_p\left(n^{-1/2}\right),$$

and the joint limiting normal distribution

$$\begin{pmatrix} U_\psi \\ U_\lambda \end{pmatrix} \overset{.}{\sim} N_{p+q}\{0, I(\psi^0, \lambda^0)\}$$

implies that

$$\frac{\partial\ell(\psi^0, \widehat{\lambda}_{\psi^0})}{\partial\psi} \overset{.}{\sim} N_p\left(0, I_{\psi\psi} - I_{\psi\lambda}I_{\lambda\lambda}^{-1}I_{\lambda\psi}\right), \qquad (4.48)$$

so

$$\frac{\partial\ell(\psi^0, \widehat{\lambda}_{\psi^0})}{\partial\psi^{\mathrm{T}}}\left(I_{\psi\psi} - I_{\psi\lambda}I_{\lambda\lambda}^{-1}I_{\lambda\psi}\right)^{-1}\frac{\partial\ell(\psi^0, \widehat{\lambda}_{\psi^0})}{\partial\psi} \overset{.}{\sim} \chi_p^2.$$

To establish (4.39), we write the likelihood ratio statistic (4.38) as

$$W_p(\psi^0) = 2\{\ell(\widehat{\psi}, \widehat{\lambda}) - \ell(\psi^0, \lambda^0)\} - 2\{\ell(\psi^0, \widehat{\lambda}_{\psi^0}) - \ell(\psi^0, \lambda^0)\},$$

and then replace $\ell(\widehat{\psi}, \widehat{\lambda})$ and $\ell(\psi^0, \widehat{\lambda}_{\psi^0})$ with second-order Taylor series expansions about (ψ^0, λ^0). The results above imply that $W_p(\psi^0)$ is approximately

$$\begin{pmatrix} \widehat{\psi} - \psi^0 \\ \widehat{\lambda} - \lambda^0 \end{pmatrix}^{\mathrm{T}} I(\psi^0, \lambda^0) \begin{pmatrix} \widehat{\psi} - \psi^0 \\ \widehat{\lambda} - \lambda^0 \end{pmatrix} - \begin{pmatrix} 0 \\ \widehat{\lambda}_{\psi^0} - \lambda^0 \end{pmatrix}^{\mathrm{T}} I(\psi^0, \lambda^0) \begin{pmatrix} 0 \\ \widehat{\lambda}_{\psi^0} - \lambda^0 \end{pmatrix},$$

and replacement of $\widehat{\lambda}_{\psi^0}$ with its expression in terms of $(\widehat{\psi}, \widehat{\lambda})$ gives

$$W_p(\psi^0) \doteq (\widehat{\psi} - \psi_0)^{\mathrm{T}}\left(I_{\psi\psi} - I_{\psi\lambda}I_{\lambda\lambda}^{-1}I_{\lambda\psi}\right)(\widehat{\psi} - \psi^0) + o_p(1). \tag{4.49}$$

But as our previous asymptotics for the maximum likelihood estimators under the full model give

$$\begin{pmatrix} \widehat{\psi} \\ \widehat{\lambda} \end{pmatrix} \overset{\cdot}{\sim} N_{p+q}\left\{\begin{pmatrix} \psi^0 \\ \lambda^0 \end{pmatrix}, I(\psi^0, \lambda^0)^{-1}\right\}, \tag{4.50}$$

the asymptotic covariance matrix of $\widehat{\psi}$ is $(I_{\psi\psi} - I_{\psi\lambda}I_{\lambda\lambda}^{-1}I_{\lambda\psi})^{-1}$, and (4.49) and (3.23) give

$$W_p(\psi^0) = 2\{\ell(\widehat{\psi}, \widehat{\lambda}) - \ell(\psi^0, \widehat{\lambda}_{\psi^0})\} \overset{\cdot}{\sim} \chi_p^2 :$$

the asymptotic distribution of the likelihood ratio statistic for comparison of two nested models is chi-squared with degrees of freedom equal to the number of parameters that are restricted by the less general model. This result applies only to nested models, and the expansions leading to it are valid only when $(\widehat{\lambda}, \widehat{\psi})$ converges to (ψ^0, λ^0). This may need checking in applications.

Exercises 4.5

1   If $Y_1, \dots, Y_n$ is a random sample from the $N(\mu, \sigma^2)$ distribution with known σ^2, show that the likelihood ratio statistic for comparing $\mu = \mu^0$ with general μ is $W(\mu^0) = n(\overline{Y} - \mu)^2/\sigma^2$. Show that $W(\mu^0)$ is a pivot, and give the likelihood ratio confidence region for μ.

2   Independent values $y_1, \dots, y_n$ arise from a distribution putting probabilities $\frac{1}{4}(1 + 2\theta)$, $\frac{1}{4}(1 - \theta), \frac{1}{4}(1 - \theta), \frac{1}{2}$ on the values 1, 2, 3, 4, where $-\frac{1}{2} < \theta < 1$. Show that the likelihood for θ is proportional to $(1 + 2\theta)^{m_1}(1 - \theta)^{m_2}$ and express m_1 and m_2 in terms of $y_1, \dots, y_n$. Find the maximum likelihood estimate of θ in terms of m_1 and m_2.
Obtain the maximum likelihood estimate and the likelihood ratio statistic for $\theta = 0$ based on data in which the frequencies of 1, 2, 3, 4 were 55, 11, 8, 26. Is it plausible that $\theta = 0$?

3   Consider Examples 4.27 and 4.33. Show that the standard error for $\eta = \beta_0 + 31\beta_1$ is $(9.289 - 2 \times 31 \times 0.142 + 31^2 \times 0.00220)^{1/2}$, and hence obtain a 95% confidence interval for η. Use this to construct an interval for $\phi = e^\eta/(1 + e^\eta)$, and compare it with the interval based on the profile log likelihood for ϕ.

4   Use (4.46) to show that $\widehat{\pi}_j = y_j/n$, and verify the contents of the corresponding observed, expected, and inverse expected information matrices.

5   Verify that the Taylor expansion $O \log(O/E) \doteq O - E + \frac{1}{2}(O - E)^2/E + \cdots$ is valid
     for small $O - E$, and hence check that provided $O_i - E_i$ is small relative to E_i, Pearson's
     statistic P is close to the likelihood ratio statistic W.

6   Let $Y_1, \ldots, Y_n$ and $Z_1, \ldots, Z_m$ be two independent random samples from the $N(\mu_1, \sigma_1^2)$
     and $N(\mu_2, \sigma_2^2)$ distributions respectively. Consider comparison of the model in which
     $\sigma_1^2 = \sigma_2^2$ and the model in which no restriction is placed on the variances, with no restriction
     on the means in either case. Show that the likelihood ratio statistic W_p to compare these
     models is large when the ratio $T = \sum(Y_j - \overline{Y})^2 / \sum(Z_j - \overline{Z})^2$ is large or small, and that
     T is proportional to a random variable with the F distribution.

7   In an experiment to assess the effectiveness of a treatment to reduce blood pressure in heart
     patients, n independent pairs of heart patients are matched according to their sex, weight,
     smoking history, initial blood pressure, and so forth. Then one of each pair is selected at
     random and given the treatment. After a set time the blood pressures are again recorded,
     and it is desired to assess whether the treatment had any effect. A simple model for this
     is that the jth pair of final measurements, (Y_{j1}, Y_{j2}) is two independent normal variables
     with means μ_j and $\mu_j + \beta$, and variances σ^2. It is desired to assess whether $\beta = 0$ or not.
     One approach is a t confidence interval based on $Z_j = Y_{j2} - Y_{j1}$. Explain this, and give
     the degrees of freedom for the t statistic. Show that the likelihood ratio statistic for $\beta = 0$
     is equivalent to $\overline{Z}^2 / \sum(Z_j - \overline{Z})^2$.

4.6 Non-Regular Models

The large-sample normal and chi-squared approximations (4.26) and (4.39) apply to
many important models. There are exceptions, however, due to failure of regularity
conditions for the parameter space, the likelihood and its derivatives, and convergence
of information quantities. A model can be non-regular in many ways, and rather
than attempt a general discussion we give some examples intended to flag possible
problems.

Parameter space

If standard asymptotics are to apply, the true parameter value must be interior to the
parameter space Θ. One way to ensure this is to insist that Θ be an open subset of
$\mathbb{R}^p$ endowed with its usual topology. If not, and if the true θ^0 lies on the edge of
the parameter space, then the maximum likelihood estimator cannot fall on 'both
sides' of θ^0, and therefore cannot have a limiting normal distribution with mean θ^0.
Alternatively, if one or more components of θ are discrete, we cannot expect the
maximum likelihood estimator to be approximately normal.

Example 4.39 (t distribution)   One model for heavy-tailed data is

$$f(y; \mu, \sigma^2, \psi) = \frac{\Gamma\{(\psi^{-1} + 1)/2\}\psi^{1/2}}{(\sigma^2\pi)^{1/2}\Gamma\{1/(2\psi)\}}\{1 + \psi(y - \mu)^2/\sigma^2\}^{-(\psi^{-1}+1)/2},$$

where $\psi, \sigma > 0$ and $-\infty < \mu, y < \infty$. This generalizes the Student t density with
$\psi^{-1} = \nu$ degrees of freedom to continuous ψ. Its tails are heavier than those of
the normal density, obtained when $\psi \to 0$; $f(y; \mu, \sigma^2, 1)$ is Cauchy. The left panel
of Figure 4.8 shows the profile log likelihood for ψ based on the $n = 15$ differences between heights of plants in the fourth column of Table 1.1; $\psi = 0$ is of
particular interest. The likelihood ratio statistic for comparing t and normal models

Figure 4.8 Likelihood
inference for t_ν
distribution. Left: profile
log likelihoods for
$\psi = \nu^{-1}$ for maize data
(solid), and for 19
simulated normal samples
(dots); $\psi = 0$ corresponds
to the $N(\mu, \sigma^2)$ density.
Right: χ_1^2 probability plot
for the 1237 positive
values of the likelihood
ratio statistic $W_p(0)$
observed in 5000
simulated normal samples
of size 15; the rest had
$W_p(0) = 0$.

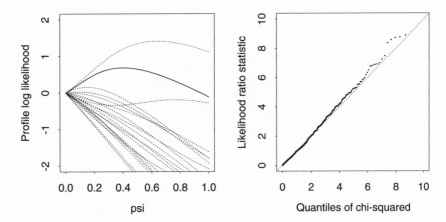

is $W_p(0) = 2\{\ell(\widehat{\mu}, \widehat{\sigma}^2, \widehat{\psi}) - \ell(\widehat{\mu}_0, \widehat{\sigma}_0^2, 0)\}$, where $\widehat{\mu}_0$ and $\widehat{\sigma}_0^2$ are maximum likelihood
estimates for the $N(\mu, \sigma^2)$ density. Its observed value of 1.366 suggests that the t fit
is only marginally better, but $\psi = 0$ is on the boundary of the parameter space and
standard asymptotics do not apply, as we see from profile log likelihoods for simulated normal samples of size 15. In many cases $\widehat{\psi} = 0$, so $W_p(0) = 0$: its distribution
cannot be χ_1^2.

To understand this, we expand $\log f(\mu, \sigma^2, \psi)$ about $\psi = 0$, giving

$$-\frac{1}{2}\{z^2 + \log(2\pi\sigma^2)\} + \frac{\psi}{4}(z^4 - 2z^2 - 1) + \frac{\psi^2}{2}\left(\frac{1}{2}z^4 - \frac{1}{3}z^6\right)$$

$$+ \frac{\psi^3}{24}(3z^8 - 4z^6 - 1) + O(\psi^4),$$

where $z = (y - \mu)/\sigma$. The first and second derivatives that involve ψ are
$\partial \log f/\partial \psi = (z^4 - 2z^2 - 1)/4$ and

$$\frac{\partial^2 \log f}{\partial \psi^2} = \frac{1}{2}z^4 - \frac{1}{3}z^6, \quad \frac{\partial^2 \log f}{\partial \psi \partial \mu} = (z - z^3)/\sigma, \quad \frac{\partial^2 \log f}{\partial \psi \partial \sigma^2} = (z^2 - z^4)/(2\sigma^2)$$

evaluated at $\psi = 0$, while Example 4.18 gives the other derivatives needed. When $\psi = 0$, $Z = (Y - \mu)/\sigma \sim N(0, 1)$, with odd moments zero and first three even moments
1, 3, and 15, so $\text{cov}(Z^4, Z^4) = 96$, $\text{cov}(Z^2, Z^4) = 12$, and $\text{var}(Z^2) = 2$. The expected
information matrix,

$$i(\mu, \sigma^2, 0) = \begin{pmatrix} \sigma^{-2} & 0 & 0 \\ 0 & \frac{1}{2}\sigma^{-4} & \sigma^{-2} \\ 0 & \sigma^{-2} & \frac{7}{2} \end{pmatrix},$$

equals the covariance matrix of the score statistic, and the third derivatives of $\log f$
are well-behaved, so the large-sample distribution of the score vector when $\psi = 0$ is
normal with mean zero and covariance matrix $ni(\mu, \sigma^2, 0)$. On setting $\lambda = (\mu, \sigma^2)$
and $\psi = 0$, (4.48) entails

$$\frac{\partial \ell(\widehat{\mu}_0, \widehat{\sigma}_0^2, 0)}{\partial \psi} \stackrel{\cdot}{\sim} N(0, 3n/2).$$

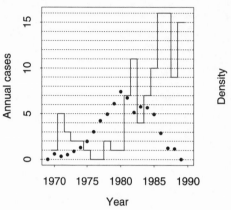
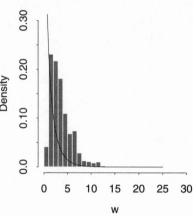

Figure 4.9 Changepoint
analysis for data on
diarrhoea-associated
haemolytic uraemic
syndrome (HUS)
(Henderson and
Matthews, 1993). Left:
counts of cases of HUS
treated in Birmingham,
1970–1989 (solid), and
scaled likelihood ratio
statistic $W_p(\tau)/10$ (blobs).
Right: density of W,
estimated from 10,000
simulations, and χ_1^2
density (solid).

In large samples this derivative is negative with probability $\frac{1}{2}$, and then $W_p(0) = 0$;
while if it is positive the usual Taylor series expansion applies and $W_p(0) \sim \chi_1^2$. Thus
the limiting distribution of $W_p(0)$ is $\frac{1}{2} + \frac{1}{2}\chi_1^2$, giving

$$\Pr\{W_p(0) \le 1.366\} = \frac{1}{2} + \frac{1}{2}\Pr(\chi_1^2 \le 1.366) = 0.88.$$

The asymptotic distribution of $\widehat{\psi}$ puts mass $\frac{1}{2}$ at $\psi = 0$, with the remaining probability
spread as a normal density confined to the positive half-line.

To assess the quality of such approximations, 5000 normal samples of size $n =$
15 were generated. Just 1237 of the $W_p(0)$ were positive, but those that were had
distribution close to χ_1^2, as the right panel of Figure 4.8 shows. Hence

$$\Pr\{W_p(0) \le 1.366\} \doteq (3763/5000) + (1237/5000)\Pr(\chi_1^2 \le 1.366) = 0.94,$$

stronger though not decisive evidence for the t model. Large-sample results are unreliable even with $n = 100$, when $\Pr\{W_p(0) = 0\} \doteq 0.37$.

Such problems also arise if the favoured model is close to the boundary. For example, despite being normal in large samples, when n is small the distribution of $\widehat{\psi}$
would have a point mass at $\psi = 0$. If several parameters lie on their boundaries, then
asymptotics become yet more cumbersome. Simulation seems preferable. ∎

Example 4.40 (HUS data) The left panel of Figure 4.9 shows annual numbers
of cases of 'diarrhoea-associated haemolytic uraemic syndrome' (HUS) treated at a
clinic in Birmingham from 1970 to 1989. HUS is a disease that can threaten the lives
of small children; physicians have speculated that it is linked to levels of *E. coli*. The
data suggest a sharp rise in incidence at about 1980.

A simple model for this increase is that the annual counts $y_1, \ldots, y_n$ are realizations
of independent Poisson variables $Y_1, \ldots, Y_n$ with positive means

$$\mathrm{E}(Y_j) = \begin{cases} \lambda_1, & j = 0, \ldots, \tau, \\ \lambda_2, & j = \tau + 1, \ldots, n. \end{cases}$$

Here the changepoint τ is a discrete parameter with possible values $0, \ldots, n$. The
simpler model of no change appears when $\tau = 0$ or n, and then λ_1 or λ_2 vanishes

from the model. Obviously these two situations are indistinguishable. Moreover, there would be no changepoint to detect if $\lambda_1 = \lambda_2$.

In terms of $s_i = y_1 + \cdots + y_i$ the log likelihood may be written

$$\ell(\tau, \lambda_1, \lambda_2) \equiv s_\tau \log \lambda_1 - \tau \lambda_1 + (s_n - s_\tau) \log \lambda_2 - (n - \tau)\lambda_2,$$

and given τ, the maximum likelihood estimates are $\widehat{\lambda}_1(\tau) = s_\tau/\tau$ and $\widehat{\lambda}_2(\tau) = (s_n - s_\tau)/(n - \tau)$. Hence the profile log likelihood for τ is

$$\ell_p(\tau) = s_\tau \log(s_\tau/\tau) + (s_n - s_\tau) \log\{(s_n - s_\tau)/(n - \tau)\} - s_n, \quad \tau = 0, \ldots, n,$$

and the likelihood ratio statistic for comparing the model of change at τ with that of constant λ is

$$W_p(\tau) = 2\left[S_\tau \log\left(\frac{S_\tau/\tau}{S_n/n}\right) + (S_n - S_\tau) \log\left\{\frac{(S_n - S_\tau)/(n - \tau)}{S_n/n}\right\}\right],$$

where S_i is the random variable corresponding to s_i. As S_i is a sum of independent Poisson variables, its distribution is Poisson. For completeness we set $W_p(0) = W_p(n) = 0$. The values of $W_p(\tau)/10$ shown in the left panel of Figure 4.9 give strong evidence of change in the rate.

If we wish to test for change at a known value of τ, the usual asymptotics will apply provided λ_1 and λ_2 can be estimated consistently from the independent Poisson variables S_τ and $S_n - S_\tau$, and this will be so if their means $\tau\lambda_1$ and $(n - \tau)\lambda_2$ both tend to infinity. Two asymptotic frameworks for this are:

- $\lambda_1, \lambda_2 \to \infty$ with n and τ fixed; and
- $n \to \infty$ and $\tau/n \to a$, with $0 < a < 1$ and λ_1, λ_2 positive and fixed.

The practical implication is that if τ is so close to one of the endpoints that $\tau\lambda_1$ or $(n - \tau)\lambda_2$ is small, a χ_1^2 approximation for the null distribution of $W_p(\tau)$ will be poor, and its quality should be checked; otherwise no new issues arise. They do, however, if τ is unknown.

The likelihood ratio statistic for existence of a changepoint, regardless of its location, is

$$W = \max\{W_p(\tau) : \tau = 1, \ldots, n - 1\}.$$

The values of $W_p(\tau)$ in the left panel of Figure 4.9 show that $\widehat{\tau} = 11$, corresponding to a change between 1980 and 1981; the observed value of W is $w = 74.14$. This seems to be the strong evidence for change that we would have anticipated from plotting the data, but can we be sure?

To find the distribution of W when $\lambda_1 = \lambda_2 = \lambda$, we first note that $Y_1, \ldots, Y_n$ are then a Poisson random sample with mean λ. For reasons given in Sections 5.2.3 it is appropriate to treat W conditional on $S_n = m$, and Example 2.36 implies that the joint distribution of $Y_1, \ldots, Y_n$ conditional on $S_n = m$ is multinomial with denominator m and probability vector $\pi = (n^{-1}, \ldots, n^{-1})^{\mathsf{T}}$. We can simulate the exact distribution of W under this setup, because no parameters are involved. The right panel of Figure 4.9 shows a histogram of 10,000 simulated values of W. Clearly W is stochastically larger than the χ_1^2 density, that is, $\Pr(W > v) > \Pr(\chi_1^2 > v)$ for any $v > 0$. Even so, $w = 74.14$ is much too large to have occurred by chance: there is overwhelming evidence for a change.

Here the maximum likelihood estimator $\widehat{\tau}$ has a discrete distribution on $0, \ldots, 20$ and normal approximation would be foolish. Other approaches have more appeal, and we revisit these data in Example 11.13.   ∎

Parameter identifiability

There must be a 1–1 mapping between models and elements of the parameter space, otherwise there may be no unique value of θ for $\widehat{\theta}$ to converge to. A model in which each θ generates a different distribution is called *identifiable*. We saw a failure of this in Example 4.40, where setting $\lambda_1 = \lambda_2$ gave the same model for any changepoint τ. A rarer possibility is that a parameter cannot be estimated from a particular set of data. In the changepoint example, for instance, the profile likelihood for τ is flat when $y_1 = \cdots = y_n$. The probability of such an event vanishes asymptotically, but such likelihoods do occasionally occur in practice; they demand a simpler model, more data or external knowledge about parameter values.

Sometimes a model has been set up in such a way that its parameters are non-identifiable from any dataset. Suppose we have data $y_1, \ldots, y_n$ with corresponding parameters $\eta_1, \ldots, \eta_n$, and that we may write both $\eta_j = \eta_j(\theta)$ and $\eta_j = \eta_j(\beta)$, where θ and $\beta = \beta(\theta)$ are $p \times 1$ and $q \times 1$ vectors of parameters, with $q < p$. Then the model with $\eta(\theta)$ is said to be *parameter redundant*. The chain rule gives

$$\frac{\partial \eta^\mathsf{T}}{\partial \theta} = \frac{\partial \beta^\mathsf{T}}{\partial \theta} \frac{\partial \eta^\mathsf{T}}{\partial \beta},$$

where both matrices on the right have rank q or lower for any θ. Hence the matrix on the left is symbolically rank-deficient: there is a $1 \times p$ vector function $\gamma(\theta)$, non-zero for all θ, such that $\gamma(\theta)\partial \eta^\mathsf{T}/\partial \theta = 0$ for all θ. It is fairly straightforward to see that the converse is true, so the model is parameter redundant if and only if $\partial \eta^\mathsf{T}/\partial \theta$ is symbolically rank-deficient. Computer algebra can be used to check the symbolic rank of $\partial \eta^\mathsf{T}/\partial \theta$ for a complex model.

Example 4.41 (Exponential density)   Let $Y_1, \ldots, Y_n$ be independent exponential variables with mean η, and set $\eta = \theta_1 \theta_2$, where $\theta_1 = \beta$ and $\theta_2 = \beta$. Evidently θ_1 and θ_2 cannot be estimated separately, and this is reflected by the $n \times 2$ matrix $\partial \eta^\mathsf{T}/\partial \theta$, which consists of a row of θ_2's above a row of θ_1's. It has symbolic rank one, as is seen on premultiplying it by $\gamma(\theta) = (\theta_1, -\theta_2)$.

The likelihood $L(\theta)$ is constant on the curves $(\theta_1, \theta_2) = (\psi\beta, \beta^{-1})$ in $\mathbb{R}_+^2$ and is maximized not at a single point but everywhere on the curve $(\theta_1, \theta_2) = (\overline{y}t, t^{-1})$, $t > 0$. A ridge such as this is a feature of parameter-redundant likelihoods.   ∎

Score and information

For regular inference the log likelihood and its derivatives must be well-behaved enough to allow Taylor series expansions and the neglect of their higher-order terms, and the score must have the asymptotic normal distribution at (4.34). For a random sample, $I(\theta^0) = ni(\theta^0)$, and so the expected information increases without limit as $n \to \infty$; in order to have a normal limit in more complicated situations we also need $I(\theta^0) \to \infty$. Furthermore the observed information must converge in probability as at (4.34).

Example 4.42 (Normal mixture) For an example of a non-smooth likelihood, let $L(\mu_1, \mu_2, \sigma_1^2, \sigma_2^2, \gamma)$ be the likelihood for a random sample $y_1, \ldots, y_n$ from the mixture of normal densities

$$\frac{\gamma}{(2\pi)^{1/2}\sigma_1} \exp\left\{-\frac{(y - \mu_1)^2}{2\sigma_1^2}\right\} + \frac{1 - \gamma}{(2\pi)^{1/2}\sigma_2} \exp\left\{-\frac{(y - \mu_2)^2}{2\sigma_2^2}\right\}, \quad 0 \le \gamma \le 1,$$

with the means and variances in their usual ranges. This corresponds to taking observations in proportions γ, $1 - \gamma$ from two normal populations, not knowing from which they come. If $\gamma \ne 0, 1$, then for each y_j

$$\lim_{\sigma_1 \to 0} L\left(y_j, \mu_2, \sigma_1^2, \sigma_2^2, \gamma\right) = \lim_{\sigma_2 \to 0} L\left(\mu_1, y_j, \sigma_1^2, \sigma_2^2, \gamma\right) = +\infty,$$

so L is a smooth surface pocked with singularities, each of which corresponds to estimating the mean and variance of one of the populations from a single observation. For large n the strong consistency result guarantees the existence of a smooth local maximum of L near the true parameter values. When finding this numerically a careful choice of starting values can help one avoid ending up at a spike instead, but it is worth asking why they occur.

The issue is rounding. As we saw in Example 4.21, the fiction that data are continuous is usually harmless and convenient. Here it is not harmless, however, because it results in infinite likelihoods. The spikes can be removed by accounting for the rounding of the y_j. If they are rounded to multiples of δ, then $\Pr(Y = k\delta) = F(k\delta + \delta/2) - F(k\delta - \delta/2)$, where

$$F(y) = \gamma\Phi\left(\frac{y - \mu_1}{\sigma_1}\right) + (1 - \gamma)\Phi\left(\frac{y - \mu_2}{\sigma_2}\right).$$

As $0 < F(y_j) < 1$, the largest possible contribution to L is then finite. See Example 5.36 for further discussion. ∎

Example 4.43 (Shifted exponential density) To see a failure of regularity conditions for the score statistic, let $y_1, \ldots, y_n$ be an exponential random sample with lower endpoint ϕ and mean $\theta + \phi$, so

$$f(y; \phi, \theta) = \theta^{-1} \exp\left\{-(y - \phi)/\theta\right\}, \quad y > \phi, \theta > 0.$$

The corresponding random variables $Y_1, \ldots, Y_n$ have the same distribution as $\phi + \theta E_1, \ldots, \phi + \theta E_n$, where $E_1, \ldots, E_n$ is a random sample from the standard exponential density. The log likelihood contribution from a single observation $y > \phi$ is $\ell(\phi, \theta) = -\log\theta - (y - \phi)/\theta$, so

$$\frac{\partial\ell(\phi, \theta)}{\partial\phi} = \begin{cases} \theta^{-1}, & y > \phi, \\ 0, & \text{otherwise.} \end{cases}$$

For a regular model this would have mean zero, but here the interchange of differentiation and integration that yields (4.32) fails because the support of the density depends on ϕ, and $E(\partial \ell / \partial \phi) = \theta^{-1}$.

The support of $g(y)$ is the set $\{y : g(y) > 0\}$.

The likelihood is $L(\phi, \theta) = \theta^{-n} \exp\{-n(\overline{y} - \phi)/\theta\}$ for $y_1, \ldots, y_n > \phi$ and $\theta > 0$, and for any θ this increases as $\phi \uparrow \min y_j$ and is zero thereafter. Thus ϕ has maximum likelihood estimate $\widehat{\phi} = y_{(1)}$, while $\widehat{\theta} = \overline{y} - \widehat{\phi} = \overline{y} - y_{(1)}$.

To find limiting distributions of $\widehat{\phi}$ and $\widehat{\theta}$, recall from Example 2.28 that the rth order statistic $E_{(r)}$ of a standard exponential random sample may be written $\sum_{j=1}^{r}(n + 1 - j)^{-1} E_j$, where $E_1, \ldots, E_n$ is an exponential random sample. As $Y_{(r)} \stackrel{D}{=} \phi + \theta E_{(r)}$, we see that $Y_{(1)} \stackrel{D}{=} \phi + n^{-1}\theta E_1$, implying that $n\theta^{-1}(\widehat{\phi} - \phi) \stackrel{D}{=} E_1$: the rescaled endpoint estimate $\widehat{\phi}$ has a non-normal limit distribution. Moreover it converges faster than usual because $\widehat{\phi} - \phi$ must be multiplied by n rather than $n^{1/2}$ in order to give a non-degenerate limit.

For the distribution of $\widehat{\theta}$, note that as $\overline{Y} - Y_{(1)} = n^{-1} \sum_{r=1}^{n} Y_{(r)} - Y_{(1)}$,

$$\widehat{\theta} \stackrel{D}{=} n^{-1} \left\{ n\phi + \theta \sum_{r=1}^{n} \sum_{j=1}^{r} \frac{E_j}{n + 1 - j} - n\phi - \theta E_1 \right\} = n^{-1}(n - 1)\theta \overline{E},$$

with $\overline{E}$ the average of $E_2, \ldots, E_n$. The central limit theorem implies that $n^{1/2}(\widehat{\theta} - \theta)/\theta \stackrel{D}{\longrightarrow} N(0, 1)$, so standard asymptotics apply to $\widehat{\theta}$ despite their failure for $\widehat{\phi}$, which converges so fast that its randomness has no impact on the limiting distribution of $\widehat{\theta}$.

In this problem exact inference is possible for any n (Exercise 4.6.4), but the general conclusion is that endpoints must be treated gingerly. ∎

Though artificial, our next example illustrates how trouble in stochastic process problems can stem from the information quantities.

Example 4.44 (Poisson birth process)   Consider a sequence $Y_0, \ldots, Y_n$ such that given the values of $Y_0, \ldots, Y_{j-1}$, the variable Y_j has a Poisson density with mean θY_{j-1}, and $E(Y_0) = \theta$. The likelihood for θ based on such data was given in Example 4.6, and the log likelihood and observed information are

$$\ell(\theta) \equiv \sum_{j=0}^{n} Y_j \log \theta - \theta \left(1 + \sum_{j=0}^{n-1} Y_j \right), \quad J(\theta) = \theta^{-2} \sum_{j=0}^{n} Y_j.$$

The expected value of Y_j, given Y_{j-1}, is θY_{j-1}, so its unconditional expectation is θ^{j+1}. Hence the expected information is $I(\theta) = \theta^{-2}(\theta + \cdots + \theta^{n+1})$. If $\theta \geq 1$, then $I(\theta) \to \infty$ as $n \to \infty$, but if not, $I(\theta)$ is asymptotically bounded. In fact, as $n \to \infty$, the process is certain to become extinct — that is, there will be an n_0 such that $Y_{n_0} = Y_{n_0+1} = \cdots = 0$ — unless $\theta > 1$, and even then there is a non-zero probability of extinction. Hence $J(\theta)$ remains finite with probability one unless $\theta > 1$, and remains finite with non-zero probability for any θ. Thus the maximum likelihood estimator $\widehat{\theta} = (Y_0 + \cdots + Y_n)/(1 + Y_0 + \cdots + Y_{n-1})$ is neither consistent nor asymptotically normal if $\theta \leq 1$.

From a practical viewpoint, this failure of standard asymptotics is less critical than it might appear. The limit (4.26) is used to obtain finite-sample approximations such as (4.27), but we can still use these if they can be justified by other means. Inference is not impossible, merely more difficult than with independent data. ∎

Wrong model

Up to now we have supposed that the model fitted to the data is correct, with only parameter values unknown. To explore some consequences of fitting the wrong model, suppose the true model is $g(y)$, but that ignorant of this we attempt to fit $f(y; \theta)$ to a random sample $y_1, \ldots, y_n$. Under mild conditions the log likelihood $\ell(\theta) = \sum \log f(y_j; \theta)$ will be maximized at $\widehat{\theta}$, say, and as $n \to \infty$ the quantity $\bar{\ell}(\widehat{\theta}) = n^{-1} \ell(\widehat{\theta})$ will tend to

$$\int \log f(y; \theta_g) g(y) \, dy,$$

where θ_g is the value of θ that minimizes the *Kullback–Leibler discrepancy*

$$D(f_\theta, g) = \int \log \left\{ \frac{g(y)}{f(y; \theta)} \right\} g(y) \, dy$$

with respect to θ. Thus θ_g is the 'least bad' value of θ given our wrong model; of course θ_g depends on g. Differentiation gives

$$0 = \int \frac{\partial \log f(y; \theta_g)}{\partial \theta} g(y) \, dy,$$

with $\widehat{\theta}$ determined by the finite-sample version of this,

$$0 = n^{-1} \sum_{j=1}^{n} \frac{\partial \log f(y_j; \widehat{\theta})}{\partial \theta}. \tag{4.51}$$

Expansion of (4.51) about θ_g yields

$$\widehat{\theta} \doteq \theta_g + \left\{ -n^{-1} \sum_{j=1}^{n} \frac{\partial^2 \log f(y_j; \theta_g)}{\partial \theta \partial \theta^{\mathsf{T}}} \right\}^{-1} \left\{ n^{-1} \sum_{j=1}^{n} \frac{\partial \log f(y_j; \theta_g)}{\partial \theta} \right\}$$

and a modification of the derivation that starts on page 124 gives

$$\widehat{\theta} \overset{\cdot}{\sim} N_p \{ \theta_g, I(\theta_g)^{-1} K(\theta_g) I(\theta_g)^{-1} \}, \tag{4.52}$$

where the *information sandwich* variance matrix depends on

$$K(\theta_g) = n \int \frac{\partial \log f(y; \theta)}{\partial \theta} \frac{\partial \log f(y; \theta)}{\partial \theta^{\mathsf{T}}} g(y) \, dy,$$

$$\tag{4.53}$$

$$I_g(\theta_g) = -n \int \frac{\partial^2 \log f(y; \theta)}{\partial \theta \partial \theta^{\mathsf{T}}} g(y) \, dy.$$

If $g(y) = f(y; \theta)$, so that the supposed density is correct, then θ_g is the true θ, the multivariate version of (4.33) gives $K(\theta_g) = I_g(\theta_g) = I(\theta)$, and (4.52) reduces to the usual approximation.

In practice $g(y)$ is of course unknown, and then $K(\theta_g)$ and $I_g(\theta_g)$ may be estimated by

$$\widehat{K} = \sum_{j=1}^{n} \frac{\partial \log f(y_j; \widehat{\theta})}{\partial \theta} \frac{\partial \log f(y_j; \widehat{\theta})}{\partial \theta^{\mathrm{T}}}, \quad \widehat{J} = -\sum_{j=1}^{n} \frac{\partial^2 \log f(y_j; \widehat{\theta})}{\partial \theta \partial \theta^{\mathrm{T}}}; \quad (4.54)$$

the latter is just the observed information matrix. We may then construct confidence intervals for θ_g using (4.52) with variance matrix $\widehat{J}^{-1} \widehat{K} \widehat{J}^{-1}$.

For future reference we give the approximate distribution of the likelihood ratio statistic. Taylor series approximation gives

$$2\{\ell(\widehat{\theta}) - \ell(\theta_g)\} \doteq (\widehat{\theta} - \theta_g)^{\mathrm{T}} \left\{ -\sum_{j=1}^{n} \frac{\partial^2 \log f(y_j; \theta_g)}{\partial \theta \partial \theta^{\mathrm{T}}} \right\} (\widehat{\theta} - \theta_g)$$

$$\doteq n(\widehat{\theta} - \theta_g)^{\mathrm{T}} I_g(\theta_g)(\widehat{\theta} - \theta_g)$$

and the normal distribution (4.52) of $\widehat{\theta}$ implies that the likelihood ratio statistic has a distribution proportional to χ_p^2, but with mean $\mathrm{tr}\{I_g(\theta_g)^{-1} K(\theta_g)\}$. If the model is correct, $I_g(\theta_g) = K(\theta_g)$, giving the previous mean, p.

Example 4.45 (Exponential and log-normal models)   Let $f(y; \theta)$ be the exponential density with mean θ, while in fact $Y = e^{\sigma Z}$, where Z is standard normal. Then Y is log-normal, with mean $e^{\sigma^2/2}$ and variance $e^{\sigma^2}(e^{\sigma^2} - 1)$.

The presumed log likelihood is $-\log \theta - y/\theta$, so that

$$\int \log f(y; \theta) g(y)\, dy = -\log \theta - \theta^{-1} \int y g(y)\, dy = -\log \theta - \theta^{-1} e^{\sigma^2/2},$$

and differentiation of this with respect to θ gives $\theta_g = e^{\sigma^2/2}$. Here the 'least bad' exponential model has the same mean as the true log-normal distribution, which must always exceed one. Further calculation gives $I(\theta_g) = \theta_g^{-2}$ and $K(\theta_g) = 1 - \theta_g^{-2}$,

The maximum likelihood estimate of θ is $\widehat{\theta} = \overline{Y}$, and either directly or using the information sandwich we see that $\mathrm{var}(\widehat{\theta}) = n^{-1}\theta_g^2(\theta_g^2 - 1)$. Note that replacement of θ_g with its estimate $\widehat{\theta}$ could result in a negative variance. This is not the case if we use the empirical variance — simple calculations give $\widehat{J} = n/\overline{y}^2$ and $\widehat{K} = \overline{y}^{-4} \sum (y_j - \overline{y})^2$, so $\widehat{J}^{-2} \widehat{K} = n^{-2} \sum (y_j - \overline{y})^2$. Reassuringly, this is a consistent estimate of the variance of the average of a random sample from any distribution with finite variance (Example 2.20).

As $I_g(\theta_g)^{-1} K(\theta_g) = e^{\sigma^2} - 1 = \theta_g^2 - 1$, the likelihood ratio statistic may be over- or under-dispersed relative to the χ_1^2 distribution.   ∎

The discussion above is too crude to be the last word. In practice the model fitted will often be elaborate enough to be reasonably close to the data, in the sense that only glaring departures from the model are certain to be detected. Thus it would be better to examine the properties of $\widehat{\theta}$ and related quantities when $f(y; \theta)$ is near $g(y)$ in a suitable sense.

Exercises 4.6

1   Data arise from a mixture of two exponential populations, one with probability π and parameter λ_1, and the other with probability $1 - \pi$ and parameter λ_2. The exponential parameters are both positive real numbers and π lies in the range $[0, 1]$, so $\Theta = [0, 1] \times \mathbb{R}_+^2$ and

$$f(y; \pi, \lambda_1, \lambda_2) = \pi \lambda_1 e^{-\lambda_1 y} + (1 - \pi) \lambda_2 e^{-\lambda_2 y}, \quad y > 0, \ 0 \leq \pi \leq 1, \lambda_1, \lambda_2 > 0.$$

Are the parameters identifiable?
Does standard likelihood theory apply when (i) using a likelihood ratio statistic to test if $\pi = 0$? (ii) estimating π when $\lambda_1 = \lambda_2$?

2   One model for outliers in a normal sample is the mixture

$$f(y; \mu, \pi) = (1 - \pi)\phi(y - \mu) + \pi g(y - \mu), \quad 0 \leq \pi \leq 1, \infty < \mu < \infty,$$

where $g(z)$ has heavier tails than the standard normal density $\phi(z)$; take $g(z) = \frac{1}{2} e^{-|z|}$, for example. Typically π will be small or zero. Show that when $\pi = 0$ the likelihood derivative for π has zero mean but infinite variance, and discuss the implications for the likelihood ratio statistic comparing normal and mixture models.

3   Show that the capture-recapture model in Example 4.13 is not parameter redundant, but that it is if different survival probabilities are allowed in each year. Why is this obvious?

4   In Example 4.43, use relations between the exponential, gamma, chi-squared and F distributions (Section 3.2.1) to show that

$$\frac{2n\widehat{\theta}}{\theta} \sim \chi_{2(n-1)}^2, \qquad \frac{n(\widehat{\phi} - \phi)}{\widehat{\theta}} \sim \frac{n}{n-1} F_{2,2(n-1)};$$

hence give exact $(1 - 2\alpha)$ confidence intervals for the parameters.

5   Show that the score statistic for a variable Y from the uniform density on $(0, \theta)$ is $U(\theta) = -\theta^{-1}$ in the range $0 < Y < \theta$ and zero otherwise, and deduce that $E\{U(\theta)\} = -1$ and $i(\theta) = -\theta^{-1}$. Why is this model non-regular?
Sketch the likelihood based on a random sample $Y_1, \ldots, Y_n$, and verify that $\widehat{\theta} = Y_{(n)}$. To find its limiting distribution, note that

$$\Pr(\widehat{\theta} \leq a) = \begin{cases} 0, & a < 0, \\ (a/\theta)^n, & 0 \leq a \leq \theta, \\ 1, & a > \theta. \end{cases}$$

Show that as $n \to \infty$, $Z_n = n(\theta - \widehat{\theta})/\theta \xrightarrow{D} E$, where E is exponential.

This requires basic knowledge of partial differential equations.

6   Suppose that $\partial \eta^{\mathrm{T}}/\partial \theta$ is symbolically rank-deficient, that is, there exist $\gamma_r(\theta)$, non-zero for all θ, such that

$$\sum_{r=1}^{p} \gamma_r(\theta) \frac{\partial \eta_j}{\partial \theta_r} = 0, \quad j = 1, \ldots, n.$$

Show that the auxiliary equations

$$\frac{d\theta_1}{\gamma_1(\theta)} = \cdots = \frac{d\theta_p}{\gamma_p(\theta)}$$

have $p - 1$ solutions given implicitly by $\beta_t(\theta) = c_t$ for constants $c_1, \ldots, c_{p-1}$. Deduce that the model is parameter redundant.
(Catchpole and Morgan, 1997)

4.7 Model Selection

Model formulation involves judgement, experience, trial, and error. Evidently models should be consistent with knowledge of the system under study, extrapolate to related sets of data, and if possible have reasonable mathematical and statistical properties. Thus, for example, we prefer discrete distributions for discrete quantities and continuous for continuous, while if a probability $\pi(x)$ depends on a quantity x, the relation $\pi(x) = e^{\beta x}/(1 + e^{\beta x})$ is preferable to $\pi(x) = \beta x$, because the latter may lie outside the interval $(0, 1)$; see Example 4.5. Often subject-matter considerations suggest a stochastic argument for a range of suitable models, which typically have primacy over purely *ad hoc* ones. Even after such principles have been applied, however, there are usually several competing models, and a basis is needed for comparing them.

A principle already used but as yet unstated is the *principle of parsimony* or Ockham's razor: 'it is vain to do with more what can be done with fewer'. According to this, given several explanations of the same phenomenon, we should prefer the simplest, or, in our terms, favour simple models over complex ones that fit our data about equally well. But what does this last phrase mean? If we have models with 1, 2, and 3 parameters and maximized log likelihoods of 0, 10, and 11, the second clearly improves on the first, but do the second and third fit 'about equally well'? For regular nested models, standard asymptotics could be applied, but more generally there are difficulties. First, model selection usually involves many fits to the same set of data, so our previous discussion focussing on comparing two prespecified models may be wildly inappropriate. Second, useful asymptotics may be unavailable, for example because the models to be compared are not nested. Third, we may wish to treat none of the models as the truth. An example is in prediction, where a fitted model is sometimes treated as a 'black box' whose contents have no intrinsic interest but are merely used to generate predictions; we should then adopt the agnostic position described at the end of Section 4.6. Here we outline how those ideas may be applied to model selection.

Suppose we have a random sample $Y_1, \ldots, Y_n$ from the unknown *true model* $g(y)$. We fit a *candidate model* $f(y; \theta)$ by maximizing $\ell(\theta) = \sum \log f(y_j; \theta)$, giving $p \times 1$ parameter estimate $\widehat{\theta}$; equivalently we could minimize $-\ell(\theta)$. The fact that the Kullback–Leibler discrepancy is positive,

$$D(f_\theta, g) = \int \log \left\{ \frac{g(y)}{f(y; \theta)} \right\} g(y) \, dy \geq 0,$$

with equality if and only if $f(y; \theta) = g(y)$, suggests that we aim to choose the candidate that minimizes $D(f_\theta, g)$. Let θ_g denote the corresponding value of θ. Unfortunately this approach to model selection is not sufficiently discriminating. The catch is that an infinity of candidate models have $D(f_{\theta_g}, g) = 0$. To see why, suppose that by a lucky chance the candidate model contains the true one. Then $f(y; \theta_g) = g(y)$ and we call f_θ *correct*. As g has fewer parameters we prefer it to f_θ, but $D(f_\theta, g) \geq 0$ with equality when $\theta = \theta_g$. Hence on this basis any correct model is indistinguishable from the true one. We want to pick out the simplest correct model, so we should favour models with few rather than many parameters, provided they fit about equally

William of Ockham or Occam (?1285–1347/1349) was an English Franciscan who studied at Oxford and Paris, was imprisoned by Pope John XXII for arguing that the Franciscan ideal of poverty was prefigured in the Gospels, and then escaped to Bavaria where he wrote in defense of Emperor Louis IV against papal claims; Eco (1984) gives some idea of these controversies. Regarded as the most important scholastic philosopher after Thomas Aquinas, his insistence that logic and human knowledge could be studied without reference to theology and metaphysics encouraged scientific research. He probably died in the Black Death of 1349.

well. For example, if g is the exponential density with unit mean, f_θ might be the Weibull density with unknown shape and scale parameters. This is correct because it reduces to g when both its parameters take value one, but given the choice we would prefer g. A example of a wrong model is the log normal density, which does not become exponential for any values of its parameters.

The expected likelihood ratio statistic for comparing g with f_θ at $\theta = \widehat{\theta}$ for another random sample $Y_1^+, \ldots, Y_n^+$ from g, independent of $Y_1, \ldots, Y_n$, is

$$
\mathrm{E}_g^+ \left[\sum_{j=1}^n \log \left\{ \frac{g(Y_j^+)}{f(Y_j^+; \widehat{\theta})} \right\} \right] = nD(f_{\widehat{\theta}}, g) \geq nD(f_{\theta_g}, g),
$$

where $\mathrm{E}_g^+(\cdot)$ denotes expectation over the density g of Y^+. If f_θ is close to g, then $nD(f_{\theta_g}, g)$ will be close to $nD(g, g)$, and we may hope that $nD(f_{\widehat{\theta}}, g)$ is close to both. But if further parameters do not give a worthwhile reduction in $D(f_{\theta_g}, \theta)$, adding degrees of freedom gives $\widehat{\theta}$ more latitude to miss θ_g, and the corresponding increase in $D(f_{\widehat{\theta}}, g)$ will tend to outweigh any decrease in $D(f_{\theta_g}, g)$. To remove dependence on $\widehat{\theta}$, we average over its distribution, giving

$$
\mathrm{E}_g \left(\mathrm{E}_g^+ \left[\sum_{j=1}^n \log \left\{ \frac{g(Y_j^+)}{f(Y_j^+; \widehat{\theta})} \right\} \right] \right) = n\mathrm{E}_g\{D(f_{\widehat{\theta}}, g)\}; \tag{4.55}
$$

the outer expectation is over the distribution of $\widehat{\theta}$, independent of Y^+. Taylor series expansion shows that $\log f(y; \widehat{\theta})$ approximately equals

$$
\log f(y; \theta_g) + (\widehat{\theta} - \theta_g)^{\mathrm{T}} \frac{\partial \log f(y; \theta_g)}{\partial \theta} + \frac{1}{2}(\widehat{\theta} - \theta_g)^{\mathrm{T}} \frac{\partial^2 \log f(y; \theta_g)}{\partial \theta \partial \theta^{\mathrm{T}}} (\widehat{\theta} - \theta_g),
$$

and as θ_g minimizes $D(f_\theta, g)$,

$$
\int \frac{\partial \log f(y; \theta_g)}{\partial \theta} g(y) \, dy = 0.
$$

Hence

$$
\begin{aligned}
nD(f_{\widehat{\theta}}, g) &= n \int \log \left\{ \frac{g(y)}{f(y; \widehat{\theta})} \right\} g(y) \, dy \\
&\doteq nD(f_{\theta_g}, g) + \frac{1}{2}\mathrm{tr}\{(\widehat{\theta} - \theta_g)(\widehat{\theta} - \theta_g)^{\mathrm{T}} I_g(\theta_g)\},
\end{aligned}
$$

where $I_g(\theta_g)$ is given at (4.53) and we have used the fact that the trace of a scalar is itself. At the end of Section 4.6 we discussed likelihood estimation under the wrong model, and saw that for regular models $\widehat{\theta}$ is asymptotically normal with mean θ_g and variance matrix $I_g(\theta_g)^{-1} K(\theta_g) I_g(\theta_g)^{-1}$, where $K(\theta_g)$ too is given at (4.53); both $I_g(\theta_g)$ and $K(\theta_g)$ are positive definite. Hence

$$
n\mathrm{E}_g\{D(f_{\widehat{\theta}}, g)\} \doteq nD(f_{\theta_g}, g) + \frac{1}{2}\mathrm{tr}\{I_g(\theta_g)^{-1} K(\theta_g)\}, \tag{4.56}
$$

where the second term penalizes the dimension p of θ. The first term here is $O(n)$, but as both $I_g(\theta)$ and $K(\theta)$ are $O(n)$, the second term is $O(p)$. When f_θ is correct and regular, $I_g(\theta_g) = K(\theta_g)$ so $\mathrm{tr}\{I_g(\theta_g)^{-1} K(\theta_g)\} = p$.

To build an estimator of (4.56), note first that the term $\int \log g(y)\, g(y)\, dy$ is constant and can be ignored. Now $\ell(\widehat{\theta}) = \ell(\theta_g) + \{\ell(\widehat{\theta}) - \ell(\theta_g)\}$, so

$$
\begin{aligned}
\mathrm{E}_g\{-\ell(\widehat{\theta})\} &= -\mathrm{E}_g\left\{\ell(\theta_g) + \frac{1}{2}W(\theta_g)\right\} \\
&\doteq n D(f_{\theta_g}, g) - \frac{1}{2}\mathrm{tr}\{I(\theta_g)^{-1}K(\theta_g)\} - n\int \log g(y)\, g(y)\, dy,
\end{aligned}
$$

where we have used the fact that under the wrong model, the likelihood ratio statistic $W(\theta_g)$ has mean approximately $\mathrm{tr}\{I(\theta_g)^{-1}K(\theta_g)\}$. Hence $-\ell(\widehat{\theta})$ tends to underestimate $n D(f_{\theta_g}, g) - n\int \log g(y)\, g(y)\, dy$. On reflection this is obvious, because $\ell(\widehat{\theta}) \geq \ell(\theta_g)$ by definition of $\widehat{\theta}$. As p increases, so will the extent of overestimation.

An estimator of (4.56) is $-\ell(\widehat{\theta}) + c$, where c estimates $\mathrm{tr}\{I(\theta_g)^{-1}K(\theta_g)\}$. Two possible choices of c are p and $\mathrm{tr}(\widehat{J}^{-1}\widehat{K})$, where $\widehat{J}$ and $\widehat{K}$ are defined at (4.54), and these lead to

$$
\text{AIC} = 2\{-\ell(\widehat{\theta}) + p\}, \quad \text{NIC} = 2\{-\ell(\widehat{\theta}) + \mathrm{tr}(\widehat{J}^{-1}\widehat{K})\}; \tag{4.57}
$$

another possibility derived in Section 11.3.1 is $\text{BIC} = -2\ell(\widehat{\theta}) + p\log n$. The model is chosen to minimize AIC, say, with the factor 2 putting differences of AIC on the same scale as likelihood ratio statistics. In practice AIC, BIC, and NIC are used far beyond random samples.

For insight into properties of AIC, suppose that by rare good fortune we fit the true and a correct model, getting maximized log likelihoods $\widehat{\ell}_g$ and $\ell(\widehat{\theta})$ with q and p parameters respectively, and $p > q$. We prefer f_θ to g if $\ell(\widehat{\theta}) - p > \widehat{\ell}_g - q$, but as g is nested within f_θ, properties of the likelihood ratio statistic give

$$
\Pr\{\ell(\widehat{\theta}) - p > \widehat{\ell}_g - q\} \doteq \Pr\{\chi^2_{p-q} > 2(p - q)\}.
$$

For every large n, and with $p - q = 1, 2, 4$ and 10, g is selected with probability 0.84, 0.86, 0.91 and 0.97. Hence model selection using AIC is inconsistent:

$$
\Pr(\text{true model selected}) \not\to 1 \quad \text{as} \quad n \to \infty.
$$

In applications many models would be fitted, and the probability of selecting the true one might be much lower than these calculations suggest.

Modification of this argument shows that NIC also gives an inconsistent procedure. For consistent model selection differences of the penalty must lie between $O(1)$ and $O(n)$ — for example, $O(\log n)$ — but in practice the true model is rarely among those fitted and finite-sample properties are more important. BIC does give consistent model selection when f_θ is correct, but in finite samples it typically leads to underfitting because it tends to suggest too parsimonious a model.

If the candidate model f_θ is not correct, then

$$
\mathrm{E}_g\{\widehat{\ell}_g - \ell(\widehat{\theta})\} \doteq n D(f_{\theta_g}, g) > 0,
$$

so the weak law of large numbers implies that $\Pr\{\widehat{\ell}_g - q > \ell(\widehat{\theta}) - p\} \to 1$ as $n \to \infty$ for fixed p. Hence with enough data we can always distinguish the true model from a fixed incorrect one.

AIC was introduced by Akaike (1973) and is known as Akaike's information criterion. Hirotugu Akaike (1927–) was educated in Tokyo and worked at the Institute of Statistical Mathematics. He has made important contributions to time series and model selection, and also to production engineering; see Findley and Parzen (1995). NIC and BIC are the *network information criterion*, and *Bayes' information criterion*. They may be modified to improve their behaviour for particular models.

Figure 4.10 Model
selection using likelihood
criteria. Upper left: $21n$
observations (blobs) with
true mean (solid) and
polynomial fits $r = 1, 2, 3$
(dots, small dashes, large
dashes); $n = 3$. Upper
right: empirical versions
of AIC, BIC and NIC for
data on left. All are
maximized with $r = 3$.
Lower left: twice expected
log likelihood $2\mathrm{E}_g\{\ell(\theta_g)\}$
(blobs) and theoretical
versions of AIC, BIC and
NIC for the panel above.
The crosses show how
$2\mathrm{E}_g\{\ell(\hat{\theta})\}$ increases with
the dimension of the fitted
model. Lower right: as
lower left panel, but with
$n = 8$ observations at
each value of x.

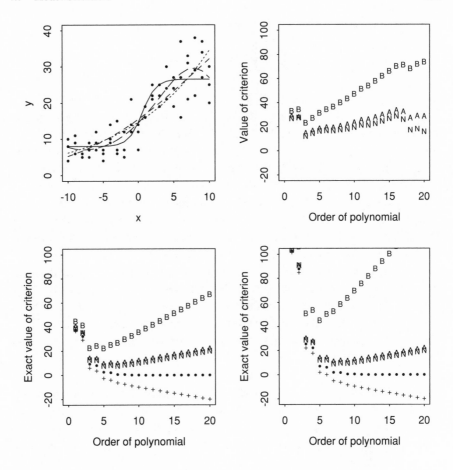

Example 4.46 (Poisson model)   We illustrate this discussion with data whose mean
$\mu(x) = 8 \exp q(x)$ is shown in the upper left panel of Figure 4.10, together with
observations generated by taking $n = 3$ independent Poisson variables with means
$\mu(-10), \mu(-9), \ldots, \mu(10)$; $21n$ variables in all. This is the true model g.

We fit candidate models f_θ with Poisson variables having means $\lambda(x) = \exp(\theta_0 +
\theta_1 x + \cdots + \theta_r x^r)$. The dimension is $p = r + 1$, and taking $r = 1, \ldots, 19$ gives in-
creasingly complex incorrect models, because $q(x) = 1.2e^x/(1 + e^x)$ is not polynomial. A polynomial with $r = 20$ terms can mimic $q(x)$ exactly at $x = -10$,
$-9, \ldots, 10$, however, so taking $r = 20$ is correct but hardly parsimonious. The difference between the linear and the quadratic fits shown in the upper left panel of
Figure 4.10 is small, but adding a cubic term seems to improve the fit.

The upper right panel shows AIC, NIC, and BIC for these data. All three suggest
the choice of $r = 3$, but BIC penalizes complexity much more drastically than the
others. In practice one should not only look at such a graph, but also examine any
models for which the chosen criterion is close to the optimum.

To see the theoretical quantities estimated by AIC, BIC, and NIC, note that the data
here comprise n variables $Y_{1,x}, \ldots, Y_{n,x}$ at each value of x. The log likelihood for an

incorrect model which takes $Y_{j,x}$ to be Poisson with mean $\lambda(x)$ is

$$\ell(\theta) \equiv \sum_{j=1}^{n} \sum_{x=-10}^{10} \{Y_{j,x} \log \lambda(x) - \lambda(x)\}.$$

Now $E_g(Y_{j,x}) = \mu(x)$, so $E_g\{\ell(\theta)\} = n \sum_x \{\mu(x) \log \lambda(x) - \lambda(x)\}$; the values of $\theta_0, \ldots, \theta_r$ that maximize this give $E_g\{\ell(\theta_g)\}$. The blobs in the lower left panel of the figure show how $-2E_g\{\ell(\theta_g)\}$ depends on r. Initially there are big decreases, but after $r = 5$ adding further parameters is barely worthwhile. The crosses show how $-2E_g\{\ell(\widehat{\theta})\}$ depends on r: not penalizing the log likelihood would lead to choosing $r = 20$. The exact values of AIC, BIC, and NIC all indicate $r = 5$. However BIC indicates fits about equally good for $r = 5$ and the simpler model $r = 3$, whereas for AIC and NIC the best fit is similar to that with the more complex model $r = 7$. The penalty applied by BIC is substantially larger than for the others, which are very similar. These functions are what is being estimated in the upper right panel.

To see the effect of increased sample size, the lower right panel of the figure shows exact values of $-E_g\{\ell(\theta_g)\}$, AIC, BIC and NIC when $n = 8$. The jumps in $-E_g\{\ell(\theta_g)\}$ are larger than with $n = 3$, and with this larger sample $r = 7$ seems appreciably better than $r = 5$: more data make it worthwhile to fit more complex models, because we can distinguish them more clearly. Enormous values of n, however, are required to separate $r = 10$ and $r = 20$ reliably: $-E_g\{\ell(\theta_g)\} \doteq -0.08$ when $n = 3$ and $r = 10$, so even a sample with $n = 100$ might indicate that $r = 10$. With $n = 8$, BIC is much more peaked than when $m = 3$, so the value $r = 5$ it indicates is better determined, even though the more complex choice $r = 7$ seems sensible on the basis of $-E_g\{\ell(\theta_g)\}$. By contrast the penalties applied by AIC and NIC are unchanged. Both indicate $r = 7$, but evidently their empirical counterparts might have minima anywhere in the range $r = 5, \ldots, 20$.

The closeness of NIC to AIC in this context leads us to ignore NIC below. ∎

Example 4.47 (Spring failure data)   To analyze the full set of spring failure data in Example 1.2, suppose that the data have Weibull densities whose parameters α and θ may depend on stress x, and consider the models:

M_1: unconnected values of α and θ at each stress, with $p = 12$ parameters;
M_2: a common value of α but unconnected θ at each stress, with $p = 7$;
M_3: a common value of α, and $\theta = (\beta x)^{-1}$, with $p = 2$; and
M_4: common values of α and θ at every stress, with $p = 2$.

The nesting structure of these models is $M_4, M_3 \subseteq M_2 \subseteq M_1$, where $\subseteq$ means 'is nested within'; neither M_3 nor M_4 is nested within the other. We anticipate from Figure 1.2 that M_4 will fit the data very poorly.

To deal with the censoring at lower stresses, note that Example 4.20 implies that the likelihood for a censored Weibull random sample $y_1, \ldots, y_n$ is

$$\prod_u \frac{\alpha}{\theta} \left(\frac{y_j}{\theta}\right)^{\alpha-1} \exp\left\{-\left(\frac{y_j}{\theta}\right)^\alpha\right\} \prod_c \exp\left\{-\left(\frac{y_j}{\theta}\right)^\alpha\right\},$$

Table 4.4 Model selection for spring failure data.

Model	p	Maximized log likelihood	AIC	BIC
M_1	12	-360.40	744.8	769.9
M_2	7	-378.90	771.8	786.5
M_3	2	-411.50	827.0	831.2
M_4	2	-460.56	925.1	929.3

Table 4.5 Parameter estimates and standard errors based on observed information for model M_1 for the spring failure data, fitting separate parameters at each stress.

Stress x_s	700	750	800	850	900	950
$\widehat{\alpha}$ (SE)	1.59 (0.82)	1.44 (0.39)	1.69 (0.39)	7.36 (1.85)	5.37 (1.23)	5.97 (2.13)
$\widehat{\theta}$ (SE)	18044 (7295)	6609 (1566)	907 (180)	372 (16.9)	232 (14.5)	181 (10.2)

where $\prod_u$ and $\prod_c$ denote products over uncensored and censored data. We regard all observations as independent, with parameters α_s and θ_s at stress x_s, and with indicator d_{sj} equalling one if the jth observation at stress x_s, y_{sj}, is uncensored and equalling zero otherwise. The overall likelihood is then

$$\prod_{s=1}^{6}\prod_{j=1}^{10}\left\{\frac{\alpha_s}{\theta_s}\left(\frac{y_{sj}}{\theta_s}\right)^{\alpha_s-1}\right\}^{d_{sj}}\exp\left\{-\left(\frac{y_{sj}}{\theta_s}\right)^{\alpha_s}\right\}.$$

Table 4.4 shows that M_4 fits much worse than any of the other models, and M_3, which has the same number of parameters, is more promising. Evidently M_1 is best by a large margin.

Table 4.5 gives estimates for M_1, with standard errors based on observed information. The values of $\widehat{\alpha}$ depend strongly on the stress, and suggest one value of α at the three lower stresses and another at the higher ones. The standard errors are useless at the lower stresses, with heavy censoring: with so little information any inference will be very uncertain.

The model with six separate values of θ_s and two values of α, one for the three upper and one for the three lower levels of x_s, has maximized log likelihood -360.92, AIC $= 737.8$, and BIC $= 754.6$, so it beats M_1. A plot of $\log\widehat{\theta}_s$ against log stress is close to a straight line, suggesting a three-parameter model with $\theta = 1/(\beta x)$ and two different levels for α, but smooth dependence of α on x is both more plausible and more useful for prediction: what value of α is suitable at stress 825 N/mm^2? Absent more knowledge about the purpose of the experiment, we proceed no further. ∎

Further discussion of model selection and the related topic of model uncertainty may be found in Sections 8.7.3 and 11.2.4.

Exercises 4.7

1   Show that both sides of (4.56) are invariant to 1–1 reparametrizations $\theta = \theta(\phi)$. Why is this important?

2   Use AIC and BIC to compare the models fitted in Example 4.34.

3   Two densities for counts $y = 0, 1, \ldots$ are the Poisson $\theta^y e^{-\theta}/y!$, $\theta > 0$ and the geometric
     $\pi(1 - \pi)^y$, $0 < \pi < 1$; their means are θ and $\pi^{-1} - 1$. Show that if the true model is one
     but the other is fitted, the 'least bad' parameter value matches the means. How easy is
     it to tell them apart when the data are Poisson with $\theta = 1, 5, 10$, and when the data are
     geometric?

4   Consider a regular penalized log likelihood $\ell(\widehat{\theta}) - c_n$, where $c_n \xrightarrow{P} c$ as $n \to \infty$, $\ell(\theta)$
     is based on a correct model, θ has dimension p, and ℓ_g is the log likelihood for the
     true model. Show that $2\{\ell(\widehat{\theta}) - c_n - \ell_g\} \xrightarrow{D} \chi_p^2 - 2c$, and deduce that the probability of
     selecting the true model is $\Pr(\chi_p^2 \leq 2c)$. Hence show that while model selection based on
     BIC is consistent, that based on AIC is not.

4.8 Bibliographic Notes

The ideas of likelihood, information, sufficiency and efficient estimation were developed in a remarkable series of papers by R. A. Fisher in the 1920s and 1930s. Most introductions to mathematical statistics contain this core material. A recent excellent account is Knight (2000). The approach here is influenced by Silvey (1970), Edwards (1972), Cox and Hinkley (1974) and Kalbfleisch (1985). See also Barndorff-Nielsen and Cox (1994) and Pace and Salvan (1997).

The literature on non-regular models is diffuse. See Self and Liang (1987), Smith (1985, 1989b, 1994) and Cheng and Traylor (1995), or Davison (2001) for a partial review. Parameter redundancy is discussed by Catchpole and Morgan (1997), with applications to capture-recapture models.

Model selection and uncertainty are topics of current research interest, with much heat generated by Chatfield (1995) and discussants. For a longer discussion, see Burnham and Anderson (2002).

4.9 Problems

1   The *logistic density* with location and scale parameters μ and σ is

$$f(y; \mu, \sigma) = \frac{\exp\{(y - \mu)/\sigma\}}{\sigma[1 + \exp\{(y - \mu)/\sigma\}]^2}, \quad -\infty < y < \infty, \quad -\infty < \mu < \infty, \sigma > 0.$$

(a) If Y has density $f(y; \mu, 1)$, show that the expected information for μ is $1/3$.
(b) Instead of observing Y, we observe the indicator Z of whether or not Y is positive.
When $\sigma = 1$, show that the expected information for μ based on Z is $e^\mu/(1 + e^\mu)^2$, and
deduce that the maximum efficiency of sampling based on Z rather than Y is $3/4$. Why is
this greatest at $\mu = 0$?
(c) Find the expected information $I(\mu, \sigma)$ based on Y when σ is unknown. Without doing
any calculations, explain why both parameters cannot be estimated based only on Z.

2   Let $\psi(\theta)$ be a 1–1 transformation of θ, and consider a model with log likelihoods $\ell(\theta)$
     and $\ell^*(\psi)$ in the two parametrizations respectively; ℓ has a unique maximum at which the
     likelihood equation is satisfied. Show that

$$\frac{\partial \ell^*(\psi)}{\partial \psi_r} = \frac{\partial \theta^{\mathrm{T}}}{\partial \psi_r} \frac{\partial \ell(\theta)}{\partial \theta}, \qquad \frac{\partial^2 \ell^*(\psi)}{\partial \psi_r \partial \psi_s} = \frac{\partial \theta^{\mathrm{T}}}{\partial \psi_r} \frac{\partial^2 \ell(\theta)}{\partial \theta \partial \theta^{\mathrm{T}}} \frac{\partial \theta}{\partial \psi_s} + \frac{\partial^2 \theta^{\mathrm{T}}}{\partial \psi_r \partial \psi_s} \frac{\partial \ell(\theta)}{\partial \theta}$$

and deduce that

$$I^*(\psi) = \frac{\partial \theta^{\mathrm{T}}}{\partial \psi} I(\theta) \frac{\partial \theta}{\partial \psi^{\mathrm{T}}},$$

but that a similar equation holds for observed information only when $\theta = \widehat{\theta}$.

3   A *location-scale model* with parameters μ and σ has density

$$f(y; \mu, \sigma) = \frac{1}{\sigma} g\left(\frac{y - \mu}{\sigma}\right), \quad -\infty < y < \infty, \ -\infty < \mu < \infty, \sigma > 0.$$

(a) Show that the information in a single observation has form

$$i(\mu, \sigma) = \sigma^{-2} \begin{pmatrix} a & b \\ b & c \end{pmatrix},$$

and express a, b, and c in terms of $h(\cdot) = \log g(\cdot)$. Show that $b = 0$ if g is symmetric about zero, and discuss the implications for the joint distribution of the maximum likelihood estimators $\widehat{\mu}$ and $\widehat{\sigma}$ when g is regular.

(b) Find a, b, and c for the normal density $(2\pi)^{-1/2} e^{-u^2/2}$ and the log-gamma density $\exp(\kappa u - e^u)/\Gamma(\kappa)$, where $\kappa > 0$ is known.

4   Let $y_1, \ldots, y_n$ be a random sample from $f(y; \mu, \sigma) = (2\sigma)^{-1} \exp(-|y - \mu|/\sigma)$, $-\infty < y, \mu < \infty, \sigma > 0$; this is the Laplace density.

means 'the number of times'.

(a) Write down the log likelihood for μ and σ and by showing that

$$\frac{d}{d\mu} \sum |y_j - \mu| = \#\{y_j < \mu\} - \#\{y_j > \mu\} = n - 2R,$$

where $R = \#\{y_j > \mu\}$, show that for any fixed $\sigma > 0$ the maximum likelihood estimate of μ is $\widehat{\mu} = \mathrm{median}\{y_j\}$, and deduce that the maximum likelihood estimate of σ is the mean absolute deviation $\widehat{\sigma} = n^{-1} \sum |y_j - \widehat{\mu}|$.

(b) Use the results of Section 2.3 to show that in large samples $\widehat{\mu} \sim N(\mu, \sigma^2/n)$ and $\widehat{\sigma} \xrightarrow{p} \sigma$. Hence give an approximate confidence interval for the difference of means based on the data in Table 1.1.

(c) Is this a regular model for maximum likelihood estimation?

5   Show that the expected information for a random sample of size n from the Weibull density in Example 4.4 is

$$I(\theta, \alpha) = n \begin{pmatrix} \alpha^2/\theta^2 & -\psi(2)/\theta \\ -\psi(2)/\theta & \{1 + \psi'(2) + \psi(2)^2\}/\alpha^2 \end{pmatrix},$$

where $\psi(z) = d \log \Gamma(z)/dz$.

Given that $\psi(2) = 0.42278$ and $\psi'(2) = 0.64493$, show that

$$I^{-1}(\theta, \alpha) = n^{-1} \begin{pmatrix} 1.108\theta^2/\alpha^2 & 0.257\theta \\ 0.257\theta & 0.608\alpha^2 \end{pmatrix}.$$

Hence find standard errors based on expected information for the estimates in the last column of Table 4.5. What problem arises in a similar calculation for the column with stress $x = 700$?

6   Persons who catch an infectious disease either die almost at once during its initial phase, or live an exponential time; denote the survival time Y and declare that $Y = 0$ if death occurs in the initial phase. Explain why the likelihood can be written as a product of terms of form

$$(1 - p)^{1-I} \times \{p\theta^{-1} \exp(-Y/\theta)\}^I, \quad 0 < p < 1, \theta > 0,$$

where I is an indicator of survival beyond the initial phase. Give interpretations of p and θ.

Table 4.6 Frequencies of eight possible sequences, with their probabilities based on a model in which the probability of a male at first birth is $\frac{1}{2}$ but the probability that the next child has the same sex is $(1 + \theta)/2$, for 6906 three-child families.

MMM	MMF	MFM	MFF
953	914	846	845
$(1+\theta)^2/8$	$(1-\theta^2)/8$	$(1-\theta)^2/8$	$(1-\theta^2)/8$

FMM	FMF	FFM	FFF
825	748	852	923
$(1-\theta^2)/8$	$(1-\theta)^2/8$	$(1-\theta^2)/8$	$(1+\theta)^2/8$

Given data $(i_1, y_1), \ldots, (i_n, y_n)$ on the survival of n persons, show that the log likelihood has form

$$\ell(p, \theta) = r \log p + (n - r) \log(1 - p) - r \log \theta - \theta^{-1} \sum_{j=1}^{n} i_j y_j,$$

where $r = \sum i_j$, and hence find the maximum likelihood estimators of p and θ, together with the observed and expected information matrices.

Comment on the form of the information matrices and give approximate 95% confidence intervals for the parameters.

7   The administrator of a private hospital system is comparing legal claims for damages against two of the hospitals in his system. In the last five years at hospital A the following 19 claims ($, inflation-adjusted) have been paid:

59	172	4762	1000	2885	1905	7094	6259	1950	1208
882	22793	30002	55	32591	853	2153	738	311	

At hospital B, in the same period, there were 16 claims settled out of court for $800 or less, and 16 claims settled in court for

36539	3556	1194	1010	5000	1370	1494	55945
19772	31992	1640	1985	2977	1304	1176	1385

The proposed model is that claims within a hospital follow an exponential distribution. How would you check this for hospital A?

Assuming that the exponential model is valid, set up the equations for calculating maximum likelihood estimates of the means for hospitals A and B. Indicate how you would solve the equation for hospital B.

The maximum likelihood estimate for hospital B is 5455.7. If a common mean is fitted for both hospitals, the maximum likelihood estimate is 5730.6. Use these results to calculate the likelihood ratio statistic for comparing the mean claims of the two hospitals, and interpret the answer.

8   Are the sexes of successive children within a family dependent? Table 4.6 gives for 6906 three-child families the frequencies of the eight possible sequences, with their probabilities based on a model in which the probability of a male at first birth is $\frac{1}{2}$ but the probability that the next child has the same sex is $(1 + \theta)/2$; here $-1 < \theta < 1$. What is special about the model in which $\theta = 0$?

(a) If y_{MMM}, y_{MMF} and so forth denote the numbers of families with orders MMM, MMF, in a sample of m families, write down the likelihood for θ and show that the numbers of consecutive pairs MM and FF is a sufficient statistic.

(b) Obtain the score statistic and observed information, and verify that for the data above the maximum likelihood estimate is $\widehat{\theta} \doteq 0.04$ with standard error 0.0085. Give a 95% confidence interval for θ. Discuss.

(c) Is it true that the probability that the first child is male is $\frac{1}{2}$? Suggest how you might generalize the model to allow for (i) this probability being unequal to $\frac{1}{2}$, and (ii) the probability that a female follows a female being unequal to the probability that a male follows a male. Write down the probabilities for Table 4.6. If you are feeling energetic, conduct a full likelihood analysis of the data.

9 Let $Y_{ij}, j = 1, \ldots, n_i, i = 1, \ldots, k$, be independent normal random variables with means μ_i and variances σ_i^2, and $n_i \geq 2$; set $\overline{Y}_{i\cdot} = n_i^{-1} \sum_j Y_{ij}$.

(a) Show that the likelihood ratio statistic for $\sigma_1^2 = \cdots = \sigma_k^2 = \sigma^2$, with no restrictions on the μ_i, is given by

$$W = \sum_{i=1}^{k} n_i \log \left(\widetilde{\sigma^2} / \widehat{\sigma}_i^2 \right), \quad \widetilde{\sigma^2} = \sum_{i=1}^{k} n_i \widehat{\sigma}_i^2 / \sum_{i=1}^{k} n_i, \quad \widehat{\sigma}_i^2 = n_i^{-1} \sum_{j=1}^{n_i} (Y_{ij} - \overline{Y}_{i\cdot})^2, \qquad (4.58)$$

and give its approximate distribution for large n_i.

(b) A modification to W to improve its behaviour in small samples replaces the n_i in (4.58) with $\nu_i = n_i - 1$. Use the modified statistic to check the homogeneity of the variances for the data in Table 1.2 at the three highest stresses, and comment.

(c) If $k = 2$ show that a test of $\sigma_1^2 = \sigma_2^2$ may be based on $\widehat{\sigma}_1^2 / \widehat{\sigma}_2^2$, and give its exact distribution.

(d) If $n_1 = \cdots = n_k = 3$, show that $\widehat{\sigma}_i^2$ may be written as $2\sigma_i^2 E_i / 3$, where the E_i are independent exponential random variables with unit means. Explain how a plot of the ordered $\widehat{\sigma}_i^2$ against exponential plotting positions can be used to check variance homogeneity and to assess the adequacy of the assumption of normality. What could be done if $n_1 = \cdots = n_k = 2$?

10 In a normal linear model through the origin, independent observations $Y_1, \ldots, Y_n$ are such that $Y_j \sim N(\beta x_j, \sigma^2)$. Show that the log likelihood for a sample $y_1, \ldots, y_n$ is

$$\ell(\beta, \sigma^2) = -\frac{n}{2} \log(2\pi\sigma^2) - \frac{1}{2\sigma^2} \sum_{j=1}^{n} (y_j - \beta x_j)^2.$$

Deduce that the likelihood equations are equivalent to $\sum x_j (y_j - \widehat{\beta} x_j) = 0$ and $\widehat{\sigma}^2 = n^{-1} \sum (y_j - \widehat{\beta} x_j)^2$, and hence find the maximum likelihood estimates $\widehat{\beta}$ and $\widehat{\sigma}^2$ for data with $x = (1, 2, 3, 4, 5)$ and $y = (2.81, 5.48, 7.11, 8.69, 11.28)$.

Show that the observed information matrix evaluated at the maximum likelihood estimates is diagonal and use it to obtain approximate 95% confidence intervals for the parameters. Plot the data and your fitted line $y = \widehat{\beta} x$. Say whether you think the model is correct, with reasons. Discuss the adequacy of the normal approximations in this example.

11 In some measurements of μ-meson decay by L. Janossy and D. Kiss the following observations were recorded from a four channel discriminator: in 844 cases the decay time was less than 1 second; in 467 cases the decay time was between 1 and 2 seconds; in 374 cases the decay time was between 2 and 3 seconds; and in 564 cases the decay time was greater than 3 seconds.

Assuming that decay time has density $\lambda e^{-\lambda t}$, $t > 0$, $\lambda > 0$, find the likelihood for λ. Find the maximum likelihood estimate, $\widehat{\lambda}$, find its standard error, and give a 95% confidence interval for λ.

Check whether the data are consistent with an exponential distribution by comparing the observed and fitted frequencies.

12 A family has two children A and B. Child A catches an infectious disease $\mathcal{D}$ which is so rare that the probability that B catches it other than from A can be ignored. Child A is infectious for a time U having probability density function $\alpha e^{-\alpha u}$, $u \geq 0$, and in any small interval of time $[t, t + \delta t]$ in $[0, U)$, B will catch $\mathcal{D}$ from A with probability $\beta \delta t + o(\delta t)$, where $\alpha, \beta > 0$. Calculate the probability ρ that B does catch $\mathcal{D}$. Show that, in a family where B is actually infected, the density function of the time to infection is $\gamma e^{-\gamma t}$, $t \geq 0$, where $\gamma = \alpha + \beta$.

An epidemiologist observes n independent similar families, in r of which the second child catches $\mathcal{D}$ from the first, at times $t_1, \ldots, t_r$. Write down the likelihood of the data as the product of the probability of observing r and the likelihood of the fixed sample $t_1, \ldots, t_r$. Find the maximum likelihood estimators $\widehat{\rho}$ and $\widehat{\gamma}$ of ρ and γ, and the asymptotic variance of $\widehat{\gamma}$.

	Round	Wrinkled
Yellow	315 (9/16)	101 (3/16)
Green	108 (3/16)	32 (1/16)

Table 4.7 Mendel's data on four kinds of pea seeds (theoretical probability) (Kendall and Stuart, 1973, p. 439).

13   Counts y_1, y_2, y_3 are observed from a multinomial density

$$\Pr(Y_1 = y_1, Y_2 = y_2, Y_3 = y_3) = \frac{m!}{y_1! y_2! y_3!} \pi_1^{y_1} \pi_2^{y_2} \pi_3^{y_3}, \ y_r = 0, \dots, m, \ \sum y_r = m,$$

where $0 < \pi_1, \pi_2, \pi_3 < 1$ and $\pi_1 + \pi_2 + \pi_3 = 1$. Show that the maximum likelihood estimate of π_r is y_r/m.
It is suspected that in fact $\pi_1 = \pi_2 = \pi$, say, where $0 < \pi < 1$. Show that the maximum likelihood estimate of π is then $\frac{1}{2}(y_1 + y_2)/m$.
Give the likelihood ratio statistic for comparing the models, and state its asymptotic distribution.

14   In experiments on cross-breeding peas, Mendel noted frequencies of seeds of different kinds when crossing plants with round yellow seeds and plants with wrinkled green seeds. His data and the theoretical probabilities according to his theory of inheritance are in Table 4.7.
Calculate the expected values under the model, and check the adequacy of the theory using the likelihood ratio and Pearson statistics W and P. How would the degrees of freedom change if the table was treated as a two-way contingency table with unknown probabilities?

Gregor Mendel (1823–1884) was the second child of farmers in Brunn, Moravia. He showed early promise but his family's poverty meant that he could continue his education only as an Augustinian monk. His work on pea plants was begun out of curiosity; it took seven years to amass enough data to formulate his theory of genetic inheritance based on discrete inheritable characteristics, which we know as genes.

15   The negative binomial density may be written

$$f(y; \mu, \psi) = \frac{\Gamma(y + \psi^{-1})}{\Gamma(\psi^{-1})y!} \frac{(\psi \mu)^y}{(1 + \psi \mu)^{y + 1/\psi}}, \quad y = 0, 1, \dots, \quad \mu, \psi > 0;$$

its limit as $\psi \to 0$ is the Poisson density. Taylor series expansion about $\psi = 0$ shows that $\log f(y; \mu, \psi)$ is

$$y \log \mu - \mu - \log y! + \frac{\psi}{2}\{(y - \mu)^2 - y\} + \frac{\psi^2}{12}\{6\mu^2 y - 4\mu^3 - y(1 - 3y + 2y^2)\}$$

$$+ \frac{\psi^3}{12}\{3\mu^4 - 4\mu^3 y + y^2(y - 1)^2\} + O(\psi^4).$$

Find the expected information $I(\mu, \psi)$ when $\psi = 0$, and show that the asymptotic distribution of the score $\partial \ell(\hat{\mu}_\psi, \psi)/\partial \psi$ based on a sample of size n is then $N(0, n\mu^2/2)$. Discuss properties of the likelihood ratio statistic for comparison of Poisson and negative binomial models.

16   A possible model for the data in Table 11.7 is that pumps are independent, and that the failures for the jth pump have the Poisson distribution with mean λx_j, where x_j is the operating hours (1000s). Find the maximum likelihood estimate of λ under this model and give its standard error. Construct the likelihood ratio statistic to compare this with the model in which all the pumps have different rates. Justifying your reasoning, say whether you expect this statistic to have an approximate χ^2 distribution.

17   If $y_1, \dots, y_n$ is a random sample with density $\sigma^{-1} f\{(y - \mu)/\sigma; \lambda\}$, where f is the skew-normal density function (Problem 3.6), write down the log likelihood for μ, σ, and λ, and investigate likelihood inference for this model.

5

Models

Chapter 4 described methods related to a central notion in inference, namely likelihood. This chapter and the next discuss how those ideas apply to some particular situations, beginning with the simplest model for the dependence of one variable on another, straight-line regression. There is then an account of exponential family distributions, which include many models commonly used in practice, such as the normal, exponential, gamma, Poisson and binomial densities, and which play a central role in statistical theory. We then briefly describe group transformation models, which are also important in statistical theory. This is followed by a description of models for data in the form of lifetimes, which are common in medical and industrial settings, and a discussion of missing data and the EM algorithm.

5.1 Straight-Line Regression

We have already met situations where we focus on how one variable depends on others. In such problems there are two or more variables, some of which are regarded as fixed, and others as random. The random quantities are known as *responses* and the fixed ones as *explanatory variables*. We shall suppose that only one variable is regarded as a response. Such models, known as regression models, are discussed extensively in Chapters 8, 9, and 10. Here we outline the basic results for the simplest regression model, where a single response depends linearly on a single covariate. We start with an example.

Example 5.1 (Venice sea level data)   Table 5.1 and Figure 5.1 show annual maximum sea levels in Venice for 1931–1981. The most obvious feature is that the maximum sea level increased by about 25 cm over that period. A simple model is of linear trend in the sea level, y, so in year j,

$$y_j = \beta_0 + \beta_1 j + \varepsilon_j, \tag{5.1}$$

where β_0 (cm) represents the expected maximum sea level in year $j = 0$, β_1 the annual increase (cm/year) , and ε_j is a random variable with mean zero and variance

103	78	121	116	115	147	119	114	89	102
99	91	97	106	105	136	126	132	104	117
151	116	107	112	97	95	119	124	118	145
122	114	118	107	110	194	138	144	138	123
122	120	114	96	125	124	120	132	166	134
138									

Table 5.1 Annual maximum sea levels (cm) in Venice, 1931–1981 (Pirazzoli, 1982). To be read across rows.

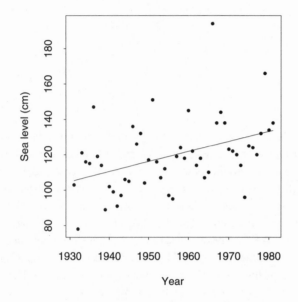

Figure 5.1 Annual maximum sea levels in Venice, 1931–1981, with fitted regression line.

σ^2 (cm^2) representing scatter about the trend. Here the response is sea level, y_j, and the year, j, is the sole explanatory variable. ∎

The simplest linear model is that independent random variables Y_j satisfy

$$Y_j = \beta_0 + \beta_1 x_j + \varepsilon_j, \quad j = 1, \dots, n, \tag{5.2}$$

where the x_j are known constants, the $\varepsilon_j \overset{\text{iid}}{\sim} N(0, \sigma^2)$, and β_0, β_1 and σ^2 are unknown parameters, Thus Y_j is normal with mean $\beta_0 + \beta_1 x_j$ and variance σ^2. The data arise as pairs $(x_1, y_1), \dots, (x_n, y_n)$, from which β_0, β_1, and σ^2 are to be estimated. In Example 5.1 the pairs are $(1931, 103), \dots, (1981, 138)$. If all the x_j are equal, we cannot estimate the slope of the dependence of y on x, so we assume that at least two x_j are distinct.

$\overset{\text{iid}}{\sim}$ means 'are independent and identically distributed as'.

A reparametrization of (5.2) is more convenient, so we consider instead

$$Y_j = \gamma_0 + \gamma_1(x_j - \overline{x}) + \varepsilon_j, \quad j = 1, \dots, n, \tag{5.3}$$

where $\overline{x} = n^{-1} \sum x_j$. In terms of the original parameters, $\gamma_1 = \beta_1$, and $\gamma_0 = \beta_0 + \beta_1 \overline{x}$. This can make better statistical sense too. In (5.1) the interpretation of β_0 as a mean sea level at the start of the Christian era — when $j = 0$ — involves a ludicrous extrapolation of the straight-line model over two millenia, whereas γ_0 concerns its level when $j = \overline{x} = 1956$; this is clearly more sensible.

Under (5.3) the Y_j are independent and normal with means and variances $\gamma_0 + \gamma_1(x_j - \bar{x})$ and σ^2, so the likelihood based on $(x_1, y_1), \ldots, (x_n, y_n)$ is

$$\prod_{j=1}^{n} \frac{1}{(2\pi\sigma^2)^{1/2}} \exp\left[-\frac{1}{2\sigma^2} \{y_j - \gamma_0 - \gamma_1(x_j - \bar{x})\}^2\right],$$
$$-\infty < \gamma_0, \gamma_1 < \infty, \sigma^2 > 0.$$

The log likelihood is

$$\ell(\gamma_0, \gamma_1, \sigma^2) \equiv -\frac{1}{2}\left[n \log \sigma^2 + \frac{1}{\sigma^2} \sum_{j=1}^{n} \{y_j - \gamma_0 - \gamma_1(x_j - \bar{x})\}^2\right]. \qquad (5.4)$$

For any σ^2, maximizing this over γ_0 and γ_1 is equivalent to minimizing the *sum of squares*

$$SS(\gamma_0, \gamma_1) = \sum_{j=1}^{n} \{y_j - \gamma_0 - \gamma_1(x_j - \bar{x})\}^2,$$

which is the sum of squared vertical deviations between the y_j and their means $\gamma_0 + \gamma_1(x_j - \bar{x})$ under the linear model. Its derivatives are

$$\frac{\partial SS}{\partial \gamma_0} = -2 \sum_{j=1}^{n} \{y_j - \gamma_0 - \gamma_1(x_j - \bar{x})\},$$

$$\frac{\partial SS}{\partial \gamma_1} = -2 \sum_{j=1}^{n} (x_j - \bar{x}) \{y_j - \gamma_0 - \gamma_1(x_j - \bar{x})\},$$

$$\frac{\partial^2 SS}{\partial \gamma_0^2} = 2n, \quad \frac{\partial^2 SS}{\partial \gamma_1^2} = 2 \sum_{j=1}^{n} (x_j - \bar{x})^2, \quad \frac{\partial^2 SS}{\partial \gamma_0 \partial \gamma_1} = 2 \sum_{j=1}^{n} (x_j - \bar{x}) = 0.$$

The solutions to the equations $\partial SS/\partial \gamma_0 = \partial SS/\partial \gamma_1 = 0$ are the *least squares estimates*,

$$\widehat{\gamma}_0 = \bar{y}, \qquad \widehat{\gamma}_1 = \frac{\sum_{j=1}^{n} y_j(x_j - \bar{x})}{\sum_{j=1}^{n} (x_j - \bar{x})^2}. \qquad (5.5)$$

As anticipated, γ_1 cannot be estimated if all the x_j are equal, for then $x_j \equiv \bar{x}$ and $\widehat{\gamma}_1$ is undefined. The matrix of second derivatives of SS is positive definite, so the estimates (5.5) minimize the sum of squares and hence maximize $\ell(\gamma_0, \gamma_1, \sigma^2)$ with respect to γ_0 and γ_1.

As the log likelihood may be written as $-\frac{1}{2}\left\{n \log \sigma^2 + SS(\gamma_0, \gamma_1)/\sigma^2\right\}$, the maximum likelihood estimate of σ^2 is

$$\widehat{\sigma}^2 = n^{-1} SS(\widehat{\gamma}_0, \widehat{\gamma}_1) = \frac{1}{n} \sum_{j=1}^{n} \{y_j - \widehat{\gamma}_0 - \widehat{\gamma}_1(x_j - \bar{x})\}^2.$$

The quantity $SS(\widehat{\gamma}_0, \widehat{\gamma}_1)$, known as the *residual sum of squares*, is the smallest sum of squares attainable by fitting (5.3) to the data.

The least squares estimators are linear combinations of normal variables, so their distributions are also normal. If we rewrite them as

$$\widehat{\gamma}_0 = n^{-1}\sum_{j=1}^{n}\{\gamma_0 + \gamma_1(x_j - \overline{x}) + \varepsilon_j\} = \gamma_0 + n^{-1}\sum_{j=1}^{n}\varepsilon_j,$$

$$\widehat{\gamma}_1 = \frac{\sum_{j=1}^{n}\{\gamma_0 + \gamma_1(x_j - \overline{x}) + \varepsilon_j\}(x_j - \overline{x})}{\sum_{j=1}^{n}(x_j - \overline{x})^2} = \gamma_1 + \frac{\sum_{j=1}^{n}(x_j - \overline{x})\varepsilon_j}{\sum_{j=1}^{n}(x_j - \overline{x})^2},$$

we see that because the ε_j are independent with means zero and variances σ^2, $\widehat{\gamma}_0$ has mean γ_0 and variance σ^2/n, and that $\widehat{\gamma}_1$ has mean γ_1 and variance $\sigma^2/\sum(x_j - \overline{x})^2$. Moreover

$$\text{cov}(\widehat{\gamma}_0, \widehat{\gamma}_1) = \text{cov}\left\{n^{-1}\sum\varepsilon_j, \frac{\sum_{j=1}^{n}(x_j - \overline{x})\varepsilon_j}{\sum_{j=1}^{n}(x_j - \overline{x})^2}\right\}$$

$$= \frac{\sum_{j=1}^{n}n^{-1}(x_j - \overline{x})\text{var}(\varepsilon_j)}{\sum_{j=1}^{n}(x_j - \overline{x})^2} = 0:$$

as $\widehat{\gamma}_0$ and $\widehat{\gamma}_1$ are uncorrelated normal random variables, they are independent.

If σ^2 is known, confidence intervals for the true values of γ_0 and γ_1 may be based on the normal distributions of $\widehat{\gamma}_0$ and $\widehat{\gamma}_1$. A $(1 - 2\alpha)$ confidence interval for γ_1, for example, is $\widehat{\gamma}_1 \pm \sigma z_\alpha/\{\sum(x_j - \overline{x})^2\}^{1/2}$.

We shall see in Chapter 8 that the residual sum of squares $SS(\widehat{\gamma}_0, \widehat{\gamma}_1) \sim \sigma^2\chi^2_{n-2}$, independent of $\widehat{\gamma}_0$ and $\widehat{\gamma}_1$. Thus when σ^2 is unknown, the estimator

$$S^2 = \frac{1}{n-2}SS(\widehat{\gamma}_0, \widehat{\gamma}_1)$$

satisfies $\text{E}(S^2) = \sigma^2$, and as S^2 is independent of $\widehat{\gamma}_0$ and $\widehat{\gamma}_1$, a $(1 - 2\alpha)$ confidence interval for γ_1 is $\widehat{\gamma}_1 \pm St_{n-2}(\alpha)/\{\sum(x_j - \overline{x})^2\}^{1/2}$, because

$$\frac{\widehat{\gamma}_1 - \gamma_1}{\left\{S^2/\sum(x_j - \overline{x})^2\right\}^{1/2}} \sim t_{n-2}.$$

Example 5.2 (Venice sea level data)   For the model $y_j = \beta_0 + \beta_1 j + \varepsilon_j$ of Example 5.1, we have $n = 51$, $x_1 = 1931, \ldots, x_n = 1981$, so $\overline{x} = 1956$. In parametrization (5.3), γ_0 is the expected annual maximum sea level in 1956 in cm, and γ_1 is the mean annual increase in maximum sea level in cm/year.

Straightforward calculation yields $\widehat{\gamma}_0 = 119.61$ cm and $\widehat{\gamma}_1 = 0.567$ cm/year, $SS(\widehat{\gamma}_0, \widehat{\gamma}_1) = 16988.1$, and $\sum(x_j - \overline{x})^2 = 11050$. The unbiased estimate of σ^2 is $s^2 = 16988.1/(51 - 2) = 346.7$, so we estimate σ by $s = 18.6$. This is very large relative to the annual increase in sea level, which as we see from Figure 5.1 is small relative to the overall vertical variation.

Standard errors for $\widehat{\gamma}_0$ and $\widehat{\gamma}_1$ are $s/n^{1/2} = 2.61$ and $s/\left\{\sum(x_j - \overline{x})^2\right\}^{1/2} = 0.177$, and a 95% confidence interval for γ_1 is $\widehat{\gamma}_1 \pm 0.177t_{49}(0.025)$, that is, $(0.213, 0.921)$. This does not include zero, confirming that the trend in Figure 5.1 is real. ∎

Linear combinations

Distributional results for linear functions of $\widehat{\gamma}_0$ and $\widehat{\gamma}_1$ are readily obtained. For example, in the original linear model (5.2) we have $\beta_0 = \gamma_0 - \gamma_1 \bar{x}$, the maximum likelihood estimator of which is $\widehat{\beta}_0 = \widehat{\gamma}_0 - \widehat{\gamma}_1 \bar{x}$. This has expected value $\gamma_0 - \gamma_1 \bar{x}$ and variance

$$\text{var}(\widehat{\gamma}_0 - \widehat{\gamma}_1 \bar{x}) = \text{var}(\widehat{\gamma}_0) - 2\bar{x}\text{cov}(\widehat{\gamma}_0, \widehat{\gamma}_1) + \bar{x}^2\text{var}(\widehat{\gamma}_1) = \sigma^2 \left\{ \frac{1}{n} + \frac{\bar{x}^2}{\sum_{j=1}^{n}(x_j - \bar{x})^2} \right\}.$$

As

$$\text{cov}(\widehat{\beta}_0, \widehat{\beta}_1) = \text{cov}(\widehat{\gamma}_0 - \widehat{\gamma}_1 \bar{x}, \widehat{\gamma}_1) = \text{cov}(\widehat{\gamma}_0, \widehat{\gamma}_1) - \bar{x}\text{var}(\widehat{\gamma}_1) = \frac{-\sigma^2 \bar{x}}{\sum_{j=1}^{n}(x_j - \bar{x})^2},$$

the normal random variables $\widehat{\beta}_0$ and $\widehat{\beta}_1$ are independent if and only if $\bar{x} = 0$.

Suppose we wish to predict the response value at x_+,

$$Y_+ = \gamma_0 + \gamma_1(x_+ - \bar{x}) + \varepsilon_+.$$

Here ε_+ represents the random variation about the expected value, which is independent of the other responses, because of our modelling assumptions. The random variable Y_+ has expected value $\gamma_0 + \gamma_1(x_+ - \bar{x})$. The maximum likelihood estimator of this, $\widehat{\gamma}_0 + \widehat{\gamma}_1(x_+ - \bar{x})$, has mean and variance

$$\gamma_0 + \gamma_1(x_+ - \bar{x}), \quad \sigma^2 \left\{ \frac{1}{n} + \frac{(x_+ - \bar{x})^2}{\sum_{j=1}^{n}(x_j - \bar{x})^2} \right\}.$$

This is the variance not of Y_+ but of $\widehat{\gamma}_0 + \widehat{\gamma}_1(x_+ - \bar{x})$: it does not account for the extra variability introduced by ε_+. The variance appropriate for the predicted response *actually observed* is

$$\text{var}(Y_+) = \text{var}\left\{\widehat{\gamma}_0 + \widehat{\gamma}_1(x_+ - \bar{x}) + \varepsilon_+\right\} = \sigma^2 \left\{ \frac{1}{n} + \frac{(x_+ - \bar{x})^2}{\sum_{j=1}^{n}(x_j - \bar{x})^2} \right\} + \sigma^2. \quad (5.6)$$

The final σ^2 is due to ε_+ and would remain even if the parameters were known.

Example 5.3 (Venice sea level data) For illustration we take $x_+ = 1993$. Our predicted value for Y_+ is $\widehat{\gamma}_0 + \widehat{\gamma}_1(x_+ - \bar{x}) = 140.59$, with estimated variance $49.75 + 346.70 = 396.45$, obtained by replacing σ^2 with s^2 in (5.6). The estimated variance of ε_+, 346.70, is much larger than the estimated variance 49.75 of the fitted value $\widehat{\gamma}_0 + \widehat{\gamma}_1(x_+ - \bar{x})$. A confidence interval for Y_+ could be obtained from the t statistic.

Our model (5.2) presupposes that the errors ε_j are normal, and that the dependence of y on x is linear. We discuss how to check these assumptions in Section 8.6.1, here noting that simple estimates of the errors ε_j are the *raw residuals* $e_j = y_j - \widehat{\beta}_0 - \widehat{\beta}_1 x_j$, which should be normal and approximately independent of x if the model is correct. We check linearity by looking for patterns in a plot of the e_j against the x_j, and check normality by a normal probability plot of the e_j; see Figure 5.2. Linearity seems justifiable, but the errors seem too skewed to be normally distributed.

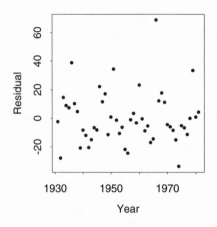
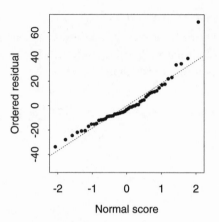

Figure 5.2 Straight-line
regression fit to annual
maximum sea levels in
Venice, 1931–1981. Left:
raw residuals plotted
against time. Right:
normal scores plot of raw
residuals; the line has
slope $\widehat{\sigma}$. The skewness of
the residuals suggests that
the errors are not normal.

The astute reader will realise that the changing sea level is due not to the rising waters of the Adriatic, but to the sinking of the marker that measures water height, along with Venice, to which it is attached.   ∎

Exercises 5.1

1   Find the observed and expected information matrices for the parameters in (5.4), and confirm that general likelihood theory gives the same variances and covariance for the least squares estimates as the direct argument on page 164.

2   Show that $(\widehat{\gamma}_0, \widehat{\gamma}_1, s^2)$ are minimal sufficient for the parameters of the straight-line regression model.

3   Consider data from the straight-line regression model with n observations and

$$x_j = \begin{cases} 0, & j = 1, \ldots, m, \\ 1, & \text{otherwise,} \end{cases}$$

where $m \leq n$. Give a careful interpretation of the parameters β_0 and β_1, and find their least squares estimates. For what value(s) of m is var($\widehat{\beta}_1$) minimized, and for which maximized? Do your results make qualitative sense?

4   Let $Y_1, \ldots, Y_n$ be observations satisfying (5.2), with not all the x_j equal. Find var($\widehat{\beta}_0 + x_+\widehat{\beta}_1$), where x_+ is fixed. Hence give exact 0.95 confidence intervals for $\beta_0 + \beta_1 x_+$ when σ^2 is known and when it is unknown.

5.2 Exponential Family Models

Exponential families include most of the models we have met so far and are widely used in applications. Densities such as the normal, gamma, Poisson, multinomial, and so forth have the same underlying structure with elegant properties giving them a central role in statistical theory. This section outlines those properties, first giving the basic ideas for scalar random variables, then extending them to more complex models, and finally considering inference.

5.2.1 Basic notions

Let $f_0(y)$ be a given probability density, discrete or continuous, under which random variable Y has support $\mathcal{Y} = \{y : f_0(y) > 0\}$ that is a subset of the real line $\mathbb{R}$. For example, $f_0(y)$ might be the uniform density on the unit interval $\mathcal{Y} = (0, 1)$, or might have probability mass function $e^{-1}/y!$ on $\mathcal{Y} = \{0, 1, \ldots\}$. Let $s(Y)$ be a function of Y, and let

When Y is discrete we interpret the integrals as sums over $y \in \mathcal{Y}$.

$$\mathcal{N} = \left\{ \theta : \kappa(\theta) = \log \int e^{s(y)\theta} f_0(y)\, dy < \infty \right\}$$

denote the values of θ for which the cumulant-generating function $\kappa(\theta)$ of $s(Y)$ is finite. Evidently $0 \in \mathcal{N}$. To avoid trivial cases we suppose that $\mathcal{N}$ has at least one other element and that $\text{var}\{s(Y)\} > 0$ under f_0, so $s(Y)$ is not a degenerate random variable. In fact the set $\mathcal{N}$ is convex, because if $\theta_1, \theta_2 \in \mathcal{N}$ and $\alpha \in [0, 1]$, then $\alpha\theta_1 + (1 - \alpha)\theta_2 \in \mathcal{N}$:

$$\int e^{s(y)\{\alpha\theta_1 + (1-\alpha)\theta_2\}} f_0(y)\, dy = \int \left\{ e^{s(y)\theta_1} \right\}^\alpha \left\{ e^{s(y)\theta_2} \right\}^{1-\alpha} f_0(y)\, dy$$
$$\leq \left\{ \int e^{s(y)\theta_1} f_0(y)\, dy \right\}^\alpha \left\{ \int e^{s(y)\theta_2} f_0(y)\, dy \right\}^{1-\alpha}$$
$$< \infty;$$

the second line follows from Hölder's inequality (Exercise 5.2.1). Moreover, as $\kappa\{\alpha\theta_1 + (1 - \alpha)\theta_2\} \leq \alpha\kappa(\theta_1) + (1 - \alpha)\kappa(\theta_2)$, the function $\kappa(\theta)$ is convex on the set $\mathcal{N}$. Equality occurs only if $\theta_1 = \theta_2$, so in fact $\kappa(\theta)$ is strictly convex.

A single fixed density f_0 is not flexible enough to be useful in practice, for which we need families of distributions. Hence we embed f_0 in the larger class

$$f(y; \theta) = \frac{e^{s(y)\theta} f_0(y)}{\int e^{s(x)\theta} f_0(x)\, dx}, \quad y \in \mathcal{Y}, \theta \in \mathcal{N},$$

by *exponential tilting*: f_0 has been tilted by multiplication by $e^{s(y)\theta}$ and then the resulting positive function has been renormalized to have unit integral. Evidently $f(y; \theta)$ has support $\mathcal{Y}$ for every θ. If $s(Y) = Y$, we have a *natural exponential family of order 1*,

$$f(y; \theta) = \exp\{y\theta - \kappa(\theta)\} f_0(y), \quad y \in \mathcal{Y}, \theta \in \mathcal{N}. \tag{5.7}$$

The family is called *regular* if the *natural parameter space* $\mathcal{N}$ is an open set.

Example 5.4 (Uniform density)   Let $f_0(y) = 1$ for $y \in \mathcal{Y} = (0, 1)$. Now

$$\kappa(\theta) = \log \int e^{y\theta} f_0(y)\, dy = \log \int_0^1 e^{y\theta}\, dy = \log\{(e^\theta - 1)/\theta\} < \infty$$

for all $\theta \in \mathcal{N} = (-\infty, \infty)$, and the natural exponential family

$$f(y; \theta) = \begin{cases} \theta e^{\theta y}/(e^\theta - 1), & 0 < y < 1, \\ 0, & \text{otherwise}, \end{cases} \tag{5.8}$$

is plotted in the left panel of Figure 5.3 for $\theta = -3, 0, 1$. For this or any natural exponential family with bounded $\mathcal{Y}$, $\mathcal{N} = (-\infty, \infty)$ and the family is regular.

168 5 · Models

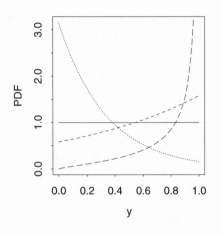

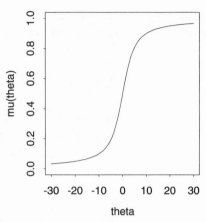

Figure 5.3 Exponential families generated by tilting the $U(0, 1)$ density. Left: original density (solid), natural exponential family when $\theta = -3$ (dots) and $\theta = 1$ (small dashes), and density generated when $s(y) = \log\{y/(1-y)\}$ when $\theta = 3/4$ (large dashes). Right: mean function $\mu(\theta)$ for the natural exponential family.

A different choice of $s(Y)$ will generate a different exponential family. With $s(Y) = \log\{Y/(1-Y)\}$, for example, the cumulant-generating function is given by

$$\int_0^1 e^{\theta \log\{y/(1-y)\}} \, dy = \int_0^1 y^{(1+\theta)-1}(1-y)^{(1-\theta)-1} \, dy$$
$$= B(1+\theta, 1-\theta)$$
$$= \frac{\Gamma(1+\theta)\Gamma(1-\theta)}{\Gamma(1+\theta+1-\theta)}, \quad |\theta| < 1,$$

For $a, b > 0$, $B(a, b) = \int_0^1 u^{a-1}(1-u)^{b-1} \, du$ is the beta function. It equals $\Gamma(a)\Gamma(b)/\Gamma(a+b)$, where $\Gamma(a) = \int_0^\infty u^{a-1}e^{-u} \, du$ is the gamma function; see Exercise 2.1.3.

and as $\Gamma(2) = 1$, we have $\kappa(\theta) = \log \Gamma(1+\theta) + \log \Gamma(1-\theta)$. Here the set $\mathcal{N} = (-1, 1)$ is open, so the resulting family is regular. Figure 5.3 shows how this family differs from the natural one, being unbounded unless $\theta = 0$. ∎

The natural exponential family of order 1 generated by a tilted version of f_0 is the same as that generated by f_0 itself. To see why, note that if $s(Y)$ has density (5.7) for some $\theta = \theta_1$, say, exponential tilting generates a density proportional to $\exp\{s(y)\theta\} \exp\{s(y)\theta_1 - \kappa(\theta_1)\}f_0(y)$ with cumulant-generating function $\kappa(\theta + \theta_1) - \kappa(\theta_1)$ for $\theta + \theta_1 \in \mathcal{N}$. The new density is $\exp\{s(y)(\theta + \theta_1) - \kappa(\theta + \theta_1)\}f_0(y)$, for $\theta + \theta_1 \in \mathcal{N}$. This is (5.7) apart from replacement of θ by $\theta + \theta_1$. Hence just one family is generated by a specific choice of f_0 and $s(Y)$, and this family is obtained by tilting any of its members.

For many purposes discussion of an exponential family is simplified if it is expressed without reference to a baseline density f_0. If a density may be written as

$$f(y; \omega) = \exp\{s(y)\theta(\omega) - b(\omega) + c(y)\}, \quad y \in \mathcal{Y}, \omega \in \Omega, \tag{5.9}$$

where $\mathcal{Y}$ is independent of the parameter ω and θ is a function of ω, it is said to be an *exponential family of order 1*. Here θ and s are called the *natural parameter* and *natural observation*.

Example 5.5 (Exponential density) The exponential density with mean ω is $f(y; \omega) = \omega^{-1} \exp(-y/\omega)$, for $y > 0$ and $\omega > 0$. Here $\Omega = \mathcal{Y} = (0, \infty)$, with natural observation and parameter $s(y) = y$ and $\theta(\omega) = -1/\omega$, and $b(\omega) = \log \omega$. The cumulant-generating function is $\kappa(\theta) = b\{\omega^{-1}(\theta)\} = -\log(-\theta)$, which has derivatives $(r-1)!(-1)^r\theta^{-r} = (r-1)!\omega^r$, the usual formula for cumulants of an exponential variable. ∎

Example 5.6 (Binomial density) If R is binomial with denominator m and probability $0 < \pi < 1$, its density is

$$\binom{m}{r}\pi^r(1-\pi)^{m-r} = \exp\left\{r\log\left(\frac{\pi}{1-\pi}\right) + m\log(1-\pi) + \log\binom{m}{r}\right\},$$

for $r \in \mathcal{Y} = \{0, 1, \ldots, m\}$. This has form (5.9) with $\omega = \pi$,

$$s(r) = r, \quad \theta(\pi) = \log\left(\frac{\pi}{1-\pi}\right), \quad b(\pi) = m\log(1-\pi), \quad c(r) = \log\binom{m}{r}.$$

The natural parameter is the *log odds* $\theta = \log\{\pi/(1-\pi)\} \in (-\infty, \infty)$. This family is regular, with cumulant-generating function $\kappa(\theta) = m\log(1 + e^\theta)$. ∎

If the function $\theta(\omega)$ in (5.9) is 1–1, the density of $S = s(Y)$ has form

$\theta(\Omega)$ denotes the set $\{\theta(\omega) : \omega \in \Omega\}$.

$$f(s; \theta) = \exp\left[s\theta - b\{\omega^{-1}(\theta)\}\right]h(s), \quad s \in s(\mathcal{Y}), \theta \in \theta(\Omega).$$

If $\Theta = \theta(\Omega) = \mathcal{N}$ for some baseline density f_0 then this is a natural exponential family with cumulant-generating function $\kappa(\theta) = b\{\omega^{-1}(\theta)\}$.

Expressed as a function of θ rather than ω, the moment-generating function of $s(Y)$ under (5.9) is, if finite,

$$\begin{aligned}
\mathrm{E}\left\{e^{ts(Y)}\right\} &= \int \exp\left\{ts(y) + \theta s(y) - \kappa(\theta) + c(y)\right\} dy \\
&= \exp\left\{\kappa(\theta+t) - \kappa(\theta)\right\} \int \exp\left\{(\theta+t)y - \kappa(\theta+t) + c(y)\right\} dy \\
&= \exp\left\{\kappa(\theta+t) - \kappa(\theta)\right\},
\end{aligned}$$

because the second integral equals unity; here $\theta = \theta(\omega)$ and $\kappa(\theta) = b\{\omega^{-1}(\theta)\}$. Hence when Y has density (5.9), the cumulant-generating function of $s(Y)$ is $\kappa(\theta+t) - \kappa(\theta)$. The cumulants result from differentiating $\kappa(\theta+t) - \kappa(\theta)$ with respect to t and then setting $t = 0$, or equivalently differentiating $\kappa(\theta)$ with respect to θ.

Mean parameter

Under (5.7) the cumulant-generating function of Y is $\kappa(\theta+t) - \kappa(\theta)$, so its mean and variance are

$$\mathrm{E}(Y) = \frac{d\kappa(\theta)}{d\theta} = \kappa'(\theta), \quad \mathrm{var}(Y) = \frac{d^2\kappa(\theta)}{d\theta^2} = \kappa''(\theta),$$

say. As Y is non-degenerate under f_0, $\mathrm{var}(Y) > 0$ for all $\theta \in \mathcal{N}$, and hence $\kappa'(\theta)$ is a strictly monotonic increasing function of θ. Thus there is a smooth 1–1 mapping between θ and the *mean parameter* $\mu = \mu(\theta) = \kappa'(\theta)$, and as θ varies in $\mathcal{N}$, μ varies in the *expectation space* $\mathcal{M}$.

The function $\mu(\theta)$ is important for likelihood inference. A natural exponential family is called *steep* if $|\mu(\theta_i)| \to \infty$ for any sequence $\{\theta_i\}$ in int $\mathcal{N}$ that converges to a boundary point of $\mathcal{N}$. Let us define the *closed convex hull* of $\mathcal{Y}$ to be $C(\mathcal{Y})$, the smallest closed set containing

The interior of a set, int $\mathcal{N}$, is what remains when its boundary is subtracted from its closure.

$$\{y : y = \alpha y_1 + (1 - \alpha)y_2, 0 \le \alpha \le 1, y_1, y_2 \in \mathcal{Y}\}.$$

Now $\mathcal{M} \subseteq C(\mathcal{Y})$, because every density (5.7) reweights elements of $\mathcal{Y}$. It can be shown that a regular natural exponential family is steep, and that for such a family, steepness is equivalent to $\mathcal{M} = \text{int } C(\mathcal{Y})$. Thus there is a duality between int $C(\mathcal{Y})$ and the expectation space $\mathcal{M}$, and hence between int $C(\mathcal{Y})$ and int $\mathcal{N}$: for every $\mu \in \text{int } C(\mathcal{Y})$ there is a unique $\theta \in \mathcal{N}$ such that $f(y; \theta)$ has mean μ. This equivalence applies widely because most natural exponential families are regular. As we shall see below, it implies that there is a unique maximum likelihood estimator of θ except for pathological samples.

Example 5.7 (Uniform density) The mean function for the natural exponential family generated by the $U(0, 1)$ density, $\mu(\theta) = (1 - e^{-\theta})^{-1} - \theta^{-1}$, is shown in the right panel of Figure 5.3. Here $\mathcal{Y} = (0, 1)$, so $C(\mathcal{Y}) = [0, 1]$ and int $C(\mathcal{Y}) = (0, 1) = \mathcal{M}$. The family is steep because the only boundary points of $\mathcal{N} = (-\infty, \infty)$ are $\pm\infty$, to which no sequence $\{\theta_i\} \subset \mathcal{N}$ can converge.

The family with $\Theta = [0, \infty)$ is not steep, because $\mu(\theta) \to 1/2$ as $\theta \downarrow 0$. ∎

Example 5.8 (Poisson density) If $\mathcal{Y} = \{0, 1, \ldots\}$ and $f_0(y) = e^{-1}/y!$, then

$$\kappa(\theta) = \log\left(\sum_{y=0}^{\infty} e^{\theta y - 1}/y!\right) = e^{\theta} - 1$$

is finite for all $\theta \in \mathcal{N} = (-\infty, \infty)$. Hence

$$f(y; \theta) = \exp(\theta y - e^{\theta})/y!, \quad y \in \mathcal{Y}, \ \theta \in \mathcal{N},$$

is a regular natural exponential family. Here $C(\mathcal{Y}) = [0, \infty)$, and the mean function is $\mu(\theta) = \kappa'(\theta) = e^{\theta}$, so $\mathcal{M} = (0, \infty) = \text{int } C(\mathcal{Y})$; the family is steep.

In terms of μ we have the familiar expression

$$f(y; \mu) = \exp(y \log \mu - \mu)/y! = \mu^y e^{-\mu}/y!, \quad y = 0, 1, \ldots, \mu > 0.$$

∎

Variance function

When Y has a natural exponential family density with cumulant-generating function $\kappa(\theta)$, its mean is $\mu(\theta) = \kappa'(\theta)$. Now $\kappa(\theta)$ is smooth and strictly convex, so the mapping between θ and $\mu = \mu(\theta) = \kappa'(\theta)$ is smooth and monotone. It follows that the density (5.7) can be reparametrized in terms of μ, setting $\theta = \theta(\mu)$. In terms of μ, $\kappa(\theta) = \kappa\{\theta(\mu)\}$, so

$$\text{var}(Y) = \kappa''(\theta) = \left.\frac{d\mu}{d\theta}\right|_{\theta=\theta(\mu)} = V(\mu), \quad \mu \in \mathcal{M},$$

say, where $V(\mu)$ is the *variance function* of the family. As we saw in Section 3.1.2, the variance function determines the variance-stabilizing transformation for Y. It plays a central role in generalized linear models, which we shall study in Section 10.3. The variance function and its domain $\mathcal{M}$ together determine their exponential family, as we shall now see.

On differentiating the identity $\mu\{\theta(\mu)\} = \mu$ with respect to μ, we obtain $\mu'\{\theta(\mu)\}d\theta/d\mu = 1$, and this implies that

$$\frac{d\theta(\mu)}{d\mu} = \frac{1}{\mu'\{\theta(\mu)\}} = \frac{1}{V(\mu)}. \tag{5.10}$$

As $\mathrm{var}(Y) > 0$, this derivative is finite for any $\mu \in \mathcal{M}$, so

$$\int_{\mu_0}^{\mu} \frac{1}{V(u)}\,du = \theta(\mu) - \theta(\mu_0),$$

and as $0 \in \mathcal{N}$ we can choose $\mu_0 \in \mathcal{M}$ to give $\theta(\mu_0) = 0$. Now

$$\kappa(\theta) = \int_0^{\theta} \kappa'(t)\,dt = \int_0^{\theta} \mu(t)\,dt = \int_{\mu_0}^{\mu} \mu\frac{dt}{d\mu}\,d\mu = \int_{\mu_0}^{\mu} \frac{u}{V(u)}\,du,$$

where we have used (5.10). Hence

$$\kappa\left\{\int_{\mu_0}^{\mu} \frac{1}{V(u)}\,du\right\} = \int_{\mu_0}^{\mu} \frac{u}{V(u)}\,du, \tag{5.11}$$

and given $\mathcal{M}$ and $V(\mu)$, we have expressed κ in terms of μ; this determines $\kappa(\theta)$ implicitly. The natural parameter space $\mathcal{N}$ is traced out by $\theta(\mu) = \int_{\mu_0}^{\mu} V(u)^{-1}\,du$ as μ varies in $\mathcal{M}$.

Example 5.9 (Linear variance function) Let Y be a random variable with $V(\mu) = \mu$ and $\mathcal{M} = (0, \infty)$. Then

$$\int_{\mu_0}^{\mu} \frac{1}{V(u)}\,du = \int_{\mu_0}^{\mu} \frac{du}{u} = \log(\mu/\mu_0), \qquad \int_{\mu_0}^{\mu} \frac{u}{V(u)}\,du = \mu - \mu_0,$$

and if $\mu_0 = 1$, (5.11) gives $\kappa(\log \mu) = \mu - 1$. On setting $\theta = \log \mu$, we have $\kappa(\theta) = e^{\theta} - 1$, and as μ varies in $\mathcal{M}$, $\theta = \log \mu$ varies in $(-\infty, \infty)$. As $e^{\theta} - 1$ is the cumulant-generating function of the Poisson density with mean e^{θ} and there is a 1–1 correspondence between cumulant-generating functions and distributions, Y is Poisson with mean $\mu = e^{\theta}$. ∎

5.2.2 Families of order p

To generalize the preceding discussion to models with several parameters, we again start from a base density $f_0(y)$, now supposing that its support $\mathcal{Y} \subseteq \mathbb{R}^d$, for $d \geq 1$, is not a subset of any space of dimension lower than d. Let the $p \times 1$ vector $s(y) = (s_1(y), \ldots, s_p(y))^{\mathrm{T}}$ consist of functions of y for which the set $\{1, s_1(y), \ldots, s_p(y)\}$ is linearly independent, and define

$$\mathcal{N} = \left\{\theta \in \mathbb{R}^p : \kappa(\theta) = \log \int e^{s(y)^{\mathrm{T}}\theta} f_0(y)\,dy < \infty\right\},$$

where $\theta = (\theta_1, \ldots, \theta_p)^T$. In general $\theta = \theta(\omega)$ may depend on a parameter ω taking values in $\Omega \subset \mathbb{R}^q$, where $\theta(\Omega) \subseteq \mathcal{N}$.

An *exponential family of order* p has density

$$f(y; \omega) = \exp \{s(y)^T \theta(\omega) - b(\omega)\} f_0(y), \quad y \in \mathcal{Y}, \omega \in \Omega, \qquad (5.12)$$

where $b(\omega) = \kappa\{\theta(\omega)\}$. This is called a *minimal representation* if the set $\{1, \theta_1(\omega), \ldots, \theta_p(\omega)\}$ is linearly independent. If there is a 1–1 mapping between $\mathcal{N}$ and Ω the family can be written as a *natural exponential family of order* p,

$$f(y; \omega) = \exp \{s(y)^T \theta - \kappa(\theta)\} f_0(y), \quad y \in \mathcal{Y}, \theta \in \mathcal{N}. \qquad (5.13)$$

Terms such as natural observation, natural parameter space, expectation space, regular model, and steep family generalize to families of order p and we shall use them below without further comment. Our proofs that the natural parameter space $\mathcal{N}$ is convex, that the family may be generated by any of its members, that $\kappa(\theta)$ is strictly convex, and that $s(Y)$ has cumulant-generating function $\kappa(\theta + t) - \kappa(\theta)$ also generalize with minor changes. The mean vector and covariance matrix of $s(Y)$ are now the $p \times 1$ vector and $p \times p$ matrix

$$E\{s(Y)\} = \frac{d\kappa(\theta)}{d\theta}, \quad \text{var}\{s(Y)\} = \frac{d^2\kappa(\theta)}{d\theta d\theta^T}.$$

Example 5.10 (Beta density) If $f_0(y)$ is uniform on $(0, 1)$ and $s(y)$ equals $(\log y, \log(1 - y))^T$, then

$$\kappa(\theta) = \log \int_0^1 \exp \{\theta_1 \log y + \theta_2 \log(1 - y)\} \, dy = \log B(1 + \theta_1, 1 + \theta_2),$$

where $B(a, b) = \Gamma(a)\Gamma(b)/\Gamma(a + b)$ is the beta function; see Example 5.4. The resulting model is usually written in terms of $a = \theta_1 + 1$ and $b = \theta_2 + 1$, giving the beta density

$$f(y; a, b) = \frac{y^{a-1}(1 - y)^{b-1}}{B(a, b)}, \quad 0 < y < 1, \quad a, b > 0. \qquad (5.14)$$

In this parametrization the natural parameter space is $\mathcal{N} = (0, \infty) \times (0, \infty)$. In Example 5.4 we took $s(y) = \log\{y/(1 - y)\}$, thereby generating the one-parameter subfamily in which $b = 2 - a$. This subfamily is also obtained by taking $s(y) = (\log y, \log(1 - y))^T$ and $\theta(\omega) = (\omega, -\omega)^T$, but this representation is not minimal because $(1, 1)\theta(\omega) = 0$.

Comparison of Figures 5.4 and 5.3 shows how tilting with two parameters broadens the variety of densities the family contains. ∎

Example 5.11 (von Mises density) Directional data are those where the observations y_j are angles — see Table 5.2, which gives the bearings of 29 homing pigeons 30, 60, and 90 seconds after release and on vanishing from sight. Another example is a wind direction, while the position of a star in the sky is an instance of directional data on a sphere.

Table 5.2 Homing
pigeon data (Artes, 1997).
Bearings (degrees) of 29
homing pigeons 30, 60
and 90 seconds after
release, with their
bearings on vanishing
from sight.

	1	2	3	4	5	6	7	8	9	10	11	12	13	14	15
30	240	300	225	285	210	265	310	330	325	290	15	330	100	35	340
60	250	290	210	325	205	240	330	315	285	335	10	305	95	65	345
90	270	305	215	295	195	210	335	315	135	10	5	325	90	70	330
van	275	285	185	290	195	225	335	285	120	30	10	85	90	80	350

	16	17	18	19	20	21	22	23	24	25	26	27	28	29
30	320	340	355	40	225	50	200	330	325	330	280	180	50	20
60	325	335	25	330	220	50	195	320	315	290	285	155	25	0
90	15	320	30	335	215	55	185	325	345	285	280	160	15	25
van	60	345	35	65	250	60	175	325	330	280	350	185	20	30

Figure 5.4 Beta
densities for different
values of a and b.
Swapping a and b reflects
the densities about
$y = 0.5$.

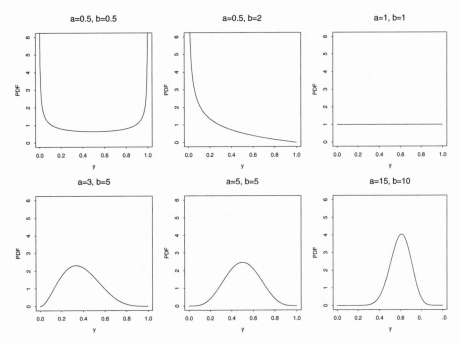

To build a class of densities for circular data we start from the uniform density on
the circle, $f_0(y) = (2\pi)^{-1}$ for $0 \le y < 2\pi$, and take

$$s(y) = (\cos y, \sin y)^{\mathrm{T}}, \quad \theta(\omega) = (\tau \cos \gamma, \tau \sin \gamma)^{\mathrm{T}},$$

where $\omega = (\tau, \gamma)$ lies in $\Omega = [0, \infty) \times [0, 2\pi)$. This choice of $s(y)$ ensures the desirable property $f(y) = f(y \pm 2k\pi)$ for all integer k. Now $s(y)^{\mathrm{T}}\theta(\omega) = \tau \cos(y - \gamma)$
and

$$\int e^{s(y)^{\mathrm{T}}\theta(\omega)} f_0(y) \, dy = \frac{1}{2\pi} \int_0^{2\pi} e^{\tau \cos(y-\gamma)} \, dy = \frac{1}{2\pi} \int_0^{2\pi} e^{\tau \cos y} \, dy = I_0(\tau),$$

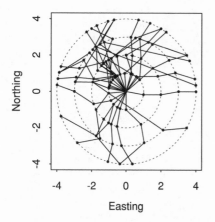

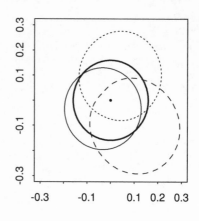

Northing

Easting

Figure 5.5 Circular data. Left: bearings of 29 homing pigeons at various intervals after release. Right: von Mises densities for different values of γ and τ. Shown are the baseline uniform density (heavy) $(2\pi)^{-1}$, and von Mises densities with $\tau = 0.3, \gamma = 5\pi/4$ (solid), $\tau = 0.7$, $\gamma = 3\pi/8$ (dots), and $\tau = 1, \gamma = 7\pi/4$ (dashes). In each case the density $f(y; \tau, \gamma)$ is given by the distance from the origin to the curve, so the areas do not integrate to one.

where $I_\nu(\tau)$ is the modified Bessel function of the first kind and order ν. The resulting exponential family is the von Mises density

$$f(y; \tau, \gamma) = \{2\pi I_0(\tau)\}^{-1} e^{\tau \cos(y-\gamma)}, \quad 0 \leq y < 2\pi, \ \tau > 0, 0 \leq \gamma < 2\pi;$$

see Figure 5.5. The *mean direction* γ gives the direction in which observations are concentrated, and the *precision* τ gives the strength of that concentration. Notice that $\tau = 0$ gives the uniform distribution on the circle, whatever the value of γ. Here interest focuses on Y rather than on $s(Y)$, which is introduced purely in order to generate a natural class of densities for y.

The estimates and standard errors for the data in Table 5.2 are $\hat{\gamma} = 320$ (15) and $\hat{\tau} = 1.08$ (0.32) at 30 seconds, with corresponding figures 316 (15) and 1.05 (0.32) at 60 seconds, 329 (21) and 0.75 (0.29) at 90 seconds, and 357 (29) and 0.52 (0.28) on vanishing. Thus as Figure 5.5 shows, the bearings of the pigeons become more dispersed as they fly away. The likelihood ratio statistics that compare the fitted two-parameter model with the uniform density are 13.80, 13.34, 7.33, and 3.75. As the mean direction γ vanishes under the uniform model, the situation is non-regular (Section 4.6), but the evidence against uniformity clearly weakens as time passes. ∎

Richard von Mises (1883–1953) was born in Lvov and educated in Vienna and Brno. He became professor of applied mathematics in Strasbourg, Dresden and Berlin, then left for Istanbul to escape the Nazis, finishing his career at Harvard. A man of wide interests, he spent the 1914–18 war as a pilot in the Austro-Hungarian army, gave the first university course on powered flight, and made contributions to aeronautics, aerodynamics and fluid dynamics as well as philosophy, probability and statistics; he was also an authority on the Austrian poet Rainer Maria Rilke. He is now perhaps best known for his frequency theory basis for probability.

Curved exponential families

In the examples above, the natural parameter $\theta = (\theta_1(\omega), \ldots, \theta_p(\omega))^{\mathrm{T}}$ is a 1–1 function of $\omega = (\omega_1, \ldots, \omega_q)^{\mathrm{T}}$, so of course $p = q$. Another possibility is that $q > p$, in which case ω cannot be identified from data. Such models are not useful in practice, and it is more interesting to consider the case $q < p$. Now $\theta(\omega)$ varies in the q-dimensional subspace $\theta(\Omega)$ of $\mathcal{N}$. If $\theta = a + B\omega$ is a linear function of ω, where a and B are a $p \times 1$ vector and a $p \times q$ matrix of constants, then $s(y)^{\mathrm{T}}\theta(\omega) = s(y)^{\mathrm{T}}a + \{s(y)^{\mathrm{T}}B\}\omega$, and the exponential family may be generated from $f_0'(y) \propto e^{a^{\mathrm{T}}s(y)} f_0(y)$ by taking $s'(y) = B^{\mathrm{T}}s(y)$. Hence it is just an exponential family of order q and no new issues arise: the original representation was not minimal. If $\theta(\omega)$ is a nonlinear function, however, and the representation is minimal, we have a (p, q) *curved exponential family*.

Example 5.12 (Multinomial density)   The multinomial density with denominator m and probability vector $\pi = (\pi_1, \ldots, \pi_p)^\mathsf{T}$ is

$$
\frac{m!}{y_1! \cdots y_p!} \pi_1^{y_1} \cdots \pi_p^{y_p} \propto \exp\{y_1 \log \pi_1 + \cdots + y_p \log \pi_p\}
$$

$$
= \exp\{y_1 \log \pi_1 + \cdots + y_{p-1} \log \pi_{p-1}
$$

$$
+ (m - y_1 - \cdots - y_{p-1}) \log(1 - \pi_1 - \cdots - \pi_{p-1})\}
$$

$$
= \exp\{y_1 \theta_1 + \cdots + y_{p-1} \theta_{p-1} - \kappa(\theta)\},
$$

where

$$
\pi_r = \frac{e^{\theta_r}}{1 + e^{\theta_1} + \cdots + e^{\theta_{p-1}}}, \quad \kappa(\theta) = m \log\left(1 + e^{\theta_1} + \cdots + e^{\theta_{p-1}}\right).
$$

This is a minimal representation of a natural exponential family of order $p - 1$ with $s(y) = (y_1, \ldots, y_{p-1})^\mathsf{T}$, $\mathcal{N} = (-\infty, \infty)^{p-1}$ and

$$
f_0(y) = \frac{p^{-m} m!}{y_1! \cdots y_p!}, \quad \mathcal{Y} = \left\{(y_1, \ldots, y_p) : y_1, \ldots, y_p \in \{0, \ldots, m\}, \sum y_r = m\right\};
$$

$\mathcal{Y}$ is a subset of the scaled p-dimensional simplex

$$
C(\mathcal{Y}) = \left\{(y_1, \ldots, y_p) : 0 \leq y_1, \ldots, y_p \leq m, \sum y_r = m\right\}.
$$

Now

$$
\mathrm{E}\{s(Y)\} = \frac{m}{1 + e^{\theta_1} + \cdots + e^{\theta_{p-1}}}(e^{\theta_1}, \ldots, e^{\theta_{p-1}}),
$$

and as $\mathrm{E}(Y_p) = m - \mathrm{E}(Y_1) - \cdots - \mathrm{E}(Y_{p-1})$, the expectation space in which $\mu(\theta) = \mathrm{E}(Y)$ varies equals int $C(\mathcal{Y})$: the model is steep.

Many multinomial models are curved exponential families. In Example 4.38, for instance, the ABO blood group data had $p = 4$ groups with

$$
\pi_A = \lambda_A^2 + 2\lambda_A \lambda_O, \quad \pi_B = \lambda_B^2 + 2\lambda_B \lambda_O, \quad \pi_O = \lambda_O^2, \quad \pi_{AB} = 2\lambda_A \lambda_B, \quad (5.15)
$$

where $\lambda_A + \lambda_B + \lambda_O = 1$. This is a $(3, 2)$ curved exponential family. In the full family of order p, the probabilities π_A, π_B and π_{AB} vary in the set

$$
\mathcal{A} = \{(\pi_A, \pi_B, \pi_{AB}) : 0 \leq \pi_A, \pi_B, \pi_{AB} \leq 1, 0 \leq \pi_A + \pi_b + \pi_{AB} \leq 1\},
$$

shown in Figure 5.6. In the sub-family given by (5.15), when λ_O is fixed we have $\lambda_A + \lambda_B = 1 - \lambda_O$, and as λ_A varies from 0 to $1 - \lambda_O$, (π_A, π_B, π_{AB}) traces a curve from $(0, 1 - \lambda_O^2, 0)$ to $(1 - \lambda_O^2, 0, 0)$ shown in the figure. As λ_O varies from 0 to 1,

$$
(\pi_A, \pi_B, \pi_{AB}) = \big(\lambda_A^2 + 2p\lambda_O, (1 - \lambda_A - \lambda_O)^2 + 2(1 - \lambda_A - \lambda_O)\lambda_O,
$$

$$
2\lambda_A(1 - \lambda_A - \lambda_O)\big)
$$

traces out the intersection of a cone with the set $\mathcal{A}$. Thus although any value of (π_A, π_B, π_{AB}) inside the tetrahedron with corners $(0, 0, 0)$, $(0, 0, 1)$, $(0, 1, 0)$ and $(1, 0, 0)$ is possible under the full model, the curved submodel restricts the probabilities to the hatched surface. ∎

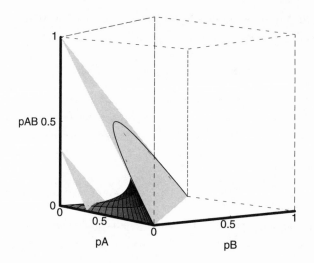

Figure 5.6 Parameter space for four-category multinomial model. The full parameter space for (π_A, π_B, π_{AB}) is the tetrahedron with corners $(0, 0, 0), (0, 0, 1), (0, 1, 0)$ and $(1, 0, 0)$, whose outer face is shaded. The other parameter $\pi_O = 1 - \pi_A - \pi_B - \pi_{AB}$. The two-parameter sub-model given by (5.15) is shown by the hatched surface.

5.2.3 Inference

Let $Y_1, \ldots, Y_n$ be a random sample from an exponential family of order p. Their joint density is

$$\prod_{j=1}^{n} f(y_j; \omega) = \exp\left\{\sum_{j=1}^{n} s(y_j)^{\mathsf{T}}\theta(\omega) - nb(\omega)\right\} \prod_{j=1}^{n} f_0(y_j), \quad \omega \in \Omega, \qquad (5.16)$$

and consequently the density of $S = \sum s(Y_j)$ is

$$f(s; \omega) = \int \prod_{j=1}^{n} f(y_j; \omega)\, dy = \exp\left\{s^{\mathsf{T}}\theta(\omega) - nb(\omega)\right\} \int \prod_{j=1}^{n} f_0(y_j)\, dy$$
$$= \exp\{s^{\mathsf{T}}\theta(\omega) - nb(\omega)\}g_0(s),$$

say, where the integral is over

$$\left\{(y_1, \ldots, y_n) : y_1, \ldots, y_n \in \mathcal{Y}, \sum_{j=1}^{n} s(y_j) = s\right\}.$$

Hence S too has an exponential family density of order p. That is, the sum of n independent variables from an exponential family belongs to the same family, with cumulant-generating function $n\kappa(\theta) = nb(\omega)$. The factorization criterion (4.15) applied to (5.16) implies that S is a sufficient statistic for ω based on $Y_1, \ldots, Y_n$, and if $f(y; \omega)$ is a minimal representation, S is minimal sufficient (Exercise 5.2.12). Thus inference for ω may be based on the density of S, while the joint density of $Y_1, \ldots, Y_n$ given the value of S is independent of ω:

$$f(y_1, \ldots, y_n; \omega) = f(y_1, \ldots, y_n \mid s)f(s; \omega). \qquad (5.17)$$

This decomposition allows us to split the inference into two parts, corresponding to the factors on its right, the first of which may be used to assess model adequacy. If satisfied of an adequate fit, we use the second term for inference on ω. We now discuss these aspects in turn.

Model adequacy

The argument for using the first factor on the right of (5.17) to assess model adequacy is that the value of ω is irrelevant to deciding if $f(y; \omega)$ fits the random sample $Y_1, \ldots, Y_n$. Hence we should assess fit using the conditional distribution of Y given S; see Example 4.10.

Example 5.13 (Poisson density) If $Y_1, \ldots, Y_n$ is a random sample from a Poisson density with mean μ, their common cumulant-generating function is $\mu(e^t - 1)$ and the natural observation is $s(y_j) = y_j$. Hence $S = \sum s(Y_j) = \sum Y_j$ has cumulant-generating function $n\mu(e^t - 1)$. The joint conditional density of $y_1, \ldots, y_n$ given that $S = s$,

$$
\begin{aligned}
f(y_1, \ldots, y_n \mid s) &= \frac{f(y_1, \ldots, y_n; \theta)}{f(s; \theta)} \\[2mm]
&= \frac{\prod_{j=1}^n \mu^{y_j} e^{-\mu} / y_j!}{(n\mu)^s e^{-n\mu} / s!} \\[2mm]
&= \begin{cases} \frac{s!}{y_1! \cdots y_n!} n^{-s}, & y_1 + \cdots + y_n = s, \\ 0, & \text{otherwise,} \end{cases}
\end{aligned}
$$

is multinomial with denominator s and $n \times 1$ probability vector $(n^{-1}, \ldots, n^{-1})$. This density is independent of μ by its construction.

The mean and variance of a Poisson variable both equal μ, so Poissonness of a random sample of counts can be assessed by comparing their average $\overline{Y}$ and sample variance $(n - 1)^{-1} \sum (Y_j - \overline{Y})^2$. A common problem with such data is overdispersion, which is suggested if $P = \sum (Y_j - \overline{Y})^2 / \overline{Y}$ greatly exceeds $n - 1$. How big is 'greatly'? As $\widehat{\mu} = \overline{Y}$ is the maximum likelihood estimate of μ, P is Pearson's statistic (Section 4.5.3) and has an asymptotic χ^2_{n-1} distribution. The argument above suggests that we assess if P is large compared to its conditional distribution given the value of $S = \sum Y_j = n\overline{Y}$, so the distribution we seek is that of P conditional on $\overline{Y}$. The conditional mean and variance of P are $(n - 1)$ and $2(n - 1)(1 - s^{-1}) \doteq 2(n - 1)$, and the conditional distribution of P is very close to χ^2_{n-1} unless s and n are both very small. Hence the *Poisson dispersion test* compares P to the χ^2_{n-1} distribution, with large values suggesting that the counts are more variable than Poisson data would be.

In Table 2.1, for example, the daily numbers of arrivals are 16, 16, 13, 11, 14, 13, 12, so P takes value 1.6, to be treated as χ^2_6, so the counts seem under- rather than overdispersed. In Example 4.40, by contrast, with counts 1, 5, 3, 2, 2, 1, 0, 0, 2, 1, 1, 7, 11, 4, 7, 10, 16, 16, 9, 15, we have $P = 99.92$, which is very large compared to the χ^2_{19} distribution; and in fact $\Pr(P \geq 99.92) \doteq 0$ to 12 decimal places. As one might expect, these data are highly overdispersed relative to the Poisson model.

Another possibility is that although all Poisson, the Y_j have different means. In Example 4.40 we compared the changepoint model under which $Y_1, \ldots, Y_\tau$ and $Y_{\tau+1}, \ldots, Y_n$ have different means with the model of equal means. The comparison involved the likelihood ratio statistic, whose exact conditional distribution was simulated under the simpler model; see Figure 4.9. ∎

Example 5.14 (Normal model)   The normal density may be written

$$f(y; \mu, \sigma^2) = \frac{1}{(2\pi)^{1/2}\sigma} \exp\left\{-\frac{1}{2\sigma^2}(y-\mu)^2\right\}$$

$$= \exp\left\{\frac{\mu}{\sigma^2}y - \frac{1}{2\sigma^2}y^2 - \frac{\mu^2}{2\sigma^2} - \log\sigma - \frac{1}{2}\log(2\pi)\right\}. \quad (5.18)$$

This is a minimal representation of an exponential family of order 2 with

$$\omega = (\mu, \sigma^2) \in \Omega = (-\infty, \infty) \times (0, \infty),$$
$$\theta(\omega)^{\mathrm{T}} = (\mu/\sigma^2, 1/(2\sigma^2)) \in \mathcal{N} = (-\infty, \infty) \times (0, \infty),$$
$$s(y)^{\mathrm{T}} = (y, -y^2),$$
$$\kappa(\theta) = \theta_1^2/(4\theta_2) - \frac{1}{2}\log(2\theta_2),$$

arising from tilting the standard normal density $(2\pi)^{-1/2}e^{-y^2/2}$.

We now consider how decomposition (5.17) applies for the normal model with $n > 2$. When $Y_1, \ldots, Y_n$ is a random sample from (5.18), our general discussion implies that $(\sum Y_j, -\sum Y_j^2)$ is minimal sufficient. As this is in 1–1 correspondence with $\overline{Y}$, $S^2 = (n-1)^{-1}\sum(Y_j - \overline{Y})^2$, our old friends the average and sample variance are also minimal sufficient. When $n > 1$ the joint distribution of $\overline{Y}$ and S^2 is nondegenerate with probability one, and (3.15) states that they are independently distributed as $N(\mu, \sigma^2/n)$ and $(n-1)^{-1}\sigma^2\chi_{n-1}^2$.

In order to compute the conditional density of $Y_1, \ldots, Y_n$ given $\overline{Y}$ and S, it is neatest to set $E_j = (Y_j - \overline{Y})/S$ and consider the conditional density of $E_1, \ldots, E_n$. As $\sum E_j = 0$ and $\sum E_j^2 = n - 1$, the random vector $(E_1, \ldots, E_n) \in \mathbb{R}^n$ lies on the intersection of the hypersphere of radius $n - 1$ and the hyperplane $\sum E_j = 0$. As this is a $(n-2)$-dimensional subset of $\mathbb{R}^n$, the joint density of $E_1, \ldots, E_n$ is degenerate but that of $E_3, \ldots, E_n$ is not.

To find the joint density of $T_3 = E_3, \ldots, T_n = E_n$ given $T_1 = \overline{Y}$ and $T_2 = S$, we need the Jacobian of the transformation from $y_1, \ldots, y_n$ to $t_1, \ldots, t_n$. In order to obtain this Jacobian, we first note that $y_j = t_1 + t_2 t_j$, for $j = 3, \ldots, n$. As $\sum e_j = 0$ and $\sum e_j^2 = n - 1$, we can write

$$e_1 + e_2 = -\sum_{j=3}^{n} t_j, \quad n - 1 - e_1^2 - e_2^2 = \sum_{j=3}^{n} t_j^2,$$

implying that there are functions h_1 and h_2 such that

$$e_1 = h_1(t_3, \ldots, t_n), \quad e_2 = h_2(t_3, \ldots, t_n),$$

which in turn gives

$$y_1 = t_1 + t_2 h_1(t_3, \ldots, t_n), \quad y_2 = t_1 + t_2 h_2(t_3, \ldots, t_n).$$

Let $h_{ij} = \partial h_i(t_3, \ldots, t_n)/\partial t_j$. The Jacobian we seek is

$$
\left| \frac{\partial(y_1, \ldots, y_n)}{\partial(t_1, \ldots, t_n)} \right| =
\begin{vmatrix}
1 & h_1 & t_2 h_{13} & t_2 h_{14} & \cdots & t_2 h_{1n} \\
1 & h_2 & t_2 h_{23} & t_2 h_{24} & \cdots & t_2 h_{2n} \\
1 & t_3 & t_2 & 0 & \cdots & 0 \\
1 & t_4 & 0 & t_2 & \cdots & 0 \\
\vdots & \vdots & \vdots & \vdots & \ddots & \vdots \\
1 & t_n & 0 & 0 & \cdots & t_2
\end{vmatrix}
= t_2^{n-2} h'(t_3, \ldots, t_n)
$$

$$
= s^{n-2} H(e), \qquad (5.19)
$$

say. Hence

$$
f(e_3, \ldots, e_n \mid \overline{y}, s) = \frac{f(y_1, \ldots, y_n; \mu, \sigma^2) s^{n-2} H(e)}{f(\overline{y}; \mu, \sigma^2) f(s; \sigma^2)} \propto H(e)
$$

after a straightforward calculation. As this depends on $e_1, \ldots, e_n$ alone, the corresponding random variables $E_1, \ldots, E_n$ are independent of $\overline{Y}$ and S^2.

Thus assessment of fit of the normal model should be based on the *raw residuals* $e_1, \ldots, e_n$. One simple tool is a normal probability plot of the e_j, which should be a straight line of unit gradient through the origin. Such plots and variants are common in regression (Section 8.6.1). Further support for use of the e_j for model checking is given in Section 5.3. ∎

Likelihood

Let $Y_1, \ldots, Y_n$ be a random sample from an exponential family of order p. Inference for the parameter may be based on the sufficient statistic $\overline{S} = n^{-1} \sum s(Y_j)$, which also belongs to a natural exponential family of order p, with support $\mathcal{S}$, say. Hence the log likelihood may be written

$$
\ell(\omega) \equiv n \{ \overline{S}^{\mathsf{T}} \theta(\omega) - b(\omega) \} = n[\overline{S}^{\mathsf{T}} \theta(\omega) - \kappa \{\theta(\omega)\}], \quad \omega \in \Omega,
$$

and the score vector and observed information matrix are given by

$$
U(\omega) = \frac{\partial \ell(\omega)}{\partial \omega} = \frac{\partial \theta^{\mathsf{T}}}{\partial \omega} n \left\{ \overline{S} - \frac{\partial \kappa(\theta)}{\partial \theta} \right\},
$$

$$
J(\omega)_{rs} = -\frac{\partial^2 \ell(\omega)}{\partial \omega_r \partial \omega_s} = -\frac{\partial^2 \theta^{\mathsf{T}}}{\partial \omega_r \partial \omega_s} n \left\{ \overline{S} - \frac{\partial \kappa(\theta)}{\partial \theta} \right\} + \frac{\partial \theta^{\mathsf{T}}}{\partial \omega_r} \left\{ n \frac{\partial^2 \kappa(\theta)}{\partial \theta \partial \theta^{\mathsf{T}}} \right\} \frac{\partial \theta}{\partial \omega_s}.
$$

The observed information is random unless the family is in natural form, in which case $\theta = \omega$ and hence $\partial^2 \theta / \partial \omega_r \partial \omega_s = 0$; then $I(\theta) = \mathrm{E}\{J(\theta)\} = J(\theta)$.

If the family is steep, there is a 1–1 relation between the interior of the closure of $\mathcal{S}$, int $C(\mathcal{S})$, the expectation space $\mathcal{M}$ of $\overline{S}$, and the natural parameter space $\mathcal{N} = \theta(\Omega)$. Thus if $\overline{S} \in$ int $C(\mathcal{S})$, there is a single value of θ such that $\overline{S} = \mu(\theta)$ and $u(\theta) = 0$, and moreover there is a 1–1 map between $\widehat{\theta}$ and $\widehat{\omega}$. Hence the maximum likelihood estimators satisfy

$$
\widehat{\mu} = \mu(\widehat{\theta}) = \mu\{\theta(\widehat{\omega})\} = \overline{S}.
$$

Thus the likelihood equation has just one solution, which maximizes the log likelihood. Moreover, as Ω is open and $\widehat{\omega} \in \Omega$, standard likelihood asymptotics will apply, so $\widehat{\omega} \;\dot{\sim}\; N\{\omega, I(\omega)^{-1}\}$ and $2\{\ell(\widehat{\omega}) - \ell(\omega)\} \;\dot{\sim}\; \chi_p^2$. If the model permits $\overline{S} \notin \mathcal{M}$, standard asymptotics will break down. The same difficulty could arise if the true parameter lies on the boundary of the parameter space.

Example 5.15 (Uniform density) The average $\overline{y}$ of a random sample from (5.8) must lie in the interval $(0, 1)$. Given $\overline{y}$, the maximum likelihood estimate $\widehat{\theta}$ is read off from the right panel of Figure 5.3 as the value of θ on the horizontal axis for which $\mu(\theta) = \overline{y}$ on the vertical axis.

As mentioned in Example 5.7, when θ is restricted to $\Theta = [0, \infty)$ the family is not steep, because $\mathcal{M} = [1/2, 1) \neq (0, 1) = \text{int } C(\mathcal{Y})$. A value $\overline{y} < 1/2$ is possible for any sample size and any $\theta \in \Theta$, and as $\widehat{\theta} = 0$ is the maximum likelihood estimate for any such $\overline{y}$, the 1–1 mapping between $\overline{y}$ and $\widehat{\theta}$ is destroyed. Furthermore, this Θ is not open, so the limiting distribution of $\widehat{\theta}$ and the likelihood ratio statistic are non-standard if $\theta = 0$; see Example 4.39. ∎

Example 5.16 (Binomial density) The binomial model with denominator m, probability $0 < \pi < 1$ and natural parameter $\theta = \log\{\pi/(1 - \pi)\} \in (-\infty, \infty)$ has $\mathcal{Y} = \{0, 1, \ldots, m\}$ and $\text{int } C(\mathcal{Y}) = \mathcal{M} = (0, m)$. The average $\overline{R}$ of a random sample $R_1, \ldots, R_n$ lies outside $(0, m)$ with probability

$$\Pr(R_1 = \cdots = R_n = 0) + \Pr(R_1 = \cdots = R_n = m) = (1 - \pi)^{mn} + \pi^{mn} > 0,$$

so the maximum likelihood estimator $\widehat{\theta} = \log\left\{\overline{R}/(m - \overline{R})\right\}$ may not be finite. As the family is steep, a unique value of θ corresponds to each $\overline{R} \in \mathcal{M}$, so the only problem that can arise is that $\widehat{\theta} = \pm\infty$ with small probability. On the other hand $\Pr(|\widehat{\theta}| = \infty) \to 0$ exponentially fast as $n \to \infty$, so infinite $\widehat{\theta}$ is rare in practice, though not unknown. It corresponds to $\widehat{\pi} = 0$ or $\widehat{\pi} = 1$.

This difficulty also arises with other discrete exponential families. ∎

Example 5.17 (Normal density) Example 4.18 gives the score and information quantities for a sample from the normal model in terms of μ and σ^2; in this parametrization the observed information is random. In Example 4.22 we saw that the log likelihood $\ell(\mu, \sigma^2)$ is unimodal and that the maximum likelihood estimators are the sole solution to the likelihood equation; this is an instance of the general result above. ∎

Derived densities

Various models derived from exponential families are themselves exponential families, and this can be useful in inference.

Consider a natural exponential family of order p with S^T and θ^T partitioned as (S_1^T, S_2^T) and (ψ^T, λ^T), where S_1 and ψ have dimension $q < p$. The marginal density of S_2, obtained by integration over the values of S_1, is

$$f(s_2; \theta) = \int \exp \left\{ s_1^{\mathsf{T}} \psi + s_2^{\mathsf{T}} \lambda - \kappa(\theta) \right\} g_0(s_1, s_2) \, ds_1$$

$$= \exp \left\{ s_2^{\mathsf{T}} \lambda - \kappa(\theta) \right\} \int \exp \left(s_1^{\mathsf{T}} \psi \right) g_0(s_1, s_2) \, ds_1$$

$$= \exp \left\{ s_2^{\mathsf{T}} \lambda - \kappa(\theta) + d_\psi(s_2) \right\},$$

say, so for fixed ψ the marginal density of S_2 is an exponential family with natural parameter λ.

The conditional density of S_1 given $S_2 = s_2$ is

$$f_{S_1 | S_2}(s_1 \mid s_2; \theta) = \frac{\exp \left\{ s_1^{\mathsf{T}} \psi + s_2^{\mathsf{T}} \lambda - \kappa(\theta) \right\} g_0(s_1, s_2)}{\exp \left\{ s_2^{\mathsf{T}} \lambda - \kappa(\theta) + d_\psi(s_2) \right\}}$$

$$= \exp \left\{ s_1^{\mathsf{T}} \psi - \kappa_{s_2}(\psi) \right\} g_{s_2}(s_1),$$

say. This is an exponential family of order q with natural parameter ψ, but the base density and cumulant-generating function depend on s_2. Such a removal of λ by conditioning is a powerful way to deal with nuisance parameters.

Example 5.18 (Gamma density) Independent gamma variables $Y_1, \ldots, Y_n$ with scale parameter λ and shape parameters $\kappa_1, \ldots, \kappa_n$ have joint density

$$\prod_{j=1}^{n} \frac{\lambda^{\kappa_j} y_j^{\kappa_j - 1}}{\Gamma(\kappa_j)} \exp(-\lambda y_j) = \lambda^{\sum \kappa_j} \exp \left(-\lambda \sum_{j=1}^{n} y_j \right) \prod_{j=1}^{n} \frac{y_j^{\kappa_j - 1}}{\Gamma(\kappa_j)}.$$

As Y_j has cumulant-generating function $-\kappa_j \log(1 - \lambda t)$, $S_1 = S = \sum Y_j$ is gamma with parameters λ and $\sum \kappa_j$. The conditional density of $Y_1, \ldots, Y_n$ given $S = s$ is

$$\frac{\Gamma\left(\sum \kappa_j \right)}{\prod_{j=1}^{n} \Gamma(\kappa_j)} s^{-n} \prod_{j=1}^{n} \left(\frac{y_j}{s} \right)^{\kappa_j - 1}, \quad y_j > 0, \sum_{j=1}^{n} y_j = s.$$

Thus the joint density of $U_1 = Y_1/S, \ldots, U_n = Y_n/S$,

$$f(u_1, \ldots, u_n; \kappa_1, \ldots, \kappa_n) = \frac{\Gamma\left(\sum \kappa_j \right)}{\prod_{j=1}^{n} \Gamma(\kappa_j)} \prod_{j=1}^{n} u_j^{\kappa_j - 1}, \quad u_j > 0, \sum_{j=1}^{n} u_j = 1, \quad (5.20)$$

lies on the simplex in n dimensions; it is called the *Dirichlet density*. Hence we may base inferences for $\kappa_1, \ldots, \kappa_n$ on the conditional density of $Y_1, \ldots, Y_n$ given their sum, or equivalently on the observed values of the U_j. ∎

The discussion above suggests that we may write

$$f(s; \theta) = f_{S_1 | S_2}(s_1 \mid s_2; \psi) f_{S_2}(s_2; \psi, \lambda). \quad (5.21)$$

If the model can be reparametrized in terms of a $(p - q) \times 1$ vector $\rho = \rho(\psi, \lambda)$ which is variation independent of ψ, in such a way that the second term on the right of (5.21) depends only on ρ, then S_2 is said to be a *cut*. The log likelihood based on (5.21) then has form $\ell_1(\psi) + \ell_2(\rho)$, maximum likelihood estimates of ρ and ψ do not depend on each other, and the observed information matrix is block diagonal. Inferences on ψ and ρ may be made separately, using the conditional density of S_1 given S_2 and the marginal density of S_2. The cut most commonly encountered in practice arises with Poisson variables; see Example 7.34 and page 501.

Exercises 5.2

1   Here is a version of Hölder's inequality: let $f(x)$ be a density supported in $[a, b]$, let $p > 1$, and let $g(y)$ and $h(y)$ be any two real functions such that the integrals

$$\int_a^b |g(y)|^p f(y)\, dy, \quad \int_a^b |h(y)|^q f(y)\, dy,$$

are finite, where $p^{-1} + q^{-1} = 1$. Then

$$\int g(y)h(y)f(y)\, dy \le \left\{ \int_a^b |g(y)|^p f(y)\, dy \right\}^{1/p} \left\{ \int_a^b |h(y)|^q f(y)\, dy \right\}^{1/q}.$$

If g and h are both non-zero, there is equality if and only if $c|g(y)|^p = d|h(y)|^q$ for positive constants c and d.
    Show strict convexity of the cumulant-generating function $\kappa(\theta)$ of an exponential family.

2   What natural exponential families are generated by (a) $f_0(y) = e^{-y}, y > 0$, and (b) $f_0(y) = \frac{1}{2}e^{-|y|}, -\infty < y < \infty$?

3   Which of Examples 4.1–4.6 are exponential families? What about the $U(0, \theta)$ density?

4   Show that the gamma density (2.7) is an exponential family. What about the inverse gamma density, for $1/Y$ when Y is gamma?

5   Show that the inverse Gaussian density

$$f(y; \mu, \lambda) = \left(\frac{\lambda}{2\pi y^3} \right)^{1/2} \exp\{-\lambda(y - \mu)^2/(2\mu^2 y)\}, \quad y > 0, \lambda, \mu > 0,$$

is an exponential family of order 2. Give a general form for its cumulants.

6   Find the exponential families with variance functions (i) $V(\mu) = a\mu(1 - \mu), \mathcal{M} = (0, 1)$, (ii) $V(\mu) = a\mu^2, \mathcal{M} = (0, \infty)$, and (iii) $V(\mu) = a\mu^2, \mathcal{M} = (-\infty, 0)$.

7   For what values of a is there an exponential family with variance function $V(\mu) = a\mu$, $\mathcal{M} = (0, \infty)$?

8   Show that the $N(\mu, \mu^2)$ model is a curved exponential family and sketch how the density changes as μ varies in $(-\infty, 0) \cup (0, \infty)$. Sketch also the subset of the natural parameter space for the $N(\mu, \sigma^2)$ distribution generated by this model.

9   Find a connection between Example 4.11 and (5.20), and hence suggest methods of checking the fit of the exponential model.

10 Explain how (5.20) may be generated as an exponential family, by showing that it generalizes (5.14).

11 Use Example 5.18 to construct a simulation algorithm for Dirichlet random variables.

12 Show that $\sum s(Y_j)$ is minimal sufficient for the parameter ω of an exponential family of order p in a minimal representation.

5.3 Group Transformation Models

Another important class of models stems from observing that many inferences should have invariance properties. If, for instance, data y are recorded in degrees Celsius, one might obtain a conclusion $s(y)$ directly from the original data, or one might transform them to degrees Fahrenheit, giving $g(y)$, say, obtain the conclusion $s\{g(y)\}$ in these terms, and then back-transform to Celsius scale, giving conclusion $g^{-1}[s\{g(y)\}]$. It is clearly essential that $g^{-1}[s\{g(y)\}] = s(y)$. The transformation from Celsius to Fahrenheit is just one of many possible invertible linear transformations that might be applied to y, however, any of which should leave the inference unchanged. More generally we might insist that inferences be invariant when any element g of a group of transformations acts on the sample space. This section explores some consequences of this requirement.

A *group* $\mathcal{G}$ is a mathematical structure having an operation $\circ$ such that:

- if $g, g' \in \mathcal{G}$, then $g \circ g' \in \mathcal{G}$;
- $\mathcal{G}$ contains an identity element e such that $e \circ g = g \circ e = g$ for each $g \in \mathcal{G}$; and
- each $g \in \mathcal{G}$ possesses an inverse $g^{-1} \in \mathcal{G}$ such that $g \circ g^{-1} = g^{-1} \circ g = e$.

A subgroup is a subset of $\mathcal{G}$ that is also a group.

A *group action* arises when elements of a group act on those of a set $\mathcal{Y}$. In the present case the group elements g_θ typically correspond to elements of a parameter space Θ and $\mathcal{Y}$ is the sample space of a random variable Y. The action of g on y, $g(y)$, say, is defined for each $y \in \mathcal{Y}$ and $g(y)$ is an element of $\mathcal{Y}$ for each $g \in \mathcal{G}$.

Setting $y \approx y'$ if and only if there is a $g \in \mathcal{G}$ such that $y = g(y')$ gives an equivalence relation, which partitions $\mathcal{Y}$ into equivalence classes called *orbits* and labelled by an index a, say. Each y belongs to precisely one orbit, and can be represented by a and its position on the orbit. Hence we can write $y = g(a)$ for some $g \in \mathcal{G}$. If this representation is unique for a given choice of index, the group action is said to be *free*.

Example 5.19 (Location model) Let $Y = \theta + \varepsilon$, where $\theta \in \Theta = \mathbb{R}$ and ε is a scalar random variable with known density $f(y)$, where $y \in \mathbb{R}$. The density of Y is $f(y - \theta) = f(y; \theta)$, say, and that of $\theta' + Y = \theta' + \theta + \varepsilon$ is $f(y; \theta + \theta')$. Thus adding θ' to Y changes the parameter of the density. Taking $\theta' = -\theta$ gives the baseline density $f(y; 0) = f(y)$ of ε.

Here group elements may be written g_θ, corresponding to the parameters θ, and the group operation is equivalent to addition. Hence $g_\theta \circ g_{\theta'} = g_{\theta + \theta'}$, the identity e is g_0 and the inverse of g_θ is $g_{-\theta}$. Each element of the group corresponds to a point in Θ, but it induces a group action $g_\theta(y) = \theta + y$ on the sample space.

For a random sample $Y_1, \ldots, Y_n$, we take $\mathcal{Y} = \mathbb{R}^n$ and interpret expressions such as $g_\theta(Y) = \theta + Y$ as vectors, with $\theta \equiv \theta 1_n$ and $Y = (Y_1, \ldots, Y_n)^\mathsf{T}$. Then y and y' belong to the same orbit if there exists a g_θ such that $g_\theta(y) = y'$, that is, there exists a θ such that $\theta + y = y'$, and this implies that y' is a location shift of y. On taking $\theta = \overline{y}' - \overline{y}$ we see that $y - \overline{y} = y' - \overline{y}'$, implying that we can represent the orbit by

1_n is the $n \times 1$ vector of ones.

the vector $a(y) = y - \bar{y}$, because this choice of index gives $a(y) = a(y')$. Thus y is equivalently written as $(y - \bar{y}, \bar{y})$, where the first term indexes the orbit and the second the position of y within it. In terms of this representation we write y as $g_{\bar{y}}(a) = \bar{y} + a = \bar{y} + y - \bar{y} = y$. The group action is free because $g_\theta(a) = y$ implies that $\theta = \bar{y}$.

In geometric terms, $a(y)$ lies on the $(n-1)$-dimensional hyperplane $\sum a_j = 0$, each point of which determines a different orbit. The orbits themselves are lines $\theta + a(y)$ passing through these points, with $\theta \in \mathbb{R}$. When $n = 2$, each point (y_1, y_2) in $\mathbb{R}^2$ is indexed by a point on the line $y_1 + y_2 = 0$, which determines the orbit, a straight line perpendicular to this.   ∎

Two points y and y' on the same orbit have the same index $a = a(y)$, which is said to be *invariant* to the action of the group because its value does not depend on whether y or $g(y)$ was observed, for any $g \in \mathcal{G}$. It is *maximal invariant* if every other invariant statistic is a function of it, or equivalently

$$a(y) = a(y') \text{ implies that } y' = g(y) \text{ for some } g \in \mathcal{G}.$$

The distribution of $A = a(Y)$ does not depend on the elements of $\mathcal{G}$. In the present context these are identified with parameter values, so the distribution of A does not depend on parameters and is known in principle; A is said to be *distribution constant*. A maximal invariant can be thought of as a reduced version of the data that represents it as closely as possible while remaining invariant to the action of $\mathcal{G}$. In some sense it is what remains of Y once minimal information about the parameter values has been extracted.

Often there is a 1–1 correspondence between the elements of $\mathcal{G}$ and the parameter space Θ, and then the action of $\mathcal{G}$ on $\mathcal{Y}$ induces a group action on Θ. If we can write g_θ for a general element of $\mathcal{G}$, then $g \circ g_\theta = g_{\theta'}$ for some $\theta' \in \Theta$. Hence g has mapped θ to θ', thereby inducing an action on Θ. In principle the action of g on Θ might be different from its action on $\mathcal{Y}$, and it is clearer to think of two related groups $\mathcal{G}$ and $\mathcal{G}^*$, the second of which acts on Θ. We use g_θ^* to denote the element of $\mathcal{G}^*$ that corresponds to $g_\theta \in \mathcal{G}$. In many cases the action of $\mathcal{G}^*$ is *transitive*, that is, each parameter can be obtained by applying an element of the group to a single baseline parameter.

Example 5.20 (Permutation group) Permutation of the indices of a random sample $Y_1, \ldots, Y_n$ should leave any inference unaffected. Hence we may consider the group of permutations π, with $g_\pi(y)$ representing the permuted version of $y \in \mathbb{R}^n$. Note that π^{-1} is also a permutation, as is the operation that leaves the indices of y unchanged. In the location model we might let $\mathcal{G}$ be the group containing all $n!$ of the g_π in addition to the g_θ. Though well-defined on the sample space, g_π has no counterpart in the parameter space, and so the enlarged group is not transitive.

To check that $a(y) = (y_{(1)} - \bar{y}, \ldots, y_{(n)} - \bar{y})^{\mathrm{T}}$ is a maximal invariant, note that if $a(y) = a(y')$, then permutations π, π' exist such that $g_\pi \circ g_{-\bar{y}}(y) = g_{\pi'} \circ g_{-\bar{y}'}(y')$. This in turn implies that $g_{-\bar{y}'}^{-1} \circ g_{\pi'}^{-1} \circ g_\pi \circ g_{-\bar{y}}(y) = y'$. Hence a is a maximal invariant.

If permutations are not included in the group, the same argument shows that $(y_1 - \bar{y}, \ldots, y_n - \bar{y})^{\mathrm{T}}$ is a maximal invariant. Thus the maximal invariant depends on the chosen group.   ∎

We shall usually ignore permutations of the order of a random sample, because the discussion below is simpler if the group considered is transitive.

Equivariance

A statistic $S = s(Y)$ defined on $\mathcal{Y}$ and taking values in the parameter space Θ is said to be *equivariant* if $s(g_\theta(Y)) = g_\theta^*(s(Y))$ for all $g_\theta \in \mathcal{G}$. Often S is chosen to be an estimator of θ, and then it is called an *equivariant estimator*. Maximum likelihood estimators are equivariant, because of their transformation property, that if $\phi = \phi(\theta)$ is a 1–1 transformation of the parameter θ, then $\widehat{\phi} = \phi(\widehat{\theta})$, where $\widehat{\theta} = s(Y)$ is the maximum likelihood estimator of θ. If the transformation ϕ corresponds to $g_\phi^* \in \mathcal{G}^*$, and $g_\phi(Y)$ is the transformation of Y whose maximum likelihood estimator is $\widehat{\phi}$, then $\widehat{\phi} = s(g_\phi(Y))$, while $\phi(\widehat{\theta}) = g_\phi^*(s(Y))$. Hence $s(g_\phi(Y)) = g_\phi^*(s(Y))$ for all such g_ϕ, which is the requirement for equivariance.

An equivariant estimator can be used to construct a maximal invariant. Note first that as $s(Y) \in \Theta$, the corresponding group elements $g_{s(Y)}^* \in \mathcal{G}^*$ and $g_{s(Y)} \in \mathcal{G}$ exist. Now consider $a(Y) = g_{s(Y)}^{-1}(Y)$. If $a(Y) = a(Y')$, then $g_{s(Y)}^{-1}(Y) = g_{s(Y')}^{-1}(Y')$, and it follows that $Y' = g_{s(Y')} \circ g_{s(Y)}^{-1}(Y)$. Hence $A = a(Y) = g_{s(Y)}^{-1}(Y)$ is maximal invariant.

Example 5.21 (Location-scale model) Let $Y = \eta + \tau\varepsilon$, where as before ε has a known density f, and the parameter $\theta = (\eta, \tau) \in \Theta = \mathbb{R} \times \mathbb{R}_+$. The group action is $g_\theta(y) = g_{(\eta,\tau)}(y) = \eta + \tau y$, so

$$g_{(\eta,\tau)} \circ g_{(\mu,\sigma)}(y) = g_{(\eta,\tau)}(\mu + \sigma y) = \eta + \tau\mu + \tau\sigma y = g_{(\eta+\tau\mu,\tau\sigma)}(y). \qquad (5.22)$$

The set of such transformations is closed with identity $g_{(0,1)}$. It is easy to check that $g_{(\eta,\tau)}$ has inverse $g_{(-\eta/\tau,\tau^{-1})}$. Therefore

$$\mathcal{G} = \left\{ g_{(\eta,\tau)} : (\eta, \tau) \in \mathbb{R} \times \mathbb{R}_+ \right\}$$

is indeed a group under the operation $\circ$ defined above.

The action of $g_{(\eta,\tau)}$ on a random sample is $g_{(\eta,\tau)}(Y) = \eta + \tau Y$, with $\eta \equiv \eta 1_n$ and Y an $n \times 1$ vector, as in Example 5.19. Expression (5.22) implies that the implied group action on Θ is

$$g_{(\eta,\tau)}^*((\mu, \sigma)) = (\eta + \tau\mu, \tau\sigma).$$

The sample average and standard deviation are equivariant, because with $s(Y) = (\overline{Y}, V^{1/2})$, where $V = (n-1)^{-1} \sum (Y_j - \overline{Y})^2$, we have

$$
\begin{aligned}
s(g_{(\eta,\tau)}(Y)) &= \left(\overline{\eta + \tau Y}, \left\{ (n-1)^{-1} \sum (\eta + \tau Y_j - \overline{\eta + \tau Y})^2 \right\}^{1/2} \right) \\
&= \left(\eta + \tau\overline{Y}, \left\{ (n-1)^{-1} \sum (\eta + \tau Y_j - \eta - \tau\overline{Y})^2 \right\}^{1/2} \right) \\
&= \left(\eta + \tau\overline{Y}, \tau V^{1/2} \right) \\
&= g_{(\eta,\tau)}^* (s(Y)).
\end{aligned}
$$

A maximal invariant is $A = g_{s(Y)}^{-1}(Y)$, and the parameter corresponding to $g_{s(Y)}^{-1}$ is $(-\overline{Y}/V^{1/2}, V^{-1/2})$. Hence a maximal invariant is the vector of residuals

$$A = (Y - \overline{Y})/V^{1/2} = \left(\frac{Y_1 - \overline{Y}}{V^{1/2}}, \ldots, \frac{Y_n - \overline{Y}}{V^{1/2}} \right)^{\mathrm{T}}, \qquad (5.23)$$

also called the *configuration*. It can be checked directly that the distribution of A depends on n and f but not on θ. Any function of A is invariant. If permutations are added to $\mathcal{G}$, a maximal invariant is $A = (Y_{(\cdot)} - \overline{Y})/V^{1/2}$, where $Y_{(\cdot)} = (Y_{(1)}, \ldots, Y_{(n)})$ represents the vector of ordered values of Y.

The orbits are determined by different values a of the statistic A, and Y has a unique representation as $g_{s(Y)}(A) = \overline{Y} + V^{1/2}A$. Hence the group action is free.

The elements of a satisfy the equations $\sum a_j = 0$ and $\sum a_j^2 = n - 1$, so A lies on an $(n - 2)$-dimensional surface in $\mathbb{R}^n$. When $n = 3$ this is easily visualized; it is the circle that forms the intersection of the sphere of radius 2 with the plane $a_1 + a_2 + a_3 = 0$. The entire space $\mathbb{R}^3$ is generated by first choosing an element of this circle, then multiplying it by a positive number to rescale it to lie on a ray passing through the origin, and finally adding the vector $\overline{y}1_3$.

Another equivariant estimator is $(Y_{(1)}, Y_{(2)} - Y_{(1)})$, where $Y_{(r)}$ is the rth order statistic, and the argument above shows that the vector $(Y - Y_{(1)})/(Y_{(2)} - Y_{(1)})$ is corresponding maximal invariant. Evidently this is just one of many possible location-scale shifts of A, which can be thought of as the 'shape' of the sample, shorn of information about its location and scale. ∎

The group-averse reader may wonder whether the generality of the discussion above is needed to deal with our motivating example of temperatures in Celsius and Fahrenheit. In fact we have not yet raised a crucial distinction between invariances intrinsic to a context and those stemming only from the mathematical structure of the model. Invariances of the first sort are more defensible than are the second, because not every mathematical expression of a statistical problem successfully preserves aspects such the interpretation of key parameters. Thus the sensible choice of group in a particular context may not be mathematically most natural. Furthermore appeal to invariance is not sensible if external information suggests that some parameter values should be favoured over others. Invariance arguments require careful thought.

Example 5.22 (Venice sea level data) The straight-line regression model (5.2) can be expressed as

$$y = X\beta + \varepsilon,$$

where

$$y = \begin{pmatrix} y_1 \\ \vdots \\ y_n \end{pmatrix}, \quad X = \begin{pmatrix} 1 & x_1 \\ \vdots & \vdots \\ 1 & x_n \end{pmatrix}, \quad \beta = \begin{pmatrix} \beta_0 \\ \beta_1 \end{pmatrix}, \quad \varepsilon = \begin{pmatrix} \varepsilon_1 \\ \vdots \\ \varepsilon_n \end{pmatrix}.$$

If the ε_j are independent normal variables then $Y \sim N_n(X\beta, \sigma^2 I_n)$. Hence $OY \sim$

An $n \times n$ orthogonal matrix of real numbers O has the properties that $O^{\mathrm{T}}O = OO^{\mathrm{T}} = I_n$.

$N_p(OX\beta, \sigma^2 I_n)$ for any $n \times n$ orthogonal matrix O that preserves the column space of X, that is, such that $X(X^{\mathrm{T}}X)^{-1}XOX = OX$. It is straightforward to check that such matrices form a group. Now $\mathrm{E}(OY) = X\gamma$, where $\gamma = (X^{\mathrm{T}}X)^{-1}X^{\mathrm{T}}OX\beta = A^{-1}\beta$, say, is the result of applying the corresponding group element in the parameter space.

The transformation giving (5.3), with

$$\begin{pmatrix} \beta_0 \\ \beta_1 \end{pmatrix} = \beta = A\gamma = \begin{pmatrix} a_{11} & a_{12} \\ a_{21} & a_{22} \end{pmatrix} \gamma = \begin{pmatrix} 1 & -\overline{x} \\ 0 & 1 \end{pmatrix} \gamma = \begin{pmatrix} \gamma_0 - \gamma_1 \overline{x} \\ \gamma_1 \end{pmatrix},$$

preserves the interpretation of $\beta_1 = a_{22}\gamma_1$ as a rate of change of $\mathrm{E}(Y)$ with respect to time, though the time origin is shifted. From a mathematical viewpoint there is no reason not to take more general invertible transformations $\beta = A\gamma$, for example with $a_{21} \neq 0$, but this makes no sense statistically. Moreover even with $a_{21} = 0$ not every choice of a_{22} makes sense: taking $a_{22} < 0$ or such that the units of γ_1 were seconds would have little appeal.   ∎

In some cases the full parameter space does not give a useful group of transformations, but subspaces of it do. If the parameter space has form $\Psi \times \Lambda$, with the same group of transformations $\mathcal{G} = \{g_\lambda : \lambda \in \Lambda\}$ acting on the sample space for each value of ψ, then we have a *composite group transformation model*.

Example 5.23 (Location-scale model)   In the previous example, suppose that the density f_ψ of ε depends on a further parameter ψ. An example is the t_ψ density. Then for each fixed ψ we have a location-scale model in terms of $\lambda = (\eta, \tau)$, with $g_\lambda(y) = \eta + \tau y$, and our previous discussion applies.

For each ψ a maximal invariant based on a random sample $Y_1, \ldots, Y_n$ is $A = (Y - \overline{Y})/V^{1/2}$, whose distribution depends on the sample size and on f_ψ but not on λ.   ∎

Exercises 5.3

1   Show that $\approx$ is an equivalence relation.

2   Suppose $Y = \tau\varepsilon$, where $\tau \in \mathbb{R}_+$ and ε is a random variable with known density f. Show that this scale model is a group transformation model with free action $g_\tau(y) = \tau y$. Show that $s_1(Y) = \overline{Y}$ and $s_2(Y) = (\sum Y_j^2)^{1/2}$ are equivariant and find the corresponding maximal invariants. Sketch the orbits when $n = 2$.

3   Suppose that ε has known density f with support on the unit circle in the complex plane, and that $Y = e^{i\theta}\varepsilon$ for $\theta \in \mathbb{R}$. Show that this is a group transformation model. Is it transitive? Is the action free?

4   Write the configuration (5.23) in terms of $\varepsilon_1, \ldots, \varepsilon_n$, where $Y_j = \mu + \sigma\varepsilon_j$, and thereby show that its distribution does not depend on the parameters.

5   Show that the gamma density with shape and scale parameters ψ and λ, is a composite transformation model under the mapping from Y to τY, where $\tau > 0$.

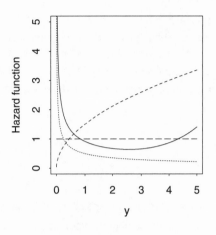 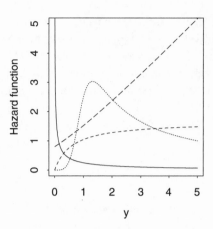

Figure 5.7 Hazard functions. Left panel: Weibull hazards with $\theta = 1$ and $\alpha = 0.5$ (dots), $\alpha = 1$ (large dashes), $\alpha = 1.5$ (dashes), and bi-Weibull hazard with $\theta_1 = 0.3$, $\alpha_1 = 0.5$, $\theta_2 = \alpha_2 = 5$ (solid). Right panel: Log-logistic hazards with $\lambda = 1$ and $\alpha = 0.5$ (solid), $\alpha = 5$ (dots), gamma hazard with $\lambda = 0.6$ and $\alpha = 2$ (dashes), and standard normal hazard (large dashes).

5.4 Survival Data

5.4.1 Basic ideas

The focus of interest in survival data is the time to an event. An important area of application is medicine, where, for example, interest may centre on whether a new treatment lengthens the life of a cancer patient, relative to those who receive existing treatments. Other common applications are in industrial reliability, where the aim may be to estimate the distribution of time to failure for a fridge, a computer program, or a pacemaker. Examples also abound in the social sciences, where for example the length of a period of unemployment may be of interest. In each case the time Y to the event is non-negative and may be censored. For example, a patient may be lost to follow-up for some reason unrelated to his disease, so that it is unknown whether or not he died from the cause under study. In general discussion we refer to the items liable to fail as *units*; these may be persons, widgets, marriages, cars, or whatever.

This section outlines some basic notions in survival analysis, concentrating on single samples. More complex models are discussed in Section 10.8.

Hazard and survivor functions

A central concept is the *hazard function* of Y, defined loosely as the probability density of failure at time y, given survival to then. If Y is a continuous random variable this is

$$h(y) = \lim_{\delta y \to 0} \frac{1}{\delta y} \Pr(y \leq Y < y + \delta y \mid Y \geq y) = \frac{f(y)}{\mathcal{F}(y)},$$

where $\mathcal{F}(y) = \Pr(Y \geq y) = 1 - F(y)$ is the *survivor function* of Y. An older term for $h(y)$ is the *force of mortality*, and it is also called the *age-specific failure rate*. Evidently $h(y) \geq 0$; some example hazard functions are shown in Figure 5.7.

The exponential density with rate λ has $\mathcal{F}(y) = \exp(-\lambda y)$ and constant hazard function $h(y) = \lambda$, and although data are rarely so simple, this model of a constant failure rate independent of the past is a natural baseline from which to develop more realistic models.

The *cumulative hazard function* is

$$H(y) = \int_0^y h(u)\,du = \int_0^y \frac{f(u)}{1 - F(u)}\,du = -\log\{1 - F(y)\},$$

as $F(0) = 0$. Thus the survivor function may be written as $\mathcal{F}(y) = \exp\{-H(y)\}$, and $f(y) = h(y)\exp\{-H(y)\}$. If $\lim_{y\to\infty} H(y) < \infty$, then $\mathcal{F}(\infty) > 0$ and the distribution is *defective*, putting positive probability on an infinite survival time. This may arise in practice if, for example, the endpoint for a study is death from a disease, but complete recovery is possible.

For a discrete distribution with probabilities f_i at $0 \le t_1 < t_2 < \cdots$, we may write $h(y) = \sum h_i \delta(y - t_i)$, where

$$h_i = \Pr(Y = t_i \mid Y \ge t_i) = \frac{f_i}{f_i + f_{i+1} + \cdots}.$$

Thus

$$\Pr(Y > t_i \mid Y \ge t_i) = 1 - h_i, \quad f_i = h_i \prod_{j=1}^{i-1}(1 - h_j), \tag{5.24}$$

and if $t_i < y \le t_{i+1}$ then

$$\mathcal{F}(y) = \Pr(Y > t_i \mid Y \ge t_i)\Pr(Y > t_{i-1} \mid Y \ge t_{i-1}) \cdots \Pr(Y > t_1)$$
$$= \prod_{i:t_i < y}(1 - h_i). \tag{5.25}$$

We define the cumulative hazard as $H(y) = -\sum_{i:t_i < y} \log(1 - h_i)$, again giving $\mathcal{F}(y) = \exp\{-H(y)\}$. The more natural definition $\sum_{i:t_i < y} h_i$ is approximately equal to $H(y)$ if the individual h_i are small.

Mixed discrete-continuous variables are important in a general treatment of survival data — for example, a patient may die so fast from complications after an operation that the survival time is effectively zero, but otherwise may live for years — but here we avoid them and the complications they bring.

Suppose that $Y = \min(Y_1, \ldots, Y_k)$, where the Y_i are continuous times to failure from k independent causes, and that their hazard functions are $h_i(y)$. Then Y exceeds y if and only if all the Y_i exceed y, so

$$\mathcal{F}(y) = \prod_{i=1}^k \Pr(Y_i \ge y) = \exp\left\{-\sum_{i=1}^k \int_0^y h_i(u)\,du\right\},$$

and it follows that Y has hazard function $h(y) = \sum h_i(y)$. That is, hazards for independent causes of failure are added.

Example 5.24 (Weibull density)   The Weibull density (4.4) has survivor function $\mathcal{F}(y) = \exp\{-(y/\theta)^\alpha\}$, so its hazard function is $\alpha\theta^{-\alpha}y^{\alpha-1}$. This is constant when $\alpha = 1$, decreasing when $\alpha < 1$, and increasing when $\alpha > 1$, as shown in the left panel of Figure 5.7. This flexibility and the tractability of its density and distribution functions make the Weibull a popular choice in reliability studies.

This density is the basis of the bi-Weibull model, which corresponds to the minimum of two independent Weibull variables, shown by the argument above to have hazard function $\alpha_1\theta_1^{-\alpha_1}y^{\alpha_1-1} + \alpha_2\theta_2^{-\alpha_2}y^{\alpha_2-1}$. If the shape parameters lie on opposite sides of unity, so $0 < \alpha_1 < 1 < \alpha_2$, say, $h(y)$ is bathtub-shaped: there is a high early failure rate during a 'burn-in period', then a flattish hazard and an eventual increase in failure rate; see Figure 5.7. If $\alpha_1 = \alpha_2$ the hazard is indistinguishable from the Weibull hazard and θ_1 and θ_2 are not identifiable. ∎

Example 5.25 (Log-logistic density) The log-logistic distribution has survivor and hazard functions

$$\mathcal{F}(y) = \{1 + (\lambda y)^\alpha\}^{-1}, \quad h(y) = \alpha\frac{\lambda^\alpha y^{\alpha-1}}{1 + (\lambda y)^\alpha}, \quad y > 0, \alpha, \lambda > 0.$$

Two examples of $h(y)$ are shown in the right panel of Figure 5.7. It is decreasing for $\alpha \leq 1$ and unimodal otherwise. The log-normal distribution, that is, the distribution of $Y = e^Z$, where Z has a normal distribution, is similar to the log-logistic, and its hazard can take similar shapes. The normal hazard, also shown, increases very rapidly due to the light tails of the normal density. ∎

Example 5.26 (Gamma density) The gamma survivor and hazard functions are

$$\mathcal{F}(y) = \int_y^\infty \frac{\lambda^\alpha u^{\alpha-1}}{\Gamma(\alpha)}e^{-\lambda u}\,du, \quad h(y) = \frac{\lambda^\alpha y^{\alpha-1}e^{-\lambda y}}{\int_y^\infty \lambda^\alpha u^{\alpha-1}e^{-\lambda u}\,du}.$$

Figure 5.7 shows an example of the gamma hazard function. ∎

Censoring

The simplest form of censoring occurs when a random variable Y is watched until a pre-determined time c. If $Y \leq c$, we observe the value y of Y, but if $Y > c$, we know only that Y survived beyond c. This is known as *Type I censoring*. *Type II censoring* arises when n independent variables are observed until there have been r failures, so the first r order statistics $0 < Y_{(1)} < \cdots < Y_{(r)}$ are observed, All that is known about the $n - r$ remaining observations is that they exceed $Y_{(r)}$. This scheme is typically used in industrial life-testing.

For simplicity we assume no ties.

Under *random censoring* we suppose that the jth of n independent units has an associated censoring time C_j drawn from a distribution G, independent of its survival time Y_j^0. The time actually observed is $Y_j = \min(Y_j^0, C_j)$, and it is known whether or not $Y_j = Y_j^0$, an event indicated by D_j. Thus a pair (y_j, d_j) is observed for each unit, with $d_j = 1$ if y_j is the survival time and $d_j = 0$ if y_j is the censoring time. This type of censoring is important in medical applications, where a patient may die of a cause unrelated to the reason they are being studied, may withdraw from the study or be lost to follow-up, or the study may end before their survival time is observed.

Figure 5.8 shows the relation between calendar time and time on trial for a medical study, with censoring both before and at the end of the trial. We assume below that failure does not depend on the calendar time at which an individual enters the study;

Figure 5.8 Lexis diagram showing typical pattern of censoring in a medical study. Each individual is shown as a line whose x coordinates run from the calendar time of entry to the trial to the calendar time of failure (blob) or censoring (circle). Censoring occurs at the end of the trial, marked by the vertical dotted line, or earlier. The vertical axis shows time on trial, which starts when individuals enter the study. The risk set for the failure at calendar time 4.5 comprises those individuals whose lines touch the horizontal dashed line; see page 543.

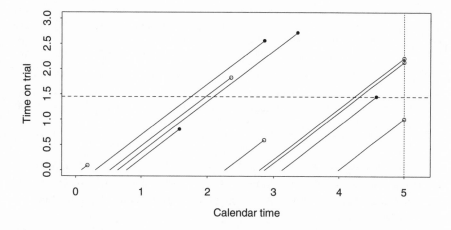

thus we study events on the vertical axis. Calendar time may be used to account for changes in medical practice over the course of a trial.

In applications the assumption that C_j and Y_j^0 are independent is critical. There would be serious bias if the illest patients drop out of a trial because the treatment makes them feel even worse, thereby inducing association between survival and censoring variables because patients die soon after they withdraw.

The examples above all involve *right-censoring*. Less common is left-censoring, where the time of origin is not known exactly, for example if time to death from a disease is observed, but the time of infection is unknown.

In practice a high proportion of the data may be censored, and there may be a serious loss of efficiency if they are ignored (Example 4.20). There will also be bias, as survival probabilities will be underestimated if censoring is not taken into account. Hence it is crucial to make proper allowance for censoring.

5.4.2 Likelihood inference

Suppose that the survival times are continuous, that data $(y_1, d_1), \ldots, (y_n, d_n)$ on n independent units are available, and that there is a parametric model for survival times, with survivor and hazard functions $\mathcal{F}(y; \theta)$ and $h(y; \theta)$. Recall that the density may be written $f(y; \theta) = h(y; \theta)\mathcal{F}(y; \theta)$ and that in terms of the integrated hazard function, $\mathcal{F}(y; \theta) = \exp\{-H(y; \theta)\}$. Under random censoring in which the censoring variables have density and distribution functions g and G, the likelihood contribution from y_j is

$$f(y_j; \theta)\{1 - G(y_j)\} \quad \text{if } d_j = 1, \quad \text{and} \quad \mathcal{F}(y_j; \theta)g(y_j) \quad \text{if } d_j = 0.$$

If the censoring distribution does not depend on θ, then $g(y_j)$ and $G(y_j)$ are constant and the overall log likelihood is

$$\ell(\theta) \equiv \sum_u \log f(y_j; \theta) + \sum_c \log \mathcal{F}(y_j; \theta),$$

0+	1+	1+	3+	3+	7	10+	11+	12+	12+	15+	18+
20+	22+	22+	24+	25+	26+	31+	36+	36+	36	38	40
47+	47+	49+	53+	53+	55+	56+	57+	61+	67+	67+	70
73	75+	77+	83+	84+	88+	89+	99	121+	122+	123+	141+
0+	0+	2+	2+	2+	2+	3	3+	4+	5+	9+	10+
11	12+	13	13+	18+	22+	22+	24+	24+	24+	25+	26+
27	28	32+	35+	36	40+	43+	50+	54			

Table 5.3
Blalock–Taussig shunt data (Oakes, 1991). The table gives survival time of shunt (months after operation) for 48 infants aged over one month at time of operation, followed by times for 33 infants aged 30 or fewer days at operation. Infants whose shunt has not yet failed are marked +.

where the sums are over uncensored and censored units. This amounts to treating the censoring pattern as fixed, and encompasses Type I censoring, for which G puts all its probability at c. In terms of the hazard function and its integral, the log likelihood is

$$\ell(\theta) = \sum_{j=1}^{n} \{d_j \log h(y_j; \theta) - H(y_j; \theta)\}. \tag{5.26}$$

Inference for θ is based on this in the usual way. As calculation of expected information involves assumptions about the censoring mechanism, standard errors for parameter estimates are based on observed information.

Example 5.27 (Exponential distribution) When $f(y; \lambda) = \lambda e^{-\lambda y}$, the hazard is $h(y; \lambda) = \lambda$, and hence the log likelihood for a random sample $(y_1, d_1), \ldots, (y_n, d_n)$ is

$$\ell(\lambda) = \sum_{j=1}^{n}(d_j \log \lambda - \lambda y_j) = \log \lambda \sum_{j=1}^{n} d_j - \lambda \sum_{j=1}^{n} y_j,$$

giving maximum likelihood estimate $\widehat{\lambda} = \sum d_j / \sum y_j$ and observed information $J(\lambda) = \sum d_j / \lambda^2$; see Example 4.20. Hence the estimate of λ is zero if there are no failures, and censored data contribute no information about λ.

The expected information $I(\lambda) = \mathrm{E}\{J(\lambda)\}$ involves $\mathrm{E}(D_j)$, where D_j indicates whether a failure or censoring time is observed for the jth observation, but this expectation cannot be obtained without some assumption about the censoring distribution G. Although this is feasible for theoretical calculations such as those in Example 4.20, in practice the inverse observed information is used to give a standard error $J(\widehat{\lambda})^{-1/2}$ for $\widehat{\lambda}$.

The mean of the exponential density is $\theta = \lambda^{-1}$, and its maximum likelihood estimate is $\widehat{\theta} = \sum y_j / \sum d_j$, with observed information $J(\widehat{\theta}) = \widehat{\theta}^2 / \sum d_j$ and maximized log likelihood $\ell(\widehat{\theta}) = -(1 + \log \widehat{\theta}) \sum d_j$. ∎

Example 5.28 (Blalock–Taussig shunt data) The Blalock–Taussig shunt is an operative procedure for infants with congenital cyanotic heart disease. Table 5.3 contains data from the University of Rochester on survival times for the shunt for 81 infants, divided into two age groups. Many of the survival times are censored, meaning that the shunt was still functioning after the given survival time; its time to failure is not known for these children, whereas it is known for the others. There are just seven failures in each group. The table suggests that the shunt fails sooner for younger children, and it is of interest to see how failure depends on age.

A simple model for these data is that the failure times are independent exponential variables, with common mean θ for both groups. Formulae from Example 5.27 show that $\widehat{\theta} = 209.1$ and the maximized log likelihood is -88.79. If the means are different, θ_1 and θ_2, say, then the maximized log likelihood is -85.98, so the likelihood ratio statistic for comparing these models is $2 \times (88.79 - 85.98) = 5.62$, to be compared with the χ_1^2 distribution. As $\Pr(\chi_1^2 \geq 5.62) \doteq 0.018$, there is strong evidence that the mean survival time is shorter for the younger group, if the exponential model is correct.

If the data were uncensored, it would be straightforward to assess the fit of this model using probability plots, but the amount of censoring is so high that this is not sensible. More specialized methods are needed, and they are discussed in Section 5.4.3.

One way to judge adequacy of the exponential model is to embed it in a larger one. A simple alternative is to suppose that the data are Weibull, with $H(y) = (y/\theta)^\alpha$. The maximized log likelihoods are -83.72 when this model is fitted separately to each group, and -83.74 when the same value of α is used for both groups. The likelihood ratio statistic for comparison of these is $2 \times (83.74 - 83.72) = 0.04$, which is negligible, but that for comparison with the best exponential model, $2 \times (85.98 - 83.74) = 4.48$, suggests that the Weibull model gives the better fit. The corresponding estimates and their standard errors are $\widehat{\theta}_1 = 181.1$ (52.7), $\widehat{\theta}_2 = 57.6$ (15.1), and $\widehat{\alpha} = 1.64$ (0.35). The value of $\widehat{\alpha}$ corresponds to an increasing hazard. ∎

Discrete data

Suppose that events could occur at pre-assigned times $0 \leq t_1 < t_2 < \cdots$, and that under a parametric model of interest the hazard function at t_i is $h_i = h_i(\theta)$. We adopt the convention that a unit censored at time t_i could have been observed to fail there, so giving likelihood contribution

$$\lim_{y \downarrow t_i} \mathcal{F}(y) = (1 - h_1) \cdots (1 - h_i),$$

from (5.25); one way to think of this is that censoring at t_i in fact takes place immediately afterwards. The contribution to the likelihood from a unit that fails at t_i is $(1 - h_1) \cdots (1 - h_{i-1})h_i$; see (5.24). Although the likelihood can be written down directly, it is more useful to express it in terms of the r_i units still in the *risk set* — that is not yet failed or censored — at time t_i and the number d_i of units who fail there. This modifies our previous notation: now d_i is the sum of the indicators of unit failures at time t_i, and can take one of values $0, 1, \ldots, r_i$. Each of the d_i failures at t_i contributes h_i to the likelihood, and the other units then still in view each contribute $1 - h_i$. It follows that the log likelihood may be written as

$$\ell(\theta) = \sum_i \left\{ d_i \log h_i + (r_i - d_i) \log (1 - h_i) \right\}, \tag{5.27}$$

with the interpretation that the probability of failure at t_i conditional on survival to t_i is h_i, and d_i of the r_i units in view at t_i fail then. Thus (5.27) is a sum of contributions from independent binomial variables representing the numbers of failures d_i at each

Age group	Hungary 900–1100	England 1640–89	Breslau 1687–91	England & Wales, 1841		England & Wales, 1980–82	
				Males	Females	Males	Females
30–35	0.0235	0.0171	0.0164	0.0108	0.0107	0.0010	0.0006
35–40	0.0291	0.0205	0.0195	0.0123	0.0118	0.0014	0.0009
40–45	0.0337	0.0195	0.0233	0.0140	0.0131	0.0024	0.0016
45–50	0.0402	0.0244	0.0282	0.0159	0.0145	0.0043	0.0028
50–55	0.0696	0.0307	0.0342	0.0181	0.0162	0.0079	0.0047
55–60	0.0814	0.0459	0.0383	0.0254	0.0220	0.0138	0.0076
60–65	0.1033	0.0513	0.0474	0.0375	0.0331	0.0227	0.0119
65–70	0.1485	0.0701	0.0630	0.0553	0.0493	0.0365	0.0187
70–75	0.1877	0.1129	0.0995	0.0815	0.0736	0.0587	0.0308
75–80	0.3008	0.1445	0.1589	0.1201	0.1097	0.0930	0.0527
80–85		0.1974		0.1771	0.1638	0.1432	0.0919
85–90				0.2617	0.2448	0.2110	0.1567
90–95				0.3884	0.3674	0.2900	0.2374
95–100						0.3894	0.3215
Deaths	2300	3133	2675	71,000	74,000	834,000	828,000

Table 5.4 Historical estimates of the force of mortality (year^{-1}), averaged for 5-year age groups (Thatcher, 1999). The bottom line gives the estimated number of deaths at age 30 years and above, on which the force of mortality is based.

time t_i, with denominators r_i and failure probabilities h_i. In fact r_i depends on the history of failures and censorings up to time t_i, so the d_i are not independent, but it turns out that for large sample inference we may proceed as if they were. This can be formalized using the theory of counting processes and martingales; see the bibliographic notes to this chapter and to Chapter 10.

Example 5.29 (Human lifetime data) The virtual elimination of many infectious diseases due to improved medical care and living conditions have led to increased life expectancy in the developed world. If the trend continues there are potentially major consequences for social security systems. Some physicians have asserted that an upper limit to the length of human life is imposed by physical constraints, and that the consequence of improved health care is that senesence will eventually be compressed into a short period just prior to death at or near this upper limit. This view is controversial, however, and there is a lively debate about the future of old age.

A natural way to assess the plausibility of the hypothesized upper limit is to examine data on mortality. Table 5.4 contains historical snapshots of the force of mortality, obtained from census data, records of births and deaths, and other sources. The earliest data were obtained by forensic examination of adult skeletons in Hungarian graveyards, using a procedure that probably underestimates ages over 60 years and overestimates those below. The table shows estimates of the average probability of dying per year, conditional on survival to then, using the following argument. For continuous-time data with survivor function $\mathcal{F}(y)$ and corresponding hazard function $h(y)$, the probability of failure in the period $[t_i, t_{i+1})$ given survival to t_i would be

$$\frac{\mathcal{F}(t_i) - \mathcal{F}(t_{i+1})}{\mathcal{F}(t_i)} = 1 - \exp\left\{-(t_{i+1} - t_i)\frac{1}{t_{i+1} - t_i}\int_{t_i}^{t_{i+1}} h(y)\,dy\right\},$$

Figure 5.9 Force of mortality for historical data, in units of deaths per person-year. Left panel, from top to bottom: data for medieval Hungary, England 1640–89, Breslau 1687–91 (dots), English and Welsh females 1841 and 1980–82. Right panel: data for England and Wales, 1980–82, males (above) and females (below) and fitted hazard functions (dots).

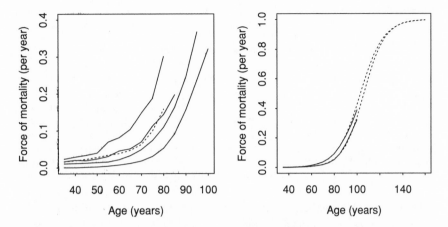

where $(t_{i+1} - t_i)^{-1} \int_{t_i}^{t_{i+1}} h(y) \, dy$ is the average hazard over the interval. Given discretized data with r_i people alive at time t_i, of whom d_i fail in $[t_i, t_{i+1})$, the corresponding empirical hazard is $-(t_{i+1} - t_i)^{-1} \log(1 - d_i/r_i)$, and this is reported in the table; the corresponding d_i and r_i are unavailable to us. For British males dying in 1980 the empirical hazard rose from about 0.001 year^{-1} at age 30 years to about 0.1 year^{-1} at 80 years to about 0.4 year^{-1} at 95 years; for females the probabilities were slightly lower. Figure 5.9 shows the force of mortality of some of the columns of the table; it is no surprise that it is lower in later than in earlier periods.

One model for such data is that

$$h(y; \theta) = \lambda + \frac{\alpha e^{\beta y}}{1 + \alpha e^{\beta y}},$$

where $\theta = (\alpha, \beta, \lambda)$, corresponding to integrated hazard and survivor functions

$$H(y; \theta) = \lambda y + \frac{1}{\beta} \log \left(\frac{1 + \alpha e^{\beta y}}{1 + \alpha} \right), \quad \mathcal{F}(y; \theta) = e^{-\lambda y} \times \left(\frac{1 + \alpha}{1 + \alpha e^{\beta y}} \right)^{1/\beta}, \quad y \geq 0.$$

One interpretation of this model is that there are two competing causes of death, one with a constant hazard, and the other with a logistic hazard.

In order to use (5.27) to fit this model to the data given in Table 5.4, we must calculate $h_i(\theta)$ and (d_i, r_i). The probability of dying in $[t_i, t_{i+1})$ conditional on survival to t_i is

$$\begin{aligned} h_i(\theta) &= \Pr(t_i \leq Y \leq t_{i+1} \mid Y \geq t_i) \\ &= \frac{\mathcal{F}(t_i; \theta) - \mathcal{F}(t_{i+1}; \theta)}{\mathcal{F}(t_i; \theta)} \\ &= 1 - \exp\{H(t_i; \theta) - H(t_{i+1}; \theta)\}, \end{aligned}$$

and this is calculated using the logistic hazard given above. The empirical values of the hazard function $h_i = d_i/r_i$, where d_i is the number of deaths among the r_i persons at risk, can be obtained from the columns of Table 5.4. Some calculation gives

$$d_1 = nh_1, \quad d_i = nh_i(1 - h_1) \cdots (1 - h_{i-1}), \quad i = 2, \ldots, k,$$

Data set	Deaths at age 30 years and over	Estimate (standard error)		
		$10^4\widehat{\alpha}$	$10^2\widehat{\beta}$	$10^2\widehat{\lambda}$
Hungary, 900–1100	2300	8.76 (3.78)	7.68 (0.65)	1.27 (0.32)
England, 1640–89	3133	1.87 (0.66)	8.65 (0.48)	1.40 (0.12)
Breslau, 1687–91	2675	1.44 (0.76)	8.88 (0.73)	1.57 (0.15)
England & Wales, 1841, males	71,000	0.50 (0.03)	10.08 (0.08)	0.97 (0.01)
England & Wales, 1841, females	74,000	0.32 (0.02)	10.50 (0.08)	0.97 (0.01)
England & Wales, 1980–82, males	834,000	0.46 (0.00)	9.93 (0.01)	−0.04 (0.00)
England & Wales, 1980–82, females	828,000	0.12 (0.00)	10.92(0.01)	0.03 (0.00)

Table 5.5 Maximum likelihood estimates for fits of logistic hazard model to the data in Table 5.4. Standard errors given as 0.00 are smaller than 0.005.

where $n = r_1$ is the number initially at risk, an estimate of which is given at the foot of the table; once the d_i are known the r_i are given by d_i / h_i. When these pieces are put together, maximum likelihood estimates of θ may be obtained by numerical maximization of (5.27), with standard errors based on the inverse observed information matrix, also obtained numerically. Table 5.5 shows that $\widehat{\alpha}$ and $\widehat{\lambda}$ decrease systematically with time, while the value of $\widehat{\beta}$ increases slightly but is broadly constant, close to 0.1. These are consistent with the overall decrease in the hazard function, but no change in its shape, that we see in the left panel of Figure 5.9. The values of $\widehat{\lambda}$ are generally similar to the observed force of mortality at age 30–35, and one interpretation is that $\widehat{\lambda}$ represents the danger from the principal risks at this age, namely infectious diseases and child-bearing, which has sharply reduced over the last 150 years.

The fits for the 1980–82 data are shown in the right panel of Figure 5.9. Although the fit is good, the extrapolation beyond the range of the data must be treated skeptically. It shows that although the model imposes no absolute upper limit on lifetimes, for a person dying in 1980–82 there was an effective limit of about 140 years, well beyond the limits of 110 or 115 years which have been suggested by physicians. In fact the longest life for which there is good documentation is that of Mme Jeanne Calment, who died in 1997 aged 122 years, and there is unlikely ever to be enough data to see if there is an upper limit well above this.

Example 5.32 gives further discussion of this model. ∎

5.4.3 Product-limit estimator

Graphical procedures are essential for initial data inspection, for suggesting plausible models and for checking their fit. One standard tool is a nonparametric estimator of the survivor function, in effect extending the empirical distribution function (Example 2.7) to censored data.

The simplest derivation of it is based on the model for failures at discrete prespecified times given above (5.25), though the estimator is useful more widely. We therefore start with expression (5.27), which gives the log likelihood for such data in terms of the hazard function $h_1, h_2, \ldots$. For parametric analysis of a discrete failure distribution the h_i are functions of a parameter θ, but for nonparametric estimation we treat each h_i as a separate parameter and estimate it by maximum likelihood.

Differentiation of (5.27) with respect to h_i gives $\widehat{h}_i = d_i/r_i$ and hence

$$\widehat{\mathcal{F}}(y) = \prod_{i:t_i<y} \left(1 - \widehat{h}_i\right) = \prod_{i:t_i<y} \left(1 - \frac{d_i}{r_i}\right).$$

This is known as the *product-limit* or *Kaplan–Meier* estimator. Note that

$$-\frac{\partial^2 \ell}{\partial h_i \partial h_j} = \begin{cases} \frac{r_i}{\widehat{h}_i(1-\widehat{h}_i)}, & i = j, \\ 0, & \text{otherwise,} \end{cases}$$

implying that that the $\widehat{h}_i$ are asymptotically independent, with diagonal variance matrix whose ith element is $\widehat{h}_i(1 - \widehat{h}_i)/r_i$.

This derivation extends to continuous failure times by supposing that the y_j are ordered and that there are no ties, giving $t_1 = y_1 < \cdots < t_n = y_n$. Then d_j simply indicates whether y_j is a failure or a censoring time, and

$$\widehat{\mathcal{F}}(y) = \prod_{j:y_j<y} \left(1 - \frac{1}{r_j}\right)^{d_j}, \tag{5.28}$$

so the estimate decreases only at those values of t_j with $d_j = 1$. This estimate is valid also when the y_j are not pre-specified, but full justification of this would take us too far afield. If there is no censoring, then $1 - \widehat{\mathcal{F}}(y)$ is the empirical distribution function.

We find the variance of $\widehat{\mathcal{F}}(y)$ by arguing that if the d_i are asymptotically independent binomial variables with denominators r_i, then

$$\begin{aligned}
\text{var}\{\log \widehat{\mathcal{F}}(y)\} &= \text{var}\left\{\sum_{i:y_i<y} \log(1 - \widehat{h}_i)\right\} \\
&\doteq \sum_{i:y_i<y} \text{var}\{\log(1 - \widehat{h}_i)\} \\
&\doteq \sum_{i:y_i<y} \frac{1}{(1 - \widehat{h}_i)^2} \frac{\widehat{h}_i(1 - \widehat{h}_i)}{r_i} \\
&= \sum_{i:y_i<y} \frac{d_i}{r_i(r_i - d_i)}, \tag{5.29}
\end{aligned}$$

where the first approximation uses the asymptotic independence of the $\widehat{h}_i$ and the second uses the delta method. As $\text{var}\{\log \widehat{\mathcal{F}}(y)\} \doteq \text{var}\{\widehat{\mathcal{F}}(y)\}/\widehat{\mathcal{F}}(y)^2$, we obtain *Greenwood's formula*,

$$\text{var}\{\widehat{\mathcal{F}}(y)\} \doteq \widehat{\mathcal{F}}(y)^2 \sum_{i:y_i<y} \frac{d_i}{r_i(r_i - d_i)},$$

variants of which are widely used to assess the uncertainty of $\widehat{\mathcal{F}}(y)$. In practice it is better to use (5.29) to compute approximate normal confidence intervals for $\log \mathcal{F}(y)$, and then to transform these intervals back to the original scale.

The cumulative hazard function can be estimated as $\widehat{H}(y) = \sum_{i:y_i<y} d_i/r_i$; this is a step function with jumps at failure times and approximate variance (5.29).

Edward Kaplan and Paul Meier were former students of John Tukey who submitted separate papers to *Journal of the American Statistical Association*. They were encouraged to merge them by the editor. Despite mixed reviews the editor decided to publish the joint paper (Kaplan and Meier, 1958), which has become one of the most-cited articles in statistics.

Major Greenwood (1880–1949) qualified as a physician before turning to statistics and epidemiology under the influence of Karl Pearson. He was the first resident statistician at any medical research institute, and worked for the British Medical Research Council and the London School of Hygiene and Tropical Medicine. He studied infant mortality, tuberculosis and hospital fatality rates, pioneered clinical trials and gradually persuaded sceptical physicians of the value of statistical thinking. Major was not his military rank but his first name.

Failure time, y_i	7	36	38	40	70	73	99
Number in view, r_i	43	29	26	25	13	12	5
Number failing, d_i	1	1	1	1	1	1	1
$1 - d_i/r_i$	0.977	0.966	0.962	0.960	0.923	0.916	0.8
$\widehat{\mathcal{F}}(y_i+)$	0.977	0.944	0.908	0.872	0.804	0.737	0.590
Standard error	0.023	0.040	0.052	0.062	0.086	0.102	0.155

Table 5.6 Product-limit estimator for older group of infants in Table 5.3.

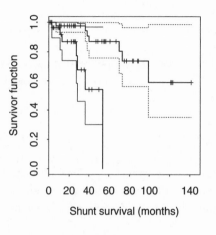
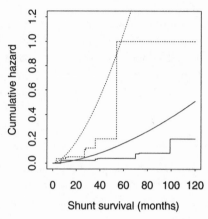

Figure 5.10 Nonparametric analysis of shunt data. Left panel: product-limit estimates of survivor function for older (upper heavy line) and younger infants (lower heavy line), with 95% confidence intervals (dots and light solid). Pluses on the product-limit estimates mark times of censored data. Right panel: estimated cumulative hazard functions for older (solid) and younger (dots) infants, using nonparametric estimate and fitted Weibull model (smooth curves).

Example 5.30 (Blalock–Taussig shunt data)   Table 5.6 illustrates the calculation of the product-limit estimator using data from Table 5.3. As the estimator changes only at times of failures, it need not be calculated at censoring times. The estimate does not approach zero for large y because the largest observation in the sample is censored.

Estimated survivor functions for both groups are shown in the left panel of Figure 5.10, together with approximate 95% confidence intervals. There is a strong effect of age, with shunts failing appreciably sooner for the younger children. The right panel compares the cumulative hazard function estimators $\widehat{H}(y) = \sum_{i:y_i \leq y} \widehat{h}_i$ with their parametric counterparts under the best Weibull model of Example 5.28. The parametric fits overstate the hazards appreciably. The apparent large difference after 60 months is largely due to a single failure in the younger group that strongly influences the analysis.   ∎

5.4.4 Other ideas

Competing risks

In some applications there may be different types of failure due to k different causes, say, and each failure time Y is accompanied by an indicator I showing which type of failure occurred. We can then define *cause-specific hazard functions*

$$h_i(y) = \lim_{\delta y \to 0} \frac{\Pr(y \leq Y \leq y + \delta y, I = i \mid Y \geq y)}{\delta y}, \quad y \geq 0, \; i = 1, \ldots, k,$$

corresponding to the rate at which failure of type i occurs, given survival to y. The overall hazard, cumulative hazard and survivor functions may be written

$$h(y) = \sum_{i=1}^{k} h_i(y), \quad H(y) = \sum_{i=1}^{k} \int_0^y h_i(u)\,du, \quad \mathcal{F}(y) = \exp\left\{\sum_{i=1}^{k} \int_0^y h_i(u)\,du\right\}.$$

If we imagine observing a population of values of (Y, I), then each of the $h_i(y)$ would be known, but we would observe no other aspect of the population. Thus without further assumptions the only estimable quantities are functions of the $h_i(y)$ such as $H(y)$ and $\mathcal{F}(y)$.

The likelihood contribution from an uncensored failure of type i is $h_i(y)\mathcal{F}(y)$, while provided censoring is independent, that from a censored failure is $\mathcal{F}(y)$, because the corresponding I is unknown. Suppose that we have independent triplets $(y_1, i_1, d_1), \ldots, (y_n, i_n, d_n)$, where y_j is the jth survival time and $d_j = 1$ if it is uncensored. If so, i_j indicates its failure type, while $i_j = 0$, say, if $d_j = 0$. The likelihood based on these data is

$$\prod_{j=1}^{n} \mathcal{F}(y_j) \prod_{i=1}^{k} h_{i_j}(y_j)^{d_j} = \prod_{i=1}^{k} \left[\prod_{j=1}^{n} \exp\left\{-\int_0^{y_j} h_i(y)\,du\right\} h_i(y_j)^{d_j I(i_j = i)} \right],$$

so it follows that to estimate $h_i(y)$ we treat any failure not of type i as a censoring. Thus, for example, the survivor function for $h_i(y)$ may be estimated by the product-limit estimator (5.28) with d_j replaced by $d_j I(i_j = i)$. Failures of types other than i are treated as censorings. Likewise for estimation of a parametric h_i.

For simplicity let $k = 2$. One way to think of competing risks is in terms of latent or potential failure times Y_1, Y_2 corresponding to the failure types. The observed quantities are $Y = \min(Y_1, Y_2)$ and $I = \{i : Y_i = Y\}$. Here Y_1 is interpreted as the time to failure that would be observed if cause 2 was removed, assuming that the failure time distribution for cause 1 when both causes of failure operate remains unchanged if cause 2 is eliminated. This assumption may be plausible in situations such as a reliability study where different types of failure are due to physically separate sub-systems and it is possible to imagine that all but one of these have been perfected, but the elimination of one failure type may alter the risk for others, particularly in medical contexts, where the assumption is often unsustainable. If it can be justified by appeal to subject-matter considerations it is very useful — the case for vaccination against infectious diseases, for example, presumes that removal of their risks increases overall survival.

An even stronger assertion is that Y_1 and Y_2 actually exist for each unit under study, with independence of causes of failure equivalent to independence of Y_1 and Y_2. In fact it is impossible to contradict this model. As mentioned above, the only observable quantities are functions of the cause-specific hazards $h_1(y)$ and $h_2(y)$. The joint survivor function

$$\mathcal{F}(y_1, y_2) = \Pr(Y_1 > y_1, Y_2 > y_2) = \exp\left\{-\int_0^{y_1} h_1(u)\,du - \int_0^{y_2} h_2(u)\,du\right\}$$

Table 5.7 Mouse data (Hoel and Walburg, 1972). Age at death (days) of RFM male mice exposed to 300 rads of x-radiation at 5–6 weeks of age. The causes of death were thymic lymphoma, reticulum cell sarcoma and other. The upper group of 95 mice were kept in a conventional environment; the lower 82 in a germ-free environment.

Lymphoma	159	189	191	198	200	207	220	235	245	250
	256	261	265	266	280	343	356	383	403	414
	428	432								
Sarcoma	317	318	399	495	525	536	549	552	554	557
	558	571	586	594	596	605	612	621	628	631
	636	643	647	648	649	661	663	666	670	695
	697	700	705	712	713	738	748	753		
Other	163	179	206	222	228	249	252	282	324	333
	341	366	385	407	420	431	441	461	462	482
	517	517	524	564	567	586	619	620	621	622
	647	651	686	761	763					
Lymphoma	158	192	193	194	195	202	212	215	229	230
	237	240	244	247	259	300	301	321	337	415
	434	444	485	496	529	537	624	707	800	
Sarcoma	430	590	606	638	655	679	691	693	696	747
	752	760	778	821	986					
Other	136	246	255	376	421	565	616	617	652	655
	658	660	662	675	681	734	736	737	757	769
	777	800	807	825	855	857	864	868	870	870
	873	882	895	910	934	942	1015	1019		

is a model for independent failures that yields cause-specific hazard functions h_1 and h_2, so whatever the form of these functions, data of form (Y, I) cannot give evidence against independent risks. Dependence can only be inferred from data in which both Y_1 and Y_2 are observed for certain units, or from subject-matter considerations. This is important because interest often focuses on the effect of eliminating failures of one type, say type 2, in which case the survivor function is $\mathcal{F}(y, 0)$. As this is not a function of h_1 and h_2 it is inestimable unless assumptions, typically unverifiable ones, are made about the relation between the risks. Some statisticians therefore insist that the only valid inferences from competing risk data concern the h_i and quantities derived from them.

Example 5.31 (Mouse data) The data in Table 5.7 are from a experiment in which two groups of RFM strain male mice were exposed to 300 rad of radiation at age 5–6 weeks. The first group lived in a conventional laboratory environment, and the second group lived in a germ-free environment. After their deaths, a pathologist ascertained whether the death was due to one of two types of cancer or to other causes. One purpose of the experiment was to assess the effect of environment on different causes of death. As irradiation took place when the mice were aged between 35 and 42 days old, it might be better to take age since irradiation as the response, but its exact value is unknown.

The panels of Figure 5.11 shows the estimated cumulative hazard functions for death from lymphoma and from other causes. Mortality from the lymphoma arises early, and seems to depend little on the environment. Deaths from other causes arise earlier in the conventional environment than in the germ-free one. See also Example 10.38. ∎

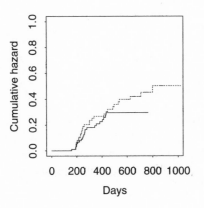 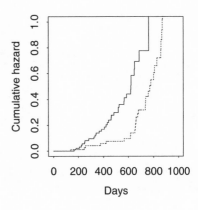

Frailty

The discussion above presupposes that all units have the same propensity to fail. In practice this is unrealistic — some cars are more reliable than others, some persons healthier than others, and so forth — and it may be important to build heterogeneity into models for survival. One reason for this is that allowing the failure rate to vary across units may greatly change the interpretation of the hazard function. It is tempting to view the population hazard function as a measure of how the risk for each unit changes as a function of time. For example, the fact that the divorce rate typically increases to a maximum a few years after marriage and thereafter decreases is sometimes interpreted as meaning that the typical marriage experiences increasing difficulties, but that if these are resolved there is eventually a more stable union. A unimodal divorce rate can be generated, however, by supposing that the hazard of failure increases with the duration of each marriage, but that the initial value of this hazard varies randomly from couple to couple. If this second interpretation is correct, then the population hazard function depends both on hazards for individual marriages and on variation across them, and reflects a selection process whereby the marriages most at risk tend to fail quickly, leaving those that were more stable to begin with. Thus the hazard rate is a more complicated quantity than it might seem at first sight.

One approach is to represent heterogeneity using the outcome of a positive random variable, Z, known as a *frailty*. We suppose that Z varies across units according to a density $f_Z(z)$, and that at time y the hazard function for a unit for whom $Z = z$ is $zh(y)$; thus the cumulative hazard to that time is $zH(y)$. Units whose z is large have high hazard functions and tend to fail sooner than those whose frailty is low. If known, the value of z could be incorporated into the analysis by modifying the likelihood, but we suppose it is unobserved, perhaps representing unobserveable genetic and environmental differences among units, and use it to model heterogeneity in the data.

As the survivor function for a unit with frailty z may be expressed as $\Pr(Y \geq y \mid Z = z) = \exp\{-zH(y)\}$, the survivor function for a unit taken randomly from the population is

$$
\begin{aligned}
\Pr(Y \geq y) &= \int_0^\infty \Pr(Y \geq y \mid Z = z) f_Z(z) \, dz \\
&= \int_0^\infty \exp\{-zH(y)\} f_Z(z) \, dz \\
&= M\{-H(y)\},
\end{aligned}
$$

where M is the moment-generating function of Z. Thus the cumulative hazard function for the population is $-\log M\{-H(y)\}$. The densities of Z conditional on failure at y and conditional on survival at least to y,

$$
f(z \mid Y = y) = \frac{z f_Z(z) \exp\{-zH(y)\}}{\int_0^\infty z f_Z(z) \exp\{-zH(y)\} \, dz},
$$

$$
f_Z(z \mid Y \geq y) = \frac{e^{-zH(y)} f_Z(z)}{\int_0^\infty \exp\{-zH(y)\} f_Z(z) \, dz}, \quad z > 0,
$$

can be used to see how frailty depends on failure and on survival.

Example 5.32 (Logistic hazard)   Let $\beta > 0$ and $H(y) = e^{\beta y} - 1$, so a unit with frailty z has hazard $z\beta e^{\beta y}$; this increases exponentially. Suppose also that Z has the gamma density with mean $\alpha\beta^{-1}/(1 + \alpha)$ and shape parameter β^{-1}. Then $M(u) = \{1 - \alpha u/(1 + \alpha)\}^{-1/\beta}$, and the population cumulative hazard function,

$$
-\log M\{-H(y)\} = \frac{1}{\beta} \log\left(\frac{1 + \alpha e^{\beta y}}{1 + \alpha}\right),
$$

is the same as that fitted to the data on old age in Example 5.29. Thus although each unit has a constant hazard, the effect of frailty is that the population hazard has an S-shaped logistic form, because of the selective effect of the early failure of the weakest units.

Simple calculations show that the density of frailties among those units failing at time y is gamma with mean $\alpha(1 + \beta^{-1})/(1 + \alpha e^{\beta y})$ and shape parameter $1 + \beta^{-1}$, while that among those units who have not failed at time y is gamma with corresponding parameters $\alpha\beta^{-1}/(1 + \alpha e^{\beta y})$ and β^{-1}. Both of these are decreasing in y, showing how the tendency for units with high frailties to fail first leads to survival of the fittest.

Information on unit hazard functions would be needed before such a model could be regarded as a serious explanation of the good fit of the logistic hazard for the data on old age. Absent such knowledge, the model is best regarded as suggesting a possible mechanism for the observed phenomenon, and as indicating the type of data needed for a more detailed investigation. ∎

Evidently frailty has the potential to greatly complicate the analysis of population phenomena. It also complicates group comparisons (Problem 5.15).

Exercises 5.4

If in doubt, think of failures of your car, fridge, computer, ...

1 Show that if there is no censoring, the product-limit estimator may be written $\widehat{\mathcal{F}}(y) = n^{-1}\#\{i : y_i > y\}$, and hence show that in this case $1 - \widehat{\mathcal{F}}(y)$ equals the empirical distribution function (2.3). Find Greenwood's formula, and comment.

2 Suggest physical phenomena that might give increasing, decreasing, and bathtub-shaped hazard functions. Sketch the corresponding survivor functions.

3 Use the relation $\mathcal{F}(y) = \exp\{-\int_0^y h(u)du\}$ between the survivor and hazard functions to find the survivor functions corresponding to the following hazards: (a) $h(y) = \lambda$; (b) $h(y) = \lambda y^\alpha$; (c) $h(y) = \alpha y^{\kappa-1}/(\beta + y^\kappa)$. In each case state what the distribution is. Show that $E\{1/h(Y)\} = E(Y)$ and hence find the means in (a), (b), and (c).

4 The *mean excess life function* is defined as $e(y) = E(Y - y \mid Y > y)$. Show that

$$e(y) = \mathcal{F}(y)^{-1} \int_y^\infty \mathcal{F}(u)\,du$$

and deduce that $e(y)$ satisfies the equation $e(y)Q'(y) + Q(y) = 0$ for a suitable $Q(y)$. Hence show that provided the underlying density is continuous,

$$\mathcal{F}(y) = \frac{e(0)}{e(y)} \exp\left\{-\int_0^y \frac{1}{e(u)}\,du\right\}.$$

As a check on this, find $e(y)$ and hence $\mathcal{F}(y)$ for the exponential density.
One approach to modelling survival is in terms of $e(y)$. For human lifetime data, let $e(y) = \gamma(1 - y/\theta)^\beta$, where θ is an upper endpoint and $\beta, \gamma > 0$. Find the corresponding survivor and hazard functions, and comment.

5 If $\mathcal{F}_1(y), \ldots, \mathcal{F}_k(y)$ are the survivor functions of independent positive random variables and $\beta_1, \ldots, \beta_k > 0$, show that $\prod \mathcal{F}_i(y)^{\beta_i}$ is also a survivor function, and find the corresponding hazard and cumulative hazard functions.
Suppose that $k = 2$ and the survivor functions are (i) log-logistic, (ii) log-normal and (iii) Weibull. Show that in the first two cases new models are obtained, but that in the third the parameters are not identifiable.

6 An empirical estimate of the survivor function $\mathcal{F}(y)$ when data $y_1, \ldots, y_n$ are not censored is given by $\widehat{\mathcal{F}}(y) = \#\{j : y_j > y\}/(n + 1)$. Suggest how plots of $\log\{-\log\widehat{\mathcal{F}}(y_j)\}$ against $\log y_j$ may be used to indicate if the data have Weibull or exponential distributions. Describe the corresponding plot for the Gumbel distribution function $F(y) = \exp[-\exp\{-(y - \eta)/\alpha\}]$.

7 Show that the log likelihood (5.26) may be expressed as

$$\ell(\theta) = \int_0^\infty \log h(y; \theta)\,dD(y) - \int_0^\infty R(y)\,dH(y; \theta),$$

where $D(y)$ is a step function with jumps of size one at the values of y that are failures and $R(y)$ is the number of units at risk of failure at time y. Establish that both integrals are over finite ranges. Such expressions are useful in a general treatment of likelihood inference for failure data.

5.5 Missing Data

5.5.1 Types of missingness

Missing observations arise in many applications, but particularly in data from living subjects, for example when frost kills a plant or the laboratory cat kills some experimental mice. They are common in data on humans, who may agree to take part in a two-year study and then drop out after six months, or refuse to answer questions about their salaries or sex-lives. They may occur by accident or by design, for example when lifetimes are censored at the end of a survival study (Section 5.4).

The central problem they pose is obvious: little can be said about unknown data, even if the pattern of missingness suggests its cause and hence indicates to what extent remaining observations can be trusted and lost ones imputed. Loss of data will clearly increase uncertainty, but a more malign effect is that inferences from the data are sharply limited unless we are prepared to make assumptions that the data themselves cannot verify. Thus, if data are missing or might be missing it is essential to consider possible underlying mechanisms and their potential effect on inferences. The discussion below is intended to focus thought about these.

Suppose that our goal is inference for a parameter θ based on data that would ideally consist of n independent pairs (X, Y), but that some values of Y are missing, as shown by an indicator variable, I. Thus the data on an individual have form $(x, y, 1)$ or $(x, ?, 0)$. We suppose that although the missingness mechanism $\Pr(I = 0 \mid x, y)$ may depend on x and y, it does not involve θ. Then the likelihood contribution from an individual with complete data is the joint density of X, Y and I, which we write as

$$\Pr(I = 1 \mid x, y) f(y \mid x; \theta) f(x; \theta),$$

while if Y is unknown we use the marginal density of X and I,

$$\int \Pr(I = 0 \mid x, y) f(y \mid x; \theta) f(x; \theta) \, dy. \tag{5.30}$$

There are now three possibilities:

- data are *missing completely at random*, that is, $\Pr(I = 0 \mid x, y) = \Pr(I = 0)$ is independent both of x and y, and (5.30) reduces to $\Pr(I = 0) f(x; \theta)$;
- data are *missing at random*, that is, $\Pr(I = 0 \mid x, y) = \Pr(I = 0 \mid x)$ depends on x but not on y, and (5.30) equals $\Pr(I = 0 \mid x) f(x; \theta)$; and
- there is *non-ignorable non-response*, meaning that $\Pr(I = 0 \mid x, y)$ depends on y and possibly also on x.

In the first two of these, which are often grouped as *ignorable non-response*, I carries no information about θ and can be omitted for most likelihood inferences. To see why, suppose that we have n independent observations of form $(x_1, y_1, I_1), \ldots, (x_n, y_n, I_n)$, let $\mathcal{M}$ be the set of j for which y_j is unobserved, and suppose that data are missing at random. Then the likelihood is

$$L(\theta) = \prod_{j \in \mathcal{M}} \Pr(I_j = 0 \mid x_j) f(x_j; \theta) \times \prod_{j \notin \mathcal{M}} \Pr(I_j = 1 \mid x_j) f(x_j, y_j; \theta)$$

$$\propto \prod_{j \in \mathcal{M}} f(x_j; \theta) \times \prod_{j \notin \mathcal{M}} f(x_j, y_j; \theta),$$

because the terms involving I_j do not depend on θ. Thus the missing data mechanism does not affect maximum likelihood estimates $\widehat{\theta}$, likelihood ratio statistics or the observed information $J(\widehat{\theta})$. It does affect the expected information, however, so standard errors for $\widehat{\theta}$ should be based on $J(\widehat{\theta})^{-1}$; see the discussion of likelihood

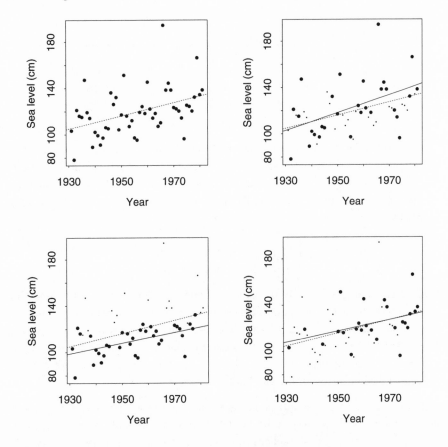

Figure 5.12 Missing data in straight-line regression for Venice sea-level data. Clockwise from top left: original data, data with values missing completely at random, data with values missing at random — missingness depends on x but not on y, and data with non-ignorable non-response — missingness depends on both x and y. Missing values are represented by a small dot. The dotted line is the fit from the full data, the solid lines those from the non-missing data.

inference in Section 5.4 and Problem 5.16. A similar argument applies if data are missing completely at random. If the non-response is non-ignorable, however, the density of I is no longer a constant of integration in (5.30). In that case, knowledge of the observed I_j is informative about θ, and likelihood inference is possible only if $\Pr(I = 0 \mid x, y)$ can be specified.

Example 5.33 (Venice sea level data)   The upper left panel of Figure 5.12 shows the data of Example 5.1. Here x represents a year in the range 1931–1981; in the absence of sea level it contains no information about any trend. The annual maximum sea level y is taken to be a normal variable with mean $\beta_0 + \beta_1(x_j - \overline{x})$ and variance σ^2; hence $\theta = (\beta_0, \beta_1, \sigma^2)$ and the full data likelihood has form $f(y \mid x; \theta) f(x)$, of which $f(x)$ is ignored.

The upper right panel of Figure 5.12 shows the effect of data missing completely at random, while in the panel below the probability that a value is unobserved depends on x but not on y; the data are missing at random, with earlier observations missing more often than later ones. The lower left panel shows non-ignorable non-response, because the probability of missingness depends on y and on x; values of y that are larger than their means are more likely to be missing. Here the fitted line differs from those in the other panels due to bias induced by the missingness mechanism.

	Truth	Full	MCAR	MAR	NIN
		Average estimate (average standard error)			
β_0	120	120 (2.79)	120 (4.02)	120 (4.73)	132 (3.67)
β_1	0.50	0.49 (0.19)	0.48 (0.28)	0.50 (0.32)	0.20 (0.25)

Table 5.8 Average estimates and standard errors for missing value simulation based on Venice data, for full dataset, with data missing completely at random (MCAR), missing at random (MAR) and with non-ignorable non-response (NIN). 1000 samples were taken. Standard errors for the averages for $\widehat{\beta}_0$ and $\widehat{\beta}_1$ are at most 0.16 and 0.01; those for their standard errors are at most 0.03 and 0.002.

To assess the extent of this bias, we generated 1000 samples from a model with parameters $\beta_0 = 120$, $\beta_1 = 0.5$ and $\sigma = 20$, close to the estimates for the Venice data and with the same covariate x. We then computed maximum likelihood estimates for the full data and for those observations that remain after applying the non-response mechanisms

$$\Pr(I = 1 \mid x, y) = \begin{cases} 0.5, \\ \Phi\left\{0.05(x - \overline{x})\right\}, \\ \Phi\left[0.05(x - \overline{x}) + \{y - \beta_0 - \beta_1(x - \overline{x})\}/\sigma\right], \end{cases}$$

to give data missing completely at random, missing at random, and with non-ignorable non-response. In each case roughly one-half of the observations are missing. Table 5.8 shows that although data loss increases the variability of the estimates, their means are unaffected, provided the probability of non-response does not depend on y. If the probability of missingness depends on the response, however, estimates based on the remaining data become entirely unreliable. ∎

The message of this example is bleak: when there is non-ignorable non-response and a non-negligible proportion of the data is missing, the only possible rescue is to specify the missingness mechanism correctly. In practice it is typically hard to tell if missingness is ignorable or not, so fully reliable inference is largely out of reach. Sensitivity analysis to assess how heavily the conclusions depend on plausible mechanisms for non-response is then useful, and we now outline one approach to this.

Publication bias

Breakthroughs in medical science are regularly reported, offering hope of a new cure or suggesting that some enjoyable activity has dire consequences. It is unwise to take them all at face value, however, as some turn out to be spurious. One reason for this is the publication process to which they are subjected. Once a study is completed, an article describing it is typically submitted to a medical journal for peer review. If the study design and analysis are found to be satisfactory, a decision is taken whether the article should be published. This decision is likely to be positive if the study reports a significant result or if it involved a large number of patients, but will often be negative if no association is found — there is no 'significant finding' — particularly if the study is small and hence deemed unreliable. The end-result of this selection process is *publication bias*, whereby studies finding associations tend to be the ones published, even if in fact there is no effect. Recommendations to change medical practice are usually based not on a single study — unless it is huge, involving many thousands of patients — but on a *meta-analysis* that combines results from all published studies.

As studies finding no effect are more likely to remain unpublished, however, wrong conclusions can be drawn.

For a simple model of this selection process, suppose that we wish to estimate a parameter μ that represents the effect of a treatment, subject to possible publication bias. A study based on n individuals produces an estimate $\widehat{\mu}$, normally distributed with mean μ and variance σ^2/n. The vagaries of the editorial process are represented by a variable Z, with the study published if Z is positive. We suppose that $\widehat{\mu}$ and Z are related by

$$\widehat{\mu} = \mu + \sigma n^{-1/2} U_1, \quad Z = \gamma_0 + \gamma_1 n^{1/2} + U_2,$$

with U_1 and U_2 standard normal variables with correlation $\rho \geq 0$. One interpretation of U_1 is as the standardized form $n^{1/2}(\widehat{\mu} - \mu)/\sigma$ of $\widehat{\mu}$, which is used to assess significance of the treatment effect. If $\rho > 0$ then publication becomes increasingly likely as U_1 increases, because Z is positively correlated with U_1. In terms of our previous discussion, Y and X correspond to $\widehat{\mu}$ and n, but now neither is observed if the study is unpublished.

The missingness indicator I equals one if $Z > 0$ and zero otherwise, so the marginal probability of publication is

$$\Pr(I = 1) = \Pr(Z > 0) = \Pr\left(U_2 > -\gamma_0 - \gamma_1 n^{1/2}\right) = \Phi\left(\gamma_0 + \gamma_1 n^{1/2}\right). \quad (5.31)$$

If $\gamma_1 > 0$ this increases with n: large studies are then more likely to be published, whatever their outcome. Conditional on the value of $\widehat{\mu}$, (3.21) implies that Z is normal with mean $\gamma_0 + \gamma_1 n^{1/2} + \rho n^{1/2}(\widehat{\mu} - \mu)/\sigma$ and variance $1 - \rho^2$. Hence the conditional probability of publication given $\widehat{\mu}$ is

$$\Pr(I = 1 \mid \widehat{\mu}) = \Pr\left(Z > 0 \mid \widehat{\mu}\right) = \Phi\left\{\frac{\gamma_0 + \gamma_1 n^{1/2} + \rho n^{1/2}(\widehat{\mu} - \mu)/\sigma}{(1 - \rho^2)^{1/2}}\right\}. \quad (5.32)$$

If $\rho > 0$, this is increasing in $\widehat{\mu}$: the probability that a study is published increases with the estimated treatment effect, at each study size n. Moreover, as $\widehat{\mu}$ appears in (5.32), non-response — non-publication of a study — is non-ignorable. If $\rho = 0$, (5.32) reduces to (5.31). Unpublished studies are then missing at random: the odds that a study is published depend on its size n but not on its outcome $\widehat{\mu}$.

Conditional on publication, the mean of $\widehat{\mu}$ is

$$E(\widehat{\mu} \mid Z > 0) = \mu + \rho \sigma n^{-1/2} \zeta\left(\gamma_0 + \gamma_1 n^{1/2}\right), \quad (5.33)$$

where $\zeta(u) = \phi(u)/\Phi(u)$ is the ratio of the standard normal density and distribution functions. If $\gamma_1, \rho > 0$, then $E(\widehat{\mu} \mid Z > 0) > \mu$, so the mean of a published $\widehat{\mu}$ is always larger than μ, but by an amount that decreases with n. For small γ_1, Taylor expansion gives

$$E(\widehat{\mu} \mid Z > 0) \doteq \mu + \rho \sigma \gamma_1 \zeta'(\gamma_0) + \rho \sigma \zeta(\gamma_0) n^{-1/2},$$

so the conditional mean of $\widehat{\mu}$ in published studies is roughly linear in $n^{-1/2}$. As just three parameters — intercept, slope and variance — can be estimated from a linear fit, simultaneous estimation of μ, ρ, σ^2, γ_0, and γ_1 is infeasible. In order to assess

Trial	Magnesium r/m	Control r/m	n	$\widehat{\mu}$	$(v/n)^{1/2}$
1	1/25	3/23	48	1.18	1.05
2	1/40	2/36	76	0.80	0.83
3	2/48	2/46	94	0.04	0.75
4	1/50	9/53	103	2.14	0.72
5	4/56	14/56	112	1.25	0.69
6	3/66	6/66	132	0.69	0.63
7	2/92	7/93	185	1.24	0.53
8	27/135	43/135	270	0.47	0.44
9	10/160	8/156	316	−0.20	0.41
10	90/1159	118/1157	2316	0.27	0.15
Meta-analysis			3652	0.41	0.11
ISIS-4	2216/29011	2103/29039	58050	−0.05	0.03

Table 5.9 Data from 11 clinical trials to compare magnesium treatment for heart attacks with control, with n patients randomly allocated to treatment and control; there are r deaths out of m patients in each group (Copas, 1999). The estimated log treatment effect $\widehat{\mu}$ will be positive if treatment is effective; $(v/n)^{1/2}$ is its standard error. The huge ISIS-4 trial is not included in the meta-analysis.

the impact of selection in the following example, we fix γ_0 and γ_1 to give plausible probabilities of publication for small and large samples, and consider inference for $\theta = (\mu, \rho, \sigma)$.

Now suppose that we wish to estimate μ based on k independent estimates $\widehat{\mu}_1, \ldots, \widehat{\mu}_k$ from published studies of sizes $n_1, \ldots, n_k$. As $\widehat{\mu}_j$ is observed only conditional on its publication, the likelihood contribution from study j is

$$f(\widehat{\mu}_j \mid Z_j > 0; \theta) = \frac{f(\widehat{\mu}_j; \theta)\Pr(Z_j > 0 \mid \widehat{\mu}_j; \theta)}{\Pr(Z_j > 0)}.$$

The marginal density of $\widehat{\mu}_j$ is normal with mean μ and variance σ^2/n_j, and on recalling (5.31) and (5.32), we see that the overall log likelihood is

$$\ell(\mu, \rho, \sigma^2) \equiv -\sum_{j=1}^{k} \left\{ \frac{1}{2}\log\sigma^2 + \frac{n_j}{2\sigma^2}(\widehat{\mu}_j - \mu)^2 + \log\Phi(a_j) - \log\Phi(b_j) \right\},$$
(5.34)

where $a_j = \gamma_0 + \gamma_1 n_j^{1/2}$ and $b_j = (1 - \rho^2)^{-1/2}\{a_j + \rho n_j^{1/2}(\widehat{\mu}_j - \mu)/\sigma\}$.

The simplest meta-analysis ignores the possibility of selection bias and amounts to setting $\rho = 0$, presuming the publication of a study to be unrelated to its result. If this is so, then $a_j = b_j$ and the log likelihood is easily maximized, the maximum likelihood estimate of μ being the weighted average

$$\frac{\sum n_j \widehat{\mu}_j}{\sum n_j}.$$
(5.35)

When $\rho = 0$, this estimator is normal with mean μ and variance $\sigma^2/\sum n_j$. If in fact $\rho > 0$, then (5.33) implies that $\widehat{\mu}_0$ will tend to exceed μ; the treatment effect will tend to be overstated by the published data.

Example 5.34 (Magnesium data) Table 5.9 shows data from clinical trials on the use of intraveneous magnesium to treat patients with suspected acute myocardial

Figure 5.13 Likelihood
analysis of magnesium
data. Left: funnel plot
showing variation of $\widehat{\mu}$
with trial size n, with 95%
confidence interval for μ
based on each trial. The
vertical dotted line is the
combined estimate of μ
from the ten small trials,
ignoring the possibility of
publication bias; the
vertical solid line shows
no treatment effect. The
solid line is the estimated
conditional mean (5.33).
Right: contours of $\widehat{\mu}$ as a
function of γ_0 and γ_1.

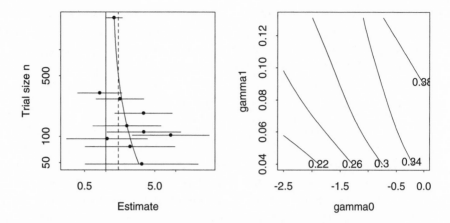

Figure 5.13 Likelihood
analysis of magnesium
data. Left: funnel plot
showing variation of $\widehat{\mu}$
with trial size n, with 95%
confidence interval for μ
based on each trial. The
vertical dotted line is the
combined estimate of μ
from the ten small trials,
ignoring the possibility of
publication bias; the
vertical solid line shows
no treatment effect. The
solid line is the estimated
conditional mean (5.33).
Right: contours of $\widehat{\mu}$ as a
function of γ_0 and γ_1.

Myocardial infarction is
the medical term for heart
attack — death of part of
the heart muscle because
of lack of oxygen and
other nutrients.

infarction. For each trial, we consider the difference in log proportion of deaths between control and treated groups, the estimated treatment effect $\widehat{\mu} = \log(r_2/m_2) - \log(r_1/m_1)$. Now $m_1 \doteq m_2$ for each trial and the proportion of deaths is small, so the delta method suggests that an approximate variance for $\widehat{\mu}$ is $4/(\widehat{\lambda}n)$, where $\widehat{\lambda} = 0.097$ is the death rate estimated from all the trials and $n = m_1 + m_2$ is the size of each trial. The combined sample is large enough to treat $\widehat{\lambda}$ and hence $\sigma^2 = 4/\widehat{\lambda}$ as constant. Although the estimated treatment effects $\widehat{\mu}$ from the ten small trials are individually inconclusive, the meta-analysis estimate (5.35) is 0.41 with standard error 0.11; this gives an estimated reduction in the probability of death by a factor $\exp(0.41) = 1.51$ with 0.95 confidence interval (1.22,1.86). A similar published meta-analysis concluded that the magnesium treatment was 'effective, safe and simple'.

For a more skeptical view, consider the *funnel plot* of n and $\exp(\widehat{\mu})$ in the left panel of Figure 5.13; note the logarithmic axes. Symmetry about the overall weighted average (5.35) would show lack of publication bias, but the visible asymmetry suggests that small studies tend to be published only if $\widehat{\mu}$ is sufficiently positive.

The right panel shows how the maximum likelihood estimate of μ from (5.34) depends on γ_0 and γ_1. The contours are very roughly parallel with slope -0.05, suggesting that the maximum likelihood estimate varies mainly as a function of $\gamma_0 + 400^{1/2}\gamma_1$, or equivalently the probability $\Phi(\gamma_0 + 400^{1/2}\gamma_1)$ that a study of size $n = 400$ is published. For example, if the selection probabilities are 0.9 and 0.1 for the largest and smallest studies in Table 5.9, then this probability is 0.32, $\widehat{\rho} = 0.5$ and the estimated treatment effect is 0.27 with standard error 0.12 from observed information. This estimate is substantially less than the value 0.41 obtained when $\rho = 0$, and the significance of the estimated treatment effect is much reduced. The estimated conditional mean (5.33) in the left panel shows how the selection due to having $\rho > 0$ affects the mean of published studies.

The sensitivity of the estimated effect to potential publication bias suggests that treatment policy conclusions cannot be based on Table 5.9. Indeed, a subsequent much larger trial — ISIS-4 — found no evidence that magnesium is effective. ∎

Publication bias is an example of selection bias, where the mechanism underlying the choice of data introduces an uncontrolled bias into the sample. This is endemic in observational studies, for example in epidemiology and the social sciences, and it can greatly weaken what conclusions may be drawn.

5.5.2 EM algorithm

The fitting of certain models is simplified by treating the observed data as an incomplete version of an ideal dataset whose analysis would have been easy. The key idea is to estimate the log likelihood contribution from the missing data by its conditional value given the observed data. This yields a very general and widely used *estimation-maximization* or EM algorithm for maximum likelihood estimation.

Let Y denote the observed data and U the unobserved variables. Our goal is to use the observed value y of Y for inference on a parameter θ, in models where we cannot easily calculate the density

$$f(y; \theta) = \int f(y \mid u; \theta) f(u; \theta) \, du$$

and hence cannot readily compute the likelihood for θ based only on y. We write the *complete-data log likelihood* based on both y and the value u of U as

$$\log f(y, u; \theta) = \log f(y; \theta) + \log f(u \mid y; \theta), \tag{5.36}$$

where the first term on the right is the *observed-data log likelihood* $\ell(\theta)$. As the value of U is unobserved, the best we can do is to remove it by taking expectation of (5.36) with respect to the conditional density $f(u \mid y; \theta')$ of U given that $Y = y$; for reasons that will become apparent we use θ' rather than θ for this expectation. This yields

$$\mathrm{E}\{\log f(Y, U; \theta) \mid Y = y; \theta'\} = \ell(\theta) + \mathrm{E}\{\log f(U \mid Y; \theta) \mid Y = y; \theta'\}, \tag{5.37}$$

which we express as

$$Q(\theta; \theta') = \ell(\theta) + C(\theta; \theta'). \tag{5.38}$$

We now fix θ' and treat $Q(\theta; \theta')$ and $C(\theta; \theta')$ as functions of θ. If the conditional distribution of U given $Y = y$ is non-degenerate and no two values of θ give the same model, then the argument at (4.31) applied to $f(y \mid u; \theta)$ shows that $C(\theta'; \theta') \geq C(\theta; \theta')$, with equality only when $\theta = \theta'$. Hence

$$Q(\theta; \theta') \geq Q(\theta'; \theta') \text{ implies } \ell(\theta) - \ell(\theta') \geq C(\theta'; \theta') - C(\theta; \theta') \geq 0. \tag{5.39}$$

Moreover under mild smoothness conditions, $C(\theta; \theta')$ has a stationary point at $\theta = \theta'$. Hence if $Q(\theta; \theta')$ is stationary at $\theta = \theta'$, so too is $\ell(\theta)$.

This leads to the *EM algorithm*: starting from an initial value θ' of θ,

1.   compute $Q(\theta; \theta') = \mathrm{E}\{\log f(Y, U; \theta) \mid Y = y; \theta'\}$; then
2.   with θ' fixed, maximize $Q(\theta; \theta')$ over θ, giving $\theta^\dagger$, say; and
3.   check if the algorithm has converged, using $\ell(\theta^\dagger) - \ell(\theta')$ if available, or $|\theta^\dagger - \theta'|$, or both. If not, set $\theta' = \theta^\dagger$ and go to 1.

Steps 1 and 2 are the expectation (E) and maximization (M) steps of the algorithm. As the M-step ensures that $Q(\theta^{\dagger}; \theta') \geq Q(\theta'; \theta')$, we see from (5.39) that $\ell(\theta^{\dagger}) \geq \ell(\theta')$: the log likelihood never decreases. Moreover, if $\ell(\theta)$ has just one stationary point, and if $Q(\theta; \theta')$ eventually reaches a stationary value at $\widehat{\theta}$, then $\widehat{\theta}$ must maximize $\ell(\theta)$. If $\ell(\theta)$ has more than one stationary point the algorithm may converge to a local maximum of the log likelihood or to a turning point. As the EM algorithm never decreases the log likelihood it is more stable than Newton–Raphson-type algorithms, which do not have this desirable property.

As one might expect, the convergence rate of the algorithm depends on the amount of missing information. If knowledge of Y tells us little about U, then $Q(\theta; \theta')$ and $\ell(\theta)$ will be very different and the algorithm slow. This may be quantified by differentiating (5.36) and taking expectations with respect to the conditional distribution of U given Y, to give

$$-\frac{\partial^2 \ell(\theta)}{\partial \theta \partial \theta^{\mathrm{T}}} = \mathrm{E}\left\{-\frac{\partial^2 \log f(y, U; \theta)}{\partial \theta \partial \theta^{\mathrm{T}}}\middle| Y = y; \theta\right\}$$
$$-\mathrm{E}\left\{-\frac{\partial^2 \log f(U \mid y; \theta)}{\partial \theta \partial \theta^{\mathrm{T}}}\middle| Y = y; \theta\right\},$$

or $J(\theta) = I_c(\theta; y) - I_m(\theta; y)$, interpreted as meaning that the observed information equals the complete-data information minus the missing information; this is sometimes called the *missing information principle*. If U is determined by Y, then the conditional density $f(u \mid y; \theta)$ is degenerate and under mild conditions the missing information will be zero. It turns out that the rate of convergence of the algorithm equals the largest eigenvalue of the matrix $I_c(\theta; y)^{-1} I_m(\theta; y)$; values of this eigenvalue close to one imply slow convergence and occur if the missing information is a high proportion of the total.

When the EM algorithm is slow it may be worth trying to accelerate it by replacing the M-step with direct maximization, assuming of course that $\ell(\theta)$ is unavailable. It turns out that (Exercise 5.5.5)

$$\frac{\partial \ell(\theta)}{\partial \theta} = \left.\frac{\partial Q(\theta; \theta')}{\partial \theta}\right|_{\theta'=\theta}, \quad \frac{\partial^2 \ell(\theta)}{\partial \theta \partial \theta^{\mathrm{T}}} = \left.\left\{\frac{\partial^2 Q(\theta; \theta')}{\partial \theta \partial \theta^{\mathrm{T}}} + \frac{\partial^2 Q(\theta; \theta')}{\partial \theta \partial \theta'^{\mathrm{T}}}\right\}\right|_{\theta'=\theta}. \quad (5.40)$$

Thus even if $\ell(\theta)$ is inaccessible, its derivatives may be obtained from those of $Q(\theta; \theta')$ and used in a generic maximization algorithm. The second of these formulae also provides standard errors for the maximum likelihood estimate $\widehat{\theta}$ when $Q(\theta; \theta')$ is known but $\ell(\theta)$ is not.

Example 5.35 (Negative binomial model) For a toy example, suppose that conditional on $U = u$, Y is a Poisson variable with mean u, and that U is gamma with mean θ and variance θ^2/ν. Inference is required for θ with the shape parameter $\nu > 0$ supposed known. Here (5.36) equals

$$y \log u - u - \log y! + \nu \log \nu - \nu \log \theta + (\nu - 1) \log u - \nu u/\theta - \log \Gamma(\nu),$$

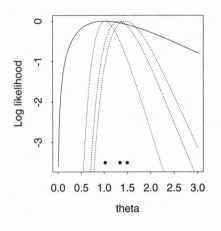

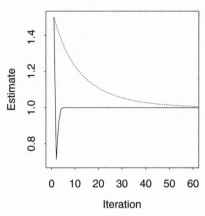

Figure 5.14 EM algorithm for negative binomial example. Left panel: observed-data log likelihood $\ell(\theta)$ (solid) and functions $Q(\theta;\theta')$ for $\theta' = 1.5, 1.347$ and 1.028 (dots, from right). The blobs show the values of θ that maximize these functions, which correspond to the first, fifth and fortieth iterations of the EM algorithm. Right: convergence of EM algorithm (dots) and Newton–Raphson algorithm (solid). The panel shows how successive EM iterations update θ' and $\widehat{\theta}$. Notice that the EM iterates always increase $\ell(\theta)$, while the Newton–Raphson steps do not.

and hence (5.37) equals

$$Q(\theta;\theta') = (y + \nu - 1)\mathrm{E}(\log U \mid Y = y; \theta') - (1 + \nu/\theta)\mathrm{E}(U \mid Y = y; \theta') - \nu \log \theta$$

plus terms that depend neither on U nor on θ.

The E-step, computation of $Q(\theta;\theta')$, involves two expectations, but fortunately $\mathrm{E}(\log U \mid Y = y; \theta')$ does not appear in terms that involve θ and so is not required. To compute $\mathrm{E}(U \mid Y = y; \theta')$, note that Y and U have joint density

$$f(y \mid u)f(u;\theta) = \frac{u^y}{y!}e^{-u} \times \frac{\nu^\nu u^{\nu-1}}{\theta^\nu \Gamma(\nu)}e^{-\nu u/\theta}, \quad y = 0, 1, \ldots, \quad u > 0, \quad \theta > 0,$$

so the marginal density of Y is

$$f(y;\theta) = \int_0^\infty f(y \mid u)f(u;\theta,\nu)\,du = \frac{\Gamma(y+\nu)\nu^\nu}{\Gamma(\nu)y!}\frac{\theta^y}{(\theta+\nu)^{y+\nu}}, \quad y = 0, 1, \ldots$$

Hence the conditional density $f(u \mid y; \theta')$ is gamma with shape parameter $y + \nu$ and mean $\mathrm{E}(U \mid Y = y; \theta') = (y + \nu)/(1 + \nu/\theta')$, and we can take

$$Q(\theta;\theta') \equiv -(1 + \nu/\theta)(y + \nu)/(1 + \nu/\theta') - \nu \log \theta,$$

where we have ignored terms independent of both θ and θ'.

The M-step involves maximization of $Q(\theta;\theta')$ over θ for fixed θ', so we differentiate with respect to θ and find that the maximizing value is

$$\theta^\dagger = \theta'(y + \nu)/(\theta' + \nu). \tag{5.41}$$

In this example, therefore, the EM algorithm boils down to choosing an initial θ', updating it to $\theta^\dagger$ using (5.41), setting $\theta' = \theta^\dagger$ and iterating to convergence.

The log likelihood based only on the observed data y is

$$\ell(\theta) = \log f(y;\theta) \equiv y \log \theta - (y + \nu)\log(\theta + \nu), \quad \theta > 0.$$

This is shown in the left panel of Figure 5.14 for $y = 1$ and $\nu = 15$. The panel also shows the functions $Q(\theta;\theta')$ on the first, fifth and fourtieth iterations starting at $\theta' = 1.5$, which gives the sequence $\theta' = 1.5, 1.45, 1.41, \ldots$. The functions $Q(\theta;\theta')$ are much more concentrated than is $\ell(\theta)$, showing that the amount of missing information is large. The difference in curvature corresponds to the information lost through not observing U.

Here the unmodified EM algorithm converges slowly. The right panel of Figure 5.14 illustrates this, as successive values of $\theta^{\dagger}$ descend gently towards the limiting value $\theta = 1$: convergence has still not been achieved after 100 iterations, at which point $\theta^{\dagger} = 1.00056$. The ratio of missing to complete-data information, $15/16$, indicates slow convergence. The Newton–Raphson algorithm (4.25) using the derivatives (5.40) converges much faster, with $\widehat{\theta} = 1$ to seven decimal places after only five iterations, so here it pays handsomely to use the derivative information in (5.40). ∎

Example 5.36 (Mixture density) Mixture models arise when an observation Y is taken from a population composed of distinct subpopulations, but it is unknown from which of these Y is taken. If the number p of subpopulations is finite, Y has a p-component mixture density

$$f(y;\theta) = \sum_{r=1}^{p} \pi_r f_r(y;\theta), \quad 0 \le \pi_r \le 1, \sum_{r=1}^{p} \pi_r = 1,$$

where π_r is the probability that Y comes from the rth subpopulation and $f_r(y;\theta)$ is its density conditional on this event. An indicator U of the subpopulation from which Y arises takes values $1, \ldots, p$ with probabilities $\pi_1, \ldots, \pi_p$. In many applications the components have a physical meaning, but sometimes a mixture is used simply as a flexible class of densities. For simplicity of notation below, let θ contain all unknown parameters including the π_r.

If the value u of U were known, the likelihood contribution from (y, u) would be $\prod_r \{f_r(y;\theta)\pi_r\}^{I(u=r)}$, giving contribution

$$\log f(y, u; \theta) = \sum_{r=1}^{p} I(u = r) \{\log \pi_r + \log f_r(y;\theta)\}$$

to the complete-data log likelihood. In order to apply the EM algorithm we must compute the expectation of $\log f(y, u; \theta)$ over the conditional distribution

$$\Pr(U = r \mid Y = y; \theta') = \frac{\pi_r' f_r(y;\theta')}{\sum_{s=1}^{p} \pi_s' f_s(y;\theta')}, \quad r = 1, \ldots, p. \tag{5.42}$$

This probability can be regarded as the weight attributable to component r if y has been observed; for compactness below we denote it by $w_r(y;\theta')$. The expected value of $I(U = r)$ with respect to (5.42) is $w_r(y;\theta')$, so the expected value of the log likelihood based on a random sample $(y_1, u_1), \ldots, (y_n, u_n)$ is

$$\begin{aligned} Q(\theta; \theta') &= \sum_{j=1}^{n} \sum_{r=1}^{p} w_r(y_j;\theta')\{\log \pi_r + \log f_r(y_j;\theta)\} \\ &= \sum_{r=1}^{p} \left\{ \sum_{j=1}^{n} w_r(y_j;\theta') \right\} \log \pi_r + \sum_{r=1}^{p} \sum_{j=1}^{n} w_r(y_j;\theta') \log f_r(y_j;\theta). \end{aligned}$$

9172	9350	9483	9558	9775	10227	10406	16084	16170	18419
18552	18600	18927	19052	19070	19330	19343	19349	19440	19473
19529	19541	19547	19663	19846	19856	19863	19914	19918	19973
19989	20166	20175	20179	20196	20215	20221	20415	20629	20795
20821	20846	20875	20986	21137	21492	21701	21814	21921	21960
22185	22209	22242	22249	22314	22374	22495	22746	22747	22888
22914	23206	23241	23263	23484	23538	23542	23666	23706	23711
24129	24285	24289	24366	24717	24990	25633	26960	26995	32065
32789	34279								

Table 5.10 Velocities (km/second) of 82 galaxies in a survey of the Corona Borealis region (Roeder, 1990). The error is thought to be less than 50 km/second.

The M step of the algorithm entails maximizing $Q(\theta; \theta')$ over θ for fixed θ'. As the π_r do not usually appear in the component density f_r, the maximizing values $\pi_r^\dagger$ are obtained from the first term of Q, which corresponds to a multinomial log likelihood; see (4.45). Thus $\pi_r^\dagger = n^{-1} \sum_j w_r(y_j; \theta')$, the average weight for component r.

Estimates of the parameters of the f_r are obtained from the weighted log likelihoods that form the second term of $Q(\theta; \theta')$. For example, if f_r is normal with mean μ_r and variance σ_r^2, simple calculations give the weighted estimates

$$\mu_r^\dagger = \frac{\sum_{j=1}^n w_r(y_j; \theta')y_j}{\sum_{j=1}^n w_r(y_j; \theta')} \qquad \sigma_r^{2\dagger} = \frac{\sum_{j=1}^n w_r(y_j; \theta')(y_j - \mu_r^\dagger)^2}{\sum_{j=1}^n w_r(y_j; \theta')}, \qquad r = 1, \dots, p.$$

Given initial values of $(\pi_r, \mu_r, \sigma_r^2) \equiv \theta'$, the EM algorithm simply involves computing the weights $w_r(y_j; \theta')$ for these initial values, updating to obtain $(\pi_r^\dagger, \mu_r^\dagger, \sigma_r^{2\dagger}) \equiv \theta^\dagger$, and checking convergence using the log likelihood, $|\theta^\dagger - \theta'|$, or both. If convergence is not yet attained, θ' is replaced by $\theta^\dagger$ and the cycle repeated.

We illustrate these calculations using the data in Table 5.10, which gives the velocities at which 82 galaxies in the Corona Borealis region are moving away from our own galaxy. It is thought that after the Big Bang the universe expanded very fast, and that as it did so galaxies formed because of the local attraction of matter. Owing to the action of gravity they tend to cluster together, but there seem also to be 'superclusters' of galaxies surrounded by voids. If galaxies are indeed super-clustered the distribution of their velocities estimated from the red-shift in their light-spectra would be multimodal, and unimodal otherwise. The data given are from sections of the northern sky carefully sampled to settle whether there are superclusters.

Cursory examination of the data strongly suggests clustering. In order to estimate the number of clusters we fit mixtures of normal densities by the EM algorithm with initial values chosen by eye. The maximized log likelihood for $p = 2$ is -220.19, found after 26 iterations. In fact this is the highest of several local maxima; the global maximum of $+\infty$ is found by centering one component of the mixture at any of the y_j and letting the corresponding $\sigma_r^2 \to \infty$; see Example 4.42. Only the local maxima yield sensible fits, the best of which is found using randomly chosen initial values. The number of iterations needed depends on these and on the number of components, but is typically less than 40. This procedure gives maximized log likelihoods -240.42, -203.48, -202.52 and -192.42 for fits with $p = 1, 3, 4$ and 5. The latter gives a single component to the two observations around 16,000 and so does not seem very

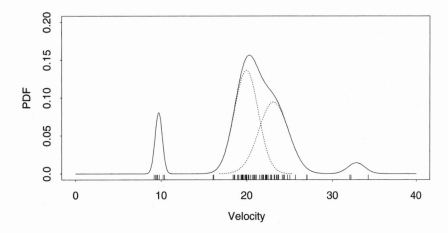

sensible. Standard likelihood asymptotics do not apply here, but evidently there is little difference between the 3- and 4-component fits, the second of which is shown in Figure 5.15. Both fits have three modes, and the evidence for clustering is very strong.

An alternative is to apply a Newton–Raphson algorithm directly to the log likelihood $\ell(\theta)$ based on the mixture density, but if this is to be reliable the model must be reparametrized so that the parameter space is unconstrained, using $\log \sigma_r^2$ and expressing $\pi_1, \ldots, \pi_p$ in terms of $\theta_1, \ldots, \theta_{p-1}$ of Example 5.12. As mentioned in Example 4.42, the effect of the spikes in $\ell(\theta)$ can be reduced by replacing $f_r(y; \theta)$ by $F_r(y + h; \theta) - F_r(y - h; \theta)$, where h is the degree of rounding of the data, here 50 km/second. ∎

Exponential family models

The EM algorithm has a particularly simple form when the complete-data log likelihood stems from an exponential family, giving

$$\log f(y, u; \theta) = s(y, u)^{\mathrm{T}}\theta - \kappa(\theta) + c(y, u).$$

The expected value of this is needed with respect to the conditional density $f(u \mid y; \theta')$. Evidently the final term will not depend on θ and can be ignored, so the M-step will involve maximizing

$$Q(\theta; \theta') = \mathrm{E}\{s(y, U)^{\mathrm{T}}\theta \mid Y = y; \theta'\} - \kappa(\theta),$$

or equivalently solving for θ the equation

$$\mathrm{E}\{s(y, U) \mid Y = y; \theta'\} = \frac{d\kappa(\theta)}{d\theta}.$$

The likelihood equation for θ based on the complete data would be $s(y, u) = d\kappa(\theta)/d\theta$, so the EM algorithm simply involves replacing $s(y, u)$ by its conditional expectation $\mathrm{E}\{s(y, U) \mid Y = y; \theta'\}$ and solving the likelihood equation. Thus a routine to fit the complete-data model can readily be adapted for missing data if the conditional expectations are available.

Example 5.37 (Positron emission tomography)   Positron emission tomography is performed by introducing a radioactive tracer into an animal or human subject. Radioactive emissions are then used to assess levels of metabolic activity and blood flow in organs of interest. Positrons emitted by the tracer annihilate with nearby electrons, giving pairs of photons that fly off in opposite directions. Some of these are counted by bands of gamma detectors placed around the subject's body, but others miss the detectors. The detected counts are used to form an image of the level of metabolic activity in the organs based on the estimated spatial concentration of isotope.

For a statistical model, the region of interest is divided into n pixels or voxels and it is assumed that the number of emissions U_{ij} from the jth pixel detected at the ith detector is a Poisson variable with mean $p_{ij}\lambda_j$; here λ_j is the intensity of emissions from that pixel and p_{ij} the probability that a single emission is detected at the ith detector. The p_{ij} depend on the geometry of the detection system, the isotope and other factors, but can be taken to be known. The U_{ij} are unknown but can plausibly be assumed independent. The counts Y_i at the d detectors are observed and have independent Poisson distributions with means $\sum_{j=1}^{n} p_{ij}\lambda_j$.

Pixels and voxels are picture and volume elements, in 2 and 3 dimensions respectively.

The complete-data log likelihood,

$$\sum_{i=1}^{d}\sum_{j=1}^{n}\{u_{ij}\log(p_{ij}\lambda_j) - p_{ij}\lambda_j\},$$

is an exponential family in which the maximum likelihood estimates of the unknown λ_j have the simple form $\widehat{\lambda}_j = \sum_i u_{ij} / \sum_i p_{ij}$. The E-step requires only the conditional expectations $E(U_{ij} \mid Y; \lambda')$. As $Y_i = U_{i1} + \cdots + U_{in}$, the conditional density of U_{ij} given $Y_i = y_i$ is binomial with denominator y_i and probability $p_{ij}\lambda'_j / \sum_h p_{ih}\lambda'_h$. Thus the M-step yields

$$\lambda_j^{\dagger} = \frac{\sum_{i=1}^{d} E(U_{ij} \mid Y_j = y_j; \lambda')}{\sum_{i=1}^{d} p_{ij}} = \frac{\sum_{i=1}^{d} y_j p_{ij}\lambda'_j / \sum_{h=1}^{n} p_{ih}\lambda'_h}{\sum_{i=1}^{d} p_{ij}}$$

$$= \lambda'_j \frac{1}{\sum_{i=1}^{d} p_{ij}} \sum_{i=1}^{d} \frac{y_i p_{ij}}{\sum_{h=1}^{n} \lambda'_h p_{ih}}, \quad j = 1, \ldots, n.$$

The algorithm converges to a unique global maximum of the observed-data log likelihood provided that $d > n$, with the positivity constraints on the λ_j satisfied at each step.

Though simple, this algorithm has the undesirable property that the resulting images are too rough if it is iterated to full convergence. The difficulty is that although we would anticipate that adjacent pixels would be similar, the model places no constraint on the λ_j and so the final image is too close to the data. Some modification is required, such as adding a smoothing step to the algorithm or introducing a roughness penalty (Section 10.7.2).   ∎

The EM algorithm is particularly attractive in exponential family problems, but is used much more widely. In more general situations both E- and M-steps may be complicated, and it often pays to break them into smaller components, perhaps involving Monte Carlo simulation to compute the conditional expectations required for the E-step. Discussion of this here would take us too far afield, but some of the recent research devoted to this is mentioned in the bibliographic notes.

Exercises 5.5

1   Data are *observed at random* if $\Pr(I = 0 \mid x, y) = \Pr(I = 0 \mid y)$, where I is the indicator that y is missing. Show that if data are observed at random and missing data are missing at random, then data are missing completely at random.

2   Show that Bayesian inference for θ is unaffected by the model for non-response if data are missing completely at random or missing at random, but not if there is non-ignorable non-response. What happens when $\Pr(I \mid x, y)$ depends on θ?

3   In Example 5.33, suppose that y is normal with mean $\beta_0 + \beta_1 x$ and variance σ^2, and that it is missing with probability $\Phi(a + by + cx)$, where a, b and c are unknown. Use (3.25) to find the likelihood contributions from pairs (x, y) and $(x, ?)$, and discuss whether the parameters are estimable.

4   When $\rho = 0$, show that (5.35) is the maximum likelihood estimate of μ and find its variance.

5   Use the fact that $\int f(u \mid y; \theta) \, du = 1$ for all y and θ to show that

$$0 = \mathrm{E}\left\{ \left. \frac{\partial \log f(U \mid Y; \theta)}{\partial \theta} \right| Y = y; \theta \right\},$$

$$0 = \mathrm{E}\left\{ \left. \frac{\partial^2 \log f(U \mid Y; \theta)}{\partial \theta \, \partial \theta^{\mathsf{T}}} + \frac{\partial \log f(U \mid Y; \theta)}{\partial \theta} \frac{\partial \log f(U \mid Y; \theta)}{\partial \theta^{\mathsf{T}}} \right| Y = y; \theta \right\}.$$

Now use (5.38) to establish (5.40).
Check this in the special case of Example 5.35, and hence give the Newton–Raphson step for maximization of the observed-data log likelihood, even though $\ell(\theta)$ itself is unknown. Write a program to compare the convergence of the EM and Newton–Raphson algorithms in that example.
(Oakes, 1999)

6   Check the forms of $\pi_r^{\dagger}$, $\mu_r^{\dagger}$ and $\sigma_r^{2\dagger}$ in Example 5.36, and verify that they respect the constraints $\sigma_r^2 > 0$, $0 \leq \pi_r \leq 1$ and $\sum \pi_r = 1$ on the parameter values.

7   Check the details of Example 5.37.

8   (a) To apply the EM algorithm to data censored at a constant c, let U denote the underlying failure time and suppose that $Y = \min(U, c)$ and $D = I(U \leq c)$ are observed. Thus the complete-data log likelihood is $\log f(u; \theta)$. Show that

$$f(u \mid y, d; \theta) = \begin{cases} \delta(u - y), & d = 1, \\ \dfrac{f(u; \theta)}{1 - F(c; \theta)}, & u > c, d = 0. \end{cases}$$

(b) If $f(u; \theta) = \theta e^{-\theta u}$, show that $\mathrm{E}(U \mid Y = y, D = d; \theta') = dy + (1 - d)(c + 1/\theta')$, and deduce that the iteration for a random sample $(y_1, d_1), \dots, (y_n, d_n)$ is

$$\theta^{\dagger} = \frac{n}{\sum_{j=1}^{n} \left\{ d_j y_j + (1 - d_j)(c + 1/\theta') \right\}}.$$

Show that the missing information is $\sum (1 - d_j)/\theta^2$ and find the rate of convergence of the algorithm. Discuss briefly.

$\delta(\cdot)$ is the Dirac delta function.

5.6 Bibliographic Notes

Linear regression is discussed in more depth in Chapter 8, and references to the enormous literature on the topic can be found in Section 8.8. Exponential family models date to work of Fisher and others in the 1930s, are widely used in applications and have been intensively studied. Chapter 5 of Pace and Salvan (1997) is a good reference, while longer more mathematical accounts are Barndorff-Nielsen (1978) and Brown (1986). The term natural exponential family was introduced by Morris (1982, 1983), who highlighted the importance of the variance function.

The roots of group transformation models go back to Pitman (1938, 1939), but owe much of their modern development to D. A. S. Fraser, summarized in Fraser (1968, 1979).

Survival analysis is a huge field with inter-related literatures on industrial and medical problems, though time-to-event data arise in many other fields also. The early literature is mostly concerned with reliability, of which Crowder *et al.* (1991) is an elementary account, while the literature on biostatistical and medical applications has grown enormously over the last 30 years. Cox and Oakes (1984), Miller (1981), Kalbfleisch and Prentice (1980), and Collett (1995) are standard accounts at about this level; see also Klein and Moeschberger (1997). Competing risks are surveyed by Tsiatis (1998); a helpful earlier account is Prentice *et al.* (1978). Their nonidentifiability was first pointed out by Cox (1959). Aalen (1994) gives an elementary account of frailty models, with further references. Keiding (1990) describes inference using the Lexis diagram.

The formal study of missing data began with Rubin (1976), though *ad hoc* procedures for dealing with missing observations in standard models were widely used much earlier. A standard reference is Little and Rubin (1987). More recently the related notion of data coarsening, which encompasses censoring, truncation and grouping as well as missingness, has been discussed by Heitjan (1994).

Although data in areas such as epidemiology and the social and economic sciences are often analyzed as if they were selected randomly from some well-defined population, the possibility that bias has entered the selection process is ever-present; publication bias is just one example of this. There is a large literature on selection bias from many points of view, much of which is mentioned by Copas and Li (1997) and its discussants. Example 5.34 is taken from Copas (1999). Molenberghs *et al.* (2001) give an example of analysis of sensitivity to missing data in contingency tables, with references to related literature.

Special cases of the EM algorithm were used well before it was crystallized and named by Dempster *et al.* (1977), who gave numerous applications and pointed the way for the substantial further work largely summarized in McLachlan and Krishnan (1997). A useful shorter account is Chapter 4 of Tanner (1996). One common criticism of the algorithm is its slowness, and Meng and van Dyk (1997) and Jamshidian and Jennrich (1997) describe some of the many approaches to speeding it up; they also contain further references. Oakes (1999) gives references to the literature on computing standard errors for EM estimates. Modern applications go far beyond the simple exponential family models used initially and may require complex E- and M-steps including Monte Carlo simulation; see for example McCulloch (1997).

Mixture models and their generalizations are widely used in applications, particularly for classification and discrimination problems; see Titterington *et al.* (1985) and Lindsay (1995). The thorny problem of selecting the number of components is given an airing by Richardson and Green (1997) and their discussants, using methods discussed in Section 11.3.3.

5.7 Problems

1 In the linear model (5.3), suppose that $n = 2r$ is an even integer and define $W_j = Y_{n-j+1} - Y_j$ for $j = 1, \ldots, r$. Find the joint distribution of the W_j and hence show that

$$\tilde{\gamma}_1 = \frac{\sum_{j=1}^{r}(x_{n-j+1} - x_j)W_j}{\sum_{j=1}^{r}(x_{n-j+1} - x_j)^2}$$

satisfies $\mathrm{E}(\tilde{\gamma}_1) = \gamma_1$. Show that

$$\mathrm{var}(\tilde{\gamma}_1) = \sigma^2 \left\{ \sum_{j=1}^{n}(x_j - \overline{x})^2 - \frac{1}{2}\sum_{j=1}^{r}(x_{n-j+1} + x_j - 2\overline{x})^2 \right\}^{-1}.$$

Deduce that $\mathrm{var}(\tilde{\gamma}_1) \geq \mathrm{var}(\widehat{\gamma}_1)$ with equality if and only if $x_{n-j+1} + x_j = c$ for some c and all $j = 1 \ldots, r$.

2 Show that the scaled chi-squared density with known degrees of freedom v,

$$f(v; \sigma^2) = \frac{v^{v/2-1}}{(2\sigma^2)^{v/2}\Gamma\left(\frac{1}{2}v\right)} \exp\left(-\frac{v}{2\sigma^2}\right), \quad v > 0, \sigma^2 > 0, v = 1, 2, \ldots,$$

is an exponential family, and find its canonical parameter and observation and cumulant-generating function.

3 Show that the geometric density

$$f(y; \pi) = \pi(1-\pi)^y, \quad y = 0, 1, \ldots, 0 < \pi < 1,$$

is an exponential family, and give its cumulant-generating function.
Show that $S = Y_1 + \cdots + Y_n$ has negative binomial density

$$\binom{n+s-1}{n-1}\pi^n(1-\pi)^s, \quad s = 0, 1, \ldots,$$

and that this is also an exponential family.

4 (a) Suppose that Y_1 and Y_2 have gamma densities (2.7) with parameters λ, κ_1 and λ, κ_2. Show that the conditional density of Y_1 given $Y_1 + Y_2 = s$ is

$$\frac{\Gamma(\kappa_1 + \kappa_2)}{s^{\kappa_1+\kappa_2-1}\Gamma(\kappa_1)\Gamma(\kappa_2)}u^{\kappa_1-1}(s-u)^{\kappa_2-1}, \quad 0 < u < s, \kappa_1, \kappa_2 > 0,$$

and establish that this is an exponential family. Give its mean and variance.
(b) Show that $Y_1/(Y_1 + Y_2)$ has the beta density.
(c) Discuss how you would use samples of form $y_1/(y_1 + y_2)$ to check the fit of this model with known v_1 and v_2.

5 If Y has density (5.7) and $\mathcal{Y}_1$ is a proper subset of $\mathcal{Y}$, show the the conditional density of Y given that $Y \notin \mathcal{Y}_1$ is also a natural exponential family.
Find the cumulant-generating function for the truncated Poisson density given by $f_0(y) \propto 1/y!$, $y = 1, 2, \ldots$, and give the likelihood equation and information quantities.
Compare with Practical 4.3.

6   Show that the two-locus multinomial model in Example 4.38 is a natural exponential family of order 2 with natural observation and parameter $s(Y) = (Y_A + Y_{AB}, Y_B + Y_{AB})^T$ and $(\theta_A, \theta_B)^T = (\log\{\alpha/(1-\alpha)\}, \log\{\beta/(1-\beta)\})$ and cumulant-generating function $m \log(1 + e^{\theta_A}) + m \log(1 + e^{\theta_B})$. Deduce that the elements of $s(Y)$ are independent. Under what circumstances will maximum likelihood estimation of θ_A, θ_B give infinite estimates?

7   Suppose that $Y_1, \ldots, Y_n$ follow (5.2). Show that the joint density of the Y_j is a linear exponential family of order three, and give the canonical statistics and parameters and the cumulant-generating function. Find the minimal representations in the cases where the x_j (i) are, and (ii) are not, all equal.
    Is the model an exponential family when $E(Y_j) = \beta_0 \exp(x_j \beta_1)$?

8   Show that the multivariate normal distribution $N_p(\mu, \Omega)$ is a group transformation model under the map $Y \mapsto a + BY$, where a is a $p \times 1$ vector and B an invertible $p \times p$ matrix. Given a random sample $Y_1, \ldots, Y_n$ from this distribution, show that

$$\overline{Y} = n^{-1} \sum_{j=1}^{n} Y_j, \quad \sum_{j=1}^{n} (Y_j - \overline{Y})(Y_j - \overline{Y})^T$$

    is a minimal sufficient statistic for μ and Ω, and give equivariant estimators of them. Use these estimators to find the maximal invariant.

9   Show that the model in Example 4.5 is an exponential family. Is it steep? What happens when $R_j = 0$ whenever $x_j < a$ and $R_j = m_j$ otherwise?
    Find its minimal representation when all the x_j are equal.

10 Independent observations $y_1, \ldots, y_n$ from the exponential density $\lambda \exp(-\lambda y)$, $y > 0$, $\lambda > 0$, are subject to Type II censoring stopping at the rth failure. Show that a minimal sufficient statistic for λ is $S = Y_{(1)} + \cdots + Y_{(r)} + (n-r)Y_{(r)}$, where $0 < Y_{(1)} < Y_{(2)} < \cdots$ are order statistics of the Y_j, and that $2\lambda S$ has a chi-squared distribution on $2r$ degrees of freedom.
    A Type II censored sample was $0.2, 0.8, 1.1, 1.4, 2.1, 2.4, 2.4+, 2.4+, 2.4+$, where + denotes censoring. On the assumption that the sample is from the exponential distribution, find a 90% confidence interval for λ. How would you check whether the data are exponential?

11 Let $X_1, \ldots, X_n$ be an exponential random sample with density $\lambda \exp(-\lambda x)$, $x > 0$, $\lambda > 0$. For simplicity suppose that $n = mr$. Let Y_1 be the total time at risk from time zero to the rth failure, Y_2 be the total time at risk between the rth and the $2r$th failure, Y_3 the total time at risk between the $2r$th and $3r$th failures, and so forth.
    (a) Let $X_{(1)} \le X_{(2)} \le \cdots \le X_{(n)}$ be the ordered values of the X_j. Show that the joint density of the order statistics is

$$f_{X_{(1)},\ldots,X_{(n)}}(x_1, \ldots, x_n) = n! f(x_1) f(x_2) \cdots f(x_n), \quad x_1 < x_2 < \cdots < x_n,$$

    and by writing $X_{(1)} = Z_1$, $X_{(2)} = Z_1 + Z_2$, $\ldots$, $X_{(n)} = Z_1 + \cdots + Z_n$, where the Z_j are the *spacings* between the order statistics $X_{(j)}$, show that the Z_j are independent exponential random variables with hazard rates $(n + 1 - j)\lambda$.
    (b) Hence show that the Y_j have independent gamma distributions with means r/λ and variances r/λ^2. Deduce that the variables $\log Y_j$ are independently distributed with constant variance.
    (c) Now suppose that the hazard rate is not constant, but is a slowly-varying smooth function of time, $\lambda(t)$. Explain how a plot of $\log Y_j$ against the midpoint of the time interval between the $(r - 1)j$th and the rjth failures can be used to estimate $\log \lambda(t)$. (Cox, 1979)

12 Let $Y_1, \ldots, Y_n$ be independent exponential variables with hazard λ subject to Type I censoring at time c. Show that the observed information for λ is D/λ^2, where D is the number of the Y_j that are uncensored, and deduce that the expected information is $i(\lambda \mid c) = n\{1 - \exp(-\lambda c)\}/\lambda^2$ conditional on c.

Now suppose that the censoring time c is a realization of a random variable C, whose density is gamma with index ν and parameter $\lambda\alpha$:

$$f(c) = \frac{(\lambda\alpha)^\nu c^{\nu-1}}{\Gamma(\nu)} \exp(-c\lambda\alpha), \qquad\qquad c > 0, \alpha, \nu > 0.$$

Show that the expected information for λ after averaging over C is

$$i(\lambda) = n\{1 - (1 + 1/\alpha)^{-\nu}\}/\lambda^2.$$

Consider what happens when (i) $\alpha \to 0$, (ii) $\alpha \to \infty$, (iii) $\alpha = 1, \nu = 1$, (iv) $\nu \to \infty$ but $\mu = \nu/\alpha$ is held fixed. In each case explain qualitatively the behaviour of $i(\lambda)$.

13   In a competing risks model with $k = 2$, write

$$\begin{aligned}\Pr(Y \leq y) &= \Pr(Y \leq y \mid I = 1)\Pr(I = 1) + \Pr(Y \leq y \mid I = 2)\Pr(I = 2)\\ &= pF_1(y) + (1 - p)F_2(y),\end{aligned}$$

say. Hence find the cause-specific hazard functions h_1 and h_2, and express F_1, F_2 and p in terms of them.
Show that the likelihood for an uncensored sample may be written

$$p^r(1 - p)^{n-r} \prod_{j=1}^r f_1(y_j) \prod_{j=r+1}^n f_2(y_j)$$

and find the likelihood when there is censoring.
If $f(y_1 \mid y_2)$ and $f(y_2 \mid y_1)$ be arbitrary densities with support $[y_2, \infty)$ and $[y_1, \infty)$, then show that the joint density

$$f(y_1, y_2) = \begin{cases} pf_1(y_1)f(y_2 \mid y_1), & y_1 \leq y_2, \\ (1 - p)f_2(y_2)f(y_1 \mid y_2), & y_1 > y_2, \end{cases}$$

produces the same likelihoods. Deduce that the joint density is not identifiable.

14   Find the cause-specific hazard functions for the bivariate survivor functions

$$\mathcal{F}(y_1, y_2) = \exp[1 - \theta_1 y_1 - \theta_2 y_2 - \exp\{\beta(\theta_1 y_1 + \theta_2 y_2)\}],$$

$$\mathcal{F}^*(y_1, y_2) = \exp\left[1 - \theta_1 y_1 - \theta_2 y_2 - \sum_{i=1}^2 \frac{\theta_i}{\theta_1 + \theta_2} \exp\{\beta(\theta_1 + \theta_2)y_i\}\right],$$

where $y_1, y_2 > 0, \theta_1, \theta_2 > 0$ and $\beta > -1$. Under what condition does $\mathcal{F}$ yield independent variables?
Write down the likelihoods based on random samples $(y_1, i_1, d_1), \ldots, (y_n, i_n, d_n)$ from these two models. Discuss the interpretation of $\widehat{\beta} \gg 0$ in the absence of external evidence for $\mathcal{F}$ over $\mathcal{F}^*$.
(Prentice *et al.*, 1978)

15   (a) Let $Z = X_1 + \cdots + X_N$, where N is Poisson with mean μ and the X_i are independent identically distributed variables with moment-generating function $M(t)$. Show that the cumulant-generating function of Z is $K_Z(t) = \mu\{M(t) - 1\}$ and that $\Pr(Z = 0) = e^{-\mu}$.
If the X_i are gamma variables, show that $K_Z(t)$ may be written as

$$\frac{\alpha}{(\alpha - 1)\delta}[\{1 - \alpha t/(\gamma\delta)\}^{1-\alpha} - 1], \qquad \gamma, \delta > 0, \qquad\qquad (5.43)$$

where $\alpha > 1$, show that $E(Z) = \gamma$ and $\text{var}(Z)/E(Z)^2 = \delta$, and find $\Pr(Z = 0)$ in terms of α, δ and γ. Show that as $\alpha \to 1$ the limiting distribution of Z is gamma, and explain why.

Z is a continuous variable for $0 < \alpha < 1$, but you need not show this.

(b) For a frailty model, set $\gamma = 1$ and suppose that an individual has hazard $Zh(y), y > 0$. Compute the population cumulative hazard $H_Y(y)$ and show that if $\alpha > 1$ then

$$\lim_{y\to\infty} H_Y(y) < \infty.$$

Give an interpretation of this in terms of the distribution of the lifetime Y. (Are all the individuals in the population liable to fail?)

(c) Obtain the population hazard rate $h_Y(y)$, take $h(y) = y^2$, and graph $h_Y(y)$ for $\delta = 0, 0.5, 1, 2.5$. Discuss this in relation to the divorce rate example on page 201.

(d) Now suppose that there are two groups of individuals, the first with individual hazards $h(y)$ and the second with individual hazards $rh(y)$, where $r > 1$. Thus the effect of transferring an individual from group 1 to group 2, if this were possible, would be to increase his hazard by a factor r. If frailties in the two groups have the same cumulant-generating function (5.43), show that the ratio of group hazard functions is

$$\frac{h_2(y)}{h_1(y)} = r \left\{ \frac{1 + \alpha^{-1}\delta H(y)}{1 + r\alpha^{-1}\delta H(y)} \right\}^{\alpha}.$$

Establish that this is a decreasing function of y, and explain why its limiting value is less than one, that is, the risk is eventually lower in group 2, if $\alpha > 1$. What difficulties does this pose for the interpretation of group differences in survival?

(Aalen, 1994; Hougaard, 1984)

16 (a) Show that when data (X, Y) are available, but with values of Y missing at random, the log likelihood contribution can be written

$$\ell(\theta) \equiv I \log f(Y \mid X; \theta) + \log f(X; \theta),$$

and deduce that the expected information for θ depends on the missingness mechanism but that the observed information does not.

(b) Consider binary pairs (X, Y) with indicator I equal to zero when Y is missing; X is always seen. Their joint distribution is given by

$$\Pr(Y = 1 \mid X = 0) = \theta_0, \quad \Pr(Y = 1 \mid X = 1) = \theta_1, \quad \Pr(X = 1) = \lambda,$$

while the missingness mechanism is

$$\Pr(I = 1 \mid X = 0) = \eta_0, \quad \Pr(I = 1 \mid X = 1) = \eta_1.$$

(i) Show that the likelihood contribution from (X, Y, I) is

$$\left[\left\{ \theta_1^Y (1 - \theta_1)^{1-Y} \right\}^X \left\{ \theta_0^Y (1 - \theta_0)^{1-Y} \right\}^{1-X} \right]^I$$
$$\times \left\{ \eta_0^I (1 - \eta_0)^{1-I} \right\}^{1-X} \left\{ \eta_1^I (1 - \eta_1)^{1-I} \right\}^X \times \lambda^X (1 - \lambda)^{1-X}.$$

Deduce that the observed information for θ_1 based on a random sample of size n is

$$-\frac{\partial^2 \ell(\theta_0, \theta_1)}{\partial \theta_1^2} = \sum_{j=1}^n I_j X_j \left\{ \frac{Y_j}{\theta_1^2} + \frac{1 - Y_j}{(1 - \theta_1)^2} \right\}.$$

Give corresponding expressions for $\partial^2 \ell(\theta_0, \theta_1)/\partial \theta_0^2$ and $\partial^2 \ell(\theta_0, \theta_1)/\partial \theta_0 \partial \theta_1$.

(ii) Statistician A calculates the expected information treating $I_1, \ldots, I_n$ as fixed and thereby ignores the missing data mechanism. Show that he gets $i_A(\theta_1, \theta_1) = M\lambda/\{\theta_1(1 - \theta_1)\}$, where $M = \sum I_j$, and find the corresponding quantities $i_A(\theta_0, \theta_1)$ and $i_A(\theta_0, \theta_0)$. If he uses this procedure for many sets of data, deduce that on average M is replaced by $n\Pr(I = 1) = n\{\lambda\eta_1 + (1 - \lambda)\eta_0\}$.

(iii) Statistician B calculates the expected information taking into account the missingness mechanism. Show that she gets $i_B(\theta_1, \theta_1) = n\lambda\eta_1/\{\theta_1(1 - \theta_1)\}$, and obtain $i_B(\theta_0, \theta_1)$ and $i_B(\theta_0, \theta_0)$.

(iv) Show that A and B get the same expected information matrices only if Y is missing completely at random. Does this accord with the discussion above?

(c) Statistician C argues that expected information should never be used in data analysis: even if the data actually observed are complete, unless it can be guaranteed that data

could not be missing at random for any reason, every expected information calculation should involve every potential missingness mechanism. Such a guarantee is impossible in practice, so no expected information calculation is ever correct. Do you agree?
(Kenward and Molenberghs, 1998)

17 (a) In Example 5.34, suppose that n patients are divided randomly into control and treatment groups of equal sizes $n_C = n_T = n/2$, with death rates λ_C and λ_T. If the numbers of deaths R_C and R_T are small, use a Poisson approximation to the binomial to show that the difference in log rates is roughly $\widehat{\mu} = \log R_C - \log R_T$. What would you conclude if $\widehat{\mu} \doteq 0$?
(b) Show that if $\lambda_C \doteq \lambda_T = \lambda$, then $\mathrm{var}(\widehat{\mu}) \doteq 4/(n\lambda)$, and use the estimates $\widehat{\lambda}_C = R_C/n_C$, $\widehat{\lambda}_T = R_T/n_T$ and $\widehat{\lambda} = (R_C + R_T)/(n_C + n_T)$ to check a few values of $\widehat{\mu}$ and the standard errors in Table 5.9.
(c) In practice the variance in (b) is typically too small, because it does not allow for inter-trial variability. Different studies are performed with different populations, in which the treatment may have different effects. We can imagine two stages: we first choose a population in which the treatment effect is $\mu + \eta$, where η is random with mean zero and variance σ^2; then we perform a trial with n subjects and produce an estimator $\widehat{\mu}$ of $\mu + \eta$ with variance v/n. Show that $\widehat{\mu}$ may be written $\mu + \eta + \varepsilon$, give the variance of ε, and deduce that when both stages of the trial are taken into account, $\widehat{\mu}$ has mean μ and variance $\sigma^2 + v/n$.
How would this affect the calculations in Example 5.34?

18 (a) Show that the t density of Example 4.39 may be obtained by supposing that the conditional density of Y given $U = u$ is $N(\mu, v\sigma^2/u)$ and that $U \sim \chi_v^2$. Show that $U \overset{D}{=} V/\{v + (y - \mu)^2/\sigma^2\}$ conditional on Y, where $V \sim \chi_{v+1}^2$, and with $\theta = (\mu, \sigma^2)$ deduce that

$$\mathrm{E}(U \mid Y; \theta) = \frac{v + 1}{v + (y - \mu)^2/\sigma^2}.$$

(b) Consider the EM algorithm for estimation of θ when v is known. Show that the complete-data log likelihood contribution from (y, u) may be written

$$-\frac{1}{2}\sigma^2 - \frac{1}{2}u(y - \mu)^2/2(v\sigma^2),$$

and hence give the M-step. Write down the algorithm in detail.
(c) Show that the result of the EM algorithm satisfies the self-consistency relation $\theta = g(\theta)$, and given the form of g when σ^2 is both known and unknown.
(d) The Cauchy log likelihood shown in the right panel of Figure 4.2 corresponds to setting $v = \sigma^2 = 1$. In this case explain why $\mu^\dagger$ converges to a local or a global maximum or a local minimum, depending on the initial value for μ.

19 Suppose that $U_1, \ldots, U_q$ have a multinomial distribution with denominator m and probabilities $\pi_1, \ldots, \pi_q$ that depend on a parameter θ, and that the maximum likelihood estimator of θ based on the U_s has a simple form. Some of the categories are indistinguishable, however, so the observed data are $Y_1, \ldots, Y_p$, where $Y_r = \sum_{s \in A_r} U_s$; $A_1, \ldots, A_p$ partition $\{1, \ldots, q\}$ and none is empty.
(a) Show that the E-step of the EM algorithm for estimation of θ involves

$$\mathrm{E}(U_s \mid Y = y; \theta') = \frac{y_r \pi_s'}{\sum_{t \in A_r} \pi_t'}, \quad s \in A_r,$$

and say how the M-step is performed.
(b) Let $(\pi_1, \ldots, \pi_5) = (1/2, \theta/4, (1 - \theta)/4, (1 - \theta)/4, \theta/4)$, and suppose that $y_1 = u_1 + u_2 = 125$, $y_2 = u_3 = 18$, $y_3 = u_4 = 20$ and $y_4 = u_5 = 34$. These data arose in a genetic linkage problem and are often used to illustrate the EM algorithm. Show that

$$\theta^\dagger = \frac{y_4 + y_1 \theta'/(2 + \theta')}{m - 2y_1/(2 + \theta')},$$

and find the maximum likelihood estimate starting with $\theta' = 0.5$.

(c) Show that the maximum likelihood estimator of $\widehat{\lambda}_A$ in the single-locus model of Example 4.38 may be written $\widehat{\lambda}_A = (2u_1 + u_2 + u_5)/m$ and establish that

$$E(U_1 \mid Y; \lambda') = y_1 \lambda'_A / (2 - 2\lambda'_B - \lambda'_A).$$

Give the corresponding expressions for $\widehat{\lambda}_B$ and $E(U_2 \mid Y; \lambda')$. Hence give the M-step for this model. Apply the EM algorithm to the data in Table 4.3, using starting-values obtained from categories with probabilities $2\lambda_A \lambda_B$ and λ_O^2.

(d) Compute standard errors for your estimates in (b) and (c).

(Rao, 1973, p. 369)

6

Stochastic Models

The previous chapter outlined likelihood analysis of some standard models. Here we turn to data in which the dependence among the observations is more complex. We start by explaining how our earlier discussion extends to Markov processes in discrete and continuous time. We then extend this to more complex indexing sets and in particular to Markov random fields, in which basic concepts from graph theory play an important role. A special case is the multivariate normal distribution, an important model for data with several responses. We give some simple notions for time series, a very widespread form of dependent data, and then turn to point processes, describing models for rare events in passing.

6.1 Markov Chains

In certain applications interest is focused on transitions among a small number of states. A simple example is rainfall modelling, where a sequence . . . 010011 . . . indicates whether or not it has rained each day. Another is in *panel studies* of employment, where many individuals are interviewed periodically about their employment status, which might be full-time, part-time, home-worker, unemployed, retired, and so forth. Here interest will generally focus on how variables such as age, education, family events, health, and changes in the job market affect employment history for each interviewee, so that there are many short sequences of state data taken at unequal intervals, unlike the single long rainfall sequence. In each case, however, the key aspect is that transitions occur amongst discrete states, even though these typically are crude summaries of reality.

Example 6.1 (DNA data)   When the double helix of deoxyribonucleic acid (DNA) is unwound it consists of two oriented linked sequences of the bases adenine (A), cytosine (C), guanine (G), and thymine (T). Just one chain determines a DNA sequence, because A in one sequence is always linked to T on the other, and likewise with C and G. An example is Table 6.1, which shows the first 150 bases from a sequence of

GTATTAAATCCGTAGTCTCGAACTAACATA
TCAATATGGTTGGAATAAAGCCTGTGAAAA
CTATGATTAGTGAATAAGGTCTCAGTAATT
TAGAATAAATATTCTGCACAATGATCAAAT
GTTTAAAGTATCCTTGTGATAAAAGCAGAC

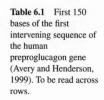

Table 6.1 First 150 bases of the first intervening sequence of the human preproglucagon gene (Avery and Henderson, 1999). To be read across rows.

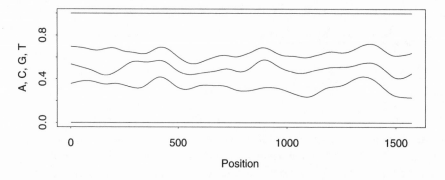

Position

Figure 6.1 Estimated proportions of bases A, C, G and T in the first intervening sequence of the human preproglucagon gene. At a point t on the x-axis, the vertical distances between the lines above correspond to the proportions of times the bases appear in a window of width 100 centred at t.

1572 bases found in the human preproglucagon gene. Figure 6.1 shows proportions of the different bases along the sequence, smoothed using a form of moving average. Roughly speaking, the number of times each base occurs in a window of width 100 centred at t has been counted, giving estimated proportions $(\widehat{\pi}_A, \widehat{\pi}_C, \widehat{\pi}_G, \widehat{\pi}_T)$. These are plotted at t, and then the procedure is repeated at $t + 1$, and so forth. Although there is local variation, the proportions seem fairly stable along the sequence, with A occurring about 30 times in every 100, C about 15 times, and so forth.

Certain sequences of bases such as CTGAC — known as *words* — are of biological interest. If they occur very often in particular stretches of the entire sequence, it may be supposed that they serve some purpose. But before trying to see what that purpose might be, it is necessary to see if they have occurred more often then chance dictates, for example by comparing the actual number of occurrences with that expected under a model. It is simplest to suppose that bases occur randomly, but the code of life is not so simple. Table 6.2 contains observed frequencies for pairs of consecutive bases. The pair AA occurs 185 times, AC 74 times, and so forth. The lower table shows corresponding proportions, obtained by dividing the frequencies by their row totals. About 80% of the bases following a C are A or T, while a G is rare; Gs occur much more frequently after A, G, or T. The sequence does not seem purely random. ∎

Example 6.2 (Breast cancer data) Table 6.3 gives data on 37 women with breast cancer treated for spinal metastases at the London Hospital. Their ambulatory status — defined as ability to walk unaided or not — was recorded before treatment began, as it started, and then 3, 6, 12, 24, and 60 months after treatment. The three states are: able to walk unaided (1); unable to walk unaided (2); and dead (3). Thus a sequence 111113 means that the patient was able to walk unaided each time she was seen, but was dead five years after the treatment began. She may have been unable to walk for periods between the times at which her state was recorded. This is illustrated in

Table 6.2 Observed frequencies of the 16 possible successive pairs of bases in the first intervening sequence of the human preproglucagon gene. There are 1571 such pairs. The lower table shows the proportion of times the second base follows the first.

| First base | Frequencies for second base | | | | |
	A	C	G	T	Total
A	185	74	86	171	516
C	101	41	6	115	263
G	69	45	34	78	226
T	161	103	100	202	566
Total	516	263	226	566	1571

| First base | Proportion for second base | | | | |
	A	C	G	T	Total
A	0.359	0.143	0.167	0.331	1.0
C	0.384	0.156	0.023	0.437	1.0
G	0.305	0.199	0.150	0.345	1.0
T	0.284	0.182	0.177	0.357	1.0
Overall	0.328	0.167	0.144	0.360	1.0

Table 6.3 Breast cancer data (de Stavola, 1988). The table gives the initial and follow-up status for 37 breast cancer patients treated for spinal metastases. The status is able to walk unaided (1), unable to walk unaided (2), or dead (3), and the times of follow-up are 0, 3, 6, 12, 24, and 60 months after treatment began. Woman 24 was alive after 6 months but her ability to walk was not recorded.

	Initial	Follow-up		Initial	Follow-up		Initial	Follow-up
1	1	111113	13	2	23	25	1	11113
2	1	1113	14	2	1113	26	2	22223
3	2	23	15	2	2	27	2	12223
4	2	121113	16	2	23	28	2	11113
5	1	111123	17	1	1113	29	2	1223
6	1	1113	18	2	223	30	2	1123
7	1	12113	19	1	13	31	2	1222
8	2	123	20	1	12223	32	1	11223
9	1	1111	21	2	23	33	2	1223
10	2	23	22	2	11111	34	1	1113
11	2	23	23	2	23	35	1	113
12	1	1113	24	1	12?3	36	2	23
						37	2	23

the left panel of Figure 6.2, which shows a possible sample path for a woman with sighting history 111223. Although there is a visit to state 1 between 12 and 24 months, it is unobserved, and the data suggest that her sojourn in state 2 is uninterrupted. The number of sightings varies from woman to woman; case 9, for example, was able to walk when seen after 12 months, but her later history is unknown.

One aspect of interest here is whether inability to walk always precedes death, while another is whether a woman's state before treatment affects her subsequent history. Although no explanatory variables are available here, their effect on the transition probabilities would often be of importance in practice. ∎

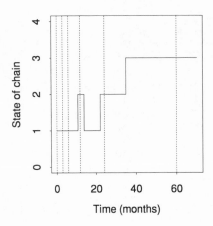

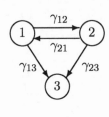

Figure 6.2 Markov chain model for breast cancer data. Left: possible sample path (solid) for a woman with states 111223 observed at 0, 3, 6, 12, 24, 60 months shown by the dotted lines. Right: parameters for possible transitions among the states.

Let X_t denote a process taking values in the *state space* $\mathcal{S} = \{1, \ldots, S\}$, where S may be infinite. For general discussion we call the quantity t on which X_t depends time, and suppose that our data have form $X_0 = s_0, X_{t_1} = s_1, \ldots, X_{t_k} = s_k$, where $0 < t_1 < \cdots < t_k$. In the DNA example t is in fact location, $k = 1571$, and $\mathcal{S} = \{1, 2, 3, 4\} \equiv \{A, C, G, T\}$. In the breast cancer example there are $S = 3$ states, $k = 5$ at most, and $t_0 = 0$, $t_1 = 3$, $t_2 = 6$, $t_3 = 12$, $t_4 = 24$, and $t_5 = 60$ months.

Let $X_{(t_j)} = s_{(j)}$ denote the composite event $X_{t_j} = s_j, \ldots, X_0 = s_0$, for $j = 0, \ldots, k - 1$. Then the joint density of the data may be written

$$\Pr\big(X_0 = s_0, \ldots, X_{t_k} = s_k\big) = \Pr(X_0 = s_0) \prod_{j=1}^{k} \Pr\big(X_{t_j} = s_j \mid X_{(t_{j-1})} = s_{(j-1)}\big);$$

using the prediction decomposition (4.7). The conditional probabilities may be complicated, but modelling is greatly simplified if the process has the *Markov property*

$$\Pr\big(X_{t_j} = s_j \mid X_0 = s_0, \ldots, X_{t_{j-1}} = s_{j-1}\big) = \Pr\big(X_{t_j} = s_j \mid X_{t_{j-1}} = s_{j-1}\big).$$

Thus the 'future' X_{t_j} is independent of the 'past' $X_{t_{j-2}}, \ldots, X_0$, given the 'present' $X_{t_{j-1}}$ — all information available about the future evolution of X_t is contained in its current state. If so, then

$$\Pr\big(X_0 = s_0, \ldots, X_{t_k} = s_k\big) = \Pr(X_0 = s_0) \prod_{j=1}^{k} \Pr\big(X_{t_j} = s_j \mid X_{t_{j-1}} = s_{j-1}\big). \quad (6.1)$$

Matters simplify further if the process is *stationary*, for then the conditional probabilities in (6.1) depend only on differences among the t_j. Thus

$$\Pr(X_t = s \mid X_u = r) = \Pr(X_{t-u} = s \mid X_0 = r),$$

and we assume this to be the case below. These simplifications yield a rich structure with many important and interesting models, in which these *transition probabilities* play a central role. They determine the likelihood (6.1), apart from the initial term

Andrei Andreyevich Markov (1856–1922) studied with Chebyshev in St Petersburg and initially worked on pure mathematics. His study of dependent sequences of variables stemmed from an attempt to understand the Central Limit Theorem.

Some authors use the term homogeneous rather than stationary.

$\Pr(X_0 = s_0)$. If k is large this term usually contains little information and can safely be dropped, but it may be important to include it when k is small; see Example 6.10.

6.1.1 Markov chains

We call a Markov model observed at discrete equally-spaced times a *Markov chain*. In this section we consider inference for simple Markov chain models, but in Section 11.3.3 we describe the use of Markov chains for inference. As the following outline of their properties serves both purposes, it is slightly more detailed than immediately required.

A stationary chain X_t on the countable set $\mathcal{S}$ of size S observed at equally-spaced times $t = 0, 1, \ldots, k$ has properties determined by the transition probabilities

If infinite matrices worry you, think of $\mathcal{S}$ as finite.

$$p_{rs} = \Pr(X_1 = s \mid X_0 = r), \quad r, s \in \mathcal{S},$$

which form the $S \times S$ *transition matrix* P whose (r, s) element is p_{rs}. The elements of P are non-negative and the fact that $\sum_s p_{rs} = 1$ implies that $P 1_S = 1_S$, so P is a *stochastic matrix*. If the rth element of the $S \times 1$ vector p is the initial probability $p_r = \Pr(X_0 = r)$, then the sth element of $p^\mathsf{T} P$ is $\Pr(X_1 = s) = \sum_r p_r p_{rs}$. Iteration shows that the density of X_n is given by $p^\mathsf{T} P^n$, so the (r, s) element of P^n is the n-step transition probability $p_{rs}(n) = \Pr(X_n = s \mid X_0 = r)$. Hence properties of X_t are governed by P. The probability of a *run* of $m \geq 1$ successive visits to state s is $p_{ss}^{m-1}(1 - p_{ss})$; this is the geometric density with mean $(1 - p_{ss})^{-1}$ (Exercise 6.1.8).

1_S is the $S \times 1$ vector of ones.

Classification of chains

It is useful to classify the states of a chain. A state s is *recurrent* if

$$\Pr(X_t = s \text{ for some } t > 0 \mid X_0 = s) = 1,$$

meaning that having started from s, eventual return is certain; s is *transient* if this probability is strictly less than one. If $T_{rs} = \min\{t > 0 : X_t = s \mid X_0 = r\}$ is the *first-passage time* from r to s, then $\mathrm{E}(T_{ss})$ is the *mean recurrence time* of state s; we set $\mathrm{E}(T_{ss}) = \infty$ if s is transient, and say that a recurrent state is *positive recurrent* if $\mathrm{E}(T_{ss}) < \infty$; otherwise it is *null recurrent*. The *period* of s is $d = \gcd\{n : p_{ss}(n) > 0\}$, the greatest common divisor of those times at which return to s is possible; s is *aperiodic* if $d = 1$, and *periodic* otherwise.

Some authors use the terms persistent and non-null rather than recurrent and positive.

We now classify chains themselves. We say that r *communicates with* s, $r \to s$, if $p_{rs}(n) > 0$ for some $n > 0$, and that r *and* s *intercommunicate*, $r \leftrightarrow s$, if $r \to s$ and $s \to r$. It may be shown that two intercommunicating states have the same period, while if one is transient so is the other, and similarly for null recurrence. A set $\mathcal{C}$ of states is *closed* if $p_{rs} = 0$ for all $r \in \mathcal{C}$, $s \notin \mathcal{C}$, and *irreducible* if $r \leftrightarrow s$ for all $r, s \in \mathcal{C}$; a closed set with just one state is called *absorbing*. It may be proved that $\mathcal{S}$ may be partitioned uniquely as $\mathcal{T} \cup \mathcal{C}_1 \cup \mathcal{C}_2 \cup \cdots$, where $\mathcal{T}$ is the set of transient states and the $\mathcal{C}_i$ are irreducible closed sets of recurrent states; if S is finite, then at least one state is recurrent, and all recurrent states are positive. A chain is called aperiodic, positive recurrent, and so forth if its states all share the corresponding property. An aperiodic irreducible positive recurrent chain is *ergodic*.

Example 6.3 (Breast cancer data)   Here $\mathcal{T} = \{1, 2\}$ contains the two transient states with the patient alive, while $\mathcal{C} = \{3\}$, death, is absorbing.   ∎

Example 6.4 (DNA data)   As transitions occur between every pair of states, $\mathcal{C} = \{A, C, G, T\}$ is an irreducible aperiodic closed set of states, all recurrent and hence all positive recurrent. This chain is ergodic.   ∎

Each of the properties of an ergodic chain is important. Irreducibility means that any state is accessible from any other. Positive recurrence implies that the chain has at least one stationary distribution with probability vector π such that $\pi^{\mathsf{T}} P = \pi^{\mathsf{T}}$, and the mean recurrence time for state s is $\mathrm{E}(T_{ss}) = \pi_s^{-1} < \infty$. There is a unique stationary distribution when the chain is both irreducible and positive recurrent. In this case each state is visited infinitely often as $t \to \infty$, but the chain need not be stationary because it might oscillate among states. Aperiodicity stops this.

When S is infinite and the chain has all three properties, the transition probabilities $p_{rs}(n) \to \pi_s$ as $n \to \infty$: the chain converges to its stationary distribution whatever the initial state. Moreover, if $m(X_t)$ is such that $\mathrm{E}_\pi \{|m(X_t)|\} = \sum_r \pi_r |m(r)| < \infty$, then

$$\Pr \left\{ n^{-1} \sum_{t=1}^{n} m(X_t) \to \sum_{r=1}^{S} \pi_r m(r) \text{ as } n \to \infty \right\} = 1 : \qquad (6.2)$$

starting from any X_0, the average of $m(X_t)$ converges almost surely to the mean $\mathrm{E}_\pi\{m(X_t)\}$ of $m(X_t)$ with respect to π. This ensures the convergence of so-called *ergodic averages* $n^{-1} \sum_{t=1}^{n} m(X_t)$ and is crucial to the use of Markov chains for inference. When S is finite, an irreducible aperiodic chain is automatically positive recurrent and hence ergodic.

If S is finite then P is an $S \times S$ matrix, whose eigenvalues $l_1, \ldots, l_S$ are roots of its characteristic polynomial $\det(P - \lambda I_S)$. If the l_r are distinct, then

$$P = E^{-1} L E, \qquad (6.3)$$

where $L = \mathrm{diag}(l_1, \ldots, l_S)$, the rth row e_r^{T} of the $S \times S$ matrix E is the left eigenvector of P corresponding to l_r and the rth column e_r' of E^{-1} is the right eigenvector of P corresponding to l_r. The l_r are complex numbers with modulus no greater than unity, but as P is real, any complex roots of its characteristic polynomial occur in conjugate pairs $a \pm ib$. For some real $r > 0$,

Here i $= \sqrt{-1}$.

$$(a \pm ib)^n = r^n \exp(\pm in\omega) = r^n(\cos n\omega \pm i \sin n\omega).$$

As P^n is a real matrix, it may be better to express its elements in terms of sines and cosines when P has complex eigenvalues.

If S is finite and the chain is irreducible with period d, then the d complex roots of unity $l_1 = \exp(2\pi i/d), \ldots, l_d = \exp\{2\pi i(d-1)/d\}$ are eigenvalues of P and $l_{d+1}, \ldots, l_S$ satisfy $|l_s| < 1$. If the chain is irreducible and aperiodic, then $l_1 = 1$, and $|l_s| < 1$ for $s = 2, \ldots, S$. Now $\pi^{\mathsf{T}} P = \pi^{\mathsf{T}}$ and $P 1_S = 1_S$, so if X_t has stationary distribution π, then π^{T} and 1_S are the left and right eigenvectors of P corresponding to $l_1 = 1$, that is, $e_1 = \pi$ and $e_1' = 1_S$. The convergence of an ergodic chain with distinct eigenvalues is obvious, because

$$P^n = (E^{-1}LE)^n = E^{-1}L^nE = \sum_{r=1}^{S} l_r^n e_r' e_r^\mathrm{T} \rightarrow e_1' e_1^\mathrm{T} = 1_S \pi^\mathrm{T} \quad \text{as } n \rightarrow \infty :$$

the (r, s) element of P^n, $p_{rs}(n)$, tends to π_s. Moreover, if $p(0)$ is the probability vector of X_0 then X_n has distribution $p(0)^\mathrm{T} P^n$, which converges to $p(0)^\mathrm{T} 1_S \pi^\mathrm{T} = \pi^\mathrm{T}$ whatever the initial vector $p(0)$.

If S is infinite and the chain ergodic, its first eigenvalue l_1 equals 1 and corresponds to the unique stationary distribution π, but the second eigenvalue l_2 need not exist. If l_2 exists and $|l_2| < 1$, then $|l_2|$ controls the rate at which the chain approaches its stationary distribution. More precisely, the chain is *geometrically ergodic* if there exists a function $V(\cdot) > 1$ such that

$$\sum_s |p_{rs}(n) - \pi_s| \leq V(r)|l_2|^n \quad \text{for all } n; \tag{6.4}$$

$|l_2|$ is then the *rate of convergence* of the chain.

An irreducible chain is *reversible* if there exists a π such that

$$\pi_r p_{rs} = \pi_s p_{sr}, \quad \text{for all } r, s \in \mathcal{S}; \tag{6.5}$$

the chain is then positive recurrent with stationary distribution π. Another way to express the *detailed balance condition* (6.5) is

$$\Pr(X_t = r, X_{t+1} = s) = \Pr(X_t = s, X_{t+1} = r), \quad \text{for all } r, s \in \mathcal{S},$$

or $\Pi P = P \Pi$, where Π is the $S \times S$ diagonal matrix whose elements are the components of the stationary distribution π.

Decomposition (6.3) applies to reversible chains, whose eigenvalues and eigenvectors l_r and e_r are real. Chains that fail to be geometrically ergodic have an infinite number of eigenvalues in any open interval containing one of ± 1, but those that are geometrically ergodic have all their eigenvalues but l_1 uniformly bounded in modulus below unity.

Example 6.5 (Two-state chain) Consider the chain for which

$$P = \begin{pmatrix} 1 - p & p \\ q & 1 - q \end{pmatrix}, \quad 0 \leq p, q \leq 1.$$

When $p = q = 0$, there are two absorbing states $\mathcal{C}_1 = \{1\}$ and $\mathcal{C}_2 = \{2\}$ and the chain is entirely uninteresting. When both p and q are positive it is clearly irreducible and $\pi^\mathrm{T} P = \pi^\mathrm{T}$, where $\pi^\mathrm{T} = (p + q)^{-1}(q, p)$. The chain is then positive recurrent with $\mathrm{E}(T_{11}) = (p + q)/q$ and $\mathrm{E}(T_{22}) = (p + q)/p$.

When $p = q = 1$, X_t takes values $\dots, 1, 2, 1, 2, 1 \dots$ and is periodic with period two, so $T_{11} = T_{22} = 2$ with probability one. If $p(0) = \pi$, then X_t has this distribution for all t, but if not, then the fact that $P^2 = I_2$ implies that $X_0, X_1, X_2, \dots$ have distributions $p(0)^\mathrm{T}, p(0)^\mathrm{T} P, p(0)^\mathrm{T}, \dots$; the chain cycles among these and never reaches stationarity.

The eigenvalues of P are $l_1 = 1, l_2 = 1 - p - q$. Its eigendecomposition is

$$\begin{pmatrix} 1 & p \\ 1 & -q \end{pmatrix} \cdot \begin{pmatrix} 1 & 0 \\ 0 & 1 - p - q \end{pmatrix} \cdot \frac{1}{p+q} \begin{pmatrix} q & p \\ 1 & -1 \end{pmatrix}.$$

If $|l_2| < 1$, then $0 < p < 1$ or $0 < q < 1$ or both, the chain is aperiodic and

$$P^n = \frac{1}{p+q} \begin{pmatrix} q + pl_2^n & p - pl_2^n \\ q - ql_2^n & p + ql_2^n \end{pmatrix} \rightarrow \frac{1}{p+q} \begin{pmatrix} q & p \\ q & p \end{pmatrix} = 1_2 \pi^{\mathsf T} \quad \text{as} \quad n \rightarrow \infty.$$

∎

Example 6.6 (Five-state chain)   The state space of the chain with

$$P = \begin{pmatrix} \frac{1}{2} & \frac{1}{2} & 0 & 0 & 0 \\ \frac{1}{4} & \frac{3}{4} & 0 & 0 & 0 \\ 0 & 0 & 0 & 1 & 0 \\ 0 & 0 & 1 & 0 & 0 \\ \frac{1}{4} & 0 & \frac{1}{4} & 0 & \frac{1}{2} \end{pmatrix}$$

decomposes as $\mathcal{C}_1 \cup \mathcal{C}_2 \cup \mathcal{T}$, where $\mathcal{C}_1 = \{1, 2\}$, $\mathcal{C}_2 = \{3, 4\}$ and $\mathcal{T} = \{5\}$. Evidently $\mathcal{C}_1$ is a special case of the previous example, so it is ergodic. The set $\mathcal{C}_2$ is closed and irreducible, but it is periodic because $X_t = X_{t+2} = X_{t+4} = \cdots$. The set $\mathcal{T}$ is transient: at each step the probability of leaving it is $\frac{1}{2}$, with equal probabilities of landing in $\mathcal{C}_1$ and $\mathcal{C}_2$. Although $\mathcal{C}_1$ is ergodic, the chain as a whole is not.

Owing to the presence of two irreducible sets, one with period two, the eigenvalues include $l_1 = 1$, $l_2 = 1$ and $l_3 = -1$. The repeated eigenvalue means that the eigendecomposition of P is not unique. One version is

$$\begin{pmatrix} 1 & 1 & 0 & 0 & -1 \\ 1 & 1 & 0 & 0 & \frac{1}{2} \\ 1 & -1 & -6 & 0 & 0 \\ 1 & -1 & 6 & 0 & 0 \\ 1 & 0 & 1 & 1 & 1 \end{pmatrix} \begin{pmatrix} 1 & 0 & 0 & 0 & 0 \\ 0 & 1 & 0 & 0 & 0 \\ 0 & 0 & -1 & 0 & 0 \\ 0 & 0 & 0 & \frac{1}{2} & 0 \\ 0 & 0 & 0 & 0 & \frac{1}{4} \end{pmatrix} \begin{pmatrix} \frac{1}{6} & \frac{1}{3} & \frac{1}{4} & \frac{1}{4} & 0 \\ \frac{1}{6} & \frac{1}{3} & -\frac{1}{4} & -\frac{1}{4} & 0 \\ 0 & 0 & -\frac{1}{12} & \frac{1}{12} & 0 \\ \frac{1}{2} & -1 & -\frac{1}{6} & -\frac{1}{3} & 1 \\ -\frac{2}{3} & \frac{2}{3} & 0 & 0 & 0 \end{pmatrix}.$$

For large n we have approximately

$$P^n = \begin{pmatrix} \frac{1}{3} & \frac{2}{3} & 0 & 0 & 0 \\ \frac{1}{3} & \frac{2}{3} & 0 & 0 & 0 \\ 0 & 0 & \frac{1}{2}\{1 + (-1)^n\} & \frac{1}{2}\{1 + (-1)^{n+1}\} & 0 \\ 0 & 0 & \frac{1}{2}\{1 + (-1)^{n+1}\} & \frac{1}{2}\{1 + (-1)^n\} & 0 \\ \frac{1}{6} & \frac{1}{3} & \frac{1}{3} & \frac{1}{6} & 2^{-n} \end{pmatrix}.$$

If $X_0 \in \mathcal{C}_1$, the stationary distribution of X_t is $(\frac{1}{3}, \frac{2}{3}, 0, 0, 0)^{\mathsf T}$ and the states have mean recurrence times 3 and $\frac{3}{2}$. If $X_0 = 3$, then $X_2 = X_4 = \cdots = 3$ and $X_1 = X_3 = \cdots = 4$, while the converse is true if $X_0 = 4$; X_t oscillates within $\mathcal{C}_2$ but has a stationary distribution only if the initial probability vector is $(0, 0, \frac{1}{2}, \frac{1}{2}, 0)^{\mathsf T}$. If $X_0 = 5$, the probability that $X_n = 5$ is essentially zero for large n and the process is equally likely to end up in $\mathcal{C}_1$ or $\mathcal{C}_2$.

∎

Likelihood inference

We now consider inference from data $s_0, s_1, \ldots, s_k$ at times $0, 1, \ldots, k$ from a stationary discrete-time Markov chain X_t with finite state space. The likelihood is

$$\Pr(X_0 = s_0, \ldots, X_k = s_k) = \Pr(X_0 = s_0) \prod_{t=0}^{k-1} \Pr(X_{t+1} = s_{t+1} \mid X_t = s_t)$$

$$= \Pr(X_0 = s_0) \prod_{t=0}^{k-1} p_{s_t s_{t+1}}$$

$$= \Pr(X_0 = s_0) \prod_{r=1}^{S} \prod_{s=1}^{S} p_{rs}^{n_{rs}}, \qquad (6.6)$$

where n_{rs} is the observed number of transitions from r to s. Apart from the first term in (6.6), the log likelihood is

$$\ell(p) = \sum_{r=1}^{S} \sum_{s=1}^{S} n_{rs} \log p_{rs}, \qquad (6.7)$$

so the $S \times S$ table of transition counts n_{rs} is a sufficient statistic; see Table 6.2. As $\sum_r p_{sr} = 1$ for each s, (6.7) sums log likelihood contributions from S separate multinomial distributions $(n_{r1}, \ldots, n_{rS})$ whose denominators $n_{r\cdot}$ equal the row sums $n_{r1} + \cdots + n_{rS}$ and whose probability vectors $(p_{r1}, \ldots, p_{rS})$ correspond to transitions out of state r; see (4.45). As $\sum_s p_{rs} = 1$ for each r, this model has $S(S-1)$ parameters.

The results of Section 4.5.3 imply that p_{rs} has maximum likelihood estimate $\widehat{p}_{rs} = n_{rs}/n_{r\cdot}$. Standard likelihood asymptotics will apply if $0 < p_{rs} < 1$ for all r and s and if the denominators $n_{r\cdot} \to \infty$ as $k \to \infty$. Now $n_{r\cdot}$ is the number of visits the chain makes to state r during the period $1, \ldots, k$, and if the chain is ergodic r is visited infinitely often as $k \to \infty$. The $\widehat{p}_{rs}$ then have an approximate joint normal distribution with covariances estimated by

$$\mathrm{cov}(\widehat{p}_{rs}, \widehat{p}_{tu}) \doteq \begin{cases} \widehat{p}_{rs}(1 - \widehat{p}_{rs})/n_{r\cdot}, & r = t, s = u, \\ -\widehat{p}_{rs}\widehat{p}_{ru}/n_{r\cdot}, & r = t, s \neq u, \\ 0, & \text{otherwise.} \end{cases}$$

The above discussion ignores the first term in (6.6). If k is large it will add only a small contribution to $\ell(p)$ and can safely be dropped, but if k is small it might be replaced by the stationary probability π_{s_0}, found from the elements of P. In general the log likelihood must then be maximized numerically.

An alternative asymptotic scenario is that m independent finite segments of Markov chains having the same parameters are observed, and $m \to \infty$. The overall information in the initial terms of the segments is then $O(m)$ and retrieval of it may be worthwhile, particularly if the segments are short. Below we continue to suppose that there is a single chain of length k.

In simpler models the p_{rs} might depend on a parameter with dimension smaller than $S(S-1)$. For instance, setting $p = q$ in Example 6.5 gives a one-parameter

	Observed frequency				Expected frequency			
First base	A	C	G	T	A	C	G	T
A	185	74	86	171	169.5	86.4	74.2	185.9
C	101	41	6	115	86.4	44.0	37.8	94.8
G	69	45	34	78	74.2	37.8	32.5	81.4
T	161	103	100	202	185.9	94.8	81.4	203.9

Table 6.4 Fit of independence model to DNA data: observed and fitted frequencies of one-step transitions.

model. If the chain is ergodic, likelihood inference for such models will be regular under the usual conditions on the parameter space.

Thus far transition probabilities have depended only on the current state, so our chains have been *first-order*. The simpler *independence model* posits transition probabilities independent of the current state, $p_{rs} \equiv p_s$; this *zeroth-order* chain has just $S - 1$ parameters. Row and column classifications in the table of counts n_{rs} are then independent, (6.7) reduces to $\sum n_{\cdot s} \log p_s$, and $\widehat{p}_s = n_{\cdot s}/n_{\cdot \cdot}$, where $n_{\cdot s} = n_{1s} + \cdots + n_{Ss}$ and $n_{\cdot \cdot} = \sum_s n_{\cdot s}$. Thus the likelihood ratio statistic for comparison of the zeroth- and first-order chains is

$$W = 2 \sum_{r,s} n_{rs} \log \left(\frac{\widehat{p}_{rs}}{\widehat{p}_s} \right) = 2 \sum_{r,s} n_{rs} \log \left(\frac{n_{rs} n_{\cdot \cdot}}{n_r \cdot n_{\cdot s}} \right) ;$$

this is the likelihood ratio statistic for testing row-column independence in the square table of counts n_{rs}. Under the zeroth-order chain the rows of P all equal $(p_1, \ldots, p_S)$, row and column classifications are independent, and W is a natural statistic to assess this; its asymptotic distribution is chi-squared with $S(S - 1) - (S - 1) = (S - 1)^2$ degrees of freedom. As we saw in Section 4.5.3, W approximately equals Pearson's statistic $P = \sum (O - E)^2/E$, where O and E denote the observed count n_{rs} and its expected counterpart $n_r \cdot n_{\cdot s}/n_{\cdot \cdot}$ under the independence model and the sum is over the cells of the table. The quantities $(O - E)/E^{1/2}$ squared give the contribution of each cell to P.

Example 6.7 (DNA data) The lowest line of Table 6.2 gives maximum likelihood estimates for the zeroth-order independence model, while the four previous lines give estimates for the first-order model. For the independence model we have $\widehat{p}_A = 516/1571 = 0.328$ and $\widehat{p}_C = 263/1571 = 0.167$, for example, while under the first-order model $\widehat{p}_{AA} = 185/516 = 0.359$, $\widehat{p}_{AC} = 74/516 = 0.143$, $\widehat{p}_{CG} = 6/263 = 0.023$ and so forth. If the independence model was correct, $W = 2 \sum_{r,s} n_{rs} \log\{n_{rs}/(n_r \cdot \widehat{p}_s)\}$ would have a χ_9^2 distribution, but the observed value $w = 64.45$ makes this highly implausible. The value of P is 50.3.

Table 6.4 shows the counts n_{rs} and the fitted values $n_r \cdot n_{\cdot s}/n_{\cdot \cdot}$ under the independence model. The largest discrepancy is for the CG cell, for which $(O - E)/E^{1/2} = -5.18$, so this cell contributes 26.79 to the value of P. The normal probability plot of

Figure 6.3 Fit of zeroth-
and first-order Markov
chains to the DNA data.
The panels show normal
probability plots of the
signed contributions
$(O - E)/E^{1/2}$ made by
the 16 cells of the
two-way table under the
independence model (left)
and the 64 cells of the
three-way table under the
first-order model (right).
The large negative value
on the left is due to the
CG cell. The dots show
the null line $x = y$.

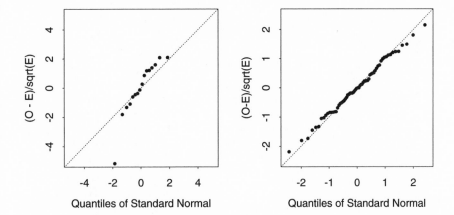

the $(O - E)/E^{1/2}$ in the left panel of Figure 6.3 shows that the other cells contribute
much less. The values of W and P remain large even if this cell is dropped from
the table, however, so it is not the sole cause of the poor fit of the independence
model. ∎

Higher-order models

First-order Markov chains extend to chains of order m, where the probability of
transition into s depends on the m preceding states. One way to think of this is that
the state of the chain is augmented from X_j to $Y_j = (X_j, X_{j-1}, \ldots, X_{j-m+1})$ and the
transition probabilities change to

$$\Pr(Y_j = y_j \mid Y_{j-1} = y_{j-1}) = \Pr(X_j = s \mid X_{j-1} = s_{j-1}, \ldots, X_{j-m} = s_{j-m})$$
$$= p_{s_{j-m}s_{j-m+1}\cdots s_{j-1}s},$$

say. Thus the 'current' state $Y_{j-1} = (s_{j-1}, \ldots, s_{j-m})$ contains information not only
from time $j - 1$ but also from the $m - 1$ previous times. Whereas with $m = 1$ the
properties of the chain were determined by the S vectors of transition probabilities
$(p_{r1}, \ldots, p_{rS})$, there are now S^m such vectors, so much more data is needed in order to
get reliable estimates of the transition probabilities. A compromise is a *variable-order*
chain, the simplest example of which is when $m = 2$ and $S = 2$, so that the chain
of order two is determined by the probabilities p_{111}, p_{121}, p_{211} and p_{221}, giving the
transition probabilities π_{sur} from (s, u) to r. A simple variable-order chain is obtained
by specifying $\pi_{111} = \pi_{211}$, that is, given that $u = 1$, the transition probabilities do not
depend on s. This chain is first-order when $u = 1$, but not when $u = 2$. In this case
the number of parameters only diminishes by one, but in general the reduction might
be much larger.

Likelihood ratio statistics or criteria such as AIC enable systematic comparison of
Markov chains of different orders, but care is needed when computing them. Suppose
that we fit models of orders up to m to a sequence of length k. There are $k - 1$
successive pairs, $k - 2$ triplets and so forth, so the fit of the mth-order model is based

First base	Second base	Frequencies for third base				
		A	C	G	T	Total
A	A	81	22	29	53	185
	C	30	7	2	35	74
	G	29	18	11	27	86
	T	54	23	33	61	171
C	A	30	20	15	36	101
	C	15	2	1	23	41
	G	2	1	0	3	6
	T	28	26	20	41	115
G	A	30	3	14	22	69
	C	18	10	1	16	45
	G	12	5	10	7	34
	T	27	11	12	27	77
T	A	44	29	28	60	160
	C	38	22	2	41	103
	G	26	21	13	40	100
	T	51	43	35	73	202

Table 6.5 Observed transition counts for second-order Markov chain for DNA data.

on the $k - m$ successive $(m + 1)$-tuples from which the transition probabilities and maximized log likelihood are computed, treating the last $k - m$ of the k observations as responses. Standard likelihood methods presuppose that the same responses are used throughout, so fits for chains of smaller order must also treat only the last $k - m$ observations as responses.

Example 6.8 (DNA data)   We compare models of order up to $m = 3$. The preceding discussion implies that as the data in Table 6.1 begin GTAT. . ., the first response is the second T, so the initial GTA, GT and G should be ignored when fitting the zeroth-, first- and second-order models respectively. The frequencies for the $k - m = 1572 - 3 = 1569$ triplets of transition counts in our sequence are shown in Table 6.5. The implied numbers of TA and GT transitions, $54 + 28 + 27 + 51 = 160$ and $27 + 3 + 7 + 40 = 77$, are smaller than the numbers 161 and 78 in Table 6.2 which include such transitions in the initial GTAT.

Estimates under the second-order model are obtained as before, by dividing each row by its total, giving $\widehat{p}_{AAA} = 81/185$, $\widehat{p}_{AAC} = 22/185$, $\widehat{p}_{ACA} = 30/74$ and so forth. Evidently estimates such as $\widehat{p}_{CGA} = 2/6$ are very unreliable.

Estimates under the first-order model are computed from the two-way table of counts obtained by collapsing the table over the first base, giving a 4×4 table whose top left (AA) element is $81 + 30 + 30 + 44 = 185$, whose CG element is $2 + 1 + 1 + 2 = 6$ and so forth. For estimates under the independence model we use the 1×4 table from a further collapse over the second base; both sets of estimates are essentially unaffected by dropping the first few bases.

The maximized log likelihoods for the zeroth-, first-, second- and third-order models are -2058.44, -2026.02, -1998.41, and -1923.25 on 3, 12, 48, and 192 degrees of freedom, so the AIC values are 4122.9, 4076.0, 4092.8, and 4230.5 and the likelihood ratio statistics for comparison of each model with the next are 64.8, 55.2, and 150.3, on 9, 36, and 144 degrees of freedom. There is strong evidence for first-order dependence compared to independence, while as $\Pr(\chi^2_{36} > 55.2) \doteq 0.02$ and $\Pr(\chi^2_{144} > 150.3) \doteq 0.34$ the evidence for second- compared to first-order dependence is weaker, and there is no suggestion of third-order dependence. The AIC values clearly indicate the first-order model.

The signed contributions $(O - E)/E^{1/2}$ to Pearson's statistic under the first-order model can be obtained using Table 6.5. The contribution for the AAA cell, for example, is $(81 - E)/E^{1/2}$, where $E = 185\widehat{p}_{AA}$, with $\widehat{p}_{AA}$ calculated under the first-order model. The value of Pearson's statistic is 52.84. The right panel of Figure 6.3 shows no highly unusual cells and apparently good fit.

The eigenvalues for the observed first-order matrix of transition probabilities $\widehat{P}$ are $1, -0.0147 \pm 0.0704i$ and 0.0524. The small absolute values of the last three suggest that the chain is close to independence, and indeed the rows of $\widehat{P}^4$ are essentially equal: four steps are (almost) enough to forget the past.

Our earlier discussion suggested that the main departures from independence occur after C, suggesting taking a model where $p_{rs} = \psi_s$ whenever $r \neq C$ and $p_{Cs} = \phi_s$. That is, for each s we have

$$\Pr(X_{t+1} = s \mid X_t = A) = \Pr(X_{t+1} = s \mid X_t = G) = \Pr(X_{t+1} = s \mid X_t = T),$$

but these do not equal p_{Cs}. This model has six independent parameters and as its log likelihood $\sum_s (\sum_{r \neq C} n_{rs}) \log \psi_s + \sum_s n_{Cs} \log \phi_s$ is of multinomial form, their estimates are readily obtained. The maximized log likelihood is -2031.0, so AIC = 4074.0 is lower than for the full first-order chain and this model seems marginally preferable. See Exercise 6.1.7 for further details. ∎

We have presumed above that X_t is stationary. If instead the transition probabilities are of form $p_{rs}(t; \theta)$, dependent on a parameter θ, then the likelihood $\Pr(X_0 = s_0; \theta) \prod_{t=0}^{k-1} p_{s_t s_{t+1}}(t; \theta)$ is found by the argument leading to (6.6). In many cases the initial probability $\Pr(X_0 = s_0; \theta)$ may be unknown, and if the series is long little will be lost by ignoring it. If the transition probabilities do not share dependence on a common θ, they can only be estimated if they are repeated. Large amounts of data will then be needed.

6.1.2 Continuous-time models

We now turn to stationary continuous-time Markov models with finite state space $\mathcal{S}$. The basic assumption is that over small intervals $[t, t + \delta t)$, transitions between states have probabilities

$o(\delta t)$ is small enough that $o(\delta t)/\delta t \to 0$ as $\delta t \to 0$.

$$\Pr(X_{t+\delta t} = s \mid X_t = r) = \begin{cases} \gamma_{rs}\delta t + o(\delta t), & s \neq r, \\ 1 + \gamma_{rr}\delta t + o(\delta t), & s = r, \end{cases} \tag{6.8}$$

where γ_{rs} is interpreted as the rate at which transitions $r \to s$ occur. The transition probabilities do not depend on t, so X_t is time homogeneous. Note that $\sum_s \gamma_{rs} = 0$, for each r, because the probabilities in (6.8) sum to one.

Let $p(t)$ denote the $S \times 1$ vector whose rth element is $p_r(t) = \Pr(X_t = r)$; note that $1_S^T p(t) = 1$ for all t. Then

$$p_s(t + \delta t) = \sum_{r=1}^{S} \Pr(X_{t+\delta t} = s \mid X_t = r) p_r(t) \doteq p_s(t) + \sum_{r=1}^{S} \gamma_{rs} p_r(t) \delta t + o(\delta t),$$

implying that

$$\frac{dp_s(t)}{dt} = \lim_{\delta t \to 0} \frac{p_s(t + \delta t) - p_s(t)}{\delta t} = \sum_{r=1}^{S} \gamma_{rs} p_r(t), \quad s = 1, \ldots, S,$$

written in matrix form as

$$\left(\frac{dp_1(t)}{dt} \quad \cdots \quad \frac{dp_S(t)}{dt} \right) = (p_1(t) \quad \cdots \quad p_S(t)) \begin{pmatrix} \gamma_{11} & \cdots & \gamma_{1S} \\ \vdots & \ddots & \vdots \\ \gamma_{S1} & \cdots & \gamma_{SS} \end{pmatrix}.$$

In terms of the *infinitesimal generator* of the chain, the matrix G whose (r, s) element is γ_{rs}, we write

$$\frac{dp(t)^T}{dt} = p(t)^T G, \tag{6.9}$$

to which the formal solution is

$$p(t)^T = p(0)^T \exp(tG),$$

where $p(0)$ is the probability vector for the states of X_0, and the matrix exponential $\exp(tG)$ is interpreted as $\sum_{m=0}^{\infty} (tG)^m / m!$, with $G^0 = I_S$. If the initial state was $X_0 = r$, $p(0)$ consists of zeros except for its rth component, implying that $\Pr(X_t = s \mid X_0 = r) = p_{rs}(t)$ is the (r, s) element of $\exp(tG)$.

Any stationary distribution π for X_t must be time-independent, so the right-hand side of (6.9) will be zero when $p(0) = \pi$. Hence π^T will be a left eigenvector of G with eigenvalue zero.

The chain is reversible if and only if there is a distribution π satisfying the detailed balance condition $\pi_r \gamma_{rs} = \pi_s \gamma_{sr}$.

If G is diagonalizable the eigendecomposition (6.3) is again useful. For if $G = E^{-1} L E$ then $G^m = E^{-1} L^m E$, so

$$\exp(tG) = E^{-1} \text{diag}\{\exp(tl_1), \ldots, \exp(tl_S)\} E.$$

Hence the sth row of E and column of E^{-1}, e_s^T and e_s', are left and right eigenvectors of $\exp(tG)$ with eigenvalue $\exp(tl_s)$. The fact that $\sum_s \gamma_{rs} = 0$ for each r implies that $G 1_S = 0$, so $e_1' = 1_S$ is a right eigenvalue of G with eigenvalue $l_1 = 0$, while $e_1 = \pi$, as we saw above. The remaining eigenvalues of G all have strictly negative real parts. Hence

$$
\exp(tG) = (e'_1 \quad \cdots \quad e'_S) \begin{pmatrix} \exp(tl_1) & & 0 \\ & \ddots & \\ 0 & & \exp(tl_S) \end{pmatrix} \begin{pmatrix} e_1^\mathsf{T} \\ \vdots \\ e_S^\mathsf{T} \end{pmatrix}
$$

$$
= \sum_{r=1}^{S} \exp(tl_r) e'_r e_r^\mathsf{T}
$$

$$
\to e'_1 e_1^\mathsf{T} = 1_S \pi^\mathsf{T} \quad \text{as} \quad t \to \infty :
$$

starting from any X_0, the (r, s) element of $\exp(tG)$, $\Pr(X_t = s \mid X_0 = r) \to \pi_s$. This transition probability may be written as a linear combination of exponentials, $c_{rs,1} e^{tl_1} + \cdots + c_{rs,S} e^{tl_S}$, where $c_{rs,v}$ is the (r, s) element of $e'_v e_v^\mathsf{T}$, that is, the product of the rth element of e'_v and the sth element of e_v.

Fully observed trajectory

If X_t had been fully observed during $[0, t_0]$, say, we would know exactly when and between which states transitions occurred. To write down the likelihood we would need probabilities for events such as $X_u = r, 0 \le u < t$, followed by transition from r to s at time t, so $X_t = s$. To obtain this we divide $[0, t)$ into m intervals of length δt and apply the Markov property to see that

$$
\Pr\left(X_{\delta t} = X_{2\delta t} = \cdots = X_{(m-1)\delta t} = r, X_{m\delta t} = s \mid X_0 = r\right)
$$

equals

$$
\Pr\left(X_{m\delta t} = s \mid X_{(m-1)\delta t} = r\right) \prod_{i=1}^{m-1} \Pr\left(X_{i\delta t} = r \mid X_{(i-1)\delta t} = r\right),
$$

and this itself is

$$
\{1 + \gamma_{rr}\delta t + o(\delta t)\}^{m-1} \{\gamma_{rs}\delta t + o(\delta t)\} = \left(1 + \frac{\gamma_{rr} t}{m}\right)^{m-1} \gamma_{rs}\delta t + o(\delta t).
$$

On dividing by δt and letting $m \to \infty$, then recalling that $\gamma_{rr} = -\sum_{v \neq r} \gamma_{rv}$, we see that the density corresponding to observing $X_u = r, 0 \le u < t$, followed by transition to $X_t = s$, is

$$
\gamma_{rs} \exp(t\gamma_{rr}) = \gamma_{rs} \exp\left(-t \sum_{v \neq r} \gamma_{rv}\right).
$$

This has the simple interpretation that the first transition out of r occurs at $T = \min\{t : X_t \neq r\} = \min_{v \neq r}\{T_{rv}\}$, where the T_{rv} are independent exponential variables with parameters γ_{rv}, that is, with means γ_{rv}^{-1}. This suggests an algorithm for simulating data from such a process (Exercise 6.1.11).

The probability of a trajectory fully observed for the period $[0, t_0]$ and with transitions at $t_1 < \cdots < t_k$ is calculated by using the Markov property to express

$$
\Pr(X_t = s_0, 0 \le t < t_1, X_t = s_1, t_1 \le t < t_2, \ldots, X_t = s_k, t_k \le t \le t_0)
$$

as

$$\Pr(X_0 = s_0) \Pr(X_t = s_0, 0 < t < t_1, X_{t_1} = s_1 \mid X_0 = s_0)$$

$$\times \prod_{j=1}^{k-1} \Pr\left(X_t = s_j, t_j < t < t_{j+1}, X_{t_{j+1}} = s_{j+1} \mid X_{t_j} = s_j\right)$$

$$\times \Pr\left(X_t = s_k, t_k < t \le t_0 \mid X_{t_k} = s_k\right).$$

Thus the likelihood for the γ_{rs} based on such data is

$$\Pr(X_0 = s_0) \times \gamma_{s_0 s_1} e^{t_1 \gamma_{s_0 s_0}} \times \prod_{j=1}^{k-1} \gamma_{s_j s_{j+1}} e^{(t_{j+1} - t_j)\gamma_{s_j s_j}} \times e^{(t_0 - t_k)\gamma_{s_k s_k}}. \tag{6.10}$$

The initial probability $\Pr(X_0 = s_0)$ might be replaced by the s_0th element of the stationary distribution of X_t, or dropped from the likelihood. In either case (6.10) may be maximized with respect to the γ_{rs}, $s \ne r$, if enough transitions have occurred — in general, no inferences can be made about transitions from r to s if none have been observed.

Partially observed trajectory

In practice trajectories may not be fully observed. One possibility is that the states $s_0, s_1, \ldots, s_k$ of X_t at times $0 < t_1 < \cdots < t_k$ are known, as are the numbers and types of transitions between the s_j, but that the times of these intervening transitions are unknown. A less informative possibility is that nothing is known about transitions, so that only the s_j and t_j are known. The likelihood is then (6.1) with $\Pr\left(X_{t_j} = s_j \mid X_{t_{j-1}} = s_{j-1}\right)$ equal to the (s_{j-1}, s_j) element of $\exp\{(t_j - t_{j-1})G\}$, that is, $p_{s_{j-1} s_j}(t_j - t_{j-1})$, and $\Pr(X_0 = s_0)$ chosen according to context.

Example 6.9 (Two-state Markov chain)   The simplest case has $S = 2$ states with transition intensities given by

$$G = \begin{pmatrix} -\gamma_{12} & \gamma_{12} \\ \gamma_{21} & -\gamma_{21} \end{pmatrix}, \quad \gamma_{12}, \gamma_{21} > 0.$$

Its eigendecomposition is

$$G = \begin{pmatrix} 1 & \gamma_{12} \\ 1 & -\gamma_{21} \end{pmatrix} \begin{pmatrix} 0 & 0 \\ 0 & -(\gamma_{21} + \gamma_{12}) \end{pmatrix} \frac{1}{\gamma_{12} + \gamma_{21}} \begin{pmatrix} \gamma_{21} & \gamma_{12} \\ 1 & -1 \end{pmatrix},$$

so the limiting distribution is $\pi^{\mathsf{T}} = (\gamma_{12} + \gamma_{21})^{-1}(\gamma_{21}, \gamma_{12})$, and

$$\exp(tG) = \frac{1}{\gamma_{12} + \gamma_{21}} \begin{pmatrix} \gamma_{21} + \gamma_{12} e^{l_2 t} & \gamma_{12}(1 - e^{l_2 t}) \\ \gamma_{21}(1 - e^{l_2 t}) & \gamma_{12} + \gamma_{21} e^{l_2 t} \end{pmatrix},$$

where $l_2 = -(\gamma_{12} + \gamma_{21}) < 0$ except in the trivial case $\gamma_{12} = \gamma_{21} = 0$, when the chain stays forever in its initial state.

The holding time in state r is exponential with parameter γ_{rs}, so the likelihood based on a trajectory fully observed on the interval $[0, t_0]$ with transitions $1 \to 2 \to 1 \to 2$ at $t_1 < t_2 < t_3$ is

$$\frac{\gamma_{21}}{\gamma_{12} + \gamma_{21}} \times \gamma_{12} e^{-t_1 \gamma_{12}} \times \gamma_{21} e^{-(t_2 - t_1)\gamma_{21}} \times \gamma_{12} e^{-(t_3 - t_2)\gamma_{12}} \times e^{-(t_0 - t_3)\gamma_{21}},$$

the first and last terms being the stationary probability $\Pr(X_0 = 1)$ and the probability that no transition occurs in $(t_3, t_0]$. Apart from the first term, the log likelihood is $n_{12} \log \gamma_{12} - \gamma_{12} t_1' + n_{21} \log \gamma_{21} - \gamma_{21} t_2'$, where n_{rs} is the number of $r \to s$ transitions and t_r' the total time spent in state r.

Each row of $\exp(tG)$ tends to π^{T} as $t \to \infty$. One effect of this is that if the process is observed so intermittently that $X_0, X_{t_1}, \ldots$ are essentially independent, the transition probabilities $p_{rs}(t_j - t_{j-1})$ will almost equal elements of π^{T}, because $\exp\{l_2(t_j - t_{j-1})\} \doteq 0$. If so, then although $\gamma_{21}/(\gamma_{12} + \gamma_{21})$ will be estimable — it will be roughly the proportion of occasions that $X_t = 1$ — the individual rates γ_{12} and γ_{21} will not. The implication for design of studies involving such models is that X_t must be observed often enough that its successive values are correlated; otherwise only the stationary distribution is estimable. If several transitions occur every week, data obtained at monthly intervals will be essentially uninformative. ∎

Example 6.10 (Breast cancer data) A model for these data has

$$G = \begin{pmatrix} -\gamma_{12} - \gamma_{13} & \gamma_{12} & \gamma_{13} \\ \gamma_{21} & -\gamma_{21} - \gamma_{23} & \gamma_{23} \\ 0 & 0 & 0 \end{pmatrix};$$

of course $\gamma_{31} = \gamma_{32} = 0$ because death is absorbing. A simpler model sets $\gamma_{13} = 0$, so a woman with the disease cannot die without first being unable to walk. Appropriate asymptotics take the number of women, rather than the number of observations on each, large; below we suppose that large-sample approximations are applicable with just 37 women. In practice it would be wise to check this by simulation.

The overall likelihood L is the product of independent contributions of form (6.1), one for each woman. Appreciable information might be lost by ignoring the terms $\Pr(X_0 = s_0)$, which comprise 37 of the 135 terms of L. Owing to the absorbing state, we cannot replace $\Pr(X_0 = s_0)$ with its stationary value

$$\lim_{t \to \infty} \Pr(X_t = 1) = \lim_{t \to \infty} \Pr(X_t = 2) = 0,$$

and we use $\lim_{t \to \infty} \Pr(X_t = s_0 \mid X_t \neq 3)$ instead, because only living women entered the study. Now for $s = 1, 2$,

$$\Pr(X_t = s \mid X_0 = r) = c_{rs,1} e^{t l_1} + c_{rs,2} e^{t l_2} + c_{rs,3} e^{t l_3},$$

where $l_3 < l_2 < l_1 = 0$, and as this probability has limit zero we must have $c_{rs,1} = 0$. As $t \to \infty$, therefore,

$$\begin{aligned}
\Pr(X_t = s \mid X_t \neq 3, X_0 = r) &= \frac{c_{rs,2} e^{l_2 t} + c_{rs,3} e^{l_3 t}}{c_{r1,2} e^{l_2 t} + c_{r1,3} e^{l_3 t} + c_{r2,2} e^{l_2 t} + c_{r2,3} e^{l_3 t}} \\
&\to \frac{c_{rs,2}}{c_{r1,2} + c_{r2,2}} \\
&= \frac{e_{2,s}}{e_{2,1} + e_{2,2}},
\end{aligned}$$

independent of r, where $e_{2,v}$ is the vth element of e_2^T, the left eigenvector of G corresponding to l_2.

The missing value complicates the likelihood contribution for woman 24, which is

$$\frac{e_{2,1}}{e_{2,1} + e_{2,2}} \times p_{12}(3) \times \{p_{21}(3)p_{13}(6) + p_{22}(3)p_{23}(6)\}.$$

The maximized log likelihoods for the three- and four-parameter models are -107.43 and -107.39. As $\gamma_{13} = 0$ lies on the boundary of the parameter space, the asymptotic distribution of the likelihood ratio statistic is $\frac{1}{2} + \frac{1}{2}\chi_1^2$; see Example 4.39. Its value, $2\{-107.39 - (-107.43)\} = 0.08$, supports the simpler model, for which maximum likelihood estimates and standard errors are $\hat{\gamma}_{12} = 0.116$ (0.025), $\hat{\gamma}_{21} = 0.057$ (0.035) and $\hat{\gamma}_{23} = 0.238$ (0.043). The transition rate γ_{21} is poorly determined, and taking the 95% confidence interval based on its profile likelihood, (0.014, 0.170), is preferable to using its standard error. The estimated mean times spent in states 1 and 2 are $\hat{\gamma}_{12}^{-1} = 8.6$ and $(\hat{\gamma}_{21} + \hat{\gamma}_{23})^{-1} = 3.4$ months, with death then occurring with estimated probability $\hat{\gamma}_{23}/(\hat{\gamma}_{21} + \hat{\gamma}_{23}) = 0.81$. Confidence intervals for these quantities should be based on profile likelihoods.

The non-zero eigenvalues of $\hat{G}$ are -0.33 and -0.08, and examination of the estimated transition matrices between the later follow-up times suggests that there is some information in the small number of later transitions.

A more thorough analysis would assess the effect of initial status, for example by seeing if the likelihood increases significantly when the three-parameter model is fitted separately to each of the two initial groups. Of particular concern is the stationarity assumption, which is hard to justify here. The data are too sparse, however, for much further modelling to be conclusive.   ■

Inhomogeneous chains

If the transition rates $\gamma_{rs}(t)$ depend on time then the fundamental equation (6.9) becomes $dp(t)^T/dt = p(t)^T G(t)$. This is a system of first-order ordinary differential equations, whose solution may be written formally as $p(t)^T = p(0)^T \exp\{\int_0^t G(s)\,ds\}$. Typically this will not be available explicitly, and the transition probabilities must be obtained using packages for solving systems of ordinary differential equations, or by discretizing time and fitting suitable models to the resulting transition probabilities.

Exercises 6.1

1   Classify the states of Markov chains with transition matrices

$$\begin{pmatrix} 0 & 1 & 0 \\ 0 & 0 & 1 \\ \frac{1}{2} & \frac{1}{2} & 0 \end{pmatrix}, \quad \begin{pmatrix} 0 & 1 & 0 & 0 \\ 0 & 0 & 0 & 1 \\ 0 & 0 & 1 & 0 \\ 1 & 0 & 0 & 0 \end{pmatrix}, \quad \begin{pmatrix} \frac{1}{2} & \frac{1}{2} & 0 & 0 & 0 & 0 \\ \frac{1}{4} & \frac{3}{4} & 0 & 0 & 0 & 0 \\ \frac{1}{4} & \frac{1}{4} & \frac{1}{4} & \frac{1}{4} & 0 & 0 \\ \frac{1}{4} & 0 & \frac{1}{4} & \frac{1}{4} & 0 & \frac{1}{4} \\ 0 & 0 & 0 & 0 & \frac{1}{2} & \frac{1}{2} \\ 0 & 0 & 0 & 0 & \frac{1}{2} & \frac{1}{2} \end{pmatrix}.$$

2 Find the eigendecomposition of

$$P = \begin{pmatrix} 0 & 1 & 0 \\ 0 & \frac{1}{2} & \frac{1}{2} \\ \frac{1}{2} & 0 & \frac{1}{2} \end{pmatrix}$$

and show that $p_{11}(n) = a + 2^{-n}\{b\cos(n\pi/2) + c\sin(n\pi/2)\}$ for some constants a, b and c. Write down $p_{11}(n)$ for $n = 0, 1$ and 2 and hence find a, b and c.

3 In Example 6.5, sketch how $p_{11}(n)$ depends on n when $l_2 < 0$, $l_2 > 0$ and $l_2 = 0$. Find $E(T_{11})$ by first showing that

$$\Pr(T_{11} = k) = \begin{cases} 1 - p, & k = 1, \\ pq(1-q)^{k-2}, & k = 2, 3, \dots \end{cases}$$

4 Say when

$$P = \begin{pmatrix} 1-p & p & 0 \\ 0 & 1-p & p \\ p & 0 & 1-p \end{pmatrix}, \quad 0 \le p \le 1,$$

has an equilibrium distribution, and write it down. Show that P has eigenvalues $1, (2 - 3p \pm i3^{1/2}p)/2$, and use them to say when the chain is ergodic.

5 Let X_t be a stationary first-order Markov chain with state space $\{1, \dots, S\}$, $S > 2$, and let I_t indicate the event $X_t = 1$. Is $\{I_t\}$ a Markov chain?

6 Consider a sequence $0100\dots10$ of variables I_j and let $\overline{I}_t = (2k+1)^{-1}\sum_{j=-k}^{k} I_{t+j}$ be the average of the $2k + 1$ variables centred at t.
(a) Verify the calculations in Example 6.5.
(b) Let the stationary first-order chain $\{I_t\}$ have state space $\{0, 1\}$ and transition probability matrix P. In the notation of Example 6.5, show that

$$\text{cov}(I_t, I_{t+j}) = \Pr(I_t = I_{t+j} = 1) - \Pr(I_t = 1)\Pr(I_{t+j} = 1) = pql_2^j/(p+q)^2,$$

and deduce that with $m = 2k + 1$,

$$\text{var}(\overline{I}_t) = \frac{2}{m^2}\sum_{j=0}^{m-1}(m - j)\text{cov}(I_0, I_j) - \frac{\text{var}(I_0)}{m}.$$

It may be useful to know that for large n,
$\sum_{j=0}^{n} jp^j \doteq p/(1-p)^2.$

Give an expression for $\text{var}(\overline{I}_t)$, and show that it is roughly $(2 - p - q)/(p + q)$ times the corresponding expression for independent I_j.

7 Check the log likelihood for the six-parameter model given at the end of Example 6.8, obtain the maximum likelihood estimates and the fitted counts, and calculate Pearson's statistic. Give its degrees of freedom and assess the fit of the model.

8 A *run of length m* of a stationary Markov chain occurs when there is a sequence of form $X_t \ne s$, $X_{t+1} = \cdots = X_{t+m} = s$, $X_{t+m+1} \ne s$. Show that this has probability $p_{ss}^{m-1}(1 - p_{ss})$ for $m = 1, 2, \dots$: the geometric density with mean $(1 - p_{ss})^{-1}$. Show that in a first-order chain the lengths of separate runs are independent. Is this true in higher-order chains? Can you construct a non-trivial 3×3 transition matrix for which it is impossible to use runs to falsify the independence model, whatever the length of the chain?

9 Recall that T_{rs} denotes the first-passage time from state r to state s. For the three-parameter model in Example 6.10, show that $E(T_{23}) = (g_{12} + g_{23})^{-1}(1 + g_{12})E(T_{13})$ and find the corresponding equation for $E(T_{13})$. Hence give expressions for $E(T_{13})$ and $E(T_{23})$ and show that their maximum likelihood estimates are 17 and 8.4 months respectively.
What additional information do you need to compute standard errors for these estimates?

10 Modify the argument from the preceding question to find the moment-generating functions of T_{13} and T_{23} in terms of γ_{12}, γ_{21}, and γ_{23}. Hence check your formulae for $E(T_{13})$ and $E(T_{23})$.

11   Let $X_1, \ldots, X_n$ be independent exponential variables with rates λ_j. Show that $Y = \min(X_1, \ldots, X_n)$ is also exponential, with rate $\lambda_1 + \cdots + \lambda_n$, and that $\Pr(Y = X_j) = \lambda_j / (\lambda_1 + \cdots + \lambda_n)$. Hence write down an algorithm to simulate data from a continuous-time Markov chain with finite state space, using exponential and multinomial random number generators.

12   Observations $s_0, \ldots, s_k$ on a discrete-time Markov chain with one-step transition matrix P are obtained at times $0 < t_1 < \ldots < t_k$, where not all the $t_j - t_{j-1}$ equal unity. Write down the likelihood in terms of elements $p_{rs}(n)$ of P^n, $n = 1, 2, \ldots$. Give explicitly the the likelihood when the states 12311 of a three-state chain with stationary distribution π are observed at times 0, 1, 3, 4, 6.
      Explain how you would calculate the likelihood L for the data in Table 6.3, with three-month transition probability matrix

$$P = \begin{pmatrix} 1 - p_{12} & p_{12} & 0 \\ p_{21} & 1 - p_{21} - p_{23} & p_{23} \\ 0 & 0 & 1 \end{pmatrix}.$$

What value has L under this model? How could P be made more plausible?   Look carefully at the data.

13   Check the eigendecomposition of G in Example 6.9. Calculate the stationary distribution when $\gamma_{12} = 0$. Is this a surprise?

6.2 Markov Random Fields

6.2.1 Basic notions

The previous section described simple models for random variables indexed by a scalar, often time, so the variables can be visualized at points along an axis. Many applications require variables associated to points in space or in space-time, however, and then more general indexing sets are needed. Think, for example, of the colours of pixels in an image, the fertility of parts of a field or the occurrence of cancer cases at points on a map. This section outlines how our earlier ideas extend to some more complex settings. There is a close connection to notions of statistical physics, from which some of the terminology is derived.

Our earlier discussion owed its relative simplicity to the Markov property — that the 'future' is independent of the 'past', conditional on the 'present' — whose importance suggests that we should seek its analogy here. The notions of 'past', 'present', and 'future' have no obvious spatial counterparts, but another formulation does generalize in a natural way. A sequence $Y_1, \ldots, Y_n$ satisfies the Markov property if

$$\Pr(Y_{j+1} = y_{j+1} \mid Y_1 = y_1, \ldots, Y_j = y_j) = \Pr(Y_{j+1} = y_{j+1} \mid Y_j = y_j)$$

for $j = 1, \ldots, n - 1$ and all y. This is equivalent to having each Y_j depend on the remaining variables $Y_{-j} = (Y_1, \ldots, Y_{j-1}, Y_{j+1}, \ldots, Y_n)$ only through the adjacent variables Y_{j-1} and Y_{j+1} (Exercise 6.2.1). To prepare for our generalization, let $\mathcal{N}_j$ denote the set of *neighbours* of j, given by $\mathcal{N}_j = \{j - 1, j + 1\}$ for $j = 2, \ldots, n - 1$, with $\mathcal{N}_1 = \{2\}$ and $\mathcal{N}_n = \{n - 1\}$; hence $Y_{\mathcal{N}_j} = (Y_{j-1}, Y_{j+1})$ for $j \neq 1, n$, while $Y_{\mathcal{N}_1} = Y_2$ and $Y_{\mathcal{N}_n} = Y_{n-1}$. Then the Markov property for variables along an axis is equivalent to

$$\Pr(Y_j = y_j \mid Y_{-j} = y_{-j}) = \Pr\bigl(Y_j = y_j \mid Y_{\mathcal{N}_j} = y_{\mathcal{N}_j}\bigr), \tag{6.11}$$

Figure 6.4 Markov random fields. Left: neighbourhood structure for first-order Markov chain and its cliques and their subsets. Right: first-order neighbourhood structure, cliques and their subsets for rectangular grid of sites.

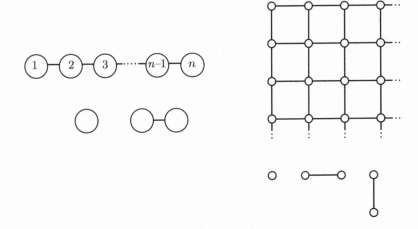

for all values of j and y. Thus Y_j depends on the other variables only through the neighbouring variables $Y_{\mathcal{N}_j}$. The probability densities on the left of (6.11) are known to statisticians as *full conditional densities*, while those on the right are called *local characteristics* in statistical physics.

For more complicated settings, let $\mathcal{J} = \{1, \ldots, n\}$ be a finite set of *sites*, each with a random variable Y_j attached. In many applications each Y_j takes the same finite number k of values, and then $Y_1, \ldots, Y_n$ may have at most k^n possible configurations; though finite, this number may be very large indeed. For any subset $A \subset \mathcal{J}$, let Y_A denote the corresponding subset of $Y \equiv Y_{\mathcal{J}}$, and let Y_{-A} indicate $Y_{\mathcal{J}-A}$, with $Y_j = Y_{\{j\}}$ and Y_{-j} defined as above. We impose a topology on $\mathcal{J}$ by defining a *neighbourhood system* $\mathcal{N} = \{\mathcal{N}_j, j \in \mathcal{J}\}$. The neighbours of j are the elements of $\mathcal{N}_j \subset \mathcal{J}$, the neighbourhoods $\mathcal{N}_j$ having the properties that

- $j \notin \mathcal{N}_j$ and
- $i \in \mathcal{N}_j$ if and only if $j \in \mathcal{N}_i$.

We visualize this as a graph $(\mathcal{J}, \mathcal{N})$ whose nodes correspond to sites, with two nodes joined by an edge if the sites are neighbours. We denote the union of $\{j\}$ and its neighbourhood by $\tilde{\mathcal{N}}_j = \mathcal{N}_j \cup \{j\}$. A subset $C \subset \mathcal{J}$ is *complete* if there are edges between all its nodes, and a maximal complete subset is a *clique* of $(\mathcal{J}, \mathcal{N})$; every pair of distinct elements of C are then neighbours, but C cannot be enlarged and retain this property. Let $\mathcal{C}$ denote the set of cliques and their subsets; in particular, $\mathcal{C}$ contains all singletons $\{j\}$ and the empty set $\emptyset$.

Some authors do not insist that cliques be maximal.

Example 6.11 (Markov chain)   For the graph on the left of Figure 6.4, each interior variable has just two neighbours, and the end variables have just one. Hence $\mathcal{C} = \{\emptyset, \{1\}, \ldots, \{n\}, \{1, 2\}, \ldots, \{n-1, n\}\}$; the cliques are the $n-1$ adjacent pairs.   ∎

Example 6.12 (Pixillated image)   Let $\mathcal{J}$ be an $m \times m$ rectangular array of sites, with neighbourhood structure shown on the right of Figure 6.4. Here $n = m^2$. Interior sites have four neighbours, while boundary sites have two or three neighbours. The cliques are horizontal or vertical pairs of adjacent sites.

This neighbourhood system is said to be first-order. It is easy to envisage enlarging the neighbourhoods, for example by adding adjacent diagonal sites to give a second-order neighbourhood system.   ∎

Having defined a neighbourhood system analogous to that implicit in a Markov chain, the extension of the Markov property is clear: a probability distribution for Y is said to be a *Markov random field* with respect to $\mathcal{N}$ if Y_j is independent of $Y_{-\tilde{\mathcal{N}}_j}$ given $Y_{\mathcal{N}_j}$, or equivalently, if (6.11) holds: the conditional distribution of Y_j depends on the other variables only through those at the neighbouring sites.

Or sometimes a Markov field or a locally dependent Markov random field.

Although the local characteristics of Y are determined by its joint density, it is not true that any collection $\Pr(Y_1 \mid Y_{\mathcal{N}_1}), \ldots, \Pr(Y_n \mid Y_{\mathcal{N}_n})$ of local characteristics yields a proper joint density. This is awkward, because in practice the local characteristics are much easier to deal with than the full joint density. Hence we ask which collections of $\Pr(Y_j \mid Y_{\mathcal{N}_j}) = \Pr(Y_j \mid Y_{-j})$ give well-defined joint distributions. It turns out that a *positivity condition* is needed, that for any $y_1, \ldots, y_n$,

$$\Pr(Y_j = y_j) > 0 \text{ for } j = 1, \ldots, n \text{ implies } \Pr(Y_1 = y_1, \ldots, Y_n = y_n) > 0: \quad (6.12)$$

if values of Y_j can occur singly they can occur together. In this case

$$\frac{\Pr(Y = y)}{\Pr(Y = y')} = \prod_{j=1}^{n} \frac{\Pr(y_j \mid y_1, \ldots, y_{j-1}, y'_{j+1}, \ldots, y'_n)}{\Pr(y'_j \mid y_1, \ldots, y_{j-1}, y'_{j+1}, \ldots, y'_n)} \quad (6.13)$$

for any two possible realizations y and y' of Y (Exercise 6.2.5). Hence (6.13) may be found for every possible y simply by taking a baseline y' and using the full conditional densities, the value of $\Pr(Y = y')$ being found by summing the ratios. Under the positivity condition, therefore, the full conditional densities determine a unique joint density for Y. This density must be unaffected by the labelling of the sites of $\mathcal{J}$, any change to which will leave (6.13) unaltered. This is a severe restriction, and we shall see at the end of this section that the joint density must have form

$$\Pr(Y = y) \propto \exp\{-\psi(y)\}, \quad (6.14)$$

where

$$\psi(y) = \sum_{C \in \mathcal{C}} \phi_C(y), \quad (6.15)$$

is a sum over all complete subsets C associated with the graph $(\mathcal{J}, \mathcal{N})$; this result, the *Hammersley–Clifford theorem*, is proved at the end of this section. Hence the only contributions to the joint density come from cliques of $(\mathcal{J}, \mathcal{N})$ and their subsets. Moreover the functions ϕ_C can be arbitrary, provided the total probability of (6.14) is finite. Many standard models have functions ϕ_C chosen so that (6.14) is an exponential family, but though convenient this is not essential. The sum in (6.15) could involve only cliques, as contributions from other complete subsets could be subsumed into those from the cliques. The collection of functions $\{\phi_C : C \in \mathcal{C}\}$ is called a *potential*.

The representation given by (6.14) and (6.15) is powerful because it enables systems whose global behaviour is very complex to be built from simple local components, namely the local characteristics determined by the ϕ_C. This is analogous to the notion that the transition probabilities of a Markov chain entirely determine its behaviour.

Example 6.13 (Markov chain) In Example 6.11 C contains the empty set, singletons, and pairs of adjacent sites, and hence

$$\psi(y) = a + \sum_{j=1}^{n} b_j(y_j) + \sum_{i \sim j} c_{ij}(y_i, y_j),$$

where the second sum is over all distinct pairs of neighbours, or equivalently all edges of the graph. The proportionality in (6.14) means that we can set $a = 0$, while setting $b_j \equiv b$ and $c_{ij}(\cdot, \cdot) = c(\cdot, \cdot)$ for all i and j gives a homogeneous field.

If the field is homogeneous and the Y_j take only values 0 and 1, we may write

$$\psi(y) = \sum_{j=1}^{n} b y_j + \sum_{j=1}^{n-1} (c_{10} y_j + c_{01} y_{j+1} + c_{11} y_j y_{j+1}),$$

and a little algebra gives

$$\Pr(Y_{j+1} = y_{j+1} \mid Y_1 = y_1, \ldots, Y_j = y_j) = \frac{e^{(\beta + \gamma y_j) y_{j+1}}}{1 + e^{(\beta + \gamma y_j)}},$$

where β and γ are functions of b, c_{10}, c_{01} and c_{11}. As expected, this conditional probability depends on $y_1, \ldots, y_j$ only through y_j and does not depend upon j directly. Hence it corresponds to a stationary first-order Markov chain with transition probabilities $\Pr(0 \mid 0) = (1 + e^{\beta})^{-1}$ and $\Pr(0 \mid 1) = (1 + e^{\beta + \gamma})^{-1}$.

If the Y_j take values in the real line and we set

$$b(y_j) = \tau \left(y_j^2 - 2\mu y_j \right)/(2\sigma^2), \quad c(y_i, y_j) = (y_i - y_j)^2/(2\sigma^2),$$

then $\psi(y) = (y^\mathsf{T} V y - 2\mu y^\mathsf{T} V 1_n)/(2\sigma^2)$, where

$$V = \begin{pmatrix} \tau + 1 & -1 & 0 & \cdots & 0 & 0 \\ -1 & \tau + 2 & -1 & \cdots & 0 & 0 \\ 0 & -1 & \tau + 2 & \cdots & 0 & 0 \\ \vdots & \vdots & \vdots & \ddots & \vdots & \vdots \\ 0 & 0 & 0 & \cdots & \tau + 2 & -1 \\ 0 & 0 & 0 & \cdots & -1 & \tau + 1 \end{pmatrix},$$

and 1_n is an $n \times 1$ vector of ones. It follows that

$$\exp\{-\psi(y)\} \propto \exp\left\{ -\frac{1}{2\sigma^2} (y - \mu 1_n)^\mathsf{T} V (y - \mu 1_n) \right\},$$

which corresponds to the multivariate normal distribution with mean vector $\mu 1_n$ and covariance matrix $V/(2\sigma^2)$.

If $\tau = 0$, the rows of V sum to zero and the distribution is degenerate. Moreover (6.14) is integrable only if $\sigma^2 > 0$. This underlines the fact that although any choice of b_j and c_{ij} yields a proper joint density when each Y_j takes only a finite number of values, restrictions may be needed to ensure this when any of the Y_j has infinite support.

Example 11.27 gives an application of this.   ∎

Example 6.14 (Ising model)   Let $\mathcal{J}$ be an $m \times m$ grid of pixels, the jth of which can take values 0 and 1, corresponding to the colours white and black. As $n = m^2$, the sample space has size 2^{m^2}, about 10^{4932} even for a small image with $m = 128$. Under a first-order neighbourhood system the cliques are horizontal and vertical pairs of adjacent pixels; see Figure 6.4. Hence if b_j and c_{ij} are homogeneous, we can take

$$\psi(y) = \sum_j b(y_j) + \sum_{i \sim j} c(y_i, y_j)$$

the second sum being over all distinct cliques. The resulting probability distribution is the *Ising model* of statistical physics, which is important in investigations of ferromagnetism.

Ernst Ising (1900–1998) was one of the generation of German scientists whose careers were destroyed by the rise of Nazism. After a period of forced labour during the war he emigrated to the USA in 1949. The Ising model described in his 1924 PhD thesis was later used to account for the phase transition between the ferromagnetic and paramagnetic states.

The conditional probability that $Y_j = 0$ given Y_{-j} is

$$\frac{\Pr(Y_j = 0, Y_{-j} = y_{-j})}{\Pr(Y_j = 0, Y_{-j} = y_{-j}) + \Pr(Y_j = 1, Y_{-j} = y_{-j})},$$

and on using (6.14) and cancelling all terms not involving y_j, we obtain

$$\frac{\exp\left\{-b(0) - \sum_{i \in \mathcal{N}_j} c(y_i, 0)\right\}}{\exp\left\{-b(0) - \sum_{i \in \mathcal{N}_j} c(y_i, 0)\right\} + \exp\left\{-b(1) - \sum_{i \in \mathcal{N}_j} c(y_i, 1)\right\}};$$

thus the full conditional densities have form

$$\Pr(Y_j = 0 \mid Y_{-j}) = \frac{1}{1 + \exp\left\{b(0) - b(1) + \sum_{i \in \mathcal{N}_j} c(y_i, 0) - c(y_i, 1)\right\}}.$$

Let n_1 denote $\sum_{i \in \mathcal{N}_j} I(Y_i = 1)$, the number of neighbours of site j that equal one, and define n_0 similarly; note that $n_0 = |\mathcal{N}_j| - n_1$. Now

$|\mathcal{A}|$ is the cardinality of the set $\mathcal{A}$.

$$\sum_{i \in \mathcal{N}_j} c(y_i, 0) - c(y_i, 1) = n_0 c(0, 0) + n_1 c(1, 0) - n_0 c(0, 1) - n_1 c(1, 1)$$

$$= n_0 \left\{c(0, 0) + c(1, 1) - c(0, 1) - c(1, 0)\right\}$$

$$+ |\mathcal{N}_j| \left\{c(1, 0) - c(1, 1)\right\},$$

from which it follows that we can write

$$\Pr(Y_j = 0 \mid Y_{-j}) = \Pr(Y_j = 0 \mid Y_{\mathcal{N}_j}) = \frac{1}{1 + \exp(\beta + \gamma |\mathcal{N}_j| + \delta n_0)}. \qquad (6.16)$$

We interpret $\beta + |\mathcal{N}_j|\gamma$ as controlling the overall size of the probability and δ its dependence on the number of its white neighbours: $\gamma = 0$ means that the colour of cell j is independent of the colours around it, while (6.16) increases to one as $\gamma \to -\infty$.

Images with more colours may be dealt with by letting Y_j take $k > 2$ values, with an analogous argument giving the local characteristics. More complex neighbourhood

Figure 6.5 A small geneology. Females are shown as circles, males as squares, and marriages leading to offspring as dots. Thus the male shown by the solid square has two parents and three children by two marriages. This would be his neighbourhood in potentially a much larger pedigree.

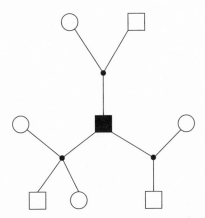

structures will introduce more parameters into the model, while these ideas can be extended to fields that allow lines, textures and other features of real images.   ∎

Example 6.15 (Genetic pedigree)   In the analysis of a genealogy, the sites typically correspond to individuals and Y_j to the genotype at a particular locus on the jth individual's DNA. Typically the genotype cannot be observed, but the phenotypes of some of the individuals are known. A simple example is in the ABO blood group system, where the observable phenotype blood group 'A' arises with genotypes AA and AO which are harder to observe; see Example 4.38. Two individuals in a pedigree are said to be *spouses* if they have mutual offspring in the pedigree, and each such pairing constitutes a *marriage*. A pedigree may be represented as a graph in which both individuals and marriages correspond to nodes, while the edges link each individual to his or her marriages and each marriage to the resulting offspring. See Figure 6.5.

The laws of genetic inheritance are Markovian. Genes are passed from parents to offspring in such a way that conditional on their parents' genotypes, individuals are independent of their earlier direct ancestors. It turns out that this dependence imposes a neighbourhood structure on the genotypes, with the neighbourhood for any individual defined to contain his parents, children and spouses. However distributions defined on this structure need not satisfy the positivity condition. A simple example is the ABO blood system: a person whose parents are both of type AB cannot be of type O. The fact that genetic models usually do not satisfy the positivity condition complicates statistical analysis of pedigree data.   ∎

Statistical inference for Markov random fields is generally based on the iterative simulation methods discussed in Section 11.3.3.

6.2.2 Directed acyclic graphs

Thus far we have supposed that all the Y_j have the same support and that the neighbourhood structure of the random field is known. The idea of expressing dependencies among variables as a graph is useful in more general settings, however, and it is then necessary to read off neighbourhoods from the joint distribution of $Y_1, \dots, Y_n$. Often

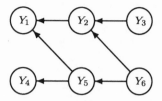 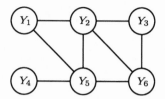

Figure 6.6 Directed acyclic and moral graphs. Left: directed acyclic graph representing (6.17). Right: moral graph, formed by moralizing the directed acyclic graph, that is, 'marrying' parents and dropping arrowheads.

the dependence structure is specified hierarchically, for example by stating the conditional distributions of Y_1 given Y_2 and of Y_2 given Y_3, Y_4 and so forth. The hierarchy may then be expressed using a directed graph, in which dependence of Y_1 on Y_2 is shown by an arrow from the *parent* Y_2 to the *child* Y_1, and Y_1 is a *descendent* of Y_3 if there is a sequence of arrows from Y_3 to Y_1. Such a graph is *directed* because each edge is an arrow, and *acyclic* if it is impossible to start from a node, traverse a path by following arrows, and return to the starting-point. The left of Figure 6.6 shows the directed acyclic graph for a model in which the joint density of $Y_1, \ldots, Y_6$ factorizes as

$$f(y) = f(y_1 \mid y_2, y_5)f(y_2 \mid y_3, y_6)f(y_3)f(y_4 \mid y_5)f(y_5 \mid y_6)f(y_6). \qquad (6.17)$$

For any directed acyclic graph we have

$$Y_j \perp \text{non-descendents of } Y_j \mid \text{parents of } Y_j, \quad \text{for all } j,$$

and (6.17) generalizes to

$$f(y) = \prod_{j \in \mathcal{J}} f(y_j \mid \text{parents of } y_j). \qquad (6.18)$$

The density is then said to be *recursive* with respect to the directed acyclic graph. Acyclicity prevents a variable from introducing a degenerate density by being its own descendent.

A directed acyclic graph does not display all the neighbourhoods of the resulting Markov random field, but its *moral graph* does. This is obtained by *moralizing* the directed acyclic graph — 'marrying' or putting edges between any parents that share a child and then cutting off the arrowheads. In Figure 6.6, for example, the directed acyclic graph on the left shows us that Y_2 and Y_5 are parents of Y_1, so they are joined in the moral graph on the right. This shows us that

$$\mathcal{N}_1 = \{2, 5\}, \ \mathcal{N}_2 = \{1, 3, 5, 6\}, \ \mathcal{N}_3 = \{2, 6\},$$
$$\mathcal{N}_4 = \{5\}, \ \mathcal{N}_5 = \{1, 2, 4, 6\}, \ \mathcal{N}_6 = \{2, 3, 5\}.$$

In general the full conditional density of y_j is

$$f(y_j \mid y_{-j}) = \frac{f(y)}{\int f(y)\,dy_j}$$

$$= \frac{\prod_{i \in \mathcal{J}} f(y_i \mid \text{parents of } y_i)}{\int \prod_{i \in \mathcal{J}} f(y_i \mid \text{parents of } y_i)\,dy_j}$$

$$\propto f(y_j \mid \text{parents of } y_j) \prod_{i: \, y_i \text{ is child of } y_j} f(y_i \mid \text{parents of } y_i),$$

$\perp$ means 'is independent of'.

Also called a conditional independence graph.

A moral graph contains no unmarried parents.

because the integral only affects terms where y_j appears. In order for the denominator to be positive for any y_{-j}, the positivity condition must hold. If so, we see that $\mathcal{N}_j$ comprises the parents and children of Y_j, and any parents of Y_j's children, precisely those variables joined to Y_j in the moral graph. Thus the distribution of Y satisfies (6.11), also called the *local Markov property*.

Consider a directed acyclic graph, let the *family* $\mathcal{F}_j$ consist of j and its parents, if any, and let $\mathcal{C}$ denote the cliques of the corresponding moral graph. Then as the families $\mathcal{F}_j$ yield cliques $C \in \mathcal{C}$, we may write

$$f(y) = \prod_j g(y_{\mathcal{F}_j}) = \prod_{C \in \mathcal{C}} h_C(y), \tag{6.19}$$

taking $g(y_{\mathcal{F}_j}) = f(y_j \mid \text{parents of } y_j)$. Thus we may write the joint density in terms of the cliques of an moral graph, analogous to (6.14) and (6.15). Let $\mathcal{A}$ and $\mathcal{B}$ be disjoint subsets of $\mathcal{J}$ that are *separated* by $\mathcal{D}$, that is, any path from an element of $\mathcal{A}$ to an element of $\mathcal{B}$ must pass through $\mathcal{D}$. Then under the positivity condition the distribution on the moral graph has the *global Markov property*, that $Y_{\mathcal{A}}$ and $Y_{\mathcal{B}}$ are independent conditional on $Y_{\mathcal{D}}$. To see this in the case where all the variables are discrete, suppose for now that $\mathcal{A} \cup \mathcal{B} \cup \mathcal{D} = \mathcal{J}$, and note that as no clique can contain elements of both $\mathcal{A}$ and $\mathcal{B}$, (6.19) implies that the joint density can be written as

$$f(y) = f(y_{\mathcal{A}}, y_{\mathcal{B}}, y_{\mathcal{D}}) = g_1(y_{\mathcal{A}}, y_{\mathcal{D}})g_2(y_{\mathcal{B}}, y_{\mathcal{D}}).$$

Thus

$$f(y_{\mathcal{A}}, y_{\mathcal{B}} \mid y_{\mathcal{D}}) = \frac{g_1(y_{\mathcal{A}}, y_{\mathcal{D}})g_2(y_{\mathcal{B}}, y_{\mathcal{D}})}{\sum_{y_{\mathcal{A}}} \sum_{y_{\mathcal{B}}} g_1(y_{\mathcal{A}}, y_{\mathcal{D}})g_2(y_{\mathcal{B}}, y_{\mathcal{D}})},$$

which factorizes in terms of $y_{\mathcal{A}}$ and $y_{\mathcal{B}}$, showing that any subset of $Y_{\mathcal{A}}$ is independent of any subset of $Y_{\mathcal{B}}$, conditional on $Y_{\mathcal{D}}$. The positivity condition ensures that the denominator here is positive for any $y_{\mathcal{D}}$. We now have only to note that if $\mathcal{A} \cup \mathcal{B} \cup \mathcal{D} \neq \mathcal{J}$, then $\mathcal{A}, \mathcal{B}$ can be enlarged to give sets $\mathcal{A}', \mathcal{B}'$ which together with $\mathcal{D}$ partition $\mathcal{J}$ such that $\mathcal{D}$ separates $\mathcal{A}', \mathcal{B}'$. Then $Y'_{\mathcal{A}} \perp Y'_{\mathcal{B}} \mid Y_{\mathcal{D}}$, implying that $Y_{\mathcal{A}} \perp Y_{\mathcal{B}} \mid Y_{\mathcal{D}}$, which is the global Markov property. The moral graph in Figure 6.6, for example, shows that

$$Y_1 \perp Y_3, Y_4 \mid Y_2, Y_5,$$

as can be verified from (6.17).

Markov properties of this sort are useful because they enable the computation of $f(y)$ or derived quantities to be broken into practicable steps. Sometimes the moral graph must be *triangulated* by adding edges to ensure that every cycle of length four or more contains an edge between two nodes that are not adjacent in the cycle itself. Triangulation can accelerate computation of $f(y)$ by making closed-form calculations possible for some model classes.

Example 6.16 (Belief network) Graphs may be used to represent supposed logical or causal relationships among variables and play an important role in probabilistic expert systems. Figure 6.7, for instance, shows a directed acyclic graph that represents

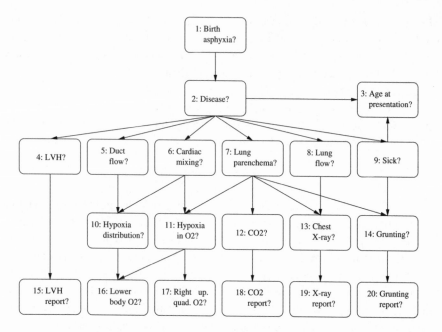

Figure 6.7 Directed acyclic graph representing the incidence and presentation of six possible diseases that would lead to a 'blue' baby (Spiegelhalter *et al.*, 1993). LVH means left ventricular hypertrophy.

the incidence and presentation of six diseases that would lead to a 'blue' baby. Early appropriate treatment is essential when such a child is born, and this expert system was developed to increase the accuracy of preliminary diagnoses. The graph shows, for example, that the level of oxygen in the lower body (node 16) is thought to be directly related to hypoxia distribution (node 10) and to its level when breathing oxygen (node 11). This last variable depends on the degree of mixing of blood in the heart (node 6) and the state of the blood vessels (parenchyma) in the lungs (node 7), and these two variables are directly influenced by which of the six possible levels the variable disease (node 2) has taken. Links such as those between nodes 6 and node 11 might be regarded as causal if poor cardiac mixing was known to contribute to hypoxia.

Each variable in such a network is typically treated as discrete, so the joint distribution of the variables is determined by a large number of multinomial distributions giving the terms on the right of (6.18). These are often obtained by eliciting opinions from experts and then updating these opinions, and perhaps the structure of the graph, as data become available. Table 6.6, for example, shows the expert view that left ventricular hypertrophy (LVH) would be present in 10% of cases of persistent foetal circulation, and that if present, it would be correctly reported in 90% of cases. The full distribution is given by specifying such tables for each of the 20 nodes of the graph, giving a sample space with more than one billion elements.

Now imagine that the LVH report for a baby is positive. In the light of this evidence the probabilities for the other variables will need updating, for example to ascribe new probabilities to the diseases or to determine which other diagnostic report will be most informative. Thus evidence must be propagated through the network to give the joint distribution of the other variables conditional on a positive LVH report. This involves

Table 6.6 Subjective expert assessments of conditional probability tables for links node 2 → node 4 and node 4 → node 15 in Figure 6.7 (Spiegelhalter *et al.*, 1993).

	Node 4: LVH	
Node 2: Disease	Yes	No
Persistent foetal circulation	0.10	0.90
Transposition of the great arteries	0.10	0.90
Teralogy of Fallot	0.10	0.90
Pulmonary atresia with intact ventricular septum	0.90	0.10
Obstructed total anomalous pulmonary venous connection	0.05	0.95
Lung disease	0.10	0.90

	Node 15: LVH report	
Node 4: LVH	Yes	No
Yes	0.90	0.10
No	0.05	0.95

the cliques of the triangulated moral graph of Figure 6.7. Details are given in the references in the bibliographic notes. ∎

Directed acyclic and their moral graphs play a useful role in the iterative simulation methods described in Section 11.3.3.

Hammersley–Clifford theorem

This can be omitted at a first reading.

We now show that if the positivity condition (6.12) holds when all the Y_j take values in $\{0, \ldots, L\}$, then the most general form that their joint density $f(y)$ can take is given by (6.14) and (6.15). Conversely these equations entail the Markov property (6.11) and positivity condition (6.12).

Let $\mathcal{Y} = \{0, \ldots, L\}^n$ denote the sample space for $Y_1, \ldots, Y_n$, and for any $y \in \mathcal{Y}$ let y_j^0 denote the vector $(y_1, \ldots, y_{j-1}, 0, y_{j+1}, \ldots, n)$. Under the positivity condition every element of $\mathcal{Y}$ occurs with positive probability, so we can define $\psi(y) = \log\{f(y)/f(0)\}$, where 0 represents a vector of n zeros. Now

$$\exp\{\psi(y) - \psi(y_j^0)\} = \frac{f(y)}{f(y_j^0)} = \frac{f(y_j \mid y_1, \ldots, y_{j-1}, y_{j+1}, \ldots, y_n)}{f(0 \mid y_1, \ldots, y_{j-1}, y_{j+1}, \ldots, y_n)} = \frac{f(y_j \mid y_{\mathcal{N}_j})}{f(0 \mid y_{\mathcal{N}_j})},$$

because the joint density satisfies the local Markov property, so knowing ψ will determine the full conditional densities and therefore the local characteristics of $f(y)$. Note that this implies that $\psi(y) - \psi(y_j^0)$ depends only on y_j and $y_{\mathcal{N}_j}$.

Now any function $\psi(y)$ has an expansion

$$\psi(y) = \sum_{j=1}^{n} y_j a_j(y_j) + \sum_{1 \leq j < k \leq n} y_j y_k a_{jk}(y_j, y_k)$$

$$+ \sum_{1 \leq j < k < l \leq n} y_j y_k y_l a_{jkl}(y_j, y_k, y_l) + \cdots + y_1 \cdots y_n a_{1\cdots n}(y_1, \ldots, y_n),$$

because we can set $y_j a_j(y_j) = \psi(0, \ldots, 0, y_j, 0, \ldots, 0) - \psi(0)$, with analogous formulae for the other a-functions. We must now show that for any subset C of $\{1, \ldots, n\}$, the corresponding a-function may be non-null if and only if C is a clique of the graph. Consider y_1 without loss of generality, and recall that owing to (6.11), $\psi(y) - \psi(y_1^0)$ depends only on y_1 and $y_{\mathcal{N}_1}$. Now $\psi(y) - \psi(y_1^0)$ equals

$$y_1 \left\{ a_1(y_1) + \sum_{2 \le j \le n} y_j a_{1j}(y_1, y_j) + \sum_{2 \le j < k \le n} y_j y_k a_{1jk}(y_1, y_j, y_k) \right.$$
$$\left. + \cdots + y_2 \cdots y_n a_{1 \cdots n}(y_1, \ldots, y_n) \right\},$$

and this must be free of y_l for any $l \notin \mathcal{N}_1$. On setting $y_j = 0$ for $j \ne 1, l$, we see that $a_{1l}(y_1, y_l) = 0$ for every possible y_l. Suitable choices of y show in like wise that every other a-function involving y_l must be identically zero. As the same is true for every other node, the Markov property (6.11) implies that the only non-zero functions $a_{j_1 \cdots j_k}$ are those in which $j_1, \ldots, j_k$ are all neighbours, that is, form a subset of a clique, and this entails (6.14) and (6.15).

For the converse, note that any set of a-functions gives a density $f(y)$ that satisfies the positivity condition (6.12). Now $\psi(y) - \psi(y_j^0)$ depends on x_l only if there is a non-null a-function containing both y_j and y_l, so the local characteristic $f(y_j \mid y_1, \ldots, y_{j-1}, y_{j+1}, \ldots, y_n)$ also depends only on those y_l that are neighbours of y_j. Thus (6.14) and (6.15) together imply (6.11) and (6.12).

Exercises 6.2

1   Show that the Markov property for a sequence $Y_1, \ldots, Y_n$ is equivalent to (6.11).

2   Give the cliques for the second-order neighbourhood system in Example 6.12.

3   Give the cliques for a second-order Markov chain, and hence write down the form of the most general density for it, under the positivity condition.

4   Consider two binary random variables with local characteristics
$$\Pr(Y_1 = 1 \mid Y_2 = 0) = \Pr(Y_1 = 0 \mid Y_2 = 1) = 1,$$
$$\Pr(Y_2 = 0 \mid Y_1 = 0) = \Pr(Y_2 = 1 \mid Y_1 = 1) = 1.$$
Show that these do not determine a joint density for (Y_1, Y_2). Is the positivity condition satisfied?

5   Let the density of a random variable $Y = (Y_1, \ldots, Y_n)$ satisfy (6.12), and let the sample space be the set $\mathcal{Y} = \{y : f(y) > 0\}$. If $(y_1, \ldots, y_n)$ and $(x_1, \ldots, x_n)$ are two elements of $\mathcal{Y}$, use the identity
$$f(y_1, \ldots, y_n) = f(y_n \mid y_1, \ldots, y_{n-1}) f(y_1, \ldots, y_{n-1})$$
to show that
$$f(y_1, \ldots, y_n) = \frac{f(y_n \mid y_1, \ldots, y_{n-1})}{f(x_k \mid y_1, \ldots, y_{n-1})} f(y_1, \ldots, y_{n-1}, x_n),$$
and then that
$$f(y_1, \ldots, y_{n-1}, x_n) = \frac{f(y_{n-1} \mid y_1, \ldots, y_{n-2}, x_n)}{f(x_{n-1} \mid y_1, \ldots, y_{n-2}, x_n)} f(y_1, \ldots, y_{n-2}, x_{n-1}, x_n),$$

Hence establish (6.13).
(Besag, 1974)

6 Use induction on the number of variables to prove (6.18).

7 Let $\mathcal{G}_m$ be the graph obtained by moralizing a finite directed acyclic graph $\mathcal{G}$. Show that every family of $\mathcal{G}$ is a clique of $\mathcal{G}_m$ but that the converse is false (consider the moral graph for Figure 6.7).

8 A subset $\mathcal{A}$ of the nodes of a graph $\mathcal{G}$ is *ancestral* if $\mathcal{A}$ contains the parents and neighbours of a whenever $a \in \mathcal{A}$. Show that if the density of Y is recursive with respect to $\mathcal{G}$, the marginal density of $Y_{\mathcal{A}}$ is recursive with respect to to subgraph induced on $\mathcal{A}$. Now consider the moral graph $\mathcal{G}'_m$ of the smallest ancestral set containing $\mathcal{A} \cup \mathcal{B} \cup \mathcal{D}$, and suppose that $\mathcal{D}$ separates $\mathcal{A}$ from $\mathcal{B}$ in $\mathcal{G}'_m$. Show that $Y_{\mathcal{A}} \perp Y_{\mathcal{B}} \mid Y_{\mathcal{D}}$.

9 Are the moral graphs for Figures 6.6 and 6.7 triangulated?

10 Write down the directed acylic and moral graphs for X, Y, and I under the missing data models described in Section 5.5. Use them to give an equation-free explanation of the differences among the models and of their consequences.

11 Give the form of the a-functions of page 253 when $L = 1$ and $n = 3$, and hence verify the Hammersley–Clifford theorem when (i) none of 1, 2, and 3 are neighbours; (ii) $1 \sim 2$ only; (iii) $1 \sim 2 \sim 3$ only; and (iv) $1 \sim 2 \sim 3 \sim 1$, where $\sim$ means 'is a neighbour of'. In each case give (6.15).

12 Suppose that the variables $U_1, \ldots, U_n$ of a Markov random field have joint density

$$f(u; \theta) = c(\theta) \exp\left[-\frac{1}{2} \left\{ \theta_1 \sum_{j \sim j'} (u_j - u_{j'})^2 + \theta_2 \sum_{j=1}^{n} u_j^2 \right\} \right], \quad \theta_1 > 0, \theta_2 \geq 0,$$

where the first sum is over all pairs of neighbours, each pair being taken only once. Let $\mathcal{N}_j$ denote the neighbours of node j.
(a) Show that their joint distribution is normal with covariance matrix determined by $\theta_1 I_n + \theta_2 A$, where the jth diagonal element of the $n \times n$ adjacency matrix A is the number of neighbours of node j, and its off-diagonal elements are

$$a_{jj'} = \begin{cases} -1, & j \sim j' \\ 0, & \text{otherwise.} \end{cases}$$

(b) Show that the density is degenerate when $\theta_2 = 0$, but that otherwise $c(\theta) = (2\pi)^{-n/2} \prod_j (\theta_2 + \theta_1 a'_j)$, where the a'_j are the eigenvalues of A.
(c) Suppose that conditional on the values u of unseen variables U, $Y_1, \ldots, Y_n$ are independent with $Y_j \mid U \sim N(u_j, \sigma^2)$. Show that the joint density of U given Y is well-defined even if $\theta_2 = 0$ provided that $\sigma^2 > 0$, and that the density of $U_j \mid Y, U_{-j}$ involves on y_j and $u_{\mathcal{N}_j}$.
(d) Work out the details using the matrix V in Example 6.13.

6.3 Multivariate Normal Data

6.3.1 Multivariate dependence

When there is a single response variable, analysis is relatively simple, the crucial aspect being how the distribution of that variable depends on any covariates. Problems with just one response variable are common in practice and typically are readily interpretable, but cases arise where relations among two or more responses are to be modelled, and to these we now briefly turn. We motivate our discussion by an example.

Mechanics (C)	Vectors (C)	Algebra (O)	Analysis (O)	Statistics (O)
77	82	67	67	81
63	78	80	70	81
75	73	71	66	81
55	72	63	70	68
63	63	65	70	63
⋮	⋮	⋮	⋮	⋮
15	38	39	28	17
5	30	44	36	18
12	30	32	35	21
5	26	15	20	20
0	40	21	9	14

Table 6.7 Marks out of 100 in five mathematics examinations for the first and last five of 88 students (Mardia *et al.*, 1979, pp. 3–4). Some of the examinations were closed-book (C), and others were open-book (O).

Example 6.17 (Maths marks data) Table 6.7 gives marks out of 100 for the first and last five students out of 88 who took five mathematics examinations. As we would anticipate, the top students tend to do best in all the exams, and the worst dismally in all, so the marks for each student are related. This is shown by the modified *scatterplot matrix* in Figure 6.8, the below-diagonal part of which contains scatterplots of the marks for each examination against those for every other one, for all 88 students. Each column of marks has been centred at zero for comparison with the panels above the diagonal, which are discussed in Example 6.20. The original marks are shown by the light histograms along the diagonal. The average marks in mechanics and in vectors were about 40 and 50, but the panel in the second row and first column shows that one candidate got about 40 more marks than this in mechanics and about 30 more in vectors; this is the first person in Table 6.7. The below-diagonal panels show generally positive association between marks for different subjects, but its strength varies. For example, algebra is strongly associated with all the other subjects, whereas the first column suggests that while mechanics is related to vectors and algebra, its association with analysis and statistics is weaker.

We could rank students by their overall averages, but this would not be useful if the question of interest is how the marks for different exams relate to one another. We would then have five response variables, and we would need to model the forms of dependence that might arise. ■

Simpson's paradox

The easiest way to deal with the multitude of potential dependencies in such situations is to ignore as many of them as possible. For example, if the response is bivariate, (Y_1, Y_2), and there is a single explanatory variable, x, one might just model Y_1 as a function of x, regardless of Y_2. Unfortunately this can be badly misleading, as Figure 6.9 illustrates. Its left panel shows how a continuous variable Y_1 depends on x for two values of the discrete variable Y_2. For each value of y_2, the mean of Y_1, $E(Y_1 \mid Y_2 = y_2, x)$, increases with x, as shown by the positive slope of the regression of Y_1 on x. However, the right panel shows that when Y_2 is ignored the mean of Y_1

Figure 6.8 Modified scatterplot matrix for the full maths marks data. Below the diagonal are scatterplots (and sample correlation coefficient) for the centred pairs of marks; for example, the lower left panel shows results for statistics plotted against those for mechanics. Above the diagonal are scatterplots of residuals (and sample partial correlation coefficients): for example, the top right panel shows the dependence remaining between mechanics and statistics after adjusting for the other variables. The diagonal shows histograms of the variables (light) and of the residuals from regression on all other variables, centred at the variable mean (dark), with the marginal and partial standard deviations.

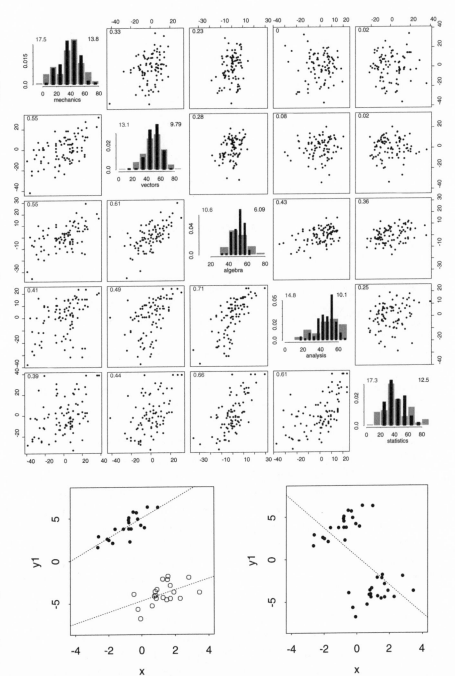

Figure 6.9 Artificial data illustrating Simpson's paradox. The left panel shows how Y_1 depends on x for each of two values of Y_2, with observations with $y_2 = 0$ shown by blobs and those with $y_2 = 1$ shown as circles. The lines are from separate straight-line regression fits of y_1 on x for each value of y_2 and show positive association. The right panel shows the fit to the data ignoring Y_2, for which the association is negative.

E. H. Simpson called attention to this effect in 1951, although it was known to G. U. Yule almost 50 years earlier.

decreases with x, that is, $E(Y_1 \mid x)$ has negative slope as a function of x. This effect — *Simpson's paradox* — is due to the fact that marginalization of the joint distribution of (Y_1, Y_2) over Y_2 has reversed the sign of the association between Y_1 and x. Here a plot at once reveals that it is a bad idea to fit a common line to both groups, but the

Age (years)	Smokers	Non-smokers
Overall	139/582 (24)	230/732 (31)
18–24	2/55 (4)	1/62 (2)
25–34	3/124 (2)	5/157 (3)
35–44	14/109 (13)	7/121 (6)
45–54	27/130 (21)	12/78 (15)
55–64	51/115 (44)	40/121 (33)
65–74	29/36 (81)	101/129 (78)
75+	13/13 (100)	64/64 (100)

Table 6.8   Twenty-year survival and smoking status for 1314 women (Appleton *et al.*, 1996). The smoker and non-smoker columns contain number dead/total (% dead).

paradox arises also in contexts where it is not obvious what to plot, and association may be strengthened or weakened as well as reversed.

Example 6.18 (Smoking and the Grim Reaper)   Table 6.8 shows data on smoking and survival for 1314 women in Whickham, near Newcastle upon Tyne in the north of England. In 1972–1974 data were collected by surveying people on the town electoral register, and among the many variables collected were age and smoking habits at that time. Twenty years later a follow-up survey was conducted and it was determined if each woman in the original survey had yet died. Just 162 women had smoked before the first survey but were non-smokers at that time, and there were only 18 whose smoking habits were not recorded; these 180 women have been excluded.

The first line of the table shows a surprising apparent health bonus of smoking: the death rate for smokers, 24%, is lower than that for non-smokers, 31%. However a breakdown of the overall figures by age shows that the death rate for smokers is higher for every age category except 25–34 and over 75 years. On inspection the reason for this discrepancy is obvious: in the early 1970s there were many more non-smokers than smokers in the age range 65–74, and as old age makes death even more likely than does smoking, most of these did not survive until the follow-up survey. To learn about the effect of smoking on health, we should compare women of the same age. Thus the appropriate analysis involves comparisons within age groups, not after merging the data across them.   ∎

These cautionary examples show that it is unwise to collapse data by ignoring variables without first examining the full dependence structure to ensure that it is safe to do so. Data with several responses are common in the social sciences, psychology, epidemiology, and public health, where outcomes may depend on numerous variables that are related in a complicated way. The modelling of such data has been extensively studied, and we only scratch its surface here, confining our discussion to the most basic models for continuous data. Discrete data such as those in Table 6.8 arise in many applications; see Chapter 10.

6.3.2 Multivariate normal distribution

Let $Y_1, \ldots, Y_n$ be a random sample from the $N_p(\mu, \Omega)$ density. That is, $Y_j = (Y_{1j}, \ldots, Y_{pj})^{\mathrm{T}}$ has the multivariate normal distribution with mean vector $\mu = (\mu_1, \ldots, \mu_p)^{\mathrm{T}}$ and $p \times p$ nonsingular covariance matrix Ω, whose (r, s) element is ω_{rs}. As each of the Y_j has density (3.20), the log likelihood is

$$\ell(\mu, \Omega) \equiv -\frac{1}{2}\left\{ n \log |\Omega| + \sum_{j=1}^n (Y_j - \mu)^{\mathrm{T}} \Omega^{-1}(Y_j - \mu) \right\}$$

$$= -\frac{1}{2}\left\{ n \log |\Omega| + \sum_{j=1}^n (Y_j - \overline{Y} + \overline{Y} - \mu)^{\mathrm{T}} \Omega^{-1}(Y_j - \overline{Y} + \overline{Y} - \mu) \right\}$$

$$= -\frac{1}{2}\left\{ n \log |\Omega| + n(\overline{Y} - \mu)^{\mathrm{T}} \Omega^{-1}(\overline{Y} - \mu) + \sum_{j=1}^n (Y_j - \overline{Y})^{\mathrm{T}} \Omega^{-1}(Y_j - \overline{Y}) \right\}$$

$$= -\frac{1}{2}\{ n \log |\Omega| + n(\overline{Y} - \mu)^{\mathrm{T}} \Omega^{-1}(\overline{Y} - \mu) + (n - 1)\mathrm{tr}(\Omega^{-1} S)\} \qquad (6.20)$$

where the $p \times 1$ vector and $p \times p$ matrix

$$\overline{Y} = n^{-1} \sum Y_j, \quad S = (n - 1)^{-1} \sum (Y_j - \overline{Y})(Y_j - \overline{Y})^{\mathrm{T}},$$

are the sample average and covariance matrix. These are minimal sufficient statistics for μ and Ω. The second and third equalities before (6.20) rest on the identities

$$\sum (Y_j - \overline{Y}) \Omega^{-1}(\overline{Y} - \mu) = 0,$$

$$\sum (Y_j - \overline{Y})^{\mathrm{T}} \Omega^{-1}(Y_j - \overline{Y}) = \sum \mathrm{tr}\{(Y_j - \overline{Y})^{\mathrm{T}} \Omega^{-1}(Y_j - \overline{Y})\}$$

$$= \sum \mathrm{tr}\{\Omega^{-1}(Y_j - \overline{Y})(Y_j - \overline{Y})^{\mathrm{T}}\}$$

$$= \mathrm{tr}\left\{ \Omega^{-1} \sum (Y_j - \overline{Y})(Y_j - \overline{Y})^{\mathrm{T}} \right\}.$$

As Ω is a positive definite matrix, so is $\Delta = \Omega^{-1}$, and it follows that for any Ω, the maximum of (6.20) with respect to μ is at $\widehat{\mu} = \overline{Y}$, the maximum likelihood estimator. In order to maximize $\ell(\widehat{\mu}, \Omega)$, we make the 1–1 transformation from Ω to Δ, in terms of which $\ell(\widehat{\mu}, \Delta) = \frac{1}{2}\{n \log |\Delta| - (n - 1)\mathrm{tr}(\Delta S)\}$. Differentiation with respect to Δ shows that the maximum likelihood estimate of $\Delta^{-1} = \Omega$ is $\widehat{\Omega} = n^{-1}(n - 1)S$; with probability one this has rank p if $n > p$. Otherwise its rank is $n - 1$.

Exercise 6.3.1 gives the details.

As in the scalar case S is unbiased. We let $\widehat{\omega}_{rs}$ denote the (r, s) element of S; this is the sample covariance between the rth and sth components of Y. The sample variances lie on the diagonal of S. The sample correlations are $\widehat{\omega}_{rs}/(\widehat{\omega}_{rr}\widehat{\omega}_{ss})^{1/2}$, the (r, s) elements of $D^{-1/2} S D^{-1/2}$, where D is the diagonal matrix $\mathrm{diag}(\widehat{\omega}_{1,1}, \ldots, \widehat{\omega}_{p,p})$ (Exercise 6.3.2).

Example 6.19 (Maths marks data) Table 6.9 shows the averages, variances, and correlations for the maths marks data. The best results are on vectors and algebra, and the worst on mechanics and statistics. The numbers below the diagonal show positive correlations among the variables, with the strongest those between algebra and the other subjects. The most variable marks are for mechanics and statistics, with

	Mechanics	Vectors	Algebra	Analysis	Statistics
Mechanics	17.5/13.8	0.33	0.23	−0.00	0.03
Vectors	0.55	13.2/9.8	0.28	0.08	0.02
Algebra	0.55	0.61	10.6/6.1	0.43	0.36
Analysis	0.41	0.49	0.71	14.8/10.1	0.25
Statistics	0.39	0.44	0.66	0.61	17.3/12.5
Average	39.0	50.6	50.6	46.7	42.3

Table 6.9 Summary statistics for maths marks data. The sample correlations between variables are below the diagonal, and the sample partial correlations are above the diagonal. The diagonal contains sample standard deviation/ sample partial standard deviation.

sample standard deviations $\widehat{\omega}_{rr}^{1/2}$ of 17.5 and 17.3 respectively, while that for algebra is smallest, at 10.6. Although the averages for mechanics and statistics are smallest, there is a wider spread of results for these subjects. The values above the diagonal are discussed in Example 6.20.   ∎

Extensions of the arguments for univariate data show that

$$\overline{Y} \sim N_p(\mu, n^{-1}\Omega), \quad \text{independent of} \quad (n-1)S \sim W_p(n-1, \Omega), \tag{6.21}$$

where $W_p(\nu, \Omega)$ denotes the *p-dimensional Wishart distribution with $p \times p$ parameter matrix Ω and ν degrees of freedom*. In fact, if $Z_1, \ldots, Z_\nu$ is a random sample from the $N_p(0, \Omega)$ distribution, then $Z_1 Z_1^\mathsf{T} + \cdots + Z_\nu Z_\nu^\mathsf{T} \sim W_p(\nu, \Omega)$; when $p = 1$ and $\Omega = 1$, the Wishart distribution reduces to the chi-squared.

The multivariate extension of the t statistic is *Hotelling's T^2 statistic*,

$$T^2 = n(\overline{Y} - \mu)^\mathsf{T} S^{-1}(\overline{Y} - \mu) \sim \frac{p(n-1)}{n-p} F_{p,n-p},$$

which can be used to test hypotheses and form confidence regions for elements of μ.

6.3.3 Graphical Gaussian models

The structure of the multivariate normal density means that variables depend on each other in a particularly simple way. Before getting into details, we need some notation. Let S be a subset of the integers $\{1, \ldots, p\}$, of cardinality $|S|$, and let Y_S and Y_{-S} be the sets of variables $\{Y_s, s \in S\}$ and $\{Y_s, s \notin S\}$. If $S = \{r\}$, we write $Y_S = Y_r$ and $Y_{-S} = Y_{-r}$. For two such subsets $\mathcal{A}$ and $\mathcal{B}$, let $\Omega_{\mathcal{A},\mathcal{B}}$ be the $|\mathcal{A}| \times |\mathcal{B}|$ matrix with elements $\omega_{ab} = \text{cov}(Y_a, Y_b)$, and let $\Omega_{\mathcal{A}|\mathcal{B}} = \text{cov}(Y_\mathcal{A} \mid Y_\mathcal{B})$ be the $|\mathcal{A}| \times |\mathcal{A}|$ conditional covariance matrix of $Y_\mathcal{A}$ given the value of $Y_\mathcal{B}$; we write its elements as $\omega_{a_1,a_2|\mathcal{B}}$.

Equation (3.21) establishes that the conditional distribution of Y_S given $Y_{-S} = y_{-S}$ is normal with mean vector and covariance matrix

$$\mu_S + \Omega_{S,-S}\Omega_{-S,-S}^{-1}(y_{-S} - \mu_{-S}), \quad \Omega_{S,S} - \Omega_{S,-S}\Omega_{-S,-S}^{-1}\Omega_{-S,S}. \tag{6.22}$$

Thus the conditional mean depends linearly on the values of the known variables, and the conditional variance is independent of them. If $S = \{r\}$ and the conditional variance of Y_r, $\omega_{rr|-r}$, is much smaller than the unconditional variance ω_{rr}, then knowing Y_{-r} is highly informative about the distribution of Y_r. Thus it will be useful to compare estimates of these variances. It is also useful to learn how knowledge of the other variables affects the covariance of Y_r and Y_s. Their 2×2 conditional covariance matrix is given by (6.22), with $\mathcal{S} = \{r, s\}$, and their *partial correlation*,

$$\rho_{rs|-\mathcal{S}} = \frac{\omega_{rs|-\mathcal{S}}}{(\omega_{rr|-\mathcal{S}} \, \omega_{ss|-\mathcal{S}})^{1/2}},$$

represents the correlation between Y_r and Y_s conditional on the remaining variables. The quantities on the right are sometimes called the *partial variances* and *partial covariance*. On page 264 we show that the partial correlation equals minus one times the (r, s) element of the correlation matrix constructed from Ω^{-1}. Thus partial variances, correlations and covariances of Y are readily computed from Ω, and we can use the transformation property of maximum likelihood estimators to estimate $\rho_{rs|-\mathcal{S}}$ and so forth by the same functions of $\widehat{\Omega}$.

Example 6.20 (Maths marks data) The second diagonal elements in Table 6.9 give the sample partial standard deviations $\widehat{\omega}_{rr|-r}^{1/2}$ for each subject. According to the normal model, our best guess of a student's mark in algebra without knowledge of his other marks would be 50.6, with standard deviation 10.6: a 95% confidence interval is $51 \pm 1.96 \times 11 = (29, 73)$, which is virtually useless. If we knew $\overline{y}$ and $\widehat{\Omega}$ and his marks y_{-r} for the other four subjects, however, we could replace the components of μ and Ω in (6.22) with $\mathcal{S} = \{r\}$ by estimates, giving estimated score $\overline{y}_r + \widehat{\Omega}_{r,-r} \widehat{\Omega}_{-r,-r}^{-1} (y_{-r} - \overline{y}_{-r})$. The estimated conditional standard deviation $\widehat{\omega}_{rr|-r}^{1/2} = 6.1$ is appreciably smaller than the unconditional value.

The above-diagonal part of Table 6.9 shows the sample partial correlations. A good mark at algebra is correlated positively with each of the other variables, given the remainder. Given the other variables, however, mechanics seems to be unrelated to analysis or statistics, and likewise for vectors: the upper right corner of the matrix is essentially zero. Thus the subjects split into three groups: vectors and mechanics; analysis and statistics; and algebra. Variables in the first two pairs are partially correlated with each other and with algebra, which itself is partially correlated with all four other variables.

This information is displayed more fully in the above-diagonal panels of Figure 6.8. Set $\mathcal{S} = \{r, s\}$, and let y denote the $n \times p$ data matrix whose jth row is y_j^{T}, y_r the rth column of y, and y_{-s} the $n \times (p - 2)$ array comprising all columns of y but the rth and sth. Then the vertical axes show the $n \times 1$ vectors of sample values

$$y_{r|-\mathcal{S}} = y_r - \overline{y}_r - \widehat{\Omega}_{r,-\mathcal{S}} \widehat{\Omega}_{-\mathcal{S},-\mathcal{S}}^{-1} (y_{-\mathcal{S}} - \overline{y}_{-\mathcal{S}})$$

of the scalar random variable

$$Y_{r|-\mathcal{S}} = Y_r - \mu_r - \Omega_{r,-\mathcal{S}} \Omega_{-\mathcal{S},-\mathcal{S}}^{-1} (Y_{-\mathcal{S}} - \mu_{-\mathcal{S}}),$$

while the horizontal axes show the $y_{s|-\mathcal{S}}$'s. The quantities $Y_{r|-\mathcal{S}}$ are normal with means zero and variances $\Omega_{r,-\mathcal{S}} \Omega_{-\mathcal{S},-\mathcal{S}}^{-1} \Omega_{-\mathcal{S},r}$, and partial correlation

$$\mathrm{corr}(Y_r, Y_s \mid Y_{-\mathcal{S}}) = \mathrm{corr}(Y_{r|-\mathcal{S}}, Y_{s|-\mathcal{S}}) = \rho_{rs|-\mathcal{S}},$$

while the correlation coefficient between the sample versions is the corresponding sample quantity $\widehat{\rho}_{rs|-\mathcal{S}}$. Thus the scatterplot in the first row and third column shows the association between mechanics on the vertical axis and algebra on the horizontal axis after adjusting for dependence on the other variables. The partial correlation of 0.23 shows that some positive correlation remains after allowing for the other variables. Summary in terms of partial correlations seems reasonable, as none of the panels shows much nonlinearity, but there is a possible outlier in the lower left corner of panels $(1, 2)$ and $(2, 1)$. This is a person whose marks $y_{81}^{\mathrm{T}} = (3, 9, 51, 47, 40)$ are dire for applied mathematics but not for pure mathematics or statistics. Dropping him makes little change to the correlations or partial correlations.

The diagonal of the scatterplot matrix compares histograms of the raw marks y_r and the marks $y_{r|-r} + \bar{y}_r$ after adjusting for all the other variables, with the sample standard deviations of these vectors.   ∎

Conditional independence graphs

As their third and higher-order joint cumulants are identically zero (Section 3.2.3), dependence among normal variables is expressed through their correlations, calculated from Ω, or equivalently their partial correlations, calculated from Ω^{-1}. Consider the graph with p nodes corresponding to the variables $Y_1, \ldots, Y_p$. Now Y_r and Y_s are independent conditional on all the other variables if and only if their partial correlation is zero, and we encode this by the absence of an edge between the corresponding nodes. Thus two nodes are neighbours — joined by an edge — if and only if the corresponding partial correlation is non-zero and hence if and only if the corresponding element of Δ is non-zero. This yields a conditional independence graph for $Y_1, \ldots, Y_p$ (Section 6.2.2).

If the density of $Y_1, \ldots, Y_p$ is non-degenerate, then the global Markov property holds. To see this, let $\mathcal{A}, \mathcal{B}$, and $\mathcal{D}$ be any disjoint nonempty subsets of $\mathcal{J} = \{1, \ldots, p\}$ such that $\mathcal{D}$ separates $\mathcal{A}$ from $\mathcal{B}$ and $\mathcal{A} \cup \mathcal{B} \cup \mathcal{D} = \mathcal{J}$. As there are no edges between $\mathcal{A}$ and $\mathcal{B}$, the density of Y has exponent

$$-\frac{1}{2}(y - \mu)^{\mathrm{T}}\Omega^{-1}(y - \mu) = -\frac{1}{2}(y - \mu)^{\mathrm{T}} \begin{pmatrix} \Delta_{\mathcal{A}\mathcal{A}} & \Delta_{\mathcal{A}\mathcal{D}} & 0 \\ \Delta_{\mathcal{D}\mathcal{A}} & \Delta_{\mathcal{D}\mathcal{D}} & \Delta_{\mathcal{D}\mathcal{B}} \\ 0 & \Delta_{\mathcal{B}\mathcal{D}} & \Delta_{\mathcal{B}\mathcal{B}} \end{pmatrix} (y - \mu),$$

with quadratic term in y_A and y_B identically zero. Hence

$$f(y) = f(y_A, y_B, y_D) = g_1(y_A, y_D)g_2(y_B, y_D),$$

for some positive functions g_1 and g_2, implying that Y_A and Y_B are conditionally independent given y_D; of course this property is inherited by any subsets of Y_A and Y_B. As any disjoint subsets of $\mathcal{J}$ separated by $\mathcal{D}$ can be augmented to give sets $\mathcal{A}, \mathcal{B}$ which are separated by $\mathcal{D}$ and which together with $\mathcal{D}$ partition $\mathcal{J}$, the global Markov property holds.

In graphical terms it is natural to restrict the degree of dependence among components of Y by deleting edges from its graph, and this means setting elements of Ω^{-1} to zero. Suppose that the inverse covariance matrix resulting from such deletions is $\Omega_0^{-1} = \Delta_0$, for which the profile log likelihood is $\ell(\widehat{\mu}, \Delta_0) \equiv \frac{1}{2}\{n \log |\Delta_0| - (n-1)\text{tr}(\Delta_0\widehat{\Omega}^{-1})\}$. For an idea of the difficulties involved in maximizing this with respect to the non-zero elements of Δ_0, we consider the simplest non-trivial case, with $p = 3$ variables and $\delta_{32} = 0$, implying that Y_2 and Y_3 are independent given Y_1. In this case the log likelihood may be written down and differentiated directly, giving five simultaneous equations to be solved for the non-zero components of $\widehat{\Delta}_0$. We lay these equations out as

$$\frac{1}{|\widehat{\Delta}_0|} \begin{pmatrix} \widehat{\delta}_{22}\widehat{\delta}_{33} & & \\ -\widehat{\delta}_{21}\widehat{\delta}_{33} & \widehat{\delta}_{11}\widehat{\delta}_{33} - \widehat{\delta}_{31}^2 & \\ -\widehat{\delta}_{31}\widehat{\delta}_{22} & ? & \widehat{\delta}_{11}\widehat{\delta}_{22} - \widehat{\delta}_{21}^2 \end{pmatrix} = \frac{n-1}{n} \begin{pmatrix} \widehat{\omega}_{11} & & \\ \widehat{\omega}_{21} & \widehat{\omega}_{22} & \\ \widehat{\omega}_{31} & ? & \widehat{\omega}_{33} \end{pmatrix},$$

where there is a missing equation ?=? corresponding to δ_{32}, which does not appear in the likelihood. The structure of these equations shows that in general we must solve a system of polynomial equations of degree p, and the properties of the graph of Δ_0 play a crucial role in determining the character of the solution. Here it turns out that if the missing equation is replaced by $\widehat{\delta}_{21}\widehat{\delta}_{31}/|\widehat{\Delta}_0| = (n-1)\widehat{\omega}_{21}\widehat{\omega}_{31}/(n\widehat{\omega}_{11})$ and the matrices are completed by symmetry, the $\widehat{\delta}_{rs}$ can be found explicitly in terms of the $\widehat{\omega}_{rs}$.

Comparisons between two nested graphical models may be based on likelihood ratio statistics, though large-sample asymptotics can be unreliable. Exact comparison of the full model with the one with a single edge missing may be based on the corresponding partial correlation coefficient (Exercise 6.3.6).

Example 6.21 (Maths marks data) The above-diagonal part of Table 6.9 suggests a graphical model in which the upper right 2×2 corner of Δ is set equal to zero. The likelihood ratio statistic for comparison of this model with the full model is 0.90, which is not large relative to the χ_4^2 distribution. This suggests strongly that the simpler model fits as well as the full one, an impression confirmed by comparing the original and fitted partial correlations,

0.33	0.23	−0.00	0.03		0.33	0.24	0.00	0.00
	0.28	0.08	0.02			0.33	0.00	0.00
		0.43	0.36				0.45	0.37
			0.25					0.26

Figure 6.10 shows the graphs for these two models. In the full model every variable is joined to every other, and there is no simple interpretation. The reduced model has a butterfly-like graph whose interpretion is that given the result for algebra, results for mechanics and vectors are independent of those for analysis and statistics. Thus a result for mechanics can be predicted from those for algebra and vectors alone, while prediction for algebra requires all four other results. ∎

The graphs described above have the drawback of taking no account of the logical status of the variables. For example, it may be known that Y_1 influences Y_2 but not vice versa, but this is not reflected in an undirected graph. In applications, therefore, it is useful to have different types of edges, with directed edges representing supposed causal effects and undirected edges linking variables that are to be put on an equal

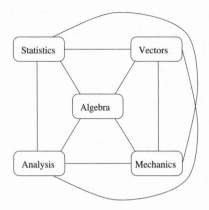 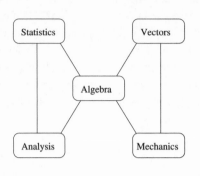

Figure 6.10 Graphs for the full model (left) and a reduced model (right) for the maths marks data. The interpretation of the reduced model is that given the result for algebra, results for vectors and mechanics are independent of those for analysis and statistics.

footing. This important topic is beyond the scope of this book; see the bibliographic notes.

Calculation of partial correlation

This may be skipped on a first reading.

Let $\mathcal{S} = \{r, s\}$, where without loss of generality $r < s$. Then the conditional variance matrix for Y_r and Y_s given $Y_{-\mathcal{S}}$ is $\Omega_{\mathcal{S},\mathcal{S}} - \Omega_{\mathcal{S},-\mathcal{S}}\Omega_{-\mathcal{S},-\mathcal{S}}^{-1}\Omega_{-\mathcal{S},\mathcal{S}}$, and hence their partial correlation is

$$
\rho_{rs|-\mathcal{S}} = \frac{\omega_{rs} - \Omega_{r,-\mathcal{S}}\Omega_{-\mathcal{S},-\mathcal{S}}^{-1}\Omega_{-\mathcal{S},s}}{\left\{\left(\omega_{rr} - \Omega_{r,-\mathcal{S}}\Omega_{-\mathcal{S},-\mathcal{S}}^{-1}\Omega_{-\mathcal{S},r}\right)\left(\omega_{ss} - \Omega_{s,-\mathcal{S}}\Omega_{-\mathcal{S},-\mathcal{S}}^{-1}\Omega_{-\mathcal{S},s}\right)\right\}^{1/2}}.
$$

The (r, s) element of Ω^{-1} is $(-1)^{r+s}\Omega_{rs}/|\Omega|$, where Ω_{rs} is the (r, s) minor of Ω. Thus the (r, s) element of the 'correlationized' version of Ω^{-1} is $(-1)^{r+s}\Omega_{rs}/(\Omega_{rr}\Omega_{ss})^{1/2}$. To show how this is related to $\rho_{rs|-\mathcal{S}}$, we use the formula

$$
\begin{vmatrix} A_{11} & A_{12} \\ A_{21} & A_{22} \end{vmatrix} = |A_{11} - A_{12}A_{22}^{-1}A_{21}| \cdot |A_{22}| \tag{6.23}
$$

for the determinant of a partitioned matrix for which A_{22}^{-1} exists. On making the row and column interchanges that bring ω_{ss} to the $(1, 1)$ position of $\Omega_{-r,-r}$, we see that

$$
\Omega_{rr} = (-1)^{2(s-1)} \begin{vmatrix} \omega_{ss} & \Omega_{s,-\mathcal{S}} \\ \Omega_{-\mathcal{S},s} & \Omega_{-\mathcal{S},-\mathcal{S}} \end{vmatrix} = \left(\omega_{ss} - \Omega_{s,-\mathcal{S}}\Omega_{-\mathcal{S},-\mathcal{S}}^{-1}\Omega_{-\mathcal{S},s}\right)|\Omega_{-\mathcal{S},-\mathcal{S}}|,
$$

with a similar expression for Ω_{ss}, while Ω_{rs} equals

$$
(-1)^{r+(s-1)} \begin{vmatrix} \omega_{sr} & \Omega_{s,-\mathcal{S}} \\ \Omega_{-\mathcal{S},r} & \Omega_{-\mathcal{S},-\mathcal{S}} \end{vmatrix} = (-1)^{r+s-1}\left(\omega_{rs} - \Omega_{s,-\mathcal{S}}\Omega_{-\mathcal{S},-\mathcal{S}}^{-1}\Omega_{-\mathcal{S},r}\right)|\Omega_{-\mathcal{S},-\mathcal{S}}|,
$$

as $\omega_{rs} = \omega_{sr}$ by symmetry of Ω. On substituting the expressions for Ω_{rr}, Ω_{ss}, and Ω_{rs} into $(-1)^{r+s}\Omega_{rs}/(\Omega_{rr}\Omega_{ss})^{1/2}$, we see that the (r, s) element of the 'correlationalized' version of Ω^{-1} equals $-\rho_{rs|-\mathcal{S}}$, as was to be proved.

Exercises 6.3

1   If A is a $p \times p$ matrix, all of whose elements are distinct and if A^{ij} denotes the cofactor of the (i, j) element a_{ij} of A, then $\partial|A|/\partial a_{ij} = A^{ij}$, whereas if A is symmetric, then

$$\frac{\partial|A|}{\partial a_{ij}} = \begin{cases} A^{ii}, & i = j, \\ 2A^{ij}, & i \neq j. \end{cases}$$

If A and B have dimensions $p \times q$ and $q \times p$, then

$$\frac{\partial \mathrm{tr}(AB)}{\partial A} = \begin{cases} B^{\mathsf{T}}, & \text{all elements of } A \text{ distinct,} \\ B + B^{\mathsf{T}} - \mathrm{diag}(B), & A \text{ symmetric.} \end{cases}$$

Use these identities to verify that $n^{-1}(n-1)S$ solves the likelihood equations for Ω for the multivariate normal model on page 259. Check that this maximizes the likelihood when $p = 2$.

2   Show that the (r, s) element of $\widehat{\Omega}$ is $\widehat{\omega}_{rs} = (n-1)^{-1} \sum_j (y_{rj} - \overline{y}_r)(y_{sj} - \overline{y}_s)$, where $\overline{y}_r$ is the rth element of the $p \times 1$ vector $\overline{y}$, and that although $\widehat{\omega}_{rs}$ is not the maximum likelihood estimate of ω_{rs}, the maximum likelihood estimate of the correlation between Y_r and Y_s equals $\widehat{\omega}_{rs}/(\widehat{\omega}_{rr}\widehat{\omega}_{ss})^{1/2}$.

3   Let Ω be the variance matrix of a p-dimensional normal variable Y. Use Cramer's rule to show that the rth diagonal element of Ω^{-1} is $\mathrm{var}(Y_r \mid Y_{-r})$.

4   Let $Y^{\mathsf{T}} = (Y_1, \ldots, Y_3)$ be a multivariate normal variable with

$$\Omega = \begin{pmatrix} 1 & m^{-1/2} & \frac{1}{2} \\ m^{-1/2} & \frac{2}{m} & m^{-1/2} \\ \frac{1}{2} & m^{-1/2} & 1 \end{pmatrix}.$$

Find Ω^{-1} and hence write down the moral graph for Y.

If $m \to \infty$, show that the distribution of Y becomes degenerate while that of (Y_1, Y_3) given Y_2 remains unchanged. Is the graph an adequate summary of the joint limiting distribution? Is the Markov property stable in the limit?

5   Suppose that $W_1, \ldots, W_n$ may be written $W_j = \mu + \sigma Z_j + \tau X$, where $Z_1, \ldots, Z_n$ and X are independent standard normal variables. Obtain the correlation matrix Ω of $Y^{\mathsf{T}} = (X, W_1, \ldots, W_n)$, write down the moral graph for Y, and hence obtain Ω^{-1}.

6   Let $y_1, \ldots, y_n$ be a $N_p(\mu, \Omega)$ random sample and let $\Delta = \Omega^{-1}$ have elements δ_{rs}. Show that apart from constants, the value of (6.20) maximized over both μ and Ω is $-\frac{1}{2}n \log |\widehat{\Omega}|$, and deduce that the likelihood ratio statistic for comparison of the full model and a sub-model obtained by constraining elements of Ω (or Δ) may be written $n \log |\widehat{\Omega}^{-1}\widehat{\Omega}_0|$, in an obvious notation.

(a) Show that the likelihood ratio statistic for testing if all the components of Y are independent is a function of the determinant of the sample correlation matrix.

(b) Use (6.23) to show that the likelihood ratio statistic to test if $\delta_{12} = 0$ may be written $-n \log(1 - \widehat{\rho}_{1,2|-S}^2)$, where $S = \{1, 2\}$, and check for what values of the partial correlation $\widehat{\rho}_{12|-S}$ this is large.

7   In the discussion on page 263, verify that if $\delta_{32} = 0$, then the likelihood equations are equivalent to

$$\widehat{\Delta}_0^{-1} = \frac{n-1}{n}\begin{pmatrix} \widehat{\omega}_{11} & & \\ \widehat{\omega}_{21} & \widehat{\omega}_{22} & \\ \widehat{\omega}_{31} & \widehat{\omega}_{21}\widehat{\omega}_{31}/\widehat{\omega}_{11} & \widehat{\omega}_{33} \end{pmatrix},$$

and hence find $\widehat{\Delta}_0$ in terms of the $\widehat{\omega}_{rs}$.

Find also the maximum likelihood estimate of $\widehat{\Delta}_0$ when $\delta_{31} = \delta_{32} = 0$ and when $\delta_{31} = \delta_{32} = \delta_{21} = 0$.

Give the graphs corresponding to each of these models.

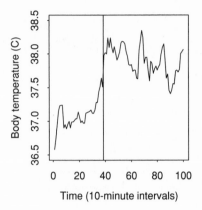

Figure 6.11 Example time series. Left: body temperatures (°C) of a female Canadian beaver measured at 10-minute intervals (Reynolds, 1994). The vertical line marks where she left her lodge. Right: FTSE closing prices, 1991–1998.

6.4 Time Series

A time series consists of data recorded in time order. Examples are monthly inflation rate, weekly demand for electricity, daily maximum temperature, number of packets of information sent per second over a communication network, and so forth. The measurements may be instantaneous, such as the daily closing prices of some stock, or may be an average, such as annual temperature averaged over the surface of the globe. Typically such data show variation on several scales. Data on internet traffic, for example, show strong diurnal variation as well as long-term upward trend. Time series are ubiquitous and their analysis is well-developed, with many techniques specific to particular areas of application. In many cases the goal of time series modelling is the forecasting of future values, while in others the intention is to control the underlying process. Here we simply introduce a few basic notions in the most common situation, where the observations are continuous and arise at regular intervals. Irregular and discrete time series also occur — see Example 6.2 — but their modelling is less well explored.

Example 6.22 (Beaver body temperature data)   The left panel of Figure 6.11 shows 100 consecutive telemetric measurements on the body temperature of a female Canadian beaver, *Castor canadensis*, taken at 10-minute intervals. The animal remains in its lodge for the first 38 recordings and then moves outside, at which point there is a sustained temperature rise. This is likely to be of main interest in such an application, with the dependence structure of the series regarded as secondary. The dependence must be accounted for, however, if confidence intervals for the rise are to be reliable.   ∎

Example 6.23 (FTSE data)   The right panel of Figure 6.11 shows the closing prices of the Financial Times Stock Exchange index of London closing prices from 1991–1998. Prices are available only for days on which the exchange was open so there are many fewer than 365 observations per year. The dominant feature is the strong upward trend. Here interest would typically focus on short-term forecasting, though portfolio managers will also wish to understand the relationship between this and other markets. In either case the dependence structure is of crucial importance. ∎

Stationarity and autocorrelation

Statistical inference cannot proceed without some assumption of stochastic regularity, and in time series this is provided by the notion of *stationarity*.

Consider data $y_1, \ldots, y_n$, supposed to be a realization of the random variables $Y_1, \ldots, Y_n$, themselves forming a contiguous stretch of a stochastic process $\{Y_t\} = \{\ldots, Y_{-1}, Y_0, Y_1, \ldots\}$. Then $\{Y_t\}$ is said to be *second-order stationary* if its first and second moments are finite and time-independent, so that the mean $E(Y_s) = \mu$ is constant and the covariances $\text{cov}(Y_s, Y_{s+t}) = \gamma_t$ do not depend on s. Finiteness of $\gamma_0 = \text{var}(Y_t)$ guarantees that $|\mu|, |\gamma_t| < \infty$ for all t. The first and second moments of a second-order stationary series do not depend on the point at which they are calculated. Neither panel of Figure 6.11 looks stationary, though it is plausible that the temperature data to the right of the vertical line are.

A series is said to be *strictly stationary* if the joint distribution of any finite subset Y_A does not depend on the origin; thus the distributions of Y_{s+A} and of Y_A are the same for any s. This is a stronger condition than second-order stationarity, because it constrains the entire distribution of the series. In particular it implies that the joint cumulants of Y_{s+A} are independent of s, if they exist. Evidently strict stationarity yields more powerful theoretical results, but as it is impossible to verify from data, they are less useful in practice. The definitions coincide if $\{Y_t\}$ has a multivariate normal distribution, as this is determined by its first and second moments. The term stationary used without qualification in this section means second-order stationary.

The second-order structure of a stationary process is summarized in its *autocorrelation function* $\rho_t = \text{corr}(Y_0, Y_t)$, $t = \pm 1, \pm 2, \ldots$, where $\rho_t = \gamma_t / \gamma_0$; $\gamma_0 = \text{var}(Y_0)$ is the marginal variance of the process $\{Y_t\}$. Note that

$$\rho_{-t} = \text{corr}(Y_s, Y_{s-t}) = \text{corr}(Y_{s+t}, Y_s) = \rho_t$$

by stationarity. A related function is the *partial autocorrelation function* $\rho_t' = \text{corr}(Y_0, Y_t \mid Y_1, \ldots, Y_{t-1})$, which summarizes any correlation between observations t lags apart after conditioning on the intervening data; see Section 6.3.3.

A *white noise process* $\{\varepsilon_t\}$ is an uncorrelated sample from some distribution with mean zero and variance σ^2; evidently it has $\rho_t = \rho_t' \equiv 0$. We shall use the term *normal white noise* when $\varepsilon_t \overset{\text{iid}}{\sim} N(0, \sigma^2)$.

Plots of estimated ρ_t and ρ_t' against positive values of t are called the *correlogram* and *partial correlogram*. Under mild conditions their ordinates are asymptotic independent $N(0, n^{-1})$ variables for a white noise series of length n, from which significance can be assessed; see Figure 6.12.

Example 6.24 (Autoregressive process)   About the simplest time series model is the autoregressive process of order one, or AR(1) model

$$Y_t - \mu = \alpha(Y_{t-1} - \mu) + \varepsilon_t, \quad t = \ldots, -1, 0, 1, \ldots, \tag{6.24}$$

(In the left margin:)

{Y_t} is also called covariance stationary, weakly stationary, or stationary in the wide sense.

where the innovation series $\{\varepsilon_t\}$ is normal white noise and ε_t is independent of $\ldots, Y_{t-2}, Y_{t-1}$. Taking variances in (6.24) yields $\gamma_0 = \alpha^2 \gamma_0 + \sigma^2$. Hence $\gamma_0 = \sigma^2/(1 - \alpha^2)$, so a necessary condition for stationarity is $|\alpha| < 1$. This condition is also sufficient, and if it is satisfied then $E(Y_t) = \mu$ and $\rho_t = \alpha^{-|t|}$ (Exercise 6.4.1).

This is a Markov process, because Y_t depends on the previous observations only through Y_{t-1}, and hence the only non-zero partial autocorrelation is $\rho_1' = \alpha$. If the ε_t are normal, then Y_t is a linear combination of normal variables and so $Y_1, \ldots, Y_n$ are jointly normal with mean vector $\mu \mathbf{1}_n$ and covariance matrix

$$\Omega = \frac{\sigma^2}{1 - \alpha^2} \begin{pmatrix} 1 & \alpha & \alpha^2 & \cdots & \alpha^{n-1} \\ \alpha & 1 & \alpha & \cdots & \alpha^{n-2} \\ \alpha^2 & \alpha & 1 & \cdots & \alpha^{n-3} \\ \vdots & \vdots & \vdots & \ddots & \vdots \\ \alpha^{n-1} & \alpha^{n-2} & \alpha^{n-3} & \cdots & 1 \end{pmatrix}.$$

One can verify directly that Ω^{-1} is the tridiagonal matrix (Example 6.13)

$$\sigma^{-2} \begin{pmatrix} 1 & -\alpha & 0 & \cdots & 0 & 0 \\ -\alpha & 1+\alpha^2 & -\alpha & \cdots & 0 & 0 \\ 0 & -\alpha & 1+\alpha^2 & \cdots & 0 & 0 \\ \vdots & \vdots & \vdots & \ddots & \vdots & \vdots \\ 0 & 0 & 0 & \cdots & 1+\alpha^2 & -\alpha \\ 0 & 0 & 0 & \cdots & -\alpha & 1 \end{pmatrix}.$$

The autoregressive process of order p or AR(p) model satisfies

$$Y_t - \mu = \sum_{j=1}^{p} \alpha_j(Y_{t-j} - \mu) + \varepsilon_t, \quad t = \ldots, -1, 0, 1, \ldots,$$

and is therefore a Markov process of order p. Constraints on $\alpha_1, \ldots, \alpha_p$ are needed for this process to be stationary, but if they are satisfied, there is a sharp cut-off in the partial autocorrelations: $\rho_t' = 0$ when $t > p$. This should be reflected in the partial correlogram of AR(p) data. The constraints are discussed after Example 6.26. ∎

Example 6.25 (Beaver body temperature data) Figure 6.12 shows the correlogram and partial correlogram for the apparently stationary observations 39–100 of the beaver temperature data. The correlogram shows positive correlations at lags 1–3. Any further evidence of structure must be treated very cautiously, as the values around lag 15 are not very significant, and as each panel of the figure shows 20 correlations estimated from only 62 observations. The partial correlogram is suggestive of an AR(1) model with $\alpha \doteq 0.75$, consistent with the geometric decrease in the correlogram at short lags.

The change in level evident in Figure 6.11 suggests that we take

$$Y_t = \begin{cases} \beta_0 + \eta_t, & t = 1, \ldots, 38, \\ \beta_0 + \beta_1 + \eta_t, & t = 39, \ldots, 100, \end{cases} \tag{6.25}$$

while the partial correlogram suggests that the η_t follow (6.24) with $\mu = 0$. This yields a Markov model with parameters $(\beta_0, \beta_1, \alpha, \sigma^2)$. If we assume normal white

Figure 6.12
Correlogram and partial correlogram for observations 39–100 of the beaver body temperature data. The dotted horizontal lines at $\pm 2n^{-1/2}$ show 95% confidence bounds for the correlation coefficients, if the data are white noise. Strong systematic departures from these are suggestive of structure in the data.

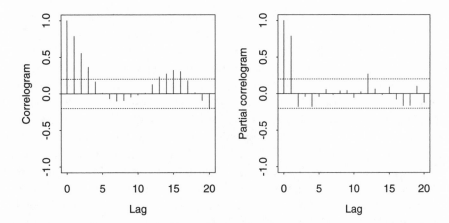

noise and initial $N\{\beta_0, \sigma^2/(1 - \alpha^2)\}$ distribution for y_1 then the log likelihood is readily obtained from (4.8); see Exercise 6.4.3. The log likelihood can be maximized numerically and standard errors obtained from the inverse observed information matrix, giving $\widehat{\beta}_0 = 37.19$ (0.119), $\widehat{\beta}_1 = 0.61$ (0.138), $\widehat{\alpha} = 0.87$ (0.068), and $\widehat{\sigma}^2 = 0.015$ (0.002). Body temperature rises by about 0.6°C when the beaver is active, and successive measurements are quite highly correlated. Treating the data as independent gives standard error 0.044 for $\widehat{\beta}_1$, so the autocorrelation greatly increases the uncertainty for β_1.

Residuals can be constructed by estimating the scaled innovations ε_t/σ. In the inactive period we define residuals $r_t = \{y_t - \widehat{\beta}_0 - \widehat{\alpha}(y_{t-1} - \widehat{\beta}_0)\}/\widehat{\sigma}$, with a similar expression in the active period. Then the correlogram, partial correlogram, and probability plots of $r_2, \ldots, r_{100}$ help assess model adequacy. Judged by these criteria, the model seems to fit well, though (6.25) does not account for the gradual rise in body temperature before the beaver left the lodge. ∎

Example 6.26 (Moving average process)   A moving average process of order q or MA(q) model satisfies the equation

$$Y_t - \mu = \sum_{j=1}^{q} \beta_j \varepsilon_{t-j} + \varepsilon_t, \quad t = \ldots, -1, 0, 1, \ldots$$

where $\{\varepsilon_t\}$ is white noise. Here $E(Y_t) = \mu$ and $\text{var}(Y_t) = \sigma^2(1 + \beta_1^2 + \cdots + \beta_q^2)$ for all t, and it is easy to check that this process is stationary and that $\rho_t = 0$ for $t > q$ (Exercise 6.4.2). Thus the correlogram of such data should show a sharp cut-off after lag q. ∎

Stationary autoregressive and moving average processes are linear processes, as the current observation Y_t may be expressed as an infinite moving average of the innovations,

$$Y_t = \sum_{j=0}^{\infty} c_j \varepsilon_{t-j}, \quad t = \ldots, -1, 0, 1, \ldots, \quad \text{with} \quad \sum_{j=0}^{\infty} |c_j| < \infty. \tag{6.26}$$

This expresses the current Y_t in terms of past innovations, provides useful models in many applications, and leads to simple computations. For example, $\text{var}(Y_t) = \sum c_j^2 < \infty$ and $\gamma_t = \sum c_j c_{j+t}$.

Evidently an MA(q) model with zero mean has a representation (6.26). To see when this is true for an AR(p) model, it is useful to introduce the *backshift operator* B such that $BY_t = Y_{t-1}$ and $B^d Y_t = Y_{t-d}$, with $B^0 = I$ the identity operator. Then an AR(p) process is expressible as $a(B)Y_t = \varepsilon_t$, where the polynomial $a(z) = 1 - \sum_{j=1}^{p} \alpha_j z^j$ corresponds to the autoregression, and we can formally write $Y_t = a(B)^{-1}\varepsilon_t = \sum_{i=0}^{\infty} c_i \varepsilon_{t-i}$, say, which is stationary if and only if $\sum c_i^2 < \infty$. Now $a(z) = \prod_{j=1}^{p}(1 - a_j z)$, where a_j^{-1} are the possibly complex roots of $a(z)$, and provided that no two of the a_j are equal, $a(z)^{-1}$ may be written using partial fractions as $\sum_{j=1}^{p} b_j/(1 - a_j z)$ for some b_j. If we take z sufficiently small then $a(z)^{-1}$ can be expressed as a sum of geometric series with coefficients $c_i = \sum_{j=1}^{p} b_j a_j^i$, giving the infinite moving average (6.26). For this to be stationary we must have $\sum c_i^2 < \infty$, which occurs if and only if $|a_j| < 1$ for each j, or equivalently all the roots of $a(z)$ lie outside the unit disk in the complex plane. Thus properties of the polynomial $a(z)$ are intimately related to those of the process $\{Y_t\}$.

Example 6.27 (ARMA process)   The autoregressive process is formed as a linear combination of previous observations, while a moving average process is based on a weighted combination of the innovations at previous steps. An obvious generalization is to combine the two, giving the *autoregressive moving average* process or ARMA(p, q) model

$$Y_t - \mu = \sum_{j=1}^{p} \alpha_j (Y_{t-j} - \mu) + \sum_{i=1}^{q} \beta_i \varepsilon_{t-i} + \varepsilon_t, \quad t = \ldots, -1, 0, 1, \ldots.$$

As in the preceding examples, the Y_t will have a joint normal distribution if the process is stationary and the ε_t represent normal white noise. Let $\mu = 0$ for simplicity.

In terms of the backshift operator we have $a(B)Y_t = b(B)\varepsilon_t$, where the polynomials $a(z) = 1 - \sum_{j=1}^{p} \alpha_j z^j$ and $b(z) = 1 + \sum_{i=1}^{q} \beta_i z^i$ represent the autoregressive and moving average components. Thus $Y_t = a(B)^{-1}b(B)\varepsilon_t = \sum_{j=-\infty}^{\infty} c_j \varepsilon_{t-j}$, where the coefficients c_j are those of the infinite series $a(z)^{-1}b(z)$. Once again, properties of these polynomials determine those of $\{Y_t\}$.

The class of ARMA processes is typically regarded as a useful 'black box' for fitting and forecasting, though fitted models sometimes have a substantive interpretation. For instance, the values of AIC when (6.25) is fitted to the beaver data and the η_t follow an ARMA(p, q) process with (p, q) equal to $(1, 1)$, $(0, 1)$, $(1, 2)$, and $(2, 0)$ are -128.34, -90.06, -126.54, and -128.78, compared with -127.55 for the AR(1) model, which therefore seems a good compromise between quality of fit and simplicity of interpretation, the latter following from its Markov structure. It is considerably harder to explain the ARMA(1,2) model in simple terms, despite its slightly better fit.   ■

Trend removal

In practice data are rarely stationary, and trends or periodic changes must be removed before fitting standard models. One simple approach to removing polynomial trends is differencing. Suppose that $Y_t = \gamma_0 + \gamma_1 t + \varepsilon_t$, so there is linear trend with possibly correlated noise superimposed. Then

$$X_t = Y_t - Y_{t-1} = (\gamma_0 + \gamma_1 t + \varepsilon_t) - \{\gamma_0 + \gamma_1(t-1) + \varepsilon_{t-1}\} = \gamma_1 + \eta_t,$$

say, where $\eta_t = \varepsilon_t - \varepsilon_{t-1}$. Thus differencing removes linear trend but complicates the error structure: if $\{\varepsilon_t\}$ had been white noise, then the differenced process $\{\eta_t\}$ follows an MA(1) model with $\beta_1 = -1$. It is straightforward to show that d-fold differencing will remove a polynomial trend of order d (Exercise 6.4.4). Over-differencing does little harm: if there had been no trend originally present then $\{X_t\}$ merely has a more complicated error structure than had $\{Y_t\}$. Differencing can also be used to remove seasonal components.

If an ARMA(p, q) model fits the d-fold difference of $\{Y_t\}$, then we have $a(B)(I - B)^d Y_t = b(B)\varepsilon_t$, and this is known as an *integrated autoregressive-moving average* or ARIMA(p, d, q) process. This generalizes the class of ARMA models to allow non-stationarity.

Example 6.28 (FTSE data) Trends such as that in the right panel of Figure 6.11 are generally removed by differencing the log closing prices, and the upper panel of Figure 6.13 shows $y_t = 100 \log(x_t/x_{t-1})$, where x_t is the original series. Thus y_t is proportional to the differences of the log x_t and represents daily percentage returns to investors. Differencing has removed the trend, but it is not clear that the y_t are stationary — their variability seems to increase from time to time. Such changes in volatility cannot be mimicked by linear processes and much effort has been expended in modelling them. Probability plots show that the y_t are somewhat asymmetric with heavier tails than the normal distribution, so the marginal distribution of $\{Y_t\}$ is non-normal.

The partial correlogram of y_t shows small but significant autocorrelation at lag one, suggestive of slight autoregressive behaviour. Its value, $\widehat{\rho}_1 = 0.09$, is too small to be of much use in predicting movements of y_t. This makes sense: high correlation could be exploited by everyone for gain, but there must be both winners and losers when shares are traded. The partial correlogram of the $(y_t - \overline{y})^2$ shows generally positive autocorrelations to about lag 20.

The y_t have average 0.043 with standard error 0.018, so if the data were independent there would be evidence that $E(Y_t) > 0$, corresponding to an average daily increase of about 0.043% in the FTSE over 1991–1998. ∎

Other approaches to trend removal can involve local smoothing by methods like those to be described in Section 10.7; very roughly the idea is to use weighted averages of the data to estimate changes in the process mean. Such averaging can be applied on different scales, for example giving separate estimates of systematic decadal, annual, and monthly variation. Robust versions of these smoothers exist and are often preferable in practice.

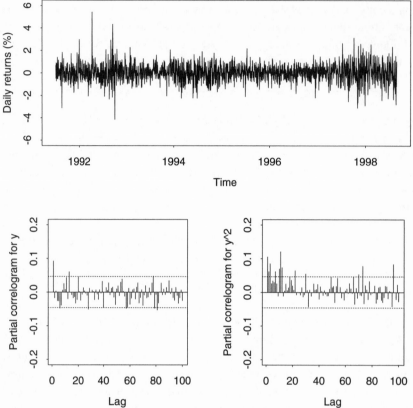

Figure 6.13 Daily returns (%) from the FTSE, 1991–1998. The lower panels show the partial correlograms of the y_t and their squares. The 95% confidence bands shown by the dotted horizontal lines are much narrower than in Figure 6.12 because there are many more data.

Volatility models

A key feature of financial time series such as that in the top panel of Figure 6.13 is their changing volatility, which leads to periods of high variability interspersed with quieter periods. A standard model for this in the financial context is the linear *autoregressive conditional heteroscedastic model of order one* or linear ARCH(1) process, which sets

$$Y_t = \sigma_t \varepsilon_t, \quad \sigma_t^2 = \beta_0 + \beta_1 Y_{t-1}^2, \quad t = \ldots, -1, 0, 1, \ldots, \quad (6.27)$$

where $\{\varepsilon_t\}$ is normal white noise with unit variance with ε_t independent of Y_{t-1}, $\beta_0 > 0$ and $\beta_1 \geq 0$. The current variance σ_t^2 is increased if the previous observation was far from zero, giving bursts of high volatility when this occurs. A necessary condition for stationarity is $E(Y_t^2) = E(\sigma_t^2)E(\varepsilon_t^2) < \infty$, implying that $\gamma_0 = \beta_0 + \beta_1 \gamma_0$ or equivalently that $\beta_1 < 1$. In this case $\{Y_t\}$ is zero-mean white noise, but as we can write $Y_t^2 = \sigma_t^2 + (Y_t^2 - \sigma_t^2) = \beta_0 + \beta_1 Y_{t-1}^2 + \eta_t$, where $\eta_t = \sigma_t^2(\varepsilon_t^2 - 1)$ has mean zero, we see that $\{Y_t^2\}$ follows an autoregressive process, albeit with non-constant variance. In order for the process $\{Y_t^2\}$ to be stationary $E(Y_t^4)$ must be finite, and this occurs when $\beta_1^2 < 1/3$. Then Y_t has fatter tails than the normal distribution.

Thus ARCH models mimic two important features of financial time series: volatility clustering and fat-tailed marginal distributions.

The assumption of normal innovations can be replaced by other distributions, a popular choice being to set $\nu\varepsilon_t/(\nu - 2) \overset{\text{iid}}{\sim} t_\nu$; the scaling ensures that $\mathrm{var}(\varepsilon_t) = 1$. ARCH models can be extended to allow dependence on $Y_{t-2}^2, \dots$ and on $\sigma_{t-1}^2, \dots$, a particularly widely-used case being the generalized ARCH or GARCH(1,1) process in which $\sigma_t^2 = \beta_0 + \beta_1 Y_{t-1}^2 + \delta\sigma_{t-1}^2$.

Example 6.29 (FTSE data)   Example 6.28 suggests that an unadorned ARCH model is unlikely to fit these data because it cannot account for the non-zero mean and non-zero correlations. Inspired by (6.27), we therefore let $Y_t - \mu = \alpha(Y_{t-1} - \mu) + \sigma_t\varepsilon_t$ with $\sigma_t^2 = \beta_0 + \beta_1(Y_{t-1} - \mu)^2$. This combines autoregressive structure for the means of the Y_t with ARCH structure for their variance. The result is a Markov process, and with normal ε_t the log likelihood contribution from the conditional density $f(y_t \mid y_{t-1})$ is

$$-\frac{1}{2} \log\{\beta_0 + \beta_1(y_{t-1} - \mu)^2\} - \frac{\{y_t - \mu - \alpha(y_{t-1} - \mu)\}^2}{2\{\beta_0 + \beta_1(y_{t-1} - \mu)^2\}}.$$

The overall log likelihood is a sum of such terms for $t = 2, \dots, n$ plus $\log f(y_1)$, but the series is so long that this initial term, which involves knowing the stationary density of Y_t, can safely be ignored.

The log likelihood is readily maximized numerically, but a correlogram suggests that structure remains in the squares of the residuals

$$r_t = \frac{y_t - \widehat{\mu} - \widehat{\alpha}(y_{t-1} - \widehat{\mu})}{\{\widehat{\beta_0} + \widehat{\beta_1}(y_{t-1} - \widehat{\mu})^2\}^{1/2}},$$

so this model is not adequate. As an alternative, we retain the AR mean structure but use GARCH structure $\sigma_t^2 = \beta_0 + \beta_1(Y_{t-1} - \mu)^2 + \delta\sigma_{t-1}^2$ for the variances. A crude way to fit this is to estimate σ_m^2 by the variance of $y_1, \dots, y_m$, and then to compute $\sigma_t^2 = \beta_0 + \beta_1(y_{t-1} - \mu)^2 + \delta_1\sigma_{t-1}^2$ for $t = m + 1, \dots, n$. The likelihood based on $f(y_{m+1}, \dots, y_n \mid y_1, \dots, y_m)$ is then readily obtained and may be maximized. Here n is large so little information is lost by conditioning on $y_1, \dots, y_m$. With $m = 30$ the maximized log likelihood is -2100.27, and both the residuals and their squares look like white noise, so the structure of the model seems correct. However a normal probability plot of the residuals suggests that slightly heavier-tailed innovations may be needed. We therefore let the ε_t have t_ν distributions, scaled so that $\mathrm{var}(\varepsilon_t) = 1$. The resulting log likelihood is -2075.64, an appreciable improvement. The maximum likelihood estimates and standard errors are $\widehat{\mu} = 0.051\ (0.018)$, $\widehat{\alpha} = 0.070\ (0.024)$, $\widehat{\beta_0} = 0.006\ (0.004)$, $\widehat{\beta_1} = 0.036\ (0.011)$, $\widehat{\delta} = 0.955\ (0.016)$ and $\widehat{\nu} = 9.7\ (1.86)$. Thus μ and α seem necessary for successful modelling. Over the period of these data the return on investment was on average $100\widehat{\mu} \doteq 5\%$ every 100 trading days, but little would be gained from using the estimated correlation $\widehat{\alpha} = 0.07$ between Y_t and Y_{t+1} for short-term prediction. The value of $\widehat{\delta}$ shows the strong dependence of σ_t^2 on σ_{t-1}^2 that leads to volatility persistence. A condition for stationarity of a GARCH process $\{Y_t\}$ is that $\beta_1 + \delta < 1$, and this is satisfied by the estimates. The value of $\widehat{\nu}$ indicates innovations somewhat heavier than normal, in agreement with the residual plot. Overall the model seems to fit surprisingly well.

∎

Time series is a large and important topic, whose surface has barely been scratched above. The bibliographic notes give some points of entry to the literature.

Exercises 6.4

1   Consider (6.24) for $t = 1, \ldots, n$, and suppose that Y_0 has a known distribution with finite variance, independent of $\varepsilon_1, \ldots, \varepsilon_n$. Deduce that

$$Y_n - \mu = \sum_{j=1}^{n} \alpha^{n-j}\varepsilon_j + \alpha^n(Y_0 - \mu)$$

and establish that a limiting distribution for Y_n as $n \to \infty$ exists only when $\lim_{n \to \infty} \sum_{j=1}^{n} \alpha^{2j} < \infty$.
Hence show that a condition for stationarity is $|\alpha| < 1$, in which case the limiting distribution for Y_n is normal with mean μ and variance $\sigma^2/(1 - \alpha^2)$. Show also that if Y_0 has this distribution, so too do all the Y_j. Show that the covariance matrix Ω of $Y_1, \ldots, Y_n$ is then that given in Example 6.24, and write down the corresponding moral graph.

2   Consider the MA(1) process; see Example 6.26. Show that its covariances are

$$\mathrm{cov}(Y_t, Y_{t+s}) = \begin{cases} \sigma^2 \left(1 + \beta_1^2\right), & s = 0, \\ \sigma^2 \beta_1, & s = 1, \\ 0 & \text{otherwise,} \end{cases}$$

find the autocorrelation function and use the matrices in Example 6.24 to deduce that there is no cut-off in the partial autocorrelations.
Generalize this to the MA(q) model.

3   Give an expression for the log likelihood in Example 6.25.

4   Suppose that $Y_t = \sum_{j=0}^{k} \xi_j t^j + \varepsilon_t$, where $\{\varepsilon_t\}$ is a stationary process. Show by induction that d-fold differencing yields a series that is stationary for any $d \geq k$.
Let $Y_t = s(t) + \varepsilon_t$, where $s(t) = s(t + kp)$, for a fixed integer p and all integers t and k. Show that $(I - B^p)Y_t$ is stationary, and discuss the implications for removal of seasonality from a monthly time series.

5   Give a formula for the residual r_t when $\sigma_t^2 = \beta_0 + \beta_1(Y_{t-1} - \mu)^2 + \delta\sigma_{t-1}^2$ in Example 6.29.

6.5 Point Processes

Data that can be summarized by points in a continuum arise in many applications. Examples are the epicentres of earthquakes, the locations of cases of leukaemia, and the times are which emails are sent. The 'point' may be merely a convenient representation of something small compared to its surroundings, and other information may be available, such as the strength of the earthquake, but here we assume that summary as a point is sensible and ignore other aspects.

6.5.1 Poisson process

The Poisson process in the line is the simplest point process and the basis for many more complex models. Suppose that we observe points in a time interval $[0, t_0]$.

Let $N(w, w + t)$ denote how many fall into the subinterval $(w, w + t]$; we write $N(t) = N(0, t)$, $t > 0$, and $N(\mathcal{A})$ for the number of points in the set $\mathcal{A}$. Let $\lambda(t)$ be a well-behaved non-negative function whose integral is finite on $[0, t_0]$, and suppose that

- events in disjoint subsets of $[0, t_0]$ are independent, that is, $N(\mathcal{A}_1)$ is independent of $N(\mathcal{A}_2)$ whenever $\mathcal{A}_1 \cap \mathcal{A}_2 = \emptyset$;

$o(\delta t)$ is small enough that $o(\delta t)/\delta t \to 0$ as $\delta t \to 0$.

- $\Pr\{N(t, t + \delta t) = 0\} = 1 - \lambda(t)\delta t + o(\delta t)$ for small δt; and
- $\Pr\{N(t, t + \delta t) = 1\} = \lambda(t)\delta t + o(\delta t)$ for small δt.

The last two properties imply that $\Pr\{N(t, t + \delta t) > 1\} = o(\delta t)$, so the process is *orderly*: multiple occurrences at the same t may not occur. The *intensity* $\lambda(t)$ is interpreted as the rate at which points occur in a small interval at t, so more points fall where $\lambda(t)$ is relatively high. Finiteness of $\int_0^{t_0} \lambda(u)\,du$ ensures that $N(t_0) < \infty$ with probability one, as we shall see below.

We find the probability that there are no points in the interval $(w, w + t]$ by dividing it into k subintervals of length $\delta t = t/k$, and then letting $\delta t \to 0$. Then the properties above imply that

$$
\begin{aligned}
\Pr\{N(w, w + t) = 0\} &= \prod_{i=0}^{k-1} \Pr\left[N\left\{w + i\delta t, w + (i + 1)\delta t\right\} = 0\right] \\
&\doteq \prod_{i=0}^{k-1} \{1 - \lambda(w + i\delta t)\delta t + o(\delta t)\} \\
&= \exp\left[\sum_{i=0}^{k-1} \log\{1 - \lambda(w + i\delta t)\delta t + o(\delta t)\}\right] \\
&= \exp\left\{-\sum_{i=0}^{k-1} \lambda(w + i\delta t)\delta t + o(k\delta t)\right\} \\
&\to \exp\left\{-\int_w^{w+t} \lambda(u)\,du\right\},
\end{aligned}
\tag{6.28}
$$

where the limit follows because as $\delta t \to 0$ with t fixed, $o(k\delta t) = t\, o(\delta t)/\delta t \to 0$. As the length of the random time T from w to the next point exceeds t if and only if $N(w, w + t) = 0$, T has probability density function

$$
f_T(t) = -\frac{d\Pr\{N(w, w + t) = 0\}}{dt} = \lambda(w + t)\exp\left\{-\int_w^{w+t} \lambda(u)\,du\right\}, \quad t > 0,
$$

and hazard function $f_T(t)/\Pr(T \geq t) = \lambda(w + t)$.

Now suppose that points in $(0, t_0]$ have been observed at times $t_1, \ldots, t_n$, where $0 < t_1 < \cdots < t_n < t_0$. As events in non-overlapping sets are independent, the joint probability density of the data is

$$
\lambda(t_1)e^{-\int_0^{t_1} \lambda(u)\,du} \times \lambda(t_2)e^{-\int_{t_1}^{t_2} \lambda(u)\,du} \times \cdots \times \lambda(t_n)e^{-\int_{t_{n-1}}^{t_n} \lambda(u)\,du} \times e^{-\int_{t_n}^{t_0} \lambda(u)\,du},
$$

where the final term is the probability of no events in $(t_n, t_0]$. This joint density reduces to

$$\exp\left\{-\int_0^{t_0} \lambda(u)\, du\right\} \prod_{j=1}^{n} \lambda(t_j), \quad 0 < t_1 < \cdots < t_n < t_0. \tag{6.29}$$

Given a parametric form for $\lambda(t)$, (6.29) gives the likelihood on which inferences may be based. In practice the integral is usually unavailable in closed form and a numerical approximation must be used.

The probability of n events occurring in the interval $[0, t_0]$ is obtained by integrating (6.29) with respect to $t_1, \ldots, t_n$ and is (Exercise 6.5.2)

$$\Pr\{N(t_0) = n\} = \frac{\Lambda(t_0)^n}{n!} \exp\{-\Lambda(t_0)\}, \quad n = 0, 1, \ldots, \tag{6.30}$$

where we have written $\Lambda(t_0) = \int_0^{t_0} \lambda(u)\, du$. Thus $N(t_0)$ is a Poisson variable with mean $\Lambda(t_0)$. As events in disjoint subsets are independent and sums of independent Poisson variables are Poisson (Example 2.35), we see that in a Poisson process, the number of events in a subset $\mathcal{A}$ is a Poisson variable whose mean $\Lambda(\mathcal{A}) = \int_{\mathcal{A}} \lambda(u)\, du$ is the integral of the rate function λ over $\mathcal{A}$. Moreover these counts are independent for disjoint subsets.

Division of (6.29) by (6.30) gives the probability that points arise at $t_1, \ldots, t_n$ conditional on there being n points, namely

$$n! \prod_{j=1}^{n} \frac{\lambda(t_j)}{\Lambda(t_0)}, \quad 0 < t_1 < \cdots < t_n < t_0.$$

This is the joint density of the order statistics of a random sample of size n with density $\lambda(t)/\Lambda(t_0)$ on the interval $[0, t_0]$; see (2.25). As we shall see, this result is useful in model-checking.

Example 6.30 (Exponential trend) Let $\lambda(t) = \exp(\beta_0 + \beta_1 t)$, so $\Lambda(t_0) = e^{\beta_0}(e^{\beta_1 t_0} - 1)/\beta_1$. When $\beta_1 = 0$ this yields a constant intensity. The log likelihood corresponding to (6.29) equals

$$\ell(\beta_0, \beta_1) = n\beta_0 + \beta_1 \sum_{j=1}^{n} t_j - e_0^{\beta}(e^{\beta_1 t_0} - 1)/\beta_1$$

and is of exponential family form.

The ratio $\lambda(t)/\Lambda(t_0)$ equals $\beta_1 e^{\beta_1 t}/(e^{\beta_1 t_0} - 1)$, corresponding to an exponential tilt of the uniform density on $[0, t_0]$, so when $\beta_1 > 0$ events tend to pile up toward the right end of the interval, and conversely. ∎

There is an intimate connection between two ways to think about such data, in terms of the counts in subsets of the region of observation and in terms of the spacings between points. Although the second approach is natural in one dimension, the count representation is generally simpler in several dimensions. To see how it extends, let $\mathcal{S}$ be a subset of $\mathbb{R}^d$ and suppose that an integrable non-negative function $\lambda(t)$ is defined such that $\Lambda(\mathcal{S}) = \int_{\mathcal{S}} \lambda(u)\, du$ is finite. Then under conditions that extend those for the univariate case, the numbers of events in disjoint subsets $\mathcal{A}_1, \ldots, \mathcal{A}_m$ of $\mathcal{S}$ have independent Poisson distributions with means $\Lambda(\mathcal{A}_1), \ldots, \Lambda(\mathcal{A}_m)$. The probability density for points observed at $\{t_1, \ldots, t_n\} \subset \mathcal{S}$ is

$$\prod_{j=1}^{n} \lambda(t_j) \times \exp\{-\Lambda(\mathcal{S})\}, \qquad (6.31)$$

from which a likelihood can again be constructed. Such models play in important role in event history and survival data, as described in Sections 5.4 and 10.8. In terms of Figure 5.8, the idea is to treat failures as events of an inhomogeneous Poisson process in the region of the plane bounded by the line $x = y$, the horizontal axis, and the vertical line marking the end of the trial; see Section 10.8.2. Another application, to statistics of extremes, will be described shortly.

Homogeneous Poisson process

The simplest situation is when the intensity function $\lambda(t)$ is a constant λ. Then $\Lambda(\mathcal{A}) = \lambda|\mathcal{A}|$ and $\Lambda(t_0) = \lambda t_0$. The number of points in $[0, t_0]$ is then Poisson with mean λt_0, and intervals between them are independent exponential variables with density $\lambda e^{-\lambda y}$. The log likelihood from (6.29) is

$$\ell(\lambda) \equiv n \log \lambda - \lambda t_0,$$

| $|\mathcal{A}|$ is the length (Lebesgue measure) of the set $\mathcal{A}$.

from which the maximum likelihood estimate $\widehat{\lambda} = n/t_0$ and information quantities may be derived; see Example 4.19.

When $\lambda(t)$ is constant, the density $\lambda(t)/\Lambda(t_0) = t_0^{-1}$ is uniform on the interval $[0, t_0]$, and hence the n points $u_j = t_j/t_0$ are distributed as order statistics of a random sample from the uniform distribution on $[0, 1]$; see Section 2.3. A graphical check of this is to plot the empirical distribution function of the u_j, $\widehat{F}(u)$. Departures from the uniform distribution $F(u) = u$, $0 \leq u \leq 1$ suggest that the intensity is not constant. Formal tests of fit using this are discussed in Section 7.3.1.

Data often exhibit clustering relative to a Poisson process. If so, there will tend to be an excess of short intervals between points, relative to the exponential distribution. Under the Poisson process model the *spacings* $y_1 = t_1 - 0$, $y_2 = t_2 - t_1, \ldots, y_{n+1} = t_0 - t_n$ form a (non-independent) sample from the exponential distribution with mean λ^{-1}, so a plot of ordered spacings against exponential order statistics should be a straight line, departures from which will suggest model failure.

Example 6.31 (Danish fire data)   Figure 6.14 shows data on the times and amounts of major insurance claims due to fire in Denmark from 1980–1990. The upper left panel shows the original 2492 claims; the original amounts have been rescaled. The data are dominated by a few large claims, shown in more detail in the upper right panel, which gives the logarithms of the 254 claims that exceed 5 units. This is a two-dimensional point process of times and log amounts, which reduces to the one-dimensional data shown as a rug at the foot of the panel if the amounts are ignored.

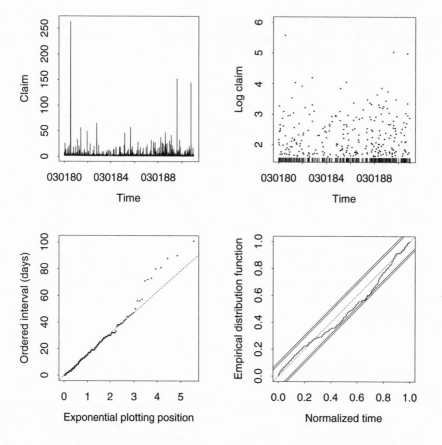

Figure 6.14 Data on major insurance claims due to fires in Denmark, 1980–1990 (Embrechts *et al.*, 1997, pp. 298–303). The upper left panel shows the original data and the upper right panel the logs of the 254 losses exceeding five units, with the rug below showing their times. The lower right panel shows the empirical distribution of the 254 $u_j = t_j/t_0$, and the lower left panel an exponential probability plot of spacings between these t_j. In each case the dotted line shows the expected pattern under a homogeneous Poisson process. The lower right panel suggests that the rate of the process may be non-uniform, with an excess of early points followed by a deficiency. The solid diagonal lines in the lower right panel show significance for a Kolmogorov–Smirnov statistic at levels 0.05 and 0.01 and are explained in Example 7.23. The lower left panel suggests that the spacings are close to exponentially distributed.

We consider only the times of these 254 largest claims. The lower right panel shows the empirical distribution function of the corresponding $u_j = t_j/t_0$, with $t_1, \ldots, t_n$ the rug in the panel above. Relative to the uniform distribution there is a slight excess of claims up to about 1983, followed by a deficiency from 1984 to 1990. Example 7.23 gives further discussion of the fit.

The exponential probability plot of the spacings in the lower left panel of the figure suggests that the times between claims are fairly close to exponential, though perhaps with a slightly longer tail. The value of $\widehat{\lambda}$ is roughly $254/(11 \times 365) = 0.063$ days^{-1}. Thus the rate of arrival of claims per day is about 0.06, corresponding to a mean time between claims of $\widehat{\lambda}^{-1} = 15.8$ days; this has standard error 1.0 calculated from the observed information.

We return to these data in Examples 6.34 and 7.23.   ■

6.5.2 Statistics of extremes

An important application of Poisson processes is to rare events — high sea levels, low temperatures, record times to run a mile, large insurance claims, and so forth. To see how, we make a detour and consider properties of the maximum of a random sample $X_1, \ldots, X_m$ from a continuous distribution function $F(x)$ with upper support

The upper support point is the smallest x_0 such that $\lim_{x \nearrow x_0} F(x) = 1$; possibly $x_0 = +\infty$.

point x_0. As $m \to \infty$, independence of the X_i implies that for any fixed $x < x_0$,

$$
\begin{aligned}
\Pr\{\max(X_1, \ldots, X_m) \leq x\} &= \Pr(X_i \leq x, i = 1, \ldots, m) \\
&= \Pr(X_1 \leq x) \times \cdots \times \Pr(X_m \leq x) \\
&= F(x)^m \to 0,
\end{aligned}
$$

so in order to obtain a non-degenerate limiting distribution for the maximum, we must rescale the X_i. We consider $Y_m = a_m^{-1}(\max_i X_i - b_m)$ for sequences of constants $\{a_m\} > 0$ and $\{b_m\}$, and ask under what conditions $Y_m \xrightarrow{D} Y$ as $m \to \infty$ for some non-degenerate random variable Y. As $m \to \infty$,

$$
\begin{aligned}
\Pr(Y_m \leq y) &= \Pr\left[a_m^{-1}\{\max(X_1, \ldots, X_m) - b_m\} \leq y\right] \\
&= F(b_m + a_m y)^m \\
&= \left[1 - \frac{m\{1 - F(b_m + a_m y)\}}{m}\right]^m
\end{aligned}
\tag{6.32}
$$

can be shown to possess a limit if and only if $\lim_{m \to \infty} m\{1 - F(b_m + a_m y)\}$ exists. As $m\{1 - F(b_m + a_m y)\}$ is the number of the $X_1, \ldots, X_m$ expected to exceed $b_m + a_m y$, suitable sequences $\{a_m\}$ and $\{b_m\}$ exist for most, but not all, continuous distributions. If they do exist, a remarkable result is that the only possible non-trivial limit is of

$(a)_+ = a$ if $a > 0$ and otherwise equals zero.

form

$$
\lim_{m \to \infty} m\{1 - F(b_m + a_m y)\} = \left(1 + \xi \frac{y - \eta}{\tau}\right)_+^{-1/\xi},
\tag{6.33}
$$

with the right-hand side taken to be $\exp\{-(y - \eta)/\tau\}$ if $\xi = 0$. The parameters τ and η control the scale and location of the limit, and account for the effect of minor changes to $\{a_m\}$ and $\{b_m\}$ — for example, replacing a_m by $\frac{1}{2}a_m$ would rescale any limit, but would not affect its existence or its shape.

On putting together (6.32) and (6.33), we see that if a limiting distribution for the maximum exists, it must be the *generalized extreme-value distribution*

Emil Julius Gumbel (1891–1966) was born and studied in Munich. His radical pacifist views and Jewish background caused conflict with his university colleagues and authorities in Heidelberg, and led to his exile in France in 1932 and later in the USA. He highlighted the importance of statistical extremes, on which he wrote an important book (Gumbel, 1958), and through his consulting strongly influenced hydrologists, meteorologists, and engineers.

$$
H(y; \eta, \tau, \xi) = \exp\left\{-\left(1 + \xi \frac{y - \eta}{\tau}\right)_+^{-1/\xi}\right\}, \quad -\infty < \xi, \eta < \infty, \tau > 0, \tag{6.34}
$$

where the range of y is such that $1 + \xi(y - \eta)/\tau > 0$. The parameter ξ controls the shape of the density, which has a heavy right tail and finite lower support point if $\xi > 0$, and a finite upper support point if $\xi < 0$. The *Gumbel distribution*

$$
H(y; \eta, \tau, 0) = \exp[-\exp\{-(y - \eta)/\tau\}], \quad -\infty < y < \infty,
$$

arises as $\xi \to 0$; see Problem 6.11. Expression (6.34) gives the only possible limiting distribution for maxima. Minima are dealt with by noting that any limit distribution for $\min_i(X_i) = -\max_i(-X_i)$ must have form $1 - H(-y; \eta, \tau, \xi)$.

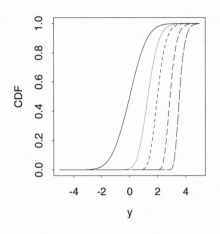

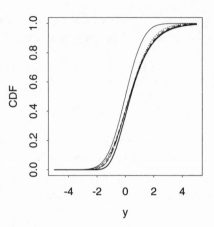

Figure 6.15
Convergence for sample
maxima. Left panel:
distributions of maxima of
$m = 1, 7, 30, 365, 3650$
standard normal variables
(from left to right). Right
panel: distributions of
renormalized maxima of
$m = 1, 7, 30, 365, 3650$
standard normal variables.
The distributions on the
right converge to the
Gumbel distribution
(heavy).

Example 6.32 (Normal distribution) For the standard normal distribution, integration by parts gives $1 - F(x) = \int_x^\infty \phi(x)\,dx \doteq \phi(x)/x$ as $x \to \infty$. Hence $m\{1 - F(b_m + a_m y)\}$ approximately equals

$$\exp\left\{-\frac{1}{2}(b_m + a_m y)^2 - \log(b_m + a_m y) + \log m - \frac{1}{2}\log 2\pi\right\}, \qquad (6.35)$$

and some tedious algebra shows that with $a_m = (2\log m)^{-1/2}$ and $b_m = a_m^{-1} - \frac{1}{2}a_m(\log\log m + \log 4\pi)$, (6.35) converges to $\exp(-y)$ as $m \to \infty$. However the convergence is very slow. With $y = 4$ the probabilities $\Phi(b_m + a_m y)^m$ are $0.9907, 0.9871$, $0.9859, 0.9855$ for $m = 30, 365, 1825, 3650$, while the target Gumbel probability is 0.9819. These values of m are chosen to correspond to random sampling of a normal distribution daily for periods of one month, and one, five, and ten years. Even with this amount of daily data the limiting probability is not attained, because the right tail of the normal distribution is so light compared to that of the Gumbel distribution that enormous samples are needed for the limit to work well.

Figure 6.15 shows the convergence graphically. The left panel shows the distributions of maxima of m standard normal variables, with $m = 1, 7, 30, 365$, and 3650, corresponding to maxima over a day, a week, a month, a year and ten years of daily normal data. The distribution becomes increasingly concentrated as m increases, and does not converge to a useful limit. The right panel shows how the distribution of $a_m^{-1}\{\max(X_1, \ldots, X_m) - b_m\}$ does converge to a limiting Gumbel distribution, given by the heavy solid line. As mentioned above, the convergence is rather slow. Fortunately the generalized extreme-value distribution usually gives a better approximation for sample maxima than this example might suggest. ∎

The upshot is that the generalized extreme-value distribution provides the natural model to fit to sample maxima or minima. For example, if a series of annual maximum sea levels $y_1, \ldots, y_n$ is available, we suppose that they are a random sample from (6.34) and fit it by maximum likelihood. Often the parameter of interest is the p quantile of the distribution, that is $y_p = \eta + \tau\{(-\log p)^{-\xi} - 1\}/\xi$, which is known in this context as the $(1-p)^{-1}$-year *return level*: it is the level exceeded once on average

$1/(1 - p)$ is known as the return period.

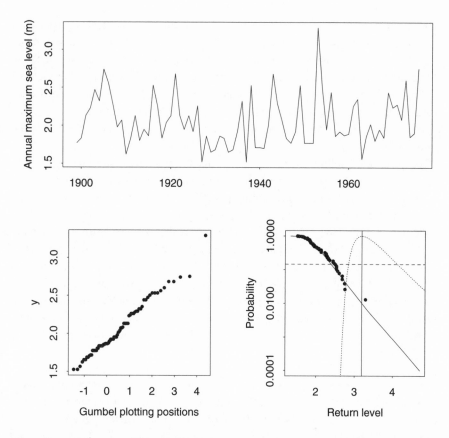

every $(1 - p)^{-1}$ years. This would be important if the data were being analyzed in order to suggest how high coastal defenses should be built. Of course quantities such as the expected insurance loss should flooding occur are also of interest.

Maximum likelihood estimation is regular if $\xi > -1/2$, as seems common in applications. When $\xi \leq -1/2$, the likelihood derivatives do not have their usual properties and Example 4.43 is relevant, as the upper support point of the density can be estimated with rate faster than the usual $n^{-1/2}$.

The return level is estimated by replacing η, τ, and ξ by their maximum likelihood estimates. Its standard error may be obtained using the delta method (page 122), though the profile log likelihood for y_p gives a more reliable confidence set. In practice n is often substantially smaller than $(1 - p)^{-1}$ and the return level is estimated well outside the range of the data. Then it is important to consider whether there are enough data underlying the $y_1, \ldots, y_n$ for the generalized extreme-value model to give a good approximate distribution for the maxima, and to check whether n is large enough for large-sample likelihood theory to be a good basis for inference. The crucial aspect is however the extent to which extrapolation to high quantiles of the distribution is sensible based on limited data, and this bears careful consideration.

Example 6.33 (Yarmouth sea level data)   The upper panel of Figure 6.16 shows a time series of annual maximum sea levels at Yarmouth on the east coast of England for

1899–1976. As is typical with such data, the largest value is considerably greater than the rest; it arose in 1953 when there was widespread flooding. The correlogram and partial correlogram show no serial dependence, so we treat the values as independent.

The lower left panel of the figure shows a probability plot of the data against Gumbel plotting positions. Upward curvature would here suggest that $\xi > 0$, and downward curvature that $\xi < 0$. In fact the plot is close to straight, indicating that $\xi \doteq 0$. The large value from 1953 does not appear outlying, because of the heavy right tail of the density.

The maximum likelihood estimates and standard errors are $\widehat{\eta} = 1.90$ (0.034), $\widehat{\tau} = 0.26$ (0.025), and $\widehat{\xi} = 0.04$ (0.096); the latter give no evidence against the Gumbel model, in agreement with the probability plot. The location and scale parameters are well determined compared to ξ.

The lower right panel of Figure 6.16 compares the estimated survivor function $\Pr(Y > y)$ with its empirical counterpart, obtained by plotting $1 - j/(n + 1)$ against $y_{(j)}$. The vertical line indicates the estimated 100-year return level, $\widehat{y}_{0.99}$, while the broken lines show the profile likelihood for $y_{0.99}$ and the corresponding 95% confidence interval. This is highly asymmetric, so this interval is much preferable to using normal approximation. In practice 1000- or even 10,000-year return levels may be needed, and then of course the statistical uncertainty is very large indeed.   ∎

Point process approximation

If more extensive data are available it is potentially wasteful to use only the annual maxima, and we now show how a Poisson process model can overcome this. Let $X_1, \ldots, X_{\lceil mt_0 \rceil}$ be a random sample from $F(x)$ and consider the pattern of points $(i/m, a_m^{-1}(X_i - b_m)), i = 1, \ldots, \lceil mt_0 \rceil$ that fall into the subset $\mathcal{S} = [0, t_0] \times [u, \infty)$ of the plane. The event $a_m^{-1}(X_i - b_m) > y$ occurs if and only if $X_i > b_m + a_m y$, so the number of points that fall into $\mathcal{A} = [t_1, t_2] \times [y, \infty)$ may be expressed as the sum of indicator random variables

$$N_m(\mathcal{A}) = \sum_{i=\lceil mt_1 \rceil}^{\lfloor mt_2 \rfloor} I\left(X_i > b_m + a_m y\right), \quad 0 \leq t_1 < t_2 \leq t_0, y \geq u.$$

[margin note:] $\lceil a \rceil$ and $\lfloor a \rfloor$ are respectively the smallest integer larger than a and the largest integer smaller than a.

The X_i are independent and identically distributed, so $N_m(\mathcal{A})$ is binomial with denominator $\lfloor mt_2 \rfloor - \lceil mt_1 \rceil + 1$ and probability $1 - F(b_m + a_m y)$ that satisfies (6.33). Hence the Poisson limit for the binomial distribution (Problem 2.3) gives

$$\lim_{m \to \infty} \Pr\{N_m(\mathcal{A}) = n\} = \frac{\Lambda(\mathcal{A})^n}{n!} \exp\{-\Lambda(\mathcal{A})\}, \quad n = 0, 1, \ldots,$$

where $\Lambda(\mathcal{A})$ equals

$$\Lambda\{[t_1, t_2] \times [y, \infty)\} = (t_2 - t_1)\left(1 + \xi \frac{y - \eta}{\tau}\right)_+^{-1/\xi}, \quad 0 \leq t_1 < t_2 \leq t_0, y \geq u,$$
(6.36)

with the second term on the right replaced by $\exp\{-(y - \eta)/\tau\}$ if $\xi = 0$. That is, $N_m(\mathcal{A}) \xrightarrow{D} N(\mathcal{A})$, where $N(\mathcal{A})$ is Poisson with mean $\Lambda(\mathcal{A})$.

Figure 6.17 Poisson
process limit for rare
events. The panels show
the values of
$a_m^{-1}(X_i - b_m)$ plotted
against i/m for random
samples of size $m = 10$,
100, 1000 and 10,000
from the exponential
distribution. The pattern
of points above the
threshold at $u = -2$ tends
to a bivariate Poisson
process with intensity
given by (6.36).

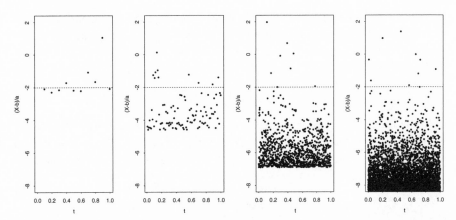

More sophisticated techniques reveal that as $m \to \infty$, the limiting joint distributions of counts $N_m(\mathcal{A}_1)$, $N_m(\mathcal{A}_2)$, … in any collection of disjoint subsets $\mathcal{A}_1, \mathcal{A}_2, \ldots$ of S is that of independent Poisson variables with means $\Lambda(\mathcal{A}_1)$, $\Lambda(\mathcal{A}_2)$, …. Hence as $m \to \infty$, the limiting positions of random values X_i, suitably rescaled, have the joint distribution of points of a Poisson process N in S with intensity (6.36), with arbitrary u. Figure 6.17 illustrates this for exponential samples.

To see the connection to extremes, suppose we have daily data for t_0 years and that $t_2 - t_1 = 1$ year. Then if we apply the Poisson limit to these data with $\mathcal{A} = [t_1, t_2] \times [y, \infty)$, effectively assuming that the limit has set in when $m = 365$ days, and let $Y^1 \geq \cdots \geq Y^r$ denote the r largest values for that year, we see that in an obvious shorthand notation,

$$\Pr(Y^1 \leq y) = \Pr\{N(\mathcal{A}) = 0\}$$
$$= \exp\left\{-\left(1 + \xi \frac{y - \eta}{\tau}\right)_+^{-1/\xi}\right\},$$
$$\Pr(Y^r \leq y^r, \ldots, Y^1 \leq y^1) = \Pr\{N(y^r, y^{r-1}) = 1, \ldots, N(y^2, y^1) = 1\}.$$

The first of these identities recovers (6.34), while the joint density of $Y^1 \geq \cdots \geq Y^r$ at $y^1 \geq \cdots \geq y^r$ is obtained either by differentiating the second identity or from (6.31), with S replaced by $[t_1, t_2] \times [y^r, \infty)$. Both routes show that the limiting joint density of the r largest values is

$$\prod_{i=1}^r \tau^{-1}\left(1 + \xi \frac{y^i - \eta}{\tau}\right)_+^{-1/\xi - 1} \times \exp\left\{-\left(1 + \xi \frac{y^r - \eta}{\tau}\right)_+^{-1/\xi}\right\}. \qquad (6.37)$$

Independence of counts in disjoint subsets implies that data for different years may be treated as independent, so an overall likelihood based on the r largest values for each year is simply the product of such terms for all t_0 years.

In many ways a more satisfactory approach to inference starts by noticing that (6.36) has form $\Lambda_1\{[t_1, t_2]\}\Lambda_2\{[y, \infty)\}$, implying that the points result from two independent Poisson processes, one giving the random 'times' T at which $X_i > u$, and the other giving the rescaled sizes $a_m^{-1}(X_i - b_m)$ of these X_i. The times of exceedances fall according to a homogeneous Poisson process of intensity $\lambda_1(t) = \{1 + \xi(u - \eta)/\tau\}_+^{-1/\xi} \equiv \lambda$, say, while their sizes follow an inhomogeneous Poisson process whose intensity is

$$\lambda_2(y) = -\frac{d\Lambda_2\{[y, \infty)\}}{dy} = \tau^{-1}\left(1 + \xi\frac{y - \eta}{\tau}\right)_+^{-1/\xi - 1}, \quad y > u.$$

This implies that the number of exceedances over level u has a Poisson distribution with mean λt_0, and conditional on n_u exceedances, their sizes $W_j = X_j - u$ are a random sample of size n_u from the *generalized Pareto distribution* (Problem 6.15)

$$G(w) = \begin{cases} 1 - (1 + \xi w/\sigma)_+^{-1/\xi}, & \xi \neq 0, \\ 1 - \exp(-w/\sigma), & \xi = 0. \end{cases} \tag{6.38}$$

The log likelihood (6.31) may be written as

$$\ell(\lambda, \sigma, \xi) \equiv n_u \log \lambda - t_0\lambda - n_u \log \sigma - \left(\frac{1}{\xi} + 1\right)\sum_{j=1}^{n_u} \log\left(1 + \xi\frac{w_j}{\sigma}\right). \tag{6.39}$$

We apply this discussion by taking a threshold u over which the Poisson approximation seems to hold; then the exceedance times should be a homogeneous Poisson process, and their sizes should follow (6.38), as typically assessed by a probability plot. If the fit is satisfactory, estimates and standard errors are obtained by our usual likelihood methods. As with the generalized extreme-value distribution, estimation of σ and ξ is not regular if $\xi \leq -1/2$, and Example 4.43 is again relevant.

We now briefly discuss the choice of u. If it is chosen so that the number of exceedances is small, then the Poisson process approximation to the extremes may be good, but the parameter estimators will have large variance. The variance can be reduced by lowering u, but at the cost of bias because the Poisson approximation for extremes cannot be expected to give good inferences when applied to the bulk of the data. Formal procedures for choosing u attempt to trade off these two aspects, but in practice graphical approaches are more common. These rest on the *threshold stability* property of a random variable W following (6.38), that is,

$$\Pr(W > w \mid W > u) = \{1 + \xi(w - u)/\sigma_u\}^{-1/\xi}, \quad w \geq u \geq 0,$$

where $\sigma_u = \sigma + \xi u$. The operation of thresholding by considering only the tail of W above u yields another random variable $W_u = W - u$, say, following (6.38) but transforms the parameters as $(\sigma, \xi) \mapsto (\sigma + \xi u, \xi)$. When $\xi = 0$ this is the lack-of-memory property of the exponential distribution.

One graphical approach uses the fact that $E(W) = \sigma/(1 - \xi)$ provided $\xi < 1$, so $E(W_u \mid W > u) = (\sigma + \xi u)/(1 - \xi)$, for $u \geq 0$. Thus if the generalized Pareto approximation is adequate for the upper tail of a random sample $X_1, \ldots, X_n$, a graph against u of the empirical version of this conditional mean, given by

$$n_u^{-1}\sum_{j=1}^{n}(X_j - u)I(X_j > u), \quad \text{where} \quad n_u = \sum_{j=1}^{n} I(X_j > u), \tag{6.40}$$

Figure 6.18 Analysis of Danish fire data. Upper left: mean residual life plot, with 95% confidence band (dots) and number of exceedances n_u at the foot of the panel. Upper right and lower left: plots of $\widehat{\sigma}_u - \widehat{\xi}_u u$ and $\widehat{\xi}_u$ against threshold u, with 95% confidence bands. Lower right: exponential probability plot of residuals $\widehat{\xi}^{-1}\log(1 + \widehat{\xi}w_j/\widehat{\sigma})$.

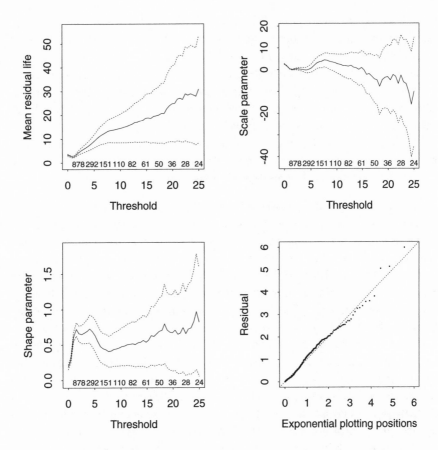

should be a straight line of gradient $\xi/(1 - \xi)$. The idea is to take the threshold to be the smallest u above which this *mean residual life plot* appears linear.

Another approach to choosing u uses the fact that if $\widehat{\xi}_u$ and $\widehat{\sigma}_u$ are maximum likelihood estimators based on the n_u positive exceedances $X_j - u$ over u, and if the generalized Pareto approximation holds, then $\widehat{\xi}_u$ and $\widehat{\sigma}_u - \widehat{\xi}_u u$ should estimate ξ and σ for all u. Thus graphs of $\widehat{\xi}_u$ and $\widehat{\sigma}_u - \widehat{\xi}_u u$ against u should be constant above a certain point, and this is the minimum threshold for which it is reasonable to apply the approximation. Interpretation of such graphs is aided by adding confidence intervals.

Example 6.34 (Danish fire data) In Example 6.31 we saw that exceedance times for the data in the upper right panel of Figure 6.14 seem to follow a homogeneous Poisson process with rate about 0.06 days^{-1}. For threshold modelling we first choose the threshold u. Figure 6.18 shows the mean residual life plot and values of $\widehat{\sigma}_u - \widehat{\xi}_u u$ and $\widehat{\xi}_u$ plotted against u. The mean residual life plot is roughly linear from $u = 7$ onwards, and its positive slope suggests that $\xi > 0$. The other two plots do not tend to constants, but in each case the confidence intervals are wide enough to contain a constant above about $u = 5$. For illustration we take $u = 5$, let $w_j = y_j - u$ denote the 254 claims that exceed $u = 5$ units, and fit the generalized Pareto distribution

(6.38) to the w_j. The maximum likelihood estimates are $\widehat{\sigma} = 3.809$ and $\widehat{\xi} = 0.632$, with standard errors 0.464 and 0.111 from observed information. The value of $\widehat{\xi}$ corresponds to a very heavy upper tail for $W = Y - u$.

The form of (6.38) shows that $\xi^{-1} \log(1 + \xi W/\sigma)$ has a standard exponential distribution, so the fit of the model for exceedances can be assessed by an exponential probability plot of the residuals $\widehat{\xi}^{-1} \log(1 + \widehat{\xi} w_j/\widehat{\sigma})$, shown in the left panel of Figure 6.18. The distribution fits fairly well but not perfectly.

Estimates and confidence regions for quantities of interest such as return levels are found in ways analogous to Example 6.33. In practice it is important to vary the threshold to see if the conclusions depend strongly on u.   ∎

In applications the underlying variables are typically neither identically distributed nor independent. For concreteness, consider using daily temperature data to model the occurrence of hot days at a site in England. These will occur in the summer months, so one way to proceed is to retain only the data for June, July, and August, to suppose that over this period the temperature distribution is roughly constant, and then to hope that about 90 rather than 365 days of data will suffice for the point process paradigm to be applicable. However, even if the summer data are roughly stationary, they will display short-term correlation owing to clustering of hot days. Some detailed mathematics establishes that if extremes far apart are asymptotically independent and the data are stationary — so that in particular all the X_i have the same marginal distribution — then the Poisson process representation with intensity (6.36) still applies, but now to the largest value in a cluster. Clusters then occur at the times of a homogeneous Poisson process, but the cluster size is random and its distribution depends on the local dependence of the X_i. This leads to the practical issues of identifying clusters from data, and of modelling their properties, which are topics of current research.

6.5.3 More general models

In a Poisson process events in disjoint intervals are independent. In practice point process data can show complex dependencies, so this property must be weakened for realistic modelling. This weakening can be done in many ways and below we merely sketch a few possibilities. We continue to suppose that the process is orderly, so events cannot coincide.

Let $\mathcal{H}_t$ denote the entire history of the process up to time t, that is, the positions of all the points in $(-\infty, t]$, and define the *complete intensity function* to be

$$\lambda_{\mathcal{H}}(t) = \lim_{\delta t \to 0} (\delta t)^{-1} \Pr\{N(t, t + \delta t) > 0 \mid \mathcal{H}_t\};$$

this is the intensity of arrival of points just after t, given the history to t. It is akin to the hazard function of Section 5.4, but here potentially dependent on the entire history of the process. The requirement of orderliness is that

$$\Pr\{N(t, t + \delta t) > 1 \mid \mathcal{H}_t\} = o(\delta t)$$

for all t and all possible $\mathcal{H}_t$. The complete intensity must be uniquely defined and well-behaved for any possible $\mathcal{H}_t$ and must moreover determine the probabilistic structure of the process. We shall take this for granted here, though a careful mathematical argument is needed in a formal discussion.

Now consider the probability of no event in $(w, w + t]$ conditional on $\mathcal{H}_w$. We divide $(w, w + t]$ into disjoint subintervals $\mathcal{I}_i = (w + i\delta t, w + (i + 1)\delta t]$, $i = 0, \ldots, k - 1$, where $\delta t = t/k$, and note that

$$\Pr\{N(w, w + t) = 0 \mid \mathcal{H}_w\} \doteq \prod_{i=0}^{k-1} \Pr\{N(\mathcal{I}_i) = 0 \mid \mathcal{H}_{w+i\delta t}\}$$

$$= \prod_{i=0}^{k-1} \{1 - \lambda_{\mathcal{H}}(w + i\delta t)\,\delta t + o(\delta t)\},$$

where $\mathcal{H}_{w+i\delta t}$ represents $\mathcal{H}_w$ followed by no events up to time $w + i\delta t$. The argument leading to (6.28) applies with $\lambda(u)$ replaced by $\lambda_{\mathcal{H}}(u)$, so

$$\Pr\{N(w, w + t) = 0 \mid \mathcal{H}_w\} = \exp\left\{-\int_w^{w+t} \lambda_{\mathcal{H}}(u)\,du\right\},$$

and the probability density that the first point subsequent to w is at t, given $\mathcal{H}_w$, is $-d\Pr\{N(w, w + t) = 0 \mid \mathcal{H}_w\}/dt$. At least in principle, this enables the likelihood for points in an interval $(0, t_0]$, conditional on $\mathcal{H}_0$, to be written down by extending our arguments for the Poisson process, giving

$$\prod_{j=1}^{n} \lambda_{\mathcal{H}}(t_j) \exp\left\{-\int_0^{t_0} \lambda_{\mathcal{H}}(u)\,du\right\} \tag{6.41}$$

as the likelihood based on events at $t_1, \ldots, t_n$ when the process is observed over $(0, t_0]$. In practice it is often hard to specify a tractable but realistic form for $\lambda_{\mathcal{H}}(t)$.

A useful implication is that if events are observed at times $0 < T_1 < \cdots < T_n < t_0$ and we write $\Lambda_{\mathcal{H}}(t) = \int_0^t \lambda_{\mathcal{H}}(u)\,du$, then the transformed times $\Lambda_{\mathcal{H}}(T_1), \ldots, \Lambda_{\mathcal{H}}(T_n)$ form a Poisson process of unit rate on $(0, \Lambda_{\mathcal{H}}(t_0)]$, the transformation $\Lambda_{\mathcal{H}}$ being random. Thus our earlier tools may be used to check the adequacy of an estimated $\widehat{\Lambda}_{\mathcal{H}}$.

Example 6.35 (Poisson process) The complete intensity function for a Poisson process may depend on t, but not on the history of the process. Thus $\lambda_{\mathcal{H}}(t) = \lambda(t)$, which is a constant λ for a homogeneous process. ∎

Example 6.36 (Renewal process) The inter-event intervals in a homogeneous Poisson process are independent exponential variables. The renewal process generalizes this to possibly non-exponential intervals and is a standard model in reliability studies, where failing components in a system may be immediately replaced by apparently identical ones, thereby renewing the system. If system failure is identified with failure of the component and the process is stationary then the complete intensity function depends only on the time since the last event. Thus if previous events have taken place at times t_i, the complete intensity at time t depends only on $v = \min(t - t_i)$

and has form $\lambda(v)$. This is the hazard function corresponding to the density of interval lengths, f. Statistical analysis for such a process is straightforward. Time series tools such as the correlogram and partial correlogram can be used to find serial dependence among successive intervals between events, though it may be clear from the context that these are independent. If independent and stationary, they can be treated as a random sample from f and inference performed in the usual way. ∎

Example 6.37 (Birth process) In a birth process the intensity at time t depends on the number of previous events. Assuming that the number n of events up to t is finite, then $\lambda_{\mathcal{H}}(t) = \beta_0 + \beta_1 n$, where $\beta_0 > 0$, $\beta_1 \geq 0$. The complete intensity function is a step function which jumps β_1 at each event; if $\beta_1 = 0$ the process is a homogeneous Poisson process. ∎

Before giving a numerical example, we briefly describe two functions useful for model checking and exploratory analysis of stationary processes.

The *variance-time curve* is defined as $V(t) = \text{var}\{N(t)\}$, for $t > 0$. A homogeneous Poisson process of intensity λ has $V(t) = \lambda t$, comparisons with which may be informative. Estimation of $V(t)$ is described in Problem 6.12.

The *conditional intensity function* is defined as

$$m_f(t) = \lim_{\delta s, \delta t \to 0} (\delta t)^{-1} \Pr\{N(t, t + \delta t) > 0 \mid N(-\delta s, 0) > 0\}, \quad t > 0,$$

which gives the intensity of events at t conditionally on there being an event at the origin. Evidently $m_f(t) = \lambda$ for a homogeneous Poisson process. An event at time t need not be the first event after that at the origin.

Example 6.38 (Japanese earthquake data) Figure 6.19 shows the times and magnitudes of earthquakes with epicentre less than 100km deep in an offshore region west of the main Japanese island of Honshū and south of the northern island of Hokkaidō. The figure shows all 483 earthquakes of magnitude 6 or more on the Richter scale in the period 1885–1980, about 5 tremors per year, in one of the most seismically active areas of Japan. A cumulative plot of the times rises fairly evenly and suggests that the data may be regarded as stationary; we shall assume this below. We take days as the units, giving $t_0 = 35,175$.

This is a *marked point process*, as in addition to the event times there is a *mark* — the magnitude — attached to each event. If we let the times be $0 < t_1 < \cdots < t_n < t_0$ and the associated magnitudes $m_1, \ldots, m_n$, their joint density may be written

$$\prod_{j=1}^{n} f(m_j \mid m_{(j-1)}, t_{(j)}) \prod_{j=1}^{n} f(t_j \mid m_{(j-1)}, t_{(j-1)}), \tag{6.42}$$

where $t_{(j-1)}$ and $m_{(j-1)}$ represent $t_1, \ldots, t_{j-1}$ and $m_1, \ldots, m_{j-1}$. Here we concentrate on inference for the times using the second term, leaving the magnitudes to Examples 10.7 and 10.31. The lower panels of Figure 6.19 show the estimated variance-time curve and conditional intensity function for the times, which are are clearly far from Poisson. The variance-time curve grows more quickly than for a Poisson process, indicating clustering of events, and this is confirmed by the

Figure 6.19 Japanese earthquake data (Ogata, 1988). The upper panel shows the times and magnitudes (Richter scale) of 483 shallow earthquakes. Lower left: estimated variance-time curve for earthquake times, with theoretical line for a Poisson process (solid) and two-sided 95% and 99% pointwise confidence limits (dots). Lower right: estimated conditional intensity, with baseline for Poisson process (solid) and two-sided 95% pointwise confidence limits (dots).

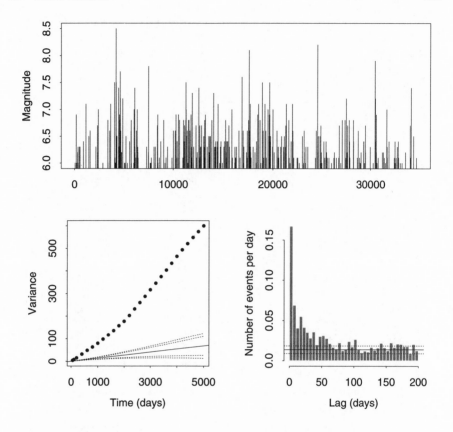

conditional intensity: for about 2–3 months after each shock the probability of another is increased.

One possible model for such data is a *self-exciting process* in which

$$\lambda_{\mathcal{H}}(t) = \mu + \sum_{j:t_j<t} w(t-t_j),$$

where μ is a positive constant and $w(u)$ is non-negative for $u > 0$ and otherwise zero. Here the intensity at any time is affected by the occurrence of previous events; often $w(u)$ is monotonic decreasing, so recent events affect the current intensity more than distant ones. This may be interpreted as asserting that events occur in clusters, whose centres occur as a Poisson process of rate μ. Subsidiary events are then spawned by the increase in intensity that occurs due to the superposition of the $w(t-t_j)$ for previous events. Seismological considerations suggest letting this function depend on m_j also, taking

$$w(t-t_j; m_j) = \frac{\kappa e^{\beta(m_j-6)}}{(t-t_j+\gamma)^\rho}, \quad t > t_j,$$

where $\rho, \gamma, \kappa, \beta, \mu > 0$, with $\beta \doteq 2$. Under this formulation the increase in intensity depends not only on the time since an event but also on its magnitude.

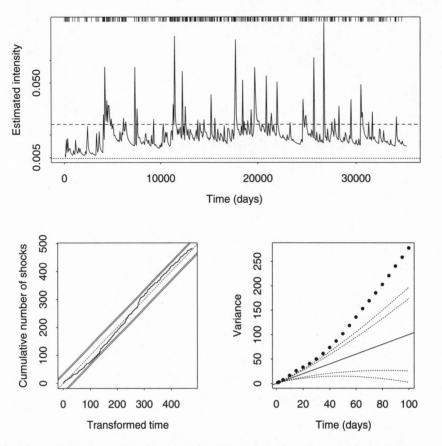

Figure 6.20 Japanese earthquake data fit. The upper panel shows the estimated intensity $\widehat{\lambda}_{\mathcal{H}}(t)$ events/day with $\widehat{\mu}$ (dots) and the mean intensity (dashes). The tick marks at the top of panel show the event times. Lower left: estimated cumulative number of events $\widehat{\Lambda}_{\mathcal{H}}(t_j)$ (solid) and two-sided 95% and 99% overall confidence limits (solid diagonal), based on the Kolmogorov–Smirnov statistic; the dotted line shows perfect fit of the model. Lower right: variance-time function for transformed process $\widehat{\Lambda}_{\mathcal{H}}(t_j)$ (blobs), with baseline for Poisson process (solid) and two-sided 95% and 99% pointwise confidence limits (dots)

The log likelihood (6.41) corresponding to the second term of (6.42) with the self-exciting model is readily obtained. Its maximized value is -2232.01, but this changes only to -2232.25 on fixing $\rho = 1$. With this restriction the estimates and standard errors are $\widehat{\mu} = 0.0049$ (0.0007) events/day, $\widehat{\kappa} = 0.020$ (0.003) events/day, $\widehat{\gamma} = 0.054$ (0.024) days, and $\widehat{\beta} = 1.61$ (0.14). These imply that after an earthquake of size $m_j = 6$, $\lambda_H(t)$ jumps by $\widehat{\kappa}/\widehat{\gamma} \doteq 0.37$ events/day, while a shock of size $m_j = 8$ induces a jump of $\widehat{\kappa}e^{2\widehat{\beta}}/\widehat{\gamma} \doteq 9.2$ events/day. The rate at which clusters arise is about $365\widehat{\mu} \doteq 1.8$ events/year, so each gives rise to a further 3.2 shocks on average.

The top panel of Figure 6.20 shows the fitted intensity $\widehat{\lambda}_{\mathcal{H}}(t)$, with the value of $\widehat{\mu}$ and the mean intensity; note the logarithmic scale. The fitted value is initially low perhaps because of the lack of data before $t = 0$, and it would be preferable to use only a portion of the likelihood, as in Example 6.29. The lower panels show the cumulative intensity for the transformed process $\widehat{\Lambda}_{\mathcal{H}}(t_j)$, which would be a straight line of unit gradient if the model fitted perfectly. The cumulative intensity lies within overall 95% confidence limits and gives no evidence against the model. However the variance-time curve of the transformed times shows clear overdispersion relative to a Poisson process. The data include an unusual series of about 25 large earthquakes in November–December 1938, all occurring in the same region. When these are removed, the remainder have variance-time curve falling within the Poisson limits and the model then seems adequate. ∎

Exercises 6.5

Recall that
$(1 + a/k)^k \to e^a$ as
$k \to \infty$.

1 For a Poisson process on $[0, t_0]$ of constant rate λ, show directly that $N(t_0)$ has a Poisson distribution of mean λt_0 by showing that

$$\Pr\{N(t_0) = m\} \doteq \frac{k!}{m!(k-m)!} \{\lambda \delta t + o(\delta t)\}^m \{1 - \lambda \delta t + o(\delta t)\}^{k-m},$$

where $\delta t = t_0/k$, and letting $k \to \infty$.

2 Check that

$$\int_0^{t_0} dt_n \int_0^{t_n} dt_{n-1} \cdots \int_0^{t_3} dt_2 \int_0^{t_2} dt_1 \, \lambda(t_1) \cdots \lambda(t_n) e^{-\Lambda(t_0)}$$

equals (6.30).

3 Consider a Poisson process of intensity λ in the plane. Find the distribution of the area of the largest disk centred on one point but containing no other points.

4 Show that the time to the rth event in a Poisson process of rate λ has the gamma distribution.

5 If T is the time to the first event in a one-dimensional Poisson process of positive intensity $\lambda(t)$, show that $\Lambda(T)$ has a standard exponential distribution.
Write down an algorithm to generate the points $0 < T_1 < \cdots < T_N < t_0$ of a Poisson process of rate $\lambda(t)$ on $[0, t_0]$. Test it.

Deletion of points of a
process is known as
thinning.

6 Over the centuries natural disasters in a particular country have occurred as a Poisson process of rate $\lambda(t)$. Any disaster at time t is known to have occurred only with probability $\pi(t)$, due to the patchiness of historical records. If records of different disasters are preserved independently, show that the point process of known disasters is Poisson with intensity $\lambda(t)\pi(t)$.

7 Find sequences $\{a_m\} > 0$ and $\{b_m\}$ such that (6.33) holds in the following cases: (i) $1 - F(x) = e^{-x}$ for $x > 0$; (ii) the distribution has a power-law upper tail, $1 - F(x) \sim x^{-\gamma}$, $\gamma > 0$, with $x_0 = \infty$; and (iii) $F(x) = x$ for $0 \le x \le 1$.
In each case give the value of κ and sketch the limiting distribution.

8 Let M_n be the maximum of the random sample $X_1, \ldots, X_n$ from a distribution F, and suppose that the limit

$$\lim_{n \to \infty} \Pr\left(\frac{M_n - b_n}{a_n} \le y\right)$$

is a nondegenerate distribution function, $H(y)$, for some sequences of constants $a_n > 0$ and b_n. Show that

$$\Pr\left(\frac{M_n - a_n}{b_n} \le y\right) = \Pr\left(\frac{M_m - a_n}{b_n} \le y\right)^l,$$

where $n = ml$, and deduce that H must be *max-stable*, that is, for any l there must exist constants c_l and d_l such that $H(y)^l = H(c_l + d_l y)$. Verify that the generalized extreme-value distribution (6.34) is max-stable.

9 Show that the Fisher information for an observation from (6.38) is

$$i(\sigma, \xi) = (1 + 2\xi)^{-1} \begin{pmatrix} 1 & (1+\xi)^{-1} \\ (1+\xi)^{-1} & 2(1+\xi)^{-1} \end{pmatrix}, \quad \xi > -1/2.$$

What happens if $\xi \le -1/2$?

10 (a) If W follows (6.38) and $u > 0$, show that conditional on $W > u$, $W - u$ follows (6.38) with parameters ξ and $\sigma_u = \sigma + \xi u$. Show also that $\mathrm{E}(W - u \mid W > u) = \sigma/(1 - \xi)$, provided $\xi < 1$. What happens if $\xi \ge 1$? And if $\xi \ge 1/2$?

(b) Derive a standard error for (6.40). For what values of ξ is it valid? Explain the saw-tooth form of the mean residual life plot.

(c) Discuss how confidence bands in plots of $\widehat{\xi}_u$ and $\widehat{\sigma}_u - \widehat{\xi}_u u$ against u might be constructed.

11 By reparametrizing (6.38) in terms of $\zeta = \xi/\sigma$ and ξ, show how to obtain maximum likelihood estimates of ξ and σ based on a random sample $w_1, \ldots, w_n$ from G, using only a one-dimensional maximization.

6.6 Bibliographic Notes

A useful general account of stochastic modelling dealing with several of the topics in this chapter is Isham (1991).

There are many books on Markov chains. Cox and Miller (1965), Grimmett and Stirzaker (2001) and Norris (1997) give standard accounts of their probabilistic aspects, while Billingsley (1961) describes inference for them. Guttorp (1995) has a nice blend of probabilistic and statistical considerations. Multi-state modelling, including the use of Markov processes, is discussed in Chapters 5 and 6 of Hougaard (2000). MacDonald and Zucchini (1997) and Künsch (2001) describe inference for hidden Markov processes. Prum *et al.* (1995) describe a systematic approach to finding words in DNA sequences, with further references to this area.

Markov random fields emerged around 1970 as a natural generalization of Markov chains to more complex phenomena, though the Ising and related models had been known to physicists since the 1920s. The key result relating Markov random fields and Gibbs distributions was proved in 1971 by J. M. Hammersley and P. Clifford but not published at that time; Clifford (1990) describes its history and some more recent ideas and gives their version of the proof. A simpler proof was given in the important paper of Besag (1974), which discusses a wide range of topics related to spatial modelling; see Smith (1997). Applications to image analysis were described in Geman and Geman (1984) and Besag (1986), which strongly influenced later work on image analysis; see for example Chellappa and Jain (1993). Applications to point processes are reviewed by Isham (1981), while Kinderman and Snell (1980) give a gentle introduction oriented towards problems of classical physics; see also Brémaud (1999). Sheehan (2000) and Thompson (2001) discuss applications in statistical genetics, with numerous further references.

Graphical models have played an increasingly important role in statistics since about 1980, though similar ideas were used in other fields decades earlier. Edwards (2000) gives an applied account of graphical models with many examples, and includes a description of the software package MIM with which certain families of models can be fitted. Lauritzen (1996) is more mathematical, with details of the necessary graph theory and its statistical application. Whittaker (1990) lies between the two, with a blend of applications and theory, while Cox and Wermuth (1996) give a general view of the subject with some substantial applications. All these books contain references to the primary literature. Those by Lauritzen and Cox and Wermuth describe graphs in which different types of edges appear; see also Wermuth and Lauritzen (1990) and Lauritzen and Richardson (2002).

Graphical representations of probabilistic expert systems are described by Lauritzen and Spiegelhalter (1988) and Spiegelhalter *et al.* (1993), from which Example 6.16 is taken. Pearl (1988), Neopolitan (1990), Almond (1995), Castillo *et al.* (1997), Cowell *et al.* (1999) and Jensen (2001) provide fuller accounts.

There are books on multivariate statistics at all levels and in all styles. Accounts of classical models for multivariate data are Anderson (1958), Mardia *et al.* (1979), and Seber (1985). Chatfield and Collins (1980) is more practical, but all predate the emergence of graphical Gaussian modelling. The bibliographic notes for Chapter 10 give references for discrete multivariate data.

Chatfield (1996), Diggle (1990) and Brockwell and Davis (1996) are standard elementary books on time series, while Brockwell and Davis (1991) is a more advanced treatment. Beran (1994) and Tong (1990) describe respectively series with long-range dependence and nonlinearity. With the growth of financial markets over the last two decades financial time series has become an area of major research effort summarized by Shephard (1996); for longer accounts see Gouriéroux (1997) and Tsay (2002). These references primarily describe modelling in the so-called time domain, in which relationships among the observations themselves are central, but a complementary approach based on frequency analysis is the main focus of Bloomfield (1976), Priestley (1981), Brillinger (1981), and Percival and Walden (1993). This second approach is particularly useful in physical applications.

The Poisson process is a fundamental stochastic model and its probabilistic aspects are described in any of the large number of excellent introductory books on stochastic processes; see for example Grimmett and Stirzaker (2001). There are also various more specialised accounts such as in Rolski *et al.* (1999). Accounts of point process theory are by Cox and Isham (1980) and Daley and Vere-Jones (1988). Cox and Lewis (1966) is a thorough account of inference for one-dimensional data, while spatial point processes are the focus of Diggle (1983). Karr (1991) gives a theoretical account of inference for point processes. Ripley (1981, 1988) and Cressie (1991) are more general accounts of the analysis of spatial data. Point processes based on notions allied to Markov random fields are reviewed by Isham (1981), and a fuller treatment is given by van Lieshout (2000).

Statistics of extremes may be said to have started with Fisher and Tippett (1928), but the first systematic book-length treatment of the subject was Gumbel (1958). Modern accounts from roughly the viewpoint taken here are Smith (1990) and Coles (2001), while Embrechts *et al.* (1997) is a systematic mathematical treatment emphasising applications in finance and insurance. The approach using point processes is described by Smith (1989a). Davison and Smith (1990) give a thorough treatment of threshold methods. Books on probabilistic aspects include Leadbetter *et al.* (1983) and Resnick (1987).

6.7 Problems

1 Dataframe `alofi` contains three-state data derived from daily rainfall over three years at Alofi in the Niue Island group in the Pacific Ocean. The states are 1 (no rain), 2 (up to

	To				To				To		
From	1	2	3	From	1	2	3	From	1	2	3
11	247	86	29	21	86	27	23	31	29	13	8
12	70	32	24	22	29	35	26	32	37	35	18
13	13	16	31	23	17	17	34	33	20	45	59

	To			To			To			To		
From	1	2	3	1	2	3	1	2	3	1	2	3
1	106	34	14	97	29	17	60	24	16	98	39	12
2	41	27	10	32	21	13	27	27	25	36	13	18
3	8	16	15	13	17	32	13	27	52	15	15	25

Table 6.10 Counts for rainfall data at Alofi (Avery and Henderson, 1999). States are 1 (no rain), 2 (up to 5mm rain) and 3 (over 5mm). Upper: transition counts for successive triplets for the entire data. Lower: transition counts for successive pairs for four sub-sequences of length 274.

5mm rain) and 3 (over 5mm). Triplets of transition counts for all 1096 observations are given in the upper part of Table 6.10; its lower part gives transition counts for successive pairs for sub-sequences 1–274, 275–548, 549–822 and 823–1096.

(a) The maximized log likelihoods for first-, second-, and third-order Markov chains fitted to the entire dataset are -1038.06, -1025.10, and -1005.56. Compute the log likelihood for the zeroth-order model, and compare the four fits using likelihood ratio statistics and using AIC. Give the maximum likelihood estimates for the best-fitting model. Does it simplify to a varying-order chain?

(b) Matrices of transition counts $\{n_{irs}\}$ are available for m independent S-state chains with transition matrices $P_i = (p_{irs})$, $i = 1, \ldots, m$. Show that the maximum likelihood estimates are $\widehat{p}_{irs} = n_{irs}/n_{i \cdot s}$, where $\cdot$ denotes summation over the corresponding index. Show that the maximum likelihood estimates under the simpler model in which $P_1 = \cdots = P_m = (p_{rs})$ are $\widehat{p}_{rs} = n_{\cdot rs}/n_{\cdot \cdot s}$. Deduce that the likelihood ratio statistic to compare these models is $2 \sum_{i,r,s} n_{irs} \log(\widehat{p}_{irs}/\widehat{p}_{rs})$ and give its degrees of freedom.

(c) Consider the lower part of Table 6.10. Explain how to use the statistic from (b) to test for equal transition probabilities in each section, and hence check stationarity of the data.

2   The nematode *Steinername feltiae* is a tiny worm used for biological control of mushroom fly larvae. Once one has found and penetrated a larva, it kills it by releasing bacteria, but death is not immediate and other nematodes may also penetrate the larva before it dies. In experiments to assess their effectiveness, m nematodes challenged a single healthy larva. Let $X_t \in \{0, \ldots, m\}$ denote the number of nematodes that have invaded the larva at time t, and let $p_r(t) = \Pr(X_t = r)$, with initial condition $p_0(0) = 1$.

(a) If the invasion process is modelled as a continuous-time Markov process with transition probabilities independent of t, explain why we may write

$$\Pr(X_{t+\delta t} = r + 1 \mid X_t = r) = \lambda_r \delta t + o(\delta t), \quad t \geq 0, \quad r = 0, \ldots, m-1,$$

where $\lambda_m = 0$, and give an interpretation of λ_r. Deduce that

$$\frac{dp_0(t)}{dt} = -\lambda_0 p_0(t), \quad \frac{dp_{r+1}(t)}{dt} = -\lambda_{r+1} p_{r+1}(t) + \lambda_r p_r(t), \quad r = 0, \ldots, m-1.$$

If $\lambda_r = (m - r)\beta$ for some $\beta > 0$, verify that these equations have solution

$$p_t(r) = \binom{m}{r} \{1 - \exp(-\beta t)\}^r \exp(-\beta t)^{m-r},$$

and give its interpretation.

Table 6.11 Numbers of nematodes invading individual fly larvae for various initial numbers of challengers (Faddy and Fenlon, 1999).

Challengers m	Number of fly larvae with $r = 0, \ldots, 10$ invading nematodes											Total
	0	1	2	3	4	5	6	7	8	9	10	
10	1	8	12	11	11	6	9	6	6	2	0	72
7	9	14	27	15	6	3	1	0				75
4	28	18	17	7	3							73
2	44	26	6									76
1	158	60										218

Table 6.12 Numbers of sites showing differences between introns of human and owl monkey insulin genes (Li, 1997, p. 83).

Human	Owl monkey			
	A	C	G	T
A	20	0	0	2
C	0	24	5	1
G	1	5	45	0
T	2	2	0	56

(b) A total of n independent experiments performed with $t = 1$ (in arbitrary units) gave data $(m_1, r_1), \ldots, (m_n, r_n)$ shown in Table 6.11. Thus, for example, of the 72 larvae challenged by 10 nematodes, 1 was not penetrated, 8 were penetrated by just one nematode, 12 were penetrated by two nematodes, and so forth. Show that the corresponding log likelihood may be written as

$$\ell(\beta) = (s_m - s_r)\beta + s_r \log(1 - e^{-\beta}),$$

and deduce that β has maximum likelihood estimate $\widehat{\beta} = \log\{s_m/(s_m - s_r)\}$ with standard error $[s_r/\{s_m(s_m - s_r)\}]^{1/2}$.

(c) Find the values of $\widehat{\beta}$ and their standard errors for models in which the value of β is (i) the same for all m and (ii) different for each m. Discuss which fits the data better, given that the likelihood ratio statistic to compare them equals 11.2.

(d) A different model has $\lambda_r = (m - r) \exp(\gamma_0 + \gamma_1 r)$, so the larva's resistance to penetration changes each time it is invaded. What feature of Table 6.11 suggests that this model might be better? What difficulties would arise in fitting it?

3   One way to estimate the evolutionary distance between species is to identify sections of their DNA which are similar and so must derive from a common ancestor species. If such sections differ at very few sites, the species are closely related and must have separated recently in the evolutionary past, but if the sections differ by more, the species are further apart. For example, data from the first introns of human and owl monkey insulin genes are in Table 6.12. The first row means that there are 20 sites with A on both genes, 0 with A on the human and C on the monkey, and so on. If all the data lay on the diagonal, this section would be identical in both species. Note that even if sites on both genes have the same base, there could have been changes such as (ancestor) A→G→T (human) and (ancestor) A→C→A→T (monkey).

Here is a (greatly simplified) model for evolutionary distance. We suppose that at a time t_0 in the past the two species we now see began to evolve away from a common ancestor species, which had a section of DNA of length n similar to those we now see. Each site on that section had one of the four bases A, C, G, or T, and for each species the base at each site has since changed according to a continuous-time Markov chain with infinitesimal generator

$$
G = \begin{pmatrix} -3\gamma & \gamma & \gamma & \gamma \\ \gamma & -3\gamma & \gamma & \gamma \\ \gamma & \gamma & -3\gamma & \gamma \\ \gamma & \gamma & \gamma & -3\gamma \end{pmatrix},
$$

independent of other sites. That is, the rate at which one base changes into, or *is substituted by*, another is the same for any pair of bases.

In fact substitutions can be of various types, but we do not distinguish them here.

(a) Check that G has eigendecomposition

$$
\frac{1}{4}\begin{pmatrix} 1 & -1 & -1 & -1 \\ 1 & -1 & -1 & 3 \\ 1 & -1 & 3 & -1 \\ 1 & 3 & -1 & -1 \end{pmatrix}\begin{pmatrix} 0 & 0 & 0 & 0 \\ 0 & -4\gamma & 0 & 0 \\ 0 & 0 & -4\gamma & 0 \\ 0 & 0 & 0 & -4\gamma \end{pmatrix}\begin{pmatrix} 1 & 1 & 1 & 1 \\ -1 & 0 & 0 & 1 \\ -1 & 0 & 1 & 0 \\ -1 & 1 & 0 & 0 \end{pmatrix},
$$

find its equilibrium distribution π, and show that the chain is reversible.

(b) Show that $\exp(tG)$ has diagonal elements $(1 + 3e^{-4\gamma t})/4$ and off-diagonal elements $(1 - e^{-4\gamma t})/4$. Use this and reversibility of the chain to explain why the likelihood for γ based on data like those above is proportional to

$$
(1 + 3e^{-8\gamma t_0})^{n-R}(1 - e^{-8\gamma t_0})^R,
$$

where R is the number of sites at which the two sections disagree. Hence find an estimate and standard error for γt_0 for the data above.

(c) Show that for each site, the probability of no substitution on either species in period t is $1 - \exp(-6\gamma t)$, deduce that substitutions occur as a Poisson process of rate 6γ, and hence show that the estimated mean number of substitutions per site for the data above is 0.120.

Discuss the fit of this model.

4   Let $Y_1, \ldots, Y_n$ represent the trajectory of a stationary two-state discrete-time Markov chain, in which

$$
\Pr(Y_j = a \mid Y_1, \ldots, Y_{j-1}) = \Pr(Y_j = a \mid Y_{j-1} = b) = \theta_{ba}, \quad a, b = 1, 2;
$$

note that $\theta_{11} = 1 - \theta_{12}$ and $\theta_{22} = 1 - \theta_{21}$, where θ_{12} and θ_{21} are the transition probabilities from state 1 to 2 and vice versa.

Show that the likelihood can be written in form $\theta_{12}^{n_{12}}(1 - \theta_{12})^{n_{11}}\theta_{21}^{n_{21}}(1 - \theta_{21})^{n_{22}}$, where n_{ab} is the number of $a \rightarrow b$ transitions in $y_1, \ldots, y_n$. Find a minimal sufficient statistic for $(\theta_{12}, \theta_{21})$, the maximum likelihood estimates $\widehat{\theta}_{12}$ and $\widehat{\theta}_{21}$, and their asymptotic variances.

5   Let $Y_{(1)} < \cdots < Y_{(n)}$ be the order statistics of a sample from the exponential density, $\lambda e^{-\lambda y}$, $y > 0$, $\lambda > 0$. Show that for $r = 2, \ldots, n$,

$$
\Pr\left(Y_{(r)} > y \mid Y_{(1)}, \ldots, Y_{(r-1)}\right) = \exp\left\{-\lambda r(y - y_{(r-1)})\right\}, \quad y > y_{(r-1)},
$$

and deduce that the order statistics from a general continuous distribution form a Markov process.

6   Let $\mathcal{G}$ denote an undirected graph with nodes $\mathcal{J}$ and for any $\mathcal{A} \subset \mathcal{J}$ let $\mathrm{cl}(\mathcal{A})$ denote the set $\bigcup_{a \in \mathcal{A}}(\{a\} \cup \mathcal{N}_a)$. Then we can write the local, global and pairwise Markov properties as

   (G) if $\mathcal{A}, \mathcal{B}, \mathcal{D}$ is a triple of disjoint sets such that $\mathcal{D}$ separates $\mathcal{A}$ from $\mathcal{B}$ in $\mathcal{G}$, then $Y_{\mathcal{A}} \perp Y_{\mathcal{B}} \mid Y_{\mathcal{D}}$;
   (L) for any node a, $Y_a \perp Y_{\mathcal{J}-\mathrm{cl}(\{a\})} \mid Y_{\mathcal{N}_a}$;
   (P) if a, b are non-adjacent nodes, then $Y_a \perp Y_b \mid Y_{\mathcal{J}-\{a,b\}}$.

(a) Show that (G) $\Rightarrow$ (L) $\Rightarrow$ (P).

(b) We say that Y satisfies (F) if the density factorizes according to (6.14) and (6.15). Show that (F) $\Rightarrow$ (G). Interpret the Hammersley–Clifford theorem as showing that if in addition (6.12) holds, then (P) $\Rightarrow$ (F).

7   Consider a rectangular grid of pixels with a first-order neighbourhood structure, and denote its random variables by u_{ij}, $i, j = 1, \ldots, m$. Suppose that the observed data are

$y_{ij} = u_{ij} + \varepsilon_{ij}$ where $\varepsilon_{ij} \overset{\text{iid}}{\sim} N(0, \sigma^2)$. Thus the u_{ij} are observed with noise. Give the moral graph for the u_{ij} and y_{ij}. Hence show that the local characteristics $f(u_{ij} \mid y, u_{-ij})$ depends on the neighbouring us and y_{ij} and find $f(u_{ij} \mid y, u_{-ij})$ when the u_{ij} follow an Ising model.

8 (a) Suppose that conditional on $U = u$, $Y \sim N_p(\mu, \nu u^{-1}\Omega)$, where $u \sim \chi_\nu^2$. Show that the marginal density of Y is multivariate t,

$$f(y; \mu, \Omega) = \frac{\Gamma\left(\frac{p+\nu}{2}\right) |\Omega|^{-1/2}}{(\pi \nu)^{p/2} \Gamma\left(\frac{\nu}{2}\right)} \{1 + (y - \mu)^{\text{T}} \Omega^{-1}(y - \mu)/\nu\}^{-(p+\nu)/2},$$

and establish that $E(U \mid Y = y) = (\nu + p)/\{\nu + (y - \mu)^{\text{T}}\Omega^{-1}(y - \mu)\}$.
(b) Use this as the basis for an EM algorithm for estimation of μ and Ω, extending that of Problem 5.18.
(c) The density of Y is called elliptical because of the shape of its contours. Other such densities may be produced by supposing that $Y \sim N_p(\mu, u^{-1}\Omega)$ conditional on $U = u$ and letting $U \sim g$, where g has support in the positive half-line. What changes to the algorithm in (b) are then needed to produce an EM algorithm for estimation of μ and Ω? (Section 5.5.2)

9 Show that the MA(1) models $Y_t = \varepsilon_t + \beta\varepsilon_{t-1}$ and $Y_t = \varepsilon_t + \beta^{-1}\varepsilon_{t-1}$ have the same correlations and deduce that they are indistinguishable from their correlograms alone. If $Y_t = (1 + \beta B)\varepsilon_t$ in terms of the backshift operator B, show that ε_t may be expressed as a linear combination of $Y_t, Y_{t-1}, \ldots$ in which the infinite past has no effect only if $|\beta| < 1$. The ARMA process $a(B)Y_t = b(B)\varepsilon_t$ is said to be invertible if the zeros of the polynomial $b(z)$ all lie outside the unit disk. Show that the MA(1) process is invertible only if $|\beta| < 1$. Compare this with the condition for stationarity of the AR(1) model. Discuss.

10 Show that strict stationarity of a time series $\{Y_j\}$ means that for any r we have

$$\text{cum}(Y_{j_1}, \ldots, Y_{j_r}) = \text{cum}(Y_0, \ldots, Y_{j_r - j_1}) = \kappa^{j_2 - j_1, \ldots, j_r - j_1},$$

This condition applies to many common models, but excludes those where variables far apart are highly correlated.

say. Suppose that $\{Y_j\}$ is stationary with mean zero and that for each r it is true that $\sum_u |\kappa^{u_1, \ldots, u_{r-1}}| = c_r < \infty$. The rth cumulant of $T = n^{-1/2}(Y_1 + \cdots + Y_n)$ is

$$\text{cum}\{n^{-1/2}(Y_1 + \cdots + Y_n)\} = n^{-r/2} \sum_{j_1, \ldots, j_r} \text{cum}(Y_{j_1}, \ldots, Y_{j_r})$$

$$= n^{-r/2} \sum_{j_1=1}^{n} \sum_{j_2, \ldots, j_r} \kappa^{j_2 - j_1, \ldots, j_r - j_1}$$

$$= n \times n^{-r/2} \sum_{j_2, \ldots, j_r} \kappa^{j_2 - j_1, \ldots, j_r - j_1}$$

$$\leq n^{1-r/2} \sum_{j_2, \ldots, j_r} |\kappa^{j_2 - j_1, \ldots, j_r - j_1}| \leq n^{1-r/2} c_r.$$

Justify this reasoning, and explain why it suggests that T has a limiting normal distribution as $n \to \infty$, despite the dependence among the Y_j.
Obtain the cumulants of T for the MA(1) model, and convince yourself that your argument extends to the MA(q) model.
Can you extend the argument to arbitrary linear combinations of the Y_j?

11 (a) Check that the Gumbel distribution arises from (6.34) in the limit as $\xi \to 0$.
(b) Derive the densities for (6.34) and the Gumbel distribution, and plot them for $\xi = -1$, $-0.5, 0, 0.5$, and 1. Which do you think is most plausible for extreme rainfall, for high tides, and for the fastest times to run a mile?
(c) Write a function that generates random samples from (6.34) by inversion.
(d) Show that the Gumbel plotting positions are $-\log[-\log\{1 - i/(n + 1)\}]$ and use these and your simulation routine to see how easy it is to detect departures from $\xi = 0$ in random samples of size $n = 40$ with $\xi = -0.3, 0.3$. Try varying ξ and n, and write a brief account of your conclusions.

12   Consider a stationary point process and denote the numbers of counts in successive inter-
      vals $(k\tau, (k + 1)\tau]$ of length τ by N_k, where $k = \ldots, -1, 0, 1, \ldots$. Let $\text{var}(N_0) < \infty$ and
      set $\gamma_j = \text{cov}(N_0, N_j)$.
      (a) Show that $\{N_j\}$ is a stationary time series and deduce that

      $$\text{var}\{N(m\tau)\} = m\gamma_0 + 2\sum_{j=1}^{m-1}(m - j)\gamma_j, \quad m = 1, 2, \ldots.$$

      Hence explain how the variance-time curve $V(t)$ for $t = \tau, 2\tau, \ldots$ may be estimated
      using the empirical covariances $\widehat{\gamma}_j$ of counts of data observed over $(0, t_0]$. Call the
      estimator $\widehat{V}(t)$.
      (b) If $k\tau = t_0$ and the data follow a Poisson process of rate λ, then

      $$\text{E}(\widehat{\gamma}_j) \doteq \begin{cases} (k - 1)\lambda\tau/k, \\ 0, \end{cases} \quad \text{var}(\widehat{\gamma}_j) = \begin{cases} \lambda\tau(2\lambda\tau + 1)/k + o(k^{-1}), & j = 0, \\ (\lambda\tau)^2/k + o(k^{-1}), & \text{otherwise}, \end{cases}$$

      while $\text{cov}(\widehat{\gamma}_i, \widehat{\gamma}_j) = o(k^{-1})$ when $i \neq j$. Hence show that in this case $\text{E}\{\widehat{V}(t)\} \doteq (1 -$
      $t/t_0)V(t)$ and

      $$\text{var}\{\widehat{V}(t)\} = \{2/3 + 4/(3m)\}(\lambda t)^2(t/t_0) + (\lambda t)(t/t_0) + o(\tau/t_0),$$

      where $t = m\tau$.
      (c) Explain the construction of the lower left panel of Figure 6.19.

13   Sampling of point processes is not straightforward. If the process is running already
      and sampling begins at an arbitrary time origin, then this origin is likely to fall into
      an interval that is longer than is typical, and this *length-biased sampling* has knock-
      on effects for subsequent intervals unless their lengths are independent. Suppose that a
      very long stretch of n intervals is available from a stationary process with mean interval
      length μ and marginal density $f(y)$ for times between events, into which the origin falls
      randomly. Of the total length $n\mu$ of the intervals, a length $nf(y) \times y$ will be taken by
      intervals of length y. Explain why the probability that the origin falls into one of these
      is $g(y)dy = nyf(y)dy/(n\mu)$, and hence show that the length of the selected interval has
      probability density g.
      Now consider the *forward recurrence time* to the next event starting from the origin. The
      origin having fallen uniformly at random into an interval of length y, the conditional
      density of its position within that interval is y^{-1}. Show that the forward recurrence time
      has density

      $$\int_x^\infty y^{-1}g(y)\,dy = \mu^{-1}\mathcal{F}(x),$$

      where $\mathcal{F}$ is the survivor function of f, and find the density of the *backward recurrence
      time* to the point before the origin.
      Show that in a homogeneous Poisson process of rate λ the interval into which the origin
      falls has density $\lambda^2 ye^{-\lambda y}$, $y > 0$, and that the forward and backward recurrence times are
      both exponential variables. Explain why these results are obvious intuitively.

14   A Poisson process of rate $\lambda(t)$ on the set $S \subset \mathbb{R}^k$ is a collection of random points with
      the following properties (among others):
      •   the number of points N_A in a subset A of S has the Poisson distribution with mean
           $\Lambda(A) = \int_A \lambda(t)\,dt$;
      •   given $N_A = n$, the positions of the points are sampled randomly from the density
           $\lambda(t)/\int_A \lambda(s)\,ds, t \in A$.
      (a) Assuming that you have reliable generators of $U(0, 1)$ and Poisson variables, show
      how to generate the points of a Poisson process of constant rate λ on the interval $[0, t_0]$.
      (b) Let $t = (x, y) \in \mathbb{R}^2, \eta, \xi \in \mathbb{R}, \tau > 0, \lambda(x, y) = \tau^{-1}\{1 + \xi(y - \eta)/\tau\}^{-1/\xi - 1}$. Give an
      algorithm to generate realisations from the Poisson process with rate $\lambda(x, y)$ on

      $$S = \{(x, y) : 0 \leq x \leq 1, y \geq u, \lambda(x, y) > 0\}.$$

Table 6.13 Times (days) between successive failures of a piece of software developed as part of a large data system (Jelinski and Moranda, 1972). The software was released after the first 31 failures. The last three failures occurred after release. The data are to be read across rows.

9	12	11	4	7	2	5	8	5	7	1	6	1	9	4	1	3
3	6	1	11	33	7	91	2	1	87	47	12	9	135	258	16	35

15 Show that the likelihood for data $(t_1, y_1), \ldots, (t_n, y_n)$ observed in $[0, t_0] \times [u, \infty)$ and with intensity (6.36) is

$$\prod_{j=1}^{n} \tau^{-1} \left(1 + \xi \frac{y_j - \eta}{\tau} \right)^{-1/\xi - 1} \times \exp \left\{ -t_0 \left(1 + \xi \frac{u - \eta}{\tau} \right)^{-1/\xi} \right\}.$$

Show that this may be reparametrized to give (6.39) and that this is the log likelihood corresponding to a decomposition

$$\Pr(N = n; \lambda) \times \prod_{j=1}^{n} g(w_j; \xi, \sigma).$$

Give the distributions of N, of the W_j, and of $Y = \max(W_1, \ldots, W_N)$. Surprised?

16 A computer program has an unknown number of bugs m. Each bug causes the program to crash, and is then located and (instantaneously!) removed. If the times at which the m failures occur are independent exponential variables with common mean β^{-1}, and if m is Poisson with mean μ/β, then show that

$$\Pr\{N(t) = 0\} = \exp\left\{-\mu(1 - e^{-\beta t})/\beta\right\}, \quad t \geq 0.$$

(a) Deduce that the times of crashes follow a Poisson process of rate $\mu e^{-\beta t}$. Show that the likelihood when failures occur at times $0 \leq t_1 < \cdots < t_n \leq t_0$ is

$$L(\mu, \beta) = \mu^n \exp\left\{-\beta \sum_{j=1}^{n} t_j - \mu\beta^{-1}\left(1 - e^{-\beta t_0}\right)\right\},$$

and that this is an exponential family model.
(b) Reliability growth occurs if $\beta > 0$. Show that a test for this may be based on the conditional distribution of $S = \sum T_j$ given that n failures have occurred in $[0, t_0]$, and that if $\beta = 0$, $E(S) = nt_0/2$ and $\text{var}(S) = nt_0^2/12$. Suggest how to perform such a test.
(c) We now treat m as a unknown parameter and aim to estimate it. Show that

$$L(m, \beta) = \frac{m!}{(m - n)!}\beta^n \exp\left\{-\beta t_0(m + s/t_0 - n)\right\}, \quad \beta > 0, m = n, n + 1, \ldots,$$

and hence find the profile log likelihood $\ell_p(m)$ for m.
(d) The code below plots $\ell_p(m)$ after the first r failures of the data in Table 6.13. Try varying r up to 30, and observe the shapes taken by the profile log likelihood.

```
y <- c(9,12,11,4,7,2,5,8,5,7,1,6,1,9,4,1,3,3,
6,1,11,33,7,91,2,1,87,47,12,9,135,258,16,35)
L <- function(m,n,s) lgamma(m+1)-lgamma(m-n+1) - n*log(m-n+s)
r <- 20   # just take data up to time of rth failure
y <- cumsum(y[1:r])
s <- sum(y)/y[r]
x <- r:(r+100)
plot(x,L(x,r,s))  # plot log likelihood
```

What problems do you see with likelihood inference for this model?
The software was released after 31 failures. Give the maximum likelihood estimate of m at that point, and its confidence interval, if possible.
(Section 6.5.1, Jelinski and Moranda, 1972)

7

Estimation and Hypothesis Testing

Chapter 4 introduced likelihood and explored associated concepts such as likelihood ratio statistics and maximum likelihood estimators, which were then extensively used for inference in Chapters 5 and 6. In this chapter we turn aside from the central theme of the book and discuss some more theoretical topics. Estimation is a fundamental statistical activity, and in Section 7.1 we consider what properties a good estimator should have, including a brief discussion of nonparametric density estimators and the mathematically appealing topic of minimum variance unbiased estimation. One of the most important approaches to constructing estimators is as solutions to systems of estimating equations. In Section 7.2 we discuss the implications of this, showing how it complements minimum variance unbiased estimation, and seeing its implications for robust estimation and for stochastic processes. We then give an account of some of the main ideas underlying another major statistical activity, the testing of hypotheses, discussing the construction of tests with good properties, and making the connection to estimation.

7.1 Estimation

7.1.1 Mean squared error

Suppose that we wish to estimate some aspect of a probability model $f(y)$. In principle we might try and estimate almost any feature of f, but we largely confine ourselves to estimation of the unknown parameter θ or a function of it $\psi(\theta)$ in a parametric model $f(y; \theta)$. Suppose that our data Y comprise a random sample $Y_1, \ldots, Y_n$ from f, and let the statistic $T = t(Y)$ be an estimator of $\psi(\theta)$. We say that T is *unbiased* for $\psi(\theta)$ if $\mathrm{E}(T) = \psi(\theta)$ for all θ, and define the *bias* of T to be $\mathrm{E}(T) - \psi(\theta)$. Large bias means that the long-run average value of T lies far from $\psi(\theta)$, and this is undesirable. The *mean squared error* of T is the expected squared distance between T and its estimand, which turns out to equal the sum of the variance and squared bias:

$$\mathrm{E}[\{T - \psi(\theta)\}^2] = \mathrm{E}[\{T - \mathrm{E}(T) + \mathrm{E}(T) - \psi(\theta)\}^2]$$
$$= \mathrm{var}(T) + \{\mathrm{E}(T) - \psi(\theta)\}^2, \tag{7.1}$$

because $E\left[\{T - E(T)\}\{E(T) - \psi(\theta)\}\right] = 0$. Mean squared error is a common measure of how well T estimates $\psi(\theta)$. The decomposition (7.1) is useful because in practice it helps to know if a large mean squared error is due to a large bias or large variance or both.

Example 7.1 (Normal variance) Let $Y_1, \ldots, Y_n$ be a random sample from the normal distribution with mean μ and variance σ^2. Then σ^2 has maximum likelihood estimator $\widehat{\sigma}^2 = n^{-1} \sum_j (Y_j - \overline{Y})^2 \stackrel{D}{=} n^{-1}\sigma^2 V$, where V has a χ^2_{n-1} distribution, by (3.15). As $E(V) = (n - 1)$ and $\text{var}(V) = 2(n - 1)$,

$$E(\widehat{\sigma}^2) = n^{-1}(n - 1)\sigma^2, \quad \text{var}(\widehat{\sigma}^2) = n^{-2}2(n - 1)\sigma^4.$$

Hence $\widehat{\sigma}^2$ has a negative bias of

$$E(\widehat{\sigma}^2) - \sigma^2 = \frac{n - 1}{n}\sigma^2 - \sigma^2 = -\frac{\sigma^2}{n}$$

and mean squared error

$$2\frac{n - 1}{n^2}\sigma^4 + \left(-\frac{\sigma^2}{n}\right)^2 = \frac{2n - 1}{n^2}\sigma^4.$$

The usual estimator is $S^2 = (n - 1)^{-1} \sum_j (Y_j - \overline{Y})^2 \stackrel{D}{=} (n - 1)^{-1}\sigma^2 V$, so

$$E(S^2) = (n - 1)^{-1}\sigma^2(n - 1) = \sigma^2, \quad \text{var}(S^2) = 2(n - 1)\sigma^4(n - 1)^{-2} = \frac{2}{n - 1}\sigma^4.$$

Hence S^2 is unbiased, but its mean squared error is greater than that of $\widehat{\sigma}^2$ because $(2n - 1)/n^2 < 2/(n - 1)$ for all $n > 1$. ∎

Both bias and variance are $O(n^{-1})$ in this example, so at least for large n the contribution that bias squared makes to mean squared error is negligible compared to that of variance. This often occurs with parametric estimators, as the following argument suggests. Suppose that the data are a random sample from a distribution with mean μ, variance σ^2, third cumulant κ_3, and well-behaved higher cumulants. Then (2.32) implies that their average $\overline{Y}$ has mean μ, variance σ^2/n, and third cumulant $\kappa_3/n^{3/2}$. Suppose also that the estimator of a scalar parameter $\psi = t(\mu)$ may be written as $\widehat{\psi} = t(\overline{Y})$, as is true for natural exponential family models, for example; see Section 5.2. Then under mild conditions, Taylor series expansion of $t(\overline{Y})$ about μ gives

$$\widehat{\psi} = t(\mu) + (\overline{Y} - \mu)t'(\mu) + \frac{1}{2}(\overline{Y} - \mu)^2 t''(\mu) + \frac{1}{6}(\overline{Y} - \mu)^3 t'''(\mu) + \cdots,$$

so

$$E(\widehat{\psi}) = t(\mu) + \frac{1}{2}n^{-1}\sigma^2 t''(\mu) + \frac{1}{6}n^{-3/2}\kappa_3 t'''(\mu) + \cdots,$$

showing that the bias of $\widehat{\psi}$ is of order n^{-1}. Thus for asymptotic purposes we can typically compare two parametric estimators in terms of their variances, as squared bias is of smaller asymptotic order. In finite samples the issue is less clear, because it is not useful to be told that $E(T) - \psi \doteq a/n$ and $\text{var}(T) \doteq b/n$ without some idea of a and b: if $a^2/n^2 \gg b/n$ for all values of n likely to be met in a particular application, then the bias term in (7.1) predominates and comparison purely in terms of variances is unhelpful. Nonetheless it is natural to ask if there is a lower bound to the variance of estimators of ψ.

Cramér–Rao lower bound

Suppose that the density $f(y;\theta)$ is regular for maximum likelihood estimation of the scalar parameter θ. If T is an unbiased estimator of a scalar $\psi = \psi(\theta)$, then under mild conditions and for all θ,

$$\text{var}(T) \geq \frac{(d\psi/d\theta)^2}{I(\theta)}, \qquad (7.2)$$

where $I(\theta)$ is the expected information in the sample. The right-hand side of (7.2) is the *Cramér–Rao lower bound*. It follows that if we can find an unbiased estimator that attains equality or near-equality in (7.2), we need search no further: no unbiased estimator could do better. We might also hope that when T has a small bias and its variance is close to the lower bound, it will be difficult to ameliorate.

To establish (7.2), note that as T is unbiased and $f(y;\theta)$ is a density,

$$E(T) = \int t(y) f(y;\theta)\, dy = \psi(\theta), \qquad \int f(y;\theta)\, dy = 1,$$

for all θ. If the order of integration and differentation can be interchanged, differentation of these equations with respect to θ gives

$$\frac{d\psi}{d\theta} = \int t(y) \frac{df(y;\theta)}{d\theta}\, dy = \int t(y) \frac{d\log f(y;\theta)}{d\theta} f(y;\theta)\, dy = E(TU),$$

where $U = d\log f(Y;\theta)/d\theta$ is the score statistic, and

$$0 = \int \frac{df(y;\theta)}{d\theta}\, dy = \int \frac{d\log f(y;\theta)}{d\theta} f(y;\theta)\, dy = E(U).$$

Hence $\text{cov}(T, U) = E(TU) - E(T)E(U) = d\psi/d\theta$. Moreover

$$\text{var}(U) = E(U^2) = E\left\{ -\frac{d^2 \log f(Y;\theta)}{d\theta^2} \right\} = I(\theta)$$

by (4.33). But the Cauchy–Schwarz inequality (Exercise 2.2.3) gives

$$\text{cov}(T, U)^2 \leq \text{var}(T)\text{var}(U),$$

or $(d\psi/d\theta)^2 \leq \text{var}(T)I(\theta)$, which entails (7.2). This will apply if maximum likelihood estimation is regular, for example. Equality in (7.2) only occurs when there is linear dependence between T and U, so that $T = b_1(\theta) + b_2(\theta)U$ for all θ and some constants $b_1(\theta)$ and $b_2(\theta) \neq 0$.

The inverse Fisher information for ψ is $I(\psi)^{-1} = (d\psi/d\theta)^2 I(\theta)^{-1}$, which equals the right-hand side of (7.2) (Problem 4.2). Hence the Cramér–Rao lower bound for estimation of ψ is $I(\psi)^{-1}$, for any sample size. However (4.26) implies that in regular cases, the large-sample distribution of the maximum likelihood estimator, $\widehat{\psi}$, is normal

Harald Cramér (1893–1985) worked all his life at the University of Stockholm, researching in biochemistry and number theory before turning to probability and statistics. He was the first Swedish professor of mathematical statistics and actuarial science, and made fundamental contributions to both subjects. His masterpiece, *Mathematical Methods of Statistics*, was written during World War II, when he was largely scientifically isolated. Calyampudi Radhakrishnan Rao (1920–) works at Pennsylvania State University. See DeGroot (1987b) for an account of his life and work.

Table 7.1 Cramér–Rao lower bound for estimation of mean of a log-normal sample of size n, when $\mu = 0$ and $\sigma = 1.5$, with properties of estimators $T = \overline{Y}$ and $\widehat{\psi} = \exp(\widehat{\sigma}^2/2)$.

n	5	10	20	40	80	160	320
Cramér–Rao lower bound	4.80	2.40	1.20	0.60	0.30	0.15	0.08
var(T)	16.11	8.05	4.03	2.01	1.01	0.50	0.25
var($\widehat{\psi}$)	296.36	7.08	1.91	0.75	0.33	0.16	0.08
Bias of $\widehat{\psi}$ ($\times 10^{-2}$)	137.73	49.66	21.83	10.30	5.01	2.47	1.23
Mean squared error of $\widehat{\psi}$	298.26	7.32	1.96	0.76	0.34	0.16	0.08

with mean ψ and variance $I(\psi)^{-1}$. Thus $\widehat{\psi}$ is asymptotically unbiased and attains the Cramér–Rao lower bound as $n \to \infty$: it has asymptotically the smallest possible variance among unbiased estimators. Such an estimator is said to be *efficient*. This suggests that the *asymptotic relative efficiency* of T be defined as

$$\frac{\text{var}(\widehat{\psi})}{\text{var}(T)} = \frac{(d\psi/d\theta)^2}{I(\theta)\text{var}(T)},$$

generalizing (4.19).

In regular cases $\text{var}(\widehat{\psi}) = O(n^{-1})$, where n is sample size, and it is impossible to find an unbiased estimator of ψ with variance of smaller order than this. If in addition the observations are identically distributed, the Cramér–Rao lower bound for θ is $1/\{ni(\psi)\}$, where $i(\psi)$ is the information in a single observation.

Example 7.2 (Log-normal mean) A log-normal random variable Y may be expressed as $\exp(\mu + \sigma Z)$, where Z is a standard normal variable; thus $X = \log Y$ has the $N(\mu, \sigma^2)$ distribution. The mean and variance of Y are (Problem 3.5)

$$\psi = e^{\mu + \sigma^2/2}, \qquad e^{2\mu + \sigma^2}\left(e^{\sigma^2} - 1\right).$$

If it is known that $\mu = 0$, then the maximum likelihood estimator of σ^2 based on a random sample $Y_1, \ldots, Y_n$ is $\widehat{\sigma}_0^2 = n^{-1}\sum X_j^2$, where $X_j = \log Y_j$, and two possible estimators for ψ are $T = \overline{Y}$ and $\widehat{\psi} = \exp(\widehat{\sigma}_0^2/2)$. Now $\widehat{\sigma}_0^2 \overset{D}{=} \sigma^2 V/n$, where V has a chi-squared distribution on n degrees of freedom, so

$$\text{E}(\widehat{\psi}^r) = \text{E}\{\exp(r\sigma^2 V/2n)\} = (1 - r\sigma^2/n)^{-n/2}, \quad n = 1, 2, \ldots,$$

because V has moment-generating function $(1 - 2t)^{-n/2}$. This enables exact calculation of the bias, variance and mean squared error of $\widehat{\psi}$.

Evidently T is an unbiased estimator of ψ with variance and mean squared error both equal to $n^{-1}\exp(\sigma^2)\{\exp(\sigma^2) - 1\}$. As $I(\sigma^2) = \sigma^4/(2n)$ and $d\psi/d\sigma^2 = \frac{1}{2}\psi$, the Cramér–Rao lower bound for ψ is $\sigma^4 \exp(\sigma^2)/(2n)$.

The first three lines of Table 7.1 give values of the lower bound, var(T) and var($\widehat{\psi}$) for various n when $\sigma = 1.5$, so $\psi = \exp(\sigma^2/2) = 9.49$. For $n \geq 80$ the bound is effectively attained by $\widehat{\psi}$, whose bias is always small compared to its variance. For $n < 80$ the bias and variance of $\widehat{\psi}$ decrease rather faster than their asymptotic rate n^{-1}. Even when $n = 5$ the contribution to mean squared error from bias is very small. The unbiased estimator T is much more efficient when $n = 5$, but otherwise is beaten by $\widehat{\psi}$; its asymptotic relative efficiency is

$$\frac{\text{var}(\widehat{\psi})}{\text{var}(T)} = \frac{\sigma^4 e^{\sigma^2}}{2e^{\sigma^2}\left(e^{\sigma^2} - 1\right)} = \frac{\sigma^4}{2\left(e^{\sigma^2} - 1\right)} \doteq 0.3$$

when $\sigma = 1.5$, close to the variance ratio in the last columns. ∎

Example 7.3 (Uniform distribution)   Let $Y_1, \ldots, Y_n$ be a random sample from the uniform distribution on $(0, \theta)$. The likelihood for θ is

$$L(\theta) = \begin{cases} \theta^{-n}, & 0 \leq Y_1, \ldots, Y_n \leq \theta, \\ 0, & \text{otherwise,} \end{cases}$$

or equivalently and in terms of the largest order statistic $Y_{(n)}$,

$$L(\theta) = \begin{cases} \theta^{-n}, & 0 \leq Y_{(n)} \leq \theta, \\ 0, & \text{otherwise,} \end{cases}$$

a sketch of which shows that $\widehat{\theta} = Y_{(n)}$. Hence $\widehat{\theta}$ has distribution

$$\text{Pr}(\widehat{\theta} \leq u) = \begin{cases} 0, & u < 0, \\ (u/\theta)^n, & 0 \leq u < 1, \\ 1, & 1 \leq u, \end{cases}$$

and it is straightforward to see that

$$\text{E}(\widehat{\theta}) = \frac{n\theta}{n+1}, \quad \text{var}(\widehat{\theta}) = \frac{n\theta^2}{(n+1)^2(n+2)}.$$

This variance is $O(n^{-2})$, suggesting a potential problem with the Cramér–Rao lower bound. In fact in this case

$$\frac{d}{d\theta}\int t(y)f(y;\theta)\,dy = \frac{d}{d\theta}\int_0^\theta \frac{t(y)}{\theta}\,dy = \frac{t(\theta)}{\theta} - \int_0^\theta \frac{t(y)}{\theta^2}\,dy \neq \int t(y)\frac{df(y;\theta)}{d\theta}\,dy,$$

because the limit of the integral depends on θ. Here the model is non-regular and the lower bound does not apply. ∎

The lower bound extends to the case where θ has dimension p and $\psi = \psi(\theta)$ has dimension $q \leq p$. Suppose that the $q \times 1$ statistic T is an unbiased estimator of ψ, with $q \times q$ covariance matrix $\text{var}(T)$. Then $\text{E}(T) = \psi(\theta)$ for all θ, so $\partial\psi/\partial\theta^{\text{T}}$ is a $q \times p$ matrix. An argument analogous to that on page 302 shows that the $q \times q$ matrix

$$\text{var}(T) - \frac{\partial\psi}{\partial\theta^{\text{T}}}I(\theta)^{-1}\frac{\partial\psi^{\text{T}}}{\partial\theta}$$

is positive semi-definite. If ψ is scalar, it follows that

$$\text{var}(T) \geq \frac{\partial\psi}{\partial\theta^{\text{T}}}I(\theta)^{-1}\frac{\partial\psi^{\text{T}}}{\partial\theta}, \tag{7.3}$$

which extends (7.2).

Example 7.4 (Log-normal mean)   Expression (4.18) implies that the inverse Fisher information for the parameters of a normal distribution with parameters $\theta = (\mu, \sigma^2)^{\text{T}}$

Figure 7.1 Unbiased
(left) and maximum
likelihood estimates
(right) of log-normal
mean based on samples of
size n when $\mu = 0$,
$\sigma = 1.5$. The horizontal
lines show the target
parameter ψ. The white
band in the centre of the
boxplots indicates the
median of the simulated
values.

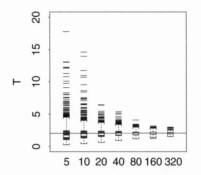
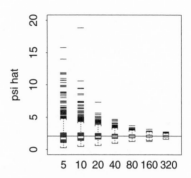

based on a random sample $X_1, \ldots, X_n$ is

$$I(\mu, \sigma^2)^{-1} = \begin{pmatrix} \sigma^2/n & 0 \\ 0 & 2\sigma^4/n \end{pmatrix}.$$

The maximum likelihood estimators are $\widehat{\mu} = \overline{X}$ and $\widehat{\sigma}^2 = n^{-1} \sum (X_j - \overline{X})^2$. Let $\psi = \exp(\mu + \sigma^2/2)$ denote the mean of the log-normal variable $Y = \exp(X)$. Then

$$\frac{\partial \psi}{\partial \theta} = \begin{pmatrix} e^{\mu + \sigma^2/2} \\ \frac{1}{2}e^{\mu + \sigma^2/2} \end{pmatrix}$$

and the Cramér–Rao lower bound (7.3) for ψ is $n^{-1}(\sigma^2 + \sigma^4/2)e^{2\mu + \sigma^2}$; this is also the asymptotic variance of the maximum likelihood estimator $\widehat{\psi} = \exp(\widehat{\mu} + \widehat{\sigma}^2/2)$. As in Example 7.2, the exact bias and variance of $\widehat{\psi}$ may be obtained explicitly (Problem 7.1).

Figure 7.1 shows 1000 simulated values of T and $\widehat{\psi}$ for various n. Their appreciable skewness when n is small shows that their distributions are then far from normal and calls into question our use of bias and variance to compare them. As $\psi > 0$, a measure of relative error such as $\mathrm{E}\{(T - \psi)^2\}/\psi^2$ might be preferable. ∎

Mean squared error is a useful measure with which to compare estimators, but it has the disadvantage of being tied to a particular scale. Thus even if T is preferable to T' as an estimator of ψ, $g(T)$ need not be better than $g(T')$ when estimating $g(\psi)$. In particular, if T is unbiased, then $g(T)$ is unbiased only when g is linear. This is not critical when interest focuses on a quantity for which transformations are irrelevant, but is awkward otherwise.

7.1.2 Kernel density estimation

The examples above suggest that estimators in parametric problems can often be compared in terms of their variances, bias being relatively unimportant. Bias and variance play more balanced roles in other contexts, as we now illustrate with a discussion of nonparametric density estimation. We also take the opportunity of introducing cross-validation, which plays a role later.

The elementary estimator of an unknown density $f(y)$, the histogram, has a number of drawbacks. In addition to often looking rather rough due to the use of bins with

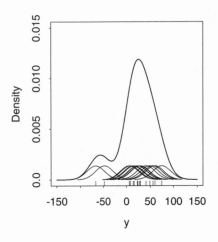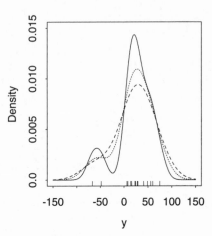

Figure 7.2 Kernel density estimates for maize data. Left: construction of kernel estimate (heavy) as sum of 15 scaled normal densities centred at the y_j, with $h = 19.5$. Right: density estimates with $h = 13.3$ (solid), $h = 23.2$ (dots) and $h = 30$ (dashes).

sharp edges, its appearance depends heavily on the placing of the bin boundaries and on bin width. Slight changes to the boundaries can give strikingly different histograms for the same data, particularly with small samples, and this is clearly undesirable.

An alternative approach is based on a kernel function $w(y)$, which is a symmetric probability density with mean zero and unit variance, descending smoothly to zero as $|y| \to \infty$. An example is the standard normal density, but there are many other possibilities. The *kernel density estimator* based on a random sample $Y_1, \ldots, Y_n$ from f is

$$\widehat{f}_h(y) = \frac{1}{nh} \sum_{j=1}^{n} w\left(\frac{y - Y_j}{h}\right),$$ (7.4)

where $h > 0$ is a bandwidth. Figure 7.2 shows how this is constructed for the data in the rightmost column of Table 1.1: the value of $\widehat{f}_h$ at y is obtained by summing contributions from densities with standard deviations h and centred at each of the Y_j, with those closest to y contributing most. The estimate depends on both bandwidth and kernel, but the choice of h is much the more important. When n is large, a smaller bandwidth can be chosen, so that the contributions are more localized. Such estimators are commonly available in statistical packages and are widely used in applications.

To find the mean and variance of $\widehat{f}_h(y)$, note that

$$E\left\{\frac{1}{nh} \sum_{j=1}^{n} w\left(\frac{y - Y_j}{h}\right)\right\} = E\left\{h^{-1} w\left(\frac{y - Y}{h}\right)\right\}.$$

Symmetry of the kernel gives

$$h^{-1} \int w\left(\frac{y - x}{h}\right) f(x)\, dx = h^{-1} \int w\left(\frac{x - y}{h}\right) f(x)\, dx = \int w(u) f(y + hu)\, du,$$

and Taylor series expansion for small h gives   Here $f'(y) = df(y)/dy$, and so forth.

$$\int w(u)\left\{f(y) + huf'(y) + \frac{1}{2}h^2 u^2 f''(y) + \cdots\right\} du = f(y) + \frac{1}{2}h^2 f''(y) + O(h^4),$$ (7.5)

because of the symmetry and moment properties of w. Thus $\widehat{f}_h(y)$ has bias $\frac{1}{2}h^2 f''(y) + O(h^4)$. The variance of $\widehat{f}_h(y)$ is

$$\mathrm{var}\{\widehat{f}_h(y)\} = \frac{1}{nh^2}\mathrm{var}\left\{w\left(\frac{Y-y}{h}\right)\right\}$$

$$= \frac{1}{nh^2}\left[\mathrm{E}\left\{w\left(\frac{Y-y}{h}\right)^2\right\} - \mathrm{E}\left\{w\left(\frac{Y-y}{h}\right)\right\}^2\right],$$

which depends on

$$\mathrm{E}\left\{h^{-2}w\left(\frac{Y-y}{h}\right)^2\right\}, \quad \mathrm{E}\left\{h^{-1}w\left(\frac{Y-y}{h}\right)\right\}.$$

An argument similar to that above shows that the first of these is

$$h^{-1}\int w(u)^2\{f(y) + huf'(y) + \cdots\}\,du = h^{-1}f(y)\int w(u)^2\,du + O(h), \quad (7.6)$$

while we have already seen that the second is $O(1)$. Thus as $h \to 0$, the variance of $\widehat{f}_h(y)$ is of order $(nh)^{-1}$.

If both the bias obtained from (7.5) and the variance obtained from (7.6) are to vanish as $n \to \infty$ we must choose h so that $h \to 0$ and $nh \to \infty$, for example taking $h \propto n^{-1/2}$. But does this give the best bias-variance tradeoff? The leading terms of the asymptotic mean squared error of $\widehat{f}_h(y)$ are

$$\frac{h^4}{4}f''(y)^2 + \frac{1}{nh}f(y)\int w(u)^2\,du. \quad (7.7)$$

Differentiation shows that this is minimized as a function of h at

$$h_{opt}(y, f) = n^{-1/5}\left\{\int w(u)^2\,du\, f(y)/f''(y)^2\right\}^{1/5},$$

giving an optimal bandwidth that varies with y. Choosing $h \propto n^{-1/5}$ gives bias of order $n^{-2/5}$ and variance of order $n^{-4/5}$, and then bias squared and variance contribute terms of equal order to (7.7). As far as variance is concerned, the effective local sample size is $n^{-4/5}$, so if $n = 1000$, say, the number of observations contributing to a local estimator will be roughly 250, considerably fewer than the sample size applicable for parametric estimation; this is typical of local problems, for which larger samples are needed.

The bandwidth plays a key role in kernel density estimation. In exploratory work a good strategy is to try several values, in the hope that different amounts of smoothing will reveal different interesting features of the data. However automatic rules for selection of h can be valuable, both to suggest starting-points for exploration and for use when density estimation plays a minor role in a larger analysis. The key difficulty is that as we saw above, an optimal choice such as $h_{opt}(y, f)$ depends on the unknown f as well as on y. Dependence on y can be removed by minimizing an overall measure of the distance between $\widehat{f}_h$ and f, such as the integral of (7.7) over y,

$$\frac{h^4}{4} \int f''(y)^2 \, dy + \frac{1}{nh} \int w(u)^2 \, du, \tag{7.8}$$

but the dependence on f remains. Many proposals have been made to sidestep it, of which we outline only two.

One simple approach is to find the bandwidth that would be optimal if f were known. If it is normal with variance σ^2, for example, and $w(u) = \phi(u)$, then the choice minimizing (7.8) is $h = 1.06\sigma n^{-1/5}$. In practice σ is replaced by its sample counterpart s, but as this is sensitive to outliers it may be better to use a rule of thumb such as

IQR is the sample
interquartile range.

$$h'_{opt} = 0.9 \min\{s, \text{IQR}/1.35\} \, n^{-1/5}.$$

This choice is very simple but tends to oversmooth for non-normal data, thereby obscuring multimodality and other features of potential interest.

A better approach is to minimize an estimate of the exact mean integrated squared error of $\widehat{f}_h(y)$,

$$\text{E} \int \{\widehat{f}_h(y) - f(y)\}^2 \, dy = \text{E} \int \widehat{f}_h(y)^2 \, dy - 2\text{E} \int \widehat{f}_h(y) f(y) \, dy + \int f(y)^2 \, dy, \tag{7.9}$$

whose true value involves expectation with respect to f and so is unknown. The third term of this does not depend on h, so it suffices to estimate the first two terms. The awkward term is the second, an obvious estimator of which may be expressed in terms of the empirical distribution function $\widehat{f}$ as

$$\int \widehat{f}_h(y) \, d\widehat{f}(y) = \frac{1}{n} \sum_{j=1}^{n} \widehat{f}_h(y_j),$$

but this is biased because y_j appears twice in $\widehat{f}_h(y_j)$, once implicitly. The bias can be removed by using the *cross-validation estimator* given by

$$\frac{1}{n} \sum_{j=1}^{n} \widehat{f}_{h,-j}(y_j), \quad \text{where} \quad \widehat{f}_{h,-j}(y) = \frac{1}{(n-1)h} \sum_{i \neq j} w\left(\frac{y - y_i}{h}\right)$$

is based on all observations except y_j and so is called a *leave-one-out* estimator. The term cross-validation means that each datum is compared with the rest. Cross-validation is widely used and has many variants. It is related to AIC; in both cases the complexity of a set of models would ideally be assessed using the accuracy of their predictions for data sets like the original. As a new dataset is unavailable, we manufacture one with just one datum, assess how well that datum can be predicted using the remainder, and average all n possible such comparisons.

It is easily shown that apart from a constant factor the first two terms of (7.9) are estimated unbiasedly by

$$\text{CV}(h) = \frac{1}{n} \sum_{j=1}^{n} \int \widehat{f}_{h,-j}(y)^2 \, dy - \frac{2}{n} \sum_{j=1}^{n} \widehat{f}_{h,-j}(y_j),$$

so it seems reasonable to hope that the value $\widehat{h}_{opt}^{CV}$ that minimizes this will be close to the value $h_{opt}(f)$ which minimizes (7.9). If the kernel is normal, then CV(h) is readily computed using results on convolutions of normal densities. Care is needed when minimizing CV(h), as it may have several local minima.

Although close to unbiased, $\widehat{h}_{opt}^{CV}$ is rather variable. Other formulations of cross-validation can be used to derive more stable bandwidth estimators, but typically at the expense of increased bias. Section 7.4 gives further reading.

Example 7.5 (Maize data) The maize data have IQR = 34 and $s = 37.74$, so $1.06sn^{-1/5} = 23.2$ and $h'_{opt} = 13.3$. The effect of these choices may be seen in Figure 7.2, where using h'_{opt} makes the two negative observations more prominent, perhaps too much so, and $h = 30$ seems rather large. Cross-validation gives $\widehat{h}_{opt}^{CV} = 19.5$, again over-emphasizing the two negative y_j.

Here $n = 15$ and it is unreasonable to hope for a useful nonparametric density estimate. ∎

7.1.3 Minimum variance unbiased estimation

Other things being equal, bias is undesirable. A badly biased estimator has expected value far from the parameter it is supposed to estimate, so it seems a good idea to minimize bias so far as possible. This has motivated a careful study of unbiased estimators, for which there is a rather complete theory in a limited class of models.

David Harold Blackwell (1919–) works at the University of California, Berkeley. See DeGroot (1986a) for an account of his life and work.

Rao–Blackwell theorem

Suppose that data Y arise from a statistical model $f(y; \theta)$ and that we want to estimate a scalar function $\psi = \psi(\theta)$ of θ. Suppose also that the statistic $S = s(Y)$ is sufficient for θ, and that $T = t(Y)$ is an unbiased estimator of ψ. Then subject to suitable regularity conditions, $W = w(S) = \mathrm{E}(T \mid S)$ is an unbiased estimator of ψ with variance no larger than that of T:

$$\mathrm{var}(T) \geq \mathrm{var}(W) \quad \text{for all } \theta. \tag{7.10}$$

This is the *Rao–Blackwell theorem*. It is a non-asymptotic result, applicable to samples of any size.

To establish (7.10), first note that W is indeed a statistic:

$$W = w(S) = \mathrm{E}(T \mid S) = \int t(y)f(y \mid s)dy$$

does not depend on θ because of the sufficiency of S. Therefore W can be calculated from Y alone. Secondly,

$$\mathrm{E}(W) = \mathrm{E}_S\{\mathrm{E}(T \mid S)\} = \mathrm{E}(T) = \psi,$$

so W is unbiased for ψ. Thirdly,

$$\begin{aligned}
\mathrm{var}(T) &= \mathrm{E}\{(T - \psi)^2\} \\
&= \mathrm{E}[\{T - \mathrm{E}(T \mid S) + \mathrm{E}(T \mid S) - \psi\}^2]
\end{aligned}$$

$$= E[\{T - E(T \mid S)\}^2] + 2E\{(T - W)(W - \psi)\} + E\{(W - \psi)^2\}$$
$$= E\{(T - W)^2\} + \text{var}(W), \tag{7.11}$$

because the middle term is

$$E_Y\{(T - W)(W - \psi)\} = E_S E_{Y \mid S}\{(T - W)(W - \psi)\}$$
$$= E_S\{(W - W)(W - \psi)\}$$
$$= 0.$$

Evidently (7.11) implies (7.10), giving $E\{(T - W)^2\} = 0$, that is, $T = W$ with probability one, so T and W are effectively the same estimator. The process of replacing an unbiased estimator T with another $E(T \mid S)$ with smaller variance is called *Rao–Blackwellization*.

Example 7.6 (Poisson mean)   Let $Y_1, \ldots, Y_n$ be a Poisson random sample whose mean θ we intend to estimate. Now $E(Y_1) = \theta$, so Y_1 is an unbiased estimator of θ. As the Poisson density is an exponential family, $S = \sum Y_j$ is minimal sufficient for θ (Section 5.2.3). Therefore $W = E(Y_1 \mid S)$ is an unbiased estimator for θ with variance at most that of Y_1.

To find W, we argue by symmetry. Evidently $E(Y_1 \mid S) = \cdots = E(Y_n \mid S)$, because the Y_j were independent and identically distributed unconditionally and the conditioning statistic S is symmetric in the Y_j. Therefore

$$E(Y_1 \mid S) = n^{-1} \sum_{j=1}^{n} E(Y_j \mid S) = E\left(n^{-1} \sum_{j=1}^{n} Y_j \; \middle| \; S \right) = E(n^{-1}S \mid S) = S/n.$$

This estimator has variance θ/n, whereas Y_1 has variance θ.   ∎

Example 7.7 (Dirac comb)   A random sample of pairs (X, Y) from a bivariate density $f(x, y)$ is available, and we wish to estimate some feature of the unknown marginal density $f(y)$ of Y. Suppose that the conditional density $f(y \mid x)$ is available. This may seem unrealistic, but in fact this situation often arises when using simulation to estimate a density; see Section 11.3.3.

The likelihood based on data $(x_1, y_1), \ldots, (x_n, y_n)$ is

$$\prod_{j=1}^{n} f(y_j \mid x_j) f(x_j),$$

and as $f(y \mid x)$ is known, $X_1, \ldots, X_n$ is sufficient for $f(x, y)$ and hence for $f(y)$.

Suppose that we wish to estimate $f(y)$ itself. We might use a kernel density estimator (7.4) with bandwidth $h > 0$, but this would be biased. We can remove the bias by letting $h \to 0$, thus obtaining the Dirac comb

$$\widehat{f}(y) = \frac{1}{n} \sum_{j=1}^{n} \delta(y - Y_j).$$

The comb, which places an infinite spike at each Y_j and can be regarded as the derivative of the empirical distribution function (2.3), gives terrible estimates of $f(y)$,

Paul Adrien Maurice Dirac (1902–1984) was born in Bristol and studied there and at Cambridge, where he later held the professorship of mathematics once held by Newton. His unifying work on the basis of quantum mechanics and relativity led to his receiving the 1933 Nobel Prize in physics. An intensely private man, he had to be persuaded that the publicity would be greater if he refused it, as he originally intended.

$\delta(u)$ is the Dirac delta function.

but it is unbiased because

$$\int \delta(y - u)g(u)\,du = g(y)$$

for any function g. To apply the Rao–Blackwell theorem, note that

$$\mathrm{E}(\widehat{f}(y) \mid X_1, \ldots, X_n) = \frac{1}{n} \sum_{j=1}^{n} \int \delta(y - y')f(y' \mid X_j)\,dy' = \frac{1}{n} \sum_{j=1}^{n} f(y \mid X_j).$$
(7.12)

This is a kernel estimator for which the kernel is the conditional density of Y given X; it is much smoother than the comb. As the first two moments of $f(y \mid X_j)$ are

$$\mathrm{E}_X \{f(y \mid X)\} = \int f(y \mid x)f(x)\,dx = f(y),$$

$$\mathrm{E}_X \{f(y \mid X)^2\} = \int f(y \mid x)^2 f(x)\,dx = \int f(y \mid x)f(x, y)\,dx, \quad (7.13)$$

we see that (7.12) is unbiased, with finite variance when (7.13) is finite, as will usually be the case.

A similar argument applies to the cumulative distribution function of Y and may be preferred by those wary of delta functions (Exercise 7.1.8). ∎

Completeness

Given an unbiased estimator T of ψ and a sufficient statistic, S, the Rao–Blackwell theorem enables us to find an unbiased estimator with variance at most that of T. However there may be many unbiased estimators, each of which could be Rao–Blackwellized. Is there one with lowest variance?

To answer this question we need the notion of *completeness*. A statistic S is *complete* if for any function h,

Strictly speaking $h = 0$ almost everywhere with respect to the density of S.

$$\mathrm{E}\{h(S)\} = 0 \quad \text{for all } \theta \text{ implies that} \quad h \equiv 0, \quad (7.14)$$

and is *boundedly complete* if (7.14) is true provided h is bounded. Evidently a complete statistic is also boundedly complete. If S is complete, we say that its density $f(s; \theta)$ is complete. As completeness must hold for all θ, it is a property of a family of densities rather than of a single density.

Completeness of minimal sufficient statistics is used to establish uniqueness of minimum variance unbiased estimators. Note the qualification here: sufficient statistics that are not minimal are not in general complete.

Example 7.8 (Poisson density) Suppose that Y is Poisson with mean $\theta > 0$, and that $h(Y)$ satisfies $\mathrm{E}\{h(Y)\} = 0$ for every value of θ. Then its expectation is proportional to a power series which is identically zero on the positive half-line:

$$0 = \mathrm{E}\{h(Y)\} = \sum_{y=0}^{\infty} h(y)\frac{\theta^y}{y!}e^{-\theta} \propto \sum_{y=0}^{\infty} \theta^y \frac{h(y)}{y!}, \quad \theta > 0.$$

Hence $h(0) = h(1) = \cdots = 0$, and Y is complete.

Now consider a Poisson sample of size n. Then $S = (Y_1, \ldots, Y_n)$ is sufficient for θ, and $h(S) = Y_1 - Y_2$ has expectation zero for all θ. This does not imply that $Y_1 = Y_2$, however, so S is not complete. The corresponding minimal sufficient statistic $\sum Y_j$ has a Poisson density, and is complete.   ■

Example 7.9 (Uniform density)   Suppose that Y is uniformly distributed on $(-\theta, \theta)$. Then $E(Y) = 0$ for every $\theta > 0$, but as $h(y) = y$ is not identically zero, Y is not complete.   ■

Example 7.10 (Exponential family)   Suppose that Y belongs to an exponential family of order p,

$$ f(y; \omega) = \exp\{s(y)^{\mathsf{T}}\theta - \kappa(\theta)\} f_0(y), \quad y \in \mathcal{Y}, \theta \in \mathcal{N}. $$

If Y is continuous and $E\{h(Y)\} = 0$, then provided that $\mathcal{N}$ contains an open set around the origin,

$$ E\{h(Y)\} = \int h(y) \exp\{s(y)^{\mathsf{T}}\theta - \kappa(\theta)\} f_0(y) \, dy = 0 $$

is proportional to the Laplace transform of $h(y)f_0(y)$. Then the uniqueness of Laplace transforms implies that $h(y)f_0(y) = 0$ except on sets of measure zero and thus $h(Y) \equiv 0$: Y is complete. When Y is discrete the corresponding argument involves series or polynomials, as in Example 7.8.

The same argument applies to any subfamily whose parameter space contains an open set around the origin, and in particular to all the standard exponential family models.   ■

To see how completeness is used, suppose that we have a parametric model $f(y; \theta)$ with complete minimal sufficient statistic S, and two unbiased estimators of $\psi = \psi(\theta)$, namely $T = t(Y)$ and $T' = t'(Y)$. Let $W = E(T \mid S)$ and $W' = E(T' \mid S)$. Now $E(W - W') = 0$ for all θ, and both W and W' are functions of the data only through S. But S is complete, so $W = W'$ except on sets of measure zero, that is, W and W' are identical for all practical purposes. Thus Rao–Blackwellization of an unbiased estimator using a complete sufficient statistic always leads to W, and no unbiased estimator of ψ has smaller variance. For suppose T' is an unbiased estimator of ψ with smaller variance than W. Then by the Rao–Blackwell theorem, $W' = E(T' \mid S)$ satisfies

$$ \mathrm{var}(W') \leq \mathrm{var}(T') < \mathrm{var}(W), $$

which is impossible because $W' \equiv W$.

Example 7.11 (Normal density)   Let $Y_1, \ldots, Y_n$ be a $N(\mu, \sigma^2)$ random sample, where $n \geq 2$. We saw in Example 5.14 that $S = (\overline{Y}, \sum(Y_j - \overline{Y})^2)$ is minimal sufficient, and as its density is an exponential family of order 2 in which we can take $\Omega = (-\infty, \infty) \times (0, \infty)$, S is complete.

Now $\overline{Y}$ is an unbiased estimator of μ that is a function of S, and therefore it is the minimum variance unbiased estimator of μ. Likewise the minimum variance unbiased estimator of σ^2 is $(n-1)^{-1} \sum(Y_j - \overline{Y})^2$.   ■

Although of theoretical interest, minimum variance unbiased estimators are not widely used in practice. One difficulty is that the restriction to exact unbiasedness can exclude every interesting estimator.

Example 7.12 (Poisson density) Let $Y_1, \ldots, Y_n$ be a Poisson random sample with mean λ, and let $\psi = \exp(-2n\lambda)$. Then an unbiased estimator $h(S)$ of ψ based on the minimal sufficient statistic $S = \sum Y_j$ must satisfy

$$\exp(-2n\lambda) = \sum_{s=0}^{\infty} h(s) \frac{(n\lambda)^s}{s!} e^{-n\lambda},$$

and completeness of S implies that the unique minimum variance unbiased estimator of ψ is the unacceptable

$$h(S) = \begin{cases} -1, & S \text{ odd,} \\ 1, & S \text{ even.} \end{cases}$$

The maximum likelihood estimator $\exp(-2S)$ is preferable despite its bias. ∎

A further difficulty is that minimum variance unbiased estimators do not transform in a simple way. Moreover, as will be evident from the discussion above, there is no easy recipe that gives unbiased estimators, and once found, it may be awkward to Rao–Blackwellize them. For these and other reasons, maximum likelihood estimators are generally preferable.

7.1.4 Interval estimation

Our focus so far has been on point estimates of a parameter and their variances. Although these are useful when estimator is approximately normal, their relevance is much less obvious when its distribution is non-normal or the sample size is small. Furthermore it is often valuable to express parameter uncertainty in terms of an interval, or more generally a region. The notion of a *pivot*, which we met in Section 3.1, then moves to centre stage.

Consider a model $f(y; \theta)$ for data Y. Then a pivot $Z = z(Y, \theta)$ is a function of Y and θ that has a known distribution independent of θ, this distribution being invertible as a function of θ for each possible value of Y. That is, given a region $\mathcal{A}$ such that $\Pr\{z(Y, \theta) \in \mathcal{A}\} = 1 - 2\alpha$, we can find a region $\mathcal{R}_\alpha(Y, \mathcal{A})$ of the parameter space such that

$$1 - 2\alpha = \Pr\{z(Y, \theta) \in \mathcal{A}\} = \Pr\{\theta \in \mathcal{R}_\alpha(Y; \mathcal{A})\}.$$

If θ is scalar then $z(Y, \mathcal{A})$ is typically a strictly monotonic function of θ for each Y. Given data y and a suitable pivot, we find a $(1 - 2\alpha)$ confidence region for the true value of θ by arguing that under repeated sampling $\mathcal{R}_\alpha(y; \mathcal{A})$ is the realization of a random region $\mathcal{R}_\alpha(Y; \mathcal{A})$ that contains the true θ with probability $(1 - 2\alpha)$. An important exact pivot is the Student t statistic, and we have extensively used an approximate pivot, the likelihood ratio statistic. For reasons to be given in Section 7.3.4, pivots such as these based on the likelihood tend to be close to optimal in the sense of providing the shortest possible confidence intervals for given α, at least in large samples.

Example 7.13 (Exponential density)   Suppose we wish to base a $(1 - 2\alpha)$ confidence interval for λ on a single observation from the exponential density $\lambda e^{-\lambda y}$, $y > 0$, $\lambda > 0$. Then $Z = Y\lambda$ is pivotal, since $\Pr(\lambda Y \leq z) = 1 - e^{-z}$, $z > 0$, independent of λ. Its upper $(1 - \alpha)$ quantile is $z_{1-\alpha} = -\log\alpha$. As

$$1 - \alpha = \Pr(Z \leq z_{1-\alpha}) = \Pr(\lambda Y \leq z_{1-\alpha}) = \Pr(\lambda \leq z_{1-\alpha}/Y),$$

an upper $(1 - \alpha)$ confidence limit is $-\log\alpha/y$. Similarly an α lower confidence limit for λ is $-\log(1 - \alpha)/y$, and an equi-tailed $(1 - 2\alpha)$ confidence interval is $(-\log(1 - \alpha)/y, -\log\alpha/y)$. This is not symmetric about the maximum likelihood estimate $\widehat{\lambda} = 1/y$, nor is it the shortest possible such interval.

To find the shortest $(1 - 2\alpha)$ confidence interval for λ based on y, we choose the upper tail probability γ, $0 < \gamma \leq 2\alpha$, to minimize the interval length

$$y^{-1}\{-\log\gamma + \log(1 - 2\alpha + \gamma)\},$$

giving $\gamma = 2\alpha$ and confidence interval $(0, -\log(2\alpha)/y)$. This is obvious from the shape of the exponential density and, not coincidentally, the likelihood.   ∎

Exercises 7.1

1   Let R be binomial with probability π and denominator m, and consider estimators of π of form $T = (R + a)/(m + b)$, for $a, b \geq 0$. Find a condition under which T has lower mean squared error than the maximum likelihood estimator R/m, and discuss which is preferable when $m = 5, 10$.

2   Let $T = a \sum(Y_j - \overline{Y})^2$ be an estimator of σ^2 based on a normal random sample. Find values of a that minimize the bias and mean squared error of T.

3   When T is a biased estimator of the scalar $\psi(\theta)$, with bias $b(\theta)$, show that under the usual regularity conditions, the mean squared error of T is no smaller than

$$\{d\psi/d\theta + db(\theta)/d\theta\}^2 /I(\theta) + b(\theta)^2.$$

If $b(\theta) = b_1(\theta)/n + b_2(\theta)/n^{3/2} + \cdots$, where $b_i(\theta)$ is $O(1)$, then show that the Cramér–Rao lower bound applies, at least in large samples.

4   Suppose that T is a $q \times 1$ unbiased estimator of $\psi = \psi(\theta)$. Show that $\mathrm{cov}(T, U) = d\psi/d\theta^{\mathrm{T}}$, and compute the variance matrix of $T - d\psi/d\theta^{\mathrm{T}} I(\theta)^{-1} U$, where U is $p \times 1$ score vector. Hence establish (7.3).

5   Consider a kernel density estimator (7.4).
(a) Verify the choice of h that minimizes (7.7). If $f(y) = \sigma^{-1}\phi\{(y - \mu)/\sigma\}$ and $w(u) = \phi(u)$, find h_{opt}. Discuss.   Note that $\phi(z)^2 = (2\pi)^{-1/2}\phi(\sqrt{2}z)$.
(b) Show that $h = 1.06\sigma n^{-1/5}$ minimises (7.8) using the densities in (a).
(c) Instead of using a constant bandwidth, we might take

$$\widehat{f}(y) = \frac{1}{nh}\sum_{j=1}^{n} \frac{1}{\lambda_j} w\left(\frac{y - y_j}{h\lambda_j}\right)$$

for local bandwidth factors $\lambda_j \propto \{\tilde{f}(y_j)\}^{-\gamma}$ based on a pilot density estimate $\tilde{f}(y)$. Show that if the pilot estimate is exact and $\gamma = -\frac{1}{2}$, then $\widehat{f}$ has bias $o(h^2)$.

6   Find the expected value of $\mathrm{CV}(h)$, and show to what extent it estimates (7.9).

7 Find minimum variance unbiased estimators of λ^2, e^λ, and $e^{-n\lambda}$ based on a random sample $Y_1, \ldots, Y_n$ from a Poisson density with mean λ. Show that no unbiased estimator of $\log \lambda$ exists.

8 In Example 7.1.3, suppose we wish to estimate $\psi = \Pr(Y \leq y)$ using the empirical distribution function $n^{-1} \sum I(Y_j \leq y)$. Show that this is unbiased and that its Rao–Blackwellized form is

$$\frac{1}{n} \sum_{j=1}^{n} \Pr(Y_j \leq y \mid X_j).$$

Hence obtain an unbiased estimator of $f(y)$.

9 Let $Y \sim N(0, \theta)$. Is Y complete? What about Y^2? And $|Y|$?

10 Let $R_1, \ldots, R_n$ be a binomial random sample with parameters m and $0 < \pi < 1$, where m is known. Find a complete minimal sufficient statistic for π and hence find the minimum variance unbiased estimator of $\pi(1 - \pi)$.

11 Let $\overline{Y}$ be the average of a random sample from the uniform density on $(0, \theta)$. Show that $2\overline{Y}$ is unbiased for θ. Find a sufficient statistic for θ, and obtain an estimator based on it which has smaller variance. Compare their mean squared errors.

7.2 Estimating Functions

7.2.1 Basic notions

Our discussion of the maximum likelihood estimator in Section 4.4.2 stressed its asymptotic properties but said little about its finite-sample behaviour. By contrast our treatment of unbiased estimators showed their finite-sample optimality under certain conditions, but suggested that the class of such estimators is often too small to be of real interest for applications. Furthermore both types of estimator can behave poorly if the data are contaminated or if the assumed model is incorrect, making it worthwhile to consider other possibilities. In this section we explore some consequences of shifting emphasis away from estimators and towards the functions that often determine them.

Suppose that we intend to estimate a $p \times 1$ parameter θ based on a random sample $Y_1, \ldots, Y_n$ from a density $f(y; \theta)$, assumed to be regular for likelihood inference. Then in most cases the maximum likelihood estimator $\widehat{\theta}$ is defined implicitly as the solution to the $p \times 1$ score equation

$$U(\theta) = u(Y; \theta) = \sum_{j=1}^{n} u(Y_j; \theta) = \sum_{j=1}^{n} \frac{\partial \log f(Y_j; \theta)}{\partial \theta} = 0.$$

Key properties of the score statistic $U(\theta)$ are

$$E\{U(\theta)\} = 0, \quad \text{var}\{U(\theta)\} = E\left\{-\frac{dU(\theta)}{d\theta^{\mathsf{T}}}\right\} = I(\theta),$$

for all θ, where the $p \times p$ Fisher information matrix $I(\theta) = ni(\theta)$ and

$$i(\theta) = \text{var}\{u(Y_j; \theta)\} = \int u(y; \theta)u(y; \theta)^{\mathsf{T}} f(y; \theta)\, dy = -\int \frac{\partial u(y; \theta)}{\partial \theta^{\mathsf{T}}} f(y; \theta)\, dy.$$

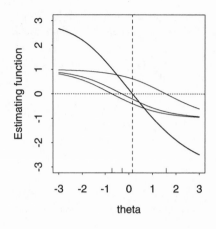 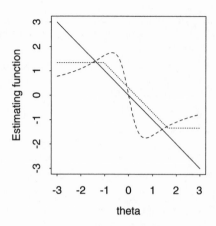

Figure 7.3 Estimating functions. Left: construction of $g(y; \theta)$ (heavy) as the sum of $g(y_j; \theta)$ for a sample of size $n = 3$ shown by the rug. The lines $g = 0$ (dots) and $\theta = \tilde{\theta}$ (dashes) are also shown. Right: estimating functions for the mean (solid), the Huber estimator (dots) and a redescending M-estimator (dashes), slightly offset to avoid overplotting.

The implicit definition of $\widehat{\theta}$ suggests that we study properties of estimators $\tilde{\theta}$ that solve a $p \times 1$ system of *estimating equations* of form

$$g(Y; \theta) = \sum_{j=1}^{n} g(Y_j; \theta) = 0. \tag{7.15}$$

We call $g(y; \theta)$ an *estimating function* and say it is unbiased if

$$\mathrm{E}\{g(Y; \theta)\} = n \int g(y; \theta) f(y; \theta) = 0 \quad \text{for all } \theta.$$

Or sometimes an inference function.

This formulation encompasses many possibilities.

Example 7.14 (Logistic density) The logistic density $e^{y-\theta}/(1 + e^{y-\theta})^2$ has score function

$$u(y; \theta) = 2e^{y-\theta}/(1 + e^{y-\theta}) - 1, \quad -\infty < y < \infty, -\infty < \theta < \infty.$$

The left panel of Figure 7.3 shows the construction of the corresponding estimating function based on a sample of size three.   ∎

Example 7.15 (Moment estimators) If $g(y; \mu) = y - \mu$, then the solution to (7.15) is the sample average $\tilde{\mu} = \overline{Y}$, which is an unbiased estimator of the mean of f, if this exists. The estimating function $y - \mu$ is shown in the right panel of Figure 7.3, with other estimating functions discussed later.

This can be extended to several parameters. The *moment estimators* of the mean and variance of Y are found by simultaneous solution of

Or method of moments estimators.

$$n^{-1} \sum_{j=1}^{n} Y_j - \mu = 0, \quad n^{-1} \sum_{j=1}^{n} Y_j^2 - \mu^2 - \sigma^2 = 0,$$

and these are of form (7.15) with $g(y; \theta) = (y - \mu, y^2 - \mu^2 - \sigma^2)^{\mathrm{T}}$ and $\theta = (\mu, \sigma^2)^{\mathrm{T}}$. Although themselves unbiased, these estimating equations produce the biased estimator $n^{-1} \sum (Y_j - \overline{Y})^2$ of σ^2.

Estimators of functions of the mean and variance may be defined similarly. For example, the Weibull density

$$f(y; \beta, \kappa) = \kappa \beta^{-1} (y/\beta)^{\kappa-1} \exp\{-(y/\beta)^\kappa\}, \quad y > 0, \beta, \kappa > 0,$$

$\Gamma(u)$ is the gamma function.

has $E(Y^r) = \beta^r \Gamma(1 + r/\kappa)$. Hence the moment estimator of $\theta = (\beta, \kappa)^{\mathsf{T}}$ can be determined as the solution to (7.15) with

$$g(y; \theta) = (y - \beta\Gamma(1 + 1/\kappa), \quad y^2 - \beta^2\Gamma(1 + 2/\kappa))^{\mathsf{T}}. \tag{7.16}$$

The parameters μ and σ^2 have the same interpretations for any model that possesses two moments, whereas (β, κ) are specific to the Weibull case. ∎

Example 7.16 (Probability weighted moment estimators) Moment estimators may be poor or even useless with data from long-tailed densities, whose moments may not exist. An alternative is use of *probability weighted moment estimators*, defined as solutions to equations of form

$$n^{-1} \sum_{j=1}^n Y^r F(Y; \theta)^s \{1 - F(Y; \theta)\}^t - \int y^r F(y; \theta)^s \{1 - F(y; \theta)\}^t f(y; \theta) \, dy = 0.$$

Even if the ordinary moments, which correspond to taking $s = t = 0$, do not exist, the integrals here may be finite for positive values of s or t or both.

An example is the generalized Pareto distribution (6.38), for which we set $\theta = (\xi, \sigma)^{\mathsf{T}}$. In this case it is convenient to take $r = 1$ and $s = 0$, giving

$$g_t(y; \theta) = y(1 + \xi y/\sigma)^{-t/\xi} - \frac{\sigma}{(t+1)(t+1-\xi)},$$

which has finite expectation provided $\xi < t + 1$. Estimators may be obtained by setting $g(y; \theta) = (g_1(y; \theta), g_2(y; \theta))^{\mathsf{T}}$ and solving (7.15) simultaneously, though equivalent more convenient forms of the equations are preferred in practice.

As with moment estimators, the choice of r, s, and t introduces an arbitrary element, because different choices will lead to different estimators. ∎

Example 7.17 (Linear model) The scalar β in the simple linear model

$$Y_j = \beta x_j + \varepsilon_j, \quad j = 1, \ldots, n,$$

where the ε_j have mean zero, can be estimated by the solution to (7.15) with $g(y; \theta) = y - \beta x$, giving $\tilde{\beta} = \sum Y_j / \sum x_j$. This estimator is unbiased whatever the distributions of the ε_j; in particular we have made no assumptions about their variances, requiring the ε_j only to have zero mean. In fact, they need not be independent, or even uncorrelated. ∎

In general discussion we shall suppose that θ is scalar and that for every value of y, we deal with an unbiased estimating function $g(y; \theta)$ that is strictly monotone decreasing in θ. It is then easy to show that $\tilde{\theta}$ is consistent for θ. Note first that $\tilde{\theta} \le a$ if and only if $g(Y; a) \le 0$. As $g(y; \theta)$ is decreasing in θ for each y, $n^{-1} g(Y; \theta - \varepsilon)$

converges to

$$n^{-1}E\{g(Y;\theta - \varepsilon)\} = n^{-1}E\{g(Y;\theta - \varepsilon) - g(Y;\theta)\} = c(\theta - \varepsilon) > 0$$

as $n \to \infty$ for any $\varepsilon > 0$, by virtue of the weak law of large numbers. Hence

$$\Pr(\tilde{\theta} \leq \theta - \varepsilon) = \Pr\{n^{-1}g(Y;\theta - \varepsilon) \leq 0\} \to 0, \quad \text{as } n \to \infty.$$

Likewise $\Pr(\tilde{\theta} > \theta + \varepsilon) \to 0$, so $\Pr(|\tilde{\theta} - \theta| \leq \varepsilon) \to 1$: $\tilde{\theta}$ is a consistent estimator.

Technical difficulties arise with non-monotone or discontinuous estimating functions, to which most of the discussion below does not apply directly. In such cases it is necessary to show that there is a consistent solution to the estimating equation, to which the arguments below can be applied.

Optimality

Having defined the class of unbiased estimating functions, the question naturally arises which of them we should use. To answer this we must find a finite-sample optimality criterion analogous to mean squared error. To motivate a suitable criterion, suppose that θ is scalar and consider its estimator $\tilde{\theta}$. Taylor series expansion of $g(Y;\tilde{\theta})$ gives

$$0 \doteq g(Y;\theta) + (\tilde{\theta} - \theta)\frac{dg(Y;\theta)}{d\theta},$$

so

$$\tilde{\theta} - \theta \doteq \frac{\sum_{j=1}^{n} g(Y_j;\theta)}{-\sum_{j=1}^{n} \frac{dg(Y_j;\theta)}{d\theta}} = \frac{\sum_{j=1}^{n} g(Y_j;\theta)}{E\{-\frac{dg(Y;\theta)}{d\theta}\}} + O_p(n^{-1}), \tag{7.17}$$

using the same argument as applied to the maximum likelihood estimator. This implies that $\tilde{\theta}$ has asymptotic variance

$$\text{var}(\tilde{\theta}) \doteq \frac{\text{var}\{g(Y;\theta)\}}{[E\{-\frac{dg(Y;\theta)}{d\theta}\}]^2} = n^{-1}\frac{\int g(y;\theta)^2 f(y;\theta)\,dy}{\{-\int \frac{dg(y;\theta)}{d\theta}f(y;\theta)\,dy\}^2}.$$

A measure of finite-sample performance of $g(y;\theta)$ should not conflict with asymptotic properties of $\tilde{\theta}$, suggesting that we regard an estimating function as optimal in the class of unbiased estimating functions if it minimizes

$$\frac{\text{var}\{g(Y;\theta)\}}{[E\{-\frac{dg(Y;\theta)}{d\theta}\}]^2} \tag{7.18}$$

for all θ. This quantity is unaffected by one-one reparametrization.

Another motivation for (7.18) rests on noting that although variance is a natural basis for comparing estimating functions, $a\,g(Y;\theta)$ is also unbiased, with variance a^2 times greater than that of $g(Y;\theta)$. Hence fair comparison is possible only after removing this arbitrary scaling. Multiplication of $g(Y;\theta)$ by a changes the slope of the estimating function, so it is natural to choose a to ensure that the expected derivative of $g(Y;\theta)$ equals one, leading to (7.18).

It can be shown that any unbiased estimating function must satisfy

$$I(\theta)^{-1} \leq \frac{\text{var}\{g(Y;\theta)\}}{\left[\text{E}\left\{-\frac{dg(Y;\theta)}{d\theta}\right\}\right]^2}, \tag{7.19}$$

so there is a lower bound on (7.18), analogous to the Cramér–Rao lower bound. If (7.18) is evaluated with $g(Y;\theta) = u(Y;\theta)$, the result is $I(\theta)^{-1}$. Hence the score function minimizes (7.18), and is in this sense optimal in finite samples. This ties in with asymptotic properties of the maximum likelihood estimator, and may be extended to the case where θ is a $p \times 1$ vector. Then

$$\text{E}\left\{-\frac{\partial g(Y;\theta)}{\partial \theta^{\text{T}}}\right\}^{-1} \text{var}\{g(Y;\theta)\} \, \text{E}\left\{-\frac{\partial g(Y;\theta)^{\text{T}}}{\partial \theta}\right\}^{-1} \geq I(\theta)^{-1} \tag{7.20}$$

in the sense that the difference of these $p \times p$ matrices is positive semi-definite, provided $\text{E}\{-\partial g(Y;\theta)/\partial \theta^{\text{T}}\}$ is invertible. The left-hand side of this inequality is the asymptotic covariance matrix of $\tilde{\theta}$, and its sandwich form generalizes that of a maximum likelihood estimator under a wrong model; see Section 4.6. Standard errors for $\tilde{\theta}$ are obtained by replacing the matrices in (7.20) by sample versions, giving

$$\left\{\sum_{j=1}^{n} \frac{\partial g(y_j;\tilde{\theta})}{\partial \theta^{\text{T}}}\right\}^{-1} \sum_{j=1}^{n} g(y_j;\tilde{\theta})g(y_j;\tilde{\theta})^{\text{T}} \left\{\sum_{j=1}^{n} \frac{\partial g(y_j;\tilde{\theta})^{\text{T}}}{\partial \theta}\right\}^{-1},$$

from which confidence sets for elements of θ may be obtained, generally by normal approximation.

Example 7.18 (Weibull model) An estimating function for the Weibull parameters β and κ is given by (7.16), for which elementary calculations give

$$\text{E}\left\{-\frac{\partial g(Y;\theta)^{\text{T}}}{\partial \theta}\right\} = n \begin{pmatrix} \Gamma(1+1/\kappa) & 2\beta\Gamma(1+2/\kappa) \\ -\beta\Gamma'(1+1/\kappa)/\kappa^2 & -2\beta^2\Gamma'(1+2/\kappa)/\kappa^2 \end{pmatrix}$$

$\Gamma'(u) = d\Gamma(u)/du$, and so forth.

and

$$I(\theta) = n \begin{pmatrix} \kappa^2/\beta^2 & -\Gamma'(2)/\beta \\ -\Gamma'(2)/\beta & \{1+\Gamma''(2)\}/\kappa^2 \end{pmatrix},$$

while $\text{var}\{g(Y;\theta)\}$ is easily found in terms of the moments $\text{E}(Y^r)$. In analogy to the discussion of efficiency on page 113, the overall efficiency of $g(Y;\theta)$ relative to the score is taken to be the square root of the ratio of the determinants of the matrices on either side of the inequality in (7.20), while the efficiency for estimation of β is the ratio of their $(1, 1)$ coefficients, with $(2, 2)$ coefficients used for κ. These efficiencies, plotted in the left panel of Figure 7.4, show that the moment estimating functions are fairly efficient when $\kappa > 2$, but are poor when κ is small. ∎

7.2.2 Robustness

Finite-sample optimality of the score function is not the whole story, for several reasons. First, we may be unwilling or unable to specify the model fully, and then the score is unavailable. Second, even if we can be fairly sure of $f(y;\theta)$, there is

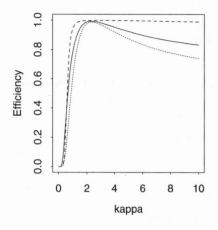 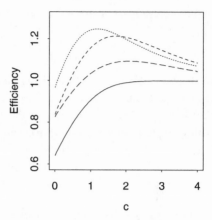

Figure 7.4 Efficiencies of estimating functions. Left: overall efficiency (solid), efficiency for β (dashes) and for κ (dots) for moment estimators of Weibull distribution. Right: finite-sample efficiency of Huber estimating function g_c relative to $g(y; \theta) = y - \theta$ for normal (solid), t_5 (dots), normal mixture $0.95N(0, 1) + 0.05N(0, 9)$ (small dashes) and logistic data (long dashes).

always the possibility of bad data — tryping errors, wild observations and so forth. In principle all data should be carefully scrutinized for these, but with big or complex datasets or where data are collected automatically this is impracticable. Estimating functions that are *robust*, that is, perform well under a wide range of potential models centred at an ideal model may be preferred, even if they are somewhat sub-optimal when that model itself holds.

Robustness entails insensitivity to departures from assumptions, but this has many aspects. Perhaps the most common usage relates to contamination by outliers. If bad values are present then we might optimistically hope to identify and delete them, or more realistically aim to downweight them. Thus we ignore or play down some 'bad' portion of the data and hope to extract useful information from the 'good' part, even if we are unsure where the boundary lies. A related usage concerns the need for procedures that perform well when assumptions underlying the ideal model are relaxed. An essential requirement is then that estimands have the same interpretation under all the potential models. In Example 7.15 the first and second moments μ and σ^2 have this property of *robustness of interpretation* but the Weibull parameters κ and β do not, because they are meaningless for models other than the Weibull.

Outliers are perhaps the most obvious form of departure from the model, but the assumed dependence structure is usually more crucial in applications. In Example 6.25, for instance, a confidence interval was three times too short when dependence was unaccounted for. Although independence is often assumed, not only is mild dependence often difficult to detect, but also it may be hard to formulate a suitable alternative. In applications independence may be assured by the design of the investigation, but often it must be checked empirically, for example using time series tools such as the correlogram.

One way to view an estimating function is that it defines a parameter $t(F)$ implicitly as the solution to the population equation

$$\int g\{y; t(F)\} \, dF(y) = 0,$$

where F is any member of the class of distributions under consideration. The requirement that $t(F)$ be robust of interpretation imposes restrictions on g. If, for instance, the density $f(y) = dF(y)/dy$ is symmetric about θ and we require $t(F) = \theta$ for any such density, then $g(y; \theta)$ must be odd as a function of $y - \theta$, with $g(\theta; \theta) = 0$. In many cases the requirement of robustness of interpretation indicates taking $t(F)$ to be a moment or related quantity, which will retain its meaning for all models possessing the necessary moments.

One approach to downweighting bad data stems from observing that (7.17) implies that the effect of Y_j on $\tilde{\theta}$ is proportional to $g(Y_j; \theta)$. If this is large, then $\tilde{\theta}$ will tend to be far from its estimand θ. This suggests that the sensitivity of $\tilde{\theta}$ to an observation y be measured by the *influence function* of $\tilde{\theta}$,

$$L(y; \theta) = \frac{g(y; \theta)}{-\int \frac{dg(u; \theta)}{d\theta} f(u; \theta) \, du};$$

this is simply a rescaling of the estimating function. Our earlier discussion implies that $\mathrm{var}(\tilde{\theta}) \doteq n^{-1} \mathrm{var}\{L(Y; \theta)\}$ in terms of a single observation Y.

Expression (7.17) suggests that the impact of outliers can be reduced by using estimating functions and hence influence functions that are bounded in y. One possibility is a redescending function such as $(y - \theta)/\{1 + (y - \theta)^2\}$, which tends to zero as $|y - \theta| \to \infty$. Another possibility is to truncate a standard function such as $y - \theta$, so that values of y distant from θ have limited impact on $\tilde{\theta}$. See Figure 7.3.

Peter Johann Huber (1934–) has been professor of statistics at ETH Zürich, Massachusetts Institute of Technology, and Harvard and Bayreuth universities, and is now retired.

Example 7.19 (Huber estimator) The effect of outliers on the estimation of a mean may be reduced by using

$$g_c(y; \theta) = \begin{cases} -c, & y \le \theta - c, \\ y - \theta, & -c < y - \theta < c, \\ c, & \theta + c \le y, \end{cases}$$

Or Huber's Proposal 2.

where the constant $c > 0$ is chosen to balance robustness and efficiency. Robustness to outliers is increased but efficiency at the normal model is reduced by decreasing c; when $c = \infty$ we have $g_\infty(y; \theta) = y - \theta$ and $\tilde{\theta} = \overline{Y}$. The estimator corresponding to $g_c(y; \theta)$ is sometimes called the *Huber estimator* of location. The parameter $t(F)$ is the centre of an underlying symmetric density and equals its mean when $c = \infty$ and its median when $c = 0$. These are not the same when the underlying density is asymmetric, and then $t(F)$ has no simple direct interpretation, though it may depend only weakly on c for certain choices of F.

The finite-sample efficiency of $g_c(y; \theta)$ as a function of c for various symmetric densities is shown in the right panel of Figure 7.4. The quantity plotted is (7.18) divided by the variance of $g_\infty(Y; \theta) = Y - \theta$, as this rather than the score function for the true density would usually be used in practice. Under the normal model the efficiency of g_c is essentially one when $c = 2$, dropping to the value $2/\pi = 0.637$ for the median when $c \to 0$. Overall a good choice seems to be $c = 1.345$, which is often the default in software packages; it has efficiency 0.95 for normal data, but beats g_∞ in the other cases shown. ∎

The discussion above presupposes that the scale of the underlying density is known, even if the location is not. In practice estimation of scale has little effect on the efficiency of location estimators, and the results above apply with little change provided scale is estimated robustly, for example using the median absolute deviation.

To illustrate optimality under weak conditions on the underlying model, suppose that we intend to estimate θ using the weighted combination of unbiased linear estimating functions

$$\sum_{j=1}^{m} w_j(\theta)\{Y_j - \mu_j(\theta)\},$$

where $\text{var}(Y_j) = V_j(\theta)$ may be a function of θ. We suppose that the mean and variance functions $\mu_j(\theta)$ and $V_j(\theta)$ for each of the Y_j are known, but make no assumption about their distributions. Notice that our argument for consistency of $\tilde{\theta}$ will apply under mild conditions on the weights and the moments. Suppose also that the Y_j are uncorrelated. Then (7.18) is

$$\frac{\sum_j w_j^2(\theta)V_j(\theta)}{\left\{\sum_j w_j(\theta)\mu_j'(\theta)\right\}^2},$$

where $\mu_j'(\theta) = d\mu_j(\theta)/d\theta$, and our earlier discussion suggests that we seek the weights $w_j(\theta)$ that minimize this. This is equivalent to the problem

$$\min_{w_1,\dots,w_n} \sum_{j=1}^{n} w_j^2 V_j \quad \text{subject to} \quad \sum_{j=1}^{n} w_j \mu_j' = c,$$

for some constant c. Use of Lagrange multipliers gives $w_j(\theta) \propto \mu_j'(\theta)/V_j(\theta)$, so the optimal estimating equation is

$$\sum_{j=1}^{n} \mu_j'(\theta)\frac{1}{V_j(\theta)}\{Y_j - \mu_j(\theta)\} = 0. \tag{7.21}$$

An exponential family variable Y_j with log likelihood contribution $y_j\theta - \kappa_j(\theta)$ has mean $\kappa_j'(\theta)$ and variance $\kappa_j''(\theta)$, so $\mu_j'(\theta) = V_j(\theta)$ and (7.21) reduces to the score equation, $\sum\{Y_j - \kappa_j'(\theta)\} = 0$, which is optimal.

Example 7.20 (Straight-line regression) Let the Y_j have means $\mu(\beta) = x_j\beta$, with x_j known. Then $\mu_j'(\beta) = x_j$, and $g(Y_j, \beta) = Y_j - x_j\beta$. If $\text{var}(Y_j) = V_j(\beta)$ is constant, (7.21) becomes $\sum x_j(Y_j - \beta x_j)$, and the corresponding estimator is $\tilde{\beta} = \sum Y_j x_j / \sum x_j^2$. This is the least squares estimator of β, corresponding to a normal distribution for Y_j, but it has much wider validity.

If $\text{var}(Y_j) = x_j\beta$, as would be the case if Y_j were Poisson with mean $x_j\beta$, then the optimal estimating function is $\sum(Y_j - \beta x_j)$, and $\tilde{\beta} = \sum Y_j / \sum x_j$. As in the normal case, $\tilde{\beta}$ is optimal more widely.   ■

Estimating equations of form similar to (7.21) are very important in the regression models encountered in Chapters 8 and 10.

7.2.3 Dependent data

This may be omitted at a first reading.

In earlier discussion, for example in Section 6.1, we used the fact that standard likelihood asymptotics also apply to some types of dependent data. For some explanation of this, consider the more general context of unbiased estimating functions for a scalar θ. Suppose that $\tilde{\theta}$ is defined as the solution to the equation

$$\sum_{j=1}^{n} g_j(Y; \theta) = 0, \tag{7.22}$$

where $g_j(Y; \theta)$ depends only on $Y_1, \ldots, Y_j$ and is such that for all θ,

$$\mathrm{E}\{g_1(Y)\} = 0, \quad \mathrm{E}\{g_j(Y; \theta) \mid Y_1, \ldots, Y_{j-1}\} = 0, \quad j = 2, \ldots, n,$$

so that the unconditional expectation $\mathrm{E}\{g_j(Y; \theta)\} = 0$ for all j. If $j > k$, then

$$
\begin{aligned}
\mathrm{cov}\{g_j(Y; \theta), g_k(Y; \theta)\} &= \mathrm{E}\{g_j(Y; \theta)g_k(Y; \theta)\} \\
&= \mathrm{E}[g_k(Y; \theta)\mathrm{E}\{g_j(Y; \theta) \mid Y_1, \ldots, Y_{j-1}\}] = 0,
\end{aligned}
$$

so

$$\mathrm{var}\left\{\sum_{j=1}^{n} g_j(Y; \theta)\right\} = \sum_{j=1}^{n} \mathrm{var}\{g_j(Y; \theta)\}.$$

The left of (7.22) is a zero-mean martingale, and under mild regularity conditions a martingale central limit theorem as $n \to \infty$ gives

$$V^{-1/2}(\tilde{\theta} - \theta) \xrightarrow{D} Z, \quad \text{where } V = \frac{\sum_{j=1}^{n} \mathrm{var}\{g_j(Y; \theta) \mid Y_1, \ldots, Y_{j-1}\}}{\left[\sum_{j=1}^{n} \mathrm{E}\{dg_j(Y; \theta)/d\theta \mid Y_1, \ldots, Y_{j-1}\}\right]^2}, \tag{7.23}$$

and Z is standard normal. Thus provided the random variable V is used to estimate the variance of $\tilde{\theta}$, confidence intervals for θ can be set in the usual way.

Two main possibilities arise for the limiting behaviour of V. In an *ergodic* model a deterministically rescaled version of V converges to a constant as $n \to \infty$, such as $nV \xrightarrow{P} v > 0$. This occurs, for instance, with independent data, ergodic Markov chains, and many time series models. Under regularity conditions the usual arguments then apply to the rescaled estimator, whose limiting distribution is normal, and the argument starting from (7.17) yields (7.18). The second possibility is that when rescaled, V converges to a nondegenerate random variable D. The model is then said to be *non-ergodic*, and as the limiting distribution of the rescaled estimator is $D^{-1/2}Z$, standard large-sample theory does not apply.

As with independent data, we can find the optimal finite-sample choice of weighting functions within the class of linear combinations of the $g_j(Y; \theta)$,

$$\sum_{j=1}^{n} W_j(\theta)g_j(Y; \theta),$$

where the $W_j(\theta)$, now random variables, can depend on $Y_1, \ldots, Y_{j-1}$ and θ. This turns out to be

$$W_j(\theta) = \frac{-\mathrm{E}\{dg_j(Y;\theta)/d\theta \mid Y_1, \ldots, Y_{j-1}\}}{\mathrm{var}\{g_j(Y;\theta) \mid Y_1, \ldots, Y_{j-1}\}}. \tag{7.24}$$

This finite-sample result is independent of the asymptotic properties of $\tilde{\theta}$.

Example 7.21 (Branching process)   The branching process was first used to model the survival of surnames, it being supposed that a surname would die out if all every male bearing it had no sons, but it has applications in epidemic modelling and elsewhere. Each of the Y_{j-1} individuals in generation $j - 1$ independently gives birth to a random number of individuals, so $Y_j = \sum_{i=1}^{Y_{j-1}} N_i$, where the N_i are independent with mean θ and variance σ^2. We take $Y_0 = 1$. Here $g_j(Y;\theta) = Y_j - \theta Y_{j-1}$ is unbiased whatever the distribution of the N_i, while

$$\mathrm{var}\{g_j(Y;\theta) \mid Y_1, \ldots, Y_{j-1}\} = Y_{j-1}\sigma^2, \quad \mathrm{E}\left\{-\frac{dg_j(Y;\theta)}{d\theta}\bigg| Y_1, \ldots, Y_{n-1}\right\} = Y_{j-1}.$$

The optimal weights are $W_j(\theta) = 1/\sigma^2$, here non-random, and the corresponding estimating equation is $\sum_{j=2}^{n}(Y_j - \theta Y_{j-1}) = 0$, whatever the distribution of the N_i. Thus $\tilde{\theta} = \sum_{j=1}^{n-1} Y_{j+1}/\sum_{j=1}^{n-1} Y_j$ is optimal and $V = \sigma^2/\sum_{j=1}^{n} Y_{j-1}$.

Extinction is certain if $\theta \leq 1$ but not if $\theta > 1$. If extinction occurs then no estimator of θ can be consistent. When $\theta > 1$ and given that extinction does not occur, (7.23) implies that $V^{-1/2}(\tilde{\theta} - \theta) \xrightarrow{D} \sigma Z$. In this case $\theta^{-n}V$ converges to a nondegenerate random variable and the asymptotics are nonstandard. Confidence intervals for θ are best constructed using V.

Other growth models such as birth processes and non-stationary diffusions can also be non-ergodic. As the discussion above suggests, inference for θ is then best performed using observed information or its generalization V^{-1}.   ∎

The argument leading to (7.23) applies in particular to maximum likelihood estimators. We write $f(y_1, \ldots, y_n;\theta) = f(y_1;\theta) \prod_{j=2}^{n} f(y_j \mid y_1, \ldots, y_{j-1};\theta)$ and express the score as

$$\frac{d\ell(\theta)}{d\theta} = \frac{d \log f(Y_1;\theta)}{d\theta} + \sum_{j=2}^{n} \frac{d \log f(Y_j \mid Y_1, \ldots, Y_{j-1};\theta)}{d\theta} = \sum_{j=1}^{n} g_j(Y;\theta).$$

Here $W_j(\theta) \equiv 1$, so the unweighted score is optimal in finite samples. In the ergodic case, Taylor series arguments establish the usual properties of maximum likelihood estimators and likelihood ratio statistics, subject to regularity conditions like those needed for independent data.

Exercises 7.2

1   Show that if an estimating function undergoes a smooth 1–1 reparametrization by writing $g(y;\theta) = g\{y;\theta(\psi)\} = g'(y;\psi)$, then $\tilde{\theta} = \theta(\tilde{\psi})$. Establish also that (7.18) is unchanged.

2   Show that the sample median of a continuous density solves (7.15) with

$$g(y;\theta) = H(y - \theta) - H(\theta - y),$$

$H(u)$ is the Heaviside function.

giving $g(Y; \theta) = \sum \{I(\theta \leq Y_j) - I(Y_j \leq \theta)\}$, a descending staircase, with a unique solution only when n is odd. Find (7.18). Surprised?

3 Find the form of estimating function for an exponential family model.

4 To verify (7.17), show that the numerator and denominator in the first ratio may be written as $n^{1/2}\varepsilon_n$ and $n\zeta + n^{1/2}\eta_n$, where $\zeta \neq 0$ and ε_n and η_n are $O_p(1)$ random variables. Deduce that the ratio is $n^{-1/2}\varepsilon_n\zeta^{-1}(1 - n^{-1/2}\eta_n\zeta^{-1} + \cdots)$, and hence find the desired result.

5 Reread the proof of the Cramér–Rao lower bound, and then establish (7.19).

6 To establish (7.20), let C and G denote the $p \times p$ matrix $E\{-\partial g(Y; \theta)^T / \partial \theta\}$ and the $p \times 1$ vector $g(Y; \theta)$, note that $C = \text{cov}\{G, U(\theta)\}$ and, assuming that C is invertible, compute the variance matrix of $C^{-1}G - I(\theta)^{-1}U(\theta)$.

7 Let F_ν represent the gamma distribution with unit mean and shape parameter ν. Investigate how the quantity $t(F_\nu)$ determined by the Huber estimating function $g_c(y; \theta)$ depends on c and ν.

8 To establish (7.24), note that (7.18) depends on

$$
E\left\{\sum_{j=1}^{n} w_j^2 E_{j-1}\left(G_j^2\right)\right\}, \quad E\left\{\sum_{j=1}^{n} w_j^2 E_{j-1}\left(\frac{dG_j}{d\theta}\right)\right\},
$$

where E_{j-1} denotes expectation conditional on $Y_1, \ldots, Y_{j-1}$ and $G_j = g_j(Y; \theta)$. Call the sums here A^2 and B, so that (7.18) has inverse $\{E(B)\}^2 / E(A^2)$.
(a) Use the fact that $E\{(B/A - cA)^2\} \geq 0$ to show that $E(B)^2/E(A^2) \leq E(B^2/A^2)$.
(b) Deduce that $E(B^2/A^2)$ is maximized by (7.24), and show that this choice gives $E(B)^2/E(A^2) = E(B^2/A^2)$.
(c) Hence show that (7.18) is minimized among the class of estimating functions $\sum w_j(\theta)g_j(Y; \theta)$ by taking (7.24).
(Godambe, 1985)

9 Find the optimal estimating function based on dependent data $Y_1, \ldots, Y_n$ with $g_j(Y; \theta) = Y_j - \theta Y_{j-1}$ and $\text{var}\{g_j(Y; \theta) \mid Y_1, \ldots, Y_{j-1}\} = \sigma^2$. Derive also the estimator $\hat{\theta}$. Find the maximum likelihood estimator of θ when the conditional density of Y_j given the past is $N(\theta y_{j-1}, \sigma^2)$. Discuss.

7.3 Hypothesis Tests

7.3.1 Significance levels

A scientific theory or hypothesis leads to assertions that are testable using empirical data. Such data may discredit the hypothesis, as when the Michelson–Morley experiment demolished the nineteenth-century notion of an aether in which the earth and planets move, or they may lead to elaboration or development of it, just as quantum theory supercedes Newtonian mechanics but does not make Newton's laws of motion useless for daily life. One way to investigate the extent to which an assertion is supported by the data Y is to choose a *test statistic*, $T = t(Y)$, large values of which cast doubt on the assertion and hence on the underlying theory. This theory, the *null hypothesis* H_0, places restrictions on the distribution of Y and is used to calculate a *significance level* or *P-value*

$$
p_{\text{obs}} = \text{Pr}_0(T \geq t_{\text{obs}}), \tag{7.25}
$$

where t_{obs} is the value of T actually observed. A distribution computed under the assumption that H_0 is true is called a *null distribution*, and then we use $\mathrm{Pr}_0, \mathrm{E}_0, \ldots$ to indicate probability, expectation and so forth. Small values of p_{obs} correspond to values t_{obs} unlikely to arise under H_0, and signal that theory and data are inconsistent. The rationale for calculating the probability that $T \geq t_{\mathrm{obs}}$ in (7.25) is that any value $t' > t_{\mathrm{obs}}$ would cast even greater doubt on H_0. A hypothesis that completely determines the distribution of Y is called *simple*; otherwise it is *composite*.

If there is a precise idea what situation will hold if the null hypothesis is false, then there is a clearly specified *alternative hypothesis*, H_1, and we can choose a test statistic that has high probability of detecting departures from H_0 in the direction of H_1. Otherwise the alternative may be very vague. In either case calculation of (7.25) involves only H_0.

For many standard tests the null distribution of T is tabulated, available in statistical packages, or readily approximated. If not, (7.25) can be estimated by generating R independent sets of data Y_r^* from the null distribution of Y, calculating the corresponding values $T_r^* = t(Y_r^*)$, and then setting

$$\widehat{p}_{\mathrm{obs}} = \frac{1 + \sum_{r=1}^R I(T_r^* \geq t_{\mathrm{obs}})}{1 + R};\tag{7.26}$$

the added 1s here arise because under H_0 the original value t_{obs} is a realization of T and trivially $t_{\mathrm{obs}} \geq t_{\mathrm{obs}}$. The indicators $I(T_r^* \geq t_{\mathrm{obs}})$ are independent Bernoulli variables with probability p_{obs} under H_0, and this enables a suitable R to be determined (Exercise 7.3.1).

Example 7.22 (Exponential density) Consider an exponential random sample $Y_1, \ldots, Y_n$ with parameter λ. We wish to test $\lambda = \lambda_0$ against the alternative $\lambda = \lambda_1$, with both λ_0 and λ_1 known, using the likelihood ratio

$$T' = \frac{\lambda_1^n \exp\left(-\lambda_1 \sum Y_j\right)}{\lambda_0^n \exp\left(-\lambda_0 \sum Y_j\right)} = \exp\left\{(\lambda_0 - \lambda_1) \sum_{j=1}^n Y_j + n \log(\lambda_1/\lambda_0)\right\}.$$

We declare that doubt is cast on λ_0 if T' or equivalently $(\lambda_0 - \lambda_1) \sum Y_j$ is large. If $\lambda_1 < \lambda_0$, the value of p_{obs} is $\mathrm{Pr}_0(\sum Y_j > t_{\mathrm{obs}})$, where $t_{\mathrm{obs}} = \sum y_j$. Under the null hypothesis, $\sum Y_j$ has a gamma distribution with index n and rate λ_0, so if $\lambda_1 < \lambda_0$, the P-value is

$$p_{\mathrm{obs}} = \int_{t_{\mathrm{obs}}}^{\infty} \frac{\lambda_0^n u^{n-1}}{\Gamma(n)} e^{-\lambda_0 u} \, du = \int_{\lambda_0 t_{\mathrm{obs}}}^{\infty} \frac{v^{n-1}}{\Gamma(n)} e^{-v} \, dv = \mathrm{Pr}(V \geq \lambda_0 t_{\mathrm{obs}}),$$

where V has a gamma distribution with index n; p_{obs} can be calculated exactly because λ_0 and t_{obs} are known. ∎

Examples of situations with a vague alternative hypothesis are given below.

Interpretation

The significance level may be written as $p_{\mathrm{obs}} = 1 - F_0(t_{\mathrm{obs}})$, where F_0 is the null distribution function of T, supposed to be continuous. One interpretation of p_{obs}

stems from the corresponding random variable, $P = 1 - F_0(T)$. For $0 \leq u \leq 1$, its null distribution is

$$\Pr_0 \{1 - F_0(T) \leq u\} = \Pr_0 \left\{ F_0^{-1}(1 - u) \leq T \right\}$$
$$= 1 - F_0 \left\{ F_0^{-1}(1 - u) \right\} = u,$$

that is, uniform on the unit interval. Hence if we regard the observed t_{obs} as being just decisive evidence against H_0, then this is equivalent to following a procedure which rejects H_0 with error rate p_{obs}: if we tested many different hypotheses and rejected them all, the same t_{obs} having arisen in each case, then a proportion p_{obs} of our decisions would be incorrect. This interpretation applies exactly if F_0 is known, and the test is then called *exact*; otherwise it will typically apply only as an approximation in large samples.

A common misinterpretation of the P-value is as the probability that the null hypothesis is true. This cannot be the case, because alternative hypotheses play no direct role in its calculation. Bayesian P-values account for alternatives and do have this more direct interpretation; see Section 11.2.2.

Hypothesis testing is very useful in certain contexts but has important limitations. A first is that statistical significance of a result may be quite different from its practical importance, because even a very small p_{obs} may correspond to an uninteresting departure from the null hypothesis. For example, a test for lack of fit of a parametric model may be highly significant even though the model is satisfactory, simply because the fit is poor only in an unimportant part of the distribution or because the sample size is so large that no simple parametric model can be expected to fit well. On the other hand a large value of p_{obs} may arise when effects of real importance are undetectable because the sample size is too small. Computer models of climate change suggest that rare weather events may be occuring more frequently, for example, but most daily temperature series are too short to detect such small changes.

A second limitation is that even a very small P-value may sometimes indicate more support for the null than for an alternative hypothesis. A simple test of the null hypothesis $\mu = 0$ based on a single $N(\mu, 1)$ random variable with value $y = 3$ against the alternative hypothesis $\mu = 20$ has significance level $1 - \Phi(y) \doteq 0.001$, but $\mu = 0$ is clearly more plausible than $\mu = 20$.

A third limitation is that a P-value simply gives evidence against the null hypothesis and does not indicate which of a family of alternatives is best supported by the data. For this reason the use of confidence intervals for model parameters is generally preferable, when it is feasible.

Goodness of fit tests

In earlier chapters we used graphs such as probability plots to assess model fit. We now briefly discuss how to supplement such informal procedures with more formal ones. Suppose initially that the null hypothesis is that a random sample $Y_1, \ldots, Y_n$ has issued from a known continuous distribution $F(y)$. Then we can compare F with the empirical distribution function

$$\widehat{F}(y) = n^{-1} \sum_{j=1}^{n} I(Y_j \le y),$$

whose mean and variance are $F(y)$ and $F(y)\{1 - F(y)\}/n$ under H_0.

Standard measures of distance between F and $\widehat{F}$ include the Kolmogorov–Smirnov, Cramér–von Mises and Anderson–Darling statistics

$$\sup_{y} |\widehat{F}(y) - F(y)| = \max_{j} \left\{ j/n - U_{(j)}, U_{(j)} - (j-1)/n \right\},$$

$$\int_{-\infty}^{\infty} \{\widehat{F}(y) - F(y)\}^2 \, dF(y) = \frac{1}{12n^2} + \frac{1}{n} \sum_{j=1}^{n} \left(U_{(j)} - \frac{2j-1}{2n} \right)^2,$$

$$n \int_{-\infty}^{\infty} \frac{\{\widehat{F}(y) - F(y)\}^2}{F(y)\{1 - F(y)\}} \, dF(y) = -n - \sum_{j=1}^{n} \frac{2j-1}{n} \log \left\{ U_{(j)}(1 - U_{(n+1-j)}) \right\},$$

where the $U_j = F(Y_j)$ have a uniform null distribution and the $U_{(j)}$ are their order statistics; see Section 2.3. The first of these is simple and widely used, while the second and third put more weight on the tails; by allowing for the dependence of the variance of $\widehat{F}(y)$ on y, the third makes it easier to detect lack of fit for extreme values of y. All three statistics converge rapidly to their limiting distributions as $n \to \infty$, but simulation can be used to estimate P-values if tables are not at hand. The Kolmogorov–Smirnov statistic has 0.95 and 0.99 quantiles $1.358n^{-1/2}$ and $1.628n^{-1/2}$ for large n; significance is declared if the empirical distribution function of the $U_{(j)}$ passes confidence bands defined in terms of these quantiles. See Figures 6.14 and 6.20.

Example 7.23 (Danish fire data)   In Section 6.5.1 we saw that the rescaled times $u_1 = t_1/t_0, \ldots, u_n = t_n/t_0$ of the events of a homogeneous Poisson process observed on $[0, t_0]$ may be regarded as the order statistics of n uniform random variables. In this case, therefore, we can take $\widehat{F}(y) = n^{-1} \sum H(y - u_j)$ and $F(y) = y$, for $0 \le y \le 1$, and use the above tests to assess the adequacy of the Poisson process.

$H(u)$ is the Heaviside function.

The lower right panel of Figure 6.14 shows $\widehat{F}(y)$ for the 254 largest Danish fire claims, for which the Kolmogorov–Smirnov, Cramér–von Mises, and Anderson–Darling statistics equal 0.095, 0.002, and 2.718 respectively. To assess the significance of these values we computed the three statistics for 10,000 samples of 254 independent variables generated from the $U(0, 1)$ distribution. Just 207 of the simulated Kolmogorov–Smirnov statistics exceeded the observed value, giving significance level 0.0208. The solid diagonal lines show the regions within which $\widehat{F}$ would have to fall in order for significance not to be achieved at the 0.05 and 0.01 levels, the inner 0.05 lines are breached but the outer 0.01 ones are not, consistent with significance at the 0.02 level. The significance levels for the Cramér–von Mises and Anderson–Darling statistics were 0.0348 and 0.0397, so the rate function for the claims does seem to vary. This illustrates one drawback of generic tests of fit such as these, which can suggest that the model is inadequate, but not how.   ∎

Figure 7.5 Analysis of maize data. Left: empirical distribution function for height differences, with fitted normal distribution (dots). Right: null density of Anderson–Darling statistic T for normal samples of size $n = 15$ with location and scale estimated. The shaded part of the histogram shows values of T^* in excess of the observed value t_{obs}.

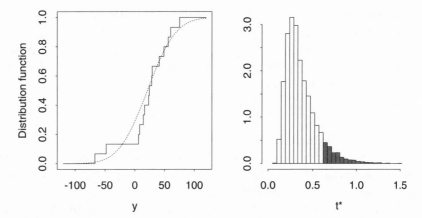

This example is atypical, because F generally depends on unknown parameters. An exact test may be available anyway, for example using the maximal invariant of a group transformation model. An observation from a location-scale model may be written as $Y = \eta + \tau \varepsilon$, where ε has known distribution G, and $F(y) = G\{(y - \eta)/\tau\}$. Most useful estimators are equivariant, with

$$\widehat{\eta}(Y_1, \ldots, Y_n) = \eta + \tau h_1(\varepsilon_1, \ldots, \varepsilon_n), \quad \widehat{\tau}(Y_1, \ldots, Y_n) = \tau h_2(\varepsilon_1, \ldots, \varepsilon_n).$$

Then the joint distribution of the residuals

$$\frac{Y_j - \widehat{\eta}}{\widehat{\tau}} = \frac{\eta + \tau \varepsilon_j - \eta + \tau h_1(\varepsilon_1, \ldots, \varepsilon_n)}{\tau h_2(\varepsilon_1, \ldots, \varepsilon_n)} = \frac{\varepsilon_j - h_1(\varepsilon_1, \ldots, \varepsilon_n)}{h_2(\varepsilon_1, \ldots, \varepsilon_n)}, \quad j = 1, \ldots, n,$$

depends only on G, h_1, and h_2 and not on the parameters. Thus the form of G may be tested by comparing the empirical and fitted distribution functions $\widehat{F}(y)$ and $G\{(y - \widehat{\eta})/\widehat{\tau}\}$.

Example 7.24 (Maize data) Under the matched pair model for the maize data of Table 1.1, the pairs of plants are independent and their height differences Y_j have mean η and variances $\tau = 2\sigma^2$. Our discussion in Section 3.2.2 presupposed that the Y_j are normally distributed, but the left panel of Figure 7.5 suggests that this may not be the case. To assess this we take $\widehat{\eta}$ and $\widehat{\tau}^2$ to be the sample average and variance, and compute the Anderson–Darling statistic based on the $(Y_j - \widehat{\eta})/\widehat{\tau}$. Its value is 0.618, with significance level $\widehat{p}_{\text{obs}} = 0.0874$ computed from the 10,000 simulations shown in the right panel of the figure. The assumption of normality seems reasonable. ∎

Similar ideas can be applied to other group transformation models. Among other goodness of fit tests are those based on the chi-squared statistics described in Section 4.5.3.

One- and two-sided tests

Often large and small values of T suggest different departures from the null hypothesis. Large values of goodness of fit statistics, for instance, imply that the model fits badly, but extremely small values might in some circumstances lead one to suspect that the data had been faked, the fit being too good to be true. With departures of two types it may be appropriate to use T^2 or equivalently $|T|$ as the test statistic, with significance level $\Pr_0(T^2 \geq t_{obs}^2)$. This is not useful in a case like Figure 7.5, however, owing to the asymmetry of the null density of T, and then we regard the test as having two possible implications, measured by

$$p_{obs}^+ = \Pr_0(T \geq t_{obs}), \quad p_{obs}^- = \Pr_0(T \leq t_{obs}),$$

corresponding to *one-sided tests*. Note that $p_{obs}^+ + p_{obs}^- = 1 + \Pr_0(T = t_{obs})$, which equals unity if the distribution of T is continuous. Let P^+ and P^- represent the random variables corresponding to these two-sided significance levels. If both large and small values of T may be regarded as evidence against H_0 we use $P = \min(P^+, P^-)$ as the overall test statistic, and take $\Pr_0\{P \leq \min(p_{obs}^+, p_{obs}^-)\}$ as the significance level. When the test is exact and T is continuous the density of P is uniform on the interval $(0, \frac{1}{2})$, and the two-sided significance level equals $2\min(p_{obs}^+, p_{obs}^-)$. This is the P-value for a *two-sided test*.

Example 7.25 (Student t test) Let $Y_1, \ldots, Y_n$ be a normal random sample with mean μ and variance σ^2. Suppose that the null hypothesis is $\mu = \mu_0$, and the two-sided alternative is that μ takes any other real value, with no restriction on σ^2 under either hypothesis. Both hypotheses are composite.

The likelihood ratio statistic is (Example 4.31)

$$W_p(\mu_0) = 2\left\{ \max_{\mu,\sigma^2} \ell(\mu, \sigma^2) - \max_{\sigma^2} \ell(\mu_0, \sigma^2) \right\} = n \log\left\{ 1 + \frac{T(\mu_0)^2}{n-1} \right\},$$

where the null distribution of $T(\mu_0) = (\overline{Y} - \mu_0)/(S^2/n)^{1/2}$ is t_{n-1}. As $W_p(\mu_0)$ is a monotone function of $T(\mu_0)^2$, the significance level is

$$p_{obs} = \Pr_0\{W_p(\mu_0) \geq w_{obs}\} = \Pr_0\left\{T^2(\mu_0) \geq t_{obs}^2\right\},$$

where w_{obs} and t_{obs} are the observed values of $W_p(\mu_0)$ and $T(\mu_0)$. Large values of w_{obs} arise when t_{obs} is distant from zero, suggesting that the population mean is not μ_0.

The results of Section 4.5 tell us that the null distribution of $W_p(\mu_0)$ is approximately χ_1^2. We could use this to approximate to p_{obs}, but an exact value is available, because

$$p_{obs} = \Pr_0\left\{T(\mu_0)^2 \geq t_{obs}^2\right\} = \Pr\left(T^2 \geq t_{obs}^2\right) = 2\Pr(T \geq |t_{obs}|), \tag{7.27}$$

where $T \sim t_{n-1}$. This is the P-value for the two-sided test.

If we suspect that $\mu > \mu_0$ but not that $\mu < \mu_0$, then large positive values of $T(\mu_0)$ will cast doubt on H_0, and the corresponding one-sided P-value is

$$p_{obs}^+ = \Pr_0\{T(\mu_0) \geq t_{obs}\} = \Pr(T \geq t_{obs}),$$

while $p_{obs}^- = \Pr(T \leq t_{obs})$ measures evidence against H_0 in the direction $\mu < \mu_0$. These differ slightly from the P-values for the one-sided likelihood ratio tests. The two-sided significance level

$$2 \min(p_{\text{obs}}^-, p_{\text{obs}}^+) = 2\Pr(|T| \geq |t_{\text{obs}}|)$$

equals (7.27). ∎

Nonparametric tests

The examples above concern tests in parametric models, where hypotheses typically determine values of the parameters, the form of the density being supposed known. *Nonparametric tests* presuppose that the data are independently sampled from an unspecified underlying model.

Example 7.26 (Sign test) A random sample $Y_1, \ldots, Y_n$ arises from an unknown distribution F. The null hypothesis H_0 asserts that F has median μ equal to μ_0, while the alternative is that $\mu > \mu_0$. Both hypotheses are composite, but neither specifies a parametric model, and we argue as follows.

If the median is μ_0, the probability that an observation Y falls on either side of μ_0 is $1/2$, and if the median is greater than μ_0, then $\Pr(Y > \mu_0) > 1/2$. This suggests that we base a test on $S = \sum_{j=1}^{n} I(Y_j > \mu_0)$, large values of which cast doubt on H_0. Under the null hypothesis, S has a binomial distribution with denominator n and probability $1/2$, so its mean and variance are $n/2$ and $n/4$. Hence the P-value is

$$p_{\text{obs}} = \Pr_0(S \geq s_{\text{obs}}) = \sum_{r=s_{\text{obs}}}^{n} \binom{n}{r} \frac{1}{2^n} \doteq 1 - \Phi\left\{ \frac{2(s_{\text{obs}} - n/2)}{n^{1/2}} \right\},$$

by normal approximation to the binomial null distribution of S. ∎

Example 7.27 (Wilcoxon signed-rank test) A random sample $Y_1, \ldots, Y_n$ has been drawn from a density that is symmetric about μ but otherwise unspecified. We wish to test the hypothesis that $\mu = 0$. The sign test is one possibility, but as it does not use the symmetry of the density, a better test can be found.

Let R_j denote the rank of $|Y_j|$ among $|Y_1|, \ldots, |Y_n|$, and let $Z_j = \text{sign}(Y_j)$. The *Wilcoxon signed-rank statistic* is $W = \sum_j Z_j R_j$. Large positive values of W suggest $\mu > 0$, while large negative values suggest $\mu < 0$.

To find the null mean and variance of W, note that when $\mu = 0$ the ranks, R_j, are independent of the signs, Z_j, by symmetry about zero, and that

$$\text{var}_0(Z_j) = (-1)^2 \frac{1}{2} + 1^2 \frac{1}{2} = 1, \quad \text{E}_0(Z_j R_j) = n^{-1} \sum_{k=1}^{n} \left\{ k \frac{1}{2} + (-k) \frac{1}{2} \right\} = 0,$$

implying that $\text{E}_0(W) = 0$. To find $\text{var}_0(W)$, we argue conditionally on the ranks $R_1, \ldots, R_n$, finding

$$\text{var}_0\left(\sum_{j=1}^{n} Z_j R_j \,\middle|\, R_1, \ldots, R_n \right) = \sum_{j=1}^{n} R_j^2 \text{var}_0(Z_j) = \sum_{j=1}^{n} R_j^2 = \sum_{j=1}^{n} j^2,$$

and this equals $n(n + 1)(2n + 1)/6$. Thus W has mean zero and variance $n(n + 1)(2n + 1)/6$ under the null hypothesis, and as its distribution is then symmetric, a normal approximation to the exact P-value may be useful. ∎

Table 7.2 Analysis of differences for maize data.

Difference d	49	−67	8	16	6	23	28	41	14	29	56	24	75	60	−48
Sign z	+	−	+	+	+	+	+	+	+	+	+	+	+	+	−
Rank r	11	14	2	4	1	5	7	9	3	8	12	6	15	13	10

Example 7.28 (Maize data) Under the model for the maize data of Table 1.1, the height differences between cross- and self-fertilized plants may be written as $D_j = \eta + \sigma(\varepsilon_{2j} - \varepsilon_{1j})$, where the ε_{ij} are independent random variables with mean zero and some common variance. If the ε_{ij} have the same distribution, the D_j will be symmetically distributed around η, while $\eta = 0$ under the null hypothesis H_0 of no difference between the effects of the different types of fertilization. If cross-fertilization increases height, then $\eta > 0$, as is suggested by the observed d_j in Table 7.2.

If the D_j were normally distributed, we would perform a Student t test based on the average and variance of the observed differences, $\bar{d} = 20.95$ and $s^2 = 1424.6$, giving $t_{\text{obs}} = n^{1/2}(\bar{d} - 0)/s = 2.15$; see Example 7.25. Under H_0 this is the realized value of a t_{14} variable, so $p_{\text{obs}} = \Pr(T \geq t_{\text{obs}}) = 0.025$, where $T \sim t_{14}$. Though low, this is not overwhelming evidence against the null hypothesis.

If we wish to avoid the assumption of normality, a nonparametric test is preferable. Under the null hypothesis, the D_j come from density symmetric about zero but not necessarily normal. Thirteen of them are positive, so the sign test statistic takes value $s_{\text{obs}} = 13$, with exact significance level

$$\Pr_0(S \geq s_{\text{obs}}) = \frac{1}{2^{15}} \sum_{r=13}^{15} \binom{15}{r} = \frac{1}{2^{15}}(1 + 15 + 105) = 0.0037;$$

normal approximation gives $1 - \Phi\{2(13 - 15/2)/\sqrt{15}\} = 0.0023$. Both give much stronger evidence against H_0 than does the t test.

Table 7.2 shows the quantities needed for the Wilcoxon signed-rank test. The observed value of $W = \sum Z_j R_j$ is 72, and its null distribution when $n = 15$ is approximately normal with mean zero and variance 1240. Therefore the P-value is roughly

$$p_{\text{obs}} = \Pr_0(W \geq 57) \doteq 1 - \Phi(57/1240^{1/2}) = 0.053,$$

to be compared with the values for the t and sign tests.   ∎

We shall see in Section 7.3.2 that likelihood considerations lead to tests that are 'best' in a certain sense when there is a parametric model. But if the model is not credible, nonparametric tests that make make fewer assumptions may be preferable, and often they perform nearly as well as parametric tests. Some situations are so ill-specified that parametric models are inappropriate, and the independence assumptions that underlie most nonparametric tests are doubtful also. Then only rough-and-ready methods can be applied and conclusions are correspondingly weaker.

7.3.2 Comparison of tests

We now consider how to compare different test statistics for the same problem. Having chosen a test statistic $T = t(Y)$ and a probability α, suppose we decide to reject the null hypothesis H_0 in favour of an alternative H_1 at level α if and only if the data Y fall into the subset $\mathcal{Y}_\alpha = \{y : t(y) \geq t_\alpha\}$ of the sample space, where t_α is chosen so that

$$\Pr_0(T \geq t_\alpha) = \Pr_0(Y \in \mathcal{Y}_\alpha) = \alpha.$$

The *size* of the test is the probability α of rejecting H_0 when it is actually true, and $\mathcal{Y}_\alpha$ is called a *size α critical region*. This construction implies that as α decreases, t_α increases and that $\mathcal{Y}_{\alpha_1} \subset \mathcal{Y}_{\alpha_2}$ whenever $\alpha_1 \leq \alpha_2$, as is essential if we are to avoid imbecilities such as 'H_0 is rejected when $\alpha = 0.01$ but not when $\alpha = 0.05$'. Choosing a test statistic and values of t_α is equivalent to specifying a system of critical regions for the different values of α, so we can discuss the test in terms of its critical regions if convenient.

By using a fixed α we have moved from regarding the significance level as a measure of evidence against H_0 to using the test to decide which of the two hypotheses is better supported by the data. Two wrong decisions are then possible, committing a *Type I* error by rejecting H_0 when it is true, or a *Type II* error by accepting H_0 when H_1 is true. The *power* of the test is the probability of detecting that H_0 is false,

$\Pr_1, E_1$ and so forth
indicate probability,
expectation and so forth
computed under H_1.

$$\Pr_1(T \geq t_\alpha) = \Pr_1(Y \in \mathcal{Y}_\alpha).$$

Example 7.29 (Normal mean) Let $Y_1, \ldots, Y_n$ be a random sample from the $N(\mu, \sigma^2)$ distribution with known σ^2, and suppose that H_0 specifies that $\mu = \mu_0$, whereas $\mu > \mu_0$ under H_1. Suppose we decide to reject H_0 if $\overline{Y}$ exceeds some constant t_α. Under H_0, $\overline{Y} \sim N(\mu_0, \sigma^2/n)$, so this test has size

$$\Pr_0(\overline{Y} \geq t_\alpha) = \Pr_0 \left\{ n^{1/2} \frac{(\overline{Y} - \mu_0)}{\sigma} \geq n^{1/2} \frac{(t_\alpha - \mu_0)}{\sigma} \right\}$$

$$= 1 - \Phi \left\{ \frac{n^{1/2}(t_\alpha - \mu_0)}{\sigma} \right\} = \Phi \left\{ \frac{n^{1/2}(\mu_0 - t_\alpha)}{\sigma} \right\},$$

using the symmetry of the normal distribution. For a test of size α, we must choose t_α such that

z_α is the α quantile of the
$N(0, 1)$ distribution.

$$\frac{n^{1/2}(\mu_0 - t_\alpha)}{\sigma} = z_\alpha,$$

giving $t_\alpha = \mu_0 - n^{-1/2}\sigma z_\alpha$. Thus the size α critical region is

$$\mathcal{Y}_\alpha = \left\{ (y_1, \ldots, y_n) : \overline{y} \geq \mu_0 - n^{-1/2}\sigma z_\alpha \right\},$$

and we can decide if Y falls into this because σ^2 and μ_0 are known under H_0.

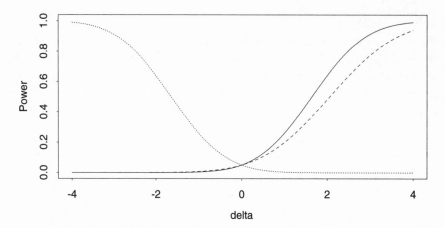

Figure 7.6 Power functions for a test of whether the mean of a $N(\mu, \sigma^2)$ random sample of size n equals μ_0 against the alternative $\mu = \mu_1$, as a function of $\delta = n^{1/2}(\mu_1 - \mu_0)/\sigma$. The test size is $\alpha = 0.05$. The solid curve is the power function for a test of $\mu_1 > \mu_0$ based on $\overline{y}$, and the dashed line is the power function for the sign test. Both critical regions are of form $\overline{y} > t_\alpha$. The dotted curve is the power function for $\overline{y}$ when the critical region is $\overline{y} < t_\alpha$.

If in fact μ equals $\mu_1 > \mu_0$, then $\overline{Y} \sim N(\mu_1, \sigma^2/n)$, and the test has power

$$\mathrm{Pr}_1\left(\overline{Y} \geq \mu_0 - \frac{\sigma z_\alpha}{n^{1/2}}\right) = \mathrm{Pr}_1\left(n^{1/2}\frac{(\overline{Y} - \mu_1)}{\sigma} \geq n^{1/2}\frac{(\mu_0 - \mu_1)}{\sigma} - z_\alpha\right)$$

$$= 1 - \Phi(-\delta - z_\alpha) = \Phi(z_\alpha + \delta), \qquad (7.28)$$

where $\delta = n^{1/2}(\mu_1 - \mu_0)/\sigma$ measures the distance between the means under the two hypotheses, standardized by $\mathrm{var}(\overline{Y})^{1/2} = \sigma/n^{1/2}$. The power is plotted in Figure 7.6, with $\alpha = 0.05$. For fixed n, σ, and μ_0, it increases with μ_1. When σ, μ_0, and μ_1 are fixed, the power increases with n.

Power can be used to choose the sample size when planning an experiment. Suppose we desire to perform a test of size α and that power of at least β is sought for detecting whether $\mu_1 = \mu_0 + \sigma\gamma$, where γ is known. Then we require $\Phi(z_\alpha + n^{1/2}\gamma) \geq \beta$ and hence $z_\alpha + n^{1/2}\gamma \geq \Phi^{-1}(\beta)$ or equivalently $n \geq (z_\beta - z_\alpha)^2/\gamma^2$.

If, for instance, $\mu_0 = 0$ and $\sigma = 1$, and we desire to detect whether a test of size 0.05 could detect $\mu_1 = 0.5$ with power 0.8 or more, then $\gamma = 0.5$, $z_\alpha = -1.645$, $z_\beta = 0.842$ and hence we would need $n \geq 24.7 \doteq 25$. ∎

Example 7.30 (Sign test) Example 7.26 describes a test for the median of a distribution to equal a specified value μ_0, using $S = \sum_{j=1}^{n} I(Y_j > \mu_0)$ as test statistic. Under H_0 the distribution of S is binomial, and if a normal approximation applies, a size α critical region is determined by the value s_α such that $\mathrm{Pr}_0(S \geq s_\alpha) = \alpha$, giving $s_\alpha = n/2 - n^{1/2}z_\alpha/2$.

For an illustrative power calculation for this test, let $Y_1, \ldots, Y_n \overset{\mathrm{iid}}{\sim} N(\mu, \sigma^2)$, with null hypothesis $\mu = \mu_0$ and alternative H_1 that $\mu = \mu_1 > \mu_0$. The normal density is symmetric, so its mean equals its median. Now

$$\mathrm{Pr}_1(Y_j \geq \mu_0) = \mathrm{Pr}_1\{(Y_j - \mu_1)/\sigma \geq (\mu_0 - \mu_1)/\sigma\} = \Phi(n^{-1/2}\delta),$$

where again $\delta = n^{1/2}(\mu_1 - \mu_0)/\sigma$. Under H_1, therefore, S is approximately normal with mean $n\Phi(n^{-1/2}\delta)$ and variance $n\Phi(n^{-1/2}\delta)\{1 - \Phi(n^{-1/2}\delta)\}$, and the probability that H_0 is rejected is

$$\Pr_1(S \geq s_\alpha) = \Pr_1\left(S \geq n/2 - n^{1/2}z_\alpha/2\right)$$

$$\doteq \Phi\left\{\frac{n\Phi\left(n^{-1/2}\delta\right) - n/2 + n^{1/2}z_\alpha/2}{\left[n\Phi\left(n^{-1/2}\delta\right)\left\{1 - \Phi\left(n^{-1/2}\delta\right)\right\}\right]^{1/2}}\right\},$$

using the normal approximation to the binomial distribution. For n large, $\Phi(n^{-1/2}\delta) \doteq \frac{1}{2} + n^{-1/2}\delta\phi(0) = \frac{1}{2} + (2\pi n)^{-1/2}\delta$, and after simplifying,

$$\Pr_1(S \geq s_\alpha) \doteq \Phi\left\{z_\alpha + \delta(2/\pi)^{1/2}\right\}. \tag{7.29}$$

As $(2/\pi)^{1/2} < 1$, the sign test has lower power than does the test using $\overline{Y}$ in Example 7.29. That test has power $\Phi(z_\alpha + \delta)$, so it requires smaller samples to attain a given power than does the test based on S. Figure 7.6 compares the power functions with $\alpha = 0.05$. Sign tests have rather low power, and better tests are almost always possible. ∎

Although power is important in planning an experiment, in giving a basis for choosing the sample size required, and in assessing the size of effects that could reasonably be detected from a given set of data, it plays no role in conducting the test itself, which simply requires a tail probability computed under the null distribution.

Neyman–Pearson lemma

Egon Sharpe Pearson (1895–1980), the second child of Karl Pearson, was very unlike his combative father. After school in Oxford and Winchester his studies in Cambridge were interrupted by illness and the 1914–18 war. He took his degree in 1920 and began work at University College London, where he stayed the rest of his life. Apart from broad contributions to statistical theory, he pioneered industrial quality control and was editor of the statistical journal *Biometrika* from 1936–1966.

Other things being equal, a test with high power is preferable to one with low power. But in order for a comparison of two tests to be fair, they must compete on an equal footing. This leads us to compare them in terms of their power for fixed size. That is, out of all possible tests with a given size, we aim to find the one with highest power.

Let $f_0(y)$ and $f_1(y)$ denote the probability densities of Y under the null and alternative hypotheses. Then the *Neyman–Pearson lemma* states that the most powerful test of size α has critical region

$$\mathcal{Y} = \left\{y : \frac{f_1(y)}{f_0(y)} \geq t_\alpha\right\}, \quad t_\alpha \geq 0,$$

determined by the likelihood ratio, if such a region exists. To explain this, suppose that such a region does exist and let $\mathcal{Y}'$ be any other critical region of size α or less. Then for any density f,

$$\int_{\mathcal{Y}} f(y)\, dy - \int_{\mathcal{Y}'} f(y)\, dy,$$

$\overline{\mathcal{Y}}$ is the complement of $\mathcal{Y}$ in the sample space.

equals

$$\int_{\mathcal{Y}\cap\mathcal{Y}'} f(y)\, dy + \int_{\mathcal{Y}\cap\overline{\mathcal{Y}'}} f(y)\, dy - \int_{\mathcal{Y}'\cap\mathcal{Y}} f(y)\, dy - \int_{\mathcal{Y}'\cap\overline{\mathcal{Y}}} f(y)\, dy,$$

and this is

$$\int_{\mathcal{Y}\cap\overline{\mathcal{Y}'}} f(y)\, dy - \int_{\mathcal{Y}'\cap\overline{\mathcal{Y}}} f(y)\, dy. \tag{7.30}$$

If $f = f_0$, this expression is non-negative, because $\mathcal{Y}'$ has size at most that of $\mathcal{Y}$. Suppose that $f = f_1$. If $y \in \overline{\mathcal{Y}}$, then $t_\alpha f_0(y) > f_1(y)$, while $f_1(y) \geq t_\alpha f_0(y)$ if $y \in \mathcal{Y}$. Hence when $f = f_1$, (7.30) is no smaller than

$$t_\alpha \left\{ \int_{\mathcal{Y} \cap \overline{\mathcal{Y}'}} f_0(y)\, dy - \int_{\mathcal{Y}' \cap \overline{\mathcal{Y}}} f_0(y)\, dy \right\} \geq 0.$$

Thus the power of $\mathcal{Y}$ is at least that of $\mathcal{Y}'$, and the result is established.

It may happen that H_0 is simple and the alternative is composite, but that the likelihood ratio critical region is most powerful for each component of the alternative hypothesis. Then $\mathcal{Y}$ is said to be *uniformly most powerful*.

Example 7.31 (Exponential family)   Consider testing the null hypothesis $\theta = \theta_0$ against the one-sided alternative $\theta = \theta_1 > \theta_0$ based on a random sample $Y_1, \ldots, Y_n$ from the one-parameter exponential family

$$f(y; \theta) = \exp\left\{ s(y)\theta - \kappa(\theta) + c(y) \right\}.$$

The likelihood ratio is

$$\exp\left\{ (\theta_1 - \theta_0) \sum_{j=1}^{n} s(Y_j) + \kappa(\theta_0) - \kappa(\theta_1) \right\},$$

so for each $\theta_1 > \theta_0$ the most powerful size α critical region is

$$\mathcal{Y}_\alpha = \left\{ (y_1, \ldots, y_n) : \sum s(y_j) \geq t_\alpha' \right\},$$

if a t_α' can be found such that $\mathrm{Pr}_0(Y \in \mathcal{Y}_\alpha) = \alpha$. This test is therefore uniformly most powerful against this one-sided alternative. When $\theta_1 < \theta_0$, the same argument shows that a uniformly most powerful critical region is obtained by replacing $\geq$ by $\leq$ in the above definition of $\mathcal{Y}_\alpha$.

A special case of this is the exponential density of Example 7.22, where the uniformly most powerful critical region of size α against one-sided alternatives $\lambda_1 < \lambda_0$ is $\mathcal{Y}_\alpha = \{(y_1, \ldots, y_n) : \sum y_j > t_\alpha'\}$, with $\lambda_0 t_\alpha'$ the $(1 - \alpha)$ quantile of the gamma distribution with unit scale and shape parameter n.

In discrete models uniformly most powerful tests of every size do not exist. In the Poisson case, for example, the null distribution of $\sum s(Y_j) = \sum Y_j$ is Poisson with mean $n\theta_0$, so $\mathcal{Y}_\alpha$ has possible sizes

$$\mathrm{Pr}_0\left(\sum_{j=1}^{n} Y_j \geq t_\alpha' \right) = \sum_{u=t_\alpha'}^{\infty} \frac{(n\theta_0)^u}{u!} \exp(-n\theta_0), \quad t_\alpha = 0, 1, \ldots.$$

Setting $n\theta_0 = 5$, for example, gives sizes $1.00, 0.993, \ldots, 0.068, 0.032, \ldots$, so a likelihood ratio critical region of size 0.05 does not exist. This does not affect the computation of a significance level, whose value is not pre-specified.   ∎

This last example shows that construction of a likelihood ratio critical region of exact size α may be impossible. If so, a *randomized test* may be used to obtain the exact size required. Suppose that critical regions of size α_1 and α_2 are available, where $\alpha_1 < \alpha < \alpha_2$. Then if I is a Bernoulli variable with success probability $p = (\alpha_2 - \alpha)/(\alpha_2 - \alpha_1)$, the test with region

$$\mathcal{Y} = \begin{cases} \mathcal{Y}_{\alpha_1}, & I = 1, \\ \mathcal{Y}_{\alpha_2}, & I = 0 \end{cases}$$

has size α. In the previous example we might take $\alpha = 0.05$, $\alpha_1 = 0.032$ and $\alpha_2 = 0.068$, giving $p = 0.5$. Then each time the test was conducted, we would flip a coin to decide whether to use $\mathcal{Y}_{\alpha_1}$ or $\mathcal{Y}_{\alpha_2}$ as the critical region. Although this trick is useful in theoretical calculations, it introduces a random element unrelated to the data. In applications it is preferable to compute a significance level and weigh the evidence accordingly.

Example 7.32 (Normal mean) In Example 7.29 the likelihood ratio for testing $\mu = \mu_0$ against $\mu = \mu_1$ with σ known is

$$\frac{f_1(Y)}{f_0(Y)} = \frac{(2\pi\sigma^2)^{-n/2} \exp\left\{ -\frac{1}{2\sigma^2} \sum_{j=1}^{n} (Y_j - \mu_1)^2 \right\}}{(2\pi\sigma^2)^{-n/2} \exp\left\{ -\frac{1}{2\sigma^2} \sum_{j=1}^{n} (Y_j - \mu_0)^2 \right\}}$$

$$= \exp\left[\frac{1}{2\sigma^2} \left\{ 2n\overline{Y}(\mu_1 - \mu_0) - \mu_1^2 + \mu_0^2 \right\} \right].$$

If $\mu_1 > \mu_0$, this is monotone increasing in $\overline{Y}$ for any fixed μ_1 and μ_0, and so the critical region rejects H_0 when $\overline{Y} \geq t'_\alpha$, with t'_α chosen to give a test of size α. Hence the size α critical region is

$$\mathcal{Y}_\alpha^+ = \left\{ (y_1, \ldots, y_n) : n^{1/2}(\overline{y} - \mu_0)/\sigma \geq z_{1-\alpha} \right\};$$

this is most powerful for any $\mu_1 > \mu_0$ and so is uniformly most powerful. The region

$$\mathcal{Y}_\alpha^- = \left\{ (y_1, \ldots, y_n) : n^{1/2}(\overline{y} - \mu_0)/\sigma \leq z_\alpha \right\}$$

is likewise uniformly most powerful against alternatives $\mu_1 < \mu_0$.

Suppose that we wish to test the same null hypothesis against the two-sided alternative that $\mu \neq \mu_0$. The null distribution of $\overline{Y}$ is symmetric about μ_0, so it is natural to use

$$\mathcal{Y}_\alpha = \left\{ (y_1, \ldots, y_n) : n^{1/2}|\overline{y} - \mu_0|/\sigma \geq z_{\alpha/2} \right\}. \tag{7.31}$$

This critical region has size α but is not uniformly most powerful against the two-sided alternative. When $\mu_1 > \mu_0$, $\mathcal{Y}_\alpha^+$ has size α and has higher power, while when $\mu_1 < \mu_0$, $\mathcal{Y}_\alpha^-$ has size α and has higher power. The power of a uniformly most powerful two-sided critical region would equal those of $\mathcal{Y}_\alpha^+$ for alternatives $\mu_1 > \mu_0$ and of $\mathcal{Y}_\alpha^-$ for $\mu_1 < \mu_0$, but its size would have to be α, whereas $\mathcal{Y}_\alpha^- \cup \mathcal{Y}_\alpha^+$ has size 2α. In fact no uniformly most powerful test exists for this two-sided alternative. This difficulty can also arise in other contexts. ∎

This last example highlights a problem with two-sided tests. One approach to dealing with it is to say that a critical region $\mathcal{Y}$ is *unbiased* if

$$\Pr_1(Y \in \mathcal{Y}) \geq \Pr_0(Y \in \mathcal{Y})$$

for all alternative hypotheses under consideration. This implies that the probability of rejecting H_0 is higher under any H_1 than under H_0, and would rule out using the critical regions $\mathcal{Y}_\alpha^+$ and $\mathcal{Y}_\alpha^-$ for two-sided tests in the previous example. If $\mu_1 < \mu_0$, for example, then $\Pr_1(Y \in \mathcal{Y}_\alpha^+) = \Phi(z_\alpha + \delta) < \alpha$ because $\delta < 0$, and hence $\mathcal{Y}_\alpha^+$ would be biased. There is a well-developed mathematical theory of such tests, but they are of little practical interest. To see why, suppose that the two-sided unbiased region $\mathcal{Y}_\alpha$ had been used in the previous example, and that doubt had been cast on the null hypothesis $\mu = \mu_0$. The test being two-sided, it would then be natural to ask whether the data suggest that $\mu > \mu_0$ or $\mu < \mu_0$, leading to use of one-sided regions such as $\mathcal{Y}_\alpha^-$ and $\mathcal{Y}_\alpha^+$. It seems more sensible to perform two one-sided tests and obtain an overall P-value by combining the individual significance levels, as outlined in Section 7.3.1. This amounts to using two one-sided tests each of size α, and in general this is not the same as an unbiased test of size 2α.

Local power

We now consider how the likelihood ratio behaves under a local alternative, when the null and alternative models $f_0(y) = f(y; \theta_0)$ and $f_1(y) = f(y; \theta_1)$ depend on a scalar parameter θ, and $\theta_1 = \theta_0 + \epsilon$ for some small ϵ. Then

$$\frac{f_1(Y)}{f_0(Y)} = \frac{f(Y; \theta_0 + \epsilon)}{f(Y; \theta_0)} = \frac{1}{f(Y; \theta_0)} \left\{ f(Y; \theta_0) + \epsilon \frac{df(Y; \theta_0)}{d\theta_0} + \cdots \right\}$$
$$\doteq 1 + \epsilon U(\theta_0),$$

where $U(\theta) = d \log f(Y; \theta)/d\theta$ is the score statistic. As $\epsilon \to 0$, this expansion shows that the likelihood ratio and score statistics are equivalent, so the Neyman–Pearson lemma implies that a locally most powerful test against H_0 may be based on large values of the score statistic. This is a *score test*.

In large samples from regular models the null distribution of $U(\theta_0)$ is approximately normal with mean zero and variance equal to the Fisher information $I(\theta_0)$, so a locally most powerful critical region has form

$$\left\{ (y_1, \ldots, y_n) : u(\theta_0) \geq I(\theta_0)^{1/2} z_{1-\alpha} \right\}.$$

Under the alternative hypothesis, $U(\theta_0)$ has mean

$$\int u(\theta_0) f(y; \theta_0 + \epsilon) \, dy = \int u(\theta_0) \left\{ f(y; \theta_0) + \epsilon u(\theta_0) f(y; \theta_0) + \cdots \right\} dy$$
$$\doteq \epsilon \int u(\theta_0)^2 f(y; \theta_0) \, dy = \epsilon I(\theta_0),$$

while its variance is $I(\theta_0) + O(n\epsilon)$. Hence the local power of the score test is

$$\Pr_1 \left\{ U(\theta_0) \geq I(\theta_0)^{1/2} z_{1-\alpha} \right\} \doteq \Phi(z_\alpha + \delta),$$

analogous to (7.28), with $\delta = I(\theta_0)^{1/2}(\theta_1 - \theta_0) = n^{1/2}(\theta_1 - \theta_0)/i(\theta_0)^{-1/2}$ playing the role of $n^{1/2}(\mu_1 - \mu_0)/\sigma$ in Example 7.29. Thus the power of the test is increased when the null Fisher information per observation $i(\theta_0)$ is large, when n is large, or when θ_1 is distant from θ_0.

Example 7.33 (Gamma density) Suppose that $Y_1, \ldots, Y_n$ is a random sample from the gamma density

$$f(y; \mu, \nu) = \frac{\nu^\nu y^{\nu-1}}{\Gamma(\nu)\mu^\nu} \exp(-\nu y/\mu), \quad y > 0, \nu, \mu > 0.$$

We consider testing if $\nu = 1$, that is, that the density is in fact exponential. Initially we suppose that μ is known. The log likelihood contribution from a single observation is $\nu \log \nu + (\nu - 1) \log y - \nu \log \mu - \nu y/\mu - \log \Gamma(\nu)$, so

$$U(\nu) = \sum_{j=1}^{n} \left\{ \log\left(\frac{Y_j}{\mu}\right) - \frac{Y_j}{\mu} + 1 - \log \nu - \frac{d \log \Gamma(\nu)}{d\nu} \right\},$$

$$I(\nu) = n \left\{ \frac{d^2 \log \Gamma(\nu)}{d\nu^2} - \frac{1}{\nu} \right\}.$$

An asymptotic test of $\nu = 1$ therefore consists in comparing $U(1)/I(1)^{1/2}$ with the standard normal distribution.

In practice an unknown μ is replaced by its maximum likelihood estimator under the null hypothesis, $\widehat{\mu} = \overline{Y}$. Then the large-sample distribution of the score is given by (4.48) with $\psi = \nu$ and $\lambda = \mu$. In this case the off-diagonal element of the Fisher information matrix is $I_{\lambda\psi} = \mathrm{E}(-\partial^2 \ell/\partial \mu \partial \nu) = 0$, so the test involves replacing μ by $\overline{Y}$. ∎

7.3.3 Composite null hypotheses

Thus far we have supposed that the null hypothesis is simple, that is, it fully specifies the null distribution of the test statistic. An exact significance level, perhaps estimated by simulation, is then in principle available. In practice exact tests are usually unobtainable because the null distribution of Y depends on unknowns. In the most common setting there is a nuisance parameter λ and a parameter of interest ψ, and the null hypothesis imposes the constraint $\psi = \psi_0$ but puts no restriction on λ. Most of the tests in preceding chapters were of this sort. The P-value may then be written

$$\mathrm{Pr}_0(T \geq t_{\mathrm{obs}}) = \mathrm{Pr}(T \geq t_{\mathrm{obs}}; \psi_0, \lambda) = \int_{\{y : t(y) \geq t_{\mathrm{obs}}\}} f(y; \psi_0, \lambda) \, dy. \tag{7.32}$$

In general this depends on λ, perhaps strongly, but sometimes a critical region $\mathcal{Y}_\alpha$ of size α can be found such that

$$\mathrm{Pr}(Y \in \mathcal{Y}_\alpha; \psi_0, \lambda) = \alpha \quad \text{for all } \lambda.$$

Such a $\mathcal{Y}_\alpha$ is called a *similar region*; it is similar to the sample space, which satisfies this equation with $\alpha = 1$. A test whose critical regions are similar is called a *similar test* and is clearly desirable if it can be found. The two main approaches to finding exact tests are use of conditioning and appeal to invariance. Before discussing these, we outline approximate ways to reduce the dependence of (7.32) on λ.

One simple idea is to replace λ by $\widehat{\lambda}_0$, the maximum likelihood estimator of λ when $\psi = \psi_0$, but this is generally unsatisfactory because the result still depends on λ, albeit to a lower order. It is better to base the test on a pivot, exact or approximate. We have already extensively used an important example of this, the likelihood ratio statistic $W_p(\psi_0) = 2\{\ell(\widehat\psi, \widehat\lambda) - \ell(\psi_0, \widehat\lambda_0)\}$. Under regularity conditions its distribution for a large sample size n is χ_p^2, where p is the dimension of ψ, and in fact as

$$\Pr\{W_p(\psi_0) \le c_p(\alpha); \psi_0, \lambda\} = \alpha\{1 + O(n^{-1})\} \quad \text{for all } \lambda, \tag{7.33}$$

tests based on $W_p(\psi_0)$ are approximately similar. In continuous models the error in (7.33) can be reduced by noting that $E_0\{W_p(\psi_0)\} \doteq p\{1 + b(\theta_0)/n\}$, where $b(\theta_0) = b(\psi_0, \lambda)$ conveys how much the null mean of $W_p(\psi_0)$ differs from its asymptotic value. Tedious calculations establish that

$$\Pr\{W_p(\psi_0)\{1 + b(\widehat\theta_0)\}^{-1} \le c_p(\alpha); \psi_0, \lambda\} = \alpha\{1 + O(n^{-2})\} \quad \text{for all } \lambda,$$

where $\widehat\theta_0 = (\psi_0, \widehat\lambda_0)$. Thus division of the likelihood ratio statistic to make its mean closer to p improves the quality of the χ^2 approximation to its entire distribution. *Bartlett adjustment* of this sort can decrease substantially the error in (7.33), and may be valuable if n is small or if the dimension of λ is appreciable.

Conditioning

When there is a minimal sufficient statistic S_0 for the unknown λ in a null distribution, it may be removed by conditioning, giving P-value

$$\Pr_0(T \ge t_{\text{obs}} \mid S_0; \psi_0) = \int_{\{y:t(y)\ge t_{\text{obs}}\}} f(y \mid s_0; \psi_o)\, dy,$$

which is independent of λ by sufficiency of S_0. If S_0 is boundedly complete, this is the only way to construct a test statistic with P-values independent of λ. To see why, let $\mathcal{Y}_\alpha$ be a critical region of size α for all λ. Then

$$0 = \Pr_0(Y \in \mathcal{Y}_\alpha; \psi_0, \lambda) - \alpha = E\{I(Y \in \mathcal{Y}_\alpha) - \alpha; \psi_0, \lambda\}$$
$$= E_{S_0}[E\{I(Y \in \mathcal{Y}_\alpha) \mid S_0; \psi_0\} - \alpha; \psi_0, \lambda],$$

for all λ, and the bounded completeness of S_0 implies that

$$E\{I(Y \in \mathcal{Y}_\alpha) \mid S_0; \psi_0\} = \Pr(Y \in \mathcal{Y}_\alpha \mid S_0; \psi_0) = \alpha.$$

Hence similar critical regions must be based on this conditional density.

Example 7.34 (Exponential family)   In Section 5.2.3 we saw that conditioning on the statistic S_2 associated with λ in the full exponential family model

$$f(s_1, s_2; \psi, \lambda) = \exp\left\{s_1^T\psi + s_2^T\lambda - \kappa(\psi, \lambda)\right\}g_0(s_1, s_2),$$

gives a density independent of λ, namely

$$f(s_1 \mid s_2; \psi) = \exp\left\{s_1^T\psi - \kappa_{s_2}(\psi)\right\}g_{s_2}(s_1). \tag{7.34}$$

If a particular value ψ_0 of ψ is fixed, then S_2 is complete and minimal sufficient for λ. Hence similar critical regions for testing $\psi = \psi_0$ must be based on (7.34).

Consider two independent Poisson variables with means μ_1 and μ_2, and suppose that we wish to test the hypothesis $\mu_1 = \mu_2$. We may equivalently set

$c_p(\alpha)$ is the α quantile of the χ_p^2 distribution.

Maurice Stevenson Bartlett (1910–2002) worked at research institutes and the universities of London, Manchester, and Oxford. Starting in the mid 1930s, he made pioneering contributions to likelihood inference, to multivariate analysis and to stochastic processes, on which he wrote a highly influential book.

$\mu_1 = \exp(\lambda + \psi)$ and $\mu_2 = \exp(\lambda)$ with $-\infty < \psi, \lambda < \infty$ and test the hypothesis $\psi = 0$ with no restriction on λ. The corresponding exponential family model is

$$\frac{\mu_1^{y_1}}{y_1!} e^{-\mu_1} \times \frac{\mu_2^{y_2}}{y_2!} e^{-\mu_2} = \frac{1}{y_1! y_2!} \exp\{y_1 \psi + (y_1 + y_2)\lambda - e^{\lambda + \psi} - e^{\lambda}\},$$

where $y_1, y_2 \in \{0, 1, \ldots\}$. Here $S_2 = Y_1 + Y_2$ has a Poisson distribution with mean $\mu_1 + \mu_2 = e^{\lambda}(1 + e^{\psi})$, so the conditional density of $S_1 = Y_1$ is binomial,

$$f(s_1 \mid s_2; \psi) = \frac{s_2!}{s_1!(s_2 - s_1)!} \left(\frac{e^{\psi}}{1 + e^{\psi}}\right)^{s_1} \left(\frac{1}{1 + e^{\psi}}\right)^{s_2 - s_1}, \qquad s_1 = 0, 1, \ldots, s_2.$$

This has denominator $s_2 = y_1 + y_2$ and so treats the total for the two variables as fixed. When $\psi = 0$ the probability equals $1/2$, so the only similar critical regions for a test of $\psi = 0$ against $\psi > 0$, that is, $\mu_1 > \mu_2$, have form

$$\Pr_0(Y_1 \geq r' \mid Y_1 + Y_2 = s_2) = \sum_{r=r'}^{s_2} \binom{s_2}{r} 2^{-r}, \qquad r' = 0, 1, \ldots, s_2.$$

Thus y_1, y_2 show evidence for $\psi > 0$ if y_1 is too close to $y_1 + y_2$.

See also Example 4.40. ∎

Example 7.35 (Permutation test) Let $Y_1, \ldots, Y_m$ and $Y_{m+1}, \ldots, Y_n$ be independent random samples with densities $g(y)$ and $g(y - \theta)$, where g is unknown. One possibility here is to base a test of $\theta = 0$ on the two-sample t statistic

$$T = \frac{\overline{Y}_2 - \overline{Y}_1}{\left[\left(\frac{1}{m} + \frac{1}{n-m}\right)\left\{(m-1)S_1^2 + (n-m)S_2^2\right\}\right]^{1/2}},$$

where $\overline{Y}_2$ and S_2^2 are the average and variance of $Y_{m+1}, \ldots, Y_n$ and $\overline{Y}_1$ and S_1^2 are the corresponding quantities for $Y_1, \ldots, Y_m$.

Under the null hypothesis $Y_1, \ldots, Y_n$ form a random sample with unknown density g, and the set of order statistics $Y_{(1)}, \ldots, Y_{(n)}$ is a minimal sufficient statistic. The conditional null distribution of $Y_1, \ldots, Y_n$ given the observed values $y_{(1)}, \ldots, y_{(n)}$ of the order statistics puts equal mass on each of the $n!$ permutations of $y_1, \ldots, y_n$, so the conditional P-value is

$$\Pr_0(T \geq t_{\text{obs}} \mid Y_{(1)}, \ldots, Y_{(n)}) = \frac{1}{n!} \sum H\{t(y_{\text{perm}}) \geq t_{\text{obs}}\}$$

where the sum is over all permutations y_{perm} of $y_1, \ldots, y_n$. ∎

Invariance

Section 5.3 describes models in which data y were transformed by the action of a group $\mathcal{G}$ on the sample space, thereby inducing a similar group action on the parameter space. In many cases it is appropriate that tests be invariant to the subgroup $\mathcal{G}_0$ of such transformations that preserves the null hypothesis. When testing the hypothesis $\mu = 0$ for a sample y from the $N(\mu, \sigma^2)$ distribution, for example, we might seek a test that is unaffected by replacing y by τy. The corresponding parameter transformation maps σ^2 to $\tau^2 \sigma^2$, thereby preserving the null hypothesis. To see some consequences of requiring such invariances, suppose that the null hypothesis splits the parameter space Θ into disjoint parts Θ_0 and Θ_1 corresponding to the null and alternative hypotheses. The problem is then said to be invariant under $\mathcal{G}_0$ if

$$\Pr\{g(Y) \in \mathcal{A}; \theta\} = \Pr\{Y \in \mathcal{A}; g^*(\theta)\}$$

for all subsets $\mathcal{A}$ of the sample space and all $g \in \mathcal{G}_0$ and corresponding $g^* \in \mathcal{G}_0^*$, where g^* satisfies $g^*(\Theta) = \Theta$, $g^*(\Theta_0) = \Theta_0$ and $g^*(\Theta_1) = \Theta_1$. Thus the action of $\mathcal{G}_0^*$ on Θ leaves Θ_0 and Θ_1 unchanged: whatever transformation is applied to Y, the null hypothesis remains equally true or false. Hence the evidence for or against the hypotheses is unaffected by observing $g(Y)$ rather than Y, for any $g \in \mathcal{G}_0$. A test with critical region $\mathcal{Y}_\alpha$ is then said to be *invariant* if

$$Y \in \mathcal{Y}_\alpha \text{ if and only if } g(Y) \in \mathcal{Y}_\alpha \text{ for all } g \in \mathcal{G}_0, \tag{7.35}$$

implying that its properties are unaffected by transformation. The hope is that appeal to invariance will simplify the problem by eliminating nuisance parameters. We can then search among invariant tests for one with high power or other good properties. As every invariant statistic is a function of a maximal invariant, we start by seeking a maximal invariant under $\mathcal{G}_0$.

Example 7.36 (Student t test)   Suppose that we wish to test $\mu = \mu_0$ against the alternative $\mu \neq \mu_0$, based on a normal random sample $Y_1, \ldots, Y_n$, with no restriction on the variance σ^2. We take $\theta = (\mu, \sigma)$, so Θ_0 is $\{\mu_0\} \times \mathbb{R}_+$ and

$$\Theta_1 = \{(-\infty, \mu_0) \cup (\mu_0, \infty)\} \times \mathbb{R}_+.$$

Let $V = (n-1)^{-1} \sum (Y_j - \overline{Y})^2$. The statistic $(\overline{Y}, V^{1/2})$ is minimal sufficient in the full model and can form the basis of our discussion. As $(\overline{Y}, V^{1/2})$ takes values in the parameter space Θ, Example 5.21 implies that an element $g_{(\eta, \tau)}$ of the group $\mathcal{G}^*$ acting on Θ transforms $(\overline{Y}, V^{1/2})$ to $(\eta + \tau \overline{Y}, \tau V^{1/2})$. This reduction to a minimal sufficient statistic taking values in Θ means that our discussion below may be expressed in terms of $\mathcal{G}^*$ rather than the group $\mathcal{G}$ acting on the original data Y.

The subset of $\mathcal{G}^*$ that preserves Θ_0 must have

$$g_{(\eta, \tau)}(\mu_0, \sigma) = (\eta + \tau \mu_0, \tau \sigma) = (\mu_0, a)$$

for some $a > 0$, and this implies that $\eta = \mu_0 - \tau \mu_0$ but imposes no restriction on τ. Hence the largest such subset is

$$\mathcal{G}_0^* = \left\{ g_{(\mu_0 - \tau \mu_0, \tau)} : \tau > 0 \right\}.$$

To verify that $\mathcal{G}_0^*$ is a subgroup of $\mathcal{G}^*$, note that it is closed, because

$$g_{(\mu_0 - \tau \mu_0, \tau)} \circ g_{(\mu_0 - \sigma \mu_0, \sigma)} = g_{(\mu_0 - \tau \mu + \tau(\mu_0 - \sigma \mu_0), \tau \sigma)} = g_{(\mu_0 - \tau \sigma \mu_0, \tau \sigma)}$$

is also an element of $\mathcal{G}_0^*$, that setting $\tau = 1$ gives the identity element $g_{(0,1)}$, and that $g_{(\mu_0 - \tau \mu_0, \tau)}$ has inverse $g_{(\mu_0 - \tau^{-1} \mu_0, \tau^{-1})}$ also an element of $\mathcal{G}_0^*$. Moreover $\mathcal{G}_0^*$ preserves Θ_1, because if $\mu \neq \mu_0$, then

$$g_{(\mu_0 - \tau \mu_0, \tau)}(\mu, \sigma) = (\mu_0 - \tau \mu_0 + \tau \mu, \tau \sigma) = (\mu_0 + \tau(\mu - \mu_0), \tau \sigma) \in \Theta_1.$$

Now $g_{(\mu_0 - \tau\mu_0, \tau)}$ maps the Student t pivot $T(\mu_0) = n^{1/2}(\overline{Y} - \mu_0)/V^{1/2}$ to

$$n^{1/2}\frac{\overline{\mu_0 - \tau\mu_0 + \tau Y} - \mu_0}{\tau V^{1/2}} = n^{1/2}\frac{\tau(\overline{Y} - \mu_0)}{\tau V^{1/2}} = T(\mu_0),$$

so $T(\mu_0)$ is invariant under $\mathcal{G}_0$. To verify that it is a maximal invariant, we find an estimator that lies in Θ_0 and is equivariant under $\mathcal{G}_0^*$, such as $s(\overline{Y}, V^{1/2}) = (\mu_0, V^{1/2})$. Then a maximal invariant is (page 185)

$$
\begin{aligned}
g^{*-1}_{(\mu_0 - \mu_0 V^{1/2}, V^{1/2})}\left(\overline{Y}, V^{1/2}\right) &= g^{*}_{(\mu_0 - \mu_0 V^{-1/2}, V^{-1/2})}\left(\overline{Y}, V^{1/2}\right) \\
&= \left(\mu_0 - \mu_0 V^{-1/2} + V^{-1/2}\overline{Y}, V^{-1/2}V^{1/2}\right) \\
&= \left(\mu_0 + (\overline{Y} - \mu_0)V^{-1/2}, 1\right),
\end{aligned}
$$

the second component of which can obviously be discarded. Under the null hypothesis μ_0 is known, so $T(\mu_0)$ is also maximal invariant, as we had anticipated. Hence any critical region based on $T(\mu_0)$ would be unaltered if a sample y was replaced by $\mu_0 - \tau\mu_0 + \tau y$, for any $\tau > 0$, because

$$n^{1/2}\frac{\overline{y} - \mu_0}{v^{1/2}} \in \mathcal{A} \quad \text{if and only if} \quad n^{1/2}\frac{\mu_0 - \tau\mu_0 + \tau\overline{y} - \mu_0}{\tau v^{1/2}} \in \mathcal{A}$$

for any set $\mathcal{A} \subset \mathbb{R}$, thus verifying (7.35). Thus any critical region based on $T(\mu_0)$ is invariant. An example is

$t_{n-1}(\alpha)$ is the α quantile of the t_{n-1} distribution.

$$\left\{(y_1, \dots, y_n) : n^{1/2}\left|\frac{\overline{y} - \mu_0}{v^{1/2}}\right| \geq t_{n-1}(1 - \alpha)\right\},$$

which has size 2α and is uniformly most powerful unbiased against two-sided alternatives, in addition to being invariant. ∎

7.3.4 Link with confidence intervals

There is a close link between tests and the construction of confidence intervals. If the density of Y depends on a scalar parameter θ, we define a *level α upper confidence limit* to be a function $T^\alpha = t^\alpha(Y)$ of Y such that

$$\Pr(\theta \leq T^\alpha; \theta) = 1 - \alpha \quad \text{for all } \theta, \tag{7.36}$$

and that $T^{\alpha_1} \leq T^{\alpha_2}$ whenever $\alpha_1 > \alpha_2$. This requirement is similar to the nesting of critical regions for tests and is imposed for the same reasons of consistency; it implies that T^α is non-increasing in α. Lower confidence limits may be defined analogously.

The random quantity in (7.36) is T^α. An equi-tailed $(1 - 2\alpha)$ confidence interval for θ is $(T^{1-\alpha}, T^\alpha)$. If the reparametrization $\psi = \psi(\theta)$ is monotonic increasing, then $\psi(T^\alpha)$ is an upper confidence limit for ψ.

In many cases confidence limits are derived from a pivot $Z(\theta)$, a function of the data and θ with the same distribution for all θ. If this distribution is continuous, we can find a z_α such that

$$\Pr\{Z(\theta) \leq z_\alpha; \theta\} = \alpha \quad \text{for all } \theta.$$

If $Z(\theta)$ is decreasing in θ for every possible value of Y, then the solution in θ to the equation $Z(\theta) = z_\alpha$ can be taken as an upper $(1 - \alpha)$ confidence limit for θ. We applied this argument to approximate normal pivots and the signed likelihood ratio statistic in Sections 3.1.1 and 4.5.2; see Figures 3.1 and 4.7.

Now suppose that $\mathcal{Y}_\alpha(\theta_0)$ is a critical region of size α constructed for tests of $\theta = \theta_0$ against lower alternatives $\theta < \theta_0$. As θ_0 increases, the critical region will vary and we can define the set

$$\{\theta : Y \notin \mathcal{Y}_\alpha(\theta)\}$$

of values of θ not rejected by the test and hence compatible with the data at level α. Under natural monotonicity conditions the supremum of this set can be taken as an upper $(1 - \alpha)$ confidence limit T^α. This inversion of a collection of critical regions to obtain a confidence interval allows us to use good tests to construct good confidence intervals. For example, the Neyman–Pearson lemma tells us that uniformly most powerful tests of simple hypotheses are commonly based on likelihood ratio statistics, which will therefore also be the basis for shortest confidence intervals.

In many cases we can express the above argument as follows. Let $G(t; \theta_0)$ denote the null distribution function of a continuous test statistic T when the null hypothesis is $\theta = \theta_0$. Then the P-value

$$p_{\mathrm{obs}}(\theta_0) = \mathrm{Pr}_0(T \geq t_{\mathrm{obs}}) = 1 - G(t_{\mathrm{obs}}; \theta_0)$$

is a realization of $P(\theta_0) = 1 - G(T; \theta_0)$, and the probability integral transform (Section 2.3) implies that the null distribution of $P(\theta_0)$ is uniform on $(0, 1)$. If the test rejects when $P(\theta_0) < \alpha$, then the set $\{\theta : \alpha \leq P(\theta)\}$ is a one-sided $(1 - \alpha)$ confidence set. In the two-sided case we take $\{\theta : \alpha \leq P(\theta) \leq 1 - \alpha\}$.

This argument applies when we can eliminate parameters other than θ by appeal to similarity or invariance; otherwise it can be sometimes be applied approximately, as with the likelihood ratio statistic. Minor complications arise when T is discrete; see Example 7.38.

Example 7.37 (Exponential density)   Let $Y_1, \ldots, Y_n$ be a random sample from the exponential density with parameter λ, and let a test of $\lambda = \lambda_0$ be conducted against the two-sided alternative $\lambda \neq \lambda_0$. We saw in Example 7.22 that the null density of $T = \sum Y_j$ is gamma with shape parameter n and scale λ_0, so the null hypothesis is rejected at level $(1 - 2\alpha)$ if

$$p_{\mathrm{obs}}(\lambda_0) = \mathrm{Pr}_0(T \geq t_{\mathrm{obs}}) = \int_{\lambda_0 t_{\mathrm{obs}}}^{\infty} \frac{v^{n-1}}{\Gamma(n)} e^{-v} \, dv$$

lies outside the interval $(\alpha, 1 - \alpha)$. For a given value of t_{obs}, this probability depends on λ_0, as shown in Figure 7.7, and a $(1 - 2\alpha)$ confidence interval can be determined as the set of values of λ for which $\alpha \leq p_{\mathrm{obs}}(\lambda) \leq 1 - \alpha$.   ■

The interpretation of two-sided confidence intervals as providing random upper and lower bounds is direct and useful for scalar parameters. Confidence regions for vector θ require a shape. It is natural to base this on likelihood, insisting that a confidence

Figure 7.7 Inversion of a two-sided test with level 0.9 to form confidence interval. Left: significance levels $p_{\text{obs}}(\lambda_0)$ for $\lambda_0 = 0.1, 0.2, 0.5, 1, 2$ (top to bottom). Horizontal lines show probabilities 0.05, 0.95 and the vertical line shows $t_{\text{obs}} = 4$. Hypotheses $\lambda_0 = 2, 0.1$ are rejected, hypotheses $\lambda_0 = 1, 0.5$ are not rejected, and $\lambda_0 = 0.2$ is just rejected. Right: significance level $p_{\text{obs}}(\lambda)$ as a function of λ. Values of λ for which $0.05 \le p_{\text{obs}}(\lambda) \le 0.95$ are contained in the 0.9 confidence interval.

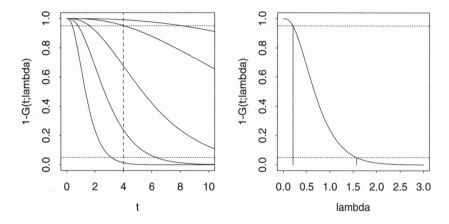

region $\mathcal{R}_\alpha$ be such that $\Pr(\theta \in \mathcal{R}_\alpha; \theta) = \alpha$ for all θ and that $L(\theta) \ge L(\theta')$ for any $\theta \in \mathcal{R}_\alpha$ and $\theta' \notin \mathcal{R}_\alpha$. This amounts to computing $\mathcal{R}_\alpha$ by inverting the likelihood ratio statistic, typically using its asymptotic distribution, perhaps with Bartlett adjustment.

Often the test inverted to obtain limits of confidence intervals is not exact. Then there is *coverage error*, defined as the difference between the actual and nominal probabilities that the confidence set contains the parameter,

$$\Pr(T^{\alpha_1} < \theta \le T^{\alpha_2}; \theta) - (\alpha_1 - \alpha_2), \quad \text{for } \alpha_1 > \alpha_2. \tag{7.37}$$

It can be helpful to know where the error occurs. The limit T^α is said to be *conservative* if it tends to be too high, that is, $\Pr(\theta \le T^\alpha; \theta) \ge 1 - \alpha$; confidence intervals for which (7.37) is positive are called conservative.

Otherwise they are called liberal.

Example 7.38 (Binomial density) An equitailed $(1 - 2\alpha)$ confidence interval for the probability π of a binomial variable Y with denominator m may be found in various ways. Exact limits may be found by inverting tests based on Y. Having observed $Y = y$, the significance level for testing the null hypothesis $\pi = \pi_0$ against the one-sided alternative $\pi < \pi_0$ is

$$\Pr_0(Y \le y) = \Pr(Y \le y; \pi_0) = \sum_{r=0}^{y} \binom{m}{r} \pi_0^r (1 - \pi_0)^{m-r},$$

so the upper α limit π^α is the solution to

$$\Pr(Y \le y; \pi) = \sum_{r=0}^{y} \binom{m}{r} \pi^r (1 - \pi)^{m-r} = \alpha,$$

and equals 1 if $y = m$. A similar argument with alternative $\pi > \pi_0$ shows that the lower α limit π_α is the solution to

$$\Pr(Y \ge y; \pi) = \sum_{r=y}^{m} \binom{m}{r} \pi^r (1 - \pi)^{m-r} = \alpha,$$

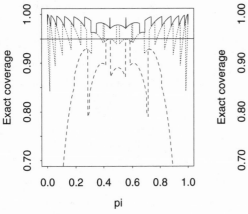

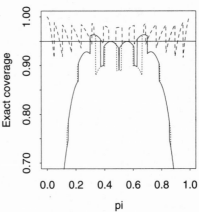

Figure 7.8 Exact coverages of equi-tailed 0.95 confidence intervals for the binomial parameter π, as functions of π, when $m = 10$. The horizontal line shows the target coverage. Left: exact (solid), score (dots) and maximum likelihood estimator (dashes). Right: signed likelihood ratio statistic (solid), modified signed likelihood ratio statistic (dots) and modified maximum likelihood estimator (dashes), obtained by replacing m and r by $m + 2$ and $r + 1$ (dashes).

$F_{\nu_1, \nu_2}(y)$ is the distribution function of an F variable with ν_1, ν_2 degrees of freedom.

but equals 0 if $y = 0$. It turns out that π^α and π_α are expressible using quantiles of the F distribution, giving $(1 - 2\alpha)$ confidence interval

$$\left(\left\{ 1 + \frac{m - y + 1}{y F^{-1}_{2y, 2(m-y+1)}(\alpha)} \right\}^{-1}, \left\{ 1 + \frac{m - y}{(y + 1) F^{-1}_{2(y+1), 2(m-y)}(1 - \alpha)} \right\}^{-1} \right),$$

with the changes mentioned above when $y = 0$ or $y = m$. This interval is exact in the sense that no approximation of binomial probabilities is involved.

Approximate intervals can be based on asymptotic standard normal distributions of the score statistic, the maximum likelihood estimator $\widehat{\pi} = Y/m$ or the signed likelihood ratio statistic,

$$Z_1(\pi) = (Y - m\pi)/\{m\pi(1 - \pi)\}^{1/2},$$
$$Z_2(\pi) = (\widehat{\pi} - \pi)/\{\widehat{\pi}(1 - \widehat{\pi})/m\}^{1/2},$$
$$Z_3(\pi) = \text{sign}(\widehat{\pi} - \pi)\left(2\left[Y \log(\widehat{\pi}/\pi) + (m - Y) \log\{(1 - \widehat{\pi})/(1 - \pi)\}\right]\right)^{1/2},$$

as well as on a quantity $Z^*(\pi) = Z_3(\pi) + Z_3(\pi)^{-1} \log\{Z_2(\pi)/Z_3(\pi)\}$ motivated in Section 12.3.3. The confidence interval based on each of these is the set of π for which $|Z(\pi)| < z_{1-\alpha}$; this must be found numerically for $Z_3(\pi)$ and $Z^*(\pi)$. Any of these intervals has coverage $\sum_{y=0}^{m} \binom{m}{y} \pi^y (1 - \pi)^{m-y} I_{1-2\alpha}(\pi, y)$, where $I_{1-2\alpha}(y, \pi)$ indicates that π lies in an interval of nominal level $(1 - 2\alpha)$ based on y.

Figure 7.8 compares the coverages for $\alpha = 0.025$ and $m = 10$. That of the exact interval always exceeds 0.975, so it is quite conservative, while that of the interval based on $Z_1(\pi)$ is fairly close to its nominal level overall. Intervals based on $Z_2(\pi)$ undercover for most π. The intervals based on $Z_3(\pi)$ and $Z^*(\pi)$ have coverage close to nominal for $0.3 < \pi < 0.7$, while perhaps the best overall performance is obtained from $Z_2(\pi)$ with m and y replaced by $m + 2$ and $y + 1$.   ∎

This example suggests that in highly discrete situations approximate confidence intervals may be preferable to exact ones. Moreover exact tests will inherit the conservatism and tend to reject too rarely. The difference decreases as the sample size increases, but even with $m = 50$ the mean exact coverage is about 0.97 in the binomial case.

Exercises 7.3

1 Show that (7.26) has mean and variance roughly p_{obs} and $p_{obs}(1 - p_{obs})/R$. Hence give minimum values of R for obtaining 5% relative error in estimation of $p_{obs} = 0.5, 0.2, 0.1, 0.05, 0.01, 0.001$. Discuss.

2 In Example 7.22, calculate the significance level for testing $H_0 : \lambda = 1$ against $H_1 : \lambda = 4$, based on the data 1.2, 3, 1.5, 0.3.

3 If $U \sim U(0, 1)$, show that $\min(U, 1 - U) \sim U(0, \frac{1}{2})$. Hence justify the computation of a two-sided significance level as $2\min(P^-, P^+)$.

4 Consider testing the hypothesis that $\mu = \mu_0$ based on a random sample $Y_1, \ldots, Y_n$ from the $N(\mu, \sigma^2)$ distribution, with two-sided alternative $\mu \neq \mu_0$. Show that the power of the region (7.31) is $\Phi(z_{\alpha/2} + \delta) + \Phi(z_{\alpha/2} - \delta)$, where $\delta = n^{1/2}(\mu - \mu_0)/\sigma$. Sketch this as a function of δ for $\alpha = 0.025$, and explain why it is invariant to the sign of $\mu - \mu_0$.

5 Check the power calculation for the sign test in Example 7.30.

6 Consider testing the hypothesis that a binomial random variable has probability $\pi = 1/2$ against the alternative that $\pi > 1/2$. For what values of α does a uniformly most powerful test exist when the denominator is $m = 5$?

7 In a random sample $Y_1, \ldots, Y_n$ from the gamma density with shape κ and scale λ, find a locally most powerful test of the null hypothesis $\kappa = 1$.

8 If I is Bernoulli with probability $p = (\alpha_2 - \alpha)/(\alpha_2 - \alpha_1)$ and $\mathcal{Y}_{\alpha_1}$ and $\mathcal{Y}_{\alpha_2}$ are critical regions of sizes α_1, α_2, show that the critical region $\mathcal{Y} = I\mathcal{Y}_{\alpha_1} + (1 - I)\mathcal{Y}_{\alpha_2}$ has size α.

9 Y_1, Y_2 are independent gamma variables with known shape parameters ν_1, ν_2 and scale parameters λ_1, λ_2, and it is desired to test the null hypothesis H_0 that $\lambda_1 = \lambda_2 = \lambda$, with λ unknown. Show that a minimal sufficient statistic for λ under H_0 is $Y_1 + Y_2$, find its distribution, and show that it is complete. Hence show that the test is based on the conditional distribution of Y_1 given $Y_1 + Y_2$ and that significance levels are computed from integrals of form

$$\frac{\Gamma(\nu_1 + \nu_2)}{\Gamma(\nu_1)\Gamma(\nu_2)} \int_0^{y_1/(y_1+y_2)} u^{\nu_1 - 1}(1 - u)^{\nu_2 - 1}\, du.$$

Explain how this argument is useful in comparison of the scale parameters of two independent exponential samples.

10 Independent data pairs $(X_1, Z_1), \ldots, (X_n, Z_n)$ arise from a joint density $f(x, z)$. The null hypothesis is that X and Z are independent, so $f(x, z) = g(x)h(z)$ for some unknown densities g and h and all x and z. Show that the order statistics $X_{(1)}, \ldots, X_{(n)}$ and $Z_{(1)}, \ldots, Z_{(n)}$ are minimal sufficient for g and h under the null hypothesis, and deduce that a similar test has P-value

$$p_{obs} = \frac{1}{n!} \sum H\{t(y_{perm}) \geq t_{obs}\},$$

where the sum is over all $y_{perm} = \{(x_1, z_{\pi(1)}), \ldots, (x_n, z_{\pi(n)})\}$ with the observed values of the zs permuted, the xs being held fixed.
If the test statistic is $T = (n^{-1}\sum X_j Z_j - \overline{X}\,\overline{Z})/(S_X^2 S_Z^2)^{1/2}$, S_X^2 and S_Z^2 being the sample variances of the X_j and the Z_j, show that it is equivalent to base the test on $\sum X_j Z_j$.

11 In a scale family, $Y = \tau\varepsilon$, where ε has a known density and $\tau > 0$. Consider testing the null hypothesis $\tau = \tau_0$ against the alternative $\tau \neq \tau_0$. Show that the appropriate group for constructing an invariant test has just one element (apart from permutations) and hence show that the test may be based on the maximal invariant $Y_{(1)}/\tau_0, \ldots, Y_{(n)}/\tau_0$.
When ε is exponential, show that the invariant test is based on $\overline{Y}/\tau_0$.

12 One natural transformation of a binomial variable R is reversal of 'success' and 'failure'. Show that this maps R to $m - R$, where m is the denominator, and that

the induced transformation on the parameter space maps π to $1 - \pi$. Which of the critical regions (a) $\mathcal{Y}_1 = \{0, 1, 20\}$, (b) $\mathcal{Y}_2 = \{0, 1, 19, 20\}$, (c) $\mathcal{Y}_3 = \{0, 1, 10, 19, 20\}$, (d) $\mathcal{Y}_4 = \{8, 9, 10, 11, 12\}$, is invariant for testing $\pi = \frac{1}{2}$ when $m = 20$? Which is preferable and why?

13   The incidence of a rare disease seems to be increasing. In successive years the numbers of new cases have been $y_1, \ldots, y_n$. These may be assumed to be independent observations from Poisson distributions with means $\lambda\theta, \ldots, \lambda\theta^n$. Show that there is a family of tests each of which, for any given value of λ, is a uniformly most powerful test of its size for testing $\theta = 1$ against $\theta > 1$.

14   A random sample $Y_1, \ldots, Y_n$ is available from the *Type I Pareto* distribution

$$F(y; \psi) = \begin{cases} 1 - y^{-\psi}, & y \geq 1, \\ 0, & y < 1. \end{cases}$$

Find the likelihood ratio statistic to test that $\psi = \psi_0$ against $\psi = \psi_1$, where ψ_0, ψ_1 are known, and show how to calculate a P-value when $\psi_0 > \psi_1$.
How does your answer change if the distribution is

$$F(y; \psi, \lambda) = \begin{cases} 1 - (y/\lambda)^{-\psi}, & y \geq \lambda, \\ 0, & y < \lambda, \end{cases}$$

with $\lambda > 0$ unspecified?

7.4 Bibliographic Notes

The main concepts described in this chapter belong to the core of statistical theory and were developed in the first half of the twentieth century by Fisher, Neyman, Pearson and others; other treatments are contained in most books on mathematical statistics. See for example the treatments of estimation in Silvey (1970), Rice (1988), Casella and Berger (1990) and Bickel and Doksum (1977), or at a more advanced level Cox and Hinkley (1974), Lehmann (1983) and Shao (1999).

Kernel density estimation has been extensively studied since it was proposed in the 1950s. Among numerous excellent expositions are Silverman (1986), Scott (1992), Wand and Jones (1995), and Bowman and Azzalini (1997). The last of these is more practical in emphasis, while Wand and Jones (1995) contains a detailed discussion of the choice of bandwidth, a topic on which there has been much progress in the 1990s. Although cross-validation is an important paradigm for selection of bandwidths and related smoothing parameters in other non- and semi-parametric contexts, other approaches to bandwidth selection give better results; see Sheather and Jones (1991). Stone (1974) is a fundamental reference on cross-validation.

Estimators based on estimating functions are widely used in practice, but there are few general expositions of them at this level. Godambe (1991) is an interesting collection of papers on the topic, with many further references, while McLeish and Small (1994) give a more abstract treatment. A fundamental reference for the role of the influence function in robust statistics is Hampel *et al.* (1986). Inference for stochastic processes is discussed in books by Hall and Heyde (1980), Basawa and Scott (1981), and Guttorp (1991), while Sørensen (1999) reviews the asymptotic theory for estimating functions.

Although the idea of significance testing goes back hundreds of years, the development of underlying theory is more recent. R. A. Fisher made extensive informal use of P-values, but resisted what he saw as the over-formalization due to Neyman and E. S. Pearson. They introduced the idea of testing as a choice between two hypotheses and introduced the notions of size, power and so forth in work that prefigured the later development of decision theory. Their joint papers are collected in Neyman and Pearson (1967). The theory of testing is explained more fully in Lehmann (1983) and in Chapters 3–6 of Cox and Hinkley (1974). Bartlett correction was first described by Bartlett (1937). Example 7.38 is based on Agresti and Coull (1998), Agresti and Caffo (2000), and Greenland (2001).

7.5 Problems

1 In Example 7.2 show that $\widehat{\psi} \overset{D}{=} \exp\{\mu + \sigma n^{-1/2} Z + \sigma^2 V/(2n)\}$. Hence give an explicit expression for $E(\widehat{\psi}^r)$ and compute the analogue of Table 7.1. Discuss your results.

2 Let $Y_1, \ldots, Y_n$ be a random sample from an unknown density f. Let I_j indicate whether or not Y_j lies in the interval $(a - \frac{1}{2}h, a + \frac{1}{2}h]$, and consider $R = \sum I_j$. Show that R has a binomial distribution with denominator n and probability

$$\int_{a-\frac{1}{2}h}^{a+\frac{1}{2}h} f(y)\, dy.$$

Hence show that $R/(nh)$ has approximate mean and variance $f(a) + \frac{1}{2}h^2 f''(a)$ and $f(a)/nh$, where f'' is the second derivative of f.
What implications have these results for using the histogram to estimate $f(a)$?

3 Suppose that the random variables $Y_1, \ldots, Y_n$ are such that

$$E(Y_j) = \mu, \quad \text{var}(Y_j) = \sigma_j^2, \quad \text{cov}(Y_j, Y_k) = 0, \quad j \neq k,$$

where μ is unknown and the σ_j^2 are known. Show that the linear combination of the Y_j's giving an unbiased estimator of μ with minimum variance is

$$\sum_{j=1}^{n} \sigma_j^{-2} Y_j \bigg/ \sum_{j=1}^{n} \sigma_j^{-2}.$$

Suppose now that Y_j is normally distributed with mean βx_j and unit variance, and that the Y_j are independent, with β an unknown parameter and the x_j known constants. Which of the estimators

$$T_1 = n^{-1} \sum_{j=1}^{n} Y_j/x_j, \quad T_2 = \sum_{j=1}^{n} Y_j x_j \bigg/ \sum_{j=1}^{n} x_j^2$$

is preferable and why?

4 In n independent food samples the bacterial counts $Y_1, \ldots, Y_n$ are presumed to be Poisson random variables with mean θ. It is required to estimate the probability that a given sample would be uncontaminated, $\pi = \Pr(Y_j = 0)$.
Show that $U = n^{-1} \sum I(Y_j = 0)$, the proportion of the samples uncontaminated, is unbiased for π, and find its variance. Using the Rao–Blackwell theorem or otherwise, show that an unbiased estimator of π having smaller variance than U is $V = \{(n-1)/n\}^{n\overline{Y}}$, where $\overline{Y} = n^{-1} \sum Y_j$. Is this a minimum variance unbiased estimator of π?
Find var(V) and hence give the asymptotic efficiency of U relative to V.

5   Let $Y_1, \ldots, Y_n$ be independent Poisson variables with means $x_1\beta, \ldots, x_n\beta$, where $\beta > 0$
    is an unknown scalar and the $x_j > 0$ are known scalars. Show that $T = \sum Y_j x_j / \sum x_j^2$ is
    an unbiased estimator of β and find its variance.
    Find a minimal sufficient statistic S for β, and show that the conditional distribution of Y_j
    given that $S = s$ is multinomial with mean $sx_j / \sum_i x_i$. Hence find the minimum variance
    unbiased estimator of β. Is it unique?

6   Given that there is a 1–1 mapping between $x_1 < \cdots < x_n$ and the sums $s_1, \ldots, s_n$, where
    $s_r = \sum x_j^r$, show that the order statistics of a random sample form a complete minimal
    sufficient statistic in the class of all continuous densities. You may find it useful to consider
    the exponential family density

    $$f(y; \theta) \propto \exp(-x^{2n} + \theta_1 x + \cdots + \theta_n x^n).$$

7   Find the maximum likelihood estimator of β based on a random sample from the shifted
    exponential density $f(y) = e^{-(y-\beta)}$ for $y \geq \beta$. Show that $\widehat{\beta}$ is biased but consistent. Does
    it satisfy the Cramér–Rao lower bound?

8   (a) Let $Y_1, \ldots, Y_n$ be a random sample from the exponential density $\lambda e^{-\lambda y}$, $y > 0$, $\lambda > 0$.
    Say why an unbiased estimator W for λ should have form a/S, and hence find a. Find
    the Fisher information for λ and show that $E(W^2) = (n-1)\lambda^2/(n-2)$. Deduce that
    no unbiased estimator of λ attains the Cramér–Rao lower bound, although W does so
    asymptotically.
    (b) Let $\psi = \Pr(Y > a) = e^{-\lambda a}$, for some constant a. Show that

    $$I(Y_1 > a) = \begin{cases} 1, & Y_1 > a, \\ 0, & \text{otherwise,} \end{cases}$$

    is an unbiased estimator of ψ, and hence obtain the minimum variance unbiased estimator.
    Does this attain the Cramér–Rao lower bound for ψ?

9   Let $X_1, \ldots, X_n$ represent the times of the first n events in a Poisson process of rate
    μ^{-1} observed from time zero; thus $0 < X_1 < \cdots < X_n$. Show that $W = 2(X_1 + \cdots +$
    $X_n)/\{n(n+1)\}$ is an unbiased estimator of μ, and establish that its Rao–Blackwellized
    form is $T = X_n/n$. Find $\text{var}(W)$ and give the asymptotic efficiency of W relative to T.

10 Show that no unbiased estimator exists of $\psi = \log\{\pi/(1-\pi)\}$, based on a binomial
    variable with probability π.

11 Let $Y_j = \eta + \tau \varepsilon_j$, where $\varepsilon_1, \ldots, \varepsilon_n$ is a random sample from a known density. Show that
    the set of order statistics $Y_{(1)}, \ldots, Y_{(n)}$ is in general minimal sufficient for η, τ (Exam-
    ple 4.12). By considering $(Y_{(2)} - Y_{(1)})/(Y_{(n)} - Y_{(1)})$ show that it is not complete.

12 Show that when the data are normal, the efficiency of the Huber estimating function
    $g_c(y; \theta)$ compared to the optimal function $g_\infty(y; \theta)$ is

    $$\frac{\{1 - 2\Phi(-c)\}^2}{1 + 2\{c^2\Phi(-c) - \Phi(-c) - c\phi(c)\}}.$$

    Hence verify that the efficiency is 0.95 when $c = 1.345$.

13 Compare the performance of the estimating function

    $$g(y; \theta) = \begin{cases} y - \theta, & |y - \theta| < c, \\ 0, & \text{otherwise,} \end{cases}$$

    with that of the Huber function $g_c(y; \theta)$ for the distributions in Example 7.19.

14 Show how (a) the Poisson birth process in Example 4.6, and (b) the Markov chain likeli-
    hood in Section 6.1.1, fall into the framework for dependent data outlined in Section 7.2.3.

15 Let $Y_1, \ldots, Y_n \overset{\text{iid}}{\sim} N(\mu, \sigma^2)$, with both parameters unknown. Suppose that we wish to test
    $\mu = \mu_0$ against the one-sided alternative $\mu > \mu_0$. By considering separately the cases

$\overline{Y} \geq \mu_0$ and $\overline{Y} < \mu_0$, show that the likelihood ratio statistic is

$$W_\mathrm{p}(\mu_0) = \begin{cases} n \log \left\{ 1 + \frac{T(\mu_0)^2}{n-1} \right\}, & \overline{Y} \geq \mu_0, \\ 0, & \overline{Y} < \mu_0. \end{cases}$$

Hence justify the one-tailed significance level described in Example 7.25.

16 Independent random samples $Y_{i1}, \ldots, Y_{in_i}$, where $n_i \geq 2$, are drawn from each of k normal distributions with means $\mu_1, \ldots, \mu_k$ and common unknown variance σ^2. Derive the likelihood ratio statistic W_p for the null hypothesis that the μ_i all equal an unknown μ, and show that it is a monotone function of

$$R = \frac{\sum_{i=1}^k n_i (\overline{Y}_{i\cdot} - \overline{Y}_{\cdot\cdot})^2}{\sum_{i=1}^k \sum_{j=1}^{n_i} (Y_{ij} - \overline{Y}_{i\cdot})^2},$$

where $\overline{Y}_{i\cdot} = n_i^{-1} \sum_j Y_{ij}$ and $\overline{Y}_{\cdot\cdot} = (\sum n_i)^{-1} \sum_{i,j} Y_{ij}$. What is the null distribution of R?

17 Let $X_1, \ldots, X_m$ and $Y_1, \ldots, Y_n$ be independent random samples from continuous distributions F_X and F_Y. We wish to test the hypothesis H_0 that $F_X = F_Y$.
Define indicator variables $I_{ij} = I(X_i < Y_j)$ for $i = 1, \ldots, m$, $j = 1, \ldots, n$ and let $U = \sum_{i,j} I_{ij}$. Assuming that H_0 is true, (i) show that $E(U) = mn/2$; (ii) find $\mathrm{cov}(I_{ij}, I_{ik})$ and $\mathrm{cov}(I_{ij}, I_{kl})$, where i, j, k, l are distinct; and (iii) hence show that $\mathrm{var}(U) = mn(m + n + 1)/12$. Why is it important that the underlying distributions are continuous?
Here are the weight gains (gms) of rats fed on low and high protein diets:

| High | 83 | 97 | 104 | 107 | 113 | 119 | 123 | 124 | 129 | 134 | 146 | 161 |
| Low | 70 | 85 | 94 | 101 | 106 | 118 | 132 | | | | | |

Use the approximate normality of U to test for a difference between diets.

18 Below are diastolic blood pressures (mm Hg) of ten patients before and after treatment for high blood pressure. Test the hypothesis that the treatment has no effect on blood pressure using a Wilcoxon signed-rank test, (a) using the exact significance level and (b) using a normal approximation. Discuss briefly.

| Before | 94 | 105 | 101 | 106 | 118 | 107 | 96 | 102 | 114 | 95 |
| After | 96 | 96 | 95 | 103 | 105 | 111 | 86 | 90 | 107 | 84 |

19 (a) A random sample of size $n = 2$ is taken from $f(y)$. For $0 < \alpha < 1/2$, find a critical region of size α for testing that $f(y)$ is

$$f_0(y) = \begin{cases} \theta^{-1}, & 0 < y < \theta, \\ 0, & \text{otherwise,} \end{cases}$$

when $\theta = 1$, against the alternative that $f(y)$ is the exponential density $f_1(y) = e^{-y}$, $y > 0$. Is there a best critical region for testing $f = f_0$ against the composite hypothesis $f(y) = \lambda \exp(-\lambda y)$, $y > 0$, for some $\lambda > 0$?
(b) Show there is no best critical region when θ is unknown.
(c) Show that the largest order statistic $Y_{(2)}$ is sufficient for θ under the null model, and deduce that there is a uniformly most powerful test based on the ratio of conditional densities of Y given $Y_{(2)}$ under the two hypotheses. Show that the most powerful conditional critical region of size α is $\mathcal{Y}_\alpha = \{(y_1, y_2) : 0 \leq y_{(1)} \leq \alpha y_{(2)})\}$.
(d) Find the conditional critical region for general n.

20 If

$$f(x; \theta) = \begin{cases} \theta^\lambda \Gamma(\lambda)^{-1} x^{\lambda-1} e^{-\theta x}, & x > 0, \\ 0, & \text{elsewhere,} \end{cases}$$

where λ is known and θ is positive, deduce that there exists a uniformly most powerful test of size α of the hypothesis $\theta = \theta_0$ against the alternative $\theta > \theta_0$, and show that when $\lambda = 1/n$ the power function of the test is $1 - (1 - \alpha)^{\theta/\theta_0}$.

21 A source at location $x = 0$ pollutes the environment. Are cases of a rare disease $\mathcal{D}$ later
    observed at positions $x_1, \ldots, x_n$ linked to the source?
    Cases of another rare disease $\mathcal{D}'$ known to be unrelated to the pollutant but with the same
    susceptible population as $\mathcal{D}$ are observed at $x_1', \ldots, x_m'$. If the probabilities of contracting
    $\mathcal{D}$ and $\mathcal{D}'$ are respectively $\psi(x)$ and ψ', and the population of susceptible individuals has
    density $\lambda(x)$, show that the probability of $\mathcal{D}$ at x, given that $\mathcal{D}$ or $\mathcal{D}'$ occurs there, is

    $$\pi(x) = \frac{\psi(x)\lambda(x)}{\psi(x)\lambda(x) + \psi'\lambda(x)}.$$

    Deduce that the probability of the observed configuration of diseased persons, conditional
    on their positions, is

    $$\prod_{j=1}^{n} \pi(x_j) \prod_{i=1}^{m} \{1 - \pi(x_i')\}.$$

    The null hypothesis that $\mathcal{D}$ is unrelated to the pollutant asserts that $\psi(x)$ is independent of
    x. Show that in this case the unknown parameters may be eliminated by conditioning on
    having observed n cases of $\mathcal{D}$ out of a total $n + m$ cases. Deduce that the null probability
    of the observed pattern is $\binom{n+m}{n}^{-1}$.
    If T is a statistic designed to detect decline of $\psi(x)$ with x, explain how permutation of
    case labels $\mathcal{D}$, $\mathcal{D}'$ may be used to obtain a significance level p_{obs}.
    Such a test is typically only conducted after a suspicious pattern of cases of $\mathcal{D}$ has been
    observed. How will this influence p_{obs}?

8

Linear Regression Models

Regression models are used to describe how one or perhaps a few *response* variables depend on other *explanatory* variables. The idea of regression is at the core of much statistical modelling, because the question 'what happens to y when x varies?' is central to many investigations. It is often required to predict or control future responses by changing the other variables, or to gain an understanding of the relation between them. There is usually a single response, treated as random. Often there are many explanatory variables, which are treated as non-stochastic. The simplest models involve linear dependence and are described in this chapter, while Chapter 9 deals with more structured situations in which the explanatory variables have been chosen by the experimenter according to a design. Chapter 10 describes some of the many extensions of regression to nonlinear dependence. Throughout we simplify our previous notation by using y to represent both the response variable and the value it takes; no confusion should arise thereby.

8.1 Introduction

If we denote the response by y and the explanatory variables by x, our concern is how changes in x affect y. In Section 5.1, for example, the key question was how the annual maximum sea level in Venice depended on the passage of time. We fitted the straight-line regression model

$$y_j = \beta_0 + \beta_1 x_j + \varepsilon_j, \quad j = 1, \ldots, n,$$

where we took y_j to be the jth annual maximum sea level and x_j to be the year in which this occurred. The parameters β_0 and β_1 represent a baseline maximum sea level and the annual rate at which sea level increases, while ε_j is a random variable that represents the difference between the underlying level, $\beta_0 + \beta_1 x_j$, and the value observed, y_j.

An immediate generalization is to increase the number of explanatory variables, setting

$$y_j = \beta_1 x_{j1} + \cdots + \beta_p x_{jp} + \varepsilon_j = x_j^T \beta + \varepsilon_j,$$

where $x_j^T = (x_{j1}, \ldots, x_{jp})$ is a $1 \times p$ vector of explanatory variables associated with the jth response, β is a $p \times 1$ vector of unknown parameters and ε_j is an unobserved error accounting for the discrepancy between the observed response y_j and $x_j^T \beta$. In matrix notation,

$$y = X\beta + \varepsilon, \tag{8.1}$$

where y is the $n \times 1$ vector whose jth element is y_j, X is an $n \times p$ matrix whose jth row is x_j^T, and ε is the $n \times 1$ vector whose jth element is ε_j. The data on which the investigation is to be based are y and X, and the aim is to disentangle systematic changes in y due to variation in X from the haphazard scatter added by the errors ε. Model (8.1) is known as a *linear regression model* with *design matrix X*.

Example 8.1 (Straight-line regression) For the straight-line regression model, (8.1) becomes

$$\begin{pmatrix} y_1 \\ y_2 \\ \vdots \\ y_n \end{pmatrix} = \begin{pmatrix} 1 & x_1 \\ 1 & x_2 \\ \vdots & \vdots \\ 1 & x_n \end{pmatrix} \begin{pmatrix} \beta_0 \\ \beta_1 \end{pmatrix} + \begin{pmatrix} \varepsilon_1 \\ \varepsilon_2 \\ \vdots \\ \varepsilon_n \end{pmatrix},$$

so X is an $n \times 2$ matrix and β a 2×1 vector of parameters. ∎

Example 8.2 (Polynomial regression) Suppose that the response is a polynomial function of a single covariate,

$$y_j = \beta_0 + \beta_1 x_j + \cdots + \beta_{p-1} x_j^{p-1} + \varepsilon_j.$$

For example, we might wish to fit a quadratic or cubic trend in the Venice sea level data, in which case we would have $p = 3$ or $p = 4$ respectively. Then

$$\begin{pmatrix} y_1 \\ y_2 \\ \vdots \\ y_n \end{pmatrix} = \begin{pmatrix} 1 & x_1 & x_1^2 & \cdots & x_1^{p-1} \\ 1 & x_2 & x_2^2 & \cdots & x_2^{p-1} \\ \vdots & \vdots & \vdots & & \vdots \\ 1 & x_n & x_n^2 & \cdots & x_n^{p-1} \end{pmatrix} \begin{pmatrix} \beta_0 \\ \beta_1 \\ \vdots \\ \beta_{p-1} \end{pmatrix} + \begin{pmatrix} \varepsilon_1 \\ \varepsilon_2 \\ \vdots \\ \varepsilon_n \end{pmatrix},$$

where X has dimension $n \times p$. ∎

A key point is that (8.1) is linear in the parameters β. Polynomial regression can be written in form (8.1) because of its linearity, not in x, but in β.

Example 8.3 (Cement data) Table 8.1 contains data on the relationship between the heat evolved in the setting of cement and its chemical composition. Data on heat evolved, y, for each of $n = 13$ independent samples are available, and for each

Table 8.1 Cement data (Woods *et al.*, 1932): y is heat evolved in calories per gram of cement, and x_1, x_2, x_3, and x_4 are percentage weight of clinkers, with x_1, $3CaO.Al_2O_3$, x_2, $3CaO.SiO_2$, x_3, $4CaO.Al_2O_3.Fe_2O_3$, and x_4, $2CaO.SiO_2$.

Case	x_1	x_2	x_3	x_4	y
1	7	26	6	60	78.5
2	1	29	15	52	74.3
3	11	56	8	20	104.3
4	11	31	8	47	87.6
5	7	52	6	33	95.9
6	11	55	9	22	109.2
7	3	71	17	6	102.7
8	1	31	22	44	72.5
9	2	54	18	22	93.1
10	21	47	4	26	115.9
11	1	40	23	34	83.8
12	11	66	9	12	113.3
13	10	68	8	12	109.4

Figure 8.1 Plots of cement data. The variables are heat evolved in calories per gram, y, percentage weight in clinkers of x_1, $3CaO.Al_2O_3$, x_2, $3CaO.SiO_2$, x_3, $4CaO.Al_2O_3.Fe_2O_3$, and x_4, $2CaO.SiO_2$.

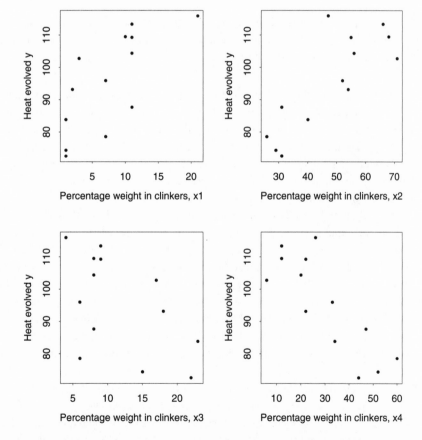

sample the percentage weight in clinkers of four chemicals, x_1, $3CaO.Al_2O_3$, x_2, $3CaO.SiO_2$, x_3, $4CaO.Al_2O_3.Fe_2O_3$, and x_4, $2CaO.SiO_2$, is recorded.

Figure 8.1 shows that although the response y depends on each of the covariates $x_1, \ldots, x_4$, the degrees and directions of the dependences differ.

In this case we might fit the model

$$y_j = \beta_0 + \beta_1 x_{1j} + \beta_2 x_{2j} + \beta_3 x_{3j} + \beta_4 x_{4j} + \varepsilon_j,$$

where Figure 8.1 suggests that β_1 and β_2 are positive, and that β_3 and β_4 are negative. The design matrix has dimension 13×5, and is

$$X = \begin{pmatrix} 1 & 7 & 26 & 6 & 60 \\ 1 & 1 & 29 & 15 & 52 \\ \vdots & \vdots & \vdots & \vdots & \vdots \\ 1 & 10 & 68 & 8 & 12 \end{pmatrix};$$

the vectors y and ε have dimension 13×1 and β has dimension 5×1.   ∎

In the examples above the explanatory variables consist of numerical quantities, sometimes called *covariates*. *Dummy variables* that represent whether or not an effect is applied can also appear in the design matrix.

Example 8.4 (Cycling data)   Norman Miller of the University of Wisconsin wanted to see how seat height, tyre pressure and the use of a dynamo affected the time taken to ride his bicycle up a hill. He decided to collect data at each combination of two seat heights, 26 and 30 inches from the centre of the crank, two tyre pressures, 40 and 55 pounds per square inch (psi) and with the dynamo on and off, giving eight combinations in all. The times were expected to be quite variable, and in order to get more accurate results he decided to make two timings for each combination. He wrote each of the eight combinations on two pieces of card, and then drew the sixteen from a box in a random order. He planned to make four widely separated runs up the hill on each of four days, first adjusting his bicycle to the setups on the successive pieces of card, but bad weather forced him to cancel the last run on the first day; he made five on the third day to make up for this. Table 8.2 gives timings obtained with his wristwatch.

The lower part of Table 8.2 shows how average time depends on experimental setup. There is a large reduction in the average time when the seat is raised and smaller reductions when the tyre pressure is increased and the dynamo is off.

The quantities that are varied in this experiment — seat height, tyre pressure, and the state of the dynamo — are known as *factors*. Each takes two possible values, known as *levels*. Here there are two types of factors: quantitative and qualitative. The two levels of seat height and tyre pressure are quantitative — other values might have been chosen, and more than two levels could have been used — but the dynamo factor has only two possible levels and is qualitative.

An experiment like this, in which data are collected at each combination of a number of factors, is known as a *factorial experiment*. Such designs and their variants

Table 8.2 Data and experimental setup for bicycle experiment (Box *et al.*, 1978, pp. 368–372). The lower part of the table shows the average times for each of the eight combinations of settings of seat height, tyre pressure, and dynamo, and the average times for the eight observations at each setting, considered separately.

Setup	Day	Run	Seat height (inches)	Dynamo	Tyre pressure (psi)	Time (secs)
1	3	2	−	−	−	51
2	4	1	−	−	−	54
3	2	2	+	−	−	41
4	2	3	+	−	−	43
5	3	3	−	+	−	54
6	2	1	−	+	−	60
7	3	1	+	+	−	44
8	4	3	+	+	−	43
9	1	1	−	−	+	50
10	4	4	−	−	+	48
11	3	5	+	−	+	39
12	4	2	+	−	+	39
13	3	4	−	+	+	53
14	1	3	−	+	+	51
15	1	2	+	+	+	41
16	2	4	+	+	+	44

	Seat height (inches from centre of crank)	Dynamo	Tyre pressure (psi)
−	26	Off	40
+	30	On	55

	Tyre pressure low		Tyre pressure high	
Dynamo	Seat low	Seat high	Seat low	Seat high
Off	52.5	42.0	49.0	39.0
On	57.0	43.5	52.0	42.5

Dynamo		Tyre pressure		Seat	
Off	On	Low	High	Low	High
45.63	48.75	48.75	45.63	52.63	41.75

are widely used; see Section 9.2.4. In this case an experimental setup with three factors each having two levels is applied twice: the design consists of two replicates of a 2^3 factorial experiment.

One linear model for the data in Table 8.2 is that at the lower seat height, with the dynamo off, and the lower tyre pressure, the mean time is μ, and the three factors act separately, changing the mean time by α_1, α_2, and α_3 respectively. This corresponds to the linear regression model

$$
\begin{pmatrix} y_1 \\ y_2 \\ y_3 \\ y_4 \\ y_5 \\ y_6 \\ y_7 \\ y_8 \\ y_9 \\ y_{10} \\ y_{11} \\ y_{12} \\ y_{13} \\ y_{14} \\ y_{15} \\ y_{16} \end{pmatrix}
=
\begin{pmatrix}
1 & 0 & 0 & 0 \\
1 & 0 & 0 & 0 \\
1 & 1 & 0 & 0 \\
1 & 1 & 0 & 0 \\
1 & 0 & 1 & 0 \\
1 & 0 & 1 & 0 \\
1 & 1 & 1 & 0 \\
1 & 1 & 1 & 0 \\
1 & 0 & 0 & 1 \\
1 & 0 & 0 & 1 \\
1 & 1 & 0 & 1 \\
1 & 1 & 0 & 1 \\
1 & 0 & 1 & 1 \\
1 & 0 & 1 & 1 \\
1 & 1 & 1 & 1 \\
1 & 1 & 1 & 1
\end{pmatrix}
\begin{pmatrix} \mu \\ \alpha_1 \\ \alpha_2 \\ \alpha_3 \end{pmatrix}
+
\begin{pmatrix} \varepsilon_1 \\ \varepsilon_2 \\ \varepsilon_3 \\ \varepsilon_4 \\ \varepsilon_5 \\ \varepsilon_6 \\ \varepsilon_7 \\ \varepsilon_8 \\ \varepsilon_9 \\ \varepsilon_{10} \\ \varepsilon_{11} \\ \varepsilon_{12} \\ \varepsilon_{13} \\ \varepsilon_{14} \\ \varepsilon_{15} \\ \varepsilon_{16} \end{pmatrix}.
$$

Table 8.2 suggests that $\mu \doteq 52.5$, that $\alpha_1 < 0$, $\alpha_2 > 0$, and $\alpha_3 < 0$. The baseline time is μ, which corresponds to the mean time at the lower level of all three factors, and the overall average time is $\bar{y} = \mu + \frac{1}{2}\alpha_1 + \frac{1}{2}\alpha_2 + \frac{1}{2}\alpha_3 + \bar{\varepsilon}$, where $\bar{\varepsilon}$ is the average of the unobserved errors.

A different formulation of the model would take the overall mean time as the baseline, leading to

$$
\begin{pmatrix} y_1 \\ y_2 \\ y_3 \\ y_4 \\ y_5 \\ y_6 \\ y_7 \\ y_8 \\ y_9 \\ y_{10} \\ y_{11} \\ y_{12} \\ y_{13} \\ y_{14} \\ y_{15} \\ y_{16} \end{pmatrix}
=
\begin{pmatrix}
1 & -1 & -1 & -1 \\
1 & -1 & -1 & -1 \\
1 & 1 & -1 & -1 \\
1 & 1 & -1 & -1 \\
1 & -1 & 1 & -1 \\
1 & -1 & 1 & -1 \\
1 & 1 & 1 & -1 \\
1 & 1 & 1 & -1 \\
1 & -1 & -1 & 1 \\
1 & -1 & -1 & 1 \\
1 & 1 & -1 & 1 \\
1 & 1 & -1 & 1 \\
1 & -1 & 1 & 1 \\
1 & -1 & 1 & 1 \\
1 & 1 & 1 & 1 \\
1 & 1 & 1 & 1
\end{pmatrix}
\begin{pmatrix} \beta_0 \\ \beta_1 \\ \beta_2 \\ \beta_3 \end{pmatrix}
+
\begin{pmatrix} \varepsilon_1 \\ \varepsilon_2 \\ \varepsilon_3 \\ \varepsilon_4 \\ \varepsilon_5 \\ \varepsilon_6 \\ \varepsilon_7 \\ \varepsilon_8 \\ \varepsilon_9 \\ \varepsilon_{10} \\ \varepsilon_{11} \\ \varepsilon_{12} \\ \varepsilon_{13} \\ \varepsilon_{14} \\ \varepsilon_{15} \\ \varepsilon_{16} \end{pmatrix}.
\qquad (8.2)
$$

In (8.2) the effect of increasing seat height from 26 to 30 inches is $2\beta_1$, the effect of switching the dynamo on is $2\beta_2$, and the effect of increasing tyre pressure is $2\beta_3$. As each column of the design matrix apart from the first has sum zero, the overall average time in this parametrization is $\beta_0 + \bar{\varepsilon}$. Although the parameter β_0 is related to the overall mean, it does not correspond to a combination of factors that can be applied to the bicycle — how can the dynamo be half on? Despite this, we shall see below that (8.2) is convenient for some purposes. ∎

Often it is better to apply a linear model to transformed data than to the original observations.

Example 8.5 (Multiplicative model) Suppose that the data consist of times to failure that depend on positive covariates x_1 and x_2 according to

$$y = \gamma_0 x_1^{\gamma_1} x_2^{\gamma_2} \eta,$$

where η is a positive random variable. Then

$$\log y = \log \gamma_0 + \gamma_1 \log x_1 + \gamma_2 \log x_2 + \log \eta,$$

which is linear in $\log \gamma_0$, γ_1, and γ_2. The variance of the transformed response $\log y$ does not depend on its mean, whereas y has variance proportional to the square of its mean, so in addition to achieving linearity, the transformation equalizes the variances. ∎

Exercises 8.1

1   Which of the following can be written as linear regression models, (i) as they are, (ii) when a single parameter is held fixed, (iii) after transformation? For those that can be so written, give the response variable and the form of the design matrix.
(a) $y = \beta_0 + \beta_1/x + \beta_2/x^2 + \varepsilon$;
(b) $y = \beta_0/(1 + \beta_1 x) + \varepsilon$;
(c) $y = 1/(\beta_0 + \beta_1 x + \varepsilon)$;
(d) $y = \beta_0 + \beta_1 x^{\beta_2} + \varepsilon$;
(e) $y = \beta_0 + \beta_1 x_1^{\beta_2} + \beta_3 x_2^{\beta_4} + \varepsilon$;

2   Data are available on the weights of two groups of three rats at the beginning of a fortnight, x, and at its end, y. During the fortnight, one group was fed normally and the other group was fed a growth inhibitor. Consider a linear model for the weights,

$$y_{jg} = \alpha_g + \beta_g x_{jg} + \varepsilon_{jg}, \quad j = 1, \ldots, 3, \quad g = 1, 2.$$

(a) Write down the design matrix for the model above.
(b) The model is to be reparametrized in such a way that it can be specialized to (i) two parallel lines for the two groups, (ii) two lines with the same intercept, (iii) one common line for both groups, just by setting parameters to zero. Give one design matrix which can be made to correspond to (i), (ii), and (iii), just by dropping columns.

8.2 Normal Linear Model

8.2.1 Estimation

Suppose that the errors ε_j in (8.1) are independent normal random variables, with means zero and variances σ^2. Then the responses y_j are independent normal random variables with means $x_j^{\mathrm{T}} \beta$ and variances σ^2, and (8.1) is the *normal linear model*. The likelihood for β and σ^2 is

$$L(\beta, \sigma^2) = \prod_{j=1}^{n} \frac{1}{(2\pi\sigma^2)^{1/2}} \exp\left\{-\frac{1}{2\sigma^2}(y_j - x_j^\mathsf{T}\beta)^2\right\},$$

and the log likelihood is

$$\ell(\beta, \sigma^2) \equiv -\frac{1}{2}\left\{n\log\sigma^2 + \frac{1}{\sigma^2}\sum_{j=1}^{n}(y_j - x_j^\mathsf{T}\beta)^2\right\}.$$

Whatever the value of σ^2, the log likelihood is maximized with respect to β at the value that minimizes the *sum of squares*

$$SS(\beta) = \sum_{j=1}^{n}(y_j - x_j^\mathsf{T}\beta)^2 = (y - X\beta)^\mathsf{T}(y - X\beta). \qquad (8.3)$$

We obtain the maximum likelihood estimate of β by solving simultaneously the equations

$$\frac{\partial SS(\beta)}{\partial\beta_r} = 2\sum_{j=1}^{n}x_{jr}(y_j - \beta^\mathsf{T}x_j) = 0, \quad r = 1,\ldots,p.$$

In matrix form these amount to the *normal equations*

$$X^\mathsf{T}(y - X\beta) = 0, \qquad (8.4)$$

which imply that the estimate satisfies $(X^\mathsf{T}X)\beta = X^\mathsf{T}y$. Provided the $p \times p$ matrix $X^\mathsf{T}X$ is of full rank it is invertible, and the *least squares estimator* of β is

$$\widehat{\beta} = (X^\mathsf{T}X)^{-1}X^\mathsf{T}y.$$

The maximum likelihood estimator of σ^2 may be obtained from the profile likelihood for σ^2,

$$\ell_\mathrm{p}(\sigma^2) = \max_\beta \ell(\beta, \sigma^2) = -\frac{1}{2}\left\{n\log\sigma^2 + \frac{1}{\sigma^2}(y - X\widehat{\beta})^\mathsf{T}(y - X\widehat{\beta})\right\}, \qquad (8.5)$$

and it follows by differentiation that the maximum likelihood estimator of σ^2 is

$$\widehat{\sigma}^2 = n^{-1}(y - X\widehat{\beta})^\mathsf{T}(y - X\widehat{\beta}) = n^{-1}\sum_{j=1}^{n}(y_j - x_j^\mathsf{T}\widehat{\beta})^2.$$

We shall see below that $\widehat{\sigma}^2$ is biased and that an unbiased estimator of σ^2 is

$$S^2 = \frac{1}{n-p}(y - X\widehat{\beta})^\mathsf{T}(y - X\widehat{\beta}) = \frac{1}{n-p}\sum_{j=1}^{n}(y_j - x_j^\mathsf{T}\widehat{\beta})^2.$$

Example 8.6 (Straight-line regression)   We write the straight-line regression model (5.3) in matrix form as

$$
\begin{pmatrix} y_1 \\ y_2 \\ \vdots \\ y_n \end{pmatrix} = \begin{pmatrix} 1 & x_1 - \overline{x} \\ 1 & x_2 - \overline{x} \\ \vdots & \vdots \\ 1 & x_n - \overline{x} \end{pmatrix} \begin{pmatrix} \gamma_0 \\ \gamma_1 \end{pmatrix} + \begin{pmatrix} \varepsilon_1 \\ \varepsilon_2 \\ \vdots \\ \varepsilon_n \end{pmatrix}.
$$

The least squares estimates are

$$
\begin{aligned}
\widehat{\beta} = \begin{pmatrix} \widehat{\gamma_0} \\ \widehat{\gamma_1} \end{pmatrix} &= \begin{pmatrix} n & \sum(x_j - \overline{x}) \\ \sum(x_j - \overline{x}) & \sum(x_j - \overline{x})^2 \end{pmatrix}^{-1} \begin{pmatrix} \sum y_j \\ \sum(x_j - \overline{x})y_j \end{pmatrix} \\
&= \begin{pmatrix} n^{-1} & 0 \\ 0 & \frac{1}{\sum(x_j - \overline{x})^2} \end{pmatrix} \begin{pmatrix} \sum y_j \\ \sum(x_j - \overline{x})y_j \end{pmatrix} \\
&= \begin{pmatrix} \overline{y} \\ \frac{\sum(x_j - \overline{x})y_j}{\sum(x_j - \overline{x})^2} \end{pmatrix}.
\end{aligned}
$$

If all the x_j are equal, $X^{\mathsf{T}}X$ is not invertible, and $\widehat{\gamma_1}$ is undetermined: any value is possible.

The unbiased estimator of σ^2 is

$$
\frac{1}{n-2} \sum_{j=1}^{n} \left\{ y_j - \overline{y} - (x_j - \overline{x}) \frac{\sum(x_k - \overline{x})y_k}{\sum(x_k - \overline{x})^2} \right\}^2.
$$

∎

Example 8.7 (Surveying a triangle)   Suppose that we want to estimate the angles α, β, and γ (radians) of a triangle ABC based on a single independent measurement of the angle at each corner. Although there are three angles, their sum is the constant $\alpha + \beta + \gamma = \pi$, and so just two of them vary independently. In terms of α and β, we have $y_A = \alpha + \varepsilon_A$, $y_B = \beta + \varepsilon_B$, and $y_C = \pi - \alpha - \beta + \varepsilon_C$, and this gives the linear model

$$
\begin{pmatrix} y_A \\ y_B \\ y_C - \pi \end{pmatrix} = \begin{pmatrix} 1 & 0 \\ 0 & 1 \\ -1 & -1 \end{pmatrix} \begin{pmatrix} \alpha \\ \beta \end{pmatrix} + \begin{pmatrix} \varepsilon_A \\ \varepsilon_B \\ \varepsilon_C \end{pmatrix}.
$$

Hence

$$
\begin{pmatrix} \widehat{\alpha} \\ \widehat{\beta} \end{pmatrix} = \frac{1}{3} \begin{pmatrix} 2 & -1 \\ -1 & 2 \end{pmatrix} \begin{pmatrix} \pi + y_A - y_C \\ \pi + y_B - y_C \end{pmatrix} = \frac{1}{3} \begin{pmatrix} \pi + 2y_A - y_B - y_C \\ \pi + 2y_B - y_A - y_C \end{pmatrix}.
$$

It is straightforward to show that $s^2 = (y_A + y_B + y_C - \pi)^2/3$.   ∎

The sum of squares $SS(\beta)$ plays a central role. Its minimum value,

$$
SS(\widehat{\beta}) = \sum_{j=1}^{n} \left(y_j - x_j^{\mathsf{T}}\widehat{\beta} \right)^2 = (y - X\widehat{\beta})^{\mathsf{T}}(y - X\widehat{\beta}),
$$

is called the *residual sum of squares* because it is the residual squared discrepancy between the observations, y, and the *fitted values*, $\widehat{y} = X\widehat{\beta}$. The vector $\widehat{y}$ is the linear combination of the columns of X that best accounts for the variation in y, in the sense of minimizing the squared distance between them. Note that

$$\widehat{y} = X\widehat{\beta} = X(X^{\mathsf{T}}X)^{-1}X^{\mathsf{T}}y = Hy,$$

say, where the *hat matrix* $H = X(X^{\mathsf{T}}X)^{-1}X^{\mathsf{T}}$ "puts hats" on y. Evidently H is a projection matrix; see Section 8.2.2.

The unobservable error $\varepsilon_j = y_j - x_j^{\mathsf{T}}\beta$ is estimated by the jth *residual* $e_j = y_j - \widehat{y}_j = y_j - x_j^{\mathsf{T}}\widehat{\beta}$. In vector terms,

Sometimes e_j is called a *raw residual.*

$$e = y - X\widehat{\beta} = y - Hy = (I_n - H)y,$$

where I_n is the $n \times n$ identity matrix.

Example 8.8 (Cycling data) For model (8.2) we find that

$$(X^{\mathsf{T}}X)^{-1} = \frac{1}{16}I_4,$$

so the least squares estimates $(X^{\mathsf{T}}X)^{-1}X^{\mathsf{T}}y$ are

$$\frac{1}{16}\begin{pmatrix} y_1 + y_2 + y_3 + y_4 + y_5 + y_6 + y_7 + y_8 + y_9 + y_{10} + y_{11} + y_{12} + y_{13} + y_{14} + y_{15} + y_{16} \\ -y_1 - y_2 + y_3 + y_4 - y_5 - y_6 + y_7 + y_8 - y_9 - y_{10} + y_{11} + y_{12} - y_{13} - y_{14} + y_{15} + y_{16} \\ -y_1 - y_2 - y_3 - y_4 + y_5 + y_6 + y_7 + y_8 - y_9 - y_{10} - y_{11} - y_{12} + y_{13} + y_{14} + y_{15} + y_{16} \\ -y_1 - y_2 - y_3 - y_4 - y_5 - y_6 - y_7 - y_8 + y_9 + y_{10} + y_{11} + y_{12} + y_{13} + y_{14} + y_{15} + y_{16} \end{pmatrix} = \begin{pmatrix} 47.19 \\ -5.437 \\ 1.563 \\ -1.563 \end{pmatrix}.$$

Thus the overall average time is 47.19 seconds, putting the seat at height 30 inches rather than 26 inches changes the time by an average of $2 \times (-5.437) = -10.87$ seconds, putting the dynamo on rather than off changes the time by an average of $2 \times 1.563 = 3.13$ seconds, and increasing the tyre pressure from 40 to 55 psi changes the time by –3.13 seconds. The largest effect is due to increasing the seat height. The model suggests that the fastest time is obtained with no dynamo, a high seat and tyres at 55 psi.

The residual sum of squares for this model is 43.25 seconds squared, the overall sum of squares is $\sum y_j^2 = 36221$ seconds squared, and therefore the sum of squares explained by the model is $36221 - 43.25 = 36177.75$ seconds squared; this is the amount of variation removed when $X\beta$ is fitted.

The fitted values are $\widehat{y} = X\widehat{\beta}$, giving $\widehat{y}_1 = \widehat{\beta}_0 - \widehat{\beta}_1 - \widehat{\beta}_2 - \widehat{\beta}_3 = 52.625$, $e_1 = y_1 - \widehat{y}_1 = 51 - 52.625 = -1.625$, and so forth. Table 8.3 gives the data, fitted values, residuals and quantities discussed in Examples 8.22 and 8.27.   ∎

8.2.2 Geometrical interpretation

Figure 8.2 shows the geometry of least squares. The n-dimensional vector space inhabited by the observation vector y is represented by the space spanned by all three axes, and the p-dimensional subspace in which $X\beta$ lies is represented by the horizontal plane through the origin. The least squares estimate $\widehat{\beta}$ minimizes $(y - X\beta)^{\mathsf{T}}(y - X\beta)$, which is the squared distance between $X\beta$ and y. We see that $(y - X\beta)^{\mathsf{T}}(y - X\beta)$ is minimized when the vector $y - X\beta$ is orthogonal to the horizontal plane spanned by the columns of X, so that for any column x of X we have $x^{\mathsf{T}}(y - X\beta) = 0$. Equivalently the normal equations $X^{\mathsf{T}}(y - X\beta) = 0$ hold, and provided $X^{\mathsf{T}}X$ is invertible

Table 8.3 Data from bicycle experiment, together with fitted values $\widehat{y}$, raw residuals e, standardized residuals, r, deletion residuals r', leverages h and Cook distances C.

Setup	Seat height	Dynamo	Tyre pressure	Time y	$\widehat{y}$	e	r	r'	h	C
1	−1	−1	−1	51	52.62	−1.625	−0.99	−0.99	0.25	0.08
2	−1	−1	−1	54	52.62	1.375	−0.84	0.83	0.25	0.06
3	1	−1	−1	41	41.75	−0.750	−0.46	−0.44	0.25	0.02
4	1	−1	−1	43	41.75	1.250	0.76	0.75	0.25	0.05
5	−1	1	−1	54	55.75	−1.750	−1.06	−1.07	0.25	0.09
6	−1	1	−1	60	55.75	4.250	2.59	3.72	0.25	0.56
7	1	1	−1	44	44.87	−0.875	−0.53	−0.52	0.25	0.02
8	1	1	−1	43	44.87	−1.875	−1.14	−1.16	0.25	0.11
9	−1	−1	1	50	49.50	0.500	0.30	0.29	0.25	0.01
10	−1	−1	1	48	49.50	−1.500	−0.91	−0.91	0.25	0.07
11	1	−1	1	39	38.62	0.375	0.23	0.22	0.25	0.00
12	1	−1	1	39	38.62	0.375	0.23	0.22	0.25	0.00
13	−1	1	1	53	52.62	0.375	0.23	0.22	0.25	0.00
14	−1	1	1	51	52.62	−1.625	−0.99	−0.99	0.25	0.08
15	1	1	1	41	41.75	−0.750	−0.46	−0.44	0.25	0.02
16	1	1	1	44	41.75	2.250	1.37	1.43	0.25	0.16

we obtain $\widehat{\beta} = (X^{\mathrm{T}}X)^{-1}X^{\mathrm{T}}y$. The fitted value $\widehat{y} = X\widehat{\beta} = X(X^{\mathrm{T}}X)^{-1}X^{\mathrm{T}}y = Hy$ is the orthogonal projection of y onto the plane spanned by the columns of X, and the matrix representing that projection is H. Notice that $\widehat{y}$ is unique whether or not $X^{\mathrm{T}}X$ is invertible.

Figure 8.2 shows that the vector of residuals, $e = y - \widehat{y} = (I_n - H)y$, and the vector of fitted values, $\widehat{y} = Hy$, are orthogonal. To see this algebraically, note that

$$\widehat{y}^{\mathrm{T}}e = y^{\mathrm{T}}H^{\mathrm{T}}(I_n - H)y = y^{\mathrm{T}}(H - H)y = 0, \tag{8.6}$$

because $H^{\mathrm{T}} = H$ and $HH = H$, that is, the projection matrix H is symmetric and idempotent (Exercise 8.2.5). The close link between orthogonality and independence for normally distributed vectors means that (8.6) has important consequences, as we shall see in Section 8.3. For now, notice that (8.6) implies that

$$y^{\mathrm{T}}y = (y - \widehat{y} + \widehat{y})^{\mathrm{T}}(y - \widehat{y} + \widehat{y}) = (e + \widehat{y})^{\mathrm{T}}(e + \widehat{y}) = e^{\mathrm{T}}e + \widehat{y}^{\mathrm{T}}\widehat{y}, \tag{8.7}$$

as is clear from Figure 8.2 by Pythagoras' theorem. That is, the overall sum of squares of the data, $\sum y_j^2 = y^{\mathrm{T}}y$, equals the sum of the residual sum of squares, $SS(\widehat{\beta}) = \sum(y_j - \widehat{y}_j)^2 = e^{\mathrm{T}}e$, and the sum of squares for the fitted model, $\sum \widehat{y}_j^2 = \widehat{y}^{\mathrm{T}}\widehat{y}$.

Such decompositions are central to analysis of variance, discussed below.

8.2.3 Likelihood quantities

Chapter 4 shows how the observed and expected information matrices play a central role in likelihood inference, by providing approximate variances for maximum likelihood estimates. To obtain these matrices for the normal linear model, note that the

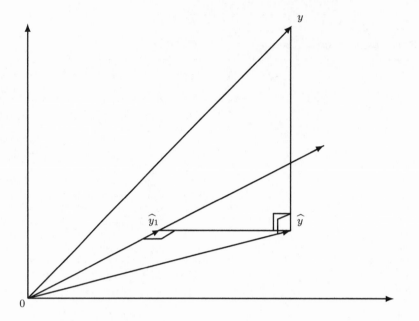

Figure 8.2 The geometry of least squares estimation. The space spanned by all three axes represents the n-dimensional observation space in which y lies. The horizontal plane through O represents the p-dimensional space in which the linear combination $X\beta$ lies, and estimation by least squares amounts to minimizing the squared distance $(y - X\beta)^T(y - X\beta)$. In the figure the value of $X\beta$ that gives the minimum lies vertically below y, which corresponds to orthogonal projection of y into the p-dimensional subspace spanned by the columns of X; the fitted value $\widehat{y} = Hy$ is the point closest to y in that subspace, and the projection matrix is $H = X(X^TX)^{-1}X^T$. The vector of residuals $e = y - \widehat{y}$ is orthogonal to the fitted value $\widehat{y}$. The line $x = z = 0$ represents the space spanned by the columns of the reduced model matrix X_1, with corresponding fitted value $\widehat{y}_1$. The orthogonality of $\widehat{y}_1, \widehat{y} - \widehat{y}_1$, and $y - \widehat{y}$ implies that when the data are normal the corresponding sums of squares are independent.

log likelihood has second derivatives

$$\frac{\partial^2 \ell}{\partial \beta_r \partial \beta_s} = -\frac{1}{\sigma^2} \sum_{j=1}^{n} x_{jr} x_{js}, \quad \frac{\partial^2 \ell}{\partial \beta_r \partial \sigma^2} = \frac{1}{\sigma^4} \sum_{j=1}^{n} x_{jr} \left(y_j - x_j^T \beta \right),$$

$$\frac{\partial^2 \ell}{\partial (\sigma^2)^2} = -\frac{1}{2} \left\{ -\frac{1}{\sigma^4} + \frac{2}{\sigma^6} \sum_{j=1}^{n} \left(y_j - x_j^T \beta \right)^2 \right\}, \quad r, s = 1, \dots, p.$$

Thus elements of the expected information matrix are

$$\mathrm{E}\left(-\frac{\partial^2 \ell}{\partial \beta_r \partial \beta_s} \right) = \frac{1}{\sigma^2} \sum_{j=1}^{n} x_{jr} x_{js}, \quad \mathrm{E}\left(-\frac{\partial^2 \ell}{\partial \beta_r \partial \sigma^2} \right) = 0, \quad \mathrm{E}\left\{ -\frac{\partial^2 \ell}{\partial (\sigma^2)^2} \right\} = \frac{n}{2\sigma^4},$$

or in matrix form

$$I(\beta, \sigma^2) = \begin{pmatrix} \sigma^{-2} X^T X & 0 \\ 0 & \frac{1}{2} n \sigma^{-4} \end{pmatrix}, \quad I(\beta, \sigma^2)^{-1} = \begin{pmatrix} \sigma^2 (X^T X)^{-1} & 0 \\ 0 & 2\sigma^4 / n \end{pmatrix}.$$

Provided that X has rank p, the matrices $I(\beta, \sigma^2)$ and $J(\widehat{\beta}, \widehat{\sigma}^2)$ are positive definite (Exercise 8.2.7).

Under mild regularity conditions on the design matrix and the errors, the general theory of likelihood estimation implies that the asymptotic distribution of $\widehat{\beta}$ and σ^2 is normal with means β and σ^2, and covariance matrix given by $I(\beta, \sigma^2)^{-1}$, the block diagonal structure of which implies that $\widehat{\beta}$ and $\widehat{\sigma}^2$ are asymptotically independent. We shall see in the next section that stronger results are true: when the errors are normal the estimates $\widehat{\beta}$ have an exact normal distribution and are independent of $\widehat{\sigma}^2$ for every value of n, while $\widehat{\sigma}^2$ has a distribution proportional to χ^2_{n-p} provided that $n > p$.

The quantities $\widehat{\beta}$ and $SS(\widehat{\beta})$ are minimal sufficient statistics for β and σ^2 (Problem 8.7).

Example 8.9 (Two-sample model) Suppose that we have two groups of normal data, the first with mean β_0,

$$y_{0j} = \beta_0 + \varepsilon_{0j}, \quad j = 1, \ldots, n_0,$$

and the second with mean $\beta_0 + \beta_1$,

$$y_{1j} = \beta_0 + \beta_1 + \varepsilon_{1j}, \quad j = 1, \ldots, n_1,$$

where the ε_{gj} are independent with means zero and variances σ^2. The matrix form of this model is

$$
\begin{pmatrix} y_{01} \\ \vdots \\ y_{0n_0} \\ y_{11} \\ \vdots \\ y_{1n_1} \end{pmatrix}
=
\begin{pmatrix} 1 & 0 \\ \vdots & \vdots \\ 1 & 0 \\ 1 & 1 \\ \vdots & \vdots \\ 1 & 1 \end{pmatrix}
\begin{pmatrix} \beta_0 \\ \beta_1 \end{pmatrix}
+
\begin{pmatrix} \varepsilon_{01} \\ \vdots \\ \varepsilon_{0n_0} \\ \varepsilon_{11} \\ \vdots \\ \varepsilon_{1n_1} \end{pmatrix}.
$$

The estimator of β is $\widehat{\beta} = (X^{\mathsf{T}}X)^{-1}X^{\mathsf{T}}y$, that is,

$$
\begin{pmatrix} \widehat{\beta}_0 \\ \widehat{\beta}_1 \end{pmatrix}
=
\begin{pmatrix} n_0 + n_1 & n_1 \\ n_1 & n_1 \end{pmatrix}^{-1}
\begin{pmatrix} n_0 \overline{y}_{0.} + n_1 \overline{y}_{1.} \\ n_1 \overline{y}_{1.} \end{pmatrix}
$$

$$
=
\begin{pmatrix} n_0^{-1} & -n_0^{-1} \\ -n_0^{-1} & n_0^{-1} + n_1^{-1} \end{pmatrix}
\begin{pmatrix} n_0 \overline{y}_{0.} + n_1 \overline{y}_{1.} \\ n_1 \overline{y}_{1.} \end{pmatrix}
$$

$$
=
\begin{pmatrix} \overline{y}_{0.} \\ \overline{y}_{1.} - \overline{y}_{0.} \end{pmatrix},
$$

where $\overline{y}_{0.} = n_0^{-1} \sum y_{0j}$ and $\overline{y}_{1.} = n_1^{-1} \sum y_{1j}$ are the group averages. One can verify directly that the elements of $\sigma^2 (X^{\mathsf{T}}X)^{-1}$ give the variances and covariance of the least squares estimators.

In this example the fitted values are $\widehat{\beta}_0 = \overline{y}_{0.}$ for the first group and $\widehat{\beta}_0 + \widehat{\beta}_1 = \overline{y}_{1.}$ for the second group, and the unbiased estimator of σ^2 is

$$
S^2 = \frac{1}{n_0 + n_1 - 2} \left\{ \sum_{j=1}^{n_0} (y_{0j} - \overline{y}_{0.})^2 + \sum_{j=1}^{n_1} (y_{1j} - \overline{y}_{1.})^2 \right\}.
$$

A minimal sufficient statistic for $(\beta_0, \beta_1, \sigma^2)$ is $(\overline{y}_{0.}, \overline{y}_{1.}, s^2)$. ∎

Example 8.10 (Maize data) The discussion in Example 1.1 suggests that a model of matched pairs better describes the experimental setup for the maize data than the two-sample model of Example 8.9. We parametrize the matched pair model so that the jth pair of observations is

$$y_{1j} = \beta_j - \beta_0 + \varepsilon_{1j}, \quad y_{2j} = \beta_j + \beta_0 + \varepsilon_{2j}, \quad j = 1, \ldots, m,$$

where we assume that the ε_{ji} are independent normal random variables with means zero and variances σ^2. We have $m = 15$. The average difference between the heights of the crossed and self-fertilized plants in a pair is $2\beta_0$, and the mean height of the pair is β_j. The matrix form of this model is

$$
\begin{pmatrix} y_{11} \\ y_{21} \\ y_{12} \\ y_{22} \\ \vdots \\ y_{1m} \\ y_{2m} \end{pmatrix}
=
\begin{pmatrix}
-1 & 1 & 0 & \cdots & 0 \\
1 & 1 & 0 & \cdots & 0 \\
-1 & 0 & 1 & \cdots & 0 \\
1 & 0 & 1 & \cdots & 0 \\
\vdots & \vdots & \vdots & & \vdots \\
-1 & 0 & 0 & \cdots & 1 \\
1 & 0 & 0 & \cdots & 1
\end{pmatrix}
\begin{pmatrix} \beta_0 \\ \beta_1 \\ \beta_2 \\ \vdots \\ \beta_m \end{pmatrix}
+
\begin{pmatrix} \varepsilon_{11} \\ \varepsilon_{21} \\ \varepsilon_{12} \\ \varepsilon_{22} \\ \vdots \\ \varepsilon_{1m} \\ \varepsilon_{2m} \end{pmatrix},
$$

so β has dimension $(m + 1) \times 1$ and $X^{\mathsf{T}}X = \mathrm{diag}(2m, 2, \ldots, 2)$ has dimension $(m + 1) \times (m + 1)$.

We see that

$$
\widehat{\beta}_0 = (y_{21} - y_{11} + y_{22} - y_{12} + \cdots + y_{2m} - y_{1m})/(2m),
$$
$$
\widehat{\beta}_j = \frac{1}{2}(y_{1j} + y_{2j}), \quad j = 1, \ldots, m,
$$

and that the estimators are independent. The unbiased estimator of σ^2 is

$$
S^2 = \frac{1}{2m - (m + 1)} \sum_{j=1}^{m} \{ (y_{1j} - \widehat{\beta}_j + \widehat{\beta}_0)^2 + (y_{2j} - \widehat{\beta}_j - \widehat{\beta}_0)^2 \},
$$

which can be written as $\{2(m - 1)\}^{-1} \sum (d_j - \overline{d})^2$, where $d_j = y_{2j} - y_{1j}$ is the difference between the heights of the crossed and self-fertilized plants in the jth pair, and $\overline{d} = m^{-1} \sum d_j$ is their average. Note that $\widehat{\beta}_0$ equals $\frac{1}{2}\overline{d}$. ∎

Likelihood ratio statistic

The likelihood ratio statistic is a standard tool for comparing nested models. In the context of the normal linear model, let

$$
y = X\beta + \varepsilon = (X_1 \quad X_2) \begin{pmatrix} \beta_1 \\ \beta_2 \end{pmatrix} + \varepsilon = X_1\beta_1 + X_2\beta_2 + \varepsilon,
$$

where X_1 is an $n \times q$ matrix, X_2 is an $n \times (p - q)$ matrix, $q < p$, and β_1 and β_2 are vectors of parameters of lengths q and $p - q$. Suppose that we wish to compare this with the simpler model in which $\beta_2 = 0$, so the mean of y depends only on X_1. Under the more general model the maximum likelihood estimators of β and σ^2 are $\widehat{\beta}$ and $\widehat{\sigma}^2 = n^{-1}SS(\widehat{\beta})$, where $SS(\beta) = (y - X\beta)^{\mathsf{T}}(y - X\beta)$, and it follows from (8.5) that the maximized log likelihood is

$$
\ell_{\mathrm{p}}(\widehat{\sigma}^2) = -\frac{1}{2}\{ n \log SS(\widehat{\beta}) + n - n \log n \},
$$

where $\ell_{\mathrm{p}}(\sigma^2) = \max_\beta \ell(\beta, \sigma^2)$ is the profile log likelihood for σ^2. When $\beta_2 = 0$, the maximum likelihood estimator of σ^2 is

$$
\widehat{\sigma}_0^2 = n^{-1}SS(\widehat{\beta}_1) = n^{-1}(y - X_1\widehat{\beta}_1)^{\mathsf{T}}(y - X_1\widehat{\beta}_1),
$$

where $\widehat{\beta}_1$ is the estimator of β_1 when $\beta_2 = 0$. Hence the likelihood ratio statistic for comparison of the models is

$$
\begin{aligned}
2\left\{\ell_{\mathrm{p}}(\widehat{\sigma}^2) - \ell_{\mathrm{p}}(\widehat{\sigma}_0^2)\right\} &= n \log\{SS(\widehat{\beta})/SS(\widehat{\beta}_1)\} \\
&= n \log\left[1 + \frac{p-q}{n-p} \frac{\{SS(\widehat{\beta}_1) - SS(\widehat{\beta})\}/(p-q)}{SS(\widehat{\beta})/(n-p)}\right] \\
&= n \log\left(1 + \frac{p-q}{n-p} F\right),
\end{aligned} \tag{8.8}
$$

say. Here $F \geq 0$, with equality only if the two sums of squares are equal. This event can occur only if the columns of X_2 are linearly dependent on those of X_1. If not, the results of Section 4.5.2 imply that the likelihood ratio statistic has an approximate χ^2 distribution, but as it is a monotonic function of F, large values of (8.8) correspond to large values of F. We shall see in Section 8.5 that the exact distribution of F is known and can be used to compare nested models, with no need for approximations.

It is instructive to express F explicitly in terms of the least squares estimators. As (8.8) is a likelihood ratio statistic for testing $\beta_2 = 0$, it is invariant to 1–1 reparametrizations that leave β_2 fixed, and we write $\mathrm{E}(y)$ as

$$
\begin{aligned}
X_1\beta_1 + X_2\beta_2 &= X_1\beta_1 + H_1X_2\beta_2 + (I - H_1)X_2\beta_2 \\
&= X_1\left\{\beta_1 + \left(X_1^{\mathrm{T}}X_1\right)^{-1}X_1^{\mathrm{T}}X_2\beta_2\right\} + Z_2\beta_2 \\
&= X_1\lambda + Z_2\psi,
\end{aligned}
$$

say, where $H_1 = X_1(X_1^{\mathrm{T}}X_1)^{-1}X_1^{\mathrm{T}}$ is the projection matrix for X_1, $Z_2 = (I - H_1)X_2$ is the matrix of residuals from regression of the columns of X_2 on those of X_1, and the new parameters are λ and $\psi = \beta_2$. Note that

$$
X_1^{\mathrm{T}}Z_2 = X_1^{\mathrm{T}}\left\{I - X_1\left(X_1^{\mathrm{T}}X_1\right)^{-1}X_1^{\mathrm{T}}\right\}X_2 = 0,
$$

and that H_1 is idempotent. In this new parametrization the parameter estimates are

$$
\begin{pmatrix}\widehat{\lambda} \\ \widehat{\psi}\end{pmatrix} = \begin{pmatrix} X_1^{\mathrm{T}}X_1 & X_1^{\mathrm{T}}Z_2 \\ Z_2^{\mathrm{T}}X_1 & Z_2^{\mathrm{T}}Z_2 \end{pmatrix}^{-1} \begin{pmatrix} X_1^{\mathrm{T}} \\ Z_2^{\mathrm{T}} \end{pmatrix} y = \begin{pmatrix} \left(X_1^{\mathrm{T}}X_1\right)^{-1}X_1^{\mathrm{T}}y \\ \left(Z_2^{\mathrm{T}}Z_2\right)^{-1}Z_2^{\mathrm{T}}y \end{pmatrix},
$$

while if $\psi = \beta_2 = 0$, the least squares estimate of λ remains $\widehat{\lambda}$. Consequently

$$
\begin{aligned}
SS(\widehat{\beta}) &= (y - X_1\widehat{\lambda} - Z_2\widehat{\psi})^{\mathrm{T}}(y - X_1\widehat{\lambda} - Z_2\widehat{\psi}) \\
&= (y - X_1\widehat{\lambda})^{\mathrm{T}}(y - X_1\widehat{\lambda}) - 2\widehat{\psi}^{\mathrm{T}}Z_2^{\mathrm{T}}(y - X_1\widehat{\lambda}) + \widehat{\psi}^{\mathrm{T}}Z_2^{\mathrm{T}}Z_2\widehat{\psi} \\
&= SS(\widehat{\beta}_1) - \widehat{\psi}^{\mathrm{T}}Z_2^{\mathrm{T}}Z_2\widehat{\psi},
\end{aligned}
$$

since

$$
\begin{aligned}
\widehat{\psi}^{\mathrm{T}}Z_2^{\mathrm{T}}(y - X_1\widehat{\lambda}) &= \widehat{\psi}^{\mathrm{T}}Z_2^{\mathrm{T}}y - \widehat{\psi}^{\mathrm{T}}Z_2^{\mathrm{T}}X_1\widehat{\lambda} \\
&= \widehat{\psi}^{\mathrm{T}}\left(Z_2^{\mathrm{T}}Z_2\right)\left(Z_2^{\mathrm{T}}Z_2\right)^{-1}Z_2^{\mathrm{T}}y \\
&= \widehat{\psi}^{\mathrm{T}}\left(Z_2^{\mathrm{T}}Z_2\right)\widehat{\psi}.
\end{aligned}
$$

Thus the F statistic in (8.8) may be written as

$$F = \frac{n-p}{p-q} \frac{\widehat{\beta}_2^\mathsf{T} X_2^\mathsf{T}(I - H_1)X_2\widehat{\beta}_2}{SS(\widehat{\beta})}$$

and this is large if $\widehat{\beta}_2$ differs greatly from zero.

If β_2 is scalar, then $p - q = 1$, the matrix $Z_2^\mathsf{T}Z_2 = X_2^\mathsf{T}(I - H_1)X_2 = v_{pp}^{-1}$ is scalar, and $F = T^2$, where

$$T = \frac{\widehat{\beta}_2 - \beta_2}{(v_{pp}s^2)^{1/2}} \tag{8.9}$$

with $s^2 = SS(\widehat{\beta})/(n - p)$ and $\beta_2 = 0$. Thus F is a monotonic function of T^2. We shall see in Section 8.3.2 that T has a t_{n-p} distribution.

8.2.4 Weighted least squares

Suppose that a normal linear model applies but that the responses have unequal variances. If the variance of y_j is σ^2/w_j, where σ^2 is unknown but the w_j are known positive quantities giving the relative precisions of the y_j, the log likelihood can be written as

$$\ell(\beta, \sigma^2) \equiv -\frac{1}{2}\left\{ n \log \sigma^2 + \frac{1}{\sigma^2}(y - X\beta)^\mathsf{T} W(y - X\beta) \right\},$$

where $W = \text{diag}\{w_1, \ldots, w_n\}$ is known as the matrix of weights. Let $W^{1/2} = \text{diag}\{w_1^{1/2}, \ldots, w_n^{1/2}\}$, and set $y' = W^{1/2}y$ and $X' = W^{1/2}X$. Then the sum of squares may be written as $(y' - X'\beta)^\mathsf{T}(y' - X'\beta)$. As this has the same form as (8.3), the estimates of β and σ^2 are

$$\widehat{\beta} = (X'^\mathsf{T}X')^{-1}X'^\mathsf{T}y' = (X^\mathsf{T}WX)^{-1}X^\mathsf{T}Wy, \tag{8.10}$$

and

$$\begin{aligned} s^2 &= (n - p)^{-1}y'^\mathsf{T}\{I - X'(X'^\mathsf{T}X')^{-1}X'^\mathsf{T}\}y' \\ &= (n - p)^{-1}y^\mathsf{T}\{W - WX(X^\mathsf{T}WX)^{-1}X^\mathsf{T}W\}y. \end{aligned} \tag{8.11}$$

These are the *weighted least squares estimates*. This device of replacing y and X with $W^{1/2}y$ and $W^{1/2}X$ allows methods for unweighted least squares models to be applied when there are weights (Exercise 8.2.9).

Example 8.11 (Grouped data) Suppose that each y_j is an average of a random sample of m_j normal observations, each with mean $x_j^\mathsf{T}\beta$ and variance σ^2, and that the samples are independent of each other. Then y_j has mean $x_j^\mathsf{T}\beta$ and variance σ^2/m_j, and the y_j are independent. The estimates of β and σ^2 are given by (8.10) and (8.11) with weights $w_j \equiv m_j$. ∎

Weighted least squares can be extended to situations where the errors are correlated but the relative correlations are known, that is, $\text{var}(y) = \sigma^2 W^{-1}$, where W is known but not necessarily diagonal. This is sometimes called *generalized least squares*. The corresponding least squares estimates of β and σ^2 are given by (8.10) and (8.11).

Weighted least squares turns out to be of central importance in fitting nonlinear models, and is used extensively in Chapter 10.

Exercises 8.2

1 Write down the linear model corresponding to a simple random sample $y_1, \ldots, y_n$ from the $N(\mu, \sigma^2)$ distribution, and find the design matrix. Verify that

$$\widehat{\mu} = (X^{\mathrm{T}}X)^{-1}X^{\mathrm{T}}y = \overline{y}, \quad s^2 = SS(\widehat{\beta})/(n - p) = (n - 1)^{-1}\sum (y_j - \overline{y})^2.$$

2 Verify the formula for s^2 given in Example 8.7, and show directly that its distribution is $\sigma^2 \chi_1^2$.

3 The angles of the triangle ABC are measured with A and B each measured twice and C three times. All the measurements are independent and unbiased with common variance σ^2. Find the least squares estimates of the angles A and B based on the seven measurements and calculate the variance of these estimates.

4 In Example 8.10, show that the unbiased estimator of σ^2 is $\{2(m - 1)\}^{-1}\sum(d_j - \overline{d})^2$.

5 Show that if the $n \times p$ design matrix X has rank p, the matrix $H = X(X^{\mathrm{T}}X)^{-1}X^{\mathrm{T}}$ is symmetric and idempotent, that is, $H^{\mathrm{T}} = H$ and $H^2 = H$, and that $\text{tr}(H) = p$. Show that $I_n - H$ is symmetric and idempotent also. By considering $H^2 a$, where a is an eigenvector of H, show that the eigenvalues of H equal zero or one. Prove also that H has rank p. Give the elements of H for Examples 8.9 and 8.10.

6 In a linear model in which $n \to \infty$ in such a way that $\widehat{\beta} \overset{P}{\longrightarrow} \beta$, show that $e_j \overset{P}{\longrightarrow} \varepsilon_j$. Generalize this to any finite subset of the residuals e. Is this true for the entire vector e? Let $y_j = \beta_0 + \beta_1 x_j + \varepsilon_j$ with $x_1 = \cdots = x_k = 0$ and $x_{k+1} = \cdots = x_n = 1$. Is $\widehat{\beta}$ consistent if $n \to \infty$ and $k = 1$? If $k = m$, for some fixed m? If $k = n/2$? Which of the ε_j can be estimated consistently in each case?

7 Show that in a normal linear model in which X has rank p, the matrices $I(\beta, \sigma^2)$ and $J(\widehat{\beta}, \widehat{\sigma}^2)$ are positive definite.

8 (a) Consider the two design matrices for Example 8.4; call them X_1 and X_2. Find the 4×4 matrix A for which $X_1 = X_2 A$, and verify that it is invertible by finding its inverse.
(b) Consider the linear models $y = X_1\beta + \varepsilon$ and $y = X_2\gamma + \varepsilon$, where $X_1 = X_2 A$, $\gamma = A\beta$, and A is an invertible matrix. Show that the hat matrices, fitted values, residuals, and sums of squares are the same for both models, and explain this in terms of the geometry of least squares.

9 (a) Consider a normal linear model $y = X\beta + \varepsilon$ where $\text{var}(\varepsilon) = \sigma^2 W^{-1}$, and W is a known positive definite symmetric matrix. Show that a inverse square root matrix $W^{1/2}$ exists, and re-express the least squares problem in terms of $y_1 = W^{1/2}y$, $X_1 = W^{1/2}X$, and $\varepsilon_1 = W^{1/2}\varepsilon$. Show that $\text{var}(\varepsilon_1) = \sigma^2 I_n$. Hence find the least squares estimates, hat matrix, and residual sum of squares for the weighted regression in terms of y, X, and W, and give the distributions of the least squares estimates of β and the residual sum of squares.
(b) Suppose that W depends on an unknown scalar parameter, ρ. Find the profile log likelihood for ρ, $\ell_{\mathrm{p}}(\rho) = \max_{\beta, \sigma^2} \ell(\beta, \sigma^2, \rho)$, and outline how to use a least squares package to give a confidence interval for ρ.

Recall that: (i) if the matrix A is square, then $\text{tr}(A) = \sum a_{ii}$; (ii) if A and B are conformable, then $\text{tr}(AB) = \text{tr}(BA)$; (iii) λ is an eigenvalue of the square matrix A if there exists a vector of unit length a such that $Aa = \lambda a$, and then a is an eigenvector of A; and (iv) a symmetric matrix A may be written as ELE^{T}, where L is a diagonal matrix of the eigenvalues of A, and the columns of E are the corresponding eigenvectors, having the property that $E^{\mathrm{T}} = E^{-1}$. If the matrix is symmetric and positive definite, then all its eigenvalues are real and positive.

8.3 Normal Distribution Theory

8.3.1 Distributions of $\widehat{\beta}$ and s^2

The derivation of the least squares estimators in the previous section rests on the assumption that the errors satisfy the *second-order assumptions*

$$\mathrm{E}(\varepsilon_j) = 0, \quad \mathrm{var}(\varepsilon_j) = \sigma^2, \quad \mathrm{cov}(\varepsilon_j, \varepsilon_k) = 0, \quad j \neq k, \qquad (8.12)$$

and in addition are normal variables. As they are uncorrelated, their normality implies they are independent. On setting $\varepsilon^{\mathrm{T}} = (\varepsilon_1, \ldots, \varepsilon_n)$, we have

$$\mathrm{E}(\varepsilon) = 0, \quad \mathrm{cov}(\varepsilon, \varepsilon) = \mathrm{E}(\varepsilon \varepsilon^{\mathrm{T}}) = \sigma^2 I_n,$$

where I_n is the $n \times n$ identity matrix. The least squares estimator equals

$$\widehat{\beta} = (X^{\mathrm{T}}X)^{-1}X^{\mathrm{T}}y = (X^{\mathrm{T}}X)^{-1}X^{\mathrm{T}}(X\beta + \varepsilon) = \beta + (X^{\mathrm{T}}X)^{-1}X^{\mathrm{T}}\varepsilon,$$

which is a linear combination of normal variables, and therefore its distribution is normal. Its mean vector and covariance matrix are

$$\mathrm{E}(\widehat{\beta}) = \beta + (X^{\mathrm{T}}X)^{-1}X^{\mathrm{T}}\mathrm{E}(\varepsilon),$$

$$\mathrm{var}(\widehat{\beta}) = \mathrm{cov}\{\beta + (X^{\mathrm{T}}X)^{-1}X^{\mathrm{T}}\varepsilon, \beta + (X^{\mathrm{T}}X)^{-1}X^{\mathrm{T}}\varepsilon\}$$

$$= (X^{\mathrm{T}}X)^{-1}X^{\mathrm{T}}\mathrm{cov}(\varepsilon, \varepsilon)X(X^{\mathrm{T}}X)^{-1},$$

so

$$\mathrm{E}(\widehat{\beta}) = \beta, \quad \mathrm{var}(\widehat{\beta}) = \sigma^2(X^{\mathrm{T}}X)^{-1}. \qquad (8.13)$$

Therefore $\widehat{\beta}$ is normally distributed with mean and covariance matrix given by (8.13). We shall see below that the residual sum of squares has a chi-squared distribution, independent of $\widehat{\beta}$. Thus the key distributional results for the normal linear model are

$$\widehat{\beta} \sim N_p\{\beta, \sigma^2(X^{\mathrm{T}}X)^{-1}\} \quad \text{independent of} \quad SS(\widehat{\beta}) \sim \sigma^2\chi^2_{n-p}. \qquad (8.14)$$

To show that the least squares estimator and residual sum of squares are independent, note that the residuals can be written as

$$e = (I_n - H)y = (I_n - H)(X\beta + \varepsilon) = (I_n - H)\varepsilon,$$

because $HX = X(X^{\mathrm{T}}X)^{-1}X^{\mathrm{T}}X = X$. Therefore the vector $e = (I_n - H)\varepsilon$ is a linear combination of normal random variables and is itself normally distributed, with mean and variance matrix

$$\mathrm{E}(e) = \mathrm{E}\{(I_n - H)\varepsilon\} = 0,$$

$$(8.15)$$

$$\mathrm{var}(e) = \mathrm{var}\{(I_n - H)\varepsilon\} = (I_n - H)\mathrm{var}(\varepsilon)(I_n - H)^{\mathrm{T}} = \sigma^2(I_n - H).$$

The covariance between $\widehat{\beta}$ and e is

$$\mathrm{cov}(\widehat{\beta}, e) = \mathrm{cov}\{\beta + (X^{\mathrm{T}}X)^{-1}X^{\mathrm{T}}\varepsilon, (I_n - H)\varepsilon\}$$

$$= (X^{\mathrm{T}}X)^{-1}X^{\mathrm{T}}\mathrm{cov}(\varepsilon, \varepsilon)(I_n - H)^{\mathrm{T}}$$

$$= (X^{\mathrm{T}}X)^{-1}X^{\mathrm{T}}\sigma^2 I_n(I_n - H)^{\mathrm{T}} = 0.$$

As both e and $\widehat{\beta}$ are normally distributed and their covariance matrix is zero, they are independent, which implies that $\widehat{\beta}$ and the residual sum of squares $SS(\widehat{\beta}) = e^{\mathsf{T}}e$ are independent.

The key to the distribution of $SS(\widehat{\beta})$ is the decomposition

$$
\begin{aligned}
\varepsilon^{\mathsf{T}}\varepsilon &= (y - X\beta)^{\mathsf{T}}(y - X\beta) \\
&= (y - X\widehat{\beta} + X\widehat{\beta} - X\beta)^{\mathsf{T}}(y - X\widehat{\beta} + X\widehat{\beta} - X\beta) \\
&= \{e + X(\widehat{\beta} - \beta)\}^{\mathsf{T}}\{e + X(\widehat{\beta} - \beta)\},
\end{aligned}
$$

which leads to

$$
\varepsilon^{\mathsf{T}}\varepsilon/\sigma^2 = e^{\mathsf{T}}e/\sigma^2 + (\widehat{\beta} - \beta)^{\mathsf{T}}X^{\mathsf{T}}X(\widehat{\beta} - \beta)/\sigma^2, \tag{8.16}
$$

because $e^{\mathsf{T}}X = y^{\mathsf{T}}(I_n - H)X = 0$. The left-hand side of (8.16) is a sum of the n independent chi-squared variables ε_j^2/σ^2, so its distribution is χ_n^2; its moment-generating function is $(1 - 2t)^{-n/2}, t < \frac{1}{2}$. It follows from applying (3.23) to the normal distribution of $\widehat{\beta}$ in (8.14) that $(\widehat{\beta} - \beta)^{\mathsf{T}}X^{\mathsf{T}}X(\widehat{\beta} - \beta)/\sigma^2 \sim \chi_p^2$. On taking moment-generating functions of both sides of (8.16) we therefore obtain

$$
(1 - 2t)^{-n/2} = \mathrm{E}\{\exp(te^{\mathsf{T}}e/\sigma^2)\} \times (1 - 2t)^{-p/2}, \quad t < \frac{1}{2},
$$

because e and $\widehat{\beta}$ are independent. Therefore $e^{\mathsf{T}}e/\sigma^2$ has moment-generating function $(1 - 2t)^{-(n-p)/2}$, showing that its distribution is χ_{n-p}^2. We need only recall that $SS(\widehat{\beta}) = e^{\mathsf{T}}e$ to establish the remaining result in (8.14): under the normal linear model, we have $SS(\widehat{\beta})/\sigma^2 \sim \chi_{n-p}^2$.

As the distribution of $SS(\widehat{\beta})$ is $\sigma^2\chi_{n-p}^2$, its mean is $\mathrm{E}\{SS(\widehat{\beta})\} = (n - p)\sigma^2$, and its variance is $\mathrm{var}\{SS(\widehat{\beta})\} = 2(n - p)\sigma^4$. Thus

$$
S^2 = \frac{1}{n - p}\sum_{j=1}^{n}(y_j - x_j\widehat{\beta})^2 = \frac{1}{n - p}SS(\widehat{\beta})
$$

is an unbiased estimator of σ^2, whereas $\widehat{\sigma}^2 = SS(\widehat{\beta})/n$ is biased.

8.3.2 Confidence and prediction intervals

Confidence intervals for components of β are based on the distributions of $\widehat{\beta}$ and S^2. Under the normal linear model the rth element of $\widehat{\beta}$ satisfies

$$
\widehat{\beta}_r \sim N(\beta_r, \sigma^2 v_{rr}),
$$

where v_{rr} is the rth diagonal element of $(X^{\mathsf{T}}X)^{-1}$, and $\widehat{\beta}$ is independent of S^2, whose distribution is $(n - p)^{-1}\sigma^2\chi_{n-p}^2$. Therefore

$$
T = \frac{\widehat{\beta}_r - \beta_r}{\sqrt{S^2 v_{rr}}} \sim t_{n-p},
$$

which makes the connection with (8.9). A $(1 - 2\alpha)$ confidence interval for β_r is $\widehat{\beta}_r \pm s v_{rr}^{1/2}t_{n-p}(\alpha)$. When σ^2 is known, we replace s by σ and $t_{n-p}(\alpha)$ by the normal quantile z_α.

$t_v(\alpha)$ is the α quantile of the t_v distribution.

Similar reasoning gives confidence intervals for linear functions of β. The maximum likelihood estimator of the linear function $x_+^T\beta$ is $x_+^T\widehat{\beta}$, which has a normal distribution with mean $x_+^T\beta$ and variance

$$\text{var}(x_+^T\widehat{\beta}) = x_+^T\text{var}(\widehat{\beta})x_+ = \sigma^2 x_+^T(X^TX)^{-1}x_+.$$

As S^2 is independent of $\widehat{\beta}$, confidence regions for $x_+^T\beta$ can be based on

$$\frac{x_+^T\widehat{\beta} - x_+^T\beta}{\{S^2 x_+^T(X^TX)^{-1}x_+\}^{1/2}} \sim t_{n-p}.$$

If σ^2 is known, the observed s is replaced in the confidence interval by σ and quantiles of the t distribution are replaced by those of the normal. Notice that the variance of a fitted value $\widehat{y}_j = x_j^T\widehat{\beta}$ is $\sigma^2 x_j^T(X^TX)^{-1}x_j$, and this equals $\sigma^2 h_{jj}$, where h_{jj} is the jth diagonal element of the hat matrix H.

A confidence interval for a function of parameters is different from a *prediction interval* for a new observation, $y_+ = x_+^T\beta + \varepsilon_+$. The presence of ε_+ would introduce uncertainty about y_+ even if β was known, and a prediction interval must take this into account. If ε_+ is normal with mean zero and variance σ^2, independent of the data from which $\widehat{\beta}$ is estimated, we have

$$E(x_+^T\widehat{\beta} + \varepsilon_+) = x_+^T\beta,$$
$$\text{var}(x_+^T\widehat{\beta} + \varepsilon_+) = \text{var}(x_+^T\widehat{\beta}) + \text{var}(\varepsilon_+) = \sigma^2\{x_+^T(X^TX)^{-1}x_+ + 1\}.$$

When σ^2 is unknown, therefore, a prediction interval for y_+ can be based on

$$\frac{y_+ - x_+^T\widehat{\beta}}{[S^2\{1 + x_+^T(X^TX)^{-1}x_+\}]^{1/2}} \sim t_{n-p},$$

with the appropriate changes if σ^2 is known.

Example 8.12 (Cycling data)   The covariance matrix for the parameter estimates in Example 8.8 is $\frac{\sigma^2}{16}I_4$. As the residual sum of squares is $SS(\widehat{\beta}) = 43.25, n = 16$ and $p = 4$, an estimate of σ^2 is $s^2 = 43.25/12 = 3.604$ on 12 degrees of freedom, and each estimate $\widehat{\beta}_r$ has standard error $(s^2/16)^{1/2} = 0.475$.

A 0.95 confidence interval for the true value of β_1 is $\widehat{\beta}_1 \pm st_{12}(0.025)/4$, and this is $-5.437 \pm 0.475 \times 2.18 = (-6.47, -4.40)$ seconds, clear evidence that the time is shorter when the seat is higher. The change due to the effect of tyre pressure is $2\widehat{\beta}_3$ seconds, for which the standard error is $2 \times s/4 = 0.95$ seconds.

A 0.95 prediction interval for a further timing y_+ made with all three factors set at their higher levels would be $41.75 \pm (1 + \frac{4}{16})^{1/2}st_{12}(0.025)$, which is $(39.49, 46.01)$. The variability introduced by ε_+ forms the bulk of the variability of y_+, whose variance is five times that of the fitted value.   ∎

Example 8.13 (Maize data)   Consider the two-sample model applied to the data in Table 1.1. If we assume that the heights of the cross-fertilized plants form a random sample with means $\beta_0 + \beta_1$, and that the heights of the self-fertilized plants form a random sample with height β_0, and that both have variance σ^2, the results of

Example 8.9 establish that the estimates are

$$\widehat{\beta}_0 = \overline{y}_{0\cdot} = 140.6, \quad \widehat{\beta}_1 = \overline{y}_{1\cdot} - \overline{y}_{0\cdot} = 161.53 - 140.6 = 20.93,$$

that the unbiased estimate of σ^2 is $s^2 = 553.19$, and that the estimated variance of $\widehat{\beta}_1$ is $s^2(n_0^{-1} + n_1^{-1}) = 73.78$. As s^2 has 28 degrees of freedom, a 0.95 confidence interval for β_1 has limits

$$\widehat{\beta}_1 \pm s\left(n_0^{-1} + n_1^{-1}\right)^{1/2} t_{28}(0.025) = 20.93 \pm 73.78^{1/2} \times 2.048 = 3.34, 38.52.$$

This does not contain zero, and is evidence that the crossed plants are significantly taller than self-fertilized plants.

For the matched pairs model of Example 8.10, there are $m = 15$ pairs, with $\widehat{\beta}_0 = 10.48$ and $s^2 = 712.36$, on $2m - (m + 1) = 14$ degrees of freedom. A 0.95 confidence interval for β_0 based on this model has limits

$$\widehat{\beta}_0 \pm \{s^2/(2m)\}^{1/2} t_{14}(0.025) = 10.48 \pm (712.36/30)^{1/2} \times 2.154 = 0.00, 20.96.$$

The corresponding interval for the height increase for crossed plants is an interval for $2\beta_0$, that is, $(0.00, 41.91)$. This is wider than the interval for the two-sample model, and just contains the value zero, giving evidence that there may be no increase due to cross-fertilization. The increase in interval width has two causes. First, the estimate of σ^2 for the matched pairs model equals 712.36, which is larger than the value 553.19 for the two-sample model. Second, there are only 14 degrees of freedom for the matched pairs estimate of variance, and $|t_{14}(0.025)| > |t_{28}(0.025)|$, which slightly inflates the matched pairs confidence interval relative to the interval from the matched analysis. ∎

Exercises 8.3

1 The following table gives the parameter estimates, standard errors and correlations, when the model $y = \beta_0 + \beta_1 x_1 + \beta_2 x_2 + \beta_3 x_3 + \varepsilon$ is fitted to the cement data of Example 8.3. The residual sum of squares is 48.11.

	Estimate	SE	Correlations of Estimates		
			(Intercept)	x1	x2
(Intercept)	48.19	3.913			
x1	1.70	0.205	x1	-0.736	
x2	0.66	0.044	x2	-0.416	-0.203
x3	0.25	0.185	x3	-0.828	0.822	-0.089

On the assumption that this normal linear model applies, compute 0.95 confidence intervals for β_0, β_1, β_2, and β_3, and test the hypothesis that $\beta_3 = 0$. Compute a 0.90 confidence interval for $\beta_2 - \beta_3$.

2 Let $\widehat{\beta}$ be a least squares estimator, and suppose that $\varepsilon_+ \sim N(0, \sigma^2)$ independent of $\widehat{\beta}$. Verify that $\mathrm{var}(x_+^T\widehat{\beta}) = \sigma^2 x_+^T(X^TX)^{-1}x_+$ and that $\mathrm{var}(x_+^T\widehat{\beta} + \varepsilon_+) = \sigma^2\{1 + x_+^T(X^TX)^{-1}x_+\}$. Assuming that a normal linear model is suitable for the cycling data, calculate a 0.90 confidence interval for the mean time to cycle up the hill when the three factors are at their lowest levels. Obtain also a 0.90 prediction interval for a future observation made with that setup.

8.4 Least Squares and Robustness

In Section 8.2.1 we established that $\widehat{\beta} = (X^{\mathrm{T}}X)^{-1}X^{\mathrm{T}}y$ is the maximum likelihood estimator of the regression parameter β under the assumption of normal responses. The model is a linear exponential family with complete minimal sufficient statistic $(\widehat{\beta}, S^2)$, and it follows that these are the unique minimum variance unbiased estimators of (β, σ^2). It is natural to ask to what optimality properties hold more generally. We shall see below that $\widehat{\beta}$ has minimum variance among all estimators linear in the responses y, under assumptions on the mean and variance structure of y alone. Thus the least squares estimator retains optimality properties even without full distributional assumptions. This has important generalizations, as we shall see in Section 10.6.

Suppose that the second-order assumptions (8.12) hold, but that the errors are not necessarily normal. Thus, although uncorrelated, they may be dependent. Then $\mathrm{E}(y) = X\beta$ and $\mathrm{var}(y) = \sigma^2 I_n$. Let $\tilde{\beta}$ denote any unbiased estimator of β that is linear in y. Then a $p \times n$ matrix A exists such that $\tilde{\beta} = Ay$, and unbiasedness implies that $\mathrm{E}(\tilde{\beta}) = AX\beta = \beta$ for any parameter vector β; this entails $AX = I_p$. Now

The $n \times n$ hat matrix $H = X(X^{\mathrm{T}}X)^{-1}X^{\mathrm{T}}$ is symmetric and idempotent and hence so is $I_n - H$.

$$
\begin{aligned}
\mathrm{var}(\tilde{\beta}) - \mathrm{var}(\widehat{\beta}) &= A\sigma^2 I_n A^{\mathrm{T}} - \sigma^2 (X^{\mathrm{T}}X)^{-1} \\
&= \sigma^2 \{ AA^{\mathrm{T}} - AX(X^{\mathrm{T}}X)^{-1}X^{\mathrm{T}}A^{\mathrm{T}} \} \\
&= \sigma^2 A(I_n - H)A^{\mathrm{T}} \\
&= \sigma^2 A(I_n - H)(I_n - H)^{\mathrm{T}}A^{\mathrm{T}}
\end{aligned}
$$

and this $p \times p$ matrix is positive semidefinite. Thus $\widehat{\beta}$ has smallest variance in finite samples among all linear unbiased estimators of β, provided that the second-order assumptions hold. This result, the *Gauss–Markov* theorem, gives further support for using $\widehat{\beta}$ if a linear estimator of β is sought, though of course nonlinear estimators may have smaller variance.

Johann Carl Friedrich Gauss (1777–1855) was born and educated in Brunswick. He studied in Göttingen and obtained a doctorate from the University of Helmstedt. His first book, published at the age of 24, contained the largest advance in geometry since the Greeks. He became director of the Göttingen observatory and invented least squares estimation for the combination of astronomical observations, though his statistical work was not published until much later. He also wrote treatises on theoretical astronomy, surveying, terrestial magnetism, infinite series, integration, number theory, and differential geometry.

Example 8.14 (Student t density)   Suppose that $y = X\beta + \sigma\varepsilon$, where the ε_j are independent and have the Student t density (3.11) with ν degrees of freedom. Now $\mathrm{var}(\varepsilon_j)$ is finite and equals $\nu/(\nu - 2)$ provided $\nu > 2$, and then the least squares estimator has variance matrix $\sigma^2 \nu/(\nu - 2) \times (X^{\mathrm{T}}X)^{-1}$.

How much efficiency is lost by using least squares rather than maximum likelihood estimation for β? To see this we must compute the expected information matrix, which gives the inverse variance of the maximum likelihood estimator. The log likelihood assuming ν and σ^2 known is

$$
\ell(\beta) \equiv -\frac{\nu + 1}{2} \sum_{j=1}^{n} \log \left\{ 1 + \left(y_j - x_j^{\mathrm{T}}\beta \right)^2 / (\nu\sigma^2) \right\},
$$

and differentiation with respect to β gives

$$
\frac{\partial \ell(\beta)}{\partial \beta} = \frac{\nu + 1}{\nu\sigma^2} \sum_{j=1}^{n} \frac{y_j - x_j^{\mathrm{T}}\beta}{1 + \left(y_j - x_j^{\mathrm{T}}\beta \right)^2 / (\nu\sigma^2)} x_j,
$$

$$
-\frac{\partial^2 \ell(\beta)}{\partial \beta \partial \beta^{\mathrm{T}}} = \frac{\nu + 1}{\nu\sigma^2} \sum_{j=1}^{n} \frac{1 - \left(y_j - x_j^{\mathrm{T}}\beta \right)^2 / (\nu\sigma^2)}{\left\{ 1 + \left(y_j - x_j^{\mathrm{T}}\beta \right)^2 / (\nu\sigma^2) \right\}^2} x_j x_j^{\mathrm{T}}.
$$

Now $\mathrm{E}\{(1 + \varepsilon^2/\nu)^{-r}\} = (\nu + 2r - 2) \cdots \nu/\{(\nu + 2r - 1) \cdots (\nu + 1)\}$, so the expected information for β is $\sigma^{-2}(\nu + 1)/(\nu + 3) \times X^\mathrm{T} X$. Thus the maximum likelihood estimator is a nonlinear function of y with large-sample variance matrix $\sigma^2(\nu + 3)/(\nu + 1) \times (X^\mathrm{T} X)^{-1}$. It follows that the least squares estimator has asymptotic relative efficiency $(\nu - 2)(\nu + 3)/\{\nu(\nu + 1)\}$, independent of the design matrix, β, or σ^2. As $\nu \to \infty$, the efficiency tends to one; for $\nu = 5, 10$, and 20 it equals 0.8, 0.95, and 0.99. Maximum likelihood estimation of β barely improves on least squares for a wide range of ν, because the t density is close to normal unless ν is small. ■

M-estimation

The least squares estimators have strong optimality properties, but because they are linear in y, they are sensitive to outliers. When data are too extensive to be carefully inspected or when bad data are present, robust or resistant estimators are more appropriate. One approach to constructing them is to replace the sum of squares with a function $\sum \rho\{(y_j - x_j^\mathrm{T}\beta)/\sigma\}$ that downweights extreme values of $(y_j - x_j^\mathrm{T}\beta)/\sigma$. The resulting estimators are called *M-estimators* because they are maximum-likelihood-like: the function ρ takes the place of a negative log likelihood. They may also be defined as the solutions of the $p \times 1$ estimating equation (Section 7.2)

$$\sigma^{-1} \sum_{j=1}^{n} x_j \rho'\{(y_j - x_j^\mathrm{T}\beta)/\sigma\} = 0, \tag{8.17}$$

where $\rho'(u) = d\rho(u)/du$, which extends the least squares estimating equation

$$X^\mathrm{T}(y - X\beta) = \sum_{j=1}^{n} x_j(y_j - x_j^\mathrm{T}\beta) = 0. \tag{8.18}$$

Many functions $\rho(u)$ have been proposed. Setting $\rho(u) = u^2/2$ gives least squares. Other possibilities include $\rho(u) = |u|$, $\rho(u) = \nu \log(1 + u^2/\nu)/2$, and

$$\rho(u) = \begin{cases} u^2, & \text{if } |u| < c, \\ c(2|u| - c), & \text{otherwise,} \end{cases}$$

corresponding to the median, a t_ν density, and a Huber estimator (Example 7.19). These have the drawback that large outliers are not downweighted to zero. This can be achieved with a redescending function such as the biweight,

$$\rho'(u) = u \max[\{1 - (u/c')^2\}^2, 0];$$

taking $c' = 4.865$ gives asymptotic efficiency 0.95 for normal data.

Notice that $\sum \rho\{(y_j - x_j^\mathrm{T}\beta)/\sigma\}$ has second derivative $\sigma^{-2} \sum x_j x_j^\mathrm{T} g'(y_j - x_j^\mathrm{T}\beta)$, whose expectation is of form $\sigma^{-2} X^\mathrm{T} X \times \mathrm{E}\{g'(\varepsilon)\}$ under a model in which $y_j = x_j^\mathrm{T}\beta + \sigma\varepsilon_j$ and the ε_j are independent and identically distributed with zero mean and unit variance. The ideas of Section 7.2 imply that the M-estimator has asymptotic variance

$$\sigma^2(X^\mathrm{T} X)^{-1} \times \mathrm{E}\{g(\varepsilon)^2\}/\mathrm{E}\{g'(\varepsilon)\},$$

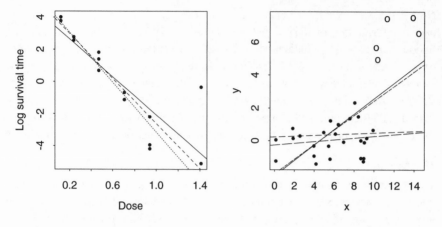

Figure 8.3 Data for
which least squares
estimation fails. Left: log
survival proportions for
rats given doses of
radiation, with lines fitted
by least squares with
(solid) and without (dots)
the outlier, and a Huber
M-estimate for the entire
data (dashes) (Efron,
1988). Right: simulated
data with a batch of
outliers (circles), and fits
by least squares to all data
(solid), least squares to
good data only (large
dash), Huber (dot-dash),
biweight (dashes), and
least trimmed squares
(medium dash). The
Huber and biweight fits
are the same to plotting
accuracy.

so its efficiency relative to least squares is simply $E\{g'(\varepsilon)\}/E\{g(\varepsilon)^2\}$. The Huber estimator for regression has efficiencies given by the right panel of Figure 7.4, for instance.

Equation (8.17) may be solved using iterative versions of least squares described in Section 10.2.2, though these may fail to converge if ρ is not convex. In practice σ too must be estimated, by the median absolute deviation of the residuals $y_j - x_j^{\mathsf{T}}\widehat{\beta}$ at each iteration, or using an M-estimator of scale.

Initial values for these fits can be found by a highly resistant procedure such as *least trimmed squares*, whereby β is chosen to minimize $\sum_{i=1}^{q}(y_j - x_j^{\mathsf{T}}\beta)^2_{(i)}$; this is the sum of the smallest $q = \lfloor n/2 \rfloor + \lfloor (p+1)/2 \rfloor$ squared residuals, found by a Monte Carlo search. Highly resistant procedures do not usually provide standard errors, which can be obtained by a data-based simulation procedure such as the bootstrap; see the bibliographic notes.

Example 8.15 (Survival data)   The left panel of Figure 8.3 shows data on batches of rats given doses of radiation. They are well fit by a straight line, apart from an apparent outlier, which strongly affects the least squares fit — note what the pattern of residuals will be. The least squares estimates of slope and its standard error with and without the outlier are $-5.91\,(1.05)$ and $-7.79\,(0.59)$, while Huber estimation gives $-7.02\,(0.46)$. Downweighting the outlier using the robust estimator gives a result intermediate between keeping it and deleting it.

This sample is small and the outlier sticks out, so robust methods are not really needed. They are more valuable for larger more complex data sets where visualization is difficult and outliers non-obvious.   ∎

Example 8.16 (Simulated data)   To illustrate and compare some robust estimators, we generated sets of 25 standard normal observations y with a single covariate x, and then added k outliers with mean 6, having the t_5 distribution. The right panel of Figure 8.3 shows one of these datasets, with $k = 5$. We then computed five estimates of slope, from least squares applied with and without the outliers, from Huber and biweight M-estimators having efficiency 0.95 at the normal model, and from

Table 8.4 Bias (standard deviation) of estimators of slope in sample of 25 good data and k outliers, estimated from 200 replications.

	Least squares		M-estimation		Least trimmed
k	No outliers	With outliers	Huber	Biweight	squares
1	0.00 (0.07)	0.17 (0.06)	0.07 (0.07)	0.01 (0.07)	−0.01 (0.13)
2	0.00 (0.07)	0.26 (0.06)	0.13 (0.07)	0.02 (0.09)	0.01 (0.14)
5	0.00 (0.07)	0.41 (0.05)	0.38 (0.06)	0.19 (0.19)	0.01 (0.14)
10	0.00 (0.06)	0.48 (0.04)	0.48 (0.04)	0.46 (0.12)	0.05 (0.20)

least trimmed squares. Table 8.4 shows the bias and standard deviation of the slope estimators for various k, computed from 200 replicate data sets.

Inclusion of just one outlier ruins the least squares estimator, which is the benchmark when outliers are excluded. The biweight gives the better of the M-estimators, but with $k \geq 5$ it is badly biased. The M-estimators perform as badly as least squares when contamination is high. Least trimmed squares is least biased overall, but is very inefficient even for $k = 1$. This suggests that a good practical data analysis strategy is to use an initial least trimmed squares fit to identify and delete outliers, and then apply M-estimation to the remaining data. ∎

Misspecified variance

Outliers are just one of many possible problems in regression. Suppose that although $E(y) = X\beta$, the variance is $\text{var}(y) = V$ rather than the assumed $\sigma^2 I_n$. Then $\widehat{\beta} = (X^\mathsf{T} X)^{-1} X^\mathsf{T} y$ has variance

$$(X^\mathsf{T} X)^{-1} (X^\mathsf{T} V X)(X^\mathsf{T} X)^{-1}. \tag{8.19}$$

If $V = \sigma^2 I_n$, then $\text{var}(\widehat{\beta}) = \sigma^2 (X^\mathsf{T} X)^{-1}$, which itself is the inverse Fisher information for β under the normal model. Thus if the variance of y is correctly supposed to equal $\sigma^2 I_n$, the least squares estimator attains the Cramér–Rao lower bound appropriate to normal responses, while (7.20) implies that $\text{var}(\widehat{\beta})$ is inflated otherwise.

Most packages use the formula $\sigma^2 (X^\mathsf{T} X)^{-1}$ and make no allowance for possible variance misspecification. If plots such as those described in Section 8.6 do not suggest a particular variance to be fitted using weighted least squares, the weights being $W = V^{-1}$, then it may be better to apply least squares but to base confidence intervals on an estimate of (8.19). One simple possibility is to replace V with $\widehat{V} = \text{diag}\{r_1^2, \ldots, r_n^2\}$, where $r_j = (y_j - \widehat{y}_j)/(1 - h_{jj})$.

Exercises 8.4

1 Check the details of Example 8.14.

2 Show that $\widehat{\beta}$ and S^2 are unbiased estimators of β and σ^2 even when the errors are not normal, provided that the second-order assumptions are satisfied.

3 Consider a linear regression model (8.1) in which the errors ε_j are independently distributed with Laplace density

$$f(u; \sigma) = (2^{3/2}\sigma)^{-1} \exp\left\{ -\left| u \big/ \left(2^{1/2}\sigma\right) \right| \right\}, \quad -\infty < u < \infty, \sigma > 0.$$

Verify that this density has variance σ^2. Show that the maximum likelihood estimate of β is obtained by minimizing the L^1 norm $\sum |y_j - x_j^\mathsf{T}\beta|$ of $y - X\beta$.

Show that if in fact the $\varepsilon_j \overset{\text{iid}}{\sim} N(0, \sigma^2)$, the asymptotic relative efficiency of the estimators relative to least squares estimators is $2/\pi$.

4   Consider a linear model $y_j = x_j\beta + \varepsilon_j$, $j = 1, \ldots, n$ in which the ε_j are uncorrelated and have means zero. Find the minimum variance linear unbiased estimators of the scalar β when (i) $\text{var}(\varepsilon_j) = x_j\sigma^2$, and (ii) $\text{var}(\varepsilon_j) = x_j^2\sigma^2$. Generalize your results to the situation where $\text{var}(\varepsilon) = \sigma^2/w_j$, where the weights w_j are known but σ^2 is not.

5   Use (8.18) to establish that (7.20) takes form

$$(X^\mathsf{T}X)^{-1}X^\mathsf{T}VX(X^\mathsf{T}X)^{-1} \geq \sigma^2(X^\mathsf{T}X)^{-1}$$

when $\text{var}(y)$ is wrongly supposed equal to $\varepsilon^2 I_n$ instead of V.

8.5 Analysis of Variance

8.5.1 *F* statistics

In most regression models a key question is whether or not the explanatory variables affect the response. For example, in the bicycle data, we were concerned how the time to climb the hill depended on the seat height and other factors. Ockham's razor suggests that we use the simplest model we can. This poses the question: which explanatory variables are needed? To be concrete, suppose that we fit a normal linear model

$$y = X\beta + \varepsilon = (X_1, X_2)\begin{pmatrix} \beta_1 \\ \beta_2 \end{pmatrix} + \varepsilon = X_1\beta_1 + X_2\beta_2 + \varepsilon, \qquad (8.20)$$

where X_1 is an $n \times q$ matrix, X_2 is an $n \times (p - q)$ matrix, $q < p$, and β_1 and β_2 are vectors with respective lengths q and $p - q$. We suppose that X has rank p and X_1 has rank q. The explanatory variables X_2 are unnecessary if $\beta_2 = 0$, in which case the simpler model $y = X_1\beta_1 + \varepsilon$ holds. How can we detect this?

In Figure 8.2, let the line $x = 0$ in the horizontal plane through the origin represent the linear subspace spanned by the columns of X_1. The fitted value $\widehat{y}_1 = X_1(X_1^\mathsf{T}X_1)^{-1}X_1^\mathsf{T}y$ is the orthogonal projection of y onto this subspace. The vector of residuals, $y - \widehat{y}_1 = \{I_n - X_1(X_1^\mathsf{T}X_1)^{-1}X_1^\mathsf{T}\}y$, resolves into the two orthogonal vectors $y - \widehat{y}$ and $\widehat{y} - \widehat{y}_1$; that is,

$$y - \widehat{y}_1 = (y - \widehat{y}) + (\widehat{y} - \widehat{y}_1),$$

where $(y - \widehat{y})^\mathsf{T}(\widehat{y} - \widehat{y}_1) = 0$. These vectors are the residual from the more complex model, $y - \widehat{y}$, and the change in fitted values when X_2 is added to the design matrix, $\widehat{y} - \widehat{y}_1$. As these vectors are orthogonal linear functions of the normally distributed vector y, they are independent. Pythagoras' theorem implies that

$$(y - \widehat{y}_1)^\mathsf{T}(y - \widehat{y}_1) = (y - \widehat{y})^\mathsf{T}(y - \widehat{y}) + (\widehat{y} - \widehat{y}_1)^\mathsf{T}(\widehat{y} - \widehat{y}_1),$$

or equivalently

$$SS(\widehat{\beta}_1) = SS(\widehat{\beta}) + \{SS(\widehat{\beta}_1) - SS(\widehat{\beta})\}. \qquad (8.21)$$

Thus the residual sum of squares for the simpler model is the sum of two independently distributed parts: the residual sum of squares for the more elaborate model, $SS(\widehat{\beta})$, and the reduction in sum of squares when the columns of X_2 are added to the design matrix, $SS(\widehat{\beta}_1) - SS(\widehat{\beta})$.

If the submodel is correct, so too is the more elaborate model, because β_2 takes the particular value zero. In this case $SS(\widehat{\beta}_1)$ has a $\sigma^2 \chi^2_{n-q}$ distribution, and $SS(\widehat{\beta})$ has a $\sigma^2 \chi^2_{n-p}$ distribution. Since $SS(\widehat{\beta}_1) - SS(\widehat{\beta})$ is independent of $SS(\widehat{\beta})$, (8.21) implies that when $\beta_2 = 0$, $SS(\widehat{\beta}_1) - SS(\widehat{\beta})$ has a $\sigma^2 \chi^2_{p-q}$ distribution, and that

$$F = \frac{\{SS(\widehat{\beta}_1) - SS(\widehat{\beta})\}/(p - q)}{SS(\widehat{\beta})/(n - p)} \sim F_{p-q,n-p};$$

recall (8.8). If β_2 is non-zero, the reduction in sum of squares due to including the columns of X_2 in the design matrix will be larger on average than if $\beta_2 = 0$. Thus if $\beta_2 \neq 0$, F will tend to be large relative to the $F_{p-q,n-p}$ distribution. We can therefore test the adequacy of the simpler model using the statistic F, large values of which suggest that $\beta_2 \neq 0$.

Exercise 8.5.3 gives the algebraic equivalent of the geometric argument above. As we saw in Section 8.2.3, F arises from the likelihood ratio statistic for comparison of the two models. When X_2 consists of a single covariate, β_2 is scalar, and tests and confidence intervals for it may be obtained by fitting the more elaborate model (8.20) and calculating $T = (\widehat{\beta}_2 - \beta_2)/(sv_{rr}^{1/2})$. Here s^2 is the estimate of σ^2 from the more elaborate model, and the null distribution of T is t_{n-p}. In this situation there is a simple connection to F: when testing $\beta_2 = 0$, $F = T^2 = \widehat{\beta}_2^2/(s^2 v_{rr})$.

Example 8.17 (Cement data)   Suppose that we want to compare the models $y = \beta_0 + x_1\beta_1 + \varepsilon$ and $y = \beta_0 + x_1\beta_1 + x_2\beta_2 + x_3\beta_3 + x_4\beta_4 + \varepsilon$. This corresponds to asking if is there any effect on y of x_2, x_3, or x_4, after allowing for the effect of x_1. Here X_1 is a 13×2 matrix whose columns are a vector of ones and x_1, and X_2 is a 13×3 matrix whose columns are x_2, x_3, and x_4; both matrices have full rank.

For the full model $p = 5$ and the residual sum of squares is $SS(\widehat{\beta}) = 47.86$, and for the simpler model $q = 2$ and the residual sum of squares is $SS(\widehat{\beta}_1) = 1265.7$. Thus the reduction in sum of squares due to the columns of X_2 after fitting X_1 is $1265.7 - 47.86 = 1217.84$ on three degrees of freedom. To test whether this is a significant reduction, we compute

$$F = \frac{(1265.7 - 47.86)/(5 - 2)}{47.86/(13 - 5)} = 67.86,$$

$F_{\nu_1,\nu_2}(\alpha)$ is the α quantile of the F distribution with ν_1 and ν_2 degrees of freedom.

which would be consistent with an $F_{3,8}$ distribution if the simpler model was adequate. As F greatly exceeds $F_{3,8}(0.95) = 4.066$, there is strong evidence that there are effects of the added covariates.

Having established that adding extra covariates helps to explain the overall variation, it is natural to ask whether this is due to a subset of them rather than to all three. Is there a more informative decomposition of the sum of squares due to adding X_2?

∎

8.5.2 Sums of squares

The interpretation of sums of squares is most useful if they can be decomposed into the reductions from successively adding different explanatory variables to the design matrix.

Suppose that we have a normal linear model

$$y = 1_n \beta_0 + X_1 \beta_1 + X_2 \beta_2 + \cdots + X_m \beta_m + \varepsilon, \tag{8.22}$$

where we call the matrices 1_n, X_1, X_2, and so forth *terms*; the constant term 1_n is a column of n ones. Usually the simplest model that might be considered sets $y = 1_n \beta_0 + \varepsilon$, in which case the fitted value is $\widehat{y}_0 = 1_n \bar{y}$, and the residual sum of squares is $SS_0 = \sum (y_j - \bar{y})^2$ with $\nu_0 = n - 1$ degrees of freedom.

We now consider the successive reductions in sum of squares due to adding the terms X_1, X_2, and so forth to the design matrix. Let $\widehat{y}_r$ be the fitted value when the terms $X_1, \ldots, X_r$ are included, and write

$$y - \widehat{y}_0 = (y - \widehat{y}_m) + (\widehat{y}_m - \widehat{y}_{m-1}) + \cdots + (\widehat{y}_1 - \widehat{y}_0).$$

This decomposition extends that leading to (8.21) and shown in Figure 8.2. The geometry of least squares implies that the quantities in parentheses on the right are mutually orthogonal. Pythagoras' theorem tells us that $(y - \widehat{y}_0)^{\mathrm{T}} (y - \widehat{y}_0)$ equals

$$(y - \widehat{y}_m)^{\mathrm{T}} (y - \widehat{y}_m) + (\widehat{y}_m - \widehat{y}_{m-1})^{\mathrm{T}} (\widehat{y}_m - \widehat{y}_{m-1}) + \cdots + (\widehat{y}_1 - \widehat{y}_0)^{\mathrm{T}} (\widehat{y}_1 - \widehat{y}_0),$$

or equivalently

$$SS_0 = SS_m + (SS_{m-1} - SS_m) + \cdots + (SS_0 - SS_1), \tag{8.23}$$

where SS_r denotes the residual sum of squares that corresponds to the fitted value $\widehat{y}_r$, on ν_r degrees of freedom. In (8.23) the difference $SS_{r-1} - SS_r$ is the reduction in residual sum of squares due to adding the term X_r when the model already contains $1_n, X_1, \ldots, X_{r-1}$. As y is normal and the vectors $\widehat{y}_r - \widehat{y}_{r-1}$ and $y - \widehat{y}_m$ are all linear functions of the data, the geometry of least squares implies that SS_m and all the $SS_{r-1} - SS_r$ are mutually independent.

As more terms are successively added to the model, the degrees of freedom of the residual sums of squares decrease, that is, $\nu_0 \geq \nu_1 \geq \cdots \geq \nu_m$, with $\nu_r = \nu_{r+1}$ when the columns of X_{r+1} are a linear combination of the columns of the matrices $1_n, X_1, \ldots, X_r$. If $\nu_r = \nu_{r+1}$, $\widehat{y}_r = \widehat{y}_{r+1}$, and $SS_r = SS_{r+1}$. The term X_{r+1} is then redundant, because its inclusion does not change the fitted model.

Analysis of variance

The sums of squares can be laid out in an *analysis of variance table*. The prototype is Table 8.5. The residual sums of squares decrease as terms are added successively to the model. Often the three leftmost columns are omitted and their bottom row is placed under the right-hand columns; SS_m is used to compute the denominator for the F statistics for inclusion of X_1, X_2 and so forth, and these may be included also, as in the examples below.

Table 8.5 Analysis of variance table.

Terms	df	Residual sum of squares	Terms added	df	Reduction in sum of squares	Mean square
1_n	$n-1$	SS_0				
$1_n, X_1$	ν_1	SS_1	X_1	$n-1-\nu_1$	$SS_0 - SS_1$	$\frac{SS_0 - SS_1}{n-1-\nu_1}$
$1_n, X_1, X_2$	ν_2	SS_2	X_2	$\nu_1 - \nu_2$	$SS_1 - SS_2$	$\frac{SS_1 - SS_2}{\nu_1 - \nu_2}$
$\vdots$	$\vdots$	$\vdots$	$\vdots$	$\vdots$	$\vdots$	$\vdots$
$1_n, X_1, \ldots, X_m$	ν_m	SS_m	X_m	$\nu_{m-1} - \nu_m$	$SS_{m-1} - SS_m$	$\frac{SS_{m-1} - SS_m}{\nu_{m-1} - \nu_m}$

Table 8.6 Analysis of variance table for the cement data, showing reductions in overall sum of squares when terms are entered in the order given.

Term	df	Reduction in sum of squares	Mean square	F
x_1	1	1450.1	1450.1	242.5
x_2	1	1207.8	1207.8	202.0
x_3	1	9.79	9.79	1.64
x_4	1	0.25	0.25	0.04
Residual	8	47.86	5.98	

Table 8.7 Models for the means of the crossed and self-fertilized plants in the pth pot and jth pair for the maize data.

Terms	Crossed	Self-fertilized
1	μ	μ
1+Fertilization	$\mu + \alpha$	μ
1+Fertilization+Pot	$\mu + \alpha + \beta_p$	$\mu + \beta_p$
1+Fertilization+Pot+Pair	$\mu + \alpha + \beta_p + \gamma_j$	$\mu + \beta_p + \gamma_j$

Example 8.18 (Cement data) Table 8.6 gives the analysis of variance when the covariates x_1, x_2, x_3, and x_4 are successively included in the design matrix. There are very large reductions due to fitting x_1 and x_2, but those due to x_3 and x_4 are smaller. The F statistics for testing the effects of x_1 and x_2 are highly significant, but once x_1 and x_2 are included the F statistic for x_3 is not large compared to the $F_{1,8}$ distribution. A similar conclusion holds for x_4. Thus once x_1 and x_2 are included, x_3 and x_4 are unnecessary in accounting for the response variation. ■

Example 8.19 (Maize data) Consider models for the maize data with means as in Table 8.7. In order, these correspond to: no differences among pairs and no difference between cross-fertilization and self-fertilization; no differences among pairs but an effect of fertilization type; differences among the pots and an effect of fertilization type; and differences among the pots and among the pairs and an effect of fertilization type. Table 8.8 gives the analysis of variance when these models are fitted successively.

Term	df	Reduction in sum of squares	Mean square	F
Fertilization	1	3286.5	3286.5	4.61
Pot	3	1053.6	351.2	0.49
Pair	11	4467.3	406.1	0.57
Residual	14	9972.5	712.3	

Table 8.8 Analysis of variance table for linear models fitted to the maize data.

There are four pot parameters β_p, but the reduction in degrees of freedom when the pots term is included is three because although the corresponding 30×4 matrix has rank four, its columns sum to a column of ones. As the design matrix already contains a column of ones, including the four columns for the pots term increases the rank of the design matrix by only three. Likewise only 11 columns of the 30×15 matrix of terms for pairs increase the rank of a design matrix that already contains the overall mean and the pots term: the remaining four columns are linear combinations of those already present.

The residual sum of squares for the eventual model is 9972.5 on 14 degrees of freedom, so the denominator for F statistics is $9972.5/14 = 712.3$. The F statistic for fertilization is just significant at the 5% level, but there seem to be no differences among pots or pairs. We can attribute to random variation the reduction in sum of squares when the pots and pairs terms are added, and obtain a better estimate of σ^2, namely

$$(9972.5 + 1053.6 + 4467.3)/(14 + 3 + 11) = 553.3$$

on 28 degrees of freedom. The F statistic for fertilization with this pooled estimate of σ^2 as denominator is 5.94 on 1 and 28 degrees of freedom and its significance level is 0.02, so the addition of the sums of squares for pots and pairs to the residual has resulted in a more sensitive analysis. ∎

8.5.3 Orthogonality

The reduction in sum of squares when a term is added depends on the terms already in the model. This can obscure the interpretation of an analysis of variance, if a term that gives a large reduction early in a sequence of fits gives a small reduction if fitted later in the sequence instead.

Suppose that a normal linear model (8.22) applies. The reductions in sum of squares due to the terms X_r are unique only if the vector spaces spanned by the columns of the X_r are all mutually orthogonal, that is, $X_r^T X_s = 0$ when $r \neq s$. Suppose that this is true, that in addition $X_r^T 1_n = 0$, and that

$$y = 1_n \beta_0 + X_1 \beta_1 + X_2 \beta_2 + \varepsilon. \tag{8.24}$$

Then the orthogonality of 1_n, X_1, and X_2 implies that the least squares estimators are

$$\begin{pmatrix} \widehat{\beta_0} \\ \widehat{\beta_1} \\ \widehat{\beta_2} \end{pmatrix} = \begin{pmatrix} 1^\mathsf{T}1 & 0 & 0 \\ 0 & X_1^\mathsf{T}X_1 & 0 \\ 0 & 0 & X_2^\mathsf{T}X_2 \end{pmatrix}^{-1} (1 \quad X_1 \quad X_2)^\mathsf{T} y,$$

so that $\widehat{\beta_0} = \overline{y}$, $\widehat{\beta_1} = (X_1^\mathsf{T}X_1)^{-1}X_1^\mathsf{T}y$, and $\widehat{\beta_2} = (X_2^\mathsf{T}X_2)^{-1}X_2^\mathsf{T}y$, with residual sum of squares

$$y^\mathsf{T}y - \widehat{\beta}^\mathsf{T}X^\mathsf{T}X\widehat{\beta} = y^\mathsf{T}y - n\overline{y}^2 - \widehat{\beta}_1^\mathsf{T}X_1^\mathsf{T}X_1\widehat{\beta}_1 - \widehat{\beta}_2^\mathsf{T}X_2^\mathsf{T}X_2\widehat{\beta}_2. \tag{8.25}$$

For the simpler models

$$y = 1_n\beta_0 + \varepsilon, \quad y = 1_n\beta_0 + X_1\beta_1 + \varepsilon \quad y = 1_n\beta_0 + X_2\beta_2 + \varepsilon,$$

a similar calculation gives residual sums of squares

$$y^\mathsf{T}y - n\overline{y}^2, \quad y^\mathsf{T}y - n\overline{y}^2 - \widehat{\beta}_1^\mathsf{T}X_1^\mathsf{T}X_1\widehat{\beta}_1, \quad y^\mathsf{T}y - n\overline{y}^2 - \widehat{\beta}_2^\mathsf{T}X_2^\mathsf{T}X_2\widehat{\beta}_2,$$

and comparison with (8.25) shows that the reductions due to X_1 and X_2 are $\widehat{\beta}_1^\mathsf{T}X_1^\mathsf{T}X_1\widehat{\beta}_1$ and $\widehat{\beta}_2^\mathsf{T}X_2^\mathsf{T}X_2\widehat{\beta}_2$ whether or not the other has been included in the design matrix. Consequently the reductions in sums of squares due to X_1 and X_2 are unique. This argument readily extends to models with more than two mutually orthogonal terms X_r. In fact (8.24) has three, as we see by writing $1_n = X_0$.

Example 8.20 (Orthogonal polynomials) Consider a normal linear model with design matrix

$$X = (1_n, x_1, x_2, x_3, x_4) = \begin{pmatrix} 1 & -2 & 2 & -1 & 1 \\ 1 & -1 & -1 & 2 & -4 \\ 1 & 0 & -2 & 0 & 6 \\ 1 & 1 & -1 & -2 & -4 \\ 1 & 2 & 2 & 1 & 1 \end{pmatrix},$$

the last four columns of which correspond to linear, quadratic, cubic, and quartic polynomials in a covariate with five values equally spaced one unit apart. The columns of X are mutually orthogonal, and it follows that the reduction due to any of them does not depend on which of the others have already been fitted.

If the values had been equally-spaced but δ units apart, the model would be $y = 1_n\beta_0 + \delta x_1\beta_1 + \cdots + \delta^4 x_4\beta_4 + \varepsilon$, and the orthogonality of the terms would be unaffected. ∎

The argument leading to (8.25) rarely applies directly, but it may do so if an overall mean, corresponding to a column of ones in the design matrix, is fitted first. Suppose that the matrices X_1 and X_2 in (8.24) are not mutually orthogonal and are not orthogonal to 1_n, but that we rewrite the model as

$$\begin{aligned} y &= 1_n\left(\beta_0 + \overline{x}_1^\mathsf{T}\beta_1 + \overline{x}_2^\mathsf{T}\beta_2\right) + \left(X_1 - 1_n\overline{x}_1^\mathsf{T}\right)\beta_1 + \left(X_2 - 1_n\overline{x}_2^\mathsf{T}\right)\beta_2 + \varepsilon \\ &= 1_n\gamma_0 + Z_1\beta_1 + Z_2\beta_2 + \varepsilon, \end{aligned}$$

say, where $\overline{x}_1^{\mathrm{T}}$ and $\overline{x}_2^{\mathrm{T}}$ are the averages of the rows of X_1 and X_2. Then Z_1 and Z_2 are centred and $Z_1^{\mathrm{T}}1_n = Z_2^{\mathrm{T}}1_n = 0$. This rearrangement of the model changes the intercept but leaves β_1 and β_2 unaffected. If the original matrices X_1 and X_2 are such that $Z_1^{\mathrm{T}}Z_2 = 0$, we can apply the argument leading to (8.25) to our new model, to obtain the successive residual sums of squares

$$SS_0 = y^{\mathrm{T}}y - n\overline{y}^2,$$
$$SS_1 = y^{\mathrm{T}}y - n\overline{y}^2 - \widehat{\beta}_1^{\mathrm{T}}Z_1^{\mathrm{T}}Z_1\widehat{\beta}_1,$$
$$SS_2 = y^{\mathrm{T}}y - n\overline{y}^2 - \widehat{\beta}_1^{\mathrm{T}}Z_1^{\mathrm{T}}Z_1\widehat{\beta}_1 - \widehat{\beta}_2^{\mathrm{T}}Z_2^{\mathrm{T}}Z_2\widehat{\beta}_2,$$

as the terms Z_1 and Z_2, or equivalently X_1 and X_2, are added to the design matrix. Since Z_1 is defined purely in terms of X_1 and 1_n, and Z_2 is defined purely in terms of X_2 and 1_n, the reduction in sum of squares due to adding X_1 after including the constant column 1_n in the design matrix is the same whether or not X_2 is present. Hence provided the constant is fitted first, the reductions in sum of squares due to X_1 and X_2 are independent of the order in which they are included. This argument extends to models with more than two X_r, provided that the centred matrices Z_r are mutually orthogonal.

Example 8.21 (3×2 **layout**)   In a 3×2 layout with no interaction the observations and their means can be written

$$
\begin{array}{cc}
y_{11} & y_{12} \\
y_{21} & y_{22}, \\
y_{31} & y_{32}
\end{array}
\qquad
\begin{array}{cc}
\mu & \mu + \alpha \\
\mu + \delta_1 & \mu + \delta_1 + \alpha\,. \\
\mu + \delta_2 & \mu + \delta_2 + \alpha
\end{array}
$$

In terms of the parameter vector $(\mu, \alpha, \delta_2, \delta_3)^{\mathrm{T}}$, the design matrix is

$$
X = \begin{pmatrix}
1 & 0 & 0 & 0 \\
1 & 1 & 0 & 0 \\
1 & 0 & 1 & 0 \\
1 & 1 & 1 & 0 \\
1 & 0 & 0 & 1 \\
1 & 1 & 0 & 1
\end{pmatrix},
$$

with X_1 the second column of X, and X_2 the third and fourth columns of X. Evidently X_1 and X_2 are not orthogonal and they are not orthogonal to the constant. On the other hand Z_1 and Z_2 in the corresponding centred matrix,

$$
\begin{pmatrix}
1 & -\frac{1}{2} & -\frac{1}{3} & -\frac{1}{3} \\
1 & \frac{1}{2} & -\frac{1}{3} & -\frac{1}{3} \\
1 & -\frac{1}{2} & \frac{2}{3} & -\frac{1}{3} \\
1 & \frac{1}{2} & \frac{2}{3} & -\frac{1}{3} \\
1 & -\frac{1}{2} & -\frac{1}{3} & \frac{2}{3} \\
1 & \frac{1}{2} & -\frac{1}{3} & \frac{2}{3}
\end{pmatrix},
$$

are orthogonal to the constant by construction and to each other because the design is balanced: δ_2 and δ_3 each occur equally often with α and without α. This balance has the consequence that provided that μ is fitted first, the reductions in sums of squares due to X_1 and X_2, or equivalently Z_1 and Z_2, are unique.   ∎

A designed experiment such as Example 8.21 can often be balanced, so that orthogonality is arranged, at least approximately, and the interpretation of its analysis of variance is relatively clear-cut. Even if the terms are not orthogonal, however, it may be possible to order them unambiguously. One example is polynomial dependence of y on x, where terms of increasing degree are added successively. Another example is when some terms represent classifications that are known to affect y but which are of secondary importance, and others correspond to the question of primary interest. For instance, it would be natural to assess the effects of different treatments on the incidence of heart disease after taking into account the effects of classifying variables such as age, sex, and previous medical history.

Exercises 8.5

1   Consider the cement data of Example 8.3, where $n = 13$. The residual sums of squares for all models that include an intercept are given below.

Model	SS	Model	SS	Model	SS
$-\,-\,-\,-$	2715.8	$1\,2\,-\,-$	57.9	$1\,2\,3\,-$	48.11
$1\,-\,-\,-$	1265.7	$1\,-\,3\,-$	1227.1	$1\,2\,-\,4$	47.97
$-\,2\,-\,-$	906.3	$1\,-\,-\,4$	74.8	$1\,-\,3\,4$	50.84
$-\,-\,3\,-$	1939.4	$-\,2\,3\,-$	415.4	$-\,2\,3\,4$	73.81
$-\,-\,-\,4$	883.9	$-\,2\,-\,4$	868.9		
		$-\,-\,3\,4$	175.7	$1\,2\,3\,4$	47.86

Compute the analysis of variance table when x_4, x_3, x_2, and x_1 are fitted in that order, and test which of them should be included in the model. Are your conclusions the same as in Example 8.18?

2   (a) Let A, B, C, and D represent $p \times p$, $p \times q$, $q \times q$, and $q \times p$ matrices respectively. Show that provided that the necessary inverses exist

$$(A + BCD)^{-1} = A^{-1} - A^{-1}B(C^{-1} + DA^{-1}B)^{-1}DA^{-1}.$$

(b) If the matrix A is partitioned as

$$A = \begin{pmatrix} A_{11} & A_{12} \\ A_{21} & A_{22} \end{pmatrix},$$

and the necessary inverses exist, show that the elements of the corresponding partition of A^{-1} are

$$A^{11} = \left(A_{11} - A_{12}A_{22}^{-1}A_{21}\right)^{-1}, \quad A^{22} = \left(A_{22} - A_{21}A_{11}^{-1}A_{12}\right)^{-1},$$
$$A^{12} = -A_{11}^{-1}A_{12}A^{22}, \quad A^{21} = -A_{22}^{-1}A_{21}A^{11}.$$

3   In (8.20), suppose that X_1 and X_2 have ranks q and $p - q$ respectively, and define $H = X(X^{\mathsf{T}}X)^{-1}X^{\mathsf{T}}$, $P = I_n - H$, $H_1 = X_1(X_1^{\mathsf{T}}X_1)^{-1}X_1^{\mathsf{T}}$ and $P_1 = I_n - H_1$. Let $\widehat{y} = Hy$, and $\widehat{y}_1 = H_1 y$.

Use the previous exercise.

(a) Show that $(y - \widehat{y})^{\mathsf{T}}(\widehat{y} - \widehat{y}_1) = 0$ if and only if $HH_1 = H_1$, and show that $H_1 H = HH_1$. Give a geometrical interpretation of the equations $H_1 H = HH_1 = H_1$.

Model	SS	Model	SS	Model	SS	Model	SS
— — —	18780	— Po —	17726	F Po —	14440	F — Pa	9972
F — —	15493	— — Pa	13259	— Po Pa	13259	F Po Pa	9972

Table 8.9 Sums of squares for models fitted to maize data.

(b) Show that

$$\left(X_1^{\mathsf{T}} P_2 X_1\right)^{-1} = \left(X_1^{\mathsf{T}} X_1\right)^{-1} - H_1 X_2 \left(X_2^{\mathsf{T}} P_1 X_2\right)^{-1} X_2^{\mathsf{T}} X_1 \left(X_1^{\mathsf{T}} X_1\right)^{-1}.$$

(c) Show that

$$H = X_1 \left(X_1^{\mathsf{T}} P_2 X_1\right)^{-1} X_1^{\mathsf{T}} - H_1 X_2 \left(X_2^{\mathsf{T}} P_1 X_2\right)^{-1} X_2^{\mathsf{T}} + X_2 \left(X_2^{\mathsf{T}} P_1 X_2\right)^{-1} X_2^{\mathsf{T}} P_1.$$

(d) Use (b) and (c) to show that $H H_1 = H_1$.

4   Under what two circumstances might one of the reductions in residual sum of squares $SS_r - SS_{r+1}$ in an analysis of variance table for a normal linear model equal zero? Does the more probable of these occur when the columns of either of the design matrices below are included successively in their models:

$$\text{(a)} \quad \begin{pmatrix} 1 & 1 & 0 & 0 \\ 1 & 1 & 0 & 1 \\ 1 & 0 & 1 & 0 \\ 1 & 0 & 1 & 1 \end{pmatrix}, \qquad \text{(b)} \quad \begin{pmatrix} 1 & 1 & 1 & 1 \\ 1 & 1 & 0 & 0 \\ 1 & 0 & 1 & 0 \\ 1 & 0 & 0 & 1 \end{pmatrix}?$$

5   Suppose that the maize data consisted of three pots each containing two pairs of plants, 12 plants in all. Using the parametrization in Example 8.19, write out the 12×11 design matrix whose first two columns are terms for the overall mean and for cross-fertilization, whose next three columns are the pots term, and whose last six columns are the pairs term. Say what the degrees of freedom for the four models in Example 8.19 would then be, and hence give the degrees of freedom in the analysis of variance table.

6   The residual sums of squares in Example 8.19 are given in Table 8.9. For which of the terms are the reductions in residual sum of squares independent of the order of fitting? Explain why adding the `Pots` term to a model that already contains the `Pairs` term does not reduce the sum of squares, even if `Fertilization` is not included.

7   Verify that the columns of the design matrix in Example 8.20 are orthogonal. Use Gram–Schmidt orthogonalization to derive the corresponding matrices for two, three, and four observations.

8   Verify that 1_n, Z_1, and Z_2 in Example 8.21 are orthogonal. Show that if one of the rows of the original design matrix is missing, the Z_r are not orthogonal.

8.6 Model Checking

8.6.1 Residuals

Discrepancies between data and a regression model may be isolated or systematic, or both. One type of isolated discrepancy is when there are outliers: a few observations that are unusual relative to the rest. Systematic discrepancies arise, for example, when a transformation of the response or a covariate is needed, when correlated errors are supposed independent, or when a term is incorrectly omitted. There are many techniques for detecting such problems. Graphs are widely used, often supplemented by more formal methods that sharpen their interpretation.

The assumptions underlying the linear regression model (8.1) are:

- *linearity* — the response depends linearly on each explanatory variable and on the error, with no systematic dependence on any omitted terms;
- *constant variance* — the responses have equal variances, which in particular do not depend on the level of the response;
- *independence* — the errors are uncorrelated, and independent if normal; and sometimes
- *normality* — in the normal linear model the errors are normally distributed.

Many graphical methods for checking these assumptions are based on the raw residuals, $e = y - \widehat{y}$. These are estimates of the unobserved errors ε, with mean vector and variance matrix

$$\mathrm{E}(e) = 0, \quad \mathrm{var}(e) = \sigma^2(I_n - H),$$

where H is the hat matrix $X(X^\mathsf{T}X)^{-1}X^\mathsf{T}$. The covariance of two different residuals, e_j and e_k, equals $-\sigma^2 h_{jk}$, so in general the residuals are correlated.

A difficulty in direct comparison of the e_j is that their variances, $\sigma^2(1 - h_{jj})$, are usually unequal. We therefore construct *standardized residuals*

$$r_j = \frac{e_j}{s(1 - h_{jj})^{1/2}} = \frac{y_j - x_j^\mathsf{T}\widehat{\beta}}{s(1 - h_{jj})^{1/2}}, \tag{8.26}$$

where $x_j^\mathsf{T}\widehat{\beta} = \widehat{y}_j$ is the jth fitted value and s^2 is the unbiased estimate of σ^2 based on the model. The r_j have means zero and approximately unit variances, and hence are comparable with standard normal variables.

The simplest check on linearity is to plot the response vector y against each column of the design matrix X. It is also useful to plot the standardized residuals r against each variable, whether or not it has been used in the model. Incorrect form of dependence on an explanatory variable, or omission of one, will show as a pattern in the corresponding plot. More formal techniques designed to detect wholesale nonlinearity are discussed below.

Constancy of variance is usually checked by a plot of the r_j or $|r_j|$ against fitted values. A common failure of this assumption occurs when the error variance increases with the level of the response; this shows as a trumpet-shaped plot. Since the raw residuals e and the fitted values $\widehat{y}$ are uncorrelated, we would expect random scatter if the model fitted adequately. This plot can also help to detect a nonlinear relation between the response and fitted value, as in Example 8.24 below.

Non-independence of the errors can be hard to detect and can have a serious effect on the standard errors of estimates, but serial correlation of time-ordered observations may show up in scatterplots of lagged r_j, or in their correlogram.

Assumptions about the distribution of the errors can be checked by probability plots of the r_j. In particular, normal scores plots are widely used.

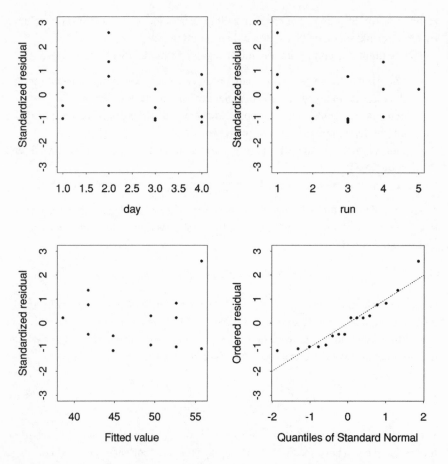

Figure 8.4 Residual plots for data on cycling up a hill. The panels showing residuals plotted against levels of day and run, and against fitted values, would show random variation if the model is adequate, as seems to be the case. The normal scores plot shows that the errors appear close to normal.

Single outliers — maybe due to mistakes in data recording, transcription, or entry — are likely to show up on any of the plots described above, while multiple outliers may lead to *masking* where each outlier is concealed by the presence of others.

Example 8.22 (Cycling data)   Figure 8.4 shows plots of the r_j for the model that includes effects of seat height, dynamo and tyre pressure. The top panels show the r_j plotted against the day on which the run took place, and the order of the run within each day. There is slight evidence of dependence on these, but we must beware of spurious patterns when there are only sixteen observations. To check whether these patterns might be genuine, we construct the F statistic for inclusion of factors corresponding to day and run after including seat height, dynamo, and tyre pressure in the model. Its value is 3.99, to be compared to $F_{7,5}(0.95) = 4.88$. Any evidence of differences among days and runs is weak, and we discount it.

The lower left panel of the figure shows residuals plotted against fitted values. There is a slight suggestion that the error variance increases as the fitted value does, but this is mostly due to the largest observation at the right of the plot.

The lower right panel of the figure shows a normal probability plot of the residuals. This is slightly upwardly curved, but not remarkably so in so small a set of data.

Inspection of Table 8.3 shows that the largest residual is for the sixth setup, of which the experimenter writes:

> Its comparison run (setup 5) was only 54 seconds. This is the largest amount of variation in the whole table. I suspect that the correct reading for setup 6 was 55 seconds, that is, I glanced at my watch and thought that it said 60 instead of 55 seconds. Since I am not sure, however, I have not changed it for the analysis. The conclusions would be the same in any case.

One reason that the conclusions would be unchanged is that a well-designed experiment like this is relatively robust to a single bad value.

To sum up: the linear model (8.2) seems to fit these data adequately. ∎

8.6.2 Nonlinearity

Linearity is usually a convenient fiction for describing how a response depends on the explanatory variables, and there are many ways it can fail. For example, a linear model may be appropriate for a transformation of the original response, so that $a(y) = x^T\beta + \varepsilon$ for some function $a(\cdot)$; then $y = a^{-1}(x^T\beta + \varepsilon)$ and error is not additive on the original scale. Another possibility is that the response is a nonlinear function of $x^T\beta$ but the error is additive, that is, $y = b(x^T\beta) + \varepsilon$ for some $b(\cdot)$. More generally we could put $a(y) = b(x^T\beta) + c(\varepsilon)$ for fairly arbitrary functions $a(\cdot)$, $b(\cdot)$ and $c(\cdot)$. Such models can be fitted, but they are beyond our scope.

For a simpler approach, we consider parametric transformation of the response, in which we assume that for some family of transformations $a(\cdot)$ indexed by a parameter λ, there is a transformation such that $a(y) = x^T\beta + \varepsilon$. In principle we might consider many possible transformations, but practical experience suggests that power and logarithmic transformations are among the most fruitful. The following example gives a general approach.

Suggested by Box and Cox (1964). George E. P. Box (1919–) was educated at London University and has held posts in industry and at Princeton and the University of Wisconsin. He has made important contributions to robust and Bayesian statistics, experimental design, time series, and to industrial statistics. Sir David Roxbee Cox (1924–) was born in Birmingham and educated in Cambridge and Leeds. He has held posts at Imperial College London, Cambridge, and Oxford where he nows works. He has made highly influential contributions across the whole of statistical theory and methods. See DeGroot (1987a) and Reid (1994).

Example 8.23 (Box–Cox transformation) Suppose that a normal linear model applies not to y, but to

$$y^{(\lambda)} = \begin{cases} \frac{y^\lambda - 1}{\lambda}, & \lambda \neq 0, \\ \log y, & \lambda = 0. \end{cases}$$

As λ varies in the range $(-2, 2)$ this encompasses the inverse transformation $(\lambda = -1)$, log $(\lambda = 0)$, cube and square roots $(\lambda = \frac{1}{3}, \frac{1}{2})$, and the original scale $(\lambda = 1)$, as well as the square transformation $(\lambda = 2)$. We assume below that all the y_j are positive. If not, the transformation must be applied to $y_j + \xi$, with ξ chosen large enough to make all the $y_j + \xi$ positive.

Now let $y^{(\lambda)}$ denote the $n \times 1$ vector of transformed responses, and assume that a normal linear model

$$y^{(\lambda)} = X\beta + \varepsilon$$

applies for some values of λ, β, and error variance σ^2. We assume that the design matrix contains a column of ones, so that using $y^{(\lambda)}$ rather than y^λ leaves the fit unchanged; it merely changes the intercept and rescales β.

To obtain the likelihood for β, σ^2, and λ, note that on taking into account the Jacobian of the transformation from $y^{(\lambda)}$ to y, the density of y_j is

$$f(y_j; \beta, \sigma^2, \lambda) = \frac{y_j^{\lambda-1}}{(2\pi\sigma^2)^{1/2}} \exp\left\{-\frac{1}{2\sigma^2}\left(y_j^{(\lambda)} - x_j^{\mathrm{T}}\beta\right)^2\right\}.$$

Consequently the log likelihood based on independent $y_1, \ldots, y_n$ is

$$\ell(\beta, \sigma^2, \lambda) \equiv -\frac{1}{2}\left\{n\log\sigma^2 + \frac{1}{\sigma^2}\sum_{j=1}^{n}\left(y_j^{(\lambda)} - x_j^{\mathrm{T}}\beta\right)^2\right\} + (\lambda-1)\sum_{j=1}^{n}\log y_j.$$

If λ is regarded as fixed, the maximum likelihood estimates of β and σ^2 are $\widehat{\beta}_\lambda = (X^{\mathrm{T}}X)^{-1}X^{\mathrm{T}}y^{(\lambda)}$ and $SS(\widehat{\beta}_\lambda)/n$, where $SS(\widehat{\beta}_\lambda)$ is the residual sum of squares for the regression of $y^{(\lambda)}$ on the columns of X. Thus the profile log likelihood for λ is

$$\ell_{\mathrm{p}}(\lambda) = \max_{\beta, \sigma^2} \ell(\beta, \sigma^2, \lambda) \equiv -\frac{n}{2}\left\{\log SS(\widehat{\beta}_\lambda) - \log g^{2(\lambda-1)}\right\},$$

where $g = (\prod y_j)^{1/n}$ is the geometric average of $y_1, \ldots, y_n$. Equivalently $\ell_{\mathrm{p}}(\lambda) = -\frac{1}{2}n\log SS_g(\widehat{\beta}_\lambda)$, where $SS_g(\widehat{\beta}_\lambda)$ is the residual sum of squares for the regression of $y^{(\lambda)}/g$ on the columns of X. Exercise 8.6.3 invites you to provide the details.

A plot of the profile log likelihood $\ell_{\mathrm{p}}(\lambda)$ summarizes the information concerning λ; a $(1 - 2\alpha)$ confidence interval is the set for which $\ell_{\mathrm{p}}(\lambda) \geq \ell_{\mathrm{p}}(\widehat{\lambda}) - \frac{1}{2}c_1(1 - 2\alpha)$. $\quad$ $c_\nu(\alpha)$ is the α quantile of the χ^2_ν distribution. The exact maximum likelihood estimate of λ is rarely used, since a nearby value is usually more easily interpreted. $\quad\blacksquare$

A different approach is to consider whether the model $y = b(x^{\mathrm{T}}\beta) + \varepsilon$ might apply. This cannot be linearized by a response transformation and if there is evidence that $b(\cdot)$ is substantially nonlinear but the variance is constant it may be necessary to fit a nonlinear normal model. The following example gives one method for detecting this sort of nonlinearity.

Example 8.24 (Non-additivity) Suppose that it is feared that $y = b(x^{\mathrm{T}}\beta) + \varepsilon$, where $b(\cdot)$ is a smooth nonlinear function. Taylor series expansion of $b(\cdot)$ about a typical value of $x^{\mathrm{T}}\beta$, η, say, gives

$$y \doteq b(\eta) + b'(\eta)(x^{\mathrm{T}}\beta - \eta) + \frac{1}{2}b''(\eta)(x^{\mathrm{T}}\beta - \eta)^2 + \varepsilon.$$

If the model contains a constant, so that $x^{\mathrm{T}}\beta = \beta_0 + x_1\beta_1 + \cdots$, then $y \doteq x^{\mathrm{T}}\gamma + \delta(x^{\mathrm{T}}\gamma)^2 + \varepsilon$, where γ is just a reparametrization of β, and $\delta \propto b''(\eta)$. A large value of δ corresponds to strong nonlinear dependence of y on $x^{\mathrm{T}}\beta$.

Let us fit the model $y = X\beta + \varepsilon$, giving fitted values $x_j^{\mathrm{T}}\widehat{\beta}$ and residual sum of squares $SS(\widehat{\beta})$. Then as $y - x^{\mathrm{T}}\gamma \doteq \delta(x^{\mathrm{T}}\gamma)^2 + \varepsilon$, non-additivity should show up as curvature in a plot of standardized residuals against fitted values.

A formal test for non-zero δ is based on refitting the model with the column $(x_j^{\mathrm{T}}\widehat{\beta})^2$ added to the design matrix. Although the residual sum of squares for this model,

Table 8.10 Poison data (Box and Cox, 1964). Survival times in 10-hour units of animals in a 3 × 4 factorial experiment with four replicates. The table underneath gives average (standard deviation) for the poison × treatment combinations.

Treatment	Poison 1	Poison 2	Poison 3
A	0.31, 0.45, 0.46, 0.43	0.36, 0.29, 0.40, 0.23	0.22, 0.21, 0.18, 0.23
B	0.82, 1.10, 0.88, 0.72	0.92, 0.61, 0.49, 1.24	0.30, 0.37, 0.38, 0.29
C	0.43, 0.45, 0.63, 0.76	0.44, 0.35, 0.31, 0.40	0.23, 0.25, 0.24, 0.22
D	0.45, 0.71, 0.66, 0.62	0.56, 1.02, 0.71, 0.38	0.30, 0.36, 0.31, 0.33

Treatment	Poison 1	Poison 2	Poison 3	Average
A	0.41 (0.07)	0.32 (0.08)	0.21 (0.02)	0.31
B	0.88 (0.16)	0.82 (0.34)	0.34 (0.05)	0.68
C	0.57 (0.16)	0.38 (0.06)	0.24 (0.01)	0.39
D	0.61 (0.11)	0.67 (0.27)	0.33 (0.03)	0.53
Average	0.62	0.55	0.28	0.48

SS_δ, depends upon the fitted values for the previous fit, the F statistic for inclusion of $(x_j^{\mathrm{T}}\widehat{\beta})^2$,

$$\frac{SS(\widehat{\beta}) - SS_\delta}{SS_\delta/(n - p - 1)}, \tag{8.27}$$

See Tukey (1949).
has an $F_{1,n-p-1}$ distribution; this is known as *Tukey's one degree of freedom for non-additivity*.   ∎

Covariates that are artificially created to help assess model fit, such as $(x_j^{\mathrm{T}}\widehat{\beta})^2$ in Example 8.24, are known as *constructed variables*.

Example 8.25 (Poisons data)   Table 8.10 contains data from a completely randomized experiment on the survival times of 48 animals. The animals were divided at random into groups of size four, and then each group was given one of three poisons and one of four treatments. Thus there are two factors, one with three and the other with four levels. The lower part of Table 8.10 and the upper panels of Figure 8.5 both show strong effects of treatment and poison: poison 3 is most potent, and treatments B and D are more efficacious than A and C. There is also evidence that the response variance depends on the mean: the standard deviations are smaller for poison × treatment combinations with smaller average response.

One model for these data is

$$y_{tpj} = \mu + \alpha_t + \beta_p + \varepsilon_{tpj}, \quad t = 1,2,3,4, \ p = 1,2,3, \ j = 1,2,3,4. \tag{8.28}$$

Here μ represents a baseline average response in the absence of treatments or poisons, α_t represents the effect of the tth treatment, β_p the effect of the pth poison and ε_{tpj} is the unobserved error for the jth replicate given the tth treatment and pth poison. We assess the fit of (8.28) initially through the plot of standardized residuals against fitted values in the upper left panel of Figure 8.6, which shows a striking increase of error

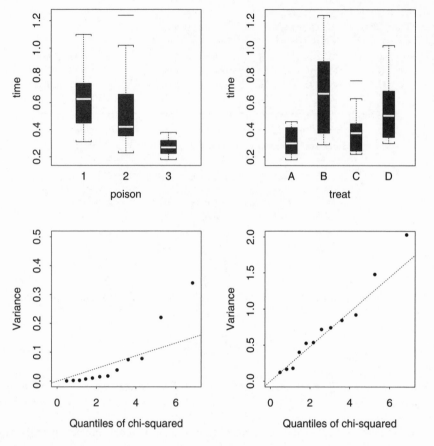

Figure 8.5 Poison data. The upper panels show how the responses depend on the factor levels. The lower left panel shows a χ^2_3 probability plots of the $3s^2_{pt}$, where s^2_{pt} is the sample variance of the four replicates y_{ptj} given the pth poison and tth treatment. The lower right panel shows the same plot for the y^{-1}_{ptj}.

variance with the mean response. The model underpredicts for the lowest responses, where $r_j > 0$ and therefore $y_j > \widehat{y}_j$, and overpredicts for the middle responses, where the residuals are mostly negative. Following Example 8.24, this suggests that the poison and treatment effects are not additive. The neighbouring panel shows that the errors are somewhat positively skewed relative to the normal distribution. The model fits the data poorly, not owing to a few bad observations, but in a systematic way, as was also suggested by the lower left panel of Figure 8.5.

Ignoring for a moment the nonconstancy of variance, we explore whether the explanatory variables act additively. The F statistic for non-additivity, (8.27), equals 14.03. This is large compared with the 0.95 quantile of the $F_{1,41}$ distribution and gives strong evidence of non-additivity.

The lower right panel of Figure 8.6 shows the profile log likelihood for the transformation parameter, λ. There is strong evidence that the original scale ($\lambda = 1$) is poor; log transformation ($\lambda = 0$) also seems inappropriate. The most readily interpretable value of λ in the 95% confidence interval seems to be -1, corresponding to fitting a linear model to the inverse response $1/y$. This can be interpreted in terms of the rate of dying, whose units are time^{-1}. The lower left panel of the figure suggests that the evidence for non-additivity has gone, and that the inverse transformation has roughly

Figure 8.6 Diagnostic plots for the two-way layout model for the poisons data. The upper left panel a plot of standardized residuals for the fit of the two-way layout model to the original data against the fitted value, while its neighbour shows the normal probability plot of these residuals. The lower right panel shows the profile log likelihood for the Box–Cox parameter λ and suggests that a linear model should be fitted to the inverse response, $1/y$. The lower left panel shows the residuals for the two-way layout model with response $1/y$ plotted against its fitted values; this does not display the non-linearity and systematic increase of variance of the panel above.

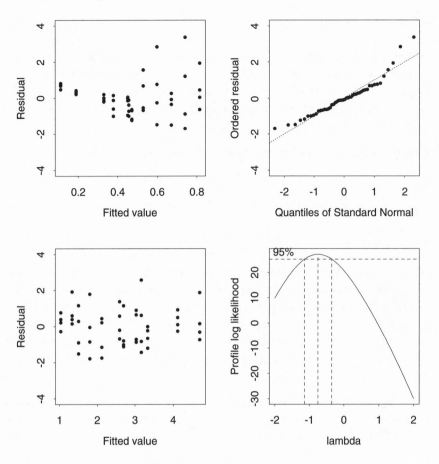

equalized the error variances. A probability plot shows that the residuals on this scale are close to normal.

To sum up, the model $y^{-1} = \mu + \alpha_t + \beta_p + \varepsilon_{tpj}$ seems to fit the data adequately, and has a direct interpretation as a linear model for the effect of poisons and treatments on the speed of dying.

We return to these data in Examples 9.6 and 9.8.   ■

8.6.3 Leverage, influence, and case deletion

We call the explanatory and response variables (x_j, y_j) the jth *case*. We have already seen how an odd y_j can arise, but there can also be effects due to unusual explanatory variables. To see how, recall that $\mathrm{var}(y_j - x_j^{\mathsf{T}}\widehat{\beta}) = \sigma^2(1 - h_{jj})$, and notice that if h_{jj} is close to one the jth fitted value must lie very close to y_j itself. Indeed, if $h_{jj} = 1$, the model is constrained so that $x_j^{\mathsf{T}}\widehat{\beta} = y_j$. This is undesirable because in effect a degree of freedom, the equivalent of one parameter, is used to fit one response value exactly. The effect on $\widehat{\beta}$ could be catastrophic if y_j were outlying.

The quantity h_{jj} is called the *leverage* of the jth case. Other things being equal, the argument above suggests that low leverage is good. But $\mathrm{tr}(H) = \sum h_{jj} = p$

(Exercise 8.2.5), so the average leverage cannot be reduced below p/n. Approximate equalization of leverage is one attribute of good design. In the factorial experiment in Table 8.3, for example, $h_{jj} = \frac{1}{4}$ for each case. A general guideline is that cases for which $h_{jj} > 2p/n$ deserve closer inspection; it may be worthwhile to repeat an analysis without them in order to assess their effect on both the values and the precision of the estimates. In itself, however, high leverage is not sufficient reason to delete a case, which if not outlying may be very informative.

Example 8.26 (Straight-line regression)   The matrix formulation of

$$y_j = \gamma_0 + (x_j - \overline{x})\gamma_1 + \varepsilon_j, \quad j = 1, \ldots, n,$$

is given in Example 8.6, and it is easily deduced that the jth leverage is

$$h_{jj} = \frac{1}{n} + \frac{(x_j - \overline{x})^2}{\sum_k (x_k - \overline{x})^2}.$$

When the constant is dropped the leverage is $(x_j - \overline{x})^2 / \sum_k (x_k - \overline{x})^2$, and when the covariate x_j is dropped the leverage is n^{-1}. Thus h_{jj} can be interpreted as a sum of contributions for each parameter. As the contribution corresponding to γ_1 is quadratic in $x_j - \overline{x}$, responses with large values of $|x_j - \overline{x}|$ will strongly affect the slope of the fitted line. All the responses have equal weight in estimating the intercept. These effects do not depend on the response values and depend purely on the design matrix. ∎

Having seen that an individual case may substantially affect least squares estimates, it is natural to ask how to measure this. One overall *influence measure* for the jth case is *Cook's distance*, defined as

$$C_j = \frac{1}{ps^2} (\widehat{y} - \widehat{y}_{-j})^{\mathsf{T}} (\widehat{y} - \widehat{y}_{-j}),$$

See Cook (1977). R. Dennis Cook is a professor of statistics at the University of Minnesota.

where $\widehat{y}_{-j} = X\widehat{\beta}_{-j}$, and subscript $-j$ denotes a quantity calculated with the jth case deleted from the model. Cook's distance measures the overall change in the fitted values when the jth case is deleted from the model, standardized by the dimension of β and the estimate of σ^2. It can be revealing to refit a model without the cases whose values of C_j are largest.

To gain some insight into C_j, note that the least squares estimate of β calculated without the jth case is

$$\widehat{\beta}_{-j} = \left(X^{\mathsf{T}} X - x_j x_j^{\mathsf{T}} \right)^{-1} (X^{\mathsf{T}} y - x_j y_j).$$

Some linear algebra shows that

$$\widehat{\beta}_{-j} = \widehat{\beta} - (X^{\mathsf{T}} X)^{-1} x_j \frac{y_j - \widehat{y}_j}{1 - h_{jj}}, \tag{8.29}$$

and it follows that (Exercise 8.6.5)

$$C_j = \frac{r_j^2 h_{jj}}{p(1 - h_{jj})}, \tag{8.30}$$

where r_j is the standardized residual. Therefore large values of C_j arise if a case has high leverage or a large standardized residual, or both. A plot of C_j against $h_{jj}/(1 - h_{jj})$ helps to distinguish between these possibilities. A crude rule is that as a residual with $|r_j| > 2$ or a case with leverage $h_{jj} > 2p/n$ deserve attention, a value of C_j greater than $8/(n - 2p)$ is worth a closer look. It is possible for the model to depend on a case whose Cook's distance is zero (Exercise 8.6.6), however, and there is no substitute for careful inspection of the data, residuals, and leverages.

As an observation with a large standardized residual can have a big effect on a fitted model, it is natural to ask whether an outlier is more easily detected by comparing y_j with its predicted value based on the other observations, $x_j^T \widehat{\beta}_{-j}$. After all, if the model is correct and y_j is not an outlier, we expect that $E(\widehat{\beta}) = E(\widehat{\beta}_{-j}) = x_j^T \beta$, although of course $\widehat{\beta}_{-j}$ will be a less precise estimate of β than $\widehat{\beta}$. On the other hand, an outlying response y_j does not affect $x_j^T \widehat{\beta}_{-j}$, so any discrepancy between them should be more obvious. There is a close connection to the idea of cross-validation. Now (8.29) implies that

$$y_k - x_k^T \widehat{\beta}_{-j} = y_k - \widehat{y}_k + x_k^T (X^T X)^{-1} x_j \frac{y_j - \widehat{y}_j}{1 - h_{jj}},$$

and since $x_k^T (X^T X)^{-1} x_j = h_{jk}$, we find that $\mathrm{var}(y_j - x_j^T \widehat{\beta}_{-j}) = \sigma^2 / (1 - h_{jj})$. This suggests that *deletion residuals* be defined as

$$r_j' = \frac{y_j - x_j^T \widehat{\beta}_{-j}}{\mathrm{var}(y_j - x_j^T \widehat{\beta}_{-j})^{1/2}} = \frac{y_j - \widehat{y}_{-j,j}}{s_{-j}(1 - h_{jj})^{1/2}},$$

where $\widehat{y}_{-j,j}$ is the jth element of the vector $\widehat{y}_{-j}$ and the estimate of σ^2 based on the data with the jth case deleted equals

$$s_{-j}^2 = \frac{1}{n - 1 - p} \left[(y - \widehat{y}_{-j})^T (y - \widehat{y}_{-j}) - \left\{ y_j - \widehat{y}_j + \frac{h_{jj}(y_j - \widehat{y}_j)}{1 - h_{jj}} \right\}^2 \right].$$

Yet more algebra shows that the deletion residual can be expressed as

$$r_j' = \left(\frac{n - p - 1}{n - p - r_j^2} \right)^{1/2} r_j,$$

which is a monotonic function of r_j that exaggerates values for which $|r_j| > 1$. As their derivation suggests, deletion residuals for outlying observations are more prominent than are the corresponding r_j.

Example 8.27 (Cycling data) Table 8.3 gives standardized residuals, deletion residuals, and measures of leverage and influence for the model with an intercept and three main effects fitted to these data. The design is balanced, and since $(X^T X)^{-1} = \frac{1}{16} I_4$, all the leverages equal $\frac{1}{4}$; consequently the standardized residuals are a simple multiple of the raw residuals. As remarked in Example 8.22, the only unusual residual is

Table 8.11 Simulated
data and case diagnostics.

Case	x_1	x_2	y	$\widehat{y}$	r	r'	h	C
1	0.02	−6.31	0.95	0.41	1.16	1.20	0.88	3.28
2	0.36	0.39	0.44	0.53	−0.08	−0.07	0.13	0.00
3	7.12	−0.64	0.27	0.38	−0.14	−0.13	0.68	0.01
4	−1.54	1.13	0.09	0.59	−0.45	−0.42	0.29	0.03
5	0.24	−1.90	−0.82	0.49	−1.07	−1.08	0.15	0.07
6	0.26	−0.06	0.03	0.53	−0.40	−0.37	0.12	0.01
7	−0.16	0.13	−0.22	0.54	−0.61	−0.59	0.14	0.02
8	0.43	0.80	0.13	0.54	−0.33	−0.31	0.15	0.01
9	−0.02	0.59	3.57	0.55	2.47	6.31	0.15	0.37
10	4.58	0.29	0.57	0.45	0.11	0.10	0.31	0.00

for setup 6, whose deletion residual is strikingly large: there is strong evidence that this is an outlier. The corresponding Cook statistic, 0.56, is by far the largest, but it is unremarkable relative to $8/(n-2p) = 1$. The belt-and-braces statistician might repeat the analysis without this datum, but it makes little difference.   ∎

Exercises 8.6

1   Show that the standardized residuals r_j have means zero and variances $(n-p)/(n-p-2)$. What can you say about their joint distribution?

2   Table 8.11 shows simulated data on the dependence of $y = \beta_0 + \beta_1 x_1 + \beta_2 x_2 + \varepsilon$ on covariates x_1 and x_2. The residual sum of squares was 12.43.
(a) Choose a case and check the relationships between $\widehat{y}$, r, r', h, and C.
(b) Discuss the fit. If it is not adequate, explain what further steps you would take in analyzing the data.

3   Provide the details for Example 8.23.

4   Compute and interpret the leverages for Examples 8.9 and 8.20.

5   Use Exercise 8.5.2(a) with $C = -1$ to show that

$$\left(X^{\mathsf{T}}X - x_j x_j^{\mathsf{T}}\right)^{-1} = (X^{\mathsf{T}}X)^{-1} + (1 - h_{jj})^{-1}(X^{\mathsf{T}}X)^{-1}x_j x_j^{\mathsf{T}}(X^{\mathsf{T}}X)^{-1};$$

it may help to note that $h_{jj} = x_j^{\mathsf{T}}(X^{\mathsf{T}}X)^{-1}x_j$. Hence show that

$$\widehat{\beta}_{-j} = \left(X^{\mathsf{T}}X - x_j x_j^{\mathsf{T}}\right)^{-1}(X^{\mathsf{T}}y - x_j y_j) = \widehat{\beta} - (1 - h_{jj})^{-1}(X^{\mathsf{T}}X)^{-1}x_j(y_j - \widehat{y}_j),$$

deduce that $\widehat{y} - \widehat{y}_{-j} = (1 - h_{jj})^{-1}X(X^{\mathsf{T}}X)^{-1}x_j(y_j - \widehat{y}_j)$, and finally that

$$C_j = \frac{(\widehat{y} - \widehat{y}_{-j})^{\mathsf{T}}(\widehat{y} - \widehat{y}_{-j})}{ps^2} = \frac{r_j^2 h_{jj}}{p(1 - h_{jj})}.$$

6   Suppose that the straight-line regression model $y = \beta_0 + \beta_1 x + \varepsilon$ is fitted to data in which $x_1 = \cdots = x_{n-1} = -a$ and $x_n = (n-1)a$, for some positive a. Show that although y_n completely determines the estimate of β_1, $C_n = 0$. Is Cook's distance an effective measure of influence in this situation?

8.7 Model Building

8.7.1 General

Once the context for a regression problem is known and the data have been scrutinized for outliers, missing values, and so forth, a model must be built. Related investigations will often suggest a form for it, the main initial questions concerning the choice of response and explanatory variables.

The purpose of the analysis determines one or perhaps more responses, which may combine several of the original variables. Once it is chosen, questions arise about whether individual responses are correlated, and if their variance is constant. If not, it may be necessary to use weighted or generalized least squares (Section 8.2.4), or to consider transformations. These may also be suggested by constraints, for example that the response is positive, but it is then also good to consider more general classes of models discussed in Chapter 10.

Scatterplots of the response against potential explanatory variables and of these variables against each another are needed to screen out bad data, to suggest which covariates are likely to be important, and perhaps also to indicate suitable transformations. Dimensional considerations or subject-matter arguments, for example that certain regression coefficients should be positive, may suggest fruitful combinations of covariates or particular relations between them and the response.

It may be clear that the response depends on a few variables, and that possible models can be fitted and compared using F and related tests. Once some suitable models have been found, the techniques of model checking outlined in Section 8.6 can be applied. Often unexpected discrepancies between a fitted model and data will lead to further thought, and then to more cycles of model-fitting, checking, and interpretation, iterated until a broadly satisfactory model has been found.

If p is much larger than n, then the design matrix must be cut down to size. One possibility is to use *principal components regression*. The basis of this is the *spectral decomposition*, which enables us to write $X^{\mathsf{T}}X = UDU^{\mathsf{T}}$, where D is the diagonal matrix $\mathrm{diag}(d_1, \ldots, d_p)$ containing the ordered eigenvalues $d_p \geq \cdots \geq d_1 \geq 0$ of $X^{\mathsf{T}}X$, and the columns of U are the corresponding eigenvectors. The matrix U can be chosen so that $UU^{\mathsf{T}} = U^{\mathsf{T}}U = I$. The idea is to form the design matrix from the columns of $Z = XU$, which are called *principal components*. The first principal component, z_1, is the linear combination $z = Xu$ of the columns of X for which $z^{\mathsf{T}}z$ is largest, the next, z_2, is the linear combination that maximizes $z_2^{\mathsf{T}}z_2$ subject to $z_1^{\mathsf{T}}z_2 = 0$, the third, z_3, maximizes $z_3^{\mathsf{T}}z_3$ subject to $z_1^{\mathsf{T}}z_2 = z_1^{\mathsf{T}}z_2 = 0$, and so forth. The hope is that much of the dependence of the response on the columns of X will be concentrated in these first few z_rs, in which case a good low-dimensional regression model may be obtainable. Sometimes it is useful to centre the columns of X by subtracting their averages, or to scale them by dividing centred columns by their standard deviations. The resulting principal components do not equal those for X.

Principal components and corresponding parameter estimates may be uninterpretable in terms of the original covariates, though this drawback is less critical when the goal of analysis is prediction.

8.7.2 Collinearity

If there is a nonzero vector c such that $Xc = 0$, the columns of the design matrix are said to be *collinear*. Then X has rank less than p and $X^\mathsf{T}X$ has no unique inverse. The simplest example of this arises in straight-line regression: if all the x_j are equal, it is impossible to find unique parameter estimates (Example 8.6). This difficulty arises more generally, because linear dependence among the columns of the design matrix means that some combinations of parameters cannot be estimated from the data; collinearity leads to indeterminable estimates with infinite variances. Related difficulties arise if the columns of X are almost collinear.

The matrix $X^\mathsf{T}X$ is invertible only if all its eigenvalues $d_p \geq \cdots \geq d_1 \geq 0$ are positive. Even if $X^\mathsf{T}X$ is invertible, however, the estimators can be very poor. The squared distance between $\widehat{\beta}$ and β is expressible as

$$(\widehat{\beta} - \beta)^\mathsf{T}(\widehat{\beta} - \beta) \overset{D}{=} \sigma^2 \sum_{r=1}^{p} Z_r^2/d_r, \quad \text{where} \quad Z_1, \ldots, Z_p \overset{\text{iid}}{\sim} N(0, 1).$$

Thus $(\widehat{\beta} - \beta)^\mathsf{T}(\widehat{\beta} - \beta)$ has mean and variance

$$\sigma^2 \sum_{r=1}^{p} d_r^{-1}, \quad 2\sigma^4 \sum_{r=1}^{p} d_r^{-2},$$

bounded below respectively by σ^2/d_1 and $2\sigma^4/d_1^2$, and $\widehat{\beta}$ may be far distant from β for small d_1. The practical implication is that parameter estimates from different but related datasets may vary greatly, giving apparently contradictory interpretations of the same phenomenon.

Diagnostics to warn of collinearity can be based on functions of the d_r such as the *condition number* $(d_p/d_1)^{1/2}$, but its statistical interpretation is not clear-cut. The condition number is sometimes reduced by replacing X with the matrix obtained on dropping the column of ones if any and centering the remaining columns, or by using the corresponding correlation matrix.

The most straightforward solution to collinearity or near collinearity is to drop columns from the design matrix until the estimates are better behaved.

A more systematic approach to dealing with weak design matrices is *ridge regression*, which starts by rewriting the original model $y = 1\beta_0 + X_1\beta_1 + \varepsilon$ as $y = 1\beta_0 + Z\gamma + \varepsilon$, where $Z^\mathsf{T}1 = 0$ and the diagonal of $Z^\mathsf{T}Z$ consists of ns. This involves centring each column of X_1 by subtracting its average, then dividing by its standard deviation, and multiplying by $n^{1/2}$. This centring and rescaling ensures that the elements of γ and of β have the same interpretations apart from a change of scale, unlike with principal components regression. Then the least squares estimates are $\widehat{\beta}_0 = \bar{y}$ and $\widehat{\gamma} = (Z^\mathsf{T}Z)^{-1}Z^\mathsf{T}y$. The idea is to replace $Z^\mathsf{T}Z$ by $Z^\mathsf{T}Z + \lambda I_{p-1}$, where $\lambda \geq 0$ is called the *ridge parameter*. The corresponding estimates, $\widehat{\gamma}_\lambda = (Z^\mathsf{T}Z + \lambda I_{p-1})^{-1}Z^\mathsf{T}y$, are biased unless $\lambda = 0$, when they are the least squares estimates of γ. Large values of λ increase the bias by shrinking the estimates towards the origin, but this decreases their variance. The value of λ is chosen empirically by minimization of a criterion

Table 8.12 Parameter estimates and their standard errors for the full model and a reduced model fitted to the cement data.

Parameter	Full model		Reduced model	
	Estimate	Standard error	Estimate	Standard error
β_0	62.41	70.07	71.64	14.14
β_1	1.55	0.74	1.45	0.12
β_2	0.51	0.72	0.42	0.19
β_3	0.10	0.75		
β_4	−0.14	0.71	−0.24	0.17

such as the *cross-validation sum of squares*

$$\mathrm{CV}(\lambda) = \sum_{j=1}^{n} (y_j - \widehat{y}_j^-)^2,$$

where $\widehat{y}_j^-$ is the fitted value for y_j predicted from the ridge regression model obtained when the jth case is deleted. Cross-validation, introduced in Section 7.1.2, is here used to assess how well the ridge regression fit would predict a new set of independent data like the original observations. A variant approach chooses λ to minimize the *generalized cross-validation sum of squares*,

$$\mathrm{GCV}(\lambda) = \sum_{j=1}^{n} \frac{(y_j - \widehat{y}_j)^2}{\{1 - \mathrm{tr}(H_\lambda)/n\}^2},$$

where $H_\lambda = n^{-1} 1_n 1_n^{\mathrm{T}} + Z(Z^{\mathrm{T}}Z + \lambda I_{p-1})^{-1} Z^{\mathrm{T}}$ is the hat matrix corresponding to the ridge regression, and the vector of fitted values $\widehat{y} = H_\lambda y$ depends on λ. We discuss these in more detail on page 523, though in another context.

Estimates such as $\widehat{\gamma}_\lambda$ that shrink towards a common value, here $\gamma = 0$, may also be derived by Bayesian arguments (Chapter 11).

Example 8.28 (Cement data) The astute reader will have realized that if the middle four columns of Table 8.1 are percentages, they may sum to 100. In fact they sum to $(99, 97, 95, 97, 98, 97, 97, 98, 96, 98, 98, 98, 98)$. As there is a column of ones in the design matrix for the full model, its columns are nearly dependent: estimation of five parameters is almost impossible. This is reflected by the standard errors in Table 8.12. The standard error for $\widehat{\beta}_0$ is vastly inflated by inclusion of x_3 because β_0 is almost impossible to estimate, whereas the other estimates are less badly affected.

The residual sum of squares for model without x_3 is 47.97, only slightly larger than that for the full model, 47.86. Thus inclusion of x_3 changes the fit of the model very little, but has a drastic effect on the precision of parameter estimation.

The eigenvalues of $X^{\mathrm{T}}X$ with all five columns of X are 44676, 5965.4, 810.0, 105.4 and 0.00012. The condition number of 6056 indicates strong ill-conditioning, and $\sum d_r^{-1} = 821$ seems very large.

The left panel of Figure 8.7 shows how the parameter estimates $\widehat{\gamma}_\lambda$ depend on the ridge parameter λ. All change fairly sharply as λ increases from zero, and are more stable for $\lambda > 0.2$. The right panel shows that $\mathrm{GCV}(\lambda)$ decreases sharply when

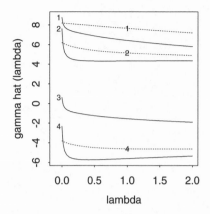
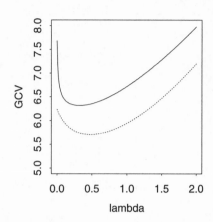

Figure 8.7 Ridge regression analysis of cement data. Left: variation of elements of $\widehat{\gamma}_\lambda$ as a function of λ, for models with all four covariates (solid) and with x_1, x_2, and x_4 only (dots). Right: generalized cross-validation criterion $GCV(\lambda)$ for these models.

λ increases from zero, and is minimized when $\lambda \doteq 0.3$. The dotted lines show that when x_3 is dropped both the $\widehat{\gamma}_\lambda$ and $GCV(\lambda)$ depend much less on λ, consistent with the discussion above.   ∎

8.7.3 Automatic variable selection

The screening and selection of many explanatory variables may be onerous. With p covariates, each to be included or not, at least 2^p possible design matrices must be fitted even before accounting for transformations, combinations of covariates, and so forth. Consequently automatic procedures for variable selection are widely used if p is large. While valuable as screening procedures, they are no substitute for careful model-building incorporating knowledge of the system under study and should be treated as a backstop; their output should always be considered critically.

Stepwise methods

Forward selection takes as baseline the model with an intercept only. Each term is added separately to this, and the base model for the next stage is taken to be the model with the intercept and the term that most reduces the sum of squares. Each of the remaining terms is added to the new base model, and the process continued, stopping if at any stage the F statistic for the largest reduction in sum of squares is not significant or if the design matrix is rank deficient.

Backward elimination starts from the model containing all terms, and then successively drops the least significant term at each stage. It stops when no term can be deleted without increasing the sum of squares significantly.

Backward elimination is generally the preferable of the two because its initial estimate of σ^2 will usually be better than that for forward selection, though at the possible expense of an unstable initial model. They may yield different final models.

In *stepwise regression* four options are considered at each stage: add a term, delete a term, swap a term in the model for one not in the model, or stop. This algorithm is often used in practice.

These three procedures have been shown to fit complicated models to completely random data, and although widely used they have no theoretical basis. This

Table 8.13 Data on light
water reactors (LWR)
constructed in the USA
(Cox and Snell, 1981,
p. 81). The covariates are
date (date construction
permit issued), T1 (time
between application for
and issue of permit), T2
(time between issue of
operating license and
construction permit),
capacity (power plant
capacity in MWe), PR (=1
if LWR already present on
site), NE (=1 if constructed
in north-east region of
USA), CT (=1 if cooling
tower used), BW (=1 if
nuclear steam supply
system manufactured by
Babcock–Wilcox), N
(cumulative number of
power plants constructed
by each
architect-engineer), PT
(=1 if partial turnkey
plant).

	cost	date	T_1	T_2	capacity	PR	NE	CT	BW	N	PT
1	460.05	68.58	14	46	687	0	1	0	0	14	0
2	452.99	67.33	10	73	1065	0	0	1	0	1	0
3	443.22	67.33	10	85	1065	1	0	1	0	1	0
4	652.32	68.00	11	67	1065	0	1	1	0	12	0
5	642.23	68.00	11	78	1065	1	1	1	0	12	0
6	345.39	67.92	13	51	514	0	1	1	0	3	0
7	272.37	68.17	12	50	822	0	0	0	0	5	0
8	317.21	68.42	14	59	457	0	0	0	0	1	0
9	457.12	68.42	15	55	822	1	0	0	0	5	0
10	690.19	68.33	12	71	792	0	1	1	1	2	0
11	350.63	68.58	12	64	560	0	0	0	0	3	0
12	402.59	68.75	13	47	790	0	1	0	0	6	0
13	412.18	68.42	15	62	530	0	0	1	0	2	0
14	495.58	68.92	17	52	1050	0	0	0	0	7	0
15	394.36	68.92	13	65	850	0	0	0	1	16	0
16	423.32	68.42	11	67	778	0	0	0	0	3	0
17	712.27	69.50	18	60	845	0	1	0	0	17	0
18	289.66	68.42	15	76	530	1	0	1	0	2	0
19	881.24	69.17	15	67	1090	0	0	0	0	1	0
20	490.88	68.92	16	59	1050	1	0	0	0	8	0
21	567.79	68.75	11	70	913	0	0	1	1	15	0
22	665.99	70.92	22	57	828	1	1	0	0	20	0
23	621.45	69.67	16	59	786	0	0	1	0	18	0
24	608.80	70.08	19	58	821	1	0	0	0	3	0
25	473.64	70.42	19	44	538	0	0	1	0	19	0
26	697.14	71.08	20	57	1130	0	0	1	0	21	0
27	207.51	67.25	13	63	745	0	0	0	0	8	1
28	288.48	67.17	9	48	821	0	0	1	0	7	1
29	284.88	67.83	12	63	886	0	0	0	1	11	1
30	280.36	67.83	12	71	886	1	0	0	1	11	1
31	217.38	67.25	13	72	745	1	0	0	0	8	1
32	270.71	67.83	7	80	886	1	0	0	1	11	1

arbitrariness is reflected in rules for deciding which terms to include, some of which use tables of the F or t distributions. Others simply drop a term from the model if its F statistic is less than a number such as 4, and otherwise include the term. Sometimes a theoretically-motivated criterion such as AIC is used.

Example 8.29 (Nuclear plant data)   Table 8.13 contains data on the cost of 32 light water reactors. The cost (in dollars $\times 10^{-6}$ adjusted to a 1976 base) is the quantity of interest, and the others are explanatory variables.

Costs are typically relative. Moreover large costs are likely to vary more than small ones, so it seems sensible to take log(cost) as the response y. For consistency we also take logs of the other quantitative covariates, fitting linear models using date, log(T1), log(T2), log(capacity), PR, NE, CT, log(N), and PT. The last of these indicates six plants for which there were partial turnkey guarantees, and some subsidies may be hidden in their costs.

	Full model		Backward		Forward	
	Est (SE)	t	Est (SE)	t	Est (SE)	t
Constant	−14.24 (4.229)	−3.37	−13.26 (3.140)	−4.22	−7.627 (2.875)	−2.66
date	0.209 (0.065)	3.21	0.212 (0.043)	4.91	0.136 (0.040)	3.38
log(T1)	0.092 (0.244)	0.38				
log(T2)	0.290 (0.273)	1.05				
log(cap)	0.694 (0.136)	5.10	0.723 (0.119)	6.09	0.671 (0.141)	4.75
PR	−0.092 (0.077)	−1.20				
NE	0.258 (0.077)	3.35	0.249 (0.074)	3.36		
CT	0.120 (0.066)	1.82	0.140 (0.060)	2.32		
BW	0.033 (0.101)	0.33				
log(N)	−0.080 (0.046)	−1.74	−0.088 (0.042)	−2.11		
PT	−0.224 (0.123)	−1.83	−0.226 (0.114)	−1.99	−0.490 (0.103)	−4.77
Residual SE (df)	0.164 (21)		0.159 (25)		0.195 (28)	

Estimates and standard errors for the full model and those found by backward elimination and forward selection are given in Table 8.14. Backward elimination starts by refitting the model without BW and then considering the t statistics for the remaining variables, dropping the next least significant, here log(T1), and so forth. The effects for the variables retained are strengthened; most are highly significant. Forward selection chooses a smaller model with larger residual sum of squares, and this results in smaller t statistics. Stepwise selection starting from this model yields the model chosen by backward elimination. Examination of residuals for this suggests no difficulty, and we are left with a model in which cost increases with capacity, though not proportionally, with presence of a cooling tower, with date, and in the north-east region of the USA, but is decreased by a partial turnkey guarantee, and with architect's experience.   ■

Likelihood criteria

A more satisfactory approach is to fit all reasonable models and adopt the one that minimizes some overall measure of discrepancy. One such measure is the residual sum of squares, but this continues to decrease as the number of parameters increases and always yields the model with all possible terms. This suggests that model complexity be penalized by balancing it against a measure of fit. We now discuss one approach to this.

Suppose that the data were generated by a true model g under which the responses Y_j are independent normal variables with means μ_j and variances σ^2 and let $E_g(\cdot)$ denote expectation with respect to this model. Following the discussion in Section 4.7, our ideal would be to choose the candidate model $f(y; \theta)$ to minimize the loss when predicting a new sample like the old one,

The scaling factor 2 is included for comparability with AIC.

$$E_g \left(E_g^+ \left[2 \sum_{j=1}^n \log \left\{ \frac{g(Y_j^+)}{f(Y_j^+; \widehat{\theta})} \right\} \right] \right). \tag{8.31}$$

Here $Y_1^+, \ldots, Y_n^+$ is another sample independent of $Y_1, \ldots, Y_n$ but with the same distribution, E_g^+ denotes expectation over $Y_1^+, \ldots, Y_n^+$, and $\widehat{\theta}$ is the maximum likelihood estimator of θ based on $Y_1, \ldots, Y_n$.

If the candidate model is normal, then θ comprises the mean responses $\mu_1, \ldots, \mu_n$ and σ^2, with maximum likelihood estimators $\widehat{\mu}_1, \ldots, \widehat{\mu}_n$ and $\widehat{\sigma}^2$. Then the sum in (8.31) equals

$$\frac{1}{2} \sum_{j=1}^n \left\{ \log \widehat{\sigma}^2 + \frac{(Y_j^+ - \widehat{\mu}_j)^2}{\widehat{\sigma}^2} - \log \sigma^2 - \frac{(Y_j^+ - \mu_j)^2}{\sigma^2} \right\},$$

and hence the inner expectation is

$$\sum_{j=1}^n \left\{ \log \widehat{\sigma}^2 + \frac{\sigma^2}{\widehat{\sigma}^2} + \frac{(\mu_j - \widehat{\mu}_j)^2}{\widehat{\sigma}^2} - \log \sigma^2 - 1 \right\}.$$

Suppose that in our earlier terminology a candidate linear model with full-rank $n \times p$ design matrix X is correct, that is, the true model is nested within it. Then the vector $\mu = (\mu_1, \ldots, \mu_n)^{\mathrm{T}}$ of true means lies in the column space of X and there is a $p \times 1$ vector β such that $\mu = X\beta$. Hence $\widehat{\mu} = (\widehat{\mu}_1, \ldots, \widehat{\mu}_n)^{\mathrm{T}}$ is normal with mean μ, from which it follows that $\sum (\mu_j - \widehat{\mu}_j)^2 = (\widehat{\mu} - \mu)^{\mathrm{T}} (\widehat{\mu} - \mu) \sim \sigma^2 \chi_p^2$ independent of $n\widehat{\sigma}^2 \sim \sigma^2 \chi_{n-p}^2$. Now the expected values of a χ_ν^2 variable and of its inverse are ν and $(\nu - 2)^{-1}$, provided $\nu > 2$, and so (8.31) equals

$$n E_g (\log \widehat{\sigma}^2) + \frac{n^2}{n - p - 2} + \frac{np}{n - p - 2} - n \log \sigma^2 - n,$$

or equivalently for our purposes,

$$n E_g (\log \widehat{\sigma}^2) + \frac{n(n + p)}{n - p - 2}.$$

This is estimated unbiasedly by the *corrected information criterion*

$$\mathrm{AIC_c} = n \log \widehat{\sigma}^2 + n \frac{1 + p/n}{1 - (p + 2)/n},$$

and the 'best' candidate model is taken to be that which minimizes this. Taylor expansion gives $\mathrm{AIC_c} \doteq n \log \widehat{\sigma}^2 + n + 2(p + 1) + O(p^2/n)$, and for large n and fixed p this will select the same model as $\mathrm{AIC} = n \log \widehat{\sigma}^2 + 2p$. When p is comparable with n, $\mathrm{AIC_c}$ penalizes model dimension more severely.

A widely used related criterion is

$$C_p = \frac{SS_p}{s^2} + 2p - n,$$

where SS_p is the residual sum of squares for the fitted model and s^2 is an estimate of σ^2; C_p can be derived as an approximation to AIC (Problem 8.16), though its original motivation was different. In some cases σ^2 can be estimated from the full model, but care is needed because the choice of s^2 is critical to successful use of C_p.

Example 8.30 (Simulation study) Twenty different $n \times 7$ design matrices X were constructed using standard normal variables, centered and scaled so that

n		1	2	3	4	5	6	7
				Number of covariates				
10	C_p		131	504	91	63	83	128
	BIC		72	373	97	83	109	266
	AIC		52	329	97	91	125	306
	AIC_c	15	398	565	18	4		
20	C_p		4	673	121	88	61	53
	BIC		6	781	104	52	30	27
	AIC		2	577	144	104	76	97
	AIC_c		8	859	94	30	8	1
40	C_p			712	107	73	66	42
	BIC			904	56	20	15	5
	AIC			673	114	90	69	54
	AIC_c			786	105	52	41	16

Table 8.15 Number of times models were selected using various model selection criteria in 50 repetitions using simulated normal data for each of 20 design matrices. The true model has $p = 3$.

each column of X had mean zero and unit variance. The parameter vector was $\beta = (3, 2, 1, 0, 0, 0, 0)^{\mathrm{T}}$, so the true model had three covariates, and the errors were taken to be independent standard normal variables. Then the models with the first p columns of X were fitted for $p = 1, \ldots, 7$, and the best of these was selected using AIC, AIC_c, the Bayesian criterion BIC, and C_p. This procedure was performed 50 times for each design matrix.

Table 8.15 shows the results of this experiment. For $n = 10$ and 20, AIC_c has the highest chance of selecting the true model, and moreover the models selected using it are the least dispersed because of the stronger penalty applied, at least for p comparable with n. For $n = 40$ the consistent criterion BIC is most likely to select the true model. In practice, however, the true model would rarely be among those fitted, and so AIC_c seems the best of the criteria considered, particularly when p is comparable with n. ∎

Example 8.31 (Nuclear plant data) When AIC_c is computed for the 2^{10} possible models in Example 8.29, the model chosen by backward elimination is selected, with $AIC_c = -71.24$. Two nearby models have AIC_c within 2 of the minimum, namely those without log(N) and without PT, but dropping these covariates together increases AIC_c sharply. The interpretation and overall fit are changed little by dropping them singly, so we retain them. ∎

Plots of the contributions to these criteria from individual observations can be useful in diagnosing whether particular cases strongly influence model choice.

There may be several different models whose values of AIC_c are similarly low. If a single model is needed the choice among them should if possible be based on subject-matter considerations. If there are several equally plausible models with quite different interpretations, then it is important to say so.

Figure 8.8 Distribution of the supposed pivot Z for inference on the slope parameter in a straight-line regression model, conditional on inclusion of slope in the model, for $\delta = 0, 1, \ldots, 5$ (left to right) and testing for inclusion at the 5% level. Conditional on inclusion, Z is near-pivotal only if $|\delta| \gg 0$.

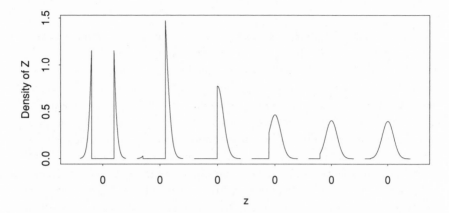

Inference after model selection

One reason that automatic variable selection should if possible be avoided is its consequence for subsequent inference. To illustrate this, consider a straight-line regression model $y = \beta_0 + x\beta_1 + \varepsilon$, based on n pairs (x_j, y_j) with $\sum x_j = 0$ and independent normal errors with mean zero and known variance σ^2. Then the least squares estimate $\widehat{\beta}_1$ is normally distributed with mean β_1 and variance $v = \sigma^2/\sum x_j$, and following the discussion in Section 8.3.2 we would base inference for β_1 on $Z = (\widehat{\beta}_1 - \beta_1)/v^{1/2}$, whose distribution is standard normal when model selection is not taken into account.

$z_{1-\alpha}$ is the $1 - \alpha$ quantile of the standard normal distribution.

Suppose, however, that before attempting to construct a confidence interval for β_1, we test for inclusion of the covariate x in the model, declaring that it should be included if $|\widehat{\beta}_1/v^{1/2}| > z_{1-\alpha}$. If not, we declare that $\beta_1 = 0$ and use the simpler model $y = \beta_0 + \varepsilon$. Now as $\widehat{\beta}_1 = \beta_1 + v^{1/2}Z$, post-model selection inference for β_1 given that x has been included will be based on the conditional density of Z given that $|Z + \beta_1/v^{1/2}| > z_{1-\alpha}$, which is

$$\phi_\delta(z) = \frac{\phi(z)\{H(z < z_\alpha - \delta) + 1 - H(z < -z_\alpha - \delta)\}}{\Phi(z_\alpha - \delta) + \Phi(z_\alpha + \delta)}, \quad -\infty < z < \infty,$$

$H(u)$ is the Heaviside function.

where $\delta = \beta_1/v^{1/2}$ is the standardized slope. Figure 8.8 displays $\phi_\delta(z)$ for $\delta = 0, 1, \ldots, 5$ and $\alpha = 0.025$, corresponding to two-sided testing at the 5% level. When $\beta_1 = 0$, for example, Z considered conditionally takes values in the tails of the standard normal distribution but not in its centre. Conditional on variable selection, Z is clearly far from pivotal unless $|\delta| \gg 0$. Hence it is only a sensible basis for inference on β_1 if the regression on x is very strong.

In practice there are three complications: the error variance σ^2 is unknown, there are typically many covariates, and the true model is not among those fitted. However the broad conclusion applies: if variables are selected automatically, the only covariates for which subsequent inference using the standard confidence intervals is reliable are those for which the evidence for inclusion is overwhelming, that is, for which it is clear that $|\delta| \gg 0$. Other covariates should be considered in the light of previous knowledge and the context of the model.

Model uncertainty

Inference is often performed after comparing different competing models, and the questions arise if, when, and how one should allow for this. Consider for example the quantity β_0 in the two models M_0 and M_1 in which $y = \beta_0 + \varepsilon$ and $y = \beta_0 + x\beta_1 + \varepsilon$, where $E(\varepsilon) = 0$. It is sometimes suggested that one should somehow average the variances of the estimators $\widehat{\beta}_0$ across the models, but this is inappropriate because the interpretation of β_0 is model-dependent. Although the same symbol is used, β_0 represents the unconditional response mean $E(Y)$ under M_0, while under M_1 it represents the conditional mean $E(Y \mid x = 0)$. Hence the meaning of β_0 depends on the context and inference for it must be conditioned on the model in which it appears: averaging is meaningless unless the quantity of interest has the same interpretation for all models considered. In particular, the interpretation of regression coefficients typically depends on the model in which they appear. Having said this, one situation in which the quantity of interest has a model-free interpretation is prediction, and below we treat the simplest example of this.

Consider using the fits of M_0 and M_1 to estimate the mean $\mu_+ = \beta_0 + x_+\beta_1$ of a future variable Y_+ with covariate $x_+ \neq 0$, assuming the error ε to be normal with mean zero and known variance σ^2; note that μ_+ has the same interpretation under both models. Suppose that n independent pairs (x_j, y_j) are available and that $\sum x_j = 0$, so that $\widehat{\beta}_0 = \bar{y}$ with variance σ^2/n under either model, independent of the slope estimate $\widehat{\beta}_1$ with variance $v = \sigma^2/\sum x_j^2$. The estimators of μ_+ and their biases, variances, and mean squared errors are

Model	Estimator	Bias	Variance	MSE
M_0:	$\widehat{\mu}_+^0 = \widehat{\beta}_0,$	$x_+\beta_1,$	$\sigma^2/n,$	$\sigma^2/n + x_+^2\beta_1^2,$
M_1:	$\widehat{\mu}_+^1 = \widehat{\beta}_0 + x_+\widehat{\beta}_1,$	$0,$	$\sigma^2/n + x_+^2 v,$	$\sigma^2/n + x_+^2 v,$

so $\widehat{\mu}_+^0$ improves on $\widehat{\mu}_+^1$ if $|\delta| < 1$, where $\delta = \beta_1/v^{1/2}$ is the standardized slope.

This suggests that it may be possible to construct a better estimator of μ_+ by choosing $\widehat{\mu}_+^0$ if an estimator of δ is close enough to zero, and otherwise taking $\widehat{\mu}_+^1$. If we decide between the models on the basis that M_1 is indicated when $|\widehat{\beta}_1|/v^{1/2} > z_{1-\alpha}$, corresponding to a two-sided test of the hypothesis that $\beta_1 = 0$ at level $(1 - 2\alpha)$, then the overall estimator is

> $I(\cdot)$ is the indicator random variable of its event.

$$\widehat{\mu}_+ = \widehat{\beta}_0 + x_+\widehat{\beta}_1 \left\{ I\left(\widehat{\beta}_1/v^{1/2} < -z_{1-\alpha}\right) + I\left(\widehat{\beta}_1/v^{1/2} > z_{1-\alpha}\right) \right\}$$
$$= \widehat{\beta}_0 + x_+ v^{1/2}(\delta + Z)\{I(Z < z_\alpha - \delta) + I(Z > z_{1-\alpha} - \delta)\},$$

where we have written $\widehat{\beta}_1 = v^{1/2}(\delta + Z)$, with $Z = (\widehat{\beta}_1 - \beta_1)/v^{1/2}$ a standard normal variable; note that $-z_{1-\alpha} = z_\alpha$. The bias and variance of $\widehat{\mu}_+$ are

$$E(\widehat{\mu}_+ - \mu_+) = x_+ v^{1/2} E(Q), \quad \text{var}(\widehat{\mu}_+) = \frac{\sigma^2}{n} + x_+^2 v \,\text{var}(Q),$$

where $Q = (\delta + Z)\{I(Z < z_\alpha - \delta) + I(Z > z_{1-\alpha} - \delta)\} - \delta$. As $v = \sigma^2/\sum x_j^2$, the bias is $O(n^{-1/2})$ and the variance is $O(n^{-1})$, while the mean squared error is $\sigma^2/n + x_+^2 v\{E(Q)^2 + \text{var}(Q)\}$. Elementary calculations give the functions $E(Q)$, var(Q), and

Figure 8.9 Properties of estimators of $\beta_0 + x_+\beta_1$ in the straight-line regression model. Left: bias (dots), variance (solid) and mean squared error (dashes) for weighted estimator $\widehat{\mu}_+^w$. Right: corresponding quantities for model-choice estimator $\widehat{\mu}_+$. The weighted estimator improves considerably on the model-choice estimator. The upper panels are for theoretical calculations, and the lower ones for the simulation experiment described in Example 8.32.

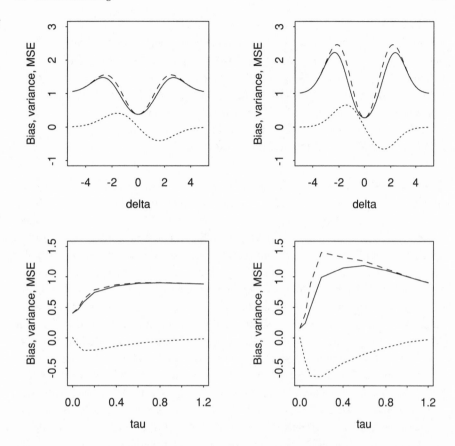

$E(Q)^2 + \text{var}(Q)$, which are shown in the upper right panel of Figure 8.9 for $\alpha = 0.025$, corresponding to choosing between the models at the two-sided 95% level. As we might have anticipated, $\widehat{\mu}_+$ is generally biased towards zero because of the possibility of using the simpler estimator $\widehat{\mu}_+^0$ even if $\beta_1 \neq 0$; its bias tends to zero when $|\delta| \gg 0$. The variance of $\widehat{\mu}_+$ is largest when $|\delta| \doteq 2$, and then decreases to the limit corresponding to use of $\widehat{\mu}_+^1$.

One difficulty with $\widehat{\mu}_+$ is that the indicator variables badly inflate its bias and variance. A simple way to avoid this is to use a weighted combination of $\widehat{\mu}_+^0$ and $\widehat{\mu}_+^1$. Take for example the estimator

$$\widehat{\mu}_+^w = (1 - W)\widehat{\mu}_+^0 + W\widehat{\mu}_+^1 = (1 - W)\widehat{\beta}_0 + W(\widehat{\beta}_0 + x_+\widehat{\beta}_1),$$

where the weight

$$W = \frac{\exp(-\text{AIC}_1/2)}{\exp(-\text{AIC}_1/2) + \exp(-\text{AIC}_0/2)}$$

depends on the information criteria AIC_0 and AIC_1 for the two models. If $\text{AIC}_1 \ll \text{AIC}_0$, then $W \doteq 1$, the data give a strong preference for M_1, and $\widehat{\mu}_+^w \doteq \widehat{\mu}_+^1$. If on the other hand $\beta_1 = 0$, then W slightly favours M_0 but the estimators under both models are unbiased.

Under our simplifying assumptions, $\mathrm{AIC}_0 - \mathrm{AIC}_1 = \widehat{\beta}_1^2/v - 2 = (\delta + Z)^2 - 2$, and as $\widehat{\mu}_+^w = \widehat{\beta}_0 + x_+ W \widehat{\beta}_1$, the quantity that corresponds to Q above is $Q^w = (\delta + Z)G\{(\delta + Z)^2/2 - 1\} - \delta$, where $G(u) = \exp(u)/\{1 + \exp(u)\}$. The bias and variance of $\widehat{\mu}_+^w$ depend on those of Q^w, which are shown in the upper left panel of Figure 8.9. Both are smaller than the values for $\widehat{\mu}_+$, and the mean squared error is considerably reduced. Evidently $\widehat{\mu}_+^w$ improves on $\widehat{\mu}_+^1$ over a wide range of values of δ, while its mean squared error is smaller than that of $\widehat{\mu}_+$. The weighted estimator $\widehat{\mu}_+^w$ clearly improves on the model-choice estimator $\widehat{\mu}_+$.

Example 8.32 (Simulation study) To assess how this approach performs in a slightly more realistic setting, we performed a small simulation study with linear model data simulated in the same way as in Example 8.30, now with $n = 15$ and $\beta^{\mathrm{T}} = \tau(0, 4, 3, 2, 1, 1, 0, 0)$; thus $p = 8$ including a constant vector. We then fitted the eight models with a constant only, constant plus the first covariate, constant plus first and second covariates, and so forth, and combined the corresponding estimators and AIC-based weights, to obtain a weighted estimator $\widehat{\theta}$ of $\theta = 1_8^{\mathrm{T}}\beta$. We compared this with the estimator $\widehat{\theta}_+$ obtained from the 'best' model, this being chosen as the model minimizing $-2\widehat{\ell}_q + 3.84q$, where $\widehat{\ell}_q$ is the log likelihood obtained when fitting the model with q parameters. This information criterion is constructed to give probability 0.05 of selecting the more complex of two nested models differing by one parameter, when in fact the simpler model is correct. This criterion is intended to mimic hypothesis testing procedures for model selection, such as backward elimination.

This experiment was repeated with 20 different response vectors for each of 250 design matrices: 5000 datasets, for $\tau = 0, 0.05, 0.1, 0.2, 0.4, \ldots, 1.2$. The lower panels of Figure 8.9 show the bias, variance, and mean squared error of $\widehat{\theta}$ and $\widehat{\theta}_+$. The results bear out the preceding toy analysis: the weighted estimator has lower mean squared error except when the regression effects are small. ∎

Although we have only considered the simplest situation, our broad conclusion generalizes to more complex settings: sharp choices among estimators from different models tends to give worse predictions than do estimators interpolating smoothly among them.

Exercises 8.7

1   Consider the cement data of Example 8.3, where $n = 13$. The residual sums of squares for all models that include an intercept are given in Exercise 8.5.1.
    (a) Use forward selection, backward elimination, and stepwise selection to select models for these data, including variables significant at the 5% level.
    (b) Use C_p to select a model for these data.

2   Another criterion for model selection is to choose the covariates that minimize the cross-validated sum of squares $\sum(y_j - x_j^{\mathrm{T}}\widehat{\beta}_{-j})^2$, where $\widehat{\beta}_{-j}$ is the estimate of β obtained when the jth case is deleted. Show this is equivalent to minimizing $\sum(y_j - x_j^{\mathrm{T}}\widehat{\beta})^2/(1 - h_{jj})^2$, and compare computational aspects of this approach with those based on AIC.

8.8 Bibliographic Notes

There are books on all aspects of the linear model. Seber (1977) and Searle (1971) give a thorough discussion of the theory, while Draper and Smith (1981), Weisberg (1985), Wetherill (1986) and Rawlings (1988) have somwhat more practical emphases; see also Sen and Srivastava (1990) and Jørgensen (1997a). Most of these books cover the central topics of this chapter in more detail. Scheffé (1959) is a classic account of the analysis of variance.

Robust approaches to regression are described by Li (1985), and in more detail in Huber (1981), Hampel *et al.* (1986), and Rousseeuw and Leroy (1987).

Davison and Hinkley (1997) and Efron and Tibshirani (1993) give accounts of bootstrap methods, which are simulation approaches to finding standard errors, confidence limits and so forth, for use with awkward estimators.

The formal analysis of transformations was discussed by Box and Cox (1964) and further developed by many others; for book-length discussions see Atkinson (1985) and Carroll and Ruppert (1988). The test for non-additivity was suggested by Tukey (1949); see also Hinkley (1985). Books on general regression diagnostics include Cook and Weisberg (1982), Belsley *et al.* (1980) and Chatterjee and Hadi (1988). Belsley (1991) focuses on problems of collinearity. Shorter accounts of aspects of model-checking are Davison and Snell (1991) and Davison and Tsai (1992). Atkinson and Riani (2000) describe how diagnostic procedures may be used to give reliable strategies for data analysis.

Stone and Brooks (1990) and their discussants give numerous references and comparison of various approaches to regression situations with fewer observations than covariates, such as principal components regression and partial least squares. Perhaps the most widespread of these is ridge regression (Hoerl and Kennard, 1970a,b; Hoerl *et al.*, 1985). Brown (1993) is a book-length treatment of these and related methods.

Variable selection for the linear model has been intensively studied. Linhart and Zucchini (1986) and Miller (1990) give useful surveys, now somewhat dated owing to the considerable amount of work in the 1990s. Model selection based on AIC was suggested by Akaike (1973) in a much-cited paper, though related criteria such as C_p were already in use (Mallows, 1973). Schwartz (1978) proposed use of BIC, and Hurvich and Tsai (1989, 1991) derive the modified AIC with improved small-sample properties. McQuarrie and Tsai (1998) give a comprehensive discussion of these and related criteria. Pötscher (1991) and Hurvich and Tsai (1990) give theoretical and numerical results on inference after model selection in linear models. More general discussion and many further references may be found in Chatfield (1995) and Burnham and Anderson (2002).

8.9 Problems

1 Consider Table 8.16. Formulate the design matrix X for the model in which E(Yield) = $\beta_i + \beta_3(z - 2)$, estimate the parameters and test whether $\beta_1 = \beta_2$.

	Level of fertilizer, z				
Variety	0	1	2	3	4
1	0.2	0.6	0.5	0.8	0.9
2	0.1	0.2	0.4	0.6	0.7

Table 8.16 Rescaled yields (tonnes/Ha) when two varieties of corn were treated with five levels of fertiliser.

2   Suppose that random variables Y_{gj}, $j = 1, \ldots, n_g$, $g = 1, \ldots, G$, are independent and that they satisfy the normal linear model $Y_{gj} = x_g^{\mathrm{T}}\beta + \varepsilon_{gj}$. Write down the covariate matrix for this model, and show that the least squares estimates can be written as $(X_1^{\mathrm{T}} W X_1)^{-1} X_1^{\mathrm{T}} W Z$, where $W = \mathrm{diag}\{n_1, \ldots, n_G\}$, and the gth element of Z is $n_g^{-1} \sum_j Y_{gj}$. Hence show that weighted least squares based on Z and unweighted least squares based on Y give the same parameter estimates and confidence intervals, when σ^2 is known. Why do they differ if σ^2 is unknown, unless $n_g \equiv 1$?
Discuss how the residuals for the two setups differ, and say which is preferable for model-checking.

3   Let $Y_1, \ldots, Y_n$ and $Z_1, \ldots, Z_m$ be two independent random samples from the $N(\mu_1, \sigma_1^2)$ and $N(\mu_2, \sigma_2^2)$ distributions respectively. Consider comparison of the model in which $\sigma_1^2 = \sigma_2^2$ and the model in which no restriction is placed on the variances, with no restriction on the means in either case. Show that the likelihood ratio statistic W_p to compare these models is large when the ratio $T = \sum(Y_j - \overline{Y})^2 / \sum(Z_j - \overline{Z})^2$ is large or small. Show that T is proportional to a random variable with the F distribution, and discuss whether the model of equal variances is plausible for the maize data of Example 1.1.

4   Find the expected information matrix for the parameters $(\beta_0, \beta_1, \sigma^2)$ of the normal straight-line regression model (5.2).

5   The usual linear model $y = X\beta + \varepsilon$ is thought to apply to a set of data, and it is assumed that the ε_j are independent with means zero and variances σ^2, so that the data are summarized in terms of the usual least squares estimates and estimate of σ^2, $\widehat{\beta}$ and S^2. Unknown to the unfortunate investigator, in fact $\mathrm{var}(\varepsilon_j) = v_j \sigma^2$, and $v_1, \ldots, v_n$ are unequal. Show that $\widehat{\beta}$ remains unbiased for β and find its actual covariance matrix.

6   Suppose that y satisfies a quadratic regression, that is,

$$y = \beta_0 + x\beta_1 + x^2\beta_2 + \varepsilon,$$

and that we can control the values of x. It is decided to choose $x = \pm a$ r times each and $x = 0$ $n - 2r$ times.
(a) Derive explicit expressions for the least squares estimates. Are they uncorrelated? If not, can they easily be made so?
(b) What value of r is best if we intend to test for the adequacy of a linear regression?
(c) What value of r is best if we intend to predict y at $x = a/2$?

7   By rewriting $y - X\beta$ as $e + X\widehat{\beta} - X\beta$ and that $e^{\mathrm{T}}X = 0$, show that

$$(y - X\beta)^{\mathrm{T}}(y - X\beta) = SS(\widehat{\beta}) + (\widehat{\beta} - \beta)^{\mathrm{T}}X^{\mathrm{T}}X(\widehat{\beta} - \beta).$$

Hence show that that the likelihood for the normal linear model equals

$$\frac{1}{(2\pi)^{n/2}\sigma^n} \exp\left\{ -\frac{SS(\widehat{\beta})}{2\sigma^2} - \frac{1}{2\sigma^2}(\widehat{\beta} - \beta)^{\mathrm{T}}X^{\mathrm{T}}X(\widehat{\beta} - \beta) \right\},$$

and use the factorization criterion to establish that $(\widehat{\beta}, SS(\widehat{\beta}))$ is a minimal sufficient statistic for (β, σ^2). The sample size n and the covariate matrix X are also needed to calculate the likelihood, so why are they not regarded as part of the minimal sufficient statistic?

8   Consider a normal linear regression $y = \beta_0 + \beta_1 x + \varepsilon$ in which the parameter of interest is $\psi = \beta_0/\beta_1$, to be estimated by $\widehat{\psi} = \widehat{\beta}_0/\widehat{\beta}_1$; let $\mathrm{var}(\widehat{\beta}_0) = \sigma^2 v_{00}$, $\mathrm{cov}(\widehat{\beta}_0, \widehat{\beta}_1) = \sigma^2 v_{01}$ and $\mathrm{var}(\widehat{\beta}_1) = \sigma^2 v_{11}$.

(a) Show that

$$\frac{\widehat{\beta}_0 - \psi\widehat{\beta}_1}{\{s^2(v_{00} - 2\psi v_{01} + \psi^2 v_{11})\}^{1/2}} \sim t_{n-p},$$

and hence deduce that a $(1 - 2\alpha)$ confidence interval for ψ is the set of values of ψ satisfying the inequality

$$\widehat{\beta}_0^2 - s^2 t_{n-p}^2(\alpha)v_{00} + 2\psi \left\{s^2 t_{n-p}^2(\alpha)v_{01} - \widehat{\beta}_0\widehat{\beta}_1\right\} + \psi^2 \left\{\widehat{\beta}_1^2 - s^2 t_{n-p}^2(\alpha)v_{11}\right\} \le 0.$$

How would this change if the value of σ was known?

(b) By considering the coefficients on the left-hand-side of the inequality in (a), show that the confidence set can be empty, a finite interval, semi-infinite intervals stretching to $\pm\infty$, the entire real line, two disjoint semi-infinite intervals — six possibilities in all. In each case illustrate how the set could arise by sketching a set of data that might have given rise to it.

(c) A government Department of Fisheries needed to estimate how many of a certain species of fish there were in the sea, in order to know whether to continue to license commercial fishing. Each year an extensive sampling exercise was based on the numbers of fish caught, and this resulted in three numbers, y, x, and a standard deviation for y, σ. A simple model of fish population dynamics suggested that $y = \beta_0 + \beta_1 x + \varepsilon$, where the errors ε are independent, and the original population size was $\psi = \beta_0/\beta_1$. To simplify the calculations, suppose that in each year σ equalled 25. If the values of y and x had been

$$\begin{array}{llllll} y: & 160 & 150 & 100 & 80 & 100 \\ x: & 140 & 170 & 200 & 230 & 260 \end{array}$$

after five years, give a 95% confidence interval for ψ. Do you find it plausible that $\sigma = 25$? If not, give an appropriate interval for ψ.

9   Over a period of $2m + 1$ years the quarterly gas consumption of a particular household may be represented by the model

$$Y_{ij} = \beta_i + \gamma j + \varepsilon_{ij}, \quad i = 1, \dots, 4, j = -m, -m + 1, \dots, m - 1, m,$$

where the parameters β_i and γ are unknown, and $\varepsilon_{ij} \overset{iid}{\sim} N(0, \sigma^2)$. Find the least squares estimators and show that they are independent with variances $(2m + 1)^{-1}\sigma^2$ and $\sigma^2/(8\sum_{i=1}^{m} i^2)$.

Show also that

$$(8m - 1)^{-1}\left[\sum_{i=1}^{4}\sum_{j=-m}^{m} Y_{ij}^2 - (2m + 1)\sum_{i=1}^{4}\overline{Y}_{i\cdot}^2 - \frac{2\sum_{j=-m}^{m} j\overline{Y}_{\cdot j}^2}{\sum_{i=1}^{m} i^2}\right]$$

is unbiased for σ^2, where $\overline{Y}_{i\cdot} = (2m + 1)^{-1}\sum_{j=-m}^{m} Y_{ij}$ and $\overline{Y}_{\cdot j} = \frac{1}{4}\sum_{i=1}^{4} Y_{ij}$.

10   A statistician travels regularly from A to B by one of four possible routes, each route crossing a river bridge at R. The times taken for the possible segments of the journey are independent random variables with means as shown in the figure, each having variance $\sigma^2/2$.

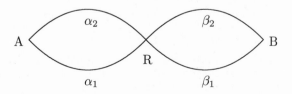

Model	SS	Model	SS	Model	SS
- - - -	11.06	1 2 - -	5.56	1 2 3 -	4.75
1 - - -	5.96	1 - 3 -	4.78	1 2 - 4	0.74
- 2 - -	10.19	1 - - 4	1.34	1 - 3 4	0.83
- - 3 -	9.96	- 2 3 -	8.09	- 2 3 4	3.05
- - - 4	9.09	- 2 - 4	7.94		
		- - 3 4	6.51	1 2 3 4	0.69

Table 8.17 Residual sums of squares for fits of linear models to output from $n = 10$ runs of a hydrological model.

He times the *complete* journey once by each route, obtaining observations y_{ij} distributed as random variables Y_{ij} having means $\mathrm{E}(Y_{ij}) = \alpha_i + \beta_j$, for $i, j = 1, 2$. Why it is not possible to estimate all the parameters from these observations?

Now define $\mu = \alpha_1 + \beta_1$, $\gamma = \alpha_2 - \alpha_1$ and $\delta = \beta_2 - \beta_1$. Obtain expressions for the least squares estimates of μ, γ and δ and also for their variance matrix.

If the observed vector of times is $(y_{11}, y_{21}, y_{12}, y_{22}) = (124, 120, 128, 136)$ minutes, determine which route has the smallest estimated mean time. Obtain a 90% confidence interval for the mean on the assumption that the times are normally distributed.

11 Suppose that we wish to construct the likelihood ratio statistic for comparison of the two linear models $y = X_1\beta_1 + \varepsilon$ and $y = X_1\beta_1 + X_2\beta_2 + \varepsilon$, where the components of ε are independent normal variables with mean zero and variance σ^2; call the corresponding residual sums of squares SS_1 and SS on ν_1 and ν degrees of freedom.

(a) Show that the maximum value of the log likelihood is $-\frac{1}{2}n(\log SS + 1 - \log n)$ for a model whose residual sum of squares is SS, and deduce that the likelihood ratio statistic for comparison of the models above is $W = n \log(SS_1/SS)$.

(b) By writing $SS_1 = SS + (SS_1 - SS)$, show that W is a monotonic function of the F statistic for comparison of the models.

(c) Show that $W \doteq (\nu_1 - \nu)F$ when n is large and ν is close to n, and say why F would usually be preferred to W.

12 Suppose that the denominator in the F statistic was replaced by $SS(\widehat{\beta}_1)/(n - q)$, giving F', say. Use the geometry of least squares to explain why F' does not have an F distribution, even if the simpler model is correct so that $SS(\widehat{\beta}_1) \sim \sigma^2\chi^2_{n-q}$. Show that F' is a monotone increasing function of F, that tends to be less than F if the simpler model is not adequate.

13 Table 8.17 gives results from $n = 10$ runs of a computer experiment to assess the accuracy of a hydrological model. The response y is the relative accuracy of predictions, and the covariates x_1, x_2, x_3, and x_4 represent parameters input to the model. The table gives the residual sums of squares for all normal linear models that include an intercept and the x_j. Taking the level of significance to be 5%, select models for the data using (a) forward selection, (b) backward elimination, (c) stepwise model selection starting from the full model, and (d) C_p. Comment briefly.

14 In the normal straight-line regression model it is thought that a power transformation of the covariate may be needed, that is, the model

$$y = \beta_0 + \beta_1 x^{(\lambda)} + \varepsilon$$

may be suitable, where $x^{(\lambda)}$ is the power transformation

$$x^{(\lambda)} = \begin{cases} \frac{x^\lambda - 1}{\lambda}, & \lambda \neq 0, \\ \log x, & \lambda = 0. \end{cases}$$

(a) Show by Taylor series expansion of $x^{(\lambda)}$ at $\lambda = 1$ that a test for power transformation can be based on the reduction in sum of squares when the constructed variable $x \log x$ is added to the model with linear predictor $\beta_0 + \beta_1 x$.

(b) Show that the profile log likelihood for λ is equivalent to $\ell_p(\lambda) \equiv -\frac{n}{2} \log SS(\widehat{\beta}_\lambda)$, where $SS(\widehat{\beta}_\lambda)$ is the residual sum of squares for regression of y on the $n \times 2$ design matrix with a column of ones and the column consisting of the $x_j^{(\lambda)}$. Why is a Jacobian for the transformation not needed in this case, unlike in Example 8.23?
(Box and Tidwell, 1962)

15 Consider model $y = X_1\beta_1 + X_2\beta_2 + \varepsilon$, which leads to least squares estimates

$$\begin{pmatrix} \widehat{\beta}_1 \\ \widehat{\beta}_2 \end{pmatrix} = \begin{pmatrix} X_1^T X_1 & X_1^T X_2 \\ X_2^T X_1 & X_2^T X_2 \end{pmatrix}^{-1} \begin{pmatrix} X_1^T y \\ X_2^T y \end{pmatrix}.$$

Let $H_1 = X_1(X_1^T X_1)^{-1} X_1^T$, $P_1 = I_n - H_1$, and define H_2 and P_2 similarly; notice that these projection matrices are symmetric and idempotent.
(a) Show that $\widehat{\beta}_2$ can be expressed as

$$\left(X_2^T P_1 X_2\right)^{-1} X_2^T y - \left(X_2^T X_2\right)^{-1} X_2^T X_1 \left(X_1^T P_2 X_1\right)^{-1} X_1^T y,$$

and use the result from Exercise 8.5.3 to deduce that $\widehat{\beta}_2 = (X_2^T P_1 X_2)^{-1} X_2^T P_1 y$, with variance matrix $\sigma^2(X_2^T P_1 X_2)^{-1}$. Note that $\widehat{\beta}_2$ is the parameter estimate from the regression of $P_1 y$ on the columns of $P_1 X_2$.
(b) Use the geometry of least squares to show that the residual sums of squares for regression of y on X_1 and X_2 is the same as for the regression of $P_1 y$ on X_1 and X_2.
(c) Suppose that in a normal linear model, X_2 is a single column that depends on y only through the fitted values from regression of y on X_1, so that X_2 is itself random. Noting that the residuals $P_1 y$ are independent of the fitted values, $H_1 y$, and arguing conditionally on $H_1 y$, show that the t statistic for $\widehat{\beta}_2$ has a distribution that is independent of X_2. Hence give the unconditional distribution of (8.27).

Recall that a model is called correct if it contains all covariates with non-zero coefficients, and called true if it contains precisely these covariates.

16 (a) Show that AIC for a normal linear model with n responses, p covariates and unknown σ^2 may be written as $n \log \widehat{\sigma}^2 + 2p$, where $\widehat{\sigma}^2 = SS_p/n$ is the maximum likelihood estimate of σ^2. If $\widehat{\sigma}_0^2$ is the unbiased estimate under some fixed correct model with q covariates, show that use of AIC is equivalent to use of $n \log\{1 + (\widehat{\sigma}^2 - \widehat{\sigma}_0^2)/\widehat{\sigma}_0^2\} + 2p$, and that this is roughly equal to $n(\widehat{\sigma}^2/\widehat{\sigma}_0^2 - 1) + 2p$. Deduce that model selection using C_p approximates that using AIC.
(b) Show that $C_p = (q - p)(F - 1) + p$, where F is the F statistic for comparison of the models with p and $q > p$ covariates, and deduce that if the model with p covariates is correct, then $E(C_p) \doteq q$, but that otherwise $E(C_p) > q$.

17 Consider the straight-line regression model $y_j = \alpha + \beta x_j + \sigma \varepsilon_j$, $j = 1, \ldots, n$. Suppose that $\sum x_j = 0$ and that the ε_j are independent with means zero, variances ε, and common density $f(\cdot)$.
(a) Write down the variance of the least squares estimate of β.
(b) Show that if σ is known, the log likelihood for the data is

$$\ell(\alpha, \beta) = -n \log \sigma + \sum_{j=1}^n \log f\left(\frac{y_j - \alpha - \beta x_j}{\sigma}\right),$$

derive the expected information matrix for α and β, and show that the asymptotic variance of the maximum likelihood estimate of β can be written as $\sigma^2/(i \sum x_j^2)$, where

$$i = E\left\{-\frac{d^2 \log f(\varepsilon)}{d\varepsilon^2}\right\}.$$

Hence show that the the least squares estimate of β has asymptotic relative efficiency $i/v \times 100\%$.

With $\Gamma(t) = \int_0^\infty u^{t-1} e^{-u} \, du$, $\Gamma''(1) - \Gamma'(1)^2 \doteq 1.64493$.

(c) Show that the cumulant-generating function of the Gumbel distribution, $f(u) = \exp\{-u - \exp(-u)\}$, $-\infty < u < \infty$, is $\log \Gamma(1 - t)$, and deduce that its variance is roughly 1.65. Find i for this distribution, and show that the asymptotic relative efficiency of least squares is about 61%.

18   Over a period of 90 days a study was carried out on 1500 women. Its purpose was to
      investigate the relation between obstetrical practices and the time spent in the delivery
      suite by women giving birth. One thing that greatly affects this time is whether or not a
      woman has previously given birth. Unfortunately this vital information was lost, giving
      the researchers three options: (a) abandon the study; (b) go back to the medical records
      and find which women had previously given birth (very time-consuming); or (c) for each
      day check how many women had previously given birth (relatively quick). The statistical
      question arising was whether (c) would recover enough information about the parameter
      of interest.

      Suppose that a linear model is appropriate for log time in delivery suite, and that the
      log time for a first delivery is normally distributed with mean $\mu + \alpha$ and variance σ^2,
      whereas for subsequent deliveries the mean time is μ. Suppose that the times for all
      the women are independent, and that for each there is a probability π that the labour
      is her first, independent of the others. Further suppose that the women are divided into
      k groups corresponding to days and that each group has size m; the overall number is
      $n = mk$. Under (c), show that the average log time on day j, Z_j, is normally distributed
      with mean $\mu + R_j\alpha/m$ and variance σ^2/m, where R_j is binomial with probability π and
      denominator m. Hence show that the overall log likelihood is

      $$\ell(\mu, \alpha) = -\frac{1}{2}k \log(2\pi\sigma^2/m) - \frac{m}{2\sigma^2} \sum_{j=1}^{k}(z_j - \mu - r_j\alpha/m)^2,$$

      where z_j and r_j are the observed values of Z_j and R_j and we take π and σ^2 to be
      known. If R_j has mean $m\pi$ and variance $m\tau^2$, show that the inverse expected information
      matrix is

      $$I(\mu, \alpha)^{-1} = \frac{\sigma^2}{n\tau^2} \begin{pmatrix} m\pi^2 + \tau^2 & -m\pi \\ -m\pi & m \end{pmatrix}.$$

      (i) If $m = 1$, $\tau^2 = \pi(1 - \pi)$, and $\pi = n_1/n$, where $n = n_0 + n_1$, show that $I(\mu, \alpha)^{-1}$
      equals the variance matrix for the two-sample regression model. Explain why.
      (ii) If $\tau^2 = 0$, show that neither μ nor α is estimable; explain why.
      (iii) If $\tau^2 = \pi(1 - \pi)$, show that μ is not estimable when $\pi = 1$, and that α is not estimable
      when $\pi = 0$ or $\pi = 1$. Explain why the conditions for these two parameters to be estimable
      differ in form.
      (iv) Show that the effect of grouping, $(m > 1)$, is that $\text{var}(\widehat{\alpha})$ is increased by a factor m
      regardless of π and σ^2.
      (v) It was known that $\sigma^2 \doteq 0.2, m \doteq 1500/90, \pi \doteq 0.3$. Calculate the standard error for $\widehat{\alpha}$.
      It was known from other studies that first deliveries are typically 20–25% longer than
      subsequent ones. Show that an effect of size $\alpha = \log(1.25)$ would be very likely to be
      detected based on the grouped data, but that an effect of size $\alpha = \log(1.20)$ would be less
      certain to be detected, and discuss the implications.

19   Suppose that model $y = X\beta + Z\gamma + \varepsilon$ holds, but that model $y = X\beta + \varepsilon$ is fitted, giving
      $\widehat{\beta} = (X^{\mathsf{T}}X)^{-1}X^{\mathsf{T}}y$ with hat matrix $H = X(X^{\mathsf{T}}X)^{-1}X^{\mathsf{T}}$ and residuals $e = y - X\widehat{\beta}$.
      (a) Show that

      $$e = (I - H)y = (I - H)Z\gamma + (I - H)\varepsilon,$$

      and hence that $\text{E}(e) = (I - H)Z\gamma$. What happens if Z lies in the space spanned by the
      columns of X?
      (b) Now suppose that Z is a single column z. Explain how an *added variable plot* of the
      residuals from the regression of y on X against the residuals from the regression of z on
      X can help in deciding whether or not to add z to the design matrix.
      (c) Discuss the interpretation of the added variable plots in Figure 8.10, bearing in mind
      the possibility of outliers and of a need to transform z before including it in the design
      matrix.

Figure 8.10 Added variable plots for four normal linear models.

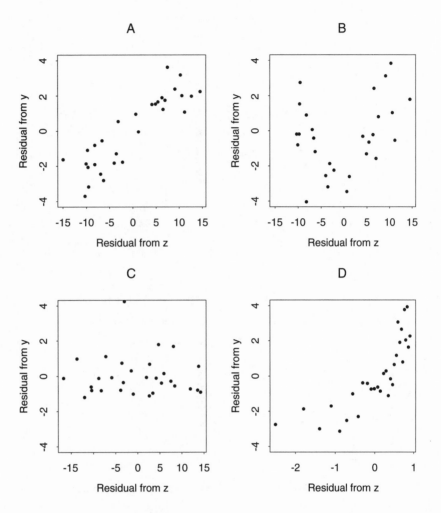

20 Figure 8.11 shows standardized residuals plotted against fitted values for linear models fitted to four different sets of data. In each case discuss the fit and explain briefly how you would try to remedy any deficiencies.

21 Data $(x_1, y_1), \ldots, (x_n, y_n)$ satisfy the straight-line regression model (5.3). In a *calibration* problem the value y_+ of a new response independent of the existing data has been observed, and inference is required for the unknown corresponding value x_+ of x.
(a) Let $s_x^2 = \sum(x_j - \overline{x})^2$ and let S^2 be the unbiased estimator of the error variance σ^2. Show that

$$T(x_+) = \frac{Y_+ - \widehat{\gamma}_0 - \widehat{\gamma}_1(x_+ - \overline{x})}{\left[S^2\left\{1 + n^{-1} + (x_+ - \overline{x})^2/s_x^2\right\}\right]^{1/2}}$$

is a pivot, and explain why the set

$$\mathcal{X}_{1-2\alpha} = \{x_+ : t_{n-2}(\alpha) \leq T(x_+) \leq t_{n-2}(1 - \alpha)\}$$

contains x_+ with probability $1 - 2\alpha$.
(b) Show that the function $g(u) = (a + bu)/(c + u^2)^{1/2}$, $c > 0$, $a, b \neq 0$, has exactly one stationary point, at $\tilde{u} = -bc/a$, that sign $g(\tilde{u}) = $ sign a, that $g(\tilde{u})$ is a local maximum if

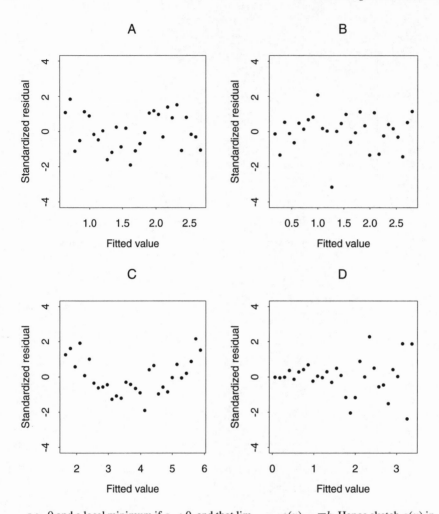

Figure 8.11
Standardized residuals
plotted against fitted
values for four normal
linear models.

$a > 0$ and a local minimum if $a < 0$, and that $\lim_{u \to \pm\infty} g(u) = \mp b$. Hence sketch $g(u)$ in the four possible cases $a, b < 0$, $a, b > 0$, $a < 0 < b$ and $b < 0 < a$.

(c) By setting $u = S(x_+ - \overline{x})/s_x$, show that $T(x_+)$ can be written in form $g(u)$. Deduce that $\mathcal{X}_{1-2\alpha}$ can be a finite interval, two semi-infinite intervals or the entire real line. Discuss.

(d) Show that if in fact $\gamma_1 = 0$, $\mathcal{X}_{1-2\alpha}$ has infinite length with probability $1 - 2\alpha$.

(e) A different approach considers x_+ to be an unknown parameter, and constructs the likelihood for β, σ^2 and x_+ based on the pairs (x_j, y_j) and y_+. Does the resulting profile log likelihood $\ell_p(x_+)$ result in confidence sets such as those in (c)?

9

Designed Experiments

A carefully planned investigation can give much more insight into the question at hand than a haphazard one, data from which may be useless. Experimental design is a highly developed subject, though its principles are not universally appreciated. In this chapter we outline some basic ideas and describe some simple designs and associated analyses. The first section discusses the importance of randomization, and shows how it can be used to justify standard linear models and how it strengthens inferences. Section 9.2 then describes some common designs and analyses. Interaction, contrasts and analysis of covariance are discussed in Section 9.3. Section 9.4 then outlines the consequences of having more than one level of variability.

9.1 Randomization

9.1.1 Randomization

The purpose of a designed experiment is to compare how *treatments* affect a response, by applying them to experimental *units*, on each of which the response is to be measured. The units are the raw material of the investigation; formally a unit is the smallest subdivision of this such that any two different units might receive different treatments. The treatments are clearly defined procedures one of which is to be applied to each experimental unit. In an agricultural field trial the treatments might be different amounts of nitrogen and potash, while a unit is a plot of land. In a medical setting, treatments might be types of operation and different therapies, with units being patients who are operated upon and then given therapy to aid recovery. In each case our concern is how the response depends on the treatment combinations and other measurable quantities. The response must be carefully defined and measured in a consistent way for every unit.

Suppose for illustration that we wish to assess the effect of a drug in reducing blood pressure, and that $n = 2m$ individuals are available. We plan to administer the drug to m of the individuals, the treatment group, and to give a placebo to the remaining m,

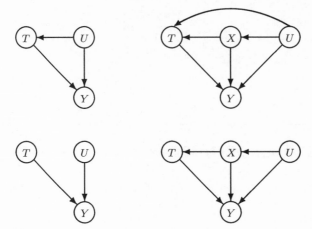

Figure 9.1 Directed acyclic graphs showing consequences of randomization. An arrow from T to Y indicates dependence of Y on T, and so forth. In general both response Y and treatment T may depend on properties U of units (upper left). Randomization (lower left) makes treatments and units independent, so any observed dependence of Y on T cannot be ascribed to joint dependence on U. The upper right graph shows the general dependence of Y, T, and covariates X on U. Randomization makes T and U independent, conditional on X (lower right), so any influence of U on T is mediated through X, for which adjustment is possible in principle. Thus having adjusted for X, dependence of Y on T cannot be due to U.

the control group. The response is to be the blood pressure of an individual measured a fixed time after the drug has first been administered. We calculate the average changes for the treated and control groups, $\overline{y}_1$ and $\overline{y}_0$, observe that $\overline{y}_1 - \overline{y}_0$ is significantly less than zero, and declare that the drug plays an effect in reducing blood pressure. Is this headline news? No!

A key difficulty is that the procedure does not avoid biased allocation of treatments to units. For example, if the control group mostly consisted of those patients with higher blood pressures at the start of the study, $\overline{y}_1$ and $\overline{y}_0$ might differ greatly even if the treatment had been ineffective. This particular source of bias could be avoided if the experimenter measured the initial blood pressures and deliberately balanced the groups with respect to them, but unknown causes of bias could not be removed in this way, and the interpretation of the results would rely on the uncheckable assertion that the experiment was also balanced with respect to these unknown factors. Any deterministic allocation scheme will have this flaw, and we turn instead to *randomization*. By allocating treatments to patients at random, we expect to equalize the effect of any factors that might affect the response, other than the treatment itself. We can then be surer that a significant difference between the groups is related to the treatment itself.

To explain randomization differently, let T represent the treatment, Y the response, and U properties of units — potential sources of bias. For example, left to their own devices physicians might be tempted to allocate a promising but untested new treatment to patients most severely affected by a disease, and an existing treatment to less severe cases. Then treatment T would depend on an attribute of the units, disease severity U; the response Y might depend on both T and U. This is shown by the directed acyclic graph in the upper left part of Figure 9.1. In general both T and Y depend on U, so any apparent relation between Y and T may be ascribed to U. Randomization induces independence between properties of the unit and any treatment allocation, making T independent of U and the lower left graph appropriate: although U may influence the response Y, it cannot entirely explain any dependence on T unless the randomization is compromised, for example by allocating all men to one group and all women to the other purely by chance. If this has not happened,

Allocation at random means that some physical device has been used, not that the experimenter has made a choice that appears haphazard.

then a highly significant effect of T implies either that treatment works, or that a rare event has occurred.

If randomization had been used and if a normal linear model was suitable, inference could be based on the two-sample model of Example 8.9, using

$$z = \frac{\overline{y}_1 - \overline{y}_0}{(2s^2/m)^{1/2}}, \tag{9.1}$$

where $s^2 = (2m - 2)^{-1} \sum_{t,j}(y_{tj} - \overline{y}_t)^2$ is the pooled estimate of error and y_{tj} is the response for the jth individual in treatment group t. In fact randomization gives a basis for the use of this and other linear models, as we shall see below.

Blocking

The design outlined above presupposes that the units are fairly homogeneous, that is, any variation among blood pressures of different patients is small enough for the design to be *completely randomized*. However, if the treatment effect was small relative to this variation, s^2 would be inflated because the division into groups made no allowance for it. The larger is s^2, the smaller is z for given $\overline{y}_1 - \overline{y}_0$, and this makes it harder to detect any treatment effect. This suggests that we should subdivide the patients into groups whose initial blood pressures are as alike as possible, and allocate the treatment randomly within these groups, a procedure known as *blocking*. As the purpose of our experiment is to compare one treatment with the control, we divide the patients into m blocks of two individuals with similar initial blood pressures, and randomly allocate one of each pair to the treatment and the other to the control, in a *paired comparison*. In the corresponding normal linear model, discussed in Example 8.10, analysis is based on the differences d_j between the treated and control individuals in the jth block, leading to confidence statements using the standardized difference given by

$$z_d = \frac{\overline{d}}{\left(s_d^2/m\right)^{1/2}}, \quad s_d^2 = (m - 1)^{-1} \sum_{j=1}^{m}(d_j - \overline{d})^2, \tag{9.2}$$

where $\overline{d} = \overline{y}_1 - \overline{y}_0$ is the average difference between pairs. The numerator of z_d is the same as that of (9.1), but the denominator may be substantially smaller if the blocking has been effective in increasing the precision of the experiment. Although here the matching is performed deliberately, randomization is still involved in the treatment allocations.

This line of reasoning suggests taking as response for each patient the difference between his initial blood pressure and that after treatment, so the comparisons are made entirely within individuals, allocated randomly to treatment or control. We ignore this design below, however, purely for purposes of exposition.

The right half of Figure 9.1 shows the effect of randomization when treatment allocation can depend on a covariate, X. For example, randomization might take into account knowledge that certain treatments should not be given to patients taking other medication. In general T might depend on unknown properties U of the unit as well as on X, so that Y and T depend on both X and U. Randomization breaks the direct

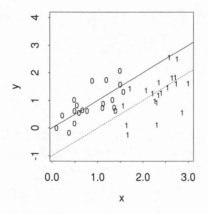 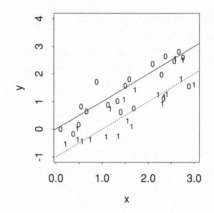

Figure 9.2 Simulated results from experiments to compare the effect of a treatment T on a response Y that varies with a covariate X. The lines show the mean response for $T = 0$ (solid) and $T = 1$ (dots). Left: the effect of T is confounded with dependence on X. Right: the experiment is balanced, with random allocation of T dependent on X.

link between U and T, so any effect of X on T is mediated through the observed X, for which adjustment is in principle possible.

To illustrate this, Figure 9.2 shows results from two simulated attempts to assess the effect of a treatment T on a response Y. Unnoticed by the virtual experimenter who obtained the data in the left panel, the mean of Y increases with a covariate X, as shown by the lines. However because all the units for which $T = 1$ also have the largest values of X, there appears to be no difference between the treatment group averages. The true treatment effect is $\delta = -1$, but the observed difference of averages is 0.2 with standard error 0.2. The 0.95 confidence interval $(-0.2, 0.6)$ does not include the true δ because of *confounding* between the effects of X and T. In practice such serious confounding would be most likely to arise due to lack of randomization, but lack of balance could occur by accident even if the treatments had been allocated at random. If so, randomization would fail to remove all possible biases due to confounders such as X.

A cannier experimenter might have formed pairs of units using values of X measured before the experiment and then randomized the treatment within pairs, leading to results like those in the right panel, where the difference of averages is -1.2 with standard error 0.3; the 0.95 confidence interval now contains δ.

In both cases the observed values of X can be used to obtain more precise estimates of δ, by fitting the model $y = \beta_0 + \beta_1 x + \delta t + \varepsilon$ to the observed triples (x, t, y), where $t = 0$ or 1. The left panel has $\widehat{\delta} = -0.7$ with standard error 0.3 and correlation $\mathrm{corr}(\widehat{\delta}, \widehat{\beta}_1) = -0.82$, while the right has $\widehat{\delta} = -1.25$, standard error 0.16 and $\mathrm{corr}(\widehat{\delta}, \widehat{\beta}_1) = -0.04$. One effect of the blocking has been to reduce the confounding of T and X by making the corresponding columns of the design matrix almost orthogonal; their parameters can then be estimated without ambiguity. There is a relation here to the discussion of collinearity in Section 8.7.2.

Although regression on x reduces the confounding between X and T in the first experiment, the lack of overlap in the values of X for the two treatment groups means that the model must be used to interpolate between them. This makes the estimate less precise and the inference less secure: an act of faith in the linearity of the model is needed, because neither of the groups has X values over the entire range.

The second experiment gives similar estimates of δ with or without adjustment for x, though the precision of $\widehat{\delta}$ is increased by making the adjustment, known as analysis of covariance; see Section 9.3.3. Moreover the data can be used to check whether the treatment effect is constant over X.

Randomization inference

In this chapter we shall assume that normal linear models are applicable. In fact the act of randomization provides a basis for inference without appealing to specific parametric assumptions, but for which the normal model often provides a good approximation. Suppose that m observations have been randomly allocated to a treatment and a further m to a control. Suppose also that *unit-treatment additivity* holds, that is there exist constants $\gamma_1, \ldots, \gamma_{2m}$, one for each unit, and δ for the treatment, such that the response on the jth unit is $\gamma_j + \delta$ when it is allocated to the treatment, and $\gamma_j - \delta$ if it is allocated to the control group, regardless of the allocation of treatments to the other units. Thus the effect of treatment is to increase the response by $\Delta = 2\delta$ relative to the control, for each unit in the experiment. Under this model the responses from the jth unit when it is allocated to treatment and to control are

$$T_j(\gamma_j + \delta), \quad (1 - T_j)(\gamma_j - \delta),$$

where T_j is an indicator of whether it has been allocated to the treatment. Therefore the difference between treatment and control averages is

$$\overline{Y}_1 - \overline{Y}_0 = \frac{1}{m} \sum_{j=1}^{2m} T_j(\gamma_j + \delta) - \frac{1}{m} \sum_{j=1}^{2m} (1 - T_j)(\gamma_j - \delta) = 2\delta + \frac{1}{m} \sum_{j=1}^{2m} (2T_j - 1)\gamma_j.$$

The properties of $\overline{Y}_1 - \overline{Y}_0$ stem from the moments of $T_1, \ldots, T_{2m}$,

$$\mathrm{E}(T_j) = \frac{1}{2}, \quad \mathrm{E}(T_j T_k) = \frac{m-1}{2(2m-1)}, \quad j \neq k. \tag{9.3}$$

Thus $\overline{Y}_1 - \overline{Y}_0$ has mean Δ and variance $2\{m(2m-1)\}^{-1} \sum_{j=1}^{2m} (\gamma_j - \overline{\gamma})^2$. Moreover the strong symmetry induced by the T_j, allied to the weak dependence among them, means that the randomization distribution of $\overline{Y}_1 - \overline{Y}_0$ is close to normal.

Example 9.1 (Shoe data) Table 9.1 shows the amount of wear in a paired comparison of materials A and B used to sole shoes. Material B is cheaper and the aim of the experiment was to see if it was less durable than A. Ten boys were chosen, material A allocated at random to one of their shoes, and material B to the other. All but two of the differences d_j are positive, suggesting that shoes soled with B wear more quickly than those with A. The average difference is $\overline{d} = 0.41$.

Suppose that there was no difference between the materials. Then A and B would simply be labels attached randomly to the shoes, and each difference might equally well have had the opposite sign. That is, each of the $2^{10} = 1024$ outcomes $\pm 0.8, \pm 0.6, \ldots, \pm 0.3$ would have been equally as likely as that actually observed. Thus the average difference $\overline{d}$ would be the observed value of $\overline{D} = m^{-1} \sum_j D_j$, where $D_j = I_j d_j$, and $I_1, \ldots, I_m$ are independent variables taking values ± 1 with

| | Material | | Difference |
Boy	A	B	d
1	13.2 (L)	14.0 (R)	0.8
2	8.2 (L)	8.8 (R)	0.6
3	10.9 (R)	11.2 (L)	0.3
4	14.3 (L)	14.2 (R)	−0.1
5	10.7 (R)	11.8 (L)	1.1
6	6.6 (L)	6.4 (R)	−0.2
7	9.5 (L)	9.8 (R)	0.3
8	10.8 (L)	11.3 (R)	0.5
9	8.8 (R)	9.3 (L)	0.5
10	13.3 (L)	13.6 (R)	0.3

Table 9.1 Shoe wear data (Box *et al.*, 1978, p. 100). The table shows the amount of shoe wear in an paired comparison experiment in which two materials A and B were randomly assigned to the soles of the left (L) or right (R) shoe of each of ten boys.

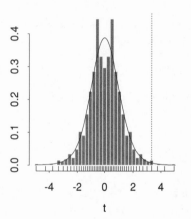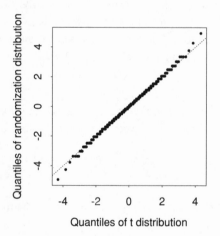

Figure 9.3 Randomization distribution of the *t* statistic for the shoes data, together with its approximating t_9 distribution. The left panel shows a histogram and rug for the randomized values of Z, with the t_9 density overlaid; the observed value is given by the vertical dotted line. The right panel shows a probability plot of the randomization distribution against t_9 quantiles.

probability $\frac{1}{2}$; here $m = 10$. In fact there are precisely three values of $\overline{D}$ that are larger than $\bar{d}$, and four values equal to it, so the exact P-value based on $\overline{D}$ is $7/1024 \doteq 0.007$.

The studentized version of $\overline{D}$, $Z = \overline{D}/[\{m(m-1)\}^{-1} \sum (D_j - \overline{D})^2]^{1/2}$, is a monotonic function of $\overline{D}$, so both Z and $\overline{D}$ give the same P-values under randomization. Figure 9.3 shows the randomization distribution of Z, with the t distribution on $m - 1 = 9$ degrees of freedom that would be used under a normal model. The agreement between the randomization distribution and the normal approximation is excellent. The observed value of Z is 3.35, with significance level 0.004 when compared to the t_9 distribution.

The pairing in this experiment could have been used to extend the validity of the results, by taking boys of different ages, with different types of shoes and so forth. As the comparisons are based only on differences between feet of the same boy, that is, within blocks, the heterogeneity of the boys themselves does not affect the comparison of A and B. If the same difference between materials was seen on a wide variety of blocks, one could be more confident that the difference in durability was general. As previously mentioned, blocking is used to ensure a generalizable result by taking blocks that are heterogeneous, while eliminating block effects by ensuring that treatment comparisons are made within blocks. ■

Although described above only in the simplest cases, the normal linear model provides approximations to randomization distributions in other settings also. Below we continue to talk of normal errors, with the understanding that these often generate approximations to randomization distributions.

9.1.2 Causal inference

In many investigations the key question is causal. Does passive smoking cause lung cancer? Does exposure to air pollution increases levels of asthma? Does applying treatment T to a unit increase its response Y by amount δ? The extensive philosophical discussion of causality is largely irrelevant here, because of its focus on deterministic relations between cause and effect. The best we can usually hope for is statements such as 'if applied to a large sample of units, T would give an average increase δ, compared with what would have been observed had they remained untreated'. This translates into probability statements for individual units.

It is important to appreciate that potential causes are aspects of units that could in principle be manipulated in the context in question. In a study of the effects of lifestyle on longevity, we can conceive of altering individuals' dietary and exercise habits, for example, but not their genders. We can imagine comparing the survival of flabby burger-loving Mr Jones with his survival as fit or vegetarian or both, but not with that of Mr Jones as female rather than male; were he a woman, he would not be Mr Jones. Here diet and exercise are potential causes, but gender is not. Intrinsic attributes of units cannot be regarded as potential causes, because to speak of a causal effect of T on Y, intervention to change the value of T must be possible.

Three types of causal statement are as follows. First, and strongest, there may be a well-understood evidence-based mechanism or set of mechanisms — biological, physical or whatever — that links a cause to its effect. This is the usual meaning of causality in so-called hard science, even though knowledge about the mechanism is invariably subject to improvement.

Second, and much weaker, is the observation that two phenomena are linked by a stable association, whose direction is established and which cannot be explained by mutual dependence on some other allowable variable. In Example 6.18, for example, ignoring age induced an apparently positive association between survival and smoking, whose direction was reversed once age was taken into account. To see this differently, consider a population of units on each of which (T, Y) may be observed, and let the association between T and Y be measured by

$$\gamma = \mathrm{E}(Y \mid T = 1) - \mathrm{E}(Y \mid T = 0) > 0,$$

say. This can be estimated by the difference in averages $\overline{Y}_1 - \overline{Y}_0$ for samples with $T = 1$ and $T = 0$. To say that the association cannot be explained away amounts to asserting that no confounding variable X exists for which

$$\gamma(x) = E(Y \mid T = 1, X = x) - E(Y \mid T = 0, X = x) \equiv 0.$$

In practice this will need to be bolstered by careful study design, often considering together studies that account for different possible confounders. The restriction to allowable variables means amongst other things that X cannot itself be a response to T.

A third interpretation of causality, intermediate between the first two, is related to experimentation and relies on the notion of a *counterfactual*. Consider a unit, and let R_0 and R_1 represent its responses on setting $T = 0$ and $T = 1$; these three variables are assumed to have a joint distribution. If in fact $T = 0$, then R_0 is observed and R_1 is counterfactual; it is the response that would have been observed had the treatment been different. Conversely R_0 is counterfactual if $T = 1$. The central difficulty of causal inference is that it is impossible to compare values of R_0 and R_1 from the same unit. Thus an assumption of homogeneity of treatment effects over units, that is, unit-treatment additivity, is essential. If unit-treatment additivity holds then the effect of T is measured by the difference of mean responses

$$\delta = E(R_1) - E(R_0),$$

but unlike γ this is not observable. In general $\delta \neq \gamma$, but if the treatment allocation is randomized, then T is independent of any property of the unit, and the consistency equation $Y = R_0(1 - T) + R_1 T$ relating the counterfactuals to the response Y entails

$$\begin{aligned}
\delta = E(R_1) - E(R_0) &= E(R_1 \mid T = 1) - E(R_0 \mid T = 0) \\
&= E(Y \mid T = 1) - E(Y \mid T = 0) = \gamma.
\end{aligned}$$

> The assumption need not apply on the original scale; it might apply to a transformed response, in which case the argument below is applied on the transformed scale.

Hence unit-treatment additivity and randomization ensure that the quantity δ we want to estimate equals γ, which we can estimate.

This argument presumes there to be no relation between treatment allocation and any property of the unit, and this is typically true only in completely randomized experiments. Suppose however that (R_1, R_0) and T are independent conditional on the value of another variable X, as in the lower right of Figure 9.1. Then

$$E(R_1) = E_X\{E(R_1 \mid X)\} = E_X\{E(R_1 \mid X, T = 1)\} = E_X\{E(Y \mid X, T = 1)\},$$

and with a parallel argument for $E(R_0)$ we have

$$\delta = E(R_1) - E(R_0) = E_X\{E(Y \mid X, T = 1) - E(Y \mid X, T = 0)\} = \gamma,$$

say. The observable effect γ, a function of the joint distribution of Y, X, and T, is now averaged over the possible values of X for the unit. The interpretation of γ and the case for a causal effect are both strengthened if in fact

$$E(Y \mid X, T = 1) - E(Y \mid X, T = 0) = \gamma(X) \equiv \gamma,$$

that is, association with T does not depend on X, as in Figure 9.2. Otherwise there is *interaction* between T and Y; see Section 9.3.1.

The use of randomization to eliminate confounding variables is a powerful tool, but it is not sufficient for causal inference. An obvious counter-example is the left panel of Figure 9.2, where it would be foolhardy to talk of a causal effect of T on Y even if the appropriate linear model had been fitted, because the observed triples (x, t, y) give no way to assess whether confounding between X and T is present despite randomization.

Even if experimentation has established that T changes the distribution of Y, it seems rash to assert causality with no idea of an underlying mechanism. In practice a combination of evidence from physical mechanisms, direct experiment, and large-scale observational data will be most compelling.

Exercises 9.1

1 (a) Show that under the two-sample model, the difference of the sample averages, $\bar{y}_2 - \bar{y}_1$, has variance $(n_1 + n_2)\sigma^2/(n_1 n_2)$. Show that subject to $n_1 + n_2 = n$, this is minimized when n_1 and n_2 are as nearly equal as possible.
(b) Suppose that n units are split into k blocks of size $m + 1$, and that one unit in each block is chosen at random to be treated, while the remaining m are controls. Suppose that the responses in the jth block are y_{j1} and $y_{j2}, \ldots, y_{j(m+1)}$, and let d_j represent the difference between the treated individual and the average of the controls. Show that the average of these differences has variance $(m + 1)\sigma^2/(km)$, and show that for fixed n this is minimized when $m = 1$.

2 Suppose a paired comparison experiment is performed, in which the jth pair satisfies the normal linear model

$$y_{0j} = \mu_j - \delta + \varepsilon_{0j}, \quad y_{1j} = \mu_j + \delta + \varepsilon_{1j}, \quad j = 1, \ldots, m,$$

but that data analysis is performed using the two-sample model. Show that the variance estimator can be written as

$$S^2 = \frac{1}{2(m-1)} \sum_{j,t} (\mu_j - \bar{\mu} + \varepsilon_{tj} - \bar{\varepsilon}_{..})^2.$$

Deduce that this has expected value $\sigma^2 + (m-1)^{-1} \sum_j (\mu_j - \bar{\mu}_.)^2$ conditional on the μ_j, and hence show that if the μ_j are normally distributed with variance τ^2, then $E(S^2) = \sigma^2 + \tau^2$.
Show that if the two-sample model is used in this situation, the length of a 95% confidence interval for 2δ is roughly $2(\sigma^2 + \tau^2)^{1/2} t_{2(m-1)}(0.025)$, whereas under the paired comparisons model the length is about $2\sigma t_{m-1}(0.025)$. For what values of τ^2/σ^2 are the two-sample intervals shorter when (a) $m = 3$, (b) $m = 11$? Discuss your results.

3 Check (9.3), find $\text{var}(T_j)$ and $\text{cov}(T_j, T_k)$ and hence verify the given formulae for the mean and variance of $\bar{Y}_1 - \bar{Y}_0$.

4 In Example 9.1, show that Z is a monotonic function of $\bar{D}$.

5 To what extent can gender be regarded as a cause in studies (a) relating longevity and lifestyle and (b) of salary differentials in employment?

6 Let $T = 0$ with probability $1 - \alpha$ and $T = 1$ otherwise, and suppose that conditional on $T = 0$, R_0 is normal with mean zero and R_1 is normal with mean δ, while conditional on $T = 1$, the corresponding means are η and $\eta + \delta$; in each case the variables have unit variances. Let $Y = R_0(1 - T) + R_1 T$ denote the observed response variable. Show that $\gamma = E(Y \mid T = 1) - E(Y \mid T = 0) = \eta + \delta$, and deduce that $\delta = E(R_1) - E(R_0)$ cannot be estimated unless (R_0, R_1) and T are independent.

Term	df	Sum of squares	Mean square
Groups	$T-1$	$\sum_{t,r}(\overline{y}_{t.} - \overline{y}_{..})^2$	$(T-1)^{-1}\sum_t(\overline{y}_{t.} - \overline{y}_{..})^2$
Residual	$T(R-1)$	$\sum_{t,r}(y_{tr} - \overline{y}_{t.})^2$	$\{T(R-1)\}^{-1}\sum_{t,r}(y_{tr} - \overline{y}_{t.})^2$

Table 9.2 Analysis of variance table for one-way layout.

9.2 Some Standard Designs

9.2.1 One-way layout

If more than two treatments are to be compared and the population is relatively homogeneous, the two-group model may be extended to a completely randomized design, known as a one-way layout. Henceforth we let T denote the number of treatments in the model under consideration.

Suppose that we wish to compare the effects of T treatments and that we have available $n = RT$ units. We divide the units at random into T groups each of size R, and apply a single treatment to all the units in each group. The corresponding linear model is

$$y_{tr} = \beta_t + \varepsilon_{tr}, \quad t = 1, \ldots, T, \quad r = 1, \ldots, R, \tag{9.4}$$

where $\varepsilon_{tr} \overset{\text{iid}}{\sim} N(0, \sigma^2)$. This assumes that the only effect of the treatment is to alter the mean response, as would be the case under a randomization distribution. Thus the observations within each group are random samples, but the groups may have different means. This explains the term one-way layout: laid out as a $T \times R$ array, only the treatment index is meaningful. In matrix terms this model is

$$
\begin{pmatrix} y_{11} \\ \vdots \\ y_{1R} \\ y_{21} \\ \vdots \\ y_{2R} \\ \vdots \\ y_{T1} \\ \vdots \\ y_{TR} \end{pmatrix}
=
\begin{pmatrix}
1 & 0 & \cdots & 0 \\
\vdots & \vdots & & \vdots \\
1 & 0 & \cdots & 0 \\
0 & 1 & \cdots & 0 \\
\vdots & \vdots & & \vdots \\
0 & 1 & \cdots & 0 \\
\vdots & \vdots & & \vdots \\
0 & 0 & \cdots & 1 \\
\vdots & \vdots & & \vdots \\
0 & 0 & \cdots & 1
\end{pmatrix}
\begin{pmatrix} \beta_1 \\ \beta_2 \\ \vdots \\ \beta_T \end{pmatrix}
+
\begin{pmatrix} \varepsilon_{11} \\ \vdots \\ \varepsilon_{1R} \\ \varepsilon_{21} \\ \vdots \\ \varepsilon_{2R} \\ \vdots \\ \varepsilon_{T1} \\ \vdots \\ \varepsilon_{TR} \end{pmatrix}.
\tag{9.5}
$$

This design matrix has full rank T and the least squares estimator of β_t it yields is the average for the tth group, $\overline{y}_{t.} = R^{-1}\sum_r y_{tr}$. If the β_t are all equal, corresponding to the model $y = 1_n\beta_0 + \varepsilon$ in our general notation, the fitted value for the entire set of data is the overall average $\overline{y}_{..}$. The sum of squares then decomposes as

$$\sum_{t,r}(y_{tr} - \overline{y}_{..})^2 = \sum_{t,r}(y_{tr} - \overline{y}_{t.})^2 + \sum_{t,r}(\overline{y}_{t.} - \overline{y}_{..})^2,$$

corresponding to (8.23), and the analysis of variance is shown in Table 9.2.

Here and below replacement of a subscript by a dot indicates averaging over the values of that subscript.

Table 9.3 Data on the teaching of arithmetic.

Group			Test result y							Average	Variance
A (Usual)	17	14	24	20	24	23	16	15	24	19.67	17.75
B (Usual)	21	23	13	19	13	19	20	21	16	18.33	12.75
C (Praised)	28	30	29	24	27	30	28	28	23	27.44	6.03
D (Reproved)	19	28	26	26	19	24	24	23	22	23.44	9.53
E (Ignored)	21	14	13	19	15	15	10	18	20	16.11	13.11

The unbiased estimator of σ^2 when (9.4) is fitted is

$$S^2 = \frac{1}{n-p}(y - \widehat{y})^{\mathrm{T}}(y - \widehat{y}) = \frac{1}{T(R-1)} \sum_{t,r}(y_{tr} - \overline{y}_{t.})^2,$$

with $T(R-1)$ degrees of freedom. If $R = 1$ it is impossible to estimate σ^2, for then there is only one observation with which to estimate β_t, and $y_{tr} \equiv \overline{y}_{t.}$. Thus *replication* of the responses for each treatment is essential unless an external estimate of σ^2 is available, for example from another experiment. A further benefit of replication is the capacity to check model assumptions, as we shall see in Examples 9.2 and 9.6.

The F statistic for assessing significance of differences among treatments,

$$F = \frac{(T-1)^{-1}\sum_t(\overline{y}_{t.} - \overline{y}_{..})^2}{S^2} \quad \sim \quad F_{T-1,T(R-1)},$$

when $\beta_1 = \cdots = \beta_T$. In applications interest generally focuses on estimation of particular differences among the β_t, however, rather than on testing for overall differences, this being merely an initial screening device.

Another possible linear model for the data is

$$y_{tr} = \alpha + \gamma_t + \varepsilon_{tr}, \quad t = 1, \ldots, T, \quad r = 1, \ldots, R,$$

in which the overall mean is represented by α, and γ_t represents the difference between the mean for treatment t and the overall mean. The design matrix for this model has $T + 1$ columns, namely the T columns of the matrix in (9.5) and a column of ones, and has rank T: the $T + 1$ parameters cannot be estimated from T groups. Although the T linear combinations $\alpha + \gamma_1, \ldots, \alpha + \gamma_T$ corresponding to the group means are estimable, the $T + 1$ parameters $\alpha, \gamma_1, \ldots, \gamma_T$ are not.

Example 9.2 (Teaching methods data) In an investigation on the teaching of arithmetic, 45 pupils were divided at random into five groups of nine. Groups A and B were taught in separate classes by the usual method. Groups C, D, and E were taught together for a number of days. On each day C were praised publicly for their work, D were publicly reproved and E were ignored. At the end of the period all pupils took a standard test, with the results given in Table 9.3 and displayed in the left panel of Figure 9.4. Groups A and B seem to have performed similarly, but the other groups have responded differently to the regimes imposed, as we see from the averages and variances in the final columns of the table. If the only differences among groups were in their means, the group variances could be expected to be independently distributed

Term	df	Sum of squares	Mean square	F
Groups	4	722.67	180.67	15.3
Residual	40	473.33	11.83	

Table 9.4 Analysis of variance for data on the teaching of arithmetic.

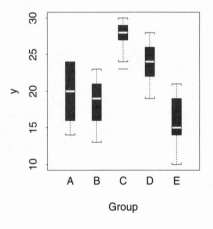 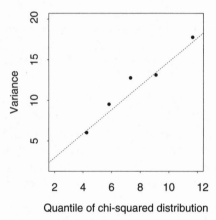

Figure 9.4 Data on teaching of arithmetic. The left panel shows the original data, and the right panel shows the ordered variances for each group plotted against plotting positions for the χ_8^2 distribution.

as $\sigma^2 \chi_8^2 / 8$. No doubt is cast on this by the corresponding probability plot, shown in the right panel of Figure 9.4; this is only available because of the replication within each group.

The analysis of variance, shown in Table 9.4, shows very strong evidence of differences among the groups, as we would expect from inspecting the data. The corresponding F statistic is 15.3, to be considered as $F_{4,40}$ under the hypothesis of no group differences, in which case the significance level is zero.

As a group average is an average of $R = 9$ observations, its variance is σ^2/R, and consequently the estimated variance for the difference between the averages for groups A and B, $\overline{y}_A - \overline{y}_B = 1.33$, is $2s^2/9 = 2.63$. The corresponding t statistic, $1.33/2.63^{1/2}$, shows no evidence of differences between the control groups, and the pooled estimate of the mean using the usual teaching method, β_U, is accordingly $\overline{y}_U = \frac{1}{2}(19.67 + 18.33) = 19$, with estimated variance $s^2/18$.

Comparisons of the usual and other methods are of interest here, and they are based on statistics such as $\overline{Y}_C - \overline{Y}_U$, each having estimated variance $s^2/18 + s^2/9 = 1.97$. Confidence intervals for the underlying differences are based on the quantities $\{\overline{Y}_C - \overline{Y}_U - (\beta_C - \beta_U)\}/\{S^2/18 + S^2/9\}^{1/2}$, each having a t_{40} distribution. Thus 95% confidence intervals are $(5.7, 11.2)$ for $\beta_C - \beta_U$, $(1.7, 7.2)$ for $\beta_D - \beta_U$, and $(-5.6, -0.14)$ for $\beta_E - \beta_U$. Giving approval and reproval improves test performance relative to the usual method, with approval working best, while ignoring pupils decreases their test scores, though by less. These conclusions are necessarily highly tentative because of the very limited scale of the experiment.   ∎

9.2.2 Randomized block design

Suppose that T treatments are to be compared, and that $n = TB$ units are available. The analogue of the paired comparisons experiment when there are more than two treatments is the *randomized block design*. The units are divided into B blocks of T units so that similar units are so far as possible in the same block. The T treatments are then applied randomly to the units, each treatment appearing precisely once in each block. A simple linear model here is that the response of the unit in block b given treatment t is

$$y_{tb} = \mu + \alpha_t + \beta_b + \varepsilon_{tb}, \quad t = 1, \ldots, T, b = 1, \ldots, B, \qquad (9.6)$$

where the ε_{tb} are a random sample of $N(0, \sigma^2)$ variables. This is the *two-way layout* model, so-called because the y_{tb} can be laid out as an array with T rows and B columns, with α_t the treatment effect for the tth row and β_b the block effect for the bth column; see Table 9.6. With $T = 4$ and $B = 3$ for definiteness, and with parameter vector $(\mu, \alpha_1, \alpha_2, \alpha_3, \alpha_4, \beta_1, \beta_2, \beta_3)^{\mathsf{T}}$, the 12×8 design matrix

$$X = \begin{pmatrix} 1 & 1 & 0 & 0 & 0 & 1 & 0 & 0 \\ 1 & 1 & 0 & 0 & 0 & 0 & 1 & 0 \\ 1 & 1 & 0 & 0 & 0 & 0 & 0 & 1 \\ 1 & 0 & 1 & 0 & 0 & 1 & 0 & 0 \\ 1 & 0 & 1 & 0 & 0 & 0 & 1 & 0 \\ 1 & 0 & 1 & 0 & 0 & 0 & 0 & 1 \\ 1 & 0 & 0 & 1 & 0 & 1 & 0 & 0 \\ 1 & 0 & 0 & 1 & 0 & 0 & 1 & 0 \\ 1 & 0 & 0 & 1 & 0 & 0 & 0 & 1 \\ 1 & 0 & 0 & 0 & 1 & 1 & 0 & 0 \\ 1 & 0 & 0 & 0 & 1 & 0 & 1 & 0 \\ 1 & 0 & 0 & 0 & 1 & 0 & 0 & 1 \end{pmatrix}$$

has rank $1 + (T - 1) + (B - 1) = 6$: all eight parameters cannot be estimated. The terms corresponding to the treatment and block effects are columns 2–5 and 6–8 of this matrix respectively. Dropping the second and sixth columns of X is equivalent to setting $\alpha_1 = \beta_1 = 0$, in which case α_t and β_b represent the mean differences in response between treatment t and treatment 1 and between block b and block 1. In this, the *corner-point* parametrization, μ is the mean response of the unit in block 1 given treatment 1, that is, in the top left corner of the two-way layout. The least squares estimates in this parametrization can be obtained from the usual formula, but their derivation is unenlightening.

Instead, let us use the original parametrization with the least squares estimates constrained so that $\sum_r \widehat{\alpha}_r = \sum_c \widehat{\beta}_c = 0$. These are constraints on the estimates, not on the parameters: nature is free to use as many parameters as she likes, but only certain linear combinations of them are estimable. These two linear restrictions ensure that our fitted model is not overparametrized, and we can use symmetry to avoid inverting a rank-deficient matrix $X^{\mathsf{T}} X$. We use Lagrange multipliers to find the values of $\widehat{\mu}, \widehat{\alpha}_t$,

$\widehat{\beta}_b$, η and ζ that minimize

$$\sum_{t,b}(y_{tb} - \widehat{\mu} - \widehat{\alpha}_t - \widehat{\beta}_b)^2 + \eta\left(\sum_t \widehat{\alpha}_t - 0\right) + \zeta\left(\sum_b \widehat{\beta}_b - 0\right).$$

On differentiating, we see that we should solve the equations

$$0 = \sum_{t,b}(y_{tb} - \widehat{\mu} - \widehat{\alpha}_t - \widehat{\beta}_b),$$

$$0 = \sum_b(y_{tb} - \widehat{\mu} - \widehat{\alpha}_t - \widehat{\beta}_b) - \eta, \quad t = 1, \ldots, T,$$

$$0 = \sum_t(y_{tb} - \widehat{\mu} - \widehat{\alpha}_t - \widehat{\beta}_b) - \zeta, \quad b = 1, \ldots, B,$$

$$0 = \sum_t \widehat{\alpha}_t, \qquad 0 = \sum_b \widehat{\beta}_b,$$

giving $\widehat{\mu} = \overline{y}_{..}$, $\widehat{\alpha}_t = \overline{y}_{t.} - \overline{y}_{..}$, and $\widehat{\beta}_b = \overline{y}_{.b} - \overline{y}_{..}$, where as before we use a dot in a subscript to indicate averaging over the corresponding index. Thus we have $\overline{y}_{..} = (TB)^{-1}\sum_{t,b} y_{tb}$, $\overline{y}_{t.} = B^{-1}\sum_b y_{tb}$, and so forth.

The fitted values are $\widehat{\mu} + \widehat{\alpha}_t + \widehat{\beta}_b = \overline{y}_{.b} + \overline{y}_{t.} - \overline{y}_{..}$, and hence the residual sum of squares is $\sum_{t,b}(y_{tb} - \overline{y}_{t.} - \overline{y}_{.b} + \overline{y}_{..})^2$; these would be the same in the corner-point parametrization, because the same subspace is spanned by the columns of the design matrix in both cases, and the fitted values — though not the parameter estimates — depend only on the column space of the design matrix; recall Figure 8.2. If the $\widehat{\beta}_b$ were not in the model, $\widehat{\mu}$ and $\widehat{\alpha}_t$ would remain the same, and the fitted values would be $\widehat{\mu} + \widehat{\alpha}_t = \overline{y}_{t.}$.

The mean difference in response between treatments r and s is estimated by the difference $\widehat{\alpha}_r - \widehat{\alpha}_s = \overline{y}_{r.} - \overline{y}_{s.}$. If we write this estimate in terms of the underlying parameters, by replacing y_{rb} by $\mu + \alpha_r + \beta_b + \varepsilon_{rb}$ and so forth, we obtain

$$B^{-1}(B\mu + B\alpha_r + \beta_1 + \cdots + \beta_B + \varepsilon_{r1} + \cdots + \varepsilon_{rB})$$
$$-B^{-1}(B\mu + B\alpha_s + \beta_1 + \cdots + \beta_B + \varepsilon_{s1} + \cdots + \varepsilon_{sB}),$$

which equals $\alpha_r - \alpha_s + \overline{\varepsilon}_{r.} - \overline{\varepsilon}_{s.}$, and this is independent of the block effects, which appeared equally often in $\overline{y}_{r.}$ and $\overline{y}_{s.}$. Thus because the design is balanced, comparisons among treatments are essentially made within blocks, and this increases precision if there are substantial block effects.

The difference between an observation and the overall average equals

$$y_{tb} - \overline{y}_{..} = (y_{tb} - \overline{y}_{t.} - \overline{y}_{.b} + \overline{y}_{..}) + (\overline{y}_{t.} - \overline{y}_{..}) + (\overline{y}_{.b} - \overline{y}_{..}),$$

and because $\sum_{t,b}(y_{tb} - \overline{y}_{..})(\overline{y}_{t.} - \overline{y}_{..}) = 0$, $\sum_{t,b}(\overline{y}_{t.} - \overline{y}_{..})(\overline{y}_{.b} - \overline{y}_{..}) = 0$, and so forth, we see that

$$\sum_{t,b}(y_{tb} - \overline{y}_{..})^2 = \sum_{t,b}(y_{tb} - \overline{y}_{t.} - \overline{y}_{.b} + \overline{y}_{..})^2 + \sum_{t,b}(\overline{y}_{t.} - \overline{y}_{..})^2 + \sum_{t,b}(\overline{y}_{.b} - \overline{y}_{..})^2,$$

which echoes (8.23). If we had set $\widehat{\beta}_b \equiv 0$, the corresponding sum of squares

Table 9.5 Analysis of variance table for two-way layout model.

Term	df	Sum of squares
Treatments	$T-1$	$\sum_{t,b}(\overline{y}_{t\cdot} - \overline{y}_{\cdot\cdot})^2$
Blocks	$B-1$	$\sum_{t,b}(\overline{y}_{\cdot b} - \overline{y}_{\cdot\cdot})^2$
Residual	$(T-1)(B-1)$	$\sum_{t,b}(y_{tb} - \overline{y}_{t\cdot} - \overline{y}_{\cdot b} + \overline{y}_{\cdot\cdot})^2$

Table 9.6 Data on weight gains in pigs.

Diet	Group 1	2	3	4	5	6	7	8	Average
I	1.40	1.79	1.72	1.47	1.26	1.28	1.34	1.55	1.48
II	1.31	1.30	1.21	1.08	1.45	0.95	1.26	1.14	1.21
III	1.40	1.47	1.37	1.15	1.22	1.48	1.31	1.27	1.33
IV	1.96	1.77	1.62	1.76	1.88	1.50	1.60	1.49	1.70
Average	1.52	1.58	1.48	1.37	1.45	1.30	1.38	1.36	1.43

decomposition would have been

$$\sum_{t,b}(y_{tb} - \overline{y}_{\cdot\cdot})^2 = \sum_{t,b}(y_{tb} - \overline{y}_{t\cdot})^2 + \sum_{t,b}(\overline{y}_{t\cdot} - \overline{y}_{\cdot\cdot})^2,$$

and it follows by symmetry that once the constant term has been fitted, the reductions in sum of squares due to treatment and block terms are respectively $\sum_{t,b}(\overline{y}_{t\cdot} - \overline{y}_{\cdot\cdot})^2$ and $\sum_{t,b}(\overline{y}_{\cdot b} - \overline{y}_{\cdot\cdot})^2$. As $\sum_{t,b}(\overline{y}_{t\cdot} - \overline{y}_{\cdot\cdot})(\overline{y}_{\cdot b} - \overline{y}_{\cdot\cdot}) = 0$, these sums of squares are independent if the errors are normal.

The analysis of variance table for a two-way layout with T rows and B columns is in Table 9.5. The residual degrees of freedom are

$$TB - 1 - (T-1) - (B-1) = (T-1)(B-1),$$

and the sums of squares are independent of the order in which terms are fitted.

Example 9.3 (Pig diet data) Twelve pigs were divided into eight groups of four, in such a way that the pigs in any one group were expected to gain weight at equal rates if fed in the same way. Four diets were compared by randomly assigning them to pigs, subject to each diet occurring once in each group. The average daily weight gains of the pigs are given in Table 9.6. The diet averages suggest that pigs on diet IV gain more weight than the others, and that any differences between II and III are small. Differences among the groups are less marked.

The analysis of variance in Table 9.7 shows strong differences among diets, but little effect of blocking into groups. The estimate of σ^2 is $s^2 = 0.024$. The diet averages are 1.48, 1.21, 1.33, and 1.70, and as the standard error for a difference of two of them is $(2s^2/8)^{1/2} = 0.077$, it is clear that diet IV leads to the fastest weight gain, with diet I second and better than diet III; it is less clear that II is worse than III. ∎

Term	df	Sum of squares	Mean square	F statistic
Diet	3	1.042	0.347	14.6
Group	7	0.247	0.035	1.48
Residual	21	0.500	0.024	

Table 9.7 Analysis of variance table for two-way layout model applied to the data of Table 9.6.

Embryo	Treat	y	Treat	y	Embryo	Treat	y	Treat	y
1	—	2.51	His–	2.15	9	His–	2.32	Lys–	2.53
2	—	2.49	Arg–	2.23	10	Arg–	2.15	Thr–	2.23
3	—	2.54	Thr–	2.26	11	Arg–	2.34	Val–	2.15
4	—	2.58	Val–	2.15	12	Arg–	2.30	Lys–	2.49
5	—	2.65	Lys–	2.41	13	Thr–	2.20	Val–	2.18
6	His–	2.11	Arg–	1.90	14	Thr–	2.26	Lys–	2.43
7	His–	2.28	Thr–	2.11	15	Val–	2.28	Lys–	2.56
8	His–	2.15	Val–	1.70					

Table 9.8 $\log_{10}$ dry weight y (μg) of chick bones after cultivation over a nutrient chemical medium, either complete (—), or with single amino acids missing (Cox and Snell, 1981, p. 95). The order of treatment pairs was randomized, but the table shows them systematically.

Balanced incomplete block design

Sometimes variation among units is large enough for blocking to be required, but a randomized block design cannot be used because the block size is smaller than the number of treatments, T. In such circumstances it may be possible to use a *balanced incomplete block design*. Suppose that there are B blocks each with K units, and that $R = BK/T$ is an integer. In the simplest such design, each treatment appears exactly once in a block, and each pair of treatments appears together λ times, in which case $R(K - 1) = (T - 1)\lambda$.

Example 9.4 (Chick bone data) Table 9.8 gives data on the growth of chick bones. Bones from 7-day-old chick embryos were cultivated over a nutrient chemical medium. Two bones were available from each chick, and the experiment was set out in a balanced incomplete block design with two units per block. The treatments were growth in the complete medium, with about 30 nutrients in carefully controlled quantities, and growth in five other media, each with a single amino acid omitted. Thus His–, Arg–, and so forth denote media without particular amino acids. This balanced incomplete block design has $T = 6$, $B = 15$, $K = 2$, $R = 5$, and $\lambda = 1$.

One way to proceed here is to let β_H, β_A, ... denote the effect of the absence of His, Arg, ..., and then regard the first pair of responses as having means $\mu_1, \mu_1 + \beta_H$, the sixth as having means $\mu_6 + \beta_H$, $\mu_6 + \beta_A$, and so forth. We then perform a linear regression with response the differences of the responses for each of the embryos, parameter vector $(\beta_H, \beta_A, \beta_T, \beta_V, \beta_L)^{\mathsf{T}}$, and a 15×5 design matrix whose first and sixth rows are $(1, 0, 0, 0, 0)$ and $(-1, 1, 0, 0, 0)$. This avoids the necessity to estimate $\mu_1, \ldots, \mu_{15}$, but gives the same estimates of the βs, shown in the first line of Table 9.9. The estimate of error variance is $s^2 = 0.013$. The initial sum of squares of 1.024 reduces to 0.132, giving overall F statistic 13.6 on 5 and 10 degrees of freedom: a

Table 9.9 Parameter estimates and standard errors for intra-block, inter-block and pooled analyses of chick data.

| Analysis | Amino acid | | | | | SE |
	His	Arg	Thy	Val	Lys	
Intra-block $\widehat{\beta}$	−0.22	−0.35	−0.35	−0.49	−0.16	0.066
Inter-block $\tilde{\beta}$	−0.55	−0.40	−0.33	−0.42	0.07	0.124
Pooled β^*	−0.29	−0.36	−0.34	−0.47	−0.11	0.058

highly significant reduction. Lack of each amino acid reduces growth; Lys has the smallest effect, but even this has the large t value of $-0.16/0.066 = -2.42$ on 10 degrees of freedom.

In this regression the terms for individual amino acids are not orthogonal, and the analysis of variance is not unique. For example, if acids are fitted in the order His, Arg, Thr, Val, Lys, the reductions in sum of squares are 0.014, 0.052, 0.078, 0.67, 0.077, while the reductions for the order Lys, Thr, Val, His, Arg are 0.074, 0.033, 0.40, 0.01, 0.38. Here balance gives equal precision for estimation of each of the βs, rather than orthogonal sums of squares.

Though simple, this so-called *intra-block* analysis uses a degree of freedom to estimate each block parameter μ_j, and if these vary little then information may be lost. To outline how *inter-block* analysis can retrieve this, we denote the responses from the jth block as y_{j1}, y_{j2} and treat these as independent normal variables with means $\mu_j + x_{j1}^T\beta$, $\mu_j + x_{j2}^T\beta$ and variances σ^2. The previous analysis was based on $y_{j1} - y_{j2}$, and this is independent of the block sum $y_{j1} + y_{j2}$. The inter-block analysis treats the μ_j as random variables with mean μ, say, and variance σ_μ^2, so the block sums have variance $2(2\sigma_\mu^2 + \sigma^2)$, perhaps much larger than the variance $2\sigma^2$ of the differences. We then fit to the 15 block sums a linear model with means $\mu 1_{15} + X\beta$, the jth row of X being $x_{j1}^T + x_{j2}^T$, thereby obtaining estimates $\tilde{\beta}$ of the amino acid effects. These estimates are independent of those obtained from the intrablock analysis; their values are given in Table 9.9, along with the standard error inflated by σ_μ^2. Both sets of estimates are unbiased, so approximate minimum variance unbiased estimates are formed as a weighted combination $\beta^* = w\widehat{\beta} + (1-w)\tilde{\beta}$, where $w = \widehat{v}^{-1}/(\widehat{v}^{-1} + \tilde{v}^{-1})$, $\widehat{v}$ and $\tilde{v}$ being the estimated variances for the intra- and inter-block estimates. As $w = 0.78$, these pooled estimates are close to the $\widehat{\beta}$, with a slightly smaller standard error. This standard error combines independent standard errors from the intra- and inter-block analyses and has approximately 13.8 degrees of freedom; see Exercise 9.2.3.

The response is $\log_{10}$ dry weight, so the effect of eliminating an amino acid is multiplicative on the original scale. A 0.95 confidence interval for the median effect of eliminating His is $10^{\widehat{\beta}_H \pm 2.15 \times 0.058} = (0.38, 0.68)$, with the estimate $10^{\widehat{\beta}_H} = 0.51$ corresponding to a 50% reduction in growth. See Practical 9.1. ∎

$t_{13.8}(0.975) = 2.15$

There are many generalizations of incomplete block designs. Their key purpose is to give good precision for estimation of the effects of interest — usually treatment effects, or perhaps a subset of them — when there are more treatments than blocks.

Higher degrees of balance are possible, in which all triples of treatments appear equally often, and sometimes constraints on the numbers of units lead to the use of partially balanced designs, which give increased precision on treatments of primary interest while sacrificing precision on those of less importance.

9.2.3 Latin square

When there are two possible blocking factors, a three-way layout could be used. In many circumstances, however, a design that requires fewer units is required, and one possibility may be a *Latin square*. Suppose that the blocking factors and the treatment have the the same number of levels, q. Then a Latin square design is constructed by laying out units in a $q \times q$ array with the blocking factors corresponding to rows and columns, and applying each treatment precisely once in each row and in each column. An example is shown in the upper left part of Table 9.11. This balanced application of treatments leads to an orthogonal decomposition of the total sum of squares. Many such layouts are possible, with randomization by choice of design and permutation of row, column and treatment labels.

The corresponding linear model treats the response in the rth row and cth column as $y_{rc} = \mu + \alpha_r + \beta_c + \gamma_{t(r,c)} + \varepsilon_{rc}$, where $t(r, c)$ is the treatment applied to that unit, and $\varepsilon_{rc} \overset{\text{iid}}{\sim} N(0, \sigma^2)$. As it stands this model contains $1 + 3q$ parameters but the design matrix would have rank $1 + 3(q - 1)$.

Least squares estimates may be obtained by extending the Lagrange multiplier argument on page 429. We minimize $\sum_{r,c}(y_{rc} - \widehat{\mu} - \widehat{\alpha}_r - \widehat{\beta}_c - \widehat{\gamma}_{t(r,c)})^2$ subject to the constraints $\sum_r \widehat{\alpha}_r = \sum_c \widehat{\beta}_c = \sum_t \widehat{\gamma}_t = 0$. This yields

$$\widehat{\mu} = \overline{y}, \quad \widehat{\alpha}_r = \overline{y}_{r.} - \overline{y}_{..}, \quad \widehat{\beta}_c = \overline{y}_{.c} - \overline{y}_{..}, \quad \widehat{\gamma}_t = \overline{y}_t - \overline{y}_{..},$$

with residual sum of squares $\sum_{r,c}(y_{rc} - \overline{y}_{r.} - \overline{y}_{.c} - \overline{y}_t + 2\overline{y}_{..})^2$.

To see this another way, suppose that we had ignored the treatment classification in the Latin square, and obtained the analysis of variance table for the row and column classifications. These would be the same as in Table 9.5, though the residual sum of squares would also contain the variation due to treatments. However, we could rewrite the table so that treatments appeared as the row classification, in which case the current row classification would take on the role of treatments and appear inside the table. The two-way analysis of variance table for the rearranged data, ignoring the new treatments (old rows), would contain sums of squares due to treatments and to columns, and its residual sum of squares would also contain the sum of squares for rows. Since the sums of squares for both two-way analyses are orthogonal, the analysis of variance table for a $q \times q$ Latin square must be as shown in Table 9.10.

Example 9.5 (Field concrete mixer data) A field concrete mixer lays down a concrete road surface while moving forward. Its efficiency is measured by the hardness of the surface it produces, as a percentage of the corresponding hardness produced under laboratory conditions. It is thought that efficiency may fall off as the speed at which the machine moves increases, and trials were performed to investigate this. On each of four days, the machine was run at four different speeds, 4, 8, 12, and

Table 9.10 Analysis of variance table for a Latin square.

Term	df	Sum of squares
Rows	$q-1$	$\sum_{r,c}(\bar{y}_{r\cdot} - \bar{y}_{\cdot\cdot})^2$
Columns	$q-1$	$\sum_{r,c}(\bar{y}_{\cdot c} - \bar{y}_{\cdot\cdot})^2$
Treatments	$q-1$	$\sum_{r,c}(\bar{y}_{t(r,c)} - \bar{y}_{\cdot\cdot})^2$
Residual	$(q-1)(q-2)$	$\sum_{r,c}(y_{rc} - \bar{y}_{r\cdot} - \bar{y}_{\cdot c} - \bar{y}_{t(r,c)} + 2\bar{y}_{\cdot\cdot})^2$

Table 9.11 Field concrete mixer data. Latin square experiment, showing application of treatments — speed in miles per hour (left) — and observed responses — machine efficiency (%) (right) — for 16 combinations of day and run. Below are average efficiencies for day, run, and speed.

	Run					Run			
Day	1	2	3	4	Day	1	2	3	4
1	8	16	4	12	1	64.2	59.8	66.2	63.6
2	16	12	8	4	2	47.5	57.3	67.7	58.6
3	4	8	12	16	3	54.2	59.9	57.1	54.1
4	12	4	16	8	4	60.1	68.4	58.7	63.7

Day	Average	Run	Average	Speed	Average
1	63.45	1	56.50	4	61.85
2	57.78	2	61.35	8	63.88
3	56.33	3	62.43	12	59.53
4	62.73	4	60.00	16	55.03

Figure 9.5 Field concrete mixer data. Left panel: efficiencies as a function of speed, with plotting symbol giving the day. Right panel: average efficiencies, with fitted quadratic curve corresponding to day 1 and run 1.

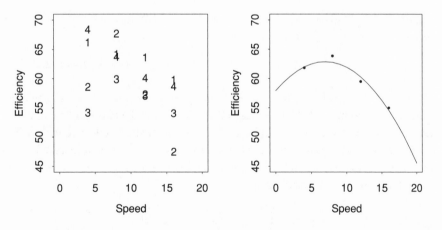

16 miles per hour, these being taken in a different order on each day. The layout in Table 9.11 gives the speeds and the response, machine efficiency. There are two blocking factors, day and run, and a quantitative treatment with four levels, speed. The averages, also in Table 9.11, show large differences among days and speeds, and smaller ones among runs, while they and the left panel of Figure 9.5 show a systematic variation of efficiency with speed.

Term	df	Sum of squares	Mean square	F statistic
Days	3	151.06	50.35	5.78
Runs	3	79.74	26.58	3.05
Speeds	3	173.58	57.86	6.64
Residual	6	52.23	8.71	

Table 9.12 Analysis of variance for Latin square fitted to field concrete mixer data.

The analysis of variance in Table 9.12 shows evidence of day and speed differences, with significance levels for their F statistics respectively 0.03 and 0.02, and weak evidence of differences among runs, at significance level 0.11.

The estimated mean response for the tth treatment is $\widehat{\mu} + \widehat{\gamma}_t = \overline{y}_t$, and the average efficiencies for the speeds are 61.85, 63.88, 59.53, and 55.03%, suggesting that the best speed is 8 mph or so. A 95% confidence interval for the true efficiency at 8 mph is $63.88 \pm t_6(0.025)(s^2/4)^{1/2}$, and since $s^2 = 8.71$ and $t_6(0.025) = -2.45$, this interval is $(60.26, 67.50)\%$. We return to these data in Example 9.12. ∎

9.2.4 Factorial design

A factorial design involves a number of treatments, each with several levels, and every combination of levels of the different factors appears together. Thus if there are two factors with two levels and one with three levels, there are $2 \times 2 \times 3 = 12$ possible treatment combinations, each of which is applied to at least one unit. A 2^3 factorial design was used in Example 8.4.

Example 9.6 (Poisons data) The data in Table 8.10 are from a 3×4 factorial experiment with four replicates. The model

$$y_{tpj} = \mu + \alpha_t + \beta_p + \gamma_{tp} + \varepsilon_{tpj}, \quad t = 1, 2, 3, 4, \ p = 1, 2, 3, \ j = 1, 2, 3, 4,$$
(9.7)

modifies (8.28) by adding terms γ_{tp} representing the *interaction* of poisons and treatments; see Section 9.3.1. If the γ_{tp} are all equal, (9.7) is the two-way layout model (9.6), except that there are four replicates at each combination of the two factors. The model (9.7) has 20 parameters, which evidently cannot be estimated separately from the 12 groups of times available. If we use Lagrange multipliers to minimize the sum of squares subject to the constraints

$$\sum_t \widehat{\alpha}_t = \sum_p \widehat{\beta}_p = \sum_t \widehat{\gamma}_{tp} = \sum_p \widehat{\gamma}_{tp} = 0,$$

then we find

$$\widehat{\mu} = \overline{y}_{...}, \ \widehat{\alpha}_t = \overline{y}_{t..} - \overline{y}_{...}, \ \widehat{\beta}_p = \overline{y}_{.p.} - \overline{y}_{...}, \ \widehat{\gamma}_{tp} = \overline{y}_{tp.} - \overline{y}_{t..} - \overline{y}_{.p.} + \overline{y}_{tp.},$$

with corresponding orthogonal decomposition

$$y_{tpj} - \overline{y}_{...} = (y_{tpj} - \overline{y}_{tp.}) + (\overline{y}_{tp.} - \overline{y}_{t..} - \overline{y}_{.p.} + \overline{y}_{tp.}) + (\overline{y}_{t..} - \overline{y}_{...}) + (\overline{y}_{.p.} - \overline{y}_{...}).$$

Table 9.13 Sums of squares for two-way layout with replication, assuming T rows, P columns, and J replicates in each cell. For the poisons data, $T = 4$, $P = 3$, and $J = 4$.

Term	df	Sum of squares
Rows	$T - 1$	$\sum_{t,p,j}(\bar{y}_{t\cdot\cdot} - \bar{y}_{\cdots})^2$
Columns	$P - 1$	$\sum_{t,p,j}(\bar{y}_{\cdot p\cdot} - \bar{y}_{\cdots})^2$
Rows×Columns	$(T - 1)(P - 1)$	$\sum_{t,p,j}(\bar{y}_{tp\cdot} - \bar{y}_{t\cdot\cdot} - \bar{y}_{\cdot p\cdot} + \bar{y}_{tp\cdot})^2$
Residual	$TP(J - 1)$	$\sum_{t,p,j}(y_{tpj} - \bar{y}_{tp\cdot})^2$

Table 9.14 Analyses of variance for the poisons data, with responses y and y^{-1}. For MS and F read 'Mean square' and 'F statistic'.

Term	df	Response y			Response y^{-1}		
		SS	MS	F	SS	MS	F
Poisons	2	1.033	0.517	23.22	34.88	17.44	72.63
Treatments	3	0.921	0.307	13.81	20.41	6.80	28.34
Treatments × Poisons	6	0.250	0.042	1.87	1.57	0.26	1.09
Residual	36	0.801	0.022		8.64	0.24	

The sums of squares and their degrees of freedom are given in Table 9.13. Notice that if $J = 1$, the residual sum of squares is zero, because $y_{tpj} \equiv \bar{y}_{tp\cdot}$, and the analysis of variance reduces to that in Table 9.5, with the interaction sum of squares used to estimate the error variance σ^2. If it was known *a priori* that an interaction was likely to be present, replication would be essential rather than merely desirable, and if replication was for some reason impossible, an external estimate of σ^2 would be required.

The left part of Table 9.14 shows the analysis of variance. There are the expected strong effects of poisons and treatments, but the interaction is less important, with a significance level of about 0.11 when treated as an $F_{6,36}$ variable. The estimate of σ^2 is $s^2 = 0.022$.

Under the model (9.7) the fitted values for each cell are $\bar{y}_{tp\cdot}$. This suggests a check on the adequacy of the model. The four observations within each combination of poison and treatment should form a random sample from the normal distribution with mean $\mu + \alpha_t + \beta_p + \gamma_{tp}$ and variance σ^2, and therefore their sample variance s^2_{tp} should have the $\sigma^2\chi^2_3/3$ distribution if the error assumption is correct. The lower left panel of Figure 8.5 shows a systematic departure from linearity, suggesting that the assumption of normal errors with constant variance is untenable. The lower right panel shows that the inverse survival times follow the normal error model more closely, suggesting that it is better to replace y_{tpj} with y_{tpj}^{-1}. The corresponding analysis of variance, shown in the right part of Table 9.14, shows that the poison and treatment effects explain a higher proportion of the response variability on the inverse scale, and that the interaction is reduced.

With response y^{-1}, the parameter estimates for the model with no interaction are $\hat{\mu} = 2.62$, $\hat{\alpha}_1 = 0.90$, $\hat{\alpha}_2 = -0.76$, $\hat{\alpha}_3 = 0.32$, $\hat{\alpha}_4 = -0.46$, $\hat{\beta}_1 = -0.82$,

$\widehat{\beta}_2 = -0.35, \widehat{\beta}_3 = 1.17$. The standard errors are 0.07 for $\widehat{\mu}$, 0.12 for the $\widehat{\alpha}$s, and 0.10 for the $\widehat{\beta}$s, all with units of $(10\text{-hours})^{-1}$. As suggested by the panels of Figure 8.5, treatments B and D prolong life best, and poison 3 shortens it most. We reconsider these data in Example 9.8.   ∎

Exercises 9.2

1   Consider the one-way layout. Show that when the model $y_{tr} = \mu + \varepsilon_{tr}$ is fitted, the residual sum of squares is $\sum_{r,t}(y_{tr} - \overline{y}_{..})^2$ on $TR - 1$ degrees of freedom, where $\overline{y}_{..}$ is the overall average. Show that

$$\sum_{r,t}(y_{tr} - \overline{y}_{..})^2 = \sum_{r,t}(y_{tr} - \overline{y}_{t.})^2 + \sum_{r,t}(\overline{y}_{t.} - \overline{y}_{..})^2,$$

and hence verify the contents of Table 9.2.
How would you form a confidence interval for $\beta_1 - \beta_2$?

2   Calculate the analysis of variance table for the data of Example 9.3, and test whether there are differences between the diets and the groups. Find the standard error of a difference between diets, and use it to give a 95% confidence interval for the mean difference in weight gain between diets IV and I.

3   Suppose that T_1 and T_2 have common mean μ and variances σ_1^2 and σ_2^2, and let S_1^2 and S_2^2 be estimators of σ_1^2, σ_2^2, independently distributed as $\chi_{\nu_1}^2, \chi_{\nu_2}^2$.
(a) Show that if σ_1^2, σ_2^2 are known, then μ has minimum variance unbiased estimator

$$T = \frac{T_1/\sigma_1^2 + T_2/\sigma_2^2}{1/\sigma_1^2 + 1/\sigma_2^2},$$

with variance $\sigma_1^2\sigma_2^2/(\sigma_1^2 + \sigma_2^2)$.
(b) Suppose that $\mathrm{var}(T)$ is estimated by V in which σ_1^2 and σ_2^2 are replaced by their estimates. Show that V has approximate mean and variance

$$\frac{\sigma_1^2\sigma_2^2}{\sigma_1^2 + \sigma_2^2}, \qquad \frac{2\sigma_1^4\sigma_2^4}{(\sigma_1^2 + \sigma_2^2)^4}\left(\frac{\sigma_1^4}{\nu_1} + \frac{\sigma_2^4}{\nu_2}\right).$$

Hence show that if V is regarded as approximately χ^2, then its degrees of freedom are $(\sigma_1^2 + \sigma_2^2)^2/(\sigma_1^4/\nu_1 + \sigma_2^4/\nu_2)$.
(c) Compute the degrees of freedom for this approximation in Example 9.4.

4   Give the analysis of variance table for a two-way layout with replication, when the numbers of replicates in the tth row and pth column, J_{tp}, are unequal.

5   Use Lagrange multipliers to verify the formulae for the estimates and fitted values given in Example 9.5, and hence check the contents of the analysis of variance table for a Latin square.

6   In Example 9.5, suppose that a confidence interval is required for the difference of the mean efficiencies between 8 and 4 mph. Show that owing to the balance of the experiment, a point estimate of this is just the difference between the average efficiencies for these speeds, and that its variance is $\frac{1}{2}\sigma^2$. Give the estimate of σ^2 ignoring day and run effects, that is, treating the data as a one-way layout with the four levels of speed as groups. How much longer is the corresponding confidence interval than when day and run effects are taken into account?

9.3 Further Notions

9.3.1 Interaction

Terms that do not act additively in a linear model are said to *interact*. An easy way to understand this is by example.

Example 9.7 (2^2 **factorial experiment**) A 2^2 factorial experiment involves a response measured at each combination of two factors each with two levels. As an illustration, consider an experiment to assess the effects of two fertilizers, in which the factors are addition or not of potash and nitrogen. If the cell means are

$$
\begin{array}{ccc}
 & \text{No potash} & \text{Potash} \\
\text{No nitrogen} & \mu & \mu + \beta \\
\text{Nitrogen} & \mu + \alpha & \mu + \alpha + \beta + \gamma
\end{array} \quad ,
$$

and $\gamma = 0$, the effects act additively because the addition of potash increases the mean response by β whether or not nitrogen is present. Similarly the effect of nitrogen does not depend on the presence of potash.

The two treatments interact if γ is non-zero, because the effect of both treatments together is not the sum of the effects of adding them separately. The difference between the effects of adding potash when there is no nitrogen present and when there is nitrogen present is

$$
\{(\mu + \alpha + \beta + \gamma) - (\mu + \alpha)\} - \{(\mu + \beta) - \mu)\} = \gamma,
$$

so we can view a non-zero interaction of the two fertilizers as a differential effect of adding potash depending on the presence or not of nitrogen. The average effect of adding potash, taken over both rows of the table, is then

$$
\frac{1}{2}\{(\mu + \alpha + \beta + \gamma) - (\mu + \alpha) + (\mu + \beta) - \mu\} = \beta + \frac{1}{2}\gamma.
$$

Thus if there is no interaction, β represents the average effect of adding potash whatever the level of nitrogen, but it loses this interpretation if γ is non-zero.

If the model is reparametized to have cell means

$$
\begin{array}{ccc}
 & \text{No potash} & \text{Potash} \\
\text{No nitrogen} & \beta_0 - \beta_1 - \beta_2 + \beta_3 & \beta_0 - \beta_1 + \beta_2 - \beta_3 \\
\text{Nitrogen} & \beta_0 + \beta_1 - \beta_2 - \beta_3 & \beta_0 + \beta_1 + \beta_2 + \beta_3
\end{array} ,
$$

the overall mean is β_0, the average effect of adding potash is

$$
\frac{1}{2}\{(\beta_0 + \beta_1 + \beta_2 + \beta_3) - (\beta_0 + \beta_1 - \beta_2 - \beta_3) + (\beta_0 - \beta_1 + \beta_2 - \beta_3)
$$
$$
- (\beta_0 - \beta_1 - \beta_2 + \beta_3)\} = 2\beta_2,
$$

and likewise the average effect of adding nitrogen is $2\beta_1$. The difference between the effects of adding potash when there is no nitrogen present and when it is present is

$$
\frac{1}{2}[\{(\beta_0 + \beta_1 + \beta_2 + \beta_3) - (\beta_0 + \beta_1 - \beta_2 - \beta_3)\} - \{(\beta_0 - \beta_1 + \beta_2 - \beta_3)
$$
$$
- (\beta_0 - \beta_1 - \beta_2 + \beta_3)\}] = 2\beta_3.
$$

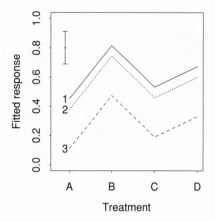
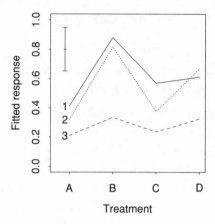

Figure 9.6 Poison data. The left panel shows how the fitted values under the model of no interaction, $\widehat{\mu} + \widehat{\alpha}_t + \widehat{\beta}_p$, for treatments A–D depend on poisons 1–3. The right panel shows the corresponding fitted values under the model of interaction, $\widehat{\mu} + \widehat{\alpha}_t + \widehat{\beta}_p + \widehat{\gamma}_{tp}$. The vertical line in each panel has length four times the standard error of a fitted value.

The difference between the two parametrizations is clear from the design matrices,

$$\begin{pmatrix} 1 & 0 & 0 & 0 \\ 1 & 1 & 0 & 0 \\ 1 & 0 & 1 & 0 \\ 1 & 1 & 1 & 1 \end{pmatrix}, \quad \begin{pmatrix} 1 & -1 & -1 & 1 \\ 1 & 1 & -1 & -1 \\ 1 & -1 & 1 & -1 \\ 1 & 1 & 1 & 1 \end{pmatrix}.$$

The first parametrization, in terms of μ, α, β and γ, represents changes in the mean response relative to the top left cell; this is the *corner-point* parametrization. In the parametrization using β_0, β_1, β_2 and β_3, the parameters can be interpreted as the overall mean, the mean effects of adding potash, nitrogen, and the effect of adding both, regardless of the other parameters. The first column corresponds to the overall mean, the middle two columns to *main effects* of potash and nitrogen, and the final column to the *first-order* or *two-factor* interaction between the two main effects. Notice that the interaction term is the product of the columns for the main effects and that the second parametrization is orthogonal, but the first is not. In practice the parametrization used would depend on the purpose of the analysis.

The order of an interaction is one less than the number of effects it involves.

If interaction is present, four parameters must be estimated from four observations. In this case σ^2 would usually be estimated by replicating the experiment.   ∎

The same ideas generalize, as the next example shows.

Example 9.8 (Poisons data)   The linear model corresponding to the data discussed in Example 9.6 is given at (9.7). If the first-order interaction parameters $\gamma_{pt} \equiv 0$, the profile of fitted values for poison p and treatments A–D may be written $(\widehat{\mu} + \widehat{\alpha}_1 + \widehat{\beta}_p, \ldots, \widehat{\mu} + \widehat{\alpha}_4 + \widehat{\beta}_p)$, and the effect of applying poison r instead of poison p is a translation of the fitted values by $\widehat{\beta}_r - \widehat{\beta}_p$. The left panel of Figure 9.6 shows the profile of fitted values for the three poisons; of course the profiles are parallel, as would also be the case for the poison profiles $(\widehat{\mu} + \widehat{\alpha}_t + \widehat{\beta}_1, \ldots, \widehat{\mu} + \widehat{\alpha}_t + \widehat{\beta}_3)$, because no interaction has been fitted.

Under the model with interaction, that is, including the γ_{tp}, the fitted values are $\widehat{\mu} + \widehat{\alpha}_t + \widehat{\beta}_p + \widehat{\gamma}_{tp} = \overline{y}_{tp\cdot}$, and the profiles are $(\overline{y}_{1p\cdot}, \ldots, \overline{y}_{4p\cdot})$; these are shown in the right panel of the figure, and are not parallel, because under the model with interaction

Table 9.15 Interactions for 2^3 factorial design.

Unit	Treatment	Intercept I	Main effects A	B	C	Two-factor interactions AB	AC	BC	Three-factor interaction ABC
1	1	+	−	−	−	+	+	+	−
2	a	+	+	−	−	−	−	+	+
3	b	+	−	+	−	−	+	−	+
4	ab	+	+	+	−	+	−	−	−
5	c	+	−	−	+	+	−	−	+
6	ac	+	+	−	+	−	+	−	−
7	bc	+	−	+	+	−	−	+	−
8	abc	+	+	+	+	+	+	+	+

the effect of changing poisons is more complex than a simple translation. The profiles are broadly similar to those in the left panel, and the evidence for interaction is weak, as shown by the F statistic for Treatments × Poisons in Table 9.14, which gives an overall test for departures from the simple pattern in the left panel. ∎

Two-factor interactions are relatively common in applications. Higher-order interactions are rarer and can indicate outliers or a poorly fitting model.

Example 9.9 (2^3 factorial experiment) A 2^3 factorial experiment has three factors, A, B, and C, each with two levels, denoted -1 and $+1$. Each of the 2^3 possible combinations of the factor levels is applied to a unit, as in the main effects columns of Table 9.15, where the signs only are given. This design, replicated twice, is used in Example 8.4.

The second column of the table shows which treatments have been applied to each unit. Under the model with main effects of A, B, and C only, no treatment is applied to unit 1 and its mean response is $\beta_0 - \beta_A - \beta_B - \beta_C$, treatment A alone is applied to unit 2 and its mean is $\beta_0 + \beta_A - \beta_B - \beta_C$, and so forth. The design matrix then corresponds to the intercept and main effects columns, and

$$\widehat{\beta}_A = \frac{1}{8}(y_a - y_1 + y_{ab} - y_b + y_{ac} - y_c + y_{abc} - y_{bc})$$

where y_a is the response for unit 2, y_{ab} is the response for unit 4, and so forth. Thus the estimate of β_A is based on contrasting the responses for units to which A was applied with those to which it was not applied. Likewise

$$\widehat{\beta}_B = \frac{1}{8}(y_b - y_1 + y_{ab} - y_a + y_{bc} - y_c + y_{abc} - y_{ac}),$$

$$\widehat{\beta}_C = \frac{1}{8}(y_c - y_1 + y_{ac} - y_a + y_{bc} - y_b + y_{abc} - y_{ab}).$$

Under the model that includes the two-factor interactions β_{AB}, β_{AC}, and β_{BC} as well as the intercept and main effects, the mean response for unit 1 is $\beta_0 - \beta_A - \beta_B - \beta_C + \beta_{AB} + \beta_{AC} + \beta_{BC}$. On including the two-factor interaction columns in the design matrix, we obtain

$$\widehat{\beta}_{AB} = \frac{1}{8}(y_{ab} - y_a - y_b + y_1 + y_{abc} - y_{ac} - y_{bc} + y_c),$$

which is based on contrasting responses for which the levels of A and B are the same with those for which the levels of A and B are different. There are similar interpretations of the other estimated two-factor interactions

$$\widehat{\beta}_{AC} = \frac{1}{8}(y_{ac} - y_a - y_c + y_1 + y_{abc} - y_{ab} - y_{bc} + y_b),$$

$$\widehat{\beta}_{BC} = \frac{1}{8}(y_{bc} - y_b - y_c + y_1 + y_{abc} - y_{ab} - y_{ac} + y_a).$$

The estimated three-factor interaction is

$$\widehat{\beta}_{ABC} = \frac{1}{8}(y_{abc} + y_a + y_b + y_c - y_{ab} - y_{ac} - y_{bc} - y_1)$$

$$= \frac{1}{8}\{(y_{abc} - y_{ac} - y_{bc} + y_c) - (y_{ab} - y_a - y_b + y_1)\},$$

which contrasts the contributions to the AB interaction when C is applied and when it is not applied. Whichever of these models is fitted, the design matrix is orthogonal, and $(X^{\mathsf{T}}X)^{-1} = \frac{1}{8}I$, so the variance of each of the estimates above is $\sigma^2/8$, as is readily verified directly.   ∎

When strong two-factor interaction is present, interpretation is often simplified by considering how responses behave separately for each level of one factor, and likewise for higher-order interactions.

Confounding

Factorial designs make it possible to assess the effects of many treatments and their interactions in a single experiment, but when there are many factors many homogeneous units must be found, and this can pose practical problems.

Example 9.10 (2^2 factorial experiment) An experiment with two two-level factors A and B is performed using two blocks each having two units. The possible treatments are $1, a, b$, and ab, and there are three possible designs depending on which treatment appears in the same block as 1. Suppose that the first block has treatments 1 and a, and the second has b and ab. The model with an intercept, a block effect, and the main effects of A and B is

$$\begin{pmatrix} y_1 \\ y_a \\ y_b \\ y_{ab} \end{pmatrix} = \begin{pmatrix} 1 & 0 & -1 & -1 \\ 1 & 0 & 1 & -1 \\ 1 & 1 & -1 & 1 \\ 1 & 1 & 1 & 1 \end{pmatrix} \begin{pmatrix} \beta_0 \\ \alpha \\ \beta_A \\ \beta_B \end{pmatrix} + \varepsilon,$$

in which the design matrix has rank three because the column for B is a linear combination of the first two columns. The design makes it impossible to distinguish these effects, which are said to be *confounded*. Evidently A would be confounded with blocks if the first block contained treatments 1 and b and the second a and ab.

Suppose instead that the experiment is set up to have 1 and ab in the same block. Then the design matrix is

$$\begin{pmatrix} 1 & 0 & -1 & -1 \\ 1 & 0 & 1 & 1 \\ 1 & 1 & -1 & 1 \\ 1 & 1 & 1 & -1 \end{pmatrix},$$

which has rank four. Then $\widehat{\beta}_0 = \frac{1}{2}(y_1 + y_{ab})$, $\widehat{\alpha} = \frac{1}{2}(y_a + y_b - y_1 - y_{ab})$, while the estimates $\widehat{\beta}_A = \frac{1}{4}(y_{ab} - y_1 + y_A - y_B)$ and $\widehat{\beta}_B = \frac{1}{4}(y_{ab} - y_1 - y_A + y_B)$ correspond to comparisons made within blocks. Thus this design does allow the estimation of the main effects of interest, though an external estimate of the error variance σ^2 is required.

The interaction between A and B would usually be estimated by

$$\widehat{\beta}_{AB} = \frac{1}{4}(y_a + y_b - y_1 - y_{ab}) = \frac{1}{2}\widehat{\alpha},$$

so use of this design entails sacrificing any information about this interaction. In examples with many factors, interactions known to be unimportant are often deliberately confounded with blocks in such a way that the effects of interest are estimated with maximum precision in the resulting fractional factorial design.

∎

If effects are confounded by accident, then further experimentation will be needed to identify the parameters, unless there is external information about their values. Some models are intrinsically non-identifiable, however; see Section 4.6.

9.3.2 Contrasts

Suppose that a model has $n \times p$ design matrix X of full rank and that we have a $p \times p$ invertible matrix A for which $XA = C$, where the first column of C is a column of ones, and the remaining columns, $c_1, \ldots, c_{p-1}$, are orthogonal to the first and to each other. In this case $C^{\mathrm{T}}C = \mathrm{diag}(n, c_1^{\mathrm{T}}c_1, \ldots, c_{p-1}^{\mathrm{T}}c_{p-1})$. Let us reparametrize the original model, $y = X\beta + \varepsilon$, by letting $\gamma = A^{-1}\beta$, thereby obtaining $y = C\gamma + \varepsilon$. The columns of C are known as *orthogonal contrasts*.

The least squares estimators for the model $y = C\gamma + \varepsilon$ are $\widehat{\gamma} = (C^{\mathrm{T}}C)^{-1}C^{\mathrm{T}}y$, with covariance matrix $\sigma^2(C^{\mathrm{T}}C)^{-1}$. As $C^{\mathrm{T}}C$ is a diagonal matrix, the estimate of γ_r is $\widehat{\gamma}_r = c_r^{\mathrm{T}}y/c_r^{\mathrm{T}}c_r$, with variance $\mathrm{var}(\widehat{\gamma}_r) = \sigma^2/c_r^{\mathrm{T}}c_r$, and different estimates $\widehat{\gamma}_r$ are uncorrelated with each other and with the overall average, $\overline{y}$. The residual sum of squares $SS(\widehat{\gamma})$ equals

$$y^{\mathrm{T}}\{I - C(C^{\mathrm{T}}C)^{-1}C^{\mathrm{T}}\}y = y^{\mathrm{T}}y - \widehat{\gamma}^{\mathrm{T}}C^{\mathrm{T}}C\widehat{\gamma}$$
$$= \sum_{j=1}^{n} y_j^2 - n\overline{y}^2 - \widehat{\gamma}_1^2 c_1^{\mathrm{T}}c_1 - \cdots - \widehat{\gamma}_{p-1}^2 c_{p-1}^{\mathrm{T}}c_{p-1}.$$

As the reduction in sum of squares due to adding c_r to the design matrix is $\widehat{\gamma}_r^2 c_r^{\mathrm{T}}c_r$, the total sum of squares can be split into the contributions from each of the columns of

Term	df	Sum of squares	Mean square	F statistic
Seat	1	473.06	473.06	112.9
Dynamo	1	39.06	39.06	9.32
Tyre	1	39.06	39.06	9.32
Seat × Dynamo	1	1.56	1.56	0.37
Seat × Tyre	1	5.06	5.06	1.21
Dynamo × Tyre	1	0.06	0.06	0.01
Seat × Dynamo × Tyre	1	3.06	3.06	0.73
Residual	8	33.50	4.19	

Table 9.16 Analysis of variance for the cycling data.

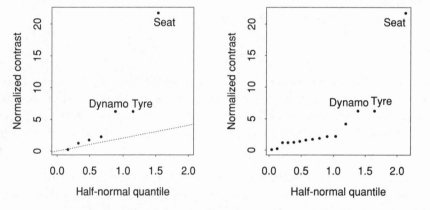

Figure 9.7 Half-normal plots of normalized contrasts $|\widehat{\gamma}_r|(c_r^{\mathrm{T}}c_r)^{1/2}$ for the data on cycling up a hill. The left panel shows the $|\widehat{\gamma}_r|(c_r^{\mathrm{T}}c_r)^{1/2}$ for the last seven columns of C_1, with the dotted line having slope $s = 4.19^{1/2}$, the residual standard error. The right panel shows the $|\widehat{\gamma}_r|(c_r^{\mathrm{T}}c_r)^{1/2}$ for the last 15 columns of C_2. In each case, only contrasts corresponding to main effects seem to be non-null. See text for details.

C, namely $n\overline{y}^2$, $\widehat{\gamma}_1^2 c_1^{\mathrm{T}}c_1$, and so forth. If γ_r equals zero, $\widehat{\gamma}_r$ has mean zero and variance $\sigma^2/c_r^{\mathrm{T}}c_r$, and if the errors are normal, $\widehat{\gamma}_r^2 c_r^{\mathrm{T}}c_r \sim \sigma^2 \chi_1^2$. A normal scores plot of the $\widehat{\gamma}_r(c_r^{\mathrm{T}}c_r)^{1/2}$, a plot of the ordered $\widehat{\gamma}_r^2 c_r^{\mathrm{T}}c_r$ against χ_1^2 plotting positions, or a half-normal plot (Practical 2.1) of the $|\widehat{\gamma}_r|(c_r^{\mathrm{T}}c_r)^{1/2}$, helps to show which of the γ_r may be non-zero.

Example 9.11 (Cycling data) The data on cycling up a hill are from a 2^3 factorial experiment, replicated twice. The design matrix D for a 2^3 experiment, in which the three main effects, the three second-order interactions, and the third-order interaction are all fitted, is obtained by adding ones to the pluses and minuses in Table 9.15. This matrix has the property that $D^{\mathrm{T}}D = 8I_8$, so its columns are orthogonal contrasts. The 16×8 design matrix for the replicated experiment may be written as

$$C_1 = \begin{pmatrix} D \\ D \end{pmatrix};$$

as $C_1^{\mathrm{T}}C_1 = 16I_8$, the columns of C_1 are also orthogonal contrasts. Table 9.16 shows the analysis of variance when this model is fitted. There are eight residual degrees of freedom, and the estimate of error is $s^2 = 4.19$. The main effects are significant, but the interactions, denoted Seat×Dynamo and so forth, are not.

The left panel of Figure 9.7 shows a half-normal plot of the quantities $|\widehat{\gamma}_r|(c_r^{\mathrm{T}}c_r)^{1/2}$ corresponding to the contrasts in the last seven columns of C_1; the dotted line has slope s. The plot confirms our impression from the analysis of variance table: only the three main effects seem to be non-zero.

The residual sum of squares can also be decomposed into its component degrees of freedom. To see how, we add eight columns to C_1, giving

$$C_2 = \begin{pmatrix} D & -D \\ D & D \end{pmatrix},$$

the last 15 columns of which are orthogonal contrasts, as $C_2^T C_2 = 16 I_{16}$. The right panel of Figure 9.7 shows a half-normal plot of the contrasts corresponding to these columns; the eight contrasts comprising s^2 — corresponding to the columns of C_2 not in C_1 — have been added to the previous seven. These eight contrast the main effects, the second- and third-order interactions between the two replicates. As no degrees of freedom remain with which to estimate σ^2, it may be estimated by pooling those contrasts that lie roughly on a straight line in the lower left corner of the graph. Here there seem to be about 12 such contrasts, the pooling of which gives error estimate 3.60 on 12 degrees of freedom.   ■

Example 9.12 (Field concrete mixer data)   In Example 9.5 we found evidence that the efficiency of the mixer depended strongly on its speed, with the best concrete produced at about 8 mph. An estimated best speed can be obtained by decomposing the sum of squares due to speed using contrasts based on orthogonal polynomials. There are three degrees of freedom for speeds. One parametrization for them gives three columns in which the rows corresponding to the four speeds 4, 8, 12, and 16 miles per hour are

$$\begin{array}{cccc} 4 & 0 & 0 & 0 \\ 8 & 1 & 0 & 0 \\ 12 & 0 & 1 & 0 \\ 16 & 0 & 0 & 1 \end{array},$$

whereas a parametrization in terms of orthogonal polynomials gives

$$\begin{array}{cccc} 4 & -3 & 1 & -1 \\ 8 & -1 & -1 & 3 \\ 12 & 1 & -1 & -3 \\ 16 & 3 & 1 & 1 \end{array},$$

where the first column is linear in speed and the other two columns are obtained by Gram–Schmidt orthogonalization of the square and cube of the first. The corresponding columns of the design matrix, s_1, s_2, and s_3, are orthogonal to the grand mean, to each other, and to the day and run effects. The parameter estimates, standard errors, and sums of squares $\widehat{\gamma}_r^2 s_r^T s_r$ for these contrasts are given in Table 9.17; note that the total sum of squares equals that for speeds in Table 9.12.

The t statistics for the linear, quadratic, and cubic effects are significant at levels about 0.01, 0.07, and 0.38 respectively, suggesting that the effect of speed on efficiency may be summarized as $\widehat{\mu} + \widehat{\gamma}_1 s_1 + \widehat{\gamma}_2 s_2$, where $\widehat{\mu}$ is the estimated grand mean in this parametrization. In terms of speed, x, $s_1 = (x - 10)/2$ and $s_2 = \{(x - 10)^2 - 20\}/16$. Since $\widehat{\gamma}_2 < 0$, efficiency is maximized as a function of speed when $\widehat{\gamma}_1 ds_1/dx + \widehat{\gamma}_2 ds_2/dx = 0$ and the estimated best speed is $10 - 4\widehat{\gamma}_1/\widehat{\gamma}_2 = 6.96$ mph.

Term	Estimate	Standard error	Sum of squares
Linear, s_1	−1.24	0.329	123.26
Quadratic, s_2	−1.63	0.737	42.58
Cubic, s_3	0.31	0.330	7.75
Total			173.58

Table 9.17 Field concrete mixer data: orthogonal decomposition of sum of squares for speed into linear, quadratic, and cubic effects.

A confidence interval for the true best speed may be obtained by the delta method or using an exact argument (Exercise 9.3.3), but the t statistic for $\widehat{\gamma}_2$ suggests that such a confidence interval will be imprecise. The right panel of Figure 9.5 shows the fitted quadratic curve as a function of speed. ∎

9.3.3 Analysis of covariance

Analysis of covariance is intended to reduce bias or increase precision when some variables cannot be controlled by design. This may arise because the importance of these variables has been recognized only after randomization, because the randomization took them partially but not fully into account, or because their values only became available after randomization.

Suppose that a model with a design matrix X from a balanced experimental setup is to be fitted, but that additional explanatory variables contained in the matrix Z have been measured that might affect the response. The design leads us to fit the model $y = X\beta + \varepsilon$, but instead we fit $y = X\beta + Z\gamma + \varepsilon$, in the hope that inclusion of Z will increase the precision of $\widehat{\beta}$. However adding Z removes balance and complicates the analysis of variance. Our interest is in the treatment effects after adjusting for Z, so for analysis of variance we fit Z before treatments and their interactions.

Example 9.13 (Cat heart data)   Table 9.18 shows the results from an experiment to determine the relative potencies of eight similar cardiac drugs, labelled A–H, where A is a standard. The method used was to infuse slowly a suitable dilution of the drug into an anaesthetized cat. The dose at which death occurred and the weight of the cat's heart were recorded. The table shows $y = 100 \times$ log dose in μgm, and, below, $z = 100 \times$ log heart weight in gm. Four observers each made two determinations on each of eight days, with a Latin square design used to eliminate observer and time differences. Here z cannot be known at the start of the experiment, but might be expected to affect comparisons among the treatments; it is assumed that heart weight is unaffected by the treatments.

The left part of Table 9.19 gives the analysis of variance without adjustment for heart weight. The seven degrees of freedom for the sum of squares between rows have been decomposed into the main effects of observer and time, and their interaction. There are clearly large differences among observers, and between times, and a smaller but substantial interaction between these terms, but there is little evidence of day-to-day variation.

Table 9.18 Data from Latin square experiment on the potencies of eight cardiac drugs given to anaesthetized cats. The table shows $y = 100 \times \log$ dose in μgm at which death occurred, and, below, $z = 100 \times \log$ heart weight in gm.

Observer	Time		Day 1		2		3		4		5		6		7		8
1	am	G	75	F	77	A	52	E	71	C	65	D	47	B	37	H	63
			91		77		102		102		84		85		73		84
1	pm	E	81	D	58	G	74	A	54	F	62	H	69	C	59	B	59
			76		90		116		87		93		79		105		71
2	am	H	94	G	86	F	104	C	66	E	94	B	72	D	82	A	58
			90		100		102		108		97		90		96		90
2	pm	B	73	A	59	E	103	F	86	G	84	C	82	H	95	D	65
			88		82		94		77		88		106		89		83
3	am	A	22	C	36	B	39	D	32	H	43	E	67	G	52	F	34
			90		81		83		94		95		101		97		66
3	pm	C	46	H	25	D	52	G	59	B	42	A	54	F	77	E	69
			90		81		91		99		90		98		106		101
4	am	F	39	E	56	H	56	B	28	D	52	G	86	A	45	C	70
			83		88		95		79		87		100		84		117
4	pm	D	87	B	82	C	72	H	92	A	58	F	92	E	99	G	89
			96		93		87		89		87		92		90		106

Table 9.19 Analysis of variance for cats data, with and without adjustment for heart weight.

Term	Without adjustment df	Sum of squares	Mean square	With adjustment df	Sum of squares	Mean square
Heart weight				1	3058	3058
Observer	3	9949	3316	3	9452	3151
Time	1	2003	2003	1	1939	1939
Observer × Time	3	2238	746	3	1890	630
Day	7	922.9	131.8	7	463	66.3
Drug	7	6098	871.1	7	5051	721.6
Residual	42	4874	116.0	41	4228	103.1

Table 9.20 Estimated differences between standard drug A and the treatments B–H, without and with adjustment for the covariate heart weight.

	B	C	D	E	F	G	H
Unadjusted	3.75 (5.4)	11.75 (5.4)	9.13 (5.4)	29.75 (5.4)	21.12 (5.4)	25.37 (5.4)	16.87 (5.4)
Adjusted	6.53 (5.2)	8.71 (5.2)	9.02 (5.1)	28.2 (5.1)	22.38 (5.1)	21.34 (5.3)	17.82 (5.1)

The estimates of the differences between the drugs and the standard, unadjusted for heart weight, are given in the upper row of Table 9.20. Their standard errors are equal because of the balance. The dose of drug B needed to cause death appears not to differ from the standard, those for C, D and H are rather larger, and those required for E, F, and G are substantially larger.

The analysis of variance with z is given in the right part of Table 9.19. Since interest centres on the drug effects, heart weight must be fitted before the term for drugs, but

	A	B	C
x	1.5, 2.2, 2.9, 4.1, 4.1	2.7, 3.8, 5.6, 6.4, 6.8	2.2, 3.5, 4.6, 5.5, 6.6
y	9.6, 11.3, 10.3, 12.5, 12.6	8.6, 7.2, 8.9, 11.6, 11.5	4.8, 5.6, 6.2, 7.5, 6.8

Table 9.21 Data from a completely randomized experiment on the comparison of diets, with initial weight x and final weight y.

otherwise it is immaterial when it is fitted; here we fit it before allowing for the experimental conditions. This results in a non-unique analysis of variance: the order of fitting is irrelevant in the left part of the table, but matters in the right part. The adjustment reduces the sums of squares for the other terms, with the reduction for days being largest. The estimate of σ^2 adjusted for heart weight, 103.1, is somewhat smaller than the unadjusted estimate, 116.0, and the precision of the comparisons between the drugs and the standard is slightly increased. In particular the adjusted estimates for B, C, and D, and for F, G, and H, are more similar — some of the variation in the unadjusted comparisons is due to heart weights.   ∎

Exercises 9.3

1   Suppose that a 2^2 factorial experiment is to be performed using eight units in four blocks of two units each. Show that the intercept, three block effects, and the main effects and interaction between the treatments can be estimated if the treatments are allocated to blocks as follows: $(1, a)$, (b, ab), $(1, ab)$, (a, b). Can they all still be estimated if an observation from the last block is lost?

2   Table 9.21 gives results from a completely randomized experiment in which five individuals were allocated at random to each of three diets.
    (a) Calculate the group averages and variances, and hence obtain the analysis of variance of the final weights, unadjusted for initial weights. Give the standard errors for differences between averages for diet A and the other two diets.
    (b) Use analysis of covariance to adjust for the initial weights. Give the new analysis of variance table, and adjusted standard errors for the differences in (a). Comment.

3   Consider the calculation of a 95% confidence interval for the speed that gives the maximum efficiency for the field concrete mixer of Example 9.12.
    (a) Use the delta method to show that

$$\mathrm{var}\left(\frac{\widehat{\gamma}_1}{\widehat{\gamma}_2}\right) \doteq \frac{\gamma_1^2}{\gamma_2^2}\left(\frac{\mathrm{var}(\widehat{\gamma}_1)}{\gamma_1^2} + \frac{\mathrm{var}(\widehat{\gamma}_2)}{\gamma_2^2}\right).$$

Use this to show that $\widehat{\gamma}_1/\widehat{\gamma}_2$ has standard error 1.595, and hence give an approximate 95% confidence interval for $10 - 4\gamma_1/\gamma_2$.
    (b) If $\psi = -\gamma_1/\gamma_2$, show that the distribution of $\widehat{\gamma}_2\psi + \widehat{\gamma}_1$ is normal with mean zero and variance $\sigma^2(\psi^2 v_{22} + v_{11})$, where v_{11} and v_{22} are the diagonal elements of the matrix $(X^{\mathrm{T}}X)^{-1}$ that correspond to γ_1 and γ_2. Deduce that as $(\widehat{\gamma}_2\psi + \widehat{\gamma}_1)^2/\{s^2(\psi^2 v_{22} + v_{11})\}$ has an $F_{1,\nu}$ distribution, an exact confidence region for ψ is the set of values such that $(\widehat{\gamma}_2\psi + \widehat{\gamma}_1)^2/\{s^2(\psi^2 v_{22} + v_{11})\} \leq F_{1,\nu}(1 - \alpha)$.
A 95% confidence set for $10 + 4\psi$ based on the calculations in Example 9.12 is $(-\infty, 9.15)$, $(38.03, \infty)$. On the same graph, plot this confidence set, the average efficiencies for the different speeds and the fitted efficiency from Example 9.12 against speed. Do you find the exact confidence set surprising?
    (c) Use part (b) to calculate the exact coverage of your delta method confidence interval.

9.4 Components of Variance

9.4.1 Basic ideas

Our models so far have involved just one level of random variation, with all the responses independent. Sometimes a more complex error structure is required.

The simplest example is the one-way layout with R units in each of T blocks. Suppose the blocking factors are of no intrinsic interest, and the block effects may be thought of as being sampled at random from a population, block means being a random sample from a normal distribution with mean μ and variance σ_b^2. Conditional on the block mean, the responses for units within a block are independent normal variables with mean zero and variance σ^2. Thus the response for the rth unit in block t is

$$y_{tr} = \mu + b_t + \varepsilon_{tr}, \tag{9.8}$$

where the b_t have zero means and variances σ_b^2, the ε_{tr} have zero means and variances σ^2, and the b_t and ε_{tr} are all mutually independent. Responses from different blocks are independent, but those within the same block are not, as $\mathrm{cov}(y_{tr}, y_{ts}) = \sigma_b^2$ for $r \neq s$. Thus the covariance matrix for the responses is block diagonal. This is called a *random effects model*, as the block effects are regarded as random variables rather than fixed parameters.

The analysis of variance for the one-way layout involves the sums of squares within and between blocks

$$SS_w = \sum_{t,r}(y_{tr} - \overline{y}_{t.})^2, \quad SS_b = \sum_{t,r}(\overline{y}_{t.} - \overline{y}_{..})^2.$$

Under the random effects model, $y_{tr} - \overline{y}_{t.} = \varepsilon_{tr} - \overline{\varepsilon}_{t.}$, and as this does not depend on the presence of the random effects, SS_w has its usual $\sigma^2 \chi^2_{T(R-1)}$ distribution. Now $\overline{y}_{t.} = \mu + b_t + \overline{\varepsilon}_{t.} \sim N(\mu, \sigma_b^2 + \sigma^2/R)$, and as the $\overline{y}_{t.}$ are independent, the distribution of SS_b is $R(\sigma_b^2 + \sigma^2/R)\chi^2_{T-1}$. Furthermore,

$$\mathrm{cov}(y_{tr} - \overline{y}_{t.}, \overline{y}_{t.} - \overline{y}_{..}) = \mathrm{cov}(b_t + \varepsilon_{tr} - b_t - \overline{\varepsilon}_{t.}, b_t + \overline{\varepsilon}_{t.} - \overline{b}. - \overline{\varepsilon}_{..}) = 0,$$

and hence the linear combinations of normal variables $y_{tr} - \overline{y}_{t.}$ and $\overline{y}_{t.} - \overline{y}_{..}$ must be independent. Thus the sums of squares SS_w and SS_b have independent chi-squared distributions with scale parameters σ^2 and $\sigma^2 + R\sigma_b^2$ respectively. Tests and confidence intervals for the ratio σ_b^2/σ^2 can be based on the $F_{T-1,T(R-1)}$ distribution of

$$\frac{\sigma^2}{\sigma^2 + R\sigma_b^2} \times \frac{SS_b/(T-1)}{SS_w/\{T(R-1)\}}. \tag{9.9}$$

An alternative derivation of the independence of SS_w and SS_b under the random effects model is to argue conditionally on the values of the b_t. Conditional on the b_t, the model is just the one-way layout described in Section 9.2.1, under which SS_w and SS_b are independent, and only the distribution of SS_b depends on the b_t. Hence SS_w and SS_b are unconditionally independent.

One aspect of interest may be statements of uncertainty for the population mean μ, which is estimated by the overall sample average, $\overline{y}_{..} = \mu + \overline{b}. + \overline{\varepsilon}_{...}$. This has

		Subject			
1	2	3	4	5	6
68	49	41	33	40	30
42	52	40	27	45	42
69	41	26	48	50	35
64	56	33	54	41	44
39	40	42	42	37	49
66	43	27	56	34	25
29	20	35	19	42	45

Table 9.22 Blood data: seven measurements from each of six subjects on a property related to the stickiness of their blood.

variance $\sigma_b^2/T + \sigma^2/(TR) = (\sigma^2 + R\sigma_b^2)/(TR)$, which is estimated unbiasedly by $SS_b/\{(T-1)TR\}$, independent of $\overline{y}_{..}$, and confidence intervals are based on the t_{T-1} distribution of $(\overline{y}_{..} - \mu)/[SS_b/\{(T-1)TR\}]^{1/2}$.

The assumptions of homogeneous variance across all blocks and of normality can be checked using probability plots.

Example 9.14 (Blood data) Six subjects were selected at random from a large population, and a property related to stickiness of samples of blood was measured seven times on each subject. The data are given in Table 9.22.

For these data, $SS_w = 4549.7$ and $SS_b = 1466.0$ on 36 and 5 degrees of freedom respectively. A point estimate of the variance for different measurements on the same subject is $SS_w/36 = 126.4$. and a point estimate of the variance of mean stickiness between subjects is $(SS_b/5 - SS_w/36)/7 = 23.83$. An equi-tailed 90% confidence interval for the ratio σ_b^2/σ^2 based on (9.9) is $(-0.01, 1.34)$; this overlaps the negative half-axis and would not usually be appropriate. ■

Nested variation

The previous example had two levels of nested variation, for subjects and for measurements. In practice data with several levels of variation arise. Consider for example comparison of the success of a surgical procedure, measured on a continuous scale. Data are available on patients, P of whom are treated by each surgeon and with S surgeons working at H hospitals. We suppose that surgeons at different hospitals are independent, and likewise for the patients, so patients are nested within surgeons within hospitals — there is no relation between the first patient of surgeon 1 at hospital 1 and the first patient of surgeon 2 at hospital 1, nor between surgeon 1 at hospital 1 and surgeon 1 at hospital 2. Put another way, labels for patients can be permuted independently within each surgeon without changing the data structure, and likewise for surgeons within each hospital. A simple model for the outcome y_{hsp} for the pth patient of the sth surgeon at the hth hospital is

$$y_{hsp} = \mu + b_h + e_{hs} + \varepsilon_{hsp}, \quad h = 1, \ldots, H, s = 1, \ldots, S, p = 1, \ldots, P,$$
$$(9.10)$$

| | | | \multicolumn{3}{c}{E(Mean square) when terms below random} |
Term	df	Sum of squares	ε	ε, e	ε, e, b
Between hospitals	$H-1$	$\sum(\overline{y}_{h\cdot\cdot} - \overline{y}_{\cdots})^2$	$PS\delta_b^2 + P\delta_e^2 + \sigma^2$	$PS\delta_b^2 + P\sigma_e^2 + \sigma^2$	$PS\sigma_b^2 + P\sigma_e^2 + \sigma^2$
Between surgeons within hospitals	$H(S-1)$	$\sum(\overline{y}_{hs\cdot} - \overline{y}_{h\cdot\cdot})^2$	$P\delta_e^2 + \sigma^2$	$P\sigma_e^2 + \sigma^2$	$P\sigma_e^2 + \sigma^2$
Between patients within surgeons	$HS(P-1)$	$\sum(y_{hsp} - \overline{y}_{hs\cdot})^2$	σ^2	σ^2	σ^2

Table 9.23 Analysis of variance table for nested model. Each sum of squares is summed over h, s and p. Mean squares are formed by dividing sums of squares by their degrees of freedom. δ_b^2 and δ_e^2 are non-centrality parameters measuring differences among the b_h and e_{hs} when they are treated as fixed.

where μ is the mean success level in a population of hospitals, from which the hth hospital departs by b_h, the e_{hs} represent surgeon effects, and the ε_{hsp} are independent normal variables with means zero and variance σ^2 corresponding to the pth patient treated by the sth surgeon at hospital h. If random, we suppose the b_h and e_{hs} to be independent normal variables with means zero and variances σ_b^2 and σ_e^2, but the decision whether they should be treated as random or as fixed depends on the context. A potential patient able to choose his surgeon would treat b_h and e_{hs} as fixed, and hope to choose h and s to optimize his prospects. If on the other hand he could choose his hospital but not his surgeon, he might treat the e_{hs} as random — in effect he will be operated upon by a randomly selected surgeon — but try and choose among hospitals, treated as fixed. A health service official hoping to estimate the national success rate for the procedure from a sample of such data would treat the b_h and the e_{hs} as random. The quantities of interest in the three cases are $\mu + b_h + e_{hs}$, $\mu + b_h$, and μ, estimated by $\overline{y}_{hs\cdot}$, $\overline{y}_{h\cdot\cdot}$, and $\overline{y}_{\cdots}$, whose variances are σ^2/P, $\sigma_e^2/S + \sigma^2/(SP)$, and $\sigma_b^2/H + \sigma_e^2/(HS) + \sigma^2/(HSP)$. In each case the analysis of variance is given by Table 9.23 and depends on

$$
\begin{aligned}
\overline{y}_{h\cdot\cdot} - \overline{y}_{\cdots} &= b_h - \overline{b}_\cdot + \overline{e}_{h\cdot} - \overline{e}_{\cdot\cdot} + \overline{\varepsilon}_{h\cdot\cdot} - \overline{\varepsilon}_{\cdots}, \\
\overline{y}_{hs\cdot} - \overline{y}_{h\cdot\cdot} &= e_{hs} - \overline{e}_{h\cdot} + \overline{\varepsilon}_{hs\cdot} - \overline{\varepsilon}_{h\cdot\cdot}, \\
y_{hsp} - \overline{y}_{hs\cdot} &= \varepsilon_{hsp} - \overline{\varepsilon}_{hs\cdot}.
\end{aligned}
$$

If all the quantities contributing to it are regarded as random, then each sum of squares has a chi-squared distribution. For example, the sum of squares between surgeons within hospitals is

$$
SS_S = \sum_{h,s,p} (\overline{y}_{hs\cdot} - \overline{y}_{h\cdot\cdot})^2 = P \sum_{h,s} (e_{hs} - \overline{e}_{h\cdot} + \overline{\varepsilon}_{hs\cdot} - \overline{\varepsilon}_{h\cdot\cdot})^2,
$$

and if e_{hs} and ε_{hsp} are random, then $e_{hs} + \overline{\varepsilon}_{hs\cdot}$ is normal with mean zero and variance $\sigma_e^2 + \sigma^2/P$. Hence

$$
\begin{aligned}
P \sum_{h,s} (e_{hs} - \overline{e}_{h\cdot} + \overline{\varepsilon}_{hs\cdot} - \overline{\varepsilon}_{h\cdot\cdot})^2 &\overset{D}{=} P(\sigma_e^2 + \sigma^2/P)(W_1 + \cdots + W_H) \\
&\sim (P\sigma_e^2 + \sigma^2)\chi^2_{H(S-1)},
\end{aligned}
$$

where the W_h are a random sample from the χ^2_{S-1} distribution. If the e_{hs} are fixed, then $e_{hs} + \bar{\varepsilon}_{hs.}$ is normal with mean e_{hs} and variance σ^2/P and hence SS_S has a non-central chi-squared distribution with $H(S-1)$ degrees of freedom and non-centrality parameter $H(S-1)\delta^2_e = P\sum_{h,s}(e_{hs} - \bar{e}_{h.})^2$ (Problem 2.12). Such calculations give the entries in Table 9.23, in which $(H-1)\delta^2_b = \sum_h (b_h - \bar{b}.)^2$. Note that $E(\delta^2_e) = \sigma^2_e$ and $E(\delta^2_b) = \sigma^2_b$.

Under the model with b_h and e_{hs} fixed, ratios of mean squares can be used to test for differences among surgeons and hospitals, for example comparing the ratio of mean squares for the last two lines of Table 9.23 with the $F_{H(S-1),HS(P-1)}$ distribution.

The assumptions underlying this model would need careful scrutiny in applications: from what populations are patients, surgeons, and hospitals drawn, and in what sense can they be treated as random samples?

Nesting is fundamentally different from the type of classification described earlier. Consider a two-way layout in which factors A and B with T and R levels respectively are applied to TR units. Then if the levels of B among $y_{11}, \ldots, y_{1R}$ were permuted, the same permutation would have to be applied to those of $y_{t1}, \ldots, y_{tR}$ for each t, because the second subscript corresponds to the same treatment for y_{2r} and y_{1r}, for example. The two classifications are then said to be *crossed*. In the random effects model described at the start of this section, however, the labelling is essentially arbitrary, $y_{t1}, \ldots, y_{tR}$ being simply replicate observations; here permutation of any or all of these groups of observations should not affect analysis. Compare Examples 9.3 and 9.14.

It is crucial that crossed and nested effects be distinguished. Typically the levels of crossed effects are of intrinsic interest and are represented by fixed parameters, while parameters associated with nested suffixes are treated as random. Different levels of nesting then correspond to different variance components. However it may be hard to write down the model appropriate to a complex design.

Split-unit experiments

Some experiments are performed with certain treatments applied to entire units and others to sub-units. As such designs originally arose in agriculture, with units being for instance plots of land sown with plant varieties, sub-plots of which were treated with different fertilisers, they are often called split-plot experiments. They also arise in industrial applications, where certain aspects of a manufacturing process may be more easily varied than others, and in medical settings where units are often patients, each with measurements taken in succession over a period, giving a series of correlated responses. Such designs are useful if it is already known that whole-unit treatments differ substantially and interest centres on sub-unit treatments and their interactions, or if physical constraints impose them; they can also arise by accident. The key idea is that there is variation within units (that is, between sub-units) as well as between units. Analysis of variance is effectively performed at two levels, as discussed below.

Suppose there are B blocks of W units, to each of which a whole-unit treatment is applied according to a randomized block design, for example. Units themselves are split into S sub-units, with a sub-unit treatment randomized to each. The corresponding linear model is

$$y_{bws} = \mu + \beta_b + \gamma_w + u_{bw} + \zeta_s + \tau_{ws} + \varepsilon_{bws}, \quad b = 1, \ldots, B, w = 1, \ldots, W,$$
$$s = 1, \ldots, S,$$

where the whole-unit effects are the overall mean μ, the block and whole-unit treatment parameters β_b and γ_w, and the whole-unit errors u_{bw}, taken to be independent normal variables with mean zero and variance σ_u^2. The sub-unit treatment effects are ζ_s, the interactions between sub- and whole-unit treatments τ_{ws} and the sub-unit errors ε_{bws}, taken to be normal with mean zero and variance σ^2 independent of each other and of the u_{bw}. Terms ξ_{bs} are not included because interaction between blocks and sub-units makes no sense.

We use the Roman letter u to indicate that the whole-unit effects are regarded as random variables rather than as parameters.

Under this model different treatments are analyzed at different levels. Whole-unit averages have variance $\sigma_u^2 + \sigma^2/S$ estimated by the residual mean square from a randomized block analysis of these averages, with $(B-1)(W-1)$ degrees of freedom. Whole-unit treatments are compared using contrasts of these averages such as

$$\overline{y}_{\cdot 2 \cdot} - \overline{y}_{\cdot 1 \cdot} = \gamma_2 - \gamma_1 + \overline{u}_{\cdot 2} - \overline{u}_{\cdot 1} + \overline{\varepsilon}_{\cdot 2 \cdot} - \overline{\varepsilon}_{\cdot 1 \cdot},$$

whose variance is $2\sigma_u^2/B + 2\sigma^2/(BS)$. Comparisons of sub-unit treatments and their interactions with whole-unit treatments use the $BW(S-1)$ remaining degrees of freedom and involve quantities such as

$$\overline{y}_{\cdot\cdot 2} - \overline{y}_{\cdot\cdot 1} = \zeta_2 - \zeta_1 + \overline{\tau}_{\cdot 2} - \overline{\tau}_{\cdot 1} + \overline{\varepsilon}_{\cdot\cdot 2} - \overline{\varepsilon}_{\cdot\cdot 1},$$
$$(\overline{y}_{\cdot 22} - \overline{y}_{\cdot 21}) - (\overline{y}_{\cdot 12} - \overline{y}_{\cdot 11}) = \tau_{22} - \tau_{21} - \tau_{12} + \tau_{11} + \overline{\varepsilon}_{\cdot 22} - \overline{\varepsilon}_{\cdot 21} - \overline{\varepsilon}_{\cdot 12} + \overline{\varepsilon}_{\cdot 11},$$

with variances respectively $2\sigma^2/(BW)$ and $4\sigma^2/B$. As there are $S-1$ degrees of freedom for sub-unit treatments and $(S-1)(W-1)$ degrees of freedom for their interactions with whole-unit treatments,

$$BW(S-1) - (S-1) - (S-1)(W-1) = W(B-1)(S-1)$$

degrees of freedom remain for estimation of σ^2. If the variability between whole units is larger than that within them, that is, $\sigma^2 < \sigma_u^2$, then comparisons among sub-unit treatments and their interactions with whole-unit treatments will be more precise than among whole-unit treatments themselves.

Example 9.15 (Cake data) Table 9.24 gives data from an experiment in which six different temperatures for cooking three recipes for chocolate cake were compared. Each time a mix was made using one of the recipes, enough batter was prepared for six cakes, which were then randomly allocated to be cooked at temperatures $175, 185, \ldots, 225°C$. Thus mixes correspond to blocks, recipes are the whole-unit treatments and baking temperatures the sub-unit treatments. We suppose that the 15 mixes of each recipe were made in order $1, \ldots, 15$, so that mix is a surrogate for time.

The response is the breaking angle, found by fixing one half of a slab of cake, then pivoting the other half about the middle until breakage occurs. Let y_{rmt} denote the response for the rth recipe, mth mixture and tth temperature, where $r = 1, \ldots, 3$,

Table 9.24 Data on breaking angles (°) of chocolate cakes (Cochran and Cox, 1959, p. 300).

Recipe	Temp °C	Mix 1	2	3	4	5	6	7	8	9	10	11	12	13	14	15
1	175	42	47	32	26	28	24	26	24	24	24	33	28	29	24	26
	185	46	29	32	32	30	22	23	33	27	33	39	31	28	40	28
	195	47	35	37	35	31	22	25	23	28	27	33	27	31	29	32
	205	39	47	43	24	37	29	27	32	33	31	28	39	29	40	25
	215	53	57	45	39	41	35	33	31	34	30	33	35	37	40	37
	225	42	45	45	26	47	26	35	34	23	33	30	43	33	31	33
2	175	39	35	34	25	31	24	22	26	27	21	20	23	32	23	21
	185	46	46	30	26	30	29	25	23	26	24	27	28	35	25	21
	195	51	47	42	28	29	29	26	24	32	24	33	31	30	22	28
	205	49	39	35	46	35	29	26	31	28	27	31	34	27	19	26
	215	55	52	42	37	40	24	29	27	32	37	28	31	35	21	27
	225	42	61	35	37	36	35	36	37	33	30	33	29	30	35	20
3	175	46	43	33	38	21	24	20	24	24	26	28	24	28	19	21
	185	44	43	24	41	25	33	21	23	18	28	25	30	29	22	28
	195	45	43	40	38	31	30	31	21	21	27	26	28	43	27	25
	205	46	46	37	30	35	30	24	24	26	27	25	35	28	25	25
	215	48	47	41	36	33	37	30	21	28	35	38	33	33	25	31
	225	63	58	38	35	23	35	33	35	28	35	28	28	37	35	25

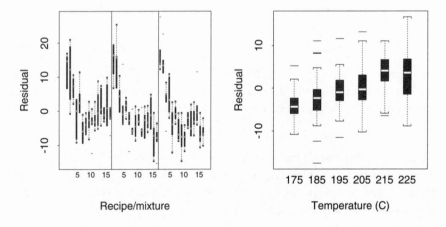

Figure 9.8 Cake data. Left: variation of $y_{rmt} - \overline{y}_{rm\cdot}$ across mixes for the three recipes. The vertical lines demarcate results for the three recipes. Right: dependence of $y_{rmt} - \overline{y}_{\cdot\cdot t}$ on temperature.

$m = 1, \ldots, 15$ and $t = 1, \ldots, 6$. The model we consider is

$$y_{rmt} = \mu + \beta_r + \gamma_m + u_{rm} + \zeta_t + \tau_{rt} + \varepsilon_{rmt}, \tag{9.11}$$

where the $u_{rm} \overset{\text{iid}}{\sim} N(0, \sigma_u^2)$ represent the whole-unit errors corresponding to the mixes made for each recipe, and the $\varepsilon_{rmt} \overset{\text{iid}}{\sim} N(0, \sigma^2)$ denote the sub-unit errors. We treat the β_r and γ_m as parameters and the u_{rm} as random variables because if the experiment was repeated, the recipes would be unchanged and the time ordering would still arise, but the mixes would be different.

The left panel of Figure 9.8 shows how $y_{rmt} - \overline{y}_{rm\cdot}$ varies across mixes for the three recipes. There is evidently a systematic effect of mix, with responses for the first few

Table 9.25 Analysis of
variance on a split-unit
basis for cakes data. The
F statistics for the upper
part are computed using
the residual sum of
squares (a) for contrasts
among whole units. Those
in the lower part are
computed using (b), the
residual sum of squares
for contrasts among split
units. There are large
differences among mixes
and temperatures, but not
among recipes. The
temperature effect is
essentially linear.

Source of variation	df	Sum of squares	Mean square	F
Mixes	14	8.159	0.583	12.15
Recipes	2	0.186	0.093	1.93
Residual (a)	28	1.343	0.048	
Temperatures	5	2.051	0.410	21.32
Linear	1	1.925	1.925	100.08
Quadratic	1	0.021	0.021	1.08
Cubic, quartic, quintic	3	0.105	0.035	1.82
Recipes × Temperatures	10	0.176	0.018	0.91
Residual (b)	210	4.040	0.019	

Table 9.26 Average log
breaking angle (degrees)
of cakes by recipe and
temperature.

	Temperature (°C)						
Recipe	175	185	195	205	215	225	Average
1	3.350	3.433	3.409	3.493	3.638	3.535	3.476
2	3.270	3.355	3.428	3.443	3.505	3.537	3.423
3	3.293	3.331	3.428	3.405	3.516	3.538	3.419
Average	3.304	3.373	3.422	3.447	3.553	3.537	3.439

mixes for each recipe substantially greater than for later ones, but any recipe differences seem small. The right panel shows roughly linear dependence on temperature of the differences $y_{rmt} - \overline{y}_{..t}$, from which whole-unit variation has been eliminated, but there is a perceptible increase in variance with mean. This suggests use of log-transformed responses, which is confirmed by a Box–Cox analysis (Example 8.23).

Table 9.25 shows the analysis of variance for the model fitted to the log responses. There are $3 \times 15 = 45$ whole units, from which a grand mean and 44 contrasts may be computed. The component of variance for these 44 degrees of freedom is shown in the upper part of the table, split into 14 degrees of freedom among mixes, 2 among recipes and 28 residuals. The mean square at (a) estimates $\sigma^2 + 6\sigma_u^2$ and is the appropriate basis for comparison of recipes and mixes. There are large differences among mixes but not among recipes.

The lower part of the table shows the $3 \times 15 \times (6 - 1) = 225$ degrees of freedom for contrasts within whole units, of which there are 5 among temperatures and $(3 - 1) \times (6 - 1)$ for the recipe × temperature interaction. The mean square for residual (b) estimates σ^2, and comparison with (a) gives estimate $(0.048 - 0.019)/6 = 0.0035$ of σ_u^2, rather small variation among mixes. A split of the overall temperature effect into linear, quadratic and remaining effects confirms the linearity of the effect of temperature on the response.

Table 9.26 shows average log breaking angles by recipe and temperature. Each average is based on 15 raw observations and has variance $\sigma_u^2 + \sigma^2/15$, but while differences between rows involve the u's, those between columns do not. Differences between two recipe averages and between two temperature averages have variances $2(\sigma_u^2 + \sigma^2/6)/15$ and $2(\sigma^2/15)/3$, estimated by $2 \times 0.048/90 = 0.033^3$ and $2 \times 0.019/45 = 0.029^2$. The difference between two temperature averages for one recipe, $\bar{y}_{r \cdot t_1} - \bar{y}_{r \cdot t_2}$, does not depend on the u_{rm}, so its variance is $2\sigma^2/15$, while the difference of two recipe averages for a given temperature, $\bar{y}_{r_1 \cdot t} - \bar{y}_{r_2 \cdot t}$, involves both u's and ε's and has variance $2(\sigma_u^2 + \sigma^2)/15$; these variances are estimated respectively by $2 \times 0.019/15 = 0.050^2$ and $2 \times \{5 \times 0.019 + 0.048)/90 = 0.056^2$.

The best summary of the results here is Table 9.26, supplemented by the standard errors for comparisons among the averages.   ∎

9.4.2 Linear mixed models

In many situations the comparison of treatments is complicated by correlations among the responses. In medical settings, for example, a common design involves repeated measures on the same individual, leading to *repeated measures* or *longitudinal* data. Related designs arise in many types of investigation. Although the notion of levels of variation underlying the classical split-plot experiment remains very useful, such data are rarely neatly balanced and their analysis and interpretation is less straightforward. In this section we briefly put such experiments in a more general context.

When confronted with a complex experiment, it is helpful to ask if it is reasonable to assume that the levels of certain factors have been selected from a population. If so, we ask if interest resides purely in the population, or also in the realized values of random variables sampled from it. When this latter is the case, then we must estimate not only properties of the population but also the underlying variables. In dairy herd breeding experiments, for example, bulls and cows are mated and the milk yield of their daughters is treated as the response. As any repetition of the experiment would involve different animals, they are regarded as randomly sampled from a population. It is useful to estimate effects for individual animals, however, in order to retain for future breeding those bulls whose daughters give the best yield. Thus although a random effects model is appropriate, estimates of the random effects are required. Similar considerations arise in many other contexts, and we now discuss inference for random effects.

We consider normal linear models of form

$$y = X\beta + Zb + \varepsilon, \tag{9.12}$$

where in addition to the usual setup the $n \times q$ matrix Z indicates how the response vector y depends on the $q \times 1$ vector of unobserved random variables b. This is called a *mixed model* because the response depends on random variables b as well as on fixed parameters β. If b is normal with mean zero and covariance matrix Ω_b, then we may write

$$y \mid b \sim N_n(X\beta + Zb, \Omega) \quad \text{and} \quad b \sim N_q(0, \Omega_b). \tag{9.13}$$

Thus the marginal density of y is normal with mean $X\beta$ and variance matrix $Z\Omega_b Z^{\mathsf{T}} + \Omega$, which does not depend on β. In most cases $\Omega = \sigma^2 I_n$, where $\sigma^2 = \text{var}(\varepsilon_j)$, and later it will be useful to write $Z\Omega_b Z^{\mathsf{T}} + \Omega = \sigma^2 \Upsilon^{-1}$, say. We use ψ to denote the vector of distinct variance ratios appearing in Υ^{-1}.

Example 9.16 (Longitudinal data) A short longitudinal study has one individual allocated to the treatment and two to the control, with observations

$$y_{1j} = \beta_0 + b_1 + \varepsilon_{1j}, \quad y_{21} = \beta_0 + b_2 + \varepsilon_{21}, \quad y_{3j} = \beta_0 + \beta_1 + b_3 + \varepsilon_{3j}, \quad j = 1, 2.$$

Thus there are two measurements on the first and third individuals, and just one on the second. The b_j represent variation among individuals and the ε_{ij} variation between measures on the same individuals. If the b's and ε's are all mutually independent with variances σ_b^2 and σ^2, then

$$\begin{pmatrix} y_{11} \\ y_{12} \\ y_{21} \\ y_{31} \\ y_{32} \end{pmatrix} = \begin{pmatrix} 1 & 0 \\ 1 & 0 \\ 1 & 0 \\ 1 & 1 \\ 1 & 1 \end{pmatrix} \begin{pmatrix} \beta_0 \\ \beta_1 \end{pmatrix} + \begin{pmatrix} 1 & 0 & 0 \\ 1 & 0 & 0 \\ 0 & 1 & 0 \\ 0 & 0 & 1 \\ 0 & 0 & 1 \end{pmatrix} \begin{pmatrix} b_1 \\ b_2 \\ b_3 \end{pmatrix} + \begin{pmatrix} \varepsilon_{11} \\ \varepsilon_{12} \\ \varepsilon_{21} \\ \varepsilon_{31} \\ \varepsilon_{32} \end{pmatrix},$$

and this fits into formulation (9.12) with $\Omega_b = \sigma_b^2 I_3$ and $\Omega = \sigma^2 I_5$. Here ψ comprises the scalar σ_b^2 / σ^2, and hence the variance matrix

$$\Omega + Z\Omega_b Z^{\mathsf{T}} = \begin{pmatrix} \sigma_b^2 + \sigma^2 & \sigma_b^2 & 0 & 0 & 0 \\ \sigma_b^2 & \sigma_b^2 + \sigma^2 & 0 & 0 & 0 \\ 0 & 0 & \sigma_b^2 + \sigma^2 & 0 & 0 \\ 0 & 0 & 0 & \sigma_b^2 + \sigma^2 & \sigma_b^2 \\ 0 & 0 & 0 & \sigma_b^2 & \sigma_b^2 + \sigma^2 \end{pmatrix}$$

may be written as

$$\sigma^2 \Upsilon^{-1} = \sigma^2 \begin{pmatrix} 1 + \psi & \psi & 0 & 0 & 0 \\ \psi & 1 + \psi & 0 & 0 & 0 \\ 0 & 0 & 1 + \psi & 0 & 0 \\ 0 & 0 & 0 & 1 + \psi & \psi \\ 0 & 0 & 0 & \psi & 1 + \psi \end{pmatrix},$$

of block diagonal form. ∎

In principle likelihood inference for the parameters of this model may be based on the marginal normal density of y, which gives log likelihood

$$\ell(\beta, \sigma^2, \psi) \equiv -\frac{1}{2\sigma^2}(y - X\beta)^T \Upsilon (y - X\beta) - \frac{n}{2} \log \sigma^2 + \frac{1}{2} \log |\Upsilon|,$$

where Υ depends on ψ. For known ψ the maximum likelihood estimators of β and σ^2 are

$$\widehat{\beta}_\psi = (X^{\mathsf{T}} \Upsilon X)^{-1} X^{\mathsf{T}} \Upsilon y, \quad \widehat{\sigma}_\psi^2 = n^{-1}(y - X\widehat{\beta})^{\mathsf{T}} \Upsilon (y - X\widehat{\beta}),$$

so the profile log likelihood for ψ is $\ell_p(\psi) \equiv -\frac{1}{2}n \log \widehat{\sigma}_\psi^2 + \frac{1}{2} \log |\Upsilon|$. We maximize this to estimate ψ, and then obtain maximum likelihood estimates $\widehat{\beta}_{\widehat{\psi}}$ and $\widehat{\sigma}_{\widehat{\psi}}^2$. Thus inference boils down to maximization of $\ell_p(\psi)$.

Unfortunately life is not so simple. One difficulty is that the maximum likelihood variance estimators can have large downward bias because no adjustment is made for the degrees of freedom lost in estimating the $p \times 1$ vector β. In such models p can be large, and then it is important to replace the divisor n in $\widehat{\sigma}^2$ by the true degrees of freedom $n - p$. Adjustment both for this and for estimation of the elements of ψ can be performed by maximizing the modified log likelihood

$$\ell(\beta, \sigma^2, \psi) + \frac{p}{2} \log \sigma^2 - \frac{1}{2} \log |X^\mathsf{T} \Upsilon X|.$$

This procedure, known as REML or restricted maximum likelihood estimation, is justified in Section 12.2. It turns out to be equivalent to use of a marginal likelihood, that is, a likelihood formed from a cunningly chosen marginal density rather than the full density of the data.

A second difficulty is that the domain for ψ is $[0, \infty)^{\dim \psi}$. If the maximum occurs on the boundary of this set, then standard likelihood theory does not apply to confidence intervals and so forth. Care must anyway be used unless the maximum lies well away from the boundary. If so, standard errors for $\widehat{\beta}$ are found from $\sigma^2(X^\mathsf{T}\Upsilon X)^{-1}$ with parameters replaced by estimates.

dim ψ is the dimension of ψ.

A third difficulty is computational: in realistic problems the matrices involved in such models can be large enough that even specially designed optimization routines converge only slowly.

Prediction of random effects

Once estimates of β, σ^2, and ψ have been obtained, the question arises how to perform inference for the random variables b. We prefer to reserve the term *estimation* for unknown parameters and to speak of *prediction* of unobserved random variables. In normal models it is natural to choose the predictor $\tilde{b} = \tilde{b}(y)$ to be the function of y that minimizes the mean squared prediction error

$$E[\{\tilde{b}(y) - b\}^\mathsf{T}\{\tilde{b}(y) - b\}],$$

where the expectation is over both b and y. It is straightforward to show that this is achieved by taking $\tilde{b}(y) = E(b \mid y)$, the conditional mean of b given y. As b and y have a joint normal distribution, we obtain (Exercise 9.4.5)

$$E(b \mid y) = \left(Z^\mathsf{T}\Omega^{-1}Z + \Omega_b^{-1}\right)^{-1} Z^\mathsf{T}\Omega^{-1}(y - X\beta), \qquad (9.14)$$

$$\mathrm{var}(b \mid y) = \left(Z^\mathsf{T}\Omega^{-1}Z + \Omega_b^{-1}\right)^{-1}. \qquad (9.15)$$

Replacement of the unknown parameters by estimates results in the predictions $\tilde{b}$ and their estimated variance. It turns out that the $\tilde{b}$ are best linear unbiased predictors (Problem 9.6). If Ω_b^{-1} was absent, then (9.14) would be the weighted least squares estimator from regressing $(y - X\beta)$ on the columns of Z with weight matrix Ω^{-1}.

They are often called BLUPs.

The presence of Ω_b^{-1} means that $\tilde{b}$ is shrunk towards zero compared to the weighted least squares estimator, and for this reason $\tilde{b}$ is known as a *shrinkage* estimator.

The residuals too are modified due to shrinkage. As

$$
\begin{aligned}
y - X\widehat{\beta} &= Z\tilde{b} + y - X\widehat{\beta} - Z\tilde{b} \\
&= Z\tilde{b} + \left\{ I_n - Z\left(Z^{\mathsf{T}}\widehat{\Omega}^{-1}Z + \widehat{\Omega}_b^{-1}\right)^{-1} Z^{\mathsf{T}}\widehat{\Omega}^{-1} \right\}(y - X\widehat{\beta}),
\end{aligned}
$$

the residuals $y - X\widehat{\beta}$ split into two parts, the first $Z\tilde{b}$ being attributable to the predicted random effects, and the second being the usual residual $y - X\widehat{\beta}$ shrunk towards zero; this estimates ε.

Example 9.17 (One-way layout) Consider the unbalanced one-way layout model

$$
y_{ij} = \mu + b_i + \varepsilon_{ij}, \quad j = 1, \ldots, n_i, \quad i = 1, \ldots, q,
$$

in which the group effects $b_i \stackrel{\text{iid}}{\sim} N(0, \sigma_b^2)$ independently of the individual errors $\varepsilon_{ij} \stackrel{\text{iid}}{\sim} N(0, \sigma^2)$. This generalizes (9.8). In terms of (9.12),

$$
\Omega = \sigma^2 I_n, \quad \Omega_b = \sigma_b^2 I_q, \quad X = 1_n, \quad Z = \begin{pmatrix} 1_{n_1} & 0 & \cdots & 0 \\ 0 & 1_{n_2} & \cdots & 0 \\ \vdots & \vdots & \ddots & \vdots \\ 0 & 0 & \cdots & 1_{n_q} \end{pmatrix},
$$

where $n = n_1 + \cdots + n_q$. Substitution into (9.14) and (9.15) reveals that the ith element of $\tilde{b}$ and its estimated variance are

$$
\tilde{b}_i = \frac{\overline{y}_{i\cdot} - \overline{y}_{\cdot\cdot}}{1 + \widehat{\sigma}^2/\left(n_i\widehat{\sigma}_b^2\right)}, \quad \frac{1}{1/\widehat{\sigma}_b^2 + n_i/\widehat{\sigma}^2},
$$

so the fixed-effects estimator $\overline{y}_{i\cdot} - \overline{y}_{\cdot\cdot}$ is shrunk towards zero by an amount that depends on the estimated variance ratio. The shrinkage will be considerable if $\widehat{\sigma}^2/n_i \gg \widehat{\sigma}_b^2$, corresponding to large variation in the group averages owing to individual variances compared to the variation between groups, as in Example 9.14. The data are then almost a simple random sample of size n, so strong shrinkage is not surprising.

The variance formula is also instructive, as $\text{var}(\tilde{b}_i \mid y) \to 0$ when $\sigma_b^2 \to 0, \sigma^2 \to 0$, or $n_i \to \infty$. In the first case, there is no variation between groups, and hence $b_i = 0$ with probability one. In the second two cases, the value of b_i is known exactly, because variation around it is negligible. The practical implication is that consistent inference for b_i is impossible when σ_b^2 and σ^2 take positive values: even if $q \to \infty$, the amount of information on any given b_i does not accumulate unless $n_i \to \infty$, and this is rarely the case. This applies to estimation of random effects more generally. ∎

Example 9.18 (Rat growth data) Table 9.27 gives the weights of $n = 30$ young rats measured for five weeks. The left panel of Figure 9.9 shows that although the weight of each rat grows roughly linearly, neither slope nor intercept appears to be common to all the animals. This is confirmed by the analysis of variance from fitting standard linear models with common intercept and slope, different intercepts, and both intercepts and slopes different: the F tests are all highly significant.

		Week						Week		
	1	2	3	4	5		1	2	3	4	5
1	151	199	246	283	320	16	160	207	248	288	324
2	145	199	249	293	354	17	142	187	234	280	316
3	147	214	263	312	328	18	156	203	243	283	317
4	155	200	237	272	297	19	157	212	259	307	336
5	135	188	230	280	323	20	152	203	246	286	321
6	159	210	252	298	331	21	154	205	253	298	334
7	141	189	231	275	305	22	139	190	225	267	302
8	159	201	248	297	338	23	146	191	229	272	302
9	177	236	285	340	376	24	157	211	250	285	323
10	134	182	220	260	296	25	132	185	237	286	331
11	160	208	261	313	352	26	160	207	257	303	345
12	143	188	220	273	314	27	169	216	261	295	333
13	154	200	244	289	325	28	157	205	248	289	316
14	171	221	270	326	358	29	137	180	219	258	291
15	163	216	242	281	312	30	153	200	244	286	324

Table 9.27 Weights (units unknown) of 30 young rats over a five-week period (Gelfand *et al.*, 1990).

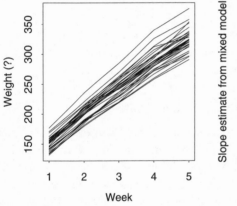 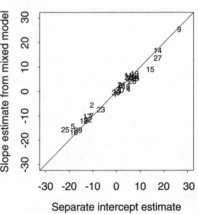

Figure 9.9 Rat growth data. Left: weekly weights of 30 young rats. Right: shrinkage of individual slope estimates towards overall slope estimate; the solid line has unit slope, and the estimates from the mixed model lie slightly closer to zero than the individual estimates.

We treat the rats as a sample from a population of similar creatures, with different initial weights and growing at different rates. To model this we express the data from the jth rat as

$$y_{jt} = \beta_0 + b_{j0} + (\beta_1 + b_{j1})x_{jt} + \varepsilon_{jt}, \quad t = 1, \dots 5,$$

where the random variables (b_{j0}, b_{j1}) have a joint normal distribution with mean vector zero and unknown variance matrix. In matrix terms we have

$$\begin{pmatrix} y_{j1} \\ \vdots \\ y_{j5} \end{pmatrix} = \begin{pmatrix} 1 & x_{j1} \\ \vdots & \vdots \\ 1 & x_{j5} \end{pmatrix} \begin{pmatrix} \beta_0 \\ \beta_1 \end{pmatrix} + \begin{pmatrix} 1 & x_{j1} \\ \vdots & \vdots \\ 1 & x_{j5} \end{pmatrix} \begin{pmatrix} b_{j0} \\ b_{j1} \end{pmatrix} + \begin{pmatrix} \varepsilon_{j1} \\ \vdots \\ \varepsilon_{j5} \end{pmatrix}, \quad j = 1, \dots, n,$$

and the overall model is obtained by stacking these expressions. Below we take $(x_{j1}, \dots, x_{j5}) = (0, \dots, 4)$, so that the intercept β_0 corresponds to the weight in week 1. There are just $p = 2$ population parameters β_0 and β_1, but $q = 60$ because

	Fixed		Random	
Parameter	Estimate	Standard error	Variance	Correlation
Intercept	156.05	2.16 (2.13)	$10.93^2\ (10.71^2)$	
Slope	43.27	0.73 (0.72)	$3.53^2\ (3.46^2)$	0.18 (0.19)

there are two random variables per rat. We assume that the within-rat errors ε_{jt} are independent normal variables with variances σ^2, independent of the b's.

Table 9.28 gives estimates from REML and maximum likelihood fits of this model. As expected, the maximum likelihood estimates for variances are smaller than the REML estimates, but here p is small and the difference is minimal. The estimated mean weight in week 1 is 156, but the variability from rat to rat has estimated standard deviation of about 11 about this. The slopes show similarly large variation. Correlation between the slope and intercept variables is small, however. The measurement error variance $\widehat{\sigma}^2 = 5.82^2$ is smaller than is the inter-rat variation in intercepts, but exceeds that for slopes.

The right panel of Figure 9.9 shows how the slope estimates from fitting separate models to each rat are shrunk towards the overall value. The amount of shrinkage is small, owing to the relatively large variation among the rats relative to $\widehat{\sigma}^2$, and as it depends on the intercepts it is not uniform.

Probability plots show that the residuals and random effects $\tilde{b}$ are reasonably close to normal, but a plot of residuals against week suggests adding quadratic terms $(\beta_2 + b_{j2})x_{jt}^2$. Their inclusion reduces AIC for REML from 1096.58 to 1013.36, a large improvement, but the resulting model involves predicting 90 b's from 150 observations, leaving only about two observations per rat for model checking. Fortunately cubic terms do not seem to be necessary as well. ∎

Sometimes it is helpful to separate b into sub-vectors corresponding to different levels of variation. For example, educational studies may involve classes of students in different schools belonging to different educational authorities, so that comparisons of outcomes must take into account different levels of random effects as well as fixed effects corresponding to types of school, socio-economic background of students, and so forth. For data with L levels of variation it may be useful to write this as a multi-level model

$$y = X\beta + Z_L b_L + \cdots + Z_0 b_0, \tag{9.16}$$

where the $q_l \times 1$ vectors b_l are all mutually independent with means zero and variance matrices Ω_l. Then the marginal mean of y is $X\beta$, while its variance is $\sum_{l=0}^{L} Z_l \Omega_l Z_l^{\mathrm{T}}$. For consistency we set $Z_0 = I_n$ and let b_0 contain the errors, so $b_0 = \varepsilon$ and $\Omega_0 = \sigma^2 I_n$. The examples above have $L = 1$, so in addition to measurement error there is just one other level of variation, corresponding to individuals. More generally $L > 1$, and var(y) is formed by adding block diagonal matrices. References to fuller discussions can be found in Section 9.5.

| | Pipette and counting chamber | | | | | | | | | |
Doctor	1	2	3	4	5	6	7	8	9	10
A	427	372	418	440	349	484	430	416	449	464
B	434	420	385	472	415	420	415	396	439	424
C	480	421	473	496	474	411	472	423	502	488
D	451	369	500	464	444	410	422	396	459	471
E	462	453	450	520	489	409	508	347	440	391

Table 9.29 The numbers of red blood cells counted by five doctors using ten sets of apparatus.

Exercises 9.4

1   Consider (9.8).
(a) Show that a confidence interval for the mean of the tth group, $\alpha_t = \mu + b_t$, may be based on $T = (\overline{y}_{t\cdot} - \alpha_t)/[SS_w/\{TR(R-1)\}]^{1/2}$, and give its distribution.
(b) Suppose we take a single observation on a randomly selected block. Show that its variance is $\sigma^2 + \sigma_b^2$, and that this is estimated unbiasedly by $\{TSS_b + (T-1)SS_w\}/\{RT(T-1)\}$.
(c) Suppose a fixed number $n = RT$ of units is available, and that it is required to minimize the variance of the population mean estimator, $\overline{y}_{\cdot\cdot}$. Show that we should take T as large as possible, that is, $R = 2$.
(d) Suppose there is a cost c_0 for measuring the response on each unit, and a cost c_1 for each group. Show that the total cost is $RTc_0 + Tc_1$, and find R to minimize $\mathrm{var}(\overline{y}_{\cdot\cdot})$ subject to a fixed total cost.

2   Discuss how to check the assumptions of the components of variance model (9.8) using (i) normal probability plots of the $y_{tr} - \overline{y}_{t\cdot}$ and of the $\overline{y}_{t\cdot}$ and (ii) chi-squared probability plots of the group sums of squares $\sum_r (y_{tr} - \overline{y}_{t\cdot})^2$. In each case give the expected slope and intercept of the plot.
Show that if R is small a normal scores plot of the $(y_{tr} - \overline{y}_{t\cdot})/\{(R-1)/R\}^{1/2}$ is preferable to one based on the $y_{tr} - \overline{y}_{t\cdot}$.
Discuss whether (9.8) is appropriate for the data of Example 9.14.

3   Table 9.29 gives the numbers of red blood cells counted by five doctors using ten sets of apparatus. Suppose that both doctors and sets of apparatus are thought of as randomly selected from suitable populations, and that the response for the rth doctor and cth set of apparatus is

$$y_{rc} = \mu + d_r + a_c + \varepsilon_{rc},$$

where d_r, a_c, and ε_{rc} are independent normal variables with zero means and variances σ_d^2, σ_a^2, and σ^2.
(a) Show that if the means of the d_r, a_c, and ε_{rc} were in fact non-zero, they could not be distinguished from μ. Give a careful interpretation of μ.
(b) By arguing conditionally on the values of the d_r and a_c, show that the sums of squares for rows, $\sum_{r,c}(y_r - \overline{y}_{\cdot\cdot})^2$, columns, $\sum_{r,c}(\overline{y}_{\cdot c} - \overline{y}_{\cdot\cdot})^2$, and residuals, $\sum_{r,c}(y_{rc} - \overline{y}_r - \overline{y}_{\cdot c} + \overline{y}_{\cdot\cdot})^2$, are independent. Obtain their distributions, and hence give formulae for unbiased estimates of the variances.
(c) The sums of squares for the analysis of variance table are 2969 for pipettes on 9 df, 2938 for doctors on 4 df, and 1176 for residual on 36 df. Obtain estimates of σ_d^2, σ_a^2, and σ^2.
(d) Now suppose that general practitioners are to perform these measurements on a routine basis, with results referred to a central laboratory. Under the assumption that the data are normal, give the standard error for a measurement taken by a particular GP (i) if the apparatus is reusable, (ii) if a new set of apparatus must be used for each measurement. What standard error is appropriate if the measurements are rarely made, so that in effect

both GP and apparatus are new? What if the average of k measurements is recorded, and (i) apparatus is reusable, (ii) apparatus is not reusable?

4 Write down the linear mixed models corresponding to (9.8) and (9.11).

5 Use (3.21) and Exercise 8.5.2 to obtain (9.14) and (9.15).

6 On page 458, let $b^\dagger$ be any predictor of b based on y. Show that

$$\text{cov}(b, \tilde{b}) = \text{var}(\tilde{b}), \quad \text{cov}(\tilde{b}, y) = \text{cov}(b, y), \quad \text{cov}(b^\dagger, b) = \text{cov}(b^\dagger, \tilde{b}),$$

and deduce that

$$\text{corr}(b^\dagger, b)^2 = \text{corr}(b^\dagger, \tilde{b})^2 \text{corr}(\tilde{b}, b)^2.$$

Hence show that $\tilde{b}$ is the predictor of b that maximizes $\text{corr}(b^\dagger, b)$.

7 Consider applying the EM algorithm (Section 5.5.2) for estimation in a normal mixed model. Show that if the random effects b are treated as unobserved data, then the complete-data log likelihood is

$$-\frac{1}{2}\log|\Omega| - \frac{1}{2}(y - X\beta - Zb)^T\Omega^{-1}(y - X\beta - Zb) - \frac{1}{2}\log|\Omega_b| - \frac{1}{2}b^T\Omega_b^{-1}b,$$

and show that the only quantity needed for the M-step for estimation of components of Ω_b is $\text{E}\left(b^T\Omega_b^{-1}b \mid y; \theta'\right)$.

In the special case $\Omega_b = \sigma_b^2 I_q$, $\Omega = \sigma^2 I_n$, show that $\text{E}(b^T b \mid y; \theta')$ equals

$$\text{tr}\left\{\sigma_b'^2 I_q - (\sigma_b'/\sigma')^2 Z^T \Upsilon' Z\right\} + (\sigma_b'/\sigma')^4 (y - X\beta')^T \Upsilon' Z Z^T \Upsilon'(y - X\beta').$$

Hence write down the form of the EM algorithm for this model.
(Searle *et al.*, 1992, Section 8.3)

8 Another approach to estimation in mixed models starts from noticing that

$$\text{E}\{(y - X\beta)(y - X\beta)^T\} = \Omega + Z\Omega_b Z^T$$

is linear in the variance parameters. Thus given an estimate $\widehat{\beta}$, we could stack the unique elements of $(y - X\widehat{\beta})(y - X\widehat{\beta})^T$ as a vector, v, say, and estimate the variance parameters by least squares regression of v on the appropriate design matrix. We then take $\widehat{\beta} = (X^T \Upsilon X)^{-1} X^T \Upsilon y$, where Υ is formed using the variance estimates, and iterate the procedure.

This is sometimes called iterative generalized least squares or IGLS estimation.

Give the details of this for Example 9.16, using as initial value $\sigma_b^2 = 0$.

What difficulties do you see with this approach in general? Say how they might be overcome.

Frank Yates (1902–1994) was born in Manchester and educated there and in Cambridge. After working on a survey in Ghana he became Fisher's assistant at Rothamsted Experimental Station, where he rapidly became head of the statistics department. He made fundamental contributions to the design and analysis of experiments and to sample surveys. He quickly saw the importance of computing: in the 1950s he and his colleagues wrote machine code programs for analysis of variance and for survey analysis.

9.5 Bibliographic Notes

Designed experiments were used in the nineteenth century and earlier, but R. A. Fisher was the first to realise the importance of randomization, and his ideas had a strong impact from the 1920s onwards. His 1935 book on design of experiments, re-issued as part of Fisher (1990), is fundamental reading. Important further developments, particularly in agricultural experimentation, were due to F. Yates, with Yates (1937) highly influential. An excellent recent account is Cox and Reid (2000), which contains a full treatment of the topics of this chapter and other topics not mentioned here, with many further references. A more elementary discussion is Cobb (1998). Older standard texts are Cochran and Cox (1959) and the excellent non-mathematical treatment of Cox (1958).

The study of causality is central to scientific thought, but has been little discussed by statisticians until fairly recently. A valuable account and excellent starting-point for further reading is Chapter 8 of Edwards (2000), while Holland (1986) is a good review making links to the philosophical study of causation. Cox (1992) and Section 8.7 of Cox and Wermuth (1996) give a somewhat different perspective. Contrasting views on the usefulness of counterfactuals are held by Dawid (2000), Lauritzen (2001), and Pearl (2000).

Scheffé (1959) is a standard account of the analysis of variance.

Box *et al.* (1978) and Fleiss (1986) respectively discuss industrial experimentation and medical studies. Atkinson and Donev (1992) give a clear discussion of optimal experimental design; see also Silvey (1980) for a more theoretical account.

Components of variance models originated in astronomy in the 1860s and have been rediscovered and renamed many times since, being also known as hierarchical or multilevel models. Chapter 2 of Searle *et al.* (1992) gives a brief history oriented towards biometry and agriculture, while Goldstein (1995) describes their use in the social sciences, using slightly different estimation techniques and with a largely disjoint set of references! Although R. A. Fisher had discussed components of variance in the 1920s and 1930s, important work by Henderson (1953) and Hartley and Rao (1967) was key in a more general reformulation, while Patterson and Thompson (1971) built on earlier work to give a general discussion of REML estimation. Robinson (1991) is a passionate advocate of best linear unbiased prediction, with an interesting and wide-ranging discussion; see particularly the contribution by T. P. Speed. McCulloch and Searle (2001) give a recent account of variance components estimation in linear and generalized linear models.

9.6 Problems

1   Example 9.6 is a *two-way layout with replication*, in which the jth replicate in row r and column c is

$$y_{rcj} = \mu + \alpha_r + \beta_c + \gamma_{rc} + \varepsilon_{rcj}, \quad r = 1, \ldots, R, \quad c = 1, \ldots, C, \quad j = 1, \ldots, k.$$

The α_r and β_c represent the main effects of rows and columns; the γ_{rc} are row×column interactions; and $\varepsilon_{rcj} \overset{\text{iid}}{\sim} N(0, \sigma^2)$.

(a) Explain why an external estimate of σ^2 is needed if the γ_{rc} are known not to be constant, and $k = 1$.

(b) A first step in the analysis of such data is to calculate the cell mean and sums of squares $\overline{y}_{rc.}$ and $\sum_j (y_{rcj} - \overline{y}_{rc.})^2$. Show that the distribution of each cell sum of squares is $\sigma^2 \chi^2_{k-1}$, and explain what you might expect to learn from a plot of $\log \sum_j (y_{rcj} - \overline{y}_{rc.})^2$ against $\log \overline{y}_{rc.}$. What does this plot show for the poisons data?

(c) The analysis of variance for this design is in Table 9.30. Show that

$$\mathrm{E}\left\{\sum_{r,c,j} (\overline{y}_{r..} - \overline{y}_{...})^2\right\} = (R-1)\sigma^2 + kC \sum_r (\alpha_r - \overline{\alpha}. + \overline{\gamma}_{r.} - \overline{\gamma}_{..})^2,$$

and write down $\mathrm{E}\{\sum_{r,c,j} (\overline{y}_{.c.} - \overline{y}_{...})^2\}$. Explain why these depend on the α_r and β_c only through $\alpha_r - \overline{\alpha}.$ and $\beta_c - \overline{\beta}.$.

Table 9.30 Analysis of variance for two-way layout with replication.

Terms	df	Sum of squares
Rows	$R - 1$	$\sum_{r,c,j}(\bar{y}_{r\cdot\cdot} - \bar{y}_{\cdots})^2$
Columns	$C - 1$	$\sum_{r,c,j}(\bar{y}_{\cdot c\cdot} - \bar{y}_{\cdots})^2$
Rows × Columns	$(R - 1)(C - 1)$	$\sum_{r,c,j}(\bar{y}_{rc\cdot} - \bar{y}_{r\cdot\cdot} - \bar{y}_{\cdot c\cdot} + \bar{y}_{\cdots})^2$
Residual	$RC(k - 1)$	$\sum_{r,c,j}(y_{rcj} - \bar{y}_{rc\cdot})^2$

(d) Show that

$$\mathrm{E}\left\{\sum_{r,c,j}(\bar{y}_{rc\cdot} - \bar{y}_{r\cdot\cdot} - \bar{y}_{\cdot c\cdot} + \bar{y}_{\cdots})^2\right\} = (R - 1)(C - 1)\sigma^2 + k\sum_{rc}(\gamma_{rc} - \bar{\gamma}_{r\cdot} - \bar{\gamma}_{\cdot c} + \bar{\gamma}_{\cdots})^2.$$

Under what circumstances does this equal $(R - 1)(C - 1)\sigma^2$?

2 Let $y_{gr}, g = 1, \ldots, G, r = 1, \ldots, R$, be independent normal random variables with means μ_{gr} and common variance σ^2.
(a) Assume the one-way analysis of variance model, namely that $\mu_{gr} = \mu_g$, so that the y_{gr} are replicate measurements with the same mean, and find the sufficient statistics for the μs and σ^2. Show that these are equivalent to

$$\bar{y}_{1\cdot}, \ldots, \bar{y}_{G\cdot}, \quad SS = \sum_{g=1}^{G}\sum_{r=1}^{R}(y_{gr} - \bar{y}_{g\cdot})^2,$$

where $\bar{y}_{g\cdot} = R^{-1}\sum_{r=1}^{R} y_{gr}$; note that

$$\sum_r(y_{gr} - \mu_g)^2 = \sum_r(y_{gr} - \bar{y}_{g\cdot})^2 + R(\bar{y}_{g\cdot} - \mu_g)^2.$$

Find the distribution of $\sum_{r=1}^{R}(y_{gr} - \bar{y}_{g\cdot})^2$.

(b) Prove that SS is independent of the group means, and that it is proportional to a chi-squared random variable on $G(R - 1)$ degrees of freedom.
(c) Let $\bar{y}_{\cdot\cdot} = G^{-1}\sum_g \bar{y}_{g\cdot}$ denote the overall mean. If $\mu_1 = \cdots = \mu_G$, show that the distribution of $SS_G = R\sum_{g=1}^{G}(\bar{y}_{g\cdot} - \bar{y}_{\cdot\cdot})^2$ is proportional to a chi-squared distribution on $G - 1$ degrees of freedom. Hence find the distribution of $G(R - 1)SS_G/(G - 1)S^2$, when the means are equal.
(d) Samples of the same material are sent to four laboratories for chemical analysis as part of a study to determine whether laboratories give the same results. The results for laboratories A–D are:

A	58.7	61.4	60.9	59.1	58.2
B	62.7	64.5	63.1	59.2	60.3
C	55.9	56.1	57.3	55.2	58.1
D	60.7	60.3	60.9	61.4	62.3

$F_{3,16}(0.95) = 3.24$.

Test the hypothesis that the means are different and comment.

3 (a) For $n = 2m + 1$ and positive integer m, suppose that $y_1, \ldots, y_n$ follow the normal linear model

$$y_j = \beta_0 + \sum_{k=1}^{m}\{\beta_k\cos(2\pi kj/n) + \gamma_k\sin(2\pi kj/n)\} + \varepsilon_j.$$

Show that the last $2m$ columns of the design matrix for this model are orthogonal contrasts, and find the least squares estimators of the parameters.
(b) Show that the overall sum of squares $\sum y_j^2$ may be split into a component $n\bar{y}^2$ corresponding to the grand mean, and m components $I_j = n(\hat{\beta}_j^2 + \hat{\gamma}_j^2)/2$ corresponding to variation with frequency $2\pi j/n$, $j = 1, \ldots, m$. Show that the I_j are independent, and

that if there is no cyclical variation, $\frac{1}{2}I_j$ has an exponential distribution with mean σ^2, whatever the value of n.

(c) Dataframe `venice` contains the annual maximum tides at Venice for the 51 years 1931–1981. It has been suggested that they may vary according to the astronomical tidal cycle, which has period 18.62 years, and that they may also be affected by the sunspot cycle, whose period is 11 years. To assess this:

```
attach(venice)
split.screen(c(1,2))
screen(1); plot(year,sea,ylab="Sea level (cm)")
n <- 51;
k1 <- 19; omega1 <- 2*pi*k1/n # roughly 18.62 years
k2 <- 11; omega2 <- 2*pi*k2/n
X19 <- cbind(sin(year*omega1),cos(year*omega1))
X11 <- cbind(sin(year*omega2),cos(year*omega2))
crossprod(cbind(rep(1,n), X11,X19)) # matrix X^\T X
venice.lm <- lm(sea~X19+X11)
anova(venice.lm,test="F")
```

Do these cycles seem to be present? How would the knowledge that the errors are non-normal affect your conclusion?

Hint: If $\omega = 2\pi p/n$, where p is an integer in the range $1, \ldots, [n/2]$, and k_1, k_2 are integers in the range $1, \ldots, n$, then

To verify this, consider the real and imaginary parts of the sums $e^{ik_1\omega} + \cdots + e^{ik_1\omega n}$ and $e^{i(k_1+k_2)\omega} + \cdots + e^{i(k_1+k_2)\omega n}$.

$$\sum_{j=1}^{n}\cos(k_1\omega j) = \sum_{j=1}^{n}\sin(k_1\omega j) = \sum_{j=1}^{n}\cos(k_1\omega j)\sin(k_2\omega j) = 0,$$

$$\sum_{j=1}^{n}\cos(k_1\omega j)\cos(k_2\omega j) = \sum_{j=1}^{n}\sin(k_1\omega j)\sin(k_2\omega j) = \begin{cases} n/2, & k_1 = k_2, \\ 0, & \text{otherwise.} \end{cases}$$

4 (a) Suppose that times had been obtained only for the odd-numbered setups Example 8.4. Further suppose that two people had been involved, and that one had ridden the bike for those setups with the seat in the low position, while the other had done so for those with the seat in the high position. Show that in this case the estimated seat and person effects are the same, and discuss what this implies about what the seat height should be.

(b) Now suppose that one person had ridden the bike for setups 1, 7, 9, and 15, while the other person had done so for setups 3, 5, 11, and 13. Show that provided there are no second- and higher-order interactions, the seat, tyre, dynamo and person effects could all be estimated from this experiment. What problem would arise if there was known to be an interaction of seat and dynamo? Explain how to modify the design to overcome this if it is known that there are no other second-order interactions.

5 The 3×3 Latin square laid out as

$$\begin{array}{ccc} A & B & C \\ B & C & A \\ C & A & B \end{array}$$

has nine observations classified by rows, columns, and the treatments A, B, and C. The corner-point parametrization for the means is

$$\begin{array}{ccc} \mu & \mu+\alpha_1+\gamma_1 & \mu+\alpha_2+\gamma_2 \\ \mu+\delta_1+\gamma_1 & \mu+\alpha_1+\delta_1+\gamma_2 & \mu+\alpha_2+\delta_1 \\ \mu+\delta_2+\gamma_2 & \mu+\alpha_1+\delta_2 & \mu+\alpha_2+\delta_2+\gamma_1 \end{array} ;$$

α_1, α_2 are column effects, δ_1, δ_2 are row effects and γ_1, γ_2 are treatment effects.

Write out the corresponding design matrix and verify that the effects of rows, columns, and treatments are not orthogonal. Check that the matrix is orthogonal when the terms for rows, columns and treatments are centred. Without doing any calculations, say whether the same is true for the *Graeco-Latin square*

Aα	Bβ	Cγ
Bγ	Cα	Aβ
Cβ	Aγ	Bα

in which there is the further set of treatments α, β, γ.

6   (a) Let A be a $q \times q$ positive definite symmetric matrix, and let b and y be two random variables with a joint distribution. Let $b^{\dagger} = b^{\dagger}(y)$ denote any predictor of b based on y, and let $\tilde{b} = E(b \mid y)$, assuming this is finite for all y. By writing

$$E\{(b^{\dagger} - b)^{\mathrm{T}} A(b^{\dagger} - b)\} = E_y E_{b|y}\{(b^{\dagger} - \tilde{b} + \tilde{b} - b)^{\mathrm{T}} A(b^{\dagger} - \tilde{b} + \tilde{b} - b) \mid y\},$$

deduce that this is minimized when $b^{\dagger} = \tilde{b}$, for any A and any joint distribution. Note that $E(\tilde{b}) = E(b)$, so in this sense $\tilde{b}$ is unbiased.

(b) Now consider the class of linear predictors $b^{\dagger}(y) = a + By$, where a and B are constants of dimensions $q \times 1$ and $q \times n$. Let $W = b - By$, and show that

$$E\{(b^{\dagger} - b)^{\mathrm{T}} A(b^{\dagger} - b)\} = \{a - E(W)\}^{\mathrm{T}} A\{a - E(W)\} + \mathrm{tr}\{A\,\mathrm{var}(W)\}.$$

Deduce that this is minimized by taking $a = -BX\beta$ and $B = \sigma^{-2}\Omega_b Z^{\mathrm{T}}\Upsilon$. Hence show that $\sigma^{-2}\Omega_b Z^{\mathrm{T}}\Upsilon(y - X\beta)$ is the best linear predictor of b whatever the distributions of b and y.

You may like to know that $|cI_k + d1_k 1_k^{\mathrm{T}}| = c^{k-1}(c + kd)$.

7   (a) Show that the log likelihood in Example 9.17 is

$$\ell(\mu, \sigma^2, \psi) \equiv -\frac{1}{2\sigma^2}\left\{ SS_w + \sum_{i=1}^{q} \frac{n_i(\bar{y}_{i.} - \mu)^2}{1 + n_i\psi} \right\} - \frac{n}{2}\log\sigma^2 - \frac{1}{2}\sum_{i=1}^{q}\log(1 + n_i\psi),$$

where $\psi = \sigma_b^2/\sigma^2$ and $SS_w = \sum_{i,j}(y_{ij} - \bar{y}_{i.})^2$.

(b) Show that using REML increases the log likelihood by

$$\frac{1}{2}\log\sigma^2 - \frac{1}{2}\log\left\{ \sum_{i=1}^{q}\frac{n_i}{1 + n_i\psi} \right\}.$$

(c) Show that if ψ is known, then

$$\widehat{\mu}(\psi) = \frac{\sum_{i=1}^{q} n_i \bar{y}_{i.}/(1 + \psi n_i)}{\sum_{i=1}^{q} n_i/(1 + \psi n_i)},$$

and deduce that $\widehat{\mu} = \bar{y}_{..}$ if the design is balanced, that is, $n_1 = \cdots = n_q$. In this case obtain $\widehat{\sigma}^2(\psi)$ and $\widehat{\psi}$ and compare with the results for (9.8). What is the effect of using REML?

10

Nonlinear Regression Models

10.1 Introduction

The regression models of Chapters 8 and 9 involve a continuous response that depends linearly on the parameters. Linear models remain the backbone of most statistical data analysis, but they have their deficiencies. In many applications, response variables are discrete, or statistical or substantive considerations suggest that covariates will appear nonlinearly. Models of this sort appeared on a somewhat *ad hoc* basis in the literature up to about 1970, since when there has been an explosion of generalizations to the linear model. Two important developments were the use of iterative weighted least squares for fitting, and the systematic use of exponential family response distributions. The iterative weighted least squares algorithm has wide applicability in nonlinear models and we outline its properties in Section 10.2, giving also a discussion of likelihood inference in this context. Exponential family response densities play a central role in generalized linear models, which we describe in Section 10.3, turning to the important special cases of binomial and Poisson responses in Sections 10.4 and 10.5. These models are widely used, but real data often display too much variation for them to be taken at face value. In Section 10.6 we outline remedies for this, based on the discussion of estimating functions in Section 7.2.

In each of these generalizations of the linear model our key notion that a few parameters summarize the entire model is retained. Section 10.7 branches out in a different direction, taking the viewpoint that the regression curve itself is more central than the parameters that summarize it. This leads to the idea of semiparametric modelling, particularly useful in exploratory analysis and assessment of model fit. Finally Section 10.8 outlines how the special features of survival data described in Section 5.4 may be dealt with in regression settings. The remainder of this section briefly motivates later developments.

Below we mostly assume that the responses $y_1, \ldots, y_n$ are independent, and that the density of y_j depends on a parameter η_j, through which systematic variation enters. The η_j depend on parameters β and explanatory variables, so we can write $\eta_j = \eta_j(\beta)$. In some models there is an incidental parameter ϕ which controls the

Time (minutes)	Calcium uptake (nmoles/mg)		
0.45	0.34170	−0.00438	0.82531
1.30	1.77967	0.95384	0.64080
2.40	1.75136	1.27497	1.17332
4.00	3.12273	2.60958	2.57429
6.10	3.17881	3.00782	2.67061
8.05	3.05959	3.94321	3.43726
11.15	4.80735	3.35583	2.78309
13.15	5.13825	4.70274	4.25702
15.00	3.60407	4.15029	3.42484

Table 10.1 Calcium uptake (nmoles/mg) of cells suspended in a solution of radioactive calcium, as a function of time suspended (minutes) (Rawlings, 1988, p. 403).

Figure 10.1 Calcium uptake (nmoles/mg) of cells suspended in a solution of radioactive calcium, as a function of time suspended (minutes).

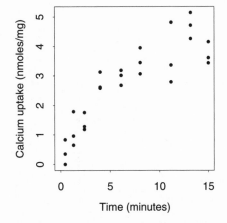

shape or spread of the distribution. Usually the aim is inference about the parameters β or prediction of future responses. For the normal linear model y_j is normally distributed with mean η_j and variance $\phi = \sigma^2$ and $\eta_j = x_j^\mathrm{T}\beta$, but much wider classes of models can be put into this framework.

Example 10.1 (Calcium data) Table 10.1 contains data on the uptake of calcium by cells that had been in "hot" calcium suspension. There are three observations at each of nine times after the start of the experiment. The data, plotted in Figure 10.1, show nonlinear dependence of calcium uptake on time.

Let y denote calcium uptake at time x after the start of the experiment. Then a differential equation that might describe how y depends on x is

$$\frac{dy}{dx} = (\beta_0 - y)/\beta_1,$$

with initial condition $y = 0$ when $x = 0$ and solution $y = \beta_0\{1 - \exp(-x/\beta_1)\}$. Allowing for measurement error, which seems to be similar at all levels of y, we might write

$$y = \beta_0\{1 - \exp(-x/\beta_1)\} + \varepsilon,$$

	Aphrissa boisduvalli N/S/E	Phoebis argante N/S/E	Dryas iulia N/S/E	Pierella luna N/S/E	Consul fabius N/S/E	Siproeta stelenes† N/S/E
Unpainted	0/0/14	6/1/0	1/0/2	4/1/5	0/0/0	0/0/1
Brown	7/1/2	2/1/0	1/0/1	2/2/4	0/0/3	0/0/1
Yellow	7/2/1	4/0/2	5/0/1	2/0/5	0/0/1	0/0/3
Blue	6/0/0	0/0/0	0/0/1	4/0/3	0/0/1	0/1/1
Green	3/0/1	1/1/0	5/0/0	6/0/2	0/0/1	0/0/3
Red	4/0/0	0/0/0	6/0/0	4/0/2	0/0/1	3/0/1
Orange	4/2/0	6/0/0	4/1/1	7/0/1	0/0/2	1/1/1
Black	4/0/0	0/0/0	1/0/1	4/2/2	7/1/0	0/1/0

Table 10.2 Response of a rufous-tailed jacamar to individuals of seven species of palatable butterflies with artifically coloured wing undersides. (N=not sampled, S = sampled and rejected, E = eaten)

† includes *Philaethria dido* also.

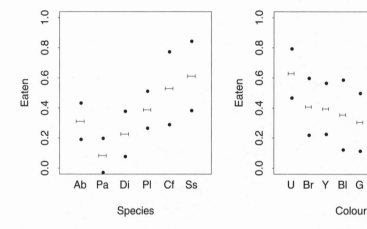

Figure 10.2 Proportion of butterflies eaten (±2SE) for diffferent species and wing colour.

where ε is normal with mean zero and variance σ^2. This fits into the general framework with $\eta(\beta) = \beta_0\{1 - \exp(-x/\beta_1)\}$, if y is normal with mean η and variance σ^2. ∎

Example 10.2 (Jacamar data)   As part of a study of the learning ability of tropical birds, Peng Chai of the University of Texas at Austin collected data on the response of a rufous-tailed jacamar, *Galbula ruficauda*, to butterflies. He used marker pens to paint the underside of the wings of eight species of butterflies, and then released each butterfly in the cage where the bird was confined. The bird responded in three ways: by not attacking the butterfly (N); by attacking the butterfly, then sampling but rejecting it (S); or by attacking and eating the butterfly, usually after removing some or all of the wings (E).

Table 10.2 gives the data in the form of an 8×6 layout, but the response is a triplet of counts. Figure 10.2 shows strong variation in the proportion of the different species and colours eaten.

If we take the number of butterflies eaten as the response, we might consider a model where the probability of the jacamar eating a butterfly would depend on its species and on its wing colour. However the low numbers in most cells make it likely that a linear model would give negative fitted probabilities, and this is clearly undesirable. In the better models described in Section 10.3 the probabilities have logistic form $\exp(\eta)/\{1 + \exp(\eta)\}$, where $\eta(\beta) = x^{\mathsf{T}}\beta$, so once again we have a model that is nonlinear in the covariates. ∎

One time-honoured way to treat examples like these is to transform the responses so that a linear model can be applied. The transformation may be chosen to stabilise the variance of the response, or better, for simplicity of interpretation. Although transformations remain useful for exploratory analysis and for plotting, they have largely been superseded by the use of likelihood estimation and more realistic modelling. In the next section we outline how this may be applied in nonlinear regression.

10.2 Inference and Estimation

10.2.1 Likelihood inference

Inference for nonlinear models is usually based on the large-sample likelihood theory described in Chapter 4. Under mild regularity conditions the asymptotic chi-squared distributions of likelihood ratio statistics and the joint normal distribution of the maximum likelihood estimates of β and ϕ form the basis for tests and confidence intervals, though the adequacy of these approximations needs to be considered in applications.

Comparisons between nested models are often based on a form of likelihood ratio statistic known as the *deviance*. Suppose that the log likelihood for a model for independent observations $y_1, \ldots, y_n$ is

$$\ell(\beta, \phi) = \sum_{j=1}^{n} \log f\{y_j; \eta_j(\beta), \phi\}, \tag{10.1}$$

and that ϕ is known; for now we suppress ϕ. If the maximum likelihood estimate of the $p \times 1$ parameter vector β is $\widehat{\beta}$, then the maximum likelihood estimate of η_j is $\widehat{\eta}_j = \eta_j(\widehat{\beta})$ and the maximized log likelihood is $\ell(\widehat{\beta})$. This is obtained by maximizing over p parameters, which connect the η_j for the different observations y_j, and thereby constrain them: larger p would lead to a larger log likelihood. The *saturated model* with no constraints on the η_j gives the largest possible log likelihood. Then β has dimension n and if the map between β and η is 1–1, the maximized log likelihood is simply the sum of the maxima of the individual terms on the right of (10.1). Let $\tilde{\eta}_j$ be the value of η_j that maximizes $\log f\{y_j; \eta_j\}$. Then the *scaled deviance* is defined as

$$D = 2 \sum_{j=1}^{n} \{\log f(y_j; \tilde{\eta}_j) - \log f(y_j; \widehat{\eta}_j)\}, \tag{10.2}$$

which is always non-negative. This will be small when the $\widehat{\eta}_j$ and $\tilde{\eta}_j$ are close, suggesting that the model fits well. Large D suggests poor fit, analogous to the sum of squares in a linear model.

Suppose we want to test whether $\beta_{q+1}, \ldots, \beta_p$ take specified values. Let A denote the model in which all p components of β vary freely and let B denote the model with $q < p$ parameters nested within A, with maximum likelihood estimates $\widehat{\eta}^A$ and $\widehat{\eta}^B$ and deviances D_A and D_B. Then the likelihood ratio statistic for comparing the models is

$$2 \sum_{j=1}^{n} \left\{ \log f\left(y_j; \widehat{\eta}_j^A\right) - \log f\left(y_j; \widehat{\eta}_j^B\right) \right\} = D_B - D_A, \tag{10.3}$$

and provided the models are regular, this has an approximate χ_{p-q}^2 distribution when model B is correct. Hence differences between scaled deviances are often used to compare nested models.

Example 10.3 (Normal deviance)   Suppose that the y_j are normal with means η_j and known variance ϕ. Then

$$\log f(y_j; \eta_j, \phi) = -\frac{1}{2} \log(2\pi\phi) - (y_j - \eta_j)^2 / \phi$$

is maximized with respect to η_j when $\tilde{\eta}_j = y_j$, giving $\log f(y_j; \tilde{\eta}_j, \phi) = -\frac{1}{2} \log(2\pi\phi)$. Therefore the scaled deviance for a model with fitted means $\widehat{\eta}_j$ is

$$D = \phi^{-1} \sum_{j=1}^{n} (y_j - \widehat{\eta}_j)^2,$$

which is just the residual sum of squares for the model, divided by ϕ. If $\eta_j = x_j^{\mathsf{T}} \beta$ is the correct normal linear model, we saw in Section 8.3 that the distribution of the residual sum of squares is $\phi \chi_{n-p}^2$, so values of D extreme relative to the χ_{n-p}^2 distribution call the model into question.

The difference between deviances for nested models A and B in which β has dimensions p and $q < p$,

$$D_B - D_A = \phi^{-1} \sum_{j=1}^{n} \left\{ \left(y_j - \widehat{\eta}_j^B\right)^2 - \left(y_j - \widehat{\eta}_j^A\right)^2 \right\} \ \dot{\sim} \ \chi_{p-q}^2$$

when model B is correct. Results from Section 8.5.1 show that this distribution is exact for linear models.   ∎

If ϕ is unknown, it is replaced by an estimate. The large-sample properties of deviance differences outlined above still apply, though in small samples it may be better to replace the approximating χ^2 distribution by an F distribution with numerator degrees of freedom equal to the degrees of freedom for estimation of ϕ.

10.2.2 Iterative weighted least squares

In order to apply large-sample likelihood approximations we must obtain the maximum likelihood estimates $\widehat{\beta}$ and their standard errors. A general procedure for this may be obtained by a variant Newton–Raphson method, an *iterative weighted least squares* algorithm widely used for nonlinear estimation. We now discuss this and some of its ramifications.

Assuming for now that ϕ is fixed, we write the log likelihood for β as

$$\ell(\beta) = \sum_{j=1}^{n} \ell_j\{\eta_j(\beta), \phi\},$$

where $\ell_j\{\eta_j(\beta), \phi\}$ is the contribution made by the jth observation. The maximum likelihood estimates $\widehat{\beta}$ usually satisfy the equations

$$\frac{\partial \ell(\widehat{\beta})}{\partial \beta_r} = 0, \qquad r = 1, \ldots, p,$$

which we put in matrix form to give the likelihood equation

$$\frac{\partial \ell(\widehat{\beta})}{\partial \beta} = \frac{\partial \eta^{\mathrm{T}}}{\partial \beta} u(\widehat{\beta}) = 0, \tag{10.4}$$

$\partial g/\partial \beta$ denotes the $p \times 1$ vector whose rth element is $\partial g/\partial \beta_r$, $\partial^2 g/\partial \beta \partial \beta^{\mathrm{T}}$ denotes the $p \times p$ matrix whose (r, s) element is $\partial^2 g/\partial \beta_r \partial \beta_s$, $\partial \eta/\partial \beta^{\mathrm{T}}$ denotes the $n \times p$ matrix with (j, r) element is $\partial \eta_j/\partial \beta_r$, and so forth. Note that $\partial \eta^{\mathrm{T}}/\partial \beta = (\partial \eta/\partial \beta^{\mathrm{T}})^{\mathrm{T}}$.

where $u(\beta)$ is the $n \times 1$ vector whose jth element is $\partial \ell(\beta)/\partial \eta_j$ because η_j only enters the log likelihood through the contribution made by the jth observation.

To find the maximum likelihood estimate $\widehat{\beta}$ starting from a trial value β, we make a Taylor series expansion in (10.4), to obtain

$$\frac{\partial \eta^{\mathrm{T}}(\beta)}{\partial \beta} u(\beta) + \left\{ \sum_{j=1}^{n} \frac{\partial \eta_j(\beta)}{\partial \beta} \frac{\partial^2 \ell_j(\beta)}{\partial \eta_j^2} \frac{\partial \eta_j(\beta)}{\partial \beta^{\mathrm{T}}} + \sum_{j=1}^{n} \frac{\partial^2 \eta_j(\beta)}{\partial \beta \partial \beta^{\mathrm{T}}} u_j(\beta) \right\} (\widehat{\beta} - \beta) \doteq 0. \tag{10.5}$$

If we denote the $p \times p$ matrix in braces on the left by the $p \times p$ matrix $-J(\beta)$, assumed invertible, we can rearrange (10.5) to obtain

$$\widehat{\beta} \doteq \beta + J(\beta)^{-1} \frac{\partial \eta^{\mathrm{T}}(\beta)}{\partial \beta} u(\beta). \tag{10.6}$$

This suggests that maximum likelihood estimates may be obtained by starting from a particular β, using (10.6) to obtain $\widehat{\beta}$, then setting β equal to $\widehat{\beta}$, and iterating (10.6) until convergence. This is the Newton–Raphson algorithm applied to our particular setting. In practice it can be more convenient to replace $J(\beta)$ by its expected value

$$I(\beta) = \sum_{j=1}^{n} \frac{\partial \eta_j(\beta)}{\partial \beta} \mathrm{E}\left(-\frac{\partial^2 \ell_j}{\partial \eta_j^2} \right) \frac{\partial \eta_j(\beta)}{\partial \beta^{\mathrm{T}}};$$

the other term vanishes because $\mathrm{E}\{u_j(\beta)\} = 0$. We write

$$I(\beta) = X(\beta)^{\mathrm{T}} W(\beta) X(\beta), \tag{10.7}$$

where $X(\beta)$ is the $n \times p$ matrix $\partial \eta(\beta)/\partial \beta^{\mathrm{T}}$ and $W(\beta)$ is the $n \times n$ diagonal matrix whose jth diagonal element is $\mathrm{E}(-\partial^2 \ell_j/\partial \eta_j^2)$.

If we replace $J(\beta)$ by $X(\beta)^{\mathrm{T}} W(\beta) X(\beta)$ and reorganize (10.6), we obtain

$$\widehat{\beta} = (X^{\mathrm{T}} W X)^{-1} X^{\mathrm{T}} W (X\beta + W^{-1} u) = (X^{\mathrm{T}} W X)^{-1} X^{\mathrm{T}} W z, \tag{10.8}$$

say, where the dependence of the terms on the right on β has been suppressed. That is, starting from β, the updated estimate $\widehat{\beta}$ is obtained by weighted linear regression of the vector $z = X(\beta)\beta + W(\beta)^{-1} u(\beta)$ on the columns of $X(\beta)$, using weight matrix $W(\beta)$. The maximum likelihood estimates are obtained by repeating this step until the log likelihood, the estimates, or more often both are essentially unchanged. The variable z plays the role of the response or dependent variable in the weighted least squares step and is sometimes called the *adjusted dependent variable*.

Often the structure of a model simplifies the estimation of an unknown value of ϕ. It may be estimated by a separate step between iterations of $\widehat{\beta}$, by including it in the step (10.6), or from the profile log likelihood $\ell_p(\phi)$.

Example 10.4 (Normal linear model)   In the normal linear model, we write $\eta_j = x_j^T\beta$. If the y_j are independently normally distributed with means η_j and variances $\phi = \sigma^2$, we have

$$\ell_j(\eta_j, \sigma^2) \equiv -\frac{1}{2}\left\{\log\sigma^2 + \frac{1}{\sigma^2}(y_j - \eta_j)^2\right\},$$

so

$$u_j(\eta_j) = \frac{\partial\ell_j}{\partial\eta_j} = \frac{1}{\sigma^2}(y_j - \eta_j), \qquad \frac{\partial^2\ell_j}{\partial\eta_j^2} = -\frac{1}{\sigma^2};$$

the jth element on the diagonal of W is the constant σ^{-2}. The (j, r) element of the matrix $\partial\eta/\partial\beta^T$ is $\partial\eta_j/\partial\beta_r = x_{jr}$, so $X(\beta)$ is simply the $n \times p$ design matrix X. We see that $z = X(\beta)\beta + W^{-1}(\beta)u(\beta) = y$, because in this situation $X(\beta)\beta = X\beta$ and $W^{-1}(\beta)u(\beta) = \sigma^2(y - X\beta)/\sigma^2$.

Here iterative weighted least squares converges in a single step.

The maximum likelihood estimate of σ^2 is obtained as in Section 8.2.1, and is $\widehat{\sigma}^2 = SS(\widehat{\beta})/n$, where $SS(\beta)$ is the sum of squares $(y - X\beta)^T(y - X\beta)$.   ∎

Example 10.5 (Normal nonlinear model)   Here the mean of the jth observation is $\eta_j = \eta_j(\beta)$. The log likelihood contribution $\ell_j(\eta_j)$ is the same as in the previous example, so u and W are the same also. However, the jth row of the matrix $X = \partial\eta/\partial\beta^T$ is $(\partial\eta_j/\partial\beta_0, \ldots, \partial\eta_j/\partial\beta_{p-1})$, and as η_j is nonlinear as a function of β, X depends on β. After some simplification, we see that the new value for $\widehat{\beta}$ given by (10.8) is

$$\widehat{\beta} \doteq (X^TX)^{-1}X^T(X\beta + y - \eta), \tag{10.9}$$

where X and η are evaluated at the current β. Here $\eta \neq X\beta$ and (10.9) must be iterated.

The log likelihood is a function of β only through the sum of squares, $SS(\beta) = \sum_{j=1}^n\{y_j - \eta_j(\beta)\}^2$. The profile log likelihood for σ^2 is

$$\ell_p(\sigma^2) = \max_\beta \ell(\beta, \sigma^2) \equiv -\frac{1}{2}\{n\log\sigma^2 + SS(\widehat{\beta})/\sigma^2\},$$

so the maximum likelihood estimator of σ^2 is $\widehat{\sigma}^2 = SS(\widehat{\beta})/n$. Although $S^2 = SS(\widehat{\beta})/(n - p)$ is not unbiased when the model is nonlinear, it turns out to have smaller bias than $\widehat{\sigma}^2$, and is preferable in applications.

In some cases the error variance depends on covariates, and we write the variance of the jth response as $\sigma_j^2 = \sigma^2(x_j, \gamma)$. Such models may be fitted by alternating iterative weighted least squares updates for β treating γ as fixed at a current value with those for γ with β fixed, convergence being attained when neither estimates nor log likelihood change materially. ∎

Iterative weighted least squares can be used for maximum likelihood estimation in linear models with non-normal errors and extends to situations with dependent responses.

Example 10.6 (Venice sea level data) In Section 5.1 the straight-line regression equation $y_j = \gamma_0 + \gamma_1(x_j - \overline{x}) + \varepsilon_j$ was fitted to data on annual maximum sea levels at Venice from 1931–1981. Fitting was by least squares, as is appropriate for normal responses, but the right-hand panel of Figure 5.2 suggests that the errors are non-normal. As the data are annual maxima, it is more appropriate to suppose that y_j has the Gumbel density

$$f(y_j; \eta_j, \tau) = \tau^{-1} \exp\left\{ -\frac{y_j - \eta_j}{\tau} - \exp\left(-\frac{y_j - \eta_j}{\tau}\right) \right\}, \tag{10.10}$$

where τ is a scale parameter and $\eta_j = \beta_0 + \beta_1(x_j - \overline{x})$; here we have replaced the γs with βs for continuity with the general discussion above. Use of this density is justified by the arguments leading to (6.34).

In this case

$$\ell_j(\eta_j, \tau) = -\log\tau - \frac{y_j - \eta_j}{\tau} - \exp\left(-\frac{y_j - \eta_j}{\tau}\right), \tag{10.11}$$

and it is straightforward to establish that

$$\frac{\partial\ell_j(\eta_j, \tau)}{\partial\eta_j} = \tau^{-1}\left\{1 - \exp\left(-\frac{y_j - \eta_j}{\tau}\right)\right\}, \quad \mathrm{E}\left\{-\frac{\partial^2\ell_j(\eta_j, \tau)}{\partial\eta_j^2}\right\} = \tau^{-2},$$

that $\partial\eta/\partial\beta^{\mathrm{T}} = X$ is the $n \times 2$ matrix whose jth row is $(1, x_j - \overline{x})$, and $W = \tau^{-2}I_n$. Hence (10.8) becomes $\widehat{\beta} \doteq (X^{\mathrm{T}}X)^{-1}(X\beta + \tau^2 u)$, where the jth element of u is $\tau^{-1}[1 - \exp\{-(y_j - \eta_j)/\tau\}]$.

Here it is simplest to fix τ, to obtain $\widehat{\beta}$ by iterating (10.8) for each fixed value of τ, and then to repeat this over a range of values of τ, giving the profile log likelihood $\ell_p(\tau)$ and hence confidence intervals for τ. Confidence intervals for β_0 and β_1 are obtained from the information matrix.

With starting value chosen to be the least squares estimates of β, and with $\tau = 5$, 19 iterations of (10.8) were required to give estimates and a maximized log likelihood whose relative change was less than 10^{-6} between successive iterations. We then took $\tau = 5.5, \ldots, 40$, using $\widehat{\beta}$ from the preceding iteration as starting-value for the next; in most cases just three iterations were needed. The left panel of Figure 10.3 shows a close-up of $\ell_p(\tau)$; its maximum is at $\widehat{\tau} = 14.5$, and the 95% confidence interval for τ is $(11.9, 18.1)$. The maximum likelihood estimates of β_0 and β_1 are 111.4 and 0.563, with standard errors 2.14 and 0.137; these compare with standard errors 2.61 and 0.177 for the least squares estimates. There is some gain in precision in using the more appropriate model. ∎

Example 10.7 (ABO blood group system) In the usual model for the ABO blood-group system (Examples 4.38, 5.12), the probabilities for the observed blood groups

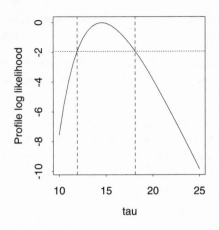
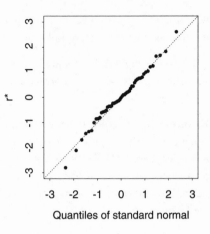

Figure 10.3 Gumbel analysis of Venice data. Left panel: profile log likelihood $\ell_p(\tau)$, with 95% confidence interval. Right panel: normal probability plot of residuals r_j^*.

A, B, AB and O are $\eta_A = \lambda_A^2 + 2\lambda_A\lambda_O$, $\eta_B = \lambda_B^2 + 2\lambda_B\lambda_O$, $\eta_{AB} = 2\lambda_A\lambda_B$, and $\eta_O = \lambda_O^2$, assuming random mating of a population in which λ_A, λ_B, and λ_O are the frequencies of alleles A, B, and O; here $\lambda_A + \lambda_B + \lambda_O = 1$. Given n independent individuals in which the groups appear with frequencies y_A, y_B, y_{AB}, and $y_O = n - y_A - y_B - y_{AB}$, the log likelihood is

$$y_A \log \eta_A + y_B \log \eta_B + y_{AB} \log \eta_{AB}$$
$$+ (n - y_A - y_B - y_{AB})\log(1 - \eta_A - \eta_B - \eta_{AB}).$$

One of the ηs is redundant and we have replaced η_O with $1 - \eta_A - \eta_B - \eta_{AB}$. If we set $\beta_1 = \log \lambda_A$ and $\beta_2 = \log \lambda_B$, we have

$$\frac{\partial \eta}{\partial \beta^{\mathrm{T}}} = 2 \begin{pmatrix} \lambda_A\lambda_O & -\lambda_A\lambda_B \\ -\lambda_A\lambda_B & \lambda_B\lambda_O \\ \lambda_A\lambda_B & \lambda_A\lambda_B \end{pmatrix}, \quad \frac{\partial \ell}{\partial \eta} = \begin{pmatrix} y_A/\eta_A - y_O/\eta_O \\ y_B/\eta_B - y_O/\eta_O \\ y_{AB}/\eta_{AB} - y_O/\eta_O \end{pmatrix}$$

and

$$W = n \begin{pmatrix} 1/\eta_A - 1/\eta_O & -1/\eta_O & -1/\eta_O \\ -1/\eta_O & 1/\eta_B - 1/\eta_O & -1/\eta_O \\ -1/\eta_O & -1/\eta_O & 1/\eta_{AB} - 1/\eta_O \end{pmatrix}.$$

Once again iterative weighted least squares can be used for maximization, although the weight matrix is not diagonal because the log likelihood contribution from y_O depends on η_A, η_B and η_{AB}.   ∎

10.2.3 Model checking

The fit of a linear model is checked using residuals and other diagnostics and by embedding it into more complex models chosen to capture particular departures of interest. These ideas extend to nonlinear models through the components of the iterative weighted least squares algorithm. For example, the leverage of the jth case, h_{jj}, is defined as the jth diagonal element of the matrix $H = W^{1/2}X(X^{\mathrm{T}}WX)^{-1}X^{\mathrm{T}}W^{1/2}$, evaluated at $\widehat{\beta}$. As the weight matrix W generally depends on $\widehat{\beta}$, the leverages depend on the responses as well as the covariates, but otherwise the h_{jj} have the same properties as in a linear model, where h_{jj} is the jth diagonal element of $X(X^{\mathsf{T}}X)^{-1}X^{\mathsf{T}}$.

There are several types of residual for nonlinear models, because no single definition plays all the roles of the standardized residual r_j defined at (8.26) for the linear model.

The residual sum of squares for a linear model can be written $\sum e_j^2$, so for more general models the analogy with the deviance suggests that we write $D = \sum d_j^2$, where d_j is the signed square root of the contribution y_j makes to the scaled deviance. This gives the *deviance residual*

$$d_j = \text{sign}(\tilde{\eta}_j - \widehat{\eta}_j)[2\{\ell_j(\tilde{\eta}_j; \phi) - \ell_j(\widehat{\eta}_j; \phi)\}]^{1/2}.$$

Analogy with the linear model suggests that the *standardized deviance residuals* $r_{Dj} = d_j/(1 - h_{jj})^{1/2}$ will be more homogeneous, and detailed calculations confirm that usually the r_{Dj} have roughly unit variances and distributions close to normal, though possibly with non-zero mean. Exceptions to this are binary data and Poisson data with small responses, whose residuals are essentially discrete.

A better general definition of residual combines the standardized deviance residual with the *standardized Pearson residual*

$$r_{Pj} = \frac{u_j(\widehat{\beta})}{\{w_j(\widehat{\beta})(1 - h_{jj})\}^{1/2}},$$

which is a standardized score statistic. Detailed calculations show that the distributions of the quantities $r_j^* = r_{Dj} + r_{Dj}^{-1} \log(r_{Pj}/r_{Dj})$ are close to normal for a wide range of models. For a normal linear model, $r_{Pj} = r_{Dj} = r_j^* = r_j$. The r_j^* or r_{Dj} may be used in the plots described in Section 8.6.1.

In a linear model, the influence of the jth case, (x_j, y_j), is proportional to $(\widehat{y} - X\widehat{\beta}_{-j})^{\mathsf{T}}(\widehat{y} - X\widehat{\beta}_{-j})$, where $\widehat{\beta}_{-j}$ is the estimate when the case is deleted from the model. For more general models a better measure is

$$2p^{-1}\{\ell(\widehat{\beta}) - \ell(\widehat{\beta}_{-j})\}, \tag{10.12}$$

where p is the dimension of β. Calculation of all n of these requires n additional fits, and it is more convenient to use the *approximate Cook statistic*

$$C_j = \frac{h_{jj}}{p(1 - h_{jj})} r_{Pj}^2. \tag{10.13}$$

This is derived by Taylor series expansion of (10.12) and reduces to (8.30) in the case of the normal linear model.

Example 10.8 (Venice sea level data) Here the weight matrix W is proportional to I_n and the matrix $\partial\eta/\partial\beta^{\mathsf{T}} = X$ is constant. It follows that leverages h_{jj} are simply the diagonal elements of $X(X^{\mathsf{T}}X)^{-1}X^{\mathsf{T}}$. If we set $z_j = (y_j - \widehat{\eta}_j)/\widehat{\tau}$, it is easy to check that the jth deviance residual is $\text{sign}(y_j - \widehat{\eta}_j)[2\{z_j + \exp(-z_j) - 1\}]^{1/2}$. The r_j^* for the fitted model are shown in the right panel of Figure 10.3. They are close to standard normal, and cast no doubt on the adequacy of the model. ∎

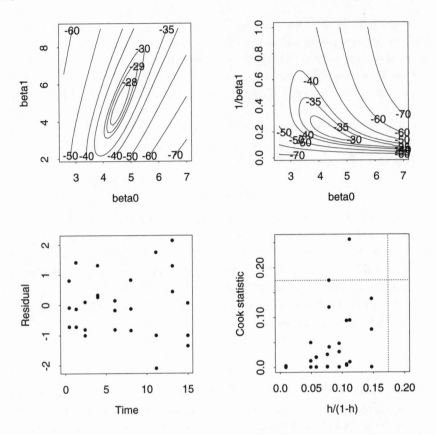

Figure 10.4 Fit of a
nonlinear model to the
calcium data. Upper left:
contours for $\ell_p(\beta_0, \beta_1)$.
Upper right: contours for
$\ell_p(\beta_0, \gamma_1)$, where
$\gamma_1 = 1/\beta_1$. Lower left:
standardized residuals
plotted against time.
Lower right: plot of Cook
statistics against
$h/(1 - h)$, where h is
leverage.

The deviance can be used to check the fit of some types of models, but in moderate samples its distribution can be far from χ^2 and plots of deviances from simulated data can be useful.

Example 10.9 (Calcium data)   The model for the calcium data of Example 10.1 sets $\eta_j(\beta) = \beta_0\{1 - \exp(-x_j/\beta_1)\}$, so $X = \partial\eta/\partial\beta^T$ has jth row

$$(\partial\eta_j/\partial\beta_0, \quad \partial\eta_j/\partial\beta_1) = \left(1 - \exp(-x_j/\beta_1), \quad -x_j\beta_0\exp(-x_j/\beta_1)/\beta_1^2\right).$$

As the initial slope of η as a function of x is β_0/β_1, and as η has asymptote β_0 for large x, we expect roughly that $\beta_0 \doteq 5$, and that $\beta_0/\beta_1 \doteq 1$. This suggests taking $\beta_0 = 5$ and $\beta_1 = 5$ as initial values for the iterative weighted least squares algorithm, which then converges rapidly to $\widehat{\beta}_0 = 4.31$ and $\widehat{\beta}_1 = 4.80$, with standard errors 0.303 and 0.905. The off-diagonal element of the inverse expected information matrix is 0.237, corresponding to $\text{corr}(\widehat{\beta}_0, \widehat{\beta}_1) \doteq 0.45$. The residual sum of squares is $SS(\widehat{\beta}) = 7.465$, and the estimate of σ^2 is $s^2 = 7.465/(27 - 2) = 0.299$.

Large-sample likelihood theory suggests that $\widehat{\beta}_0$ and $\widehat{\beta}_1$ have approximately a bivariate normal distribution. This rests on quadratic approximation to the log likelihood, which seems reasonable from the upper left panel of Figure 10.4. The upper right panel contains contours of $\ell_p(\beta_0, \gamma_1) = \max_{\sigma^2} \ell(\beta_0, \gamma_1, \sigma^2)$, where $\gamma_1 = 1/\beta_0$, for which quadratic approximation would be poor.

The standardized Pearson residuals in the lower left panel suggests possible poor fit at the three longest times, although the normal scores plot is good. The lower right panel shows that one of the approximate Cook statistics is somewhat large, but due to an outlier rather than a high leverage point.

To check model fit more formally, we allow unconnected means at each time, giving nine parameters and residual sum of squares 4.797 on 18 degrees of freedom. The F statistic that compares this and the fitted nonlinear model is equal to $\{(7.645 - 4.797)/7\}/(4.797/18) = 1.53$, and as the 0.95 quantile of the $F_{7,18}$ distribution is 2.58, there is no evidence of poor fit overall. Under the unconnected model the sums of squares between the three observations at each time have distribution $\sigma^2 \chi_2^2$, so a chi-squared probability plot of the ordered sums of squares should be straight. This plot suggests difficulties at $x = 11.15$ minutes, as was evident from the original data. ∎

Exercises 10.2

1   Show that the scaled deviance contribution for a binomial response with probability density $\binom{m}{r}\pi^r(1-\pi)^{m-r}, 0 < \pi < 1, r = 0, \ldots, m$, and $\eta = \pi$ is

$$2\{r \log\{r/(m\widehat{\pi})\} + (m-r)\log[(m-r)/\{m(1-\widehat{\pi})\}]\}.$$

2   Show that the scaled deviance contribution for a Poisson response with density $\eta^y e^{-\eta}/y!$, $\eta > 0, y = 0, 1, \ldots$, is $2\{y \log(y/\widehat{\eta}) - y + \widehat{\eta}\}$.

3   Consider a linear model with non-normal errors, in which the jth response is $y_j = \eta_j + \tau\varepsilon_j$, where $\eta_j = x_j^T\beta$, and ε_j has density $\exp u(\varepsilon)$, for $j = 1, \ldots, n$.
(a) Show that the log likelihood contribution from y_j may be written as $u(z_j) - \log \tau$, and hence express in terms of $u(\cdot)$ and τ the quantities needed to obtain the maximum likelihood estimates of β by iterative weighted least squares.
(b) Let $x_j^T = (1, z_j^T)$, so that the covariate matrix equals $X = (1_n, Z)$ with $1_n^T Z = 0$. Show that the expected information matrix may be written

$$\tau^{-2}\begin{pmatrix} Z^T Z & 0 \\ 0 & nA \end{pmatrix},$$

where A is a 2×2 matrix that does not depend on the parameters. Show further that A is diagonal if the density of ε_j is symmetric about zero.
(c) Give the matrix A for the t density (3.11) and for the Gumbel density (10.10).

4   Suppose that n years of daily data are available, for each of which the r largest observations are known. Find the score and observed information when the model (6.37) is fitted to these, with $\eta = x_j^T\beta$ but constant τ and ξ. Hence write down the steps needed to apply weighted least squares to estimation of β, when τ and ξ are known. How would you estimate τ and ξ?

5   Show that for a normal linear model in which ϕ is replaced by $\widehat{\phi} = SS(\widehat{\beta})/(n-p)$, the standardized deviance and Pearson residuals both equal the usual standardized residual r_j, and hence verify that (10.13) reduces to (8.30).

6   Verify the contents of the matrices X and W in Example 10.7.

7   In a nonlinear normal regression model, suppose that $\eta = \beta_0(1 - e^{-\beta_1 x})$. Let $SS(\beta_1)$ be the residual sum of squares when β_1 is known, that is, when the single covariate $1 - e^{-\beta_1 x}$

is fitted. Show that the profile log likelihood for β_1 can be written as

$$\ell_p(\beta_1) = \max_{\beta_0, \sigma^2} \ell(\beta_0, \beta_1, \sigma^2) \equiv -\frac{n}{2} \log s^2(\beta_1),$$

and give the form of a $(1 - 2\alpha)$ confidence interval for β_1 based on ℓ_p.

10.3 Generalized Linear Models

10.3.1 Density and link functions

Linear models play a central role in regression. In their simplest form they apply when the response variable is continuous, takes values on the real line, and has constant variance. Transformations and weighting broaden their applicability but can give fits that are awkward to interpret, as well as being unsatisfactory with discrete responses. In this section we describe how the key features of the linear model may be extended to situations where the response comes from any of a wide class of distributions. Such models are widely used in practice and the main ideas form the basis for much further development, in which the iterative weighted least squares algorithm plays an important role.

Three aspects of the normal linear model for a continuous response y are:

- the *linear predictor* $\eta = x^T\beta$ through which $\mu = \mathrm{E}(y)$ depends on the $p \times 1$ vectors x of explanatory variables and β of parameters;
- the density of the response y, which is normal with mean μ and variance σ^2; and
- the fact that the mean response equals the linear predictor, $\mu = \eta$.

In a generalized linear model the second and third of these are extended to:

- the response y has density

$$f(y; \theta, \phi) = \exp\left\{\frac{y\theta - b(\theta)}{\phi} + c(y; \phi)\right\}, \qquad (10.14)$$

  where θ depends on the linear predictor, and the *dispersion parameter* ϕ is often known; and
- the linear predictor and the mean of y are related by a monotone *link function* g, with

$$\eta = g(\mu). \qquad (10.15)$$

If ϕ is known then (10.14) is a linear exponential family with natural parameter θ/ϕ. Thus it contains old friends such as the normal, gamma, binomial, and Poisson densities, but it includes also more casual acquaintances such as the inverse Gaussian and negative binomial models. This broadens the applicability of linear model ideas to data where the responses are positive, counts or proportions, without the need for transformations.

If Y has density (10.14), its moment-generating function is (Exercise 10.3.3)

$$M(t) = \exp\{b(\theta + t\phi) - b(\theta)\},$$

so the response has mean and variance

' denotes differentiation
with respect to θ.

$$\mathrm{E}(Y) = b'(\theta) = \mu, \quad \mathrm{var}(Y) = \phi b''(\theta) = \phi b''\{b'^{-1}(\mu)\} = \phi V(\mu), \qquad (10.16)$$

say, provided that a function inverse to $b'(\theta)$ exists. We met the variance function $V(\mu)$ in Section 5.2.1.

Example 10.10 (Poisson density)   The Poisson density may be written as

$$f(y; \mu) = \exp(y \log \mu - \mu - \log y!), \quad y = 0, 1, \ldots, \quad \mu > 0,$$

which has form (10.14) with $\theta = \log \mu$, $b(\theta) = e^\theta$, $\phi = 1$, and $c(y; \phi) = -\log y!$. The mean of y is $\mu = b'(\theta) = e^\theta = \mu$, and its variance is $b''(\theta) = e^\theta = \mu$, so the variance function is linear: $V(\mu) = \mu$; see Example 5.9. ∎

Example 10.11 (Normal density)   The normal density with mean μ and variance σ^2 may be written

$$f(y; \mu, \sigma^2) = \exp\left\{ -\frac{(y^2 - 2y\mu + \mu^2)}{2\sigma^2} - \frac{1}{2}\log(2\pi\sigma^2) \right\},$$

so

$$\theta = \mu, \quad \phi = \sigma^2, \quad b(\theta) = \frac{1}{2}\theta^2, \quad c(y; \phi) = -\frac{1}{2\phi}y^2 - \frac{1}{2}\log(2\pi\phi).$$

As the first and second derivatives of $b(\theta)$ are θ and 1, we have $V(\mu) = 1$; the variance function is constant. ∎

Example 10.12 (Binomial density)   We write the binomial density

$$f(r; \pi) = \binom{m}{r}\pi^r(1 - \pi)^{m-r}, \quad 0 < \pi < 1, \quad r = 0, \ldots, m,$$

in the form

$$\exp\left[m\left\{ \frac{r}{m}\log\left(\frac{\pi}{1 - \pi}\right) + \log(1 - \pi) \right\} + \log\binom{m}{r} \right],$$

so

$$y = \frac{r}{m}, \quad \phi = \frac{1}{m}, \quad \theta = \log\left(\frac{\pi}{1 - \pi}\right), \quad b(\theta) = \log(1 + e^\theta), \quad c(y; \phi) = \log\binom{m}{r}.$$

The mean and variance of y are

$$\mu = b'(\theta) = \frac{e^\theta}{1 + e^\theta}, \quad \phi b''(\theta) = \frac{e^\theta}{m(1 + e^\theta)^2};$$

the variance function is $V(\mu) = \mu(1 - \mu)$. ∎

By allowing nonlinear relations between the mean response and the covariates, the link function (10.15) permits the relationship between the linear part of the model and the mean response to be chosen on subject-matter or statistical grounds. Restrictions on μ can be imposed through the choice of g, which maps the domain of μ to the set inhabited by η, usually the real line. One particular choice is the canonical link, which is obtained when $\eta = \theta = b'^{-1}(\mu)$. When ϕ is known and the canonical link is used, the model is a natural exponential family and there is a p-dimensional minimal sufficient statistic for β. This is attractive from the vantage of statistical theory, but substantive considerations are more important.

Example 10.13 (Poisson link functions)   The Poisson mean is positive, so the most common link function is the log, for which $\log \mu = \eta$. As $\theta = \log \mu$, this is the canonical link.

Suppose that data n_1 and n_2 are gathered from two independent Poisson processes, with means $t_1\gamma_1$ and $t_2\gamma_2$, where t_1 and t_2 are known, but that the aggregated count $y = n_1 + n_2$ only is known. Then $E(y) = t_1\gamma_1 + t_2\gamma_2$ and the identity link function would be appropriate.   ∎

Example 10.14 (Normal link functions)   The usual link function for the normal density is the identity, yielding the normal linear model. Suppose instead that $y = \mu + \varepsilon$, but

$$\mu = \alpha_0 \frac{x}{\alpha_1 + x}.$$

On rewriting μ^{-1} as $\eta = \beta_0 + \beta_1 z$, where $\beta_0 = \alpha_0^{-1}$, $\beta_1 = \alpha_1/\alpha_0$, and $z = x^{-1}$, we see that this model fits into our general setup with normal distribution for the response and the inverse link function, $\eta = \mu^{-1}$.   ∎

10.3.2 Estimation and inference

The log likelihood of independent responses $y_1, \ldots, y_n$ from density (10.14) is

$$\ell(\beta) = \sum_{j=1}^{n} \left\{ \frac{y_j\theta_j - b(\theta_j)}{\phi_j} + c(y_j; \phi_j) \right\}, \tag{10.17}$$

where $\theta_j = \theta(\eta_j)$ with $\eta_j = x_j^{\mathrm{T}}\beta$. Maximum likelihood estimates $\widehat{\beta}$ are obtained by solving the score equation (10.4) by iterative weighted least squares. Some differentiation shows that

$$\frac{\partial\ell(\beta)}{\partial\beta} = \frac{\partial\eta^{\mathrm{T}}}{\partial\beta} \frac{\partial\theta}{\partial\eta^{\mathrm{T}}} \frac{\partial\ell}{\partial\theta^{\mathrm{T}}} = X^{\mathrm{T}}u(\beta), \tag{10.18}$$

where the design matrix $\partial\eta/\partial\beta^{\mathrm{T}} = X$ does not depend on β. The components of the score statistic $u(\beta)$ and the weight matrix $W(\beta)$ may be expressed in terms of components μ_j of the mean vector μ as

$$u_j = \frac{\partial\theta_j}{\partial\eta_j} \frac{\partial\ell_j(\theta_j)}{\partial\theta_j} = \frac{y_j - \mu_j}{g'(\mu_j)\phi_j V(\mu_j)},$$

$$w_j = \left(\frac{\partial\theta_j}{\partial\eta_j}\right)^2 \frac{\partial^2\ell_j(\theta_j)}{\partial\theta_j^2} = \frac{1}{g'(\mu_j)^2\phi_j V(\mu_j)}, \tag{10.19}$$

where $g'(\mu_j) = dg(\mu_j)/d\mu_j$. Thus $\widehat{\beta}$ is obtained by iterative weighted least squares regression of response $z = X\beta + g'(\mu)(y - \mu)$ on the columns of X using weights (10.19). By using y as an initial value for μ and $g(y)$ as an initial value for $\eta = X\beta$, we avoid needing an initial value for β.

Most generalized linear models in which the ϕ_j are unknown have $\phi_j = \phi a_j$, where the a_j are known and only ϕ must be estimated; this is analogous to weighted least squares, with a_j^{-1} playing the role of the weight attached to y_j. When ϕ is unknown, the scaled deviance is replaced by the *deviance*, defined as ϕ times the scaled deviance.

Under the usual regularity conditions, $\widehat{\beta}$ has a large-sample normal distribution with mean β and variance matrix $X^{\mathsf{T}}WX)^{-1}$.

The maximum likelihood estimator of ϕ can behave poorly, but another estimator is suggested by noting that if the regression parameters β were known, an unbiased estimator of $\phi = \mathrm{var}(y_j)/\{a_j V(\mu_j)\}$ would be

$$\frac{1}{n} \sum_{j=1}^{n} \frac{(y_j - \mu_j)^2}{a_j V(\mu_j)}.$$

This motivates the use of

$$\widehat{\phi} = \frac{1}{n-p} \sum_{j=1}^{n} \frac{(y_j - \widehat{\mu}_j)^2}{a_j V(\widehat{\mu}_j)}, \qquad (10.20)$$

where the divisor $n - p$ allows for estimation of β, analogous to the unbiased estimate of variance in a normal linear model. When ϕ is known, fit can be measured using *Pearson's statistic*,

$$P = \frac{1}{\phi} \sum_{j=1}^{n} \frac{(y_j - \widehat{\mu}_j)^2}{V(\widehat{\mu}_j)/a_j}, \qquad (10.21)$$

analogous to the scaled residual sum of squares in a normal model.

When the dispersion parameter is known, overall tests of fit are provided by Pearson's statistic and the scaled deviance, but their distributions depend on the situation. For gamma or binomial data with small dispersion, that is small ϕ, the distribution of D or P when the model fits is roughly χ^2_{n-p}. This corresponds to large ν for gamma data and to large m for binomial data. For Poisson data χ^2 approximations are useful unless all the fitted means are small, $\widehat{\mu} < 5$, say. Empirical evidence suggests that although such approximations are better for P than for D, they are poor if the data are sparse. The problem is most acute for the deviance of binary data, for which the large-sample approximation is useless (Exercise 10.4.1).

Example 10.15 (Jacamar data)   For the data of Example 10.2, we treat the number of butterflies of species s painted the cth colour and eaten, r_{cs}, as binomial with denominator m_{cs} and probability

$$\pi_{cs} = \frac{\exp(\alpha_c + \gamma_s)}{1 + \exp(\alpha_c + \gamma_s)}, \qquad c = 1, \ldots, 8, s = 1, \ldots, 6.$$

Terms	df	Deviance
1	43	134.24
1+Species	38	114.59
1+Colour	36	108.46
1+Species+Colour	31	67.28

Table 10.3 Deviances and analysis of deviance for models fitted to jacamar data. The lower part of the analysis of deviance table shows results for the reduced data, without two outliers.

Terms	df	Deviance reduction	Terms	df	Deviance reduction
Species (unadj. for Colour)	5	19.64	Species (adj. for Colour)	5	41.18
Colour (adj. for Species)	7	47.31	Colour (unadj. for Species)	7	25.78
Species (unadj. for Colour)	4	27.63	Species (adj. for Colour)	4	35.18
Colour (adj. for Species)	7	18.03	Colour (unadj. for Species)	7	10.48

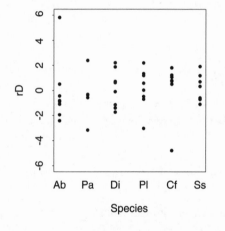 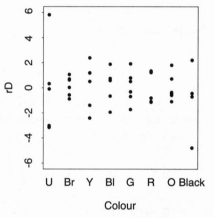

Figure 10.5 Standardized deviance residuals r_D for binomial two-way layout fitted to jacamar data.

From Example 10.12 we see that this is equivalent to a generalized linear model with binomial errors and response $y_{cs} = r_{cs}/m_{cs}$ whose mean $\mu_{cs} = \pi_{cs}$ is related to the linear predictor $\eta_{cs} = \alpha_c + \gamma_s$ by the *logit* link function

$$\eta = \log\left(\frac{\pi}{1 - \pi}\right).$$

Colour and species effects are represented by the α_c and γ_s respectively.

Table 10.3 contains the analysis of deviance. As there are four cells with zero counts there are 44 degrees of freedom in total. The reductions in deviance depend on the order of fitting, but are highly significant compared to the appropriate chi-squared distributions. The residual deviance, 67.28, is large compared to the χ^2_{31} distribution and suggests poor fit.

Figure 10.5 shows at least two outliers, corresponding to the first cell in the top row and the penultimate cell in the last row of Table 10.2. Deletion of the second of

Table 10.4 Estimated parameters and standard errors for the reduced jacamar data.

Aphrissa boisduvalli	*Phoebis argante*	*Dryas iulia*	*Pierella luna*	*Consul fabius*	*Siproeta stelenes*
−1.99 (0.79)	−2.22 (0.85)	−0.56 (0.67)	0.16 (0.54)	−	1.50 (0.78)

Brown	Yellow	Blue	Green	Red	Orange	Black
0.16 (0.73)	0.33 (0.68)	−0.53 (0.81)	−0.83 (0.75)	−1.93 (0.88)	−1.94 (0.85)	−1.26 (0.86)

Table 10.5 Times in minutes taken by four chimpanzees to learn ten words (Brown and Hollander, 1977, p. 257).

Chimpanzee	Word									
	1	2	3	4	5	6	7	8	9	10
1	178	60	177	36	225	345	40	2	287	14
2	78	14	80	15	10	115	10	12	129	80
3	99	18	20	25	15	54	25	10	476	55
4	297	20	195	18	24	420	40	15	372	190

Incidentally when only the last cell in this column was dropped, neither of the two packages used signalled a failure of $\widehat{\gamma}_s$ to converge.

these necessitates dropping its entire column, because all the remaining *Consul fabius* butterflies were eaten and the corresponding $\widehat{\gamma}_s$ is infinite. In this context a case is a single butterfly, so deletion of a cell involves deleting a number of cases.

The lower part of Table 10.3 shows the analysis of deviance for the reduced data. The overall deviance drops from 73.68 on 35 degrees of freedom to 28.02 on 24 degrees of freedom. The significance of the colours depends on whether or not species is fitted first. Overall significance for each term is assessed by its significance after adjusting for the other. Thus Colour is significant at about the 0.01 level, when treated as χ^2_7. The residual deviance is not large compared to the χ^2_{24} distribution, and casts no doubt on the model.

Table 10.4 shows the estimates and standard errors. The odds that an unpainted *Aphrissa boisduvalli* is eaten are $e^{-1.99} = 0.14$ and the correponding probability is $e^{-1.99}/(1 + e^{-1.99}) = 0.12$. Painting the underside of its wings green multiplies the odds of its being eaten by $e^{-0.83} = 0.44$. The most marked reductions are when red, orange, or black paint is used; locally these colours are associated with unpalatable butterflies.

Pearson's statistic is 25.58 on 24 degrees of freedom. The standardized deviance residuals lie between −2.03 and 1.96, and these and other case diagnostics show that the model fits the reduced data well. ∎

Example 10.16 (Chimpanzee learning data) Table 10.5 gives times in minutes taken by four chimpanzees to learn each of ten words. The data are a two-way layout, but the responses are positive and vary over two orders of magnitude, suggesting that a linear model is inappropriate. When a linear model with mean $\alpha_c + \gamma_w$ is fitted,

Term	df	Deviance reduction	Term	df	Deviance reduction
Chimp (unadj. for Word)	3	6.95	Chimp (adj. for Word)	3	6.22
Word (adj. for Chimp)	9	38.46	Word (unadj. for Chimp)	9	39.19

Table 10.6 Analysis of deviance for models fitted to chimpanzee data.

where α_c and γ_w correspond to chimpanzee and words effect, the F statistic (8.27) for non-additivity strongly indicates a change of scale.

We fit a model with gamma errors and the log link function, in which the mean time taken by the cth chimpanzee to learn the wth word is μ_{cw}, where $\log \mu_{cw} = \eta_{cw} = \alpha_c + \gamma_w$. If the gamma density with mean μ and shape parameter ν is written in form

$$f(y; \mu, \nu) = \frac{1}{\Gamma(\nu)} y^{\nu-1} \left(\frac{\nu}{\mu}\right)^{\nu} \exp(-\nu y/\mu), \quad y > 0, \quad \nu, \mu > 0, \quad (10.22)$$

we see (Exercise 10.3.4) that the dispersion parameter $\phi = 1/\nu$, which must be estimated.

The deviances for models 1, 1+Chimp, 1+Word, and 1+Chimp+Word are 60.38, 53.43, 21.19, and 14.97, with 39, 36, 30, and 27 degrees of freedom. Table 10.6 shows the analysis of deviance. The two-way layout is balanced and the order of fitting matters less than in Example 10.15. Use of (10.20) gives $\widehat{\phi} = 0.432$, and $\widehat{\nu} = \widehat{\phi}^{-1} = 2.31$. The significance of the deviance reductions for chimps and words is gauged by F tests using $\widehat{\phi}$ as the denominator. For example, we test for differences between chimps by comparing $(6.22/3)/0.432 = 4.78$ with the $F_{3,27}$ distribution, giving significance level about 0.01.

An alternative is to fit a normal two-way layout to the log data, but the residuals suggest that the gamma model is preferable. The largest residual is for the first chimpanzee and fifth word, which is very large compared to the other times for that word. This is also the most influential value for the gamma model, but we shall not pursue this.

A different approach uses gamma errors and the inverse link $\eta_{cw} = 1/\mu_{cw}$, giving a linear model for the speed with which a word is learnt. The residual deviance for this model is 17.08. To test whether this link is suitable we extend the nonadditivity test outlined in Example 8.24 by adding the constructed variable $\widehat{\eta}^2$ to the linear predictor, where $\widehat{\eta}$ is the fitted linear predictor for the model with inverse link. The deviance drops by 1.26 on one degree of freedom, and since $\widehat{\phi} = 0.47$ for the extended model, the test statistic is $(1.26/1)/0.47 = 2.68$, to be compared to the $F_{1,26}$ distribution. The significance level, 0.11, gives only weak evidence against the inverse link. ■

Exercises 10.3

1 Suppose that y is the number of events in a Poisson process of rate λ observed for a period of length T. Show that y has a generalized linear model density and give θ, $b(\theta)$, ϕ and $c(y; \phi)$.

2 Use the identities $E(\partial\ell/\partial\theta) = 0$ and $\text{var}(\partial\ell/\partial\theta) = E(-\partial^2\ell/\partial\theta^2)$ to derive the mean and variance of the density (10.14).

3   Check that the moment-generating function that corresponds to (10.14) is $\exp\{b(\theta + t\phi) - b(\theta)\}$, give the corresponding cumulant-generating function, and hence verify the mean and variance in (10.16).

4   Show that the gamma density (10.22) can be put in form (10.14) with canonical parameter $\theta = -\mu^{-1}, b(\theta) = -\log(-\theta)$ and dispersion parameter $\phi = 1/\nu$. Give the canonical link function, and use $b(\theta)$ to show that (10.22) has mean μ, variance function $V(\mu) = \mu^2$, and variance μ^2/ν.

5   Verify that the inverse Gaussian density

$$f(y; \lambda, \mu) = \left(\frac{\lambda}{2\pi y^3}\right)^{1/2} \exp\left\{-\frac{\lambda(y-\mu)^2}{2\mu^2 y}\right\}, \qquad y > 0, \quad \lambda > 0, \mu > 0,$$

can be written in form (10.14) by giving $\theta, b(\theta), \phi$, and $c(y; \phi)$, and show that its variance function is $V(\mu) = \mu^3$.

6   In (10.17), suppose that $\phi_j = \phi a_j$, where the a_j are known constants, and that ϕ is functionally independent of β. Show that the likelihood equations for β are independent of ϕ, and deduce that the profile log likelihood for ϕ is

$$\ell_p(\phi) = \phi^{-1} \sum_{j=1}^{n} \left\{\frac{y_j \widehat{\theta}_j - b(\widehat{\theta}_j)}{a_j} + c(y_j; \phi a_j)\right\}.$$

Hence show that for gamma data the maximum likelihood estimate of ν solves the equation $\log \nu - \psi(\nu) = n^{-1} \sum_{(} z_j - \log z_j - 1)$, where $z_j = y_j/\widehat{\mu}_j$ and $\psi(\nu)$ is the digamma function $d \log \Gamma(\nu)/d\nu$.

7   Suppose that the canonical link is used with the log likelihood (10.17), so that $\theta_j = \eta_j$, and that it is required to check this link. Let $\theta_j = \theta(\eta_j)$, where $\theta(\cdot)$ is a potentially nonlinear function. By quadratic Taylor series expansion of $\theta(.)$ about a suitable η^0, show that provided η contains an intercept term, we can write $\theta(\eta) \doteq \eta' + \delta\eta'^2$, where δ is proportional to $b''(\eta^0)$ and η' is a linear function of η. Hence verify that when the fit of a model with the canonical link has given linear predictor $\widehat{\eta}$, a constructed variable test of the canonical link is based on the change in fit when $\widehat{\eta}^2$ is added to the linear predictor. Discuss Example 8.24 in this light.

8   Show that in any generalized linear model, the Fisher information $\mathrm{E}(-\partial^2 l/\partial\beta_r \partial\phi)$ for the dispersion parameter ϕ and any regression parameter β_r is zero; ϕ and β are said to be *orthogonal*. What is the implication for the maximum likelihood estimates $\widehat{\phi}$ and $\widehat{\beta}_r$? Prove also the stronger result that, whatever the value of ϕ, the joint observed information $-\sum \partial^2 \log f(y_j; \beta, \phi)/\partial\phi\partial\beta$ is zero, evaluated at $\widehat{\beta}$.

10.4 Proportion Data

10.4.1 Binary data

A binary response Y takes values 1 and 0 with probabilities π and $1 - \pi$, denoting a dichotomous outcome such as success/failure, won/lost, or well/ill. Such data are common in applications. The simplest relation between $\mathrm{E}(Y) = \pi$ and a linear predictor is $\pi = x^\mathsf{T}\beta$. This is unsuitable for general use because π may then lie outside the unit interval. It is usually better to force $0 < \pi < 1$ by taking it to be a nonlinear monotone function of $x^\mathsf{T}\beta$, whose inverse is the link function of the corresponding generalized linear model.

One way to derive link functions for dichotomous variables is to suppose that Y is a binary version of an underlying continuous response Z. Let $Z = x^\mathsf{T}\gamma + \sigma\varepsilon$, and

	Distribution		Link function	
Logistic	$e^u/(1+e^u)$	Logit	$\eta = \log\{d\pi/(1-\pi)\}$	
Normal	$\Phi(u)$	Probit	$\eta = \Phi^{-1}(\pi)$	
Log Weibull	$1 - \exp(-\exp(u))$	Log-log	$\eta = -\log\{-\log(\pi)\}$	
Gumbel	$\exp\{-\exp(-u)\}$	Complementary log-log	$\eta = \log\{-\log(1-\pi)\}$	

Table 10.7 Tolerance distributions and corresponding link functions for binary data.

suppose that ε has continuous distribution function F. The mean of Z increases with $x^\mathsf{T}\gamma$, and $Y = 1$ if $Z > 0$, with probability

$$\pi = \Pr(Y = 1) = 1 - F(-x^\mathsf{T}\gamma/\sigma) = 1 - F(-x^\mathsf{T}\beta),$$

say. The ratio $\beta = \gamma/\sigma$ is estimable from the binary data, but γ and σ are not. If F is symmetric about zero, then π equals $F(x^\mathsf{T}\beta)$ and the corresponding link function (10.15) is $\eta = x^\mathsf{T}\beta = F^{-1}(\pi)$. Some standard choices of the so-called *tolerance distribution* F and corresponding link functions are shown in Table 10.7. The logit and probit functions are symmetric and usually hard to distinguish in practice, while the log-log and complementary log-log functions are asymmetric in opposite directions. Numerous other links have been proposed, but those in the table usually suffice in applications.

Much information may be lost by splitting and it is generally better to work with the original responses if they are available. Otherwise less information is lost by taking several categories. Difficulties in the binary case are illustrated in the following example.

Example 10.17 (Dichotomization) Suppose independent observations $Z_j = x_j^\mathsf{T}\beta + \varepsilon_j$ are dichotomized by setting $Y_j = 1$ if $Z_j > 0$ and $Y_j = 0$ otherwise, and let F and f denote the distribution and density of the ε_j. If the original Z_j were available, the jth log likelihood contribution would be $\log f(z_j - x_j^\mathsf{T}\beta)$ and the expected information matrix would be (10.7), with X the constant matrix whose jth row is x_j^T and $W = kI_n$, where $k = -\int d^2 \log f(\varepsilon)/d\varepsilon^2 f(\varepsilon)\, d\varepsilon$. Thus if the Z_j are available the asymptotic covariance matrix of the maximum likelihood estimator $\widehat{\beta}_Z$ is $k^{-1}(X^\mathsf{T}X)^{-1}$.

Suppose now that only the binary variables $Y_1, \ldots, Y_n$ are known. As Y_j has success probability $\pi_j = 1 - F(-\eta_j)$, where $\eta_j = x_j^\mathsf{T}\beta$, its log likelihood contribution is $\ell_j = Y_j \log \pi_j + (1 - Y_j) \log(1 - \pi_j)$, and the Fisher information matrix is (10.7) with the same X as before but with W the diagonal matrix whose jth element is $\mathrm{E}(-d^2\ell_j/d\eta_j^2) = f(-\eta_j)^2/[F(-\eta_j)\{1 - F(-\eta_j)\}]$. The asymptotic variance of the maximum likelihood estimator $\widehat{\beta}_Y$ based on $Y_1, \ldots, Y_n$ is thus $(X^\mathsf{T}WX)^{-1}$.

The efficiency of large-sample inferences based on Z and Y may be compared through the asymptotic variance matrices $k^{-1}(X^\mathsf{T}X)^{-1}$ and $(X^\mathsf{T}WX)^{-1}$ of the corresponding maximum likelihood estimators. Rather than attempt a general discussion, we illustrate this numerically. Let $\eta_j = \beta_0 + \beta_1 x_j$, with x_j taking $n = 21$ values equally spaced from -1 to 1. The left panel of Figure 10.6 shows data simulated

The asymptotics here arise if we imagine m replicate observations at each x_j, and let $m \to \infty$.

Figure 10.6 Efficiency loss due to reducing continuous variables to binary ones. Left panel: simulated data. Blobs above the dotted line are counted as successes, with zeros below it as failures; the solid line is $0.5 + 2x$. Right panel: Comparison of asymptotic t statistics when continuous data are dichotomized, for normal error distribution, when $\beta_0 = 0.5, 1, 1.5$ (solid, dots, dashes).

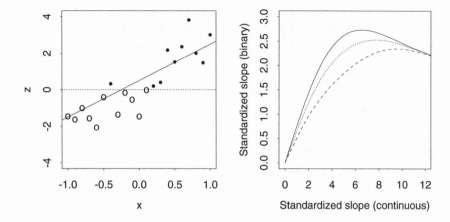

from this model, with $\beta_0 = 0.5$, $\beta_1 = 2$, and standard normal errors. The right panel plots $\beta_1 / v_Y^{1/2}$ against $\beta_1 / v_Z^{1/2}$, where v_Y and v_Z are the large-sample variances of the maximum likelihood estimates of β_1 based on the Ys and the Zs, for three values of β_0. The quantity $\beta_1 / v_Z^{1/2}$ is the limiting value of the t statistic for testing whether $\beta_1 = 0$, based on the full data, while $\beta_1 / v_Y^{1/2}$ is the corresponding quantity for the binary data. The ratio v_Z / v_Y is the asymptotic efficiency for estimating β_1 from the binary data, relative to the original data, and in the graph this has largest value of about $(3/4)^2 \doteq 0.56$ when $\beta_1 = 0$, decreasing to about $(2/12)^2 \doteq 0.03$. For the data in the left panel, the t-statistic is $\widehat{\beta}_1 / v_Z^{1/2} = 2.39/0.36 = 6.6$, whereas the corresponding quantity for binary data is $3.15/1.20 = 2.6$, which is much weaker — though still strong — evidence of non-zero slope.

An argument analogous to that giving (7.28) shows that the power of a size α two-sided test of $\beta_1 = 0$ using the asymptotic normal distribution of $\widehat{\beta}_1 / v_Y^{1/2}$ is $\Phi(z_{\alpha/2} + \delta_Y) + \Phi(z_{\alpha/2} - \delta_Y)$, where $\delta_Y = \beta_1 / v_Z^{1/2}$ is replaced by $\delta_Z = \beta_1 / v_Y^{1/2}$ in the corresponding power from the full data. Use of the binary data can sharply reduce the power. When $\beta_1 / v_Z^{1/2} \doteq 2$, for example, $\beta_1 / v_Y^{1/2} \doteq 1.4$, and with $\alpha = 0.025$ the power is reduced from 0.52 to 0.29.

A peculiarity of binary regression is the decrease in $\beta_1 / v_Y^{1/2}$ as $\beta_1 \to \infty$, because the information $f(-\eta)^2 / [F(-\eta)\{1 - F(-\eta)\}]$ tends to zero so quickly that $v_Y^{1/2} \to \infty$ faster than $\beta_1 \to \infty$. Thus the reduced efficiency for estimating β_1 becomes extreme when $|\beta_1|$ is large; to put this another way, as $|\beta_1| \to \infty$, the power for testing for zero slope based on $\widehat{\beta}_1$ tends to zero. The explanation for this is that most information in binary data is contributed by those responses whose variances are largest, for which π is not too close to zero or one, but as $\beta_1 \to \pm\infty$, the variances of all the observations tend to zero and β_1 cannot be reliably estimated.

Complete separation of successes and failures can occur. To see how, note that the estimate of β_1 from the binary data in the left panel of Figure 10.6 depends crucially on the value y_7 at $x_7 = -0.4$. If y_7 had equalled zero, then a perfect fit would have been obtained by setting $\widehat{\beta}_1 = +\infty$ and choosing $\widehat{\beta}_0$ so that $\widehat{\pi} = 0$ for $x \leq 0.1$ and $\widehat{\pi} = 1$ otherwise. This is harder to spot when there are several covariates, but a good model-fitting routine will signal convergence problems as $|\widehat{\beta}| \to \infty$. Near-complete separation will be indicated by regression diagnostics, which here suggest that the pair (y_7, x_7) is an outlier, as it has a large residual and is highly influential. ∎

Logistic regression

The most common choice of function F is the logistic distribution, which gives the canonical, logit, link function. Then

$$\Pr(Y = 1) = \pi = \frac{\exp(x^{\mathrm{T}}\beta)}{1 + \exp(x^{\mathrm{T}}\beta)}, \quad \Pr(Y = 0) = 1 - \pi = \frac{1}{1 + \exp(x^{\mathrm{T}}\beta)},$$

and the resulting logistic regression model is a linear model for the logarithm of the odds of success,

$$\frac{\Pr(Y = 1)}{\Pr(Y = 0)} = \frac{\pi}{1 - \pi} = \exp(x^{\mathrm{T}}\beta).$$

The likelihood for independent binary observations $y_1, \ldots, y_n$ with covariate vectors $x_1, \ldots, x_n$ is

$$L(\beta) = \prod_{j=1}^{n} \left\{ \frac{\exp\left(x_j^{\mathrm{T}}\beta\right)}{1 + \exp\left(x_j^{\mathrm{T}}\beta\right)} \right\}^{y_j} \left\{ \frac{1}{1 + \exp\left(x_j^{\mathrm{T}}\beta\right)} \right\}^{1-y_j} = \frac{\exp\left(\sum_j y_j x_j^{\mathrm{T}}\beta\right)}{\prod_j \left\{1 + \exp\left(x_j^{\mathrm{T}}\beta\right)\right\}}.$$

$$(10.23)$$

This is a linear exponential family model in which $S = \sum Y_j x_j$ is minimal sufficient for β. If any of the covariate vectors are repeated, S may be written as $\sum_d x_d R_d$, where the distinct covariate vectors are labelled x_d and $R_d = \sum_j Y_{dj}$, the total number of successes for responses with covariates x_d, is a binomial variable. Apart from a constant, (10.23) is the same likelihood as would be obtained from responses aggregated by covariate vectors.

If R is binomial with denominator m, then the log odds may be estimated by the *empirical logistic transform* $\log\{(R + \frac{1}{2})/(m - R + \frac{1}{2})\}$, whose estimated variance is $(R + \frac{1}{2})^{-1} + (m - R + \frac{1}{2})^{-1}$, and this is sometimes useful for plotting.

Many model-checking procedures break down for unaggregated binary data. For a given fitted probability $\widehat{\pi}$, a residual takes just two values, so comparison with a normal distribution is not useful. Moreover the deviance for a binary logistic model is a function of the data through $\widehat{\beta}$ alone, and hence it provides no information about fit in any absolute sense (Exercise 10.4.1). Pearson's statistic is strongly correlated with the deviance and shares this difficulty.

Example 10.18 (Nodal involvement data)   Table 10.8 summarizes data on 53 patients with prostate cancer. There are five binary explanatory variables: age in years $(0 = \text{less than } 60, 1 = 60 \text{ or more})$; `stage`, a measure of the seriousness of the tumour $(0 = \text{less serious}, 1 = \text{more serious})$; `grade`, a measure of the pathology of the tumour $(0 = \text{less serious}, 1 = \text{more serious})$; `xray` $(0 = \text{less serious}, 1 = \text{more serious})$; and `acid`, the level of serum acid phosphatase $(0 = \text{less than } 0.6, 1 = 0.6 \text{ or more})$. The response, nodal involvement, indicates whether the cancer has spread to neighbouring lymph nodes. The first row of the table shows that for five out of six patients

Table 10.8 Data on nodal involvement (Brown, 1980).

m	r	age	stage	grade	xray	acid
6	5	0	1	1	1	1
6	1	0	0	0	0	1
4	0	1	1	1	0	0
4	2	1	1	0	0	1
4	0	0	0	0	0	0
3	2	0	1	1	0	1
3	1	1	1	0	0	0
3	0	1	0	0	0	1
3	0	1	0	0	0	0
2	0	1	0	0	1	0
2	1	0	1	0	0	1
2	1	0	0	1	0	0
1	1	1	1	1	1	1
1	1	1	1	0	1	1
1	1	1	0	1	1	1
1	1	1	0	0	1	1
1	0	1	0	1	0	0
1	1	0	1	1	1	0
1	0	0	1	1	0	0
1	1	0	1	0	1	0
1	1	0	0	1	0	1
1	0	0	0	0	1	1
1	0	0	0	0	1	0

aged less than 60 and with high levels of the other explanatory variables, there was nodal involvement. A case is an individual patient rather than a row of the table. The explanatory variables are relatively easily collected and the aim of analysis was to predict nodal involvement from them.

Table 10.9 contains the deviances for all 2^5 combinations of explanatory variables when a binary logistic model is fitted to the data. The model with terms for stage, xray, and acid has deviance 19.64 on 49 degrees of freedom and the smallest AIC; it seems best overall, though it has several close competitors. The fitted linear predictor for this model is $-3.05 + 1.65I_{\text{stage}} + 1.91I_{\text{xray}} + 1.64I_{\text{acid}}$, where I_{stage} indicates that stage takes its higher level, and so forth. The fitted odds of nodal involvement when all the explanatory variables take their lower levels are a low $e^{-3.05} \doteq 0.047$, though this must be viewed with caution as there are no such cases in the data. The odds increase by a factor $e^{1.91} \doteq 6.75$ when acid takes its higher level, and are $e^{-3.05+1.91+1.65+1.64} \doteq 8.6$ at the higher levels of stage, acid, and xray.

The residual scaled deviance of 19.64 on 49 degrees of freedom suggests that the model fits well, but the binomial denominators are too small for confidence in χ^2 asymptotics. The deviance does not measure model fit for binary data: it is a function of $\widehat{\beta}$ alone and hence does not contrast the data with the fitted model. If the data in Table 10.8 had been analyzed as written there, that is as 23 binomial rather than 53 binary observations, the degrees of freedom for the best-fitting model would be

age	stage	grade	xray	acid	df	Deviance	age	stage	grade	xray	acid	df	Deviance
					52	40.71	+	+	+			49	29.76
+					51	39.32	+	+		+		49	23.67
	+				51	33.01	+	+			+	49	25.54
		+			51	35.13	+		+	+		49	27.50
			+		51	31.39	+		+		+	49	26.70
				+	51	33.17	+			+	+	49	24.92
+	+				50	30.90		+	+	+		49	23.98
+		+			50	34.54		+	+		+	49	23.62
+			+		50	30.48		+		+	+	49	19.64
+				+	50	32.67			+	+	+	49	21.28
	+	+			50	31.00	+	+	+	+		48	23.12
	+		+		50	24.92	+	+	+		+	48	23.38
	+			+	50	26.37	+	+		+	+	48	19.22
		+	+		50	27.91	+		+	+	+	48	21.27
		+		+	50	26.72		+	+	+	+	48	18.22
			+	+	50	25.25	+	+	+	+	+	47	18.07

Table 10.9 Scaled deviances for 32 logistic regression models for nodal involvement data. A plus denotes a term included in the model.

$23 - 4 = 19$ rather than 49, but the deviance of 19.64 would be unchanged. This ambiguity is another reason not to rely on the deviance to measure fit for binary data.

Figure 10.7 illustrates difficulties with binary residuals. The left panel shows the 53 residuals for the unaggregated data. The linear predictors are slightly jittered to prevent over-plotting. The upper and lower bands, corresponding to ones and zeros respectively, are typical of data with only a few response values. The right panel shows the 23 residuals for the aggregated data; the fitted values are the same as on the left. Banding remains but is much less obvious, and the apparent outliers are gone. There is little useful information in either plot.

We reconsider these data in Example 12.18. ■

10.4.2 2 × 2 table

A very common data structure classifies individuals by two sets of binary categories. In a medical setting, for example, we may observe success or failure for patients randomly allocated to be either a case — receiving some treatment — or a control. The resulting data may be laid out as in Table 10.10. The simplest model for this regards the numbers of successes R_1 and R_0 as independent binomial variables with probabilities

$$\pi_1 = \frac{e^{\lambda+\psi}}{1 + e^{\lambda+\psi}}, \quad \pi_0 = \frac{e^\lambda}{1 + e^\lambda}$$

and denominators m_1 and m_0. Then the joint density of R_1 and R_0 is

$$\Pr(R_1 = r_1, R_0 = r_0; \psi, \lambda) = \binom{m_1}{r_1}\binom{m_0}{r_0} \frac{e^{(\lambda+\psi)r_1}}{(1 + e^{\lambda+\psi})^{m_1}} \frac{e^{\lambda r_0}}{(1 + e^\lambda)^{m_0}}, \quad (10.24)$$

Table 10.10 Notation for 2 × 2 table.

	Success	Failure	Total
Case	R_1	$m_1 - R_1$	m_1
Control	R_0	$m_0 - R_0$	m_0
Total	$R_1 + R_0$	$m_1 + m_0 - R_1 - R_0$	$m_1 + m_0$

Figure 10.7 Standardized deviance residuals for nodal involvement data, for ungrouped responses (left) and grouped responses (right).

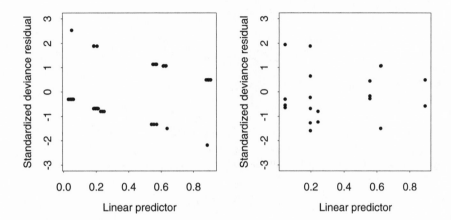

which is an exponential family of order two with natural parameter (ψ, λ) and natural observation $(R_1, R_1 + R_0)$ (Section 5.2.2). This is a generalized linear model with binomial errors and logit link function.

The usual purpose of analysis is to compare π_1 with π_0. Although quantities such as the difference $\pi_1 - \pi_0$ are sometimes of interest, we focus here on the difference in log odds,

$$\psi = \log\left(\frac{\pi_1}{1 - \pi_1}\right) - \log\left(\frac{\pi_0}{1 - \pi_0}\right).$$

This is a natural parameter of the exponential family, but more importantly its interpretation does not depend on whether the data are obtained prospectively or retrospectively. To appreciate this, suppose that a prospective study is performed: an individual is allocated randomly to cases $(T = 1)$ or controls $(T = 0)$ and then followed until a binary outcome Y is observed. Then

$$\Pr(Y = 1 \mid T = 1) = \frac{e^{\lambda+\psi}}{1 + e^{\lambda+\psi}}, \quad \Pr(Y = 1 \mid T = 0) = \frac{e^{\lambda}}{1 + e^{\lambda}} \tag{10.25}$$

and as T is allocated and then Y observed, the scheme fixes the vertical margin in Table 10.10. The drawback is that it may be costly and difficult to follow up enough individuals to obtain a precise estimate of ψ.

In a retrospective study the treatment status T is determined only after the outcomes Y are known; the scheme fixes the horizontal margin in Table 10.10. Often the treatment undergone can be ascertained from medical records, so large samples can be assembled more easily and cheaply than a prospective study, though the lack of randomization weakens subsequent inferences. Let $Z = 1$ indicate that an individual is chosen for the retrospective study, and suppose that this occurs with probabilities

$$\Pr(Z = 1 \mid Y = 1) = p_1, \quad \Pr(Z = 1 \mid Y = 0) = p_0,$$

independent of treatment status T. Then the success probability for an individual who was treated, conditional on their being chosen for inclusion in the study is $\Pr(Y = 1 \mid Z = 1, T = 1)$. This equals

$$\frac{\Pr(Z = 1 \mid Y = 1)\Pr(Y = 1 \mid T = 1)}{\Pr(Z = 1 \mid Y = 1)\Pr(Y = 1 \mid T = 1) + \Pr(Z = 1 \mid Y = 0)\Pr(Y = 0 \mid T = 1)}$$

by Bayes' theorem, so

$$\Pr(Y = 1 \mid Z = 1, T = 1) = \frac{p_1 e^{\lambda+\psi}}{p_1 e^{\lambda+\psi} + p_0} = \frac{e^{\lambda'+\psi}}{1 + e^{\lambda'+\psi}},$$

where $\lambda' = \lambda + \log(p_1/p_0)$. A similar argument gives

$$\Pr(Y = 1 \mid Z = 1, T = 0) = \frac{e^{\lambda'}}{1 + e^{\lambda'}},$$

so although retrospective sampling alters λ, the difference of log odds ψ is unchanged. This gives a strong motivation for using ψ to summarize the treatment effect, particularly if estimates from both types of study will ultimately be combined.

This argument applies also if ψ is replaced by $x^T\beta$, where x contains covariates as well as an indicator of treatment status. The key point is that the selection probabilities p_1 and p_0 must be independent of x.

Example 10.19 (Smoking and the Grim Reaper) Table 6.8 contains seven 2×2 tables, containing a prospective observational, that is, non-randomized, study on outcomes for women smokers and non-smokers. The simplest model ignores age by using only the overall data in the first line of Table 6.8, and gives parameter estimates for (10.25) of $\widehat{\lambda} = 0.78$ (0.08) and $\widehat{\psi} = 0.38$ (0.13). The significant positive value of $\widehat{\psi}$ shows an unlikely preservative effect of smoking. The deviance is 632.3 on 12 degrees of freedom, however, so the model is evidently inadequate.

When different values of λ are fitted to each table, the deviance drops to 2.38 on 6 degrees of freedom, and $\widehat{\psi} = -0.43$ (0.18): smoking significantly increases the death rate. There are 14 residuals, but as they arise in negatively correlated pairs and have only 6 degrees of freedom, it is better to examine one residual for each 2×2 table. They show nothing untoward.   ∎

Small sample analysis

The discussion above relies on large-sample likelihood results. Special techniques are needed for 2×2 tables with small counts. As (10.24) is an exponential family, the nuisance parameter λ may be eliminated by conditioning on its associated statistic $A = R_1 + R_0$, whose density is

$$\sum_{u=r_-}^{r_+} \binom{m_1}{u}\binom{m_0}{a-u} \frac{e^{\lambda a}}{(1+e^\lambda)^{m_0}} \frac{e^{\psi u}}{(1+e^{\lambda+\psi})^{m_1}}, \quad a = 0, \ldots, m_1 + m_0,$$

where $r_- = \max(0, a - m_0)$, $r_+ = \min(m_1, a)$. The conditional density of R_1 given $A = a$ is the *non-central hypergeometric density*

$$f(r \mid a; \psi) = \frac{\binom{m_1}{r}\binom{m_0}{a-r}e^{\psi r}}{\sum_{u=r_-}^{r_+} \binom{m_1}{u}\binom{m_0}{a-u}e^{\psi u}}, \quad r = r_-, \ldots, r_+, \tag{10.26}$$

on which exact inferences for ψ may be based; this amounts to conditioning on both margins of Table 10.10. Tests of $\pi_1 = \pi_0$ compare the observed value of R_1 with its null distribution, obtained by setting $\psi = 0$ in (10.26). To test $\psi = 0$ against the one-sided alternative $\psi > 0$ we use the P-value

$$\Pr(R_1 \geq r_1 \mid A = a; 0) = \sum_{r=r_1}^{r_+} f(r \mid a; 0), \tag{10.27}$$

and take $\Pr(R_1 \leq r_1 \mid A = a; 0)$ when testing $\psi < 0$. Exact confidence intervals are obtained by inverting these tests, solving for ψ_α, ψ^α the equations

$$\Pr(R_1 \geq r_1 \mid A = a; \psi_\alpha) = \alpha, \quad \Pr(R_1 \leq r_1 \mid A = a; \psi^\alpha) = \alpha.$$

When the margins of the table are small, the conditional distribution of R_1 is very discrete and the difficulties seen in Example 7.38 arise: exact conditional confidence intervals are quite conservative and it is preferable to replace (10.27) by the *mid-p significance level*

$$p_{+,\mathrm{mid}} = \frac{1}{2}\Pr(R_1 = r_1 \mid a; 0) + \Pr(R_1 > r_1 \mid a; 0). \tag{10.28}$$

When exact significance levels for testing $\psi = 0$ are unavailable, approximate ones may be obtained by treating

$$Z = \frac{R_1 - \frac{1}{2} - \mathrm{E}(R_1 \mid A = a; 0)}{\mathrm{var}(R_1 \mid A = a; 0)^{1/2}}$$

as standard normal, where the $\frac{1}{2}$ is a continuity correction, and

$$\mathrm{E}(R_1 \mid A = a; 0) = \frac{m_1 a}{m_0 + m_1}, \quad \mathrm{var}(R_1 \mid A = a; 0) = \frac{m_0 m_1 a(m_0 + m_1 - a)}{(m_0 + m_1)^2(m_0 + m_1 - 1)}.$$

Example 10.20 (Ulcer data) In a trial to compare two treatments for stomach ulcer, 28 persons with ulcers were divided randomly into two groups, one of size $m_1 = 15$ who were given a new surgical treatment, and the other of size $m_0 = 13$ who were given an existing one; see Table 10.11, which also contains data from other trials. The numbers in these groups without an adverse outcome, recurrent bleeding, were $r_1 = 8$ and $r_0 = 2$. Does the new treatment reduce the number of adverse outcomes? Here the null hypothesis is $\psi = 0$, with alternative $\psi > 0$. The attainable significance levels for the conditional test are in Table 10.12, and $p_+ = 0.0434$ and $p_{+,\mathrm{mid}} = 0.0243$. There is some evidence that the new treatment improves on the old.

Table 10.11 Data from 40 independent experiments to compare a new surgery for stomach ulcer with an older surgery; data from Efron (1996) corrected from original articles. Shown are the number of persons given the new treatment, m_1, of whom r_1 did not have recurrent bleeding, and the number given the old treatment, m_0, of whom r_0 did not have recurrent bleeding.

	r_1	m_1	r_0	m_0		r_1	m_1	r_0	m_0
1	8	15	2	13	21	34	40	8	21
2	11	19	8	16	22	14	18	34	39
3	29	34	35	39	23	54	68	61	74
4	14	20	13	21	24	20	24	19	27
5	9	12	12	12	25	6	6	0	6
6	3	7	0	4	26	9	10	10	15
7	13	17	11	24	27	12	17	10	15
8	15	16	3	16	28	10	10	2	14
9	11	14	15	22	29	22	22	16	24
10	36	38	20	32	30	16	18	11	21
11	6	12	0	8	31	14	15	6	13
12	5	7	2	9	32	9	12	2	9
13	12	21	17	24	33	20	20	18	23
14	14	21	20	25	34	13	17	14	16
15	22	25	21	32	35	30	40	8	20
16	7	11	4	10	36	13	16	14	16
17	8	10	2	10	37	30	34	14	19
18	30	31	23	27	38	31	38	22	37
19	24	28	16	31	39	34	34	0	34
20	36	43	27	43	40	9	9	16	16

Table 10.12
Significance probabilities for a test of no treatment effect in the first 2×2 table of the ulcer data. Here $p_+ = \Pr(R_1 \geq r_1 \mid A = a; 0)$, and z' and z are the standardized forms of R_1 without and with continuity correction. Note how closely $1 - \Phi(z)$ and $1 - \Phi(z')$ match p_+ and $p_{+,\text{mid}}$ respectively.

r_1	0	1	2	3	4	5	6	7	8	9	10
p_+	1	1	0.999	0.989	0.929	0.751	0.456	0.184	0.043	0.005	0
$1 - \Phi(z)$	1	1	0.999	0.987	0.925	0.747	0.456	0.187	0.048	0.007	0
$p_{+,\text{mid}}$	1	1	0.994	0.959	0.840	0.604	0.320	0.114	0.024	0.003	0
$1 - \Phi(z')$	1	1	0.995	0.966	0.854	0.609	0.309	0.101	0.020	0.002	0

The left panel of Figure 10.8 shows the conditional and unconditional distributions of Z; for the unconditional distribution $\lambda = 0$. Though both are discrete, the unconditional distribution is much more nearly continuous.

The right panel shows summaries for likelihood analysis of the data. The difference between the conditional likelihood based on (10.26) and the profile likelihood for ψ is small, but we should be wary of using large-sample likelihood approximations, because the sample is rather small. The panel also shows slices through the likelihood corresponding to various values of λ; evidently the likelihood depends strongly on both parameters.   ■

The history of the 2×2 table has been dogged by controversy, partly because of the effect of the discreteness of the conditional distribution of R_1 on confidence intervals for ψ. The unconditional distribution is more nearly continuous, so it yields shorter confidence intervals and more powerful tests. Hence some authors believe that inference should be based on the unconditional rather than on the conditional distribution. The drawback is that as the unconditional distribution depends on λ it does not give exact tests and confidence intervals, whereas the conditional approach does.

Figure 10.8 Analysis for first 2 × 2 table of ulcer data. Left: conditional (bold) and unconditional distribution of standardized R_1. Right: relative likelihoods based on conditional distribution of R_1 given A (heavy), profile likelihood (solid), and slices through likelihood based on R_1 and R_2 for fixed values of λ, equal to $-0.5, -1, -1.5, 2, -2.5$, from left to right (dots).

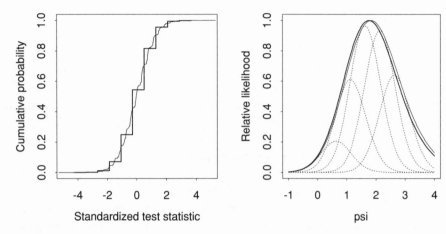

Cumulative probability — Standardized test statistic

Relative likelihood — psi

Exercises 10.4

1   Data $y_1, \ldots, y_n$ are assumed to follow a binary logistic model in which y_j takes value 1 with probability $\pi_j = \exp(x_j^{\mathrm T}\beta)/\{1 + \exp(x_j^{\mathrm T}\beta)\}$ and value 0 otherwise, for $j = 1, \ldots, n$.
(a) Show that the deviance for a model with fitted probabilities $\widehat\pi_j$ can be written as

$$D = -2\left\{ y^{\mathrm T} X\widehat\beta + \sum_{j=1}^{n} \log(1 - \widehat\pi_j) \right\}$$

and that the likelihood equation is $X^{\mathrm T} y = X^{\mathrm T}\widehat\pi$. Hence show that the deviance is a function of the $\widehat\pi_j$ alone.
(b) If $\pi_1 = \cdots = \pi_n = \pi$, then show that $\widehat\pi = \bar y$, and verify that

$$D = -2n\left\{ \bar y \log \bar y + (1 - \bar y)\log(1 - \bar y) \right\}.$$

Comment on the implications for using D to measure the discrepancy between the data and fitted model.
(c) In (b), show that Pearson's statistic (10.21) is identically equal to n. Comment.

2   (a) Show that the parametric link function

$$g(\pi; \gamma) = \log\left[\gamma^{-1}\{(1 - \pi)^{-\gamma} - 1\} \right], \quad \gamma \neq 0,$$

gives the logit and complementary log-log links when $\gamma = 1$ and when $\gamma \to 0$.
Give a similar function containing the logit and log-log link functions.
(b) Show that the link function

$$g(\pi; \gamma) = 2\gamma^{-1}\frac{\pi^\gamma - (1 - \pi)^\gamma}{\pi^\gamma + (1 - \pi)^\gamma}, \quad \gamma \neq 0,$$

is symmetric for all γ and gives the logit and identity functions when $\gamma \to 0$ and when $\gamma = 1$.

3   If X is a Poisson variable with mean $\mu = \exp(x^{\mathrm T}\beta)$ and Y is a binary variable indicating the event $X > 0$, find the link function between $\mathrm E(Y)$ and $x^{\mathrm T}\beta$.

10.5 Count Data

10.5.1 Log-linear models

The basic model for count data treats the response Y as a Poisson variable with mean μ. With the canonical, log, link, $\mu = \exp(x^T\beta)$; this is a log-linear model. In certain applications Y may be thought of as the number of events in a Poisson process of rate $\exp(x^T\beta)$ observed for a period T, in which case $\mu = T \exp(x^T\beta) = \exp(x^T\beta + \log T)$. This is a log-linear model with linear predictor $\eta' = x^T\beta + \log T$; the *offset* term $\log T$ is a fixed part of the linear predictor.

The connection between the Poisson and binomial distributions induces a relationship between log-linear and logistic models. Let Y_1 and Y_2 be independent Poisson variables with means μ_1 and μ_2. Then the conditional distribution of Y_2 given that $Y_1 + Y_2 = m$ is binomial with probability $\pi = \mu_2/(\mu_1 + \mu_2)$ and denominator m. If $\mu_1 = \exp(\gamma + x_1^T\beta)$ and $\mu_2 = \exp(\gamma + x_2^T\beta)$, then $\pi = \exp\{(x_2 - x_1)^T\beta\}/[1 + \exp\{(x_2 - x_1)^T\beta\}]$, so β may be estimated either by a log-linear model based on both observations, or by a logistic model using the conditional distribution of the second given their sum; in this second case γ cannot be estimated.

Example 10.21 (Premier League data)   We consider the numbers of goals scored in the 380 soccer matches played in the English Premier League in the 2000–2001 season. The data are the home and away scores, y_{ij}^h and y_{ij}^a, when team i is at home to team j, treated as independent Poisson variables with means

$$\mu_{ij}^h = \exp(\Delta + \alpha_i - \beta_j), \quad \mu_{ij}^a = \exp(\alpha_j - \beta_i),$$

where Δ represents the home advantage and α_i and β_i the offensive and defensive strengths of team i. We expect to find $\Delta > 0$, corresponding to better performance for teams playing at home.

Table 10.13 contains the analysis of deviance for this log-linear model. There are large home and offensive effects and weaker but still very significant defensive effects. Although the residual deviance is substantially larger than its degrees of freedom, only 36 of the individual scores exceed three goals, so asymptotics based on large counts are suspect. For the same reason residual analysis is not very useful, and we assess model fit by Monte Carlo methods. Simulation from the fitted model gave 999 deviances with average value of 826, of which 748 exceeded the observed value. Thus the observed residual deviance is not unusual, suggesting that the model is broadly adequate. Under this model, $\widehat{\Delta} = 0.37$ (0.07), so the mean score of a team playing at home is increased by a substantial multiplier of $\exp(\widehat{\Delta}) = 1.45$. Estimates of the other parameters are given in the lower part of Table 10.13. The fitted mean scores are readily computed; for example when Manchester United is at home to Coventry the fitted means are $\exp\{0.37 + 0.22 - (-0.52)\} = 3.03$ and $\exp(-0.53 - 0.15) = 0.51$. In fact this match was a 4–2 win for the home team, Coventry doing better than expected but losing anyway.

A different analysis models the home score, given the total score m for each match. The paragraph preceding this example shows that if the log-linear model is correct,

Table 10.13 Log-linear and logistic models fitted to Premier League data. The upper part shows the analysis of deviance for log-linear models with parameters for home advantage, offense and defense. The lower part shows a league table based on the overall strengths estimated from the binomial model, with estimated offensive and defensive capabilities from the log-linear model. The baseline team is Arsenal, some of whose parameters are aliased. Individual standard errors are not shown, but they are within ±0.02 of the values at the foot of the table.

Log-linear model			Logistic model		
Terms	df	Deviance reduction	Terms	df	Deviance reduction
Home	1	33.58	Home	1	33.58
Defense	19	39.21	Team	19	79.63
Offense	19	58.85			
Residual	720	801.08	Residual	332	410.65

	Overall (δ)	Offensive (α)	Defensive (β)
Manchester United	0.39	0.22	0.15
Liverpool	0.13	0.12	−0.08
Arsenal	—	0.04	—
Chelsea	−0.09	0.08	−0.22
Leeds	−0.10	0.02	−0.17
Ipswich	−0.16	−0.10	−0.13
Sunderland	−0.33	−0.31	−0.10
Aston Villa	−0.48	−0.31	−0.15
West Ham	−0.53	−0.33	−0.30
Middlesborough	−0.53	−0.35	−0.17
Charlton	−0.55	−0.21	−0.43
Tottenham	−0.58	−0.28	−0.38
Newcastle	−0.59	−0.35	−0.30
Southampton	−0.60	−0.45	−0.25
Everton	−0.75	−0.32	−0.46
Leicester	−0.77	−0.47	−0.31
Manchester City	−0.90	−0.40	−0.56
Coventry	−0.93	−0.53	−0.52
Derby	−0.93	−0.51	−0.45
Bradford	−1.29	−0.71	−0.62
SEs	0.29	0.20	0.20

the distribution of the number of goals scored when team i plays at home to team j is binomial with denominator m and probability

$$\frac{\mu_{ij}^h}{\mu_{ij}^h + \mu_{ij}^a} = \frac{\exp(\Delta + \alpha_i - \beta_j)}{\exp(\Delta + \alpha_i - \beta_j) + \exp(\alpha_j - \beta_i)}$$
$$= \frac{\exp(\Delta + \delta_i - \delta_j)}{1 + \exp(\Delta + \delta_i - \delta_j)}, \tag{10.29}$$

where $\delta_i = \alpha_i + \beta_i$ represents the overall strength of team i. Under this logistic model, no-score draws contribute no information, as the conditional distribution of Y_2 is degenerate when $m = 0$, and if there was no home advantage and no differences among the teams, the number of goals scored by the home side would be binomial with denominator m and probability $\frac{1}{2}$. This analysis will give no information on the absolute goal-scoring abilities of the teams, merely their relative strengths. As an arbitrary constant may be added to the δ_i, they cannot all be estimated; we deal with this by declaring that $\delta = 0$ for Arsenal.

When this model is fitted to the 352 matches with at least one goal scored, we obtain the deviances in the right part of Table 10.13. There are strong differences among the teams. The home advantage remains $\widehat{\Delta} = 0.37$ (0.07), and the estimated δs are given in the lower part of the table. The broad pattern is the same as in the log-linear model, though the ordering of clubs in the middle of the league is different. However, the standard errors for the team effects are larger than those for the log-linear model, because information is lost when the logistic model is fitted; see Exercise 10.5.2.

The logistic model gives an overall ranking of the teams similar to the official ranking, though with differences of detail: Arsenal and Liverpool are interchanged, and so are Derby and Manchester City. The centre of the table has further differences, but as the standard error for each $\widehat{\delta}_j - \widehat{\delta}_i$ is about 0.3, it is dangerous to read much into them. One reason for the differences is that the ranking here is based on numbers of goals, while the official ranking gives 2 for a win, 1 for a draw, and 0 for a loss.   ■

As the lowest three sides were relegated to the first division, their supporters might not regard this as a matter of detail!

10.5.2 Contingency tables

Count data often arise in the form of contingency tables that cross-classify individuals according to their attributes. The appropriate class of models for such a table depends on the sampling scheme. Suppose that an $R \times C$ table arises by randomly sampling a population over a fixed period and then classifying the resulting individuals. For example, a researcher interested in the association of gender (rows) and voting intentions (columns) might stand on a street corner for an hour recording data from anyone willing to talk to him. There are then no constraints on the row and column totals, and a simple model is that the count in the (r, c) cell, y_{rc}, has a Poisson distribution with mean μ_{rc}. The resulting likelihood is

This would not be a good way to proceed because of likely bias due to non-random sampling.

$$\prod_{r,c} \left\{ \frac{\mu_{rc}^{y_{rc}}}{y_{rc}!} e^{-\mu_{rc}} \right\} ;$$

this is simply the Poisson likelihood for the counts in the RC groups.

Our hapless researcher may set out with the intention of interviewing a fixed number m of individuals, stopping only when $\sum_{rc} y_{rc} = m$. In this case the data are multinomially distributed, with likelihood

$$\frac{m!}{\prod_{r,c} y_{rc}!} \prod_{r,c} \pi_{rc}^{y_{rc}}, \quad \sum_{r,c} \pi_{rc} = 1,$$

with $\pi_{rc} = \mu_{rc} / \sum_{s,t} \mu_{st}$ the probability of falling into the (r, c) cell.

A third scheme is to interview fixed numbers of men and of women, thus fixing the row totals $m_r = \sum_c y_{rc}$ in advance. In effect this treats the row categories as subpopulations, and the column categories as the response. This yields independent multinomial distributions for each row, and product multinomial likelihood

$$\prod_r \left\{ \frac{m_r!}{\prod_c y_{rc}!} \prod_c \pi_{rc}^{y_{rc}} \right\}, \quad \sum_c \pi_{1c} = \cdots = \sum_c \pi_{Rc} = 1,$$

in which $\pi_{rc} = \mu_{rc} / \sum_t \mu_{rt}$. See Table 10.2, in which the response is the fate of a fixed number of butterflies for each combination of species and colour; the appropriate product multinomial model fixes the total for each triplet.

These three set-ups can all be fitted as log-linear models, provided the appropriate baseline terms are included in the linear predictor. To see this, we arrange our data as a two-way layout, with row totals fixed: the multinomial sampling scheme gives just one row, whereas we would arrange the data in Table 10.2 as a 48×3 table. Suppose that the cell counts y_{rc} are independent Poisson variables with means $\mu_{rc} = \exp(\gamma_r + x_{rc}^T \beta)$, where γ_r corresponds to the overall count in the rth row; interest focuses on the parameter β. The multinomial model has fixed row totals $\sum_c y_{rc} = m_r$ and probabilities

$$\pi_{rc} = \frac{\mu_{rc}}{\sum_d \mu_{rd}} = \frac{\exp\left(\gamma_r + x_{rc}^T \beta\right)}{\sum_d \exp\left(\gamma_r + x_{rd}^T \beta\right)} = \frac{\exp\left(x_{rc}^T \beta\right)}{\sum_d \exp\left(x_{rd}^T \beta\right)},$$

so the corresponding log likelihood is

$$\begin{aligned}
\ell_{\text{Mult}}(\beta; y \mid m) &\equiv \sum_{rc} y_{rc} \log \pi_{rc} \\
&= \sum_r \left\{ \sum_c y_{rc} x_{rc}^T \beta - m_r \log \left(\sum_c e^{x_{rc}^T \beta} \right) \right\}, \quad (10.30)
\end{aligned}$$

where we have emphasized the fact that the likelihood is based on the conditional distribution of the counts y given the row totals m.

For the Poisson model there is no conditioning, so the log likelihood is

$$\begin{aligned}
\ell_{\text{Poiss}}(\beta, \gamma) &\equiv \sum_{r,c} (y_{rc} \log \mu_{rc} - \mu_{rc}) \\
&= \sum_r \left(m_r \gamma_r + \sum_c y_{rc} x_{rc}^T \beta - e^{\gamma_r} \sum_c e^{x_{rc}^T \beta} \right).
\end{aligned}$$

As the γ_r are not of central concern, we express this log likelihood as a function of the row totals $\tau_r = \sum_c \mu_{rc} = e^{\gamma_r} \sum_c e^{x_{rc}^T \beta}$ and the parameter of interest, β. In terms of β and the τ_r, we have $\gamma_r = \log \tau_r - \log\{\sum_c \exp(x_{rc}^T \beta)\}$, giving

$$\begin{aligned}
\ell_{\text{Poiss}}(\beta, \tau) &\equiv \sum_r (m_r \log \tau_r - \tau_r) + \sum_r \left\{ \sum_c y_{rc} x_{rc}^T \beta - m_r \log \left(\sum_c e^{x_{rc}^T \beta} \right) \right\}, \\
&= \ell_{\text{Poiss}}(\tau; m) + \ell_{\text{Mult}}(\beta; y \mid m),
\end{aligned}$$

say. The first term on the right of this decomposition is the log likelihood that corresponds to the Poisson distribution of the row total m_r — a sum of independent Poisson variables — while the second is the multinomial log likelihood (10.30). Thus the m_r form a cut, and the maximum likelihood estimates of β and τ based on $\ell_{\text{Poiss}}(\beta, \tau)$ are the same as those based on separate maximizations of $\ell_{\text{Poiss}}(\tau; m)$ and $\ell_{\text{Mult}}(\beta; y \mid m)$; see (5.21). Hence $\widehat{\beta}$ equals the maximum likelihood estimate for the multinomial log likelihood (10.30), and $\widehat{\tau}_r = m_r$. Moreover, the observed and expected information matrices for the model are block diagonal, with blocks corresponding to $-\partial^2 \ell_{\text{Poiss}}(\tau; m)/\partial \tau \partial \tau^{\text{T}}$ and $-\partial^2 \ell_{\text{Mult}}(\beta; y \mid m)/\partial \beta \partial \beta^{\text{T}}$.

To see that the standard errors for $\widehat{\beta}$ based on the multinomial and Poisson models are equal, note that $\partial^2 \ell_{\text{Poiss}}(\beta, \tau)/\partial \beta \partial \beta^{\text{T}}$ depends on the data only through the m_r. Therefore the expected information for β under the multinomial model, in which the m_r are fixed, equals the observed information for β under the Poisson model. Under the Poisson model, the expected information for β is

$$\sum_r \text{E}(m_r) \frac{\partial^2 \log\left(\sum_c e^{x_{rc}^{\text{T}}\beta}\right)}{\partial \beta \partial \beta^{\text{T}}} = \sum_r \tau_r \frac{\partial^2 \log\left(\sum_c e^{x_{rc}^{\text{T}}\beta}\right)}{\partial \beta \partial \beta^{\text{T}}},$$

and the standard errors for $\widehat{\beta}$ are obtained by replacing τ_r and β with their estimates, and inverting the resulting matrix. But as $\widehat{\tau}_r = m_r$, the resulting standard errors will equal those obtained by inverting the expected information matrix obtained from (10.30). It follows that the numerical values of standard errors and maximum likelihood estimates for β under the Poisson model are the same as those under the multinomial model, provided that the parameters associated with the margin fixed under the multinomial model, the γ_r, are included in the fit. The log linearity is important here, as it ensures that second derivatives of both log likelihoods with respect to β involve the counts y_{rc} only through their row totals m_r.

Example 10.22 (Jacamar data) Let y_{csf} denote the number of butterflies of the cth colour and sth species suffering the fth fate, where $c = 1, \ldots, 8, s = 1, \ldots, 6$, and $f = 1, 2, 3$. If we treat fate as the response, any model should fix the total count for each of the 48 combinations of species and colour, giving 48 trinomial variables. Any Poisson model should have a term α_{cs} in the linear predictor. For example, $\log \mu_{csf} = \alpha_{cs}$ corresponds to equal probability of each of the three fates, whatever the colour and species, because

$$(\pi_{cs1}, \pi_{cs2}, \pi_{cs3}) = \left(\frac{\mu_{cs1}}{\sum_f \mu_{csf}}, \frac{\mu_{cs2}}{\sum_f \mu_{csf}}, \frac{\mu_{cs3}}{\sum_f \mu_{csf}}\right) = \left(\frac{1}{3}, \frac{1}{3}, \frac{1}{3}\right),$$

and this is independent of c and s, while $\log \mu_{csf} = \alpha_{cs} + \gamma_f$ corresponds to probabilities

$$(\pi_{cs1}, \pi_{cs2}, \pi_{cs3}) = \left(\frac{\mu_{cs1}}{\sum_f \mu_{csf}}, \frac{\mu_{cs2}}{\sum_f \mu_{csf}}, \frac{\mu_{cs3}}{\sum_f \mu_{csf}}\right)$$

$$= \frac{1}{e^{\gamma_1} + e^{\gamma_2} + e^{\gamma_3}} (e^{\gamma_1}, e^{\gamma_2}, e^{\gamma_3}),$$

also independent of colour and species. Linear predictor $\log \mu_{csf} = \alpha_{cs} + \gamma_{cf}$

Table 10.14 Deviances for log-linear models fitted to jacamar data.

Terms	df	Deviance
C⋆S	22	259.42
C⋆S+F	86	173.86
C⋆S+C⋆F	72	139.62
C⋆S+S⋆F	76	148.23
C⋆S+C⋆F+S⋆F	62	90.66
C⋆S⋆F	0	0

corresponds to probability vector

$$(\pi_{cs1}, \pi_{cs2}, \pi_{cs3}) = \left(\frac{\mu_{cs1}}{\sum_f \mu_{csf}}, \frac{\mu_{cs2}}{\sum_f \mu_{csf}}, \frac{\mu_{cs3}}{\sum_f \mu_{csf}} \right)$$

$$= \frac{1}{e^{\gamma_{c1}} + e^{\gamma_{c2}} + e^{\gamma_{c3}}} (e^{\gamma_{c1}}, e^{\gamma_{c2}}, e^{\gamma_{c3}}),$$

in which the probabilities of the different fates depend on colour, but not on species. Let C⋆S+F denote the terms of the linear predictor $\alpha_{cs} + \gamma_f$. Then the terms for the three models above are C⋆S, C⋆S+F, and C⋆S+C⋆F. Any model that treats F as the response must contain a term C⋆S, which fixes the row totals. The term C⋆F indicates that the response probabilities depend on colour.

Table 10.14 contains the deviances for the models with C⋆S, with degrees of freedom adjusted for triplets with zero totals. The best-fitting model is the full model C⋆S⋆F. The best reasonable model is C⋆S+C⋆F+S⋆F, which extends to trinomial responses the binomial two-way layout model of Example 10.15, but its deviance is large compared to its asymptotic χ^2_{62} distribution.

If categories N and S were merged there would be 96 observations and two possible fates. In this case the linear predictor $\log \mu_{csf} = \alpha_{cs} + \gamma_{cf}$ corresponds to probabilities

$$(\pi_{cs1}, \pi_{cs2}) = \frac{1}{e^{\gamma_{c1}} + e^{\gamma_{c2}}} (e^{\gamma_{c1}}, e^{\gamma_{c2}}) = \frac{1}{1 + e^{\gamma_{c2}-\gamma_{c1}}} (1, e^{\gamma_{c2}-\gamma_{c1}}),$$

which is the binomial logistic model with terms 1+Colour fitted in Example 10.15. When the response classification has two categories, it simplifies matters to fit the model as binomial rather than Poisson, although identical inferences are drawn about its parameters. ∎

Example 10.23 (Lung cancer data) Example 1.4 gives data on the lung cancer mortality of cigarette smokers among British male physicians. The response is the number of deaths in each cell of the table, which also gives the total number of man-years of exposure T in each category.

We initially fit a log-linear model with factors for both margins and offset log man-years at risk in each cell. Thus $T_{rc} \exp(\alpha_r + \beta_c)$ is the mean number of deaths in the (r, c) cell. This model has deviance 51.47 on 48 degrees of freedom and appears to fit well. Figure 10.9 shows the coefficients for this model; the first level of each factor is taken to have coefficient zero. The figure suggests that there is a linear effect of dose d on the cancer rate, but that the increase with age is faster. However the standard

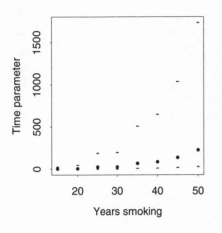
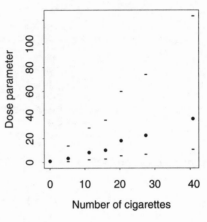

Figure 10.9 Results for two-way layout model, plotted against age and cigarette consumption. Shown are exponentials of coefficients, plus/minus two standard errors.

errors for the individual parameters are very large, reflecting the small numbers of deaths in most cells.

For a more concise model that is not log-linear, let the death rate for those smoking d cigarettes per day after t years of smoking be

$$\lambda(d, t) = \left(\beta_0 + \beta_1 d^{\beta_2}\right) t^{\beta_3}, \tag{10.31}$$

deaths per 100,000 man-years at risk; here β_0 and β_1 are non-negative, and β_2 and β_3 are real. We take t to be the midpoint of each group, divided by 42.5, so that β_0 represents the background rate of cancer for non-smokers aged 62.5 years, for whom the rescaled $t = 1$. The broadly exponential pattern in the left panel of Figure 10.9 suggests that $\beta_3 > 1$. The term $\beta_1 d^{\beta_2}$ describes the effect of smoking on death rates; we expect $\beta_2 \doteq 1$, corresponding to the linear increase seen in the right panel of Figure 10.9.

A likelihood ratio test of the effect of smoking on death rate would be non-regular, because setting either $\beta_1 = 0$ or $\beta_2 = 0$ eliminates both of these parameters; moreover $\beta_1 = 0$ is a boundary hypothesis. In either case the resulting model is log-linear, with deviance 180.8 on 61 degrees of freedom; the fit seems poor, despite the low counts and hence the likely inapplicability of chi-squared deviance asymptotics.

To fit the full model it is better to recast (10.31) as

$$\lambda(d, t) = \{e^{\gamma_0} + \exp(\gamma_1 + \beta_2 \log d)\} \exp(\beta_3 \log t),$$

so that all the parameters are unconstrained; the term $\exp(\gamma_1 + \beta_2 \log d)$ is omitted for the non-smokers. In this form it is straightforward to maximize the log likelihood by iterative weighted least squares, giving deviance 59.58 on 59 degrees of freedom, so marked an improvement on the model without smoking that the non-regularity of the asymptotics is immaterial.

Table 10.15 shows that the precision of $\widehat{\gamma}_0$ depends heavily on the data for non-smokers. The background non-smoker death-rate from cancer at age 62.5 is $e^{\widehat{\gamma}_0} = 18.9$ per 100,000 years at risk.

With the restriction $\beta_2 = 1$ the deviance increases to 61.84 on 60 degrees of freedom, a deviance difference of 2.26 on 1 degree of freedom. Linear dependence of

Table 10.15 Parameter estimates (standard errors) for lung cancer data.

	γ_0	γ_1	β_2	β_3
Smokers only	0.96 (25.4)	2.15 (1.45)	1.20 (0.40)	4.50 (0.34)
All data	2.94 (0.58)	1.82 (0.66)	1.29 (0.20)	4.46 (0.33)
All data ($\beta_2 = 1$)	2.75 (0.56)	2.72 (0.09)	—	4.43 (0.33)

Table 10.16 Joint distribution of visual impairment on both eyes by race and age Liang *et al.* (1992). Combination $(0, 0)$ means neither eye is visually impaired.

Eye		Prevalence for whites aged				Prevalence for blacks aged			
Left	Right	40–50	51–60	61–70	70+	40–50	51–60	61–70	70+
0	0	602	541	752	606	729	551	452	307
1	0	11	15	31	60	19	24	22	29
0	1	15	16	37	67	21	23	21	37
1	1	4	9	11	79	10	14	28	56

death rate on d appears plausible, in which case the background death rate drops somewhat to 15.6 deaths per 100,000 man-years at risk, but rises by an additional $e^{\widehat{\gamma_1}} \doteq 15.2$ for every cigarette smoked daily. Case analysis shows no residuals out of line, and the model appears to fit well. It is both more parsimonious than the log-linear model and motivated by substantive considerations, so it seems preferable. ∎

Marginal models

Although mathematically elegant and simple to fit, log-linear models have some awkward statistical properties because their parameters have interpretations that depend on other terms in the model, as we shall now see.

Example 10.24 (Eye data) Table 10.16 gives data from the Baltimore Eye Study Survey. Drivers are classified by age, race and visual impairment, defined as vision less than 20/60; in the original data their level of education is also available, and is treated as a surrogate for socioeconomic status. The aim of the original analysis was to see how visual impairment depends on age and race, controlling for education, but we shall simply consider dependence on age and race.

We treat the data as eight 2×2 tables corresponding to columns 3–10 of the table, that is one for each different combination of race and age, given by the covariate vector x. Each table has elements $(y_{00}, y_{01}; y_{10}, y_{11})$, where y_{00} is the number of men without visual impairment, y_{01} is the number whose right eye only is poor, and so forth. The total number of men with covariate combination x is $m = y_{00} + y_{01} + y_{10} + y_{11}$, which we treat as fixed. The corresponding probabilities $(\pi_{00}, \pi_{01}; \pi_{10}, \pi_{11})$ depend on x.

A natural preliminary to joint analysis of data for both eyes is to fit logistic regression models and estimate the probability of impairment in each eye separately. For the left eye we would treat $r_L = y_{10} + y_{11}$ as a binomial response with denominator m and probability

$$\pi_{10} + \pi_{11} = \pi_L = \exp(x^T \beta_L) / \{1 + \exp(x^T \beta_L)\},$$

say. For the right eye the response would be $r_R = y_{01} + y_{11}$ with denominator m and probability $\pi_{01} + \pi_{11} = \pi_R = \exp(x^{\mathsf{T}}\beta_R)/\{1 + \exp(x^{\mathsf{T}}\beta_R)\}$. Here β_L and β_R summarize the effect of x on the marginal distributions of r_L and r_R. Our earlier arguments show that these logistic models are also log-linear.

In generalizing these marginal models to allow for the anticipated dependence between the eyes, it is natural to augment π_L and π_R by adding further parameters. One possibility is to write the odds ratio as

$$\frac{\pi_{11}\pi_{00}}{\pi_{10}\pi_{01}} = \frac{\pi_{11}(1 - \pi_L - \pi_R + \pi_{11})}{(\pi_L - \pi_{11})(\pi_R - \pi_{11})} = \exp(x^{\mathsf{T}}\beta_{LR}).$$

If $x^{\mathsf{T}}\beta_{LR} = \gamma$ was independent of x, there would be constant association between the eyes after adjusting for marginal effects of age and race, with more complicated models indicating more complex patterns of association. As $0 < \pi_{00}, \pi_{01}, \pi_{10}, \pi_{11} < 1$, the probability π_{11} must lie in the interval $(\max(0, \pi_L + \pi_R - 1), \min(\pi_L, \pi_R))$, and a little algebra shows that π_{11} may be expressed as the root of a quadratic equation whose coefficients depends on π_L, π_R, and $x^{\mathsf{T}}\beta_{LR}$, thereby enabling us to express the probabilities in each 2×2 table in terms of the marginal probabilities and the odds ratio.

The log-linear model for the joint density of $(y_{00}, y_{01}; y_{10}, y_{11})$ has probabilities

$$(\pi_{00}, \pi_{01}; \pi_{10}, \pi_{11}) = \frac{1}{1 + e^{\gamma_R} + e^{\gamma_L} + e^{\gamma_R + \gamma_L + \gamma_{LR}}}(1, e^{\gamma_R}, e^{\gamma_L}, e^{\gamma_R + \gamma_L + \gamma_{LR}}),$$

where $\gamma_L = x^{\mathsf{T}}\delta_L$, $\gamma_R = x^{\mathsf{T}}\delta_R$, $\gamma_{LR} = x^{\mathsf{T}}\delta_{LR}$. Under this model the marginal probability of an unimpaired left eye is

$$\pi_L' = \frac{e^{\gamma_L} + e^{\gamma_R + \gamma_L + \gamma_{LR}}}{1 + e^{\gamma_R} + e^{\gamma_L} + e^{\gamma_R + \gamma_L + \gamma_{LR}}},$$

which has logistic form $e^{\gamma_L}/(1 + e^{\gamma_L})$ only when $\gamma_{LR} = 0$, that is conditional on x visual impairment occurs independently in each eye. Otherwise the marginal probability of an impaired left eye depends on γ_R and γ_{LR}, implying that the initial logistic fits shed no light on γ_L.

To put this another way, note that γ_L may be written as

$$\log\left(\frac{\pi_{10}}{\pi_{00}}\right) = \log\frac{\Pr(L = 1 \mid R = 0, x)}{\Pr(L = 0 \mid R = 0, x)},$$

$L = 1$ denotes visual
impairment in the left eye,
etc.

with a similar expression for γ_R, and that

$$\gamma_{LR} = \log\left\{\frac{\Pr(L = 1 \mid R = 1, x)}{\Pr(L = 0 \mid R = 1, x)}\right\} - \log\left\{\frac{\Pr(L = 1 \mid R = 0, x)}{\Pr(L = 0 \mid R = 0, x)}\right\}.$$

Thus the parameters of the log-linear model have interpretations in terms of contrasts of log odds for one eye *conditional* on the state of the other, and these do not yield marginal probabilities with simple interpretations. Therefore the log-linear model for the joint outcomes is not upwardly compatible with the logistic models for the marginal outcomes. This poses problems in applications where marginal properties of the variables are of interest.   ∎

Inference for marginal models is awkward because complete specification of their likelihoods is ordinarily neither possible nor desirable. An alternative is to base inference on systems of estimating equations, and we now sketch how this is done.

Suppose that the jth of n individuals contributes a $q \times 1$ response vector Y_j and a $p \times 1$ vector of explanatory variables x_j, and let Z_j denote the $q(q-1)/2 \times 1$ vector containing the distinct products of pairs of elements of Y_j. Now $\mathrm{E}(Y_j) = \mu(x_j^{\mathsf{T}}\beta)$ is specified by the marginal model, while the covariance structure among the responses is given by $\mathrm{E}(Z_j) = \xi(x_j^{\mathsf{T}}\beta, \gamma)$, where β represents the parameters of the marginal model, and γ additional parameters that account for association among elements of Y_j.

In the preceding example $q = 2$ and Y_j^{T} equals $(0,0)$, $(0,1)$, $(1,0)$, or $(1,1)$, indicating the state of the left and right eyes, while

$$\mathrm{E}\big(Y_j^{\mathsf{T}}\big) = \mu\big(x_j^{\mathsf{T}}\beta\big)^{\mathsf{T}} = (\pi_L, \pi_R) = \left(\frac{\exp(x^{\mathsf{T}}\beta_L)}{1 + \exp(x^{\mathsf{T}}\beta_L)}, \frac{\exp(x^{\mathsf{T}}\beta_R)}{1 + \exp(x^{\mathsf{T}}\beta_R)} \right),$$

and $\xi(x_j^{\mathsf{T}}\beta, \gamma)$, the probability of visual impairment in both eyes for individual j, depends both on x_j and on the degree of association between the eyes.

Ideas from Section 7.2 suggest that consistent estimators of β and γ may be obtained by combining the unbiased estimating functions $Y_j - \mu(x_j^{\mathsf{T}}\beta)$ and $Z_j - \xi(x_j^{\mathsf{T}}\beta, \gamma)$, and the form of (7.21) suggests that the estimators that solve the *generalized estimating equations*

$$\sum_{j=1}^{n} \frac{\partial\big\{\mu\big(x_j^{\mathsf{T}}\beta\big)^{\mathsf{T}}, \xi\big(x_j^{\mathsf{T}}\beta, \gamma\big)^{\mathsf{T}}\big\}}{\partial(\beta, \gamma)} \left(\begin{array}{cc} \mathrm{var}(Y_j) & \mathrm{cov}(Y_j, Z_j) \\ \mathrm{cov}(Z_j, Y_j) & \mathrm{var}(Z_j) \end{array} \right)^{-1}$$
$$\times \left(\begin{array}{c} Y_j - \mu\big(x_j^{\mathsf{T}}\beta\big) \\ Z_j - \xi\big(x_j^{\mathsf{T}}\beta, \gamma\big) \end{array} \right) = 0$$

will have smallest asymptotic variance. The presence of $\mathrm{cov}(Y_j, Z_j)$ means that third-order moments of Y_j must in principle be specified, and one way to avoid this is to replace this term by a zero matrix. The resulting estimators $\widehat{\beta}$ and $\widehat{\gamma}$ are consistent but the variance of $\widehat{\gamma}$ can be much larger than when the correct covariance matrix is used. If γ is of interest then some of the lost efficiency can be retrieved by assuming a simple form for $\mathrm{cov}(Y_j, Z_j)$. Standard errors for $\widehat{\beta}$ and $\widehat{\gamma}$ are based on a sandwich covariance matrix; see Section 7.2. In many applications it is important to be able to accomodate missing data, and this is achieved by allowing the length of Y_j to vary with the individual; no essentially new points arise.

10.5.3 Ordinal responses

Discrete data often arise in which the response comprises numbers in ordered categories that may be labelled $1, \dots, k$. Examples are individuals undergoing some treatment and asked to say if they experience one of {no pain, slight pain, moderate pain, extreme pain}, or where curries are classified as {bland, mild, ..., volcanic}. The goal is then typically to assess how these *ordinal responses* depend on explanatory variables x. Sometimes the response is a discretized version of an underlying continuous variable, though this interpretation is not always plausible. In either case suitable models are based on the multinomial distribution. If there are n independent individuals whose responses are $I_1, \ldots, I_n$, and $I_j = l$ indicates that the jth response falls in category l, then $\Pr(I_j = l) = \pi_l$ for $l = 1, \ldots, k$, and the corresponding cumulative probabilities are $\gamma_l = \Pr(I_j \leq l) = \pi_1 + \cdots + \pi_l$ for $l = 1, \ldots, k$; of course $\gamma_k = 1$. Individual responses with common explanatory variables x can be merged to give a multinomial variable $(Y_1, \ldots, Y_k)$, where Y_l represents the number in category l; thus $Y_1 + \cdots + Y_k = n$. Typically the joint distribution of $(Y_1, \ldots, Y_k)$ depends on x through a linear predictor $x^T \beta$.

In many applications it is appropriate to require that the interpretation of the model parameters remains unchanged when adjacent categories are merged. One class of models with this property may be motivated by positing the existence of an underlying continuous variable ε with distribution function F, with I indicating into which of the k intervals

$$(-\infty, \zeta_1], \ (\zeta_1, \zeta_2], \ldots, \ (\zeta_{k-2}, \zeta_{k-1}], \ (\zeta_{k-1}, \infty), \quad \zeta_1 < \cdots < \zeta_{k-1},$$

$x^T \beta + \varepsilon$ falls. For convenience let $\zeta_0 = -\infty$ and $\zeta_k = \infty$. Then

$$\pi_l(x^T \beta) = \Pr(I = l; x^T \beta) = \Pr(\zeta_{l-1} < x^T \beta + \varepsilon \leq \zeta_l)$$
$$= F(\zeta_l - x^T \beta) - F(\zeta_{l-1} - x^T \beta),$$

and $\gamma_l(x^T \beta) = F(\zeta_l - x^T \beta)$, for $l = 1, \ldots, k$. Thus large $x^T \beta$ leads to higher probabilities for the higher categories. A natural choice is the logistic distribution function $F(u) = \exp(u)/\{1 + \exp(u)\}$, which leads to the *proportional odds model*, so-called because the odds ratio of appearing in category l or lower for two individuals with explanatory variables x_1 and x_2,

$$\frac{\Pr(I \leq l; x_2)/\Pr(I > l; x_2)}{\Pr(I \leq l; x_1)/\Pr(I > l; x_1)} = \frac{\exp\left(\zeta_l - x_2^T \beta\right)}{\exp\left(\zeta_l - x_1^T \beta\right)} = \exp\left\{-(x_2 - x_1)^T \beta\right\},$$

is independent of l. Another possibility that often works well in practice is $F(u) = 1 - \exp\{-\exp(u)\}$. Whatever the choice of F, interest focuses on how the response depends on the covariates, summarized in β; typically $\zeta_1, \ldots, \zeta_{k-1}$ are of little concern. Any overall intercept term in $x^T \beta$ is aliased with the ζ_l.

Although this model is motivated by arguing as if an underlying continuous variable exists, this is not essential in order for it to be applied — the model may be useful even when ε clearly does not exist. As Examples 4.21 and 10.17 show, the loss of information due to categorizing continuous data can be substantial if the number of categories is very small.

The log likelihood based on independent responses $i_1, \ldots, i_n$ with corresponding vectors of explanatory variables $x_1, \ldots, x_n$ may be written as

$$\ell(\beta, \zeta) = \sum_{j=1}^{n} \sum_{l=1}^{k} I(i_j = l) \log \pi_l(x_j^T \beta),$$

a multinomial log likelihood to which by now familiar methods can be applied.

Example 10.25 (Pneumoconiosis data)   The data in Table 10.17 concern the period x in years of work at a coalface and the degree of pneumoconiosis in a group of miners. The response consists of counts $\{y_1, y_2, y_3\}$ in $k = 3$ categories {Normal,

Table 10.17 Period of exposure x and prevalence of pneumoconiosis amongst coalminers (Ashford, 1959).

	Period of exposure (years)							
	5.8	15	21.5	27.5	33.5	39.5	46	51.5
Normal	98	51	34	35	32	23	12	4
Present	0	2	6	5	10	7	6	2
Severe	0	1	3	8	9	8	10	5

Figure 10.10 Pneumoconiosis data analysis. The left panel shows how empirical logistic transformations z_2 and z_3 depend on exposure x. The right panel shows how the implied fitted logistic distributions depend on x. The vertical lines show $\widehat{\zeta}_1$ and $\widehat{\zeta}_2$; the areas lying left of, between, and right of them equal the fitted probabilities $\widehat{\pi}_1(x)$, $\widehat{\pi}_2(x)$, and $\widehat{\pi}_3(x)$.

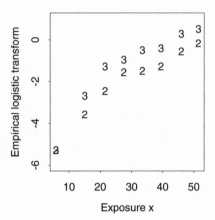
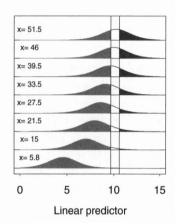

Present, Severe} assessed radiologically and is qualitative. As the period of exposure increases, the proportion of miners with the disease present or in severe form increases sharply.

A simple analysis starts by combining categories, either as {Normal or Present, Severe} or as {Normal, Present or Severe}, to which models for binomial responses may be fitted. The plot of the empirical logistic transforms

$$z_2 = \log\left(\frac{y_3 + \frac{1}{2}}{y_1 + y_2 + \frac{1}{2}}\right), \quad z_3 = \log\left(\frac{y_2 + y_3 + \frac{1}{2}}{y_1 + \frac{1}{2}}\right),$$

in the left panel of Figure 10.10 shows that the linear predictor should contain $\log x$ rather than x. The logistic regression model with linear predictor $\beta_0 + \beta_1 \log x$ and response $y_2 + y_3$ gives $\widehat{\beta}_0 = -9.6$ and $\widehat{\beta}_1 = 2.58$. The corresponding model with response y_3 yields $\widehat{\beta}_0 = -10.9$ and $\widehat{\beta}_1 = 2.69$. Both models fit well, and the similarity of the slope estimates suggests that fitting the proportional odds model will be worthwhile.

Maximum likelihood fitting of the proportional odds model with linear predictor $\beta_1 \log x$ gives $\widehat{\beta}_1 = 2.60$ (0.38), $\widehat{\zeta}_1 = 9.68$ (1.32), and $\widehat{\zeta}_2 = 10.58$ (1.34), entirely consistent with the binomial fits. Pearson's statistic is 4.7 on 13 degrees of freedom, so the fit seems good. The interpretation of $\widehat{\beta}_1$ is that every doubling of exposure increases the linear predictor by $2.6 \times \log 2 \doteq 1.8$ and hence the odds of having the disease by a factor 6 or so. The same increase applies to the odds of having the disease in severe form. The right panel of Figure 10.10 illustrates how the fitted logistic distribution implied by the model changes with x. ∎

Such models can be broadened by taking an underlying variable $x^T\beta + \sigma\varepsilon$, with σ dependent on explanatory variables.

Continuation ratio models may be based on the decomposition of the multinomial distribution of $(Y_1, \ldots, Y_k)$ as

$$Y_1 \sim B(n, \pi_1),$$

$$Y_2 \mid Y_1 \sim B\left(n - y_1, \frac{\pi_2}{1 - \pi_1}\right),$$

$$\vdots$$

$$Y_{k-1} \mid Y_1, \ldots, Y_{k-2} \sim B\left(n - y_1 - \cdots - y_{k-2}, \frac{\pi_{k-1}}{1 - \pi_1 - \cdots - \pi_{k-2}}\right);$$

(10.32)

of course Y_k is constant conditional on $Y_1, \ldots, Y_{k-1}$. At each stage the number of individuals in category l, given the numbers in categories $1, \ldots, l-1$, is treated as a binomial variable with response probability $\pi_l/(1 - \gamma_{l-1})$, to which a logistic regression or other suitable binomial response model may be fitted. Thus the original k-nomial response is broken into $k - 1$ separate binomial responses. Unlike in the proportional odds model there is no necessity that the same explanatory variables be used in each of the $k - 1$ fits, nor that their link functions be the same; this would depend on the scientific context.

Exercises 10.5

1   Consider the $2 \times n$ table of independent Poisson variables

$$\begin{array}{ccccc} Y_{11} & \cdots & Y_{1j} & \cdots & Y_{1n} \\ Y_{21} & \cdots & Y_{2j} & \cdots & Y_{2n} \end{array},$$

where

$$\eta_{1j} = \log \mathrm{E}(Y_{1j}) = x_{1j}^T\beta, \quad \eta_{2j} = \log \mathrm{E}(Y_{2j}) = x_{2j}^T\beta.$$

Show that the conditional density of Y_{1j} given that $Y_{1j} + Y_{2j} = m_j$ is binomial with denominator m_j and probability π_j satisfying $\log\{\pi_j/(1 - \pi_j)\} = x_j^T\beta$, where $x_j = x_{1j} - x_{2j}$. This implies that a contingency table in which a single, binary, classification is regarded as the response can be analyzed using logistic regression. What advantages are there to doing so in terms of model-fitting and the examination of residuals?

2   In light of the preceding exercise and the discussion on page 501, reconsider the models fitted in Example 10.21. Say why Table 10.13 contains much larger standard errors for the logistic than for the log-linear model.

3   For a 2×2 contingency table with probabilities

$$\begin{array}{cc} \pi_{00} & \pi_{01} \\ \pi_{10} & \pi_{11} \end{array},$$

the maximal log-linear model may be written as

$$\eta_{00} = \alpha + \beta + \gamma + (\beta\gamma), \quad \eta_{01} = \alpha + \beta - \gamma - (\beta\gamma),$$
$$\eta_{10} = \alpha - \beta + \gamma - (\beta\gamma), \quad \eta_{11} = \alpha - \beta - \gamma + (\beta\gamma),$$

where $\eta_{jk} = \log \mathrm{E}(Y_{jk}) = \log(m\pi_{jk})$ and $m = \sum_{j,k} y_{jk}$. Show that the 'interaction' term $(\beta\gamma)$ may be written $(\beta\gamma) = \frac{1}{4}\log\Delta$, where Δ is the odds ratio $(\pi_{00}\pi_{11})/(\pi_{01}\pi_{10})$, so that $(\beta\gamma) = 0$ is equivalent to $\Delta = 1$.

4   Give the matrices needed for iterative weighted least squares for the nonlinear model (10.31) in Example 10.23. How might starting-values be obtained?

5   In Example 10.24, discuss whether a marginal model or a log-linear model is preferable for (a) a white man aged 43 with a visually impaired left eye, who wants to assess his probability of having visual impairment in the other eye at the age of 65, and (b) a scientist comparing how visual impairment deveops with age for men of different races.

6   Give the form of the proportional odds model obtained when an underlying continuous variable $x^T\beta + \exp(x^T\gamma)\varepsilon$ is categorized; ε has the logistic density $e^u/(1+e^u)^2$, $-\infty < u < \infty$. Derive the iterative weighted least squares algorithm for estimation of β when it is known that $\gamma = 0$. Explain how you would need to change your algorithm to deal with $\gamma \neq 0$.

7   Establish (10.32).

10.6 Overdispersion

Thus far we have supposed that our data are well-described by a model with a simple error distribution. Nature is not usually so obliging, however, and in practice it is common to find that count and proportion data are more variable than would be expected under the Poisson and binomial models. Other types of data may also exhibit such *overdispersion*, manifested by models with over-large deviances and residuals, but otherwise showing no systematic lack of fit. Structure in the data is obscured by additional noise, so overdispersion increases uncertainty. Underdispersion also arises but is much rarer.

Two approaches to dealing with overdispersion are explicit parametric modelling of the heterogeneity, and the use of quasi-likelihood and associated estimating functions.

Parametric models

Suppose that the response Y has a standard distribution conditional on the unobserved variable ε, but that ε induces extra variation in Y. Here ε might represent unobserved — perhaps unobserveable — covariates that affect the response. Let ε have unit mean and variance $\xi > 0$, and to be concrete suppose that conditional on ε, Y has the Poisson distribution with mean $\mu\varepsilon$. Then (3.12) and (3.13) give

$$\mathrm{E}(Y) = \mathrm{E}_\varepsilon\left\{\mathrm{E}(Y \mid \varepsilon)\right\}, \quad \mathrm{var}(Y) = \mathrm{var}_\varepsilon\left\{\mathrm{E}(Y \mid \varepsilon)\right\} + \mathrm{E}_\varepsilon\left\{\mathrm{var}(Y \mid \varepsilon)\right\},$$

so the response has mean and variance

$$\mathrm{E}(Y) = \mathrm{E}_\varepsilon(\mu\varepsilon) = \mu, \quad \mathrm{var}(Y) = \mathrm{var}_\varepsilon(\mu\varepsilon) + \mathrm{E}_\varepsilon(\mu\varepsilon) = \mu(1 + \xi\mu).$$

If on the other hand the variance of ε is ξ/μ, then $\mathrm{var}(Y) = (1 + \xi)\mu$. In both cases the variance of Y is greater than its value under the standard Poisson model, for which $\xi = 0$. In the first case the variance function is quadratic, and in the second it is linear.

Table 10.18 illustrates the difference between these variance functions under modest overdispersion. Large amounts of data will be needed to detect overdispersion when the counts are small. The variances are equal when $\mu = 15$, but evidently a lot of data over a limited range of values of μ or alternatively a large range of mean responses would be needed to discriminate well between the two variance functions. This is one reason to consider a more robust approach, rather than to model the

μ	1	2	5	10	15	20	30	40	60
Linear	1.5	3.0	7.5	15.0	22.5	30	45	60	90
Quadratic	1.0	2.1	5.8	13.3	22.5	33	60	93	180

Table 10.18
Comparison of variance functions for overdispersed count data. The linear and quadratic variance functions are $V_L(\mu) = (1 + \xi_L)\mu$ and $V_Q(\mu) = \mu(1 + \xi_Q\mu)$, with $\xi_L = 0.5$ and ξ_Q chosen so that $V_L(15) = V_Q(15)$.

overdispersion in detail. If a full likelihood analysis is desired regardless, one can proceed as in the following example.

Example 10.26 (Negative binomial model)   In the discussion above, suppose that ε has the gamma distribution with unit mean and variance $1/\nu$. Then Y has the negative binomial density (Exercise 10.6.1)

$$f(y; \mu, \nu) = \frac{\Gamma(y + \nu)}{\Gamma(\nu)y!} \frac{\nu^\nu \mu^y}{(\nu + \mu)^{\nu + y}}, \quad y = 0, 1, \ldots, \quad \mu, \nu > 0, \quad (10.33)$$

and quadratic and linear variance functions are obtained on setting $\nu = 1/\xi$ and $\nu = \mu/\xi$ respectively. The first leads to simpler likelihood equations and so is preferable in purely numerical terms. When independent responses y_j have associated covariates x_j, it is natural to take the log link, giving means $\mu_j = \exp(x_j^T \beta)$. The value of ξ may be estimated from its profile log likelihood or by equating the Pearson statistic and its expected value; see Example 10.28.   ∎

A similar analysis applies to proportions. Suppose that conditional on ε, $R = mY$ is binomial with denominator m and success probability $\pi\varepsilon$, and that ε has unit mean and variance ξ. Then calculations like those above give

$$\mathrm{E}(Y) = \pi, \quad \mathrm{var}(Y) = m^{-1}\{\pi(1 - \pi) + \xi\pi^2(m - 1)\}. \quad (10.34)$$

Hence overdispersion increases with m if ξ is constant. Heterogeneity is undetectable in pure binary data, for which $m = 1$. When $m > 1$ and $\gamma > 0$, the choice $\xi = \gamma(1 - \pi)/\{\pi(m - 1)\}$ gives $\mathrm{var}(Y) = (1 + \gamma)\pi(1 - \pi)/m$, corresponding to uniform overdispersion. This is explored further in Exercise 10.6.4.

Quasi-likelihood

In all but the simplest cases the modelling of overdispersion by integrating out an unobserved variable leads to use of numerical integration. This can be awkward, but a more serious difficulty is that inferences might depend strongly on the unobserved component, which can be validated only indirectly. Hence it is often preferable to modify standard methods to accommodate overdispersion, in analogy with the use of least squares estimation when responses are non-normal (Section 8.4). We shall see below that provided the mean and variance functions are correctly specified, the estimators obtained by fitting standard models retain their large-sample normal distributions, but with an inflated variance matrix. This is very convenient, because standard software can then be used for fitting, with minor modification to the output.

Unrecognised overdispersion is a form of model misspecification, so one starting-point is to apply the ideas of Section 7.2, treating the generalized linear model score statistic (10.18) as an estimating function $g(Y; \beta)$ for β. An estimator $\tilde{\beta}$ is obtained

X is the $n \times p$ matrix whose *j*th row is x_j^T.

by solving the quasi-likelihood equation

$$g(Y; \beta) = X^\mathrm{T} u(\beta) = \sum_{j=1}^{n} x_j u_j(\beta) = \sum_{j=1}^{n} x_j \frac{Y_j - \mu_j}{g'(\mu_j)\phi_j V(\mu_j)} = 0, \qquad (10.35)$$

where the link function gives $g(\mu_j) = \eta_j = x_j^\mathrm{T}\beta$. Now if the mean structure has been chosen correctly, then $\mathrm{E}(Y_j) = \mu_j$ and the estimating function is unbiased, that is $\mathrm{E}\{g(Y; \beta)\} = 0$ for all β. Then the quasi-likelihood estimator $\tilde{\beta}$ is consistent under mild regularity conditions.

In large samples $\tilde{\beta}$ is normal with variance matrix (Section 7.2.1)

$$\mathrm{E}\left\{-\frac{\partial g(Y; \beta)}{\partial \beta^\mathrm{T}}\right\}^{-1} \mathrm{var}\{g(Y; \beta)\} \, \mathrm{E}\left\{-\frac{\partial g(Y; \beta)^\mathrm{T}}{\partial \beta}\right\}^{-1}. \qquad (10.36)$$

In order to compute this we require $\mathrm{E}\{-\partial g(Y; \beta)/\partial \beta^\mathrm{T}\}$ and $\mathrm{var}\{g(Y; \beta)\}$. Now

$$\frac{\partial u_j(\beta)}{\partial \beta^\mathrm{T}} = \frac{\partial \eta_j}{\partial \beta^\mathrm{T}} \frac{\partial \mu_j}{\partial \eta_j} \frac{\partial u_j(\beta)}{\partial \mu_j}$$

$$= x_j^\mathrm{T} \frac{1}{g'(\mu_j)} \left\{ -\frac{g''(\mu_j)}{g'(\mu_j)} u_j(\beta) - \frac{V'(\mu_j)}{V(\mu_j)} u_j(\beta) - \frac{1}{g'(\mu_j)\phi_j V(\mu_j)} \right\},$$

and as $\mathrm{E}\{u_j(\beta)\} = 0$, it follows that

$$\mathrm{E}\left\{-\frac{\partial g(Y; \beta)}{\partial \beta^\mathrm{T}}\right\} = -\sum_{j=1}^{n} x_j \mathrm{E}\left\{\frac{\partial u_j(\beta)}{\partial \beta^\mathrm{T}}\right\}$$

$$= \sum_{j=1}^{n} x_j x_j^\mathrm{T} \frac{1}{g'(\mu_j)^2 \phi_j V(\mu_j)} = X^\mathrm{T} W X,$$

where W is the $n \times n$ diagonal matrix with *j*th element $\{g'(\mu_j)^2 \phi_j V(\mu_j)\}^{-1}$. Moreover if in addition the variance function has been correctly specified, then $\mathrm{var}(Y_j) = \phi_j V(\mu_j)$, and hence

$$\mathrm{var}\{g(Y; \beta)\} = X^\mathrm{T} \mathrm{var}\{u(\beta)\} X = \sum_{j=1}^{n} x_j x_j^\mathrm{T} \frac{\mathrm{var}(Y_j)}{g'(\mu_j)^2 \phi_j^2 V(\mu_j)^2} = X^\mathrm{T} W X.$$

Thus (10.36) equals $(X^\mathrm{T} W X)^{-1}$. Had the variance function been wrongly specified, the variance matrix of $\tilde{\beta}$ would have been of sandwich form $(X^\mathrm{T} W X)^{-1}$ $(X^\mathrm{T} W' X)(X^\mathrm{T} W X)^{-1}$, where W' is a diagonal matrix involving the true and assumed variance functions. Only if the variance function has been chosen very badly will this sandwich matrix differ greatly from $(X^\mathrm{T} W X)^{-1}$, which therefore provides useful standard errors unless a plot of absolute residuals against fitted means is markedly non-random. In that case the choice of variance function should be reconsidered.

Quasi-likelihood estimates and standard errors are easily obtained using software that fits generalized linear models. Usually $\phi_j = a_j \phi$, where the a_j are known constants and $\phi = 1$ corresponds to a model such as the Poisson or binomial, for which the software finds estimates and standard errors by solving (10.35) with $\phi = 1$. As ϕ cancels from (10.35), the quasi-likelihood estimate $\tilde{\beta}$ equals the maximum likelihood estimate. Software that sets $\phi = 1$ will yield a variance matrix that is too small by a factor ϕ, however, so the usual standard errors must be multiplied by $\widehat{\phi}^{1/2}$, where $\widehat{\phi}$ is defined at (10.20).

Under an exponential family model, the quantity $g(Y; \beta)$ in (10.35) is the score statistic, so estimators based upon it are asymptotically optimal. Even if that model is false, inference based on $g(Y; \beta)$ is valid provided the mean and its relation with the variance $V(\mu)$ have been correctly specified. Moreover the argument on page 322 shows that $\tilde{\beta}$ is optimal among estimators based on linear combinations of the $Y_j - \mu_j$, in analogy with the Gauss–Markov theorem. The essential requirement for this is that the $u_j(\beta)$ satisfy the two key properties

$$\mathrm{E}(\partial\ell/\partial\mu) = 0, \quad \mathrm{var}(\partial\ell/\partial\mu) = \mathrm{E}(-\partial^2\ell/\partial\mu^2)$$

of a log likelihood derivative. In fact, $g(Y; \beta)$ is the derivative with respect to β of the *quasi-likelihood* function

Strictly $Q(\beta; Y)$ is a quasi-log likelihood.

$$Q(\beta; Y) = \sum_{j=1}^{n} \int_{Y_j}^{\mu_j} \frac{Y_j - u}{\phi a_j V(u)}\, du,$$

and we can define a deviance as $-2\phi Q(\beta; Y)$. This is positive by construction and can be used to compare nested models under overdispersion.

Example 10.27 (Weighted least squares)   The simplest example of quasi-likelihood estimation arises when $V(\mu) = 1$, $\phi_j = \phi a_j$, and the mean of Y_j is $\mu_j = x_j^{\mathrm{T}}\beta$. Then (10.35) becomes

$$\sum_{j=1}^{n} x_j \frac{Y_j - x_j^{\mathrm{T}}\beta}{\phi a_j} = X^{\mathrm{T}}W(Y - X\beta) = 0,$$

where W is the diagonal matrix $\phi^{-1}\mathrm{diag}(1/a_1, \ldots, 1/a_n)$, and

$$\tilde{\beta} = (X^{\mathrm{T}}WX)^{-1}X^{\mathrm{T}}WY$$

is the weighted least squares estimator of β, found using weights a_j^{-1}. This estimator is the maximum likelihood estimator only if the Y_j are independent and normal, but even if not, $\tilde{\beta}$ is the minimum variance unbiased estimator linear in the Y_j (Section 8.4).

Integration shows that the deviance $Q(\beta; Y)$ equals the weighted sum of squares $(Y - X\beta)^{\mathrm{T}}W(Y - X\beta)$, while

$$\widehat{\phi} = \frac{1}{n-p} \sum_{j=1}^{n} \frac{\left(Y_j - x_j^{\mathrm{T}}\tilde{\beta}\right)^2}{a_j} = \frac{1}{n-p}(Y - X\tilde{\beta})^{\mathrm{T}}W(Y - X\tilde{\beta});$$

see Section 8.2.4.   ∎

Example 10.28 (Cloth fault data)   The left panel of Figure 10.11 shows the numbers of flaws in $n = 32$ cloth samples of various lengths. A plausible model is that the number of faults y in a sample of length x has a Poisson distribution with mean βx. A maximum likelihood fit of this model gives $\widehat{\beta} = 1.51$ with standard error 0.09. However the deviance of 64.5 on 31 degrees of freedom and the right panel of the figure suggest that the data are more variable than the Poisson model might indicate.

Figure 10.11 Cloth data analysis (Bissell, 1972). The left panel shows the numbers of flaws in 32 cloth samples of various lengths (m). The dotted line shows the fitted mean number of faults under the model. The right panel shows that absolute residuals for the fit are overdispersed relative to the standard normal distribution appropriate under the Poisson assumption.

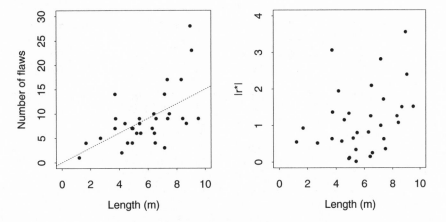

On reflection this is not surprising, as the rate β is likely to vary from one sample to another.

For quasi-likelihood estimation with $\text{var}(Y) = \phi\mu$, (10.35) is given by

$$\sum_{j=1}^{n} x_j \frac{y_j - \mu_j}{\phi\mu_j} = 0, \quad \mu_j = x_j\beta,$$

and as $\sum(y_j - \widehat{\mu}_j)^2/\widehat{\mu}_j = 68.03$ on 31 degrees of freedom, $\widehat{\phi} = 68.03/31 = 2.19$. The standard error for $\widehat{\beta}$ is then $0.09\widehat{\phi}^{1/2} = 0.13$, appreciably larger than under the Poisson model.

When the negative binomial model with variance function $\mu(1 + \xi\mu)$ is fitted, the maximum likelihood estimates of β and ξ are 1.51 and 0.115 with standard errors 0.13 and 0.056. The maximized log likelihood is -87.73, and $\sum(y_j - \widehat{\mu}_j)^2/\{\widehat{\mu}_j(1 + \widehat{\xi}\widehat{\mu}_j)\} = 32.57$ on 31 degrees of freedom, giving no evidence of poor fit. For the alternative negative binomial model with variance function $(1 + \xi)\mu$, the maximized log likelihood is -88.63, so the fit is slightly worse. This is borne out by the right panel of Figure 10.11, which suggests that the variance of the residuals increases with x, as would be the case if the linear variance function was fitted when the quadratic was more appropriate. ∎

The discussion above shows how standard errors should be modified in the presence of overdispersion. A similar adjustment applies when using deviance differences for model selection. Let model A be nested within a more complicated model B, with deviances $D_A > D_B$ and parameters $p_A < p_B$. For binomial and Poisson data, the usual approach is to compare $D_A - D_B$ with the $\chi^2_{p_B - p_A}$ distribution. In the presence of overdispersion this is modified by analogy with F tests in linear regression: if $\widehat{\phi}_B$ is the estimate of ϕ under the more complex model, then the adequacy of model A relative to model B is assessed by referring $\{(D_A - D_B)/(p_B - p_A)\}/\widehat{\phi}_B$ to the $F_{p_B - p_A, n - p_B}$ distribution.

Example 10.29 (Toxoplasmosis data) Table 10.19 gives data on the relation between rainfall and the proportions of people with toxoplasmosis for 34 cities in

Table 10.19
Toxoplamosis data:
rainfall (mm) and the
numbers of people testing
positive for
toxoplasmosis, r, our of m
people tested, for 34 cities
in El Salvador (Efron,
1986).

City	Rain	r/m	City	Rain	r/m	City	Rain	r/m	City	Rain	r/m
1	1735	2/4	11	2050	7/24	21	1756	2/12	31	1780	8/13
2	1936	3/10	12	1830	0/1	22	1650	0/1	32	1900	3/10
3	2000	1/5	13	1650	15/30	23	2250	8/11	33	1976	1/6
4	1973	3/10	14	2200	4/22	24	1796	41/77	34	2292	23/37
5	1750	2/2	15	2000	0/1	25	1890	24/51			
6	1800	3/5	16	1770	6/11	26	1871	7/16			
7	1750	2/8	17	1920	0/1	27	2063	46/82			
8	2077	7/19	18	1770	33/54	28	2100	9/13			
9	1920	3/6	19	2240	4/9	29	1918	23/43			
10	1800	8/10	20	1620	5/18	30	1834	53/75			

Table 10.20 Analysis of
deviance for polynomial
logistic models fitted to
the toxoplasmosis data.

Terms	df	Deviance
Constant	33	74.21
Linear	32	74.09
Quadratic	31	74.09
Cubic	30	62.63

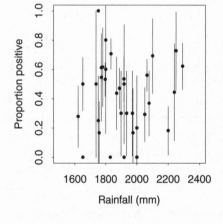
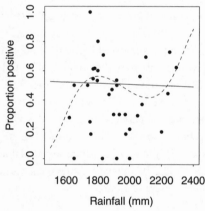

Figure 10.12
Toxoplasmosis data. The
left panel shows the
proportion of people
testing positive, r/m,
plus/minus
$2\{r(m-r)/m^3\}^{1/2}$, as a
function of rainfall in 34
cities in El Salvador. The
right panel shows the data
and linear (solid),
quadratic (dots, almost
identical to the linear fit),
and cubic (dashes)
polynomial models fitted
on the logistic scale.

El Salvador. There is wide variation in the numbers tested, as well as in the proportions testing positive, and the left panel of Figure 10.12 indicates a possible nonlinear relation between rainfall and toxoplasmosis incidence.

The right panel shows fitted proportions for logistic regression models in which the linear predictor contains terms linear, quadratic, and cubic in rainfall. Table 10.20 contains the analysis of deviance when the polynomial terms are included successively. The residual deviance of 62.63 on 30 degrees of freedom indicates overdispersion by a factor of roughly two.

Under the binomial assumption, the cubic model is tested against the constant model by comparing the deviance difference $74.21 - 62.63 = 11.58$ with the χ_3^2

distribution, giving significance level 0.009. This overstates the significance of the test because it makes no allowance for overdispersion. Under quasi-likelihood with $\text{var}(R) = \phi m \pi (1 - \pi)$ we obtain $\tilde{\phi} = 1.94$, and our general discussion suggests that we should compare the F statistic $(11.58/3)/\tilde{\phi} = 1.99$ with the $F_{3,30}$ distribution. This gives significance level 0.14, only weak evidence of a relationship between rainfall and incidence. We return to these data in Example 10.32. ∎

If the responses are dependent, the above discussion can be extended by taking as estimating function $X^{\mathsf{T}} V(\mu)^{-1}(Y - \mu)$, where $V(\mu)$ is an $n \times n$ covariance matrix for Y; see page 507. This is a common technique for modelling longitudinal data, in which short, often irregular time series are available on independent individuals. In such cases there may be no function whose derivatives with respect to β give the estimating function, and then no quasi-likelihood exists.

In some cases the response variance may be expressed as $\phi(\gamma) V(\mu; \xi)$, with γ and ξ unknown. An example is the quadratic variance function $\mu + \xi \mu^2$ in Example 10.26. The definition of the deviance depends on ξ, so models with different values of ξ cannot be compared using differences of deviances. An extended quasi-likelihood can be defined as the sum of the contributions

$$-\frac{1}{2} \log\{\phi_j(\gamma) V(\mu_j; \xi)\} - \frac{1}{2} \int_{Y_j}^{\mu_j} \frac{Y_j - u}{\phi_j(\gamma) a_j V(u; \xi)} \, du,$$

however, and used for inference about the unknown parameters. Unfortunately this definition is ambiguous: for example $\mu + \xi \mu^2$ can be written as $\phi(\gamma) = 1$, $V(\mu; \xi) = \mu + \xi \mu^2$ or as $\phi(\gamma) = \mu$, $V(\mu; \xi) = 1 + \xi \mu$, and these give different extended quasi-likelihoods. Uniqueness can be imposed by insisting that $\phi(\gamma)$ not involve μ or that $V(\mu) = 1$, leading to two different systems of estimating equations. The first system gives inconsistent estimators and the second gives consistent estimators. However simulation shows that for sample sizes of most interest the second estimators are worse than the first. Thus in practice the solutions to the first system are preferable, though neither is really satisfactory.

Exercises 10.6

1   Use (2.8) to establish (10.33). Give formulae for the corresponding deviance residuals when $\nu = 1/\xi$ and when $\nu = \mu/\xi$.
    Suppose that independent counts $y_1, \ldots, y_n$ arise with means $\mu_j = \exp(x_j^{\mathsf{T}} \beta)$. Under the model with constant $\nu = 1/\xi$, write down the negative binomial log likelihood for β and ξ. Explain why the likelihood equations become more complicated if the shape parameter changes for each observation, so $\nu_j = \mu_j/\xi$.
    If we estimate ξ by equating the Pearson statistic $\sum(y_j - \hat{\mu}_j)^2/V(\hat{\mu}_j)$ to $n - p$, where $V(\mu_j) = \text{var}(Y_j)$, discuss how to obtain the estimate under the above two variance functions.

2   Let I be a binary variable with success probability π, and suppose that π is given a density h. Show that I remains a binary variable whatever the choice of h, and hence explain the form of the variance in (10.34).
    Against what variable should the squared Pearson residual be plotted if it is desired to assess if (10.34) gives a suitable fit to data?

3   Find $Q(\beta; Y)$ when $u_j(\beta) = (Y_j - \mu_j)/\{\phi g'(\mu_j)V(\mu_j)\}$ and $V(\mu)$ equals μ, $\mu(1 - \mu)$, and μ^2.

4   One standard model for over-dispersed binomial data assumes that R is binomial with denominator m and probability π, where π has the beta density

$\Gamma(a)$ is the gamma function.

$$f(\pi; a, b) = \frac{\Gamma(a + b)}{\Gamma(a)\Gamma(b)}\pi^{a-1}(1 - \pi)^{b-1}, \quad 0 < \pi < 1, a, b > 0.$$

(a) Show that this yields the *beta-binomial* density

$$\Pr(R = r; a, b) = \frac{\Gamma(m + 1)\Gamma(r + a)\Gamma(m - r + b)\Gamma(a + b)}{\Gamma(r + 1)\Gamma(m - r + 1)\Gamma(a)\Gamma(b)\Gamma(m + a + b)}, \quad r = 0, \ldots, m.$$

(b) Let μ and σ^2 denote the mean and variance of π. Show that in general,

$$\mathrm{E}(R) = m\mu, \quad \mathrm{var}(R) = m\mu(1 - \mu) + m(m - 1)\sigma^2,$$

and that the beta density has $\mu = a/(a + b)$ and $s^2 = ab/\{(a + b)(a + b + 1)\}$. Deduce that the beta-binomial density has mean and variance

$$\mathrm{E}(R) = ma/(a + b), \quad \mathrm{var}(R) = m\mu(1 - \mu)\{1 + (m - 1)\delta\}, \quad \delta = (a + b + 1)^{-1}.$$

Hence re-express $\Pr(R = r; a, b)$ as a function of μ and δ. What is the condition for uniform overdispersion?

5   Conditional on ε, the observation Y has a generalized linear model density with canonical parameter $\eta + \tau\varepsilon$, where $\tau > 0$. If ε is standard normal, show that the marginal density of Y can be written

$$f(y; \eta, \tau) = \frac{1}{(2\pi)^{1/2}} \int_{-\infty}^{\infty} \exp\left\{\frac{y\eta + y\tau\varepsilon - b(\eta + \tau\varepsilon)}{a(\phi)} + c(y; \phi) - \varepsilon^2/2\right\} d\varepsilon.$$

By second-order Taylor series expansion of $b(\eta + \tau\varepsilon)$ for small τ, or otherwise, show that $f(y; \eta, \tau)$ equals

$$f(y; \eta, 0)\exp\left[\frac{\tau^2}{2}\frac{\{y - b'(\eta)\}^2}{a(\phi)^2\{1 + \tau^2 b''(\eta)/a(\phi)\}}\right]\{1 + \tau^2 b''(\eta)/a(\phi)\}^{-1/2} + o_p(\tau^2).$$

Prove that this approximation is exact when the conditional density of Y given ε is normal, and then find the unconditional mean and variance of Y.

10.7 Semiparametric Regression

Our earlier regression models have involved responses that depend on explanatory variables x through simple parametric functions such as $\beta_0 + \beta_1 x + \beta_2 x^2$. Their conciseness and direct interpretation gives such formulations great appeal, but they are not flexible enough to cater for all the situations met in practice and more general approaches are desirable, especially for exploratory analysis, model-checking, and other situations where the data should be allowed to 'speak for themselves'. Many ways to do this have been proposed in recent years, under the heading of *nonparametric* or *semiparametric* models, the aim typically being to extract a smooth curve from the data. An algorithm that does this is often termed a *smoother*. In fact smoothing operations typically do involve parameters, but in less prescriptive ways than before, and the results are best understood graphically. There are many approaches to semiparametric modelling, and below we merely sketch the possibilities by extending our previous discussion in two directions.

The adjective 'smoother' has become a noun in this context.

Figure 10.13
Earthquake magnitudes
plotted against fitted
intensity just before the
earthquake shock and time
since the preceding shock.
Note the log scales. The
magnitudes have been
jittered to reduce
overplotting.

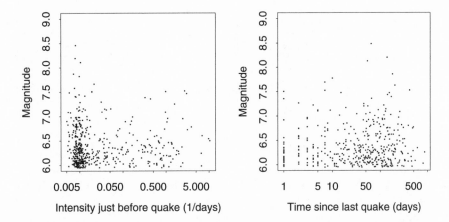

Example 10.30 (Japanese earthquake data) Figure 6.19 shows data on 483 earthquake shocks, of magnitude at least 6 on the Richter scale, offshore from Japan from 1885–1980. In Example 6.38 a self-exciting point process model was fitted to the data, in which the intensity at time t was given by

$$\lambda_{\mathcal{H}}(t) = \mu + \kappa \sum_{j:t_j<t} \frac{e^{\beta(m_j-6)}}{t - t_j + \gamma},$$

where m_j is the magnitude and t_j the time of the jth earthquake. This fits the times adequately, but $\lambda_{\mathcal{H}}(t)$ models only the time of the next shock and not its its magnitude. It is natural to ask if the magnitude of a shock depends on the past, for example on the value of $\lambda_{\mathcal{H}}(t)$ just before the shock occurs, or on the time elapsed since the previous shock.

Figure 10.13 contains scatterplots of m_i against $\lim_{\delta\to 0} \widehat{\lambda}_{\mathcal{H}}(t_i - \delta)$ and against $t_i - t_{i-1}$. The lack of pattern in both panels suggests that magnitude is unrelated to these other quantities, but the clustering of points at the left of the left panel makes it hard to be sure, and a smooth curve would sharpen our judgement. Fitting a particular parametric model seems difficult to justify, so the curve should be defined flexibly. ∎

10.7.1 Local polynomial models

Suppose that the response y equals $g(x) + \varepsilon$, where g is a smooth function of the scalar x, and ε has mean zero and variance σ^2. Although we assume that g has as many derivatives as we need, it will not usually have a simple form. The data consist of pairs $(x_1, y_1), \ldots, (x_n, y_n)$, and initially we shall suppose that we wish to make inferences about $g(x)$ at a single point $x = x_0$ in the interval spanned by the design points x_j. One approach is to fit a polynomial to all n pairs and then read off its value at x_0. However, such fits are often unconvincing — see the right panel of Figure 10.12, for example — and furthermore they can be sensitive to observations distant from x_0. Rather than treat all data pairs equally, it seems natural to attach more importance to observations close to x_0. This is closely related to kernel density estimation (Section 7.1.2).

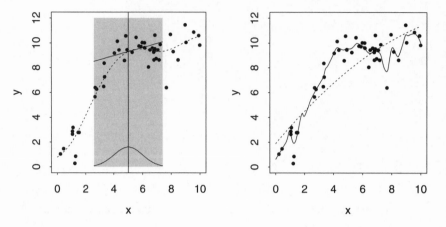

Figure 10.14
Construction of a local
linear smoother. Left
panel: observations in the
shaded part of the panel
are weighted using the
kernel shown at the foot,
with $h = 0.8$, and the
solid straight line is fitted
by weighted least squares.
The local estimate is the
fitted value when $x = x_0$,
shown by the vertical line.
Two hundred local
estimates formed using
equi-spaced x_0 were
interpolated to give the
dotted line, which is the
estimate of $g(x)$. Right
panel: local linear
smoothers with $h = 0.2$
(solid) and $h = 5$ (dots).

Recall that a kernel function $w(u)$ is a unimodal density function symmetric about $u = 0$ and with unit variance. One choice of w is the standard normal density. Another is a rescaled form of the *tricube* function

$$w(u) = \begin{cases} (1 - |u|^3)^3, & |u| \leq 1, \\ 0, & \text{otherwise,} \end{cases} \tag{10.37}$$

and there are many others.

When estimating $g(x)$ at $x = x_0$, we attach weight $w_j = h^{-1}w\{h^{-1}(x_j - x_0)\}$ to (x_j, y_j), where h is a bandwidth, and fit the polynomial

$$\beta_0 + \beta_1(x - x_0) + \cdots + \beta_k(x - x_0)^k$$

to the responses by weighted least squares (Section 8.2.4). The weights decrease the influence of points for which $|x_j - x_0|/h$ is big: for large h points far from x_0 will affect the fit, whereas as $h \to 0$ the regression becomes ever more local. The number of data included will vary as x_0 changes, with fewer points when x_0 is near the limits of its range or where the x_j are sparse.

The estimate of $g(x_0)$ is obtained by fitting the linear model

$$\begin{pmatrix} y_1 \\ y_2 \\ \vdots \\ y_n \end{pmatrix} = \begin{pmatrix} 1 & (x_1 - x_0) & \cdots & (x_1 - x_0)^k \\ \vdots & \vdots & & \vdots \\ 1 & (x_n - x_0) & \cdots & (x_n - x_0)^k \end{pmatrix} \begin{pmatrix} \beta_0 \\ \beta_1 \\ \vdots \\ \beta_k \end{pmatrix} + \begin{pmatrix} \varepsilon_1 \\ \varepsilon_2 \\ \vdots \\ \varepsilon_n \end{pmatrix},$$

that is $y = X\beta + \varepsilon$, with weight matrix $W = \text{diag}(w_1, \ldots, w_n)$. This results in the weighted least squares estimate $\widehat{\beta} = (X^\mathsf{T}WX)^{-1}X^\mathsf{T}Wy$, of which the component of interest is $\widehat{\beta}_0$, the value of the fitted polynomial when $x = x_0$.

In practice a smooth estimate $\widehat{g}(x)$ of the entire function $g(x)$ is usually required. It is obtained by interpolating the estimates $\widehat{\beta}_0$ for different values of x_0, as in the left panel of Figure 10.14. The right panel shows how the curve estimate depends on h. When h is large, $\widehat{g}$ is too smooth to capture the pattern of the data, and hence it will be biased. When h is small, $\widehat{g}$ follows the data better but wiggles implausibly and so has a high variance. The intermediate estimate shown in the left panel balances bias and variance more satisfactorily. The choice of h is important and we discuss it below.

The dips in $\widehat{g}$ towards the right of both panels suggest that local polynomial fits are sensitive to outliers. This is no surprise because they involve least squares estimation, which is non-robust. Some implementations robustify the fit, for example using the Huber estimator or discounting outliers by making the weights depend on the residuals $y_j - \widehat{g}(x_j)$ from an initial fit. The popular iterative algorithm *lowess* takes this approach, though it uses nearest neighbours, weighting the proportion p of the data nearest to x_0 instead of using a fixed bandwidth. Typically $p = 2/3$ in standard implementations. Nearest neighbour fitting automatically allows for changes in the density of the x_j and hence reduces fluctuations like those at $x = 2$ in the right panel of Figure 10.14; it amounts to taking a locally-varying bandwidth.

Lowess stands for locally weighted scatterplot smoother.

Local polynomial estimators can be studied using ideas from least squares. The estimator of $g(x_0)$ is a weighted average of the y_j, that is

$$\widehat{g}(x_0) = \widehat{\beta}_0 = \sum_{j=1}^{n} S(x_0; x_j, h)y_j,$$

where the elements of the *effective kernel* $S(x_0; x_1, h), \ldots, S(x_0; x_n, h)$ form the first row of the $(k + 1) \times n$ matrix $(X^{\mathsf{T}}WX)^{-1}X^{\mathsf{T}}W$. This depends on the design matrix, the bandwidth, and x_0, but not on y, so

$$\mathrm{E}\{\widehat{g}(x_0)\} = \sum_{j=1}^{n} S(x_0; x_j, h)g(x_j), \quad \mathrm{var}\{\widehat{g}(x_0)\} = \sigma^2 \sum_{j=1}^{n} S(x_0; x_j, h)^2. \quad (10.38)$$

The second expression here gives a finite-sample variance for $\widehat{g}(x_0)$, provided that σ^2 is replaced by an estimate. One natural choice is

$$s^2(h) = \frac{1}{n - 2\nu_1 + \nu_2} \sum_{j=1}^{n} \{y_j - \widehat{g}(x_j)\}^2, \quad (10.39)$$

where ν_1 and ν_2 are defined below. The corresponding estimator is unbiased when g is a polynomial of degree k, but otherwise is biased upwards. A simple way to reduce the bias is to construct $s^2(h)$ using a smaller h than that used for the curve estimate.

For theoretical and practical purposes it is fruitful to represent the smoothing operation in matrix form. The fitted values $\widehat{g}(x_i)$ may be obtained by setting $x_0 = x_1, \ldots, x_n$, and to each of these corresponds an effective kernel $S_{i1}(h), \ldots, S_{in}(h)$, where $S_{ij}(h)$ is an abbreviation of $S(x_i; x_j, h)$. Let us stack these as an $n \times n$ *smoothing matrix* S_h. Then the vector of fitted values $\widehat{g} = (\widehat{g}(x_1), \ldots, \widehat{g}(x_n))^{\mathsf{T}}$ may be written as $\widehat{g} = S_h y$. There is an analogy with the hat matrix $H = X(X^{\mathsf{T}}X)^{-1}X^{\mathsf{T}}$ for a linear regression with fitted values $\widehat{y} = Hy$. Unlike a hat matrix, however, S_h is not idempotent and it may be asymmetric. Any sensible smoothing operation will leave a constant unchanged, so $S_h 1_n = 1_n$. Equivalently any effective kernel sums to one.

Although kernel smoothers are most conveniently defined in terms of their bandwidth, it is useful to know their degrees of freedom when comparing their output.

In a standard linear model the trace of the hat matrix equals the number of parameters, suggesting by analogy that we define the degrees of freedom of the smoother to be $\nu_1 = \text{tr}(S_h)$. The analogy is not perfect, however, and alternatives such as $\nu_2 = \text{tr}(S_h^T S_h)$ have been proposed. Both ν_1 and ν_2 decrease as h increases, and $k \leq \nu_2 \leq \nu_1 \leq n$.

Three things must be chosen when implementing a local polynomial fit: the kernel function w, the degree k of the polynomial, and the bandwidth h. Experience shows that the choice of w is rarely important, though the best-looking curves are obtained when w descends smoothly to zero. Sharp-edged kernels such as the uniform density give rough and visually unappealing fits.

Choice of polynomial

It turns out to be sensible to take k odd, at least in theory. To see this in qualitative terms, consider the simplest form of local smoothing, with $k = 0$. Then the local polynomial is a constant and the least squares estimator of $g(x_0)$ is a weighted average, the *Nadaraya–Watson estimator*

$$\widehat{\beta}_{0,NW} = \frac{\sum_{j=1}^{n} w_j y_j}{\sum_{j=1}^{n} w_j}. \tag{10.40}$$

This is a simple estimator, but not a very good one, because of its bias at the ends of the data. Figure 10.15 shows what happens when local constant and local linear estimators are fitted to data, shown noiseless for clarity. The left panels show that local linear and local constant fits give similar estimates when x_0 is central; the effective kernel is almost the same. In the right panels x_0 lies at the right-hand edge of the data, and as $\widehat{\beta}_{0,NW}$ does not allow for $g'(x_0)$ it is badly biased upwards. The effective kernel for the local linear fit is smaller away from the boundary values of x_j, however, and so its bias is smaller. The fact that local polynomial fits adapt automatically to the presence of a boundary gives them a large practical advantage over many other approaches.

An asymptotic argument given at the end of this section, under which $n \to \infty$ and $h \to 0$ with $nh \to \infty$, shows that when x_0 is away from a boundary, both local constant and local linear fits have approximate bias and variance

$$\frac{1}{2}h^2 g''(x_0), \quad \frac{\sigma^2}{nh f(x_0)} \int w(u)^2 \, du, \tag{10.41}$$

where $f(x)$ represents the limiting density of design points. The bias increases if $g(x)$ has high curvature or if h is large; see Figure 10.14. The bias has the desirable property of not depending on the pattern of design points, at least asymptotically. One interpretation of the variance formula is that the effective number of observations that contribute to the weighted average $\widehat{\beta}_0$ is $nh f(x_0)/\int w(u)^2 \, du$, and this explains why we must have $nh \to \infty$ for a sensible asymptotic framework, while a small value of $f(x_0)$ — that is, few points near x_0 — decreases the precision with which $g(x_0)$ may be estimated.

When x_0 is near a boundary it may be shown that the bias of the local constant estimator increases to $O(h)$. The local linear estimator retains its $O(h^2)$ bias, however,

Geoffrey Stuart Watson (1921–1998) was educated in Melbourne, North Carolina and Cambridge, and held posts in Australia and North America, the last being at Princeton. He made important contributions to time series, to directional data analysis, and to mathematical biology. See Beran and Fisher (1998). E. A. Nadaraya (1935–) worked at Tbilisi State University. Both men published papers describing this estimator in 1964.

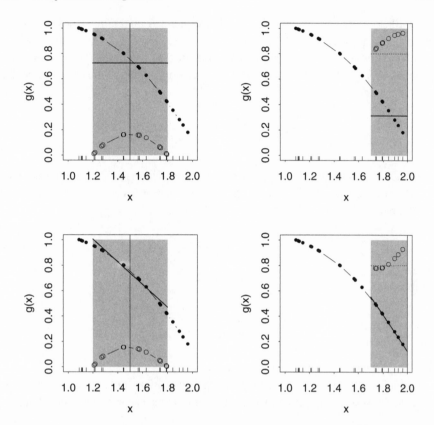

Figure 10.15 Local polynomial fitting by least squares. In each panel the function $g(x)$ is shown by a line joining the solid blobs (x_j, y_j), shown without error for clarity, and the target value x_0 at which g is to be estimated is given by the vertical line; $x_0 = 1.5$ for the left panels and $x_0 = 2$ for the right panels. Only observations falling inside the shaded region contribute to the fit, and the effective kernel is shown by the circles; in the right panels the effective kernel has been shifted upwards by 0.8. The heavy solid lines show the local polynomial fits, which are constant in the upper panels and linear in the lower panels. The local constant fit is more biased than the local linear fit, especially at the edge $x_0 = 2$.

confirming what is suggested by Figure 10.15. Near a boundary both variances remain of order $(nh)^{-1}$. Thus in asymptotic terms the variance of the local linear estimator is no worse than that of the local constant estimator, while its bias is of smaller order. The same argument applies in general: whenever k is even, the variance of $\widehat{\beta}_0$ is not increased asymptotically by fitting a polynomial of order $k + 1$, and the bias of $\widehat{\beta}_0$ is reduced thereby.

Asymptotic arguments are useful backstops, but the theoretical benefits of higher-order fitting are outweighed by a finite-sample increase in variance. In practice local linear and quadratic polynomials are commonest, but local cubic curves are also sometimes fitted, particularly to data showing high-frequency variation.

Choice of smoothing parameter

One possibility is to choose the bandwidth h to minimize the asymptotic mean squared error

$$\frac{h^4}{4} g''(x_0)^2 + \frac{\sigma^2}{nhf(x_0)} \int w(u)^2 \, du$$

of $\widehat{g}(x_0)$ with respect to h, giving a local bandwidth at x_0. An overall bandwidth could be found using the integrated asymptotic mean squared error; see (7.7) and (7.8). Unfortunately both local and overall choices of h involve the unknowns g'' and f, and further bandwidths are needed to estimate them. A quagmire stands before us, and we must skirt it if we can.

A better, finite-sample, approach is to trade off the bias and variance when each observation is predicted from the rest, choosing h to minimize the cross-validation sum of squares

$$\text{CV}(h) = \sum_{j=1}^{n} \{y_j - \widehat{g}_{-j}(x_j)\}^2,$$

where $\widehat{g}_{-j}(x)$ is obtained by applying the smoother with bandwidth h but dropping the jth case from the data. At first sight it appears that n fits are needed to compute $\text{CV}(h)$, but as (Exercise 10.7.5)

$$y_j - \widehat{g}_{-j}(x_j) = \frac{y_j - \widehat{g}(x_j)}{1 - S_{jj}(h)}, \tag{10.42}$$

it turns out that

$$\text{CV}(h) = \sum_{j=1}^{n} \left\{ \frac{y_j - \widehat{g}(x_j)}{1 - S_{jj}(h)} \right\}^2$$

may be computed from components of the overall fit.

In practice $\text{CV}(h)$ must be minimized over a grid of values of h, so fast computation of the $S_{jj}(h)$ is essential. An alternative is to minimize the generalized cross-validation criterion

$$\text{GCV}(h) = \sum_{j=1}^{n} \left\{ \frac{y_j - \widehat{g}(x_j)}{1 - \text{tr}(S_h)/n} \right\}^2$$

instead. This replaces the individual $S_{jj}(h)$ by their average, which is found more speedily.

Cross-validation and information criteria are closely related, so it is not surprising that the values of h found by minimizing $\text{CV}(h)$ and $\text{GCV}(h)$ tend to be too small, analogous to the overfitting seen when AIC is used. In Section 8.7.3 we saw that for linear models this may be remedied by using the corrected information criterion AIC_c. For semiparametric models this suggests that we choose h to minimize $\text{AIC}_c(h)$, given by

$$\text{AIC}_c(h) = n \log \widehat{\sigma}^2(h) + n \frac{1 + \text{tr}(S_h)/n}{1 - \{\text{tr}(S_h) + 2\}/n}, \quad \widehat{\sigma}^2(h) = n^{-1} \sum \{y_j - \widehat{g}(x_j)\}^2,$$

where the number of regressors p in the linear model is replaced by $\text{tr}(S_h)$. Simulation shows that this procedure reduces the overfitting.

Semiparametric versions of other criteria also exist, but in practice an automatic choice of h should typically be used as a starting-point for investigation, rather than as a black box procedure.

Inference

At first sight it seems that approximate pointwise confidence intervals for $g(x_0)$ can be constructed using the mean and variance (10.38), simply by supposing that

$$\frac{\widehat{g}(x_0) - \mathrm{E}\{\widehat{g}(x_0)\}}{\widehat{\mathrm{var}}\{\widehat{g}(x_0)\}^{1/2}} \overset{.}{\sim} N(0, 1).$$

Unfortunately, however, this leads to confidence statements for $\mathrm{E}\{\widehat{g}(x_0)\}$ rather than for the usual quantity of interest $g(x_0)$. If the bias $\mathrm{E}\{\widehat{g}(x_0)\} - g(x_0)$ is substantial, the confidence interval may be centred in quite the wrong place. One way to deal with this is to correct the interval using a bias estimate, for example taken from (10.41) or found by comparing $\widehat{g}(x_0)$ and a less biased estimate obtained using a smaller bandwidth or higher k. These solutions tend to be complicated and can work poorly in small samples. A simpler approach ignores the bias issue and constructs a $(1 - 2\alpha)$ *variability band*, in which $\pm z_\alpha \widehat{\mathrm{var}}\{\widehat{g}(x_0)\}^{1/2}$ is added to $\widehat{g}(x_0)$ to give an idea of its variability. In effect this is a confidence band for $\mathrm{E}\{\widehat{g}(x_0)\}$.

Sometimes it is useful to construct an overall $(1 - 2\alpha)$ confidence band for g over the set $\mathcal{A}$ of x-values. This requires two curves $L(x)$ and $U(x)$ such that

$$\Pr\{L(x) \leq g(x) \leq U(x), x \in \mathcal{A}\} = 1 - 2\alpha,$$

which may be found by probability approximation to the distribution of $\sup_{x \in \mathcal{A}} |\widehat{g}(x) - g(x)|/\widehat{\mathrm{var}}\{\widehat{g}(x)\}^{1/2}$. References to this are given in the bibliographic notes.

It is useful to have a means of testing the overall significance of an apparent departure from a given parametric form. Suppose that a non-local linear model has yielded the fitted values $\widehat{y} = Hy$, where H is the corresponding hat matrix, and that a more comprehensive, smooth, model has smoothing matrix S_h. Then it is natural to compute a P-value for departures from the non-local model using the ratio of the corresponding residual sums of squares

$$R(h) = \frac{\sum_{j=1}^{n}(y_j - \widehat{y}_j)^2}{\sum_{j=1}^{n}\{y_j - \widehat{g}(x_j)\}^2} = \frac{y^{\mathrm{T}}(I_n - H)y}{y^{\mathrm{T}}(I_n - S_h)^{\mathrm{T}}(I_n - S_h)y}.$$

If the observations are normal it turns out to be possible to compute good approximations to the corresponding P-value

$$p_{\mathrm{obs}}(h) = \Pr_0\{R(h) \geq r_{\mathrm{obs}}\}, \tag{10.43}$$

where r_{obs} is the observed value of $R(h)$ (Problem 10.13). It is useful to plot a *significance trace* of $p_{\mathrm{obs}}(h)$ as a function of h, to assess the strength of evidence against the parametric model at different bandwidths.

Example 10.31 (Japanese earthquake data) The left panel of Figure 10.13 suggests that the magnitude of an earthquake might depend on the conditional intensity $\widehat{\lambda}_{\mathcal{H}}(t)$ just before it, with a larger release of energy and hence a bigger shock when the $\widehat{\lambda}_{\mathcal{H}}(t)$ is small. To assess this we fit a local linear smoother with tricube kernel

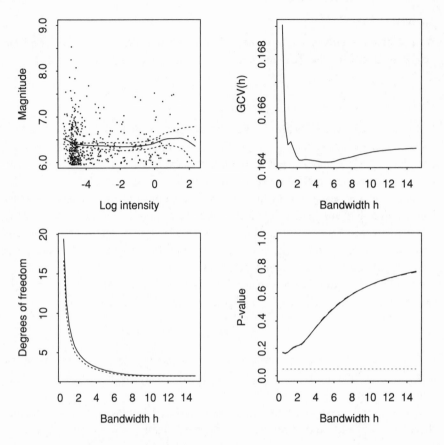

Figure 10.16 Smooth analysis of earthquake data. Upper left: local linear regression of magnitude on log intensity just before quake (solid), with 0.95 pointwise confidence bands (dots). Upper right: generalized cross-validation criterion GCV(h) as a function of bandwidth h. Lower left: relation between degrees of freedom ν_1 (solid), ν_2 (dots), and h. Lower right: significance traces for test of no relation between magnitude and log intensity, based on chi-squared approximation (dots) and saddlepoint approximation (solid). The horizontal line shows the conventional 0.05 significance level.

(10.37) and bandwidth $h = 2$. The upper left panel of Figure 10.16 shows this fit, with a 0.95 pointwise confidence interval. The width of the interval increases at the boundaries and at the right of the panel, owing to the lower density of design points x_j. There is a suggestion of an increased magnitude at very low and high intensities, but the evidence is not compelling.

The upper right panel shows that although the cross-validation criterion GCV(h) is minimized when $h \doteq 5.5$, the minimum changes little for $h > 2$. In the lower left panel we see that $h > 2$ corresponds to fitting curves with at most 4 or so equivalent degrees of freedom, while for $h > 8$ there are essentially two degrees of freedom, corresponding to straight-line regression. The corresponding plots for AIC(h) and AIC$_c$(h) decrease sharply and then tail off slowly, and also suggest that large bandwidths are appropriate.

The lower right panel of the figure shows two approximations to the significance trace for an overall test of no relation between log intensity and magnitude. The values of $p_{obs}(h)$ suggest that the evidence for such a relation varies from weak to non-existent. The approximations rest on the assumption that the data are normal. The large number of observations should mitigate the fact that this assumption is plainly incorrect, and it is unlikely to be critical, at least in this case.

These data show no relation between the fitted intensity just prior to an earthquake and its magnitude. This conclusion is of course very tentative, because seismological knowledge has not been incorporated. ∎

Extensions

The locally weighted polynomial fit arises naturally from a modified form of likelihood. For if the ε_j were independent and normal, the contribution from (x_j, y_j) to the overall log likelihood for a polynomial fit of degree k centred at x_0 would be

$$\ell_j(\beta, \sigma; x_0) \equiv -\frac{1}{2\sigma^2} \{y_j - \beta_0 - \beta_1(x_j - x_0) - \cdots - \beta_k(x_j - x_0)^k\}^2 - \frac{1}{2} \log \sigma^2,$$

and $\widehat{\beta}$ maximizes the *local log likelihood*

$$\ell(\beta, \sigma; x_0) = \sum_{j=1}^{n} \frac{1}{h} w \left(\frac{x_j - x_0}{h} \right) \ell_j(\beta, \sigma; x_0).$$

This idea extends fairly directly, for example to generalized linear and stochastic process models, using the appropriate log likelihood contribution and estimating $\widehat{\beta}$ by iterative weighted least squares. The ideas described above then go through largely unchanged, though for a generalized linear model $\mathrm{AIC}_c(h)$ must be changed to

$$\mathrm{AIC}_c(h) = \sum_{j=1}^{n} d_j\{y_j; \widehat{\mu}_j(h)\} + n \frac{1 + \mathrm{tr}(S_h)/n}{1 - \{\mathrm{tr}(S_h) + 2\}/n}, \tag{10.44}$$

where $d_j\{y_j; \widehat{\mu}_j(h)\}$ is the deviance contribution from y_j when the fitted value is $\widehat{\mu}_j(h)$. This is a large topic, to which the bibliographic notes give some entry points. The key ideas are summarized in the following example.

Example 10.32 (Toxoplasmosis data) Example 10.29 described how allowance might be made for the overdispersion of the data in Table 10.19, to which a logistic regression model with cubic dependence on rainfall was fitted. In view of the implausibility of the cubic model shown in the right panel of Figure 10.12, we consider local fitting with binomial probability $\pi(x) = \exp\{\theta(x)\}/[1 + \exp\{\theta(x)\}]$ depending on a local log odds $\theta(x)$. We fit a Taylor series expansion, $\theta(x) \doteq \beta_0 + \beta_1(x - x_0) + \cdots + \beta_k(x - x_0)^k/k!$, and take $\widehat{\beta}_0$ as the estimate of $\theta(x_0)$.

The local log likelihood is

$$\ell(\beta; x_0, h) \equiv \sum_{j=1}^{n} \frac{w_j m_j}{\phi} \{y_j x_j^{\mathsf{T}} \beta - \log (1 + e^{x_j^{\mathsf{T}} \beta})\},$$

where m_j is the binomial denominator, $y_j = r_j/m_j$ is the observed proportion positive, $x_j^{\mathsf{T}} = (1, x_j - x_0, \ldots, (x_j - x_0)^k)$, and taking $\phi > 0$ will allow for overdispersion relative to the binomial model. The kernel function reduces the effective value of m_j to $m_j w_j$, so towns whose rainfall x_j is far from x_0 count for less in the estimation of β.

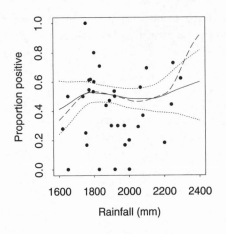 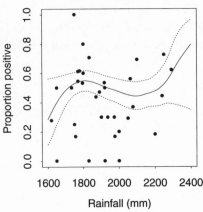

Figure 10.17 Local fits to the toxoplasmosis data. The left panel shows fitted probabilities $\widehat{\pi}(x)$, with the fit of local linear logistic model with $h = 400$ (solid) and 0.95 pointwise confidence bands (dots). Also shown is the local linear fit with $h = 300$ (dashes). The right panel shows the local quadratic fit with $h = 400$ and its 0.95 confidence band. Note the increased variability due to the quadratic fit, and its stronger curvature at the boundaries.

The local score function may be written $X^{\mathsf{T}} W u(\beta)$, to be compared with (10.18), and $\widehat{\beta}$ is obtained by applying iterative weighted least squares to the binomial model with artificial denominators $w_j m_j$. A sandwich variance matrix $(X^{\mathsf{T}} W_1 X)^{-1} X^{\mathsf{T}} W_2 X (X^{\mathsf{T}} W_1 X)^{-1}$ is required, where the jth elements of the diagonal matrices W_1 and W_2 are $w_j m_j \widehat{\pi}_j (1 - \widehat{\pi}_j)$ and $w_j^2 m_j \widehat{\pi}_j (1 - \widehat{\pi}_j)$, with $\widehat{\pi}_j$ the fitted probabilities. The dispersion parameter ϕ does not appear in the local score equation, and plays no role in the estimation of β, in the effective kernel, in the smoothing matrix or the degrees of freedom. An estimator $\widehat{\phi}$ is obtained by replacing the divisor $n - p$ in (10.20) by its counterpart $n - 2\nu_1 + \nu_2$, hence generalizing (10.39) to accommodate the binomial variance function.

Figure 10.17 shows linear and quadratic local fits and their 0.95 pointwise confidence bands, obtained with $h = 400$; the left panel also shows the fit with $h = 300$. The confidence bands for the quadratic fit are appreciably wider, and the fit itself is more curved, particularly at the boundaries. As might be expected, taking $h = 300$ gives a more locally adapted fit, whose effect is similar to increasing the order of the polynomial. All the fits are more plausible than the polynomial shown in Figure 10.12.

The confidence bands are appreciably narrower when no allowance is made for the overdispersion, and they suggest that the probability depends on rainfall. Overdispersion makes this much less plausible, and indeed a horizontal line would lie inside the bands in both panels of Figure 10.17. Any evidence for a relation between the probability and rainfall seems weak, though an analogue of (10.43) would be required for a more definite conclusion.   ∎

In some applications trigonometric or other expansions may be more appropriate than polynomial expansions; they too may be fitted locally using kernel or nearest neighbour weighting.

Similar ideas may be applied for smoothing in several dimensions, though the curse of dimensionality can then become heavy. It is useful to scale the covariates so that a common bandwidth can be used for them all, for example by using bandwidth $h s_r$ on the rth axis, where s_r is the standard deviation of the rth covariate.

This can be omitted on a first reading.

Computation of bias and variance

To express the lessons of Figure 10.15 in algebraic terms, we compute the mean and variance of $\widehat{\beta}$. Taylor series expansion gives

$$
\begin{aligned}
g(x) &= g(x_0) + (x - x_0)g'(x_0) + \cdots + \frac{1}{k!}(x - x_0)^k g^{(k)}(x_0) \\
&\quad + \frac{1}{(k+1)!}(x - x_0)^{k+1} g^{(k+1)}(x) + \cdots \\
&= \beta_0 + (x - x_0)\beta_1 + \cdots + (x - x_0)^k \beta_k + b(x),
\end{aligned}
$$

say, where the final term is the remainder. Consequently

$$
\begin{pmatrix} g(x_1) \\ g(x_2) \\ \vdots \\ g(x_n) \end{pmatrix} = \begin{pmatrix} 1 & (x_1 - x_0) & \cdots & (x_1 - x_0)^k \\ \vdots & \vdots & & \vdots \\ 1 & (x_n - x_0) & \cdots & (x_n - x_0)^k \end{pmatrix} \begin{pmatrix} \beta_0 \\ \beta_1 \\ \vdots \\ \beta_k \end{pmatrix} + \begin{pmatrix} b(x_1) \\ b(x_2) \\ \vdots \\ b(x_n) \end{pmatrix},
$$

or equivalently $g = X\beta + b$, where b is the $n \times 1$ vector whose jth element is $b(x_j)$. Let y, g, and ε represent the $n \times 1$ vectors whose jth elements are y_j, $g(x_j)$, and ε_j, and recall that the ε_j are independent with mean zero and variance σ^2. Then $y = g + \varepsilon = X\beta + b + \varepsilon$, giving

$$
\mathrm{E}(\widehat{\beta}) = \mathrm{E}\{(X^\mathrm{T} W X)^{-1} X^\mathrm{T} W (X\beta + b + \varepsilon)\} = \beta + (X^\mathrm{T} W X)^{-1} X^\mathrm{T} W b,
$$

$$
\mathrm{var}(\widehat{\beta}) = \sigma^2 (X^\mathrm{T} W X)^{-1} X^\mathrm{T} W^2 X (X^\mathrm{T} W X)^{-1},
$$

Hence $\widehat{\beta}$ has a bias that depends on the polynomial terms of degree $k + 1$ and higher. If $g(x)$ is indeed a polynomial of degree k or lower then $b = 0$ and $\widehat{\beta}$ is unbiased.

For the local linear fit, $k = 1$, and the bias of $\widehat{\beta}$ is

$$
(X^\mathrm{T} W X)^{-1} X^\mathrm{T} W b = \begin{pmatrix} \sum w_j & \sum w_j(x_j - x_0) \\ \sum w_j(x_j - x_0) & \sum w_j(x_j - x_0)^2 \end{pmatrix}^{-1} \begin{pmatrix} \sum w_j b_j \\ \sum w_j(x_j - x_0)b_j \end{pmatrix}.
$$

Hence the bias of $\widehat{\beta}_0$ is

$$
\frac{\sum w_j(x_j - x_0)^2 \sum w_j b_j - \sum w_j(x_j - x_0) \sum w_j(x_j - x_0)b_j}{\sum w_j \sum w_j(x_j - x_0)^2 - \left\{\sum w_j(x_j - x_0)\right\}^2}.
$$

To approximate this, we suppose that the x_j are sufficiently dense to have a well-behaved smooth density, $f(x)$, let $n \to \infty$ and $h \to 0$ in such a way that $nh \to \infty$, and replace the sums by integrals. We then see, for example, that

$$
\begin{aligned}
\sum w_j(x_j - x_0)^2 &\doteq n \int \frac{1}{h} w\left(\frac{x - x_0}{h}\right)(x - x_0)^2 f(x)\, dx \\
&= nh^2 \int w(u)u^2 f(x_0 + hu)\, du \\
&\doteq nh^2 \int w(u)u^2 \{f(x_0) + huf'(x_0) + \cdots\}\, du \\
&\doteq nh^2 f(x_0) + O(nh^4),
\end{aligned}
$$

on changing the variable of integration to $u = (x - x_0)/h$ and recalling that w has
unit variance and is symmetric. This calculation presupposes that x_0 is sufficiently
far from the boundary relative to h that the range of integration for integrals such as
$\int w(u)u\,du$ is effectively infinite; otherwise odd powers of h do not vanish and the
result is $anh^2 f(x_0) + O(nh^3)$, with $0 < a < 1$. Provided the odd terms do cancel,
similar calculations give

$$\sum w_j b_j \doteq \frac{1}{2}nh^2 f(x_0)g''(x_0), \quad \sum w_j(x_j - x_0) \doteq nh^2 f'(x_0),$$
$$\sum w_j(x_j - x_0)b_j \doteq O(nh^4), \qquad \sum w_j \doteq nf(x_0) \tag{10.45}$$

and on putting the pieces together we find that $\widehat{\beta}_0$ has bias whose leading term is
$\frac{1}{2}h^2 g''(x_0)$. It turns out that the bias has order h^2 even when x_0 is near the boundary,
but a similar calculation for the Nadaraya–Watson estimator (10.40) shows that its
bias near the boundary is $O(h)$ (Exercise 10.7.6).

To get a handle on the variance of $\widehat{\beta}_0$, it is simplest to orthogonalize X by replacing
the jth element of its second column with $x_j - \overline{x}_w$, where $\overline{x}_w = \sum w_j x_j / \sum w_j$. In
this parametrization the weighted least squares estimators are

$$\widehat{\gamma} = \begin{pmatrix} \sum w_j & 0 \\ 0 & \sum w_j(x_j - \overline{x}_w)^2 \end{pmatrix}^{-1} \begin{pmatrix} \sum w_j y_j \\ \sum w_j(x_j - \overline{x}_w)y_j \end{pmatrix},$$

and $\widehat{\beta}_0 = \widehat{\gamma}_0 + (x_0 - \overline{x}_w)\widehat{\gamma}_1$. This gives a simple explicit formula for $\widehat{\beta}_0$, useful for
numerical work. Its variance is

$$\mathrm{var}(\widehat{\beta}_0) = \sigma^2 \left[\frac{\sum w_j^2}{\left(\sum w_j\right)^2} + (x_0 - \overline{x}_w)^2 \frac{\sum w_j^2(x_j - \overline{x}_w)^2}{\left\{\sum w_j(x_j - \overline{x}_w)^2\right\}^2} \right],$$

the first term of which equals the variance of the local constant estimator (10.40).
It turns out that $x_0 - \overline{x}_w \doteq h^2 f'(x_0)/f(x_0)$ away from the boundary, and is $O(h)$
otherwise, and calculations like those above show that away from the boundary, both
local linear and constant estimators have the approximate variance given in (10.41).

10.7.2 Roughness penalty methods

Local polynomial fitting is brought under the likelihood umbrella by local weighting
of the log likelihood contributions. A different approach to curve estimation is based
on fitting a family of flexible functions to the data, with the most appropriate of these
specified indirectly by penalizing the roughness of the result. The idea is to fit a model
with potentially as many parameters as there are observations, but to constrain these
parameters to the extent desired.

To see how this might be done, we first consider suitably parametrized families of
smooth functions. Let the data consist of pairs $(t_1, y_1), \ldots, (t_n, y_n)$, where $a = t_0 <$
$t_1 < \cdots < t_n < t_{n+1} = b$. We seek a smooth summary $g(t)$ of how the response y
depends on t over the interval $[a, b]$.

One approach is to use a *natural cubic spline* $g(t)$ with knots $t_1, \ldots, t_n$. Such a
function consists of separate cubic polynomials on each of the intervals $[t_1, t_2], \ldots,$

We denote the covariate
by t rather than x for ease
of generalization below.

Figure 10.18 Natural
cubic spline fits to $n = 15$
data pairs simulated from
the model $y = 8x^2 + \varepsilon$.
Left panel: fit with 15
degrees of freedom (solid)
that interpolates the data,
with values of t_j shown by
the vertical dashed lines.
Right panel: fits with
degrees of freedom 2
(solid), 7 (dashes), and 3.7
(dots); the latter is chosen
by cross-validation.

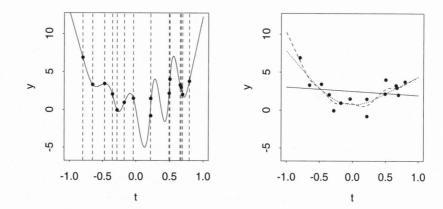

$[t_{n-1}, t_n]$, constrained to be continuous and to have continuous first and second derivatives at each knot. The spline is linear on the extreme intervals $[a, t_1]$ and $[t_n, b]$.

As there are $2 + 4(n - 1) + 2$ coefficients for these polynomial pieces and $3n$ constraints, just n numbers specify the spline. It turns out to be convenient to express it both in terms of its values $g^{\mathrm{T}} = (g_1, \ldots, g_n)$ at the knots and the second derivatives $\gamma^{\mathrm{T}} = (\gamma_2, \ldots, \gamma_{n-1})$ at $t_2, \ldots, t_{n-1}$, where $g_j = g(t_j)$ and $\gamma_j = g''(t_j)$. The second derivatives at t_1 and t_n are zero, so $\gamma_1 = \gamma_n = 0$. In fact there exist $n \times (n - 2)$ and $(n - 2) \times (n - 2)$ matrices Q and R, depending only on $t_1, \ldots, t_n$, such that $Q^{\mathrm{T}}g = R\gamma$; R is positive definite and hence invertible, and both Q and R have simple structure that makes numerical work with them very efficient (Problem 10.14). Note that $\gamma = R^{-1}Q^{\mathrm{T}}g$, so $g(t)$ is completely determined by its values at the knots.

Here and below, $g''(t)$ is
the second derivative of
$g(t)$.

An example is shown in the left panel of Figure 10.18. As outlined above, the spline $g(t)$ is linear outside (t_1, t_n) and cubic between the vertical lines that show the t_j, with smooth joins between cubic portions. One way to imagine this is that the spline adjusts to pass smoothly through beads — the y_j — that move on vertical wires fixed at the t_j.

Penalized log likelihood

Although perhaps a useful numerical summary of the data in Figure 10.18, the spline in the left panel is a poor statistical summary: we need a smoother fit. Suppose that $y_j = g(t_j) + \varepsilon_j$, where the ε_j are independent normal errors with common variance σ^2, and $g(t)$ is a natural cubic spline with its n parameters $g^{\mathrm{T}} = (g_1, \ldots, g_n)$, where $g_j = g(t_j)$; let y denote the $n \times 1$ vector of observed responses. Maximization of the likelihood over g boils down to minimization of the sum of squares $(y - g)^{\mathrm{T}}(y - g)$. This is achieved when $g_j = y_j$, clearly overfitted.

When a similar difficulty arose in our discussion of model selection (Section 4.7), we dealt with it by penalizing the log likelihood to account for model complexity, and we apply this idea here as well. If we judge that a straight line is the acme of simplicity, then one measure of the complexity of $g(t)$ over the interval $[a, b]$ is $\int_a^b \{g''(t)\}^2 \, dt$, which would be zero for a linear fit. Rather than maximize the usual log likelihood, therefore, we maximize the *penalized log likelihood*

$$\ell_\lambda(g, \sigma^2) = \sum_{j=1}^n \log f\{y_j; g(t_j), \sigma^2\} - \frac{\lambda}{2\sigma^2} \int_a^b \{g''(t)\}^2 \, dt, \quad \lambda \geq 0. \quad (10.46)$$

This trades off the increase in the log likelihood term for more complex g against the second term, which penalizes nonlinearity. The extent of the trade-off is controlled by λ, a dimensionless quantity related to the degrees of freedom of the maximizing curve $\widehat{g}_\lambda(t)$. When $\lambda = 0$, no penalty is applied and there are n degrees of freedom, corresponding to unconstrained variation of each element of the vector g. As $\lambda \to \infty$, the penalty becomes so large that $g(t)$ becomes a straight line, which has two degrees of freedom. Intermediate values give curves lying between these extremes. For now we suppose that λ is fixed, deferring discussion of how to choose it.

It turns out that when $g(t)$ is a natural cubic spline, the integral in (10.46) may be expressed as $\gamma^T Q^T g = g^T Q R^{-1} Q^T g = g^T K g$, say, where K is a $n \times n$ symmetric matrix of rank $n - 2$ (Problem 10.14). For normal errors the jth log likelihood contribution is

$$\log f\{y_j; g(t_j), \sigma^2\} \equiv -\frac{1}{2}\sigma^{-2}\{y_j - g(t_j)\}^2 - \frac{1}{2}\log \sigma^2,$$

so $\widehat{g}_\lambda(t)$ is determined by the vector $\widehat{g}$ that minimizes the *penalized sum of squares*

$$(y - g)^T(y - g) + \lambda g^T K g \quad (10.47)$$

with respect to g. On completing the square we find that $\widehat{g}$ minimizes

$$\left\{(I + \lambda K)^{-1}y - g\right\}^T (I + \lambda K)\left\{(I + \lambda K)^{-1}y - g\right\}, \quad (10.48)$$

which differs from (10.47) only by a constant independent of g. It is straightforward to see that $\widehat{g} = (I + \lambda K)^{-1}y$ is the unique natural cubic spline that minimizes (10.48); furthermore, as it turns out that it does so among all functions that are differentiable on $[a, b]$ and have absolutely continuous first derivative, $\widehat{g}$ is optimal in a large class of smooth functions.

The structure of the matrix K can be exploited to give a fast algorithm for fitting the spline. Recall that $\gamma = R^{-1}Q^T g$ and $K = QR^{-1}Q^T$. Hence $\widehat{g}$ is the solution to $(I + \lambda QR^{-1}Q^T)g = y$. Equivalently the corresponding $\widehat{\gamma}$ solves the system $(R + \lambda Q^T Q)\gamma = Q^T y$, which can be solved in $O(n)$ operations because both R and Q are band matrices; their only non-zero elements lie on the diagonal or just above and below it.

Like a local polynomial fit, the spline is a linear smoother. Its smoothing matrix is $S_\lambda = (I + \lambda K)^{-1}$. Note that $Kg = 0$ for vectors of form $g = \beta_0 1_n + \beta_1(t_1, \ldots, t_n)^T$, because the roughness penalty is zero when $g(t)$ is linear, and it follows that $S_\lambda g = g$ for such vectors. Once again there are several definitions of the degrees of freedom for the smooth fit, the most obvious being $\text{tr}(S_\lambda)$.

The right panel of Figure 10.18 shows three fits, of which the linear fit is evidently too smooth, and the one with seven degrees of freedom too rough. The fit with

3.7 degrees of freedom, chosen by cross-validation as described below, seems more plausible.

For later development we must deal with two complications. The first arises when weights w_j are attached to the cases (t_j, y_j). Then $(y - g)^\mathsf{T}(y - g)$ must be replaced by $(y - g)^\mathsf{T} W(y - g)$, where $W = \text{diag}\{w_1, \ldots, w_n\}$; see Section 8.2.4. The second occurs when some of the t_j are tied. If so, we let $s_1 < \cdots < s_q$ denote the ordered distinct values among $t_1, \ldots, t_n$ and denote by N the $n \times q$ *incidence matrix* whose (j, k) element indicates whether $t_j = s_k$; obviously $q \geq 2$, and $N = I$ if the t_j are distinct and ordered. With these changes (10.47) alters to

$$(y - Ng)^\mathsf{T} W(y - Ng) + \lambda g^\mathsf{T} K g,$$

which is minimized at $\widehat{g} = N(N^\mathsf{T} W N + \lambda K)^{-1} N^\mathsf{T} W y$. The smoothing matrix $S_\lambda = N(N^\mathsf{T} W N + \lambda K)^{-1} N^\mathsf{T} W$ reduces to the previous expression when $W = N = I$.

How much smoothing?

The smoothing parameter λ plays the same role as the bandwidth in a local polynomial model, and it too is typically chosen by minimizing information or cross-validation criteria such as

$$\text{CV}(\lambda) = \sum_{j=1}^n \left\{ \frac{y_j - \widehat{g}(t_j)}{1 - S_{jj}(\lambda)} \right\}^2, \quad \text{GCV}(\lambda) = \sum_{j=1}^n \left\{ \frac{y_j - \widehat{g}(t_j)}{1 - \text{tr}(S_\lambda)/n} \right\}^2,$$

where both the diagonal elements $S_{jj}(\lambda)$ of the smoothing matrix S_λ and the fitted values $g(t_j)$ depend on λ. As with other applications of smoothing, the goal is to trade off fidelity to the data against smoothness of the fit. Once again a caveat is needed: the results from an automatic procedure cannot always be trusted, and it is often valuable to apply different levels of smoothing. As mentioned above, it is useful to know the degrees of freedom of a smooth fit.

Example 10.33 (Spring barley data) Table 10.21 gives standardized yields from an agricultural field trial in which three blocks of long narrow plots were sown with 75 varieties of spring barley in a random order within each block. The yield from variety 27 in the third block is missing, but otherwise there are three replicates for each variety. The plot of the yields in the left panel of Figure 10.19 shows strong spatial patterns owing to fertility trends within each block, in addition to the variety effects. For the moment we ignore differences among the varieties, and illustrate how fitting a natural cubic spline can account for the fertility gradient in the first block.

The left panel of Figure 10.19 shows some of the disadvantages of polynomial fitting. The lower curve, for example, wiggles implausibly compared to a spline fit with the same degrees of freedom, shown in the upper right panel. The lower right panel shows how $\text{CV}(\lambda)$ and $\text{GCV}(\lambda)$ vary with the equivalent degrees of freedom, for the three blocks. The fit to block 2 seems fairly reasonable, but block 3 is evidently overfitted with 40 degrees of freedom, and block 1 is probably also overfitted. We reconsider these data in Example 10.35. ∎

Location t	Block 1		Block 2		Block 3	
	Variety	Yield y	Variety	Yield y	Variety	Yield y
1	57	9.29	49	7.99	63	11.77
2	39	8.16	18	9.56	38	12.05
3	3	8.97	8	9.02	14	12.25
4	48	8.33	69	8.91	71	10.96
5	75	8.66	29	9.17	22	9.94
6	21	9.05	59	9.49	46	9.27
7	66	9.01	19	9.73	6	11.05
8	12	9.40	39	9.38	30	11.40
9	30	10.16	67	8.80	16	10.78
10	32	10.30	57	9.72	24	10.30
11	59	10.73	37	10.24	40	11.27
12	50	9.69	26	10.85	64	11.13
13	5	11.49	16	9.67	8	10.55
14	23	10.73	6	10.17	56	12.82
15	14	10.71	47	11.46	32	10.95
16	68	10.21	36	10.05	48	10.92
17	41	10.52	64	11.47	54	10.77
18	1	11.09	63	10.63	37	11.08
19	64	11.39	33	11.03	21	10.22
20	28	11.24	74	10.85	29	10.59
21	46	10.65	13	11.35	62	11.35
22	73	10.77	43	10.25	5	11.39
23	37	10.92	3	10.08	70	10.59
24	55	12.07	53	10.25	13	11.26
25	19	11.03	23	9.57	11	11.79
26	10	11.64	62	11.34	44	12.25
27	35	11.37	52	10.19	36	12.23
28	26	10.34	12	10.80	52	10.84
29	17	9.52	2	10.04	60	10.92
30	71	8.99	32	9.69	68	10.41
31	8	8.34	22	9.36	3	10.96
32	62	9.25	42	9.43	19	9.94
33	44	9.86	72	11.46	67	11.27
34	53	9.90	73	9.29	59	11.79
35	74	11.04	25	10.10	2	11.51
36	20	10.30	45	9.53	75	11.64
37	56	11.56	15	10.55	27	—
38	29	9.69	35	11.34	43	9.78
39	2	10.68	66	11.36	51	8.86
40	47	10.91	5	10.88	10	10.28
41	11	10.05	56	11.61	35	12.15
42	38	10.80	46	10.33	74	10.36
43	65	10.06	71	10.53	66	9.59
44	13	10.04	51	8.67	34	10.53
45	31	10.50	21	9.56	18	11.26
46	40	9.51	1	9.95	50	10.37
47	4	9.20	31	11.10	42	10.10
48	67	9.74	11	10.11	1	9.95
49	22	8.84	41	9.36	58	9.80
50	49	9.33	61	10.23	26	10.58
51	58	9.51	55	11.38	41	9.31
52	43	9.35	14	11.30	25	9.29

Table 10.21 Spring barley data (Besag *et al.*, 1995). Spatial layout and plot yield at harvest y (standardized to have unit crude variance) in a final assessment trial of 75 varieties of spring barley. The varieties are sown in three blocks, with each variety replicated thrice in the design. The yield for variety 27 is missing in the third block.

Table 10.21   (*cont.*)

Location t	Block 1		Block 2		Block 3	
	Variety	Yield y	Variety	Yield y	Variety	Yield y
53	7	9.01	44	10.90	33	10.03
54	25	10.58	34	10.97	9	9.49
55	61	11.03	54	12.22	17	11.52
56	16	9.89	24	10.10	57	12.24
57	52	11.39	4	11.22	65	11.64
58	70	11.24	65	10.01	49	10.74
59	34	12.18	75	10.29	73	10.29
60	42	10.21	38	10.95	7	10.25
61	24	11.08	17	9.66	23	11.39
62	33	11.05	68	9.31	72	13.34
63	51	10.29	7	8.84	55	12.73
64	60	10.57	27	10.64	31	12.62
65	69	10.42	58	9.45	39	10.19
66	15	10.49	48	9.66	47	11.61
67	6	10.00	28	9.85	15	10.52
68	63	9.23	60	9.24	20	9.07
69	54	10.57	30	10.11	61	10.76
70	18	10.27	70	9.63	28	9.91
71	45	8.86	20	9.04	53	10.17
72	72	9.45	9	8.43	69	8.68
73	9	8.03	40	10.97	45	8.74
74	36	9.22	50	8.98	12	9.15
75	27	8.70	10	9.88	4	9.39

10.7.3 More general models

We now consider how the discussion above should be modified when there are explanatory variables as well as a smooth variable, treating certain covariates nonparametrically and others not, and allowing the response to have a density other than the normal.

Let the data consist of independent triples $(x_1, t_1, y_1), \ldots, (x_n, t_n, y_n)$, with jth log likelihood contribution $\ell_j(\eta_j, \kappa)$, where $\eta_j = x_j^{\mathrm{T}}\beta + g(t_j)$; for now we suppress dependence on κ. Then the analogue of (10.47) is the *penalized log likelihood*

$$\ell_\lambda(\beta, g) = \sum_{j=1}^{n} \ell_j(\eta_j) - \frac{1}{2}\lambda \int_a^b \{g''(t)\}^2 \, dt, \quad \lambda > 0, \qquad (10.49)$$

where a and b are chosen so that $a < t_1, \ldots, t_n < b$. If all the t_j are distinct and $\lambda = 0$, the maximum is obtained by choosing $g_j = g(t_j)$ to maximise the jth log likelihood contribution, but this is not useful because the resulting model has n parameters and is too rough. The integral in (10.49) penalizes roughness of $g(t)$, so λ has the same interpretation as before.

If the ordered distinct values of $t_1, \ldots, t_n$ are $s_1 < \cdots < s_q$ and if $g(t)$ is a natural cubic spline with knots at the s_i, then the integral in (10.49) may be written $g^{\mathrm{T}}Kg$, where the $q \times 1$ vector g has ith element $g_i = g(s_i)$. Given a value of λ, our aim

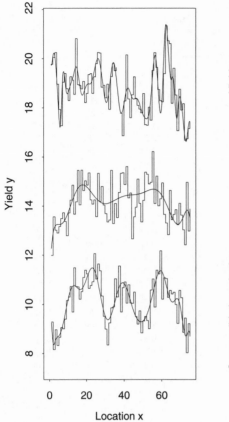

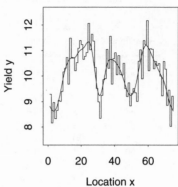

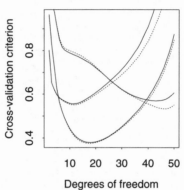

Figure 10.19 Spring barley data analysis. Left panel: yield y as a function of location x for the three blocks. Yields for blocks 2, 3 have been offset by adding 4, 8 respectively. The smooth solid lines are the fits of polynomials of degree 20, 10 and 40 to the data from blocks 1, 2 and 3. Upper right: yields for block 1, with smoothing spline fit with 18 degrees of freedom. Lower right: cross-validation (solid) and generalized cross-validation (dots) criteria for smoothing spline fits to blocks 1, 2 and 3, with minima at roughly 20, 10 and 40 equivalent degrees of freedom.

is to find values of β and g that maximize $\ell_\lambda(\beta, g)$. As the $n \times 1$ vector of linear predictors may be written $\eta = X\beta + Ng$, where N is the $n \times q$ incidence matrix for the elements of g and the $n \times p$ matrix X has jth row x_j^T, the score equations are

$$\begin{pmatrix} \partial\ell_\lambda(\widehat{\beta}, \widehat{g})/\partial\beta \\ \partial\ell_\lambda(\widehat{\beta}, \widehat{g})/\partial g \end{pmatrix} = \begin{pmatrix} X^\mathsf{T} u(\widehat{\eta}) \\ N^\mathsf{T} u(\widehat{\eta}) - \lambda K \widehat{g} \end{pmatrix} = 0, \qquad (10.50)$$

where the $n \times 1$ vector $u(\eta)$ has jth element $\partial\ell_j(\eta_j)/\partial\eta_j$. The usual Taylor series expansion (Sections 4.4.1, 10.2.2) then gives

$$\begin{pmatrix} X^\mathsf{T} W X & X^\mathsf{T} W N \\ N^\mathsf{T} W X & N^\mathsf{T} W N + \lambda K \end{pmatrix} \begin{pmatrix} \widehat{\beta} \\ \widehat{g} \end{pmatrix} \doteq \begin{pmatrix} X^\mathsf{T} W z \\ N^\mathsf{T} W z \end{pmatrix}, \qquad (10.51)$$

where $W = \text{diag}\{w_1, \ldots, w_n\}$, with $w_j = \mathrm{E}\{-\partial^2\ell_j(\eta_j)/\partial\eta_j^2\}$ and where $z = W^{-1}u(\eta) + \eta$ is the $n \times 1$ adjusted dependent variable. Fisher scoring would solve (10.51) iteratively starting from suitable initial values of β and g, but here there are $p + q$ regressors, where q is typically comparable to n, and an approach known as *backfitting* is generally used instead. The idea is to alternate between the two matrix equations in (10.51), rewritten as

$$(X^\mathsf{T} W X)\widehat{\beta} \doteq X^\mathsf{T} W(z - N\widehat{g}), \qquad (10.52)$$

$$(N^\mathsf{T} W N + \lambda K)\widehat{g} \doteq N^\mathsf{T} W(z - X\widehat{\beta}). \qquad (10.53)$$

Given initial values β_0 and g_0 of β and g, we calculate η, W, and z, replace $\widehat{g}$ in (10.52) by g_0, and then obtain an approximate value of $\widehat{\beta}$ by regressing $z - N g_0 = W^{-1} u + X \beta_0$ on the columns of X with weights W. We then recalculate $\eta = X\widehat{\beta} + N g_0$, W, and z, and solve (10.53) by applying the matrix $(N^\mathsf{T} W N + \lambda K)^{-1} N^\mathsf{T} W$ to $z - X\widehat{\beta} = W^{-1} u + N g_0$, thus obtaining an approximate value of $\widehat{g}$. We then set β_0 and g_0 equal to $\widehat{\beta}$ and $\widehat{g}$ and iterate the cycle to convergence.

Example 10.34 (Partial spline model)   The case where y_j is normal with mean η_j and variance σ^2 is known as a *partial spline model*. Here $u(\eta) = \sigma^{-2}(y - \eta)$ and $W = \sigma^{-2} I_n$, so $z = y$ (Example 10.4).

The first backfitting step is least squares regression of $y - N g_0$ on the columns of X. The second applies the linear smoother $(N^\mathsf{T} N + \sigma^2 \lambda K)^{-1} N^\mathsf{T}$ to the residual $y - X\widehat{\beta}$ from the first step; the effective penalty is thus $\lambda' = \sigma^2 \lambda$. At each step of the iteration either least squares or linear smoothing is applied to the residual from the previous operation, continuing until any systematic structure has been removed from both $y - X\widehat{\beta}$ and $y - N\widehat{g}$.   ∎

Example 10.36 gives a further illustration of such fitting.

Backfitting yields parameter estimates, but unless the fit is purely exploratory other quantities are required for inference. The deviance is defined in the usual way, and the error degrees of freedom of a fit are

$$n - \nu_\lambda = n - \text{tr}(S_\lambda) - \text{tr}[\{X^\mathsf{T} W(I - S_\lambda)X\}^{-1} X^\mathsf{T} W(I - S_\lambda)^2 X],$$

where this and the smoothing matrix

$$S_\lambda = N(N^\mathsf{T} W N + \lambda K)^{-1} N^\mathsf{T} W$$

are computed at convergence. The usual chi-squared theory is often used for approximate comparison of nested models, even though it has no firm theoretical basis. Standard errors for elements of $\widehat{\beta}$ and the fitted $\widehat{g}(t)$ are generally obtained using the approximate linearization entailed by (10.51).

As usual in semiparametric modelling, the degree of smoothing is critical. Various forms of the cross-validation and generalized cross-validation criteria for choice of λ have been suggested. One approach takes

$$\text{CV}(\lambda) = \sum_{j=1}^{n} \frac{(y_j - \widehat{y}_j)^2}{v_j \{1 - h_{jj}(\lambda)\}^2}, \quad \text{GCV}(\lambda) = \sum_{j=1}^{n} \frac{(y_j - \widehat{y}_j)^2}{v_j \{1 - \text{tr}(H_\lambda)/n\}^2},$$

where v_j is the estimated variance of y_j, $\widehat{y}_j$ is the jth fitted value, and $h_{jj}(\lambda)$ is the partial derivative of $\widehat{y}_j$ with respect to y_j, analogous to the earlier role of $S_{jj}(h)$. Another possibility is to cross-validate the approximating linear problem obtained at each step of the fitting algorithm.

When several covariates might be treated nonparametrically, say t, u, and v, we can take the linear predictor to be $x^T\beta + g_1(t) + g_2(u) + g_3(v)$ and fit smoothing splines or other nonparametric curves with a version of backfitting in which β and the gs are iteratively estimated in succession. Such a setup is known as a *generalized additive model*, to which the same ideas apply as outlined above. The degrees of smoothing are controlled by separate penalties for each of the gs, and the corresponding λs may be estimated by minimizing a cross-validation or similar criterion. Surfaces $g(t, u)$ can also be fitted using similar ideas.

In some cases it is necessary to diminish the computational burden by reducing the number of knots and hence the number of parameters q that specify the fitted curve. Although the resulting fit no longer has the optimality properties of the natural cubic spline, this is typically unimportant in practice.

Example 10.35 (Spring barley data)   In addition to their strong spatial dependence, the spring barley yields in Table 10.21 depend on variety effects. The simplest model that would accomodate these is a two-way layout with variety and block effects, in which the response is

$$y_{vb} = \tau_b + \beta_v + \varepsilon_{vb}, \quad v = 1, \ldots, 75, b = 1, 2, 3, \quad \text{where} \quad \varepsilon_{vb} \overset{\text{iid}}{\sim} N(0, \sigma^2).$$

This has residual sum of squares 94.87 on 147 degrees of freedom, giving $\widehat{\sigma}^2 = 0.645$, while the standard error for a difference of variety effects $\widehat{\beta}_{v_1} - \widehat{\beta}_{v_2}$ is 0.655.

As the two-way layout ignores the spatial variation, it greatly overestimates σ^2, thereby decreasing the sensitivity of comparisons among the varieties. Moreover the variety effect estimators may be biased if all three replicates of a particular variety happen to fall where the fertilities are higher than average. It seems more sensible to fit a model in which the yield for the vth variety in the bth block depends on its location t_{vb} through

$$y_{vb} = g_b(t_{vb}) + \beta_v + \varepsilon_{vb}, \quad v = 1, \ldots, 75, b = 1, 2, 3,$$

where $g_b(t)$ is a smooth function that determines how the fertility pattern in block b depends on location t. When this model is fitted using smoothing splines with 40 knots for each of the g_b, 77 degrees of freedom are needed to account for the variety and block effects, the degrees of freedom that minimize the generalized cross-validation criterion are 16.4, 8.3, and 25.2 for $b = 1, 2$, and 3, the residual sum of squares is 16.85 on $224 - 77 - 16.4 - 8.3 - 25.3 = 97$ degrees of freedom, and $\widehat{\sigma}^2 = 0.174$, about one quarter of the value for the two-way layout. The standard errors for differences of variety effects are roughly 0.41, so more precise comparisons are possible than in the simpler fit. Fewer degrees of freedom are needed to model the spatial variation here than the 20, 10, and 40 required for the three blocks in Example 10.33, because allowance for variety effects enables smoother fertility trends to be used.

Figure 10.20 shows how this model decomposes the original data into fertility trends, variety effects, and residuals. As their degrees of freedom would suggest, the estimate $\widehat{g}_2(t)$ is appreciably smoother than the fertility trends in blocks 1 and 3. The best-yielding varieties in decreasing order of $\widehat{\beta}_v$ are 35, 56, 31, 54, 72, 55, 47, 18,

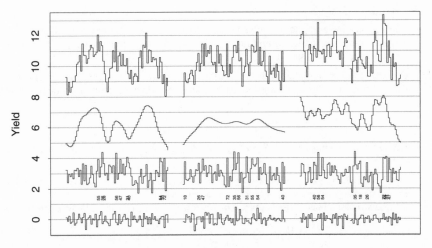

Figure 10.20 Spring barley data analysis. Block 1 is shown on the left and block 3 on the right. The panel shows, from the top, the original yields y, the fertility trend and variety effect estimates $\widehat{g}_b(t)$ and $\widehat{\beta}_v$, both offset for display, and the crude residuals. The varieties with the ten largest $\widehat{\beta}_v$ are marked.

40, and 26, but as a 0.95 confidence interval for differences of two variety effects has width 1.61, there is no clear-cut best variety.

A probability plot of the residuals shows nothing to undermine normality of the errors, but the correlograms and partial correlograms of residuals from blocks 1 and 3 show slight negative correlations, suggesting that $\widehat{g}_1(t)$ and $\widehat{g}_3(t)$ may be overfitted. A more complete analysis would try and remedy this by refitting the model with fewer degrees of freedom for the fertility trends in blocks 1 and 3. ∎

Exercises 10.7

1 Explain how the derivatives $g'(x_0), \ldots, g^{(k)}(x_0)$ may be estimated using the least squares estimator $\widehat{\beta}$ from a local polynomial fit of degree k.

2 What is the bias of the local fit of a polynomial of degree k to a function that is polynomial of degree $l \leq k$? How would you measure the disadvantage of this relative to an unweighted fit?

3 By writing $\sum \{y_j - \widehat{g}(x_j)\}^2 = (y - \widehat{g})^{\mathrm{T}}(y - \widehat{g})$ and recalling that $y = g + \varepsilon$ and $\widehat{g} = Sy$, where S is a smoothing matrix, show that

$$\mathrm{E}\left[\sum_{j=1}^{n}\{y_j - \widehat{g}(x_j)\}^2\right] = \sigma^2(n - 2\nu_1 + \nu_2) + g^{\mathrm{T}}(I - S)^{\mathrm{T}}(I - S)g.$$

Hence explain the use of $s^2(h)$ as an estimator of σ^2. Under what circumstances is it unbiased?

4 (a) If $S_{11}(h), \ldots, S_{nn}(h)$ are the *leverages* of a smoothing matrix S_h, establish

$$\nu_1 = \sum_{j=1}^{n} S_{jj}(h), \quad \nu_2 = \sum_{i,j=1}^{n} S_{ij}(h)^2 = \sigma^{-2} \sum_{j=1}^{n} \mathrm{var}\left\{\widehat{g}(x_j)\right\}.$$

(b) Show that $S(x_j; x_j, h)$ is proportional to the $(1, 1)$ element of $(X^{\mathrm{T}}WX)^{-1}$, and let the *influence function* of the smoother be $I(x) = w(0)e_1^{\mathrm{T}}(X^{\mathrm{T}}WX)^{-1}e_1$, where $e_1^{\mathrm{T}} = (1, 0, \ldots, 0)$ has length $k + 1$. Show that

$$\sigma^{-2}\mathrm{var}\{\widehat{g}(x)\} \leq I(x),$$

and deduce that $v_2 \leq v_1$.

(c) Let $I_1(x)$ and $I_2(x)$ be the influence functions corresponding to bandwidths $h_1 < h_2$, and let the corresponding weight matrices be W_1 and W_2. Show that $W_1 \leq W_2$ componentwise and deduce that $I_2(x) \geq I_1(x)$.

5   Consider a linear smoother with $n \times n$ smoothing matrix S_h, so $\widehat{g} = S_h y$, and show that the function $a_j(u)$ giving the fitted value at x_j as a function of the response u there satisfies

$$a_j(u) = \begin{cases} \widehat{g}(x_j), & u = y_j, \\ \widehat{g}_{-j}(x_j), & u = \widehat{g}_{-j}(x_j). \end{cases}$$

Explain why this implies that $S_{jj}(h)\{y_j - \widehat{g}_{-j}(x_j)\} = \widehat{g}(x_j) - \widehat{g}_{-j}(x_j)$, and hence obtain (10.42).

6   (a) Check (10.45), and hence verify that $\mathrm{E}(\widehat{\beta}_0) - g(x_0) \doteq \frac{1}{2}h^2 g''(x_0)$ far from a boundary. Do the corresponding calculation for x_0 near a boundary.

(b) Show that the bias of the Nadarayah–Watson estimator (10.40) may be expressed as

$$\frac{\sum w_j \left\{ (x_j - x_0)g'(x_0) + \frac{1}{2}(x_j - x_0)^2 g''(x_0) + \cdots \right\}}{\sum w_j},$$

and deduce that this is approximately $h g'(x_0)a$ near a boundary, where $a \neq 0$, and $\frac{1}{2}h^2\{g''(x_0) + f'(x_0)g'(x_0)\}$ elsewhere.

7   Develop the details of local likelihood smoothing when a linear polynomial is fitted to Poisson data, using link function $\log \mu = \beta_0 + \beta_1(x - x_0)$.

10.8 Survival Data

10.8.1 Introduction

Survival or event history analysis concerns the times of events. Such data are particularly common in medicine and the social sciences, but also arise in many other domains. As we saw in Section 5.4, the responses may be incompletely observed owing to censoring or truncation. Here we give an introduction to regression analysis of such data in the simplest and most common situation, with just one event per individual. Throughout we use the term 'failure' to describe the event of interest, and refer to the time to failure as a survival time.

Let the data available on the jth of n independent individuals be (x_j, y_j, d_j), where x_j is a $p \times 1$ vector of explanatory variables, $d_j = 1$ indicates that y_j is an observed survival time, and $d_j = 0$ indicates that the survival time is right-censored at y_j. Consider a parametric model under which the survival time has density $f(y; x, \beta)$, survivor function $\mathcal{F}(y; x, \beta) = 1 - F(y; x, \beta)$, and hazard and cumulative hazard functions $h(y; x, \beta)$ and $H(y; x, \beta) = -\log \mathcal{F}(y; x, \beta)$. Assume that the censoring mechanism is uninformative, that is, independent of the failure time and uninformative about β. Then the discussion in Section 5.4 implies that the log likelihood may be written as

$$\ell(\beta) = \sum_{j=1}^{n} \{d_j \log h(y_j; x_j, \beta) - H(y_j; x_j, \beta)\}, \tag{10.54}$$

from which maximum likelihood estimates $\widehat{\beta}$ may be obtained by iterative weighted least squares. Any additional parameters ϕ may be estimated by interleaving updates to $\widehat{\beta}$ and $\widehat{\phi}$. As usual with incomplete data, confidence intervals should be based on observed information, as in Example 4.47.

Residuals are important for model checking. The relation $F(y; x, \beta) = 1 - \exp\{-H(y; x, \beta)\}$ implies that if Y is continuous and uncensored with distribution function $F(y; x, \beta)$, then $H(Y; x, \beta)$ is exponentially distributed with unit mean.

After Cox and Snell (1968). This suggests that for diagnostic purposes the *Cox–Snell residuals* $H(y_j; x, \widehat{\beta})$ can be regarded as an exponential random sample. For observations censored at c, we argue that as $E\{H(Y; x, \beta) \mid Y > c\} = H(c; x, \beta) + 1$, an appropriate residual is $H(c; x, \widehat{\beta}) + 1$. This yields modified Cox–Snell residuals

$$r_j = H(y_j; x, \widehat{\beta}) + 1 - d_j. \tag{10.55}$$

Other residuals may be defined for the proportional hazards model, described below. Case diagnostics discussed in Section 10.2.3 may also be useful.

Accelerated life models

One notion used particularly in reliability studies is that time to failure may be accelerated or retarded relative to some baseline. Let Y and Y_0 denote failure times for individuals with covariates x and $x = 0$. Then the accelerated life model posits the existence of a positive function $\tau(\beta; x)$ such that Y and $\tau(\beta; x)Y_0$ have the same distribution; equivalently $Y/\tau(\beta; x) \overset{D}{=} Y_0$, a baseline random variable. An individual with $\tau(\beta; x) < 1$ will 'wear out' at a faster rate than the baseline, and conversely. If Y_0 has survivor, density, and hazard functions $\mathcal{F}_0(y)$, $f_0(y)$, and $h_0(y)$, the corresponding functions for Y are

$$\mathcal{F}_0\{y/\tau(\beta; x)\}, \quad \tau(\beta; x)^{-1} f_0\{y/\tau(\beta; x)\}, \quad \tau(\beta; x)^{-1} h_0\{y/\tau(\beta; x)\}. \tag{10.56}$$

This is a scale model, so obvious possibilities are to let Y_0 be an exponential, gamma, Weibull, log-normal, or log-logistic variable. If $\tau(\beta; x) = \exp(x^\mathsf{T}\beta)$, then $\log Y \overset{D}{=} x^\mathsf{T}\beta + \varepsilon$, where $\varepsilon = \log Y_0$. The regression-scale model $x^\mathsf{T}\beta + \sigma\varepsilon$ is equivalent to taking $(Y/\tau)^{1/\sigma} \overset{D}{=} Y_0$. Any of these gives a linear model for the log responses, and if there is no censoring this can be fitted using least squares, though typically information will be lost by doing so. However there is no special difficulty with maximum likelihood estimation by iterative weighted least squares, if the density of ε or equivalently of Y_0 is known.

Example 10.36 (Leukaemia data) Table 10.22 contains data on the survival of acute leukaemia victims. The covariate x is $\log_{10}$ white blood cell count at time of diagnosis, and the patients are grouped according to the presence or not of a morphologic characteristic of their white blood cells. Within each group suppose that survival time Y is exponential with mean $\tau = \exp(\eta)$, where $\eta = \beta_0 + \beta_1 I(\text{Group} = 1) + \beta_2 x$. This is a generalized linear model with gamma errors, log link function, and dispersion parameter $\phi = 1$.

	Group 1					Group 2			
	x	y		x	y		x	y		x	y
1	3.36	65	10	3.85	143	18	3.64	56	27	4.45	3
2	2.88	156	11	3.97	56	19	3.48	65	28	4.49	8
3	3.63	100	12	4.51	26	20	3.60	17	29	4.41	4
4	3.41	134	13	4.54	22	21	3.18	7	30	4.32	3
5	3.78	16	14	5.00	1	22	3.95	16	31	4.90	30
6	4.02	108	15	5.00	1	23	3.72	22	32	5.00	4
7	4.00	121	16	4.72	5	24	4.00	3	33	5.00	43
8	4.23	4	17	5.00	65	25	4.28	4			
9	3.73	39				26	4.43	2			

Table 10.22 Survival times y (weeks) for two groups of acute leukaemia patients, together with $x = \log_{10}$ white blood cell count at time of diagnosis (Feigl and Zelen, 1965). Patients in group 1 had Auer rods and/or significant granulation of the leukaemic cells in the bone marrow at the time of diagnosis; those in group 2 did not.

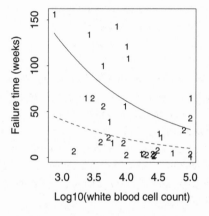
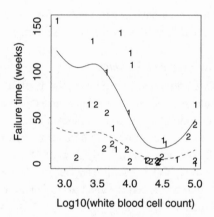

Figure 10.21 Plots of data and fitted means for generalized linear (left) and generalized additive (right) models fitted to two groups of survival times for leukaemia patients: group 1 (solid); group 2 (dashed).

When this model is fitted the deviance drops by 17.82 to 40.32, and the degrees of freedom drop from 32 to 30. The parameter estimates and standard errors are $\widehat{\beta_0} = 5.81$ (1.29), $\widehat{\beta_1} = 1.02$ (0.35), and $\widehat{\beta_2} = -0.70$ (0.30). The mean survival time drops rapidly with x, but is increased in group 1; both effects are significant. The fitted means $\widehat{\tau}$ and data are shown in the left panel of Figure 10.21. An exponential probability plot of the residuals $y/\widehat{\tau}$ casts no doubt on the model. This can be verified more formally by fitting gamma, Weibull, and log-logistic distributions, none of which improves on the exponential.

Inspection of the left panel of Figure 10.21 suggests some lack of fit of the systematic part of the model and we use this to illustrate the fitting of a generalized additive model with linear predictor $\eta = \beta_0 + \beta_1 I(\text{Group} = 1) + s(x)$, where the smooth function $s(x)$ is a natural cubic spline. The right panel shows smooth dependence on x with 3.36 degrees of freedom, found by generalized cross-validation starting from a cubic spline with 6 knots equi-spaced along the range of x. With this model $\widehat{\beta_0} = 2.80$ (0.24) and $\widehat{\beta_1} = 1.15$ (0.31); the value of $\widehat{\beta_0}$ changes because $s(x)$ is parametrized to be orthogonal to a constant. The deviance is 31.68 with 27.63 equivalent degrees of freedom, so the F statistic for comparison of this with the fully parametric model is $(40.32 - 31.68363)/2.37/(31.68/27.63) = 3.18$ on 2.37 and

27.63 degrees of freedom, giving significance level 0.05. Here chi-squared asymptotics are of dubious relevance, and as simulation from the parametric model gives a rather larger significance level, there is no reason to choose the more complex generalized additive model, particularly as increased mean survival time at the highest white blood cell counts seems implausible. ∎

Often called the Cox model, because introduced by Cox (1972).

10.8.2 Proportional hazards model

In medical applications the focus of interest is typically on how treatments or classifications of the units affect survival, the form of the survival distribution being of secondary importance. This suggests that we seek inferences that will be valid for any such distribution. This is difficult for accelerated life models, and instead we let the covariates act directly on the hazard. Suppose that an individual with baseline

We can take $x_0 = 0$ without loss of generality.

covariate x_0 has hazard function $h_0(y)$ after a time y on trial, while an individual with covariate x has hazard function $h(y) = \xi(\beta; x)h_0(y)$, where $\xi(\beta; x)$ is a positive function sometimes called a risk score; usually $\xi(\beta; x) = \exp(x^\mathsf{T}\beta)$. The ratio $h(y)/h_0(y)$ does not involve h_0. This *proportional hazards* assumption turns out to be crucial, but it is strong and must be checked in practice.

The basic relationship between the survivor and hazard functions, $\mathcal{F}(y) = \exp\{-H(y)\}$, where $H(y) = \int_0^y h(u)\,du$, implies that the survival time for an individual with covariate x has survivor and density functions

$$\mathcal{F}_0(y)^{\xi(\beta;x)}, \quad \xi(\beta;x)h_0(y)\mathcal{F}_0(y)^{\xi(\beta;x)}.$$

Thus whereas accelerated life models scale the axis of a baseline survivor function, $\mathcal{F}_0$, proportional hazards raise the baseline survivor function to a power.

The action of the covariates being of primary interest, we seek a likelihood on which to base inference for β, regardless of $h_0(y)$. To motivate the argument below, note that if the hazard function was entirely arbitrary, then inference could only be based on events where failures actually occurred, because the hazard might in principle be zero at every other time. Thus it suffices to estimate the baseline cumulative hazard function by a step function $H_0(y) = \sum_{j:y_j \le y} h_j$, where $h_j = h_0(y_j) > 0$ only at observed failure times.

Suppose there are no ties, take $0 < y_1 < \cdots < y_n$ without loss of generality, and let $\mathcal{R}_j$ denote the *risk set* of individuals still available to fail at the instant before y_j, that is, all except those who have previously failed or been censored; see Figure 5.8. For brevity set $\xi_j = \xi(\beta; x_j)$. Then the log likelihood is

$$\sum_{j=1}^n \{d_j \log(\xi_j h_j) - \xi_j H_0(y_j)\} = \sum_{j=1}^n \left\{ d_j \log(\xi_j h_j) - \xi_j \sum_{i=1}^j h_i \right\}$$

$$= \sum_{j=1}^n \left(d_j \log \xi_j + d_j \log h_j - h_j \sum_{i \in \mathcal{R}_j} \xi_i \right).$$

With β fixed the h_j have maximum likelihood estimators $\widehat{h}_j = d_j / \sum_{i \in \mathcal{R}_j} \xi_i$, positive only when $d_j = 1$, so the profile log likelihood for β is

$$\ell_p(\beta) = \max_{h_1,\ldots,h_n} \ell(\beta, h_1, \ldots, h_n) \equiv \sum_{j=1}^{n} d_j \log\left(\frac{\xi_j}{\sum_{i \in \mathcal{R}_j} \xi_i}\right). \qquad (10.57)$$

The corresponding profile likelihood is

$$\prod_{j=1}^{n} \left\{\frac{\xi(\beta; x_j)}{\sum_{i \in \mathcal{R}_j} \xi(\beta; x_i)}\right\}^{d_j} = \prod_{\text{failures}} \frac{\xi(\beta; x_j)}{\sum_{i \in \mathcal{R}_j} \xi(\beta; x_i)}. \qquad (10.58)$$

Alternatively we may reason that the probability of the particular failure observed to occur at y_j, conditional on a failure occurring then, is

$$\frac{\xi_j h_0(y_j)}{\sum_{i \in \mathcal{R}_j} \xi_i h_0(y_j)} = \frac{\xi_j}{\sum_{i \in \mathcal{R}_j} \xi_i}, \qquad (10.59)$$

and hence (10.58) is the probability that failures occur in the observed order, conditional on their occurrence times and margining over times of censoring. Thus (10.58) is the product of a nested sequence of multinomial variables. There is a close connection to the discussion of Poisson variables and log-linear models on page 501.

Expression (10.58) is known as a *partial likelihood*. In Section 12.2 it is derived as a marginal likelihood based on the observed ranking of failure times. Although a mathematically complete derivation is beyond our scope, it turns out that despite the maximization over n nuisance parameters, (10.58) can be treated as an ordinary likelihood: the maximum partial likelihood estimator $\widehat{\beta}$ is consistent for β under mild conditions, and standard errors can be based on the inverse observed information matrix.

Information contained in the failure times is lost, because they are treated as fixed in constructing the partial likelihood. The loss of information compared to using the correct parametric model turns out to be small in most cases, however, so standard errors from partial likelihood are close to those obtained under the true model. Partial likelihood inferences make essentially no assumptions about h_0, and are in this sense semiparametric.

Tied failure times have probability zero for continuous distributions, but nevertheless they arise in data due to rounding. Three possible modifications of the partial likelihood to adjust for the simultaneous failure of a elements of the risk set $\mathcal{R}_j$ at time y_j are to include a term corresponding to each failure occurring first, to compute the exact probability of a failures, and to use an approximation to this. Thus (10.59) is replaced by one of

$$\prod_{i=1}^{a} \frac{\xi_i}{\sum_{k \in \mathcal{R}_j} \xi_k}, \quad \frac{\prod_{i=1}^{a} \xi_i}{\sum \prod_{k=1}^{a} \xi_{l_k}}, \quad \prod_{i=1}^{a} \frac{\xi_i}{\sum_{k \in \mathcal{R}_j} \xi_k - \frac{i-1}{a}\sum_{k=1}^{a}\xi_k}, \qquad (10.60)$$

where the sum in the exact central formula is over all subsets of $\mathcal{R}_j$ of size a. The first of these arises from applying the profile likelihood argument above to tied data. In practice these corrections often give similar results.

Example 10.37 (Leukaemia data) Consider the data in Table 10.22 with $\xi = \exp\{\beta_0 + \beta_1 I(\text{Group} = 1) + \beta_2 x\}$. Now β_0 cancels from the partial likelihood, maximization of which gives $\widehat{\beta}_1 = -1.07\,(0.43)$ and $\widehat{\beta}_2 = 0.85\,(0.31)$. These are similar to the values for the exponential model, apart from the sign change because the hazard and mean survival time are inversely related. Note in particular that the standard errors barely differ, confirming our comments about the efficiency of partial likelihood estimation.

These data have 17 ties. The estimates above result from using the third, approximate, term in (10.60), while the second, exact, formula gives $\widehat{\beta}_1 = -1.08\,(0.45)$ and $\widehat{\beta}_2 = 0.90\,(0.34)$, and the simple first approximation gives $\widehat{\beta}_1 = -1.02\,(0.42)$ and $\widehat{\beta}_2 = 0.83\,(0.31)$. There is little to choose among these, but rather more to choose among the likelihood ratio statistics for inclusion of the two covariates, which are 15.6 using the second and third terms, and 14.6 using the first. The third term in (10.60) thus seems preferable to the first, as both require the same computational effort.

It may be useful to contrast two types of semiparametric procedure. Partial likelihood inference requires no assumptions about the baseline hazard and distribution function, and in a sense relaxes the vertical axis of Figure 10.21. Use of splines or other smoothing procedures can also be described as semiparametric, but it relaxes the horizontal axis of the figure, which relates the covariate and response. Spline terms can be introduced into the linear predictor of the proportional hazards model, replacing $\beta_2 x$ by $s(x)$, but this does not improve significantly on the exponential model. ∎

The baseline cumulative hazard and survivor functions may be estimated by

$$\widehat{H}_0(y) = \sum_{j:y_j \le y} \frac{d_j}{\sum_{i \in \mathcal{R}_j} \widehat{\xi}_i}, \quad \widehat{\mathcal{F}}_0(y) = \prod_{j:y_j \le y} \left(1 - \frac{d_j}{\sum_{i \in \mathcal{R}_j} \widehat{\xi}_i}\right), \tag{10.61}$$

where $\widehat{\xi}_j = \xi(\widehat{\beta}; x_j)$. These are needed to assess fit and to predict survival probabilities for individuals from a fitted model. The estimated survivor function for an individual with covariates x_+ is $\mathcal{F}_+(y) = \exp\{-\xi(\widehat{\beta}; x_+)\widehat{H}_0(y)\}$, from which the probability of survival beyond a given point can be read off, with standard errors found using the delta method.

The construction above extends to stratified data, with the baseline hazard varying between strata but the parameter being common to all strata. This is useful in checking proportionality of hazards.

Log rank test

We now briefly discuss use of the proportional hazards model to construct tests for equality of survival distributions. When $\xi(\beta; x) = \exp(x^{\mathsf{T}}\beta)$, the log partial likelihood (10.57) equals

$$\ell_{\mathrm{p}}(\beta) = \sum_{\text{failures}} \left[x_j^{\mathsf{T}}\beta - \log\left\{\sum_{i \in \mathcal{R}_j} \exp\left(x_i^{\mathsf{T}}\beta\right)\right\}\right] = \sum_{\text{failures}} \left\{x_j^{\mathsf{T}}\beta - \log A_j(\beta)\right\},$$

say, with first derivative

$$U(\beta) = \sum_{\text{failures}} \left\{ x_j - \frac{\sum_{i \in \mathcal{R}_j} x_i \exp\left(x_i^{\mathsf{T}}\beta\right)}{A_j(\beta)} \right\} = \sum_{\text{failures}} \left\{ x_j - \frac{B_j(\beta)}{A_j(\beta)} \right\}, \qquad (10.62)$$

say, and negative second derivative

$$J(\beta) = \sum_{\text{failures}} \left\{ \frac{\sum_{i \in \mathcal{R}_j} x_i x_i^{\mathsf{T}} \exp\left(x_i^{\mathsf{T}}\beta\right)}{A_j(\beta)} - \frac{B_j(\beta)B_j(\beta)^{\mathsf{T}}}{A_j(\beta)^2} \right\}. \qquad (10.63)$$

Suppose the data fall into two groups with respective hazard functions $h_0(y)$ and $h(y) = e^{\beta} h_0(y)$. Then a score test for $\beta = 0$, that is, equality of survival distributions, is obtained by letting x_j be scalar, with $x_j = 1$ or 0 indicating that failure j belongs to groups 1 or 0, and taking $U(0) \overset{\cdot}{\sim} N\{0, J(0)\}$ or equivalently $U(0)^2/J(0) \overset{\cdot}{\sim} \chi_1^2$. This is known as the *log rank test*.

Now $A_j(0)$ and $B_j(0)$ respectively equal

$$\sum_{i \in \mathcal{R}_j} \exp\left(x_i^{\mathsf{T}}\beta\right)\Bigg|_{\beta=0} = m_{0j} + m_{1j}, \qquad \sum_{i \in \mathcal{R}_j} x_i \exp\left(x_i^{\mathsf{T}}\beta\right)\Bigg|_{\beta=0} = m_{1j},$$

the total number of individuals and the number of group 1 individuals available to fail at time y_j. Thus

$$U(0) = \sum_{\text{failures}} \left(R_j - \frac{m_{1j}}{m_{0j} + m_{1j}} \right), \qquad J(0) = \sum_{\text{failures}} \frac{m_{0j}m_{1j}}{(m_{0j} + m_{1j})^2},$$

where $R_j = 1$ if the individual failing at time y_j belongs to group 1 and $R_j = 0$ otherwise. Hence the score statistic is a sum of centred binary variables. These are not independent but under mild conditions the normal limiting distribution above will nonetheless hold.

An alternative argument proceeds by cross-classifying the risk set at each failure time by group membership and failure/survival, and using the hypergeometric distribution for the number of group 1 failures conditional on the row and column totals in the resulting 2×2 table. This applies also when there are ties, and yields

$$U(0) = \sum_{\text{failures}} \left(R_j - \frac{m_{1j}a_j}{m_{0j} + m_{1j}} \right),$$

$$J(0) = \sum_{\text{failures}} \frac{m_{0j}m_{1j}a_j(m_{0j} + m_{1j} - a_j)}{(m_{0j} + m_{1j})^2(m_{0j} + m_{1j} - 1)},$$

with R_j now the number of group 1 failures at y_j among the total number of failures a_j at the jth failure time; see the discussion after (10.28). This reduces to the previous version when there are no ties, that is, $a_j \equiv 1$.

Example 10.38 (Mouse data) Figure 5.11 compares cumulative hazard functions for subsets of the data of Table 5.7. The values of $U(0)^2/J(0)$ for the left and right panels are 3.3 and 40.1, each to be treated as χ_1^2. The first has significance level 0.07, weak evidence that the distributions differ. The second strongly supports the visual impression of quite different distributions. ∎

The log rank test generalizes to quantitative covariates x, to multiple survival distributions, and to weighted sums

$$\sum_{\text{failures}} w_j \left(R_j - \frac{m_{1j} a_j}{m_{0j} + m_{1j}} \right),$$

where the w_j can depend on the failure times, on m_{0j}, m_{1j}, and on a_j. Such statistics can give better power against alternatives other than proportional hazards. Their variances may be found using ideas from Section 7.2.3.

Time-dependent covariates

Thus far we have supposed that the covariate vector x_j takes the same value throughout the period over which the jth individual is observed. This is appropriate for variables such as age on entry, sex, and summaries of medical history prior to entry to the study, but it is also necessary to be able to accommodate explanatory variables that vary during the study. Quantities such as a patient's blood pressure may be available at various points over the observation period, for example, or a treatment may be not allocated until well after the study has begun, or changed during the trial. In reliability trials the key explanatory variable may be cumulative stress, or perhaps instantaneous stress, both of which may change during the experiment.

The interpretation of effects of covariates that may be influenced by the treatments demands careful thought. Consider for example a study in which treatments for hypertension are compared, blood pressure being an explanatory variable. Use of initial blood pressure as a covariate should increase the precision with which the treatment effects can be estimated, but interest would focus on the treatments, the estimated effect of blood pressure being of little direct interest. Use of blood pressure monitored after treatment allocation, by contrast, would allow the analyst to assess the extent to which treatments affect survival by influencing blood pressure; the estimate might then be of prime concern.

Time-varying covariates may also be constructed for technical reasons, for instance to check adequacy of the proportional hazards assumption by including y or $\log y$ in the linear predictor.

Whatever the interpretation, use of time-dependent covariates leads to replacement of the p elements $x_{j1}, \ldots, x_{jp}$ of x_j by functions $x_{j1}(y), \ldots, x_{jp}(y)$, $0 \leq y \leq y_j$. These may be indicator variables, for example showing the treatment being applied at time y. The covariates are typically measured only at certain times, so the function $x_{jr}(y)$ is usually obtained by interpolation. Let $x_j(y)$ denote the $p \times 1$ vector $(x_{j1}(y), \ldots, x_{jp}(y))^{\text{T}}$. The hazard function $\xi\{\beta; x_j\} h_0(y)$ becomes $\xi\{\beta; x_j(y)\} h_0(y)$, and our previous argument shows that the log partial likelihood is

$$\ell_{\text{p}}(\beta) = \sum_{j=1}^{n} d_j \left(\xi\{\beta; x_j(y_j)\} - \log \left[\sum_{i \in \mathcal{R}_j} \xi\{\beta; x_i(y_j)\} \right] \right),$$

the outer sum being over failure times y_j. Thus rather than x_j, the covariates needed for the jth individual are $\{x_j(y_i) : y_i \leq y_j\}$, where the failure times y_i are those at which case j lies in the risk set.

Standard large-sample likelihood results may be used for inference on β.

Model checking

When the data contain two groups, a graphical check on proportional hazards may be based on their estimated cumulative hazard functions. If the cumulative hazard functions for the two groups are $H_0(y)$ and $H(y)$, then proportional hazards asserts that $H(y) = \xi(\beta; x)H_0(y)$. Thus $\log \widehat{H}(y) - \log \widehat{H}_0(y)$ should appear independent of y.

Various residuals can be defined. The modified Cox–Snell residuals (10.55) equal $r_j = \widehat{H}_j(y_j) + 1 - d_j$, where the estimated cumulative hazard function for the jth individual under a proportional hazards model with constant covariates is $\widehat{H}_j(y) = \xi(\widehat{\beta}; x_j)\widehat{H}_0(y_j)$ and $\widehat{H}_0$ is given at (10.61). Plots of the r_j for subsets of the observations may cast light on interactions, but are not useful for assessing distributional assumptions in the proportional hazards model because $H_0(y)$ is not specified parametrically.

If Y_j is a continuous random variable with censoring indicator D_j and cumulative hazard function $H_j(y)$, then $I(Y_j \leq y, D_j = 1) - H_j\{\min(y, Y_j)\}$ is a zero-mean continuous-time martingale; see page 552. With $y = \infty$ this gives $D_j - H_j(Y_j)$, and implies that a *martingale residual* may be constructed as $d_j - \widehat{H}_j(y_j) = 1 - r_j$, just the residual above apart from a location and sign change. The functional form of a covariate in a proportional hazards model with $\xi(\beta; x) = \exp(x^{\mathsf{T}}\beta)$ can be checked by plotting $1 - r_j$ computed with the covariate omitted against the covariate itself.

The strong negative skewness of martingale residuals can be reduced by transformation, giving *deviance residuals*

$$\operatorname{sign}\{d_j - \widehat{H}_j(y_j)\}[2\{\widehat{H}_j(y_j) - d_j - d_j \log \widehat{H}_j(y_j)\}]^{1/2},$$

which are useful for checking for outliers; they are formally equivalent to treating the D_j as Poisson variables with means $H_j(Y_j)$.

An approach based on (10.62) uses components of the contributions

$$x_j - \frac{\sum_{i \in \mathcal{R}_j} x_i \exp\left(x_i^{\mathsf{T}}\widehat{\beta}\right)}{\sum_{i \in \mathcal{R}_j} \exp\left(x_i^{\mathsf{T}}\widehat{\beta}\right)},$$

to the score vector, thus giving a residual for each covariate and for each individual seen to fail. These $p \times 1$ vectors can be scaled by pre-multiplication by $J_j(\widehat{\beta})$, where the $p \times p$ matrix $J_j(\beta)$ is the contribution to (10.63) from the jth failure. They are closely related to the influence measures (8.29) and (10.13), and plots of their components help to determine which of the observations are influential for elements of $\widehat{\beta}$. They are also useful for assessing adequacy of proportional hazards. A natural way in which hazards might not be proportional is $h(y) = h_0(y) \exp\{x^{\mathsf{T}}\beta(y)\}$, that is, the coefficient of x depends on time. If this is the case, and there are no tied failures, then $\mathrm{E}(S_j) + \widehat{\beta} \doteq \beta(y_j)$, where S_j is a standardized version of the score contributions computed using only the risk set at time y_j. A non-constant plot of observed S_j against y_j suggests this type of model failure.

These and other diagnostics for the proportional hazards model can be extended to time-dependent covariates.

Example 10.39 (PBC data) Primary biliary cirrhosis (PBC) is a chronic fatal disease of the liver, with an incidence of about 50 cases per million. Controlled clinic trials are hard to perform with very rare diseases, so the double-blinded randomized trial conducted at the Mayo Clinic from 1974–1984 is a valuable resource for liver specialists. A total of 424 patients were eligible for the trial, and the 312 who consented to take part were randomized to be treated either with the drug D-penicillamine or with a placebo. Although basic data are available on all 424 patients, we consider only these 312 individuals. Covariates available on each of them at recruitment include the demographic variables sex and age; clinical variables, namely presence or absence of ascites, hepatomegaly, spiders, and a ternary varable edtrt whose values 0, 1/2, 1 indicate no, mild, and severe edema; and biochemical variables, namely levels of serum bilirubin (mg/dl), serum cholesterol (mg/dl), albumin (gm/dl), urine copper (μg/day), alkaline phosphatase (U/ml), SGOT (U/ml), and triglycerides (mg/dl), platelet count (coded), prothombin time (seconds), and the histologic stage of the disease (1–4). There are 28 missing values of serum cholesterol and 30 of triglycerides, and we ignore these covariates. Four missing values of platelets and two of urine copper were replaced by the medians of the remaining values; this should have little effect on the analysis. At the time at which the data considered here became available, 125 patients had died, with just 11 deaths not due to PBC, eight patients had been lost to follow-up, and 19 had undergone a liver transplant. As the response is time to death, these patients are regarded as censored.

The upper left panel of Figure 10.22 shows that estimated survivor functions for the patients with the drug and the placebo are very close, and it is no surprise that the log-rank statistic has value 0.1, insignificant when treated as χ_1^2. This is borne out by the estimated treatment effect of -0.057 (0.179) for a fit of the proportional hazards model with treatment effect only. Analysis stratified by sex gives an estimate of -0.045 (0.179). Neither differs significantly from zero. The corresponding baseline survival function estimates in the upper right panel of Figure 10.22 suggest no need to stratify.

Similar analyses for subgroups of the data and the corresponding log-rank statistics also show no significant treatment effects.

Having established that treatment has no effect on survival, we try constructing a model for prediction of survivor functions for new patients. This should be useful in assessing for whom liver transplant is a priority. The first step is to see which readily accessible covariates are highly predictive of survival. We exclude histologic stage, which requires a liver biopsy, and urine copper and SGOT, which are frequently unmeasured. The product-limit estimates and log rank statistics show strong dependence of failure on the other variables individually, so we fit a proportional hazards model

Edema is the accumulation of fluids in body tissues.

| Variable | Estimate (SE) | | | |
	Full	Reduced	Transformed	Final
age	0.028 (0.009)	0.030 (0.009)	0.033 (0.009)	0.041 (0.009)
alb	−0.97 (0.027)	−1.09 (0.24)	−3.06 (0.72)	−3.07 (0.72)
alkphos	0.015 (0.035)			
ascites	0.29 (0.31)			
bili	0.11 (0.02)	0.11 (0.02)	0.88 (0.10)	0.88 (0.10)
edtrt	0.69 (0.32)	0.77 (0.31)	0.79 (0.30)	0.69 (0.30)
hepmeg	0.49 (0.22)	0.50 (0.22)	0.25 (0.22)	
platelet	−0.61 (1.02)			
protime	0.24 (0.08)	0.25 (0.08)	3.01 (1.02)	3.57 (1.13)
sex	−0.48 (0.26)	−0.55 (0.25)		
spiders	0.29 (0.21)	0.30 (0.21)		

Table 10.23 Parameter estimates and standard errors for proportional hazards models fitted to the PBC data. The full fit is reduced by backwards elimination. In the last two columns log transformation is applied to alb, bili, and protime.

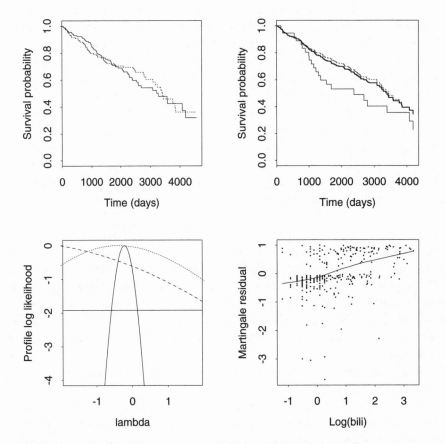

Figure 10.22 PBC data analysis (Fleming and Harrington, 1991). Top left: product-limit estimates for control (solid) and treatment (dots) groups. Top right: estimates of baseline survivor function for data stratified by sex, men (dots), women (solid). The heavy line shows the unstratified estimate. Lower left: profile likelihood for Box–Cox transformations of bilirubin (solid), albumin (dots), and prothrombin time (dashes); the horizontal line indicates 95% confidence limits for the transformation parameter. Lower right: martingale residuals from the model with terms age, log(alb), edtrt, log(protime) against log bilirubin, and lowess smooth with $p = 2/3$.

with all but the excluded covariates. Table 10.23 suggests that serum bilirubin is most significant and that several other covariates can be dropped. Backward selection based on AIC leads to the reduced model in the table. The likelihood ratio statistic for comparison of the two models is 1.22, plainly insignificant. Dropping sex and `spiders` also leads to a likelihood ratio statistic of 7.29, with significance level 0.20 when treated as χ_5^2. Bearing in mind the tendency of AIC to overfit, we now ignore these covariates.

To investigate whether transformation is worthwhile we apply the Box–Cox approach (Example 8.23) to `alb`, `bili`, and `protime`. The lower left panel of Figure 10.22 clearly indicates log transformation of `bili`, but not of the other variables. The need for transformation of `bili` can also be assessed through the plot of martingale residuals obtained when it is dropped, given in the lower right panel of the figure. Note the strong negative skewness of the residuals. The near-linearity of the lowess smooth shows the appropriateness of the transformation. The corresponding plot against `bili` itself is harder to read because the points are bunched towards zero. The plots for `alb` and `protime` are more ambiguous. If we take logs of all three variables, then the maximized log partial likelihood increases by 13.8 and `hepmeg` can be dropped; see Table 10.23.

A model with terms `age+log(alb)+log(bili)+edtrt+log(protime)` is medically plausible. As the disease progresses, the liver's ability to produce albumin decreases, leading to the negative coefficient for `alb`, while damage to the bile ducts reduces excretion of bilirubin and so increases its level in the body. Edema is often associated with the later stages of the disease, while prothrombin is decreased, leading to slower clotting of the blood. Finally and unsurprisingly, risk increases with age.

The upper panels of Figure 10.23 show deviance residuals plotted against age and prothrombin time. Inspection of those in the left panel lying outside the 0.01 and 0.99 normal quantiles reveals an error in the data coding; case 253 has residual -2.55 but his age should be 54.4 rather than 78.4. The right panel shows an unusually high prothrombin time of 17.1, which should have been 10.7. The estimates after these corrections are shown in the final column of Table 10.23.

The lower left panel of Figure 10.23 shows the scaled scores plotted against prothrombin time. There is some suggestion of non-proportionality, but it is too limited to suggest model failure. Such plots for the other variables cast no doubt on proportionality of hazards, and we accept the model.

To illustrate prediction, consider an individual with `age=60`, `alb=4`, `bili=1`, `edtrt=0`, and `protime=8`, for whom $x^{\mathsf{T}}\widehat{\beta} = -1.618$ and whose hazard is reduced by a factor $\exp(x^{\mathsf{T}}\widehat{\beta}) = 0.20$ compared to baseline. Setting `edtrt=1` and `bili=20` gives estimated risk scores of 0.4 and 2.8. The lower right panel of Figure 10.23 shows how the survivor functions then vary. The median estimated lifetime in each case can be found by solving for y the equation $\widehat{\mathcal{F}}_0(y)^{\exp(x^{\mathsf{T}}\widehat{\beta})} = 0.5$. ∎

The proportional hazards model has been broadened in many directions. Suppose, for instance, that individuals move between states 1 and 2 and back again, baseline time-dependent transition rates $\gamma_{12}(y)$ and $\gamma_{21}(y)$ being modified to $\gamma_{12}(y)\xi_{12}(\beta; x)$ and $\gamma_{21}(y)\xi_{21}(\beta; x)$ for an individual with explanatory variables x. The partial likelihood for β is a product of terms corresponding to each of the observed transitions between states. For instance, the contribution from transition $1 \to 2$ at time y by an

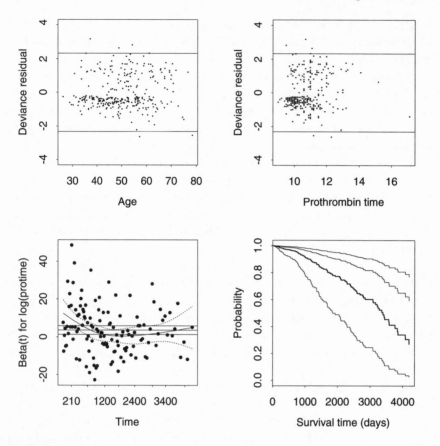

Figure 10.23 PBC data analysis. Upper panels: deviance residuals plotted against age and prothrombin time, with horizontal lines showing 0.01 and 0.99 standard normal quantiles. Lower left: scaled scores S_j^* plotted against prothrombin time, with lowess smooth and approximate 0.95 pointwise confidence bands (curved lines). Also shown are overall estimate and 0.95 confidence interval (horizontal lines). Lower right: baseline survivor function estimate (heavy), with predicted survivor functions for individuals with risk factors 0.2, 0.4, and 2.8 (top to bottom).

individual with covariates x_j is

$$\frac{\gamma_{12}(y)\xi_{12}(\beta;x_j)}{\sum \gamma_{12}(y)\xi_{12}(\beta;x_k)} = \frac{\xi_{12}(\beta;x_j)}{\sum \xi_{12}(\beta;x_k)},$$

the sum being over individuals in state 1 at time y. Individuals unobserved at y, or not in state 1, do not appear in the sum. Such extensions of partial likelihood enable inference for many types of partially observed and censored multi-state data, but details cannot be given here.

Counting processes and martingale residuals

Consider a random variable Y with censoring indicator D and hazard function $h(y)$, and let

This can be skipped at a first reading.

$$V(y) = I(Y \geq y), \quad N(y) = I(Y \leq y, D = 1),$$

be random variables that indicate whether Y is in view at time y, and whether failure has been observed by y. As $V(y)$ is left-continuous, its value at time y can be predicted the moment before, y^-, whereas the counting process $N(y)$ is right-continuous and so is not predictable. Let $\{\mathcal{H}_y : y \geq 0\}$ denote the history of the process up to time y. This is known as a filtration or increasing collection of sigma-algebras: $\mathcal{H}_x \subset \mathcal{H}_y$ for

$x \leq y$; knowledge accumulates. Define also $dN(y) = N\{(y + dy)^-\} - N(y)$, which equals 1 if failure is observed to occur at y, and otherwise equals 0. Then

$$E\{dN(y) \mid \mathcal{H}_{y^-}\} = \Pr\{dN(y) = 1 \mid \mathcal{H}_{y^-}\} = h(y)V(y), \quad y \geq 0; \quad (10.64)$$

the mean failure rate at y can be predicted from the history to y^-. However potential dependence on $\mathcal{H}_{y^-}$ makes $h(y)V(y)$ a random variable.

Now (10.64) implies that $dM(y) = dN(y) - h(y)V(y)$ is a zero-mean continuous-time martingale with respect to $\mathcal{H}_{y^-}$, for all $y > 0$. This implies that

$$M(y) = \int_0^y dM(y) = N(y) - \int_0^y h(u)V(u) \, du = N(y) - H\{\min(y, Y)\},$$

has the property that for any $y \geq x$,

$$E\{M(y) \mid \mathcal{H}_x\} - M(x) = E\left\{\int_x^y dM(u) \,\middle|\, \mathcal{H}_x\right\}$$

$$= \int_x^y E[E\{dM(u) \mid \mathcal{H}_{u^-}\} \mid \mathcal{H}_x] = 0,$$

and is therefore a martingale. Thus $E\{M(y)\} = 0$ for all y, and in particular $E\{M(\infty)\} = E\{D - H(Y)\} = 0$.

Let independent variables $Y_1, \ldots, Y_n$ with cumulative hazard functions $H_j(y)$ and censoring indicators $D_1, \ldots, D_n$ be observed, and set $V_j(y) = I(Y_j \geq y)$ and $N_j(y) = I(Y_j \leq y, D_j = 1)$. The corresponding continuous-time martingale is $M_j(y) = N_j(y) - H_j\{\min(y, Y_j)\}$, from which martingale residuals are obtained by setting $y = \infty$ and replacing unknowns with estimates.

Developments of the above formulation are central to mathematical treatments of time-to-event data, references to which are given in Section 10.9.

Exercises 10.8

1 Show that if Y is continuous with cumulative hazard function $H(y)$, then $H(Y)$ has the unit exponential distribution. Hence establish that $E\{H(Y) \mid Y > c\} = 1 + H(c)$, and explain the reasoning behind (10.55).

2 Let Y be a positive continuous random variable with survivor and hazard functions $\mathcal{F}(y)$ and $h(y)$. Let $\psi(x)$ and $\chi(x)$ be arbitrary continuous positive functions of the covariate x, with $\psi(0) = \chi(0) = 1$. In a proportional hazards model, the effect of a non-zero covariate is that the hazard function becomes $h(y)\psi(x)$, whereas in an accelerated life model, the survivor function becomes $\mathcal{F}\{y\chi(x)\}$. Show that the survivor function for the proportional hazards model is $\mathcal{F}(y)^{\psi(x)}$, and deduce that this model is also an accelerated life model if and only if

$$\log \psi(x) + G(\tau) = G\{\tau + \log \chi(x)\},$$

where $G(\tau) = \log\{-\log \mathcal{F}(e^\tau)\}$. Show that if this holds for all τ and some non-unit $\chi(x)$, we must have $G(\tau) = \kappa\tau + \alpha$, for constants κ and α, and find an expression for $\chi(x)$ in terms of $\psi(x)$. Hence or otherwise show that the classes of proportional hazards and accelerated life models coincide if and only if Y has a Weibull distribution.

3   In the usual notation for a linear regression model, $X^T(y - X\widehat{\beta}) = 0$. By writing the partial likelihood corresponding to (10.62) as $\sum_{j=1}^{n} d_j\{x_j^T\beta - \log A_j(\beta)\}$, show that

$$\sum_{j=1}^{n} x_j\left\{d_j - \exp\left(x_j^T\widehat{\beta}\right)\widehat{H}_0(y_j)\right\} = 0.$$

Which type of residual for a proportional hazards model is analogous to the raw residual in a linear model?

4   Suppose that survival data data consist of independent observations $(Y_j, C_j), j = 1 \ldots, n$, where Y_j is an exponential random variable with mean $\exp(x_j^T\beta)$, censored at random, and the censoring indicator is C_j, which equals 0 if Y_j is censored and equals 1 otherwise. Show that the likelihood for these data is the same as if the counts C_j had Poisson distributions with means $y_j \exp(-x_j^T\beta)$. Hence show that maximum likelihood estimates for the censored data model, and their standard errors based on observed information, can be obtained by regarding the censoring variable as having the Poisson distribution with log link function and offset $\log y_j$.

5   Write down the partial likelihood contributions from failure times $y = 1, 2$, for the data in Table 10.22, using the model $\xi = \exp\{\beta_0 + \beta_1 I(\text{Group} = 1) + \beta_2 x\}$.

6   Suppose that the continuous-time proportional hazards model holds, but that the failure times are grouped into intervals $0 = u_0 < u_1 < \cdots < u_m = \infty$. Show that the corresponding grouped hazards

$$h_i(x) = \Pr(Y < u_i \mid Y \geq u_{i-1}; x), \quad i = 1, \ldots, m,$$

satisfy

$$\log\{1 - h_i(x)\} = \xi(\beta; x)\log\{1 - h_i(0)\},$$

and write down the corresponding log likelihood when $\xi(\beta; x_j) = \exp(x_j^T\beta)$. Hence find the maximum likelihood estimator of β when the $h_i(0)$ are treated as nuisance parameters. Does this have the usual properties if $n \to \infty$ and m is fixed?
(Prentice and Gloeckler, 1978)

10.9 Bibliographic Notes

Driven by the needs of applications, the literature on regression models has expanded hugely over the last 30 years, and most of the development has been in nonlinear modelling. Generalized linear models were first explicitly formulated by Nelder and Wedderburn (1972), though others had previously suggested special cases. The resulting conceptual unification of apparently disparate models has had a major influence on subsequent developments, not least because of the part played by the iterative weighted least squares algorithm, for which Green (1984) is a standard reference. McCullagh and Nelder (1989) give an excellent account of generalized linear models and their ramifications, while Dobson (1990) is more elementary. Shorter accounts of generalized linear models and corresponding diagnostics are Firth (1991), ?), and Davison and Tsai (1992). Jørgensen (1997b) describes more general classes of exponential family-like distributions.

Data with binary responses are discussed by Collett (1991) and by Cox and Snell (1989).

Bishop *et al.* (1975) and Fienberg (1980) are standard references to log-linear models, though their approach is rather different to that adopted here. Log-linear and marginal models are discussed in Chapter 6 of McCullagh and Nelder (1989), with more recent work by Liang *et al.* (1992), Glonek and McCullagh (1995), and others. Generalized estimating equations and marginal modelling are of great importance in longitudinal data, a good discussion of which is given in Diggle *et al.* (1994). Agresti (1984) discusses models for ordinal data.

Quasi-likelihood was introduced by Wedderburn (1974) in a seminal article. For subsequent developments see McCullagh (1991), the useful survey by Firth (1993), and Davison (2001). Heyde (1997) gives a longer more theoretical account. See also the bibliographic notes for Chapter 7.

There are now many books on semiparametric regression. Bowman and Azzalini (1997) give an elementary account of kernel methods, with an applied emphasis, while Wand and Jones (1995) contains a more theoretical treatment, and Simonoff (1996) gives an excellent general discussion. Fan and Gijbels (1996) describe the theory of local polynomial modelling in detail, while the more practical account by Loader (1999) includes references to purpose-written software for local fitting. Hastie and Loader (1993) give a shorter more intuitive account of the properties of these methods. The account of spline methods in Section 10.7.2 is based on Green and Silverman (1994). Hastie and Tibshirani (1990) give a book-length account of generalized additive models. Wood (2000) gives a recent account of smoothing parameter selection for penalized likelihood procedures.

Survival data analysis has developed very rapidly over the last three decades. A major impetus was given by the introduction of the proportional hazards model by Cox (1972), which led to greatly increased interest in the area, the use of point process methods by Aalen (1978), and a flood of subsequent work. Fleming and Harrington (1991) and Andersen *et al.* (1993) are standard references to this topic, the latter also treating event history analysis in other areas. Therneau and Grambsch (2000) is an excellent recent book highlighting computation for proportional hazards models and their extensions, while Hougaard (2000) is an account of more advanced topics such as frailty and multistate models. See also Klein and Moeschberger (1997). Although most developments have centred on the proportional hazards model, it is not always suitable in practice, and many other possibilities have been suggested. See also the bibliographic notes to Chapter 5.

10.10 Problems

1 Suppose that Y has a density with generalized linear model form

$$f(y; \theta, \phi) = \exp \left\{ \frac{y\theta - b(\theta)}{a(\phi)} + c(y; \phi) \right\},$$

where $\theta = \theta(\eta)$ and $\eta = \beta^{\mathrm{T}} x$.
(a) Show that the weight for iterative weighted least squares based on expected information is

$$w = b''(\theta)(d\theta/d\eta)^2 / a(\phi),$$

and deduce that $w^{-1} = V(\mu)a(\phi)\{dg(\mu)/d\mu\}^2$, where $V(\mu)$ is the variance function, and that the adjusted dependent variable is $\eta + (y - \mu)dg(\mu)/d\mu$.

Note that initial values are not required for β, since w and z can be determined in terms of η and μ; initial values can be found from y as $\mu^1 = y$ and $\eta^1 = g(y)$.

(b) Give explicit formulae for the weight and adjusted dependent variable when $R = mY$ is binomial with denominator m and probability $\pi = e^\eta/(1 + e^\eta)$.

2   The independent observations Y_j, $j = 1, \ldots, n$, have Poisson distributions with means μ_j, where $g(\mu_j) = \eta_j$, $g(\cdot)$ is the link function, and η_j is the linear predictor $x_j^\mathrm{T}\beta$. The x_j are $p \times 1$ vectors of known covariates such that the matrix X whose jth row is x_j^T has rank p. Show that the likelihood equation for the maximum likelihood estimator $\widehat{\beta}$ of β can be written

$$X^\mathrm{T}s(\widehat{\beta}) = 0,$$

and hence derive the iterative weighted least squares algorithm for estimation of β, giving explicit formulae for the weight matrix and the adjusted dependent variable.

In a set of data on faults in lengths of textile, there were y faults in independent samples of length x. Five pairs (y, x) were $(6, 5.5)$, $(4, 6.5)$, $(17, 8.3)$, $(9, 3.8)$, and $(14, 7.2)$. Suppose that the Y_j are independent Poisson variables with means η_j, and $\eta_j = \beta_0 + \beta_1 x_j$. Give the link function for this model, verify that the maximum likelihood estimates are $\widehat{\beta}_0 = 1.006$ and $\widehat{\beta}_1 = 1.437$, and calculate their asymptotic covariance matrix. Is there evidence that $\beta_0 \neq 0$?

3   For a generalized linear model with known dispersion parameter ϕ and canonical link function, write the deviance as $\sum_{j=1}^n d_j^2$, where d_j^2 is the contribution from the jth observation. Also let

$$u_j(\beta) = \partial \log f(y_j; \eta_j, \phi)/\partial \eta_j, \quad w_j = -\partial^2 \log f(y_j; \eta_j, \phi)/\partial \eta_j^2,$$

denote the elements of the score vector and observed information, let X denote the $n \times p$ matrix whose jth row is x_j^T, where $\eta_j = \beta^\mathrm{T}x_j$, and let H denote the matrix $W^{1/2}X(X^\mathrm{T}WX)^{-1}X^\mathrm{T}W^{1/2}$, where $W = \mathrm{diag}\{w_1, \ldots, w_n\}$.   Recall Exercise 8.5.2.

(a) Let $\widehat{\beta}_{(k)}$ be the solution of the likelihood equation when case k is deleted,

$$\sum_{j \neq k} x_j u_j\left(\widehat{\beta}_{(k)}\right) = 0, \tag{10.65}$$

and let $\widehat{\beta}$ be the maximum likelihood estimate based on all n observations. Use first-order Taylor series expansion of (10.65) about $\widehat{\beta}$ to show that

$$\widehat{\beta}_{(k)} \doteq \widehat{\beta} - (X^\mathrm{T}WX)^{-1}x_k \frac{u_k(\widehat{\beta})}{1 - h_{kk}}.$$

Express $\widehat{\beta}_{(k)}$ in terms of the standardized Pearson residual $r_{Pk} = u_k/\{w_k(1 - h_{kk})\}^{1/2}$.

(b) Use a second order Taylor series expansion of the deviance to show that the change in the deviance when the kth case is deleted is approximately

$$r_{Gk}^2 = (1 - h_{kk})r_{Dk}^2 + h_{kk}r_{Pk}^2,$$

$r_{Gk} = \mathrm{sign}(y_k - \widehat{\mu}_k)\sqrt{r_{Gk}^2}$ is called a jackknifed deviance residual.

where r_{Dk} is the standardized deviance residual $d_k/(1 - h_{kk})^{1/2}$.

(c) Suppose models A and B have deviances D_A and D_B. Use (b) to find an expression for the change in the likelihood ratio statistic $D_A - D_B$, when the kth case is deleted.

(d) Show that your results (a)–(c) are exact in models with normal errors.

4   In a study on the relation between social class, education, and income, m independently sampled individuals are classified according to the social class of their parents, their income group, and their level of education. m is fixed in advance. The number of individuals with parents in class j, income group k, and with educational level l is y_{jkl}, where $j = 1, \ldots, J$, $k = 1, \ldots, K$ and $l = 1, \ldots, L$. Show that the joint multinomial distribution for the y_{jkl} which is appropriate to this sampling scheme is equivalent to that derived by treating the y_{jkl} as independent Poisson random variables with means μ_{jkl}, conditional on $\sum_{kjl} y_{jkl} = m$, and give the multinomial probabilities in terms of the μ_{jkl}.

One possible model for such data would be that the multinomial probabilities π_{jkl} may be written in the form $\alpha_j(\beta\gamma)_{kl}$, where $\sum_j \alpha_j = \sum_{kl}(\beta\gamma)_{kl} = 1$. Show that the maximum likelihood estimate for α_j is then $y_{j\cdot\cdot}/m$, where a dot indicates summation over the corresponding subscript, and find the maximum likelihood estimates of the $(\beta\gamma)_{kl}$. Derive the deviance statistic to test the adequacy of this model, and show that for large m it is equivalent to

$$\frac{1}{m}\sum_{jkl} \frac{(my_{jkl} - y_{j\cdot\cdot}y_{\cdot kl})^2}{y_{j\cdot\cdot}y_{\cdot kl}},$$

when the model is correct.

5 The rate of growth of an epidemic such as AIDS for a large population can be estimated fairly accurately and treated as a known function $g(t)$ of time t. In a smaller area where few cases have been observed the rate is hard to estimate because data are scarce. However predictions of the numbers of future cases in such an area must be made in order to allocate resources such as hospital beds. A simple assumption is that cases in the area arise in a non-homogeneous Poisson process with rate $\lambda g(t)$, for which the mean number of cases in period (t_1, t_2) is $\lambda \int_{t_1}^{t_2} g(t)dt$. Suppose that $N_1 = n_1$ individuals with the disease have been observed in the period $(-\infty, 0)$, and that predictions are required for the number N_2 of cases to be observed in a future period (t_1, t_2).
(a) Find the conditional distribution of N_2 given $N_1 + N_2$, and show it to be free of λ. Deduce that a $(1 - 2\alpha)$ prediction interval (n_-, n_+) for N_2 is found by solving approximately the equations

$$\alpha = \Pr(N_2 \leq n_- | N_1 + N_2 = n_1 + n_-),$$
$$\alpha = \Pr(N_2 \geq n_+ | N_1 + N_2 = n_1 + n_+).$$

(b) Use a normal approximation to the conditional distribution in (a) to show that for moderate to large n_1, n_- and n_+ are the solutions to the quadratic equation

$$(1 - p)^2 n^2 + p(p - 1)\left(2n_1 + z_\alpha^2\right)n + n_1 p\left\{n_1 p - (1 - p)z_\alpha^2\right\} = 0,$$

where $\Phi(z_\alpha) = \alpha$ and

$$p = \int_{t_1}^{t_2} g(t)dt \Big/ \left\{\int_{t_1}^{t_2} g(t)dt + \int_{-\infty}^{0} g(t)dt\right\}.$$

(c) Find approximate 0.90 prediction intervals for the special case where $g(t) = 2^{t/2}$, so that the doubling time for the epidemic is two years, $n_1 = 10$ cases have been observed until time 0, and $t_1 = 0$, $t_2 = 1$ (next year) (Cox and Davison, 1989).

6 Let R_0 and R_1 be independent binomial variables with denominators m_0 and m_1 and probabilities π_0 and π_1, and let $\Delta = \{\pi_1(1 - \pi_0)\}/\{\pi_0(1 - \pi_1)\}$ be the odds ratio for the 2×2 table $(R_0, m_0 - R_0; R_1, m_1 - R_1)$. Let $A = R_0 + R_1$, and let $Y^{(s)} = Y(Y - 1)\cdots(Y - s + 1) = Y!/(Y - s)!$, with $Y^{(s)} = 0$ if $Y + 1 \leq s$.
(a) Show that $\mathrm{E}\{R_1^{(s)}(m_0 - R_0)^{(s)} \mid A = a\} = \Delta^s \mathrm{E}\{R_0^{(s)}(m_1 - R_1)^{(s)} \mid A = a\}$, and that when $\Delta = 1$, $\mathrm{E}(R_1^{(s)} \mid A = a) = m_1^{(s)} a^{(s)}/(m_0 + m_1)^{(s)}$.
(b) When $\Delta = 1$, show that

$$\mathrm{E}(R_1 \mid A = a) = \frac{m_1 a}{m_0 + m_1}, \quad \mathrm{var}(R_1 \mid A = a) = \frac{m_0 m_1 a(m_0 + m_1 - a)}{(m_0 + m_1)^2(m_0 + m_1 - 1)}.$$

(c) Show that unconditionally $\{\mathrm{E}(R_1)\mathrm{E}(m_0 - R_0)\}/\{\mathrm{E}(R_0)\mathrm{E}(m_1 - R_1)\} = \Delta$, whereas conditionally on A,

$$\{\mathrm{E}(R_1)\mathrm{E}(m_0 - R_0) + \mathrm{var}(R_1)\}/\{\mathrm{E}(R_0)\mathrm{E}(m_1 - R_1) + \mathrm{var}(R_1)\} = \Delta.$$

What does this indicate about the conditional maximum likelihood estimate of Δ relative to the unconditional one?

(d) Show that conditional on A, R_1 has a generalized linear model density with

$$b(\theta) = \log\left\{\sum_{u=u_-}^{u_+} \binom{m_1}{u}\binom{m_0}{a-u}e^{u\theta}\right\}, \quad u_- = \max\{0, a-m_0\}, \ u_+ = \min\{m_1, a\}.$$

Deduce that a score test of $\Delta = 1$ based on data from n independent 2×2 tables $(R_{0j}, m_{0j} - R_{0j}; R_{1j}, m_{1j} - R_{1j})$ is obtained by treating $\sum R_{1j}$ as approximately normal with mean and variance

$$\sum_{j=1}^{n} \frac{m_{1j}a_j}{m_{0j} + m_{1j}}, \quad \sum_{j=1}^{n} \frac{m_{0j}m_{1j}a_j(m_{0j} + m_{0j} - a_j)}{(m_{0j} + m_{1j})^2(m_{0j} + m_{1j} - 1)};$$

when continuity-corrected this is the *Mantel–Haenszel test*.
(Mantel and Haenszel, 1959)

7   Suppose that the cumulant-generating function of X can be written in the form $m\{b(\theta + t) - b(\theta)\}$. Let $E(X) = \mu = mb'(\theta)$ and let $\kappa_2(\mu)$ and $\kappa_3(\mu)$ be the variance and third cumulant respectively of X, expressed in terms of μ; $\kappa_2(\mu)$ is the variance function $V(\mu)$.
(a) Show that

$$\kappa_3(\mu) = \kappa_2(\mu)\kappa_2'(\mu) \quad \text{and} \quad \frac{\kappa_3}{\kappa_2^2} = \frac{d}{d\mu}\log\kappa_2(\mu).$$

Verify that the binomial cumulants have this form with $b(\theta) = \log(1 + e^\theta)$.
(b) Show that if the derivatives of $b(\theta)$ are all $O(1)$, then $Y = g(X)$ is approximately symmetrically distributed if g satisfies the second-order differential equation

$$3\kappa_2^2(\mu)g''(\mu) + g'(\mu)\kappa_3(\mu) = 0.$$

Show that if $\kappa_2(\mu)$ and $\kappa_3(\mu)$ are related as in (a), then

$$g(x) = \int^x \kappa_2^{-1/3}(\mu)d\mu.$$

(c) Hence find symmetrizing transformations for Poisson and binomial variables.
(McCullagh and Nelder, 1989, Section 4.8)

8   Show that the chi-squared density with known degrees of freedom ν,

$$\frac{y^{\nu/2-1}}{2^{\nu/2}\sigma^\nu\Gamma(\nu/2)}\exp\left(-\frac{y}{2\sigma^2}\right), \quad y > 0, \sigma > 0, \nu = 1, 2, \dots,$$

can be written in generalized linear model form (10.14), where θ and ϕ are functions, to be found, of ν and σ^2. Hence derive an expression for its rth cumulant, $r \geq 1$.
The yield of an industrial process was measured r_i times independently at m different temperatures t_i. The resulting yields $Z_{ij}, i = 1, \dots, m, j = 1, \dots, r_i$ may be assumed to be independent and normally distributed with both means ζ_i and variances τ_i dependent on t_i. Explain how the sums of squares $Y_i = \sum_{j=1}^{r_i}(Z_{ij} - \overline{Z}_i)^2$, where $\overline{Z}_i = r_i^{-1}\sum_{j=1}^{r_i}Z_{ij}$, may be used to assess the dependence of variance on temperature in a suitable generalized linear model. Briefly discuss the advantages and disadvantages of the canonical link function of your model.

9   At each of the doses $x_1 < x_2 < \cdots < x_n$ of a drug, m animals are tested. At dose x_i, r_i animals respond. Derive the maximum likelihood equation when the linear predictor takes the form $\eta = \beta x$ when a probit link function is used. If only one dosage $x_0 > 0$ is used, show that

$$\widehat{\beta} = \frac{1}{x_0}\Phi^{-1}(r/m), \quad \text{var}(\widehat{\beta}) \doteq \frac{\Phi(\beta x_0)\{1 - \Phi(\beta x_0)\}}{mx_0^2\{\phi(\beta x_0)\}^2},$$

where ϕ and Φ are the standard normal density and distribution functions. Plot the function $\Phi(\eta)\{1 - \Phi(\eta)\}/\phi(\eta)^2$ for η in the range $-3 \leq \eta \leq 3$, and comment on the implications for the choice of x_0 if there is some prior knowledge of the likely value of β.

Table 10.24 Simulated data with two covariates, binary response, and fitted values.

Case	x_1	x_2	y	$\hat{y}$
1	3.7	0.83	1	0.999
2	3.5	1.09	1	0.999
3	1.25	2.50	1	0.875
4	0.75	1.50	1	0.066
5	0.8	3.2	1	0.886
6	0.7	3.5	1	0.921
7	0.6	0.75	0	0.005
8	1.1	1.70	0	0.320
9	0.9	0.75	0	0.017
10	0.9	0.45	0	0.008

10   Let Y be binomial with probability $\pi = e^\lambda/(1 + e^\lambda)$ and denominator m.
(a) Show that $m - Y$ is binomial with $\lambda' = -\lambda$. Consider

$$\tilde{\lambda} = \log\left(\frac{Y + c_1}{m - Y + c_2}\right)$$

as an estimator of λ. Show that in order to achieve consistency under the transformation $Y \to m - Y$, we must have $c_1 = c_2$.
(b) Write $Y = m\pi + \sqrt{m\pi(1 - \pi)}Z$, where $Z = O_p(1)$ for large m. Show that

$$\mathrm{E}\{\log(Y + c)\} = \log(m\pi) + \frac{c}{m\pi} - \frac{1 - \pi}{2m\pi} + O(m^{-3/2}).$$

Find the corresponding expansion for $\mathrm{E}\{\log(m - Y + c)\}$, and with $c_1 = c_2 = c$ find the value of c for which $\tilde{\lambda}$ is unbiased for λ to order m^{-1}.
What is the connection to the empirical logistic transform?
(Cox, 1970, Section 3.2)

11   Arcturian society is surprisingly similar to ours, the main differences being that Arcturians have three eyes (left, centre, and right) and are better at quantum physics. Their statistics is relatively rudimentary. On a recent study visit to our planet an Arcturian statistician encountered marginal models and decided to use one for visual impairment data similar to those in Table 10.16. He set up a $2 \times 2 \times 2$ table of probabilities $(\pi_{000}, \pi_{001}, \pi_{010}, \pi_{011}, \pi_{100}, \pi_{101}, \pi_{110}, \pi_{111})$ and used logistic models with marginal probabilities $\pi_L = \pi_{100} + \pi_{101} + \pi_{110} + \pi_{111}$ and so forth, and odds ratios

He usually comes disguised as Elvis, but attends statistical congresses in the guise of an eminent statistician; this may account for the other-worldly discussion.

$$\gamma_{LC} = \frac{\Pr(L = C = 1)\Pr(L = C = 0)}{\Pr(L = 1, C = 0)\Pr(L = 0, C = 1)},$$

$$\gamma_{LR} = \frac{\Pr(L = R = 1)\Pr(L = R = 0)}{\Pr(L = 1, R = 0)\Pr(L = 0, R = 1)}.$$

Show that the corresponding odds ratio γ_{CR} may be expressed as

$$\frac{\Pr(C = R = 1)\{1 - \pi_C - \pi_R + \Pr(C = R = 1)\}}{\{\pi_C - \Pr(C = R = 1)\}\{\pi_R - \Pr(C = R = 1)\}},$$

and that $\Pr(C = R = 1)$ lies between $\min(\pi_C, \pi_R)$ and $\max(0, \Pr(L = C = 1) + \Pr(L = R = 1) - \pi_L)$. Deduce that if $\pi_L, \pi_C, \pi_R, \gamma_{LC}$ and γ_{LR} are fixed, then the range of values that γ_{CR} can take is limited by the other parameters. What problems of fitting and interpretation might be encountered with such a model? Compare this with the corresponding log-linear model.

12   The data in Table 10.24 are from an experiment with a binary response in which two covariates are fitted. The parameter estimates and their standard errors are $\hat{\beta}_0 = -9.530(3.224)$, $\hat{\beta}_1 = 3.882(1.425)$, and $\hat{\beta}_2 = 2.649(0.9121)$. The table gives the data and fitted values for the first ten of the 39 observations. The overall deviance for the model is 29.77.

Give a careful interpretation of the effect of the covariates on the response.

Verify that the fitted value for case 8 is correct.

Is the deviance a useful guide to the fit of the model?

13 (a) In (10.43), show that $p_{obs}(h) = Pr_0(y^T A y \geq 0)$, where A is a $n \times n$ real symmetric matrix.

(b) Let $y \sim N_n(\mu, \Omega)$, where $\Omega = LL^T$ and L is a non-singular lower triangular matrix. If $\mu = 0$, then show that $y^T A y \overset{D}{=} \sum_{j=1}^{n} \lambda_j U_j^2$, where $U_1, \ldots, U_n \overset{iid}{\sim} N(0, 1)$ and $\lambda_1 \geq \cdots \geq \lambda_n$ are the eigenvalues of $L^T A L$. Deduce that the rth cumulant of $y^T A y$ equals $\kappa_r = 2^{r-1}(r - 1)!\mathrm{tr}\{(\Omega A)^r\}$. Show that the same is true whenever μ lies in the null space of A.

(c) For a simple approximation to the distribution of $y^T A y$, we match its first three cumulants with those of the random variable $aW + c$, where $W \sim \chi_b^2$. Show that this gives $a = |\kappa_3|/(4\kappa_2)$, $b = 8\kappa_2^3/\kappa_3^2$ and $c = \kappa_1 - ab$. Outline how this can be used to approximate $p_{obs}(h)$.

(d) Compute the cumulant-generating function of $y^T A y$, and develop a saddlepoint approximation to its distribution.

(Azzalini et al., 1989; Azzalini and Bowman, 1993; Kuonen, 1999)

14 In the penalized least squares setup of Section 10.7.2, with $t_0 < t_1 < \cdots < t_n < t_{n+1}$, set $h_j = t_{j+1} - t_j$ for each $j = 1, \ldots, n - 1$, let $g_1, \ldots, g_n$ and $\gamma_1, \ldots, \gamma_n$ be arbitrary real numbers, and define

$$g(t) = \frac{(t - t_j)g_{j+1} + (t_{j+1} - t)g_j}{h_j}$$
$$-\frac{1}{6}(t - t_j)(t_{j+1} - t)\left\{\left(1 + \frac{t - t_j}{h_j}\right)\gamma_{j+1} + \left(1 + \frac{t_{j+1} - t}{h_j}\right)\gamma_j\right\}$$

on each interval $t_j \leq t \leq t_{j+1}$, $j = 1, \ldots, n - 1$.

(a) Show that on each such interval $g(t)$ is a cubic function with $g(t_j) = g_j$.

(b) Show that

$$\lim_{t \downarrow t_j} g'(t) = \frac{g_{j+1} - g_j}{h_j} - \frac{1}{6}h_j(2\gamma_j + \gamma_{j+1}),$$

$$\lim_{t \uparrow t_{j+1}} g'(t) = \frac{g_{j+1} - g_j}{h_j} + \frac{1}{6}h_j(\gamma_j + 2\gamma_{j+1}),$$

and

$$g''(t) = \frac{(t - t_j)\gamma_{j+1} + (t_{j+1} - t)\gamma_j}{h_j}, \quad g'''(t) = h_j^{-1}(\gamma_{j+1} - \gamma_j), \quad t_j \leq t \leq t_{j+1},$$

and hence deduce that $g''(t_j) = \gamma_j$.

(c) Show that

$$g'(t_1) = \frac{g_2 - g_1}{t_2 - t_1} - \frac{1}{6}(t_2 - t_1)\gamma_2, \quad g'(t_n) = \frac{g_n - g_{n-1}}{t_n - t_{n-1}} - \frac{1}{6}(t_n - t_{n-1})\gamma_{n-1},$$

and deduce that if $g(t)$ is to be a natural cubic spline, then we must define

$$g(t) = \begin{cases} g_1 - (t_1 - t)g'(t_1), & t \leq t_1, \\ g_n + (t - t_n)g'(t_n), & t \geq t_n, \end{cases}$$

independent of the values of t_0 and t_{n+1}. Deduce that $\gamma_1 = \gamma_n = 0$.

(d) We have seen that $g(t)$ is continuous, with continuous first derivative at t_1 and t_n, and that $\lim_{t \uparrow t_j} g''(t) = \lim_{t \downarrow t_j} g''(t) = \gamma_j$ for each j. If $g(t)$ is to be a natural cubic spline, it must also satisfy $\lim_{t \uparrow t_j} g'(t) = \lim_{t \downarrow t_j} g'(t)$ for each $j = 2, \ldots, n - 1$. Show that this implies that

$$\frac{g_{j+1} - g_j}{h_j} - \frac{g_j - g_{j-1}}{h_{j-1}} = \frac{1}{6}h_{j-1}\gamma_{j-1} + \frac{1}{3}(h_{j-1} + h_j)\gamma_j + \frac{1}{6}h_j\gamma_{j+1},$$
$$j = 2, \ldots, n-1,$$

Recall that $\gamma_1 = \gamma_n = 0$.

and that this system of equations may be rewritten as $Q^{\mathsf T} g = R\gamma$, where $g^{\mathsf T} = (g_1, \ldots, g_n)$, $\gamma^{\mathsf T} = (\gamma_2, \ldots, \gamma_{n-1})$, and Q and R have dimensions $n \times (n-2)$ and $(n-2) \times (n-2)$; it is necessary to label the columns of Q from 2 to $n-2$ and both rows and columns of R likewise, so their top left elements are respectively q_{12} and r_{22}.

(e) Use integration by parts to show that the integral in (10.46) may be written

$$\int_{t_0}^{t_{n+1}} \{g''(t)\}^2 \, dt = \sum_{j=1}^{n-1} \frac{\gamma_{j+1} - \gamma_j}{h_j}(g_j - g_{j+1}),$$

and deduce that the integral may be written $\gamma^{\mathsf T} Q^{\mathsf T} g = g^{\mathsf T} K g$.

(f) Write down Q and R when $n = 5$ and $h_1 = \cdots = h_{n-1} = 1$. Show that R is then invertible, and give $K = QR^{-1}Q^{\mathsf T}$.

(Green and Silverman, 1994, pp. 22–25)

15 (a) Let $U_1, \ldots, U_n$ be independent exponential variables with parameters $\lambda_1, \ldots, \lambda_n$, and let $H_0(u)$ be a differentiable monotone increasing function of $u > 0$, with derivative $h_0(u)$. Show that $Y_1 = H_0(U_1), \ldots, Y_n = H_0(U_n)$ have joint density

$$\prod_{j=1}^{n} \lambda_j h_0(y_j) \exp\{-\lambda_j H_0(y_j)\}.$$

(b) Show that the joint density of $U_1, \ldots, U_n$ may be written as

$$\prod_{j=1}^{n} \frac{\lambda_{(j)}}{\sum_{i=j}^{n} \lambda_{(i)}} \times \prod_{j=1}^{n} \left(\sum_{i=j}^{n} \lambda_{(i)}\right) \exp(-\lambda_{(j)} u_{(j)}), \tag{10.66}$$

where the elements of the rank statistic $R = \{(1), \ldots, (n)\}$ are determined by the ordering on the failure times, $U_{(1)} < \cdots < U_{(n)}$. Establish that the first term of this product is invariant to transformations $Y = H_0(U)$ but that the second is not.

(c) Suppose that $\lambda_j = \exp(x_j^{\mathsf T}\beta)$. Give an argument why inference for β should be based on the first term of (10.66) only.

16 In Figure 5.8, let y represent time on trial and t calendar time, and suppose that the hazard function for an individual with covariates x has form $h_0(y)h_0^\dagger(t)\exp(x^{\mathsf T}\beta)$, where $h_0(y)$ represents a baseline hazard for time on trial and $h_0^\dagger(t)$ a baseline hazard for calendar time. Discuss how partial likelihood inference might be generalized to account for inclusion of $h_0^\dagger(t)$, which is included to allow for changes in medical practice during the course of the trial.

17 Consider independent exponential variables Y_j with densities $\lambda_j \exp(-\lambda_j y_j)$, where $\lambda_j = \exp(\beta_0 + \beta_1 x_j)$, $j = 1, \ldots, n$, where x_j is scalar and $\sum x_j = 0$ without loss of generality.

(a) Find the expected information for β_0, β_1 and show that the maximum likelihood estimator $\widehat{\beta}_1$ has asymptotic variance $(nm_2)^{-1}$, where $m_2 = n^{-1}\sum x_j^2$.

(b) Under no censoring, show that the partial log likelihood for β_1 equals

$$-\sum_{j=1}^{n} \log\left\{\sum_{i=j}^{n} \exp\left(\beta_1 x_{(i)}\right)\right\},$$

where the elements of the rank statistic $R = \{(1), \ldots, (n)\}$ are determined by the ordering on the failure times, $y_{(1)} < \cdots < y_{(n)}$. Deduce the information in the partial likelihood is

$$I_R(\beta_1) = \sum_{j=1}^{n} \mathrm{E}_R\{m_{2,j}(\beta_1) - m_{1,j}(\beta_1)^2\},$$

where the expectation is over the distribution of R and

$$m_{k,j}(\beta_1) = \frac{\sum_{i=j}^n x_{(i)}^k \exp\left(\beta_1 x_{(i)}\right)}{\sum_{i=j}^n \exp\left(\beta_1 x_{(i)}\right)}.$$

Show that when $\beta_1 = 0$,

$$\mathrm{E}_R\{m_{2,j}(\beta_1)\} = m_2, \quad \mathrm{E}_R\{m_{1,j}(\beta_1)^2\} = \frac{m_2}{n-1}\sum_{i=1}^n \frac{i-1}{n-i+1},$$

and hence find the efficiency of partial likelihood estimation of β_1 relative to maximum likelihood estimation. Compute this for $n = 2, 5, 10, 20, 50, 100$, and comment.
(c) It can be shown that as $n \to \infty$ for small β_1, the relative efficiency equals $\exp(-m_2\beta_1^2)$. Show that in the two-sample problem with equal numbers of observations in each group and $x_j = \pm 1/2$, the relative efficiency exceeds 0.75 when $|\beta_1| < 1.07$, corresponding to a ratio of failure rates between the two groups in the range $(1/3, 3)$. Discuss.
(Kalbfleisch, 1974)

18 Suppose that n independent Poisson processes of rates $\lambda_j(y)$ are observed simultaneously, and that the m events occur at $0 < y_1 < \cdots < y_m < y_0$, in processes $j_1, \ldots, j_m$.
(a) Show that the probabilities that the first event occurs at y_1 and that given this it has type j_1 are respectively

$$\left\{\sum_{j=1}^n \lambda_j(y_1)\right\} \exp\left\{-\sum_{j=1}^n \int_0^{y_1} \lambda_j(u)\,du\right\}, \quad \frac{\lambda_{j_1}(y_1)}{\sum_{j=1}^n \lambda_j(y_1)}.$$

Hence interpret the quantities

$$\exp\left\{-\sum_{j=1}^n \int_0^{y_0} \lambda_j(u)\,du\right\} \prod_{i=1}^m \left\{\sum_{j=1}^n \lambda_j(y_i)\right\}, \quad \prod_{i=1}^m \frac{\lambda_{j_i}(y_i)}{\sum_{j=1}^n \lambda_j(y_i)}. \tag{10.67}$$

(b) Now suppose that $\lambda_j(y) = h_0(y)\xi\{\beta; x_j(y)\}V_j(y)$, where $h_0(y)$ is a baseline hazard function, $\xi\{\beta; x_j(y)\}$ depends on parameters β and time-varying covariates $x_j(y)$, and $V_j(y)$ is a predictable process, with $V_j(y) = 1$ if the jth process is in view at time y, and $V_j(y) = 0$ if not. Thus if the jth process is censored at time c_j, $V_j(y) = 0$, $y > c_j$. If $\mathcal{R}_i$ is the set $\{j : V_j(y_i) = 1\}$, show that the second term in (10.67) equals

$$\prod_{i=1}^m \frac{\xi\{\beta; x_{j_i}(y_i)\}}{\sum_{j \in \mathcal{R}_i} \xi\{\beta; x_j(y_i)\}}.$$

How does this specialize for time-varying explanatory variables in the proportional hazards model?

19 Two individuals with cumulative hazard functions $uH_1(y_1)$ and $uH_2(y_2)$ are independent conditional on the value u of a frailty U whose density is $f(u)$.
(a) For this shared frailty model, show that

$$\mathcal{F}(y_1, y_2) = \Pr(Y_1 > y_1, Y_2 > y_2) = \int_0^\infty \exp\left\{-uH_1(y_1) - uH_2(y_2)\right\} f(u)\,du.$$

If $f(u) = \lambda^\alpha u^{\alpha-1} \exp(-\lambda u)/\Gamma(\alpha)$, for $u > 0$ is a gamma density, then show that

$$\mathcal{F}(y_1, y_2) = \frac{\lambda^\alpha}{\{\lambda + H_1(y_1) + H_2(y_2)\}^\alpha}, \quad y_1, y_2 > 0,$$

and deduce that in terms of the marginal survivor functions $\mathcal{F}_1(y_1)$ and $\mathcal{F}_2(y_2)$ of Y_1 and Y_2,

$$\mathcal{F}(y_1, y_2) = \left\{\mathcal{F}_1(y_1)^{-1/\alpha} + \mathcal{F}_2(y_2)^{-1/\alpha} - 1\right\}^{-\alpha}, \quad y_1, y_2 > 0.$$

What happens to this joint survivor function as $\alpha \to \infty$?

(b) Find the likelihood contributions when both individuals are observed to fail, when one is censored, and when both are censored.

(c) Extend this to k individuals with parametric regression models for survival.

20 A positive stable random variable U has $E(e^{-sU}) = \exp(-\delta s^\alpha/\alpha)$, $0 < \alpha \leq 1$.

(a) Show that if Y follows a proportional hazards model with cumulative hazard function $u \exp(x^T\beta)H_0(y)$, conditional on $U = u$, then Y also follows a proportional hazards model unconditionally. Are β, α, and δ estimable from data with single individuals only?

(b) Consider a shared frailty model, as in the previous question, with positive stable U. Show that the joint survivor function may be written as

$$\mathcal{F}(y_1, y_2) = \exp\left(-\left[\{-\log \mathcal{F}_1(y_1)\}^{1/\alpha} + \{-\log \mathcal{F}_2(y_2)\}^{1/\alpha}\right]^\alpha\right), \quad y_1, y_2 > 0,$$

in terms of the marginal survivor functions $\mathcal{F}_1$ and $\mathcal{F}_2$. Show that if the conditional cumulative hazard functions are Weibull, $uH_r(y) = u\xi_r y^\gamma$, $\gamma > 0$, $r = 1, 2$, then the marginal survivor functions are also Weibull. Show also that the time to the first event has a Weibull distribution.

21 Consider individuals arising in k independent clusters of sizes $n_1, \ldots, n_k$, and such that conditional on the values $u_1, \ldots, u_k$ of unobserved frailties $U_1, \ldots, U_k$, the individuals in the ith cluster have survival times independently distributed according to a proportional hazards model with cumulative hazards $u_i\xi_{ij}H_0(y_{ij})$, for $j = 1, \ldots, n_i$, where ξ_{ij} is short-hand for $\xi(\beta; x_{ij})$, x_{ij} being a vector of explanatory variables. Let $h_0(y)$ be the derivative of $H_0(y)$, and suppose that the U_i are independent gamma variables with unit means and shape parameter θ.

(a) If the survival times are subject to non-informative censoring, show that the joint density of U_i and the (survival time, censoring indicator) pairs (Y_{ij}, D_{ij}) for the ith cluster is

$$\prod_{j=1}^{n_i} \{u_i\xi_{ij}h_0(y_{ij})\}^{d_{ij}} \times \exp\left\{-\sum_{j=1}^{n} u_i\xi_{ij}H_0(y_{ij})\right\} \times \frac{\theta^\theta u_i^{\theta-1}}{\Gamma(\theta)} \exp(-\theta u_i),$$

and deduce that the conditional means of U_i and of $\log U_i$ given the observed data are $w_i(\theta, \beta) = A_i/B_i$ and $\psi(A_i) - \log B_i$, where

$$A_i = \theta + d_{i\cdot}, \quad B_i = \theta + \sum_{j=1}^{n_i} \xi_{ij}H_0(y_{ij}), \quad d_{i\cdot} = \sum_{j=1}^{n_i} d_{ij}, \quad \psi(\alpha) = d\log\Gamma(\alpha)/d\alpha.$$

Discuss the merits and demerits (if any) of inference in terms of $\psi = \theta^{-1}$: what happens as $\psi \to 0$?

(b) Show that a step of the EM algorithm for estimation of (θ, β) involves updating (θ', β') by maximization of $\ell_1(\beta, H_0) + \ell_2(\theta)$ over β, H_0, and θ, where

$$\ell_1(\beta, H_0) = \sum_{i=1}^{k} \sum_{j=1}^{n_i} [d_{ij}\{\log \xi_{ij} + \log h_0(y_{ij})\} - w_i'\xi_{ij}H_0(y_{ij})],$$

$$\ell_2(\theta) = \sum_{i=1}^{k} \{(\theta + d_{i\cdot} - 1)(\psi(A_i') - \log B_i') - A_i\theta/B_i'\} + k\{\theta \log \theta - \log \Gamma(\theta)\},$$

$w_i' = w_i(\theta', \beta')$ and A_i' and B_i' are evaluated at (θ', β'). Extend the argument leading to (10.57) to establish that the step for β involves maximizing the partial likelihood that is a product over individuals of terms

$$\left\{\frac{w_i'\xi(\beta; x_{ij})}{\sum_{k \in \mathcal{R}_{ij}} w_i'\xi(\beta; x_k)}\right\}^{d_{ij}},$$

with the risk set $\mathcal{R}_{ij}$ containing those individuals from every cluster available to fail at failure time y_{ij}. When $\xi(\beta; x) = \exp(x^T\beta)$, show that this amounts to using an offset in the proportional hazards model. Find the form of $\widehat{H}_0(y)$, and give an algorithm for estimation of β and θ.

(c) Show that the joint survivor function for the individuals in a cluster is

$$\Pr(Y_{i1} > y_{i1}, \ldots, Y_{in_i} > y_{in_i}) = \theta^\theta \left\{ \theta + \sum_{j=1}^{n_i} \xi_{ij} H_0(y_j) \right\}^{-\theta},$$

and hence give the log likelihood contribution from (y_{ij}, d_{ij}), for $j = 1, \ldots, n_i$. Explain how to use this to obtain the observed information matrix for θ and β based on the estimates obtained in (b).
(Klein, 1992)

22 Let $Y_1, \ldots, Y_n$ be independent exponential variables with hazards $\lambda_j = \exp(\beta^T x_j)$.
(a) Show that the expected information for β is $X^T X$, in the usual notation.
(b) Now suppose that Y_j is subject to uninformative right censoring at time c_j, so that y_j is a censoring time or a failure time as the case may be. Show that the log likelihood is

$$\ell_U(\beta) = \sum_f \beta^T x_j - \sum_{j=1}^n \exp(\beta^T x_j) y_j,$$

where $\sum_f$ denotes a sum over observations seen to fail. If the jth censoring-time is exponentially distributed with rate κ_j, show that the expected information for β is $X^T X - X^T C X$, where $C = \mathrm{diag}\{c_1, \ldots, c_n\}$, and $c_j = \kappa_j/(\kappa_j + \lambda_j)$ is the probability that the jth observation is censored. What is the implication for estimation of β if the c_j are constant?
(c) Sometimes a variable W_j has been measured which can act as a surrogate response variable for censored individuals. We formulate this as $W_j = Z_j/U_j$, where Z_j is the unobserved remaining life-time of the jth individual from the moment of censoring, and U_j is a noise component which has a fixed distribution independent of the censoring time and of x_j. Owing to the exponential assumption, the excess life Z_j is independent of Y_j if censoring occurred. If U_j has gamma density

$$\alpha^\kappa u^{\kappa-1} \exp(-\alpha u)/\Gamma(\kappa), \quad \alpha, \kappa > 0, u > 0,$$

show that W_j has density

$$\lambda_j \kappa \alpha^\kappa/(\alpha + \lambda_j w)^{\kappa+1}, \quad w > 0.$$

Show that the log likelihood for the data, including the additional information in the W_j, is

$$\ell(\beta) = L_U(\beta) + \sum_c \left\{ \beta^T x_j + \log \kappa + \kappa \log \alpha - (\kappa + 1) \log \left(\alpha + e^{\beta^T x_j} w_j \right) \right\},$$

where $\sum_c$ denotes a sum over censored individuals, and we have assumed that α and κ are known. Show that the expected information for β is

$$X^T X - 2/(\kappa + 2) X^T C X,$$

and compare this with (b). Explain qualitatively in terms of the variability of the distribution of U why the loss of information decreases as κ increases.
(Cox, 1983)

11

Bayesian Models

Every statistical investigation takes place in a context. Information about what question is to be addressed will suggest what data are needed to give useful answers. Before the data are available, one role for this information is to suggest suitable probability models. There may also be information about the values of unknown parameters, and if this can be expressed as a probability density, an approach to inference based on Bayes' theorem is possible. Many statisticians make the stronger claim that this theorem provides the only entirely consistent basis for inference, and insist on its use.

This chapter outlines some aspects of the Bayesian approach to modelling. We first give an account of basic uses of Bayes' theorem and of the role and construction of prior densities. We then turn to inference, dealing with analogues of confidence intervals, tests, approaches to model criticism, and model uncertainty. Until recently computational difficulties placed realistic Bayesian modelling largely out of reach, but over the last 20 years there has been rapid progress and complex models can now be fitted routinely. Section 11.3 gives an account of Bayesian computation, first of analytical approaches based on integral approximations, and then of Monte Carlo methods. The chapter concludes with brief introductions to hierarchical and empirical Bayesian procedures.

11.1 Introduction

11.1.1 Bayes' theorem

Let $A_1, \ldots, A_k$ be events that partition a sample space, and let B be an arbitrary event on that space for which $\Pr(B) > 0$. Then Bayes' theorem is

$$\Pr(A_j \mid B) = \frac{\Pr(B \mid A_j)\Pr(A_j)}{\sum_{i=1}^{k} \Pr(B \mid A_i)\Pr(A_i)}.$$

This reverses the order of conditioning by expressing $\Pr(A_j \mid B)$ in terms of $\Pr(B \mid A_j)$ and the marginal probability $\Pr(B)$ in the denominator. For continuous random variables Y and Z,

$$f_{Z|Y}(z \mid y) = \frac{f_{Y|Z}(y \mid z) f_Z(z)}{\int f_{Y|Z}(y \mid z) f_Z(z) \, dz}, \tag{11.1}$$

provided the marginal density $f(y) > 0$, with integration replaced by summation for discrete variables.

Inference

To see how Bayes' theorem is used for inference, suppose that there is a probability model $f(y \mid \theta)$ for data y. In earlier chapters we have written $f(y \mid \theta) = f(y; \theta)$, but here we use the conditional notation to emphasize that the probability model is a density for the data given the value of θ. Suppose also that we are able to summarize our beliefs about θ in a *prior density*, $\pi(\theta)$, constructed separately from the data y. This implies that we think of the unknown value θ that underlies our data as the outcome of a random variable whose density is $\pi(\theta)$, just as our probability model is that the data y are the observed value of a random variable Y with density $f(y \mid \theta)$. Once the data have been observed, our beliefs about θ are contained in its conditional density given that $Y = y$,

$$\pi(\theta \mid y) = \frac{\pi(\theta) f(y \mid \theta)}{\int \pi(\theta) f(y \mid \theta) \, d\theta}. \tag{11.2}$$

This is the *posterior density* for θ given y. Note that $f(y \mid \theta)$ is the likelihood for θ based on y, so that in terms of θ, we have *posterior $\propto$ prior $\times$ likelihood*.

Frequentist inference treats θ as an unknown constant, whereas the Bayesian approach treats it as a random variable. We make this distinction explicit by using π to denote a density for θ, which thus has prior and posterior densities $\pi(\theta)$ and $\pi(\theta \mid y)$, rather than $f(\theta)$ and $f(\theta \mid y)$.

It is useful to note that any quantity that does not depend on θ cancels from the denominator and numerator of (11.2). This implies that if we can recognise which density is proportional to (11.2), regarded solely as a function of θ, we can read off the posterior density of θ. Furthermore, the factorization criterion (4.15) implies that the posterior density depends on the data solely through any minimal sufficient statistic for θ.

Example 11.1 (Bernoulli trials)   Suppose that conditional on θ, the data $y_1, \ldots, y_n$ are a random sample from the Bernoulli distribution, for which $\Pr(Y_j = 1) = \theta$ and $1 - \Pr(Y_j = 0) = -\theta$, where $0 < \theta < 1$. The likelihood is

$$L(\theta) = f(y \mid \theta) = \prod_{j=1}^{n} \theta^{y_j} (1-\theta)^{1-y_j} = \theta^r (1-\theta)^{n-r}, \quad 0 < \theta < 1,$$

where $r = \sum y_j$.

A natural prior here is the beta density with parameters a and b,

$$\pi(\theta) = \frac{1}{B(a,b)} \theta^{a-1} (1-\theta)^{b-1}, \quad 0 < \theta < 1, \quad a, b > 0, \tag{11.3}$$

where $B(a, b)$ is the beta function $\Gamma(a)\Gamma(b)/\Gamma(a+b)$. Figure 5.4 shows (11.3) for various values of a and b.

$\Gamma(a) = \int_0^\infty u^{a-1} e^{-u} \, du$ is the gamma function; see Exercise 2.1.3.

The posterior density of θ conditional on the data is given by (11.2), and is

$$\pi(\theta \mid y) = \frac{\theta^{r+a-1}(1-\theta)^{n-r+b-1}/B(a,b)}{\int_0^1 \theta^{r+a-1}(1-\theta)^{n-r+b-1}\,d\theta/B(a,b)}$$

$$\propto \theta^{r+a-1}(1-\theta)^{n-r+b-1}, \quad 0 < \theta < 1. \tag{11.4}$$

As (11.3) has unit integral for all positive a and b, the constant normalizing (11.4) must be $B(a+r, b+n-r)$. Therefore

$$\pi(\theta \mid y) = \frac{1}{B(a+r, b+n-r)}\theta^{r+a-1}(1-\theta)^{n-r+b-1}, \quad 0 < \theta < 1.$$

Thus the posterior density of θ has the same form as the prior: acquiring data has the effect of updating (a, b) to $(a + r, b + n - r)$. As the mean of the $B(a, b)$ density is $a/(a + b)$, the posterior mean is $(r + a)/(n + a + b)$, and this is roughly r/n in large samples. Hence the prior density inserts information equivalent to having seen a sample of $a + b$ observations, of which a were successes. If we were very sure that $\theta \doteq 1/2$, for example, we might take $a = b$ very large, giving a prior density tightly concentrated around $\theta = 1/2$, whereas taking smaller values of a and b would increase the prior uncertainty.

To illustrate this, suppose that $a = b = 1$, so that the initial density of θ is the uniform prior shown in the upper right panel of Figure 5.4, representing ignorance about θ. Then data with $n = 23$ and $r = \sum y_j = 14$ update the prior density to the posterior density in the lower right panel. ∎

The use of the beta density as prior for a model whose likelihood is proportional to $\theta^r(1 - \theta)^s$ leads to a posterior density that is also beta. This is an example of a *conjugate prior*, an idea discussed in Section 11.1.3.

When the parameter takes one of a finite number of values, labelled $1, \ldots, k$, with prior probabilities $\pi_1, \ldots, \pi_k$, the posterior density is the probability mass function

$$\Pr(\theta = j \mid y) = \frac{\pi_j f(y \mid \theta = j)}{\sum_{i=1}^k \pi_i f(y \mid \theta = i)}. \tag{11.5}$$

Example 11.2 (Diagnostic tests) A disease occurs with prevalence γ in a population, and θ indicates that an individual has the disease. Hence $\Pr(\theta = 1) = \gamma$, $\Pr(\theta = 0) = 1 - \gamma$. A diagnostic test gives a result Y, whose distribution is $F_1(y)$ for a diseased individual and $F_0(y)$ otherwise. The commonest type of test declares that a person is diseased if $Y > y_0$, say, where y_0 is fixed on the basis of past data. The probability that a person is diseased, given a positive test result, is

$$\Pr(\theta = 1 \mid Y > y_0) = \frac{\gamma\{1 - F_1(y_0)\}}{\gamma\{1 - F_1(y_0)\} + (1 - \gamma)\{1 - F_0(y_0)\}};$$

this is sometimes called the *positive predictive value* of the test. Its *sensitivity* and *specificity* are $1 - F_1(y_0)$ and $F_0(y_0)$. These are the probabilities of correct classification of diseased and non-diseased persons, while the false negative and false positive ratios are $F_1(y_0)$ and $1 - F_0(y_0)$. One aims to construct tests whose sensitivity and specificity are as high as possible. ∎

Prediction

Prediction of the value of a future random variable, Z, is straightforward when there is a prior density for the parameters. The joint density of Z and the data Y may be written

$$f(y, z) = \int f(z \mid y, \theta) f(y \mid \theta) \pi(\theta) \, d\theta,$$

and hence once Y has taken the value y, inference for Z is based on its *posterior predictive density*,

$$f(z \mid y) = \int f(z \mid y, \theta) \pi(\theta \mid y) \, d\theta = \frac{\int f(z \mid y, \theta) f(y \mid \theta) \pi(\theta) \, d\theta}{\int f(y \mid \theta) \pi(\theta) \, d\theta}. \qquad (11.6)$$

This is (11.1) expanded to make explicit the integration over the posterior density of θ.

Example 11.3 (Bernoulli trials) Heads occurs r times among the first n tosses in a sequence of independent throws of a coin. What is the probability of a head on the next throw?

Let θ be the unknown probability of a head and let $Z = 1$ indicate the event that the next toss yields a head. Conditional on θ, $\Pr(Z = 1 \mid y, \theta) = \theta$ independent of the data y so far. If the prior density for θ is beta with parameters a and b, then

$$\begin{aligned}
\Pr(Z = 1 \mid y) &= \int_0^1 \Pr(Z = 1 \mid \theta, y) \pi(\theta \mid y) \, d\theta \\
&= \int_0^1 \theta \, \frac{\theta^{a+r-1}(1-\theta)^{b+n-r-1}}{B(a+r, b+n-r)} \, d\theta \\
&= \frac{B(a+r+1, b+n-r)}{B(a+r, b+n-r)} = \frac{a+r}{a+b+n},
\end{aligned}$$

on using results for beta functions; see Example 11.1 and Exercise 2.1.3. As $n, r \to \infty$, this tends to the sample proportion of heads r/n, so the prior information is drowned by the sample. ∎

11.1.2 Likelihood principle

There have been many attempts to justify the use of Bayes' theorem as a basis for inference. One line of argument rests on axioms that individuals can use to make optimal decisions in the face of uncertain events, and leads to the view that probability is a measure of personal belief about the world, to be updated by additional knowledge using Bayes' theorem. An account of this would take us too far afield, and instead we outline another argument, which centres on principles intended to guide inference. The force of this is that two basic principles — the sufficiency and conditionality principles — together imply a third — the likelihood principle — which is difficult to apply except through Bayes' theorem. Many statisticians do subscribe to the first two, at least implicitly, thus setting them on the path to Bayesian inference.

We begin by introducing the notion of an experiment E, which yields data y, on which we wish to base inference about θ through the *evidence* $\mathrm{Ev}(E, y)$. The form of this function need not be specified; we merely suppose that it exists and contains all the information about θ based on E and y.

Sufficiency and conditionality principles

The form of the sufficiency principle we shall use is that if an experiment E could give rise to y_1 and y_2, but that there is a statistic $s(\cdot)$ sufficient for θ such that $s(y_1) = s(y_2)$, then any inference for θ should be the same whether y_1 or y_2 is observed, that is $\mathrm{Ev}(E, y_1) = \mathrm{Ev}(E, y_2)$. This is widely accepted, as the factorization criterion (4.15) implies that given the sufficient statistic, the data contain no further information about θ.

A second principle can be motivated by the following classic example.

Example 11.4 (Measuring machines) Suppose that a physical quantity θ can be measured by two machines, both giving normal measurements Y with mean θ. A measurement from the first machine has unit variance, but one from the second has variance 100. The more precise machine is often busy, while the second is used only if the first is unavailable; the upshot is that each is equally likely to be used. Thus if A takes value 1 or 2 depending on the machine used, $\Pr(A = 1) = \Pr(A = 2) = \frac{1}{2}$.

Suppose that an observation obtained is from machine 1. Then clearly any inference about θ should not take into account that machine 2 might have been used, when it is known that it was not. Mathematically this is expressed by saying that the revelant distribution for inference about θ is the conditional distribution of Y given A, rather than the unconditional distribution of Y. For example, the conditional 95% confidence interval for θ given that $A = 1$ is $y \pm 1.96$, whereas the unconditional interval is $y \pm 16.45$, which is clearly much too long if it is known that y came from the $N(\theta, 1)$ distribution. ∎

The lesson of this is formalized as follows. Suppose that an experiment E can be thought of as arising in two stages. In the first stage we observe that a random variable A with known distribution independent of θ takes value a, and in the second stage we observe y_a from a component experiment E_a. This is a mixture experiment, for which the data are (a, y_a). Then one form of the *conditionality principle* says that $\mathrm{Ev}\{E, (a, y_a)\} = \mathrm{Ev}(E_a, y_a)$: the evidence concerning θ based on the compound experiment E is equal to the evidence from the component experiment E_a *actually performed*, the results of other possible components being irrelevant. The key point is that since the distribution of A does not depend on θ, conditioning on A does not lead to a loss of information about θ, but selects the relevant component of the mixture experiment. This principle is widely, even if sometimes unconsciously, accepted; we discuss its implications in more detail in Chapter 12.

Likelihood principle

Suppose that two experiments relating to θ, E_1 and E_2, give rise to data y_1 and y_2 such that the corresponding likelihoods are proportional, that is, for all θ,

$$L(\theta; y_1, E_1) = c L(\theta; y_2, E_2).$$

Then according to one expression of the *likelihood principle*, $\mathrm{Ev}(E_1, y_1) = \mathrm{Ev}(E_2, y_2)$: inference should be based on the observed likelihood alone. Full acceptance of this means rejecting frequentist tools such as significance tests, as the following example shows.

Example 11.5 (Bernoulli trials)   Suppose that E_1 consists of observing the number y_1 of successes in a fixed number n_1 of independent Bernoulli trials. The likelihood is then

$$L_1(\theta) = \binom{n_1}{y_1} \theta^{y_1}(1 - \theta)^{n_1 - y_1}, \quad 0 < \theta < 1,$$

corresponding to the binomial number of successful trials.

Experiment E_2 consists of conducting Bernoulli trials independently until y_2 successes occur, at which point there have been n_2 trials. Here the likelihood,

$$L_2(\theta) = \binom{n_2 - 1}{y_2 - 1} \theta^{y_2}(1 - \theta)^{n_2 - y_2}, \quad 0 < \theta < 1,$$

corresponds to the negative binomial number of trials up to y_2 successes.

Now suppose that it happens that $n_1 = n_2 = n$ and $y_1 = y_2 = y$, giving $L_1(\theta) \propto L_2(\theta)$. Then according to the likelihood principle, inferences based on the two experiments should be the same. But consider testing the hypothesis $H_0 : \theta = \frac{1}{2}$ against the alternative that $\theta < \frac{1}{2}$. In E_1, the test statistic would be the random number of successes, Y, and the P-value would be

$$\Pr\left(Y \le y \mid \theta = \frac{1}{2}\right) = \sum_{r=0}^{y} \binom{n}{r} 2^{-n}, \tag{11.7}$$

while in E_2 the test statistic would be the total number of trials, N, with P-value

$$\Pr\left(N \ge n \mid \theta = \frac{1}{2}\right) = \sum_{m=n}^{\infty} \binom{m-1}{y-1} 2^{-m}. \tag{11.8}$$

The catch is that (11.7) and (11.8) need not be equal. For example, if $y = 3$ and $n = 12$, the P-values are respectively 0.073 and 0.033, conveying different evidence against H_0. In particular, use of the fixed significance level 0.05 would lead to acceptance or rejection of H_0 depending on the experiment performed. The reason for this is that (11.7) and (11.8) involve summation over portions of two different sample spaces. This conflicts with the likelihood principle, according to which only the data actually observed should contribute to the inference.   ∎

Construction of tail probabilities such as (11.7) or (11.8), or of confidence intervals, involves consideration of data not actually observed, and thereby disobeys the likelihood principle. This poses a problem for frequentist procedures, because a rational statistician who rejects the likelihood principle should also reject one of the apparently reasonable sufficiency and conditionality principles, which together entail the likelihood principle.

To see this, suppose that we accept the sufficiency and conditionality principles, and that experiments E_1 and E_2 have yielded data y_1 and y_2 such that $L(\theta; y_1, E_1) = cL(\theta; y_2, E_2)$ for some $c > 0$ and all θ. Consider the mixture experiment E that consists of observing (E_a, y_a), where a is the observed value of the binary random variable such that

$$\Pr(A = 1) = \frac{1}{c+1}, \quad \Pr(A = 2) = \frac{c}{c+1};$$

the distribution of A is independent of θ. The outcomes for E are (E_1, y_1) and (E_2, y_2), and the decomposition $\Pr(E_a, y_a; \theta) = \Pr(y_a \mid E_a; \theta)\Pr(E_a)$ shows that the corresponding likelihoods,

$$\frac{1}{c+1}L(\theta; y_1, E_1), \quad \frac{c}{c+1}L(\theta; y_2, E_2),$$

are equal for all θ. Since the likelihood function is itself a minimal sufficient statistic for θ (Exercise 4.2.11), the sufficiency principle implies

$$\mathrm{Ev}\{E, (E_1, y_1)\} = \mathrm{Ev}\{E, (E_2, y_2)\}. \tag{11.9}$$

But the conditionality principle implies

$$\mathrm{Ev}\{E, (E_1, y_1)\} = \mathrm{Ev}(E_1, y_1), \quad \mathrm{Ev}\{E, (E_2, y_2)\} = \mathrm{Ev}(E_2, y_2),$$

and combined with (11.9) we get $\mathrm{Ev}(E_1, y_1) = \mathrm{Ev}(E_2, y_2)$. Thus acceptance of the sufficiency and conditionality principles implies acceptance of the likelihood principle. The converse is also true (Problem 11.6). In fact it can be shown that a stronger version of the conditionality principle on its own implies the likelihood principle.

Statisticians attempting to weaken the force of this argument have criticized its central notions of evidence and mixture experiments, or have insisted that the sufficiency and conditionality principles apply only in a more limited way. They can then accept some form of these principles but not the conclusion of the argument, and continue to use such tools as confidence intervals and P-values. Others deny the validity of the argument on the grounds that it applies only to models known to be true, and this is rare in practice.

Statisticians who embrace the likelihood principle find themselves in an awkward position: their inference should be based on the observed likelihood, $L(\theta)$, but how should it be expressed? In particular, what can be inferred about a scalar component of vector θ? The obvious solution of profiling over the other components of θ can go badly awry, as we shall see in Chapter 12, and the alternative of integrating them out does not give a unique answer (Problem 11.7). Thus the idea of multiplying $L(\theta)$ by a prior density and applying the simple recipe of Bayes' theorem starts to appear very attactive. Moreover, we see from (11.2) that given a particular prior $\pi(\theta)$, Bayesian inference for θ does conform to the likelihood principle, because any constants in $f(y \mid \theta)$ do not appear in the posterior density.

11.1.3 Prior information

Despite its conformity to the likelihood principle, inference based on Bayes' theorem has often been seen as controversial. This is not due to the result itself, which simply states mathematically how the probability density of one random variable changes when another has been observed, but because its use in statistical inference for θ requires the investigator to treat θ as a random variable, and to specify a prior density $\pi(\theta)$ separate from the data. A key issue is the interpretation and choice of π.

In some circumstances it is uncontroversial to treat θ as random. At one extreme the data at hand may be the latest in a stream of similar datasets, each having an underlying parameter that may be supposed to be drawn from a distribution. For example, an accountant may wish to estimate the level of errors in a company's books, θ, based on a sample of transactions that reveals y errors. It will be sensible to treat θ as randomly chosen from a density $\pi(\theta)$ of error rates based on experience with previous firms. Then inference on θ will use both y and $\pi(\theta)$. An example in the use of forensic evidence is when there is a close match between DNA profile data from the scene of a crime and a suspect. Then a database of prior profiles may help to establish whether DNA found at the scene of the crime could plausibly have come from someone else. In these applications the prior information has a frequentist basis, so new issues of interpretation do not arise.

Despite this, the London Court of Appeal (Regina vs. Adams, 1996, 1997) ruled that 'introducing Bayes' theorem ... into a criminal trial plunges the jury into inappropriate and unnecessary realms of complexity, deflecting them from their proper task'.

At the other end of the range of possibilities is the situation where the data are to be used to make subjective decisions such as 'should I bet on the outcome of this race?' Although likely to depend on how facts such as 'Flatfoot has not won a race this season' are viewed, both model and prior information here reflect a personal judgement. Here Bayes' theorem provides the mechanism for updating prior beliefs in the light of whatever data is available, but the inference is a personal assessment of the evidence and has no claim to objective force.

The debate arises when the prior information does not have a frequency interpretation, but the inference required is not purely personal. Many statisticians regard the information in data as being qualitatively different from their prior beliefs about model parameters, and hence find it unacceptable to use Bayes' theorem to combine the two. They argue that although the choice of probability model is usually a matter of individual judgement, that judgement can be checked by comparing the data and fitted model, while by definition prior information cannot be checked directly. To which a Bayesian might reply that the epistemological distinction between data, model, and prior is unclear, because collection of any data must be based on some prior belief, which will often include information about possible models and the likely values of their parameters. Furthermore Bayes' theorem provides a single recipe for inference about unknowns, while frequentist notions such as confidence intervals can violate what seem reasonable principles of inference. Much has been written on this, but we shall avoid getting embroiled, simply noting that in many situations the Bayesian approach is simpler and more direct than frequentist alternatives, and that when they can be compared, the inferences produced by Bayesian and good frequentist procedures are often rather similar, so that the practical consequences of choosing between them are usually not critical. When a frequentist inference differs strongly from any conceivable Bayesian one, it seems wise to pause and reflect awhile.

Whatever its interpretation, a prior must be specified in order for Bayesian analysis to proceed. We now consider aspects of this.

Conjugate densities

In Example 11.1 the combination of a beta prior density for a probability and the likelihood for several Bernoulli trials led to a beta posterior density. Although too inflexible to encompass the range of prior knowledge that arises in applications, such conjugate combinations of prior and likelihood are useful because of their simple closed forms. They are closely tied to exponential family models.

Example 11.6 (Exponential family) Suppose that $y_1, \ldots, y_n$ is a random sample from the exponential family (5.12)

$$f(y \mid \omega) = \exp\{s(y)^{\mathrm{T}}\theta(\omega) - b(\omega)\} f_0(y),$$

so that in terms of $s = \sum s(y_j)$, the likelihood is proportional to

$$\exp\{s^{\mathrm{T}}\theta(\omega) - nb(\omega)\}. \tag{11.10}$$

If the prior density for ω depends on the quantities ξ and ν and has form

$$\pi(\omega) = \exp\{\xi^{\mathrm{T}}\theta(\omega) - \nu b(\omega) + c(\xi, \nu)\},$$

then the posterior density is proportional to

$$\exp\{(\xi + s)^{\mathrm{T}}\theta(\omega) - (\nu + n)b(\omega)\}.$$

Provided this is integrable the posterior density therefore must be

$$\pi(\omega \mid y) = \exp\{(\xi + s)^{\mathrm{T}}\theta(\omega) - (\nu + n)b(\omega) + c(\xi + s, \nu + n)\}.$$

Thus the prior parameters (ξ, ν) are updated to $(\xi + s, \nu + n)$ by the data. One interpretation of the *hyperparameters* ξ and ν is that the prior information is equivalent to ν prior observations summing to ξ.

For example, the Poisson density with mean ω has kernel $\exp(y \log \omega - \omega)$, so the conjugate prior must have kernel $\exp(\xi \log \omega - \nu\omega)$. For $\xi, \nu > 0$, this is proportional to the gamma density with mean ξ/ν, whose density is

$$\pi(\omega) = \frac{\nu^{\xi}\omega^{\xi-1}}{\Gamma(\xi)} e^{-\nu\omega}, \quad \omega > 0,$$

and which is therefore the conjugate prior for the Poisson mean. As the data update (ξ, ν) to $(\xi + s, \nu + n)$, the posterior density

$$\pi(\omega \mid y) = \frac{(\nu + n)^{\xi+s}\omega^{\xi+s-1}}{\Gamma(\xi + s)} e^{-(\nu+n)\omega}, \quad \omega > 0,$$

also has gamma form. ∎

Example 11.7 (Normal distribution) Let $y_1, \ldots, y_n$ be a normal random sample with mean μ and known variance σ^2. The likelihood is

$$\frac{1}{(2\pi\sigma^2)^{n/2}} \exp\left\{-\frac{1}{2\sigma^2}\sum_{j=1}^{n}(y_j - \mu)^2\right\} \propto \exp\left(\mu\frac{n\overline{y}}{\sigma^2} - \frac{n}{\sigma^2}\frac{1}{2}\mu^2\right),$$

$\overline{y}$ is the sample average $n^{-1}\sum y_j$.

which is of form (11.10) with $s = n\overline{y}/\sigma^2$, $k = n/\sigma^2$, $a(\mu) = \mu$, and $\kappa(\mu) = \frac{1}{2}\mu^2$. Therefore the conjugate prior is proportional to

$$\exp\left(\mu\frac{\mu_0}{\tau^2} - \frac{1}{\tau^2}\frac{1}{2}\mu^2\right),$$

and must be the normal density with mean μ_0 and variance τ^2. The effect of the data is to update $(\mu_0\tau^{-2}, \tau^{-2})$ to $(\mu_0\tau^{-2} + s\sigma^{-2}, \tau^{-2} + n\sigma^{-2})$, so the posterior density for μ is normal with mean and variance

$$\frac{n\overline{y}/\sigma^2 + \mu_0/\tau^2}{n/\sigma^2 + 1/\tau^2}, \quad \frac{1}{n/\sigma^2 + 1/\tau^2}. \tag{11.11}$$

On writing the mean in (11.11) as

$$\frac{n\overline{y} + (\sigma^2/\tau^2)\mu_0}{n + \sigma^2/\tau^2},$$

we see that the prior injects information equivalent to σ^2/τ^2 observations with mean μ_0, and shrinks the sample average, $\overline{y}$, towards the prior mean by an amount that depends on the ratio of τ^2 to σ^2/n. As $n \to \infty$ or $\tau^2 \to \infty$, corresponding to increasing information in the data relative to the prior, the posterior density becomes normal with mean $\overline{y}$ and variance σ^2/n, so the effect of the prior withers away. As $\tau^2 \to 0$, corresponding to more definite prior knowledge, the posterior approaches the normal density with mean μ_0 and variance τ^2, which is the prior. ∎

Conjugate priors are often too restrictive for expression of realistic prior information, but it is straightforward to establish that mixtures of conjugate densities are also conjugate, and this considerably broadens the class of priors with closed-form posterior densities (Problem 11.3).

Ignorance

Sometimes the prior density must express prior ignorance about a parameter. One reason for this may be the need for a 'baseline' analysis as a basis for discussion. Another is the belief that a non-informative prior will allow the data 'to speak for themselves', though it seems optimistic to think that they will spill their secrets without careful interrogation. Nevertheless it is important to weigh how much an inference depends on the prior compared to the data. One way to do this is to contrast inferences from a minimally informative prior with those from the prior actually used.

When θ has bounded support, as in Example 11.1, a uniform prior density, with $\pi(\theta) \propto 1$, seems an obvious choice. When the support of θ is unbounded, such a prior has infinite integral and so is *improper*. An improper prior may nevertheless lead to a proper posterior density. In Example 11.7, for example, we can represent complete ignorance about the prior value of μ by letting $\tau^2 \to \infty$, in which case the prior is $\pi(\mu) \propto 1$ with support on the entire real line, and the posterior density of μ is normal with mean $\overline{y}$ and variance σ^2/n, which is proper. Prior ignorance about σ in models where the density of the data is of form $\sigma^{-1}g(u/\sigma)$, $u > 0$, $\sigma > 0$, is usually represented by the improper prior $\pi(\sigma) \propto \sigma^{-1}$, $\sigma > 0$. Non-informative priors of this sort exist for more general situations, but there is a fundamental difficulty in representing ignorance in a way that is independent both of the data to be collected and the parametrization of the model (Problem 11.4). The key question is: ignorance about what? The following classic example illustrates this.

Example 11.8 (Bernoulli probability) The probability of success in a Bernoulli trial lies in the interval $[0, 1]$, so if we are completely ignorant of its true value, the obvious prior to use is uniform on the unit interval: $\pi(\theta) = 1$, $0 \leq \theta \leq 1$. But if we are completely ignorant of θ, we are also completely ignorant of $\psi = \log\{\theta/(1-\theta)\}$, which takes values in the real line. The density implied for ψ by the uniform prior for θ is

$$\pi(\psi) = \pi\{\psi(\theta)\} \times \left|\frac{d\theta}{d\psi}\right| = \frac{e^\psi}{(1+e^\psi)^2}, \quad -\infty < \psi < \infty :$$

the standard logistic density. Far from expressing ignorance about ψ, this density asserts that the prior probability of $|\psi| < 3$ is about 0.9. ∎

Jeffreys priors

Sir Harold Jeffreys (1891–1989) studied first in Newcastle and then in Cambridge, where he remained for the rest of his life, becoming Plumian Professor of Astronomy. During World War I he worked in the Cavendish Laboratory, and thereafter studied and taught hydrodynamics and geophysics, being the first to claim that the core of the earth is liquid. In an important series of books he championed objective Bayesian inference long before it became popular (Jeffreys, 1961), and also wrote important works on geophysics and mathematical physics. His unassuming character inspired deep affection.

Apparent paradoxes like that of Example 11.8 led to a widespread rejection of Bayesian inference in the early twentieth century. The key difficulty is that the representation of ignorance is not invariant under reparametrization. A solution to this is to seek invariant priors. For scalar θ the best-known of these is the *Jeffreys prior*

$$\pi(\theta) \propto |i(\theta)|^{1/2}, \tag{11.12}$$

where $i(\theta) = -\mathrm{E}\{d^2\ell(\theta)/d\theta^2\}$ is the expected information for θ based on the log likelihood $\ell(\theta)$; $i(\theta)$ is positive in a regular statistical model. For a smooth reparametrization $\theta = \theta(\psi)$ in terms of ψ, the expected information for ψ is

$$i(\psi) = -\mathrm{E}\left[\frac{d^2\ell\{\theta(\psi)\}}{d\psi^2}\right] = -\mathrm{E}\left\{\frac{d^2\ell(\theta)}{d\theta^2}\right\} \times \left|\frac{d\theta}{d\psi}\right|^2 = i(\theta) \times \left|\frac{d\theta}{d\psi}\right|^2.$$

Consequently $|i(\theta)|^{1/2}d\theta = |i(\psi)|^{1/2}d\psi$: with the choice (11.12), prior information does behave consistently under reparametrization; furthermore such priors give widely-accepted solutions in some standard problems. When θ is vector, $|i(\theta)|$ is taken to be the determinant of $i(\theta)$.

This prior was initially proposed with the aim of giving an 'objective' basis for inference, but after further paradoxes emerged its use was suggested for convenience, a matter of scientific convention rather than as a logically unassailable expression of ignorance about the parameter.

Example 11.9 (Bernoulli probability)   The log likelihood for a single Bernoulli trial with success probability θ is $y \log \theta + (1 - y) \log(1 - \theta)$, and the Fisher information is $i(\theta) = \theta^{-1}(1 - \theta)^{-1}$. Thus the Jeffreys prior is proportional to $\theta^{-1/2}(1 - \theta)^{-1/2}$, and so equals the beta density (11.3) shown in the top left panel of Figure 5.4, which while proper does not look uninformative. It can be interpreted as carrying information equivalent to one trial, in which one-half of a success was observed. As the prior information for n independent trials is $ni(\theta)$, the Jeffreys prior is the same because the constant of proportionality is independent of θ.   ∎

Example 11.10 (Location-scale model)   Suppose that $y_1, \ldots, y_n$ is a random sample from a location model $f(y; \eta) = g(y - \eta)$, for real y and η. Then the log likelihood is $\ell(\eta) = \sum \log g(y_j - \eta)$, so

$$i(\eta) = -n \int_{-\infty}^{\infty} \frac{d^2 \log g(y - \eta)}{d\eta^2} g(y - \eta) \, dy.$$

The substitution $u = y - \eta$ shows that $i(\eta)$ is independent of η, and therefore the Jeffreys prior is the constant non-informative prior $\pi(\eta) \propto 1$ for all η.

A modification of this argument (Problem 11.2) shows that the Jeffreys prior for $f(y; \tau) = \tau^{-1} g(y/\tau)$, $y, \tau > 0$, is $\pi(\tau) \propto \tau^{-1}$, which is also widely accepted as non-informative. Both $\pi(\tau)$ and $\pi(\eta)$ are improper.

A difficulty with this approach appears when we consider the location-scale model $f(y; \eta, \tau) = \tau^{-1} g\{(y - \eta)/\tau\}$. Its information matrix has form $i(\eta, \tau) = n\tau^{-2} A$, where the 2×2 matrix A is free of parameters, so $\pi(\eta, \tau) = |i(\eta, \tau)|^{1/2} \propto \tau^{-2}$. This does not equal the prior τ^{-1} arising from taking independent Jeffreys priors for η and τ separately.

The approach is here unsatisfactory because the prior τ^{-2} is not widely accepted as a non-informative statement of uncertainty about τ. More generally this example shows that a non-informative inference for a parameter of interest, η, say, may depend on the model in which η is embedded, in the sense that the inference may depend on the prior chosen for nuisance parameters, even when these are *a priori* independent of η.   ∎

Jeffreys' general solution to the difficulty raised in Example 11.10 was to treat location parameters as fixed when computing $i(\theta)$. Let $\theta = (\mu_1, \ldots, \mu_p, \psi)$, where the μ_r are location parameters and ψ contains all other parameters in the problem. Then the prior he recommended is

$$\pi(\mu_1, \ldots, \mu_p, \psi) \propto \left| \mathrm{E} \left\{ -\frac{\partial^2 \ell(\mu_1, \ldots, \mu_p, \psi)}{\partial \psi \partial \psi^{\mathrm{T}}} \right\} \right|^{1/2},$$

which produces $\pi(\theta) \propto \tau^{-1}$ in the location-scale model.

Numerous other approaches to representing prior ignorance have been proposed, based for example on notions of invariance, of minimal information, or of matching the coverage of Bayesian and frequentist confidence intervals. To a large extent these are regarded as useful to the extent that they yield Jeffreys priors, and we shall not consider them in detail. To be more explicit about links with the frequentist approach, however, note that if a uniform prior is taken in (11.11), corresponding to $\tau \to \infty$, and we define $\mathcal{A}_y$ to be the interval with limits $\bar{y} \pm z_\alpha n^{-1/2}\sigma$, then the posterior probability $\Pr(\theta \in \mathcal{A}_y \mid y) = 1 - 2\alpha$. Thus $\mathcal{A}_y$ has posterior coverage $(1 - 2\alpha)$. But $\mathcal{A}_y$ also has the same coverage for any fixed θ unconditional on y, so the uniform prior yields an interval justifiable from both Bayesian and frequentist viewpoints. Exact results such as this are unobtainable in more general settings, but nonetheless it can be helpful to consider the extent to which Bayesian and frequentist procedures agree.

Some further aspects of Jeffreys priors are outlined in Problem 11.4.

Exercises 11.1

1   In Example 11.3, calculate the predictive probability for k future heads out of m tosses based on r heads observed in n tosses, using a beta prior density.

2   Show that the limits of an unconditional confidence interval of level $(1 - 2\alpha)$ in Example 11.4 involve the solutions to the equation

$$\frac{1}{2}\Phi\{(y - \theta)/10\} + \frac{1}{2}\Phi(y - \theta) = \alpha, 1 - \alpha.$$

Hence justify the approximate 0.95 interval given in the example.

3   (a) Let $y_1, \ldots, y_n$ be a Poisson random sample with mean θ, and suppose that the prior density for θ is gamma,

$$\pi(\theta) = g(\theta; \alpha, \lambda) = \frac{\lambda^\alpha \theta^{\alpha-1}}{\Gamma(\alpha)} \exp(-\lambda\theta), \quad \theta > 0, \ \lambda, \alpha > 0.$$

Show that the posterior density of θ is $g(\theta; \alpha + \sum y_j, \lambda + n)$, and find conditions under which the posterior density remains proper as $\alpha \downarrow 0$ even though the prior density becomes improper in the limit.
(b) Show that $\int \theta g(\theta; \alpha, \lambda) d\theta = \alpha/\lambda$. Find the prior and posterior means $E(\theta)$ and $E(\theta \mid y)$, and hence give an interpretation of the prior parameters.
(c) Let Z be a new Poisson variable independent of $Y_1, \ldots, Y_n$, also with mean θ. Find its posterior predictive density. To what density does this converge as $n \to \infty$? Does this make sense?

4   How would you express prior ignorance about an angle? About the position of a star in the firmament?

5   If $Y_{ij} \sim N(\mu_i, \sigma^2)$ independently for $i = 1, \ldots, k$ and $j = 1, \ldots, m$, show that the Jeffreys prior for $\mu_1, \ldots, \mu_k, \sigma$ equals $\sigma^{-(k+1)}$. Discuss the form of posterior inferences on σ^2 when $m = 2$. Is this prior reasonable? If not, suggest a better alternative.

6   According to the *principle of insufficient reason* probabilities should be ascribed uniformly to finite sets unless there is some definite reason to do otherwise. Thus the most natural way to express prior ignorance for a parameter θ that inhabits a finite parameter space $\theta_1, \ldots, \theta_k$ is to set $\pi(\theta_1) = \cdots = \pi(\theta_k) = 1/k$. Let $\pi_i = \pi(\theta_i)$.
Consider a parameter space $\{\theta_1, \theta_2\}$, where θ_1 denotes that there is life in orbit around the star Sirius and θ_2 that there is not. Can you see any reason not to take $\pi_1 = \pi_2 = 1/2$?
Now consider the parameter space $\{\omega_1, \omega_2, \omega_3\}$, where ω_1, ω_2, and ω_3 denote the events that there is life around Sirius, that there are planets but no life, and that there are no planets. With this parameter space the principle of insufficient reason gives $\Pr(\text{life around Sirius}) = 1/3$.
Discuss this partitioning paradox. What solutions do you see?
(Schafer, 1976, pp. 23–24)

7   Compute the prior and posterior means and variances for exponential family data with the conjugate prior distribution, and discuss their interpretation.

$f(y \mid \theta)$	Parameter	Prior
Binomial	success probability	beta
Poisson	mean	gamma
Exponential	mean	gamma
Normal	mean (known variance)	normal
Normal	variance (known mean)	inverse gamma
Multinomial	probabilities	Dirichlet

Table 11.1 Conjugate prior densities for exponential family samling distributions.

8   Use Example 11.6 to verify the contents of Table 11.1.

9   Let θ be a randomly chosen physical constant. Such constants are measured on an arbitrary scale, so transformations from θ to $\psi = c\theta$ for some constant c should leave the density $\pi(\theta)$ of θ unchanged. Show that this entails $\pi(c\theta) = c^{-1}\pi(\theta)$ for all $c, \theta > 0$, and deduce that $\pi(\theta) \propto \theta^{-1}$.

Let $\tilde{\theta}$ be the first significant digit of θ in some arbitrary units. Show that

$$\Pr(\tilde{\theta} = d) \propto \int_{d10^a}^{(d+1)10^a} u^{-1}\, du, \quad d = 1, \ldots, 9,$$

and hence verify that $\Pr(\tilde{\theta} = d) = \log_{10}(1 + d^{-1})$. Check whether some set of physical 'constants' (e.g. sizes of countries or of lakes) fits this distribution.

11.2 Inference

11.2.1 Posterior summaries

If the information regarding θ is contained in its posterior density given the data y, $\pi(\theta \mid y)$, how do we get at it? In principle this is easy: we simply use the posterior density to calculate the probability of any event of interest. But some summary quantities may be useful. For example, if $\theta = (\psi, \lambda)$ is a vector, and we are interested in ψ, the *marginal posterior density*

$$\pi(\psi \mid y) = \int \pi(\psi, \lambda \mid y)\, d\lambda,$$

contains the marginal information in the model and prior concerning ψ. It is most useful when ψ has dimension one or two, in which case it can be plotted. It condenses further to moments, quantiles, or the mode of $\pi(\psi \mid y)$.

Normal approximation

One simple approximate summary of a unimodal posterior rests on quadratic series expansion of the log posterior density, analogous to expansion of the log likelihood. In terms of $\tilde{\ell}(\theta) = \log L(\theta) + \log \pi(\theta)$ and the posterior mode $\tilde{\theta}$, we have

$$\tilde{\ell}(\theta) \doteq \tilde{\ell}(\tilde{\theta}) + (\theta - \tilde{\theta})^{\mathrm{T}}\frac{\partial \tilde{\ell}(\tilde{\theta})}{\partial \theta} + \frac{1}{2}(\theta - \tilde{\theta})^{\mathrm{T}}\frac{\partial^2 \tilde{\ell}(\tilde{\theta})}{\partial \theta \partial \theta^{\mathrm{T}}}(\theta - \tilde{\theta})$$

$$= \tilde{\ell}(\tilde{\theta}) - \frac{1}{2}(\theta - \tilde{\theta})^{\mathrm{T}}\tilde{J}(\tilde{\theta})(\theta - \tilde{\theta}),$$

Table 11.2 Mortality rates r/m from cardiac surgery in 12 hospitals (Spiegelhalter *et al.*, 1996b, p. 15). Shown are the numbers of deaths r out of m operations.

A	0/47	B	18/148	C	8/119	D	46/810	E	8/211	F	13/196
G	9/148	H	31/215	I	14/207	J	8/97	K	29/256	L	24/360

provided the mode lies inside the parameter space. Here $\tilde{J}(\theta)$ is the second derivative matrix of $-\tilde{\ell}(\theta)$. This expansion corresponds to a posterior multivariate normal density for θ, with mean $\tilde{\theta}$ and variance matrix $\tilde{J}(\tilde{\theta})^{-1}$, based on which an equitailed $(1 - 2\alpha)$ confidence interval for the rth component θ_r of θ is $\tilde{\theta}_r \pm z_\alpha \tilde{v}_{rr}^{1/2}$, where $\tilde{v}_{rr}$ is the rth diagonal element of $\tilde{J}(\tilde{\theta})^{-1}$.

In large samples the log likelihood contribution is typically much greater than that from the prior, so $\tilde{\theta}$ and $\tilde{J}(\tilde{\theta})$ are essentially indistinguishable from the maximum likelihood estimate $\widehat{\theta}$ and observed information $J(\widehat{\theta})$. Thus likelihood-based confidence intervals may be interpreted as giving approximate Bayesian inferences, if the sample is large. This approximation will usually be better if applied to the marginal posterior of a low-dimensional subset of θ, because of the averaging effect of integration over the other parameters. The same caveats apply when using this approximation as to use of normal approximations for the maximum likelihood estimator; in particular, it may be more suitable for a transformed parameter. We describe a more refined approach in Section 11.3.1.

Other distributions may be used to approximate posterior densities, for example by matching first and second moments.

Posterior confidence sets

The mean and mode of the posterior density are point summaries of $\pi(\theta \mid y)$, but confidence regions or intervals are usually more useful. The Bayesian analogue of a $(1 - 2\alpha)$ confidence interval is a $(1 - 2\alpha)$ *credible set*, defined to be a set, C, of values of θ, whose posterior probability content is at least $1 - 2\alpha$. When θ is continuous this is

$$1 - 2\alpha = \Pr(\theta \in C \mid y) = \int_C \pi(\theta \mid y)\, d\theta.$$

When θ is discrete, the integral is replaced by $\sum_{\theta \in C} \pi(\theta \mid y)$. For scalar θ, such a set is equi-tailed if it has form (θ_L, θ_U), where θ_L and θ_U are the posterior α and $1 - \alpha$ quantiles of θ, that is, $\Pr(\theta < \theta_L \mid y) = \Pr(\theta > \theta_U \mid y) = \alpha$.

Often C is chosen so that the posterior density for any θ in C is higher than for any θ not in C. That is, if $\theta \in C$, $\pi(\theta \mid y) \geq \pi(\theta' \mid y)$ for any $\theta' \notin C$. Such a region is called a *highest posterior density credible set*, or more concisely a *HPD credible set*.

Example 11.11 (Cardiac surgery data) Table 11.2 contains data on the mortality levels for cardiac surgery on babies at 12 hospitals. A simple model treats the number of deaths r as binomial with mortality rate θ and denominator m. At hospital A, for example, $m = 47$ and $r = 0$, giving maximum likelihood estimate $\widehat{\theta}_A = 0/47 = 0$, but it seems too optimistic to suppose that θ_A could be so small when the other rates are evidently larger. If we take a beta prior density with $a = b = 1$, the posterior density is beta with parameters $a + r = 1$ and $b + m - r = 48$, as shown in the

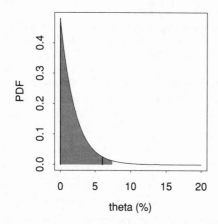

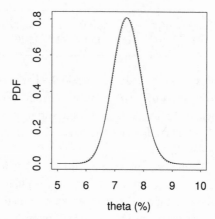

Figure 11.1 Cardiac surgery data. Left panel: posterior density for θ_A, showing boundaries of 0.95 highest posterior credible interval (vertical lines) and region between posterior 0.025 and 0.975 quantiles of $\pi(\theta_A \mid y)$ (shaded). Right panel: exact posterior beta density for overall mortality rate θ (solid) and normal approximation (dots).

left panel of Figure 11.1. The 0.95 HPD credible interval is $(0, 6.05)\%$, while the equitailed credible interval uses the 0.025 and 0.975 quantiles of $\pi(\theta_A \mid y)$ and is $(0.05, 7.40)\%$.

The right panel of Figure 11.1 shows the posterior density for the overall mortality rate θ, obtaining by merging all the data, giving $r = 208$ deaths in $m = 2814$ operations. Here the prior parameters a and b have essentially no effect on the posterior, and hence

$$\tilde{\theta} = \frac{a+r-1}{a+b+m-2} \doteq \frac{r}{m}, \quad \tilde{J}(\tilde{\theta})^{-1} = \frac{(a+r-1)(b+m-r-1)}{(a+b+m-2)^3} \doteq \frac{r(m-r)}{m^3}.$$

The figure shows the corresponding normal approximation to $\pi(\theta \mid y)$. Evidently inferences from exact and approximate posterior densities will be equivalent for practical purposes.

Both separate and pooled analyses of mortality rates seem unsatisfactory, because although some variation among hospitals is plausible they are likely also to have elements in common. Example 11.26 describes an approach intermediate between those used here. ∎

Example 11.12 (Normal distribution) Consider a normal random sample $y_1, \ldots, y_n$ with mean μ and variance σ^2 both unknown. We shall give them independent prior densities. As the posterior for (μ, σ^2) depends on y only through the minimal sufficient statistic $(\overline{y}, s^2)$, we have

$\overline{y} = n^{-1} \sum y_j$ and $s^2 = (n-1)^{-1} \sum (y_j - \overline{y})^2$ are the sample average and variance.

$$\begin{aligned}
\pi(\mu, \sigma^2 \mid \overline{y}, s^2) &\propto f(\overline{y}, s^2 \mid \mu, \sigma^2)\pi(\mu, \sigma^2) \\
&= f(\overline{y} \mid \mu, \sigma^2)f(s^2 \mid \mu, \sigma^2)\pi(\mu, \sigma^2) \\
&= f(\overline{y} \mid \mu, \sigma^2)f(s^2 \mid \sigma^2)\pi(\mu)\pi(\sigma^2) \\
&\propto \pi(\mu \mid \overline{y}, \sigma^2)f(s^2 \mid \sigma^2)\pi(\sigma^2), \quad (11.13)
\end{aligned}$$

where the first step follows from Bayes' theorem, the second from the conditional independence of $\overline{y}$ and σ^2 given μ and σ^2, the third from the prior independence of μ and σ^2 and the independence of s^2 and μ, and the fourth on using Bayes' theorem to get the posterior density for μ conditional on $\overline{y}$ and σ^2. Integration of (11.13) with respect to μ shows that $\pi(\sigma^2 \mid \overline{y}, s^2) \propto f(s^2 \mid \sigma^2)\pi(\sigma^2)$: the marginal posterior density of σ^2 depends only on s^2. However, as σ^2 appears in all three terms, integration of (11.13) with respect to σ^2 shows that the marginal posterior for μ depends on both $\overline{y}$ and s^2.

Let us use the improper priors $\pi(\mu) \propto 1$, $\pi(\sigma^2) \propto \sigma^{-2}$. Example 11.7 shows that the posterior density for μ when σ^2 is known is $N(\overline{y}, \sigma^2/n)$. Conditional on σ^2, the distribution of $(n-1)s^2$ is $\sigma^2\chi^2_{n-1}$, so our choice of prior gives

$$\pi(\sigma^2 \mid s^2) \propto \pi(\sigma^2)f(s^2 \mid \sigma^2)$$

$$\propto (\sigma^2)^{-1}(\sigma^2)^{-(n-1)/2} \exp\left\{-\frac{1}{2}(n-1)s^2/\sigma^2\right\}, \quad \sigma^2 > 0.$$

Thus the marginal posterior density of σ^2 is *inverse gamma*,

$$\frac{\beta^\alpha}{\Gamma(\alpha)x^{\alpha+1}} \exp(-\beta/x), \quad x > 0, \quad \alpha, \beta > 0, \tag{11.14}$$

with $x = \sigma^2$, $\alpha = \frac{1}{2}(n-1)$ and $\beta = \frac{1}{2}(n-1)s^2$; a useful shorthand for (11.14) is $IG(\alpha, \beta)$. Its mean and variance are $\beta/(\alpha-1)$ and $\beta^2/\{(\alpha-1)^2(\alpha-2)\}$, provided that $\alpha > 2$. Equivalently, the posterior distribution of σ^2 given s^2 is that of $(n-1)s^2/V$, where $V \sim \chi^2_{n-1}$. The joint posterior density for (μ, σ^2),

$$\pi(\mu, \sigma^2 \mid \overline{y}, s^2) \propto \pi(\mu \mid \overline{y}, \sigma^2)\pi(\sigma^2 \mid s^2).$$

is proportional to

$$(\sigma^2)^{-1/2} \exp\left\{-\frac{n}{2\sigma^2}(\mu - \overline{y})^2\right\} \times (\sigma^2)^{-(n-1)/2-1} \exp\left\{-\frac{(n-1)s^2}{2\sigma^2}\right\}, \tag{11.15}$$

integration of which over σ^2 yields the marginal posterior density for μ,

$$\pi(\mu \mid \overline{y}, s^2) = \frac{\Gamma\left(\frac{n}{2}\right)}{\Gamma\left(\frac{n-1}{2}\right)} \left\{\frac{n}{(n-1)s^2\pi}\right\}^{1/2} \left\{1 + \frac{n(\mu - \overline{y})^2}{(n-1)s^2}\right\}^{-n/2}.$$

Therefore $n^{1/2}(\mu - \overline{y})/s \sim t_{n-1}$ *a posteriori*. The corresponding frequentist result treats $\overline{y}$ and s^2 as random and μ as fixed; here the random variable is μ, with $\overline{y}$ and s^2 regarded as constants.

Figure 11.2 shows posterior densities for μ and σ^2 based on the height differences for the 15 pairs of plants in Table 1.1; here $\overline{y} = 20.93$ and $s^2 = 1424.64$. Evidently the posterior densities are not independent. While the HPD credible set for μ is equi-tailed, that for σ^2 is not. ∎

A credible set may contain the same values of θ as a confidence interval, but its interpretation is different. In the Bayesian framework the data are regarded as fixed and the parameter as random, so the endpoints of the credible set are fixed and the probability statement concerns the parameter, regarded as a random variable. The frequentist approach treats the parameter as an unknown constant and the confidence interval endpoints as random variables; the probability statement concerns their behaviour in repeated sampling from the model.

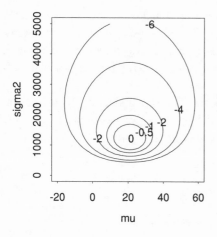 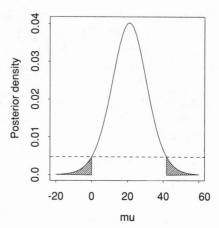

Figure 11.2 Posterior densities of (μ, σ^2) of normal model for maize data. Left: contours of the normalized log joint posterior density. Right: marginal posterior density for μ, showing 95% HPD credible set, which is the set of values of μ whose values of the posterior density $\pi(\mu \mid y)$ lie above the dashed line. The shaded region has area 0.05.

11.2.2 Bayes factors

The frequentist approach to hypothesis testing compares a null hypothesis H_0 with an alternative H_1 through a test statistic T that tends to be larger under H_1 than under H_0, and rejects H_0 for small values of the significance probability $p_{\text{obs}} = \text{Pr}_0(T \geq t_{\text{obs}})$, where t_{obs} is the value of T actually observed and the probability is computed as if H_0 were true.

The Bayesian approach attaches prior probabilities to the models corresponding to H_0 and H_1 and compares their posterior probabilities

$$\text{Pr}(H_i \mid y) = \frac{\text{Pr}(y \mid H_i)\text{Pr}(H_i)}{\text{Pr}(y \mid H_0)\text{Pr}(H_0) + \text{Pr}(y \mid H_1)\text{Pr}(H_1)}, \quad i = 0, 1.$$

An obvious distinction between this and the frequentist approach is that $\text{Pr}(H_0 \mid y)$ is the probability of H_0 conditional on the data, whereas the P-value may not be interpreted in this way. In Bayesian settings increasing amounts of data may lead to increasing support for one hypothesis relative to the alternatives. This differs from the frequentist approach, where non-rejection of H_0 does not indicate increasing support for it in large samples. A further important difference is that the P-value does not depend on the particular alternative H_1 under discussion. Indeed, whereas frequentist testing does not require H_1 to be fully specified, this is essential for Bayesian testing, which is in this sense more restrictive.

For some purposes it is valuable to use the odds in favour of H_1,

$$\frac{\text{Pr}(H_1 \mid y)}{\text{Pr}(H_0 \mid y)} = \frac{\text{Pr}(y \mid H_1)}{\text{Pr}(y \mid H_0)} \times \frac{\text{Pr}(H_1)}{\text{Pr}(H_0)}. \tag{11.16}$$

The change in prior to posterior odds for H_1 relative to H_0 depends on data only through the *Bayes factor*

$$B_{10} = \frac{\text{Pr}(y \mid H_1)}{\text{Pr}(y \mid H_0)}. \tag{11.17}$$

Thus analogous to the updating rule for inference on θ, we update evidence comparing the models by the rule *posterior odds = Bayes factor × prior odds*.

Table 11.3
Interpretation of Bayes
factor B_{10} in favour of H_1
over H_0. Since
$B_{10} = B_{01}^{-1}$, negating the
values of $2 \log B_{10}$ gives
the evidence against H_1.

B_{10}	$2 \log B_{10}$	Evidence against H_0
1–3	0–2	Hardly worth a mention
3–20	2–6	Positive
20–150	6–10	Strong
> 150	> 10	Very strong

The simplest situation is when both hypotheses are simple, in which case B_{10} equals the likelihood ratio in favour of H_1. Usually, however, both hypotheses involve parameters, say θ_0 and θ_1, and

$$\Pr(y \mid H_i) = \int f(y \mid H_i, \theta_i)\pi(\theta_i \mid H_i)\, d\theta_i, \quad i = 0, 1,$$

where $\pi(\theta_i \mid H_i)$ is the prior for θ_i under H_i. In this case the Bayes factor is a ratio of weighted likelihoods. By analogy with the likelihood ratio statistic, the quantity $2 \log B_{10}$ is often used to summarize the evidence for H_1 compared to H_0, with the rough interpretation shown in Table 11.3. This contrasts with the interpretation of a likelihood ratio statistic, whose null χ^2 distribution for nested models would depend on the difference in their degrees of freedom. The log Bayes factor $\log B_{10}$ is sometimes called the *weight of evidence*.

Example 11.13 (HUS data)   Example 4.40 introduced data on the numbers of cases of haemolytic uraemic syndrome (HUS) treated at a clinic in Birmingham from 1970 to 1989. The data suggest a sharp rise in incidence around 1980. In that example it was supposed that the annual counts $y_1, \ldots, y_n$ are realizations of independent Poisson variables with means $E(Y_j) = \lambda_1$ for $j = 1, \ldots, \tau$ and $E(Y_j) = \lambda_2$ for $j = \tau + 1, \ldots, n$. Here the changepoint τ can take values $1, \ldots, n - 1$.

Suppose that our baseline model H_0 is that $\lambda_1 = \lambda_2 = \lambda$, that is, no change, and consider the alternative H_τ of change after year τ. Under H_τ we suppose that λ_1 and λ_2 have independent gamma prior densities with parameters γ and δ. This density has mean γ/δ and variance γ/δ^2. Then $\Pr(y \mid H_\tau)$ equals

$$\int_0^\infty \prod_{j=1}^\tau \frac{\lambda_1^{y_j}}{y_j!} e^{-\lambda_1} \times \frac{\delta^\gamma \lambda_1^{\gamma-1}}{\Gamma(\gamma)} e^{-\delta\lambda_1}\, d\lambda_1 \int_0^\infty \prod_{j=\tau+1}^n \frac{\lambda_2^{y_j}}{y_j!} e^{-\lambda_2} \times \frac{\delta^\gamma \lambda_2^{\gamma-1}}{\Gamma(\gamma)} e^{-\delta\lambda_2}\, d\lambda_2,$$

or equivalently

$$\frac{\delta^{2\gamma}}{\Gamma(\gamma)^2 \prod_{j=1}^n y_j!} \frac{\Gamma(\gamma + s_\tau)\Gamma(\gamma + s_n - s_\tau)}{(\delta + \tau)^{\gamma + s_\tau}(\delta + n - \tau)^{\gamma + s_n - s_\tau}},$$

where $s_\tau = y_1 + \cdots + y_\tau$.

Under H_0 we assume that λ also has the gamma density with parameters γ and δ. Then the Bayes factor for a changepoint in year τ is

$$B_{\tau 0} = \frac{\Gamma(\gamma + s_\tau)\Gamma(\gamma + s_n - s_\tau)\delta^\gamma(\delta + n)^{\gamma + s_n}}{\Gamma(\gamma)\Gamma(\gamma + s_n)(\delta + \tau)^{\gamma + s_\tau}(\delta + n - \tau)^{\gamma + s_n - s_\tau}}, \quad \tau = 1, \ldots, n - 1.$$

For completeness we set $B_{n0} = 1$.

	1970	1971	1972	1973	1974	1975	1976	1977	1978	1979
y	1	5	3	2	2	1	0	0	2	1
$2 \log B_{\tau 0}, \gamma = \delta = 1$	4.9	−0.5	0.6	3.9	7.5	13	24	35	41	51
$2 \log B_{\tau 0}, \gamma = \delta = 0.01$	−1.3	−5.9	−4.5	−1.0	3.0	9.7	20	32	39	51
$2 \log B_{\tau 0}, \gamma = \delta = 0.0001$	−10	−15	−14	−10	−6.1	0.6	11	23	30	42

	1980	1981	1982	1983	1984	1985	1986	1987	1988	1989
y	1	7	11	4	7	10	16	16	9	15
$2 \log B_{\tau 0}, \gamma = \delta = 1$	63	55	38	42	40	31	11	−2.9	−5.3	0
$2 \log B_{\tau 0}, \gamma = \delta = 0.01$	64	57	40	47	46	38	18	1.8	1.2	0
$2 \log B_{\tau 0}, \gamma = \delta = 0.0001$	55	48	31	38	37	29	8.8	−7.1	−7.7	0

Table 11.4 Bayes factors for comparison of model of change in Poisson parameter after τ years, H_τ, with model of no change H_0, for HUS data y. There is very strong evidence of change in any year from 1976–86.

Table 11.4 gives $2 \log B_{\tau 0}$ for $\tau = 1, \ldots, 19$, for values of γ and δ such that the prior density for λ has unit mean and variances respectively $1, 10^2, 10^4$, corresponding to increasing prior uncertainty. Negative values of $2 \log B_{\tau 0}$ correspond to evidence in favour of H_0. There is very strong evidence for change in any year from 1976 to 1986, but the most plausible changepoint is just after 1980. The evidence for change is overwhelming for all the priors chosen. See Practical 11.6. ∎

Example 11.14 (Forensic evidence)   The following situation can arise when forensic evidence is used in criminal trials: material y found on a suspect is similar to other material, x, at the scene of the crime, and it is desired to know how this affects our view of the case. For simplicity we shall suppose that if x and y come from the same source, the suspect is guilty, an event we shall denote by G. Let E denote any other evidence. Then the odds of guilt, conditional on E and the data, are

$$\frac{\Pr(G \mid x, y, E)}{\Pr(\overline{G} \mid x, y, E)} = \frac{\Pr(x, y \mid G, E)}{\Pr(x, y \mid \overline{G}, E)} \frac{\Pr(G \mid E)}{\Pr(\overline{G} \mid E)}$$

$$= \frac{\Pr(x, y \mid G)}{\Pr(x \mid \overline{G})\Pr(y \mid \overline{G})} \times \frac{\Pr(G \mid E)}{\Pr(\overline{G} \mid E)}, \qquad (11.18)$$

where we have supposed that x and y are independent of E, and that they are independent given the event $\overline{G}$ that the suspect is not guilty. The first ratio on the right of (11.18) is the Bayes factor due to the forensic evidence.

Let y and x represent single measurements on the refractive index of glass fragments found on a suspect and at the scene of a burglary. We model the corresponding random variables as

$$X \mid \theta_1 \sim N(\theta_1, \sigma^2), \quad Y \mid \theta_2 \sim N(\theta_2, \sigma^2),$$

where θ_1 and θ_2 are the true refractive indexes and σ^2 is known. If the suspect is guilty, then $\theta_1 = \theta_2 = \theta$, say. We model natural variation among refractive indexes by supposing that θ is drawn from a population of types of glass whose true refractive indexes are $N(\mu, \tau^2)$, where μ and $\tau^2 \gg \sigma^2$ both known. Thus under G,

$$X, Y \mid \theta \overset{\text{iid}}{\sim} N(\theta, \sigma^2), \quad \theta \sim N(\mu, \tau^2),$$

while under $\overline{G}$, the true indexes θ_1 and θ_2 are independent, giving

$$X \mid \theta_1 \sim N(\theta_1, \sigma^2), \quad Y \mid \theta_2 \sim N(\theta_2, \sigma^2), \quad \theta_1, \theta_2 \overset{\text{iid}}{\sim} N(\mu, \tau^2).$$

It turns out to be easier to work in terms of transformed observations $u = x - y$ and $z = \frac{1}{2}(x + y)$, and to write the corresponding random variables as

$$U = \theta_1 - \theta_2 + \varepsilon_1 - \varepsilon_2, \quad Z = \frac{1}{2}(\theta_1 + \theta_2 + \varepsilon_1 + \varepsilon_2), \quad \varepsilon_1, \varepsilon_2 \overset{\text{iid}}{\sim} N(0, \sigma^2).$$

Then U and Z are independent and normal both conditionally on θ_1, θ_2 and unconditionally. Under G, $\theta_1 = \theta_2$, so

$$U \sim N(0, 2\sigma^2), \quad Z \sim N\left(\mu, \tau^2 + \frac{1}{2}\sigma^2\right),$$

while under $\overline{G}$,

$$U \sim N(0, 2\tau^2 + 2\sigma^2), \quad Z \sim N\left(\mu, \frac{1}{2}\tau^2 + \frac{1}{2}\sigma^2\right).$$

As the Jacobian of the transform from (x, y) to (u, z) equals one under both G and $\overline{G}$, and $\tau^2 \gg \sigma^2$, the Bayes factor is roughly

$$\frac{(2\sigma^2)^{-1/2} \exp\{-u^2/(4\sigma^2)\}(\tau^2)^{-1/2} \exp\{-(z - \mu)^2/(4\tau^2)\}}{(2\tau^2)^{-1/2} \exp\{-u^2/(4\tau^2)\}(\tau^2/2)^{-1/2} \exp\{-(z - \mu)^2/\tau^2\}},$$

which equals

$$\left(\frac{\tau^2}{2\sigma^2}\right)^{1/2} \times \exp\left(-\frac{u^2}{4\sigma^2}\right) \times \exp\left\{\frac{(z - \mu)^2}{2\tau^2}\right\}.$$

The interpretation of the second term is that if the difference $u = x - y$ is large relative to its variance $2\sigma^2$, there is strong evidence that θ_1 and θ_2 differ, and this favours $\overline{G}$. The third term measures how typical x and y are. If $z = \frac{1}{2}(x + y)$ is far from its mean, μ, compared to its variance $\frac{1}{2}\tau^2$ under $\overline{G}$, both x and y have similar but unusual refractive indexes, and this strengthens the evidence for G. With $\tau/\sigma = 100$, $u/(2\sigma^2)^{1/2} = 2$, and $(z - \mu)/(\frac{1}{2}\tau^2)^{1/2} = 2$, for example, these factors are respectively 0.135 and 2.718, and the overall Bayes factor is 26.01. Under G a frequentist test for a difference between θ_1 and θ_2 based on u would suggest that $\theta_1 \neq \theta_2$ at the 5% level, but the Bayes factor gives strong evidence in favour of guilt, as the values of x and y correspond to similar, unusual, types of glass.

A more realistic model would account for non-normality of the distribution of θ. Other forms of evidence, such as DNA fingerprints or cloth samples, require more complex likelihoods in the Bayes factor and use prior information from specially tailored databases. Moreover when the probabilities being modelled are very small, it is important to allow for the possibility of events such as mistakes at the forensic laboratory.   ∎

We often wish to test nested hypotheses. In a typical example $\theta = (\psi, \lambda)$ for real ψ, and λ varies in an open subset of $\mathbb{R}^p$, with $H_0 : \psi = \psi_0$ and $H_1 : \psi \neq \psi_0$. Then if the same proper continuous prior $\pi(\psi, \lambda)$ is used under both hypotheses, the prior odds in favour of H_1 are infinite because

$$\Pr(H_0) = \int \pi(\psi_0, \lambda) \, d\lambda = 0$$

is an integral over a set of prior probability zero. Thus the posterior odds in favour of H_1 are infinite, whatever the data. This vexation can be eliminated by using different prior densities, weighted according to prior belief in H_0 and H_1, giving overall prior

$$\pi(\psi, \lambda) = \delta(\psi - \psi_0)\pi(\psi_0, \lambda \mid H_0)\Pr(H_0) + \pi(\psi, \lambda \mid H_1)\Pr(H_1),$$

$\delta(\cdot)$ is the Dirac delta function.

where

$$\int \pi(\psi_0, \lambda \mid H_0) \, d\lambda = \int \pi(\psi, \lambda \mid H_1) \, d\psi d\lambda = 1.$$

One result of this is that Bayes factors are more sensitive to the prior than are posterior densities. In particular, improper priors cannot be used, as the Bayes factor depends on the ratio of the two arbitrary constants of proportionality that appear in the priors. One way to remove the arbitrariness is to fix the ratio of these constants using some external argument.

A further difficulty is that when a large number of models must be compared, prior probabilities and proper priors must be assigned to each. This can be hard in practice, and the results may depend strongly on how it is done. This contrasts with frequentist hypothesis testing, where such difficulties do not arise. An apparently even more striking contrast is provided by the following example.

Example 11.15 (Jeffreys–Lindley paradox)   Consider testing $H_0 : \mu = 0$ against $H_1 : \mu \neq 0$ based on a normal random sample $y_1, \ldots, y_n$ with mean μ and known variance σ^2. The usual test is based on the normal distribution of $n^{1/2}\overline{Y}/\sigma$ under H_0, and gives P-value $p = \Phi(-n^{1/2}|\overline{y}|/\sigma)$. In the Bayesian framework, we write $\pi_0 = \Pr(H_0)$, and suppose that under H_1, μ is normal with mean zero and variance τ^2. Then the posterior probabilities are

Dennis Victor Lindley (1923–) was educated at Cambridge, and held academic posts there, in Aberwystwyth, and in London. He is a strong advocate of Bayesian statistics. See Smith (1995).

$\overline{Y}$ is the average of the random variables $Y_1, \ldots, Y_n$; its observed value is $\overline{y}$.

$$\Pr(H_0 \mid y) = \frac{\pi_0}{(2\pi\sigma^2)^{n/2}} \exp\left(-\frac{1}{2\sigma^2} \sum_{j=1}^{n} y_j^2\right),$$

$$\Pr(H_1 \mid y) = \frac{1 - \pi_0}{(2\pi\sigma^2)^{n/2}(2\pi\tau^2)^{1/2}} \int \exp\left\{-\frac{1}{2\sigma^2} \sum_{j=1}^{n}(y_j - \mu)^2 - \frac{\mu^2}{2\tau^2}\right\} d\mu,$$

leading to Bayes factor

$$B_{01} = \left(1 + n\frac{\tau^2}{\sigma^2}\right)^{1/2} \exp\left\{-\frac{n\overline{y}^2}{2\sigma^2(1 + n^{-1}\sigma^2/\tau^2)}\right\}$$

Table 11.5 Dependence of Bayes factor B_{01} on sample size n for a test with significance level 0.01.

n	1	10	100	1000	10^4	10^6	10^8
B_{01}	0.269	0.163	0.376	1.15	3.63	36.2	362

in favour of H_0. Now suppose that $n\overline{y}^2/\sigma^2 = z_{\alpha/2}^2$. The significance level of the conventional test is α, but as $n \to \infty$ we see that $B_{01} \doteq n^{1/2}\tau\sigma^{-1}\exp(-z_{\alpha/2}^2/2)$, giving increasingly strong evidence in favour of H_0. Hence the paradox: although with $\overline{y}$ corresponding to $\alpha = 10^{-6}$ we would reject H_0 decisively, the Bayes factor gives increasingly strong support for H_0, because as $n \to \infty$, the weight of the alternative distribution is more and more widely spread compared to the distance from $\overline{y}$ to the null hypothesis value of μ. Table 11.5 gives some values of B_{01} when $\tau^2 = \sigma^2$.

One resolution of this hinges on noticing that a fixed alternative is not appropriate as $n \to \infty$. A test is used when there is doubt as to its outcome — when the data do not evidently contradict the null hypothesis. Mathematically, this means that sensible alternatives are $O(n^{-1/2})$ distant from the null hypothesis. In this case we take $\tau^2 = n^{-1}\delta\sigma^2$, so that as $n \to \infty$ the range of alternatives is fixed relative to the null; sensible values for δ might be in the range 5–20. Then the Bayes factor corresponding to significance level α, $B_{01} = (1 + \delta)^{1/2}\exp\{-\frac{1}{2}z_{\alpha/2}^2/(1 + \delta^{-1})\}$, does not increase with n. If we take $\delta = 10$ and $\alpha = 0.05, 0.01, 0.001$, and 0.0001, B_{10} equals 1.73, 6.2, 41.4, and 293. According to Table 11.3 these correspond respectively to evidence against H_0 that is hardly worth mentioning, positive, strong, and very strong, broadly agreeing with the usual interpretation of the P-values. ■

11.2.3 Model criticism

The prior density $\pi(\theta)$ introduces further information into the model, with the benefit of directness of inference for θ. The corresponding disbenefit is the need to assess the appropriateness of $\pi(\theta)$ and the sensitivity of posterior conclusions to the prior, added to the usual concerns about the sampling model $f(y \mid \theta)$. Sensitivity analysis is generally performed simply by comparing posterior inferences based on a range of priors and models. The problems this poses are mainly computational, and we discuss them briefly in Section 11.3.

When just a few parametrized alternative models are in view, the ideas for model comparison outlined in Section 11.2.2 can be applied, supplemented with suitable graphs. In practice, however, consideration of all possible models is usually infeasible, not least because data can spring surprises on the investigator, and so we turn to model-checking when the alternatives are not explicit.

Marginal inference

From a Bayesian viewpoint all information concerning the data and model is contained in the joint density

$$f(y, \theta) = \pi(\theta \mid y)f(y). \tag{11.19}$$

and this suggests that $f(y)$ should be used to check the model. It is relatively clear how to do this when there is a sufficient statistic s and $s = (t, a)$, where a is a function of s whose distribution does not depend on θ; a is an *ancillary statistic*, a notion explored in Section 12.1. Then we can write

$$f(y) = f(y \mid s) f(a) \int f(t \mid a, \theta) \pi(\theta) \, d\theta, \qquad (11.20)$$

the first two components of which do not depend on the prior, and hence can be used to give information about the sampling model. The third component of (11.20), $f(t \mid a)$, can be regarded as carrying information about agreement between data and prior. In simple models, consideration of the first two terms can yield standard model-checking tools.

Example 11.16 (Location-scale model)   Let $y_1, \ldots, y_n$ be a random sample from the location-scale model $y_j = \eta + \tau \varepsilon_j$, where the ε_j have density g. In general, the order statistics $s = (y_{(1)}, \ldots, y_{(n)})$ form a minimal sufficient statistic for $\theta = (\eta, \tau)$ based on $y_1, \ldots, y_n$. They may be re-expressed as

$$t = \widehat{\theta} = (\widehat{\eta}, \widehat{\tau}), \quad a = \left(\frac{y_{(1)} - \widehat{\eta}}{\widehat{\tau}}, \ldots, \frac{y_{(n)} - \widehat{\eta}}{\widehat{\tau}} \right),$$

where t consists of the maximum likelihood estimators of θ, and the joint distribution of the maximal invariant a is degenerate but independent of η and τ. The suitability of g can be checked by probability plots of a against quantiles of g. Similar ideas extend to regression models.

Given a particular choice of g, agreement between the prior and data would be assessed through the conditional density of $\widehat{\theta}$ given a.

When g is normal, the minimal sufficient statistic is $(\overline{y}, s^2)$ and the assumption of normality is checked using the distribution of y given $\overline{y}$ and s^2. Example 5.14 established that the raw residuals $((y_1 - \overline{y})/s, \ldots, (y_n - \overline{y})/s)$ are independent of $\overline{y}$ and s^2.

The marginal joint distribution of $\overline{y}$ and s^2 enables the prior to be criticized. For instance, suppose that a joint conjugate prior is used for μ and σ^2, with

$IG(\cdot, \cdot)$ denotes the inverse gamma distribution.

$$\mu \mid \sigma^2 \sim N(\mu_0, \sigma^2/k_0), \quad \sigma^2 \sim IG\left(\frac{1}{2}\nu_0, \frac{1}{2}\nu_0\sigma_0^2 \right).$$

Then integration shows that the marginal densities of $\overline{y}$ and s^2 are given by

$$d_1 = \frac{\overline{y} - \mu_0}{\sigma_0 \left(n^{-1} + k_0^{-1} \right)^{1/2}} \sim t_{\nu_0}, \quad d_2 = \frac{s^2}{\sigma_0^2} \sim F_{n-1, \nu_0}.$$

Values of d_1 and d_2 that are unusual relative to the distributions of the corresponding random variables D_1 and D_2 can cast doubt on both prior and sampling models. For example, if a probability plot cast no doubt on the assumption of normality, and $d_1 = 100$ nevertheless, the relevance of the prior values μ_0 and σ_0^2 would be called into question. But if the data were not normal but Cauchy, then $\overline{y}$ would have the same distribution as y_1 and very large values of d_1 could arise even if the prior and data agreed about μ.

Consider again the data of Example 11.12, for which the model was normal. Suppose that our prior is that conditional on σ^2, $\mu \sim N(0, \sigma^2)$, and that the prior distribution for σ^2 is $IG(3, 3 \times 100^2)$. Then $d_1 = 0.202$ and $d_2 = 0.1424$. The first is close enough to zero to cast no doubt on the prior mean, but d_2 is rather small relative to the $F_{14,6}$ distribution, and casts some doubt on the prior variance. The corresponding Bayesian P-values are $\Pr(|D_1| > |d_1|) = 0.75$ and $\Pr(D_2 < d_2) = 0.045$; the data are rather more precise than our prior information would suggest. ∎

One overall measure of the plausibility of the data under the model is the probability $\Pr\{f(Y_+) \leq f(y)\}$, where $f(y)$ is the marginal density of the data actually observed, and Y_+ is a set of data that might have been observed (Problem 11.12). Some controversy surrounds this test and the P-values calculated in the previous example, as they flout the likelihood principle. One view is that the essence of Bayesian inference is to use Bayes' theorem to update prior belief in light of the data. This entails using posterior probabilities or equivalently Bayes factors to compare competing models, and leaves no place for tail probability calculations. A contrary argument is that a Bayes factor measures the relative support for two hypotheses and therefore requires prior specification of each, while some model-checking techniques do not require explicit alternatives: if my prior belief is that $y_1, \ldots, y_{20} \overset{iid}{\sim} N(0, 1)$, I am surprised to learn that the smallest value is -10, even before considering how this could have arisen. Furthermore, a strict interpretation of the argument for Bayes factors requires the specification of a proper prior distribution over all reasonable alternatives, which seems infeasible in practice. Finally, the argument for the likelihood principle assumes that the model is correct and the case for strict adherence to the principle seems weaker when assessing fit than when performing inference for a parameter.

Prediction diagnostics

Most models do not have a useful reduction in terms of exact minimal sufficient or ancillary statistics, so the ideas outlined above cannot usually be applied. Moreover, $\pi(\theta)$ is often improper in practice and then $f(y)$ is typically improper also, though this need not undercut diagnostic use of $f(y \mid s)f(a)$ if there is a useful sufficient reduction. When $\pi(\theta)$ is improper, posterior predictive distributions can be used to diagnose both problems with individual cases and more general model failures. The idea is to assess the posterior plausibility of suitable functions of the data.

One way to detect single outliers compares observations with their predicted values conditional on the remaining data through the *conditional predictive ordinates* $f(y_j \mid y_{-j})$, where y_{-j} consists of all the data except y_j. Since these quantities may be written in terms of ratios of densities, they depend less on the propriety of priors. There is a close link to cross-validation.

Example 11.17 (Normal linear model) In the normal linear model with known $n \times p$ design matrix X of rank $p < n$, the distribution of the $n \times 1$ response vector y conditional on the $p \times 1$ vector of parameters β and the error variance σ^2 is normal with mean $X\beta$ and covariance matrix $\sigma^2 I_n$, and the least squares estimates and residual estimate of error

$$\widehat{\beta} = (X^{\mathrm{T}}X)^{-1}X^{\mathrm{T}}y, \quad s^2 = (n-p)^{-1}y^{\mathrm{T}}\{I - X(X^{\mathrm{T}}X)^{-1}X^{\mathrm{T}}\}y,$$

are independent and minimal sufficient for β and σ^2.

It would be alarming if the usual standardized residuals r_j had no Bayesian justification. Fortunately they do, as we now see. The simplest argument is that the joint distribution of $a = (r_1, \ldots, r_n)$ is free of the parameters $\theta = (\beta, \sigma^2)$, for which $\widehat{\theta} = (\widehat{\beta}, s^2)$ form a complete minimal sufficient statistic. Basu's theorem (page 649) implies that a is independent of $\widehat{\theta}$, so we infer from (11.20) that the sampling model can be checked by comparing a to its joint distribution. This justifies residual plots and other tricks of the trade.

Concentrationally-challenged readers may want to jump to (11.23).

For a longer more tedious argument for Bayesian use of deletion residuals and hence of the r_j, we compute the conditional predictive ordinate $f(y_j \mid y_{-j})$ under the conjugate prior distribution for β and σ^2,

$$\beta \mid \sigma^2 \sim N(\gamma, \sigma^2 V), \quad \sigma^2 \sim IG\left(\frac{1}{2}v, \frac{1}{2}v\tau^2\right),$$

where the hyperparameters are the $p \times 1$ vector γ, the $p \times p$ positive definite symmetric matrix V, and the scalars v and τ^2; these are all regarded as known. An argument analogous to that leading to (11.13) gives

$$\pi(\beta, \sigma^2 \mid y) \propto \pi(\beta \mid \widehat{\beta}, \sigma^2)\pi(\sigma^2 \mid s^2),$$

so we need only find the posterior distributions of β given $\widehat{\beta}$ and σ^2 and of σ^2 given s^2. As the joint distribution of $(\beta^{\mathrm{T}}, \widehat{\beta}^{\mathrm{T}})^{\mathrm{T}}$ given σ^2 is

$$N_{2p}\left\{\begin{pmatrix}\gamma\\\gamma\end{pmatrix}, \sigma^2\begin{pmatrix}V & V\\V & V+(X^{\mathrm{T}}X)^{-1}\end{pmatrix}\right\},$$

(3.21) and Exercise 8.5.2 shows that the posterior distribution of β given $\widehat{\beta}$ and σ^2 is normal with mean and variance matrix

$$(X^{\mathrm{T}}X + V^{-1})^{-1}(X^{\mathrm{T}}X\widehat{\beta} + V^{-1}\gamma), \quad \sigma^2(X^{\mathrm{T}}X + V^{-1})^{-1}, \tag{11.21}$$

which generalizes (11.11). As prior uncertainty about γ increases, $V^{-1} \to 0$, and then we see from (11.21) that the posterior mean and variance of β approach $\widehat{\beta}$ and $\sigma^2(X^{\mathrm{T}}X)^{-1}$. Direct calculation shows that the posterior distribution of σ^2 given s^2 is $IG[(v+n)/2, \{v\tau^2 + (n-p)s^2\}/2]$. If the constant prior $\pi(\beta) \propto 1$ is used, then the posterior mean and variance of β given σ^2 are $\widehat{\beta}$ and $\sigma^2(X^{\mathrm{T}}X)^{-1}$, but the posterior density for σ^2 is $IG[(v+n-p)/2, \{v\tau^2 + (n-p)s^2\}/2]$; letting $v \to 0$ gives the effect of taking $\pi(\beta, \sigma^2) \propto \sigma^{-2}$.

For future reference we note that the distribution of y conditional on σ^2 is normal with mean $X\gamma$ and variance $\sigma^2(I + XVX^{\mathrm{T}})$, and that on integrating over the prior distribution for σ^2, we find that the marginal density $f(y)$ has a multivariate t form

$$\frac{\Gamma\left(\frac{n+\nu}{2}\right)(\nu\tau^2)^{\nu/2}}{\pi^{n/2}\Gamma\left(\frac{\nu}{2}\right)|I+XVX^{\mathrm{T}}|^{1/2}}\{\nu\tau^2+(y-X\gamma)^{\mathrm{T}}(I+XVX^{\mathrm{T}})^{-1}(y-X\gamma)\}^{-(n+\nu)/2}.$$

(11.22)

To find the posterior predictive density of another observation y_+ with $p \times 1$ covariate vector x_+, assumed independent of y conditional on β and σ^2, we write

$$
\begin{aligned}
f(y_+ \mid y) &= \int f(y_+ \mid \theta)\pi(\theta \mid y)\,d\theta \\
&= \int\int f(y_+ \mid \beta,\sigma^2)\pi(\beta \mid \widehat{\beta},\sigma^2)\pi(\sigma^2 \mid s^2)\,d\beta\,d\sigma^2 \\
&= \int \pi(\sigma^2 \mid s^2)\int f(y_+ \mid \beta,\sigma^2)\pi(\beta \mid \widehat{\beta},\sigma^2)\,d\beta\,d\sigma^2.
\end{aligned}
$$

Now

$$
\begin{aligned}
y_+ \mid \beta,\sigma^2 &\sim N(x_+^{\mathrm{T}}\beta,\sigma^2), \\
\beta \mid \widehat{\beta},\sigma^2 &\sim N\{(X^{\mathrm{T}}X+V^{-1})^{-1}(X^{\mathrm{T}}X\widehat{\beta}+V^{-1}\gamma),\sigma^2(X^{\mathrm{T}}X+V^{-1})^{-1}\},
\end{aligned}
$$

from which it follows that conditional on $\widehat{\beta}$ and σ^2, the distribution of y_+ is normal with mean and variance

$$x_+^{\mathrm{T}}(X^{\mathrm{T}}X+V^{-1})^{-1}(X^{\mathrm{T}}X\widehat{\beta}+V^{-1}\gamma),\quad \sigma^2\{1+x_+^{\mathrm{T}}(X^{\mathrm{T}}X+V^{-1})^{-1}x_+\}.$$

Integration over the posterior distribution of σ^2 shows that the posterior predictive distribution of y_+ conditional on y is given by

$$\frac{y_+ - x_+^{\mathrm{T}}(X^{\mathrm{T}}X+V^{-1})^{-1}(X^{\mathrm{T}}X\widehat{\beta}+V^{-1}\gamma)}{\left[\{\frac{(n-p)s^2+\nu\tau^2}{n+\nu}\}\{1+x_+^{\mathrm{T}}(X^{\mathrm{T}}X+V^{-1})^{-1}x_+\}\right]^{1/2}} \sim t_{n+\nu}.$$

(11.23)

For prediction of y_j given the other observations y_{-j}, based on the improper prior $\pi(\beta,\sigma^2) \propto \sigma^{-2}$, we set $V^{-1}=0$ and $\nu=0$ and replace y_+ with y_j, x_+ with x_j, $n+\nu$ with $n-p-1$, and $\widehat{\beta}$, s^2 and X with the corresponding quantities $\widehat{\beta}_{-j}$, s^2_{-j} and X_{-j} based on y_{-j}. Then (11.23) becomes

$$\frac{y_j - x_j^{\mathrm{T}}\widehat{\beta}_{-j}}{\left[s^2_{-j}\{1+x_j^{\mathrm{T}}(X_{-j}^{\mathrm{T}}X_{-j})^{-1}x_j\}\right]^{1/2}} \sim t_{n-p-1}.$$

A straightforward calculation reveals that the term in braces in the denominator here is $(1-h_j)^{-1}$, where h_j is the jth leverage based on the full model. Hence prediction of y_j given y_{-j} may be based on the t_{n-p-1} distribution of the deletion residual

$$r_j^* = \frac{(y_j - x_j^{\mathrm{T}}\widehat{\beta}_{-j})(1-h_j)^{1/2}}{s_{-j}}.$$

Thus outlier detection based on the conditional predictive ordinate is conducted using the usual deletion residuals r_j^*. As these are monotonic functions of the standardized residuals r_j, this supports Bayesian use of the r_j. ∎

More general diagnostics can be based on measures of discrepancy between data and the model, $d = d(y, \theta)$, compared to data Y_+ that might have been generated by the model. Posterior predictive checks are based on comparison of $D_+ = d(Y_+, \theta)$ with its predictive distribution, via

$$\Pr\{d(Y_+, \theta) \geq d(y, \theta) \mid y\}, \tag{11.24}$$

where the averaging is over both Y_+ and the posterior distribution of θ. Since Y_+ is independent of y given θ, we can write

$$\int \Pr\{D_+ \geq d(y, \theta) \mid y, \theta\} \pi(\theta \mid y) d\theta = \int \Pr\{D_+ \geq d(y, \theta) \mid \theta\} \pi(\theta \mid y) d\theta.$$

Thus a simple way to evaluate (11.24) is to calculate $\Pr\{D_+ \geq d(y, \theta) \mid \theta\}$ for fixed θ, and then to average this probability over the posterior density of θ. One omnibus measure of discrepancy is the analogue of Pearson's statistic,

$$d(y, \theta) = \sum_{j=1}^{n} \frac{\{y_j - E(Y_j \mid \theta)\}^2}{\text{var}(Y_j \mid \theta)},$$

but this may be inappropriate, and typically D_+ is chosen with key aspects of the model in mind. As mentioned above, authors differ over whether (11.24) should be used, though unlike the use of the marginal density of y, inference based on (11.24) does condition on the data.

11.2.4 Prediction and model averaging

In the Bayesian framework prediction is performed through the posterior predictive density (11.6). In practice this is not as simple as it appears, because there may be a number of possible models $M_1, \ldots, M_k$ on which to the base the prediction. Conditional on M_i, the predictive density for z based on y is $f(z \mid y, M_i)$, but this ignores any uncertainty concerning the selection of M_i. This uncertainty can be incorporated by averaging over the posterior distribution of the model selected, to give the *model-averaged prediction*

$$f(z \mid y) = \sum_{i=1}^{k} f(z \mid y, M_i) \Pr(M_i \mid y) \tag{11.25}$$

which is an average of the posterior distributions of z under the different models, weighted according to their posterior probabilities

$$\Pr(M_i \mid y) = \frac{f(y \mid M_i) \Pr(M_i)}{\sum_{l=1}^{k} f(y \mid M_l) \Pr(M_l)}, \tag{11.26}$$

where

$$f(y \mid M_i) = \int f(y \mid \theta_i, M_i) \pi(\theta_i \mid M_i) d\theta_i,$$

$$f(z \mid M_i, y) = \frac{\int f(z \mid y, \theta_i, M_i) f(y \mid \theta_i, M_i) \pi(\theta_i \mid M_i) d\theta_i}{f(y \mid M_i)}.$$

Here θ_i is the parameter for model M_i, under which the prior is $\pi(\theta_i \mid M_i)$ and the prior probability of M_i is $\Pr(M_i)$. Formally, (11.25) is just a re-expression of (11.6) in which the parameter splits into two parts, one a model indicator, M_i, and the other the parameters conditional on M_i. In using (11.25) it is crucial that z is the same quantity under all models considered, rather than one whose interpretation depends on the model.

In practice the main obstacle to model averaging is computational. For each model, the integrations involved must usually be done numerically using ideas described in Section 11.3. Furthermore there can be many models in some applications — for example, selecting among 15 covariates in a regression problem gives $2^{15} = 32,768$ models, corresponding to inclusion or exclusion of each covariate separately, without considering outliers, transformations, and so forth. Thus it may be difficult to find the most plausible models, quite apart from the calculations conditional on each model and the difficulties of specifing a prior over model space — giving the same weight to all combinations of covariates will rarely be sensible.

Example 11.18 (Cement data) We fit linear models to the data in Table 8.1 with $n = 13$ observations and four covariates. There are 2^4 possible subsets of the covariates, giving us models $M_1, \ldots, M_{16}$, which for sake of illustration we regard as equally probable *a priori*, though in practice we should hope that a small number of covariates is more likely than a large number. The models are on different parameter spaces, so the discussion in Section 11.2.2 implies that proper, preferably weak, priors should be used. We use the conjugate prior described in Example 11.17, and without loss of generality centre and scale each covariate vector to have average zero and unit variance. We then set V to be the 5×5 matrix with diagonal elements $\phi^2(v, 1, 1, 1, 1)$, where v is the sample variance of y, $\gamma^{\mathrm{T}} = (\overline{y}, 0, 0, 0, 0)$, $\nu = 2.58$, $\tau^2 = 0.28$, and $\phi = 2.85$. This choice implies that the elements of β are independent *a priori*, and should give a weak but proper prior that is consistent between different models and invariant to location and scale changes of the response and explanatory variables.

The marginal density of y under this model is (11.22); for each subset of covariates we use the corresponding submatrix of V. Table 11.6 shows the quantities $2 \log B_{10}$, where $B_{10} = \Pr(y \mid M_1)/\Pr(y \mid M_0)$ is the Bayes factor in favour of a subset of covariates relative to the model with none, the posterior probabilities of each subset, and, for comparison, the residual sums of squares under the usual linear models, which are broadly in line with the probabilities.

Let us try and predict the value of a new response y_+ with covariates $x_+^{\mathrm{T}} = (1, 10, 40, 20, 30)$. Conditional on a particular subset of covariate vectors, the predictive distribution for y_+ is given by (11.23). Figure 11.3 shows these densities for the six models shown in Table 11.6 to have non-negligible support, and the model-averaged predictive density. ∎

A different approach to dealing with model uncertainty is to find a plausible model, $f(y \mid \psi)\pi(\psi)$, and then add further parameters λ whose variation allows for the most uncertain aspects of the model, together with a prior that expresses belief about them.

Model	RSS	$2 \log B_{10}$	$\Pr(M \mid y)$	a	b
– – – –	2715.8	0.0	0.0000		
1 – – –	1265.7	7.1	0.0000		
– 2 – –	906.3	12.2	0.0000		
– – 3 –	1939.4	0.6	0.0000		
– – – 4	883.9	12.6	0.0000		
1 2 – –	57.9	45.7	0.2027	93.77	2.31
1 – 3 –	1227.1	4.0	0.0000		
1 – – 4	74.8	42.8	0.0480	99.05	2.58
– 2 3 –	415.4	19.3	0.0000		
– 2 – 4	868.9	11.0	0.0000		
– – 3 4	175.7	31.3	0.0002		
1 2 3 –	48.11	43.6	0.0716	95.96	2.80
1 2 – 4	47.97	47.2	0.4344	95.88	2.45
1 – 3 4	50.84	44.2	0.0986	94.66	2.89
– 2 3 4	73.81	33.2	0.0004		
1 2 3 4	47.86	45.0	0.1441	95.20	2.97

Table 11.6 Bayesian prediction using model averaging for the cement data. For each of the 16 possible subsets of covariates, the table shows the log Bayes factor in favour of that subset compared to the model with no covariates and gives the posterior probability of each model. The values of the posterior mean and scale parameters a and b are also shown for the six most plausible models; $(y_+ - a)/b$ has a posterior t density. For comparison, the residual sums of squares are also given.

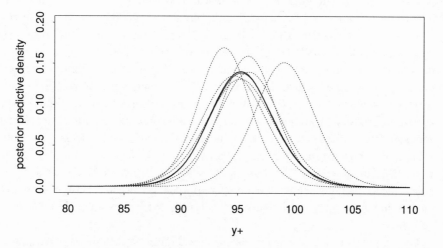

Figure 11.3 Posterior predictive densities for cement data. Predictive densities for y_+ based on individual models are given as dotted curves, and the heavy curve is the averaged prediction from all 16 models.

This gives an expanded model $f(y \mid \psi, \lambda)\pi(\psi, \lambda)$, to which (11.6) is then applied with $\theta = (\psi, \lambda)$.

Exercises 11.2

1   Find elements $\tilde{\theta}$ and $\tilde{J}(\tilde{\theta})$ of the normal approximation to a beta density, and hence check the formulae in Example 11.11. Find also the posterior mean and variance of θ. Give an approximate 0.95 credible interval for θ. How does this differ from a 0.95 confidence interval? Comment.

2   Let $Y_1, \ldots, Y_n$ be a random sample from the uniform distribution on $(0, \theta)$, and take as prior the Pareto density with parameters β and λ,

$$\pi(\theta) = \beta \lambda^\beta \theta^{-\beta-1}, \quad \theta > \lambda, \quad \beta, \lambda > 0.$$

(a) Find the prior distribution function and quantiles for θ, and hence give prior one- and two-sided credible intervals for θ. If $\beta > 1$, find the prior mean of θ.

(b) Show that the posterior density of θ is Pareto with parameters $n + \beta$ and $\max\{Y_1, \ldots, Y_n, \lambda\}$, and hence give posterior credible intervals and the posterior mean for θ.

(c) Interpret λ and β in terms of a prior sample from the uniform density.

3 Check the details of Example 11.7.

4 Two independent samples $Y_1, \ldots, Y_n \overset{\text{iid}}{\sim} N(\mu, \sigma^2)$ and $X_1, \ldots, X_m \overset{\text{iid}}{\sim} N(\mu, c\sigma^2)$ are available, where $c > 0$ is known. Find posterior densities for μ and σ based on prior $\pi(\mu, \sigma) \propto 1/\sigma$.

5 Verify (11.21), (11.22), and (11.23). How do (11.21) and (11.22) change when $\text{var}(y_j \mid \beta, \sigma^2) = \sigma^2/w_j$, the w_j being known weights?

6 Travelling in a foreign country, you arrive at midnight in a town you have never heard of. You have no idea of its size. The first thing you see is a bus with the number $y = 100$. What is a reasonable estimate of the total number θ of buses in the town, assuming that they are numbered $1, \ldots, \theta$?

(a) Explain why it is sensible to use the improper prior $\pi(\theta) \propto \theta^{-1}$, $\theta = 1, 2, \ldots$. Assuming that $f(y \mid \theta)$ is uniform on $1, \ldots, \theta$, show that θ has posterior density

$$\pi(\theta \mid y) = \frac{\theta^{-2}}{\sum_{u=y}^{\infty} u^{-2}}, \qquad \theta = y, y+1, \ldots.$$

(b) Show that the posterior mean of θ is infinite. Show also that the posterior distribution function is approximately

$$\Pr(\theta \leq v \mid y) \doteq \frac{\int_{y-1/2}^{v+1/2} u^{-2}\, du}{\int_{y-1/2}^{\infty} u^{-2}\, du},$$

and that the posterior median is approximately $2y - 3/2$. Give an equi-tailed 95% posterior confidence interval and a 95% HPD interval for θ.

(c) What would you conclude if you saw two buses, numbered 100 and 30?

7 In Example 11.12, calculate the Bayes factor for $H_0 : \mu \leq 0$ and $H_1 : \mu > 0$.

8 A forensic laboratory assesses if the DNA profile from a specimen found at a crime scene matches the DNA profile of a suspect. The technology is not perfect, as there is a (small) probability ρ that a match occurs by chance even if the suspect was not present at the scene, and a (larger) probability γ that a match is reported even if the profiles are different; this can arise due to laboratory error such as cross-contamination or accidental switching of profiles.

(a) Let R, S, and M denotes the events that a match is reported, that the specimen does indeed come from the suspect, and that there is a match between the profiles, and suppose that

$\overline{M}$ denotes the complement of M, and $\cap$ means 'and'.

$$\Pr(R \mid M \cap S) = \Pr(R \mid M \cap \overline{S}) = \Pr(R \mid M) = 1, \ \Pr(\overline{M} \mid S) = 0, \ \Pr(R \mid S) = 1.$$

Show that the posterior odds of the profiles matching, given that a match has been reported, depend on

$$\frac{\Pr(R \mid S)}{\Pr(R \mid \overline{S})} = \frac{\Pr(R \mid M \cap S)\Pr(M \mid S) + \Pr(R \mid \overline{M} \cap S)\Pr(\overline{M} \mid S)}{\Pr(R \mid M \cap \overline{S})\Pr(M \mid \overline{S}) + \Pr(R \mid \overline{M} \cap \overline{S})\Pr(\overline{M} \mid \overline{S})},$$

and establish that this equals $\{\rho + \gamma(1 - \rho)\}^{-1}$.

(b) Tabulate $\Pr(R \mid S)/\Pr(R \mid \overline{S})$ when $\rho = 0, \ 10^{-9}, \ 10^{-6}, \ 10^{-3}$ and $\gamma = 0, \ 10^{-4}, \ 10^{-3}, 10^{-2}$.

(c) At what level of posterior odds would you be willing to convict the suspect, if the only evidence against them was the DNA analysis, and you should only convict if convinced of their guilt 'beyond reasonable doubt'? Would your chosen odds level depend on the likely sentence, if they are found guilty? How does your answer depend on the prior odds of the profiles matching, $\Pr(S)/\Pr(\overline{S})$?

9   One way to set the ratio of arbitrary constants that appears when two models are compared using Bayes factors and improper priors is by *imaginary observations*: we imagine the smallest experiment that would enable the models to be discriminated but maximizes evidence in favour of H_0, and then choose the constants so that the Bayes factor equals one for these data.

Consider data from a Poisson process observed on $[0, t_0]$, and let H_0 and H_1 represent the models with rates $\lambda(t) = \rho$ and $\lambda(t) = \mu\beta^{-1}\{1 - \exp(-\beta t)\}$, where $\rho, \mu, \beta > 0$. Take improper priors $\pi(\rho) = c_0\rho^{-1}$ and $\pi(\mu, \beta) = c_1\mu^{-2}$, with $c_1, c_0 > 0$.

(a) Explain why the smallest experiment that enables the models to be discriminated must have two events, and show that it gives $\Pr(y \mid H_0) = c_0/t_0^2$. Find $\Pr(y \mid H_1)$ and show that it is minimized when both events occur at t_0, with

$$\Pr(y \mid H_1) = c_1 \int_0^\infty \frac{\beta e^{-2\beta t_0}}{1 - e^{-\beta t_0}} \, d\beta = c_1 t_0^{-2}\left(\frac{\pi^2}{6} - 1\right).$$

Deduce that the device of imaginary observations gives $c_0/c_1 = \pi^2/6 - 1$.

(b) Compute the Bayes factor when these two models are compared using the data in Table 6.13. Discuss.

(Section 6.5.1; Raftery, 1988; Spiegelhalter and Smith, 1982)

10 A random sample $y_1, \ldots, y_n$ arises either from a log-normal density, with $\log Y_j \sim N(\mu, \sigma^2)$, or from an exponential density $\rho^{-1}e^{-y/\rho}$. The improper priors chosen are $\pi(\rho) = c_0/\rho$ and $\pi(\mu, \sigma) = c_1/\sigma$, for $\rho, \sigma > 0$ and $c_0, c_1 > 0$. Use imaginary observations to give a value for c_1/c_0.

11.3 Bayesian Computation

11.3.1 Laplace approximation

The goal of Bayesian data analysis is posterior inference for quantities of interest, and this involves integration over one or more of the parameters. Usually the integrals cannot be obtained in closed form and numerical approximations must be used. Deterministic integration procedures such as Gaussian quadrature can sometimes be applied, but they are typically useful only for low-dimensional integrals, and have the drawback of requiring information about the position and width of any modes of the integrand that unavailable in practice. The most powerful tool for approximate calculation of posterior densities is numerical integration by Monte Carlo simulation, to which we turn after describing an analytical approach known as *Laplace's method*.

Consider the one-dimensional integral

$$I_n = \int_{-\infty}^{\infty} e^{-nh(u)} \, du, \tag{11.27}$$

where $h(u)$ is a smooth convex function with minimum at $u = \tilde{u}$, at which point $dh(\tilde{u})/du = 0$ and $d^2h(\tilde{u})/du^2 > 0$. For compactness of notation we write $h_2 = d^2h(\tilde{u})/du^2$, $h_3 = d^3h(\tilde{u})/du^3$, and so forth. Close to $\tilde{u}$ a Taylor series expansion gives $h(u) \doteq h(\tilde{u}) + \frac{1}{2}h_2(u - \tilde{u})^2$, so

$$I_n \doteq e^{-nh(\tilde{u})} \int_{-\infty}^{\infty} e^{-nh_2(u-\tilde{u})^2/2} \, du$$

$$= e^{-nh(\tilde{u})} \int_{-\infty}^{\infty} e^{-z^2/2} \frac{du}{dz} \, dz$$

$$= \left(\frac{2\pi}{nh_2}\right)^{1/2} e^{-nh(\tilde{u})},$$

where the first and second equalities use the substitution $z = (nh_2)^{1/2}(u - \tilde{u})$ and the fact that the normal density has unit integral. A more detailed accounting (Exercise 11.3.2) gives

$$I_n = \left(\frac{2\pi}{nh_2}\right)^{1/2} e^{-nh(\tilde{u})} \times \left\{1 + n^{-1}\left(\frac{5h_3^2}{24h_2^3} - \frac{h_4}{8h_2^2}\right) + O(n^{-2})\right\}. \qquad (11.28)$$

The leading term on the right of (11.28) is known as the *Laplace approximation* to I_n, and we denote it by $\tilde{I}_n$.

There are several points to note about (11.28). First, as $I_n/\tilde{I}_n = 1 + O(n^{-1})$, the error is relative, and $\tilde{I}_n$ is often remarkably accurate. Second, $\tilde{I}_n$ involves only h and its second derivative at $\tilde{u}$, so it is relatively easy to obtain, numerically if necessary. Third, the right-hand side of (11.28) is an asymptotic series for I_n, implying that its partial sums need not converge, and that the approximation may not be improved by including further terms of the series. And fourth, because the bulk of the normal probability integral lies within three standard deviations of its centre, the limits of the integral will not affect $\tilde{I}_n$ provided they lie outside the interval with endpoints $\tilde{u} \pm 3(nh_2)^{-1/2}$ or so.

In the multivariate case, with $h(u)$ again a smooth convex function but u a vector of length p, the same argument but using the multivariate normal density shows that the Laplace approximation to (11.27) is

$$\left(\frac{2\pi}{n}\right)^{p/2} |h_2|^{-1/2} e^{-nh(\tilde{u})}, \qquad (11.29)$$

where $\tilde{u}$ solves the $p \times 1$ system of equations $\partial h(u)/\partial u = 0$ and $|h_2|$ is the determinant of the $p \times p$ matrix of second derivatives $\partial^2 h(u)/\partial u \partial u^{\mathsf{T}}$, evaluated at $u = \tilde{u}$, at which point the matrix is positive definite.

In applications an approximation is often required to an integral of form

$$J_n(u_0) = \left(\frac{n}{2\pi}\right)^{1/2} \int_{-\infty}^{u_0} a(u) e^{-ng(u)} \{1 + O(n^{-1})\} \, du, \qquad (11.30)$$

where u is scalar, $a(u) > 0$, and in addition to possessing the properties of $h(u)$ above, g is such that $g(\tilde{u}) = 0$. The first step in approximating (11.30) is to change the variable of integration from u to $r(u) = \text{sign}(u - \tilde{u})\{2g(u)\}^{1/2}$; that is, $r^2/2 = g(u)$. Then $g'(u) = dg(u)/du$ and $r(u)$ have the same sign, and $r \, dr/du = g'(u)$, so

$$J_n(u_0) = \left(\frac{n}{2\pi}\right)^{1/2} \int_{-\infty}^{r_0} a(u) \frac{r}{g'(u)} e^{-nr^2/2} \{1 + O(n^{-1})\} \, dr$$

$$= \left(\frac{n}{2\pi}\right)^{1/2} \int_{-\infty}^{r_0} e^{-nr^2/2 + \log b(r)} \{1 + O(n^{-1})\} \, dr,$$

where the positive quantity $b(r) = a(u)r/g'(u)$ is regarded as a function of r.

We now change variables again, from r to $r^* = r - (rn)^{-1} \log b(r)$, so

$$-nr^{*2} = -nr^2 + 2\log b(r) - n^{-1}r^{-2}\{\log b(r)\}^2.$$

The Jacobian of the transformation and the third term in $-nr^{*2}$ contribute only to the error of $J_n(u_0)$, so

$$J_n(u_0) = \left(\frac{n}{2\pi}\right)^{1/2} \int_{-\infty}^{r_0^*} e^{-nr^{*2}/2}\{1 + O(n^{-1})\}\, dr^*$$
$$= \Phi\left(n^{1/2}r_0^*\right) + O(n^{-1}), \tag{11.31}$$

where

$$r_0^* = r_0 + (r_0 n)^{-1} \log\left(\frac{v_0}{r_0}\right), \quad r_0 = \text{sign}(u_0 - \tilde{u})\{2g(u_0)\}^{1/2}, \quad v_0 = \frac{g'(u_0)}{a(u_0)}.$$

Variants on this expression play an important role in Chapter 12.

Here is a further approximation for later use. Let $u = (u_1, u_2)$, where u_1 is scalar and u_2 a $p \times 1$ vector, and consider

$$(2\pi)^{-(p+1)/2}c \int_{-\infty}^{u_1^0} du_1 \int du_2 \exp\{-nh(u_1, u_2)\}, \tag{11.32}$$

where c is constant, the inner integral being over $\mathbb{R}^p$. Here h has its previous smoothness properties, is maximized at $(\tilde{u}_1, \tilde{u}_2)$, and in addition $h(\tilde{u}_1, \tilde{u}_2) = 0$. We fix u_1 and apply Laplace approximation to the inner integral, obtaining

$$(2\pi)^{-1/2}c \int_{-\infty}^{u_1^0} |nh_{22}(u_1, \tilde{u}_{21})|^{-1/2} \exp\{-nh(u_1, \tilde{u}_{21})\}\{1 + O(n^{-1})\}\, du_1,$$

where $\tilde{u}_{21} = \tilde{u}_2(u_1)$ maximizes $h(u_1, u_2)$ with respect to u_2 when u_1 is fixed, and $h_{22}(u_1, u_2) = \partial^2 h(u_1, u_2)/\partial u_2 \partial u_2^{\mathsf{T}}$ is the $p \times p$ Hessian matrix of h with respect to u_2. Apart from multiplicative constants, this integral has form (11.30), and so (11.31) may be used to approximate to (11.32), with

$$r_0 = \text{sign}\left(u_1^0 - \tilde{u}_1\right)\left\{2h\left(u_1^0, \tilde{u}_{20}\right)\right\}^{1/2}, \quad v_0 = c^{-1}\frac{\partial h\left(u_1^0, \tilde{u}_{20}\right)}{\partial u_1}\left|h_{22}\left(u_1^0, \tilde{u}_{20}\right)\right|^{1/2},$$

where $\tilde{u}_{20}$ is the maximizing value of u_2 when $u_1 = u_1^0$.

Although the formulation of (11.27), (11.30), and (11.32) in terms of n and the $O(1)$ functions h and g simplifies the derivation of (11.29) and (11.31) by clarifying the orders of the various terms, for applications it is equivalent and usually simpler to set $n = 1$ and allow h and g and their derivatives to be $O(n)$.

Inference

One application of Laplace approximation is to the Bayes factor (11.17). For one of the hypotheses we write $\Pr(y) = \int f(y \mid \theta)\pi(\theta)\, d\theta$, with integrand expressed as $\exp\{-h(\theta)\}$, where $h(\theta) = -\ell_m(\theta)$ and

$$\ell_m(\theta) = \log f(y \mid \theta) + \log \pi(\theta)$$

is the log likelihood modified by addition of the log prior. Typically the first term of ℓ_m is $O(n)$, and the second is $O(1)$. The value $\tilde{\theta}$ that minimizes $h(\theta)$ is the maximum *a posteriori* estimate of θ — the value that maximizes the modified log likelihood — and we can apply (11.29). The result is

$$\log \Pr(y) \doteq \log f(y \mid \tilde{\theta}) + \log \pi(\tilde{\theta}) - \frac{1}{2} p \log n + \frac{1}{2} p \log(2\pi) - \frac{1}{2} \log \left| -\frac{\partial^2 \ell_m(\tilde{\theta})}{\partial \theta \partial \theta^{\mathrm{T}}} \right|,$$

where p is the dimension of θ. To further simplify this, note that in large samples the log prior is negligible relative to the log likelihood and $\tilde{\theta}$ is roughly the maximum likelihood estimate $\widehat{\theta}$, and if p is fixed we can drop terms that are $O(1)$. Crudely speaking, therefore,

$$-2 \log \Pr(y) \doteq \mathrm{BIC} = -2 \log f(y \mid \widehat{\theta}) + p \log n.$$

This *Bayes information criterion*, which we met in Section 4.7, is used for rough comparison of competing models.

For a more sophisticated application we write a vector parameter θ as $(\psi, \lambda^{\mathrm{T}})^{\mathrm{T}}$ and approximate the marginal posterior density for the scalar ψ,

$$\pi(\psi \mid y) = \frac{\int f(y \mid \psi, \lambda) \pi(\psi, \lambda) \, d\lambda}{\int f(y \mid \psi, \lambda) \pi(\psi, \lambda) \, d\lambda \, d\psi}, \tag{11.33}$$

by applying Laplace's method to each integral. The discussion above gives the approximation to the denominator. For the numerator we take $h_\psi(\lambda) = -\ell_m(\psi, \lambda)$, where the notation emphasises that the approximation is applied only to the integral over λ, for a fixed value of ψ. The resulting approximation may be written as

$$\pi(\psi \mid y) \doteq \left(\frac{n}{2\pi} \right)^{1/2} \left\{ \frac{\left| -\frac{\partial^2 \ell_m(\tilde{\psi}, \tilde{\lambda})}{\partial \theta \partial \theta^{\mathrm{T}}} \right|}{\left| -\frac{\partial^2 \ell_m(\psi, \tilde{\lambda}_\psi)}{\partial \lambda \partial \lambda^{\mathrm{T}}} \right|} \right\}^{1/2} \frac{f(y \mid \psi, \tilde{\lambda}_\psi) \pi(\psi, \tilde{\lambda}_\psi)}{f(y \mid \tilde{\psi}, \tilde{\lambda}) \pi(\tilde{\psi}, \tilde{\lambda})}, \tag{11.34}$$

where $\tilde{\lambda}_\psi$ is the maximum *a posteriori* estimate of λ for fixed ψ and the denominator and numerator determinants are of Hessian matrices of sides $(p-1)$ and p respectively.

The posterior marginal cumulative distribution for ψ may be approximated by applying (11.31) to the integral of (11.34) over the range (∞, ψ_0). We take $u_0 = \psi_0$,

$$g(\psi) = \ell_m(\tilde{\psi}, \tilde{\lambda}) - \ell_m(\psi, \tilde{\lambda}_\psi), \quad a(\psi) = \left\{ \frac{\left| -\frac{\partial^2 \ell_m(\tilde{\psi}, \tilde{\lambda})}{\partial \theta \partial \theta^{\mathrm{T}}} \right|}{\left| -\frac{\partial^2 \ell_m(\psi, \tilde{\lambda}_\psi)}{\partial \lambda \partial \lambda^{\mathrm{T}}} \right|} \right\}^{1/2},$$

and set $r_0^* = r_0 + r_0^{-1} \log(v_0/r_0)$, where

$$r_0 = \mathrm{sign}(\psi_0 - \tilde{\psi}) [2\{\ell_m(\tilde{\psi}, \tilde{\lambda}) - \ell_m(\psi_0, \tilde{\lambda}_{\psi_0})\}]^{1/2},$$

$$v_0 = -\frac{\partial \ell_m(\psi_0, \tilde{\lambda}_{\psi_0})}{\partial \psi} \left\{ \frac{\left| -\frac{\partial^2 \ell_m(\psi_0, \tilde{\lambda}_{\psi_0})}{\partial \lambda \partial \lambda^{\mathrm{T}}} \right|}{\left| -\frac{\partial^2 \ell_m(\tilde{\psi}, \tilde{\lambda})}{\partial \theta \partial \theta^{\mathrm{T}}} \right|} \right\}^{1/2};$$

here $\tilde{\lambda}_{\psi_0}$ is the maximum *a posteriori* estimate of λ when ψ is fixed at ψ_0. It is often convenient to find the derivatives numerically.

			Rate estimate ($\times 10^2$)	
Case	x	y	Crude	Empirical Bayes
1	94.320	5	5.3	6.1
2	15.720	1	6.4	10.7
3	62.880	5	8.0	9.1
4	125.760	14	11.1	11.7
5	5.240	3	57.3	58.8
6	31.440	19	60.4	60.6
7	1.048	1	95.4	80.0
8	1.048	1	95.4	80.0
9	2.096	4	190.8	143.7
10	10.480	22	209.9	194.4

Table 11.7 Numbers of failures y of ten pumps in x thousand operating hours, with the crude rate estimate y/x (Gaver and O'Muircheartaigh, 1987). The final column gives empirical Bayes rate estimates derived in Problem 11.26.

Numerous variant approaches are possible. For example, the ratio of priors in the integral of (11.34) may be included in the function $a(u)$ of (11.30), which case ℓ_m is simply the log likelihood, $\tilde{\theta}$ and $\tilde{\lambda}_\psi$ are maximum likelihood estimates, the Hessians are observed information matrices, and r_0 is the directed likelihood ratio statistic for testing the hypothesis $\psi = \psi_0$. The prior then appears only in v_0. The resulting approximation is generally poorer than that described above, but this idea does suggest a quick way to assess sensitivity to the prior density. The key is to notice that the approximate effect on (11.34) of taking a different prior, $\pi_1(\psi, \lambda)$, say, would be to multiply (11.34) by the ratio $c(\psi) = \{\pi_1(\psi, \tilde{\lambda}_\psi)/\pi(\psi, \tilde{\lambda}_\psi)\}/\{\pi_1(\tilde{\psi}, \tilde{\lambda})/\pi(\tilde{\psi}, \tilde{\lambda})\}$; the effect is approximate because Laplace approximation based on π_1 would not lead to integrals maximized at $\tilde{\lambda}_\psi$ and $(\tilde{\psi}, \tilde{\lambda})$. On the other hand, the effect on these maximizing values of changing the prior is often relatively small. Thus the effect of modifying the prior from π to π_1 may be gauged by changing v_0 to $v_0/c(\psi_0)$, and recalculating r_0^* and $\Phi(r_0^*)$. This involves no further maximization or numerical differentation.

Example 11.19 (Pump failure data) Table 11.7 contains the numbers of failures y_j of $n = 10$ pumps in operating periods of x_j thousands of hours. The pumps are from several systems in the nuclear plant Farley 1; pumps 1, 3, 4, and 6 operate continuously, while the rest operate only intermittantly or on standby. For now we suppose that the pumps may be expected to have similar rates of failure, with the jth pump having failure rate λ_j, and that conditional on λ_j, the numbers of failures y_j have independent Poisson distributions with means $\lambda_j x_j$. We further suppose that the λ_j are independent realizations of a gamma variable with parameters α and β, and that β itself has a prior gamma distribution with parameters ν and ϕ. Thus

$$f(y \mid \lambda) = \prod_{j=1}^{n} \frac{(x_j \lambda_j)^{y_j}}{y_j!} e^{-x_j \lambda_j}, \quad \pi(\lambda \mid \beta) = \prod_{j=1}^{n} \frac{\beta^\alpha \lambda_j^{\alpha-1}}{\Gamma(\alpha)} e^{-\beta \lambda_j},$$

$$\pi(\beta) = \frac{\phi^\nu \beta^{\nu-1}}{\Gamma(\nu)} e^{-\phi \beta},$$

$$(11.35)$$

Table 11.8 Integrals of two approximate posterior densities for β for the pumps data. The first, $\tilde{I}_1$, involves a one-dimensional Laplace approximation to (11.36), while $\tilde{I}_{10}$ involves ten-dimensional Laplace approximation. The table shows how the integral changes when the curvature of the likelihood is increased by a.

a	1	2	3	4	5	10	20
$\tilde{I}_1$	1.022	1.017	1.014	1.012	1.011	1.009	1.007
$\tilde{I}_{10}$	1.782	1.309	1.183	1.127	1.096	1.042	1.019

so that the joint density of the data y, the rates λ, and β is

$$f(y \mid \lambda) f(\lambda \mid \beta) \pi(\beta) = c \prod_{j=1}^{n} \left\{ \lambda_j^{y_j + \alpha - 1} e^{-\lambda_j (x_j + \beta)} \right\} \times \beta^{n\alpha + \nu - 1} e^{-\phi\beta}, \tag{11.36}$$

where c is a constant of proportionality.

To find the conditional density of β, we integrate over the λ_j, to obtain

$$f(y, \beta) = c \prod_{j=1}^{n} \left\{ (x_j + \beta)^{-(y_j + \alpha)} \Gamma(y_j + \alpha) \right\} \times \beta^{n\alpha + \nu - 1} e^{-\phi\beta}, \tag{11.37}$$

from which the marginal density of y is obtained by further integration to give

$$f(y) = c \prod_{j=1}^{n} \Gamma(y_j + \alpha) \times \int_0^{\infty} e^{-h(\beta)} \, d\beta,$$

where $h(\beta) = \phi\beta - (n\alpha + \nu - 1) \log \beta + \sum (y_j + \alpha) \log(x_j + \beta)$; we use I to denote the integral in this expression.

For sake of illustration we take a proper but fairly uninformative prior for β, with $\nu = 0.1$ and $\phi = 1$, and take $\alpha = 1.8$. Application of Laplace's method to I then results in the approximate posterior density for β, $\tilde{\pi}(\beta \mid y) = \tilde{I}^{-1} \exp\{-h(\beta)\}$, which has integral 1.022.

The accuracy of Laplace's method can be tested by taking a different approach, in which we first integrate (11.36) over β, and then apply the multivariate version of Laplace's method to the resulting ten-dimensional integral with respect to the λ_j. In this case the density approximation has integral 1.782, because the ten-dimensional integral approximation, $\tilde{I}_{10}$, is less accurate than $\tilde{I}_1$. To compare the two approaches we recalculate the approximations for data (ax_j, ay_j) and various values of a. This leaves unchanged the failure rates y_j/x_j, but increases by a factor a the Fisher information for each of the λ_j, thereby increasing the curvature of the log likelihood and the accuracy of the approximation. The results in Table 11.8 show that $\tilde{I}_{10}$ rapidly improves as a increases, and that with counts about 4–5 times as large as those observed, Laplace's method gives adequately accurate answers, even in ten dimensions. In practice, of course, $\tilde{I}_1$ would be used.

To calculate approximate posterior densities for λ_j, we integrate (11.36) over λ_i, $i \neq j$, and then apply Laplace's method to the numerator and denominator integrals of

$$\pi(\lambda_j \mid y) = \frac{\lambda_j^{y_j + \alpha - 1} e^{-\lambda_j x_j} \int_0^{\infty} e^{-h_j(\beta)} \, d\lambda}{\Gamma(y_j + \alpha) \int_0^{\infty} e^{-h(\beta)} \, d\beta},$$

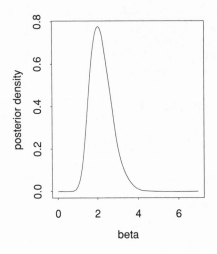 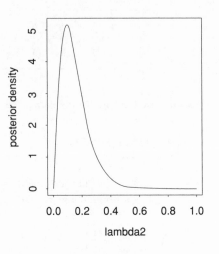

Figure 11.4
Approximate posterior
densities for β and λ_2 for
the pumps data, based on
Laplace approximation.

where

$$h_j(\beta) = (\phi + \lambda_j)\beta - (n\alpha + \nu - 1)\log\beta + \sum_{i \neq j}(y_i + \alpha)\log(x_i + \beta).$$

The resulting denominator is again $\tilde{I}_1$, while the numerator must be recalculated at each of a range of values of λ_j. Figure 11.4 shows these approximate densities for β and for λ_2. That for λ_2 has integral 1.0004 and is presumably closer to one because it is based on a ratio of Laplace approximations. ∎

The ideal situation for Laplace approximation is when the posterior density is strongly unimodal. When the posterior is multimodal, the approximation can be applied separately to each mode — provided they can all be found. Different approximations apply when the posterior is peaked at the end of its range (Exercise 11.3.5).

11.3.2 Importance sampling

Many Monte Carlo techniques may be applied in Bayesian computation. In this section we discuss ideas based on importance sampling, and in the next section we turn to iterative methods based on simulating Markov chains. Importance sampling gives independent samples, and so measures of uncertainty for estimators are usually fairly readily obtained, but it applies to a limited range of problems. Iterative methods are more widely applicable but it can be difficult to assess their convergence and to give statements of uncertainty for their output.

Suppose we wish to calculate an integral of form

$$\mu = \int m(\theta, y, z)\pi(\theta \mid y)\, d\theta.$$

If we take $m(\theta, y, z) = I(\theta \leq a)$, for example, then $\mu = \Pr(\theta \leq a \mid y)$, while taking $m(\theta, y, z) = f(z \mid y, \theta)$ gives $\mu = f(z \mid y)$, the posterior predictive density for z given the data. Suppose that direct computation of μ is awkward, but that it is straightforward both to generate a sample $\theta_1, \ldots, \theta_S$ from a density $h(\theta)$ whose support includes that of $\pi(\theta \mid y)$, and to calculate $m(\theta, y, z)$ and $f(y \mid \theta)$. We can then apply importance sampling for estimation of μ, obtaining the unbiased estimator (Section 3.3.2)

$$\widehat{\mu} = S^{-1} \sum_{s=1}^{S} m(\theta_s, y, z) \frac{\pi(\theta_s \mid y)}{h(\theta_s)} = S^{-1} \sum_{s=1}^{S} m(\theta_s, y, z) w(\theta_s), \tag{11.38}$$

say, where $w(\theta) = \pi(\theta \mid y)/h(\theta)$ is an importance sampling weight. An important advantage of $\widehat{\mu}$ over the iterative procedures to be disussed later is that its variance is readily obtained (Exercise 11.3.6).

In practice the importance sampling ratio estimator of μ,

$$\widehat{\mu}_{\text{rat}} = \frac{\sum_{s=1}^{S} m(\theta_s, y, z) w(\theta_s)}{\sum_{s=1}^{S} w(\theta_s)},$$

is more commonly used. This is typically less variable than $\widehat{\mu}$; indeed it performs perfectly if $m(\theta, y, z)$ is constant, as is clear from its variance, given by (Example 2.25)

$$\widehat{\text{var}}(\widehat{\mu}_{\text{rat}}) = \frac{1}{S(S-1)} \sum_{s=1}^{S} \frac{\{m(\theta_s, y, z) - \widehat{\mu}_{\text{rat}}\}^2 w(\theta_s)^2}{\overline{w}^2}, \quad \overline{w} = S^{-1} \sum_{s=1}^{S} w(\theta_s).$$

As usual with importance sampling, a good choice of $h(\theta)$ is crucial if the simulation is to be useful. One possibility is a normal approximation to the posterior density of θ, taking $h(\theta)$ to be $N\left\{\widehat{\theta}, J(\widehat{\theta})^{-1}\right\}$, where $\widehat{\theta}$ and $J(\widehat{\theta})$ are the maximum likelihood estimate and the observed information. Normal approximation may be better if applied to a transformed parameter $\psi = \psi(\theta)$, however, while the light-tailed normal distribution typically gives too few simulations in the tail of the posterior density. Hence it is usually better to generate the θ_s from a shifted and rescaled t_ν density.

Example 11.20 (Challenger data)   Table 1.3 gives data on launches of the space shuttle, including the ill-fated Challenger launch. In Examples 1.3, 4.5 and 4.33 we saw how these data may be modelled using a logistic regression model, under which the number of O-rings suffering thermal distress when a launch takes place at temperature $x_1°$F is binomial with denominator $m = 6$ and probability $\pi(\beta + \beta_1 x_1) = \exp(\beta_0 + \beta_1 x_1)/\{1 + \exp(\beta_0 + \beta_1 x_1)\}$. The likelihood (4.6) for this model is shown in Figure 4.3. Let us represent the data for the 23 successful launches by y, with likelihood $f(y \mid \theta)$; here $\theta = (\beta_0, \beta_1)$.

One aspect of interest when deciding whether to launch the Challenger should have been the number Z of distressed O-rings at its launch temperature of $x_1 = 31°$F. We suppose that, conditional on θ, $f(z \mid \theta)$ is binomial with denominator $m = 6$ and probability $\pi(\beta_0 + 31\beta_1)$, independent of other launches. Then in the Bayesian framework we should calculate the posterior predictive density for Z,

$$\frac{\int f(z \mid \theta) f(y \mid \theta) \pi(\theta) \, d\theta}{\int f(y \mid \theta) \pi(\theta) \, d\theta},$$

where $\pi(\theta)$ is the prior density on (β_0, β_1).

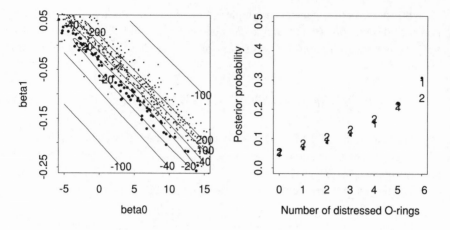

Figure 11.5 Importance sampling applied to shuttle data. Left: pairs (β_0, β_1) simulated from a prior density, with log likelihood contours superimposed. Pairs whose weight w_s exceeds $(100S)^{-1}$ are shown as blobs. The other pairs have very low likelihoods and hence essentially zero posterior probabilities w_s. Right: posterior predictive density for the number of distressed O-rings for a launch at 31°F, using beta prior with $a = b = 0.5$ (blobs), $a = b = 1$ (1) and $a = 1, b = 4$ (2), estimated by importance sampling with $S = 10,000$.

The parameters β_0 and β_1 are difficult to interpret directly, and instead we consider the probabilities $\pi_1 = \pi(\beta + 60\beta_1)$ and $\pi_2 = \pi(\beta + 80\beta_1)$ that a single O-ring will be distressed at 60 and 80°F. In practice specification of the joint prior density of π_1 and π_2 would require engineering expertise, but in default of this we simply suppose that they have independent beta densities (11.3) with $a = b = 1/2$. For the initial step of the importance sampling algorithm we generate 10,000 independent pairs (π_1, π_2) and then set

$$\beta_1 = \frac{1}{80 - 60} \log \left\{ \frac{\pi_2(1 - \pi_1)}{\pi_2(1 - \pi_1)} \right\}, \quad \beta_0 = \log \left\{ \frac{\pi_1}{1 - \pi_1} \right\} - 60\beta_1.$$

The left panel of Figure 11.5 shows some of the resulting pairs $\theta_s = (\beta_0, \beta_1)$, superimposed on contours of the log likelihood. Pairs whose weight w_s exceeds one-hundredth of its average are shown by blobs. About 30% of the simulated values fall into this category, for which $\sum w_s = 0.9996$, so just 4/10,000ths of the posterior probability is placed on the other 7000 pairs. This occurs both because the prior is much more dispersed than the likelihood, and because they are mismatched, in the sense that the prior value of β_1 for a given β_0 is generally too large — the mode of $f(\beta_1 \mid \beta_0)$ lies to the right of that of $f(y \mid \beta_1, \beta_0)$, considered as a function of β_1 for fixed β_0.

The right panel of Figure 11.5 shows the posterior probabilities of $z = 0, \ldots,$ 6 distressed rings. There is appreciable probability of damage to most of the rings, as $\Pr(Z \geq 4 \mid y) \doteq 0.65$, with little dependence on the prior. ∎

This examples show both the strengths and weaknesses of importance sampling. It is simple to apply, and because $\theta_1, \ldots, \theta_S$ are independent it is easy to obtain a standard error for $\hat{\mu}$, and then to increase S if necessary. On the other hand the prior is sometimes so overdispersed relative to the likelihood that S must be huge before an appreciable number of the w_s are non-zero, and a better importance sampling distribution must be found. This problem becomes acute when the dimension of θ is large and the curse of dimensionality bites. There are clever ways to improve importance sampling in such situations, but Markov chain methods apply readily to many high-dimensional problems, and to these we now turn.

11.3.3 Markov chain Monte Carlo

The idea of Markov chain Monte Carlo simulation is to construct a Markov chain that will, if run for an infinitely long period, generate samples from a posterior distribution π, specified implicitly and known only up to a normalizing constant. Although it has roots in areas such as statistical physics, its application in mainstream Bayesian statistics is relatively recent and the discussion below is merely a snapshot of a topic in full spate of development. The reader whose memory of Markov chains is hazy may find it useful to review the early pages of Section 6.1.1.

Gibbs sampler

The term Gibbs sampling comes from an analogy with statistical physics, where similar methods are used to generate states from Gibbs distributions. In that context it is called the *heat bath* algorithm.

Let $U = (U_1, \ldots, U_k)$ be a random variable of dimension k whose joint density $\pi(u)$ is unknown. Our goal is to estimate aspects of $\pi(u)$, such as joint or marginal densities and their quantiles, moments such as $\mathrm{E}(U_1)$ and $\mathrm{var}(U_1)$, and so forth. Although $\pi(u)$ itself is unknown, we suppose that we can simulate observations from the *full conditional* densities $\pi(u_i \mid u_{-i})$, where $u_{-i} = (u_1, \ldots, u_{i-1}, u_{i+1}, \ldots, u_k)$. Often in practice the constant normalizing $\pi(u)$ is unknown, but as it does not appear in the $\pi(u_i \mid u_{-i})$, this causes no difficulty. If $\pi(u)$ is proper, then the Hammersley–Clifford theorem implies that under mild conditions $\pi(u)$ is determined by these densities; this does not imply that any set of full conditional densities determines a proper joint density. *Gibbs sampling* is successive simulation from the $\pi(u_i \mid u_{-i})$ according to the algorithm:

1. initialize by taking arbitrary values of $U_1^{(0)}, \ldots, U_k^{(0)}$.
2. Then for $i = 1, \ldots, I$,
   - (a) generate $U_1^{(i)}$ from $\pi\left(u_1 \mid u_2 = U_2^{(i-1)}, \ldots, u_k = U_k^{(i-1)}\right)$,
   - (b) generate $U_2^{(i)}$ from $\pi\left(u_2 \mid u_1 = U_1^{(i)}, u_3 = U_3^{(i-1)}, \ldots, u_k = U_k^{(i-1)}\right)$,
   - (c) generate $U_3^{(i)}$ from

$$\pi\left(u_3 \mid u_1 = U_1^{(i)}, u_2 = U_2^{(i)}, u_4 = U_4^{(i-1)}, \ldots, u_k = U_k^{(i-1)}\right),$$

   $\vdots$

   - (d) generate $U_k^{(i)}$ from $\pi\left(u_k \mid u_1 = U_1^{(i)}, \ldots, u_{k-1} = U_{k-1}^{(i)}\right)$.

Here we update each of the U_j in turn, basing each value generated on the $k - 1$ previous simulations. This gives a stream of random variables

$$U_1^{(1)}, \ldots, U_k^{(1)}, U_1^{(2)}, \ldots, U_k^{(2)}, \quad \ldots, \quad U_1^{(I-1)}, \ldots, U_k^{(I-1)}, U_1^{(I)}, \ldots, U_k^{(I)},$$

so for the jth component of U we have a sequence $U_j^{(1)}, \ldots, U_j^{(I)}$.

To see why we might hope that $(U_1^{(I)}, \ldots, U_k^{(I)})$ is approximately a sample from $\pi(u)$, suppose that $k = 2$ and that U_1 and U_2 take values in the finite sets $\{1, \ldots, n\}$ and $\{1, \ldots, m\}$. We write their joint and marginal densities as

$$\Pr(U_1 = r, U_2 = s) = \pi(r, s),$$

$$\Pr(U_1 = r) = \pi_1(r) = \sum_{s=1}^{m} \pi(r, s), \quad r = 1, \ldots, n,$$

$$\Pr(U_2 = s) = \pi_2(s) = \sum_{r=1}^{n} \pi(r, s), \quad s = 1, \ldots, m,$$

with $\pi_1(r), \pi_2(s) > 0$ for all r and s. The conditional densities are

$$p_{sr} = \Pr(U_1 = r \mid U_2 = s) = \frac{\pi(r, s)}{\pi_2(s)}, \quad q_{rs} = \Pr(U_2 = s \mid U_1 = r) = \frac{\pi(r, s)}{\pi_1(r)},$$

which we express as an $m \times n$ matrix P_{21} with (s, r) element p_{sr} and an $n \times m$ matrix P_{12} with (r, s) element q_{rs}. These transition matrices give the probabilities of going from the m possible values of U_2 to the n possible values of U_1 and back again. As they are ratios, p_{rs} and q_{rs} do not involve the normalizing constant for π.

If f_0 is an $m \times 1$ vector containing the distribution of $U_2^{(0)}$, the distributions of $U_1^{(1)}, U_2^{(1)}, U_1^{(2)}, \ldots$, are $f_0^{\mathrm{T}} P_{21}, f_0^{\mathrm{T}} P_{21} P_{12}, f_0^{\mathrm{T}} P_{21} P_{12} P_{21}, \ldots$. Thus each iteration of step 2 of the algorithm corresponds to postmultiplying the current distribution of $U_2^{(i)}$ by the $m \times m$ matrix $H = P_{21} P_{12}$. Hence $U_2^{(I)}$ has distribution $f_0^{\mathrm{T}} H^I$. Conditional on $U_2^{(i)}$, $U_2^{(i+1)}$ is independent of earlier values, so the sequence $U_2^{(1)}, \ldots, U_2^{(I)}$ is a Markov chain with transition matrix H. If the chain is ergodic, then $U_2^{(I)}$ has a unique limiting distribution f as $I \to \infty$, satisfying the equation $f^{\mathrm{T}} H = f^{\mathrm{T}}$. As this limit is unique, we need only show that f is the marginal distribution of U_2 to see that the algorithm ultimately produces a variable with density π_2. Now the rth element of $\pi_2^{\mathrm{T}} H = \pi_2^{\mathrm{T}} P_{21} P_{12}$ equals

$$\sum_{t=1}^{n} \sum_{s=1}^{m} \pi_2(t) p_{ts} q_{sr} = \sum_{t=1}^{n} \sum_{s=1}^{m} \pi_2(t) \frac{\pi(s, t)}{\pi_2(t)} \frac{\pi(r, s)}{\pi_1(s)} = \pi_2(r),$$

so π_2 is indeed the unique solution to the equation $f^{\mathrm{T}} H = f^{\mathrm{T}}$. By symmetry, $U_1^{(1)}, \ldots, U_1^{(I)}$ is a Markov chain with transition matrix $P_{12} P_{21}$ and limiting distribution π_1. Moreover the fact that $\pi_2^{\mathrm{T}} P_{21} = \pi_1^{\mathrm{T}}$ ensures that the joint distribution of $(U_1^{(I)}, U_2^{(I)})$ converges to $\pi(r, s)$ as $I \to \infty$. Generalization to $k > 2$ works in an obvious way.

Most of the densities $\pi(u)$ met in applications are continuous, so this argument is not directly applicable. However any continuous density can be closely approximated by one with countable support, for which essentially the same results hold, so it is not surprising that the ideas apply more widely, and from now on we shall assume that they are applicable to our problems.

Such a simulation will only be useful if convergence to the stationary distribution is not too slow. In discrete cases like that above, the convergence rate is determined by the modulus of the second largest eigenvalue l_2 of H, where $1 = l_1 \geq |l_2| \geq \cdots$. If $|l_2| < 1$, then convergence is geometrically ergodic; see (6.4).

In the continuous case it can occur that $|l_2| = 1$ or that l_2 does not exist, either of which will spell trouble. A reversible chain has real eigenvalues and satisfies the detailed balance condition (6.5). Hence it can be useful to make the chain reversible, for example by generating variables in order $1, \ldots, k, k-1, \ldots, 2, \ldots$ or by choosing the next update at random. Either involves modifying step 2 of the algorithm.

Output analysis

The only sure way to know how long a Markov chain simulation algorithm should be run is by theoretical analysis to determine its rate of convergence. This requires knowledge of the stationary distribution being estimated, however, and is possible only in very special cases. A more pragmatic approach is to declare that the algorithm has converged when its output satisfies tests of some sort. Such *convergence diagnostics* can at best detect non-convergence, however; they cannot guarantee that the output will be useful. Both empirically- and theoretically-based diagnostics have been proposed, and references to them are given in the bibliographic notes. Empirical approaches include contrasting output from the start and the end of a run, and comparing results from parallel independent runs whose initial values have been chosen to be overdispersed relative to the target distribution. Theoretical approaches generally assess whether the output satisfies known properties of stationary chains. In practice it is sensible to use several diagnostics but also to scrutinize time series plots of the output. As different parameters may converge at different rates, it is important to examine all parameters of interest and also global quantities such as the current log likelihood, prior, and posterior.

If stationarity seems to have been attained, then it is useful to examine correlograms and partial correlograms of output. If the autocorrelations are high, then the statistical efficiency of the algorithm will be low. A chain with low correlations will yield estimators with smaller variance, and is more likely to visit all regions of significant probability mass. The algorithm may need modification to reduce high autocorrelations, for example by reparametrization; see Example 11.24.

Multimodal target densities are awkward because it can be hard to know if all significant modes have been visited. Use of widely separated starting values may then be useful, and so too may be occasional insertion of large random jumps into the algorithm, so that it effectively restarts from a location unrelated to its previous position.

Suppose that the chain seems to have converged after B iterations and is run for a total of $I \gg B$ iterations. In general discussion below we suppose that I is so much larger than B that inference can safely be based on all I iterations, but in practice we use only output from iterations $B + 1, \ldots, I$. Let the quantity of interest be $\mu = \int m(u)\pi(u)\,du$, where $\int |m(u)|\pi(u)\,du < \infty$. Unless there is qualitative knowledge about $\pi(u)$ this may involve an act of faith. For example, taking $m(u) = u_1$ gives $\mu = E(U_1)$, which could be infinite although $\pi(u)$ is proper. Hence unless properties of the posterior density are known it is safer to base inferences on density and quantile estimates than on moments. If μ is finite then it can be estimated by the ergodic average

$$\widehat{\mu} = I^{-1} \sum_{i=1}^{I} m(U^{(i)}), \qquad (11.39)$$

where $U^{(i)}$ denotes $(U_1^{(i)}, \ldots, U_k^{(i)})$. The ergodic theorem (6.2) implies that $\widehat{\mu}$ converges almost surely to μ as $I \to \infty$, and under further conditions

$$I^{1/2}(\widehat{\mu} - \mu) \xrightarrow{D} N(0, \sigma_m^2), \quad \text{where} \quad 0 < \sigma_m^2 < \infty, \qquad (11.40)$$

so $\widehat{\mu}$ is approximately normal for large I. In that case

$$I \times \text{var}(\widehat{\mu}) = I^{-1} \sum_{i=-I+1}^{I-1} (I - |i|)\gamma_i \sim \sigma_m^2 = \sum_{i=-\infty}^{\infty} \gamma_i = \gamma_0 \sum_{i=-\infty}^{\infty} \rho_i,$$

where $\gamma_i = \text{cov}\{m(U^{(0)}), m(U^{(i)})\}$ depends on π and on the construction of the chain, and $\rho_i = \gamma_i/\gamma_0$ is the ith autocorrelation. The marginal variance of $m(U)$ is $\gamma_0 = \text{var}_\pi\{m(U)\}$, which depends only on m and π. The effect of using correlated output is to inflate $\text{var}(\widehat{\mu})$ by a factor $\tau = \sum_{-\infty}^{\infty} \rho_i$ relative to an independent sample of size I, so an estimate $\widehat{\tau}$ from a pilot run may suggest how large I should be. The obvious estimator of τ based on the correlogram is inconsistent, but better ones exist. One simple possibility is $\widehat{\tau} = \sum_{i=-M}^{M} \widehat{\rho}_i$, where $M = \lfloor 3\widehat{\tau} \rfloor$ is found by iteration.

$\lfloor x \rfloor$ is the smallest integer greater than or equal to x.

Another approach splits the output into b blocks of k successive iterations, with k taken so large that the block averages of the $m(U^{(i)})$ have correlations lower than 0.05, say, and gives the standard error for $\widehat{\mu}$ as if the block averages were a simple random sample.

The density of U_1 at u_1 may be estimated by a kernel method (Section 7.1.2), or by the unbiased estimator (7.12), written in this context as

$$I^{-1} \sum_{i=1}^{I} \pi(u_1 \mid U_{-1}^{(i)}). \qquad (11.41)$$

The discussion above presupposes a single long run of the chain. An alternative is S independent parallel runs of length I, leading ultimately to S independent values $U^{(I)}$ from $\pi(u)$. An estimate based on these may be less variable than one based on SI dependent samples from a single chain, and its variance is more easily estimated. Roughly SB iterations must be disregarded, however, compared to B when there is only one chain. From this viewpoint a single run is preferable, but it is then harder to detect lack of convergence.

Example 11.21 (Bivariate normal density) If (U_1, U_2) are bivariate normal with means zero, variances one and correlation ρ, then

ϕ denotes the standard normal density.

$$\pi(u_1 \mid u_2) = \frac{1}{(1 - \rho^2)^{1/2}} \phi\left\{ \frac{u_1 - \rho u_2}{(1 - \rho^2)^{1/2}} \right\},$$

with a symmetric result for $\pi(u_2 \mid u_1)$, and we can use the marginal standard normal densities of U_1 and U_2 to assess convergence. The upper left panel of Figure 11.6 shows the contours of the joint density when $\rho = 0.75$, together with a sample path of the process starting from an initial value generated uniformly on the

Figure 11.6 Gibbs
sampler for bivariate
normal density. Top left:
contours of the bivariate
normal density with
$\rho = 0.75$, with the first
five iterations of a Gibbs
sampler; the blobs are at
$(u_1^{(i)}, u_2^{(i)})$, for
$i = 0, \ldots, 5$, starting
from the top left of the
panel. Top right: sample
paths of $U_1^{(i)}$ and $U_2^{(i)}$ for
$i = 1, \ldots, 100$. Bottom
left: kernel density
estimates of $\pi_1(u_1)$ (heavy
solid) based on 100
parallel chains after I
iterations, with $I = 0$
(solid), 2 (dots), 5
(dashes), 10 (large
dashes), and 100 (largest
dashes); the bandwidth is
chosen by uniform
cross-validation. Bottom
right: estimates (dots) of
$\pi_1(u_1)$ (heavy solid) after
100 iterations of 5
replicate chains, based on
(11.41).

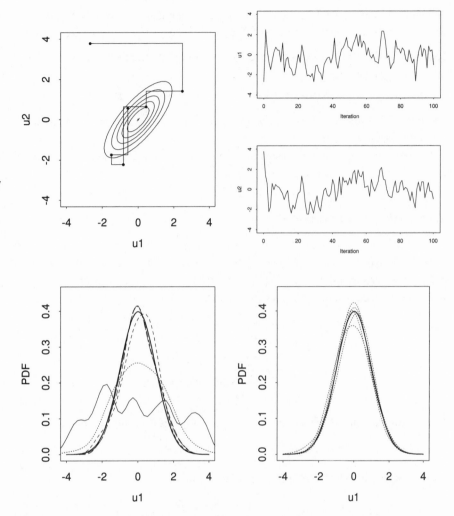

square $(-4, 4) \times (-4, 4)$. The updating scheme forces the sample path to consist of steps parallel to the coordinate axes. The upper right panel shows that the sample paths of the Markov chains appear to converge rapidly to their limit distributions, as the calculations in Problem 11.20 show will be the case. This is confirmed by the estimated variance inflation factor $\widehat{\tau} \doteq 3$. The lower left panel shows rapid convergence of the kernel density estimates to their target, based on $S = 100$ parallel chains. The lower right panel illustrates the variability of (11.41), which here performs better than the kernel estimator. ∎

Bayesian application

The essence of Bayesian inference is to treat all unknowns as random variables, and to compute their posterior distributions given the data y. The Gibbs sampler is applied by taking $U_1, \ldots, U_k$ to be the unknowns, usually parameters, and simulating conditional on y. The full conditional densities $\pi(u_i \mid u_{-i})$ are typically of form $\pi(\theta_i \mid \theta_{-i}, y)$

and must be obtained before the algorithm can be applied. Fortunately this is often possible for 'nice' models, where the full conditional densities have conjugate forms.

Example 11.22 (Random effects model)   The sampling model in the simplest normal one-way layout is

$$y_{tr} = \theta_t + \varepsilon_{tr}, \quad t = 1, \ldots, T, \ r = 1, \ldots, R,$$

where $\theta_1, \ldots, \theta_T \stackrel{\text{iid}}{\sim} N(\nu, \sigma_\theta^2)$ and independent of this $\varepsilon_{tr} \stackrel{\text{iid}}{\sim} N(0, \sigma^2)$. The focus of interest is usually σ^2 and σ_θ^2.

Bayesian analysis requires prior information, which we suppose to be expressed through the conjugate densities

$$\mu \sim N(\mu_0, \tau^2), \quad \sigma^2 \sim IG(\alpha, \beta), \quad \sigma_\theta^2 \sim IG(\alpha_\theta, \beta_\theta).$$

The full posterior density is then

$$\pi\left(\mu, \theta, \sigma^2, \sigma_\theta^2 \mid y\right) \propto f(y \mid \theta, \sigma^2) f\left(\theta \mid \mu, \sigma_\theta^2\right) \pi(\mu) \pi(\sigma^2) \pi\left(\sigma_\theta^2\right). \quad (11.42)$$

We now take $(U_1, U_2, U_3, U_4) = (\sigma_\theta^2, \sigma^2, \mu, \theta)$, and calculate the full conditional densities needed for Gibbs sampling, always treating the data y as fixed. Each calculation requires integration over just one parameter. For example,

$$
\begin{aligned}
\pi\left(\sigma_\theta^2 \mid \sigma^2, \mu, \theta, y\right) &= \frac{f(y \mid \theta, \sigma^2) f\left(\theta \mid \mu, \sigma_\theta^2\right) \pi(\mu) \pi(\sigma^2) \pi\left(\sigma_\theta^2\right)}{\int f(y \mid \theta, \sigma^2) f\left(\theta \mid \mu, \sigma_\theta^2\right) \pi(\mu) \pi(\sigma^2) \pi\left(\sigma_\theta^2\right) d\sigma_\theta^2} \\
&= \frac{f\left(\theta \mid \mu, \sigma_\theta^2\right) \pi(\mu) \pi\left(\sigma_\theta^2\right)}{\int f\left(\theta \mid \mu, \sigma_\theta^2\right) \pi(\mu) \pi\left(\sigma_\theta^2\right) d\sigma_\theta^2} \\
&= \pi\left(\sigma_\theta^2 \mid \mu, \theta\right).
\end{aligned}
$$

Similar calculations reveal that $\pi(\theta \mid \sigma_\theta^2, \sigma^2, \mu, y)$ does not simplify, but that

$$\pi\left(\sigma^2 \mid \sigma_\theta^2, \mu, \theta, y\right) = \pi(\sigma^2 \mid \theta, y), \quad \pi\left(\mu \mid \sigma_\theta^2, \sigma^2, \theta, y\right) = \pi\left(\mu \mid \sigma_\theta^2, \theta\right). \quad (11.43)$$

Arguments paralleling those in Example 11.12 lead to

$$\sigma_\theta^2 \mid \mu, \theta \sim IG\left(\alpha_\theta + \frac{1}{2}T, \beta_\theta + \frac{1}{2}\sum_{t=1}^{T}(\theta_t - \mu)^2\right), \quad (11.44)$$

$$\sigma^2 \mid \theta, y \sim IG\left(\alpha + \frac{1}{2}TR, \beta + \frac{1}{2}\sum_{t=1}^{T}\sum_{r=1}^{R}(y_{tr} - \theta_t)^2\right), \quad (11.45)$$

$$\mu \mid \sigma_\theta^2, \theta \sim N\left(\frac{\sigma_\theta^2 \mu_0 + \tau^2 \sum_{t=1}^{T} \theta_t}{\sigma_\theta^2 + T\tau^2}, \frac{\sigma_\theta^2 \tau^2}{\sigma_\theta^2 + T\tau^2}\right). \quad (11.46)$$

The conditional density $\pi(\theta \mid \sigma_\theta^2, \sigma^2, \mu, y)$ is most readily calculated by noting that given μ, σ_θ^2 and σ^2, the statistic $\overline{y}_t$ is sufficient for θ_t, with distribution $N(\theta_t, \sigma^2/R)$, while the prior density for θ_t given σ_θ^2, σ^2, and μ is $N(\mu, \sigma_\theta^2)$. Hence the posterior density for θ_t is

$$\theta_t \mid \sigma_\theta^2, \sigma^2, \mu, y \sim N\left(\frac{R\sigma_\theta^2 \overline{y}_t + \sigma^2 \mu}{R\sigma_\theta^2 + \sigma^2}, \frac{\sigma_\theta^2 \sigma^2}{R\sigma_\theta^2 + \sigma^2}\right), \quad t = 1, \ldots, T, \quad (11.47)$$

and the θ_t are conditionally independent.

Table 11.9 Estimated posterior means and standard deviations for the model fitted to the blood data, and simple frequentist estimates from analysis of variance.

	σ_θ^2	σ^2	μ	θ_1	θ_2	θ_3	θ_4	θ_5	θ_6
Estimate	23.8	126.4	41.9	53.9	43.0	34.9	39.9	41.3	38.6
Posterior mean	17.1	138.0	41.9	45.8	42.3	39.6	41.2	41.7	40.8
Posterior SD	30.3	33.8	2.4	4.1	2.9	3.4	2.9	2.9	3.0

Figure 11.7 Graphs for random effects model of Example 11.22. Left: directed acyclic graph showing dependence of random variables (circles) on themselves and on fixed quantities (rectangles). Right: conditional independence graph, formed by moralizing the directed acyclic graph, that is, joining parents and dropping arrowheads.

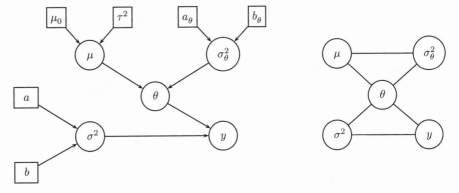

Expressions (11.44)–(11.47) give the steps required for an iteration of the Gibbs sampler. As the T updates in (11.47) are independent, they may all be performed at once, if the programming language used permits simultaneous generation of several non-identically-distributed normal variates.

Ideas from Section 6.2.2 render the structure of the full conditional densities more intelligible. Figure 11.7 shows the directed acyclic graph and the corresponding conditional independence graph for the present model. Each of μ, σ_θ^2, and σ^2 has two hyperparameters, considered fixed, and μ and σ_θ^2 are parents of $\theta_1, \ldots, \theta_T$. Each iteration of the Gibbs sampler traverses the parameter nodes in the conditional independence graph, simulating from the full conditional distribution corresponding to each node with remaining parameters set at their current values. The data y are held fixed throughout.

We applied this algorithm to the data in Table 9.22 on the stickiness of blood. For illustration we took $\alpha = \alpha_\theta = 0.5, \beta = \beta_\theta = 1, \mu = 0$, and $\tau^2 = 1000$, and generated starting-values for the parameters from the uniform distribution on $(0, 100)$. We ran 25 independent chains with $I = 1000$.

Figure 11.8 shows simulated series for three parameters and estimates of their posterior densities. The burn-in period seems to last for about $B = 100$ iterations, after which the chains seem stable. The chain for σ_θ^2 makes some large positive excursions, but the others seem fairly homogeneous, though they both show fairly strong autocorrelations. Estimated variance inflation factors are about 10 for σ_θ^2 and μ, but only 1–2.5 for the other parameters, consistent with the top left panels of the figure.

Table 11.9 shows the posterior means and standard deviations for the parameters, with their frequentist estimates. The posterior mean for μ is essentially equal to the overall average $\bar{y}$, but the posterior densities of the θ_t are strongly shrunk towards it, because there is evidence that σ_θ^2 is small; its posterior 0.1, 0.5, and 0.9 quantiles

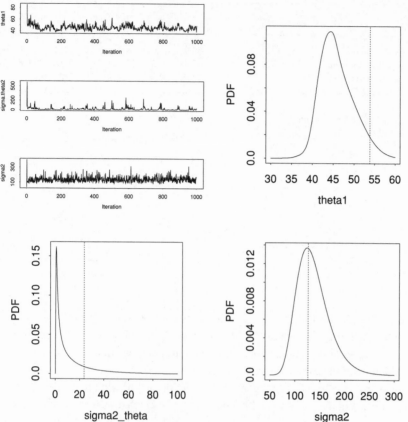

Figure 11.8 Gibbs sampler for normal components of variance model and blood data. Top left: time plots of θ_1, σ_θ^2, and σ^2. The other panels show estimated posterior densities for these parameters, based on applying analogues of (11.41) to the last 200 estimates from each of 25 parallel chains of length 1000. Frequentist estimates are shown as the dotted vertical lines.

are 0.46, 7.1, and 42.1. The variability mostly comes from measurement error, not inter-subject variation.   ∎

Metropolis–Hastings algorithm

The Gibbs sampler is easy to program, but if the full conditional densities it involves are unavailable or too nasty then a more general algorithm may be needed. A powerful approach known as the *Metropolis–Hastings algorithm* works as follows. In order to update the current value u of a Markov chain, a new value u' is generated using a proposal density $q(u' \mid u)$. Any density q can be used provided $q(u' \mid u) > 0$ if and only if $q(u \mid u') > 0$ and the resulting chain has the properties desired. Having generated u', a move from u to u' is accepted with probability

$$a(u, u') = \min \left\{ 1, \frac{\pi(u')q(u \mid u')}{\pi(u)q(u' \mid u)} \right\},$$

but otherwise the chain remains at u. Hence the probability density for a move to u', given that the chain has current value u, is

$$p(u' \mid u) = q(u' \mid u)a(u, u') + r(u)\delta(u - u'),$$

δ denotes the Dirac delta function.

where

$$r(u) = 1 - \int q(v \mid u) a(u, v) \, dv.$$

The first and second terms of $p(u' \mid u)$ are the probability density for a move from u to u' being proposed and accepted, and the probability that a move away from u is rejected.

The Metropolis–Hastings update step satisfies the detailed balance condition (6.5), because

$$\pi(u)p(u' \mid u) = \pi(u)q(u' \mid u) \min \left\{ 1, \frac{\pi(u')q(u \mid u')}{\pi(u)q(u' \mid u)} \right\} + \pi(u)r(u)\delta(u - u')$$

$$= \pi(u')q(u \mid u') \min \left\{ \frac{\pi(u)q(u' \mid u)}{\pi(u')q(u \mid u')}, 1 \right\} + \pi(u')r(u')\delta(u' - u)$$

$$= \pi(u')p(u \mid u').$$

Hence the corresponding Markov chain is reversible with equilibrium distribution π, provided it is irreducible and aperiodic. As π appears only in a ratio $\pi(u')/\pi(u)$ in the acceptance probability $a(u, u')$, the algorithm requires no knowledge of the constant that normalizes π.

If $q(u' \mid u) = q(u \mid u')$, the kernel is called symmetric, and $a(u, u') = \min\{1, \pi(u')/\pi(u)\}$. This occurs in particular if $u' = u + \varepsilon$, where ε is symmetric with density g; then $q(u' \mid u) = g(u' - u) = g(u - u') = q(u \mid u')$. This is called *random walk Metropolis* sampling. It is often applied to transformations of u, or to subsets of its elements, using a different proposal distribution for each subset.

The Gibbs sampler is a form of Metropolis–Hastings algorithm, the proposal density at the ith step of an iteration being

$$q(u' \mid u) = \begin{cases} \pi(u_i' \mid u_{-i}), & u_{-i}' = u_{-i}, \\ 0, & \text{otherwise.} \end{cases}$$

It then follows that

$$\frac{\pi(u')q(u \mid u')}{\pi(u)q(u' \mid u)} = \frac{\pi(u')/\pi(u_i' \mid u_{-i})}{\pi(u)/\pi(u_i \mid u_{-i}')} = \frac{\pi(u')/\pi(u_i' \mid u_{-i}')}{\pi(u)/\pi(u_i \mid u_{-i})} = \frac{\pi(u_{-i}')}{\pi(u_{-i})} = 1,$$

because $u_{-i}' = u_{-i}$. Here the proposals always have $u_{-i}' = u_{-i}$ and are always accepted, because $a(u, u') = \min[1, \pi(u')q(u \mid u')/\{\pi(u)q(u' \mid u)\}] = 1$.

Although there are few theoretical restrictions on the choice of q, practical constraints intervene. For example, if $q(u' \mid u)$ is so chosen that the acceptance probability $a(u, u')$ is essentially zero, the chain will spend long periods without moving and its output will be useless, and if the acceptance probability is close to one at each step but the chain barely moves, the state space will be traversed too slowly. Hence it is important to balance a reasonably high acceptance probability $a(u, u')$ with a chain that moves around its state space quickly enough. This can demand creativity and patience from the programmer.

Example 11.23 (Normal density) For illustration we take the toy problem of using the Metropolis–Hastings algorithm to simulate from the standard normal density

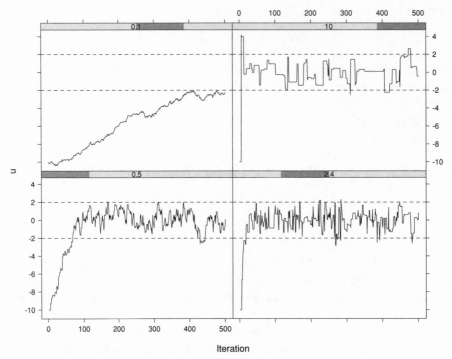

Figure 11.9 Sample paths for Metropolis–Hastings algorithm. The stationary density is standard normal and the proposal density $q(u' \mid u)$ is $N(u, \sigma^2)$, with $\sigma = 0.1, 0.5, 2.4$ and 10. The initial value is $u_0 = -10$ and the same seed is used for the random number generator in each case. Note the dependence of the acceptance rate and convergence to stationarity on σ. The horizontal dashed lines show the 'usual' range for u.

$\phi(u) = \pi(u)$. The proposal density, $q(u' \mid u) = \sigma^{-1}\phi\{(u' - u)/\sigma\}$, depends on σ. We take initial value $u_0 = -10$ far from the centre of the stationary distribution. As $q(u' \mid u) = q(u \mid u')$, the acceptance probability is $a(u, u') = \min\{1, \phi(u')/\phi(u)\}$.

Figure 11.9 shows sample paths $u_0, \ldots, u_{500}$ for four values of σ. When $\sigma = 0.1$, only small steps occur but they are accepted with high probability because $\phi(u')/\phi(u) \doteq 1$. Although u changes at almost every step, it moves so little that the chain has not reached equilibrium after 500 iterations. When $\sigma = 0.5$ it takes 100 or so iterations to reach convergence and the chain then appears to mix fairly fast. When $\sigma = 2.4$ convergence is almost immediate but as the acceptance probability is lower the chain tends to get stuck for slightly longer. When $\sigma = 10$ the acceptance probability is low and although the chain jumps to its stationary range almost at once, it spends long periods without moving.

For comparison the experiment above was repeated 50 times, and the estimated means of $\pi(u)$ were compared. The estimator was the average of the last half of $u_0, \ldots, u_I$, with $I = 500$ iterations; that is, (11.39) with $m(u) = u$ and $B = 250$. Each of the 50 replicates used the same seed and initial value u_0 for each σ; the values of u_0 were generated from the t_5 density. The estimated values of σ_m^2 in (11.40) were 170, 17.7, 6.2, and 8.0 for $\sigma = 0.1, 0.5, 2.4$, and 10; the larger values of σ are preferable, but there is a large efficiency loss relative to the value $\sigma_m^2 = 1$ for independent sampling. This is because of the serial correlations of $u_{B+1}, \ldots, u_I$, which were roughly 0.97, 0.89, 0.62, and 0.83 for $\sigma = 0.1, 0.5, 2.4$, and 10.

Exercise 11.3.11 sheds more light on this example.   ■

Table 11.10 Motorette data (Nelson and Hahn, 1972). Censored failure times are denoted by +.

x (° F)	Failure time (hours)									
150	8064+	8064+	8064+	8064+	8064+	8064+	8064+	8064+	8064+	8064+
170	1764	2772	3444	3542	3780	4860	5196	5448+	5448+	5448+
190	408	408	1344	1344	1440	1680+	1680+	1680+	1680+	1680+
220	408	408	504	504	504	528+	528+	528+	528+	528+

Example 11.24 (Motorette data) Table 11.10 contains failure times y_{ij} from an accelerated life trial in which ten motorettes were tested at each of four temperatures, with the objective of predicting lifetime at 130°F. We analyse these data using a Weibull model with

$$\Pr(Y_{ij} \leq y; x_i) = 1 - \exp\left\{(y/\theta_i)^\gamma\right\}, \quad \theta_i = \exp\left(\beta_0 + \beta_1 x_i\right), \tag{11.48}$$

for $i = 1, \ldots, 4$, $j = 1, \ldots, 10$, where failure time is taken in units of hundreds of hours and x_i is log(temperature/100).

Here we describe a simple Bayesian analysis using the Metropolis–Hastings algorithm. For illustration we take independent priors on the parameters, $N(0, 100)$ on β_0 and β_1 and exponential with mean 2 on γ. Then the log posterior is

$$\ell_m(\beta_0, \beta_1, \gamma) \equiv -\left(\beta_0^2 + \beta_1^2\right)/200 - \gamma/2$$
$$+ \sum_{i=1}^{4} \sum_{j=1}^{10} d_{ij}\{\log \gamma + \gamma \log(y_{ij}/\theta_i)\} - (y_{ij}/\theta_i)^\gamma,$$

where $d_{ij} = 0$ for uncensored y_{ij}.

For proposal distribution we update all three parameters simultaneously, by taking $(\beta_0', \beta_1', \log \gamma') = (\beta_0, \beta_1, \log \gamma) + c(s_1 Z_1, s_2 Z_2, s_3 Z_3)$, where the s_r are the standard errors of the corresponding maximum likelihood estimates, $Z_r \overset{\text{iid}}{\sim} N(0, 1)$, and c can be chosen to balance the acceptance probability and the size of the move. The ratio $q(u \mid u')/q(u' \mid u)$ reduces to γ'/γ, so the acceptance probability equals

$$a\left\{(\beta_0', \beta_1', \gamma'), (\beta_0, \beta_1, \gamma)\right\} = \min\left[1, \exp\left\{\ell_m(\beta_0, \beta_1, \gamma) - \ell_m(\beta_0', \beta_1', \gamma')\right\} \gamma'/\gamma\right].$$

The chain is clearly irreducible and aperiodic, so the ergodic theorem applies.

We take initial values near the maximum likelihood estimates, and run the chain for 5000 iterations with $c = 0.5$. The sample path for β_1 in the upper left panel of Figure 11.10 shows that despite its acceptance probability of about 0.3, the chain is not moving well over the parameter space. This is confirmed by the correlogram and partial correlogram for successive values of β_1, which suggest that the chain is essentially an AR(1) process with $\rho_1 \doteq 0.99$. In this case the variance inflation factor is $\widehat{\tau} = 199$, so 5000 successive observations from the chain are worth about 25 independent observations. Sample paths for the other parameters are similar, and varying c does not improve matters. One reason for this is that β_0 and β_1 have correlation about -0.97 *a posteriori*, and the proposal distribution does not respect this. It is better to

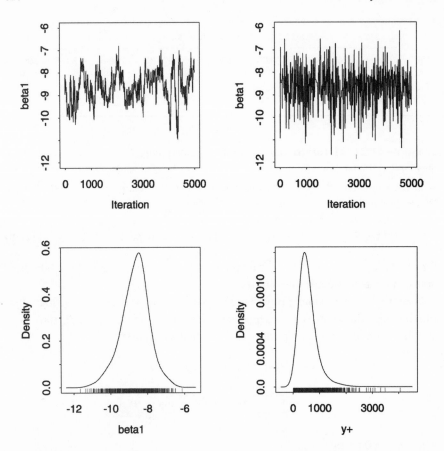

Figure 11.10 Bayesian
analysis of motorette data
using
Metropolis–Hastings
algorithm. Upper panels:
sample paths for β_1 using
two parametrizations, the
right one more nearly
orthogonal. Lower left:
kernel density estimates of
$\pi(\beta_1 \mid y)$ and of
$\pi(Y_+ \mid y)$, where Y_+ is
failure time predicted for
130°F.

reduce this correlation by replacing x by $x - \overline{x}$, after which $\mathrm{corr}(\beta_0, \beta_1 \mid y) \doteq -0.4$. The sample path for β_1 from a run of the algorithm starting near the new maximum likelihood estimates, with the new s_r and with $c = 2$, is shown in the upper right panel of Figure 11.10. This chain mixes much better, though its acceptance probability is about 0.2. The usual plots suggest that β_1 follows an AR(1) process with $\rho \doteq 0.9$, and likewise for the other parameters, whose chains show similar good behaviour. Here $\widehat{\tau}$ has the more acceptable value 19, though 5000 iterations would remain too small in practice.

The lower panels of the figure show kernel density estimates of the posterior densities for β_1 and for a predicted failure time Y_+ for temperature 130°F. Once convergence has been verified, it is easy to obtain values for Y_+, simply by simulating a Weibull variable from (11.48) using the current parameter values at each iteration. Quantiles of the simulated distributions may be used to obtain posterior confidence intervals for the corresponding quantities.

The Metropolis–Hastings update described above changes all three parameters on each iteration, or none of them. Alternatively we may attempt to update one parameter, chosen at random. The resulting chain is also ergodic, but it does not improve on the second approach described above.   ∎

Table 11.11 Accuracy of Stirling's formula and related approximations.

α	0.5	1	2	3	4	5
$I_{\alpha+1}$	0.8862	1	2	6	24	120
$\tilde{I}_{\alpha+1}/I_{\alpha+1}$	0.8578	0.9221	0.9595	0.9727	0.9794	0.9834
$\tilde{I}'_{\alpha+1}/I_{\alpha+1}$	0.9905	0.9960	0.9987	0.9994	0.9996	0.9998

Metropolis–Hastings updates using an appropriate proposal distribution can be used when the full conditional densities needed for particular steps of the Gibbs sampler are not available. Generalizations can be constructed to jump between spaces of differing dimensions, and these are valuable in applications where averaging over various spaces or choosing among them is important. More details are given in the bibliographic notes.

Exercises 11.3

1 Show that Laplace approximation to the gamma function

$$I_{\alpha+1} = \Gamma(\alpha+1) = \int_0^\infty u^\alpha e^{-u} \, du$$

gives *Stirling's formula*, $\Gamma(\alpha+1) \doteq \tilde{I}_{\alpha+1} = (2\pi)^{1/2} \alpha^{\alpha+1/2} e^{-\alpha}$, and verify that the $O(\alpha^{-1})$ term in (11.28) is $(12\alpha)^{-1}$. Show that this can be incorporated by modifying $\tilde{I}_{\alpha+1}$ to $\tilde{I}'_{\alpha+1} = (2\pi)^{1/2}(\alpha + \frac{1}{6})^{1/2} \alpha^\alpha e^{-\alpha}$, and check some of the numbers in Table 11.11.

2 Use the facts that if Z is a standard normal variable, $E(Z^4) = 3$ and $E(Z^6) = 15$, to check (11.28). Use properties of normal moments to explain why (11.28) is an expansion with terms in increasing powers of n^{-1} rather than $n^{-1/2}$.

3 Let $f(y; \theta)$ be a unimodal density with mode at $\tilde{y}_\theta$. Show that $\int_{-\infty}^y f(u; \theta) \, du$ may be approximated by (11.31), with

$$g(u) = \log f(\tilde{y}_\theta; \theta) - \log f(u; \theta), \quad a(u) = (2\pi)^{1/2} f(\tilde{y}_\theta; \theta),$$

and verify that the approximation is exact for the $N(\theta, \sigma^2)$ density. Investigate its accuracy numerically for the gamma density with shape parameter $\theta > 1$, and for the t_ν density.

4 Consider predicting the outcome of a future random variable Z on the basis of a random sample $Y_1, \ldots, Y_n$ from density $\lambda^{-1} e^{-u/\lambda}$, $u > 0$, $\lambda > 0$. Show that $\pi(\lambda) \propto \lambda^{-1}$ gives posterior predictive density

$$f(z \mid y) = \frac{\int f(z, y \mid \lambda) \pi(\lambda) \, d\lambda}{\int f(y \mid \lambda) \pi(\lambda) \, d\lambda} = n s^n / (s+z)^{n+1}, \quad z > 0,$$

where $s = y_1 + \cdots + y_n$.
Show that when Laplace's method is applied to each integral in the predictive density the result is proportional to the exact answer, and assess how close the approximation is to a density when $n = 5$.

5 Consider the integral

$$I_n = \int_{u_1}^{u_2} e^{-nh(u)} \, du,$$

where $h(u)$ is a smooth increasing function with minimum at u_1, at which point its derivatives are $h_1 = h'(u_1) > 0$, $h_2 = h''(u_1)$ and so forth. Show that

$$I_n = \frac{1}{nh_1} e^{-nh(u_1)} \left\{ 1 - e^{-nh_1(u_2-u_1)} + O(n^{-1}) \right\},$$

and deduce that

$$\int_{u_1}^{u_2} e^{-nh(u)}\, du \Big/ \int_{u_1}^{\infty} e^{-nh(u)}\, du \doteq 1 - e^{-nh_1(u_2-u_1)}.$$

A posterior density has form $\pi(\theta \mid y) \propto \theta^{-m-1}$, for $\theta > \theta_1$ (Exercise 11.2.2). Find the approximate and exact posterior density and distribution functions of θ, and compare them numerically when $m = 5, 10, 20$ and $\theta_1 = 1$. Discuss.

Investigate how the approximation will change if $h_1 = 0$.

6   Give an approximate variance for the importance sampling estimator (11.38), and verify the formula for $\mathrm{var}(\widehat{\mu}_{\mathrm{rat}})$.

7   *Sampling-importance resampling* (SIR) works as follows: instead of using (11.38) as an estimator of μ, an independent sample $\theta_1^*, \ldots \theta_Q^*$ of size $Q \ll S$ is taken from $\theta_1, \ldots, \theta_S$ with probabilities proportional to $w(\theta_1), \ldots, w(\theta_S)$. The estimator of μ is $\widehat{\mu}^* = Q^{-1} \sum \theta_q^*$.
(a) Discuss SIR critically when the initial sample is taken from the prior $\pi(\theta)$; this is sometimes called the *Bayesian bootstrap*. Give an explicit discussion in the case of an exponential family model and conjugate prior.
(b) Show that $\mathrm{E}^*(\widehat{\mu}^*) = \widehat{\mu}_{\mathrm{rat}}$, and find its variance. Use the Rao–Blackwell theorem to show that the variance of $\widehat{\mu}^*$ exceeds that of $\widehat{\mu}_{\mathrm{rat}}$.
Under what circumstances would it be sensible to use SIR anyway?
(Rubin, 1987; Smith and Gelfand, 1992; Ross, 1996)

8   Show that the Gibbs sampler with $k > 2$ components updated in order

$$1, \ldots, k, 1, \ldots, k, 1, \ldots, k, \ldots$$

is not reversible. Are samplers updated in order $1, \ldots, k, k - 1, \ldots, 1, 2, \ldots$, or in a random order reversible?

9   Show that the acceptance probability for a move from u to u' when random walk Metropolis sampling is applied to a transformation $v = v(u)$ of u is

$$\min\left\{ 1, \frac{\pi(u')|dv/du|}{\pi(u)|dv'/du'|} \right\}.$$

Hence verify the form of $q(u \mid u')/q(u' \mid u)$ given in Example 11.24.
Find the acceptance probability when a component of u takes values in (a, b), and a random walk is proposed for $v = \log\{(u - a)/(b - u)\}$.

10 Suppose that $Y_1, \ldots, Y_n$ are taken from an AR(1) process with innovation variance σ^2 and correlation parameter ρ such that $|\rho| < 1$. Show that

$$\mathrm{var}(\overline{Y}) = \frac{\sigma^2}{n^2(1 - \rho^2)} \left\{ n + 2 \sum_{j=1}^{n-1} (n - j)\rho^j \right\},$$

and deduce that as $n \to \infty$ for any fixed ρ, $n\,\mathrm{var}(\overline{Y}) \to \sigma^2/(1 - \rho)^2$.
What happens when $|\rho| = 1$?
Discuss estimation of $\mathrm{var}(\overline{Y})$ based on $(n - 1)^{-1} \sum (Y_j - \overline{Y})^2$ and an estimate $\widehat{\rho}$.

11 In Example 11.23, show that the probability of acceptance of a move starting from $u > 0$ equals

$$\frac{1}{2} + (1 + \sigma^2)^{-1/2} \exp(a^2/2) \{\Phi(a) + \Phi(b)\} - \Phi(-2u/\sigma),$$

where

$$a = -\frac{\sigma u}{\sqrt{1 + \sigma^2}}, \qquad b = \frac{-(2 + \sigma^2)u}{\sqrt{\sigma^2(1 + \sigma^2)}}.$$

Show that the expected move size may be written as

$$\exp\left(\frac{a^2}{2}\right)\left[\frac{\sigma}{1+\sigma^2}\{\phi(a)-\phi(b)\}-\frac{\sigma^2 u}{(1+\sigma^2)^{3/2}}\{\Phi(a)+\Phi(b)\}\right]$$
$$+\sigma\left\{\phi\left(\frac{-2u}{\sigma}\right)-\phi(0)\right\}.$$

Plot these functions over the range $0 \le u \le 15$ for $\sigma = 0.1, 1, 2.4, 10$, and also with $0 \le \sigma \le 10$ for $u = 0, 1, 2, 3, 10$. What light do these plots cast on the behaviour of the chains in Figure 11.9?

11.4 Bayesian Hierarchical Models

Hierarchical models are useful when data have layers of variation. The incidence of a disease may vary from region to region of a country, for instance, while within regions there is variation due to differences in poverty, pollution, or other factors. If the regional and local incidence rates are regarded as random, we can imagine a hierarchy in which the numbers of diseased persons depend on random local rates, which themselves depend on random regional rates. Such models were discussed briefly from a frequentist viewpoint in Section 9.4. Here we outline the Bayesian approach, using the notion of exchangeability.

The random variables $U_1, \ldots, U_n$ are called *finitely exchangeable* if their density has the property

$$f(u_1, \ldots, u_n) = f\left(u_{\xi(1)}, \ldots, u_{\xi(n)}\right)$$

for any permutation ξ of the set $\{1, \ldots, n\}$. Then the density is completely symmetric in its arguments and in probabilistic terms the $U_1, \ldots, U_n$ are indistinguishable; this does not mean that they are independent. An infinite sequence $U_1, U_2, \ldots$, is called *infinitely exchangeable* if every finite subset of it is finitely exchangeable.

A key result in this context is *de Finetti's theorem*, whose simplest form says that if $U_1, U_2, \ldots$, is an infinitely exchangeable sequence of binary variables, taking values $u_j = 0, 1$, then for any n there is a distribution G such that

$$f(u_1, \ldots, u_n) = \int_0^1 \prod_{j=1}^n \theta^{u_j}(1-\theta)^{1-u_j}\, dG(\theta) \qquad (11.49)$$

where

$$G(\theta) = \lim_{m\to\infty} \Pr\{m^{-1}(U_1 + \cdots + U_m) \le \theta\}, \quad \theta = \lim_{m\to\infty} m^{-1}(U_1 + \cdots + U_m).$$

This is justified at the end of this section. It implies that any set of exchangeable binary variables $U_1, \ldots, U_n$ may be modelled as if they were independent Bernoulli variables, conditional on their success probability θ, this having distribution G and being interpretable as the long-run proportion of successes. More general versions of (11.49) hold for real U_j, for example. The upshot is that a judgement that certain quantities are exchangeable implies that they may be represented as a random sample conditional on a variable that itself has a distribution. This provides the basis of a

Bruno de Finetti (1906–1985) was born in Innsbruck and studied in Milan and Rome, where he eventually became professor, after working in Trieste as an actuary and at the University of Padova. His main contribution to statistics was to develop personalistic probability, teaching that 'probability does not exist'. (You may think this should have been made clear on page 1 of the book!) He argued that probability distributions express a person's view of the world, with no objective force. His ideas have strongly influenced Bayesian thought.

case in favour of Bayesian inference, because it implies that the conditional density $\Pr(U_{n+1} \mid U_1, \ldots, U_n)$ for a future variable U_{n+1} given the outcomes of $U_1, \ldots, U_n$, may be represented as a ratio of two integrals of form (11.49), and this is formally equivalent to Bayesian prediction using a prior density on θ.

The essence of hierarchical modelling is to treat not data but particular sets of parameters as exchangeable. For if our model contains parameters $\theta_1, \ldots, \theta_n$, and if we believe *a priori* that these are to be treated completely symmetrically, then they are exchangeable and may be thought of as a random sample from a distribution that is itself unknown. In principle that distribution might be anything, but in practice a tractable one is often chosen.

Example 11.25 (Normal hierarchical model)   A prototypical case is the normal model under which $y_1, \ldots, y_n$ satisfy

$$y_j \mid \theta_j \overset{\text{ind}}{\sim} N(\theta_j, v_j), \quad \theta_1, \ldots, \theta_n \mid \mu \overset{\text{iid}}{\sim} N(\mu, \sigma^2), \quad \mu \sim N(\mu_0, \tau^2),$$

where $v_1, \ldots, v_n$, σ^2, μ_0 and τ^2 are known; the last two are *hyperparameters* that control the uncertainty injected at the top level of the hierarchy. The y_j have different variances, but their means θ_j are supposed indistinguishable and hence are modelled as exchangeable, being normal with unknown mean μ. As the joint density of $(\mu, \theta^{\mathrm T}, y^{\mathrm T})^{\mathrm T}$ is multivariate normal of dimension $2n + 1$, with mean vector and covariance matrix

$$\mu_0 1_{2n+1}, \quad \begin{pmatrix} \tau^2 & \tau^2 1_n^{\mathrm T} & \tau^2 1_n^{\mathrm T} \\ \tau^2 1_n & \tau^2 1_n 1_n^{\mathrm T} + \sigma^2 I_n & \tau^2 1_n 1_n^{\mathrm T} + \sigma^2 I_n \\ \tau^2 1_n & \tau^2 1_n 1_n^{\mathrm T} + \sigma^2 I_n & V + \tau^2 1_n 1_n^{\mathrm T} + \sigma^2 I_n \end{pmatrix}, \tag{11.50}$$

where $V = \mathrm{diag}(v_1, \ldots, v_n)$, the posterior density of $(\mu, \theta^{\mathrm T})^{\mathrm T}$ given y is also normal. Unenlightening matrix calculations give

$$\mathrm{E}(\mu \mid y) = \frac{\mu_0/\tau^2 + \sum y_j/(\sigma^2 + v_j)}{1/\tau^2 + \sum 1/(\sigma^2 + v_j)}, \quad \mathrm{var}(\mu \mid y) = \frac{1}{1/\tau^2 + \sum 1/(\sigma^2 + v_j)},$$

and

$$\mathrm{E}(\theta_j \mid y) = \mathrm{E}(\mu \mid y) + \frac{\sigma^2}{\sigma^2 + v_j}\{y_j - \mathrm{E}(\mu \mid y)\}.$$

The posterior mean of μ is a weighted average of its prior mean μ_0 and of the y_j, weighted according to their precisions conditional on μ. Typically τ^2 is very large, and then $\mathrm{E}(\mu \mid y)$ is essentially a weighted average of the data. Even when $v_j \to 0$ for all j there is still posterior uncertainty about μ, whose variance is σ^2/n because $y_1, \ldots, y_n$ is then a random sample from $N(\mu, \sigma^2)$.

The posterior mean of θ_j is a weighted average of y_j and $\mathrm{E}(\mu \mid y)$, showing shrinkage of y_j towards $\mathrm{E}(\mu \mid y)$ by an amount that depends on v_j. As $v_j \to 0$, $\mathrm{E}(\theta_j \mid y) \to y_j$, while as $v_j \to \infty$, $\mathrm{E}(\theta_j \mid y) \to \mathrm{E}(\mu \mid y)$. This is a characteristic feature of hierarchical models, in which there is a 'borrowing of strength' whereby all the data combine to estimate common parameters such as μ, while estimates of individual parameters such as the θ_j are shrunk towards common values by amounts

that depend on the precision of the corresponding observations, here represented by
the v_j. ∎

Example 11.26 (Cardiac surgery data) Table 11.2 contains data on mortality of
babies undergoing cardiac surgery at 12 hospitals. Although the numbers of operations
and the death rates vary, we have no further knowledge of the hospitals and hence no
basis for treating them other than entirely symmetrically, suggesting the hierarchical
model

$$r_j \mid \theta_j \overset{\text{ind}}{\sim} B(m_j, \theta_j), \quad j = A, \ldots, L, \quad \theta_A, \ldots, \theta_L \mid \zeta \overset{\text{iid}}{\sim} f(\theta \mid \zeta), \quad \zeta \sim \pi(\zeta).$$

Conditional on θ_j, the number of deaths r_j at hospital j is binomial with probability
θ_j and denominator m_j, the number of operations, which plays the same role as v_j^{-1}
in Example 11.25: when m_j is large then a death rate is relatively precisely known.
Conditional on ζ, the θ_j are a random sample from a distribution $f(\theta \mid \zeta)$, and ζ itself
has a prior distribution that depends on fixed hyperparameters.

One simple formulation is to let $\beta_j = \log\{\theta_j/(1 - \theta_j)\} \sim N(\mu, \sigma^2)$, conditional
on $\zeta = (\mu, \sigma^2)$, thereby supposing that the log odds of death have a normal distribution, and to take $\mu \sim N(0, c^2)$ and $\sigma^2 \sim IG(a, b)$, where $a, b,$ and c express proper
but vague prior information. For sake of illustration we let $a = b = 10^{-3}$, so σ^2 has
prior mean one but variance 10^3, and $c = 10^3$, giving μ prior variance 10^6. The joint
density then has form

$$\prod_j \binom{m_j}{r_j} \frac{e^{r_j \beta_j}}{(1 + e^{\beta_j})^{m_j}} \frac{1}{(2\pi\sigma^2)^{1/2}} \exp\left\{-\frac{1}{2\sigma^2}(\beta_j - \mu)^2\right\} \times \pi(\mu)\pi(\sigma^2),$$

so the full conditional densities for μ and σ^2 are normal and inverse gamma. Apart
from a constant, the full conditional density for β_j has logarithm

$$r_j \beta_j - m_j \log(1 + e^{\beta_j}) - \frac{(\beta_j - \mu)^2}{2\sigma^2},$$

and as this is a sum of two functions concave in β_j, adaptive rejection sampling may
be used to simulate β_j given μ, σ^2, and the data; see Example 3.22.

This model was fitted using the Gibbs sampler with 5500 iterations, of which the
first 500 were discarded. Convergence appeared rapid.

Figure 11.11 compares results for the hierarchical model with the effect of treating
each hospital separately using uniform prior densities for the θ_j. Shrinkage due to the
hierarchical fit is strong, particularly for the smaller hospitals; the posterior mean of
θ_A, for example, has changed from about 2% to over 5%. Likewise the posterior means
of θ_H and θ_B have decreased considerably towards the overall mean. By contrast, the
posterior mean of θ_D barely changes because of the large value of m_D. Posterior
credible intervals for the hierarchical model are only slightly shorter but they are
centred quite differently. The posterior mean rate is about 7.3%, with 0.95 credible
interval $(5.3, 9.4)\%$. ∎

In some cases the hierarchical element is merely a component of a more complex
model, as the following example illustrates.

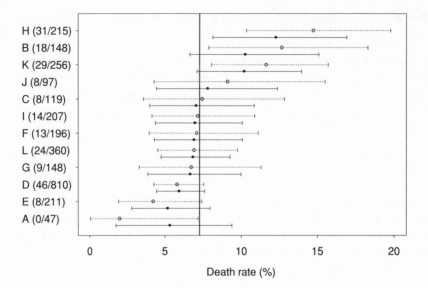

Figure 11.11 Posterior
summaries for mortality
rates for cardiac surgery
data. Posterior means and
0.95 equitailed credible
intervals for separate
analyses for each hospital
are shown by hollow
circles and dotted lines,
while blobs and solid lines
show the corresponding
quantities for a
hierarchical model. Note
the shrinkage of the
estimates for the
hierarchical model
towards the overall
posterior mean rate,
shown as the solid vertical
line; the hierarchical
intervals are slightly
shorter than those for the
simpler model.

Example 11.27 (Spring barley data) Table 10.21 contains data on a field trial intended to compare the yields of 75 varieties of spring barley allocated randomly to plots in three long narrow blocks. The data were analysed in Example 10.35 using a generalized additive model to accommodate the strong fertility trends over the blocks. In the absence of detailed knowledge about the varieties it seems natural to treat them as exchangeable, and we outline a Bayesian hierarchical approach. We also show how the fertility patterns may be modelled using a simple Markov random field.

Let $y = (y_1, \ldots, y_n)^{\mathrm{T}}$ denote the yields in the $n = 225$ plots and let ψ_j denote the unknown fertility of plot j. Let X denote the $n \times p$ design matrix that shows which of the $p = 75$ variety parameters $\beta = (\beta_1, \ldots, \beta_p)^{\mathrm{T}}$ have been allocated to the plots. Then a normal linear model for the yields is

$$y \mid \beta, \psi, \lambda_y \sim N_n(\psi + X\beta, I_n/\lambda_y), \qquad (11.51)$$

where ψ is the $n \times 1$ vector containing the fertilities and λ_y is the unknown precision of the ys.

We take the prior density of λ_y to be gamma with shape and scale parameters a and b, $G(a, b)$, so that its prior mean and variance are a/b and a/b^2, where a and b are specified. As there is no special treatment structure, we take for the β_r the exchangeable prior $\beta \sim N_p(0, I_p/\lambda_\beta^{-1})$, with $\lambda_\beta \sim G(c, d)$ and c, d specified. For the fertilities we take the normal Markov chain of Example 6.13, for which

$$\pi(\psi \mid \lambda_\psi) \propto \lambda_\psi^{n/2} \exp\left\{ -\frac{1}{2}\lambda_\psi \sum_{i \sim j}(\psi_i - \psi_j)^2 \right\}, \quad \lambda_\psi > 0, \qquad (11.52)$$

the summation being over pairs of neighbouring plots and λ_ψ^{-1} being the variance of differences between fertilities. Each ψ_j occurs in n_j terms, where $n_j = 1$ or 2 is the number of plots adjacent to plot j. The sum in (11.52) equals $\psi^{\mathrm{T}} W \psi$, where W is the $n \times n$ tridiagonal matrix with elements

$$
w_{ij} = \begin{cases} n_i, & i = j, \\ -1, & i \sim j, \\ 0, & \text{otherwise.} \end{cases}
$$

Thus W is block diagonal, with three blocks like the matrix V in Example 6.13 with $\tau = 0$, corresponding to the three physical blocks of the experiment. We take $\lambda_\psi \sim G(g, h)$, with g and h specified.

With these conjugate prior densities, the joint posterior density is

$$
\pi(\beta, \psi, \lambda) \propto \lambda_y^{n/2} \exp\left\{ -\frac{1}{2} \lambda_y (y - \psi - X\beta)^{\mathrm{T}} (y - \psi - X\beta) \right\}
$$

$$
\times \lambda_\beta^{p/2} \exp\left(-\frac{1}{2} \lambda_\beta \beta^{\mathrm{T}} \beta \right) \times \lambda_\psi^{p/2} \exp\left(-\frac{1}{2} \lambda_\psi \psi^{\mathrm{T}} W \psi \right)
$$

$$
\times \lambda_y^{a-1} \exp(-b\lambda_y) \times \lambda_\beta^{c-1} \exp(-c\lambda_\beta) \times \lambda_\psi^{g-1} \exp(-h\lambda_\psi),
$$

where $\lambda = (\lambda_y, \lambda_\beta, \lambda_\psi)^{\mathrm{T}}$. The full conditional densities turn out to be

$$
\beta \mid \psi, \lambda, y \sim N\left\{ \lambda_y Q_\beta^{-1} X^{\mathrm{T}} (y - \psi), Q_\beta^{-1} \right\}, \tag{11.53}
$$

$$
\psi \mid \beta, \lambda, y \sim N\left\{ \lambda_y Q_\psi^{-1} (y - X\beta), Q_\psi^{-1} \right\}, \tag{11.54}
$$

$$
\lambda_y \mid \psi, \beta, y \sim G\{ a + n/2, b + (y - X\beta - \psi)^{\mathrm{T}} (y - X\beta - \psi)/2 \}, \tag{11.55}
$$

$$
\lambda_\beta \mid \psi, \beta, y \sim G(c + p/2, d + \beta^{\mathrm{T}} \beta/2), \tag{11.56}
$$

$$
\lambda_\psi \mid \psi, \beta, y \sim G(g + n/2, h + \psi^{\mathrm{T}} W \psi/2), \tag{11.57}
$$

where

$$
Q_\beta = \lambda_y X^{\mathrm{T}} X + \lambda_\beta I_p, \quad Q_\psi = \lambda_y I_n + \lambda_\psi W.
$$

The elements of λ are independent conditional on the remaining variables. The relatively simple form of the densities in (11.53)–(11.57) suggests using a time-reversible Gibbs sampler, in which β, ψ, and λ are updated in a random order at each iteration. The most direct approach to simulation in (11.53) and (11.54) is through Cholesky decomposition of Q_β and Q_ψ: in (11.53), for example, we find the lower triangular matrix L such that $LL^{\mathrm{T}} = Q_\beta^{-1}$, generate $\varepsilon \sim N_p(0, I_p)$, and let $\beta = \lambda_y Q_\beta^{-1} X^{\mathrm{T}} (y - \psi) + L\varepsilon$. The block diagonal structure of W means that the ψs for different blocks can be updated separately, so the largest Cholesky decomposition needed is that of a 75×75 matrix. An alternative is to update individual ψ_js in a random order, but although the computational burden is smaller, the algorithm then converges more slowly than with direct use of (11.54).

Note the strong resemblance of (11.53) and (11.54) to the steps of the backfitting algorithm for the corresponding generalized additive model.

The missing response in block 3 is simply a further unknown whose value may be simulated using the relevant marginal density of (11.51). This adds a fourth component to the simulation in random order of β, ψ, and λ at each iteration; there are no other changes to the algorithm.

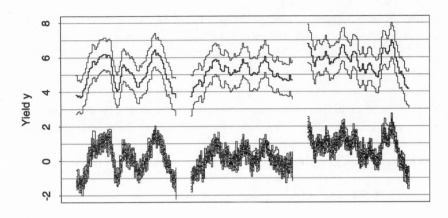

Yield y

Location

Figure 11.12 Posterior summaries for fertility trend ψ for the three blocks of spring barley data, shown from left to right. Above: median trend (heavy) and overall 0.9 posterior credible bands. Below: 20 simulated trends from Gibbs sampler output.

If the matrix $X^{\mathrm{T}}X$ is diagonal, then the full conditional density for the rth variety effect has form

$$\beta_r \mid \psi, \lambda, y \sim N\left(\frac{\lambda_y m_r \bar{z}_r}{\lambda_\beta + \lambda_y m_r}, \frac{1}{\lambda_\beta + \lambda_y m_r}\right),$$

where $\bar{z}_r$ is the current average of $y_j - \psi_j$ for the m_r plots receiving variety r. Thus the β_r are shrunk towards zero by an amount that depends on the ratio λ_β/λ_y; with $\lambda_\beta = 0$ the mean for β in (11.53) is the least squares estimate computed by regressing $y - \psi$ on the columns of X. Unlike in Example 11.25, however, the normal distributions of the β_r are here averaged over the posterior densities of ψ, λ_y and λ_β.

The algorithm described above was run with random initial values for 10,500 iterations. Time series plots of the parameters and log likelihood suggested that it had converged after 500 iterations, and inferences below are based on the final 10,000 iterations. The variance inflation factors $\hat{\tau}$ were less than 4 for ψ and β, about 44, 6 and 30 for λ_y, λ_τ and λ_ψ, and about 6 for y_{187}. Thus estimation for λ_y is least reliable, being based on a sample equivalent to about 220 independent observations. A longer run of the algorithm would seem wise in practice. Based on this run, the posterior 0.9 credible intervals for λ_y, λ_ψ and λ_β were (5.2, 12.4), (5.0, 11.5) and (2.7, 5.7) respectively, and differences of two variety effects have posterior densities very close to normal with typical standard deviation of 0.35. The corresponding standard error for the generalized additive model was 0.41, so use of a hierarchical model and injection of prior information has increased the precision of these comparisons.

Figure 11.12 shows some simulated values of ψ and pointwise 0.90 credible envelopes for the true ψ. These envelopes are constructed by joining the 0.05 quantiles of the fertilities simulated from the posterior density, for each location, and likewise with the 0.95 quantiles. By contrast with the analysis in Example 10.35, the effective degrees of freedom for ψ, controlled by λ_ψ, are here equal for each block, leading to apparent overfitting of the fertilities for block 2 compared to the generalized additive model. A difference between the models is that the current model corresponds

Table 11.12 Posterior probabilities that a variety is ranked among the best r varieties, estimated from 10,000 iterations of Gibbs sampler.

	Variety									
r	56	35	72	31	55	47	54	18	38	40
1	0.327	0.182	0.149	0.129	0.075	0.055	0.019	0.015	0.012	0.006
2	0.518	0.357	0.299	0.270	0.174	0.136	0.050	0.042	0.035	0.020
5	0.814	0.690	0.643	0.621	0.486	0.416	0.234	0.183	0.153	0.106
10	0.959	0.908	0.887	0.871	0.795	0.743	0.560	0.497	0.429	0.344

to first differences of ψ being a normal random sample, while in the earlier model the second differences are a normal random sample, giving a smoother fit.

The posterior probabilities that certain varieties rank among the r best are given in Table 11.12. The ordering is somewhat different from that in Example 10.35, perhaps due to the slightly different treatment of fertility effects. As mentioned previously, no single variety strongly outperforms the rest, and future field experiments would have to include several of those included in this trial. This type of information is difficult to obtain using frequentist procedures, but is readily found by manipulating the output of the simulation algorithm described above.

This analysis is relatively easily modified when elements of the model are changed. Indeed the priors and other components chosen largely for convenience should be varied in order to assess the sensitivity of the conclusions to them; see Exercise 11.3.6. Metropolis–Hastings steps would then typically replace the Gibbs updates in the algorithm. ∎

As mentioned above, more complicated hierarchies involve several layers of nested variation. Such models are widely used in certain applications, but their assessment and comparison can be difficult. For instance, shrinkage makes it unclear just how many parameters a hierarchical model has. Hierarchical modelling is an active area of current research.

Justification of (11.49)

To establish (11.49), suppose that r lies in $0, \ldots, n$ and that $m > n$. Then exchangeability of $U_1, \ldots, U_m$ implies that the conditional probability

$$\Pr(U_1 + \cdots + U_n = r \mid U_1 + \cdots + U_m = s)$$

equals the probability of seeing r 1's in n draws without replacement from an urn containing s 1's and $m - s$ 0's, which is $\binom{m}{n}^{-1}\binom{s}{r}\binom{m-s}{n-r}$ for $s = r, \ldots, m - (n - r)$ and zero otherwise. Hence

$$\Pr(U_1 + \cdots + U_n = r) = \sum_{s=r}^{m-(n-r)} \binom{m}{n}^{-1} \binom{s}{r}\binom{m-s}{n-r}\Pr(U_1 + \cdots + U_m = s)$$

$$= \binom{n}{r} \sum_{s=r}^{m-(n-r)} \frac{s^{(r)}(m-s)^{(n-r)}}{m^{(n)}}\Pr(U_1 + \cdots + U_m = s),$$

where $s^{(r)} = s(s-1)\cdots(s-r+1)$ and so forth. If $G_m(\theta)$ denotes the distribution putting mass $\Pr(U_1 + \cdots + U_m = s)$ at s/m, for $s = 0, \ldots, m$, then

$$\Pr(U_1 + \cdots + U_n = r) = \binom{n}{r} \int_0^1 \frac{(m\theta)^{(r)}\{m(1-\theta)\}^{(n-r)}}{m^{(n)}} \, dG_m(\theta).$$

As $m \to \infty$,

$$\frac{(m\theta)^{(r)}\{m(1-\theta)\}^{(n-r)}}{m^{(n)}} \to \theta^r(1-\theta)^{n-r},$$

and in fact there is an infinite subsequence of values of m such that G_m converges to a limit G that is a distribution function. To establish (11.49) we simply note that

$$\binom{n}{r} f\left(u_{\xi(1)}, \ldots, u_{\xi(n)}\right) = \Pr(U_1 + \cdots + U_n = r)$$

for any permutation ξ of $\{1, \ldots, n\}$ such that $u_{\xi(1)} + \cdots + u_{\xi(n)} = r$, giving

$$f(u_1, \ldots, u_n) = \int_0^1 \theta^r(1-\theta)^{n-r} \, dG(\theta) = \int_0^1 \prod_{j=1}^n \theta^{u_j}(1-\theta)^{1-u_j} \, dG(\theta)$$

as desired.

Exercises 11.4

1   Two balls are drawn successively without replacement from an urn containing three white and two red balls. Are the outcomes of the first and second draws independent? Are they exchangeable?

2   Under what conditions are the Bernoulli random variables Y_1 and $Y_2 = 1 - Y_1$ exchangeable? What about $Y_1, \ldots, Y_n$ given that $Y_1 + \cdots + Y_n = m$?

3   Establish (11.50), and use it and (3.21) to verify the given formulae for the posterior mean and variance for μ.

4   Describe how a Metropolis–Hastings update could be used to avoid adaptive rejection sampling from the full conditional density for β in Example 11.26. Compare and contrast the two approaches.

5   In a variant on the hierarchical Poisson model in Example 11.19, let $Y_1, \ldots, Y_n$ be independent Poisson variables with means $\theta_1, \ldots, \theta_n$, let $\theta_1, \ldots, \theta_n$ be a random sample from the density $\beta e^{-\theta\beta}$, $\theta > 0$, and let the prior density of β be uniform on the positive half-line. Find $E(\theta_j \mid y, \beta)$, and show that if $n\overline{y} > 1$ then the posterior distribution of $\gamma = 1/(1+\beta)$ is Beta with parameters $n\overline{y} - 1$ and $n + 1$. Hence show that the posterior mean of θ_j is $(y_j + 1)(n\overline{y} - 1)/(n\overline{y} + n)$. Under what condition is this greater than the estimate $\widehat{\theta}_j = y_j$ obtained under the classical model with no link among the θs? Explain.

6   (a) Give the directed acyclic and conditional independence graphs for the model in Example 11.27, and verify (11.53)–(11.57).
    (b) What changes to the algorithm are needed if (11.52) is replaced by

$$\pi(\psi \mid \lambda_\psi) \propto \lambda_\psi^{n/2} \exp\left\{-\frac{1}{2}\lambda_\psi \sum_{i \sim j} |\psi_i - \psi_j|\right\}, \qquad \lambda_\psi > 0?$$

What changes are needed if (11.51) specifies that the y_j have independent t_ν densities, for some known ν?
    (c) How would you allow different degrees of smoothing for the different blocks?
    (Besag *et al.*, 1995)

11.5 Empirical Bayes Inference

11.5.1 Basic ideas

The borrowing of strength achieved by hierarchical Bayes models increases the precision of parameter estimation at the cost of specifying prior distributions at two levels. This can be bothersome in practice, because priors on hyperparameters are difficult to verify and it is natural to worry about their effect on subsequent inferences. Sensitivity analysis, comparing results from different priors, is valuable, but another possibility in some cases is to estimate the hyperparameters from the data. Many Bayesians deprecate this empirical Bayes approach as essentially frequentist; we shall skirt this issue and simply sketch the main ideas.

Consider the model

$$y_1, \ldots, y_n \mid \theta_1, \ldots, \theta_n \overset{\text{ind}}{\sim} f(y_1 \mid \theta_1), \ldots, f(y_n \mid \theta_n), \quad \theta_1, \ldots, \theta_n \overset{\text{iid}}{\sim} \pi(\theta \mid \gamma).$$

A fully Bayesian specification would add a prior density $\pi(\gamma)$ for γ, with inference for the θ_j based on the marginal posterior densities $\pi(\theta_j \mid y)$. If we do not add this further level of complexity, then the data have marginal density

$$f(y_1, \ldots, y_n \mid \gamma) = \prod_{j=1}^{n} \int f(y_j \mid \theta_j) \pi(\theta_j \mid \gamma) \, d\theta_j$$

from which we might estimate γ. An obvious approach is to use the maximum likelihood estimator $\widehat{\gamma}$ found from this density, and then to base inferences on the posterior densities $\pi(\theta_j \mid y, \widehat{\gamma})$, for example computing posterior moments

$$\text{E}(\theta_j^r \mid y, \widehat{\gamma}) = \left. \frac{\int \theta_j^r f(y_j \mid \theta_j) \pi(\theta_j \mid \gamma) \, d\theta_j}{\int f(y_j \mid \theta_j) \pi(\theta_j \mid \gamma) \, d\theta_j} \right|_{\gamma = \widehat{\gamma}}.$$

Numerical methods are generally needed to evaluate the integrals. Full Bayesian analysis would integrate out γ with respect to its prior density, thereby accounting for uncertainty about γ rather than simply setting it to $\widehat{\gamma}$.

Example 11.28 (Normal distribution) Consider the model

$$y_1, \ldots, y_n \mid \theta_1, \ldots, \theta_n \overset{\text{ind}}{\sim} N(\theta_j, v_j), \quad \theta_1, \ldots, \theta_n \overset{\text{iid}}{\sim} N(\mu, \tau^2),$$

where the v_j are known positive constants, and suppose initially that $\tau^2 > 0$ is also known. The conditional distribution of θ_j given y is

$$N(\xi_j \mu + (1 - \xi_j) y_j, (1 - \xi_j) v_j), \quad \text{with } \xi_j = \frac{v_j}{v_j + \tau^2}, \quad j = 1, \ldots, n, \quad (11.58)$$

and the y_j are marginally independent with $N(\mu, v_j + \tau^2)$ densities. The maximum likelihood estimate of μ is therefore

$$\widehat{\mu} = \widehat{\mu}(\tau^2) = \frac{\sum_{j=1}^{n} y_j / (v_j + \tau^2)}{\sum_{j=1}^{n} 1 / (v_j + \tau^2)},$$

and the empirical Bayes estimate of θ_j is found by substituting this into $\mathrm{E}(\theta_j \mid y)$, to give

$$\tilde{\theta}_j = \xi_j \widehat{\mu} + (1 - \xi_j) y_j = \widehat{\mu} + (1 - \xi_j)(y_j - \widehat{\mu}). \tag{11.59}$$

When $\xi_j = 0$ then $\tilde{\theta}_j = y_j$ is unbiased for θ_j. Taking $\xi_j > 0$ gives non-zero shrinkage and biased estimation of $\tilde{\theta}_j$, but the hope is that the borrowing of strength induced by shrinkage towards a common mean will reduce overall mean squared error. The degree of shrinkage towards $\widehat{\mu}$ depends on v_j / τ^2. This is disquieting because the amount of shrinkage bears no relation to the data. Thus if the y_j were very different doubt would be cast on the model, but the formulation pays no heed to this.

When τ^2 is unknown, its profile log likelihood is

$$\ell_{\mathrm{p}}(\tau^2) \equiv -\frac{1}{2} \sum_{j=1}^{n} \log(v_j + \tau^2) - \frac{1}{2} \sum_{j=1}^{n} \{y_j - \widehat{\mu}(\tau^2)\}^2 / (v_j + \tau^2), \quad \tau^2 \geq 0,$$

from which the maximum likelihood estimate $\widehat{\tau}^2$ can be obtained. If $\widehat{\tau}^2 = 0$ then the data give no evidence of variation in the θ_j, all the y_j have mean μ, and all the $\tilde{\theta}_j$ are shrunk to $\widehat{\mu}$. If $\widehat{\tau}^2 > 0$, then ξ_j is replaced by $v_j / (v_j + \widehat{\tau}^2)$ in (11.59). As $0 \leq v_j / (v_j + \widehat{\tau}^2) \leq 1$, $\tilde{\theta}_j$ lies between y_j and $\widehat{\mu}$.

Confidence intervals for the θ_j may be computed by replacing μ and τ^2 in (11.58) by estimates, but their coverage will be lower than the nominal level because the variability of $\widehat{\mu}$ and $\widehat{\tau}^2$ is unaccounted for. Approaches to overcoming this have been proposed, but we shall not treat them here.   ∎

Example 11.29 (Toxoplasmosis data)   Example 10.29 discusses estimation of levels of toxoplasmosis in 34 cities in El Salvador. For a simple analysis of these data, we let $y_j = \log\{(r_j + 1/2)/(m_j - r_j + 1/2)\}$ represent empirical logistic transformations of the binomial responses giving the level of toxoplasmosis, with approximate variances $v_j = (r_j + 1/2)^{-1} + (m_j - r_j + 1/2)^{-1}$ treated as known. We generalize Example 11.28 to encompass regression by taking

$$y_1, \ldots, y_n \mid \theta_1, \ldots, \theta_n \stackrel{\mathrm{ind}}{\sim} N(\theta_j, v_j), \quad \theta_j \mid \beta \stackrel{\mathrm{ind}}{\sim} N(x_j^{\mathrm{T}} \beta, v_j'), \quad j = 1, \ldots, n,$$

so that the θ_j vary around means $x_j^{\mathrm{T}} \beta$. Then

$$\theta_j \mid y, \beta, v_j' \stackrel{\mathrm{ind}}{\sim} N\left\{(1 - \xi_j) y_j + \xi_j x_j^{\mathrm{T}} \beta, v_j (1 - \xi_j)\right\}, \quad \xi_j = v_j / (v_j + v_j'),$$

and marginally $y_j \stackrel{\mathrm{ind}}{\sim} N(x_j^{\mathrm{T}} \beta, v_j + v_j')$, for $j = 1, \ldots, n$. Maximum likelihood yields the weighted least squares estimator $\widehat{\beta} = (X^{\mathrm{T}} W X)^{-1} X^{\mathrm{T}} W y$, where W is the diagonal matrix with elements $w_j = (v_j + v_j')^{-1}$, leading to shrinkage estimators $\tilde{\theta}_j = (1 - \xi_j) y_j + \xi_j x_j^{\mathrm{T}} \widehat{\beta}$ of the θ_j, with estimated variances $v_j (1 - \xi_j)$.

The v_j' typically depend on unknown parameters that may be estimated from the profile likelihood. Here we take $v_1' = \cdots = v_n' = \tau^2$. If $x^{\mathrm{T}} \beta$ equals a constant, then $\widehat{\tau}^2 = 0.17$, but it is better to let $x^{\mathrm{T}} \beta$ be a cubic function of rainfall, leading to $\widehat{\tau}^2 = 0.1$. Figure 11.13 shows strong shrinkage of the individual estimates y_j towards their regression counterparts $x_j \widehat{\beta}$. The average variance reduces by a factor of almost ten,

Table 11.13
Shakespeare's word type frequencies (Efron and Thisted, 1976; Thisted and Efron, 1987). Entry r is n_r, the number of word types used exactly r times. There are 846 word types which appear more than 100 times, for a total of 31,534 word types.

r	1	2	3	4	5	6	7	8	9	10	Total
0+	14376	4343	2292	1463	1043	837	638	519	430	364	26305
10+	305	259	242	223	187	181	179	130	127	128	1961
20+	104	105	99	112	93	74	83	76	72	63	881
30+	73	47	56	59	53	45	34	49	45	52	513
40+	49	41	30	35	37	21	41	30	28	19	331
50+	25	19	28	27	31	19	19	22	23	14	227
60+	30	19	21	18	15	10	15	14	11	16	169
70+	13	12	10	16	18	11	8	15	12	7	122
80+	13	12	11	8	10	11	7	12	9	8	101
90+	4	7	6	7	10	10	15	7	7	5	78

Figure 11.13 Shrinkage of individual estimates (lower blobs) towards regession estimates (upper blobs) for toxoplasmosis data.

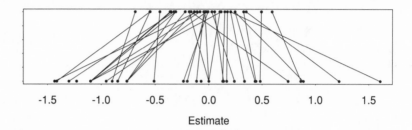

from $\overline{v} = 0.68$ to $\overline{v(1 - \widehat{\xi})} = 0.07$, and one would expect a large decrease in overall mean squared error.

The empirical Bayes estimates of the toxoplasmosis levels themselves are obtained by inverse logistic transformation, with standard errors from the delta method. A more detailed analysis, or simulation, would be needed to account for the uncertainty in $\widehat{\beta}$ and $\widehat{\tau}^2$. ∎

The previous examples illustrate parametric empirical Bayes inference, in which the prior for θ is taken from a parametrized family of distributions. In practice an alternative is to try and estimate the prior nonparametrically. The resulting estimators are generally unstable if the data are not extensive, and some form of smoothing may be needed.

Example 11.30 (Shakespeare's vocabulary data) The canon of Shakespeare's accepted works contains 884,647 words, with 31,534 distinct word types. A word type is a distinguishable arrangement of letters, so 'king' is different from 'kings' and 'alehouse' different from both 'ale' and 'house'. Table 11.13 shows how many word types occurred once, twice, and so on in the canon: 14,376 appear just once, 4343 appear twice, and so forth. If n_r is the number of word types appearing r times, then $\sum_{r=1}^{\infty} n_r = 31,534$.

If a new body of work containing $884,647t$ words was found, how many new word types might it contain? Taking $t = 1$ corresponds to finding a new set of works the same size as the canon, while setting $t = \infty$ enables us to estimate Shakespeare's total vocabulary.

Finding a new word type in a body of work is analogous to finding a new species of animal among those caught in a trap. Suppose that there are S species in total, and that after trapping over the period $[-1, 0]$ we have y_s members of species s. We assume that they enter the trap according to a Poisson process of rate λ_s per unit of time, so y_s is Poisson with mean λ_s, and let $n_r = \sum_s I(y_s = r)$ be the number of species observed exactly r times in the trapping period $[-1, 0]$. Let $G(\lambda)$ be the unknown distribution function of $\lambda_1, \dots, \lambda_S$. Then the expected number of species seen in $(0, t]$ that were seen exactly r times in the previous interval $[-1, 0]$ is

$$
\begin{aligned}
v_r(t) &= S \int_0^\infty e^{-\lambda} \frac{\lambda^r}{r!} (1 - e^{-\lambda t}) dG(\lambda) \\
&= S \int_0^\infty e^{-\lambda} \frac{\lambda^r}{r!} \left\{ \lambda t - \frac{(\lambda t)^2}{2!} + \frac{(\lambda t)^3}{3!} - \cdots \right\} dG(\lambda) \\
&= \sum_{k=1}^\infty (-1)^{k+1} \binom{r+k}{k} t^k \eta_{r+k},
\end{aligned}
\tag{11.60}
$$

where

$$
\eta_r = E(n_r) = S \int_0^\infty \frac{\lambda^r}{r!} e^{-\lambda} dG(\lambda), \quad r = 1, 2, \dots.
$$

The convergence of (11.60) will depend on t, but if it does converge, then an unbiased nonparametric empirical Bayes estimator $\tilde{v}_r(t)$ is obtained by replacing the η_r by estimates $\tilde{\eta}_r = n_r$ obtained from the marginal distribution across the species. If the S Poisson processes are independent, then the n_r will be approximately independent Poisson variables with means η_r. Thus for example,

$$
\mathrm{var}\{\tilde{v}_0(t)\} = \mathrm{var}(n_1 t - n_2 t^2 + n_3 t^3 - \cdots) \doteq \sum_{r=1}^\infty \eta_r t^{2r} \doteq \sum_{r=1}^\infty n_r t^{2r}
$$

provides a standard error for $\tilde{v}_0(t)$.

For the data in Table 11.13, $\tilde{v}_0(1) = 11{,}430$ with standard error 178. It turns out not to be possible to give an upper bound for the size of Shakepeare's vocabulary, but a fairly realistic lower bound can be established of about 35,000 word types that he knew but which do not appear in the canon.

Parametric empirical Bayes models employ parametric distributions for G, one candidate being gamma with mean and variance ξ/β and ξ/β^2. Then

$$
\eta_r = \eta_1 \frac{\Gamma(r+\xi)}{r!\Gamma(1+\xi)} \left(\frac{\beta}{1+\beta} \right)^{r-1}, \quad r = 1, 2, \dots,
$$

proportional to the negative binomial density truncated so that $r > 0$. In the negative binomial case $\xi > 0$, but here any value of $\xi > -1$ is possible; $\xi = 0$ gives the logarithmic series distribution, the first to be fitted to species abundance data. The parameters can be estimated by maximum likelihood fitting of the multinomial distribution of $n_1, \dots, n_{r_0}$, for some suitable r_0. Taking $r_0 = 40$ yields $\widehat{\eta}_1 = 14{,}376$, $\widehat{\xi} = -0.3954$ and $\widehat{\beta} = 104.3$. The fit to Table 11.13 is then remarkably good, giving $\tilde{v}_0(1) \doteq 11{,}483$, very close to the nonparametric empirical Bayes estimate.

In 1985 a previously unknown nine-stanza poem was found in the Bodleian Library in Oxford. It consists of 429 words with 258 word types, of which nine do not appear in the canon. The empirical counts can be compared with the values $\tilde{v}_r(t)$ with $t = 429/884,647$; for example $\tilde{v}_0(t) = 6.97$ is in fair agreement with the observed number of nine new words. Detailed work suggests that at least on the basis of the word counts, the poem might be attributable to Shakespeare. Scholarly debate continues, however, as word usage in the new poem differs from that in the canon. ∎

Shrinkage improves estimators in many models. Before discussing an unexpected consequence of this, we outline some key notions of decision theory.

11.5.2 Decision theory

Sometimes data are gathered in order to decide among decisions whose payoffs are known explicitly. The decision chosen will depend on the data y, and the choice is made according to a *decision rule* $\delta(y)$, which takes a value in a decision space $\mathcal{D}$. Thus δ is a mapping from the sample space $\mathcal{Y}$ to $\mathcal{D}$.

The fact that some decisions have better consequences than others is quantified through a *loss function* $l(d, \theta)$, which represents the loss due to making decision d when the true state of nature is θ. A bad decision incurs a big loss, a better decision a smaller one.

At the time a decision is taken its loss is unknown because of uncertainty about θ. Nevertheless, provided we have prior information on θ, we can calculate the posterior expected loss,

$$\mathrm{E}\left\{l(d, \theta) \mid y\right\} = \int l(d, \theta)\pi(\theta \mid y)\, d\theta = \frac{\int l(d, \theta)f(y \mid \theta)\pi(\theta)\, d\theta}{\int f(y \mid \theta)\pi(\theta)\, d\theta}.$$

This is a function of d and y. If we want to make a decision leading to as small a loss as possible, one strategy is to choose the decision d that minimizes the posterior expected loss for the particular y that has been observed. Thus $\delta(y) = d$, where $\mathrm{E}\{l(d', \theta) \mid y\} \geq \mathrm{E}\{l(d, \theta) \mid y\}$ for every $d' \in \mathcal{D}$. This is called the *Bayes rule* for loss function l with respect to prior π.

Example 11.31 (Discrimination) Suppose we must decide whether or not a patient with measurements y has a disease that has prevalence γ in the population. Let $\theta = 1$ indicate the event that he is diseased. Then

$$\mathrm{Pr}(\theta = 1) = \gamma, \quad \mathrm{Pr}(\theta = 0) = 1 - \gamma,$$

and y has densities $f_1(y)$ and $f_0(y)$ according to the unknown value of θ, which represents the state of nature. The possible decisions are

$$d_0 = \text{'patient is not diseased'}, \quad d_1 = \text{'patient is diseased'},$$

and a decision rule $\delta(y)$ is a procedure that chooses one of these.

Let l_{ij} denote the loss made when $\theta = i$ and decision d_j is made. We set $l_{00} = l_{11} = 0$, so there is no loss when a decision is correct, and assume that $l_{10}, l_{01} > 0$. The posterior expected losses associated with d_0 and d_1 are

$$\mathrm{E}\{l(d_0, \theta) \mid y\} = \frac{l_{00}(1 - \gamma)f_0(y) + l_{10}\gamma f_1(y)}{(1 - \gamma)f_0(y) + \gamma f_1(y)} = \frac{l_{10}\gamma f_1(y)}{(1 - \gamma)f_0(y) + \gamma f_1(y)}$$

and

$$\mathrm{E}\{l(d_1, \theta) \mid y\} = \frac{l_{01}(1 - \gamma)f_0(y) + l_{11}\gamma f_1(y)}{(1 - \gamma)f_0(y) + \gamma f_1(y)} = \frac{l_{01}(1 - \gamma)f_0(y)}{(1 - \gamma)f_0(y) + \gamma f_1(y)}.$$

The posterior expected loss is minimized by d_0 if $l_{10}\gamma f_1(y) < l_{01}(1 - \gamma)f_0(y)$ and otherwise by d_1; we are indifferent if $l_{10}\gamma f_1(y) = l_{01}(1 - \gamma)f_0(y)$.

This Bayes rule can be expressed in more familiar terms: choose d_0 if

$$\frac{f_0(y)}{f_1(y)} > \frac{l_{10}\gamma}{l_{01}(1 - \gamma)},$$

and otherwise choose d_1. This is reminiscent of the Neyman–Pearson lemma, though here the value determining the decision involves γ and the loss function rather than a null distribution for y. ∎

The set-up described thus far applies to decisions to be made once the data are known. But actions must sometimes be taken before any data are available — for example, an experimental design should be chosen to maximize the information in future data. It then seems wise to average the loss incurred over the future data. The expected loss due to using decision rule $\delta(y)$ when the true state of nature is θ is called the *risk function* of δ,

$$R_\delta(\theta) = \int l\{\delta(y), \theta\} f(y \mid \theta)\, dy.$$

If we have prior density $\pi(\theta)$ for θ, the overall expected loss due to using δ is the *Bayes risk*,

$$\int R_\delta(\theta)\pi(\theta)\, d\theta = \int \pi(\theta) \int l\{\delta(y), \theta\} f(y \mid \theta)\, dy\, d\theta$$

$$= \int f(y) \int l\{\delta(y), \theta\}\pi(\theta \mid y)\, d\theta\, dy.$$

For any given y this is minimized by the decision $\delta(y)$ minimizing the inner integral, and this choice of δ is the Bayes rule for the prior $\pi(\theta)$. Thus the Bayes rule minimizes expected loss for both post-data and pre-data decisions.

If we view estimation as a decision problem, then a decision is a choice of the value $\tilde{\theta}$ to be used to estimate θ, and the loss depends on θ and $\tilde{\theta}$. A common choice is squared error loss, $l(\tilde{\theta}, \theta) = (\tilde{\theta} - \theta)^2$. The Bayes rule then uses as estimator the posterior mean of θ,

$$m(y) = \int \theta\pi(\theta \mid y)\, d\theta.$$

To see why, let $\tilde{\theta}(y)$ be any other estimator, and note that as

$$\{\tilde{\theta}(y) - \theta\}^2 = \{\tilde{\theta}(y) - m(y)\}^2 + 2\{\tilde{\theta}(y) - m(y)\}\{m(y) - \theta\} + \{m(y) - \theta\}^2,$$

the posterior expected loss

$$\int \{\tilde{\theta}(y) - \theta\}^2 \pi(\theta \mid y)\,d\theta = \{\tilde{\theta}(y) - m(y)\}^2 + \int \{m(y) - \theta\}^2 \pi(\theta \mid y)\,d\theta \tag{11.61}$$

is minimized by choosing $\tilde{\theta}(y) = m(y)$.

Admissible decision rules

We saw above that if a prior density for θ is available, one should choose the decision that minimizes the posterior expected loss with respect to that prior. But if no prior is available then we must attempt to make a good decision whatever the value of θ. We can compare two decision rules δ and δ' through their risk functions. If $R_{\delta'}(\theta) \geq R_\delta(\theta)$ for all θ, with strict inequality for some θ, then we say that δ' is *inadmissible* — it is beaten by another rule. If no such rule can be found, δ' is said to be *admissible*. Provided the decision formulation is accepted and considerations such as robustness may be ignored, we should clearly restrict attention to admissible decision rules.

The Bayes rule δ_B corresponding to a proper prior $\pi(\theta)$ is always admissible. For if not, there is a rule δ' such that $R_{\delta'}(\theta) \leq R_{\delta_B}(\theta)$, with strict inequality for some set of values of θ to which π attaches positive probability. The corresponding Bayes risks satisfy

$$\int \pi(\theta) R_{\delta'}(\theta)\,d\theta < \int \pi(\theta) R_{\delta_B}(\theta)\,d\theta,$$

contradicting the fact that δ_B minimizes the Bayes risk with respect to $\pi(\theta)$.

In a particular setting there may be many admissible decision rules. We can choose among them by minimizing $\sup_\theta R_\delta(\theta)$. This generally very conservative choice is called a *minimax rule*. An admissible decision rule δ with constant risk is minimax. For otherwise there exists a rule δ' such that for all θ,

$$R_{\delta'}(\theta) \leq \sup_\theta R_{\delta'}(\theta) < \sup_\theta R_\delta(\theta).$$

But if δ has constant risk, then the right-hand side of this expression is constant, and δ must be inadmissible, which is a contradiction.

Example 11.32 (Normal distribution) Suppose that $Y_1, \ldots, Y_n$ is a random sample from the $N(\mu, \sigma^2)$ distribution with known σ^2 and that we wish to choose an estimator $\tilde{\mu}$ of μ among

1. $\delta_1(Y) = \overline{Y}$, the sample average;
2. $\delta_2(Y)$ is the median of $Y_1, \ldots, Y_n$; and
3. $\delta_3(Y) = (n\overline{Y}/\sigma^2 + \mu_0/\tau^2)/(n/\sigma^2 + 1/\tau^2)$, the posterior mean for μ under the prior $N(\mu_0, \tau^2)$; see (11.11).

We take loss function $(\tilde{\mu} - \mu)^2$, so $\delta(Y)$ has risk $R_\delta(\mu)$ equal to its mean squared error, $E[\{\delta(Y) - \mu\}^2]$, the expectation being over Y for fixed μ.

The average $\delta_1(Y)$ has mean and variance μ and σ^2/n, while the median $\delta_2(Y)$ has approximate mean and variance μ and $\pi\sigma^2/(2n)$. Their risks are

$$R_{\delta_1}(\mu) = \sigma^2/n, \quad R_{\delta_2}(\mu) \doteq \pi\sigma^2/(2n).$$

The posterior mean $\delta_3(Y)$ has bias and variance

$$\frac{n\mu/\sigma^2 + \mu_0/\tau^2}{n/\sigma^2 + 1/\tau^2} - \mu, \quad \frac{n/\sigma^2}{(n/\sigma^2 + 1/\tau^2)^2},$$

and so

$$R_{\delta_3}(\mu) = \frac{n/\sigma^2 + (\mu - \mu_0)^2/\tau^2}{(n/\sigma^2 + 1/\tau^2)^2}.$$

As $R_{\delta_2}(\mu) > R_{\delta_1}(\mu)$ for all μ, δ_2 is inadmissible. It can be shown that δ_1 is admissible, and as it has constant risk it is minimax. The rule δ_3 is Bayes and hence admissible. If τ^2 is small, δ_3 will be greatly preferable to δ_1 for values of μ close to the prior mean μ_0. Contrariwise if τ^2 is large, corresponding to weak prior information, then $R_{\delta_3}(\mu) < R_{\delta_1}(\mu)$ over a wide range, but the improvement is small. When $\tau \to \infty$, we see that $\delta_3 \to \delta_1$. ∎

Shrinkage and squared error loss

Having set up machinery for the comparison of estimators using risk, we investigate the gains due to shrinkage when using empirical Bayes estimation.

Let $Y_1, \ldots, Y_n$ be independent normal variables with means $\theta_1, \ldots, \theta_n$ and unit variance. We consider estimation of $\theta_1, \ldots, \theta_n$ by $\tilde{\theta}_1, \ldots, \tilde{\theta}_n$ using as risk function the sum of squared errors

$$R_{\tilde{\theta}}(\theta) = \mathrm{E}\left\{\sum_{j=1}^{n}(\tilde{\theta}_j - \theta_j)^2\right\}, \tag{11.62}$$

the expectation being over Y with θ fixed. At first sight this formulation seems highly artificial, but in fact it is paradigmatic of many situations, one being the semiparametric models discussed in Section 10.7. The maximum likelihood estimators arise when $\tilde{\theta}_j = Y_j$ and have risk $R_{\tilde{\theta}}(\theta) = n$. Are better estimators available?

One possibility stems from taking (11.59) when $v_1 = \cdots = v_n$. Then $\widehat{\mu} = \overline{Y}$ does not depend on τ^2, whose maximum likelihood estimator is given by

$$\widehat{\tau}_+^2 = \max(n^{-1}W - 1, 0), \quad W = \sum_{j=1}^{n}(Y_j - \overline{Y})^2.$$

The eventual conclusion is unchanged but the computations below simplify if we replace $\widehat{\tau}_+^2$ by W/b, where we choose b to minimize the risk. Substitution into (11.59) gives the shrinkage estimators

$$\tilde{\theta}_j = \overline{Y} + (1 - b/W)(Y_j - \overline{Y}), \quad j = 1, \ldots, n. \tag{11.63}$$

These are more appealing than (11.59), because the degree of shrinkage depends on the data, being small if the Y_j are widely separated and W is large. 'Overshrinkage' occurs if $b/W > 1$, so in practice one would use a non-negative estimator such as $\widehat{\tau}_+$.

We show below that the risk of (11.63) using squared error loss is

$$R_{\tilde{\theta}}(\theta) = n + b\,\{b - 2(n-3)\}\,\mathrm{E}(W^{-1}).\tag{11.64}$$

This has minimum value $n - (n-3)^2\mathrm{E}(W^{-1})$ when $b = n - 3$, and as $\mathrm{E}(W^{-1}) > 0$ this risk is uniformly less than n when $n > 3$. That is, when means of four or more normal variables are estimated simultaneously using (11.63) and squared error loss, the maximum likelihood estimator is inadmissible: the paragon of point estimation should not be used. This risk improvement is often called the *Stein effect* after its chief discoverer.

Charles Stein (1920–) studied at Chicago and Columbia universities and since 1953 has worked at Stanford University. He has made important contributions to mathematical statistics. See DeGroot (1986b).

This striking result rests on the cumulation of risk across observations; the chosen risk function would not be sensible if interest focused on a single θ_j. The extent to which shrinkage reduces the risk depends on the distribution of W, which is non-central chi-squared with non-centrality parameter $\rho = \sum(\theta_j - \bar{\theta})^2$. If $\rho = 0$, that is, all the θ_j are equal, then $\mathrm{E}(W^{-1}) = (n-3)^{-1}$ and $R_{\tilde{\theta}}(\theta) = 3$ independent of n. In this case shrinkage yields a dramatically improved estimator. If ρ is large, then the means of the Y_j are widely separated and $\mathrm{E}(W^{-1})$ is small, so $R_{\tilde{\theta}}(\theta)$ is only slightly less than n: the gain from shrinkage is then small. When $\bar{Y}$ in (11.63) and in W is replaced by a fixed prior value μ, then essentially the same result applies, with the maximum likelihood estimator then inadmissible when $n > 2$. The amount of shrinkage then depends on the distance from θ to the prior mean μ, and is large if this distance is small.

Similar results apply more generally, for example to regression and to multivariate situations. The broad lesson is that frequentist estimation of related quantities may be improved by using shrinkage procedures.

Derivation of (11.64)

Note first that with $\tilde{\theta}_j$ given in (11.63), $\sum(\tilde{\theta}_j - \theta_j)^2$ equals

$$\sum_{j=1}^{n}\{\bar{Y} + (1 - b/W)(Y_j - \bar{Y}) - \theta_j\}^2 = \sum_{j=1}^{n}\{Y_j - \theta_j - b(Y_j - \bar{Y})/W\}^2$$

and this equals

$$\sum_{j=1}^{n}(Y_j - \theta_j)^2 - 2bW^{-1}\sum_{j=1}^{n}(Y_j - \theta_j)(Y_j - \bar{Y}) + b^2 W^{-1}.\tag{11.65}$$

The first term has expectation n and the last appears in (11.64), so we must deal with the middle term.

Consider $\mathrm{E}\left\{(Y_j - \theta_j)h_j(Y)\right\}$, where $h_j(y)$ is a sufficiently well-behaved function. Integration by parts, recalling that $Y_j \overset{\text{ind}}{\sim} N(\theta_j, 1)$, and that $d\phi(z)/dz = -z\phi(z)$, implies that $\mathrm{E}\{(Y_j - \theta_j)h_j(Y)\} = \mathrm{E}\{\partial h_j(Y)/\partial Y_j\}$. Setting

$$h_j(Y) = \frac{Y_j - \bar{Y}}{W} = \frac{Y_j - \bar{Y}}{\sum_i(Y_i - \bar{Y})^2}$$

yields

$$\frac{\partial h_j(Y)}{\partial Y_j} = \frac{1 - n^{-1}}{W} - 2\frac{(Y_j - \overline{Y})^2}{W^2},$$

and a little algebra establishes that the central term in (11.65) has expectation $-2b(n-3)\mathrm{E}(W^{-1})$. Expression (11.64) follows directly.

Exercises 11.5

1   In Example 11.29, suppose that $v'_j = \tau^2 v_j$. Show that an unbiased estimator of τ^2 is then $SS/(n-p) - 1$, where SS is the residual sum of squares and p is the dimension of β, and explain why a better estimator is $\max\{SS/(n-p) - 1, 0\}$.
Find also the profile log likelihood when $v'_j = \tau^2$.

2   Consider estimating the success probability θ for a binomial variable R with denominator m, using a beta prior distribution with parameters $a, b > 0$.
(a) Show that the marginal probability $\Pr(R = r \mid \mu, \nu)$ has beta-binomial form

$$\frac{\Gamma(\nu)}{\Gamma(\nu\mu)\Gamma\{\nu(1-\mu)\}}\binom{m}{r}\frac{\Gamma(r+\nu\mu)\Gamma\{m-r+\nu(1-\mu)\}}{\Gamma(m+\nu)}, \quad r = 0, \ldots, m,$$

where $\mu = a/(a+b)$ and $\nu = a + b$, and deduce that

$$\mathrm{E}(R/m) = \mu, \quad \mathrm{var}(R/m) = \frac{\mu(1-\mu)}{m}\left(1 + \frac{m-1}{\nu+1}\right).$$

(b) Show that methods of moments estimators based on a random sample $R_1, \ldots, R_n$ all with denominator m are

$$\widehat{\mu} = \overline{R}, \quad \widehat{\nu} = \frac{\widehat{\mu}(1-\widehat{\mu}) - S^2}{S^2 - \widehat{\mu}(1-\widehat{\mu})/m},$$

where $\overline{R}$ and S^2 are the sample average and variance of the R_j.
(c) Find the mean and variance of the conditional distribution of θ given R, and show that the mean can be written as a shrinking of R/m towards μ. Hence give the empirical Bayes estimates of the θ_j.

3   Consider a logistic regression model for Example 11.29. Show that the marginal log likelihood for β, τ^2 may be written as

$$\sum_{j=1}^{n} \log \int \frac{e^{r_j\theta}}{(1+e^\theta)^{m_j}}\phi\left(\frac{\theta - x_j^{\mathrm{T}}\beta}{\tau}\right) d\theta - \log\tau.$$

Use Laplace approximation to remove the integrals, and outline how you would then estimate β and τ^2. Give also a Laplace approximation for the posterior mean of θ_j given the data, β and τ.

4   Consider the exponential family density $f(y \mid \theta) = \theta^y e^{-\kappa(\theta)} f_0(y)$ for integer y, where $f_0(y)$ is known. If $\pi(\theta)$ is any prior on θ, show that

$$\mathrm{E}(\theta \mid y) = \frac{\int \theta^{y+1} e^{-\kappa(\theta)}\pi(\theta)\,d\theta}{\int \theta^y e^{-\kappa(\theta)}\pi(\theta)\,d\theta} = \frac{\Pr_\pi(Y = y + 1)f_0(y)}{\Pr_\pi(Y = y)f_0(y+1)},$$

where $\Pr_\pi(Y = y)$ is the marginal probability that $Y = y$, averaged over π. Given a sample $y_1, \ldots, y_n$ from the corresponding empirical Bayes model, explain why $\mathrm{E}(\theta_j \mid y_j)$ may be estimated by

$$\frac{f_0(y_j)\sum_{i=1}^{n} I(y_i = y_j + 1)}{f_0(y_j + 1)\sum_{i=1}^{n} I(y_i = y_j)}.$$

Do you think this estimator will be numerically stable? Check by simulating some data and trying it out.

5 Let $X_1, \ldots, X_n$ be a Poisson random sample with mean μ. Previous experience suggests prior density

$$\pi(\mu) = \frac{1}{\Gamma(\nu)} \mu^{\nu-1} e^{-\mu}, \quad 0 < \mu < \infty, \nu > 0.$$

If the loss function for an estimator $\tilde{\mu}$ of μ is $(\tilde{\mu} - \mu)^2$, determine an estimator that minimizes the expected loss and compare its bias and variance with those of the maximum likelihood estimator.

6 The proportion θ of defective items from a production process varies because of fluctuations in the the the raw material. Records show that the prior density for θ is proportional to $\theta(1 - \theta)^4$. A hundred items are inspected from a large batch all made from a homogeneous batch of raw material, and six are found to be defective.
Find the posterior density function for the proportion θ of defectives in the batch. The cost of estimating θ by $\widehat{\theta}$ is $\theta^2(\widehat{\theta} - \theta)^2$. Find also the value of $\widehat{\theta}$ which minimizes the expected cost, and the value of the minimum expected cost.

7 The loss when the success probability θ in Bernoulli trials is estimated by $\tilde{\theta}$ is $(\tilde{\theta} - \theta)^2 \theta^{-1}(1 - \theta)^{-1}$. Show that if the prior distribution for θ is uniform and m trials result in r successes then the corresponding Bayes estimator for θ is r/m. Hence show that r/m is also a minimax estimator for θ.

8 A population consists of k classes $\theta_1, \ldots, \theta_k$ and it is required to classify an individual on the basis of an observation Y having density $f_i(y \mid \theta_i)$ when the individual belongs to class $i = 1, \ldots, k$. The classes have prior probabilities $\pi_1, \ldots, \pi_k$ and the loss in classifying an individual from class i into class j is l_{ij}.
(a) Find the posterior probability $\pi_i(y) = \Pr(\text{class } i \mid y)$ and the posterior risk of allocating the individual to class i.
(b) Now consider the case of 0–1 loss, that is, $l_{ij} = 0$ if $i = j$ and $l_{ij} = 1$ otherwise. Show that the risk is the probability of misclassification.
(b) Suppose that $k = 3$, that $\pi_1 = \pi_2 = \pi_3 = 1/3$ and that Y is normally distributed with mean i and variance 1 in class i. Find the Bayes rule for classifying an observation. Use it to classify the observation $y = 2.2$.

9 Let $Y_j \overset{\text{ind}}{\sim} N(\theta_j, 1), j = 1, \ldots, n$, let $\mu^{\mathsf{T}} = (\mu_1, \ldots, \mu_n)$ be a constant vector, and consider the estimator of $\theta_1, \ldots, \theta_n$ given by

$$\tilde{\theta}_j = \mu + \left\{ 1 - b \left/ \sum (Y_i - \mu_i)^2 \right. \right\} (Y_j - \mu), \quad j = 1, \ldots, n.$$

Show that the risk under squared error loss, (11.62), reduces to (11.64) with $n - 3$ replaced by $n - 2$. Discuss the consequences of this.

11.6 Bibliographic Notes

The Bayesian approach to statistics, then called the inverse probability approach, played a central role in the early and middle parts of the nineteenth century, and was central to Laplace's work. It then fell into disrepute after strong attacks were made on the principle of insufficient reason and remained there for many years. During the 1920s and 1930s R. A. Fisher strongly criticised the use of prior distributions to represent ignorance. The publication in 1939 of the first edition of the influential Jeffreys (1961) marked the start of a resurgence of interest in Bayesian inference, which was consolidated by further important advocacy in the 1950s, particularly after difficulties with frequentist procedures emerged. Interest has mounted especially strongly since serious Bayesian computation became routinely possible.

Introductory books on the Bayesian approach are O'Hagan (1988), Lee (1997), and Robert (2001), while the excellent Carlin and Louis (2000) and Gelman *et al.* (1995) are more oriented towards applications; see also Box and Tiao (1973), and Leonard and Hsu (1999). More advanced accounts are Berger (1985) and Bernardo and Smith (1994), while De Finetti (1974, 1975) is *de rigeur* for the serious reader. The likelihood principle and its relation to the Bayesian approach is discussed at length by Berger and Wolpert (1988). Bayesian model averaging is described by Hoeting *et al.* (1999), who give other references to the topic.

The role and derivation of prior information has been much debated. For some flavour of this, see Lindley (2000) and its discussion. A valuable review of arguments for non-subjective representations of prior ignorance is given by Kass and Wasserman (1996). The elicitation of priors is extensively discussed by Kadane and Wolfson (1998), O'Hagan (1998), and Craig *et al.* (1998).

Laplace approximation is a standard tool in asymptotics, with close links to saddlepoint approximation. A statistical account is given by Barndorff-Nielsen and Cox (1989), which gives further references. It has been used sporadically in Bayesian contexts at least since the 1960s. Tierney and Kadane (1986) and Tierney *et al.* (1989) raised its profile for modern readers. The same idea can be applied to other distributions; see for example Leonard *et al.* (1994).

Markov chain Monte Carlo methods originated in statistical physics. The original algorithm of Metropolis *et al.* (1953) was broadened to what is now called the Metropolis–Hastings algorithm by Hastings (1970), a paper astonishingly overlooked for two decades, though known to researchers in spatial statistics and image analysis (Geman and Geman, 1984; Ripley, 1987, 1988). The last decade has made up for this oversight, with rapid progress being made in the 1990s following Gelfand and Smith (1990)'s adoption of the Gibbs sampler for mainstream Bayesian application. Valuable books on Bayesian use of such procedures are Gilks *et al.* (1996), Gamerman (1997), and Robert and Casella (1999), while Brooks (1998) and Green (2001) give excellent shorter accounts. Example 11.27 is taken from Besag *et al.* (1995), while further interesting applications are contained in Besag *et al.* (1991) and Besag and Green (1993). Tanner (1996) describes a number of related algorithms, including variants on the EM algorithm and data augmentation. Green (1995) and Stephens (2000) describe procedures that may be applied when the parameter space has varying dimension.

Spiegelhalter *et al.* (1996a) describe software for Bayesian use of Gibbs sampling algorithms, with many examples in the accompanying manuals (Spiegelhalter *et al.*, 1996b,c). Cowles and Carlin (1996) and Brooks and Gelman (1998) review numerous convergence diagnostics for Markov chain Monte Carlo output.

Decision theory is treated by Lindley (1985), Smith (1988), Raiffa and Schlaifer (1961), and Ferguson (1967). Hierarchical modelling is discussed in many of the above references. Carlin and Louis (2000) give a modern account of empirical Bayes methods, while the more theoretical Maritz and Lwin (1989) predates modern computational developments. The discovery of the inadmissibility of the maximum likelihood estimator by Stein (1956) and the effects of shrinkage spurred much work; see Morris (1983) for a review.

11.7 Problems

1   Show that the integration in (11.6) is avoided by rewriting it as

$$f(z \mid y) = \frac{f(z \mid y, \theta)\pi(\theta \mid y)}{\pi(\theta \mid y, z)}.$$

Note that the terms on the right need be calculated only for a single θ.
Use this formula to give a general expression for the density of a future observation in an exponential family with a conjugate prior, and check your result using Example 11.3. (Besag, 1989)

2   (a) Consider a scale model with density $f(y) = \tau^{-1}g(y/\tau)$, $y > 0$, depending on a positive parameter τ. Show that this can be written as a location model in terms of $\log y$ and $\log \tau$, and infer that the non-informative prior for τ is $\pi(\tau) \propto \tau^{-1}$, for $\tau > 0$.
(b) Verify that the expected information matrix for the location-scale model $f(y; \eta, \tau) = \tau^{-1}g\{(y - \eta)/\tau\}$, for real η and positive τ, has the form given in Example 11.10, and hence check the Jeffreys prior for η and τ given there.

3   Show that if $y_1, \ldots, y_n$ is a random sample from an exponential family with conjugate prior $\pi(\theta \mid \lambda, m)$, any finite mixture of conjugate priors,

$$\sum_{j=1}^{k} p_j \pi(\theta, \lambda_j, m_j), \quad \sum_j p_j = 1, \, p_j \geq 0,$$

is also conjugate. Check the details when $y_1, \ldots, y_n$ is a random sample from the Bernoulli distribution with probability θ.

4   Inference for a probability θ proceeds either by observing a single Bernoulli trial, X, with probability θ, or by observing the outcome of a geometric random variable, Y, with density $\theta(1 - \theta)^{y-1}$, $y = 1, 2, \ldots,$. Show that the corresponding Jeffreys priors are $\theta^{-1/2}(1 - \theta)^{-1/2}$ and $\theta^{-1}(1 - \theta)^{-1/2}$, and deduce that although the likelihoods for X and Y are equal, subsequent inferences may differ. Does this make sense to you?

5   Let y_1, y_2 be the observed value of a random variable from the bivariate density

$$f(y_1, y_2; \theta) = \pi^{-3/2} \frac{\exp\{-(y_1 + y_2 - 2\theta)^2/4\}}{1 + (y_1 - y_2)^2}, \quad -\infty < y_1, y_2, \theta < \infty.$$

Show that the likelihood for θ is the same as for two independent observations from the $N(\theta, 1)$ density, but that confidence intervals for θ based the average $\overline{y}$ are not the same under both models, in contravention of the likelihood principle.

6   Show that acceptance of the likelihood principle implies acceptance of the sufficiency and conditionality principles.

7   Consider a likelihood $L(\psi, \lambda)$, and suppose that in order to respect the likelihood principle we base inferences for ψ on the integrated likelihood

$$\int L(\psi, \lambda) \, d\lambda.$$

(a) Compare what happens when X and Y have independent exponential distributions with means (i) λ^{-1} and $(\lambda\psi)^{-1}$, (ii) λ and λ/ψ. Discuss.
(b) Suppose that the parameters in (i) are given prior density $\pi(\psi, \lambda)$ and that we compute the marginal posterior density for ψ. Establish that if the corresponding prior density is used in the parametrization in (ii), the problems in (a) do not arise.

8   Obtain expressions for the mean, variance, and mode of the inverse gamma density (11.14), and express its quantiles in terms of those of the gamma density. Use your results to summarize the posterior density of σ^2 in Example 11.12. Calculate also 95% HPD and equi-tailed credible sets for σ^2.

9   (a) Let y be Poisson with mean θ and gamma prior $\lambda^\nu \theta^{\nu-1} \exp(-\lambda\theta)/\Gamma(\nu)$, for $\theta > 0$. Show that if $\nu = \frac{1}{2}$ and $y = 0$, the posterior density for θ has mode zero, and that a HPD credible set for θ has form $(0, \theta_U)$.

> You may like to check that for $b > 0$, the function $g(u) = au - be^u$ is concave with a maximum at a finite u if $a > 0$, but that if $a < 0$, it is monotonic decreasing.

(b) Show that a HPD credible set for $\phi = \log\theta$ has form (ϕ_L, ϕ_U), with both endpoints finite. How does this compare to the interval transformed from (a)? Why does the difference arise?

(c) Compare the intervals in (a) and (b) with the use of quantiles of $\pi(\theta \mid y)$ to construct an equi-tailed credible set for θ, and with confidence intervals based on the likelihood ratio statistic.

10   Use (11.15) to show that the joint conjugate density for the normal mean and variance has $\mu \sim N(\mu_0, \sigma^2/k)$ conditional on σ^2, with σ^2 having an inverse gamma density. Give interpretations of the hyperparameters, and investigate under what conditions the conjugate prior approaches the improper prior in which $\pi(\mu, \sigma^2) \propto \sigma^{-2}$.
Consider instead replacing the prior variance σ^2/k of μ by a known quantity τ^2. Is the resulting joint prior conjugate?

11   Two competing models for a random sample of count data $y_1, \ldots, y_n$ are that they are independent Poisson variables with mean θ, or independent geometric variables with density $\theta(1-\theta)^{y-1}$, for $y = 0, 1, \ldots$, with $0 < \theta < 1$; this density has mean θ^{-1}. Give the posterior odds and Bayes factor for comparison of these models, using conjugate priors for θ in both cases.
What are your prior mean and variance for the numbers of seedlings per five foot square quadrat in a fir plantation? Use them to deduce the corresponding parameters of the conjugate priors for the Poisson and geometric models. Calculate your prior odds and Bayes factor for comparison of the two models applied to the data in Table 11.14. Investigate their sensitivity to other choices of prior mean and variance.

12   Consider a random sample $y_1, \ldots, y_n$ from the $N(\mu, \sigma^2)$ distribution, with conjugate prior $N(\mu_0, \sigma^2/k)$ for μ; here σ^2 and the hyperparameters μ_0 and k are known. Show that the marginal density of the data

$$f(y) \propto \sigma^{-(n+1)}(\sigma^2 n^{-1} + \sigma^2 k^{-1})^{1/2} \exp\left[-\frac{1}{2}\left\{ \frac{(n-1)s^2}{\sigma^2} + \frac{(\bar{y} - \mu_0)^2}{\sigma^2/n + \sigma^2/k} \right\} \right]$$

$$\propto \exp\left\{ -\frac{1}{2}d(y) \right\},$$

say. Hence show that if Y_+ is a set of data from this marginal density, $\Pr\{f(Y_+) \leq f(y)\} = \Pr\{\chi_n^2 \geq d(y)\}$. Evaluate this for the sample 77, 74, 75, 78, with $\mu_0 = 70$, $\sigma^2 = 1$, and $k_0 = \frac{1}{2}$. What do you conclude about the model?
Do the corresponding development when σ^2 has an inverse gamma prior.
(Box, 1980)

13   Suppose that $y_1, \ldots, y_n$ is a random sample from the Poisson distribution with mean θ, and that the prior information for θ is gamma with scale and shape parameters λ and ν. Show that the marginal density of y is

$$f(y) = \frac{s!}{\prod_{j=1}^n y_j!} n^{-s} \times \frac{\Gamma(s+\nu)}{\Gamma(\nu)s!} \frac{\lambda^\nu n^s}{(\lambda + n)^{\nu+s}}, \quad y_1, \ldots, y_n \geq 0,$$

where $s = \sum_j y_j$, and give an interpretation of it.
Suppose that the data in Table 11.14 are treated as Poisson variables, and that prior information suggests that $\lambda = 1$ and $\nu = \frac{1}{2}$. Is this compatible with the data? Do the data seem Poisson, regardless of the prior?

Table 11.14 Counts of of balsam-fir seedlings in five feet square quadrats.

0	1	2	3	4	3	4	2	2	1
0	2	0	2	4	2	3	3	4	2
1	1	1	1	4	1	5	2	2	3
4	1	2	5	2	0	3	2	1	1
3	1	4	3	1	0	0	2	7	0

14 In the usual normal linear regression model, $y = X\beta + \varepsilon$, suppose that σ^2 is known and that β has prior density

$$\pi(\beta) = \frac{1}{|\Omega|^{1/2}(2\pi)^{p/2}} \exp\{-(\beta - \beta_0)^{\mathrm{T}}\Omega^{-1}(\beta - \beta_0)/2\},$$

where Ω and β_0 are known. Find the posterior density of β.

15 Show that the $(1 - 2\alpha)$ HPD credible interval for a continuous unimodal posterior density $\pi(\theta \mid y)$ is the shortest credible interval with level $(1 - 2\alpha)$.

16 An autoregressive process of order one with correlation parameter ρ is stationary only if $|\rho| < 1$. Discuss Bayesian inference for such a process. How might you (a) impose stationarity through the prior, (b) compute the probability that the process underlying data y is non-stationary, (c) compare the models of stationarity and non-stationarity?

17 Study the derivation of BIC for a random sample of size n. Investigate the sizes of the neglected terms for nested normal linear models with known variance. Suggest a better model comparison criterion that is almost equally simple.

18 The lifetime in months, y, of an individual with a certain disease is thought to be exponential with mean $1/(\alpha + \beta x)$, where $\alpha, \beta > 0$ are unknown parameters and x a known covariate. Data (x_j, y_j) are observed for n independent individuals, some of the lifetimes being right-censored. The prior density for α and β is

$$\pi(\alpha, \beta) = ab \exp(-\alpha a - \beta b), \quad \alpha, \beta > 0,$$

where $a, b > 0$ are specified. Show that an approximate predictive density for the uncensored lifetime, z, of a future individual with covariate t is

$$\widehat{f}(z|t, y_1, \ldots, y_n) = (\widehat{\alpha} + \widehat{\beta}t) \exp\{-(\widehat{\alpha} + \widehat{\beta}t)z\}, \quad z > 0,$$

where $\widehat{\alpha}$ and $\widehat{\beta}$ satisfy the equations

$$b + \sum_{j=1}^{n} x_j y_j = \sum_{j \in \mathcal{U}} \frac{x_j}{\alpha + \beta x_j}, \quad a + \sum_{j=1}^{n} y_j = \sum_{j \in \mathcal{U}} \frac{1}{\alpha + \beta x_j},$$

and $\mathcal{U}$ denotes the set of uncensored individuals.

19 Suppose that (U_1, U_2) lies in a product space, of form $\mathcal{U}_1 \times \mathcal{U}_2$.
(a) Show that

$$\pi(u_1) = \frac{\pi(u_1 \mid u_2)}{\pi(u_2 \mid u_1)}\pi(u_2), \quad \text{for any } u_1 \in \mathcal{U}_1, u_2 \in \mathcal{U}_2,$$

and deduce that for each $u_2 \in \mathcal{U}_2$ and an arbitrary $u_1' \in \mathcal{U}_1$,

$$\pi(u_2) = \left\{\int \frac{\pi(u_1 \mid u_2)}{\pi(u_2 \mid u_1)} du_1\right\}^{-1} = \frac{\pi(u_2 \mid u_1')}{\pi(u_1' \mid u_2)}\left\{\int \frac{\pi(u_2 \mid u_1')}{\pi(u_1' \mid u_2)} du_2\right\}^{-1}.$$

(b) If $U_2^1, \ldots, U_2^S$ is a random sample from $\pi(u_2 \mid u_1')$, show that

$$\widehat{\pi}(u_2) = \frac{\pi(u_2 \mid u_1')}{\pi(u_1' \mid u_2)}\left\{S^{-1}\sum_{s=1}^{S} \pi\left(u_1' \mid U_2^s\right)^{-1}\right\}^{-1} \xrightarrow{P} \pi(u_2) \text{ as } S \to \infty.$$

(c) Verify that the code below applies this approach to the bivariate normal model in Example 11.21.

```
S <- 1000; rho <- 0.75; u1p <- -2  # u1p is u1prime
z <- seq(from=-4,to=4,length=200)
plot(z,dnorm(z),type="l",ylim=c(0,1.5))
for (r in 1:20)          # 20 replicates of the simulation
{ u2.sim <- rnorm(S, rho*u1p, sqrt(1-rho^2))
  if (r==1) rug(u2.sim)  # rug with one of the u2 samples
  const <- mean( 1/dnorm(u1p,rho*u2.sim,sqrt(1-rho^2)) )
  dz <- dnorm(z,rho*u1p,sqrt(1-rho^2))/dnorm(u1p,rho*z,sqrt(1-rho^2))
  lines(z, dz/const) }
```

Does this work well? Why not? Try with $u_1' = -2, -1, 0$.
What lesson does this example suggest for the use of this approach in general?

20 (a) Let (U_1, U_2) have a joint density π, marginal densities π_1 and π_2, and conditional densities $\pi_{1|2}$ and $\pi_{2|1}$. Show that π_1 satisfies the integral equation

$$\pi_1(u) = \int h(u, v)\pi_1(v)\,dv, \quad \text{where} \quad h(u, v) = \int \pi_{1|2}(u \mid w)\pi_{2|1}(w \mid v)\,dw.$$

(b) In Example 11.21, establish that the conditional distributions of $U_2^{(i+1)} \mid U_1^{(i)} = v$, $U_1^{(i+1)} \mid U_2^{(i+1)} = w$, and $U_1^{(i+1)} \mid U_1^{(i)} = v, i = 1, \ldots, I - 1$, are those of

$$\rho v + (1 - \rho^2)^{1/2}\varepsilon_1, \quad \rho w + (1 - \rho^2)^{1/2}\varepsilon_2, \quad \rho^2 v + (1 - \rho^4)^{1/2}\varepsilon_3,$$

where $\varepsilon_j \overset{\text{iid}}{\sim} N(0, 1)$. Hence write down $h(u, v)$ for this problem.

(c) Show by induction that the conditional distribution of $U_1^{(I+1)} \mid U_1^{(1)} = v$ is the same as that of $\rho^{2I}v + (1 - \rho^{4I})^{1/2}\varepsilon_4$, and hence show that (i) the Markov chain $U_1^{(1)}, U_1^{(2)}, \ldots$ is in equilibrium when $U_2^{(0)}$ has the standard normal density, and (ii) the chain will reach equilibrium provided $U_2^{(0)}$ may not equal $\pm\infty$.

21 The unmodified Gibbs sampler can be a poor way to generate values from a posterior density with several widely separated modes. Let $U = (U_1, U_2)^{\mathsf{T}}$ and consider

$$\pi(u) = \gamma\phi(u_1 - \delta)\phi(u_2 - \delta) + (1 - \gamma)\phi(u_1 + \delta)\phi(u_2 + \delta),$$

where $u = (u_1, u_2)^{\mathsf{T}}$, $0 < \gamma < 1$ and $\delta > 0$; this is a mixture of two bivariate normal densities whose separation depends on δ and whose relative sizes depend on γ.
(a) When $\gamma = 1/2$, sketch contours of π and the conditional density of U_1 given $U_2 = u_2$ for $u_2 = -2\delta, \delta, 0, \delta, 2\delta$. Sketch also some sample paths for a Gibbs sampling algorithm. What problem do you foresee if $\delta > 4$, say?
(b) Show that the conditional density of U_1 given $U_2 = u_2$ may be written

$$\alpha(u_2)\phi(u_1 - \delta) + \{1 - \alpha(u_2)\}\phi(u_1 + \delta), \quad \text{where} \quad \alpha(u_2) = \frac{\gamma e^{2\delta u_2}}{1 - \gamma + \gamma e^{2\delta u_2}},$$

and write down a Gibbs sampling algorithm for π.
(c) If $c > 0$ is large enough that $\Phi(-c)$ is negligible, show that the probability that the sampler stays in the same mode during R iterations of the sampler is bounded below by

$$\exp\{-2R(\gamma^{-1} - 1)e^{-2\delta c}\},$$

and compute this for $\delta = 2, 3$ and some suitable values of c. Comment.
(d) Find the joint distribution of $V = (V_1, V_2)^{\mathsf{T}} = 2^{-1/2}(U_1 + U_2, U_1 - U_2)^{\mathsf{T}}$ and show that if simulation is performed in terms of V, convergence is immediate. Comment on the implications for implementing the Gibbs sampler.

22 Table 5.9 gives data from k clinical trials as 2×2 tables $(R_{Tj}, m_{Tj}; R_{Cj}, m_{Cj})$, where R_{Tj} is the number of deaths in the treatment group of m_{Tj} patients and similarly in the control group, for $j = 1, \ldots, k$. As a model for such data, ignoring publication bias, assume that R_{Cj} and R_{Tj} are independent binomial variables with denominators m_{Cj} and m_{Tj} and probabilities

$$\frac{\exp(\mu_j)}{1 + \exp(\mu_j)}, \quad \frac{\exp(\mu_j + \delta_j)}{1 + \exp(\mu_j + \delta_j)}, \quad j = 1, \ldots, k,$$

where $\delta_j \overset{\text{iid}}{\sim} N(\gamma, \tau^2)$ represent the treatment effects. Suitable prior densities are assumed for $\mu_1, \ldots, \mu_k, \gamma$ and τ^2.

(a) Write down the directed acyclic graph for this model, derive its conditional independence graph, and hence give steps of a Markov chain Monte Carlo algorithm to sample from the posterior density of $\mu_1, \ldots, \mu_k, \gamma$ and τ^2. If any steps require Metropolis–Hastings sampling, suggest how you would implement it and give the acceptance probabilities.

(b) How does your sampler change if one of the R_Cs is missing?

(c) How should your sampler be modified to generate from the posterior predictive density of δ_+, the value of δ for a new trial?

(d) How should your algorithm be modified if an hierarchical model is used for the μ_j?

23 A Poisson process with rate

$$\lambda(t) = \begin{cases} \lambda_0, & 0 < t \leq \tau, \\ \lambda_1, & \tau < t \leq t_0, \end{cases}$$

where τ is known, is observed on the interval $(0, t_0]$. Let n_0 and n_1 denote the numbers of events seen before and after τ, and suppose that λ_0 and λ_1 are independent gamma variables with parameters ν and β, where ν is specified and β has a gamma prior density with specified parameters a and b.

(a) Check that the joint density of $n_0, n_1, \lambda_0, \lambda_1$, and β is

$$\frac{(\lambda_0 \tau)^{n_0}}{n_0!} e^{-\lambda_0 \tau} \frac{\{\lambda_1(t_0 - \tau)\}^{n_1}}{n_1!} e^{-\lambda_1(t_0-\tau)} \frac{\lambda_0^{\nu-1}\beta^{\nu}}{\Gamma(\nu)} e^{-\lambda_0\beta} \frac{\lambda_1^{\nu-1}\beta^{\nu}}{\Gamma(\nu)} e^{-\lambda_1\beta} \frac{\beta^{a-1}b^a}{\Gamma(a)} e^{-b\beta}.$$

Show that λ_0, λ_1, and β have gamma full conditional densities, and hence give a reversible Gibbs sampler algorithm for simulating from their joint posterior density. Extend this to a process with known multiple change points $\tau_1, \ldots, \tau_k$, for which

$$\lambda(t) = \begin{cases} \lambda_0, & 0 < t \leq \tau_1, \\ \lambda_1, & \tau_1 < t \leq \tau_2, \\ \cdots & \cdots \\ \lambda_k, & \tau_k < t \leq t_0. \end{cases}$$

(b) Now suppose that ν is unknown, with prior gamma density with specified parameters c and d. Show that a random walk Metropolis–Hastings move to update $\log \nu$ to $\log \nu'$ has acceptance probability

$$\min\left[1, \left\{\frac{\Gamma(\nu)}{\Gamma(\nu')}\right\}^{k+1} \left(\frac{\nu'}{\nu}\right)^c \left(e^{-d}\beta^{k+1}\prod \lambda_j\right)^{\nu'-\nu}\right].$$

How would you add this to the algorithm in (a) to retain reversibility?

(c) Now suppose that although k is known, $\tau_1, \ldots, \tau_k$ are not. Show that the joint density of the even order statistics from a random sample of size $2k + 1$ from the uniform density on $(0, t_0)$ is proportional to

$$\tau_1(\tau_2 - \tau_1)\cdots(\tau_k - \tau_{k-1})(t_0 - \tau_l), \quad 0 < \tau_1 < \cdots < \tau_k < t_0.$$

Suppose that this is taken as the prior for the positions of the k changepoints, and that these are updated singly with proposals in which τ_i' is drawn uniformly from (τ_{i-1}, τ_{i+1}), with obvious changes for τ_1 and τ_k. Find the acceptance probabilities for these moves.

24 In a Bayesian formulation of Problem 6.16, we suppose that the computer program is one of many to be debugged, and that the mean number of bugs per program has a Poisson distribution with mean μ/β, where $\mu, \beta > 0$. The actual number of bugs in a particular program is m, and each gives rise to a failure after an exponential time with mean β^{-1}, independent of the others. On failure, the corresponding bug is found and removed at once.

(a) Debugging takes place over the interval $[0, t_0]$ and failures are seen to occur at times $0 < t_1 < \cdots < t_n < t_0$. Show that

$$f(y \mid m, \beta) = \frac{m!}{(m-n)!} \beta^n \exp\{-\beta t_0(m + s/t_0 - n)\}, \quad \beta > 0, m = n, n+1, \ldots,$$

where y represents the failure times and $s = \sum_{j=1}^{n} t_j$.

(b) We take prior $\pi(\mu, \beta) \propto \mu^{-2}$, $\mu > 0$. Show that

$$\pi(y, m) = \int_0^\infty \int_0^\infty f(y \mid m, \beta) f(m \mid \beta, \mu) \pi(\beta, \mu) \, d\beta d\mu$$

$$\propto (m - n + s/t_0)^{-n} \prod_{i=1}^{n-2} (m - n + i), \quad m = n, n+1, \ldots,$$

and give expressions for the posterior probabilities (i) that the program has been entirely debugged and (ii) that there are no failures in $[t_0, t_0 + u]$.

(c) Use the data in Table 6.13 to give a 95% HPD credible interval for the number of bugs remaining after 31 failures. Compute the probability that the program had been entirely debugged (i) after 31 failures and (ii) after 34 failures. Should the program have been released when it was?

(d) Discuss how the appropriateness of the model might be checked.

(Example 2.28; Raftery, 1988)

25 Let $Y_1, \ldots, Y_n$ be independent normal variables with means $\mu_1, \ldots, \mu_n$ and common variance σ^2. Show that if the prior density for μ_j is δ is the Dirac delta function.

$$\pi(\mu_j) = \gamma \tau^{-1} \phi(\mu_j/\tau) + (1 - \gamma)\delta(\mu_j), \quad \tau > 0, 0 < \gamma < 1,$$

with all the μ_j independent *a priori*, then $\pi(\mu_j \mid y_j)$ is also a mixture of a point mass and a normal density, and give an interpretation of its parameters.

(a) Find the posterior mean and median of μ_j when σ is known, and sketch how they vary as functions of y_j. Which would you prefer if the signal is sparse, that is, many of the μ_j are known *a priori* to equal zero but it is not known which?

(b) How would you find empirical Bayes estimates of τ, γ, and σ?

(c) In applications of the tails of the normal density might be too light to represent the distribution of non-zero μ_j well. How could you modify π to allow for this?

26 Suppose that $y_1, \ldots, y_n$ are independent Poisson variables with means $\lambda_j x_j$, where the x_j are known constants, and that the λ_j are a random sample from the gamma density with mean ξ/ν.

(a) Show that the marginal density of y_j is

$$f(y_j; \xi, \nu) = \frac{\Gamma(y_j + \xi)}{\Gamma(\xi) y_j!} \frac{x_j^{y_j} \nu^\xi}{(x_j + \xi)^{y_j + \xi}}, \quad y_j = 0, 1, \ldots, \quad \xi, \nu > 0,$$

and give its mean. Say how you would estimate ξ and ν based on $y_1, \ldots, y_n$.

(b) Establish that

$$E(\lambda_j \mid y, \xi, \nu) = \frac{y_j + \xi}{x_j + \nu}, \quad \text{var}(\lambda_j \mid y, \xi, \nu) = \frac{y_j + \xi}{(x_j + \nu)^2},$$

and give an interpretation of this.

(c) Check that the code below computes the maximum likelihood estimates $\widehat{\xi}$ and $\widehat{\nu}$, and yields the empirical Bayes estimates in Table 11.7. Discuss.

```
x <- c(94.32,15.72,62.88,125.76,5.24,31.44,1.048,1.048,2.096,10.48)
y <- c(5,1,5,14,3,19,1,1,4,22)
L <- function(p, y, x)
        -sum(dnbinom(y, size=p[1], prob=p[2]/(x+p[2]), log=T))
fit <- nlm(L, p=c(1,1), y=y, x=x) # marginal maximum likelihood
xi <- fit$estimate[1]
nu <- fit$estimate[2]
ests <- (y+xi)/(x+nu)
vars <- (y+xi)/(x+nu)^2
cbind(ests,vars)
```

12

Conditional and Marginal Inference

In most models the parameter vector can be split into two parts, $\theta = (\psi, \lambda)$, where ψ is the *interest parameter* or *parameter of interest* and λ is a *nuisance* or *incidental parameter*. The former is the focus of enquiry, while the latter summarizes aspects of the model that are not of central concern, but which are nevertheless essential to realistic modelling. Usually ψ has small dimension, often being scalar, while λ may be of high dimension. Different elements of θ may be nuisance or interest parameters at different stages of an investigation. We suppose throughout the discussion below that ψ and λ are variation independent, that is, the parameter space has form $\Psi \times \Lambda$, so knowledge about the range of λ imparts no information about ψ. In many cases it is desirable that inferences be invariant to *interest-preserving reparametrizations*, that is, one-one maps between (ψ, λ) and $(\eta(\psi), \zeta(\psi, \lambda))$.

Example 12.1 (Log-normal mean)   If $X \sim N(\mu, \sigma^2)$, then the log-normal variable $Y = \exp(X)$ has mean and variance

$$\mathrm{E}(Y) = \exp(\mu + \sigma^2/2) = \psi, \quad \mathrm{var}(Y) = \exp(2\mu + \sigma^2)\{\exp(\sigma^2) - 1\} = \lambda,$$

say. A confidence interval (ψ_-, ψ_+) for ψ should transform to $(\log \psi_-, \log \psi_+)$ if the model is expressed in terms of $\eta = \log \psi$ and $\zeta = \zeta(\psi, \lambda) = \sigma^2$.   ∎

In many important cases the density of data Y may be factorized as

$$f(y; \psi, \lambda) \propto f(t_1 \mid a; \psi) f(t_2 \mid t_1, a; \psi, \lambda),$$

or as

$$f(y; \psi, \lambda) \propto f(t_1 \mid t_2, a; \psi) f(t_2 \mid a; \psi, \lambda),$$

where terms free of the parameters have been neglected. The quantity a may not be present, but if it is, it is usually chosen to be an ancillary statistic, a notion discussed in Section 12.1. If such a factorization holds, it is natural to base inference for ψ on the leading term on its right side. Information about ψ can then be extracted from the data without needing to estimate or otherwise account for λ. Leaving aside the presence of a, these terms are respectively marginal and conditional densities of T_1, and can be viewed as a *marginal likelihood* and a *conditional likelihood* for ψ. One reason for considering them is that when λ is high-dimensional, the standard likelihood methods described in Chapter 4 and used throughout the book can become unreliable or fail utterly, as in the following classic example.

Example 12.2 (Neyman–Scott problem)   The log likelihood based on data y_{ij}, $i = 1, \ldots, m$, $j = 1, \ldots, k$, supposed to be realized values of independent normal variables with means μ_i and common variance σ^2, is

So-called because first pointed out by Neyman and Scott (1948).

$$
\begin{aligned}
\ell(\mu_1, \ldots, \mu_m, \sigma^2) &\equiv -\tfrac{1}{2} \sum_{i=1}^{m} \left\{ k \log \sigma^2 + \frac{1}{\sigma^2} \sum_{j=1}^{k} (y_{ij} - \mu_i)^2 \right\} \\
&= -\tfrac{1}{2} \sum_{i=1}^{m} \left\{ k \log \sigma^2 + \frac{1}{\sigma^2} \sum_{j=1}^{k} (y_{ij} - \bar{y}_{i.})^2 + \frac{k}{\sigma^2} (\bar{y}_{i.} - \mu_i)^2 \right\}.
\end{aligned}
$$

As the maximum likelihood estimate of μ_i is $\bar{y}_{i.} = k^{-1} \sum_j y_{ij}$, the maximum likelihood estimate of σ^2 is $\widehat{\sigma}^2 = (mk)^{-1} \sum_{ij} (y_{ij} - \bar{y}_{i.})^2$. The sums of squares $\sum_j (y_{ij} - \bar{y}_{i.})^2$ are independently distributed as $\sigma^2 \chi^2_{k-1}$, so

$$
\mathrm{E}(\widehat{\sigma}^2) = \sigma^2 \frac{(k-1)}{k}, \quad \mathrm{var}(\widehat{\sigma}^2) = 2\sigma^4 \frac{(k-1)}{mk^2}.
$$

Thus if k is fixed and m increases, $\widehat{\sigma}^2$ converges to the wrong value. When $k = 2$, for example, $\widehat{\sigma}^2 \xrightarrow{P} \sigma^2/2$, so $\widehat{\sigma}^2$ is both biased in small samples and inconsistent as $m \to \infty$.

The problem here is that the number of parameters increases with the sample size, but the information about each of the μ_i stays fixed. Moreover, even though the expected information for σ^2 does increase with m, it gives a misleading impression of the precision with which σ^2 can be estimated. This is an extreme situation, but similar difficulties arise in many other models. Here they could be eliminated by replacing k by $k - 1$ in the denominator of $\widehat{\sigma}^2$, but such a repair is impossible in general. It turns out, however, that use of marginal likelihood rescues the situation.   ∎

This chapter gives an introduction to conditional and marginal inference. In Section 12.1 we define ancillary statistics and discuss their properties, and in Sections 12.2 and 12.3 describe marginal and conditional likelihoods, which have their major applications for group transformation and exponential family models respectively. In many cases these likelihoods are difficult to construct exactly. Highly accurate approximations to them may be based on saddlepoint procedures, which are closely related to the Bayesian use of Laplace approximation, and these are described in Section 12.3. Finally Section 12.4 gives a brief account of modifications intended to improve inferences based on the profile likelihood.

12.1 Ancillary Statistics

Conditioning arguments have several important roles in statistics. One is to restrict the sample space under consideration, thereby ensuring that repeated sampling comparisons are made within a so-called *relevant subset* of the full sample space. The conditionality principle (Section 11.1.2) suggests that this will often result in inferences that are close to Bayesian ones, but without the need to specify a prior density that may be difficult to justify. A second role of conditioning is the elimination of nuisance parameters; see Sections 5.2.3, 7.3.3, and 12.3. In this section we focus on the use of conditioning for ensuring relevance of tests and confidence intervals.

Consider a statistical model for data Y that depends on a parameter θ, and suppose that the minimal sufficient statistic for θ is $S = s(Y)$. The factorization theorem implies that inference for θ may be based on the density of S. If we can write $S = (T, A)$, where the distribution of A does not depend on θ, then A is said to be *ancillary for θ*, or, more loosely, an *ancillary statistic*. Note our requirement that A be a function of the minimal sufficient statistic. Some authors do not impose this, and say that any function of Y whose distribution does not depend on θ is ancillary, but we reserve the term *distribution constant* for such statistics; see Section 5.3. We shall see below that in a particular sense ancillary statistics determine the amount of information in the data, and as a result they play a central role in conditional inference.

The likelihood for θ may be written

$$L(\theta) \propto f(s; \theta) = f(t, a; \theta) = f(a) f(t \mid a; \theta),$$

suggesting that inference for θ be based on the conditional distribution of T given the observed value a of A. One argument for this is that, in principle at least, the experiment that generates S may be regarded as having two stages. The first stage consists of observing a value a from the marginal density of A. In the second stage a value for T is observed from the conditional density of T given that $A = a$. As the distribution of A does not depend upon θ, the conditionality principle implies that inference for θ should be based only on the second stage, which contributes $f(t \mid a; \theta)$ to the likelihood. According to this argument the subset of the sample space relevant to inference for θ is the set $\{y : a(y) = a_{\text{obs}}\}$, where a_{obs} denotes the value of A actually observed.

Example 12.3 (Uniform density) Let $Y_1, \ldots, Y_n$ be a random sample from the density uniform on $(\theta - 1/2, \theta + 1/2)$, with θ unknown. Then the likelihood,

$$L(\theta) = \begin{cases} 1, & \theta - \frac{1}{2} \le y_1, \ldots, y_n \le \theta + \frac{1}{2}, \\ 0, & \text{otherwise}, \end{cases}$$

can be re-expressed as

$$L(\theta) = \begin{cases} 1, & \max(y_j) - \frac{1}{2} \le \theta \le \min(y_j) + \frac{1}{2}, \\ 0, & \text{otherwise}, \end{cases}$$

so the minimal sufficient statistic consists of the smallest and largest order statistics, $S = (Y_{(1)}, Y_{(n)})$. The joint density of $V = Y_{(n)}$ and $U = Y_{(1)}$ is

$$f(u, v) = n(n-1)(v-u)^{n-2}, \quad \theta - \frac{1}{2} \le u < v < \theta + \frac{1}{2},$$

and on changing variables from (v, u) to (a, t), where $a = v - u = y_{(n)} - y_{(1)}$ and $t = (y_{(1)} + y_{(n)})/2$, we obtain

$$f(t, a) = n(n - 1)a^{n-2}, \quad \theta - \tfrac{1-a}{2} \leq t \leq \theta + \tfrac{1-a}{2}, \ 0 \leq a \leq 1;$$

the random variable $T = (Y_{(1)} + Y_{(n)})/2$ is an unbiased estimator of θ. The support of the joint density for (t, a) is a triangle with vertices $(\theta, 1)$ and $(\theta \pm 1/2, 0)$. Straightforward calculations give

$$f(t) = n (1 - 2|t - \theta|)^{n-1}, \quad \theta - \tfrac{1}{2} \leq t \leq \theta + \tfrac{1}{2}, \tag{12.1}$$

$$f(a) = n(n - 1)(1 - a)a^{n-2}, \quad 0 \leq a \leq 1,$$

$$f(t \mid a; \theta) = \frac{1}{1 - a}, \quad \theta - \tfrac{1-a}{2} \leq t \leq \theta + \tfrac{1-a}{2}. \tag{12.2}$$

Thus A is ancillary, as is clear from its location-invariance. Conditional on $A = a$, the density of T is uniform on an interval of length $1 - a$ centred at θ. From (12.2) we see that the conditional likelihood for θ equals $(1 - a)^{-1}$ on the interval with endpoints $t \pm (1 - a)/2$ and is zero elsewhere. Thus $a \doteq 0$ conveys little information about θ, whereas $a \doteq 1$ pins down θ very precisely.

For example, a sample of size $n = 2$ with $y_{(1)} = -0.25$ and $y_{(2)} = 0.45$ has $t = 0.1$ and $a = 0.7$. The unconditional equi-tailed 0.9 confidence interval for θ based on (12.1) is $(-0.242, 0.442)$ (Exercise 12.1.2), while the 0.9 conditional interval based on (12.2) is $(-0.035, 0.235)$; their respective lengths are 0.68 and 0.27. The unconditional interval is logically inconsistent with the data, because it includes values such as $\theta = 0.3$ for which $y_{(1)} = -0.25$ could not be observed. The conditional interval takes the observed value of A into account and eliminates such absurdities. ∎

Example 12.4 (Regression model) In many regressions the observed explanatory variables x and responses y could both be regarded as realizations of random variables X and Y, in the sense that different values (x, y) might have occurred. This is clearest in an observational study, where individuals each have a number of variables recorded, some being later regarded as explanatory and others as responses. Unless the values of the Xs are restricted, the data collection scheme implies that they may be modelled as random variables. If, however, the joint density of (X, Y) factorizes as

$$f(y \mid x; \psi)f(x),$$

and properties of the marginal distribution of X are not of interest, it will be appropriate to treat X as fixed throughout the analysis, and this is what is generally done in practice. Notice that ψ is a parameter of the conditional density of Y given X; analysis based on $f(y \mid x; \psi)$ alone would not be appropriate if ψ also entered the distribution of X.

Consider the linear model $Y = x\beta + \sigma\varepsilon$, where Y and x are the vector of responses and the observed design matrix, respectively. The least squares estimator of β is $(x^{\mathsf{T}}x)^{-1}x^{\mathsf{T}}Y$, and conditional on $X = x$, it has variance $\sigma^2(x^{\mathsf{T}}x)^{-1}$. Thus the estimator and its variance both depend on the explanatory variables actually observed, and not on those that might have been observed but in fact were not. For example, if certain

In this example (only) we depart from our convention that X is the observed design matrix.

parameters cannot be estimated from the design matrix actually used, it is irrelevant that with another design they might have been estimable. ∎

See Basu (1955, 1958).

Basu's theorem

If S is a complete minimal sufficient statistic, then it is independent of any distribution constant statistic C. To see this in the discrete case, note that for any c and θ the marginal density of C may be written as

$$f_C(c) = \sum_s f_{C|S}(c \mid s) f_S(s; \theta),$$

where the sum is over all possible values of s. This implies that for all θ

$$\sum_s \left\{ f_C(c) - f_{C|S}(c \mid s) \right\} f_S(s; \theta) = 0,$$

and completeness of S gives $f_C(c) = f_{C|S}(c \mid s)$ for every c and s. Hence C and S are independent. This result is useful, because it assures independence without the effort of computing the joint density of C and S. The argument in the continuous case is analogous.

Example 12.5 (Normal linear model) The minimal sufficient statistic in the normal linear model $y = X\beta + \varepsilon$ consists of $\widehat{\beta} = (X^{\mathsf{T}}X)^{-1}X^{\mathsf{T}}y$ and the rescaled sum of squares $S^2 = (y - X\widehat{\beta})^{\mathsf{T}}(y - X\widehat{\beta})/(n - p)$. That the pair $(\widehat{\beta}, S^2)$ is complete and minimal sufficient follows from properties of the exponential family; see Example 7.10. The standardized residuals

$$\frac{y_j - x_j^{\mathsf{T}}\widehat{\beta}}{S(1 - h_j)^{1/2}}, \quad j = 1, \ldots, n,$$

have a (degenerate) joint distribution that does not depend on β and σ^2. Basu's theorem implies that this distribution is independent of $\widehat{\beta}$ and S^2; see Example 5.14. ∎

Location model

Suppose that $y_1, \ldots, y_n$ is a random sample from the continuous density $f(y; \theta) = g(y - \theta)$, where $-\infty < \theta < \infty$ and the density g is known. Here θ determines the location of f, and we can write a random variable from f as $Y = \theta + \varepsilon$, where ε has density g. Except for certain special choices of g, such as the normal and uniform densities, the minimal sufficient statistic for θ consists of the order statistics $Y_{(1)} < \cdots < Y_{(n)}$, whose joint density is

$$f(y_{(1)}, \ldots, y_{(n)}; \theta) = n! \prod_{j=1}^{n} g(y_{(j)} - \theta), \quad y_{(1)} < \cdots < y_{(n)}.$$

We now discuss conditional inference for θ based on the maximum likelihood estimator $\widehat{\theta}$. Although rarely of interest in applications, the simplicity of this model brings out the main ideas without unnecessary complication, and helps to motivate later developments.

We assume that $\widehat{\theta}$ is the sole solution to the score equation

$$0 = \sum_{j=1}^{n} \frac{\partial \log g(Y_{(j)} - \widehat{\theta})}{\partial \theta} = \sum_{j=1}^{n} \frac{\partial \log g(A_j)}{\partial \theta}, \qquad (12.3)$$

where $A_j = Y_{(j)} - \widehat{\theta}$. As $\widehat{\theta}$ is equivariant, the random variables $A_1, \ldots, A_n$ form a maximal invariant, which is of course distribution constant. Moreover the *configuration* $A = (A_1, \ldots, A_n)$ is a function of the minimal sufficient statistic. Hence it is ancillary. The discussion above suggests that inference be based on the conditional distribution of $\widehat{\theta}$ given the value a of A. Equivariance of $\widehat{\theta}$ means that $Z(\theta) = \widehat{\theta} - \theta$ is a pivot, so if we let $z_\alpha(a)$ denote the α quantile of the conditional distribution of $Z(\theta)$, a $(1 - 2\alpha)$ conditional confidence interval for θ will have limits

$$\widehat{\theta} - z_{1-\alpha}(a), \quad \widehat{\theta} - z_\alpha(a)$$

that depend on the observed a.

Another possibility is to use the unconditional distribution of $Z(\theta)$, whose quantiles z_α and $z_{1-\alpha}$ do not depend on a. These quantiles are readily estimated by simulation, but the $z_\alpha(a)$ are not. In fact the conditional distributions of $\widehat{\theta}$ and hence of $Z(\theta)$ are fairly easily found, as we now see.

Owing to the constraint (12.3), the density of A is degenerate, and there exists a function h such that $a_1 = h(a_2, \ldots, a_n)$; this is assumed sufficiently smooth for the development below. Thus

$$y_{(1)} = \widehat{\theta} + h(a_2, \ldots, a_n), \quad y_{(j)} = \widehat{\theta} + a_j, \quad j = 2, \ldots, n.$$

The Jacobian for transformation from $(y_{(1)}, \ldots, y_{(n)})$ to $(\widehat{\theta}, a_2, \ldots, a_n)$ is

$$\begin{vmatrix} \frac{\partial y_{(1)}}{\partial \theta} & \frac{\partial y_{(2)}}{\partial \theta} & \cdots & \frac{\partial y_{(n)}}{\partial \theta} \\ \frac{\partial y_{(1)}}{\partial a_2} & \frac{\partial y_{(2)}}{\partial a_2} & \cdots & \frac{\partial y_{(n)}}{\partial a_2} \\ \vdots & \vdots & \cdots & \vdots \\ \frac{\partial y_{(1)}}{\partial a_n} & \frac{\partial y_{(2)}}{\partial a_n} & \ddots & \frac{\partial y_{(n)}}{\partial a_n} \end{vmatrix} = \begin{vmatrix} 1 & 1 & \cdots & 1 \\ h_1 & 1 & \cdots & 0 \\ \vdots & \vdots & \ddots & \vdots \\ h_n & 0 & \cdots & 1 \end{vmatrix} = H(a),$$

say, where $h_j = \partial h(a_2, \ldots, a_n)/\partial a_j$. Hence the joint density of $\widehat{\theta}$ and A is

$$f(\widehat{\theta}, a; \theta) = n! H(a) \prod_{j=1}^{n} g(y_{(j)} - \theta)\Big|_{y_{(j)} = \widehat{\theta} + a_j} = n! H(a) \prod_{j=1}^{n} g(a_j + \widehat{\theta} - \theta),$$

so for $a_1 \leq \cdots \leq a_n$ and $\widehat{\theta}$ real,

$$f(\widehat{\theta} \mid a; \theta) = \frac{f(\widehat{\theta}, a; \theta)}{f(a)} = \frac{\prod_{j=1}^{n} g(a_j + \widehat{\theta} - \theta)}{\int_{-\infty}^{\infty} \prod_{j=1}^{n} g(a_j + u)\, du}. \qquad (12.4)$$

This conditional density contains all the information in the data concerning θ, appropriately conditioned. Changing variables to $z(\theta) = \widehat{\theta} - \theta$ and integrating gives

$$\Pr\{Z(\theta) \leq z \mid A = a\} = \frac{\int_{-\infty}^{z} \prod_{j=1}^{n} g(a_j + u)\, du}{\int_{-\infty}^{\infty} \prod_{j=1}^{n} g(a_j + u)\, du}.$$

Figure 12.1 Conditional
inference for location in
two samples of size $n = 5$
from the t_3 density.
Conditional density of
pivot $Z(\theta) = \widehat{\theta} - \theta$ given
configuration ancillary
(solid), and unconditional
density of $Z(\theta)$ (dots).
The blobs show the
observed configuration,
and the rug the expected
configuration.

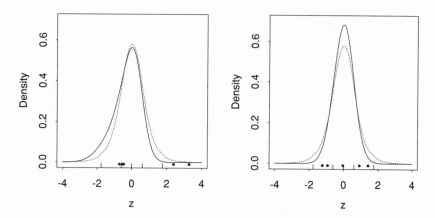

Figure 12.1 Conditional inference for location in two samples of size $n = 5$ from the t_3 density. Conditional density of pivot $Z(\theta) = \widehat{\theta} - \theta$ given configuration ancillary (solid), and unconditional density of $Z(\theta)$ (dots). The blobs show the observed configuration, and the rug the expected configuration.

This function can be obtained by evaluating the integrand on a grid of values, or by more sophisticated methods. Once it has been approximated the quantiles $z_\alpha(a)$ are easily found.

Example 12.6 (Student t distribution) Figure 12.1 shows the conditional densities of $Z(\theta)$ for two samples of size five from the t_3 density. In the left panel the configuration is very asymmetric, and hence so is the conditional density of $\widehat{\theta}$. In the right panel the configuration is quite symmetric but underdispersed relative to a typical configuration. Hence the conditional density of $Z(\theta)$ is more peaked than the unconditional density, thereby giving more precise inferences.

In the left panel the conditional 0.025 and 0.975 quantiles are -2.11 and 1.03, so if $\widehat{\theta} = 0$ the 0.95 confidence interval, $(-1.03, 2.11)$, is asymmetric in the same direction as the configuration. The corresponding confidence interval for the right panel would be $(-1.13, 1.18)$, to be compared to the unconditional interval $(-1.61, 1.61)$. ∎

It is suggestive to express $f(\widehat{\theta} \mid a; \theta)$ using the log likelihood for the model, which we write as $\ell(\theta; \widehat{\theta}, a)$ to stress its dependence on the data through $\widehat{\theta}$ and a. Setting $u = \theta + \widehat{\theta} - v$ we see that (12.4) becomes

$$f(\widehat{\theta} \mid a; \theta) = \frac{\exp\left\{\ell(\theta; \widehat{\theta}, a)\right\}}{\int_{-\infty}^{\infty} \exp\left\{\ell(v; \widehat{\theta}, a)\right\} dv}. \tag{12.5}$$

Laplace approximation to the integral (Section 11.3.1) gives

$$f(\widehat{\theta} \mid a; \theta) = (2\pi)^{-1/2} |J(\widehat{\theta}; \widehat{\theta}, a)|^{1/2} \exp\left\{\ell(\theta; \widehat{\theta}, a) - \ell(\widehat{\theta}; \widehat{\theta}, a)\right\} \left\{1 + O(n^{-1})\right\}. \tag{12.6}$$

where $J(\widehat{\theta}; \widehat{\theta}, a)$ equals $-\partial^2 \ell(\theta; \widehat{\theta}, a)/\partial\theta^2$ evaluated at $\theta = \widehat{\theta}$. In fact the quantity $J(\widehat{\theta}; \widehat{\theta}, a)$ does not depend on $\widehat{\theta}$, because the log likelihood may be written as $\ell(\widehat{\theta} - \theta)$. By (12.5) we see that renormalizing (12.6) to have unit integral will recover the exact expression, giving

This is called the p^* formula or Barndorff-Nielsen's formula.

$$f(\widehat{\theta} \mid a; \theta) = (2\pi)^{-1/2} \overline{c}(a) |J(\widehat{\theta}; \widehat{\theta}, a)|^{1/2} \exp\left\{\ell(\theta; \widehat{\theta}, a) - \ell(\widehat{\theta}; \widehat{\theta}, a)\right\}. \tag{12.7}$$

Detailed analysis of the Laplace expansion shows that $\overline{c}(a) = 1 + O(n^{-1})$.

Formula (12.7) expresses the conditional density of $\widehat{\theta}$ in terms of the log likelihood and its derivatives and is the basis for much further development. It is exact for group transformation models and approximately true more generally, though we shall prove neither of these assertions; see instead the bibliographic notes.

If we accept that (12.7) is correct, how may it be used for inference? One possibility is to obtain conditional confidence intervals for θ by transforming $\widehat{\theta}$ to a normally-distributed pivot. To do this, we write the integral of (12.6) in form (11.30), taking $a(u) = |J(u; u, a)|^{1/2}$ and $g(u) = \ell(u; u, a) - \ell(\theta; u, a)$. Then the argument leading to (11.31) gives

$$\Pr(\widehat{\theta} \leq t \mid A = a; \theta) = \Phi\left\{r^*(\theta)\right\} + O(n^{-1}), \tag{12.8}$$

where $r^*(\theta) = r(\theta) + r(\theta)^{-1} \log\{v(\theta)/r(\theta)\}$, with

$$r(\theta) = \text{sign}(t - \theta) \left[2\left\{\ell(t; t, a) - \ell(\theta; t, a)\right\}\right]^{1/2}, \quad v(\theta) = \frac{\ell_{;\widehat{\theta}}(t; t, a) - \ell_{;\widehat{\theta}}(\theta; t, a)}{|J(t; t, a)|^{1/2}}.$$

The last of these involves a *sample space derivative* of ℓ,

$$\ell_{;\widehat{\theta}}(\theta; \widehat{\theta}, a) = \frac{\partial \ell(\theta; \widehat{\theta}, a)}{\partial \widehat{\theta}}.$$

In order to obtain this, ℓ must be expressed as a function of $\widehat{\theta}$ and a so that it can be differentiated partially with respect to $\widehat{\theta}$, holding a fixed. Usually this re-expression is difficult, and approximations to such derivatives are needed. For the location model, however, $\ell_{;\widehat{\theta}}(t; t, a) = 0$ for any t, and

$$\ell_{;\widehat{\theta}}(\theta; \widehat{\theta}, a) = -\frac{\partial \ell(\theta; \widehat{\theta}, a)}{\partial \theta} = -\ell_{\theta;}(\theta; \widehat{\theta}, a),$$

say. Thus $v(\theta) = \ell_{\theta;}(\theta; t, a)/|J(t; t, a)|^{1/2}$ is the score statistic.

More detailed computations show that in a *moderate deviation region*, where $n^{1/2}(t - \widehat{\theta}) = O(1)$, the error in (12.8) is relative and of order $O(n^{-3/2})$: the right-hand side is in fact $\Phi\left\{r^*(\theta)\right\}\left\{1 + O(n^{-3/2})\right\}$. In a *large deviation region*, for which $t - \widehat{\theta} = O(1)$, the error remains relative but becomes $O(n^{-1})$. The relative error property helps explain why the extraordinary accuracy of this type of approximation persists far into the tails of the distribution of $\widehat{\theta}$.

Expression (12.8) shows that the random quantity $R^*(\theta)$ corresponding to $r^*(\theta)$ is an approximate pivot, conditional on $A = a$, because its distribution is almost standard normal. As this distribution is essentially independent of a, $R^*(\theta)$ is also approximately normal unconditionally. The limits of a $(1 - 2\alpha)$ confidence interval for θ may be found as those θ for which $\Phi\{r^*(\theta)\} = \alpha, 1 - \alpha$; equivalently we may solve for θ the equations $r^*(\theta) = \Phi^{-1}(\alpha)$ and $r^*(\theta) = \Phi^{-1}(1 - \alpha)$.

Owing to the structure of the location model, $R^*(\theta)$ is a monotone function of $Z(\theta) = \widehat{\theta} - \theta$, so both $Z(\theta)$ and $R^*(\theta)$ yield the same conditional confidence intervals; $R^*(\theta)$ can be regarded as a transformation of $\widehat{\theta} - \theta$ to normality. Such transformations have the same effect in much greater generality. There is a close link to the Bayesian approximations of Section 11.3.1.

Figure 12.2 Use of
normal approximation to
pivotal quantities $r^*(\theta)$
(solid), $r(\theta)$ (large
dashes), and $z_1(\theta)$ (small
dashes) to construct 0.95
confidence intervals for
location based on two
samples of size 5. The
dotted horizontal lines
shows the 0.025 and 0.975
quantiles of the standard
normal distribution, and
the dotted vertical lines
mark the limits of the
confidence interval based
on $r^*(\theta)$. Note how the
confidence intervals based
on $r^*(\theta)$ and $r(\theta)$ take
account of the skewness
of the configuration
ancillary.

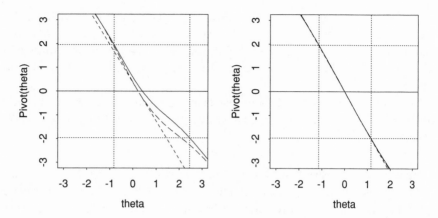

The approximate pivot $R^*(\theta) = R(\theta) + R_{\mathrm{INF}}(\theta)$, where

$$R_{\mathrm{INF}}(\theta) = R(\theta)^{-1} \log \{V(\theta)/R(\theta)\}, \tag{12.9}$$

is a modified form of the signed likelihood ratio statistic

$$R(\theta) = \mathrm{sign}(\widehat{\theta} - \theta) \left[2 \left\{\ell(\widehat{\theta}; \widehat{\theta}, a) - \ell(\theta; \widehat{\theta}, a)\right\}\right]^{1/2}.$$

As $R(\theta)$ has an approximate normal distribution with absolute error of $O(n^{-1/2})$ rather than relative error of $O(n^{-3/2})$, $R_{\mathrm{INF}}(\theta)$ can be regarded as a correction to $R(\theta)$ that gives an improved normal approximation; although $R^*(\theta)$ is slightly more complicated, it yields inferences that are much more accurate. It can be useful to compute values $r_{\mathrm{INF}}(\theta)$ of $R_{\mathrm{INF}}(\theta)$ for the plausible range of values of θ.

Example 12.7 (Student t distribution) The log likelihood contribution for the t_ν density is $g(u) = -\frac{1}{2}(\nu + 1) \log(1 + u^2/\nu)$, so the log likelihood and its derivatives are

$$\ell(\theta; \widehat{\theta}, a) = -\frac{\nu + 1}{2} \sum_{j=1}^{n} \log \left\{1 + (a_j + \widehat{\theta} - \theta)^2/\nu\right\},$$

$$\ell_{\theta;}(\theta; t, a) = (\nu + 1) \sum_{j=1}^{n} (a_j + t - \theta)/\left\{\nu + (a_j + t - \theta)^2\right\},$$

$$J(t; t, a) = (\nu + 1) \sum_{j=1}^{n} (\nu - a_j^2)/(\nu + a_j^2)^2,$$

from which $r(\theta)$, $v(\theta)$, and $r^*(\theta)$ may be found.

The data used in the left panel of Figure 12.1 have $n = 5, \widehat{\theta} = 0.22$, and configuration $a = (-0.71, -0.54, -0.47, 2.39, 3.28)$. The left panel of Figure 12.2 compares the values of $R^*(\theta)$, $R(\theta)$ and the approximate pivot $Z_1(\theta) = J(\widehat{\theta}; \widehat{\theta}, a)^{1/2}(\widehat{\theta} - \theta)$ derived from the large-sample $N(\theta, J(\widehat{\theta}; \widehat{\theta}, a)^{-1})$ distribution of $\widehat{\theta}$. All three decrease as functions of θ, but $r^*(\theta)$ and $r(\theta)$ are curved for positive θ and shifted upwards relative to $z_1(\theta)$, reflecting the asymmetry of the configuration. The limits of confidence intervals based on $R^*(\theta)$ are read off as those values for which $r^*(\theta) = z_\alpha, z_{1-\alpha}$, and likewise with $r(\theta)$ and $z_1(\theta)$. Thus 0.95 confidence intervals based on $r^*(\theta)$, $r(\theta)$, and $z_1(\theta)$ are $(-0.83, 2.45)$, $(-0.87, 2.08)$, and $(-1.00, 1.44)$. These intervals reflect the upward shift and curvature in $r^*(\theta)$, as the corresponding interval is moved to the right and quite asymmetric.

The right panel of Figure 12.2 shows that for the symmetric configuration in the right of Figure 12.1, confidence intervals based on $Z(\theta)$, $R(\theta)$, and $R^*(\theta)$ are essentially identical.   ■

Applications are usually much more complicated than this, and though it nicely illustrates how conditioning can affect inference, one might wonder how relevant the above toy example is to a real world in which models contain nuisance parameters and have no exact ancillary statistics. In practice a key aspect is not the number of observations, but the information they contain. If ten parameters are estimated from 50 observations, say, then one may worry that inference for one key parameter is weakened by having to deal with the other nine: the proportion of parameters to observations is the same as in the example above. Then small-sample procedures may in some cases provide reassurance that standard inferences are adequate, while giving improvements when they are not.

Difficulties with ancillaries

One difficulty in a general treatment of conditional inference is that some statistical models admit no exact ancillary statistic, while others have several. An instance of this is the location model, where any subset of the configuration is ancillary, but there it is clear that the sample space is partitioned most finely by conditioning on the entire configuration, which is a maximal ancillary statistic. Such arguments may not always be applied, however.

Example 12.8 (Multinomial distribution) Consider the multinomial distribution with denominator m and responses $R_1, \dots, R_4$ corresponding to cells with probabilities

$$\pi_1 = \tfrac{1}{6}(1 + \theta), \ \pi_2 = \tfrac{1}{6}(1 - \theta), \ \pi_3 = \tfrac{1}{6}(2 + \theta), \ \pi_4 = \tfrac{1}{6}(2 - \theta), \quad -1 < \theta < 1.$$

The minimal sufficient statistic is $S = (R_1, R_2, R_3)$, but $A_1 = R_1 + R_2$ and $A_2 = R_2 + R_3$ are binomial with success probabilities $\tfrac{1}{3}$ and $\tfrac{1}{2}$. As A_1 and A_2 are functions of S whose distributions are free of θ, both are ancillary. They are not jointly ancillary, however (Problem 12.6).

Conditioning on $A_1 = a_1$ splits the multinomial likelihood into two binomial contributions, one from R_1 with denominator a_1 and probability $\tfrac{1}{2}(1 + \theta)$, and the other from R_3 with denominator $m - a_1$ and probability $\tfrac{1}{4}(2 + \theta)$, so the expected information is the sum of the information from the two binomials, that is $I_1(\theta) = 4a_1/(1 - \theta^2) + 16(m - a_1)/(2 - \theta^2)$. Conditioning on A_2 gives a similar split, but the conditional expected information is

$$I_2(\theta) = 9a_2/\{(1-\theta)(2+\theta)\} + 9(m - a_2)/\{(1+\theta)(2-\theta)\}.$$

Evidently there is no reason that $I_1(\theta)$ should equal $I_2(\theta)$, and therefore conditional large-sample confidence intervals for θ would depend on whether A_1 or A_2 was used. ∎

This lack of uniqueness is annoying in principle but turns out not to be crucial in applications. More awkward difficulties arise when there is no exact ancillary. Then it is necessary to construct approximate ancillaries, and several approaches to this have been proposed. One conceptually simple notion can be used when the model under study can be regarded as nested within a density $f(y; \theta, \lambda)$, with $\lambda = \lambda_0$ corresponding to the model under study. Suppose that the likelihood for the larger model may be expressed as a function of the data through $(\widehat\theta, \widehat\lambda, a_0)$, where $\widehat\theta$ and $\widehat\lambda$ are maximum likelihood estimators and a_0 is ancillary for (θ, λ). Let $\widehat\theta_0$ be the maximum likelihood estimator of θ when $\lambda = \lambda_0$. Then if the model under study is correct, the likelihood ratio statistic $2\{\ell(\widehat\theta, \widehat\lambda; a_0) - \ell(\widehat\theta_0, \lambda_0; a_0)\}$ has an approximate chi-squared distribution whatever the value of θ and so is approximately ancillary. More elaborate variants of this argument are required to produce ancillary statistics on which conditioning can readily be performed in practice, for example by decomposing the ancillary into a sum of squared approximate standard normal variables, and conditioning on these. It may also be necessary to modify these variables to have a joint normal distribution to higher order. The details are tedious and since explicit forms of these ancillaries are not needed below we shall not delve into them. Although there are different forms of approximate ancillary, most of the approximations in later sections do not depend on the particular one used.

Exercises 12.1

1   Let $Y_1, \ldots, Y_n \overset{\text{iid}}{\sim} N(\mu, c^2\mu^2)$, with c known. Show that $\overline{Y}/S$ is ancillary for μ.

2   In Example 12.3, show that $(1 - 2\alpha)$ equi-tailed conditional and unconditional confidence intervals are $t \pm (1/2 - \alpha)(1 - a)$ and $t \pm \{(2\alpha)^{1/n} - 1\}/2$. Hence verify the intervals given in the example.

3   If $Y_1, \ldots, Y_n$ is an exponential random sample, show that the joint distribution of $Y_1/S, \ldots, Y_n/S$ is independent of that of $S = \sum Y_j$, without computing it. Is this true when S is replaced by any other equivariant estimator of scale?

4   Let $Y_1, \ldots, Y_n \overset{\text{iid}}{\sim} N(\mu, \sigma^2)$ with σ^2 known. Show that $(Y_1 - \overline{Y}, \ldots, Y_n - \overline{Y})$ is distribution constant, and deduce that $\overline{Y}$ and $\sum(Y_j - \overline{Y})^2$ are independent.

5   When is the configuration $(Y_1 - \widehat\eta)/\widehat\tau, \ldots, (Y_n - \widehat\eta)/\widehat\tau$ in a location-scale model ancillary as well as maximal invariant?

6   Show that conditional and unconditional inference for the mean of the normal location-scale model are the same.

7   Give expressions for the quantities needed to compute $r^*(\theta)$ for the location model when the data have the Gumbel density, so $g(u) = -u - e^{-u}$. Hence find confidence intervals based on $r(\theta)$, $r^*(\theta)$, and $z_1(\theta)$ with the data -0.32, -0.49, -0.25, 2.61, 3.50 from Example 12.6.

12.2 Marginal Inference

Consider a model for data with minimal sufficient statistic (T_1, T_2, A), where A is ancillary, and for which

$$f(y; \psi, \lambda) \propto f(t_1, t_2, a; \psi, \lambda) = f(a) f(t_1 \mid a; \psi) f(t_2 \mid t_1, a; \psi, \lambda).$$

As the density of T_1 conditional on A depends only on ψ, inference for ψ may be based on the *marginal likelihood* $L_m(\psi) = f(t_1 \mid a; \psi)$. In many cases there is no ancillary, and then $L_m(\psi)$ is simply the marginal density of T_1. There may be some loss of information about ψ due to neglecting the third term in the decomposition above, but the complications in retrieving it are presumed to outweigh the benefits.

We have already encountered marginal likelihood and maximum marginal likelihood estimators in several contexts. Such estimators and derived statistics will have the usual large-sample distributions provided that the usual regularity conditions apply to $f(t_1 \mid a; \psi)$.

Example 12.9 (Normal linear model)   Consider the usual normal linear model $y = X\beta + \sigma\varepsilon$, where the ε_j are independent standard normal variables and β has dimension $p \times 1$. The minimal sufficient statistics for (β, σ^2) are

$$T_1 = S^2 = (n - p)^{-1}(y - X\widehat{\beta})^{\mathsf{T}}(y - X\widehat{\beta}), \quad T_2 = \widehat{\beta} = (X^{\mathsf{T}}X)^{-1}X^{\mathsf{T}}y,$$

and these are independent, with $(n - p)S^2/\sigma^2$ having a χ_{n-p}^2 distribution. If the interest parameter is σ^2, and β is the nuisance parameter, we have

$$f(y; \sigma^2, \beta) = f(y \mid \widehat{\beta}, s^2) f(\widehat{\beta}; \beta, \sigma^2) f(s^2; \sigma^2).$$

As the density of S^2 depends only on σ^2, we may base inference for σ^2 on the corresponding marginal likelihood

$$L_m(\sigma^2) = f(s^2; \sigma^2) = \left(\frac{n - p}{2\sigma^2}\right)^{(n-p)/2} \frac{(s^2)^{(n-p)/2-1}}{\Gamma\left(\frac{n-p}{2}\right)} \exp\left\{-\frac{(n - p)s^2}{2\sigma^2}\right\},$$

where $\sigma^2 > 0$, $s^2 > 0$. The marginal maximum likelihood estimate of σ^2 is s^2, with expected information $(n - p)/(2\sigma^4)$ from the marginal likelihood. These give inferences for σ^2 essentially equivalent to those described in Chapter 8.

This argument applies to any normal linear model, and in particular to Example 12.2, where $p = k$ and

$$S^2 = \frac{1}{n - p}(y - X\widehat{\beta})^{\mathsf{T}}(y - X\widehat{\beta})$$

$$= \frac{1}{m(k - 1)} \sum_{i=1}^{m} \sum_{j=1}^{k} (y_{ij} - \overline{y}_i)^2 \sim \frac{\sigma^2}{m(k - 1)} \chi_{m(k-1)}^2.$$

Here marginal likelihood automatically produces corrected inferences for σ^2.   ■

Example 12.10 (Partial likelihood)   The partial likelihood used with the proportional hazards model (Section 10.8.2) is a marginal likelihood. Suppose that survival time Y has hazard $\xi h(y)$, so its density may be written $\xi h(y) \exp\{-\xi H(y)\}$, where $H(y) = \int_0^y h(u)\,du$ is the baseline cumulative hazard. Observe that

$$\int_u^\infty h(s)e^{-\gamma H(s)}\,ds = \gamma^{-1}e^{-\gamma H(u)}.$$

To illustrate the argument, consider $n = 4$ continuous observations that fall in the order $0 < Y_2 < Y_3^+ < Y_1 < Y_4$, where Y_3^+ is right-censored. With nothing known about the baseline hazard $h(y)$, a minimal sufficient statistic is the set of failure times and censoring indicators $(Y_1, 1), (Y_2, 1), (Y_3, 0), (Y_4, 1)$; these are in 1–1 correspondence with the order statistics $(Y_{(1)}, Y_{(2)}, Y_{(3)}, Y_{(4)})$ and inverse ranks $(2, 3^+, 1, 4)$, the jth of the inverse ranks giving the original index of $Y_{(j)}$ and its censoring status. We shall compute the probability of seeing this particular realization of inverse ranks. If Y_3 had been observed, the joint density of the Y_j would be

$$\xi_2 h(y_2)e^{-\xi_2 H(y_2)} \times \xi_3 h(y_3)e^{-\xi_3 H(y_3)} \times \xi_1 h(y_1)e^{-\xi_1 H(y_1)} \times \xi_4 h(y_4)e^{-\xi_4 H(y_4)},$$

so the probability that $0 < Y_2 < Y_1 < Y_4$ with Y_3 censored somewhere to the right of Y_2 is

$$\xi_2 h(y_2)e^{-\xi_2 H(y_2)} \times e^{-\xi_3 H(y_2)} \times \xi_1 h(y_1)e^{-\xi_1 H(y_1)} \times \xi_4 h(y_4)e^{-\xi_4 H(y_4)}.$$

Hence the probability that the uncensored observations fail in the order observed, with Y_3 censored to the right of Y_2, is

$$\xi_2 \xi_1 \xi_4 \int_0^\infty dy_2 \int_{y_2}^\infty dy_1 \int_{y_1}^\infty dy_4\, h(y_2)h(y_1)h(y_4)e^{-(\xi_2+\xi_3)H(y_2)-\xi_1 H(y_1)-\xi_4 H(y_4)},$$

and this equals

$$\frac{\xi_2}{\xi_1 + \xi_2 + \xi_3 + \xi_4} \times \frac{\xi_1}{\xi_1 + \xi_4} = \prod_j \frac{\xi_j}{\sum_{i \in \mathcal{R}_j} \xi_i},$$

where the product is over those j for which Y_j is uncensored and $\mathcal{R}_j$ denotes the risk set of individuals available to fail at the jth failure time. This last expression is simply the partial likelihood (10.58) in this case. Plainly this argument generalizes to n failures, the only complication being notational. Thus partial likelihood is a marginal likelihood based on the inverse ranks of the failure and censoring times. ∎

Restricted maximum likelihood

An important application of marginal likelihood is to normal mixed models, which we met in Section 9.4.2. There the response vector may be written

$$y = X\beta + Zb + \varepsilon, \tag{12.10}$$

where the $n \times p$ and $n \times q$ matrices X and Z are known, β is an unknown $p \times 1$ parameter vector, and the $q \times 1$ and $n \times 1$ random vectors b and ε are independent with respective $N_q(0, \Omega_b)$ and $N_n(0, \sigma^2 I_n)$ distributions. Suppose that the variance matrix $\sigma^2 \Upsilon^{-1} = \sigma^2 I_n + Z\Omega_b Z^{\mathsf{T}}$ exists and that Υ depends on parameters ψ but not on β. We aim to construct a marginal likelihood for σ^2 and ψ, eliminating the nuisance parameter β.

We can write $y = X\beta + \zeta$, where $\zeta \sim N_n(0, \sigma^2 \Upsilon^{-1})$. If ψ and hence Υ were known, the maximum likelihood estimator of β would be

$$\widehat{\beta}_\psi = (X^\mathsf{T} \Upsilon X)^{-1} X^\mathsf{T} \Upsilon y = \beta + (X^\mathsf{T} \Upsilon X)^{-1} X^\mathsf{T} \Upsilon \zeta,$$

whose distribution is $N_p(\beta, \sigma^2 (X^\mathsf{T} \Upsilon X)^{-1})$.

Let H denote the $n \times n$ matrix $X(X^\mathsf{T} X)^{-1} X^\mathsf{T}$; it satisfies $HX = X$. A natural basis for building a marginal likelihood is the vector of residuals $(I_n - H)y$, whose distribution does not depend on β, but as $I_n - H$ has rank $(n - p)$ this distribution is degenerate and it seems better to take just $(n - p)$ linearly independent residuals. Consider therefore the $(n - p) \times 1$ random variable $U = B^\mathsf{T} y$, where B is any $n \times (n - p)$ matrix with $BB^\mathsf{T} = I_n - H$ and $B^\mathsf{T} B = I_{n-p}$ (Exercise 12.2.3); B has rank $(n - p)$. Now $B^\mathsf{T} y = B^\mathsf{T} BB^\mathsf{T} y = B^\mathsf{T}(I_n - H)y$ is a linear combination of the residuals. Furthermore

$$B^\mathsf{T} X = B^\mathsf{T} BB^\mathsf{T} X = B^\mathsf{T}(I_n - H)X = 0,$$

giving

$$U = B^\mathsf{T} y = B^\mathsf{T}(X\beta + \zeta) = B^\mathsf{T}\zeta,$$

whose distribution does not depend on β. Moreover U and $\widehat{\beta}_\psi$ are normal variables, with covariance

$$\begin{aligned}
\mathrm{E}\left\{U(\widehat{\beta}_\psi - \beta)\right\} &= B^\mathsf{T} \mathrm{E}(\zeta\zeta^\mathsf{T}) \Upsilon X (X^\mathsf{T} \Upsilon X)^{-1} \\
&= \sigma^2 B^\mathsf{T} \Upsilon^{-1} \Upsilon X (X^\mathsf{T} \Upsilon X)^{-1} = 0,
\end{aligned}$$

because $B^\mathsf{T} X = 0$. Hence U and $\widehat{\beta}_\psi$ are independent, and therefore

$$f(u; \psi) = \frac{f(u; \psi) f(\widehat{\beta}_\psi; \beta, \psi)}{f(\widehat{\beta}_\psi; \beta, \psi)} = \frac{f(u, \widehat{\beta}_\psi; \beta, \psi)}{f(\widehat{\beta}_\psi; \beta, \psi)} = \frac{f(y; \beta, \psi)}{f(\widehat{\beta}_\psi; \beta, \psi)} \left| \frac{\partial y}{\partial(u, \widehat{\beta}_\psi)} \right|, \tag{12.11}$$

while the Jacobian for the change of variable from $(u, \widehat{\beta}_\psi)$ to y is

$$\begin{aligned}
\left| \frac{\partial(u, \widehat{\beta}_\psi)}{\partial y} \right| &= |\, B \quad X(X^\mathsf{T} X)^{-1} \,| \\
&= \left| \begin{pmatrix} B^\mathsf{T} \\ (X^\mathsf{T} X)^{-1} X^\mathsf{T} \end{pmatrix} (B \quad X(X^\mathsf{T} X)^{-1}) \right|^{1/2} \\
&= \left| \begin{pmatrix} B^\mathsf{T} B & B^\mathsf{T} X(X^\mathsf{T} X)^{-1} \\ (X^\mathsf{T} X)^{-1} X^\mathsf{T} B & (X^\mathsf{T} X)^{-1} \end{pmatrix} \right|^{1/2} \\
&= \left| \begin{pmatrix} I_{n-p} & 0 \\ 0 & (X^\mathsf{T} X)^{-1} \end{pmatrix} \right|^{1/2} = |X^\mathsf{T} X|^{-1/2}.
\end{aligned}$$

On substituting this and the normal densities of y and $\widehat{\beta}_\psi$ into (12.11), we find

$$f(u; \psi) = \frac{|X^\mathsf{T} X|^{1/2} |\Upsilon|^{1/2}}{(2\pi\sigma^2)^{(n-p)/2} |X^\mathsf{T} \Upsilon X|^{1/2}} \exp\left\{ -\frac{1}{2\sigma^2} (y - X\widehat{\beta}_\psi)^\mathsf{T} \Upsilon (y - X\widehat{\beta}_\psi) \right\}, \tag{12.12}$$

in which B does not appear. Hence (12.12) is independent of the choice of B and therefore of the linear combination of elements of $(I_n - H)y$ used.

Expression (12.12) is the marginal likelihood on which inference for ψ is based.

It is also called the residual likelihood. Restricted maximum likelihood estimation is often abbreviated to REML.

It is also known as the *restricted likelihood*, because its parameter space involves σ^2 and ψ alone. The log restricted likelihood is

$$\ell_m(\psi, \sigma^2) \equiv \tfrac{1}{2} \log |\Upsilon| - \tfrac{1}{2} \log |X^\mathsf{T} \Upsilon X| - \frac{1}{2\sigma^2}(y - X\widehat{\beta}_\psi)^\mathsf{T} \Upsilon (y - X\widehat{\beta}_\psi)$$
$$- \frac{n-p}{2} \log \sigma^2, \tag{12.13}$$

where Υ and $\widehat{\beta}_\psi$ depend on ψ. This can be maximized with respect to ψ by a

Or RIGLS algorithm.

Newton–Raphson procedure, leading to a *restricted iterative generalized least squares* algorithm, or using the slower but more stable EM algorithm. Computational considerations are often important in practice, because the matrices involved can be very large.

The profile log likelihood for ψ and σ^2 based directly on the density of y is

$$\ell_p(\psi, \sigma^2) \equiv \frac{1}{2} \log |\Upsilon| - \frac{1}{2\sigma^2}(y - X\widehat{\beta}_\psi)^\mathsf{T} \Upsilon (y - X\widehat{\beta}_\psi) - \frac{n}{2} \log \sigma^2,$$

from which $\ell_m(\psi)$ differs by the addition of the term $\frac{p}{2} \log \sigma^2 - \frac{1}{2} \log |X^\mathsf{T} \Upsilon X|$, of order p. Thus if the dimension of β is large, the residual maximum likelihood estimator obtained from $\ell_m(\psi)$ may differ substantially from the usual maximum likelihood estimator from $\ell_p(\psi)$. When the term Zb does not appear in (12.10), $\Upsilon = I_n$ and the estimator of σ^2 obtained by maximizing (12.13) is the usual unbiased quantity (Exercise 12.2.4). Thus the argument leading to (12.13) generalizes that in Example 12.9.

We now illustrate how restricted likelihood can lead to modified inferences; see also Example 9.18.

Example 12.11 (Short time series) Consider m independent individuals on each of which k measurements y_{ij} are available from the distribution

$$\begin{pmatrix} Y_{i1} \\ \vdots \\ Y_{ik} \end{pmatrix} \sim N_k \left\{ \begin{pmatrix} \mu_i \\ \vdots \\ \mu_i \end{pmatrix}, \frac{\sigma^2}{1 - \rho^2} \begin{pmatrix} 1 & \rho & \cdots & \rho^{k-1} \\ \rho & 1 & \cdots & \rho^{k-2} \\ \vdots & \vdots & \ddots & \vdots \\ \rho^{k-1} & \rho^{k-2} & \cdots & 1 \end{pmatrix} \right\}, \quad i = 1, \ldots, m.$$

Denote the matrix here by Ω_0. Its determinant is $(1 - \rho^2)^{k-1}$ and its inverse has tridiagonal form

$$\frac{1}{1 - \rho^2} \begin{pmatrix} 1 & -\rho & 0 & \cdots & 0 & 0 \\ -\rho & 1 + \rho^2 & -\rho & \ddots & & 0 \\ 0 & -\rho & \ddots & \ddots & \ddots & \vdots \\ \vdots & \ddots & \ddots & \ddots & -\rho & 0 \\ 0 & & \ddots & -\rho & 1 + \rho^2 & -\rho \\ 0 & 0 & \cdots & 0 & -\rho & 1 \end{pmatrix}.$$

This formulation is typical of that arising in longitudinal data, where individuals are measured at successive time points. Correlation among measurements on a single individual is represented by Ω_0, here corresponding to a first-order Markov chain; see Example 6.13. To illustrate the argument above, we construct the profile and marginal log likelihoods for the correlation parameter ρ, with $m + 1$ nuisance parameters $\mu_1, \ldots, \mu_m, \sigma^2$.

The matrices X and $\sigma^2 \Upsilon^{-1}$ are of side $mk \times m$ and $mk \times mk$ respectively and

$$X = \begin{pmatrix} 1_k & 0 & \cdots & 0 \\ 0 & 1_k & \cdots & 0 \\ \vdots & \vdots & \ddots & \vdots \\ 0 & 0 & \cdots & 1_k \end{pmatrix}, \quad \sigma^2 \Upsilon^{-1} = \sigma^2 (1 - \rho^2)^{-1} \mathrm{diag}\left\{\Omega_0, \ldots, \Omega_0\right\},$$

and some algebra gives

$$\tfrac{1}{2} \log |\Upsilon| = \frac{m}{2} \log(1 - \rho^2), \quad -\tfrac{1}{2} \log |X^{\mathsf{T}} \Upsilon X| = -\frac{m}{2} \log \left[(1 - \rho)\left\{k - (k - 2)\rho\right\}\right].$$

More algebra establishes that $\widehat{\mu}_i = \bar{y}_i$, so

$$(y - X\widehat{\beta}_\psi)^{\mathsf{T}} \Upsilon (y - X\widehat{\beta}_\psi) = \sum_{i=1}^{m} T_i(\bar{y}_i, \rho),$$

where

$$T_i(\mu, \rho) = \sum_{j=1}^{k}(y_{ij} - \mu)^2 + \rho^2 \sum_{j=2}^{k-1}(y_{ij} - \mu)^2 - 2\rho \sum_{j=1}^{k-1}(y_{ij} - \mu)(y_{i,j+1} - \mu).$$

Thus the profile log likelihood for ρ is given by

$$\ell_{\mathrm{p}}(\rho) = \frac{m}{2} \log(1 - \rho^2) - \frac{mk}{2} \log \widehat{\sigma}_{\mathrm{p}}^2(\rho), \quad \widehat{\sigma}_{\mathrm{p}}^2(\rho) = \frac{1}{mk} \sum_{i=1}^{m} T_i(\bar{y}_i, \rho),$$

while the marginal log likelihood (12.13) when profiled gives

$$\max_{\sigma^2} \ell_{\mathrm{m}}(\rho, \sigma^2) = -\frac{m}{2} \log \left\{\frac{k - (k - 2)\rho}{1 + \rho}\right\} - \frac{m(k - 1)}{2} \log \widehat{\sigma}_{\mathrm{m}}^2(\rho),$$

where $\widehat{\sigma}_{\mathrm{m}}^2(\rho) = k\widehat{\sigma}_{\mathrm{p}}^2(\rho)/(k - 1)$. As $\widehat{\sigma}_{\mathrm{m}}^2(\rho) > \widehat{\sigma}_{\mathrm{p}}^2(\rho)$, there is an upward bias correction like that in Example 12.9. Let $\widehat{\rho}$ and $\widehat{\rho}_{\mathrm{m}}$ denote the estimators from maximizing the usual and marginal likelihoods for ρ.

Figure 12.3 shows the profile log likelihood and the profile marginal log likelihood for 20 samples with $m = 10$, $k = 5$ and $\rho = 0.7$. Most of the ordinary maximum likelihood estimates $\widehat{\rho}$ are much smaller than ρ, but the marginal maximum likelihood estimator $\widehat{\rho}_{\mathrm{m}}$ is much less downwardly biased, as is confirmed by further simulation. Some values of $\widehat{\rho}_{\mathrm{m}}$ equal the upper limit of $\rho = 1$, so the small-sample distributions of $\widehat{\rho}_{\mathrm{m}}$ and the marginal likelihood ratio statistic will be poorly approximated by asymptotic results. ∎

Despite its artificiality, this example illustrates how estimation using restricted likelihood can give substantial bias corrections, and it is generally preferable to ordinary likelihood for variance estimation in complex models.

Figure 12.3 Inference
for correlation parameter
in short time series with
$\rho = 0.7$. Left: profile log
likelihoods for 20 samples
consisting of $m = 10$
series of length $k = 5$.
Right: profile marginal log
likelihoods for ρ, for the
same datasets. Maximum
marginal likelihood
estimators are much less
biased than are ordinary
maximum likelihood
estimators.

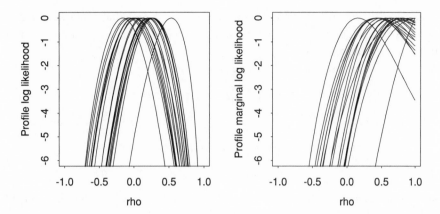

Figure 12.3 Inference
for correlation parameter
in short time series with
$\rho = 0.7$. Left: profile log
likelihoods for 20 samples
consisting of $m = 10$
series of length $k = 5$.
Right: profile marginal log
likelihoods for ρ, for the
same datasets. Maximum
marginal likelihood
estimators are much less
biased than are ordinary
maximum likelihood
estimators.

Regression-scale model

Consider the linear regression model $Y = X\beta + e^{\tau}\varepsilon$, where the errors $\varepsilon_1, \ldots, \varepsilon_n$ are independent, the $n \times p$ design matrix X has rank p and rows $x_1^{\mathsf{T}}, \ldots, x_n^{\mathsf{T}}$, τ is scalar and β is a $p \times 1$ vector of parameters. Exact results for this linear regression model were discussed in Chapters 8 and 9 for normal errors, but are typically unavailable otherwise. Accurate approximations to distributions needed for inference may be found, however. Let d be the negative log density of ε, and note that the log likelihood is

$$\ell(\beta, \tau; y) = -n\tau - \sum_{j=1}^{n} d\left\{e^{-\tau}(y_j - x_j^{\mathsf{T}}\beta)\right\}.$$

The minimal sufficient statistic is typically the full set of responses $Y_1, \ldots, Y_n$. Differentiation of the log likelihood shows that the maximum likelihood estimators $\widehat{\beta}$ and $\widehat{\tau}$ are determined by the score equations

$$\sum_{j=1}^{n} x_j d'\left(A_j\right) = 0, \quad n - \sum_{j=1}^{n} A_j d'\left(A_j\right) = 0, \tag{12.14}$$

where $d'(u)$ is the derivative of $d(u)$ with respect to u, and $A_j = e^{-\widehat{\tau}}(Y_j - x_j^{\mathsf{T}}\widehat{\beta})$, for $j = 1, \ldots, n$. Evidently $A = (A_1, \ldots, A_n)^{\mathsf{T}}$ is distribution constant, and as $\widehat{\beta}$ and $\widehat{\tau}$ are functions of the minimal sufficient statistic, A is in general also ancillary. Its distribution is degenerate, with $p + 1$ constraints imposed by (12.14), so just $n - p - 1$ of the A_j are needed to determine them all. Let A_0 denote such a subset, say, $A_0 = (A_1, \ldots, A_{n-p-1})$.

We now consider how conditional tests and confidence intervals may be derived for an element of β, say β_1 without loss of generality. The idea is to construct a marginal density that depends only on β_1, by using the joint density of $\widehat{\beta}$ and $\widehat{\tau}$ conditional on A to obtain

$$f(u_1, u_2 \mid a; \beta, \tau) = f(\widehat{\beta}, \widehat{\tau} \mid a; \beta, \tau)\left|\frac{\partial(\widehat{\beta}, \widehat{\tau})}{\partial(u_1, u_2)}\right|,$$

where $U_1 = e^{-\widehat{\tau}}(\widehat{\beta}_1 - \beta_1)$, U_2 denotes the $p \times 1$ vector with elements $\widehat{\tau} - \tau$ and $e^{-\widehat{\tau}}(\widehat{\beta}_{-1} - \beta_{-1})$, and β_{-1} denotes $(\beta_2, \ldots, \beta_p)^{\mathsf{T}}$. We shall see that U_1 and U_2 are pivots, and that the distribution of U_1 conditional on the ancillary statistic A forms the basis of conditional tests and confidence intervals for β_1. These are readily obtained using p^*-type approximations.

As a preliminary step, we write the log likelihood as

$$\ell(\beta, \tau; y) = \ell\left\{e^{-\widehat{\tau}}(\beta - \widehat{\beta}), \tau - \widehat{\tau}; a\right\} - n\widehat{\tau}; \qquad (12.15)$$

its maximized value is $\ell(\widehat{\beta}, \widehat{\tau}; a) = -n\widehat{\tau} - \sum_{j=1}^{n} d(a_j)$. For later use define

$$
\begin{aligned}
h_a(u_1, u_2) &= \ell(0, 0; a) - \ell(u_1, u_2; a) \\
&= \ell(0, 0; a) - \ell\left\{e^{-\widehat{\tau}}(\beta - \widehat{\beta}), \tau - \widehat{\tau}; a\right\},
\end{aligned}
$$

and observe that its second derivative matrix with respect to (u_1, u_2) has determinant of form $e^{-2\widehat{\tau}} H_a(u_1, u_2)$, and

$$\frac{\partial h_a(u_1, u_2)}{\partial u_1} = \frac{\partial \ell(\beta, \tau; y)}{\partial \beta_1}.$$

We now compute the density of $(\widehat{\beta}, \widehat{\tau})$ conditional on A. An extension of the argument giving (5.19) shows that the Jacobian for the transformation $y \mapsto (\widehat{\beta}, \widehat{\tau}, a_0)$ may be written $e^{(n-p)\widehat{\tau}} J(a)$, so

$$
\begin{aligned}
f(\widehat{\beta}, \widehat{\tau}, a_0; \beta, \tau) &= J(a) \exp\left\{(n - p)\widehat{\tau} + \ell(\beta, \tau; y)\right\} \\
&= J(a) \exp\left[-p\widehat{\tau} + \ell\left\{e^{-\widehat{\tau}}(\beta - \widehat{\beta}), \tau - \widehat{\tau}; a\right\}\right].
\end{aligned}
$$

Hence

$$f(\widehat{\beta}, \widehat{\tau} \mid a; \beta, \tau) = \frac{\exp\left[-p\widehat{\tau} + \ell\left\{e^{-\widehat{\tau}}(\beta - \widehat{\beta}), \tau - \widehat{\tau}; a\right\}\right]}{\int \exp\left[-p\widehat{\tau} + \ell\left\{e^{-\widehat{\tau}}(\beta - \widehat{\beta}), \tau - \widehat{\tau}; a\right\}\right] d\widehat{\beta} d\widehat{\tau}},$$

the domain of integration being $\mathbb{R}^{p+1}$. We now change variables in the integral from $(\widehat{\beta}, \widehat{\tau})$ to (u_1, u_2), with Jacobian $\exp(p\widehat{\tau})$. Thus the integral may be written

$$\int \exp\left\{\ell(0, 0; a) - h_a(u_1, u_2)\right\} \, du_1 du_2$$

and this equals

$$(2\pi)^{(p+1)/2} \left|\frac{\partial^2 h_a(0, 0)}{\partial u \partial u^{\mathsf{T}}}\right|^{-1/2} e^{\ell(0,0;a)} \left\{1 + O(n^{-1})\right\}$$

by Laplace approximation, the $O(n^{-1})$ term depending only on a, and the conditional density $f(\widehat{\beta}, \widehat{\tau} \mid a; \beta, \tau)$ equals

$$\bar{c}(a)(2\pi)^{-(p+1)/2} \left|\frac{\partial^2 h_a(0, 0)}{\partial u \partial u^{\mathsf{T}}}\right|^{1/2} \widehat{\tau}^{-p} \exp\left[\ell\left\{e^{-\widehat{\tau}}(\beta - \widehat{\beta}), \tau - \widehat{\tau}; a\right\} - \ell(0, 0; a)\right],$$

where $\bar{c}(a) = 1 + O(n^{-1})$ depends only on a. On changing variables to u_1 and u_2, we have

$$f(u_1, u_2 \mid a; \beta, \tau) = \bar{c}(a)(2\pi)^{-(p+1)/2} \left|\frac{\partial^2 h_a(0, 0)}{\partial u \partial u^{\mathsf{T}}}\right|^{1/2} \exp\left\{-h_a(u_1, u_2)\right\},$$

and as the the right-hand side does not depend on the parameters, $U_1(\beta_1)$ and $U_2(\beta_{-1}, \tau)$ are pivots. Note that the inferences below are not parametrization invariant, because the accuracy of Laplace approximation depends on the closeness of h to quadratic; taking $\widehat{\tau} - \tau$ should give better results than the more obvious pivot $\widehat{\sigma}/\sigma$.

Consider testing the hypothesis that β_1 takes value β_1^0. If so, the observed value of U_1 is $u_1^0 = e^{-\widehat{\tau}}(\widehat{\beta}_1 - \beta_1^0)$, with corresponding tail probability

$$\Pr\left(U_1 \le u_1^0 \mid A = a\right) = \bar{c}(a)\frac{\left|\partial^2 h_a(0, 0)/\partial u \partial u^{\mathsf{T}}\right|^{1/2}}{(2\pi)^{(p+1)/2}} \int_{-\infty}^{u_1^0} du_1 \int du_2 e^{-h_a(u_1, u_2)},$$

the inner integral being over $\mathbb{R}^p$. This expression has form (11.32), so

$$\Pr\left(U_1 \le u_1^0 \mid A = a\right) = \Phi\left\{r^*(\beta_1^0)\right\}\left\{1 + O(n^{-1})\right\}, \tag{12.16}$$

where $r^*(\beta_1^0) = r(\beta_1^0) + r(\beta_1^0)^{-1}\log\{v(\beta_1^0)/r(\beta_1^0)\}$, and

$$r(\beta_1^0) = \mathrm{sign}(u_1^0)\left\{2h_a(u_1^0, \tilde{u}_{20})\right\}^{1/2},$$

$$v(\beta_1^0) = \frac{\partial h_a(u_1^0, \tilde{u}_{20})}{\partial u_1}\left|\frac{\partial^2 h_a(u_1^0, \tilde{u}_{20})}{\partial u_2 \partial u_2^{\mathsf{T}}}\right|^{1/2}\left|\frac{\partial^2 h_a(0, 0)}{\partial u \partial u^{\mathsf{T}}}\right|^{-1/2},$$

where $\tilde{u}_{20}$ is the value of u_2 that maximizes $h_a(u_1^0, u_2)$ with $u_1^0 = e^{-\widehat{\tau}}(\widehat{\beta}_1 - \beta_1^0)$ fixed. It is straightforward to check that in terms of the log likelihood and its derivatives,

$$r(\beta_1^0) = \mathrm{sign}(\widehat{\beta}_1 - \beta_1^0)\left[2\left\{\ell(\widehat{\beta}, \widehat{\tau}; y) - \ell(\widehat{\beta}^0, \widehat{\tau}^0; y)\right\}\right]^{1/2},$$

$$v(\beta_1^0) = \frac{\partial \ell(\widehat{\beta}^0, \widehat{\tau}^0; y)}{\partial \beta_1}\frac{\left|J_{\lambda\lambda}(\widehat{\beta}^0, \widehat{\tau}^0; y)\right|^{1/2}}{\left|J(\widehat{\beta}^0, \widehat{\tau}^0; y)\right|^{1/2}} \times \frac{\left|J(\widehat{\beta}^0, \widehat{\tau}^0; y)\right|^{1/2}}{\left|J(\widehat{\beta}, \widehat{\tau}; y)\right|^{1/2}}, \tag{12.17}$$

say, where $\widehat{\beta}^0 = (\beta_1^0, \widehat{\beta}_{-1}^0)$, $\widehat{\beta}_{-1}^0$ and $\widehat{\tau}^0$ are the maximum likelihood estimates of β_{-1} and τ when $\beta_1 = \beta_1^0$, $J(\beta, \tau; y)$ is the observed information matrix for β and τ, and $J_{\lambda\lambda}(\beta, \tau; y)$ is the observed information matrix corresponding to $\lambda = (\beta_{-1}, \tau)$ only. Note that $r(\beta_1^0) = v(\beta_1^0) = 0$ when $\beta_1^0 = \widehat{\beta}_1$; it turns out that $r^*(\beta_1^0)$ has a finite limit as $\beta_1^0 \to 0$.

Our argument has established that the conditional tail probability (12.16) depends on the signed likelihood ratio statistic $r(\beta_1^0)$ and a modified score statistic $v(\beta_1^0)$ for testing the hypothesis $\beta_1 = \beta_1^0$. Expression (12.17) may be written as a product, γC, say, where the first term is a score statistic standardized by its standard error and the second contains information matrix determinants. Then $r^*(\beta_1^0)$ may be decomposed as

$$r^* = r + r^{-1}\log(\gamma/r) + r^{-1}\log C = r + r_{\mathrm{INF}} + r_{\mathrm{NP}}, \tag{12.18}$$

say, where dependence on β_1^0 has been suppressed. The discussion around (12.9) suggests that r_{INF} should be regarded as a correction that improves the normal approximation to the distribution of the signed likelihood ratio statistic. The presence of λ introduces the term r_{NP}, large values of which indicate that strong correction for nuisance parameters is required. Although r_{INF} is typically small in applications, the nuisance parameter correction r_{NP} can be substantial.

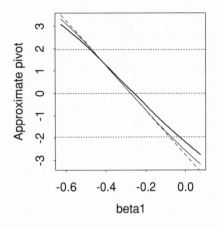
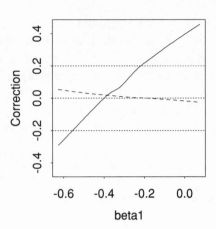

Figure 12.4
Small-sample inference
on parameter β_1 for
partial turnkey guarantee
for nuclear plant data,
using regression-scale
model with t_5 errors. Left:
approximate pivots $r(\beta_1)$
(solid), $r^*(\beta_1)$ (heavy),
and $(\widehat{\beta}_1 - \beta_1)/v_{11}^{1/2}$
(dashes) as functions of
β_1. Horizontal dotted lines
are at 0 and ± 1.96, so
their intersections with the
diagonal lines show the
limits of approximate 0.95
confidence intervals.
Right: small-sample
corrections r_{NP} (solid) and
r_{INF} (dashes) as functions
of β_1. The horizontal lines
at ± 0.2 give a crude rule
of thumb for 'large'
corrections.

The limits of an equi-tailed $(1 - 2\alpha)$ conditional confidence interval for β_1 are the values β_1^+, β_1^- that satisfy

$$\Pr\left\{ U_1 \le e^{-\widehat{\tau}}(\widehat{\beta}_1 - \beta_1)\big| a \right\} \doteq \Phi\left\{ r^*(\beta_1) \right\} = \alpha, 1 - \alpha.$$

The likelihood ratio confidence limits are obtained by replacing $r^*(\beta_1)$ with $r(\beta_1)$ in this expression, so computation of $v(\beta_1)$ for use in $r^*(\beta_1)$ is the only addition needed for small-sample conditional inference. Routines for large-sample likelihood inference on the model using $r(\beta_1)$ compute most of the quantities involved for small-sample conditional inferences using $r^*(\beta_1)$, so rather little extra work is needed.

Example 12.12 (Nuclear plant data)   To illustrate the effect of these small-sample modifications, we fit the model with six covariates chosen in Example 8.31 to the data in Table 8.13. We focus on the effect of the partial turnkey guarantee, PT, the coefficient of which under the normal model is -0.266 with standard error 0.144, giving an exact 0.95 confidence interval $(-0.46, 0.01)$. The evidence for inclusion of this effect is somewhat marginal under the normal model, and it is interesting to see the effect of using a t_5 error distribution instead. Computations that generalize those in Example 12.7 lead to the results shown in Figure 12.4, the left panel of which shows how $r(\beta_1)$ and $r^*(\beta_1)$ depend on the parameter β_1 for the partial turnkey effect. The 0.95 confidence intervals based on these pivots are respectively $(-0.47, -0.07)$ and $(-0.48, -0.03)$, while that based on normal approximation to the maximum likelihood estimator $\widehat{\beta}_1$ is $(-0.47, -0.08)$. The more accurate interval based on the small-sample approximation is rather wider than the others, and the right panel of the figure shows that this is primarily due to the nuisance parameter adjustment.

None of the intervals differs much from that found with normal errors, however, giving some reassurance that conclusions found using the normal model are valid more broadly, at least for this example.   ∎

Exercises 12.2

1   Suppose that $Y_1, \ldots, Y_n$ are independent Poisson variables with means $\psi \pi_j$, where $0 < \pi_j < 1$ and $\sum \pi_j = 1$. Find a marginal likelihood for ψ based on $Y_1, \ldots, Y_n$, and show that no information about ψ is lost by using the marginal likelihood rather than the full likelihood.

2   Verify the argument in Example 12.10 by writing down the partial likelihood for observations $0 < Y_4 < Y_1^+ < Y_3^+ < Y_5 < Y_2^+$ from the proportional hazards model, and then deriving it by the reasoning in the example.

3   Use the spectral decomposition to show that an $n \times (n - p)$ matrix B exists such that $B^{\mathrm{T}} B = I_n - H$, and $B B^{\mathrm{T}} = I_{n-p}$, where $H = X(X^{\mathrm{T}} X)^{-1} X^{\mathrm{T}}$ is the projection matrix for a linear model with $n \times p$ design matrix X of rank p.

4   Show that the maximum likelihood estimator of σ^2 based on (12.13) is given by

$$\widehat{\sigma}_{\mathrm{m}}^2 = \frac{1}{n - p}(y - X\widehat{\beta}_\psi)^{\mathrm{T}} \Upsilon (y - X\widehat{\beta}_\psi), \quad \widehat{\beta}_\psi = \left(X^{\mathrm{T}} \Upsilon X\right)^{-1} X^{\mathrm{T}} \Upsilon y.$$

Deduce that when Zb does not appear in (12.10), the restricted maximum likelihood estimator of σ^2 is $(y - X\widehat{\beta})^{\mathrm{T}}(y - X\widehat{\beta})/(n - p)$.

5   Find marginal likelihoods for σ^2 and for $\psi = (\beta_0, \sigma^2)$ in Example 8.10.

6   Check the details of Example 12.11.

7   Verify the p^* formula (12.7) for the regression-scale model, replacing $(2\pi)^{-1/2}$ by $(2\pi)^{-(p+1)/2}$.

8   (a) Express the quantities $r(\beta_1^0)$ and $v(\beta_1^0)$ defined after (12.16) in terms of the log likelihood and its derivatives.
     (b) Compute the modified signed likelihood ratio statistic for inference on τ.
     (c) Find $r^*(\beta_1)$ for the regression-scale model when $d(u) = u^2/2$.

12.3 Conditional Inference

12.3.1 Exact conditioning

Thus far this chapter has focused on the use of conditioning to ensure the relevance of tests and confidence intervals. In exponential family models there is typically no ancillary statistic and conditioning has the somewhat different purpose of removing a nuisance parameter from consideration; see Sections 10.4.2, 10.5, and 10.8.2. If the model density may be factorized as

$$f(y; \psi, \lambda) \propto f(t_1 \mid t_2; \psi) f(t_2; \psi, \lambda), \tag{12.19}$$

then although both terms on the right contain information on ψ, it may not be worthwhile to try and extract it from the second term. Moreover in Section 7.3.3 we saw that similar critical regions for tests on ψ must be based on the conditional density of T_1 given T_2. These considerations suggest that we restrict consideration to this density, which we treat as a conditional likelihood for ψ. Exact inference is typically infeasible, however.

Example 12.13 (Logistic regression) Let $Y_1, \ldots, Y_n$ be independent binary variables having $p \times 1$ covariate vectors x_j and satisfying a logistic regression model.

The minimal sufficient statistic $\sum x_j Y_j = S = (S_1, \ldots, S_p)^{\mathsf{T}}$ has joint density

$$\Pr(S_1 = s_1, \ldots, S_p = s_p; \beta) = \frac{c(s_1, \ldots, s_p) \exp\left(s_1 \beta_1 + \cdots + s_p \beta_p\right)}{\sum_{j=1}^{n} \left\{1 + \exp(x_j^{\mathsf{T}} \beta)\right\}},$$

found by summing the joint density (10.23) of the Y_j over all $c(s_1, \ldots, s_p)$ binary sequences of length n that yield the same value of S as did the data. If β_p is taken as the parameter of interest and the other β_j are treated as nuisance parameters, then they do not appear in the conditional density

$$\Pr(S_p = s_p \mid S_1 = s_1, \ldots, S_{p-1} = s_{p-1}; \beta_p) = \frac{c(s_1, \ldots, s_p) e^{s_p \beta_p}}{\sum c(s_1, \ldots, s_{p-1}, u) e^{u \beta_p}},$$

where the sum is over the possible values of u for which $s_1, \ldots, s_{p-1}$ are fixed. Exact tests and confidence intervals may be obtained by adapting the argument on page 495, but rely on ready computation of the combinatorial coefficients. In principle these may be obtained by considering the coefficients of $w_1^{s_1} \cdots w_p^{s_p}$ in the expansion of the generating function

$$\prod_{j=1}^{n} \left(1 + w_1^{x_{j1}} \cdots w_p^{x_{jp}}\right),$$

where x_{jr} is the rth element of x_j. Recourse to computer algebra or special algorithms is needed for all but the simplest problems, however, and typically the computational burden puts exact inference out of reach.   ∎

In subsequent sections we discuss accurate analytical approximations to exact conditional inferences, but we first illustrate how Markov chain Monte Carlo simulation (Section 11.3.3) may be used to explore a conditional sample space.

Example 12.14 (Several 2×2 tables)   Consider n independent 2×2 tables containing the numbers of successes R_{1j} and R_{0j} in m_{1j} and m_{0j} independent Bernoulli trials with success probabilities

$$\pi_{1j} = \frac{e^{\lambda_j + \psi_j}}{1 + e^{\lambda_j + \psi_j}} \quad \text{and} \quad \pi_{0j} = \frac{e^{\lambda_j}}{1 + e^{\lambda_j}}, \quad j = 1, \ldots, n.$$

The overall joint density is

$$\prod_{j=1}^{n} \binom{m_{1j}}{r_{1j}} \binom{m_{0j}}{r_{0j}} \frac{e^{r_{1j}(\lambda_j + \psi_j)}}{\left(1 + e^{\lambda_j + \psi_j}\right)^{m_{1j}}} \frac{e^{r_{0j} \lambda_j}}{\left(1 + e^{\lambda_j}\right)^{m_{0j}}}.$$

Suppose we wish to test whether the log odds ratios are equal for each table, that is $\psi_1 = \cdots = \psi_n = \psi$, where ψ is unknown. Let $W = w(R)$ be a statistic constructed to test this against the alternative of unequal ψ_j. Under the null hypothesis the statistics $S_j = R_{1j} + R_{0j}$ and $T = R_{11} + \cdots + R_{1n}$ are associated with λ_j and ψ, and as we wish to test homogeneity of the log odds ratios regardless of the values of $\lambda_1, \ldots, \lambda_n, \psi$, we should use the conditional distribution of W given the values of

$S_1, \ldots, S_n, T$ that were actually observed, namely $s_1, \ldots, s_n, t$. It is straightforward to establish that

$$\Pr(R \mid S, T) = \frac{\prod_{j=1}^{n} \binom{m_{1j}}{r_{1j}} \binom{m_{0j}}{s_j - r_{1j}}}{\sum \prod_{j=1}^{n} \binom{m_{1j}}{u_j} \binom{m_{2j}}{s_j - u_j}} = C^{-1} \prod_{j=1}^{n} c_j(r_{1j}), \qquad (12.20)$$

say, where the sum is over the set

$$\mathcal{U} = \left\{ (u_1, \ldots, u_n) : u_{j,-} \leq u_j \leq u_{j,+}, u_j \in \mathbb{Z}, u_1 + \cdots + u_n = t \right\},$$

with $u_{j,-} = \max(0, s_j - m_{0j})$ and $u_{j,+} = \min(m_{1j}, s_j)$. If the 2×2 tables were stacked on top of one another, then this conditioning would amount to fixing all three margins of the stack. We would like to use (12.20) to calculate the tail probability $\Pr\{w(R) \geq w \mid S = s, T = t\}$, but enumeration of $\mathcal{U}$ in order to find the normalising constant in (12.20) is typically difficult. A simplifying feature is that any 2×2 table for which s_j equals zero or $m_{0j} + m_{1j}$ is conditionally constant and can be ignored.

One possibility is to use the Metropolis–Hastings algorithm. The idea is to update the stack of 2×2 tables with contents r to have contents r', while holding its margins constant. Although it is hard to simulate directly from (12.20), it is straightforward to generate from the conditional distribution of R_{1i} given $R_{1j} + R_{1i}$, which is univariate with density

$$\Pr(R_{1i} = r \mid R_{1j} + R_{1i} = v) = \frac{c_i(r) c_j(v - r)}{\sum c_i(u) c_j(v - u)},$$

where the sum is over values of $R_{1i} = u$ for which $R_{1j} + R_{1i} = v$ is possible. The algorithm starts with original data r and repeats the following steps for $k = 1, \ldots, K$.

1. For $l = 1, \ldots, L$,
   - select two tables in the current stack at random, say i and j;
   - generate proposal data r' by leaving the remaining 2×2 tables intact and generating a value r'_{1i} from the conditional distribution of R_{1i} given $R_{1i} + R_{1j}$; then set $r'_{1j} = r_{1i} + r_{1j} - r'_{1i}$.
   - Set $r \leftarrow r'$ with probability

$$p = \min \left\{ 1, \frac{\pi(r') q(r \mid r')}{\pi(r) q(r' \mid r)} \right\}. \qquad (12.21)$$

2. Calculate $w_k^* = w(r)$ based on the current data r.

The estimated P-value is then $\{\sum I(w_k^* \geq w_{\text{obs}}) + 1\}/(K + 1)$, where w_{obs} is the observed value of $w(R)$.

The target density is

$$\pi(r) \propto \prod_j c_j(r_{1j})$$

and the transition density for the Markov chain is

$$q(r' \mid r) \propto c_i(r'_{1i}) c_j(r_{1i} + r_{1j} - r'_{1i}) \sum_u c_i(u) c_j(r_{1i} + r_{1j} - u),$$

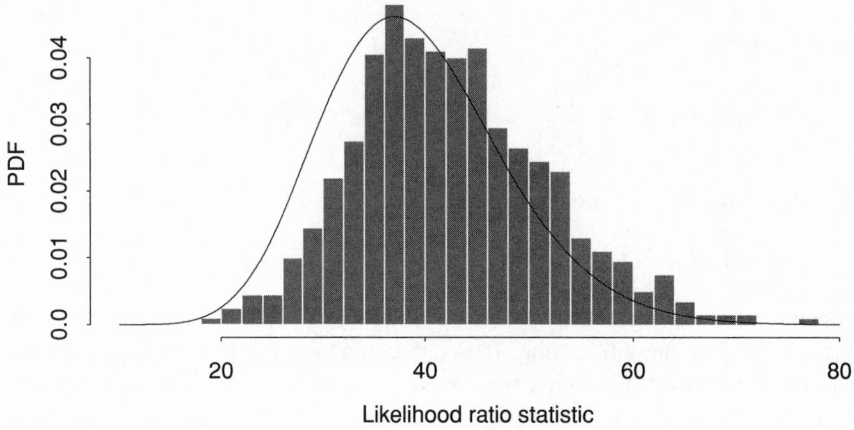

Figure 12.5 Simulation approximation to conditional distribution of test statistic W for homogeneity of parameters ψ_j for the ulcer data (Table 10.11). The line is the asymptotic χ^2_{39} density.

so the ratio in (12.21) is

$$\frac{\pi(r')q(r \mid r')}{\pi(r)q(r' \mid r)} = \frac{c_i(r'_{1i})c_j(r'_{1j})c_i(r_{1i})c_j(r_{1j})}{c_i(r_{1i})c_j(r_{1j})c_i(r'_{1i})c_j(r'_{1j})} = 1.$$

Thus the proposal r' is always accepted.

In order to assess whether the parameters ψ_j for recurrent bleeding vary across the 41 studies in Table 10.11, we apply this algorithm to the 40 non-degenerate tables; the last one is conditionally constant. We take W to be the likelihood ratio statistic for testing the null hypothesis. Its observed value is $w = 177.6$ on 39 degrees of freedom, which is highly significant relative to its asymptotic χ^2_{39} distribution. Values w^* of W^* calculated every $L = 100$ steps using the algorithm above were uncorrelated, and are compared with the asymptotic distribution in Figure 12.5. The evidence that the ψ_j vary across the tables is overwhelming relative to either distribution, but as the distributions are rather different, conflict between asymptotic and exact P-values could arise for other such sets of data; then the P-value based on W^* would be preferred. Such a P-value is called *Monte Carlo exact*: it would be exact if the number of Monte Carlo replicates was infinite. ■

Algorithms of this type provide powerful tools for testing hypotheses in contingency tables with low counts, and can be adapted to provide Monte Carlo approximations to conditional likelihoods. A difficulty with exact conditioning, however, is that in some cases the inference depends very sharply on the conditioning event, giving likelihoods, significance levels and so forth that are highly sensitive to the exact data values. This raises the question whether some form of approximate conditioning would be more stable. Below we outline how conditional inferences in exponential families may be approximated using saddlepoint methods.

12.3.2 Saddlepoint approximation

Saddlepoint approximation to density and distribution functions is the basis of many highly accurate small-sample procedures. In this section we describe informally the

Figure 12.6 Saddlepoint approximations to the density and distribution functions of an average of n $U(-1, 1)$ variables. Left panel: exact density functions (solid) and their basic (dots) and normalized (dashes) approximations for $n = 2, 5$; the peakier curves are for $n = 5$. Right panel: exact distribution function (solid) and its approximation (dots) for $n = 2$.

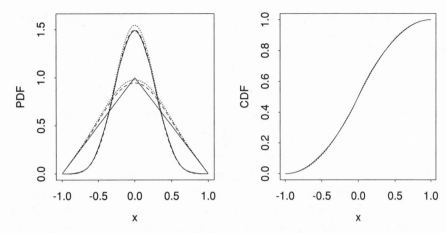

underlying ideas. Initially we state the basic approximations, leaving their justification to the end of the section.

Let $\overline{X}$ denote the average of a random sample of continuous scalar random variables $X_1, \ldots, X_n$, each having cumulant-generating function $K(u)$. Then the *saddlepoint approximation* to the density of $\overline{X}$ at x is

$$f_{\overline{X}}(x) \doteq \left\{ \frac{n}{2\pi K''(\tilde{u})} \right\}^{1/2} \exp\left[n \left\{ K(\tilde{u}) - \tilde{u}x \right\} \right], \qquad (12.22)$$

where $\tilde{u} = \tilde{u}(x)$, known as the saddlepoint, is the unique value of u satisfying the *saddlepoint equation* $K'(u) = x$, and $K'(u)$ and $K''(u)$ are the first and second derivatives of $K(u)$ with respect to u. There is a corresponding approximation to the cumulative distribution function of $\overline{X}$, namely

$$\Pr\left(\overline{X} \leq x \right) \doteq \Phi\left\{ r^*(x) \right\}, \qquad (12.23)$$

where $r^*(x) = r(x) + r(x)^{-1} \log\{v(x)/r(x)\}$ depends on

$$r(x) = \text{sign}(\tilde{u}) \left[2n \left\{ \tilde{u}x - K(\tilde{u}) \right\} \right]^{1/2}, \quad v(x) = \tilde{u} \left\{ n K''(\tilde{u}) \right\}^{1/2}.$$

Calculation of (12.22) and (12.23) requires knowledge of $K(u)$ and computation of $\tilde{u}$ for each x of interest, but such approximations are often extremely accurate far into the tails of the density of $\overline{X}$. A more accurate density approximation can be obtained by renormalizing (12.22), that is, dividing by its integral, which is usually obtained numerically.

Example 12.15 (Uniform distribution) If X has the uniform distribution on $(-1, 1)$, then $K(u) = \log\{\sinh(u)/u\}$. The left panel of Figure 12.6 compares the exact density function of $\overline{X}$ with its saddlepoint approximations when $n = 2$ and $n = 5$. Numerical integrals of the density approximations are 1.097 and 1.034, and the renormalized densities are also shown; that for $n = 5$ seems essentially exact. The distribution function approximation (12.23) is very accurate even in the extreme situation $n = 2$, although (12.22) fails to capture the cusp in the density at the origin. ∎

These basic approximations can be generalized in various ways. For our purpose the most useful is to the situation where the X_j are vectors of length p, in which case u is a $p \times 1$ vector. Then (12.22) extends to

$$f_{\overline{X}}(x) \doteq \left\{ \frac{n^p}{(2\pi)^p |K''(\tilde{u})|} \right\}^{1/2} \exp\left[n \left\{ K(\tilde{u}) - \tilde{u}^{\mathsf{T}} x \right\} \right], \tag{12.24}$$

where now the $p \times 1$ saddlepoint $\tilde{u}$ solves the $p \times 1$ system of equations

$$K'(\tilde{u}) = \frac{\partial K(\tilde{u})}{\partial u} = x,$$

and $K''(u) = \partial^2 K(u)/\partial u \partial u^{\mathsf{T}}$ is the $p \times p$ matrix of second derivatives of $K(u)$.

Let $X^{\mathsf{T}} = (X_1, X_2^{\mathsf{T}})$, where X_2 has dimension $(p-1) \times 1$, and split u^{T} into (u_1, u_2^{T}) conformably with X^{T}; both X_1 and u_1 are scalar. Then the marginal density of $\overline{X}_2$ at x_2 is obtained by saddlepoint approximation using the marginal cumulant-generating function $K(0, u_2)$ of X_2, and is

$$f_{\overline{X}_2}(x_2) \doteq \left\{ \frac{n^{p-1}}{(2\pi)^{p-1} |K_{22}''(\tilde{u}_0)|} \right\}^{1/2} \exp\left[n \left\{ K(\tilde{u}_0) - (0, x_2^{\mathsf{T}})\tilde{u}_0 \right\} \right], \tag{12.25}$$

where $\tilde{u}_0^{\mathsf{T}} = (0, \tilde{u}_2^{\mathsf{T}})$ is the solution to the $(p-1) \times 1$ system of equations $\partial K(0, u_2)/\partial u_2 = x_2$, and K_{22}'' is the $(p-1) \times (p-1)$ corner of K'' corresponding to u_2. Division of (12.24) by (12.25) gives an approximation to the conditional density $f_{\overline{X}_1 | \overline{X}_2}(x_1 \mid x_2)$, the *double saddlepoint approximation*

$$\left(\frac{n}{2\pi} \right)^{1/2} \frac{|K_{22}''(\tilde{u}_0)|^{1/2}}{|K''(\tilde{u})|^{1/2}} \exp\left[n \left\{ K(\tilde{u}) - \tilde{u}^{\mathsf{T}} x - K(\tilde{u}_0) + (0, x_2^{\mathsf{T}})\tilde{u}_0 \right\} \right], \tag{12.26}$$

corresponding to which is the distribution function approximation

$$\Pr(\overline{X}_1 \leq x_1 \mid \overline{X}_2 = x_2) \doteq \Phi\left\{ r^*(x_1) \right\}, \tag{12.27}$$

where again $r^*(x_1)$ equals $r(x_1) + r(x_1)^{-1} \log\{v(x_1)/r(x_1)\}$, but now with

$$r(x_1) = \text{sign}(\tilde{u}_1) \left[2n \left\{ K(\tilde{u}_0) - (0, x_2^{\mathsf{T}})\tilde{u}_0 \right\} - n \left\{ K(\tilde{u}) - \tilde{u}^{\mathsf{T}} x \right\} \right]^{1/2},$$
$$v(x_1) = \tilde{u}_1 n^{1/2} |K''(\tilde{u})|^{1/2} / |K_{22}''(\tilde{u}_0)|^{1/2}.$$

These formulae break down when $x_1 = \mathrm{E}(X_1)$ and a different more complicated approximation is then available. In practice it is simplest to evaluate $r^*(x_1)$ for at a grid of values of x_1 excluding any too close to $\mathrm{E}(X_1)$, and then to interpolate them using a spline or other numerical method.

Example 12.16 (Poisson distribution)   Let V_1 and V_2 be independent Poisson variables with means λ_1 and λ_2, and set $X_1 = V_1, X_2 = V_1 + V_2$. Then the joint cumulant-generating function of X_1 and X_2 is $K(u) = \lambda_1 \exp(u_1 + u_2) + \lambda_2 \exp(u_2)$, and the exact density of X_1 given $X_2 = x_2$ is binomial with probability $\pi = \lambda_1/(\lambda_1 + \lambda_2)$ and denominator x_2.

Figure 12.7 shows the conditional density and distribution of X_1, when $\lambda_1 = 1$, $\lambda_2 = 2.5$, and $x_2 = 15$. The saddlepoint density gives a continuous approximation to the discrete binomial density of X_1 given X_2, which it closely matches on the support

Figure 12.7 Double saddlepoint approximations to the density and distribution functions of $X_1 = V_1$ given $X_2 = V_1 + V_2 = 15$, where V_1 and V_2 are Poisson with means 1 and 2.5. Left panel: exact density function and its approximation. Right panel: exact distribution function (heavy) and its approximation without (solid) and with (dots) a continuity correction.

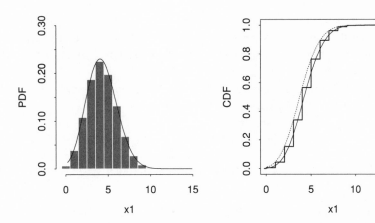

of X_1. For $x_1 = 0, \ldots, 15$, the distribution function approximation $\Phi\{r^*(x_1)\}$ agrees closely with

$$\Pr(X_1 \le x_1 - 1 \mid X_2) + \tfrac{1}{2}\Pr(X_1 = x_1 \mid X_2),$$

a quantity akin to the mid-p significance level (10.28). Continuity correction applied by taking $\Phi\{r^*(x_1 + 1/2)\}$ approximates $\Pr(X_1 \le x_1 \mid X_2)$ well. ∎

In applications saddlepoint formulae are sometimes used simply by taking the cumulant-generating function for a variable of interest, and formally setting $n = 1$, as we shall do below.

The rest of this section sketches derivations of the approximations above and can be skipped on a first reading. No attempt is made at a rigorous treatment; the intention is to give some justification for the approximations, and the flavour of the manipulations involved.

Edgeworth series

Francis Ysidro Edgeworth (1845–1926) was born in Longford, Ireland, studied in Dublin and Oxford, where he took a first-class degree in classics, and then became a London barrister. In 1880 he began a university career as a mathematician, eventually holding professorships of political economy in London and Oxford. His pioneering articles systematically applied ideas from the theory of errors to social science and economics.

Although they have practical disadvantages compared to saddlepoint approximations, Edgeworth series play a central role in theoretical discussions of small-sample inference. As we shall see below, they allow a relatively simple derivation of saddlepoint approximations.

Let $X_1, \ldots, X_n$ be a random sample of continuous variables with cumulant-generating function $K(u)$ and finite cumulants κ_r, let $\rho_r = \kappa_r/\kappa_2^{r/2}$ denote the rth standardized cumulant, and let $Z_n = (S_n - n\kappa_1)/(n\kappa_2)^{1/2}$ denote the standardized version of $S_n = X_1 + \cdots + X_n$. The large-sample distribution of Z_n may be found by noting that its cumulant-generating function is

$$u^2/2 + n^{-1/2}\rho_3 u^3/3! + n^{-1}\rho_4 u^4/4! + \cdots.$$

Thus its moment-generating function is

$$\exp(u^2/2)\left\{ 1 + \tfrac{1}{6}\rho_3 n^{-1/2} u^3 + \left(\frac{\rho_3^2}{72} + \frac{\rho_4}{24} \right) n^{-1} u^4 + O(n^{-3/2}) \right\}. \qquad (12.28)$$

All that remains when $n \to \infty$ is the leading term, which is the moment-generating function of the standard normal density. The continuity theorem and uniqueness of moment-generating functions then yield the limiting standard normal distribution of Z_n.

To find better finite-sample approximations to the density and distribution functions of Z_n, note that integration by parts gives

$$\int_{-\infty}^{\infty} e^{uz}(-1)^r \frac{d^r \phi(z)}{dz^r} \, dz = u^r e^{u^2/2}, \quad r = 0, 1, \ldots.$$

Letting $(-1)^r d^r \phi(z)/dz^r = \phi(z) H_r(z)$ determine the rth-order Hermite polynomial $H_r(z)$ gives

$$H_1(z) = z, \quad H_2(z) = z^2 - 1, \quad H_3(z) = z^3 - 3z, \quad H_4(z) = z^4 - 6z^2 + 3,$$

$$H_5(z) = z^5 - 10z^3 + 15z, \quad H_6(z) = z^6 - 15z^4 + 45z^2 - 15.$$

Term by term inversion of (12.28) shows that the corresponding density is

$$f_{Z_n}(z) = \phi(z) \left[1 + \frac{\rho_3}{6n^{1/2}} H_3(z) + \frac{1}{n} \left\{ \frac{\rho_4}{24} H_4(z) + \frac{\rho_3^2}{72} H_6(z) \right\} + O(n^{-3/2}) \right],$$
(12.29)

inversion of which gives the first three terms of the cumulant-generating function of Z_n. Integration of (12.29) gives the corresponding distribution function,

$$F_{Z_n}(z) = \Phi(z) - \phi(z) \left[\frac{\rho_3}{6n^{1/2}} H_2(z) + \frac{1}{n} \left\{ \frac{\rho_4}{24} H_3(z) + \frac{\rho_3^2}{72} H_5(z) \right\} + O(n^{-3/2}) \right].$$
(12.30)

The leading terms of the *Edgeworth expansions* (12.29) and (12.30) give the standard normal approximation for Z_n, with subsequent terms giving improvements, based respectively on the skewness of Z_n and on a combination of its skewness and kurtosis. Higher-order terms depend on further cumulants of the X_j.

The expansions (12.29) and (12.30) are more useful for theoretical development than for applications, because there is no reason for (12.29) to remain positive or for (12.30) to be increasing in z, or indeed even for it to lie between zero and one, whereas the saddlepoint formula (12.22) is guaranteed to give a positive density approximation. On the other hand saddlepoint approximation requires the entire cumulant-generating function, while (12.29) and (12.30) use only first few cumulants and so are more readily calculated.

When the density for Z_n is evaluated at $z = 0$, the series in (12.29) contains only powers of n^{-1}, because the odd powers of $n^{-1/2}$ depend on odd Hermite polynomials, all of which vanish at $z = 0$.

Derivation of saddlepoint approximation

To derive (12.22) we first embed the density f_X of X in the exponential family

$$f_X(x; u) = \exp\{xu - K(u)\} f_X(x),$$

where K is the cumulant-generating function of X. Here u plays the role of a parameter, and the new density $f_X(x; u)$ is an exponential tilt of f_X. Under this new model the density of S_n may be written

$$f_{S_n}(s; u) = \exp\{su - nK(u)\} f_{S_n}(s), \tag{12.31}$$

and its cumulant-generating function as $K_{S_n}(t) = n\{K(t + u) - K(u)\}$. Thus with parameter u the mean and variance of S_n under the new density are $nK'(u)$ and $nK''(u)$. Expression (12.31) gives

$$f_{S_n}(s) = \exp\{nK(u) - su\} f_{S_n}(s; u)$$

for all u. We now replace $f_{S_n}(s; u)$ in this expression by its Edgeworth expansion, but with the value of u chosen so that the tilted density $f_{S_n}(s; u)$ has mean s. Then $nK'(\tilde{u}) = s$, and the first term of the Edgeworth series is $\{2\pi nK''(\tilde{u})\}^{-1/2}$. The resulting approximation to the density of S_n is

$$f_{S_n}(s) = \{2\pi nK''(\tilde{u})\}^{-1/2} \exp\{nK(\tilde{u}) - s\tilde{u}\} \left\{1 + O(n^{-1})\right\},$$

where the order of the error follows from the fact that at its mean the Edgeworth series contains only powers of n^{-1}. A change of variables $s \mapsto x = s/n$ gives the leading term of (12.22), with subsequent terms found by retaining more terms of the Edgeworth series.

The argument leading to (12.23) starts by integrating (12.22), giving

$$\Pr(\overline{X} \le x_0) \doteq \int_{-\infty}^{x_0} \left\{\frac{n}{2\pi K''(\tilde{u})}\right\}^{1/2} \exp\left[n\{K(\tilde{u}) - \tilde{u}x\}\right] dx,$$

where $\tilde{u}$ is a function of the variable of integration through $K'(\tilde{u}) = x$. A change of variable from x to $\tilde{u}$, using $dx/d\tilde{u} = K''(\tilde{u})$, gives

$$\Pr(\overline{X} \le x_0) \doteq \int_{-\infty}^{u_0} \left\{\frac{nK''(\tilde{u})}{2\pi}\right\}^{1/2} \exp\left[n\{K(\tilde{u}) - \tilde{u}K'(\tilde{u})\}\right] d\tilde{u},$$

which is of form (11.30); here u_0 satisfies $K'(u_0) = x_0$. In this case the approximation (11.31) takes the form (12.23). Detailed accounting shows that under broad conditions the error in (12.23) is relative, of size $O(n^{-1})$ in large deviation regions, where $x_0 - E(X)$ is $O(1)$, and is only $O(n^{-3/2})$ for moderate deviation regions, in which $x_0 - E(X)$ is $O(n^{-1/2})$. The large-deviation property leads to the extraordinary accuracy of the approximations, which have low relative error far into the tails of the distribution of $\overline{X}$.

Discrete distributions are also important in practice. Suppose that the X_j take values in the lattice $a + bk$, where a and b are constants and $k = 0, 1, \ldots$. Then $\overline{X}$ takes values in the lattice $a + bk/n$. The saddlepoint approximation for the density of $\overline{X}$ at these values is unchanged, and its error remains relative and of $O(n^{-1})$, but in the cumulative distribution function approximation $v(x) = \tilde{u}\left\{nK''(\tilde{u})\right\}^{1/2}$ is replaced by $b^{-1}\{1 - \exp(-b\tilde{u})\}\left\{nK''(\tilde{u})\right\}^{1/2}$; note that this produces the continuous version when $b \to 0$. The error in this approximation is $O(n^{-1})$. A continuity-corrected approximation to the quantity $\Pr\{\overline{X} \le x_0 + 1/(2n)\}$ replaces $1 - \exp(-b\tilde{u})$ by $2\sinh(\tilde{u}/2)$.

12.3.3 Approximate conditional inference

To see how saddlepoint approximation may be applied for inference, we consider initially the case where a random sample $Y_1, \ldots, Y_n$ from the continuous exponential family $\exp\{\theta y - \kappa(\theta)\} f_0(y)$ depends on a scalar θ. The maximum likelihood estimator $\widehat{\theta}$ solves the likelihood equation $\overline{Y} = \kappa'(\widehat{\theta})$, so the log likelihood and observed information may be expressed as

$$\ell(\theta; \widehat{\theta}) \equiv n\left\{\theta \kappa'(\widehat{\theta}) - \kappa(\theta)\right\}, \quad J(\theta; \widehat{\theta}) = -\partial^2 \ell(\theta; \widehat{\theta})/\partial\theta^2 = n\kappa''(\theta).$$

Now $\overline{Y}$ is a minimal sufficient statistic for θ, so no information is lost by considering its density or equivalently that of $\widehat{\theta}$, which is a 1–1 function of $\overline{Y}$. The cumulant-generating function of Y_j is $K(u) = \kappa(\theta + u) - \kappa(\theta)$, so the saddlepoint equation is $\kappa'(\theta + \tilde{u}) = \overline{y}$. This implies that $\tilde{u} = \widehat{\theta} - \theta$; furthermore the second derivative $K''(\tilde{u}) = \kappa''(\theta + \tilde{u}) = n^{-1} J(\widehat{\theta}; \widehat{\theta})$. Thus the density approximation (12.22) for $\overline{Y}$ is

$$f_{\overline{Y}}(\overline{y}; \theta) \doteq \left\{\frac{n}{2\pi K''(\tilde{u})}\right\}^{1/2} \exp\left[n\left\{K(\tilde{u}) - \tilde{u}\overline{y}\right\}\right]$$

$$= \left\{\frac{n}{2\pi \kappa''(\widehat{\theta})}\right\}^{1/2} \exp\left[n\left\{\kappa(\widehat{\theta}) - \kappa(\theta) - (\widehat{\theta} - \theta)\overline{y}\right\}\right].$$

Now $\partial\overline{y}/\partial\widehat{\theta} = n^{-1} J(\widehat{\theta}; \widehat{\theta})$, and hence the density of $\widehat{\theta}$ may be written as

$$f(\widehat{\theta}; \theta) = c|J(\widehat{\theta}; \widehat{\theta})|^{1/2} \exp\left\{\ell(\theta; \widehat{\theta}) - \ell(\widehat{\theta}; \widehat{\theta})\right\}\left\{1 + O(n^{-1})\right\}, \tag{12.32}$$

where $c = (2\pi)^{-1/2}$. Thus (12.7) again arises as an approximation to the density of a maximum likelihood estimator, here with no ancillary statistic.

Saddlepoint approximation to the cumulative distribution function of $\widehat{\theta}$ or equivalently of $\overline{Y}$ gives

$$\Pr(\overline{Y} \leq \overline{y}; \theta) \doteq \Phi\left\{r^*(\theta)\right\}, \tag{12.33}$$

where $r^*(\theta) = r(\theta) + r(\theta)^{-1} \log\{v(\theta)/r(\theta)\}$, and

$$r(\theta) = \text{sign}(\widehat{\theta} - \theta)[2\{\ell(\widehat{\theta}) - \ell(\theta)\}]^{1/2}, \quad v(\theta) = (\widehat{\theta} - \theta)J(\widehat{\theta})^{1/2}; \tag{12.34}$$

recall that $\kappa'(\widehat{\theta}) = \overline{y}$. An approximate confidence interval for θ may be obtained by finding the values of θ for which $\Phi\{r^*(\theta)\} = \alpha, 1 - \alpha$ or equivalently for which $r^*(\theta) = z_\alpha, z_{1-\alpha}$. This improves on the likelihood ratio limits, which are found by solving $r(\theta) = z_\alpha, z_{1-\alpha}$.

Example 12.17 (Gamma distribution) Let $Y_1, \ldots, Y_n$ be a gamma random sample with mean θ and known shape parameter v. The log likelihood is then $\ell(\theta; \widehat{\theta}) \equiv -nv(\log\theta + \widehat{\theta}/\theta)$, with $\widehat{\theta} = \overline{Y}$ and observed information $J(\theta; \widehat{\theta}) = nv\theta^{-2}$. Thus the approximate density for $\widehat{\theta}$ is

$$c|J(\widehat{\theta}; \widehat{\theta})|^{1/2} \exp\left\{\ell(\theta; \widehat{\theta}) - \ell(\widehat{\theta}; \widehat{\theta})\right\} = \left(\frac{nv}{2\pi}\right)^{1/2} \frac{1}{\widehat{\theta}} \left(\frac{\widehat{\theta}}{\theta}\right)^{nv} \exp(nv - nv\widehat{\theta}/\theta).$$

Table 12.1 Approximate tail probabilities ($\times 10^2$) corresponding to quantiles of the maximum likelihood estimator of the mean of an exponential variable.

x_p	0.001001	0.01005	0.02532	0.05129	2.996	3.689	4.605	6.908
Exact, $100p$	0.1	1	2.5	5	95	97.5	99	99.9
$100\Phi\{r^*(\theta)\}$	0.104188	1.02937	2.55549	5.08002	94.923	97.454	98.978	99.897
$100\Phi\{r(\theta)\}$	0.029354	0.36040	1.00503	2.21778	90.997	95.189	97.926	99.760
$100\Phi\{v(\theta)\}$	15.88975	16.1099	16.4859	17.1385	97.702	99.642	99.984	100.000

The exact density of $\widehat{\theta}$ is gamma with mean θ and shape parameter $n\nu$, so the approximation is here exact after renormalization; it merely substitutes Stirling's formula for the gamma function that appears in the exact density.

In this example,

$$r(\theta) = \text{sign}(\widehat{\theta} - \theta)[2n\nu\{\widehat{\theta}/\theta - 1 - \log(\widehat{\theta}/\theta)\}]^{1/2}, \quad v(\theta) = (n\nu)^{1/2}(\widehat{\theta} - \theta)/\widehat{\theta},$$

and Table 12.1 shows that (12.33) is essentially exact for this model when $n\nu = 1$. The tail probability approximation based on the signed likelihood ratio statistic $r(\theta)$ is also fairly accurate, while that based on the standardized maximum likelihood estimate $v(\theta)$ is very poor. The accuracy of the tail approximation for $r^*(\theta)$ should carry over to the corresponding confidence intervals. ∎

Conditional inference

Consider now conditional inference for the scalar parameter ψ in the exponential family

$$f(t_1, t_2; \psi, \lambda) = \exp\left\{t_1\psi + t_2^{\mathrm{T}}\lambda - \kappa(\psi, \lambda)\right\} m(t_1, t_2), \tag{12.35}$$

where t_1 is scalar and t_2 has dimension $(p - 1) \times 1$, and λ is treated as a $(p - 1) \times 1$ nuisance parameter. We obtain an approximate conditional density for T_1 given T_2 by double saddlepoint approximation of

$$f(t_1 \mid t_2; \psi) = \frac{f(t_1, t_2; \psi, \lambda)}{f(t_2; \psi, \lambda)}. \tag{12.36}$$

The cumulant-generating function of (T_1, T_2) is

$$K(u) = \kappa(\psi + u_1, \lambda + u_2) - \kappa(\psi, \lambda),$$

where $u^{\mathrm{T}} = (u_1, u_2^{\mathrm{T}})$, u_1 is scalar, and u_2 is a $(p - 1) \times 1$ vector. The saddlepoint equation corresponding to the numerator of (12.36) is

For brevity we let κ_λ denote $\partial\kappa/\partial\lambda$, $\kappa_{\psi\lambda}$ denote $\partial^2\kappa/\partial\psi\partial\lambda^{\mathrm{T}}$, and so forth.

$$t_2 = \frac{\partial K(\psi, \lambda + \tilde{u}_2)}{\partial u_2} = \frac{\partial\kappa(\psi, \widehat{\lambda}_\psi)}{\partial\lambda} = \kappa_\lambda(\widehat{\theta}_\psi),$$

say, where $\widehat{\theta}_\psi^{\mathrm{T}} = (\psi, \widehat{\lambda}_\psi^{\mathrm{T}})$ and $\widehat{\lambda}_\psi$ is the maximum likelihood estimate of λ with ψ fixed, based on t_2; note that $\tilde{u}_2 = \widehat{\lambda}_\psi - \lambda$. The matrix of second derivatives

$$\frac{\partial^2 K(\psi, \lambda + \tilde{u}_2)}{\partial u_2 \partial u_2^{\mathrm{T}}} = \frac{\partial^2\kappa(\psi, \widehat{\lambda}_\psi)}{\partial\lambda\partial\lambda^{\mathrm{T}}} = \kappa_{\lambda\lambda}(\widehat{\theta}_\psi) = J_{\lambda\lambda}(\widehat{\theta}_\psi)$$

is the observed information for λ when ψ is fixed. The saddlepoint equation corresponding to the denominator of (12.36) is

$$\begin{pmatrix} t_1 \\ t_2 \end{pmatrix} = \begin{pmatrix} \frac{\partial K(\psi + \tilde{u}_1, \lambda + \tilde{u}_2)}{\partial u_1} \\ \frac{\partial K(\psi + \tilde{u}_1, \lambda + \tilde{u}_2)}{\partial u_2} \end{pmatrix} = \begin{pmatrix} \kappa_\psi(\widehat{\theta}) \\ \kappa_\lambda(\widehat{\theta}) \end{pmatrix},$$

while the matrix of second derivatives is

$$\begin{pmatrix} \frac{\partial^2 K(\psi + \tilde{u}_1, \lambda + \tilde{u}_2)}{\partial u_1^2} & \frac{\partial^2 K(\psi + \tilde{u}_1, \lambda + \tilde{u}_2)}{\partial u_1 \partial u_2^T} \\ \frac{\partial^2 K(\psi + \tilde{u}_1, \lambda + \tilde{u}_2)}{\partial u_1 \partial u_2} & \frac{\partial^2 K(\psi + \tilde{u}_1, \lambda + \tilde{u}_2)}{\partial u_2 \partial u_2^T} \end{pmatrix} = \begin{pmatrix} \kappa_{\psi\psi}(\widehat{\theta}) & \kappa_{\psi\lambda}(\widehat{\theta}) \\ \kappa_{\lambda\psi}(\widehat{\theta}) & \kappa_{\lambda\lambda}(\widehat{\theta}) \end{pmatrix} = J(\widehat{\theta}),$$

where $\widehat{\theta}^T = (\widehat{\psi}, \widehat{\lambda}^T)$ is the overall maximum likelihood estimate. Substitution of these formulae into the double saddlepoint approximation (12.26) and then reorganization along the lines that leads to (12.32) gives

$$f(t_1 \mid t_2; \psi) \doteq \left\{ \frac{|J_{\lambda\lambda}(\widehat{\theta}_\psi)|}{2\pi |J(\widehat{\theta})|} \right\}^{1/2} \exp\left\{ \ell(\widehat{\theta}_\psi) - \ell(\widehat{\theta}) \right\}, \qquad (12.37)$$

where $\ell(\theta) = \ell(\psi, \lambda) = \log f(t_1, t_2; \psi, \lambda)$ is the overall log likelihood.

Approximate tail probabilities associated with particular values of ψ can be formed in the way outlined after (12.33), and are used to construct confidence intervals. The basis of the tail probability approximation is (12.27), which here depends on ψ, and is

$$\Pr(T_1 \leq t_1 \mid T_2 = t_2; \psi) \doteq \Phi\{r^*(\psi)\}, \qquad (12.38)$$

where $r^*(\psi) = r(\psi) + r(\psi)^{-1} \log\{r(\psi)/v(\psi)\}$, with

$$r(\psi) = \text{sign}(\widehat{\psi} - \psi)[2\{\ell(\widehat{\theta}) - \ell(\widehat{\theta}_\psi)\}]^{1/2}, \quad v(\psi) = (\widehat{\psi} - \psi) \left\{ \frac{|J(\widehat{\theta})|}{|J_{\lambda\lambda}(\widehat{\theta}_\psi)|} \right\}^{1/2}.$$

Thus the improved approximation again involves modifying the signed likelihood ratio statistic, here using a standardized maximum likelihood estimate rather than the score statistic that appeared with the regression-scale model. We can write

$$v(\psi) = (\widehat{\psi} - \psi) \left\{ \frac{|J(\widehat{\theta}_\psi)|}{|J_{\lambda\lambda}(\widehat{\theta}_\psi)|} \right\}^{1/2} \times \left\{ \frac{|J(\widehat{\theta})|}{|J(\widehat{\theta}_\psi)|} \right\}^{1/2} = \gamma \times C, \qquad (12.39)$$

say, yielding a three-part decomposition of $r^*(\psi)$ like (12.18). It is an exercise to show that $r^*(\psi)$ is invariant to interest-preserving reparametrization.

Example 12.18 (Nodal involvement data)   A central issue for the data in Table 10.8 is how nodal involvement depends on the five binary covariates. For purpose of illustration we consider setting confidence intervals for the parameter ψ associated with acid, and regard the parameters for other covariates as incidental. The left panel of Figure 12.8 shows $r(\psi)$ and $r^*(\psi)$ as functions of ψ, when acid only is included, and when all five covariates are included. The small-sample modification is appreciably larger when several nuisance parameters are eliminated, and the standard error, corresponding roughly to the slope, is larger. The right panel shows information and nuisance parameter corrections r_{INF} and r_{NP} for models with 1+acid, 1+stage+xray+acid, and with all five covariates. All three information corrections

Table 12.2 Estimate, standard errors, and 0.95 confidence intervals for the coefficient of acid, ψ, for the nodal involvement data with all the other covariates fitted. The continuity corrected version of $r^*(\psi)$ is obtained by multiplying $v(\psi)$ by $(e^{\widehat{\psi}-\psi} - 1)/(\widehat{\psi} - \psi)$.

Method	Estimate (SE)	Interval
Normal approximation to $\widehat{\psi}$	1.68 (0.79)	(0.136, 3.232)
Normal approximation to $\widehat{\psi}_a$	1.49 (0.74)	(0.048, 2.930)
Directed deviance $r(\psi)$	—	(0.209, 3.378)
Modified directed deviance $r^*(\psi)$	—	(0.086, 2.998)
Modified directed deviance $r^*(\psi)$ with continuity correction	—	(-0.131, 3.330)

Figure 12.8 Conditional inference for coefficient of acid, ψ, for nodal involvement data. Left: signed likelihood ratio statistic $r(\psi)$ (dashes) and modified version $r^*(\psi)$ (solid) for models with five nuisance parameters (left pair of curves) and with one nuisance parameter (right pair of curves, shifted right by one unit). The dotted horizontal lines are at 0, ±1.96. Right: information corrections r_{INF} (dashes) and nuisance parameter corrections r_{NP} (solid) for models with one, three, and five nuisance parameters. The r_{NP} increase in size with the number of nuisance parameters eliminated. The dotted horizontal lines at ±0.2 show a rule of thumb for substantial corrections.

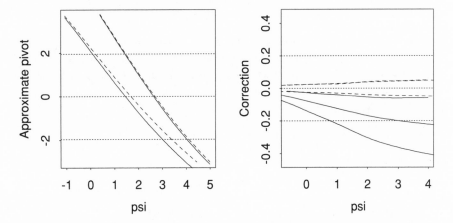

are small, but the nuisance parameter correction increases sharply as more parameters are eliminated.

An approximate conditional maximum likelihood estimate $\widehat{\psi}_a$ is obtained as the solution to $r^*(\psi) = 0$, with standard error given by the slope of $r^*(\psi)$ at that point. Together these yield a confidence interval for ψ based on normal approximation. Table 12.2 shows that $\widehat{\psi}_a$ has smaller magnitude and standard error than has the ordinary maximum likelihood estimate $\widehat{\psi}$. Small-sample adjustment shortens confidence intervals and moves them closer to zero. In such cases a continuity correction can be applied to give results closer to those from exact conditioning. This gives appreciably wider confidence intervals, as the discussion in Example 7.38 would suggest. ∎

Curved exponential family

Our previous discussion has applied to linear exponential families, but curved exponential family models are also important in applications. The key difficulty in developing small-sample inferences is then to find an analogue of the quantities (12.17) and (12.39) that modify the signed likelihood ratio statistic. An exact expression is unavailable but one of several possible approximations is

$$v(\psi) \doteq |I(\widehat{\theta})|^{-1} |C(\widehat{\theta}, \widehat{\theta}_\psi)| \left\{ |J(\widehat{\theta})| / |J_{\lambda\lambda}(\widehat{\theta}_\psi)| \right\}^{1/2}, \qquad (12.40)$$

with $I(\theta)$ the expected information and $C(\theta, \theta_0)$ denoting the $p \times p$ matrix

$$\begin{pmatrix} \mathrm{cov}_0 \left\{ \ell(\theta_0) - \ell(\theta), \ell_\theta(\theta_0) \right\} \\ \mathrm{cov}_0 \left\{ \ell_\lambda(\theta), \ell_\theta(\theta_0) \right\} \end{pmatrix}.$$

Here $\ell_\theta(\theta) = \partial\ell(\theta)/\partial\theta$, and so forth, and cov_0 denotes covariance taken with respect to the model with parameter value θ_0. Expression (12.40) admits a decomposition into two parts like that at (12.39). It requires neither that an ancillary statistic be specified nor does it involve sample space derivatives, but it does entail a loss of precision relative to (12.39), use of which in continuous models gives relative error of sizes $O(n^{-3/2})$ in moderate deviation regions and $O(n^{-1/2})$ in large deviation regions. By contrast use of (12.40) reduces the relative error in moderate deviation regions to $O(n^{-1})$. The key point, however, is that the relative error properties, which give highly accurate approximations far into the tails of the distribution of the modified signed likelihood ratio statistic $R^*(\psi)$, are preserved.

Example 12.19 (Nonlinear model)   Consider a model in which the $n \times 1$ vector of responses $Y \sim N_n(\eta, \sigma^2 I_n)$ and the $n \times 1$ mean vector $\eta = \eta(\beta)$ depends on a $p \times 1$ parameter vector β. Let the parameter of interest ψ be an element of β, β_1, say, let λ denote the $p \times 1$ vector comprising the remaining elements of β and σ^2, and let $\theta^{\mathrm{T}} = (\psi, \lambda^{\mathrm{T}})$. The log likelihood,

$$\ell(\theta) \equiv -\tfrac{1}{2}\left\{ n\log\sigma^2 + (y - \eta)^{\mathrm{T}}(y - \eta)/\sigma^2 \right\},$$

has score vector given by

$$\ell_\theta(\theta)^{\mathrm{T}} = \left(\sigma^{-2}(y - \eta)^{\mathrm{T}}\eta_\beta, \left\{ (y - \eta)^{\mathrm{T}}(y - \eta) - n\sigma^2 \right\}/(2\sigma^4) \right)$$

where η_β is the $n \times p$ matrix $\partial\eta/\partial\beta^{\mathrm{T}}$. On using the decomposition

$$\ell(\theta) = -\tfrac{1}{2}\left\{ n\log\sigma^2 + (y - \eta_0 + \eta_0 - \eta)^{\mathrm{T}}(y - \eta_0 + \eta_0 - \eta)/\sigma^2 \right\}$$
$$= -\tfrac{1}{2}\left\{ (y - \eta_0)^{\mathrm{T}}(y - \eta_0) + 2(y - \eta_0)^{\mathrm{T}}(\eta_0 - \eta) \right\}/\sigma^2 + d,$$

where d depends only on the parameters and subscript 0 indicates that a quantity is evaluated at $\theta_0 = (\beta_0, \sigma_0^2)$, we find

$$\mathrm{cov}_0\left\{ \ell(\theta), \ell_\theta(\theta_0) \right\} = -\sigma^{-2}\left((\eta_0 - \eta)^{\mathrm{T}}\eta_{\beta 0}, n/2 \right),$$

from which we see that the first row of $C(\theta, \theta_0)$ equals

$$\left((\eta_0 - \eta)^{\mathrm{T}}\eta_{\beta 0}/\sigma^2, n(\sigma^{-2} - \sigma_0^{-2})/2 \right),$$

and a similar calculation shows that the remaining $p \times (p + 1)$ submatrix is

$$\mathrm{cov}_0\left\{ \ell_\lambda(\theta), \ell_\theta(\theta_0) \right\} = \begin{pmatrix} \sigma^{-2}\eta_{\beta_2}^{\mathrm{T}}\eta_{\beta 0} & 0 \\ \sigma^{-4}(\eta_0 - \eta)^{\mathrm{T}}\eta_{\beta 0} & n/(2\sigma^4) \end{pmatrix}.$$

Thus the approximation to $v(\psi)$ involves this matrix, evaluated with $\theta_0 = \widehat{\theta}$ and $\theta = \widehat{\theta}_\psi$, and the other information matrices. In this case $I(\widehat{\theta}) = J(\widehat{\theta})$.   ∎

Example 12.20 (Calcium data)   For numerical illustration we consider the data of Example 10.1, to which we fit a normal model with constant variance σ^2 and mean $\beta_0\{1 - \exp(-x/\beta_1)\}$, where x represents time in minutes. First-order inference was discussed in Example 10.9.

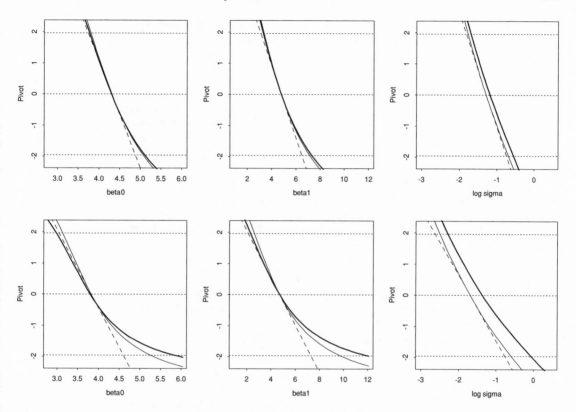

Figure 12.9 Conditional inference for parameters of nonlinear regression model for calcium data. Each panel shows how the signed likelihood ratio statistic (solid), the modified signed likelihood ratio statistic (heavy), and the standardized maximum likelihood estimate (dashes) depend on the corresponding parameter. The dotted horizontal lines are at 0, ±1.96. Upper panels: full data; lower panels: just one observation at each time.

The result of applying the computations in Example 12.19 are given in Figure 12.9, which shows, for each parameter of interest ψ, the quantities $r(\psi)$, $r^*(\psi)$ and the pivot $(\widehat{\psi} - \psi)/s_\psi$ from normal approximation to the maximum likelihood estimator; s_ψ is the standard error for $\widehat{\psi}$ based on the observed information matrix. Results are given for the full dataset, and for a reduced dataset comprising just the nine responses in the final column of Table 10.1. The shallower slopes in the lower panels show the effect of reducing the sample size.

Normal approximation is clearly inadequate, particularly at the upper limits of confidence intervals, but the ordinary unmodified signed likelihood ratio statistic yields intervals for β_0 and β_1 that are reasonably close to those from the modified version, at least for the full dataset. The unmodified estimates of $\log \sigma$ are strongly biased downwards, and the higher-order procedures give worthwhile improvements. Much of this bias may be removed by dividing the residual sum of squares by denominator $n - p$ rather than n.

It seems wise to use the modified statistics for all three parameters with the smaller dataset. The 0.95 confidence intervals based on $r(\beta_1)$ and $r^*(\beta_1)$ are (2.63, 9.75) and (2.31, 12.02), for instance, while the corresponding intervals for σ are (0.08, 0.55) and (0.10, 0.94), whose right tails differ substantially. ∎

Variants have been proposed that replace the expectations by averages in order to give a purely empirical version of (12.40); see the bibliographic notes.

Exercises 12.3

1   Let Y and X be independent exponential variables with means $1/(\lambda + \psi)$ and $1/\lambda$. Find the distribution of Y given $X + Y$ and show that when $\psi = 0$ it has mean $s/2$ and variance $s^2/12$. Construct an exact conditional test of the hypothesis $E(Y) = E(X)$.

2   A discrete exponential family for independent pairs $(S_1, T_1), \ldots, (S_n, T_n)$ has form

$$\exp\left\{s_j\lambda_j + t_j\psi_j + \kappa(\lambda_j, \psi_j)\right\} c_j(s_j, t_j), \quad j = 1, \ldots, n.$$

It is intended to test the hypothesis $\psi_1 = \cdots = \psi_n$ regardless of the values of the λ_j. Generalize Example 12.14 to explain how to perform Monte Carlo exact significance tests. Give the acceptance probability for the Metropolis–Hastings algorithm, and write out the algorithm when

$$c_j(s_j, t_j) = \{(s_j - t_j)!t_j!\}^{-1}, \quad s_j = t_j, t_j + 1, \ldots, \quad t_j = 0, 1, \ldots.$$

How does the argument change if the variables are continuous?

3   Show that the saddlepoint approximations for the density and distribution of the average of a normal random sample are exact.

4   Show that saddlepoint approximation for the inverse Gaussian density

$$f(y; \mu, \lambda) = \left(\frac{\lambda}{2\pi y^3}\right)^{1/2} \exp\left\{-\lambda(y - \mu)^2/(2\mu^2 y)\right\}, \quad y > 0, \lambda, \mu > 0,$$

is exact after renormalization. Investigate the accuracy of the distribution function approximation (12.23) when $n = 1$.

5   Consider independent exponential observations Y_j with means $(x_j^{\mathsf{T}}\beta)^{-1}$, for $j = 1, \ldots, n$, where the x_j are $p \times 1$ vectors of explanatory variables and the $p \times 1$ parameter vector β is unknown. Find the ingredients of (12.37) and (12.38) when inference is required for $\psi = \beta_1$, with $\lambda = (\beta_2, \ldots, \beta_p)$ treated as incidental.
    Give the exact conditional likelihood for ψ when $Y_1, \ldots, Y_{n-1}$ have mean λ^{-1} and Y_n has mean $(\lambda + \psi)^{-1}$, and investigate the accuracy of (12.38) when $n = 2$.

6   Show that (12.40) retrieves (12.39) for a linear exponential family (12.35).

7   Check the details of Examples 12.19 and 12.20. Find (12.40) when σ^2 is of interest.

12.4 Modified Profile Likelihood

12.4.1 Likelihood adjustment

The profile log likelihood is a standard tool for inference in large-sample situations, and it is natural to consider if and how it may be modified for use in small-sample problems. We saw at (12.13), for instance, that a log marginal likelihood for parameters ψ controlling the variance of a normal distribution could be obtained by adding a term to the profile log likelihood, while (12.37) suggests that an approximate conditional likelihood for a linear exponential family model has form

$$\ell(\psi, \widehat{\lambda}_\psi) + \tfrac{1}{2} \log|J_{\lambda\lambda}(\psi, \widehat{\lambda}_\psi)|,$$

where $\widehat{\lambda}_\psi$ is the maximum likelihood estimate of λ for fixed ψ and $J_{\lambda\lambda}(\psi, \lambda)$ is the corner of the observed information matrix corresponding to λ. This amounts to a penalization of the log likelihood by an amount that depends on the information available for λ; when this is large, the profile log likelihood is more strongly penalized than when it is small.

The form of these expressions suggests that in general settings we multiply the profile likelihood

$$L_\mathrm{p}(\psi) = \exp\left\{\ell_\mathrm{p}(\psi)\right\} = \exp\left\{\ell(\psi, \widehat{\lambda}_\psi)\right\}$$

by a cunningly chosen function of ψ, giving a *modified profile likelihood*

$$L_\mathrm{mp}(\psi) = \exp\left\{\ell_\mathrm{mp}(\psi)\right\} = M(\psi)L_\mathrm{p}(\psi). \tag{12.41}$$

It is natural to try and choose $M(\psi)$ so that $L_\mathrm{mp}(\psi)$ gives inferences equivalent to using a marginal or conditional likelihood for ψ, if such is available. It is a remarkable fact that taking

$$M(\psi) = \left| J_{\lambda\lambda}(\psi, \widehat{\lambda}_\psi) \right|^{-1/2} \left| \frac{\partial\widehat{\lambda}}{\partial\widehat{\lambda}_\psi^\mathrm{T}} \right| \tag{12.42}$$

achieves this to a high degree of accuracy in some generality. The first term of (12.42) involves the part of the observed information matrix mentioned above, while the second term is a Jacobian needed if the modified profile likelihood is to be invariant to interest-preserving transformations: typically this term depends on the parameter ψ. A derivation of (12.42) is given after Example 12.18. Here is a toy example.

Example 12.21 (Normal linear model) Let the parameter of interest in the normal linear model be the variance σ^2, with the $p \times 1$ vector β treated as incidental. The log likelihood is

$$\ell(\beta, \sigma^2) \equiv -\frac{n}{2}\log\sigma^2 - \frac{1}{2\sigma^2}(y - X\beta)^\mathrm{T}(y - X\beta),$$

and the maximum likelihood estimator of β for σ^2 fixed is $\widehat{\beta}_{\sigma^2} = (X^\mathrm{T}X)^{-1}X^\mathrm{T}y$. As this is independent of σ^2, we have $\widehat{\beta} = \widehat{\beta}_{\sigma^2}$. Thus

$$J_{\beta\beta}(\beta, \sigma^2) = \sigma^{-2}X^\mathrm{T}X, \quad \frac{\partial\widehat{\beta}_{\sigma^2}^\mathrm{T}}{\partial\widehat{\beta}} = I_p, \quad M(\sigma^2) = (\sigma^2)^{p/2}|X^\mathrm{T}X|^{-1/2},$$

and the modified profile log likelihood is

$$\begin{aligned}
\ell_\mathrm{mp}(\sigma^2) &\equiv -\frac{n - p}{2}\log\sigma^2 - \frac{1}{2\sigma^2}(y - X\widehat{\beta})^\mathrm{T}(y - X\widehat{\beta}) \\
&= -\frac{n - p}{2}\left(\log\sigma^2 - S^2/\sigma^2\right),
\end{aligned}$$

where S^2 is the unbiased estimator of σ^2. Hence (12.41) produces inferences identical to those based on the marginal distribution of the residual sum of squares, or equivalently the marginal likelihood for σ^2; see Example 12.9.

In fact $|\partial\widehat{\beta}_{\sigma^2}^\mathrm{T}/\partial\widehat{\beta}| = 1$ in all regression-scale models (Exercise 12.4.2). ∎

Computation of (12.41) demands knowledge of the two terms of (12.42). In applications the first is easily found by numerical or analytical differentation of $\ell(\psi, \lambda)$, but the second is a sample space derivative and approximations to it are usually needed. When the log likelihood can be written in terms of the maximum likelihood estimates $\widehat{\psi}$ and $\widehat{\lambda}$ and an ancillary statistic a, however, the equation determining $\widehat{\lambda}_\psi$ may be expressed as

$$\frac{\partial \ell(\psi, \widehat{\lambda}_\psi; \widehat{\psi}, \widehat{\lambda}, a)}{\partial \lambda} = 0,$$

and partial differentiation with respect to $\widehat{\lambda}$, holding ψ, $\widehat{\psi}$, and a fixed, yields

$$\frac{\partial^2 \ell(\psi, \widehat{\lambda}_\psi; \widehat{\psi}, \widehat{\lambda}, a)}{\partial \lambda \partial \lambda^{\mathsf{T}}} \frac{\partial \widehat{\lambda}_\psi^{\mathsf{T}}}{\partial \widehat{\lambda}} + \frac{\partial^2 \ell(\psi, \widehat{\lambda}_\psi; \widehat{\psi}, \widehat{\lambda}, a)}{\partial \lambda \partial \widehat{\lambda}^{\mathsf{T}}} = 0.$$

This gives the alternative and often more convenient expression

$$\left| \frac{\partial \widehat{\lambda}_\psi^{\mathsf{T}}}{\partial \widehat{\lambda}} \right| = \left| J_{\lambda\lambda}(\psi, \widehat{\lambda}_\psi; \widehat{\psi}, \widehat{\lambda}, a) \right|^{-1} \left| \frac{\partial^2 \ell(\psi, \widehat{\lambda}_\psi; \widehat{\psi}, \widehat{\lambda}, a)}{\partial \lambda \partial \widehat{\lambda}^{\mathsf{T}}} \right|. \tag{12.43}$$

Although the ancillary must be held fixed when differentiating with respect to $\widehat{\lambda}$, it is not needed explicitly, and in some cases (12.43) can be obtained without the burden of specifying a.

Example 12.22 (Linear exponential family)   In a linear exponential family with log likelihood expressed as

$$\ell(\psi, \lambda) \equiv t_1^{\mathsf{T}} \psi + t_2^{\mathsf{T}} \lambda - \kappa(\psi, \lambda)$$

there is no ancillary statistic and the maximum likelihood estimates $\widehat{\psi}$ and $\widehat{\lambda}$ are solutions of the equations

$$t_1 = \kappa_\psi(\widehat{\psi}, \widehat{\lambda}), \quad t_2 = \kappa_\lambda(\widehat{\psi}, \widehat{\lambda}).$$

Thus the log likelihood may be written

$$\ell(\psi, \lambda; \widehat{\psi}, \widehat{\lambda}) = \kappa_\psi(\widehat{\psi}, \widehat{\lambda})^{\mathsf{T}} \psi + \kappa_\lambda(\widehat{\psi}, \widehat{\lambda})^{\mathsf{T}} \lambda - \kappa(\psi, \lambda),$$

and (12.43) implies that

$$\left| \frac{\partial \widehat{\lambda}_\psi}{\partial \widehat{\lambda}} \right| = \left| \kappa_{\lambda\lambda}(\psi, \widehat{\lambda}_\psi) \right|^{-1} \left| \kappa_{\lambda\lambda}(\widehat{\psi}, \widehat{\lambda}) \right| = \left| J_{\lambda\lambda}(\psi, \widehat{\lambda}_\psi; \widehat{\psi}, \widehat{\lambda}) \right|^{-1} \left| J_{\lambda\lambda}(\widehat{\psi}, \widehat{\lambda}; \widehat{\psi}, \widehat{\lambda}) \right|.$$

Thus the modified profile log likelihood for ψ is

$$\ell_{\mathrm{mp}}(\psi) \equiv \ell_{\mathrm{p}}(\psi) + \tfrac{1}{2} \log \left| J_{\lambda\lambda}(\psi, \widehat{\lambda}_\psi; \widehat{\psi}, \widehat{\lambda}) \right|,$$

where terms independent of ψ have been neglected. Hence (12.41) and (12.42) do indeed retrieve the approximate conditional likelihood (12.37), apart from constants of proportionality.   ∎

Inference on ψ is performed by treating (12.41) as a likelihood. Its maximizing value $\widehat{\psi}_{\mathrm{mp}}$ yields a confidence interval for ψ, for example by normal approximation with variance based on the second derivative of $\ell_{mp}(\psi)$ at $\widehat{\psi}_{mp}$, or by chi-squared approximation to the distribution of $2\{\ell_{mp}(\widehat{\psi}_{mp}) - \ell_{mp}(\psi)\}$. The maximum likelihood estimator based on L_{mp} typically has better properties than $\widehat{\psi}$, as is the case in Example 12.21, but likelihood modification is not a universal panacea. One difficulty is that the underlying inferential basis is first-order distributional results such as normal approximation to the distribution of the maximum modified likelihood estimator rather than more accurate forms such as (12.38). Another difficulty is that modification may not be enough to remove inconsistency of maximum likelihood estimators.

Example 12.23 (Binary matched pairs) Consider matched pairs of binary observations R_{0j}, R_{1j}, with success probabilities

$$\pi_{0j} = \frac{e^{\lambda_j}}{1 + e^{\lambda_j}}, \quad \pi_{1j} = \frac{e^{\lambda_j + \psi}}{1 + e^{\lambda_j + \psi}}, \quad j = 1, \ldots, n.$$

This logistic regression model is a linear exponential family with minimal sufficient statistic $S_1, \ldots, S_n, T$, where $S_j = R_{0j} + R_{1j}$ is associated with λ_j and $T = \sum R_{1j}$ with ψ. We suppose that ψ is finite, and compare its maximum likelihood estimators based on the conditional, modified profile, and usual likelihoods as $n \to \infty$.

Pairs for which $S_j = 0$ or 2 are uninformative (Problem 12.11), and we suppose that they have already been dropped from the analysis, so all n pairs are discordant, that is, $S_1 = \cdots = S_n = 1$.

The exact conditional likelihood for ψ is obtained from the conditional density of $T = \sum R_{1j}$ given that $S_1 = \cdots = S_n = 1$. Conditional on $S_j = 1$, R_{1j} is a Bernoulli variable with success probability $\pi = e^\psi / (1 + e^\psi)$, so T has a binomial distribution with denominator n and probability π. Thus the conditional log likelihood is $\ell_c(\psi) \equiv T\psi - n\log(1 + e^\psi)$, which is maximized at $\widehat{\psi}_c = \log\{T/(n - T)\} \xrightarrow{P} \psi$ as $n \to \infty$: hence $\widehat{\psi}_c$ is consistent.

The overall log likelihood is

$$\ell(\psi, \lambda) \equiv T\psi + \sum_{j=1}^{n} \left\{ \lambda_j - \log(1 + e^{\lambda_j}) - \log(1 + e^{\psi + \lambda_j}) \right\},$$

and the values of λ_j that maximise $\ell(\psi, \lambda)$ for fixed ψ all equal $\widehat{\lambda}_\psi = -\frac{1}{2}\psi$. Thus the profile log likelihood is

$$\ell_p(\psi) = T\psi - 2n\log(1 + e^{\psi/2}).$$

It is straightforward to see that the maximum likelihood estimator $\widehat{\psi} \xrightarrow{P} 2\psi$ as $n \to \infty$ (Problem 12.11). Thus $\widehat{\psi}$ is inconsistent.

Differentiation of $\ell(\psi, \lambda)$ establishes that $J_{\lambda\lambda}(\psi, \widehat{\lambda}_\psi)$ is a $n \times n$ matrix with elements $2\gamma/(1 + \gamma)^2$ on the diagonal and zeros elsewhere, where $\gamma = \exp(\psi/2)$. Hence the modified profile log likelihood is

$$\ell_{mp}(\psi) \equiv \tfrac{1}{4}(n + 4T)\psi - 3n\log(1 + e^{\psi/2}),$$

ψ	0	0.5	1	1.5	2	2.5	3	4	5
Limit of $\widehat{\psi}_c$	0	0.5	1	1.5	2	2.5	3	4	5
Limit of $\widehat{\psi}_{mp}$	0	0.66	1.27	1.81	2.23	2.56	2.79	3.05	3.16
Limit of $\widehat{\psi}$	0	1	2	3	4	5	6	8	10

Table 12.3 Probability limits of ordinary, conditional, and maximum modified likelihood estimators of log odds ratio ψ in binary matched pairs.

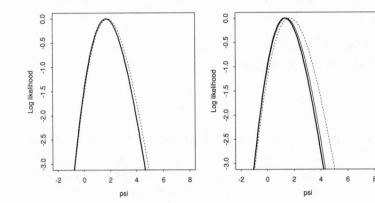
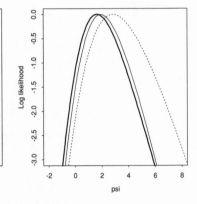

which is maximized at $\widehat{\psi}_{mp} = 2\log\{(1 + 4T/n)/(5 - 4T/n)\}$. Now $T/n \xrightarrow{P} \pi$ as $n \to \infty$, so $\widehat{\psi}_{mp} \xrightarrow{P} 2\log\{(1 + 5e^{\psi})/(5 + e^{\psi})\}$.

Table 12.3 compares these limiting values. Although $\widehat{\psi}_{mp}$ is inconsistent, it is not horribly biased in the range $|\psi| < 3$ usually met in applications: the modification is partly but not wholly successful in eliminating bias. ∎

Figure 12.10 Likelihood inference for nodal involvement data. Left panel: conditional (heavy), modified profile (solid) and profile log likelihoods (dots) for model with terms 1+acid. Centre and right panels: corresponding log likelihoods with terms 1+stage+xray+acid and with all five covariates.

Example 12.24 (Nodal involvement data) For numerical illustration of likelihood modification we fit logistic regression models to the rows of Table 10.8 with $m = 2$ or $m = 1$: this gives 17 binary responses in all. We let ψ be the parameter corresponding to acid, and fit models with terms 1+acid, 1+stage+xray+acid, and with all five covariates; thus the nuisance parameter λ has dimensions $q = 1$, 3, and 5.

Figure 12.10 shows the conditional, modified profile, and profile log likelihoods for these three models. The conditional likelihoods were obtained by symbolic computation of the necessary generating functions; see Example 12.13. The profile log likelihoods are quite different from the conditional and modified profile log likelihoods. The effect of modification depends on the number of nuisance parameters; it works almost perfectly when $q = 1$ but less well otherwise. In the right panel five nuisance parameters are eliminated from a likelihood based on just 17 binary observations, and it is perhaps surprising that modification works so well.

Table 12.4 compares the corresponding estimates and confidence intervals. The difference between results based on the profile likelihood and conditional and modified profile likelihoods increases sharply with the number of nuisance parameters, but the difference between these two last sets of results remains modest. ∎

Table 12.4 Estimates, standard errors, and 0.95 confidence intervals based on profile, modified profile and conditional likelihoods for reduced nodal data having 17 binary responses, with $q = 1, 3$ and 5 nuisance parameters.

	Estimate (SE)		Confidence interval		
q	Profile	Modified	Profile	Modified	Conditional
1	1.79 (1.08)	1.68 (1.04)	$(-0.21, 4.15)$	$(-0.28, 3.86)$	$(-0.26, 3.92)$
3	1.72 (1.13)	1.43 (1.02)	$(-0.41, 4.21)$	$(-0.54, 3.51)$	$(-0.53, 3.48)$
5	2.83 (1.61)	1.91 (1.15)	$(0.13, 6.90)$	$(-0.39, 4.57)$	$(-0.44, 4.71)$

Derivation of (12.42)

Consider a model with interest and nuisance parameters ψ and λ, and for which the data $y \mapsto (\widehat{\psi}, \widehat{\lambda}, a)$, where a is ancillary. If a factorization

$$f(\widehat{\psi}, \widehat{\lambda} \mid a; \psi, \lambda) = f(\widehat{\psi} \mid a; \psi) f(\widehat{\lambda} \mid \widehat{\psi}, a; \psi, \lambda) \qquad (12.44)$$

holds, then a marginal likelihood for ψ may be based on the first term on the right. We now apply the p^* formula to the left-hand density, giving

$$(2\pi)^{-p/2} \overline{c}_1(\psi, \lambda, a) \left| J(\widehat{\psi}, \widehat{\lambda}) \right|^{1/2} \exp \left\{ \ell(\psi, \lambda) - \ell(\widehat{\psi}, \widehat{\lambda}) \right\}, \qquad (12.45)$$

where (ψ, λ) has dimension p and $\overline{c}_1(\psi, \lambda, a) = 1 + O_p(n^{-1})$. To approximate to $f(\widehat{\lambda} \mid \widehat{\psi}, a; \psi, \lambda)$, we note that with ψ, $\widehat{\psi}$, and a held fixed we have

$$f(\widehat{\lambda} \mid \widehat{\psi}, a; \psi, \lambda) = f(\widehat{\lambda}_\psi \mid \widehat{\psi}, a; \psi, \lambda) \left| \frac{\partial \widehat{\lambda}_\psi}{\partial \widehat{\lambda}} \right|,$$

and apply the p^* formula to the density on the right, giving

$$(2\pi)^{-q/2} \overline{c}_2(\psi, \lambda, a) \left| J_{\lambda\lambda}(\psi, \widehat{\lambda}_\psi) \right|^{1/2} \exp \left\{ \ell(\psi, \lambda) - \ell(\psi, \widehat{\lambda}_\psi) \right\} \left| \frac{\partial \widehat{\lambda}_\psi}{\partial \widehat{\lambda}} \right|, \qquad (12.46)$$

where q is the dimension of λ and $\overline{c}_2(\psi, \lambda, a) = 1 + O_p(n^{-1})$. On substituting (12.45) and (12.46) into (12.44) and rearranging we find that $f(\widehat{\psi} \mid a; \psi)$ equals

$$(2\pi)^{-(p-q)/2} \overline{c}(\psi, \lambda, a) \frac{\left| J(\widehat{\psi}, \widehat{\lambda}) \right|^{1/2}}{\left| J_{\lambda\lambda}(\psi, \widehat{\lambda}_\psi) \right|^{1/2}} \exp \left\{ \ell(\psi, \widehat{\lambda}_\psi) - \ell(\widehat{\psi}, \widehat{\lambda}) \right\} \left| \frac{\partial \widehat{\lambda}}{\partial \widehat{\lambda}_\psi} \right|,$$

where $\overline{c}(\psi, \lambda, a) = \overline{c}_1(\psi, \lambda, a)/\overline{c}_2(\psi, \lambda, a) = 1 + O(n^{-1})$. Thus

$$f(\widehat{\psi} \mid a; \psi) \propto \exp \left\{ \ell_p(\psi) \right\} M(\psi) \left\{ 1 + O(n^{-1}) \right\}, \qquad (12.47)$$

as a function of ψ, with $M(\psi)$ given by (12.42). It can be shown that the error term is $O(n^{-3/2})$ in a moderate deviation region, that is, when $\widehat{\psi}$ differs from the true ψ by only $O(n^{-1/2})$.

Exercise 12.4.4 derives (12.47) as an approximate conditional likelihood.

12.4.2 Parameter orthogonality

This section demands a nodding acquaintance with partial differential equations.

In applications it is rarely possible to find an explicit expression for the second term of (12.42). Approximations to it are available in certain cases, but when they are not, it is natural to seek to reduce the importance of that term. If $\widehat{\lambda}_\psi$ is independent of ψ, then $\widehat{\lambda}_\psi = \widehat{\lambda}$ for all ψ, and $|\partial\widehat{\lambda}/\partial\widehat{\lambda}_\psi|$ plays no part in the likelihood modification. Although this occurs only in particular models, it suggests that we seek to reduce the dependence of $\widehat{\lambda}_\psi$ on ψ more generally. One approach to this is through orthogonal parameters.

To motivate subsequent discussion, let $\bar{\ell} = n^{-1}\ell$ denote the log likelihood, standardized to be of order one, and note that $\widehat{\lambda}_\psi$ is determined by the equation $\bar{\ell}_\lambda(\psi, \widehat{\lambda}_\psi) = 0$. If we suppose that $\widehat{\psi} - \psi = O(n^{-1/2})$, then Taylor series expansion around the overall maximum likelihood estimator gives

$$0 = \bar{\ell}_\lambda(\widehat{\psi}, \widehat{\lambda}) + \bar{\ell}_{\lambda\psi}(\widehat{\psi}, \widehat{\lambda})(\psi - \widehat{\psi}) + \bar{\ell}_{\lambda\lambda}(\widehat{\psi}, \widehat{\lambda})(\widehat{\lambda}_\psi - \widehat{\lambda}) + O_p(n^{-1}),$$

where second and higher derivatives such as $\bar{\ell}_{\lambda\psi}(\widehat{\psi}, \widehat{\lambda})$ are $O_p(1)$. Hence

$$\begin{aligned}
\widehat{\lambda}_\psi - \widehat{\lambda} &= \bar{\ell}_{\lambda\lambda}(\widehat{\psi}, \widehat{\lambda})^{-1}\bar{\ell}_{\lambda\psi}(\widehat{\psi}, \widehat{\lambda})(\widehat{\psi} - \psi) + O_p(n^{-1}) \\
&= J_{\lambda\lambda}(\widehat{\psi}, \widehat{\lambda})^{-1}J_{\lambda\psi}(\widehat{\psi}, \widehat{\lambda})(\widehat{\psi} - \psi) + O_p(n^{-1}) \\
&= I_{\lambda\lambda}(\psi, \lambda)^{-1}I_{\lambda\psi}(\psi, \lambda)(\widehat{\psi} - \psi) + O_p(n^{-1}),
\end{aligned}$$

where $I_{\lambda\lambda}$ and $I_{\lambda\psi}$ are components of the expected information matrix. In regular models these and the corresponding observed information quantities are of order n, so $\widehat{\lambda}_\psi - \widehat{\lambda}$ will be of precise order $n^{-1/2}$ unless the model is set up so that $J_{\lambda\psi}(\widehat{\psi}, \widehat{\lambda})$ or $I_{\lambda\psi}(\psi, \lambda)$ vanishes. If this can be arranged, then $\widehat{\lambda}_\psi$ differs from $\widehat{\lambda}$ by less than $O_p(n^{-1/2})$ and the asymptotic dependence of $\widehat{\lambda}_\psi$ on ψ is reduced.

To be more explicit, suppose $I_{\lambda\psi}(\psi, \lambda) = 0$ at the true parameter value. Then $\widehat{\lambda} = \widehat{\lambda}_\psi + O_p(n^{-1})$, and it follows that the term $|\partial\widehat{\lambda}/\partial\widehat{\lambda}_\psi^{\mathrm{T}}|$ that appears in (12.42) equals $1 + O_p(n^{-1})$, in contrast to the value $1 + O_p(n^{-1/2})$ typically obtained. It then seems reasonable to hope that little damage will be done by dropping the Jacobian term from (12.42) and approximating $\ell_{\mathrm{mp}}(\psi)$ by

$$\ell(\psi, \widehat{\lambda}_\psi) - \tfrac{1}{2}\log\left|J_{\lambda\lambda}(\psi, \widehat{\lambda}_\psi)\right|. \tag{12.48}$$

This argument has a serious drawback, because knowledge that a term is $O_p(n^{-1})$ gives no notion of its actual size: $c\psi/n$ is $O(n^{-1})$ both when $c = 100$ and $c = 0.01$, but the numerical values are very different. Asymptotic arguments are valuable heuristics but cannot ensure accuracy in applications. With this in mind, we nevertheless press forward with vigour.

If $I_{\lambda\psi}(\psi, \lambda) = 0$ for all (ψ, λ), then the parameters λ and ψ are said to be *orthogonal*. Among the consequences of this is that the inverse information matrix is block diagonal, so the maximum likelihood estimators $\widehat{\psi}$ and $\widehat{\lambda}$ are asymptotically independent, and the asymptotic standard error for $\widehat{\psi}$ when λ is unknown is the same as when it is known. Another advantage of is that likelihood maximization may be numerically more stable in the orthogonal parametrization. Below we briefly consider the implications of parameter orthogonality for likelihood modification, first outlining how to obtain parameters orthogonal to a given interest parameter.

Consider a model with log likelihood $\ell^*(\psi, \gamma)$ in terms of the scalar interest parameter ψ and an arbitrary nuisance parameter $\gamma = (\gamma_1, \ldots, \gamma_q)^{\mathrm{T}}$. We seek $\lambda = \lambda(\psi, \gamma)$

such that λ is orthogonal to ψ. Writing $\gamma = \gamma(\psi, \lambda)$, we have

$$\ell(\psi, \lambda) = \ell^* \{\psi, \gamma(\psi, \lambda)\},$$

and differentiation with respect to ψ and λ yields

$$\frac{\partial^2 \ell}{\partial \lambda \partial \psi} = \frac{\partial \gamma^\mathsf{T}}{\partial \lambda} \frac{\partial^2 \ell^*}{\partial \gamma \partial \psi} + \frac{\partial \gamma^\mathsf{T}}{\partial \lambda} \frac{\partial^2 \ell^*}{\partial \gamma \partial \gamma^\mathsf{T}} \frac{\partial \gamma}{\partial \psi} + \frac{\partial^2 \gamma^\mathsf{T}}{\partial \lambda \partial \psi} \frac{\partial \ell^*}{\partial \gamma}.$$

If λ and ψ are to be orthogonal, this expression must have expectation zero. Hence

$$0 = \frac{\partial \gamma^\mathsf{T}}{\partial \lambda} I^*_{\gamma\psi} + \frac{\partial \gamma^\mathsf{T}}{\partial \lambda} I^*_{\gamma\gamma} \frac{\partial \gamma}{\partial \psi},$$

where $I^*_{\gamma\psi}$ and $I^*_{\gamma\gamma}$ are components of the expected information matrix in the non-orthogonal parametrization. Thus provided that the $q \times q$ matrix $\partial \gamma^\mathsf{T} / \partial \lambda$ is invertible and taking for granted the necessary regularity conditions, we see that $\lambda(\psi, \gamma)$ is a solution of the system of q partial differential equations

$$\frac{\partial \gamma}{\partial \psi} = -I^{*-1}_{\gamma\gamma}(\psi, \gamma) I^*_{\gamma\psi}(\psi, \gamma). \tag{12.49}$$

There is latitude in the choice of λ, because if λ and ψ are orthogonal, then any smooth functions of ψ and of λ are orthogonal. Thus λ can in some cases be chosen to have a desirable property such as directness of interpretation.

When the data are a random sample of size n, the expected information equals $I^*(\psi, \gamma) = ni^*(\psi, \gamma)$, and the partial differential equation (12.49) may be expressed using the information matrix $i^*(\psi, \gamma)$ for a single observation.

It is not necessary to find λ in terms of ψ and γ in order to obtain (12.48). To see this, write $\gamma = \gamma(\psi, \lambda)$ and note that

$$\frac{\partial \ell(\psi, \lambda)}{\partial \lambda} = \frac{\partial \gamma^\mathsf{T}}{\partial \lambda} \frac{\partial \ell^*(\psi, \gamma)}{\partial \gamma},$$

$$\frac{\partial^2 \ell(\psi, \lambda)}{\partial \lambda \partial \lambda^\mathsf{T}} = \frac{\partial^2 \gamma^\mathsf{T}}{\partial \lambda \partial \lambda^\mathsf{T}} \frac{\partial \ell^*(\psi, \gamma)}{\partial \gamma} + \frac{\partial \gamma^\mathsf{T}}{\partial \lambda} \frac{\partial^2 \ell^*(\psi, \gamma)}{\partial \gamma \partial \gamma^\mathsf{T}} \frac{\partial \gamma}{\partial \lambda^\mathsf{T}}.$$

If the maximum likelihood estimates of λ and γ for fixed ψ are $\widehat{\lambda}_\psi$ and $\widehat{\gamma}_\psi$, then

$$J_{\lambda\lambda}(\psi, \widehat{\lambda}_\psi) = \frac{\partial \gamma(\psi, \widehat{\lambda}_\psi)^\mathsf{T}}{\partial \lambda} J^*_{\gamma\gamma}(\psi, \widehat{\gamma}_\psi) \frac{\partial \gamma(\psi, \widehat{\lambda}_\psi)}{\partial \lambda^\mathsf{T}}.$$

Provided $|\partial \gamma / \partial \lambda^\mathsf{T}| \neq 0$, we have

$$\left| \frac{\partial \gamma(\psi, \widehat{\lambda}_\psi)}{\partial \lambda^\mathsf{T}} \right| = \left| \frac{\partial \lambda(\psi, \widehat{\gamma}_\psi)}{\partial \gamma^\mathsf{T}} \right|^{-1},$$

and so (12.48) equals

$$\ell^*(\psi, \widehat{\gamma}_\psi) - \tfrac{1}{2} \log \left| J^*_{\gamma\gamma}(\psi, \widehat{\gamma}_\psi) \right| + \log \left| \frac{\partial \lambda(\psi, \widehat{\gamma}_\psi)}{\partial \gamma^\mathsf{T}} \right|, \tag{12.50}$$

which can be computed without writing λ explicitly in terms of ψ and γ.

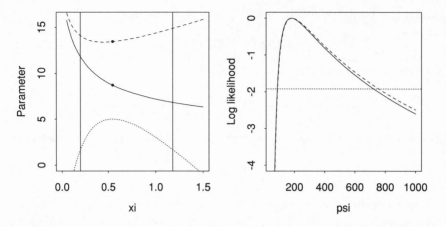

Figure 12.11
Likelihood analysis of
Danish fire data. Left:
variation of $\widehat{\sigma}_\xi$ (solid) and
$\widehat{\lambda}_\xi$ (dashes) as a function
of the shape parameter ξ.
The blobs show the
maximum likelihood
estimates. The dotted line
is the profile log
likelihood $\ell_p(\xi)$ and the
vertical lines mark the
limits of a 0.99 confidence
interval for ξ. The
orthogonal parameter $\widehat{\lambda}_\xi$
varies much less than does
the non-orthogonal
parameter $\widehat{\sigma}_\xi$ over the
range of ψ considered.
Right: profile log
likelihood (solid) and
modified profile log
likelihood (dashes) for
0.99 quantile ψ of
generalized Pareto
distribution. The
horizontal line determines
the limits of a 0.95
confidence interval for ψ.

Example 12.25 (Generalized Pareto distribution)   The expected information matrix for an observation with distribution function (6.38) is

$$i^*(\xi, \sigma) = \frac{1}{\sigma^2(1 + \xi)(1 + 2\xi)} \begin{pmatrix} 2\sigma^2 & \sigma \\ \sigma & 1 + \xi \end{pmatrix}.$$

Hence a parameter $\lambda = \lambda(\xi, \sigma)$ orthogonal to the shape parameter ξ satisfies the partial differential equation

$$\frac{\partial \sigma}{\partial \xi} = -i_{\sigma\sigma}^{*-1}(\xi, \sigma)i_{\sigma\xi}^*(\xi, \sigma) = -\frac{\sigma}{1 + \xi}.$$

It is straightforward to check that this implies that $g(\lambda) = \sigma(1 + \xi)$ for a suitably smooth function g. With the choice $\lambda = \sigma(1 + \xi)$ we find that

$$i(\xi, \lambda) = \frac{1}{\lambda^2(1 + \xi)^2(1 + 2\xi)} \begin{pmatrix} \lambda^2(1 + 2\xi) & 0 \\ 0 & (1 + \xi)^2 \end{pmatrix}.$$

We illustrate numerically the effect of parameter orthogonality using the Danish fire insurance claim data described in Examples 6.31 and 6.34. We apply a threshold $u = 15$ to the claim sizes and fit the generalized Pareto distribution to the resulting 60 exceedances. The left panel of Figure 12.11 shows that $\widehat{\sigma}_\xi$ varies more over the range of appreciable likelihood for ξ than does $\widehat{\lambda}_\xi = \widehat{\sigma}_\xi(1 + \xi)$, as our general discussion anticipates.

Now suppose that interest focuses on the $(1 - p)$ quantile of the distribution,

$$\psi = \begin{cases} \sigma(p^{-\xi} - 1)/\xi, & \xi \neq 1, \\ -\sigma \log p, & \xi = 0, \end{cases}$$

a return level, to which we seek an orthogonal parameter $\lambda = \lambda(\psi, \xi)$. As

$$i^*(\xi, \psi) = \frac{\partial(\xi, \sigma)^{\mathrm{T}}}{\partial(\xi, \psi)} i^*(\xi, \sigma) \frac{\partial(\xi, \sigma)}{\partial(\xi, \psi)^T} \bigg|_{\sigma = \sigma(\xi, \psi)},$$

a tedious calculation shows that $i^*(\xi, \psi)$ may be written as

$$\begin{pmatrix} \psi^2 a(\xi) & -\psi a(\xi)/b(\xi) \\ -\psi a(\xi)/b(\xi) & c(\xi) \end{pmatrix},$$

and so $\lambda = \lambda(\psi, \xi)$ solves the equation $\psi b(\xi) \partial \xi / \partial \psi = 1$. Thus we may take

$$\lambda = g \left\{ \log \psi + \int^\xi b(u) \, du \right\}$$

for any suitably smooth function g. With the choice $g(u) = u$, we have $\partial \lambda / \partial \xi = b(\xi)$, and this may be used in (12.50).

The right panel of Figure 12.11 shows the effect of likelihood modification when setting a confidence interval for the 0.99 quantile of the generalized Pareto model using the 60 exceedances. Application of standard chi-squared asymptotics to the profile log likelihood yields the highly asymmetric 0.95 confidence interval $(86.1, 723.8)$, while the corresponding confidence interval based on the modified profile log likelihood is $(87.6, 757.4)$.

Although modification changes the right-hand limit of the confidence interval appreciably, of overwhelmingly greater concern in applications would be the representativeness of the largest few observations in the sample, on which inferences will hinge. Moreover in practice the quantile also depends on the Poisson rate of exceedance times, and hence must be orthogonalized with respect to two parameters; see Section 6.5.2. ∎

Although (12.49) gives a basis for orthogonalizing γ with respect to a scalar ψ, parameters orthogonal to a vector ψ cannot be found in general. For if ψ contains ψ_1 and ψ_2, say, then λ must simultaneously satisfy both systems of equations

$$\frac{\partial \gamma}{\partial \psi_1} = -I_{\gamma\gamma}^{*-1}(\psi, \gamma) I_{\gamma\psi_1}^*(\psi, \gamma), \quad \frac{\partial \gamma}{\partial \psi_2} = -I_{\gamma\gamma}^{*-1}(\psi, \gamma) I_{\gamma\psi_2}^*(\psi, \gamma),$$

for all γ, ψ_1 and ψ_2. However there is no guarantee that the compatibility condition $\partial^2 \gamma / \partial \psi_1 \partial \psi_2 = \partial^2 \gamma / \partial \psi_2 \partial \psi_1$ will hold; if not, a simultaneous joint solution does not exist. When a solution exists, it can produce familiar results.

Example 12.26 (Linear exponential family) In a linear exponential family with log likelihood

$$\ell^*(\psi, \gamma) \equiv t_1^{\mathsf{T}} \psi + t_2^{\mathsf{T}} \gamma - \kappa(\psi, \gamma),$$

the parameters $\lambda = \lambda(\psi, \gamma)$ orthogonal to ψ are determined by

$$\frac{\partial \gamma}{\partial \psi^{\mathsf{T}}} = -\kappa_{\gamma\gamma}^{-1}(\psi, \gamma) \kappa_{\gamma\psi}(\psi, \gamma). \tag{12.51}$$

If we reparametrize in terms of ψ and $\lambda = \kappa_\gamma(\psi, \gamma) = \partial \kappa(\psi, \gamma) / \partial \gamma$, then in this new parametrization, γ is a function of ψ and λ, and

$$0 = \frac{\partial \lambda^{\mathsf{T}}}{\partial \psi} = \frac{\partial \gamma^{\mathsf{T}}}{\partial \psi} \kappa_{\gamma\gamma}(\psi, \gamma) + \kappa_{\psi\gamma}(\psi, \gamma),$$

so $\lambda = \kappa_\gamma(\psi, \gamma)$ is a solution to (12.51). That is, the parameter orthogonal to ψ is the so-called complementary mean parameter $\lambda(\psi, \gamma) = \mathrm{E}(T_2; \psi, \gamma)$. By symmetry, $\mathrm{E}(T_1; \psi, \gamma)$ is orthogonal to γ.

The normal distribution with mean μ and variance σ^2 has canonical parameter $(\mu/\sigma^2, -1/(2\sigma^2))$. The canonical statistic (Y, Y^2) has expectation $(\mu, \mu^2 + \sigma^2)$, so μ is orthogonal to $-1/(2\sigma^2)$, and hence to σ^2, while μ/σ^2 is orthogonal to $\mu^2 + \sigma^2$.

Independent Poisson variables Y_1 and Y_2 with means $\exp(\gamma)$ and $\exp(\gamma + \psi)$ have log likelihood

$$\ell^*(\psi, \gamma) \equiv (y_1 + y_2)\gamma + y_2\psi - e^\gamma - e^{\gamma+\psi}.$$

The discussion above suggests that

$$\lambda = \mathrm{E}(Y_1 + Y_2) = \exp(\gamma) + \exp(\gamma + \psi) = e^\gamma(1 + e^\psi)$$

is orthogonal to ψ, so $\gamma = \log\lambda - \log(1 + e^\psi)$ and

$$\ell(\psi, \lambda) \equiv y_2\psi - (y_1 + y_2)\log(1 + e^\psi) + (y_1 + y_2)\log\lambda - \lambda.$$

The separation of ψ and λ implies that the profile and modified profile likelihoods for ψ are proportional. They correspond to the conditional likelihood obtained from the density of Y_2 given $Y_1 + Y_2$.   ■

Example 12.27 (Restricted likelihood) If $Y \sim N_n(X\beta, \sigma^2\Upsilon^{-1})$, where the parameter of interest ψ appears in the $n \times n$ matrix Υ but not in β, the log likelihood is

$$\ell(\beta, \sigma^2, \psi) \equiv -\frac{n}{2}\log\sigma^2 + \tfrac{1}{2}\log|\Upsilon| - \frac{1}{2\sigma^2}(y - X\beta)^\mathsf{T}\Upsilon(y - X\beta),$$

differentiation of which yields $\partial\ell/\partial\beta = \sigma^{-2}\Upsilon(y - X\beta)$. It follows that β is orthogonal to both σ^2 and ψ. Now $J_{\beta\beta}(\beta, \sigma^2, \psi) = \sigma^{-2}X^\mathsf{T}\Upsilon X$, so apart from the term $|\partial\widehat{\beta}_{\psi,\sigma^2}/\partial\widehat{\beta}|^{-1}$, the modified profile log likelihood for ψ and σ^2 equals the marginal log likelihood (12.13). Note that $\widehat{\beta}_{\psi,\sigma^2} = (X^\mathsf{T}\Upsilon X)^{-1}X^\mathsf{T}\Upsilon y$ depends on ψ but not on σ^2.

This argument also applies when the mean of Y is a nonlinear function of β, provided that no parameter appears in both mean and variance.   ■

The notion of parameter orthogonality is useful, but the resulting modified likelihoods can be viewed as unsatisfactory, partly because the arbitrariness of the choice of orthogonal parameter results in a lack of uniqueness. A second difficulty is that the partial differential equation and hence its solution will change if there is a minor change to the model, such as the introduction of censoring, so any statistical interpretation of the orthogonal parameter is then compromised. A third is that inferences are typically based on first-order distributional approximations, as mentioned above.

Exercises 12.4

1   Let $Y_1, \ldots, Y_n \overset{\text{iid}}{\sim} N(\mu, \sigma^2)$ and let μ be the interest parameter. Show that

$$\ell(\mu, \sigma^2; \widehat{\mu}, \widehat{\sigma}^2) \equiv -\frac{n}{2}\left\{\log\sigma^2 + \frac{\widehat{\sigma}^2 + (\widehat{\mu} - \mu)^2}{\sigma^2}\right\},$$

where $\widehat{\mu}$ and $\widehat{\sigma}^2$ are the maximum likelihood parameter estimators, and hence find the modified profile likelihood for μ. Compare this with the marginal likelihood based on the t_{n-1} density of $(\overline{Y} - \mu)/(S^2/n)^{1/2}$, where $\overline{Y}$ and S^2 are the unbiased estimators of μ and σ^2. Discuss.

2   Compute (12.43) for (12.15). Hence show that $M(\tau) = |J_{\beta\beta}(\tau, \widehat{\beta}_\tau)|^{-1/2}$ for any regression-scale model.

3   In Example 12.23, find the asymptotic relative efficiencies of $\widehat{\psi}$ and $\widehat{\psi}_{\mathrm{mp}}$ when $\psi = 0$.

4   Suppose that $y \mapsto (\widehat{\psi}, \widehat{\lambda}, a)$, that a is ancillary, and that

$$f(\widehat{\psi}, \widehat{\lambda} \mid a; \psi, \lambda) = f(\widehat{\psi} \mid \widehat{\lambda}, a; \psi) f(\widehat{\lambda} \mid a; \psi, \lambda).$$

By modifying the argument on page 685 show that the first term on the right is proportional to $M(\psi) \exp\{\ell_{\mathrm{p}}(\psi)\}$, with $M(\psi)$ given by (12.42), apart from a relative error of size n^{-1}.

5   Independent Poisson variables Y_1 and Y_2 have means $\exp(\gamma)$ and $\exp(\gamma + \psi)$. Find the profile and modified profile log likelihoods for ψ in this parametrization. Comment.

6   A Poisson variable Y has mean μ, which is itself a gamma random variable with mean θ and shape parameter ν. Find the marginal density of Y, and show that $\mathrm{var}(Y) = \theta + \theta^2/\nu$ and that ν and θ are orthogonal. Hence show that ν is orthogonal to β for any model in which $\theta = \theta(x^\mathrm{T}\beta)$, x being a covariate vector. Is the same true for the model in which $\nu = \theta/\kappa$, so that $\mathrm{var}(Y) = (1 + \kappa)\mu$? Discuss the implications for inference on β when the variance function is unknown.

7   Let X and Y be independent exponential variables with means γ^{-1} and $(\gamma\psi)^{-1}$. Show that the parameter $\lambda(\gamma, \psi)$ orthogonal to ψ is the solution to the equation $\partial\gamma/\partial\psi = -\gamma/(2\psi)$, and verify that taking $\lambda = \gamma/\psi^{-1/2}$ yields an orthogonal parametrization.
   Investigate how this solution changes when X and Y are subject to Type I censoring at c.

8   Consider n pairs of independent binomial variables with denominators m_{0j} and m_{1j} and success probabilities

$$\frac{\exp(\lambda_j)}{1 + \exp(\lambda_j)}, \quad \frac{\exp(\lambda_j + \psi)}{1 + \exp(\lambda_j + \psi)}, \quad j = 1, \ldots, n.$$

Find a parameter orthogonal to ψ and hence obtain a modified profile likelihood for ψ. How does it compare to that in Example 12.23?

12.5 Bibliographic Notes

R. A. Fisher (1934) suggested that conditioning plays a central role in inference, building on earlier work, and subsequently developed this notion largely through examples; see Fisher (1922, 1925, 1935b, 1956, 1990). Although many of these examples are convincing, a fully satisfactory systematic development remains elusive, owing among other things to the non-existence of exact ancillary statistics in some important models and their non-uniqueness in others. However very substantial progress has been made over the past two decades, initially stemming from Efron and Hinkley (1978) and Barndorff-Nielsen and Cox (1979). The p^* formula was crystallized as being central by Barndorff-Nielsen (1983), building on earlier work going back to Fisher (1934), though a fully general proof is hard to establish; perhaps the most satisfactory is given by Skovgaard (1990). Reid (1995, 2003) gives excellent reviews of the roles in inference of conditioning and asymptotics, while Barndorff-Nielsen and Cox (1994) and Severini (2000) give more extended accounts of modern likelihood asymptotics and many further references.

Conditional and marginal likelihoods were introduced by Bartlett (1937) and are now widely used in applications. Restricted maximum likelihood estimation was first described by Patterson and Thompson (1971), though restricted likelihood itself had been obtained by Hartley and Rao (1967) and has earlier roots. Diggle *et al.* (1994) discuss its use in the context of longitudinal data. Example 12.11 is based on Cruddas *et al.* (1989). Kalbfleisch and Prentice (1973) show the link between marginal and partial likelihoods for the proportional hazards model.

The use of Monte Carlo simulation for conditional testing is described by Besag and Clifford (1989, 1991), and applied extensively in contingency tables by Forster *et al.* (1996) and Smith *et al.* (1996). See also Section 4.2 of Davison and Hinkley (1997).

Saddlepoint approximation was introduced into statistics in the pioneering paper of Daniels (1954), which was largely ignored until interest in small-sample asymptotics revived in the 1970s, focussed particularly by Barndorff-Nielsen and Cox (1979). Reid (1988) surveys statistical aspects of saddlepoint approximation, while Barndorff-Nielsen and Cox (1989) is a standard reference to these and related procedures. Jensen (1995) is a thorough mathematical treatment. Distribution function approximations are described by Daniels (1987), Skovgaard (1987), and Barndorff-Nielsen (1986). The approximation (12.40) was proposed by Skovgaard (1996) and has itself been approximated by Severini (1999). Numerous other approaches have been suggested; see for example the account of likelihoods for component parameters in Fraser (2002) or Chapter 7 of Severini (2000).

Cox and Reid (1987) and its discussion give a systematic treatment of parameter orthogonality and its consequences for conditional inference.

The complexity of many likelihood expressions has led authors such as McCullagh (1987) and Andrews and Stafford (2000) to develop powerful tools for analytic work with or symbolic computation applied to asymptotic expansions.

Brazzale (1999, 2000), and Bellio (1999) describe systematic attempts to implement small-sample likelihood asymptotics for practical use.

12.6 Problems

1   Find the Fisher information matrix for Example 12.2, and show that it gives the wrong asymptotic variances for all the parameters.

2   Does the argument of Example 12.5 apply to the location-scale model with any error distribution?

3   Show that a missingness indicator (Section 5.5.1) is not generally an ancillary statistic, and discuss the implications for inference on θ.

4   A normal random variable Y has mean θ and variance determined by Table 12.5, where k_1 and k_2 are chosen so that the values of π lie in the unit interval. Show that A_1 is exactly ancillary but is not informative about the precision of Y, while A_2 is not exactly ancillary but is indicative of the precision of Y.
Discuss briefly the merits of exact and approximate conditional inference here. (Lloyd, 1992)

Table 12.5 Conditional distributions of normal variables.

var(Y)	100	100	1	1
π	$\frac{1}{2}(1 - k_1\theta)$	$\frac{1}{2}(1 + k_2\theta)$	$\frac{1}{2}(1 + k_1\theta)$	$\frac{1}{2}(1 - k_2\theta)$
A_1	1	0	1	0
A_2	0	0	1	1

5   Show that if Y has the Cauchy density

$$f(y; \mu, \sigma) = \frac{\sigma}{\pi \left\{\sigma^2 + (y - \mu)^2\right\}}, \quad -\infty < y, \mu < \infty, \quad \sigma > 0,$$

then $1/Y$ has density $f(y; \mu', \sigma')$, where $\mu' = \mu/(\mu^2 + \sigma^2)$ and $\sigma' = \sigma/(\mu^2 + \sigma^2)$. Deduce that a random sample $Y_1, \ldots, Y_n$ of Cauchy variables yields two distinct sets of maximal ancillary statistics. Discuss inference for μ, σ, and for μ/σ.
(McCullagh, 1992)

6   In Example 12.8, show that A_1 and A_2 are not jointly ancillary for θ.
Conditioning on an ancillary statistic is intended to divide the sample space into relevant subsets according to their information content, so one basis for choice among competing ancillaries is to take that whose the conditional Fisher information has largest variance. Show that $\mathrm{E}\{i_1(\theta)\} = \mathrm{E}\{i_2(\theta)\}$, but that $\mathrm{var}\{i_1(\theta)\} > \mathrm{var}\{i_2(\theta)\}$ for $0 < \theta < 1$, and deduce that A_1 is preferable.
Discuss critically this idea.
(Cox, 1971)

7   Consider two independent exponential random samples, $Y_1, \ldots, Y_n$ having rate $\psi > 0$ and $X_1, \ldots, X_n$ having rate λ/ψ, where $\lambda > 0$.
(a) Suppose it is required to find an approximate ancillary for ψ when $\lambda = 1$. Find the likelihood ratio statistic for testing $\lambda = 1$ against the alternative putting no restriction on λ, and show that it is a function only of $\overline{X}\,\overline{Y}$. Hence give an exact ancillary statistic for ψ.
(b) Give explicit expressions for the p^* formula (12.7) and for $r^*(\psi)$ when $\lambda = 1$. Investigate the numerical accuracy of (12.8) when $n = 1$.

8   A Poisson process of rate $\lambda e^{\psi t}$ is observed on the interval $[0, 1]$, over which events occur at times $0 < t_1 < \cdots < t_n < 1$. Show that the likelihood is

$$\exp\left\{-\lambda \int_0^1 e^{\psi u}\, du\right\} \prod_{j=1}^n \lambda e^{\psi t_j},$$

and deduce that inference for ψ may be based on the conditional density

$$n! \prod_{j=1}^n \frac{e^{\psi t_j}}{\int_0^1 e^{\psi u}\, du}$$

of the times of events $T_1 < \cdots < T_N$ given that $N = n$. Show that this is the joint density of the order statistics of a random sample of n variables with density $e^{\psi t}/\int_0^1 e^{\psi u}\, du$ on $(0, 1)$, and derive a conditional test of the hypothesis $\psi = 0$ against $\psi > 0$, giving the null mean and variance of your test statistic.
When $\psi = 0$, how might you test the hypothesis that the T_j are clustered relative to a Poisson process?
(Cox and Lewis, 1966, pp. 45–51)

9   Under the model of Example 1.4, the number of deaths due to lung cancer in the (i, j) cell of Table 1.4, Y_{ij}, is a Poisson variable with mean expressible as

$$x_{ij}\, g(t_i, \phi)(1 + \psi_1 d_j^{\psi_2}),$$

where the notation reflects our interest in the effect of smoking.

Show that the marginal density of $M_i = Y_{i1} + \cdots + Y_{ic}$ is Poisson with mean

$$\lambda_i = g(t_i; \phi) \sum_{j=1}^{c} x_{ij}(1 + \psi_1 d_j^{\psi_2}), \quad i = 1, \ldots, r,$$

while the conditional distribution of $Y_{i1}, \ldots, Y_{ic}$ given $M_i = m_i$ is multinomial with denominator $m_i = y_{i1} + \cdots + y_{ic}$ and probabilities

$$\pi_{ij} = \frac{x_{ij}(1 + \psi_1 d_j^{\psi_2})}{\sum_{k=1}^{c} x_{ik}(1 + \psi_1 d_k^{\psi_2})}, \quad j = 1, \ldots, c.$$

Outline how this computation may be used as a basis for inference on ψ, and in particular how evidence for $\psi_2 = 1$ may be assessed. Do the usual likelihood asymptotics apply when testing the hypothesis that $\psi_1 = 0$, regardless of ψ_2?

10   In an exponential family density

$$f(t_1, t_2; \psi, \lambda) = \exp\left\{t_1^{\mathrm{T}}\psi + t_2^{\mathrm{T}}\lambda - \kappa(\psi, \lambda) + c(t_1, t_2)\right\},$$

show that the conditional distribution of T_1 given T_2 is unchanged if λ is randomly taken from a density $g(\lambda)$.
Independent pairs of observations $(x_1, y_1), \ldots, (x_n, y_n)$ are supposed to have independent Poisson distributions with means $(\mu_j, \beta\mu_j)$, for $j = 1, \ldots, n$. Does your inference for β depend on the knowledge that $\mu_1, \ldots, \mu_n \overset{\text{iid}}{\sim} g$?
If the density $g(\mu) = g(\mu; \gamma)$ is known up to the value of a parameter γ, say, suggest how to retrieve any information on ψ in the marginal density of the y_j.

11   Independent pairs of binary observations $(R_{01}, R_{11}), \ldots, (R_{0n}, R_{1n})$ have success probabilities $(e^{\lambda_j}/(1 + e^{\lambda_j}), e^{\psi+\lambda_j}/(1 + e^{\psi+\lambda_j}))$, for $j = 1, \ldots, n$.
(a) Show that the maximum likelihood estimator of ψ based on the conditional likelihood is $\widehat{\psi}_c = \log(R^{01}/R^{10})$, where R^{01} and R^{10} are respectively the numbers of $(0,1)$ and $(1,0)$ pairs. Does $\widehat{\psi}_c$ tend to ψ as $n \to \infty$?
(b) Write down the unconditional likelihood for ψ and λ, and show that the likelihood equations are equivalent to

$$r_{0j} + r_{1j} = \frac{e^{\widehat{\lambda}_j}}{1 + e^{\widehat{\lambda}_j}} + \frac{e^{\widehat{\lambda}_j + \widehat{\psi}}}{1 + e^{\widehat{\lambda}_j + \widehat{\psi}}}, \quad j = 1, \ldots, n, \qquad (12.52)$$

$$\sum_{j=1}^{n} r_{1j} = \sum_{j=1}^{n} \frac{e^{\widehat{\lambda}_j + \widehat{\psi}}}{1 + e^{\widehat{\lambda}_j + \widehat{\psi}}}.$$

(i) Show that the maximum likelihood estimator of λ_j is ∞ if $r_{0j} = r_{1j} = 1$ and $-\infty$ if $r_{0j} = r_{1j} = 0$; such pairs are not informative. (ii) Use (12.52) to show that $\widehat{\lambda}_j = -\widehat{\psi}/2$ for those pairs for which $r_{0j} + r_{1j} = 1$. (iii) Hence deduce that the unconditional maximum likelihood estimator of ψ is $\widehat{\psi}_u = 2\log(R^{01}/R^{10})$. What is the implication for unconditional estimation of ψ?

12   Consider two independent Poisson random samples $X_1, \ldots, X_n$ and $Y_1, \ldots, Y_n$, the first having mean λ and the second having mean λ^ψ, where $\lambda, \psi > 0$.
(a) Show that $(T_1, T_2) = (X_1 + \cdots + X_n, Y_1 + \cdots + Y_n)$ is minimal sufficient, and for any fixed value of ψ establish that λ may be eliminated by conditioning on $T_\psi = T_1 + \psi T_2$.
(b) Let $(t_{1,\text{obs}}, t_{2,\text{obs}})$ denote the observed value of (T_1, T_2). Sketch the sample space for (T_1, T_2), and consider how the relevant subset

$$\left\{(t_1, t_2) : t_1 + \psi t_2 = t_{1,\text{obs}} + \psi t_{2,\text{obs}}\right\}$$

What happens if ψ is rational? What if ψ is irrational?

varies with ψ. Hence explain how an exact significance level for a test of $\psi = \psi_0$ against the alternative $\psi > \psi_0$ will depend on ψ_0. Do you find this satisfactory?

13   Adapt the argument giving inference for the regression-scale model to the location-scale model $Y = \mu + e^\tau \varepsilon$, and outline how to make small-sample inferences for μ and for τ.

Compare the resulting confidence intervals for μ with the the posterior credible intervals found by Bayesian inference using prior density $\pi(\mu, \sigma) \propto \sigma^{-1}$ and distribution function approximation (11.31). Discuss.

14   (a) Consider a location-scale model in which $y = \eta + \sigma \varepsilon$, where ε has a known density g. Find parameters orthogonal to η and to σ, and give conditions under which η and σ are themselves orthogonal.

(b) Consider a regression model $y = x^{\mathsf{T}} \beta + \sigma \varepsilon$, where ε again has density g. Find a parameter orthogonal to the first component of β, and compare your result with the discussion following (8.8).

15   Independent exponential variables $Y_1, \ldots, Y_n$ have means $\mathrm{E}(Y_j) = \lambda e^{x_j \psi}$, where $\sum x_j = 0$. Show that λ and ψ are orthogonal parameters.

(a) Show that $\widehat{\lambda}_\psi = n^{-1} \sum e^{x_j \psi}$, and deduce that the likelihood ratio statistic for testing $\psi = 0$ can be written as $Wp(0) = 2n \log(\widehat{\lambda}_0 / \widehat{\lambda}_{\widehat\psi})$.

(b) Let $\gamma = \log \lambda$. By writing the model in linear regression form, show that $\partial \widehat{\gamma} / \partial \widehat{\gamma}_\psi = 1$, and deduce that $\partial \widehat{\lambda} / \partial \widehat{\lambda}_\psi = \widehat{\lambda} / \widehat{\lambda}_\psi$. Hence find the modified profile likelihood both with and without this term, and compare the resulting likelihood ratio statistics with $Wp(0)$. (Cox and Reid, 1987)

16   The Michaelis–Menton model of nonlinear regression is usually specified as

$$Y_j = \frac{\beta_0 x_j}{\beta_1 + x_j} + \varepsilon_j, \quad \varepsilon_1, \ldots, \varepsilon_n \overset{\text{iid}}{\sim} N(0, \sigma^2);$$

we assume that σ^2 is known. Show that the log likelihood is

$$\ell^*(\beta_0, \beta_1) = -\frac{1}{2\sigma^2} \sum_{j=1}^{n} \left\{ y_j - \beta_0 x_j / (\beta_1 + x_j) \right\}^2,$$

and find the expected information matrix.

(a) Show that the parameter $\lambda = \lambda(\beta_1, \beta_0)$ orthogonal to β_1 is determined by

$$g(\lambda) = \beta_0^2 \sum_{j=1}^{n} \frac{x_j^2}{(\beta_1 + x_j)^2}$$

for an appropriate smooth function g. Choose g suitably, and write the log likelihood explicitly in terms of λ and β_1.

(b) Show that the parameter $\lambda = \lambda(\beta_1, \beta_0)$ orthogonal to β_0 is determined by

$$\beta_0 \sum_{j=1}^{n} \frac{x_j^2}{(\beta_1 + x_j)^4} \frac{\partial \beta_1}{\partial \beta_0} = \sum_{j=1}^{n} \frac{x_j^2}{(\beta_1 + x_j)^3}$$

and check that its solution is

$$g_1(\lambda) = \beta_0^3 \sum_{j=1}^{n} \frac{x_j^2}{(\beta_1 + x_j)^3}.$$

Can the log likelihood be expressed explicitly in terms of λ and β_0?

(c) The orthogonal parametrizations above depend on the design points x_j. Do you find this satisfactory?

(Hills, 1987)

APPENDIX A

Practicals

The list below gives key words for practicals written in the statistical language S and intended to accompany the chapters of the book. The practicals themselves may be downloaded from

`http://statwww.epfl.ch/people/~davison/SM`

together with a library of functions and data.

2. Variation

1. Speed of light data. Exploratory data analysis.
2. Maths marks data. Brush and spin plots.
3. Probability plots for simulated data.
4. Illustration of central limit theorem using simulated data.
5. Data on air-conditioning failures. Exponential probability plots.

3. Uncertainty

1. Properties of half-normal distribution. Half-normal plot.
2. Simulation of Student t statistic, following original derivation.
3. Simulation of Wiener process and Brownian bridge.
4. Normal random number generation by summing uniform variables.
5. Implementation and assessment of a linear congruential generator.
6. Coverage of Student t confidence interval under various scenarios.

4. Likelihood

1. Loss of information due to rounding of normal data.
2. Birth data. Maximum likelihood estimation for Poisson and gamma models. Assessment of fit.
3. Data on sizes of groups of people. Maximum likelihood fit of truncated Poisson distribution. Pearson's statistic.
4. α-particle data. Maximum likelihood fit of Poisson process model.

5. Blood group data. Maximum likelihood fit of multinomial model.
6. Generalized Pareto distribution. Nonregular estimation of endpoint.

5. Models

1. Boiling point of water data. Straight-line regression.
2. Survival data on leukaemia. Exponential and Weibull models.
3. HUS data. EM algorithm for mixture of Poisson distributions.
4. EM algorithm for mixture of normal distributions.

6. Stochastic Models

1. Markov chain fitting to Alofi rainfall data. Assessment of fit.
2. Multivariate normal fit to data on head sizes.
3. Time series analysis of Manaus river height data. ARMA modelling.
4. Inhomogeneous Poisson process fitted to freezes of Lake Constance.
5. Extreme-value analysis of FTSE return data.

7. Theory

1. Neurological data. Kernel density estimation. Test of unimodality.
2. Mean integrated squared error of kernel density estimator applied to mixtures of normal densities.
3. Test for spatial Poisson process. Beetle data.
4. Coverage of confidence intervals for Poisson mean.

8. Linear Models

1. Cherry tree data. Linear model.
2. Salinity data. Linear model.
3. Data on IQs of identical twins. Linear model.
4. Cement data. Simulation of collinear data.
5. Simulation to assess properties of stepwise model selection procedures.
6. Data on pollution and mortality. Linear model. Ridge regression.

9. Designed Experiments

1. Chick bone data. Inter- and intra-block recovery of information.
2. Millet plant data. Latin square. Outliers. Orthogonal polynomials.
3. Data on marking of examination scripts. Analysis of variance.
4. Teak plant data. 2×3 factorial experiment.

10. Nonlinear Models

1. Space shuttle data. Logistic regression model.
2. Beetle data. Regression models for binary data.
3. Stomach ulcer data. Logistic regression for 2×2 tables. Overdispersion.
4. Speed limit data. Log-linear model. Logistic regression model.
5. Lizards data. Log-linear model. Logistic regression model.
6. Titanic survivor data. Log-linear model.

7. Seed germination data. Overdispersion. Quasi-likelihood. Beta-binomial model.
8. Coal-mining disaster data. Inhomogeneous Poisson process. Generalized additive model.
9. Urine crystal data. Logistic regression.
10. Survival data on leukaemia. Proportional hazards model.
11. Motorette data. Survival data analysis.
12. PBC data. Survival data analysis. Proportional hazards model.

11. Bayesian Models

1. Coin spun on edge. Updating individual and group priors.
2. Cloth data. Hierarchical Poisson model. Laplace approximation.
3. Gibbs sampler for bivariate truncated exponential distribution.
4. Random walk Metropolis–Hastings algorithm with Cauchy proposals.
5. Pump failure data. Gibbs sampler for hierarchical Poisson model.
6. HUS data. Gibbs sampler. Changepoint in Poisson variables.
7. Beaver body temperature data. Gibbs sampler. Changepoint in normal variables.
8. Data augmentation algorithm with multinomial data.

12. Marginal and Conditional Likelihood

1. Ancillary statistic. Simulation with Cauchy data.
2. Saddlepoint approximation. Laplace distribution.
3. Urine data. Logistic regression. Approximate conditional inference.

Bibliography

Aalen, O. O. (1978) Nonparametric inference for a family of counting processes. *Annals of Statistics* **6**, 701–726.

Aalen, O. O. (1994) Effects of frailty in survival analysis. *Statistical Methods in Medical Research* **3**, 227–243.

Agresti, A. (1984) *Analysis of Ordinal Categorical Data*. New York: Wiley.

Agresti, A. and Caffo, B. (2000) Simple and effective confidence intervals for proportions and differences of proportions result from adding two successes and two failures. *The American Statistician* **54**, 280–288.

Agresti, A. and Coull, B. A. (1998) Approximate is better than "exact" for interval estimation of binomial proportions. *The American Statistician* **52**, 119–126.

Akaike, H. (1973) Information theory and an extension of the maximum likelihood principle. In *Second International Symposium on Information Theory*, eds B. N. Petrov and F. Czáki, pp. 267–281. Budapest: Akademiai Kiadó. Reprinted in *Breakthroughs in Statistics*, volume 1, eds S. Kotz and N. L. Johnson, pp. 610–624. New York: Springer.

Almond, R. (1995) *Graphical Belief Modelling*. New York: Chapman & Hall.

Andersen, P. K., Borgan, Ø., Gill, R. D. and Keiding, N. (1993) *Statistical Models Based on Counting Processes*. New York: Springer.

Anderson, T. W. (1958) *Introduction to Multivariate Statistical Analysis*. New York: Wiley.

Andrews, D. F. and Stafford, J. E. (2000) *Symbolic Computation for Statistical Inference*. Oxford: Clarendon Press.

Appleton, D. R., French, J. M. and Vanderpump, M. P. J. (1996) Ignoring a covariate: An example of Simpson's paradox. *The American Statistician* **50**, 340–341.

Arnold, B. C., Balakrishnan, N. and Nagaraja, H. N. (1992) *A First Course in Order Statistics*. New York: Wiley.

Artes, R. (1997) *Extensões da Teoria das Equações de Estimação Generalizadas a Dados Circulares e Modelos de Dispersão*. Ph.D. thesis, University of São Paulo.

Ashford, J. R. (1959) An approach to the analysis of data for semi-quantal responses in biological assay. *Biometrics* **15**, 573–581.

Atkinson, A. C. (1985) *Plots, Transformations, and Regression*. Oxford: Clarendon Press.

Atkinson, A. C. and Donev, A. N. (1992) *Optimum Experimental Designs*. Oxford: Clarendon Press.

Atkinson, A. C. and Riani, M. (2000) *Robust Diagnostic Regression Analysis*. New York: Springer.

Avery, P. J. and Henderson, D. A. (1999) Fitting Markov chain models to discrete state series such as DNA sequences. *Applied Statistics* **48**, 53–61.

Azzalini, A. and Bowman, A. W. (1993) On the use of nonparametric regression for checking linear relationships. *Journal of the Royal Statistical Society series B* **55**, 549–557.

Azzalini, A., Bowman, A. W. and Härdle, W. (1989) On the use of nonparametric regression for model-checking. *Biometrika* **76**, 1–11.

Barndorff-Nielsen, O. E. (1978) *Information and Exponential Families in Statistical Theory*. New York: Wiley.

Barndorff-Nielsen, O. E. (1983) On a formula for the distribution of the maximum likelihood estimator. *Biometrika* **70**, 343–365.

Barndorff-Nielsen, O. E. (1986) Inference on full or partial parameters based on the standardized signed log likelihood ratio. *Biometrika* **73**, 307–322.

Barndorff-Nielsen, O. E. and Cox, D. R. (1979) Edgeworth and saddle-point approximations with statistical applications (with Discussion). *Journal of the Royal Statistical Society series B* **41**, 279–312.

Barndorff-Nielsen, O. E. and Cox, D. R. (1989) *Asymptotic Techniques for Use in Statistics*. London: Chapman & Hall.

Barndorff-Nielsen, O. E. and Cox, D. R. (1994) *Inference and Asymptotics*. London: Chapman & Hall.

Bartlett, M. S. (1936a) Statistical information and properties of sufficiency. *Proceedings of the Royal Society of London, series A* **154**, 124–137.

Bartlett, M. S. (1936b) The information available in small samples. *Proceedings of the Cambridge Philosophical Society* **32**, 560–566.

Bartlett, M. S. (1937) Properties of sufficiency and statistical tests. *Proceedings of the Royal Society of London, series A* **160**, 268–282.

Basawa, I. V. and Scott, D. J. (1981) *Asymptotic Optimal Inference for Non-Ergodic Models*. Volume 17 of *Lecture Notes in Statistics*. New York: Springer.

Basu, D. (1955) On statistics independent of a complete sufficient statistic. *Sankhyā* **15**, 377–380.

Basu, D. (1958) On statistics independent of sufficient statistics. *Sankhyā* **20**, 223–226.

Bellio, R. (1999) *Likelihood Asymptotics: Applications in Biostatistics*. Ph.D. thesis, Department of Statistical Science, University of Padova.

Belsley, D. A. (1991) *Conditioning Diagnostics: Collinearity and Weak Data in Regression*. New York: Wiley.

Belsley, D. A., Kuh, E. and Welsch, R. E. (1980) *Regression Diagnostics: Identifying Influential Data and Sources of Collinearity*. New York: Wiley.

Beran, J. (1994) *Statistics for Long-Memory Processes*. London: Chapman & Hall.

Beran, R. J. and Fisher, N. I. (1998) A conversation with Geoff Watson. *Statistical Science* **13**, 75–93.

Berger, J. O. (1985) *Statistical Decision Theory and Bayesian Analysis*. Second edition. New York: Springer.

Berger, J. O. and Wolpert, R. L. (1988) *The Likelihood Principle*. Second edition, volume 6 of *Lecture Notes — Monograph Series*. Hayward, California: Institute of Mathematical Statistics.

Bernardo, J. M. and Smith, A. F. M. (1994) *Bayesian Theory*. New York: Wiley.

Besag, J. E. (1974) Spatial interaction and the statistical analysis of lattice systems (with Discussion). *Journal of the Royal Statistical Society series B* **34**, 192–236.

Besag, J. E. (1986) On the statistical analysis of dirty pictures (with Discussion). *Journal of the Royal Statistical Society series B* **48**, 259–302.

Besag, J. E. (1989) A candidate's formula: a curious result in Bayesian prediction. *Biometrika* **76**, 183.

Besag, J. E. and Clifford, P. (1989) Generalized Monte Carlo significance tests. *Biometrika* **76**, 633–642.

Besag, J. E. and Clifford, P. (1991) Sequential Monte Carlo p-values. *Biometrika* **78**, 301–304.

Besag, J. E. and Green, P. J. (1993) Spatial statistics and Bayesian computation. *Journal of the Royal Statistical Society series B* **55**, 25–37.

Besag, J. E., Green, P. J., Higdon, D. and Mengersen, K. (1995) Bayesian computation and stochastic systems (with Discussion). *Statistical Science* **10**, 3–66.

Besag, J. E., York, J. and Mollié, A. (1991) Bayesian image restoration, with two applications in spatial statistics (with Discussion). *Annals of the Institute of Statistical Mathematics* **43**, 1–59.

Bickel, P. J. and Doksum, K. A. (1977) *Mathematical Statistics: Basic Ideas and Selected Topics*. San Francisco: Holden-Day.

Billingsley, P. (1961) *Statistical Inference for Markov Processes*. Chicago: Chicago University Press.

Bishop, Y. M., Fienberg, S. E. and Holland, P. W. (1975) *Discrete Multivariate Analysis*. Cambridge, Massachussetts: MIT Press.

Bissell, A. F. (1972) A negative binomial model with varying element sizes. *Biometrika* **59**, 435–441.

Bloomfield, P. (1976) *Fourier Analysis of Time Series: An Introduction*. New York: Wiley.

Bowman, A. W. and Azzalini, A. (1997) *Applied Smoothing Techniques for Data Analysis: The Kernel Approach with* S-Plus *Illustrations*. Oxford: Clarendon Press.

Box, G. E. P. (1980) Sampling and Bayes inference in scientific modelling and robustness (with Discussion). *Journal of the Royal Statistical Society series A* **143**, 383–430.

Box, G. E. P. and Cox, D. R. (1964) An analysis of transformations (with Discussion). *Journal of the Royal Statistical Society series B* **26**, 211–246.

Box, G. E. P., Hunter, W. G. and Hunter, J. S. (1978) *Statistics for Experimenters*. New York: Wiley.

Box, G. E. P. and Tiao, G. C. (1973) *Bayesian Inference in Statistical Analysis*. Second edition. Reading, Massachussetts: Addison–Wesley.

Box, G. E. P. and Tidwell, P. W. (1962) Transformation of the independent variables. *Technometrics* **4**, 531–550.

Brazzale, A. R. (1999) Approximate conditional inference in logistic and loglinear models. *Journal of Computational and Graphical Statistics* **8**, 653–661.

Brazzale, A. R. (2000) *Practical Small-Sample Parametric Inference*. Ph.D. thesis, Department of Mathematics, Swiss Federal Institute of Technology, Lausanne, Switzerland.

Brémaud, P. (1999) *Markov Chains: Gibbs Fields, Monte Carlo Simulation, and Queues*. New York: Springer.

Brillinger, D. R. (1981) *Time Series: Data Analysis and Theory*. Expanded edition. San Francisco: Holden-Day.

Brockwell, P. J. and Davis, R. A. (1991) *Time Series: Theory and Methods*. Second edition. New York: Springer.

Brockwell, P. J. and Davis, R. A. (1996) *Introduction to Time Series and Forecasting*. New York: Springer.

Brooks, S. P. (1998) Markov chain Monte Carlo and its application. *The Statistician* **47**, 69–100.

Brooks, S. P. and Gelman, A. (1998) General methods for monitoring convergence of iterative simulations. *Journal of Computational and Graphical Statistics* **7**, 434–455.

Brown, B. W. (1980) Prediction analysis for binary data. In *Biostatistics Casebook*, eds R. G. Miller, B. Efron, B. W. Brown and L. E. Moses, pp. 3–18. New York: Wiley.

Brown, B. W. and Hollander, M. (1977) *Statistics: A Biomedical Introduction*. New York: Wiley.

Brown, L. D. (1986) *Fundamentals of Statistical Exponential Families, with Applications in Statistical Decision Theory*. Volume 9 of *Lecture Notes — Monograph Series*. Hayward, California: Institute of Mathematical Statistics.

Brown, P. J. (1993) *Measurement, Regression, and Calibration*. Oxford: Clarendon Press.

Burnham, K. P. and Anderson, D. R. (2002) *Model Selection and Multi-Model Inference: A Practical Information Theoretic Approach*. Second edition. New York: Springer.

Carlin, B. P. and Louis, T. A. (2000) *Bayes and Empirical Bayes Methods for Data Analysis*. Second edition. London: Chapman & Hall.

Carroll, R. J. and Ruppert, D. (1988) *Transformation and Weighting in Regression*. London: Chapman & Hall.

Casella, G. and Berger, R. L. (1990) *Statistical Inference*. Belmont, California: Wadsworth & Brooks/Cole.

Castillo, E., Gutiérrez, J. M. and Hadi, A. S. (1997) *Expert Systems and Probabilistic Network Models*. New York: Springer.

Catchpole, E. A. and Morgan, B. J. T. (1997) Detecting parameter redundancy. *Biometrika* **84**, 187–196.

Chatfield, C. (1988) *Problem-Solving: A Statistician's Guide*. London: Chapman & Hall.

Chatfield, C. (1995) Model uncertainty, data mining and statistical inference (with Discussion). *Journal of the Royal Statistical Society series A* **158**, 419–466.

Chatfield, C. (1996) *The Analysis of Time Series*. Fifth edition. London: Chapman & Hall.

Chatfield, C. and Collins, A. J. (1980) *Introduction to Multivariate Analysis*. London: Chapman & Hall.

Chatterjee, S. and Hadi, A. S. (1988) *Sensitivity Analysis in Linear Regression*. New York: Wiley.

Chellappa, R. and Jain, A. (eds) (1993) *Markov Random Fields: Theory and Application*. New York: Academic Press.

Cheng, R. C. H. and Traylor, L. (1995) Non-regular maximum likelihood problems (with Discussion). *Journal of the Royal Statistical Society series B* **57**, 3–44.

Cleveland, W. S. (1993) *Vizualizing Data*. New Jersey: Hobart Press.

Cleveland, W. S. (1994) *The Elements of Graphing Data*. Revised edition. New Jersey: Hobart Press.

Clifford, P. (1990) Markov random fields in statistics. In *Disorder in Physical Systems: A Volume in Honour of John M. Hammersley*, eds G. R. Grimmett and D. J. A. Welsh, pp. 19–32. Oxford: Clarendon Press.

Cobb, G. W. (1998) *Introduction to Design and Analysis of Experiments*. New York: Springer.

Cochran, W. G. and Cox, G. M. (1959) *Experimental Designs*. Second edition. New York: Wiley.

Coles, S. G. (2001) *An Introduction to the Statistical Modeling of Extreme Values*. New York: Springer.

Collett, D. (1991) *Modelling Binary Data*. London: Chapman & Hall.

Collett, D. (1995) *Modelling Survival Data in Medical Research*. London: Chapman & Hall.

Cook, R. D. (1977) Detection of influential observations in linear regression. *Technometrics* **19**, 15–18.

Cook, R. D. and Weisberg, S. (1982) *Residuals and Influence in Regression*. London: Chapman & Hall.

Copas, J. B. (1999) What works?: Selectivity models and meta-analysis. *Journal of the Royal Statistical Society series A* **162**, 96–109.

Copas, J. B. and Li, H. G. (1997) Inference for non-random samples (with Discussion). *Journal of the Royal Statistical Society series B* **59**, 55–95.

Cowell, R. G., Dawid, A. P. and Lauritzen, S. L. (1999) *Probabilistic Networks and Expert Systems*. New York: Springer.

Cowles, M. K. and Carlin, B. P. (1996) Markov chain Monte Carlo convergence diagnostics: A comparative review. *Journal of the American Statistical Association* **91**, 883–904.

Cox, D. R. (1958) *Planning of Experiments*. New York: Wiley.

Cox, D. R. (1959) The analysis of exponentially distributed life-times with two types of failure. *Journal of the Royal Statistical Society series B* **21**, 411–421.

Cox, D. R. (1970) *Analysis of Binary Data*. London: Chapman & Hall.

Cox, D. R. (1971) The choice between alternative ancillary statistics. *Journal of the Royal Statistical Society series B* **33**, 251–255.

Cox, D. R. (1972) Regression models and life tables (with Discussion). *Journal of the Royal Statistical Society series B* **34**, 187–220.

Cox, D. R. (1978) Some remarks on the role in statistics of graphical methods. *Applied Statistics* **27**, 4–9.

Cox, D. R. (1979) A note on the graphical analysis of survival data. *Biometrika* **66**, 188–190.

Cox, D. R. (1983) A remark on censoring and surrogate response variables. *Journal of the Royal Statistical Society series B* **45**, 391–393.

Cox, D. R. (1990) Role of models in statistical analysis. *Statistical Science* **5**, 169–174.

Cox, D. R. (1992) Causality: Some statistical aspects. *Journal of the Royal Statistical Society series A* **155**, 291–301.

Cox, D. R. and Davison, A. C. (1989) Prediction for small subgroups. *Philosophical Transactions of the Royal Society of London, series B* **325**, 185–187.

Cox, D. R. and Hinkley, D. V. (1974) *Theoretical Statistics*. London: Chapman & Hall.

Cox, D. R. and Isham, V. (1980) *Point Processes*. London: Chapman & Hall.

Cox, D. R. and Lewis, P. A. W. (1966) *The Statistical Analysis of Series of Events*. London: Chapman & Hall.

Cox, D. R. and Miller, H. D. (1965) *The Theory of Stochastic Processes*. London: Chapman & Hall.

Cox, D. R. and Oakes, D. (1984) *Analysis of Survival Data*. London: Chapman & Hall.

Cox, D. R. and Reid, N. (1987) Parameter orthogonality and approximate conditional inference (with Discussion). *Journal of the Royal Statistical Society series B* **49**, 1–39.

Cox, D. R. and Reid, N. (2000) *The Theory of the Design of Experiments*. London: Chapman & Hall.

Cox, D. R. and Snell, E. J. (1968) A general definition of residuals (with Discussion). *Journal of the Royal Statistical Society series B* **30**, 248–275.

Cox, D. R. and Snell, E. J. (1981) *Applied Statistics: Principles and Examples*. London: Chapman & Hall.

Cox, D. R. and Snell, E. J. (1989) *Analysis of Binary Data*. Second edition. London: Chapman & Hall.

Cox, D. R. and Wermuth, N. (1996) *Multivariate Dependencies: Models, Analysis and Interpretation*. London: Chapman & Hall.

Craig, P. S., Goldstein, M., Seheult, A. H. and Smith, J. A. (1998) Constructing partial prior specifications for models of complex physical systems (with Discussion). *The Statistician* **47**, 37–53.

Cressie, N. A. C. (1991) *Statistics for Spatial Data*. New York: Wiley.

Crowder, M. J., Kimber, A. C., Smith, R. L. and Sweeting, T. J. (1991) *Statistical Analysis of Reliability Data*. London: Chapman & Hall.

Cruddas, A. M., Reid, N. and Cox, D. R. (1989) A time series illustration of approximate conditional likelihood. *Biometrika* **76**, 231–237.

Dalal, S. R., Fowlkes, E. B. and Hoadley, B. (1989) Risk analysis of the space shuttle: Pre-Challenger prediction of failure. *Journal of the American Statistical Association* **84**, 945–957.

Daley, D. J. and Vere-Jones, D. (1988) *An Introduction to the Theory of Point Processes*. New York: Springer.

Daniels, H. E. (1954) Saddlepoint approximations in statistics. *Annals of Mathematical Statistics* **25**, 631–650.

Daniels, H. E. (1987) Tail probability approximations. *International Statistical Review* **54**, 37–48.

Davison, A. C. (2001) *Biometrika* centenary: Theory and general methodology. *Biometrika* **88**, 13–52. Reprinted in *Biometrika: One Hundred Years*, edited by D. M. Titterington and D. R. Cox. Oxford University Press, [11]–[50].

Davison, A. C. and Hinkley, D. V. (1997) *Bootstrap Methods and Their Application*. Cambridge: Cambridge University Press.

Davison, A. C. and Smith, R. L. (1990) Models for exceedances over high thresholds (with Discussion). *Journal of the Royal Statistical Society series B* **52**, 393–442.

Davison, A. C. and Snell, E. J. (1991) Residuals and diagnostics. In *Statistical Theory and Modelling: In Honour of Sir David Cox, FRS*, eds D. V. Hinkley, N. Reid and E. J. Snell, pp. 83–106. London: Chapman & Hall.

Davison, A. C. and Tsai, C.-L. (1992) Regression model diagnostics. *International Statistical Review* **60**, 337–353.

Dawid, A. P. (2000) Causality without counterfactuals (with Discussion). *Journal of the American Statistical Association* **95**, 407–448.

De Finetti, B. (1974) *Theory of Probability: Volume 1.* New York: Wiley.

De Finetti, B. (1975) *Theory of Probability: Volume 2.* New York: Wiley.

de Stavola, B. L. (1988) Testing departures from time homogeneity in multistate Markov processes. *Applied Statistics* **37**, 242–250.

DeGroot, M. H. (1986a) A conversation with David Blackwell. *Statistical Science* **1**, 40–53.

DeGroot, M. H. (1986b) A conversation with Charles Stein. *Statistical Science* **1**, 454–462.

DeGroot, M. H. (1987a) A conversation with George Box. *Statistical Science* **2**, 239–258.

DeGroot, M. H. (1987b) A conversation with C. R. Rao. *Statistical Science* **2**, 53–67.

Dempster, A. P., Laird, N. M. and Rubin, D. B. (1977) Maximum likelihood from incomplete data via the EM algorithm (with Discussion). *Journal of the Royal Statistical Society series B* **39**, 1–38.

Desmond, A. and Moore, J. (1991) *Darwin.* London: Penguin.

Diggle, P. J. (1983) *Statistical Analysis of Spatial Point Patterns.* London: Academic Press.

Diggle, P. J. (1990) *Time Series: A Biostatistical Introduction.* Oxford: Clarendon Press.

Diggle, P. J., Liang, K.-Y. and Zeger, S. L. (1994) *Analysis of Longitudinal Data.* Oxford: Clarendon Press.

Dobson, A. J. (1990) *An Introduction to Generalized Linear Models.* London: Chapman & Hall.

Draper, N. R. and Smith, H. (1981) *Applied Regression Analysis.* Second edition. New York: Wiley.

Eco, U. (1984) *The Name of the Rose.* London: Pan Books.

Edwards, A. W. F. (1972) *Likelihood.* Cambridge: Cambridge University Press.

Edwards, D. (2000) *Introduction to Graphical Modelling.* Second edition. New York: Springer.

Efron, B. (1986) Double exponential families and their use in generalized linear regression. *Journal of the American Statistical Association* **81**, 709–721.

Efron, B. (1988) Computer-intensive methods in statistical regression. *SIAM Review* **30**, 421–449.

Efron, B. (1996) Empirical Bayes methods for combining likelihoods (with Discussion). *Journal of the American Statistical Association* **91**, 538–565.

Efron, B. and Hinkley, D. V. (1978) Assessing the accuracy of the maximum likelihood estimator: Observed versus expected Fisher information. *Biometrika* **65**, 457–481.

Efron, B. and Thisted, R. (1976) Estimating the number of unseen species: How many words did Shakespeare know? *Biometrika* **63**, 435–448.

Efron, B. and Tibshirani, R. J. (1993) *An Introduction to the Bootstrap.* New York: Chapman & Hall.

Embrechts, P., Klüppelberg, C. and Mikosch, T. (1997) *Modelling Extremal Events for Insurance and Finance.* Berlin: Springer.

Faddy, M. J. and Fenlon, J. S. (1999) Stochastic modelling of the invasion process of nematodes in fly larvae. *Applied Statistics* **48**, 31–37.

Fan, J. and Gijbels, I. (1996) *Local Polynomial Modelling and Its Applications.* London: Chapman & Hall.

Feigl, P. and Zelen, M. (1965) Estimation of exponential survival probabilities with concomitant information. *Biometrics* **21**, 826–838.

Ferguson, T. S. (1967) *Mathematical Statistics: A Decision-Theoretic Approach.* New York: Academic Press.

Fernholz, L. T. and Morgenthaler, S. (2000) A conversation with John W. Tukey and Elizabeth Tukey. *Statistical Science* **15**, 79–94.

Fienberg, S. E. (1980) *The Analysis of Cross-Classified Categorical Data.* Second edition. Cambridge, Massachussetts: MIT Press.

Findley, D. F. and Parzen, E. (1995) A conversation with Hirotugu Akaike. *Statistical Science* **10**, 104–117.

Firth, D. (1991) Generalized linear models. In *Statistical Theory and Modelling: In Honour of Sir David Cox, FRS*, eds D. V. Hinkley, N. Reid and E. J. Snell, pp. 55–82. London: Chapman & Hall.

Firth, D. (1993) Recent developments in quasi-likelihood methods. *Bulletin of the 49th Session of the International Statistical Institute* pp. 341–358.

Fisher, R. A. (1922) On the mathematical foundations of theoretical statistics. *Philosophical Transactions of the Royal Society of London, series A* **222**, 309–368.

Fisher, R. A. (1925) Theory of statistical estimation. *Proceedings of the Cambridge Philosophical Society* **22**, 700–725.

Fisher, R. A. (1934) Two new properties of mathematical likelihood. *Proceedings of the Royal Society of London, series A* **144**, 285—307.

Fisher, R. A. (1935a) *The Design of Experiments*. Edinburgh: Oliver and Boyd.

Fisher, R. A. (1935b) The logic of inductive inference. *Journal of the Royal Statistical Society* **98**, 39–54.

Fisher, R. A. (1956) *Statistical Methods and Scientific Inference*. Edinburgh: Oliver and Boyd.

Fisher, R. A. (1990) *Statistical Methods, Experimental Design, and Scientific Inference*. Oxford: Clarendon Press.

Fisher, R. A. and Tippett, L. H. C. (1928) Limiting forms of the frequency distributions of the largest or smallest member of a sample. *Proceedings of the Cambridge Philosophical Society* **24**, 180–190.

Fisher Box, J. (1978) *R. A. Fisher: The Life of a Scientist*. New York: Wiley.

Fishman, G. S. (1996) *Monte Carlo Concepts, Algorithms, and Applications*. New York: Springer.

Fleiss, J. L. (1986) *The Design and Analysis of Clinical Experiments*. New York: Wiley.

Fleming, T. R. and Harrington, D. P. (1991) *Counting Processes and Survival Analysis*. New York: Wiley.

Forster, J. J., McDonald, J. W. and Smith, P. W. F. (1996) Monte Carlo exact conditional tests for log-linear and logistic models. *Journal of the Royal Statistical Society series B* **58**, 445–453.

Fraser, D. A. S. (1968) *The Structure of Inference*. New York: Wiley.

Fraser, D. A. S. (1979) *Inference and Linear Models*. New York: McGraw Hill.

Fraser, D. A. S. (2003) Likelihood for component parameters. *Biometrika* **90**, to appear.

Frome, E. L. (1983) The analysis of rates using Poisson regression models. *Biometrics* **39**, 665–674.

Gamerman, D. (1997) *Markov Chain Monte Carlo: Stochastic Simulation for Bayesian Inference*. London: Chapman & Hall.

Gaver, D. P. and O'Muircheartaigh, I. G. (1987) Robust empirical Bayes analysis of event rates. *Technometrics* **29**, 1–15.

Gelfand, A. E., Hills, S. E., Racine-Poon, A. and Smith, A. F. M. (1990) Illustration of Bayesian inference in normal data models using Gibbs sampling. *Journal of the American Statistical Association* **85**, 972–985.

Gelfand, A. E. and Smith, A. F. M. (1990) Sampling-based approaches to calculating marginal densities. *Journal of the American Statistical Association* **85**, 398–409.

Gelman, A., Carlin, J. B., Stern, H. S. and Rubin, D. B. (1995) *Bayesian Data Analysis*. London: Chapman & Hall.

Geman, S. and Geman, D. (1984) Stochastic relaxation, Gibbs distributions, and the Bayesian restoration of images. *IEEE Transactions on Pattern Analysis and Machine Intelligence* **6**, 721–741.

Gilks, W. R., Richardson, S. and Spiegelhalter, D. J. (eds) (1996) *Markov Chain Monte Carlo in Practice*. London: Chapman & Hall.

Gilks, W. R. and Wild, P. (1992) Adaptive rejection sampling for Gibbs sampling. *Applied Statistics* **41**, 337–348.

Glonek, G. F. V. and McCullagh, P. (1995) Multivariate logistic models. *Journal of the Royal Statistical Society series B* **57**, 533–546.

Godambe, V. P. (1985) The foundations of finite sample estimation in stochastic processes. *Biometrika* **72**, 419–428.

Godambe, V. P. (ed.) (1991) *Estimating Functions*. Oxford: Clarendon Press.

Goldstein, H. (1995) *Multilevel Statistical Methods*. Second edition. London: Edward Arnold.

Gouriéroux, C. (1997) *ARCH Models and Financial Applications*. New York: Springer.

Green, P. J. (1984) Iteratively reweighted least squares for maximum likelihood estimation and some robust and resistant alternatives (with Discussion). *Journal of the Royal Statistical Society series B* **46**, 149–192.

Green, P. J. (1995) Reversible jump Markov chain Monte Carlo computation and Bayesian model determination. *Biometrika* **82**, 711–732.

Green, P. J. (2001) A primer on Markov chain Monte Carlo. In *Complex Stochastic Systems*, eds C. Klüppelberg, O. E. Barndorff-Nielsen and D. R. Cox, pp. 1–62. London: Chapman & Hall.

Green, P. J. and Silverman, B. W. (1994) *Nonparametric Regression and Generalized Linear Models: A Roughness Penalty Approach*. London: Chapman & Hall.

Greenland, S. (2001) Letter to the editor. *The American Statistician* **55**, 172.

Grimmett, G. R. and Stirzaker, D. R. (2001) *Probability and Random Processes*. Third edition. Oxford: Clarendon Press.

Grimmett, G. R. and Welsh, D. J. A. (1986) *Probability: An Introduction*. Oxford: Clarendon Press.

Gumbel, E. J. (1958) *Statistics of Extremes*. New York: Columbia University Press.

Guttorp, P. (1991) *Statistical Inference for Branching Processes*. New York: Wiley.

Guttorp, P. (1995) *Stochastic Modelling of Scientific Data*. London: Chapman & Hall.

Hall, P. G. and Heyde, C. C. (1980) *Martingale Limit Theory and its Application*. New York: Academic Press.

Hampel, F. R., Ronchetti, E. M., Rousseeuw, P. J. and Stahel, W. A. (1986) *Robust Statistics: The Approach Based on Influence Functions*. New York: Wiley.

Hartley, H. O. and Rao, J. N. K. (1967) Maximum-likelihood estimation for the mixed analysis of variance model. *Biometrika* **54**, 93–108.

Hastie, T. J. and Loader, C. (1993) Local regression: automatic kernel carpentry (with Discussion). *Statistical Science* **8**, 120–143.

Hastie, T. J. and Tibshirani, R. J. (1990) *Generalized Additive Models*. London: Chapman & Hall.

Hastings, W. K. (1970) Monte Carlo sampling methods using Markov chains and their applications. *Biometrika* **57**, 97–109.

Heitjan, D. F. (1994) Ignorability in general incomplete-data models. *Biometrika* **81**, 701–708.

Henderson, C. R. (1953) Estimation of variance and covariance components. *Biometrics* **9**, 226–252.

Henderson, R. and Matthews, J. N. S. (1993) An investigation of changepoints in the annual number of cases of haemolytic uraemic syndrome. *Applied Statistics* **42**, 461–471.

Heyde, C. C. (1997) *Quasi-likelihood and its Application: A General Approach to Optimal Parameter Estimation*. New York: Springer.

Heyde, C. C. and Seneta, E. (eds) (2001) *Statisticians of the Centuries*. New York: Springer.

Hills, S. E. (1987) Contribution to the discussion of Cox, D. R. and Reid, N., Parameter orthogonality and approximate conditional inference. *Journal of the Royal Statistical Society series B* **49**, 23–24.

Hinkley, D. V. (1985) Transformation diagnostics for linear models. *Biometrika* **72**, 487–496.

Hoaglin, D. C., Mosteller, F. and Tukey, J. W. (eds) (1983) *Understanding Robust and Exploratory Data Analysis*. New York: Wiley.

Hoaglin, D. C., Mosteller, F. and Tukey, J. W. (eds) (1985) *Exploring Data Tables, Trends, and Shapes*. New York: Wiley.

Hoaglin, D. C., Mosteller, F. and Tukey, J. W. (eds) (1991) *Fundamentals of Exploratory Analysis of Variance*. New York: Wiley.

Hoel, D. G. and Walburg, H. E. (1972) Statistical analysis of survival experiments. *Journal of the National Cancer Institute* **49**, 361–372.

Hoerl, A. E. and Kennard, R. W. (1970a) Ridge regression: Biased estimation for nonorthogonal problems. *Technometrics* **12**, 661–676.

Hoerl, A. E. and Kennard, R. W. (1970b) Ridge regression: Applications to nonorthogonal problems. *Technometrics* **12**, 69–82.

Hoerl, A. E., Kennard, R. W. and Hoerl, R. W. (1985) Practical use of ridge regression: A challenge met. *Applied Statistics* **34**, 114–120.

Hoeting, J. A., Madigan, D., Raftery, A. E. and Volinsky, C. T. (1999) Bayesian model averaging: A tutorial (with Discussion). *Statistical Science* **14**, 382–417.

Holland, P. W. (1986) Statistics and causal inference (with Discussion). *Journal of the American Statistical Association* **81**, 945–970.

Hougaard, P. (1984) Life table methods for heterogeneous populations: Distributions describing the heterogeneity. *Biometrika* **71**, 75–83.

Hougaard, P. (2000) *Analysis of Multivariate Survival Data*. New York: Springer.

Huber, P. J. (1981) *Robust Statistics*. New York: Wiley.

Hurvich, C. M. and Tsai, C.-L. (1989) Regression and time series model selection in small samples. *Biometrika* **76**, 297–307.

Hurvich, C. M. and Tsai, C.-L. (1990) The impact of model selection on inference in linear regression. *The American Statistician* **44**, 214–217.

Hurvich, C. M. and Tsai, C.-L. (1991) Bias of the corrected AIC criterion for underfitted regression and time series models. *Biometrika* **78**, 499–509.

Isham, V. (1981) An introduction to spatial point processes and Markov random fields. *International Statistical Review* **49**, 21–43.

Isham, V. S. (1991) Modelling stochastic phenomena. In *Statistical Theory and Modelling: In Honour of Sir David Cox, FRS*, eds D. V. Hinkley, N. Reid and E. J. Snell, pp. 177–203. London: Chapman & Hall.

Jamshidian, M. and Jennrich, R. I. (1997) Acceleration of the EM algorithm by using quasi-Newton methods. *Journal of the Royal Statistical Society series B* **59**, 569–587.

Jeffreys, H. (1961) *Theory of Probability*. Third edition. Oxford: Clarendon Press.

Jelinski, Z. and Moranda, P. B. (1972) Software reliability research. In *Statistical Computer Performance Evaluation*, ed. W. Freiberger, pp. 465–484. London: Academic Press.

Jensen, F. V. (2001) *Bayesian Networks and Decision Graphs*. New York: Springer.

Jensen, J. L. (1995) *Saddlepoint Approximations*. Oxford: Clarendon Press.

Jørgensen, B. (1997a) *The Theory of Linear Models*. New York: Chapman & Hall.

Jørgensen, B. (1997b) *The Theory of Dispersion Models*. New York: Chapman & Hall.

Kadane, J. B. and Wolfson, L. J. (1998) Experiences in elicitation (with Discussion). *The Statistician* **47**, 3–19.

Kalbfleisch, J. D. (1974) Some efficiency calculations for survival distributions. *Biometrika* **61**, 31–38.

Kalbfleisch, J. D. and Prentice, R. L. (1973) Marginal likelihoods based on Cox's regression and life model. *Biometrika* **60**, 267–278.

Kalbfleisch, J. D. and Prentice, R. L. (1980) *The Statistical Analysis of Failure Time Data*. New York: Wiley.

Kalbfleisch, J. G. (1985) *Probability and Statistical Inference*. Second edition, volume 2. New York: Springer.

Kaplan, E. L. and Meier, P. (1958) Nonparametric estimation from incomplete observations. *Journal of the American Statistical Association* **53**, 457–481.

Karr, A. F. (1991) *Point Processes and their Statistical Inference*. Second edition. New York: Marcel Dekker.

Kass, R. E. and Wasserman, L. (1996) The selection of prior distributions by formal rules. *Journal of the American Statistical Association* **91**, 1343–1370.

Keiding, N. (1990) Statistical inference in the Lexis diagram. *Philosophical Transactions of the Royal Society of London, series A* **332**, 487–509.

Kendall, M. G. and Stuart, A. (1973) *The Advanced Theory of Statistics, Volume 2: Inference and Relationship*. Third edition. London: Griffin.

Kendall, M. G. and Stuart, A. (1976) *The Advanced Theory of Statistics, Volume 3: Design and Analysis, and Time Series*. Third edition. London: Griffin.

Kendall, M. G. and Stuart, A. (1977) *The Advanced Theory of Statistics, Volume 1: Distribution Theory*. Fourth edition. London: Griffin.

Kenward, M. G. and Molenberghs, G. (1998) Likelihood based frequentist inference when data are missing at random. *Statistical Science* **13**, 236–247.

Kinderman, R. and Snell, J. L. (1980) *Markov Random Fields and their Applications*. Volume 1 of *Contemporary Mathematics*. Providence, Rhode Island: American Mathematical Society.

Klein, J. P. (1992) Semiparametric estimation of random effects using the Cox model based on the EM algorithm. *Biometrics* **48**, 175–806.

Klein, J. P. and Moeschberger, M. L. (1997) *Survival Analysis: Techniques for Censored and Truncated Data*. New York: Springer.

Knight, K. (2000) *Mathematical Statistics*. New York: Chapman & Hall.

Kullback, S. and Leibler, R. A. (1951) On information and sufficiency. *Annals of Mathematical Statistics* **22**, 79–86.

Künsch, H. R. (2001) State space and hidden Markov models. In *Complex Stochastic Systems*, eds C. Klüppelberg, O. E. Barndorff-Nielsen and D. R. Cox, pp. 109–173. London: Chapman & Hall.

Kuonen, D. (1999) Saddlepoint approximations for distributions of quadratic forms in normal variables. *Biometrika* **86**, 929–935.

Lauritzen, S. L. (1996) *Graphical Models*. Oxford: Clarendon Press.

Lauritzen, S. L. (2001) Causal inference from graphical models. In *Complex Stochastic Systems*, eds C. Klüppelberg, O. E. Barndorff-Nielsen and D. R. Cox, pp. 63–107. London: Chapman & Hall.

Lauritzen, S. L. and Richardson, T. S. (2002) Chain graph models and their causal interpretations (with Discussion). *Journal of the Royal Statistical Society series B* **64**, 321–361.

Lauritzen, S. L. and Spiegelhalter, D. J. (1988) Local computations with probabilities on graphical structures and their application to expert systems (with Discussion). *Journal of the Royal Statistical Society series B* **50**, 157–224.

Leadbetter, M. R., Lindgren, G. and Rootzén, H. (1983) *Extremes and Related Properties of Random Sequences and Processes*. New York: Springer.

Lee, P. M. (1997) *Bayesian Statistics: An Introduction*. Second edition. London: Edward Arnold.

Lehmann, E. L. (1983) *Theory of Point Estimation*. New York: Wiley.

Lehmann, E. L. (1990) Model specification: The views of Fisher and Neyman, and later developments. *Statistical Science* **5**, 160–168.

Leonard, T. and Hsu, J. S. J. (1999) *Bayesian Methods: An Analysis for Statisticians and Interdisciplinary Researchers*. Cambridge University Press.

Leonard, T., Hsu, J. S. J. and Ritter, C. (1994) The Laplacian t-approximation in Bayesian inference. *Statistica Sinica* **4**, 127–142.

Li, G. (1985) Robust regression. In *Exploring Data Tables, Trends, and Shapes*, eds F. M. D. C. Hoaglin and J. W. Tukey, pp. 281–343. New York: Wiley.

Li, W.-H. (1997) *Molecular Evolution*. Sunderland, MA: Sinauer.

Liang, K., Zeger, S. L. and Qaqish, B. (1992) Multivariate regression analyses for categorical data (with Discussion). *Journal of the Royal Statistical Society series B* **54**, 3–40.

Lindley, D. V. (1985) *Making Decisions*. New York: Wiley.

Lindley, D. V. (2000) The philosophy of statistics (with Comments). *The Statistician* **49**, 293–337.

Lindley, D. V. and Scott, W. F. (1984) *New Cambridge Elementary Statistical Tables*. Cambridge: Cambridge University Press.

Lindsay, B. G. (1995) *Mixture Models: Theory, Geometry, and Applications*. Number 5 in NSF-CBMS Regional Conference Series in Probability and Statistics. Hayward, CA: Institute for Mathematical Statistics.

Linhart, H. and Zucchini, W. (1986) *Model Selection*. New York: Wiley.

Little, R. J. A. and Rubin, D. B. (1987) *Statistical Analysis with Missing Data*. New York: Wiley.

Lloyd, C. J. (1992) Effective conditioning. *Australian Journal of Statistics* **34**, 241–260.

Loader, C. (1999) *Local Regression and Likelihood*. New York: Springer.

MacDonald, I. L. and Zucchini, W. (1997) *Hidden Markov and Other Models for Discrete-valued Time Series*. London: Chapman & Hall.

Mallows, C. L. (1973) Some comments on C_p. *Technometrics* **15**, 661–675.

Mantel, N. and Haenszel, W. (1959) Statistical aspects of the analysis of data from retrospective studies of disease. *Journal of the National Cancer Institute* **22**, 719–748.

Mardia, K. V., Kent, J. T. and Bibby, J. M. (1979) *Multivariate Analysis*. London: Academic Press.

Maritz, J. S. and Lwin, T. (1989) *Empirical Bayes Methods*. Second edition. London: Chapman & Hall.

McCullagh, P. (1987) *Tensor Methods in Statistics*. London: Chapman & Hall.

McCullagh, P. (1991) Quasi-likelihood and estimating functions. In *Statistical Theory and Modelling: In Honour of Sir David Cox, FRS*, eds D. V. Hinkley, N. Reid and E. J. Snell, pp. 265–286. London: Chapman & Hall.

McCullagh, P. (1992) Conditional inference and Cauchy models. *Biometrika* **79**, 247–259.

McCullagh, P. and Nelder, J. A. (1989) *Generalized Linear Models*. Second edition. London: Chapman & Hall.

McCulloch, C. E. (1997) Maximum likelihood algorithms for generalized linear mixed models. *Journal of the American Statistical Association* **92**, 162–170.

McCulloch, C. E. and Searle, S. R. (2001) *Generalized, Linear, and Mixed Models*. New York: Wiley.

McLachlan, G. J. and Krishnan, T. (1997) *The EM Algorithm and Extensions*. New York: Wiley.

McLeish, D. and Small, C. G. (1994) *Hilbert Space Methods in Probability and Statistical Inference*. New York: Wiley.

McQuarrie, A. D. R. and Tsai, C.-L. (1998) *Regression and Time Series Model Selection*. Singapore: World Scientific.

Meng, X.-L. and van Dyk, D. (1997) The EM algorithm — an old folk-song sung to a fast new tune (with Discussion). *Journal of the Royal Statistical Society series B* **59**, 511–567.

Metropolis, N., Rosenbluth, A. W., Rosenbluth, M. N., Teller, A. H. and Teller, E. (1953) Equations of state calculations by fast computing machines. *Journal of Chemical Physics* **21**, 1087–1091.

Miller, A. J. (1990) *Subset Selection in Regression*. London: Chapman & Hall.

Miller, R. G. (1981) *Survival Analysis*. New York: Wiley.

Molenberghs, G., Kenward, M. G. and Goetghebeur, E. (2001) Sensitivity analysis for incomplete contingency tables: the Slovenian plebiscite case. *Applied Statistics* **50**, 15–29.

Morgan, B. J. T. (1984) *Elements of Simulation*. London: Chapman & Hall.

Morris, C. N. (1982) Natural exponential families with quadratic variance functions. *Annals of Statistics* **10**, 65–80.

Morris, C. N. (1983) Parametric empirical Bayes inference: Theory and applications. *Journal of the American Statistical Association* **78**, 47–65.

Mosteller, F. and Tukey, J. W. (1977) *Data Analysis and Regression*. Reading, Massachussetts: Addison–Wesley.

Nelder, J. A. and Wedderburn, R. W. M. (1972) Generalized linear models. *Journal of the Royal Statistical Society series A* **135**, 370–384.

Nelson, W. D. and Hahn, G. J. (1972) Linear estimation of a regression relationship from censored data. Part 1 — simple methods and their application (with Discussion). *Technometrics* **14**, 247–276.

Neopolitan, E. (1990) *Probabilistic Reasoning in Expert Systems*. New York: Wiley.

Neyman, J. and Pearson, E. S. (1967) *Joint Statistical Papers*. Cambridge University Press.

Neyman, J. and Scott, E. L. (1948) Consistent estimates based on partially consistent observations. *Econometrica* **16**, 1–32.

Norris, J. R. (1997) *Markov Chains*. Cambridge: Cambridge University Press.

Oakes, D. (1991) Life-table analysis. In *Statistical Theory and Modelling: In Honour of Sir David Cox, FRS*, eds D. V. Hinkley, N. Reid and E. J. Snell, pp. 107–128. London: Chapman & Hall.

Oakes, D. (1999) Direct calculation of the information matrix via the EM algorithm. *Journal of the Royal Statistical Society series B* **61**, 479–482.

Ogata, Y. (1988) Statistical models for earthquake occurrences and residual analysis for point processes. *Journal of the American Statistical Association* **83**, 9–27.

O'Hagan, A. (1988) *Probability: Methods and Measurement*. London: Chapman & Hall.

O'Hagan, A. (1998) Eliciting expert beliefs in substantial practical applications (with Discussion). *The Statistician* **47**, 21–35.

Pace, L. and Salvan, A. (1997) *Principles of Statistical Inference from a Neo-Fisherian Perspective*. Singapore: World Scientific.

Patterson, H. D. and Thompson, R. (1971) Recovery of inter-block information when block sizes are unequal. *Biometrika* **58**, 545–554.

Pearl, J. (1988) *Probabilistic Reasoning in Intelligent Systems: Networks of Plausible Inference*. San Francisco: Morgan Kaufmann.

Pearl, J. (2000) *Causality: Models, Reasoning and Inference*. Cambridge: Cambridge University Press.

Pearson, E. S. and Hartley, H. O. (1976) *Biometrika Tables for Statisticians*. Third edition, volumes 1 and 2. London: Biometrika Trust: University College.

Percival, D. B. and Walden, A. T. (1993) *Spectral Analysis for Physical Applications: Multitaper and Conventional Univariate Techniques*. Cambridge: Cambridge University Press.

Pirazzoli, P. A. (1982) Maree estreme a Venezia (periodo 1872–1981). *Acqua Aria* **10**, 1023–1039.

Pitman, E. J. G. (1938) The estimation of location and scale parameters of a continuous population of any given form. *Biometrika* **30**, 391–421.

Pitman, E. J. G. (1939) Tests of hypotheses concerning location and scale parameters. *Biometrika* **31**, 200–215.

Pötscher, B. M. (1991) Effects of model selection on inference. *Econometric Theory* **7**, 163–185.

Prentice, R. L. and Gloeckler, L. A. (1978) Regression analysis of grouped survival data with application to breast cancer data. *Biometrics* **34**, 57–67.

Prentice, R. L., Kalbfleisch, J. D., Peterson, A. V., Flournoy, N., Farewell, V. T. and Breslow, N. E. (1978) The analysis of failure times in the presence of competing risks. *Biometrics* **34**, 541–554.

Priestley, M. B. (1981) *Spectral Analysis and Time Series*. London: Academic Press.

Prum, B., Rodolphe, F. and de Turckheim, E. (1995) Finding words with unexpected frequencies in deoxyribonucleic acid sequences. *Journal of the Royal Statistical Society series B* **57**, 205–220.

Raftery, A. E. (1988) Analysis of a simple debugging model. *Applied Statistics* **37**, 12–22.

Raiffa, H. and Schlaifer, R. (1961) *Applied Statistical Decision Theory*. Cambridge, Mass: MIT Press.

Rao, C. R. (1973) *Linear Statistical Inference and its Applications*. Second edition. New York: Wiley.

Rawlings, J. O. (1988) *Applied Regression Analysis: A Research Tool*. Pacific Grove, California: Wadsworth & Brooks/Cole.

Reid, N. (1988) Saddlepoint methods and statistical inference (with Discussion). *Statistical Science* **3**, 213–238.

Reid, N. (1994) A conversation with Sir David Cox. *Statistical Science* **9**, 439–455.

Reid, N. (1995) The roles of conditioning in inference (with Discussion). *Statistical Science* **10**, 138–199.

Reid, N. (2003) Asymptotics and the theory of inference. *Annals of Statistics*, to appear.

Resnick, S. I. (1987) *Extreme Values, Point Processes and Regular Variation*. New York: Springer.

Reynolds, P. S. (1994) Time-series analyses of beaver body temperatures. In *Case Studies in Biometry*, eds N. Lange, L. Ryan, L. Billard, D. R. Brillinger, L. Conquest and J. Greenhouse, pp. 211–228. New York: Wiley.

Rice, J. A. (1988) *Mathematical Statistics and Data Analysis*. Belmont, California: Wadsworth & Brooks/Cole.

Richardson, S. and Green, P. J. (1997) On Bayesian analysis of mixtures with an unknown number of components (with Discussion). *Journal of the Royal Statistical Society series B* **59**, 731–792.

Ripley, B. D. (1981) *Spatial Statistics*. New York: Wiley.

Ripley, B. D. (1987) *Stochastic Simulation*. New York: Wiley.

Ripley, B. D. (1988) *Statistical Inference for Spatial Processes*. Cambridge: Cambridge University Press.

Robert, C. P. (2001) *The Bayesian Choice*. Second edition. New York: Springer.

Robert, C. P. and Casella, G. (1999) *Monte Carlo Statistical Methods*. New York: Springer.

Robinson, G. K. (1991) That BLUP is a good thing: The estimation of random effects (with Discussion). *Statistical Science* **3**, 15–51.

Roeder, K. (1990) Density estimation with confidence sets exemplified by superclusters and voids in galaxies. *Journal of the American Statistical Association* **85**, 617–624.

Rolski, T., Schmidli, H., Schmidt, V. and Teugels, J. (1999) *Stochastic Processes for Insurance and Finance*. Chichester: Wiley.

Ross, S. M. (1996) Bayesians should not resample a prior sample to learn about the posterior. *American Statistician* **50**, 116.

Rousseeuw, P. J. and Leroy, A. M. (1987) *Robust Regression and Outlier Detection*. New York: Wiley.

Rubin, D. B. (1976) Inference and missing data (with Discussion). *Biometrika* **63**, 581–592.

Rubin, D. B. (1987) *Multiple Imputation for Nonresponse in Surveys*. New York: Wiley.

Rubinstein, R. Y. (1981) *Simulation and the Monte Carlo Method*. New York: Wiley.

Schafer, G. (1976) *A Mathematical Theory of Evidence*. Princeton, NJ: Princeton University Press.

Scheffé, H. (1959) *Analysis of Variance*. New York: Wiley.

Schervish, M. J. (1995) *Theory of Statistics*. New York: Springer.

Schwartz, G. (1978) Estimating the dimension of a model. *Annals of Statistics* **6**, 461–464.

Scott, D. W. (1992) *Multivariate Density Estimation: Theory, Practice, and Visualization*. New York: Wiley.

Searle, S. R. (1971) *Linear Models*. New York: Wiley.

Searle, S. R., Casella, G. and McCulloch, C. E. (1992) *Variance Components*. New York: Wiley.

Seber, G. A. F. (1977) *Linear Regression Analysis*. New York: Wiley.

Seber, G. A. F. (1985) *Multivariate Observations*. New York: Wiley.

Self, S. G. and Liang, K.-Y. (1987) Asymptotic properties of maximum likelihood estimators and likelihood ratio tests under nonstandard conditions. *Journal of the American Statistical Association* **82**, 605–610.

Sen, A. and Srivastava, M. (1990) *Regression Analysis: Theory, Methods, and Applications*. New York: Springer.

Severini, T. A. (1999) An empirical adjustment to the likelihood ratio statistic. *Biometrika* **86**, 235–247.

Severini, T. A. (2000) *Likelihood Methods in Statistics*. Oxford: Clarendon Press.

Shao, J. (1999) *Mathematical Statistics*. New York: Springer.

Sheather, S. J. and Jones, M. C. (1991) A reliable data-based bandwidth selection method for kernel density estimation. *Journal of the Royal Statistical Society series B* **53**, 683–690.

Sheehan, N. A. (2000) On the application of Markov chain Monte Carlo methods to genetic analyses on complex pedigrees. *International Statistical Review* **68**, 83–110.

Shephard, N. G. (1996) Statistical aspects of ARCH and stochastic volatility. In *Time Series Models In Econometrics, Finance and Other Fields*, eds D. R. Cox, D. V. Hinkley and O. E. Barndorff-Nielsen, pp. 1–67. London: Chapman & Hall.

Silverman, B. W. (1986) *Density Estimation for Statistics and Data Analysis*. London: Chapman & Hall.

Silvey, S. D. (1970) *Statistical Inference*. London: Chapman & Hall.

Silvey, S. D. (1980) *Optimal Design*. London: Chapman & Hall.

Simonoff, J. S. (1996) *Smoothing Methods in Statistics*. New York: Springer.

Skovgaard, I. M. (1987) Saddlepoint expansions for conditional distributions. *Journal of Applied Probability* **24**, 875–887.

Skovgaard, I. M. (1990) On the density of minimum contrast estimators. *Annals of Statistics* **18**, 779–789.

Skovgaard, I. M. (1996) An explicit large-deviation approximation to one-parameter tests. *Bernoulli* **2**, 145–166.

Smith, A. F. M. (1995) A conversation with Dennis Lindley. *Statistical Science* **10**, 305–319.

Smith, A. F. M. and Gelfand, A. E. (1992) Bayesian statistics without tears: A sampling-resampling perspective. *American Statistician* **46**, 84–88.

Smith, J. Q. (1988) *Decision Analysis: A Bayesian Approach*. London: Chapman & Hall.

Smith, P. W. F., Forster, J. J. and McDonald, J. W. (1996) Monte Carlo exact tests for square contingency tables. *Journal of the Royal Statistical Society series A* **159**, 309–321.

Smith, R. L. (1985) Maximum likelihood estimation in a class of non-regular cases. *Biometrika* **72**, 67–92.

Smith, R. L. (1989a) Extreme value analysis of environmental time series: An example based on ozone data (with Discussion). *Statistical Science* **4**, 367–393.

Smith, R. L. (1989b) A survey of nonregular problems. *Bulletin of the International Statistical Institute* **53**, 353–372.

Smith, R. L. (1990) Extreme value theory. In *Handbook of Applicable Mathematics, Supplement*, eds W. Ledermann, E. Lloyd, S. Vajda and C. Alexander, chapter 14. Chichester: Wiley.

Smith, R. L. (1994) Nonregular regression. *Biometrika* **81**, 173–183.

Smith, R. L. (1997) Introduction to Besag (1974) Spatial interaction and the statistical analysis of lattice systems. In *Breakthroughs in Statistics, Volume 3*, eds S. Kotz and N. L. Johnson, pp. 285–291. New York: Springer.

Sørensen, M. (1999) On asymptotics of estimating functions. *Brazilian Journal of Probability and Statistics* **13**, 111–136.

Spiegelhalter, D. J., Dawid, A. P., Lauritzen, S. L. and Cowell, R. G. (1993) Bayesian analysis in expert systems. *Statistical Science* **8**, 219–283.

Spiegelhalter, D. J. and Smith, A. F. M. (1982) Bayes factors for linear and log-linear models with vague prior information. *Journal of the Royal Statistical Society series B* **44**, 377–387.

Spiegelhalter, D. J., Thomas, A., Best, N. G. and Gilks, W. R. (1996a) *BUGS 0.5: Bayesian Inference Using Gibbs Sampling (Version ii)*. Cambridge: MRC Biostatistics Unit.

Spiegelhalter, D. J., Thomas, A., Best, N. G. and Gilks, W. R. (1996b) *BUGS 0.5 Examples Volume 1 (Version ii)*. Cambridge: MRC Biostatistics Unit.

Spiegelhalter, D. J., Thomas, A., Best, N. G. and Gilks, W. R. (1996c) *BUGS 0.5 Examples Volume 2 (Version ii)*. Cambridge: MRC Biostatistics Unit.

Stein, C. (1956) Inadmissibility of the usual estimator for the mean of a multivariate normal distribution. In *Proceedings of the 3rd Berkeley Symposium on Mathematical Statistics and Probability*, volume 1, pp. 197–206. University of California Press: Berkeley, CA.

Stephens, M. (2000) Bayesian analysis of mixture models with an unknown number of components: An alternative to reversible jump methods. *Annals of Statistics* **28**, 40–74.

Stigler, S. M. (1986) *The History of Statistics: The Measurement of Uncertainty Before 1900*. Cambridge, MA: Belknap Press.

Stirzaker, D. R. (1994) *Elementary Probability*. Cambridge: Cambridge University Press.

Stone, M. (1974) Cross-validatory choice and assessment of statistical predictions (with Discussion). *Journal of the Royal Statistical Society series B* **36**, 111–147.

Stone, M. and Brooks, R. J. (1990) Continuum regression: Cross-validated sequentially constructed prediction embracing ordinary least squares, partial least squares and principal components regression (with Discussion). *Journal of the Royal Statistical Society series B* **52**, 237–269.

Tanner, M. A. (1996) *Tools for Statistical Inference: Methods for the Exploration of Posterior Distributions and Likelihood Functions*. Third edition. New York: Springer.

Taylor, G. L. and Prior, A. M. (1938) Blood groups in England. *Annals of Eugenics* **8**, 343–355.

Thatcher, A. R. (1999) The long-term pattern of adult mortality and the highest attained age (with Discussion). *Journal of the Royal Statistical Society series A* **162**, 5–43.

Therneau, T. M. and Grambsch, P. M. (2000) *Modeling Survival Data: Extending the Cox Model*. New York: Springer.

Thisted, R. and Efron, B. (1987) Did Shakespeare write a newly-discovered poem? *Biometrika* **74**, 445–455.

Thompson, E. A. (2001) Monte Carlo methods on genetic structures. In *Complex Stochastic Systems*, eds C. Klüppelberg, O. E. Barndorff-Nielsen and D. R. Cox, pp. 175–218. London: Chapman & Hall.

Tierney, L. and Kadane, J. B. (1986) Accurate approximations for posterior moments and marginal densities. *Journal of the American Statistical Association* **81**, 82–86.

Tierney, L., Kass, R. E. and Kadane, J. B. (1989) Approximate marginal densities of nonlinear functions. *Biometrika* **76**, 425–433.

Titterington, D. M., Smith, A. F. M. and Makov, U. E. (1985) *Statistical Analysis of Finite Mixture Distributions*. New York: Wiley.

Tong, H. (1990) *Non-linear Time Series: A Dynamical System Approach*. Oxford: Clarendon Press.

Tsay, R. S. (2002) *Analysis of Financial Time Series*. New York: Wiley.

Tsiatis, A. A. (1998) Competing risks. In *Encyclopedia of Biostatistics*, eds P. Armitage and T. Colton, volume 1, pp. 824–834. New York: Wiley.

Tufte, E. R. (1983) *The Visual Display of Quantitative Information*. Cheshire, Connecticut: Graphics Press.

Tufte, E. R. (1990) *Envisioning Information*. Cheshire, Connecticut: Graphics Press.

Tukey, J. W. (1949) One degree of freedom for non-additivity. *Biometrics* **5**, 232–242.

Tukey, J. W. (1977) *Exploratory Data Analysis*. Reading, Massachussetts: Addison–Wesley.

van der Vaart, A. W. (1998) *Asymptotic Statistics*. Cambridge University Press.

van Lieshout, M. N. M. (2000) *Markov Point Processes and their Applications*. Singapore: World Scientific.

Wand, M. P. and Jones, M. C. (1995) *Kernel Smoothing*. London: Chapman & Hall.

Wedderburn, R. W. M. (1974) Quasi-likelihood functions, generalized linear models, and the Gauss–Newton method. *Biometrika* **61**, 439–447.

Weisberg, S. (1985) *Applied Linear Regression*. Second edition. New York: Wiley.

Welsh, A. H. (1996) *Aspects of Statistical Inference*. New York: Wiley.

Wermuth, N. and Lauritzen, S. L. (1990) On substantive research hypotheses, conditional independence graphs and graphical chain models (with Discussion). *Journal of the Royal Statistical Society series B* **52**, 21–72.

Wetherill, G. B. (1986) *Regression Analysis with Applications*. London: Chapman & Hall.

Whittaker, J. (1990) *Graphical Models in Applied Multivariate Statistics*. New York: Wiley.

Wild, P. and Gilks, W. R. (1993) Algorithm AS 287: Adaptive rejection sampling from log-concave density functions. *Applied Statistics* **42**, 701–709.

Wood, S. N. (2000) Modelling and smoothing parameter estimation with multiple quadratic penalties. *Journal of the Royal Statistical Society series B* **62**, 413–428.

Woods, H., Steinour, H. H. and Starke, H. R. (1932) Effect of composition of Portland cement on heat evolved during hardening. *Industrial Engineering and Chemistry* **24**, 1207–1214.

Yates, F. (1937) The Design and Analysis of Factorial Experiments. Technical report, Imperial Bureau of Soil Science, Harpenden. Technical communication **35**.

Name Index

Example Index

Index